Series in Mathematical Analysis and Applications

Edited by Ravi P. Agarwal and Donal O'Regan

T0133774

VOLUME 8

NONSMOOTH CRITICAL POINT THEORY AND NONLINEAR BOUNDARY VALUE PROBLEMS

SERIES IN MATHEMATICAL ANALYSIS AND APPLICATIONS

Series in Mathematical Analysis and Applications (SIMAA) is edited by Ravi P. Agarwal, Florida Institute of Technology, USA and Donal O'Regan, National University of Ireland, Galway, Ireland.

The series is aimed at reporting on new developments in mathematical analysis and applictions of a high standard and or current interest. Each volume in the series is devoted to a topic in analysis that has been applied, or is potentially applicable, to the solutions of scientific, engineering and social problems.

Series in Mathematical Analysis and Applications

Edited by Ravi P. Agarwal and Donal O'Regan

VOLUME 8

NONSMOOTH CRITICAL POINT THEORY AND NONLINEAR BOUNDARY VALUE PROBLEMS

Leszek Gasiński

Nikolaos S. Papageorgiou

CRC Press

Taylor & Francis Group

Boca Raton London New York

CRC Press is an imprint of the
Taylor & Francis Group, an **informa** business

A CHAPMAN & HALL BOOK

CRC Press
Taylor & Francis Group
6000 Broken Sound Parkway NW, Suite 300
Boca Raton, FL 33487-2742

First issued in paperback 2019

© 2005 by Taylor & Francis Group, LLC
CRC Press is an imprint of Taylor & Francis Group, an Informa business

No claim to original U.S. Government works

ISBN-13: 978-1-58488-485-9 (hbk)
ISBN-13: 978-0-367-39405-9 (pbk)

Library of Congress Cataloging-in-Publication Data

Gasiński, Leszek.
    Nonsmooth critical point theory and nonlinear boundary value problems / Leszek Gasiński and Nikolaos S. Papageorgiou.
        p. cm. — (Mathematical analysis and applications ; 8)
    Includes bibliographical references and index.
    ISBN 1-58488-485-1 (alk. paper)
    1. Critical point theory (Mathematical analysis) 2. Nonlinear boundary value problems. I. Papageorgiou, Nikolaos Socrates.
    II. Title. III. Series: Series in mathematical analysis and applications ; v. 8.

    QA614.7.G39 2005
    514'.74—dc22   2004050153

Library of Congress Card Number 2004050153
Visit the Taylor & Francis Web site at
http://www.taylorandfrancis.com

and the CRC Press Web site at
http://www.crcpress.com

To Krystyna, Halszka and Krystyna (LG)
To my brother A.S. Papageorgiou (NSP)

Contents

x

Preface

Variational methods have turned out to be a very effective analytical tool in the study of nonlinear problems. The idea behind them is to try to find solutions of a given boundary value problem by looking for critical (stationary) points of a suitable "energy" functional defined on an appropriate function space dictated by the data of the problem. Then the boundary value problem under consideration is the Euler-Lagrange equation satisfied by a critical point. In many cases of interest, the energy functional is unbounded (from both above and below; indefinite functional) and so we cannot hope for a global maximum or minimum. Therefore we must look for local extrema and for saddle points obtained by minimax arguments.

One useful technique in obtaining critical points is based on deformations along the paths of steepest descent of the energy functionals. Another approach can be based on the Ekeland variational principle. The classical critical point theory was developed in the sixties and seventies for C^1-functionals. The needs of specific applications (such as nonsmooth mechanics, nonsmooth gradient systems, mathematical economics, etc.) and the impressive progress in nonsmooth analysis and multivalued analysis led to extensions of the critical point theory to nondifferentiable functions, in particular locally Lipschitz and even continuous functions. The resulting theory succeeded in extending a big part of the smooth (C^1) theory.

In this book, we present the existing nonsmooth critical point theories (Chapter 2) and use them to study nonlinear boundary value problems of ordinary and partial (elliptic) differential equations, which are in variational form. We also investigate nonlinear boundary value problems (BVPs) in nonvariational form, using a great variety of methods and techniques which involve upper-lower solutions, fixed point and degree theories, nonlinear operator theory, nonsmooth analysis, and multivalued analysis (Chapter 3 and Chapter 4). The necessary mathematical background to understand these methods is developed in Chapter 1 (see also the Appendix). This way we present a large part of the methods used today in the study of nonlinear boundary value problems with nonsmooth and multivalued terms.

Acknowledgments

In preparing this book, we have received help and encouragement from a number of friends and colleagues. In particular we wish to thank Prof. Z. Denkowski and Prof. S. Migórski (Jagiellonian University), who did everything possible to make this job easier. The second author wishes also to thank Prof. F.S. De Blasi (University of Roma II), Prof. S. Hu (Southwest Missouri State University) and Prof. F. Papalini (University of Ancona) for their help and support.

We are also grateful to Prof. D. O'Regan, who recommended our book to CRC Press. For his continuous support and interest in our work we thank him warmly.

Finally, we wish to thank the people of CRC Press for the very effective and pleasant cooperation.

Chapter 1

Mathematical Background

In this chapter, we review the basic mathematical material that we need in the development of the nonsmooth critical point theories and in the study of the nonlinear boundary value problems (ordinary and partial) that follow. So in the first section we outline the basic facts about Sobolev spaces. Sobolev spaces provide the appropriate functional framework for the analysis of the ordinary and partial differential equations problems that we consider in this volume. The subdifferential of a nonsmooth (nondifferentiable) function is a multivalued map. So the resulting nonsmooth critical point theories and the corresponding boundary value problems are of multivalued nature, since the potential function is nonsmooth. Moreover, in our formulation of the problems we allow the nonlinear perturbation term to be set-valued. Therefore, to handle such problems we need to know a few basic facts about Set-Valued Analysis. In Section 1.2 we review from the theory the main items that will be helpful in what follows. Since one of our goals in this volume is to present the main facts about the existing nonsmooth critical point theories, we need the notions and results of Nonsmooth Analysis. In Section 1.3, we review the main items of Nonsmooth Analysis, which are needed for what follows. Nonsmooth Analysis is closely related to Set-Valued Analysis and to the theory of nonlinear operators. Set-Valued Analysis has already been covered in Section 1.2. So in Section 1.4 we deal with nonlinear operators, with particular emphasis on operators of monotone type. We also discuss briefly the Nemytskii (superposition) operator and present various forms of the Ekeland Variational Principle. Finally in Section 1.5, we present some basic facts about semilinear and nonlinear elliptic equations. Our starting point is the derivation of the spectra of the ordinary and partial Laplacian and p-Laplacian differential operators under Dirichlet and periodic boundary conditions. We also consider certain weighted eigenvalue problems driven by a strongly elliptic linear partial differential operator. We establish the existence of eigenvalues, provide variational characterizations of them (via the Rayleigh quotient) and examine the corresponding eigenfunctions. This analysis is based on some regularity results and maximum principles that we also present.

1.1   Sobolev Spaces

For the reader's convenience, in this section we present a quick review of the theory of Sobolev spaces. The results that we present here are standard and their proofs as well as a more detailed and deeper analysis can be found in several classical textbooks on the subject such as Adams (1975), Brézis (1983) and Kufner, John & Fučik (1977).

1.1.1   Basic Definitions and Properties

Let $\Omega \subseteq \mathbb{R}^N$ be a nonempty open set. By $\partial\Omega$ we denote the **boundary** of Ω, i.e. $\partial\Omega \stackrel{df}{=} \overline{\Omega} \cap \Omega^c = \overline{\Omega} \setminus \Omega$. Also we say that another open set Ω' is **strongly included** in Ω, denoted by $\Omega' \subset\subset \Omega$, if Ω' is bounded and $\overline{\Omega'} \subseteq \Omega$. For a **multi-index** $\alpha = (\alpha_1, \ldots, \alpha_N) \in \mathbb{N}_0^N$, by $|\alpha|$ we denote the **length of the multi-index**, defined by

$$|\alpha| \stackrel{df}{=} \sum_{k=1}^{N} \alpha_k$$

and by $D^\alpha u$ we denote the **weak derivative** of u of order α, i.e.

$$D^\alpha u \stackrel{df}{=} \frac{\partial^{|\alpha|} u}{\partial z_1^{\alpha_1} \ldots \partial z_N^{\alpha_N}}.$$

By $C_c^\infty(\Omega)$ we denote the space of functions $\vartheta \in C^\infty(\Omega)$ for which their **support**, defined by

$$\operatorname{supp} \vartheta \stackrel{df}{=} \overline{\{x \in \Omega : \vartheta(x) \neq 0\}},$$

is a compact set contained in Ω. We furnish $C_c^\infty(\Omega)$ with a convergence notion according to which $\{\vartheta_n\}_{n\geq 1} \subseteq C_c^\infty(\Omega)$ converges to 0 if and only if there exists a compact set $K \subseteq \Omega$, such that

$$\bigcup_{n\geq 1} \operatorname{supp} \vartheta_n \subseteq K$$

and the sequence $\{D^\alpha \vartheta_n\}_{n\geq 1}$ converges uniformly to 0 for all $\alpha \in \mathbb{N}_0^N$. Usually $C_c^\infty(\Omega)$ equipped with this convergence notion is denoted by $\mathcal{D}(\Omega)$ and is known as the **space of test functions**. Recall that $C_c^\infty(\Omega)$ is dense in $L^p(\Omega)$ for all $p \in [1, +\infty)$. By $\mathcal{D}'(\Omega)$ we denote the **space of distributions**, i.e. the space of all linear maps $L : \mathcal{D}(\Omega) \longrightarrow \mathbb{R}$, such that $L(\vartheta_n) \longrightarrow 0$ for all $\{\vartheta_n\}_{n\geq 1} \subseteq \mathcal{D}(\Omega)$, such that $\vartheta_n \longrightarrow 0$. For a given distribution $L \in \mathcal{D}'(\Omega)$ and for all $\alpha \in \mathbb{N}_0^N$, we define the distribution $D^\alpha L$ by

$$D^\alpha L(\vartheta) \stackrel{df}{=} (-1)^{|\alpha|} L(D^\alpha \vartheta) \qquad \forall\, \vartheta \in \mathcal{D}(\Omega).$$

For every $u \in L^1_{\mathrm{loc}}(\Omega)$, we can introduce the so-called **regular distribution** L_u by

$$L_u(\vartheta) \stackrel{df}{=} \int_\Omega u(x)\vartheta(x)dx \qquad \forall\, \vartheta \in \mathcal{D}(\Omega).$$

We have $L_u = L_v$ if and only if $u(x) = v(x)$ for almost all $x \in \Omega$. For given $u, v \in L^1_{\mathrm{loc}}(\Omega)$ and $\alpha \in \mathbb{N}_0^N$ we write $v = D^\alpha u$ to express the equality $L_v = D^\alpha L_u$. So it is equivalent to saying that

$$\int_\Omega v(x)\vartheta(x)dx = (-1)^{|\alpha|}\int_\Omega u(x)D^\alpha\vartheta(x)dx \qquad \forall\, \vartheta \in \mathcal{D}(\Omega).$$

We say that $D^\alpha u \in L^1_{\mathrm{loc}}(\Omega)$, if we can find $v \in L^1_{\mathrm{loc}}(\Omega)$, such that $D^\alpha u = v$. We say that $D^\alpha u \in L^p(\Omega)$ (with $1 \leq p \leq +\infty$), if we can find $v \in L^p(\Omega)$, such that $D^\alpha u = v$. Note that, if $u \in C^{|\alpha|}(\Omega)$, then this generalized derivative coincides with the usual (classical) partial derivative.

DEFINITION 1.1.1   *For $m \in \mathbb{N}_0 \stackrel{df}{=} \mathbb{N} \cup \{0\}$ and $1 \leq p \leq +\infty$, we define the **Sobolev space***

$$W^{m,p}(\Omega) \stackrel{df}{=} \left\{ u \in L^p(\Omega) : D^\alpha u \in L^p(\Omega) \text{ for all } \alpha \in \mathbb{N}_0^N \text{ with } |\alpha| \leq m \right\}.$$

For every $u \in W^{m,p}(\Omega)$, we define

$$\|u\|_{W^{m,p}(\Omega)} \stackrel{df}{=} \left(\sum_{|\alpha| \leq m} \|D^\alpha u\|_p^p \right)^{\frac{1}{p}} \qquad \text{if } 1 \leq p < +\infty,$$

where $\|\cdot\|_p$ is the norm of $L^p(\Omega)$, and

$$\|u\|_{W^{m,\infty}(\Omega)} \stackrel{df}{=} \sum_{|\alpha| \leq m} \|D^\alpha u\|_\infty,$$

where $\|\cdot\|_\infty$ is the norm of $L^\infty(\Omega)$. We also set

$$W^{m,p}_0(\Omega) \stackrel{df}{=} \overline{\mathcal{D}(\Omega)}^{\|\cdot\|_{W^{m,p}(\Omega)}}.$$

REMARK 1.1.1   The space $\left(W^{m,p}(\Omega), \|\cdot\|_{W^{m,p}(\Omega)}\right)$ is a Banach space, which is reflexive and uniformly convex if $p \in (1, +\infty)$ and separable if $p \in [1, +\infty)$. $\left(W^{m,p}_0(\Omega), \|\cdot\|_{W^{m,p}(\Omega)}\right)$ is a closed subspace of $\left(W^{m,p}(\Omega), \|\cdot\|_{W^{m,p}(\Omega)}\right)$. If $p = 2$, we write

$$H^m(\Omega) \stackrel{df}{=} W^{m,2}(\Omega) \quad \text{and} \quad H^m_0(\Omega) \stackrel{df}{=} W^{m,2}_0(\Omega).$$

These spaces are Hilbert spaces with inner product given by

$$(u, v)_{H^m(\Omega)} \stackrel{df}{=} \sum_{|\alpha| \le m} (D^\alpha u, D^\alpha v)_2 = \sum_{|\alpha| \le m} \int_\Omega D^\alpha u(x) D^\alpha v(x) dx.$$

$\Box$

The next theorem is known as the Meyers-Serrin Theorem and it says that Sobolev functions can be approximated by smooth ones.

THEOREM 1.1.1 (Meyers-Serrin Theorem)
If $\Omega \subseteq \mathbb{R}^N$ is open, $m \in \mathbb{N}_0$ and $p \in [1, +\infty)$,
then $C^\infty(\Omega) \cap W^{m,p}(\Omega)$ is dense in $W^{m,p}(\Omega)$.

REMARK 1.1.2   Note that in Theorem 1.1.1 we do not claim that the approximating sequence of smooth functions belongs in $C^\infty(\overline{\Omega})$. To be able to approximate Sobolev functions by functions which are smooth all the way up to the boundary, we need to strengthen our hypotheses about the geometry of Ω. $\Box$

DEFINITION 1.1.2   *We say that the boundary $\partial\Omega$ of an open set $\Omega \subseteq \mathbb{R}^N$ is **Lipschitz**, if for each $x = (x_1, \ldots, x_N) \in \partial\Omega$, there exist $r > 0$ and a Lipschitz continuous map $\gamma \colon \mathbb{R}^{N-1} \longrightarrow \mathbb{R}$ which, after rotation and relabelling of the coordinate axes if necessary, satisfies*

$$\Omega \cap C_r(x) = \big\{ (y_1, \ldots, y_N) \in \mathbb{R}^N : \gamma(y_1, \ldots, y_{N-1}) < y_N \big\} \cap C_r(x),$$

where

$$C_r(x) \stackrel{df}{=} \big\{ (y_1, \ldots, y_N) \in \mathbb{R}^N : |x_k - y_k| < r \text{ for } k \in \{1, \ldots, N\} \big\}.$$

REMARK 1.1.3   So $\partial\Omega$ is Lipschitz, if locally it is the graph of a Lipschitz continuous function. By Rademacher's theorem (see Theorem A.2.4), the outer unit normal $n(z)$ to Ω exists for almost all $z \in \partial\Omega$ (on $\partial\Omega$ we consider the $(N-1)$-dimensional Hausdorff (surface) measure; see Definition A.2.3). $\Box$

Using this notion we can have a stronger approximation result by smooth functions.

THEOREM 1.1.2
If $\Omega \subseteq \mathbb{R}^N$ is a bounded open set with Lipschitz boundary $\partial\Omega$ and $u \in W^{1,p}(\Omega)$ with $p \in [1, +\infty)$,

<u>*then*</u> *we can find a sequence* $\{u_n\}_{n \geq 1} \subseteq W^{1,p}(\Omega) \cap C^\infty\left(\overline{\Omega}\right)$, *such that* $u_n \longrightarrow u$ *in* $W^{1,p}(\Omega)$.

The next theorem (known as the ***Trace Theorem***), for every $u \in W^{m,p}(\Omega)$ assigns a meaning to expressions like $u|_{\partial\Omega}$ and $\frac{\partial u}{\partial n}$ (the normal derivative on $\partial\Omega$). Because in general the N-dimensional Lebesgue measure of $\partial\Omega$ is zero, it is not meaningful to talk *a priori* of $u|_{\partial\Omega}$ when $u \in W^{1,p}(\Omega)$, unless u is at least continuous. So we have to generalize the meaning of boundary values for Sobolev functions.

THEOREM 1.1.3 (Trace Theorem)
If $\Omega \subseteq \mathbb{R}^N$ *is a bounded open set with Lipschitz boundary and* $p \in [1, +\infty)$, <u>*then*</u> *there exists a unique continuous linear operator*

$$\gamma_0 \colon W^{1,p}(\Omega) \longrightarrow L^p(\partial\Omega),$$

such that $\gamma_0(u) = u|_{\partial\Omega}$ *for all* $u \in C\left(\overline{\Omega}\right)$. *We say that* $\gamma_0(u)$ *is the* ***trace*** *of* $u \in W^{1,p}(\Omega)$ *on* $\partial\Omega$.

REMARK 1.1.4   For a bounded open set $\Omega \subseteq \mathbb{R}^N$ with Lipschitz boundary, we have
$$\ker \gamma_0 \;=\; W_0^{1,p}(\Omega).$$
The range of γ_0 is less than $L^p(\partial\Omega)$. There are functions $v \in L^p(\partial\Omega)$ which are not the trace of an element $u \in W^{1,p}(\Omega)$. More precisely

$$\gamma_0\left(W^{1,p}(\Omega)\right) = W^{1-\frac{1}{p},p}(\partial\Omega),$$

where $v \in W^{1-\frac{1}{p},p}(\partial\Omega)$ if and only if $v \in L^p(\partial\Omega)$ and $\|v\|_{W^{1-\frac{1}{p},p}(\partial\Omega)} < +\infty$, with

$$\|v\|_{W^{1-\frac{1}{p},p}(\partial\Omega)} \overset{df}{=} \left(\int\limits_{\partial\Omega} |v(x)|^p d\sigma(x) + \int\limits_{\partial\Omega \times \partial\Omega} \frac{|v(x) - v(x')|}{|x - x'|^{N+p-2}} d\sigma(x) d\sigma(x') \right)^{\frac{1}{p}}.$$

$\Box$

Clearly a function $u \in W_0^{1,p}(\Omega)$ can be extended by zero to a Sobolev function on all $\mathbb{R}^N$. Can we do this for any Sobolev function $u \in W^{1,p}(\Omega)$?

THEOREM 1.1.4 (Extension Theorem)
If $\Omega \subseteq \mathbb{R}^N$ *is a bounded open set and* $\partial\Omega$ *is Lipschitz,* <u>*then*</u> *there exists a bounded linear operator*

$$E \colon W^{1,p}(\Omega) \longrightarrow W^{1,p}\left(\mathbb{R}^N\right),$$

such that $E(u)|_\Omega = u$ for all $u \in W^{1,p}(\Omega)$. This operator is called **extension operator** and it is also continuous from $L^p(\Omega)$ into $L^p(\mathbb{R}^N)$.

1.1.2   Embedding Theorems

By their definition (see Definition 1.1.1), the elements of $W^{m,p}(\Omega)$ belong in the Lebesgue space $L^p(\Omega)$. Can we have more regularity properties for the Sobolev functions? This is achieved via the so-called **Sobolev Embedding Theorems**, which are one of the main tools in the analysis of boundary value problems.

Let $(X, \|\cdot\|_X)$ and $(Y, \|\cdot\|_Y)$ be two Banach spaces, such that $X \subseteq Y$. We say that the embedding $X \subseteq Y$ is **continuous** if there exists a constant $c > 0$, such that $\|x\|_Y \le c \|x\|_X$ for all $x \in X$. We say that the embedding $X \subseteq Y$ is **compact** if it is continuous and maps bounded sets into relatively compact sets.

In what follows by $C^k(\overline{\Omega})$ we denote the space of all functions $u \in C^k(\Omega)$, such that $D^\alpha u$ is bounded and uniformly continuous on Ω for all $|\alpha| \le k$ (hence it possesses a unique bounded continuous extension on $\overline{\Omega}$). This is a Banach space with the norm

$$\|u\|_{C^k(\overline{\Omega})} \stackrel{df}{=} \sum_{|\alpha| \le k} \|D^\alpha u\|_\infty .$$

The next theorem is usually called the **Sobolev Embedding Theorem**.

THEOREM 1.1.5 (Sobolev Embedding Theorem)
 If $\Omega \subseteq \mathbb{R}^N$ is a bounded open set with Lipschitz boundary, $p \in [1, +\infty)$ and $k, m \in \mathbb{N}_0$,
then

(a) *if $mp < N$, then the embedding*

$$W^{k+m,p}(\Omega) \subseteq W^{k,r}(\Omega)$$

is continuous for $r \in \left[1, \frac{Np}{N-mp}\right]$ and compact for $r \in \left[1, \frac{Np}{N-mp}\right)$.
In particular if $k = 0$, we have that the embedding

$$W^{m,p}(\Omega) \subseteq L^r(\Omega)$$

is continuous for $r \in \left[1, \frac{Np}{N-mp}\right]$ and compact for $r \in \left[1, \frac{Np}{N-mp}\right)$;

(b) *if $mp = N$, then the embedding*

$$W^{k+m,p}(\Omega) \subseteq W^{k,r}(\Omega)$$

is compact for $r \in [1, +\infty)$.
In particular if $k = 0$, we have that the embedding

$$W^{m,p}(\Omega) \subseteq L^r(\Omega)$$

is compact for $r \in [1, +\infty)$;

(c) *if $mp > N$, then the embeddings*

$$W^{k+m,p}(\Omega) \subseteq C^k(\overline{\Omega})$$

and

$$W^{k+m,p}(\Omega) \subseteq W^{k,r}(\Omega)$$

are compact for $r \in [1, +\infty]$.
In particular if $k = 0$, we have that the embedding

$$W^{m,p}(\Omega) \subseteq L^r(\Omega)$$

is compact for $r \in [1, +\infty]$.

1.1.3 Poincaré Inequality

The next result is known as the ***Poincaré inequality*** and is extremely useful in the analysis of Dirichlet boundary value problems.

THEOREM 1.1.6 (Poincaré Inequality)
If $\Omega \subseteq \mathbb{R}^N$ is a bounded open set and $p \in [1, +\infty)$,
then there exists a constant $c > 0$, such that

$$\|u\|_{L^p(\Omega)} \leq c \|Du\|_{L^p(\Omega;\mathbb{R}^N)} \qquad \forall\, u \in W_0^{1,p}(\Omega)$$

REMARK 1.1.5 This theorem implies that $\|Du\|_{L^p(\Omega;\mathbb{R}^N)}$ is an equivalent norm for the Sobolev space $W_0^{1,p}(\Omega)$. If Ω is a bounded, open and connected set (i.e. a ***bounded domain***) with Lipschitz boundary, then Theorem 1.1.6 can be generalized as follows:

"If V is a closed subspace of $W^{1,p}(\Omega)$, such that the only constant function belonging to V is the zero function, then there exists $c > 0$, such that

$$\|u\|_{L^p(\Omega)} \leq c \|Du\|_{L^p(\Omega;\mathbb{R}^N)} \qquad \forall\, u \in V."$$

For example V can be the space

$$V = \left\{ u \in W^{1,p}(\Omega) : \gamma_0(u)(x) = 0 \text{ for all } x \in \Gamma_1 \right\},$$

where Γ_1 is a subset of $\partial\Omega$ with strictly positive $(N-1)$-dimensional Hausdorff (surface) measure (see Definition A.2.3). Another possibility is to take

$$V = \left\{ u \in W^{1,p}(\Omega) : \int_\Omega u(x)dx = 0 \right\}.$$

With this choice we are led to the second basic inequality for Sobolev functions, known as the **Poincaré-Wirtinger inequality**, which is very important for periodic and Neumann problems. $\Box$

THEOREM 1.1.7 (Poincaré-Wirtinger Inequality)
  If $\Omega \subseteq \mathbb{R}^N$ *is a bounded domain and* $p \in [1, +\infty)$,
<u>*then*</u> *there exists* $c > 0$, *such that*

$$\|u - \overline{u}\|_{L^p(\Omega)} \leq c \|Du\|_{L^p(\Omega;\mathbb{R}^N)} \qquad \forall\, u \in W^{1,p}(\Omega) \text{ with } \overline{u} \stackrel{df}{=} \int_\Omega u(x)\,dz.$$

REMARK 1.1.6   By virtue of Theorem 1.1.5(a), for $p < N$ we have

$$\|u - \overline{u}\|_{L^{p^*}(\Omega)} \leq c \|Du\|_{L^p(\Omega;\mathbb{R}^N)} \qquad \forall\, u \in W^{1,p}(\Omega),$$

where

$$p^* \stackrel{df}{=} \frac{Np}{N-p}$$

is the so-called **critical Sobolev exponent**. $\Box$

1.1.4   Dual Space

By $W^{-1,p'}(\Omega)$ we denote the dual of $W_0^{1,p}(\Omega)$, i.e.

$$W^{-1,p'}(\Omega) \stackrel{df}{=} \left(W_0^{1,p}(\Omega) \right)^*,$$

where $\frac{1}{p} + \frac{1}{p'} = 1$. If $p = 2$, then we put

$$H^{-1}(\Omega) \stackrel{df}{=} \left(H_0^1(\Omega) \right)^*.$$

By the Riesz Representation Theorem (see Theorem A.3.13), we identify $L^2(\Omega)$ with its dual, but we do not identify $H_0^1(\Omega)$ with its dual. If $\Omega \subseteq \mathbb{R}^N$ is bounded and open, then the embeddings

$$W_0^{1,p}(\Omega) \subseteq L^2(\Omega) \subseteq W^{-1,p'}(\Omega)$$

are continuous provided that $\frac{2N}{N+2} \leq p < +\infty$ (see Theorem 1.1.5) and both embeddings are dense. We have the following characterization for the dual space $W^{-1,p'}(\Omega)$.

THEOREM 1.1.8

If $\Omega \subseteq \mathbb{R}^N$ *is a bounded open set and* $p \in [1, +\infty)$,
<u>*then*</u> $L \in W^{-1,p'}(\Omega)$ *if and only if there exists* $\widehat{g} = (g_1, \ldots, g_N) \in L^{p'}(\Omega; \mathbb{R}^N)$, *such that*

$$L = -\sum_{k=1}^{N} D_k g_k,$$

where $D_k \stackrel{df}{=} \dfrac{\partial}{\partial x_k}$, $k = 1, \ldots, N$, *i.e.*

$$\langle L, u \rangle_{W_0^{1,p}(\Omega)} = \sum_{k=1}^{N} \int_{\Omega} g_k D_k u \, dx \qquad \forall \, u \in W_0^{1,p}(\Omega),$$

where by $\langle \cdot, \cdot \rangle_{W_0^{1,p}(\Omega)}$ *we denote the duality brackets for the following pair* $\left(W_0^{1,p}(\Omega), W^{-1,p'}(\Omega) \right)$.

1.1.5 Green Formula

We present an identity known as the **Green Formula**. We present here Green's formula for quasilinear elliptic operators. The usual Green's formula is a special case of this. So let $\Omega \subseteq \mathbb{R}^N$ be a bounded open set with Lipschitz boundary. We consider the following space:

$$V^q(\mathrm{div}, \Omega) \stackrel{df}{=} \left\{ v \in L^q(\Omega; \mathbb{R}^N) : \mathrm{div}\, v \in L^q(\Omega) \right\}, \qquad q \in (1, +\infty).$$

On $V^q(\mathrm{div}, \Omega)$ we consider the norm

$$\|v\|_{V^q(\mathrm{div}, \Omega)} \stackrel{df}{=} \left(\|v\|_{L^q(\Omega; \mathbb{R}^N)}^q + \|\mathrm{div}\, v\|_{L^q(\Omega)}^q \right)^{\frac{1}{q}}.$$

Furnished with this norm, $V^q(\mathrm{div}, \Omega)$ becomes a separable reflexive Banach space and $C_c^\infty(\Omega; \mathbb{R}^N)$ is dense in it. In Remark 1.1.4, we introduced the "boundary" Sobolev space $W^{\frac{1}{p'}, p}(\partial\Omega) = W^{1-\frac{1}{p}, p}(\partial\Omega)$ by $W^{\frac{1}{p'}, p}(\partial\Omega) \stackrel{df}{=} \gamma_0\left(W^{1,p}(\Omega) \right)$, with γ_0 being the trace operator. We denote by $W^{-\frac{1}{p'}, p'}(\partial\Omega)$ the dual of the space $W^{\frac{1}{p'}, p}(\partial\Omega)$ and by $\langle \cdot, \cdot \rangle_{\partial\Omega}$ the duality brackets for the pair $\left(W^{\frac{1}{p'}, p}(\partial\Omega), W^{-\frac{1}{p'}, p'}(\partial\Omega) \right)$.

THEOREM 1.1.9 (Green Formula)

If $\Omega \subseteq \mathbb{R}^N$ *is a bounded open set with Lipschitz boundary and* $p \in (1, +\infty)$, <u>*then*</u> *there exists a unique continuous linear operator*

$$\gamma_n \colon V^{p'}(\mathrm{div}, \Omega) \longrightarrow W^{-\frac{1}{p'}, p'}(\partial\Omega),$$

such that

$$\gamma_n(v) = (v, n)_{L^p(\partial\Omega)} \qquad \forall\, v \in C_c^\infty(\Omega; \mathbb{R}^N)$$

and

$$\int_\Omega (Du(x), v(x))_{\mathbb{R}^N}\, dx + \int_\Omega u(x) \operatorname{div} v(x) dx$$

$$= \langle \gamma_0(u), \gamma_n(v) \rangle_{\partial\Omega} \qquad \forall\, u \in W^{1,p}(\Omega),\ v \in V^{p'}(\operatorname{div}, \Omega).$$

REMARK 1.1.7   If $p = p' = 2$ and $v = Dy$ for some $y \in H_0^1(\Omega)$, such that $\Delta y \in L^2(\Omega)$, then we recover the usual Green formula. ⬚

1.1.6   One Dimensional Sobolev Spaces

Next let us specialize to the case where $N = 1$. In this case we take $\Omega = T$ to be an open interval in $\mathbb{R}$. The following result distinguishes the one-dimensional ($N = 1$) case from the multidimensional one ($N > 1$).

THEOREM 1.1.10
If T is an open interval in $\mathbb{R}$ and $1 \leq p \leq +\infty$,
then $u \in W^{1,p}(T)$ if and only if $u \in L^p(T)$, u is absolutely continuous and $\frac{du}{dt} \in L^p(T)$, where by $\frac{du}{dt}$ we denote the classical derivative of u (which by the Lebesgue Theorem exists for almost all $t \in T$; see Theorem A.2.3).

From this theorem it follows that for every open interval $T \subseteq \mathbb{R}$ and every $1 \leq p \leq +\infty$, we have $W^{1,p}(T) \subseteq C(\overline{T})$. More precisely we have the next theorem.

THEOREM 1.1.11
Let $T \subseteq \mathbb{R}$ be an open interval.

(a) *The embedding*
$$W^{1,p}(T) \subseteq L^\infty(T)$$
is continuous for all $1 \leq p \leq +\infty$;

(b) *If T is a bounded open interval,*
then the embedding
$$W^{1,p}(T) \subseteq C(\overline{T})$$
is compact for all $1 < p \leq +\infty$;

(c) *If T is a bounded open interval,*
then the embedding
$$W^{1,1}(T) \subseteq L^r(T)$$
is compact for all $1 \leq r < +\infty$.

REMARK 1.1.8   The embedding

$$W^{1,1}(T) \subseteq C(\overline{T})$$

is always continuous but never compact, even if T is bounded. Using Theorem 1.1.11(a) and the fact that $C_c^\infty(\mathbb{R})|_T \cap W^{1,p}(T)$ is dense in $W^{1,p}(T)$ for $p \in [1, +\infty)$, we can show that if T is not bounded and $u \in W^{1,p}(T)$ with $p \in [1, +\infty)$, then

$$\lim_{\substack{t \in T \\ |t| \to +\infty}} u(t) = 0.$$

Note that the just mentioned density result is not generally true for the multidimensional case. More precisely, if $\Omega \subseteq \mathbb{R}^N$ ($N > 1$) is an open set, $p \in [1, +\infty)$ and $u \in W^{1,p}(\Omega)$, we can find $\{u_n\}_{n \geq 1} \subseteq C_c^\infty(\mathbb{R}^N)$, such that $u_n|_\Omega \longrightarrow u$ in $L^p(\Omega)$ and $D(u_n|_{\Omega'}) \longrightarrow D(u|_{\Omega'})$ in $L^p(\Omega'; \mathbb{R}^N)$ for all $\Omega' \subset\subset \Omega$. ∎

In the one-dimensional case, the ***Poincaré inequality*** (see Theorem 1.1.6 and Remark 1.1.5) takes the following form.

THEOREM 1.1.12 (One Dimensional Poincaré Inequality)
If $T = (a, b)$ *is a bounded open interval,* $1 \leq p \leq +\infty$, $t_0 \in [a, b]$ *and*

$$V \overset{df}{=} \left\{ v \in W^{1,p}(T) : \ v(t_0) = 0 \right\},$$

then there exists $c > 0$, *such that*

$$\|v\|_p \leq c \|v'\|_p \qquad \forall \, v \in V.$$

REMARK 1.1.9   If $T = (0, b)$ and

$$V \overset{df}{=} \left\{ v \in W^{1,p}(T) : \ \int_0^b v(t) \, dt = 0 \right\},$$

then

$$\|v\|_\infty \ \leq \ b^{\frac{1}{p'}} \|v'\|_p \quad \text{for a.a. } v \in V.$$

If $p = 2$, then

$$\|v\|_2^2 \ \leq \ \frac{b^2}{4\pi^2} \|v'\|_\infty^2 \quad \text{for a.a. } v \in V$$

and

$$\|v\|_\infty^2 \ \leq \ \frac{b}{12} \|v'\|_2^2 \quad \text{for a.a. } v \in V.$$

∎

Returning to the general case ($N \geq 1$), let us mention the following powerful chain rule for Sobolev functions.

THEOREM 1.1.13 (Chain Rule for Sobolev Functions)
If $\Omega \subseteq \mathbb{R}^N$ is an open set, $f \colon \mathbb{R} \longrightarrow \mathbb{R}$ is a Lipschitz continuous function and $u \in W^{1,p}(\Omega)$, with $p \in [1, +\infty)$,
then

(a) $f \circ u \in W^{1,p}(\Omega)$;

(b) $D(f \circ u)(x) = \widehat{f}(u(x)) Du(x)$ *for almost all $x \in \Omega$, where $\widehat{f} \colon \mathbb{R} \longrightarrow \mathbb{R}$ is any Borel function, such that $\widehat{f} = f'$ almost everywhere on $\mathbb{R}$;*

(c) *the **Nemytskii operator** $N_f \colon W^{1,p}(\Omega) \longrightarrow W^{1,p}(\Omega)$, defined by*

$$N_f(u)(\cdot) \stackrel{df}{=} f(u(\cdot))$$

is continuous.

REMARK 1.1.10   If $u \in W^{1,p}(\Omega)$, $f \colon \mathbb{R} \longrightarrow \mathbb{R}$ is a locally Lipschitz function and C is a Borel null subset of $\mathbb{R}$, then $Du(x) = 0$ for almost all $x \in f^{-}(C)$. So the chain rule in Theorem 1.1.13 is not affected if $\widehat{f}$ is modified on a null-set. Hence we can always assume that $\widehat{f}$ is a bounded function. As a consequence of the above chain rule we have that if $u \in W^{1,p}(\Omega)$ (respectively $u \in W_0^{1,p}(\Omega)$) then $u^+, u^-, |u| \in W^{1,p}(\Omega)$ (respectively $u^+, u^-, |u| \in W_0^{1,p}(\Omega)$), for $p \in [1, +\infty)$. This is not true for higher order Sobolev spaces. Moreover, we have

$$Du^+(x) = \begin{cases} Du(x) & \text{for a.a. } x \in \{u > 0\}, \\ 0 & \text{for a.a. } x \in \{u \leq 0\}, \end{cases}$$

$$Du^-(x) = \begin{cases} 0 & \text{for a.a. } x \in \{u \geq 0\}, \\ -Du(x) & \text{for a.a. } x \in \{u < 0\} \end{cases}$$

and

$$D|u|(x) = \begin{cases} Du(x) & \text{for a.a. } x \in \{u > 0\}, \\ 0 & \text{for a.a. } x \in \{u = 0\}, \\ -Du(x) & \text{for a.a. } x \in \{u < 0\}. \end{cases}$$

$\square$

Let us conclude this section by giving some equivalent norms for the Sobolev spaces $W^{1,p}(\Omega)$.

THEOREM 1.1.14

If $\Omega \subseteq \mathbb{R}^N$ *is a bounded domain (i.e. a bounded open connected set) with Lipschitz boundary* $\partial\Omega$ *and* $p \in [1, +\infty)$,
then the following are equivalent norms for the Sobolev spaces $W^{1,p}(\Omega)$:

(a)

$$\|u\| = \|Du\|_p + \left| \int_\Omega u(x)dx \right|$$

or

$$\|u\| = \left(\|Du\|_p^p + \left| \int_\Omega u(x)dx \right|^p \right)^{\frac{1}{p}};$$

(b)

$$\|u\| = \|Du\|_p + \left| \int_{\partial\Omega} u(x)d\sigma \right|$$

or

$$\|u\| = \left(\|Du\|_p^p + \left| \int_{\partial\Omega} u(x)d\sigma \right|^p \right)^{\frac{1}{p}};$$

(c)

$$\|u\| = \|Du\|_p + \|u\|_{L^p(\partial\Omega)}$$

or

$$\|u\| = \left(\|Du\|_p^p + \|u\|_{L^p(\partial\Omega)}^p \right)^{\frac{1}{p}};$$

(d)

$$\|u\| = \|Du\|_p + \|u\|_r,$$

with r, such that

$$\begin{cases} 1 \le r \le p^* & \text{if } p < N, \\ 1 \le r < +\infty & \text{if } p = N, \\ 1 \le r \le +\infty & \text{if } p > N. \end{cases}$$

If $N = 1$ *and* $\Omega = (a, b)$, *with* $-\infty < a < b < +\infty$, *then in* **(b)** *we put*

$$\int_{\partial\Omega} u(x)d\sigma = u(a) + u(b)$$

and **(d)** *becomes*
(d)'

$$\|u\| = \|u'\|_p + \|u\|_r \qquad \text{with any } 1 \le r \le +\infty.$$

REMARK 1.1.11   For higher order Sobolev spaces $W^{m,p}(\Omega)$, **(d)** becomes

$$\|u\| = \sum_{|\alpha| \le m} \|D^\alpha u\|_p + \|u\|_r,$$

with the same restriction on r. $\square$

The next proposition provides a simple characterization of functions in $W^{1,p}(\Omega)$.

PROPOSITION 1.1.1
If $p \in (1, +\infty]$ *and* $u \in L^p(\Omega)$,
then the following properties are equivalent:

(a) $u \in W^{1,p}(\Omega)$;

(b) $\int_{\Omega} (u D_i \vartheta) \, dz \leq c \|\vartheta\|_{p'}$ *for some* $c > 0$ *and all* $\vartheta \in C_c(\overline{\Omega})$, $i \in \{1, \ldots, N\}$.

We state a **change of variables formula** useful in the study of boundary value problems.

THEOREM 1.1.15 (Change of variables)
If $f \colon [0, b] \longrightarrow [c, d]$ *is absolutely continuous,* $g \in L^1(c, d)$ *and* $(g \circ f) f' \in L^1(0, b)$,
then

$$\int_{f(0)}^{f(b)} g(t) \, dt \;=\; \int_0^b g\big(f(t)\big) f'(t) \, dt.$$

1.2   Set-Valued Analysis

The aim of this section is to give some basic definitions and results from Set-Valued Anàlysis (Multivalued Analysis). More precisely we present certain basic results about the continuity and the measurability of multifunctions (set-valued functions) and also state the main results on the existence of continuous and measurable selections. We also discuss decomposable sets and mention a few basic things about limits of sequences of sets. Again the results are presented without proofs. This section is based on the books of Aubin & Frankowska (1990), Hu & Papageorgiou (1997, 2000), Kisielewicz (1991) and Klein & Thompson (1984), where the reader can find the proofs and an in depth discussion of these and related issues.

First let us fix our notation. For a Hausdorff topological space X we introduce the following spaces:

$$P_f(X) \overset{df}{=} \{A \subseteq X : \ A \text{ is nonempty and closed}\},$$
$$P_k(X) \overset{df}{=} \{A \subseteq X : \ A \text{ is nonempty and compact}\}.$$

If X is a normed space, we can also consider the following spaces:

$$P_{fc}(X) \stackrel{df}{=} \{A \subseteq X : A \text{ is nonempty, closed and convex}\},$$
$$P_{kc}(X) \stackrel{df}{=} \{A \subseteq X : A \text{ is nonempty, compact and convex}\},$$
$$P_{wkc}(X) \stackrel{df}{=} \{A \subseteq X : A \text{ is nonempty, weakly compact and convex}\},$$
$$P_{bf(c)}(X) \stackrel{df}{=} \{A \subseteq X : A \text{ is nonempty, bounded, closed (and convex)}\}.$$

For a Hausdorff topological space X and $x \in X$, by $\mathcal{N}(x)$ we denote the filter of neighbourhoods of x (see Definition A.1.2(a)). Also, if (X, d_x) is a metric space and $x \in X$, $r > 0$, then by $B_r(x)$ (or $B_r^X(x)$) we denote the **open r-ball in X centered at x**, i.e.

$$B_r(x) \stackrel{df}{=} \{y \in X : d_x(x,y) < r\}.$$

In addition, if X is a normed space, B_r denotes the open r-ball centered at the origin, i.e.

$$B_r = \{y \in X : \|y\|_X < r\}.$$

For sets X, Y, a multifunction $F: X \longrightarrow 2^Y$ and a subset $A \subseteq Y$, we define

$$F^+(A) \stackrel{df}{=} \{x \in X : F(x) \subseteq A\},$$
$$F^-(A) \stackrel{df}{=} \{x \in X : F(x) \cap A \neq \emptyset\}.$$

Of course $F^+(A) \subseteq F^-(A) \subseteq X$.

1.2.1 Upper and Lower Semicontinuity

In what follows X, Y are Hausdorff topological spaces.

DEFINITION 1.2.1   *Let $F: X \longrightarrow 2^Y$ be a multifunction.*

(a) *We say that F is **upper semicontinuous at x_0**, if for any open subset $V \subseteq Y$ with $F(x_0) \subseteq V$, there exists $U \in \mathcal{N}(x_0)$, such that $F(U) \subseteq V$. If F is upper semicontinuous at every $x_0 \in X$, we say that F is **upper semicontinuous**. In the sequel we shall use the abbreviation **usc**.*

(b) *We say that F is **lower semicontinuous at x_0**, if for any open subset $V \subseteq Y$ with $F(x_0) \cap V \neq \emptyset$, there exists $U \in \mathcal{N}(x_0)$, such that $F(x) \cap V \neq \emptyset$ for all $x \in U$. If F is lower semicontinuous at every $x_0 \in X$, we say that F is **lower semicontinuous**. In the sequel we shall use the abbreviation **lsc**.*

(c) *We say that F is **continuous** (or **Vietoris continuous**) **at x_0**, if it is both upper semicontinuous and lower semicontinuous at x_0. If F is continuous at every $x_0 \in X$, we say that F is **continuous** (or **Vietoris continuous**).*

Directly from these definitions we get some alternative equivalent descriptions of these notions.

PROPOSITION 1.2.1
For a multifunction $F \colon X \longrightarrow 2^Y$ the following statements are equivalent:

(a) *F is upper semicontinuous;*

(b) *for every closed set $C \subseteq Y$, the set $F^-(C)$ is closed in X;*

(c) *if $x \in X$, $\{x_\alpha\}_{\alpha \in J}$ is a net in X (see Definition A.1.5(b)), $x_\alpha \longrightarrow x$, $V \subseteq Y$ is open such that $F(x) \subseteq V$, then there exists $\alpha_0 \in J$, such that for all $\alpha \in J$, $\alpha \geq \alpha_0$ we have $F(x_\alpha) \subseteq V$.*

PROPOSITION 1.2.2
For a multifunction $F \colon X \longrightarrow 2^Y$ the following statements are equivalent:

(a) *F is lower semicontinuous;*

(b) *for every closed set $C \subseteq Y$, the set $F^+(C)$ is closed in X;*

(c) *if $x \in X$, $\{x_\alpha\}_{\alpha \in J}$ is a net in X, $x_\alpha \longrightarrow x$, $V \subseteq Y$ is open such that $F(x) \cap V \neq \emptyset$, then there exists $\alpha_0 \in J$, such that for all $\alpha \in J$, $\alpha \geq \alpha_0$ we have $F(x_\alpha) \cap V \neq \emptyset$;*

(d) *if $x \in X$, $\{x_\alpha\}_{\alpha \in J}$ is a net in X, $x_\alpha \longrightarrow x$ and $y \in F(x)$, then for every $\alpha \in J$ we can find $y_\alpha \in F(x_\alpha)$, such that $y_\alpha \longrightarrow y$.*

PROPOSITION 1.2.3
For a multifunction $F \colon X \longrightarrow 2^Y$ the following statements are equivalent:

(a) *F is continuous;*

(b) *for every closed set $C \subseteq Y$, the sets $F^+(C)$ and $F^-(C)$ are closed in X;*

(c) *if $x \in X$, $\{x_\alpha\}_{\alpha \in J}$ is a net in X, $x_\alpha \longrightarrow x$, $V, W \subseteq Y$ are open such that $F(x) \subseteq V$ and $F(x) \cap W \neq \emptyset$, then there exists $\alpha_0 \in J$, such that for all $\alpha \in J$, $\alpha \geq \alpha_0$ we have $F(x_\alpha) \subseteq V$ and $F(x_\alpha) \cap W \neq \emptyset$.*

DEFINITION 1.2.2   *For a given multifunction $F \colon X \longrightarrow 2^Y$, by the* **graph** *of F we mean the set*

$$\operatorname{Gr} F \overset{df}{=} \{(x, y) \in X \times Y : y \in F(x)\}.$$

We say that F is **closed**, *if $\operatorname{Gr} F$ is closed in $X \times Y$.*

PROPOSITION 1.2.4
If Y is a regular topological space (see Definition A.1.2(b)) and the multi-

function $F\colon X \longrightarrow P_f(Y)$ *is upper semicontinuous,*
then F is closed.

REMARK 1.2.1   If F is $P_k(Y)$-valued, then we can drop the hypothesis
that Y is a regular topological space. Also if $F\colon X \longrightarrow P_k(Y)$ is upper
semicontinuous and $K \subseteq X$ is compact, then $F(K)$ is compact.   ▯

The converse of Proposition 1.2.4 is not in general true. We need to impose
additional conditions on F.

PROPOSITION 1.2.5
If $F\colon X \longrightarrow P_f(Y)$ *is closed and locally compact (see Definition A.4.3),*
then F is upper semicontinuous.

If (X, d_X) is a metric space and $A \subseteq X$, for every $x \in X$ we define

$$d_X(x, A) \stackrel{df}{=} \inf_{a \in A} d_X(x, a)$$

(as usual we adopt the convention $\inf_{\emptyset} = +\infty$). The distance function $X \ni$
$x \longmapsto d_X(x, A) \in \mathbb{R}$ is a contraction (see Definition A.1.13).

DEFINITION 1.2.3   *If X is a normed space (see Definition A.3.1(c)),*
then for every $x^* \in X^*$ *we define*

$$\sigma_X(x^*, A) \stackrel{df}{=} \sup_{a \in A} \langle x^*, a \rangle_X$$

(here by $\langle \cdot, \cdot \rangle_X$ *we denote the duality brackets for the pair* (X, X^*)*). The func-*
tion $\sigma_X(\cdot, A)\colon X^* \longrightarrow \mathbb{R}^* \stackrel{df}{=} \mathbb{R} \cup \{\pm\infty\}$ *is known as the* **support function**
of the set A.

PROPOSITION 1.2.6
Let $F\colon X \longrightarrow 2^Y \setminus \{\emptyset\}$ *be a multifunction.*

(a) *If Y is a metric space,*
    then F is lower semicontinuous if and only if for all $y \in Y$*, the function*
    $x \longmapsto d_Y(y, F(x))$ *is upper semicontinuous;*

(b) *If Y is a metric space and F is upper semicontinuous*
    then for all $y \in Y$*, the function* $x \longmapsto d_Y(y, F(x))$ *is lower semicontinu-*
    ous. The converse is true if F is locally compact (see Definition A.4.3);

(c) *If Y is a normed space furnished with the weak topology and F is upper*
    semicontinuous,

<u>then</u> for all $y^* \in Y^*$, the function $x \longmapsto \sigma_Y(y^*, F(x))$ is upper semicontinuous;

Let X, Y be Hausdorff topological spaces, $u \colon X \times Y \longrightarrow \mathbb{R}^*$ and $F \colon Y \longrightarrow 2^X \setminus \{\emptyset\}$. We consider the following parametric optimization problem:

$$\sup_{x \in F(y)} u(x, y) = v(y).$$

Also let $S(y)$ be the **solution multifunction**, defined by

$$S(y) \stackrel{df}{=} \{ x \in F(y) \colon u(x, y) = v(y) \}.$$

PROPOSITION 1.2.7 (Berge Maximum Theorem)
Let $X, Y, F, v, S(y)$ be as above.

(a) <u>*If*</u> *u is a lower semicontinuous function and F is a lower semicontinuous multifunction,*
<u>*then*</u> *the function $v \colon Y \longrightarrow \overline{\mathbb{R}} \stackrel{df}{=} \mathbb{R} \cup \{+\infty\}$ is lower semicontinuous.*

(b) <u>*If*</u> *u is an upper semicontinuous function and F is an upper semicontinuous and $P_k(X)$-valued multifunction,*
<u>*then*</u> *the function $v \colon Y \longrightarrow \overline{\mathbb{R}}$ is upper semicontinuous.*

(c) <u>*If*</u> *$u \colon X \times Y \longrightarrow \mathbb{R}$ is a continuous function and F is a continuous and $P_k(X)$-valued multifunction,*
<u>*then*</u> *the function $v \colon Y \longrightarrow \mathbb{R}$ is continuous and the multifunction $S \colon Y \longrightarrow P_k(X)$ is upper semicontinuous.*

1.2.2 h-Lower and h-Upper Semicontinuity

When X is a metric space, then we can introduce a metric structure on certain subspaces of 2^X. To do this we introduce the following quantities.

DEFINITION 1.2.4   *Let (X, d_X) be a metric space and $A, C \in 2^X$. We define*

$$h_X^*(A, C) \stackrel{df}{=} \sup \{ d_X(a, C) \colon a \in A \} \qquad \text{(the "excess" of A over C)}.$$

*The **Hausdorff distance** of A and C, we define by:*

$$h_X(A, C) \stackrel{df}{=} \max \{ h_X^*(A, C), h_X^*(C, A) \}.$$

REMARK 1.2.2   It is easy to see that $h_X(A, C) = 0$ if and only if $\overline{A} = \overline{C}$ and so $(P_f(X) \cup \{\emptyset\}, h_X)$ is a (generalized) metric space. The empty set is an isolated point in this metric space. We call h_X the **Hausdorff metric**. We have

$$h_X(A, C) = \sup\left\{ \left| d_X(x, A) - d_X(x, C) \right| : x \in X \right\}$$

and if X is a normed space and $A, C \in P_{bfc}(X)$, then the following **Hörmander's formula** holds:

$$h_X(A, C) = \sup\left\{ \left| \sigma_X(x^*, A) - \sigma_X(x^*, C) \right| : x^* \in X^*, \|x^*\|_{X^*} \leq 1 \right\}.$$

If X is a complete metric space, then so is $(P_f(X), h_X)$ and both spaces $(P_{bf}(X), h_X)$ and $(P_k(X), h_X)$ are closed subsets of it. Moreover, if X is a normed space, then $P_{kc}(X) \subseteq P_{bfc}(X) \subseteq P_{fc}(X)$ and all are closed subspaces of $(P_f(X), h_X)$. $\quad\Box$

DEFINITION 1.2.5 *Let X be a Hausdorff topological space, Y a metric space and $F: X \longrightarrow 2^Y \setminus \{\emptyset\}$ a multifunction.*

(a) *We say that F is h-**upper semicontinuous at** $x_0 \in X$, if the function $X \ni x \longmapsto h_X^*(F(x), F(x_0)) \in \mathbb{R}_+$ is continuous at x_0. If F is h-upper semicontinuous at every $x \in X$, we say that F is h-**upper semicontinuous**. In the sequel we shall use the abbreviation h-**usc**.*

(b) *We say that F is h-**lower semicontinuous at** $x_0 \in X$, if the function $X \ni x \longmapsto h_X^*(F(x_0), F(x)) \in \mathbb{R}_+$ is continuous at x_0. If F is h-lower semicontinuous at every $x \in X$, we say that F is h-**lower semicontinuous**. In the sequel we shall use the abbreviation h-**lsc**.*

(c) *We say that F is h-**continuous at** $x_0 \in X$, if it is both h-upper semicontinuous and h-lower semicontinuous at x_0. If F is h-continuous at every $x \in X$, we say that F is h-**continuous**.*

A natural question that arises is how are the notions of Definition 1.2.5 related to those introduced in Definition 1.2.1. The next proposition gives an answer to this problem.

PROPOSITION 1.2.8
Let X be a Hausdorff topological space, Y a metric space and let $F: X \longrightarrow 2^Y \setminus \{\emptyset\}$ be a multifunction.

(a) *If F is upper semicontinuous, then F is h-upper semicontinuous;*

(b) *If F is h-lower semicontinuous, then F is lower semicontinuous;*

(c) *If F is $P_k(Y)$-valued, then the converse of (a) and (b) hold. So $F: X \longrightarrow P_k(Y)$ is continuous if and only if it is h-continuous.*

Before passing to the measurability of multifunctions, let us see how these continuity notions behave with respect to some basic operations.

PROPOSITION 1.2.9

Let X, Y be Hausdorff topological spaces and let $F, F_1, F_2 \colon X \longrightarrow 2^Y \setminus \{\emptyset\}$ be multifunctions.

(a) *F is lower semicontinuous if and only if $\overline{F}$ is lower semicontinuous, where $\overline{F}$ denotes the multifunction $X \ni x \longmapsto \overline{F(x)} \in 2^Y \setminus \{\emptyset\}$;*

(b) *If Y is normal and F is upper semicontinuous, then $\overline{F}$ is upper semicontinuous;*

(c) *If F_1 is lower semicontinuous, F_2 has open graph and*

$$F_1(x) \cap F_2(x) \neq \emptyset \qquad \forall \, x \in X,$$

then the multifunction $X \ni x \longmapsto (F_1 \cap F_2)(x) \stackrel{df}{=} F_1(x) \cap F_2(x) \in 2^Y \setminus \{\emptyset\}$ is lower semicontinuous;

(d) *If Y is normal, $F_1, F_2 \colon X \longrightarrow P_f(Y)$ are upper semicontinuous and*

$$F_1(x) \cap F_2(x) \neq \emptyset \qquad \forall \, x \in X,$$

then the multifunction $X \ni x \longmapsto (F_1 \cap F_2)(x) \in 2^Y \setminus \{\emptyset\}$ is upper semicontinuous;

(e) *If Y is a normed space and F is lower semicontinuous, then so are the multifunctions $X \ni x \longmapsto \operatorname{conv} F(x) \in 2^Y \setminus \{\emptyset\}$ and $X \ni x \longmapsto \overline{\operatorname{conv}}\, F(x) \in 2^Y \setminus \{\emptyset\}$;*

(f) *If Y is a Banach space and $F \colon X \longrightarrow P_k(Y)$ is upper semicontinuous, then so is $X \ni x \longmapsto \overline{\operatorname{conv}}\, F(x) \in 2^Y \setminus \{\emptyset\}$;*

(g) *If Y is a metric space, then F is h-lower semicontinuous (respectively h-upper semicontinuous, h-continuous) if and only if $\overline{F}$ is;*

(h) *If Y is a normed space and F is h-lower semicontinuous (respectively h-upper semicontinuous, h-continuous) then so are the multifunctions $X \ni x \longmapsto \operatorname{conv} F(x) \in 2^Y \setminus \{\emptyset\}$ and $X \ni x \longmapsto \overline{\operatorname{conv}}\, F(x) \in 2^Y \setminus \{\emptyset\}$;*

(i) *If Y is a normed space, $F_1, F_2 \colon X \longrightarrow P_{bfc}(Y)$ are h-lower semicontinuous and*

$$\operatorname{int}\big(F_1(x) \cap F_2(x)\big) \neq \emptyset \qquad \forall \, x \in X,$$

then $X \ni x \longmapsto (F_1 \cap F_2)(x) \in 2^Y \setminus \{\emptyset\}$ is h-lower semicontinuous;

(j) *If $F_1 \colon X \longrightarrow P_f(Y)$ and $F_2 \colon X \longrightarrow P_k(Y)$ are h-upper semicontinuous and*

$$F_1(x) \cap F_2(x) \neq \emptyset \qquad \forall \, x \in X,$$

then $X \ni x \longmapsto (F_1 \cap F_2)(x) \in 2^Y \setminus \{\emptyset\}$ *is h-upper semicontinuous (in fact upper semicontinuous because it is $P_k(Y)$-valued).*

1.2.3 Measurability of Multifunctions

We can pass to the measurability of multifunctions.

DEFINITION 1.2.6 *Let (Ω, Σ) be a measurable space and let (X, d_X) be a separable metric space (see Definition A.1.4(e)). For a multifunction $F: \Omega \longrightarrow 2^X$ we say that:*

(a) *F is **measurable** if and only if for any open set $U \subseteq X$, we have*

$$F^-(U) \overset{df}{=} \{\omega \in \Omega : F(\omega) \cap U \neq \emptyset\} \in \Sigma;$$

(b) *F is **graph measurable** if and only if*

$$\mathrm{Gr}\, F \overset{df}{=} \{(\omega, x) \in \Omega \times X : x \in F(\omega)\} \in \Sigma \times \mathcal{B}(X),$$

with $\mathcal{B}(X)$ being the Borel σ-field of X;

(c) *F is **scalarly measurable** if and only if for every $x^* \in X^*$, the function $\Omega \ni \omega \longmapsto \sigma_X(x^*, F(\omega)) \in \mathbb{R}$ is measurable.*

The next theorem summarizes the situation about the measurability of multifunctions.

THEOREM 1.2.1
Let (Ω, Σ) be a measurable space, X a separable Banach space and $F: \Omega \longrightarrow P_f(X)$. If we consider the following properties:

(i) for every $D \in \mathcal{B}(X)$, $F^-(D) \in \Sigma$;

(ii) for every closed $C \subseteq X$, $F^-(C) \in \Sigma$;

(iii) F is measurable;

(iv) for every $x \in X$, the $\mathbb{R}_+$-valued function $\omega \longmapsto d_X(x, F(\omega))$ is Σ-measurable;

(v) F is graph measurable,

then the following implications hold among the above properties:

(a) *(i) $\implies$ (ii) $\implies$ (iii) $\iff$ (iv) $\implies$ (v);*

(b) *if X is σ-compact (see Definition A.1.9(b)), then (ii) $\iff$ (iii);*

(c) *if* $\Sigma = \widehat{\Sigma}$ *(where $\widehat{\Sigma}$ is the universal σ-field; i.e. Σ is complete) and X is complete (i.e. X is a Polish space; see Definition A.1.12(b)), then all conditions from (i) to (v) are equivalent.*

REMARK 1.2.3   Recall that the ***universal σ-field*** $\widehat{\Sigma}$ corresponding to Σ is defined by

$$\widehat{\Sigma} \overset{df}{=} \bigcap_{\mu} \overline{\Sigma}_{\mu},$$

where μ ranges over all finite measures on Σ and $\overline{\Sigma}_{\mu}$ denotes the μ-completion of Σ. If μ is a σ-finite measure on (Ω, Σ), then there exists a finite measure with the same null-sets. That is why it suffices to take μ to be a finite measure. Finally, if μ is a σ-finite measure on (Ω, Σ) and Σ is a μ-complete σ-field, then $\Sigma = \widehat{\Sigma}$.   ☐

1.2.4   Measurable Selections

DEFINITION 1.2.7   *For a given multifunction $F: \Omega \longrightarrow 2^X \setminus \{\emptyset\}$, we say that a function $f: \Omega \longrightarrow X$ is a **measurable selection** or **measurable selector** of F, if f is Σ-measurable and*

$$f(\omega) \in F(\omega) \qquad \forall \; \omega \in \Omega.$$

There are two basic theorems concerning the existence of measurable selections.

THEOREM 1.2.2 (Kuratowski-Ryll Nardzewski Selection Theorem)
If (Ω, Σ) *is a measurable space, X is a Polish space (Definition A.1.12(b)) and $F: \Omega \longrightarrow P_f(X)$ is measurable,*
then F admits a measurable selection.

REMARK 1.2.4   In fact it can be shown that $F: \Omega \longrightarrow P_f(X)$ is measurable if and only if there exists a sequence $\{f_n\}_{n \geq 1}$ of measurable selections of F, such that
$$F(\omega) = \overline{\{f_n(\omega)\}}_{n \geq 1} \qquad \forall \, \omega \in \Omega.$$

☐

The second theorem on the existence of measurable selections is graph conditioned and is known as the Yankov-von Neumann-Aumann Selection Theorem. First let us give a definition.

DEFINITION 1.2.8   *A Hausdorff topological space X is said to be a **Souslin space**, if there exists a Polish space Y and a continuous surjection from Y to X.*

REMARK 1.2.5   Evidently a Souslin space is always separable but need not be metrizable. For example an infinite dimensional separable Banach space furnished with the weak topology, denoted by X_w, is a nonmetrizable Souslin space. Closed, open subsets of a Souslin space are Souslin and so are countable products, countable unions and countable intersections of Souslin spaces. In particular then if X is an infinite dimensional separable Banach space and $X_{w^*}^*$ is its dual equipped with the w^*-topology, then $X_{w^*}^*$ is another nonmetrizable Souslin space. Finally the continuous image of a Souslin space is Souslin space.   ⬚

THEOREM 1.2.3 (Yankov-von Neumann-Aumann Selection Theorem)
If (Ω, Σ, μ) is a complete measure space, X is a Souslin space (see Definition 1.2.8) and $F \colon \Omega \longrightarrow 2^X \setminus \{\emptyset\}$ is a multifunction, such that $\mathrm{Gr}\, F \in \Sigma \times \mathcal{B}(X)$,
then F admits a measurable selection.

REMARK 1.2.6   If the σ-field Σ is not μ-complete, then we can say that there exists a Σ-measurable function $f \colon \Omega \longrightarrow X$, such that

$$f(\omega) \in F(\omega) \quad \text{for } \mu\text{-a.a. } \omega \in \Omega.$$

As was the case with the Kuratowski-Ryll Nardzewski Selection Theorem (see Remark 1.2.4), we can conclude the existence of a whole sequence of measurable selections $\{f_n\}_{n \geq 1}$ of F, such that

$$F(\omega) = \overline{\bigcup_{n \geq 1} \{f_n(\omega)\}} \quad \forall\, \omega \in \Omega.$$

⬚

We can describe the measurability of a multifunction using its support function.

THEOREM 1.2.4
If (Ω, Σ) is a measurable space, X is a separable Banach space and $F \colon \Omega \longrightarrow P_{wkc}(X)$,
then F is measurable if and only if it is scalarly measurable.

1.2.5   Continuous Selections

What about continuous selections of a multifunction $F \colon X \longmapsto 2^Y \setminus \{\emptyset\}$ (where X, Y are two Hausdorff topological spaces)? So we are looking for a continuous function $f \colon X \longrightarrow Y$, such that

$$f(x) \in F(x) \qquad \forall\, x \in X.$$

The basic result in this direction is the celebrated **Michael Selection Theorem**.

THEOREM 1.2.5 (Michael Selection Theorem)
If X is a paracompact space (see Definition A.1.10(b)), Y is a Banach space and $F \colon X \longrightarrow P_{fc}(Y)$ is lower semicontinuous,
<u>*then*</u> *F admits a continuous selection.*

REMARK 1.2.7   If X is a metric space and Y is a separable Banach space, then as with the measurable selections, we can find a whole sequence $\{f_n\}_{n \geq 1}$ of continuous selections, such that

$$F(x) = \overline{\bigcup_{n \geq 1} \{f_n(x)\}} \qquad \forall\, x \in X.$$

Also if X and Y are as above, $F \colon X \longrightarrow 2^Y$ is a lower semicontinuous multifunction with nonempty convex values such that $\operatorname{int} F(x) \neq \emptyset$ for all $x \in X$, then we can find a continuous map $f \colon X \longrightarrow Y$, such that

$$f(x) \in \operatorname{int} F(x) \qquad \forall\, x \in X.$$

$\square$

Simple counterexamples in $\mathbb{R}$ show that Theorem 1.2.5 fails if F is upper semicontinuous instead of lower semicontinuous. In that case we can speak of continuous ε-approximate selections.

THEOREM 1.2.6
If X is a metric space, Y is a Banach space and $F \colon X \longrightarrow 2^Y \setminus \{\emptyset\}$ is upper semicontinuous with convex values,
<u>*then*</u> *for any given $\varepsilon > 0$ we can find a locally Lipschitz function $f_\varepsilon \colon X \longrightarrow Y$, such that*

$$f_\varepsilon(X) \subseteq \operatorname{conv} F(X) \qquad \text{and} \qquad h^*_{X \times Y}(\operatorname{Gr} f_\varepsilon, \operatorname{Gr} F) < \varepsilon.$$

If we strengthen the conditions on F, we can improve the conclusion of this theorem and produce a locally Lipschitz selection for F.

THEOREM 1.2.7
If X is a metric space, Y is a normed space and $F\colon X \longrightarrow 2^Y \setminus \{\emptyset\}$ has convex values and for every $x \in X$ there exists $U \in \mathcal{N}(x)$, such that

$$\bigcap_{z \in U} F(z) \neq \emptyset,$$

then F admits a locally Lipschitz selection.

In many occasions an effective substitute for convexity is the notion of decomposability. Let (Ω, Σ, μ) be a σ-finite measure space and X a separable Banach space.

DEFINITION 1.2.9　*A set $K \subseteq L^0(\Omega; X)$ is said to be **decomposable**, if for every $(A, f_1, f_2) \in \Sigma \times K \times K$, we have*

$$\chi_A f_1 + \chi_{A^c} f_2 \in K.$$

REMARK 1.2.8　Since $\chi_{A^c} = 1 - \chi_A$, we see that the notion of decomposability formally looks like convexity. Only now the "coefficients" in the "convex combination" are themselves $\mathbb{R}$-valued functions. ⬚

The typical decomposable set is

$$S_F \stackrel{df}{=} \{f \in L^0(\Omega; X) : f(\omega) \in F(\omega) \text{ for } \mu\text{-a.a. } \omega \in \Omega\}$$

for a multifunction $F\colon \Omega \longrightarrow 2^X \setminus \{\emptyset\}$ and the set

$$S_F^p \stackrel{df}{=} S_F \cap L^p(\Omega; X), \qquad p \in [1, +\infty].$$

In fact within closure these are all the decomposable sets.

The next theorem gives us a remarkable consequence of decomposability and will be a basic tool in what follows. For a given integrand $u\colon \Omega \times X \longrightarrow \overline{\mathbb{R}}$, we introduce the integral functional

$$I_u(f) \stackrel{df}{=} \int_{\Omega} u\big(\omega, f(\omega)\big) \, d\mu, \qquad f \in L^p(\Omega; X),$$

whenever the integral is defined, i.e. $u\big(\cdot, f(\cdot)\big)^+$ or $u\big(\cdot, f(\cdot)\big)^-$ is integrable.

THEOREM 1.2.8
If $u\colon \Omega \times X \longrightarrow \overline{\mathbb{R}}$ is measurable, $F\colon \Omega \longrightarrow 2^X \setminus \{\emptyset\}$ is graph measurable, $I_u(f)$ is defined (maybe is $+\infty$ or $-\infty$) for every $f \in S_F^p$, $1 \leq p \leq +\infty$ and

there exists at least one $f_0 \in S_F^p$, such that $I_u(f_0) > -\infty$,
<u>then</u>

$$\sup\left\{I_u(f) : f \in S_F^p\right\} = \int_\Omega \sup\left\{u(\omega, x) : x \in F(\omega)\right\} d\mu.$$

The next proposition gives some properties of the set S_F^p.

PROPOSITION 1.2.10
Let (Ω, Σ, μ) be a σ-finite measure space, X a separable Banach space and $F : \Omega \longrightarrow 2^X \setminus \{\emptyset\}$.

(a) *If F is graph measurable and $S_F^p \neq \emptyset$, $p \in [1, +\infty)$,*
<u>then</u>

$$\overline{\mathrm{conv}}\, S_F^p = S_{\overline{\mathrm{conv}}\, F}^p \quad \text{and} \quad \overline{S_F^p} = S_{\overline{F}}^p,$$

where by $\overline{F}$ we denote the multifunction $\Omega \ni \omega \longrightarrow \overline{F(\omega)} \in 2^X \setminus \{\emptyset\}$;

(b) *If μ is nonatomic (see Remark A.2.2), F is graph measurable and $S_F^p \neq \emptyset$, $p \in [1, +\infty)$,*
<u>then</u>

$$\overline{S_F^p}^{\,w} = S_{\overline{\mathrm{conv}}\, F}^p.$$

The next theorem is a useful weak compactness theorem for $S_F^1 \subseteq L^1(\Omega; X)$. First a definition.

DEFINITION 1.2.10
*A multifunction $F : \Omega \longrightarrow 2^X \setminus \{\emptyset\}$ is said to be L^p-**integrably bounded** (for $1 \leq p \leq +\infty$), if there exists $h \in L^p(\Omega)$ such that*

$$|F(\omega)| = \sup\left\{\|x\|_X : x \in F(\omega)\right\} \leq h(\omega) \qquad \text{for } \mu\text{-a.a. } \omega \in \Omega.$$

*If $p = 1$ we say simply that F is **integrably bounded**.*

THEOREM 1.2.9
If $F : \Omega \longrightarrow P_{wkc}(X)$ is graph measurable and integrably bounded, <u>then</u> *S_F^1 is a nonempty, convex and weakly compact subset of $L^1(\Omega; X)$.*

Of interest to us for what follows in the subsequent chapters is the extremal structure of the set S_F^p, $1 \leq p \leq +\infty$. The basic result in this direction is the following theorem.

THEOREM 1.2.10
If (Ω, Σ, μ) is a σ-finite measure space and $F : \Omega \longrightarrow P_{wkc}(X)$ is scalarly measurable,

<u>*then*</u>

$$\text{ext } S_F^p = S_{\text{ext } F}^p \qquad \forall\, 1 \le p \le +\infty.$$

When we started our discussion of decomposable sets, we said that decomposability can be used as a substitute for convexity, when we deal with multifunctions which have values in the Lebesgue-Bochner space $L^1(\Omega; X)$. So the Michael Selection Theorem (see Theorem 1.2.5) has a counterpart for decomposable-valued (but not necessary convex-valued) multifunctions. This result is useful in the study of nonconvex multivalued boundary value problems (see Chapter 3).

THEOREM 1.2.11
<u>*If*</u> *Z is a separable metric space and the multifunction* $F\colon Z \longrightarrow P_f\big(L^1(\Omega; X)\big)$ *is lower semicontinuous and has decomposable values (see Definition 1.2.9),*
<u>*then*</u> *F admits a continuous selection.*

There is another such selection theorem which is useful when we look for extremal solutions and we want to prove strong relaxation theorems (see Section 3.3). To state this theorem we need to introduce some auxiliary material. So let $T = [0, b]$ be furnished with the Lebesgue measure λ. Also let Y be a Polish space and as before let X be a separable Banach space. On $L^1(T; X)$ in addition to the usual norm, we also consider the following one.

DEFINITION 1.2.11 *The **weak norm** on $L^1(T; X)$ is defined by*

$$\|f\|_w \overset{df}{=} \sup\left\{ \left\| \int_t^{t'} f(s)\,ds \right\|_X : 0 \le t \le t' \le b \right\} \qquad \forall\, f \in L^1(T; X).$$

The space $\big(L^1(T; X), \|\cdot\|_w\big)$ is denoted by $L_w^1(T; X)$.

REMARK 1.2.9 Clearly an equivalent norm to the weak one can be obtained by setting

$$\|f\|_w \overset{df}{=} \sup\left\{ \left\| \int_0^t f(s)\,ds \right\|_X : 0 \le t \le b \right\} \qquad \forall\, f \in L^1(T; X).$$

$\square$

Let $F\colon T \times Y \longrightarrow 2^X \setminus \{\emptyset\}$ be a multifunction which satisfies the following hypotheses:

<u>$H(F)$</u> $F\colon T \times Y \longrightarrow P_{wkc}(X)$ is a multifunction, such that:

(i) for every $y \in Y$, the multifunction $t \longmapsto F(t, y)$ is measurable;

(ii) for almost all $t \in T$, the multifunction $y \longmapsto F(t, y)$ is h continuous;

(iii) for every $C \in P_k(Y)$, there exists $a_C \in L^1(T)$, such that for almost all $t \in T$ and all $y \in C$ we have

$$|F(t, y)| \stackrel{df}{=} \sup_{u \in F(t,y)} \|u\|_X \leq a_C(t).$$

Let $F: T \times Y \longrightarrow 2^X$ be a multifunction and let $K \subseteq C(Y)$ be a nonempty and compact set. Let $\widetilde{F}_K : K \longrightarrow P_{wkc}(L^1(T; X))$ be defined by

$$\widetilde{F}_K(y) \stackrel{df}{=} S^1_{F(\cdot, y(\cdot))}.$$

In what follows by $CS^w_{\widetilde{F}_K}$ (respectively $CS^w_{\text{ext }\widetilde{F}_K}$), we denote the selections of $\widetilde{F}_K$ (respectively ext $\widetilde{F}_K$) which are continuous from K into $L^1_w(T; X)$.

THEOREM 1.2.12
If F satisfies $H(F)$ and $K \subseteq C(Y)$ is a nonempty and compact set,
then
$$CS^w_{\widetilde{F}_K} = \overline{CS^w_{\text{ext }\widetilde{F}_K}}^{\|\cdot\|_w}.$$

A useful byproduct of the proof of this theorem (see Hu & Papageorgiou (1997, Theorem 8.31, p. 260)) is the following lower semicontinuity result that we shall need in our discussion of multivalued boundary value problems (see Chapter 3).

PROPOSITION 1.2.11
If (Ω, Σ, μ) is a finite measure space, Z is a metric space, $G: Z \longrightarrow P_f(L^1(\Omega; X))$ is lower semicontinuous and has decomposable values, $g: Z \longrightarrow L^1(\Omega; X)$, $\varphi: Z \longrightarrow L^1(\Omega)$ are continuous maps and for every $z \in Z$, we have

$$H(z) \stackrel{df}{=} \{u \in G(z) : \|u(\omega) - g(z)(\omega)\|_X < \varphi(z)(\omega) \ \mu\text{-a.e. on } \Omega\} \neq \emptyset,$$

then $z \longmapsto H(z)$ is lower semicontinuous with decomposable values.

1.2.6   Convergence in the Kuratowski Sense

Finally let us define a basic mode of set convergence and mention a useful result on the pointwise behaviour of weak sequences in $L^p(\Omega; X)$ ($p \in [1, +\infty)$) which will be used repeatedly in what follows.

DEFINITION 1.2.12   *Let (X, τ) be a Hausdorff topological space (τ denotes the topology) and let $\{A_n\}_{n \geq 1} \subseteq 2^X \setminus \{\emptyset\}$. We define*

$$\tau\text{-}\liminf_{n \to +\infty} A_n \overset{df}{=} \left\{ x \in X : \ x = \tau\text{-}\lim_{n \to +\infty} x_n, \ x_n \in A_n, \ n \geq 1 \right\}$$

and

$$\tau\text{-}\limsup_{n \to +\infty} A_n \overset{df}{=} \left\{ x \in X : x = \tau\text{-}\lim_{k \to +\infty} x_{n_k}, \ x_{n_k} \in A_{n_k}, \ n_k < n_{k+1}, \ k \geq 1 \right\}.$$

*The set $\tau\text{-}\liminf\limits_{n \to +\infty} A_n$ is called the τ-**Kuratowski limit inferior** of the sequence $\{A_n\}_{n \geq 1}$ and the set $\tau\text{-}\limsup\limits_{n \to +\infty} A_n$ is called the τ-**Kuratowski limit superior** of the sequence $\{A_n\}_{n \geq 1}$. If*

$$A \ = \ \tau\text{-}\liminf_{n \to +\infty} A_n \ = \ \tau\text{-}\limsup_{n \to +\infty} A_n,$$

*then we say that the sequence $\{A_n\}_{n \geq 1}$ **converges in the Kuratowski sense** to A and we write*

$$A \ = \ \tau\text{-}\lim_{n \to +\infty} A_n \quad \text{or} \quad A_n \ \overset{K}{\longrightarrow} \ A.$$

REMARK 1.2.10   Clearly we always have that

$$\tau\text{-}\liminf_{n \to +\infty} A_n \ \subseteq \ \tau\text{-}\limsup_{n \to +\infty} A_n$$

and the inclusion may be strict. If X is a metric space with a metric d_X generating the topology τ, then

$$\tau\text{-}\liminf_{n \to +\infty} A_n = \left\{ x \in X : \ \lim_{n \to +\infty} d_X(x, A_n) = 0 \right\}$$

and

$$\tau\text{-}\limsup_{n \to +\infty} A_n = \left\{ x \in X : \ \liminf_{n \to +\infty} d_X(x, A_n) = 0 \right\}.$$

Also if X is first countable (see Definition A.1.4(c))), then

$$\tau\text{-}\limsup_{n \to +\infty} A_n \ = \ \bigcap_{k \geq 1} \overline{\bigcup_{n \geq k} A_n}^{\tau}.$$

Finally if the topology τ is clear from the context, then we shall drop the letter τ. $\quad\square$

Also there is another mode of convergence more suitable to deal with sequences of sets and functions defined in a Banach space X.

DEFINITION 1.2.13　 *Let X be a Banach space and let $\{A_n\}_{n\geq 1} \subseteq 2^X \setminus \{\emptyset\}$. We say that the sets $\{A_n\}_{n\geq 1}$ converge to A in the **Mosco sense**, denoted by*

$$A_n \xrightarrow{M} A,$$

if and only if

$$s\text{-}\liminf_{n\to +\infty} A_n = w\text{-}\limsup_{n\to +\infty} A_n = A,$$

where $s\text{-}\liminf\limits_{n\to +\infty} A_n$ and $w\text{-}\limsup\limits_{n\to +\infty} A_n$ are as in Definition 1.2.12 with the strong and weak topologies respectively.

There is a corresponding mode of convergence for functions $\varphi_n \in \Gamma_0(X)$, (see Definition 1.3.1) $n \geq 1$.

DEFINITION 1.2.14　 *We say that the sequence of functions $\{\varphi_n\}_{n\geq 1} \subseteq \Gamma_0(X)$ converges to $\varphi \in \Gamma_0(X)$ in the **Mosco sense**, denoted by*

$$\varphi_n \xrightarrow{M} \varphi,$$

if and only if

$$\operatorname{epi} \varphi_n \xrightarrow{M} \operatorname{epi} \varphi \quad \text{in } X \times \mathbb{R}.$$

REMARK 1.2.11　 The above notion of convergence in the Mosco sense is equivalent to the following two properties:

(a) for each $x \in X$, there exists a sequence $x_n \longrightarrow x$ in X such that $\varphi_n(x_n) \longrightarrow \varphi(x)$;

(b) for all $x \in X$ and all sequences $\{x_n\}_{n\geq 1} \subseteq X$ such that $x_n \xrightarrow{w} x$ in X, we have

$$\varphi(x) \leq \liminf_{n\to +\infty} \varphi_n(x_n).$$

▯

Related to this mode of functional convergence is the so-called *G-convergence* of subdifferentials. Here $\Gamma_0(X)$ is as in Definition 1.3.1.

DEFINITION 1.2.15　 *We say that the sequence of subdifferentials $\{\partial\varphi_n\}_{n\geq 1}, \{\varphi_n\}_{n\geq 1} \subseteq \Gamma_0(X)$ G-**converges** to $\partial\varphi, \varphi \in \Gamma_0(X)$, denoted by*

$$\partial\varphi_n \xrightarrow{G} \partial\varphi,$$

if and only if

$$\mathrm{Gr}\,\partial\varphi \subseteq \underset{n\to+\infty}{s\text{--}\liminf}\,\mathrm{Gr}\,\partial\varphi_n \quad in\ X \times X^*.$$

REMARK 1.2.12   Due to the maximal monotonicity of the subdifferential, G-convergence of $\{\partial\varphi_n\}_{n\geq 1}$ implies that

$$w \times s\text{--}\underset{n\to+\infty}{\limsup}\,\mathrm{Gr}\,\partial\varphi_n \subseteq \mathrm{Gr}\,\partial\varphi.$$

Also the convergence $\varphi_n \xrightarrow{M} \varphi$ implies $\partial\varphi_n \xrightarrow{G} \partial\varphi$. ⬚

The next result provides information about the pointwise behaviour of a weakly convergent sequence in the Lebesgue-Bochner space $L^p(\Omega; X)$, $p \in [1, +\infty)$.

PROPOSITION 1.2.12
If (Ω, Σ, μ) *is a σ-finite measure space, X is a Banach space, $\{f_n\}_{n\geq 1} \subseteq \overline{L^p}(\Omega; X)$, $f \in L^p(\Omega; X)$, $p \in [1, +\infty)$,*

$$f_n \xrightarrow{w} f \quad in\ L^p(\Omega; X)$$

and

$$f_n(\omega) \in G(\omega) \qquad for\ \mu\text{-a.a.}\ \omega \in \Omega\ and\ all\ n \geq 1,$$

for some $G\colon \Omega \ni \omega \longmapsto G(\omega) \in P_{wk}(X)$,
then

$$f(\omega) \in \overline{\mathrm{conv}}\,\underset{n\to+\infty}{w\text{--}\liminf}\{f_n(\omega)\} \qquad for\ \mu\text{-a.a.}\ \omega \in \Omega.$$

The final results of this section are two important **Projection Theorems** for measurable sets. The starting point for such results is the well known fact that a Borel set in $\mathbb{R}^2$ does not necessarily project to a Borel set in $\mathbb{R}$.

THEOREM 1.2.13 (Yankov-von Neumann-Aumann Projection Theorem)
If (Ω, Σ) *is a measurable space, X is a Souslin space (see Definition 1.2.8) and $G \in \Sigma \times \mathcal{B}(X)$,*
then $\mathrm{proj}_\Omega\, G \in \widehat{\Sigma}$.

THEOREM 1.2.14
If T, X *are Polish spaces, $G \in \mathcal{B}(T \times X)$ and for every $t \in T$ the section $G(t) \subseteq X$ is σ-compact,*
then $\mathrm{proj}_T\, G \in \mathcal{B}(X)$.

1.3   Nonsmooth Analysis

In this section we present some aspects of Nonsmooth Analysis that we shall need in what follows. We do not aim at a complete presentation of today's subdifferential theory, but we only want to outline those parts of the theory that give us analytical tools for the study of the problems that we shall be investigating in the next chapters. For this reason we limit ourselves to convex functions (convex subdifferential), to locally Lipschitz functions (Clarke's theory) and to continuous functions (theory of weak slope).

1.3.1   Convex Functions

Let X be a Banach space and let $\varphi \colon X \longrightarrow \overline{\mathbb{R}} \stackrel{df}{=} \mathbb{R} \cup \{+\infty\}$. We define the *effective domain* of φ by

$$\operatorname{dom} \varphi \stackrel{df}{=} \{x \in X : \ \varphi(x) < +\infty\}$$

and the *epigraph* of φ by

$$\operatorname{epi} \varphi \stackrel{df}{=} \{(x, \lambda) \in X \times \mathbb{R} : \ \varphi(x) \leq \lambda\}.$$

The function φ is **convex** if and only if $\operatorname{epi} \varphi$ is convex in $X \times \mathbb{R}$ and of course this definition is equivalent to the usual one which requires that for all $x, y \in \operatorname{dom} \varphi$ and all $\lambda \in [0, 1]$, we have

$$\varphi\big(\lambda x + (1 - \lambda)y\big) \leq \lambda \varphi(x) + (1 - \lambda)\varphi(y).$$

If this inequality is strict for $x, y \in \operatorname{dom} \varphi$ with $x \neq y$ and for $\lambda \in (0, 1)$, then we say that φ is **strictly convex**. Another equivalent definition of a convex function says that φ is convex, if for all $N \geq 1$, all $y_0, \ldots, y_N \in X$ and all $\lambda_0, \ldots, \lambda_N \in \mathbb{R}_+$, such that $\sum\limits_{k=1}^{N} \lambda_N = 1$, we have

$$\varphi\left(\sum_{k=1}^{N} \lambda_k y_k\right) \leq \sum_{k=1}^{N} \lambda_k \varphi(y_k).$$

Since we are dealing with $\overline{\mathbb{R}}$-valued functions, to avoid trivial situations we focus on **proper functions**, i.e. functions φ, such that $\operatorname{dom} \varphi \neq \emptyset$. Convex functions have remarkable continuity properties as it is evident in the next theorem, which essentially says that for proper convex functions continuity follows from local upper boundedness.

THEOREM 1.3.1
If $\varphi \colon X \longrightarrow \overline{\mathbb{R}}$ *is a proper convex function,*
then the following statements are equivalent:

(a) φ is bounded above in a neighbourhood of some point $x_0 \in X$;

(b) φ is continuous at some point $x_0 \in X$;

(c) $\text{int epi}\,\varphi \neq \emptyset$;

(d) $\text{int dom}\,\varphi \neq \emptyset$ and $\varphi|_{\text{int dom}\,\varphi}$ is continuous.

Moreover, if the above statements hold, then

$$\text{int epi}\,\varphi \;=\; \{(x,\lambda) \in X \times \mathbb{R} : \; x \in \text{int dom}\,\varphi, \; \varphi(x) < \lambda\}.$$

PROOF **(a)$\Longrightarrow$(b):** Let $U \in \mathcal{N}(x_0)$ be such that $\varphi|_U$ is bounded above, i.e. there exists $c > 0$, such that

$$\varphi(y) \leq c \qquad \forall\, y \in U.$$

Replacing if necessary U by $U - x$ and $\varphi(y)$ by $\varphi(y + x_0) - \varphi(x_0)$, we may assume that $x_0 = 0$ and that $\varphi(0) = 0$. We will show that φ is continuous at $x_0 = 0$. Let $0 < \varepsilon \leq c$ and set

$$V_\varepsilon \stackrel{df}{=} \left(\frac{\varepsilon}{c}U\right) \cap \left(-\frac{\varepsilon}{c}U\right) \in \mathcal{N}(0).$$

We shall show that

$$|\varphi(y)| \leq \varepsilon \qquad \forall\, y \in V_\varepsilon \tag{1.1}$$

(which implies the continuity of φ at $x_0 = 0$). So let $y \in V_\varepsilon$. We have $\frac{c}{\varepsilon}y \in U$ and because φ is convex,

$$\varphi(y) \;\leq\; \frac{\varepsilon}{c}\varphi\left(\frac{c}{\varepsilon}y\right) + \left(1 - \frac{\varepsilon}{c}\right)\varphi(0) \;\leq\; \frac{\varepsilon}{c}c \;=\; \varepsilon.$$

Also $-\frac{c}{\varepsilon}y \in U$ and so

$$0 \;=\; \varphi(0) \;=\; \varphi\left(\frac{1}{1+\frac{\varepsilon}{c}}y + \frac{\frac{\varepsilon}{c}}{1+\frac{\varepsilon}{c}}\left(-\frac{c}{\varepsilon}y\right)\right)$$

$$\leq\; \frac{1}{1+\frac{\varepsilon}{c}}\varphi(y) + \frac{\frac{\varepsilon}{c}}{1+\frac{\varepsilon}{c}}\varphi\left(-\frac{c}{\varepsilon}y\right) \;\leq\; \frac{1}{1+\frac{\varepsilon}{c}}\varphi(y) + \frac{\varepsilon}{1+\frac{\varepsilon}{c}}$$

and so $-\varepsilon \leq \varphi(y)$. So finally we obtain (1.1) and this proves the continuity of φ at the origin.

(b)$\Longrightarrow$(a): Obvious.

(a)$\Longrightarrow$(c): By hypothesis, there exists $U \in \mathcal{N}(x_0)$, such that

$$\varphi(y) \leq c \qquad \forall\, y \in U.$$

So $U \subseteq \text{int dom}\,\varphi$ and

$$\{(y,\lambda) \in X \times \mathbb{R} : y \in U, \ \lambda > c\} \subseteq \text{epi}\,\varphi,$$

which implies that $\text{int epi}\,\varphi \neq \emptyset$.

(c)$\Longrightarrow$(d): If $(x_0, \lambda_0) \in \text{int epi}\,\varphi$, then φ is bounded above in a neighbourhood of x and so from the equivalence of **(a)** and **(b)** established earlier, we infer that φ is continuous at x_0. Also note that

$$\{x \in X : \exists\,\lambda \in \mathbb{R} : (x, \lambda) \in \text{int epi}\,\varphi\} \subseteq \text{int dom}\,\varphi$$

and this proves the implication.

(d)$\Longrightarrow$(b): Obvious.

Finally let us show that

$$\text{int epi}\,\varphi \ = \ \{(x,\lambda) \in X \times \mathbb{R} : x \in \text{int dom}\,\varphi, \ \varphi(x) < \lambda\}.$$

Let us denote the right hand side set by W. Clearly

$$\text{int epi}\,\varphi \subseteq W.$$

On the other hand let $\overline{x} \in \text{int dom}\,\varphi$ and $\varphi(\overline{x}) < \overline{\lambda}$. Let $\widehat{\lambda} \in (\varphi(\overline{x}), \overline{\lambda})$. Because $\varphi|_{\text{int epi}\,\varphi}$ is continuous, there exists $U \in \mathcal{N}(\overline{x})$, such that $U \subseteq \text{int dom}\,\varphi$ and

$$\varphi(y) < \widehat{\lambda} \qquad \forall\, y \in U.$$

So $(\overline{x}, \overline{\lambda}) \in U \times (\widehat{\lambda}, +\infty) \subseteq \text{int epi}\,\varphi$, and so

$$W \subseteq \text{int epi}\,\varphi.$$

$\square$

In fact we can improve our conclusion that $\varphi|_{\text{int dom}\,\varphi}$ is continuous. This will be done with the help of the following preliminary result.

LEMMA 1.3.1
If $\varphi \colon X \longrightarrow \mathbb{R}$ *is convex and continuous at* $x_0 \in \text{dom}\,\varphi$,
<u>*then*</u> *we can find* $r > 0$, *such that* $\varphi|_{\overline{B}_r(x_0)}$ *is Lipschitz continuous.*

PROOF   Since φ is continuous at x_0, we have $x_0 \in \text{int dom}\,\varphi$. So there exist $r, \varepsilon > 0$ and $m, M \in \mathbb{R}$, such that $B_{r+\varepsilon}(x_0) \subseteq \text{dom}\,\varphi$ and $m \leq \varphi(y) \leq M$ for all $y \in B_{r+\varepsilon}(x_0)$. Let $y_1, y_2 \in B_r(x_0)$ with $y_1 \neq y_2$. We set

$$z \ = \ y_1 + \varepsilon \frac{y_1 - y_2}{\|y_1 - y_2\|_X} \qquad \text{and} \qquad \lambda \ = \ \frac{\|y_1 - y_2\|_X}{\|y_1 - y_2\|_X + \varepsilon}.$$

Evidently $z \in B_{r+\varepsilon}(x)$, $\lambda \in (0,1)$ and $y_1 = \lambda z + (1-\lambda)y_2$. Because of the convexity of φ we have

$$\varphi(y_1) \leq \lambda\varphi(z) + (1-\lambda)\varphi(y_2) = \lambda(\varphi(z) - \varphi(y_2)) + \varphi(y_2)$$

and so

$$\varphi(y_1) - \varphi(y_2) \leq \frac{\|y_1 - y_2\|_X}{\|y_1 - y_2\|_X + \varepsilon}(M - m) \leq \frac{M - m}{\varepsilon}\|y_1 - y_2\|_X.$$

Interchanging the roles of y_1 and y_2 we obtain

$$|\varphi(y_1) - \varphi(y_2)| \leq \frac{M - m}{\varepsilon}\|y_1 - y_2\|_X \qquad \forall\, y_1, y_2 \in \overline{B}_r(x),$$

i.e. $\varphi|_{\overline{B}_r(x)}$ is Lipschitz continuous. $\qquad\qquad\qquad\qquad\qquad\qquad\quad \Box$

Combining Theorem 1.3.1 with Lemma 1.3.1 we are led to the following precise characterization of the continuity of $\varphi|_{\text{int dom}\,\varphi}$, which is the starting point of Clarke's theory that we shall examine later in this section.

THEOREM 1.3.2
<u>*If*</u> $\varphi\colon X \longrightarrow \overline{\mathbb{R}}$ *is proper, convex and lower semicontinuous,*
<u>*then*</u> $\varphi|_{\text{int dom}\,\varphi}$ *is locally Lipschitz.*

PROOF Note that

$$\operatorname{dom}\varphi = \bigcup_{n \geq 1}\{\varphi \leq n\}.$$

Let $x_0 \in \operatorname{int dom}\varphi$. Because φ is lower semicontinuous, for every $n \geq 1$ the set $\{\varphi \leq n\}$ is closed. From the Baire Category Theorem (see Theorem A.1.9), it follows that there exists $n_0 \geq 1$, such that

$$\operatorname{int}\{\varphi < n_0\} \neq \emptyset \qquad \text{and} \qquad \varphi(x_0) < n_0.$$

Let $y \in \operatorname{int}\{\varphi < n_0\}$ and let us define $\xi\colon \mathbb{R} \longrightarrow \mathbb{R}$, by

$$\xi(t) \stackrel{df}{=} \varphi(x_0 + t(y - x_0)) \qquad \forall\, t \in \mathbb{R}.$$

Of course ξ is convex and proper. Since $x_0 \in \operatorname{int dom}\varphi$, we can find $\rho > 0$, such that

$$B_{\rho\|y - x_0\|_X}(x_0) \subseteq \operatorname{dom}\varphi.$$

Therefore

$$[-\rho, \rho] \subseteq \operatorname{dom}\xi \quad \text{and} \quad 0 \in \operatorname{int dom}\xi,$$

which implies that ξ is continuous at 0 (see Theorem 1.3.1). Recall that $\xi(0) < n_0$. So we can find $\eta \in (0,1)$, such that

$$\xi(t) < n_0 \qquad \forall\, t \in [-\eta, 0].$$

Set $z = x_0 - \eta(y - x_0)$. We have $\varphi(z) = \xi(-\eta) < n_0$ and so $z \in \{\varphi < n_0\}$. Because

$$x_0 \in [y, z] \overset{df}{=} \{ty + (1-t)z : t \in [0,1]\},$$

we have that $x_0 \in \text{int}\,\{\varphi < n_0\}$. From Theorem 1.3.1 and Lemma 1.3.1, it follows that φ is locally Lipschitz at x_0. Because $x_0 \in \text{int}\,\text{dom}\,\varphi$ was arbitrary, we conclude that $\varphi|_{\text{int}\,\text{dom}\,\varphi}$ is locally Lipschitz. $\qquad\Box$

When X is finite dimensional, then in the above theorem the lower semi-continuity hypothesis can be dropped.

THEOREM 1.3.3
If X is a finite dimensional Banach space and the function $\varphi \colon X \longrightarrow \overline{\mathbb{R}}$ is proper and convex,
then $\varphi|_{\text{int}\,\text{dom}\,\varphi}$ *is locally Lipschitz.*

PROOF   Let $x_0 \in \text{int}\,\text{dom}\,\varphi$. We can find $u_0, \ldots, u_N \in X$ ($N = \dim X$) and $r > 0$, such that

$$B_r(x_0) \subseteq \text{conv}\,\{u_0, \ldots, u_N\} \subseteq \text{dom}\,\varphi.$$

Let $M \overset{df}{=} \max\{\varphi(u_0), \ldots, \varphi(u_N)\}$. For every $y \in B_r(x_0)$, we can find $\lambda_0, \ldots, \lambda_N \in \mathbb{R}_+$, such that

$$y = \sum_{k=0}^{N} \lambda_k u_k \qquad \text{and} \qquad \sum_{k=0}^{N} \lambda_k = 1.$$

Because φ is convex, we have

$$\varphi(y) \le \sum_{k=0}^{N} \lambda_k \varphi(u_k) \le \left(\sum_{k=0}^{N} \lambda_k\right) \max_{0 \le k \le N} \varphi(u_k) = M.$$

Note that $M \in \mathbb{R}$ is not depending on $y \in B_r(x_0)$. Thus by virtue of Theorem 1.3.1, φ is continuous at x_0 and from Lemma 1.3.1, $\varphi|_{\text{int}\,\text{dom}\,\varphi}$ is locally Lipschitz. $\qquad\Box$

The next theorem is a major tool in the execution of the so-called "Direct Method of the Calculus of Variations."

THEOREM 1.3.4

If X *is a reflexive Banach space,* $\varphi\colon X \longrightarrow \overline{\mathbb{R}}$ *is a proper, convex, lower semicontinuous function (or a proper, weakly lower semicontinuous function) and*

$$\lim_{\|x\|_X \to +\infty} \varphi(x) = +\infty,$$

then there exists $x_0 \in X$, *such that* $\varphi(x_0) = \inf_X \varphi$.

PROOF Because φ is proper, we can find $\eta \in \mathbb{R}$, such that $\inf_X \varphi < \eta$. Also since $\varphi(x) \longrightarrow +\infty$ as $\|x\|_X \to +\infty$, we can find $r > 0$, such that

$$\varphi(x) > \eta \qquad \forall\, x \in X \text{ with } \|x\|_X > r.$$

Therefore

$$\{\varphi \leq \eta\} \subseteq \overline{B}_r \qquad \text{and so} \qquad \inf_{\overline{B}_r} \varphi = \inf_X \varphi.$$

Let $\{x_n\}_{n \geq 1} \subseteq \overline{B}_r$ be a minimizing sequence of φ. Due to the reflexivity of X, $\overline{B}_r$ is weakly compact and so by the Eberlein-Smulian Theorem (see Theorem A.3.5) and by passing to a subsequence if necessary, we may assume that

$$x_n \xrightarrow{w} x_0 \in \overline{B}_r.$$

The function φ being convex, from Mazur's lemma (see Theorem A.3.7), it is also weakly lower semicontinuous. Hence

$$\varphi(x_0) \leq \liminf_{n \to +\infty} \varphi(x_n) = \inf_X \varphi$$

and so $\varphi(x_0) = \inf_X \varphi$.

The proof is similar when φ is a proper, weakly lower semicontinuous function. $\square$

From the previous results it is clear that the proper, convex and lower semicontinuous functions form an important set (actually cone) of functions. For this reason we introduce a special notation for this set.

DEFINITION 1.3.1 *Let X be a locally convex space. By $\Gamma_0(X)$ we denote the space (cone) of all proper, convex and lower semicontinuous functions* $\varphi\colon X \longrightarrow \overline{\mathbb{R}}$.

The next proposition explains why the family $\Gamma_0(X)$ is important in convex analysis.

PROPOSITION 1.3.1

If X *is a locally convex space and* $\varphi \in \Gamma_0(X)$,
then φ *is bounded below by an affine function.*

PROOF   Since $\varphi \in \Gamma_0(X)$, the set $\operatorname{epi}\varphi \subseteq X \times \mathbb{R}$ is closed and convex and

$$(x, \varphi(x) - \varepsilon) \notin \operatorname{epi}\varphi \qquad \forall\, x \in \operatorname{dom}\varphi,\ \varepsilon > 0.$$

Let us fix any $\varepsilon > 0$ and $x_0 \in \operatorname{dom}\varphi$. Apply the Strong Separation Theorem (see Theorem A.3.4) to obtain $(x^*, \beta) \in (X^* \times \mathbb{R}) \setminus \{(0,0)\}$ and $\delta > 0$, such that

$$\sup\left\{ \langle x^*, y\rangle_X + \beta\lambda : (y, \lambda) \in \operatorname{epi}\varphi\right\} \;\leq\; \langle x^*, x_0\rangle_X + \beta \cdot \big(\varphi(x_0) - \varepsilon\big) - \delta.$$

Since

$$\big(x_0, \varphi(x_0) + n\big) \in \operatorname{epi}\varphi \qquad \forall\, n \geq 1,$$

it follows that $\beta < 0$. On the other hand, we have that

$$\big(y, \varphi(y)\big) \in \operatorname{epi}\varphi \qquad \forall\, y \in \operatorname{dom}\varphi$$

and so

$$\langle x^*, y\rangle_X + \beta\varphi(y) \;\leq\; \langle x^*, x_0\rangle_X + \beta\varphi(x_0) \qquad \forall\, y \in \operatorname{dom}\varphi,$$

hence

$$-\frac{1}{\beta} \langle x^*, y\rangle_X + \frac{1}{\beta} \langle x^*, x_0\rangle_X + \varphi(x_0) \;\leq\; \varphi(y) \qquad \forall\, y \in X.$$

The function

$$X \ni y \longmapsto -\frac{1}{\beta} \langle x^*, y\rangle_X + \frac{1}{\beta} \langle x^*, x_0\rangle_X + \varphi(x_0) \in \mathbb{R}$$

is the desired affine function.   ⬚

1.3.2   Fenchel Transform

The Fenchel transform (convex conjugate) that we are about to introduce extends the classical Legendre transform to functions which are not necessarily smooth, by using affine minorants (Proposition 1.3.1) instead of tangent hyperplanes.

DEFINITION 1.3.2   *Let X be a Banach space and $\varphi\colon X \longrightarrow \mathbb{R}^* \stackrel{df}{=} \mathbb{R} \cup \{\pm\infty\}$. The **Fenchel transform** or **conjugate** of φ is the function $\varphi^*\colon X^* \longrightarrow \mathbb{R}^*$, defined by*

$$\varphi^*(x^*) \stackrel{df}{=} \sup\left\{ \langle x^*, x\rangle_X - \varphi(x) : x \in \operatorname{dom}\varphi\right\}.$$

*We can repeat this process to φ^* and obtain the **second conjugate** $\varphi^{**} = (\varphi^*)^*\colon X \longrightarrow \mathbb{R}^*$, defined by*

$$\varphi^{**}(x) \stackrel{df}{=} \sup\left\{ \langle x^*, x\rangle_X - \varphi^*(x^*) : x^* \in X^*\right\}.$$

REMARK 1.3.1   From the definition we see that φ^* being the supremum of continuous affine functions is itself convex and w^*-lower semicontinuous, while φ^{**} is convex and lower semicontinuous (also weakly lower semicontinuous). Moreover,

$$\varphi^{**} \leq \varphi$$

and

$$\text{if } \varphi_1 \leq \varphi_2, \text{ then } \varphi_2^* \leq \varphi_1^*.$$

In addition, from Proposition 1.3.1, we see that

$$\varphi \in \Gamma_0(X) \text{ if and only if } \varphi^* \in \Gamma_0(X_{w^*}^*).$$

Here $X_{w^*}^*$ stands for the dual space X^* furnished with the weak*-topology. Finally the inequality

$$\langle x^*, x \rangle_X \leq \varphi(x) + \varphi^*(x^*) \qquad \forall \, x \in X, \, x^* \in X^*,$$

which is an immediate consequence of Definition 1.3.2, is usually called the ***Young-Fenchel inequality***.   ⏹

PROPOSITION 1.3.2
If X, Y are two Banach spaces, $A \in \mathcal{L}(X; Y)$ is an isomorphism, $g \colon Y \longrightarrow \overline{\mathbb{R}}$ is proper and

$$\varphi(x) = \lambda_0 g(Ax + y_0) + \langle x_0^*, x \rangle_X + \vartheta_0 \qquad \forall \, x \in X,$$

with $y_0 \in Y$, $x_0^* \in X^*$, $\vartheta_0 \in \mathbb{R}$ and $\lambda_0 > 0$,
then

$$\varphi^*(x^*) = \lambda_0 g^* \left(\frac{1}{\lambda_0} (A^*)^{-1} (x^* - x_0^*) \right) - \langle x^* - x_0^*, A^{-1} y_0 \rangle_X - \vartheta_0.$$

PROOF   We have

$$
\begin{aligned}
\varphi^*(x^*) &= \sup_{x \in X} \left\{ \langle x^*, x \rangle_X - \lambda_0 g(Ax + y_0) - \langle x_0^*, x \rangle_X - \vartheta_0 \right\} \\
&= \lambda_0 \sup_{y \in Y} \left\{ \frac{1}{\lambda_0} \langle (A^*)^{-1}(x^* - x_0^*), y \rangle_X - g(y) \right\} \\
&\quad - \langle x^* - x_0^*, A^{-1} y_0 \rangle_X - \vartheta_0 \\
&= \lambda_0 g^* \left(\frac{1}{\lambda_0} (A^*)^{-1}(x^* - x_0^*) \right) - \langle x^* - x_0^*, A^{-1} y_0 \rangle_X - \vartheta_0.
\end{aligned}
$$

   ⏹

As an immediate consequence of this proposition, we have the following conjugation rules.

COROLLARY 1.3.1

If $g\colon X \longrightarrow \overline{\mathbb{R}}$ *a is proper function and* $x_0 \in X$, $x_0^* \in X^*$, $\lambda_0 > 0$, $\vartheta_0 \in \mathbb{R}$, *then*

(a) *for* $\varphi(x) = g(x + x_0)$, $\varphi^*(x^*) = g^*(x^*) - \langle x^*, x_0 \rangle_X$;
(b) *for* $\varphi(x) = g(x) + \langle x_0^*, x \rangle_X$, $\varphi^*(x^*) = g^*(x^* - x_0^*)$;
(c) *for* $\varphi(x) = \lambda_0 g(x)$, $\varphi^*(x^*) = \lambda_0 g^*\left(\frac{1}{\lambda_0} x^*\right)$;
(d) *for* $\varphi(x) = \lambda_0 g\left(\frac{1}{\lambda_0} x\right)$, $\varphi^*(x^*) = \lambda_0 g^*(x^*)$;
(e) *for* $\varphi(x) = g(\lambda_0 x)$, $\varphi^*(x^*) = g^*\left(\frac{1}{\lambda_0} x^*\right)$;
(f) *for* $\varphi(x) = \lambda_0 g(\lambda_0 x + x_0) + \vartheta_0$, $\varphi^*(x^*) = \lambda_0 g^*(x^*) - \lambda_0 \langle x^*, x_0 \rangle_X - \vartheta_0$.

We already said that $\varphi^{**} \leq \varphi$. The next theorem says when we have equality.

THEOREM 1.3.5

If $\varphi\colon X \longrightarrow \overline{\mathbb{R}}$ *is a proper function,*
then

$$\varphi^{**} = \varphi \text{ if and only if } \varphi \in \Gamma_0(X).$$

PROOF $\Longrightarrow$: Recall that $\varphi^{**} \in \Gamma_0(X)$ and since $\varphi^{**} = \varphi$, we conclude that $\varphi \in \Gamma_0(X)$.

$\Longleftarrow$: We know that we have $\varphi^{**} \leq \varphi$. So we need to show that the opposite inequality also holds. To this end let $x \in X$ and $\mu \in \mathbb{R}$ be such that $\mu < \varphi(x)$. Then $(x, \mu) \notin \operatorname{epi} \varphi$ and so we can apply the Strong Separation Theorem for convex sets (see Theorem A.3.4) and find $(x^*, \beta) \in X^* \times \mathbb{R}$, $(x^*, \beta) \neq (0, 0)$ and $\delta > 0$, such that

$$\langle x^*, y \rangle_X + \beta \lambda \leq \langle x^*, x \rangle_X + \beta \mu - \delta \qquad \forall (y, \lambda) \in \operatorname{epi} \varphi.$$

Since λ can go to $+\infty$, from this inequality it follows that $\beta \leq 0$.
   First suppose that $\beta < 0$. We have

$$\langle x^*, y \rangle_X + \beta \varphi(y) < \langle x^*, x \rangle_X + \beta \mu \qquad \forall y \in X,$$

from which it follows that

$$(-\beta \varphi)^*(x^*) \leq \langle x^*, x \rangle_X + \beta \mu.$$

Using Corollary 1.3.1(c), we obtain

$$-\beta \varphi^* \left(\frac{x^*}{-\beta} \right) \leq \langle x^*, x \rangle_X + \beta \mu$$

and thus

$$\mu \leq \left\langle -\frac{x^*}{\beta}, x \right\rangle_X - \varphi^*\left(-\frac{x^*}{\beta}\right) \leq \varphi^{**}(x).$$

Because $\mu < \varphi(x)$ was arbitrary, we infer that $\varphi(x) \le \varphi^{**}(x)$ as desired.

Next assume that $\beta = 0$. We have

$$\langle x^*, y \rangle_X \;\le\; \langle x^*, x \rangle_X - \delta \qquad \forall\, y \in \mathrm{dom}\,\varphi$$

and so, we see that $x \notin \mathrm{dom}\,\varphi$ and $\varphi(x) = +\infty$. It is enough to show that also $\varphi^{**}(x) = +\infty$. Let $\eta \in \mathbb{R}$ be such that

$$\langle x^*, y \rangle \;<\; \eta \;<\; \langle x^*, x \rangle_X \qquad \forall\, y \in \mathrm{dom}\,\varphi.$$

As φ is bounded below by an affine function (see Proposition 1.3.1), we have that there exist $y^* \in X^*$ and $\vartheta \in \mathbb{R}$, such that

$$\langle y^*, y \rangle_X - \vartheta \;\le\; \varphi(y) \qquad \forall\, y \in X.$$

So for all $\gamma > 0$, we have

$$\langle y^*, y \rangle_X - \vartheta + \gamma\big(\langle x^*, y \rangle_X - \eta \big) \;\le\; \varphi(y) \qquad \forall\, y \in X.$$

Then

$$\langle y^* + \gamma x^*, y \rangle_X - \varphi(y) \;\le\; \vartheta + \gamma\eta \qquad \forall\, y \in X$$

and so

$$\varphi^*(y^* + \gamma x^*) \;\le\; \vartheta + \gamma\eta.$$

Therefore

$$\langle y^*, x \rangle_X - \vartheta + \gamma\big(\langle x^*, x \rangle_X - \eta \big) \;\le\; \langle y^* + \gamma x^*, x \rangle_X - \varphi^*(y^* + \gamma x^*)$$
$$\le\; \varphi^{**}(x).$$

Since $\eta < \langle x^*, x \rangle_X$ and $\gamma > 0$ was arbitrary, we see that the left hand side is arbitrarily large and so $\varphi^{**}(x) = +\infty$. Thus $\varphi(x) \le \varphi^{**}(x)$. $\quad\square$

If by $\overline{\varphi}$ we denote the **closure** or **lower semicontinuous hull** of the proper function $\varphi\colon X \longrightarrow \overline{\mathbb{R}}$, i.e. $\mathrm{epi}\,\overline{\varphi} = \overline{\mathrm{epi}\,\varphi}$, then we deduce the following corollary of Theorem 1.3.5.

COROLLARY 1.3.2
If $\varphi\colon X \longrightarrow \overline{\mathbb{R}}$ *is a proper and convex function,*
then $\varphi^{**} = \overline{\varphi}$.

Another consequence of Theorem 1.3.5 is the following result.

COROLLARY 1.3.3
If $\varphi\colon X \longrightarrow \overline{\mathbb{R}}$ *is a proper function,*
then $\varphi \in \Gamma_0(X)$ *if and only if it is an upper envelope of all continuous affine functions majorized by* φ.

1.3.3   Subdifferential of Convex Functions

Now we shall introduce the subdifferential of a convex function. This notion generalizes the concept of gradient.

DEFINITION 1.3.3   *Let $\varphi\colon X \longrightarrow \overline{\mathbb{R}}$ be a proper and convex function. The **subdifferential** of φ at x is the multivalued map $\partial\varphi\colon X \longrightarrow 2^{X^*}$ defined by*

$$\partial\varphi(x) \stackrel{df}{=} \{x^* \in X^* : \ \langle x^*, y - x\rangle_X \leq \varphi(y) - \varphi(x) \ \ \forall y \in \mathrm{dom}\,\varphi\}.$$

*The elements of $\partial\varphi(x)$ are called **subgradients** of φ at x. Also we define the **domain** of $\partial\varphi$ by*

$$\mathrm{Dom}\,(\partial\varphi) \stackrel{df}{=} \{x \in X : \ \partial\varphi(x) \neq \emptyset\}$$

and we say that φ is subdifferentiable at $x \in X$, if $x \in \mathrm{Dom}\,(\partial\varphi)$.

PROPOSITION 1.3.3
*If $\varphi\colon X \longrightarrow \overline{\mathbb{R}}$ is a proper convex function,
then the following properties are equivalent:*

(a) $x^* \in \partial\varphi(x)$;

(b) $\varphi(x) + \varphi^*(x^*) = \langle x^*, x\rangle_X$.

If additionally $\varphi \in \Gamma_0(X)$, then properties **(a)** *and* **(b)** *are also equivalent to:*

(c) $x \in \partial\varphi^*(x^*)$.

PROOF   **(a)$\Longrightarrow$(b):** By definition we have

$$\langle x^*, y - x\rangle_X \ \leq \ \varphi(y) - \varphi(x) \qquad \forall\, y \in X.$$

So

$$\langle x^*, y\rangle_X - \varphi(y) \ \leq \ \langle x^*, x\rangle_X - \varphi(x),$$

from which it follows that

$$\varphi^*(x^*) \ \leq \ \langle x^*, x\rangle_X - \varphi(x).$$

Since the opposite inequality is always true (Young-Fenchel inequality; see Remark 1.3.1), we conclude that

$$\varphi(x) + \varphi^*(x^*) \ = \ \langle x^*, x\rangle_X.$$

(b)$\Longrightarrow$(a): From **(b)** and Definition 1.3.2, we have that

$$\langle x^*, x \rangle_X = \varphi(x) + \varphi^*(x^*) \geq \varphi(x) + \langle x^*, y \rangle_X - \varphi(y) \qquad \forall \, y \in X,$$

hence

$$\langle x^*, y - x \rangle_X \leq \varphi(y) - \varphi(x) \qquad \forall \, y \in X$$

and so $x^* \in \partial\varphi(x)$.

Finally from the established equivalence of **(a)** and **(b)**, we see that **(c)** can be equivalently expressed as

$$\varphi^*(x^*) + (\varphi^*)^*(x) = \langle x^*, x \rangle_X,$$

with $(\varphi^*)^*: X^{**} \longrightarrow \overline{\mathbb{R}}$ being the conjugate of φ^*. Since

$$(\varphi^*)^*|_X = \varphi^{**} = \varphi$$

(see Theorem 1.3.5), we obtain **(b)**. $\qquad \qquad \qquad \square$

From the definition of the subdifferential we see that $\partial\varphi(x)$ is a closed, convex set possibly empty. With additional assumptions we can strengthen this observation.

PROPOSITION 1.3.4
If $\varphi: X \longrightarrow \overline{\mathbb{R}}$ is a convex function which is continuous at $x_0 \in X$,
<u>*then*</u> *$x_0 \in \mathrm{Dom}\,(\partial\varphi)$ and $\partial\varphi(x_0)$ is a bounded and w^*-compact subset of X^*.*

PROOF   From Theorem 1.3.1 we know that $\mathrm{int}\,\mathrm{epi}\,\varphi \neq \emptyset$ and $(x_0, \varphi(x_0))$ belongs to the boundary of $\mathrm{epi}\,\varphi$. So by the Weak Separation Theorem for convex sets (see Theorem A.3.3), we can find $x^* \in X^*$ and $\beta \in \mathbb{R}$, $(x^*, \beta) \neq (0,0)$, such that

$$\langle x^*, x_0 \rangle_X + \beta\varphi(x_0) \leq \langle x^*, y \rangle_X + \beta\lambda \qquad \forall \, (y, \lambda) \in \mathrm{epi}\,\varphi.$$

If $\beta = 0$, then

$$\langle x^*, y - x_0 \rangle_X \geq 0 \qquad \forall \, y \in \mathrm{dom}\,\varphi.$$

But $x_0 \in \mathrm{int}\,\mathrm{dom}\,\varphi$ (being a point of continuity of φ) and so we deduce that $x^* = 0$, a contradiction since $(x^*, \beta) \neq (0,0)$. Since λ can increase up to $+\infty$, we infer that $\beta > 0$. So

$$\langle -x^*, y - x_0 \rangle_X \leq \beta\,(\varphi(y) - \varphi(x_0)) \qquad \forall \, y \in X,$$

i.e. $-\frac{1}{\beta}x^* \in \partial\varphi(x_0) \neq \emptyset$.

From Lemma 1.3.1 we know that there exists $r > 0$, such that $\varphi|_{\overline{B}_r(x_0)}$ is Lipschitz. Then for some $k > 0$ we have

$$\langle x^*, y \rangle_X \leq \varphi(x_0 + y) - \varphi(x_0) \leq k\,\|y\|_X \qquad \forall \, y \in \overline{B}_r,$$

hence
$$\|x^*\|_{X^*} \leq k \qquad \forall\, x^* \in \partial\varphi(x_0).$$

Therefore $\partial\varphi(x_0)$ is w^*-compact (Alaoglu theorem; see Theorem A.3.1). ⬜

COROLLARY 1.3.4
If $\varphi \in \Gamma_0(X)$, *then* $\operatorname{int} \operatorname{dom} \varphi \subseteq \operatorname{Dom}(\partial\varphi) \subseteq \operatorname{dom} \varphi$.

A useful tool for the study of the subdifferential is provided by the directional derivative.

DEFINITION 1.3.4   *Let* $\varphi\colon X \longrightarrow \overline{\mathbb{R}}$ *be a proper and convex function. Then for every* $x \in \operatorname{dom} \varphi$, *the* **directional derivative** *of* φ *at* $x \in X$ *in the direction* $h \in X$ *is defined by*

$$\varphi'(x;h) \overset{df}{=} \lim_{\lambda \searrow 0} \frac{\varphi(x + \lambda h) - \varphi(x)}{\lambda} = \inf_{\lambda > 0} \frac{\varphi(x + \lambda h) - \varphi(x)}{\lambda}.$$

REMARK 1.3.2   It is clear from this definition that the function $X \ni h \longmapsto \varphi'(x;\cdot)$ is sublinear. ⬜

PROPOSITION 1.3.5
If $\varphi\colon X \longrightarrow \overline{\mathbb{R}}$ *is a proper and convex function which is continuous at* $x_0 \in \operatorname{dom} \varphi$,
then

$$\varphi'(x_0;h) = \sigma_{X^*}\big(h, \partial\varphi(x_0)\big) \qquad \forall\, h \in X$$

(see Definition 1.2.3).

PROOF   Let
$$\psi(h) = \varphi'(x_0;h) \qquad \forall\, h \in X.$$
From Proposition 1.3.4, we have that $\partial\varphi(x_0) \neq \emptyset$. Let $x^* \in \partial\varphi(x_0)$. From the definition of $\partial\varphi$, we have that

$$\langle x^*, h \rangle_X \leq \frac{\varphi(x_0 + \lambda h) - \varphi(x_0)}{\lambda} \qquad \forall\, \lambda > 0$$

and so

$$\langle x^*, h \rangle_X \leq \psi(h) \leq \frac{\varphi(x_0 + \lambda h) - \varphi(x_0)}{\lambda} \qquad \forall\, \lambda > 0.$$

From this it follows that ψ is everywhere finite and so it is continuous on X. Using Corollary 1.3.1(f), we see that the conjugate function of

$$h \longmapsto \frac{\varphi(x_0 + \lambda h) - \varphi(x_0)}{\lambda}$$

is the function
$$h^* \longmapsto \frac{\varphi^*(h^*) - \langle h^*, x_0 \rangle_X + \varphi(x_0)}{\lambda}.$$

Hence since $\psi(h) = \inf\limits_{\lambda > 0} \dfrac{\varphi(x_0 + \lambda h) - \varphi(x_0)}{\lambda}$, we have

$$\psi^*(h^*) = \sup\limits_{\lambda > 0} \frac{\varphi(x_0) + \varphi^*(h^*) - \langle h^*, x_0 \rangle_X}{\lambda}.$$

By virtue of Proposition 1.3.3, we have

$$\partial \varphi(x_0) = \left\{ h^* \in X^* : \; \varphi(x_0) + \varphi^*(h^*) = \langle h^*, x_0 \rangle_X \right\}$$

and so

$$\psi^*(h^*) = \begin{cases} 0 & \text{if } h^* \in \partial \varphi(x_0), \\ +\infty & \text{otherwise.} \end{cases}$$

But $\psi \in \Gamma_0(X)$ and so, for all $h \in X$, we have

$$\begin{aligned} \psi(h) = \psi^{**}(h) &= \sup\limits_{x^* \in X^*} \left(\langle x^*, h \rangle_X - \psi^*(x^*) \right) \\ &= \sup\limits_{x^* \in \partial\varphi(x_0)} \langle x^*, h \rangle_X = \sigma_{X^*}(h, \partial\varphi(x_0)). \end{aligned}$$

□

The case of Gâteaux differentiability corresponds to the situation where the subdifferential is a singleton.

PROPOSITION 1.3.6
If $\varphi \colon X \longrightarrow \overline{\mathbb{R}}$ is a proper and convex function,
then

(a) *if φ is Gâteaux differentiable at x_0, then $\partial\varphi(x_0) = \{\varphi'_G(x_0)\}$;*

(b) *if φ is continuous at x_0 and $\partial\varphi(x_0)$ is a singleton, then φ is Gâteaux differentiable at x_0 and $\partial\varphi(x_0) = \{\varphi'_G(x_0)\}$.*

PROOF **(a)** From the convexity of φ, for all $\lambda \in (0, 1)$ and all $h \in X$, we have

$$\begin{aligned} \varphi(x_0 + \lambda h) &= \varphi\big(\lambda(x_0 + h) + (1 - \lambda)x_0\big) \\ &\leq \lambda\varphi(x_0 + h) + (1 - \lambda)\varphi(x_0) \\ &= \lambda\big(\varphi(x_0 + h) - \varphi(x_0)\big) + \varphi(x_0), \end{aligned}$$

so we obtain

$$\frac{\varphi(x_0 + \lambda h) - \varphi(x_0)}{\lambda} \leq \varphi(x_0 + h) - \varphi(x_0) \qquad \forall \lambda \in (0, 1), \; h \in X,$$

hence

$$\langle \varphi'_G(x_0), h \rangle_X \leq \varphi(x_0 + h) - \varphi(x_0) \qquad \forall\, h \in X,$$

which implies that $\varphi'_G(x) \in \partial\varphi(x_0)$. On the other hand, if $x^* \in \partial\varphi(x_0)$, we have

$$\langle x^*, y - x_0 \rangle_X \leq \varphi(y) - \varphi(x_0) \qquad \forall\, y \in X.$$

Let $y = x_0 + \lambda h$. We obtain

$$\langle x^*, h \rangle_X \leq \frac{\varphi(x_0 + \lambda h) - \varphi(x_0)}{\lambda} \qquad \forall\, \lambda > 0,\ h \in X$$

and so

$$\langle \varphi'_G(x_0) - x^*, h \rangle_X \geq 0 \qquad \forall\, h \in X.$$

Thus $x^* = \varphi'_G(x)$, i.e. $\partial\varphi(x) = \{\varphi'_G(x)\}$.

(b) Let $\partial\varphi(x_0) = \{x_0^*\}$. From Proposition 1.3.5, we have that

$$\varphi'(x_0; h) = \sigma_{X^*}(h, \partial\varphi(x_0)) = \langle x^*, h \rangle_X \qquad \forall\, h \in X.$$

In particular this implies that φ is Gâteaux differentiable at $x_0 \in X$ and $\varphi'_G(x_0) = x_0^*$, so $\partial\varphi(x_0) = \{\varphi'_G(x_0)\}$. ∎

We close our discussion of the convex subdifferential with two calculus rules. The first concerns the sum of two convex functions.

THEOREM 1.3.6
If $\varphi, \psi \colon X \longrightarrow \mathbb{R}$ are proper and convex functions and there exists $x_0 \in \operatorname{dom}\varphi \cap \operatorname{dom}\psi$ where one of the functions is continuous,
<u>*then*</u>
$$\partial(\varphi + \psi)(y) = \partial\varphi(y) + \partial\psi(y) \qquad \forall\, y \in X.$$

PROOF   Assume that φ is continuous at x_0. It is an immediate consequence of the definition of the subdifferential that we have

$$\partial(\varphi + \psi)(y) \supseteq \partial\varphi(y) + \partial\psi(y) \qquad \forall\, y \in X.$$

Let us fix any $y_0 \in X$ and $y^* \in \partial(\varphi + \psi)(y_0)$ and consider the following two convex sets in $X \times \mathbb{R}$:

$$C_1 \stackrel{df}{=} \{(z, \lambda) \in X \times \mathbb{R} \colon \varphi(z) - \varphi(y_0) - \langle y^*, z - y_0 \rangle_X \leq \lambda\}$$

and

$$C_2 \stackrel{df}{=} \{(v, \mu) \in X \times \mathbb{R} \colon \mu \leq \psi(y_0) - \psi(v)\}.$$

Note that if we put

$$g(z) \stackrel{df}{=} \varphi(z) - \varphi(y_0) - \langle y^*, z - y_0 \rangle_X \qquad \forall \, z \in X,$$

then g is convex and continuous at x_0 and moreover $C_1 = \mathrm{epi} \, g$. Therefore $\mathrm{int} \, C_1 \neq \emptyset$ (see Theorem 1.3.1) and $\mathrm{int} \, C_1 \cap C_2 = \emptyset$. So we can apply the Weak Separation Theorem (see Theorem A.3.3) and obtain $(z^*, \gamma) \in X^* \times \mathbb{R} \setminus \{(0,0)\}$ and $\xi \in \mathbb{R}$, such that

$$\langle z^*, z \rangle_X + \gamma \lambda \;<\; \xi \;\leq\; \langle z^*, v \rangle_X + \gamma \mu \qquad (1.2)$$
$$\forall \, (z, \lambda) \in \mathrm{int} \, C_1, \; (v, \mu) \in C_2.$$

Because $y^* \in \partial(\varphi + \psi)(y_0)$, it follows that $\gamma < 0$. Note that $(y_0, 0) \in C_1 \cap C_2$ and so $\xi = \langle z^*, y_0 \rangle_X$. From (1.2), we have

$$\langle z^*, z \rangle_X + \gamma \left(\varphi(z) - \varphi(y_0) - \langle y^*, z - y_0 \rangle_X \right) \;\leq\; \langle z^*, y_0 \rangle_X$$
$$\leq \; \langle z^*, v \rangle_X + \gamma \big(\psi(y_0) - \psi(v) \big) \qquad \forall \, z \in \mathrm{dom} \, g, \; v \in \mathrm{dom} \, \psi.$$

From these inequalities, it follows that

$$\frac{1}{\gamma} z^* \in \partial \psi(y_0) \qquad \text{and} \qquad y^* - \frac{1}{\gamma} z^* \in \partial \varphi(y_0).$$

Therefore

$$y^* \;=\; \left(y^* - \frac{1}{\gamma} z^* \right) + \frac{1}{\gamma} z^* \;\in\; \partial \varphi(y_0) + \partial \psi(y_0)$$

and we have proved that

$$\partial(\varphi + \psi)(y) \;\subseteq\; \partial \varphi(y) + \partial \psi(y) \qquad \forall \, y \in X.$$

Since the opposite inclusion is always true, we have proved the theorem. $\qquad \square$

The other calculus rule concerns the composition $\varphi \circ A$ of a proper convex function φ and of a continuous linear operator A.

THEOREM 1.3.7
If X, Y are two Banach spaces, $A \in \mathcal{L}(X; Y)$ and $\varphi \colon Y \longrightarrow \overline{\mathbb{R}}$ is a proper and convex function which is continuous at some point $y_0 \in R(A)$, then

$$A^* \big(\partial \varphi(Ax) \big) \;=\; \partial(\varphi \circ A)(x) \qquad \forall \, x \in X.$$

PROOF Let $x^* \in \partial(\varphi \circ A)(x)$. By definition we have

$$\langle x^*, z - x \rangle_X + (\varphi \circ A)(x) \;\leq\; (\varphi \circ A)(z) \qquad \forall \, z \in X.$$

Let

$$V \overset{df}{=} \{(y,\mu) \in Y \times \mathbb{R} : y = Az,$$
$$\mu = \langle x^*, z - x \rangle_X + (\varphi \circ A)(x), \; z \in X\}.$$

Evidently $V \cap \text{int epi}\,\varphi = \emptyset$. So we can apply the Weak Separation Theorem (see Theorem A.3.3) and obtain $(y^*, \gamma) \in (Y^* \times \mathbb{R}) \setminus \{(0,0)\}$, such that

$$\langle y^*, u \rangle_X + \gamma\lambda \;\leq\; \langle y^*, y \rangle_X + \gamma\mu \qquad \forall\, (y,\mu) \in V, \; (u,\lambda) \in \text{epi}\,\varphi.$$

Because of the nature of the epigraph we have that $\gamma \leq 0$. If $\gamma = 0$, then $y^* \neq 0$ and

$$\langle y^*, u \rangle_X \;\leq\; \langle y^*, y \rangle_X \qquad \forall\, u \in \text{dom}\,\varphi, \; y \in R(A),$$

which contradicts the hypothesis that $\text{int epi}\,\varphi \cap R(A) \neq \emptyset$. So $\gamma < 0$ and we can always assume that $\gamma = -1$. So

$$\langle y^*, u \rangle_X - \varphi(u) \;\leq\; \langle y^*, y \rangle_Y - \mu \qquad \forall\, u \in \text{dom}\,\varphi, \; (y,\mu) \in V.$$

Therefore
$$\langle y^*, A(z - x) \rangle_X \;=\; \langle x^*, z - x \rangle_X \qquad \forall\, z \in X$$

and so $x^* = A^* y^*$. Then

$$\langle y^*, u - Ax \rangle_X + \varphi(Ax) \;\leq\; \varphi(u) \qquad \forall\, u \in \text{dom}\,\varphi$$

and so $y^* \in \partial\varphi(Ax)$, hence $A^* y^* = x^* \in A^* \partial\varphi(Ax)$. Thus we have proved that $\partial(\varphi \circ A)(x) \subseteq A^* \partial\varphi(Ax)$. Since the opposite inclusion is always true, we obtain the thesis. $\qquad\Box$

1.3.4   Clarke Subdifferential

Now we pass to the nonconvex part of Nonsmooth Analysis, starting with an outline of the basic aspects of Clarke's theory. Again X is a Banach space.

DEFINITION 1.3.5   *A function $\varphi\colon X \longrightarrow \mathbb{R}$ is said to be **locally Lipschitz**, if for every $x \in X$ there exists $U \in \mathbb{N}(x)$ and a constant $k_U > 0$, such that*

$$|\varphi(y) - \varphi(z)| \;\leq\; k_U \, \|y - z\|_X \qquad \forall\, y, z \in U.$$

REMARK 1.3.3   From Theorem 1.3.2 we know that if $\varphi \in \Gamma_0(X)$, then $\varphi|_{\text{int dom}\,\varphi}$ is locally Lipschitz. $\qquad\Box$

The next notion generalizes the directional derivative of convex functions.

DEFINITION 1.3.6 *For a given locally Lipschitz function* $\varphi \colon X \longrightarrow \mathbb{R}$, *the **generalized directional derivative** of* φ *at* $x \in X$ *in the direction* $h \in X$ *is defined by*

$$\varphi^0(x; h) \overset{df}{=} \limsup_{\substack{y \to x \\ \lambda \searrow 0}} \frac{\varphi(y + \lambda h) - \varphi(y)}{\lambda}$$

$$= \inf_{\substack{\varepsilon, \delta > 0}} \sup_{\substack{\|x - y\|_X < \varepsilon \\ 0 < \lambda < \delta}} \frac{\varphi(y + \lambda h) - \varphi(y)}{\lambda}.$$

REMARK 1.3.4 This definition involves only the behaviour of φ near x and the limit supremum is finite. □

Based on Definition 1.3.6 one can easily verify the following facts.

PROPOSITION 1.3.7
If $\varphi \colon X \longrightarrow \mathbb{R}$ *is locally Lipschitz,*
<u>*then*</u>

(a) *for every* $x \in X$, *the function* $h \longmapsto \varphi^0(x; h)$ *is sublinear and continuous;*

(b) *the function* $X \times X \ni (x, h) \longmapsto \varphi^0(x; h) \in \mathbb{R}$ *is upper semicontinuous;*

(c) *for all* $(x, h) \in X \times X$, *we have* $\varphi^0(x; -h) = (-\varphi)^0(x; h)$.

Proposition 1.3.7(a) and Hahn-Banach theorem lead to the following definition:

DEFINITION 1.3.7 *Let* $\varphi \colon X \longrightarrow \mathbb{R}$ *be a locally Lipschitz function. The **generalized subdifferential** of* φ *at* $x \in X$ *is the nonempty set* $\partial \varphi(x) \subseteq X^*$ *defined by*

$$\partial \varphi(x) \overset{df}{=} \left\{ x^* \in X^* \colon \langle x^*, h \rangle_X \leq \varphi^0(x; h) \text{ for all } h \in X \right\}.$$

REMARK 1.3.5 Note that for the generalized subdifferential of φ (see Definition 1.3.7), we use the same notation $\partial \varphi(x)$ as for the subdifferential of a convex function φ (see Definition 1.3.3). This is justified by the fact that if φ is a convex function, then both definitions are equivalent. This fact will be shown in Proposition 1.3.11.

If $X = \mathbb{R}^N$, then $\varphi'(x)$ exists for all $\xi \in \mathbb{R}^N \setminus N_f$, where N_f is a Lebesgue-null set in $\mathbb{R}^N$. Suppose that S is any set of Lebesgue measure zero in $\mathbb{R}^N$. Then $\partial \varphi(x) = \text{conv} \left\{ \lim_{n \to +\infty} \varphi'(x_n) \colon x_n \longrightarrow x, \ x_n \notin S \cup N_f \right\}.$ □

From the last definition we deduce at once the following facts about the set $\partial\varphi(x)$.

PROPOSITION 1.3.8

If $\varphi\colon X \longrightarrow \mathbb{R}$ is a locally Lipschitz function,
<u>*then*</u> *for every $x \in X$, the set $\partial\varphi(x) \subseteq X^*$ is nonempty, w^*-compact and for every $y \in U \in \mathcal{N}(x)$ and every $x^* \in \partial\varphi(y)$ we have $\|x^*\|_{X^*} \leq k_U$ and*

$$\sigma_{X^*}\big(h, \partial\varphi(x)\big) \;=\; \varphi^0(x;h) \qquad \forall\, h \in X.$$

In the next proposition we give a useful property of this multifunction $X \ni x \longmapsto \partial\varphi(x) \in 2^{X^*} \setminus \{\emptyset\}$.

PROPOSITION 1.3.9

If $\varphi\colon X \longrightarrow \mathbb{R}$ is a locally Lipschitz function,
<u>*then*</u> *the multifunction $x \longmapsto \partial\varphi(x)$ is upper semicontinuous from X into $X^*_{w^*}$.*

PROOF   From Proposition 1.3.8 and from the Alaoglu theorem (see Theorem A.3.1) it follows that $\partial\varphi$ is locally compact into $X^*_{w^*}$. So according to Proposition 1.2.5 it suffices to show that $\operatorname{Gr}\partial\varphi \subseteq X \times X^*_{w^*}$ is closed. To this end suppose that $x_\alpha \longrightarrow x$ in X, $x^*_\alpha \xrightarrow{w^*} x^*$ in X^* and $x^*_\alpha \in \partial\varphi(x_\alpha)$. We have

$$\langle x^*_\alpha, h \rangle_X \;\leq\; \varphi^0(x_\alpha; h) \qquad \forall\, h \in X.$$

Passing to the limit we obtain

$$\langle x^*, h \rangle_X \;\leq\; \varphi^0(x; h) \qquad \forall\, h \in X,$$

hence $x^* \in \partial\varphi(x)$.   ◻

In general differentiability at $x \in X$ is not enough to guarantee that $\partial\varphi(x)$ is a singleton.

EXAMPLE 1.3.1   The function $\varphi\colon \mathbb{R} \longrightarrow \mathbb{R}$ defined by

$$\varphi(x) \;\overset{df}{=}\; \begin{cases} x^2 \sin\frac{1}{x} & \text{if } x \neq 0, \\ 0 & \text{if } x = 0 \end{cases}$$

is locally Lipschitz near 0 and $\varphi^0(0; h) = |h|$. So $\partial\varphi(0) = [-1, 1] \supseteq \varphi'(0) = \{0\}$.   ◻

To have a singleton subdifferential, we need stronger conditions on φ. We present such a result. A more general version of it can be found in Clarke (1983, p. 31).

PROPOSITION 1.3.10
If $\varphi \in C^1(X)$, *then* $\partial\varphi(x) = \{\varphi'(x)\}$.

How do the notions of generalized directional derivative and generalized subdifferential compare with the corresponding notions for convex functions? In the next proposition we show that in fact Clarke's theory is an extension of the subdifferential theory for continuous, convex functions.

PROPOSITION 1.3.11
If $\varphi\colon X \longrightarrow \mathbb{R}$ *is a continuous convex function (hence locally Lipschitz), then Clarke's subdifferential* $\partial\varphi(x)$ *coincides with the convex subdifferential introduced in Definition 1.3.3 and for all* $x \in X$ *we have* $\varphi^0(x; \cdot) = \varphi'(x; \cdot)$.

PROOF For the needs of this proof, by $\partial_c\varphi$ we denote the subdifferential of φ in the sense of convex analysis. By virtue of Proposition 1.3.5 to prove the result it suffices to show that $\varphi^0(x; h) = \varphi'(x; h)$ for all $h \in X$. Since φ is convex, we know that

$$\sup_{0 < \lambda < \varepsilon} \frac{\varphi(y + \lambda h) - \varphi(y)}{\lambda} = \frac{\varphi(y + \varepsilon h) - \varphi(y)}{\varepsilon}.$$

Because φ is locally Lipschitz, for $\delta > 0$ small enough and $y \in B_{\varepsilon\delta}(x)$ we have

$$\left| \frac{\varphi(y + \lambda h) - \varphi(y)}{\lambda} - \frac{\varphi(x + \lambda h) - \varphi(x)}{\lambda} \right| \leq 2\delta k,$$

where $k > 0$ is the local Lipschitz constant. Hence

$$\varphi^0(x; h) \leq 2\delta k + \varphi'(x; h).$$

Let $\delta \searrow 0$ to conclude that $\varphi^0(x; h) \leq \varphi'(x; h)$. Since the opposite inequality is always true, we conclude that $\varphi^0(x; h) = \varphi'(x; h)$ for all $h \in X$ and so $\partial\varphi(x) = \partial_c\varphi(x)$. $\qquad\Box$

DEFINITION 1.3.8 *If for a locally Lipschitz (not necessarily convex) function, we have that for all* $h \in X$, $\varphi'(x; h)$ *exists and equals* $\varphi^0(x; h)$, *then we say that* φ *is* **regular** *at* $x \in X$.

The second statement in the next proposition generalizes Fermat's principle to the present nonsmooth and nonconvex setting.

PROPOSITION 1.3.12
If $\varphi\colon X \longrightarrow \mathbb{R}$ *is a locally Lipschitz function, then*

(a) $\partial(\lambda\varphi)(x) = \lambda\partial\varphi(x)$ *for all* $\lambda \in \mathbb{R}$;

(b) *if* $x \in X$ *is a local extremum of* φ, *then* $0 \in \partial\varphi(x)$.

PROOF   **(a)** If $r \geq 0$, then $(r\varphi)^0 = r\varphi^0$ and so $\partial(r\varphi) = r\partial\varphi$. If $r < 0$ it suffices to assume that $r = -1$. From Proposition 1.3.7(c) we know that $(-\varphi)^0(x; h) = \varphi^0(x; -h)$. So for $x^* \in \partial(-\varphi)(x)$ we have

$$\langle x^*, h\rangle_X \leq (-\varphi)^0(x; h) = \varphi^0(x; -h)$$

and so $-x^* \in \partial\varphi(x)$. Since the argument is reversible, we conclude that $\partial(-\varphi)(x) = -\partial\varphi(x)$.

(b) Since from **(a)**, we have that $\partial(-\varphi)(x) = -\partial\varphi(x)$, it suffices to assume that x is a local minimum. We have

$$0 \leq \varphi^0(x; h) \qquad \forall\, h \in X$$

and so $0 \in \partial\varphi(x)$.   ▯

Next we shall present some useful calculus rules for the generalized subdifferential.

PROPOSITION 1.3.13
If $\varphi_k \colon X \longrightarrow \mathbb{R}$ *with* $k \in \{1, \ldots, N\}$ *are locally Lipschitz functions,* <u>*then*</u>

$$\partial\left(\sum_{k=1}^{N} \varphi_k\right)(x) \subseteq \sum_{k=1}^{N} \partial\varphi_k(x)$$

and equality holds if the functions are regular (see Definition 1.3.8).

PROOF   It suffices to assume that $N = 2$ and the general case follows by induction. We have

$$\sigma_{X^*}\big(h; \partial(\varphi_1 + \varphi_2)(x)\big) = (\varphi_1 + \varphi_2)^0(x; h)$$
$$\leq \varphi_1^0(x; h) + \varphi_2^0(x; h) = \sigma_{X^*}\big(h; \partial\varphi_1(x)\big) + \sigma_{X^*}\big(h; \partial\varphi_2(x)\big) \qquad \forall\, h \in X$$

and so
$$\partial(\varphi_1 + \varphi_2)(x) \subseteq \partial\varphi_1(x) + \partial\varphi_2(x).$$

If φ_1 and φ_2 are regular, then so is $\varphi_1 + \varphi_2$ and so

$$(\varphi_1 + \varphi_2)^0(x; h) = (\varphi_1 + \varphi_2)'(x; h)$$
$$= \varphi_1'(x; h) + \varphi_2'(x; h) = \varphi_1^0(x; h) + \varphi_2^0(x; h) \qquad \forall\, h \in X.$$

Therefore

$$\partial(\varphi_1 + \varphi_2)(x) \;=\; \partial\varphi_1(x) + \partial\varphi_2(x).$$

$\square$

The following ***mean value property*** will be a very useful tool in our study of "nonsmooth" boundary value problems.

PROPOSITION 1.3.14 (Mean Value Theorem for Locally Lipschitz Functions)

If $x, y \in X$ *and* $\varphi \colon X \longrightarrow \mathbb{R}$ *is a locally Lipschitz function on an open set containing* $[x, y] \overset{df}{=} \{(1 - t)x + ty : \; 0 \leq t \leq 1\}$,
then there exists $z \in (x, y) \overset{df}{=} \{(1 - t)x + ty : \; 0 < t < 1\}$ *and* $z^* \in \partial\varphi(z)$, *such that* $\varphi(y) - \varphi(x) = \langle z^*, y - x \rangle_X$.

PROOF   Consider the function $\xi \colon [0, 1] \longrightarrow \mathbb{R}$ defined by

$$\xi(t) \overset{df}{=} \varphi(x_t) \qquad \text{with } x_t \overset{df}{=} x + t(y - x).$$

Evidently ξ is locally Lipschitz and we have

$$\xi^0(t; r) \overset{df}{=} \underset{\substack{s \to t \\ \lambda \searrow 0}}{\lim\sup} \frac{\xi(s + \lambda r) - \xi(s)}{\lambda}$$

$$= \underset{\substack{s \to t \\ \lambda \searrow 0}}{\lim\sup} \frac{\varphi\big(x + (s + \lambda r)(y - x)\big) - \varphi\big(x + s(y - x)\big)}{\lambda}$$

$$\leq \underset{\substack{y \to x_t \\ \lambda \searrow 0}}{\lim\sup} \frac{\varphi\big(y + \lambda r(y - x)\big) - \varphi(y)}{\lambda} \;=\; \varphi^0\big(x_t; r(y - x)\big).$$

Taking $r = \pm 1$ we obtain

$$\partial\xi(t) \subseteq \big\{ \langle u^*, y - x \rangle_X : \; u^* \in \partial\varphi(x_t) \big\}.$$

Now consider the function $\vartheta \colon [0, 1] \longrightarrow \mathbb{R}$ defined by

$$\vartheta(t) \overset{df}{=} \varphi(x_t) + t\big(\varphi(x) - \varphi(y)\big).$$

We have that $\vartheta(0) = \vartheta(1) = \varphi(x)$. So there exists $t \in (0, 1)$ where ϑ has a local extremum and so from Proposition 1.3.13 we have

$$0 \in \partial\vartheta(t) \;=\; \partial\xi(t) + \varphi(x) - \varphi(y).$$

Therefore

$$0 \in \langle \partial \varphi(u), y - x \rangle_X + \varphi(x) - \varphi(y) \qquad \forall \, u \in (x, y).$$

□

The **chain rules** that follow will be useful when we look for positive or negative solutions for boundary value problems.

PROPOSITION 1.3.15

If X, Y are Banach spaces, $g \colon X \longrightarrow Y$ is a continuously Gâteaux differentiable function and $\psi \colon Y \longrightarrow \mathbb{R}$ is a locally Lipschitz function,
<u>*then*</u>

(a) *$\varphi = \psi \circ g \colon X \longrightarrow \mathbb{R}$ is locally Lipschitz;*

(b) *$\partial \varphi(x) \subseteq \partial \psi(g(x)) \circ g'(x) \stackrel{df}{=} \{ y^* \circ g'(x) \colon y^* \in \partial \psi(g(x)) \}$.*

Moreover, if ψ (or $-\psi$) is regular at $g(x)$, then φ (or $-\varphi$) is regular at x and

$$\partial \varphi(x) \; = \; \partial \psi(g(x)) \circ g'(x).$$

PROOF   Clearly φ is locally Lipschitz. We need to show that

$$\varphi^0(x; h) \; \leq \; \max \{ \langle y^*, g'(x) \rangle_X \colon y^* \in \partial \psi(g(x)) \} \qquad \forall \, h \in X. \qquad (1.3)$$

From Proposition 1.3.14 we know that for some $y \in (g(v), g(v + \lambda h))$ and some $y^* \in \partial \psi(y)$ we have

$$\frac{|\varphi(v + \lambda h) - \varphi(v)|}{\lambda} \; = \; \frac{|\langle y^*, g(v + \lambda h) - g(v) \rangle_X|}{\lambda} \in \operatorname{conv} g'([v, v + \lambda h]),$$

where the last inclusion is a consequence of the classical Mean Value Theorem. Let $v \to x$ in X and $\lambda \searrow 0$. Then by the Alaoglu theorem (see Theorem A.3.1) and by passing to a suitable subnet, we may assume that $y^* \xrightarrow{w^*} y_0^* \in \partial \psi(g(x))$ (see Proposition 1.3.9). Thus we have that

$$\varphi^0(x; h) \; \leq \; \langle y^*, g'(x)h \rangle_X \qquad \forall \, h \in X.$$

So (1.3) holds.

Suppose that ψ is regular at $g(x)$. We have

$$\max \{ \langle y^*, g'(x)h \rangle_X \colon y^* \in \partial \psi(g(x)) \} \; = \; \psi^0(g(x); g'(x)h)$$

$$= \; \psi'(g(x); g'(x)h) \; = \; \lim_{\lambda \searrow 0} \frac{\psi(g(x) + \lambda g'(x)h) - \psi(g(x))}{\lambda}$$

$$= \; \lim_{\lambda \searrow 0} \frac{(\psi \circ g)(x + \lambda h) - (\psi \circ g)(x)}{\lambda} \; = \; \varphi'(x; h) \; \leq \; \varphi^0(x; h). \qquad (1.4)$$

From (1.3) and (1.4) we infer that

$$\partial\varphi(x) \;=\; \{y^* \circ g'(x): \; y^* \in \partial\psi(g(x))\}.$$

⧠

REMARK 1.3.6   This chain rule is usually expressed as

$$\partial\varphi(x) \;\subseteq\; g'(x)^*\partial\psi(g(x)).$$

⧠

The second **chain rule** is proved in a similar way. We omit the proof here.

PROPOSITION 1.3.16
If $\psi\colon X \longrightarrow \mathbb{R}^N$, with $\psi = (\psi_1,\ldots,\psi_N)$ and $f\colon \mathbb{R}^N \longrightarrow \mathbb{R}$ are locally Lipschitz functions and $\varphi = f \circ \psi$,
then φ is locally Lipschitz and

$$\partial\varphi(x) \;\subseteq\; \overline{\mathrm{conv}}^{\,w^*}\left\{\sum_{k=1}^N \xi_k x_k^* : \; x_k^* \in \partial\psi_k(x), \; \xi = (\xi_1,\ldots,\xi_N) \in \partial f(\psi(x))\right\}.$$

Moreover, equality holds under any of the following conditions:

(i) *f is regular at $\psi(x)$, each ψ_k is regular at x and $\partial f(\psi(x)) \subseteq \mathbb{R}_+^N$ (then also φ is regular at x);*

(ii) *$N = 1$ and $f \in C^1(\mathbb{R})$ (in this case $\partial\varphi(x) = f'(\psi(x))\partial\psi(x)$);*

(iii) *f is regular at $\psi(x)$ and ψ is C^1 at x (then φ is regular at x and $\partial\varphi(x) = \partial f(\psi(x)) \circ \psi'(x)$).*

We have three useful consequences of the above chain rule.

COROLLARY 1.3.5
If $\varphi_k\colon X \longrightarrow \mathbb{R}$ are locally Lipschitz functions for $k \in \{1,\ldots,N\}$ and $\varphi = \max\limits_{1\leq k\leq N} \varphi_k$,
then φ is locally Lipschitz and

$$\partial\varphi(x) \;\subseteq\; \mathrm{conv}\left\{\partial\varphi_k(x): \; k \in J(x)\right\}$$

where

$$J(x) \stackrel{df}{=} \{k \in \{1,\ldots,N\} : \varphi_k(x) = \varphi(x)\}.$$

Moreover, if φ_k is regular at x for every $k \in J(x)$, then φ is regular at x and equality holds.

PROOF   Apply Proposition 1.3.16 with $\psi = (\varphi_1, \ldots, \varphi_N)$ and the function $f \colon \mathbb{R}^N \longrightarrow \mathbb{R}$ defined by

$$f(y) \stackrel{df}{=} \max_{1 \leq k \leq N} y_k \qquad \forall\, y = (y_1, \ldots, y_N) \in \mathbb{R}^N.$$

$\square$

COROLLARY 1.3.6
If $\varphi_1, \varphi_2 \colon X \longrightarrow \mathbb{R}$ are locally Lipschitz functions and $\varphi = \varphi_1 \varphi_2$, then φ is locally Lipschitz and

$$\partial(\varphi_1 \varphi_2)(x) \subseteq \varphi_2(x) \partial \varphi_1(x) + \varphi_1(x) \partial \varphi_2(x) \qquad \forall\, x \in X.$$

Moreover, if $\varphi_1(x) \geq 0$, $\varphi_2(x) \geq 0$ and φ_1, φ_2 are both regular at x, then φ is regular at x and equality holds.

PROOF   Apply Proposition 1.3.16 with

$$\psi(x) \stackrel{df}{=} \big(\varphi_1(x), \varphi_2(x)\big) \quad \text{and} \quad f(y_1, y_2) \stackrel{df}{=} y_1 y_2.$$

$\square$

Similarly we also have the following quotient rule.

COROLLARY 1.3.7
If $\varphi_1, \varphi_2 \colon X \longrightarrow \mathbb{R}$ are locally Lipschitz near $x \in X$ and $\varphi_2(x) \neq 0$, then $\frac{\varphi_1}{\varphi_2}$ is locally Lipschitz near x and

$$\partial \left(\frac{\varphi_1}{\varphi_2}\right)(x) \subseteq \frac{\varphi_2(x) \partial \varphi_1(x) - \varphi_1(x) \partial \varphi_2(x)}{\varphi_2(x)^2}.$$

Moreover, if $\varphi_1(x) \geq 0$, $\varphi_2(x) > 0$ and $\varphi_1, -\varphi_2$ are regular at x, then $\frac{\varphi_1}{\varphi_2}$ is regular at x and equality holds.

Next let us present a consequence of the first chain rule, which is useful in the study of nonsmooth boundary value problems.

PROPOSITION 1.3.17
If X, Y are Banach spaces, the embedding $X \subseteq Y$ is continuous and dense, $\psi \colon Y \longrightarrow \mathbb{R}$ is locally Lipschitz and $\varphi = \psi|_X$, then φ is locally Lipschitz and $\partial \varphi(x) = \partial \psi(x)$ for all $x \in X$.

PROOF   Let $e: X \longrightarrow Y$ be the continuous embedding. Then $\varphi = \psi \circ e$. Using Proposition 1.3.15 we obtain

$$\partial \varphi(x) \ = \ e^* \partial \psi(x) \ = \ \partial \psi(x) \qquad \forall \, x \in X$$

(since $e^* \in \mathcal{L}(Y^*; X^*)$ is the restriction on $Y^* \subseteq X^*$ operator). ∎

PROPOSITION 1.3.18
If $g \in C^1\big([0,1]; X\big)$ and $\varphi: X \longrightarrow \mathbb{R}$ is locally Lipschitz,
then $\xi = \varphi \circ g: [0,1] \longrightarrow \mathbb{R}$ is differentiable almost everywhere and

$$\xi'(t) \ \leq \ \sigma\big(g'(t); \partial\varphi(g(t))\big).$$

PROOF   Evidently ξ is locally Lipschitz and so differentiable almost everywhere. Let t be such a point of differentiability of ξ. We have

$$\xi'(t) \ = \ \lim_{s \to 0} \frac{1}{s}\big[\varphi\big(g(t+s)\big) - \varphi\big(g(t)\big)\big]$$

$$= \ \lim_{s \to 0} \frac{1}{s}\big[\varphi\big(g(t) + g'(t) + o(s)\big) - \varphi\big(g(t)\big)\big]$$

(where $\frac{o(s)}{s} \longrightarrow 0$ as $s \to 0$)

$$= \ \lim_{s \to 0} \frac{1}{s}\big[\varphi\big(g(t) + g'(t) + o(s)\big) - \varphi\big(g(t) + g'(t)s\big)\big]$$

$$+ \lim_{s \to 0} \frac{1}{s}\big[\varphi\big(g(t) + g'(t)s\big) - \varphi(g(t))\big]$$

$$\leq \ \lim_{s \to 0} \frac{1}{s}\big[\varphi\big(g(t) + g'(t)s\big) - \varphi(g(t))\big]$$

(since φ is locally Lipschitz)

$$\leq \ \limsup_{\substack{h \,\to\, 0 \\ s \,\searrow\, 0}} \frac{1}{s}\big[\varphi\big(g(t) + h + g'(t)s\big) - \varphi\big(g(t) + h\big)\big]$$

$$= \ \varphi^0\big(g(t); g'(t)\big) \ = \ \sigma\big(g'(t); \partial\varphi(g(t))\big).$$

∎

Let $H: \mathbb{R}^N \times \mathbb{R}^k \longrightarrow \mathbb{R}^k$ be a locally Lipschitz function and consider the equation

$$H(y, z) \ = \ 0.$$

We wish to solve this equation for z as a function of y, around a point (y_0, z_0) at which the equation holds. The notation $p_z \partial H(y_0, z_0)$ signifies the set of all

$k \times k$ matrices M such that for some $k \times N$ matrix L, the $k \times (k + N)$ matrix $[L, M]$ belongs to $\partial H(y_0, z_0)$. We can state the following theorem.

THEOREM 1.3.8 (Nonsmooth Implicit Function Theorem)

If $\mathrm{proj}_z\, \partial H(y_0, z_0)$ *is of maximal rank,*

then there exists a neighbourhood U of y_0 and a Lipschitz map $h\colon U \longrightarrow \mathbb{R}^k$ such that $h(y_0) = z_0$ and $H\big(y, h(y)\big) = 0$ for all $y \in U$.

Next we shall state two theorems concerning the generalized subdifferential of certain **integral functionals**. For the proofs of these results we refer to Clarke (1983).

Let (Ω, Σ, μ) be a σ-finite measure space, X a separable Banach space and $\varphi\colon \Omega \times X \longrightarrow \mathbb{R}$ an integrand which satisfies the following hypotheses:

$\underline{H(\varphi)_1}$ $\varphi\colon \Omega \times X \longrightarrow \mathbb{R}$ is a function, such that:

(i) for every $x \in X$, the function $\Omega \ni \omega \longmapsto \varphi(\omega, x) \in \mathbb{R}$ is Σ-measurable;

(ii) for any bounded subset $B \subseteq X$, there exists $k_B \in L^1(\Omega)$, such that for almost all $\omega \in \Omega$ and all $x, y \in B$ we have

$$|\varphi(\omega, x) - \varphi(\omega, y)| \ \leq\ k_B(\omega)\, \|x - y\|_X\,.$$

We consider the integral functional $\Phi\colon X \longrightarrow \mathbb{R}$ defined by

$$\Phi(x) \overset{df}{=} \int_\Omega \varphi(\omega, x)\, d\mu.$$

THEOREM 1.3.9

If $\varphi\colon \Omega \times X \longrightarrow \mathbb{R}$ *satisfies hypotheses $H(\varphi)_1$ and Φ is finite at some point $x \in X$,*

then Φ is finite, it is Lipschitz on every bounded subset of X and

$$\partial\Phi(y) \ \subseteq\ \int_\Omega \partial\varphi(\omega, y)\, d\mu \qquad \forall\, y \in X.$$

Moreover, if $\varphi(\omega, \cdot)$ is μ-almost everywhere regular at $y_0 \in X$, then so is Φ and

$$\partial\Phi(y_0) \ =\ \int_\Omega \partial\varphi(\omega, y_0)\, d\mu.$$

REMARK 1.3.7 In the above theorem by $\int\limits_{\Omega} \partial\varphi(\omega, y)\, d\mu$ we mean all $u^* \in X^*$, such that

$$\langle u^*, h \rangle_X = \int\limits_{\Omega} \langle x^*(\omega), h \rangle_X \, d\mu \qquad \forall\, h \in X,$$

with $x^* \colon \Omega \longrightarrow X^*$ being w^*-measurable and $x^*(\omega) \in \partial\varphi(\omega, x)$ for almost all $\omega \in \Omega$ (set-valued Gelfand or w^*-integral of $\partial\varphi(\omega, x)$). ☐

The second theorem is about integral functionals on the Lebesgue-Bochner space $L^p(\Omega; X)$, $p \in [1, +\infty)$. Let (Ω, Σ, μ) be a complete finite measure space and X a separable Banach space. The hypotheses on the integrand $\varphi(\omega, x)$ are the following:

$\underline{H(\varphi)_2}$ $\varphi \colon \Omega \times X \longrightarrow \mathbb{R}$ is a function, such that:

(i) for every $x \in X$, the function $\Omega \ni \omega \longmapsto \varphi(\omega, x) \in \mathbb{R}$ is Σ-measurable;

(ii) there exists $k \in L^{p'}(\Omega)$ $(\frac{1}{p} + \frac{1}{p'} = 1)$, such that for μ-almost all $\omega \in \Omega$ and all $x, y \in X$ we have

$$\big|\varphi(\omega, x) - \varphi(\omega, y)\big| \leq k(\omega)\, \|x - y\|_X\,.$$

$\underline{H(\varphi)_3}$ $\varphi \colon \Omega \times X \longrightarrow \mathbb{R}$ is a function, such that:

(i) for every $x \in X$, the function $\Omega \ni \omega \longmapsto \varphi(\omega, x) \in \mathbb{R}$ is Σ-measurable;

(ii) for μ-almost all $\omega \in \Omega$, the function $X \ni x \longmapsto \varphi(\omega, x) \in \mathbb{R}$ is locally Lipschitz;

(iii) for μ-almost all $\omega \in \Omega$, all $x \in X$ and all $x^* \in \partial\varphi(\omega, x)$ we have $\|x^*\|_{X^*} \leq a(\omega) + c\, \|x\|_X^{p-1}$ with $a \in L^\infty(\Omega)$, $c > 0$.

The integral functional I_φ on $L^p(\Omega; X)$ is defined by

$$I_\varphi(y) \overset{df}{=} \int\limits_{\Omega} \varphi\big(\omega, y(\omega)\big)\, d\mu \qquad \forall\, y \in L^p(\Omega; X).$$

THEOREM 1.3.10
If $\varphi \colon \Omega \times X \longrightarrow \mathbb{R}$ satisfies hypotheses $H(\varphi)_2$ or $H(\varphi)_3$, then I_φ is Lipschitz on every bounded set and

$$\partial I_\varphi(x) \subseteq S^{p'}_{\partial\varphi(\cdot, x(\cdot))} \subseteq L^{p'}(\Omega; X^*_{w^*}).$$

Moreover, if for μ-almost all $\omega \in \Omega$, $\varphi(\omega, \cdot)$ is regular at $x(\omega)$, then I_φ is regular at x and

$$\partial I_\varphi(x) = S^{p'}_{\partial\varphi(\cdot,x(\cdot))}.$$

We conclude our discussion of Clarke's theory with a few basic things about tangent cones, normal cones and the constrained minimizer of a locally Lipschitz function.

Let X be a Banach space and $C \subseteq X$ a nonempty subset. The ***distance function***

$$X \ni x \longmapsto d(x,C) \stackrel{df}{=} \inf\left\{ \|x - c\|_X : c \in C \right\} \in \mathbb{R}$$

is Lipschitz. Using this fact we introduce the following definition.

DEFINITION 1.3.9   *Let $x \in X$. The **tangent cone** to C at x is the set*

$$T_C(x) \stackrel{df}{=} \{h \in X : d^0(x,C;h) \leq 0\}.$$

*The **normal cone** to C at x is the **polar cone** of $T_C(x)$, i.e.*

$$N_C(x) \stackrel{df}{=} (T_C(x))^* = \{x^* \in X^* : \langle x^*, h\rangle_X \leq 0 \ \ \forall\, h \in T_C(x)\}.$$

REMARK 1.3.8   Because the function $X \ni h \longmapsto d^0(x,C;h) \in \mathbb{R}$ is positively homogeneous and continuous, the tangent cone $T_C(x)$ is always closed and convex. Of course the same is true for $N_C(x)$. If C is smooth manifold and $x \in C$, then $T_C(x)$ is the usual ***tangent space*** to C at x. If C is convex, then $N_C(x)$ coincides with the normal cone of convex analysis, namely

$$N_C(x) \stackrel{df}{=} \left\{ x^* \in X^* : \langle x^*, x\rangle_X = \sup_{u \in C} \langle x^*, u\rangle_X = \sigma_X(x^*,C) \right\}.$$

Also for any nonempty $C \subseteq X$, we have

$$N_C(x) = \overline{\bigcup_{\lambda \geq 0} \lambda \partial d(x,C)}^{w^*}.$$

$\square$

The next result is useful because it replaces a ***constrained minimization problem*** by an unconstrained one.

PROPOSITION 1.3.19
If $\varphi\colon X \longrightarrow \mathbb{R}$ is Lipschitz with constant $k_0 > 0$ on $U \subseteq X$, $x_0 \in C \subseteq U$ and $\varphi(x_0) = \inf_C \varphi$,

then *for any* $k \geq k_0$, *the function*

$$X \ni y \longmapsto \psi(y) \stackrel{df}{=} \varphi(y) + kd(y, C) \in \mathbb{R}$$

attains its minimum over U *at* x_0. *If* $k > k_0$ *and* C *is closed, then any other minimizer of* ψ *over* U *also lies in* C.

PROOF   Suppose that x_0 does not minimize ψ over U. Then we can find $y \in U$ and $\varepsilon > 0$, such that

$$\varphi(y) + kd(y, C) < \varphi(x_0) - k\varepsilon.$$

Let $c \in C$ be such that $\|y - c\|_X \leq d(y, C) + \varepsilon$. We have

$$\varphi(c) \leq \varphi(y) + k\|y - c\|_X \leq \varphi(y) + k\big(d(y, C) + \varepsilon\big) < \varphi(x_0),$$

which contradicts the hypothesis that x_0 minimizes φ on C. Next let $k > k_0$ and let x_1 be a minimizer of ψ on U. From the first assertion we have

$$\varphi(x_1) + kd(x_1, C) = \varphi(x_0) \leq \varphi(x_1) + \frac{1}{2}(k_0 + k)d(x_1, C)$$

and so $d(x_1, C) = 0$, i.e. $x_1 \in C$.   ▯

COROLLARY 1.3.8
If φ *is Lipschitz on* $U \in \mathcal{N}(x_0)$ *and* x_0 *minimizes* φ *over* C,
then $0 \in \partial\varphi(x_0) + N_C(x_0)$.

PROOF   We may assume that $C \subseteq U$ (since $N_C(x_0) = N_{C \cap U}(x_0)$). Using Proposition 1.3.19 we have that

$$0 \in \partial\big(\varphi + kd(\cdot, C)\big)(x_0) \subseteq \partial\varphi(x_0) + k\partial d(x_0, C) \subseteq \partial\varphi(x_0) + N_C(x_0)$$

(see Remark 1.3.8).   ▯

We conclude our discussion of the subdifferential theory of locally Lipschitz functions with a multiplier rule. Suppose that X is a Banach space and $f, g \colon X \longrightarrow \mathbb{R}$ are locally Lipschitz functions. We consider the constrained minimization problem:

$$\inf_{\substack{x \in X \\ g(x) = 0}} f(x). \tag{1.5}$$

THEOREM 1.3.11
If $x_0 \in X$ *solves* (1.5),
then there exist $r, \xi \geq 0$ *not both equal to zero such that*

$$0 \in r\partial f(x_0) + \xi\partial g(x_0).$$

Moreover, if g is also convex, then $r > 0$ (and so it can be taken to be equal to 1).

1.3.5   Weak Slope

In this last subsection we introduce a generalized notion of the derivative which in Chapter 2 will be used to develop a critical point theory for continuous functions on a metric space (X, d_X). In the following on $X \times \mathbb{R}$ we consider the metric

$$d_{X \times \mathbb{R}}\big((x, \lambda), (y, \mu)\big) \stackrel{df}{=} \big(d_X(x, y)^2 + (\lambda - \mu)^2\big)^{\frac{1}{2}}$$

and for a given function $\varphi\colon X \longrightarrow \overline{\mathbb{R}}$ we introduce the continuous function $E_\varphi\colon \operatorname{epi}\varphi \longrightarrow \mathbb{R}$ defined by $E_\varphi(x, \lambda) \stackrel{df}{=} \lambda$.

DEFINITION 1.3.10   *For every $x \in \operatorname{dom}\varphi$, by $|d\varphi|(x)$, we denote the supremum of all $\xi \geq 0$, such that there exist $\delta > 0$ and a map*

$$H\colon [0, \delta] \times \big(B_\delta(x, \varphi(x)) \cap \operatorname{epi}\varphi\big) \longrightarrow X$$

satisfying

$$d\big(H\big(t, (u, \lambda)\big), u\big) \leq t$$

and

$$\varphi\big(H\big(t, (u, \lambda)\big)\big) \leq \mu - \xi t$$

*for all $(u, \lambda) \in B_\delta\big(x, \varphi(x)\big) \cap \operatorname{epi}\varphi$ and $t \in [0, \delta]$. The quantity $|d\varphi|(x) \in \overline{\mathbb{R}}$ is called **weak slope** of φ at x.*

In the case of continuous φ, we can make this definition more precise.

PROPOSITION 1.3.20
If $\varphi\colon X \longrightarrow \mathbb{R}$ is continuous,
then the weak slope $|d\varphi|(x)$ is the supremum of all $\xi \geq 0$, such that there exist $\delta > 0$ and a continuous map

$$H\colon [0, \delta] \times B_\delta(x) \longrightarrow X,$$

such that

$$\begin{cases} d\big(H(t, u), u\big) \leq t \\ \varphi\big(H(t, u)\big) \leq \varphi(u) - \xi t \end{cases} \quad \forall\, (t, u) \in [0, \delta] \times B_\delta(x).$$

PROOF   Let $\delta' > 0$ and $H_1\colon [0, \delta'] \times \big(B_\delta(x, \varphi(x)) \cap \operatorname{epi}\varphi\big) \longrightarrow X$ be as postulated by Definition 1.3.10. Exploiting the continuity of φ, we can find $\delta > 0$ small enough and define $H\colon [0, \delta] \times B_\delta(x) \longrightarrow X$ by

$$H(t, u) \stackrel{df}{=} H_1\big(t, (u, \varphi(u))\big).$$

It is easy to see that this is the desired function.

Conversely, let $H\colon [0,\delta] \times B_\delta(x) \longrightarrow X$ be as in the proposition. Then $H_1\colon [0,\delta] \times \big(B_\delta(x,\varphi(x)) \cap \operatorname{epi}\varphi\big) \longrightarrow X$, defined by

$$H_1\big(t,(u,\lambda)\big) \overset{df}{=} H(t,u),$$

has the properties assumed in Definition 1.3.10. $\qquad$ ⬜

The weak slope for a proper function $\varphi\colon X \longrightarrow \overline{\mathbb{R}}$ can be defined in terms of the corresponding notion for the map

$$E_\varphi\colon \operatorname{epi}\varphi \ni (x,\lambda) \longmapsto \lambda \in \mathbb{R}.$$

Evidently E_φ is Lipschitz with constant 1. So

$$|dE_\varphi|(x,\lambda) \leq 1 \qquad \forall\, (x,\lambda) \in \operatorname{epi}\varphi.$$

PROPOSITION 1.3.21
If $\varphi\colon X \longrightarrow \overline{\mathbb{R}}$ *is proper and* $x \in \operatorname{dom}\varphi$,
then

$$|d\varphi|(x) \overset{df}{=} \begin{cases} \dfrac{|dE_\varphi|(x,\varphi(x))}{\sqrt{1 - |dE_\varphi|(x,\varphi(x))^2}} & \text{if } |dE_\varphi|(x,\varphi(x)) < 1, \\[2ex] +\infty & \text{if } |dE_\varphi|(x,\varphi(x)) = 1. \end{cases}$$

PROOF First we prove that $|d\varphi|(x)$ is bigger or equal to the right hand side. If $|dE_\varphi|(x,\varphi(x)) = 0$, the inequality is trivial. So suppose that $0 < \xi < |dE_\varphi|(x,\varphi(x))$. Because of the continuity of E_φ, from Proposition 1.3.20 we can find $\delta > 0$ and continuous map

$$H\colon [0,\delta] \times \big(B_\delta(x,\varphi(x)) \cap \operatorname{epi}\varphi\big) \longrightarrow \operatorname{epi}\varphi,$$

with the properties stated there. Let $0 < \delta' < \delta\sqrt{1 - \xi^2}$ and let

$$K\colon [0,\delta'] \times \big(B_{\delta'}(x,\varphi(x)) \cap \operatorname{epi}\varphi\big) \longrightarrow X$$

be defined by

$$K\big(t,(u,\lambda)\big) \overset{df}{=} H_1\left(\frac{t}{\sqrt{1 - \xi^2}},(u,\lambda)\right),$$

with H_1 being the first component of H. Clearly K is continuous and

$$d\big(K\big(t,(u,\lambda)\big),u\big)^2 \leq \frac{t^2}{1 - \xi^2} - \left|H_2\left(\frac{t}{\sqrt{1 - \xi^2}},(u,\lambda)\right) - \lambda\right|^2$$

$$\leq \frac{t^2}{1 - \xi^2} - \frac{\xi^2 t^2}{1 - \xi^2} = t^2,$$

with H_2 being the second component of H. Also we have

$$\varphi\big(K\big(t,(u,\lambda)\big)\big) \leq H_2\left(\frac{t}{\sqrt{1-\xi^2}},(u,\lambda)\right)$$

$$\leq E_\varphi(u,\lambda) - \frac{\xi}{\sqrt{1-\xi^2}}t = \lambda - \frac{\xi}{\sqrt{1-\xi^2}}t.$$

Thus from Definition 1.3.10 it follows that

$$|d\varphi|(x) \geq \frac{\xi}{\sqrt{1-\xi^2}}$$

and from the arbitrariness of $\xi < |dE_\varphi|(x,\varphi(x))$ we conclude that indeed $|d\varphi|(x)$ is bigger or equal to the right hand side of the equality of the proposition.

Next we show that the opposite inequality also holds. If $|d\varphi|(x) = 0$ or $|dE_\varphi|(x,\varphi(x)) = 1$ the desired inequality is clear. So suppose that $0 < \xi < |d\varphi|(x)$. Let

$$H\colon [0,\delta] \times \big(B_\delta\big(x,\varphi(x)\big) \cap \operatorname{epi}\varphi\big) \longrightarrow X$$

be as in Definition 1.3.10 and let

$$K\colon [0,\delta] \times \big(B_\delta\big(x,\varphi(x)\big) \cap \operatorname{epi}\varphi\big) \longrightarrow \operatorname{epi}\varphi$$

be defined by

$$K\big(t,(u,\lambda)\big) \overset{df}{=} \left(\lambda - \frac{\xi}{\sqrt{1+\xi^2}}t, H\left(\frac{t}{\sqrt{1+\xi^2}},(u,\lambda)\right)\right).$$

Evidently K is continuous and

$$d_{X\times\mathbb{R}}\big(K\big(t,(u,\lambda)\big),(u,\lambda)\big)$$

$$= \left[d\left(H\left(\frac{t}{\sqrt{1+\xi^2}},(u,\lambda)\right),u\right)^2 + \left(\frac{\xi}{\sqrt{1+\xi^2}}t\right)^2\right]^{\frac{1}{2}}$$

$$\leq \left(\frac{t^2}{1+\xi^2} + \frac{\xi^2 t^2}{1+\xi^2}\right)^{\frac{1}{2}} = t.$$

We also have

$$E_\varphi\big(t,K(u,\lambda)\big) = \lambda - \frac{\xi}{\sqrt{1+\xi^2}}t = E_\varphi(u,\lambda) - \frac{\xi}{\sqrt{1+\xi^2}}t,$$

hence

$$|dE_\varphi|(x,\varphi(x)) \geq \frac{\xi}{\sqrt{1+\xi^2}}.$$

Therefore

$$\xi \leq \frac{|dE_\varphi|(x, \varphi(x))}{\sqrt{1 - |dE_\varphi|(x, \varphi(x))^2}}.$$

Because $0 < \xi < |d\varphi|(x)$ was arbitrary, we have the opposite inequality and the proof of the proposition is complete. □

An immediate consequence of Proposition 1.3.20 is the following property of the weak slope.

PROPOSITION 1.3.22

_If_ $\varphi\colon X \longrightarrow \mathbb{R}$ _is continuous,_
<u>_then_</u> _the function_ $X \ni x \longmapsto |d\varphi|(x) \in \mathbb{R}$ _is lower semicontinuous._

The next proposition will permit comparisons with the C^1 (smooth) case and the locally Lipschitz case.

PROPOSITION 1.3.23

_If_ X _is a Banach space,_ $\varphi\colon X \longrightarrow \mathbb{R}$ _is continuous and_

$$\overline{D}_+\varphi(y)(h) \overset{df}{=} \limsup_{t \to 0^+} \frac{\varphi(y + th) - \varphi(y)}{t} \qquad \forall\, y \in X,$$

<u>_then_</u> _we have_

$$\|h\|_X \, |d\varphi|(x) \geq -\limsup_{y \to x} \overline{D}_+\varphi(y)(h).$$

PROOF If $-\limsup\limits_{y \to x} \overline{D}_+\varphi(y)(h) \leq 0$, then the inequality is automatically true. So suppose that $-\limsup\limits_{y \to x} \overline{D}_+\varphi(y)(h) > 0$. We may assume that $\|h\|_X = 1$. Let $0 < \xi < -\limsup\limits_{y \to x} \overline{D}_+\varphi(y)(h)$. Thus we can find $\delta > 0$, such that

$$-\overline{D}_+\varphi(u)(h) > \xi \qquad \forall\, u \in B_{2\delta}(x).$$

Let $H\colon [0, \delta] \times B_\delta \longrightarrow X$ be a map defined by $H(t, y) \overset{df}{=} y + th$. We have $\|H(t, y) - y\|_X \leq t$ and

$$\limsup_{\tau \to t^+} \frac{\varphi\big(H(\tau, y)\big) - \varphi\big(H(t, y)\big)}{\tau - t} < -\xi \qquad \forall\, t \in [0, \delta]$$

and so $\varphi\big(H(t, y)\big) \leq \varphi(y) - \xi t$. This implies that $\xi \leq |d\varphi|(x)$ and so finally from the arbitrariness of ξ we obtain the desired inequality. □

COROLLARY 1.3.9
If X *is a Banach space and* $\varphi \in C^1(X)$,
then

$$|d\varphi|(x) \;=\; \|\varphi'(x)\|_X \qquad \forall \, x \subset X$$

PROOF   Let $H(t,x)$ be as in Proposition 1.3.20. We have

$$\xi \;\leq\; \liminf_{t \to 0^+} \frac{\varphi(x) - \varphi\big(H(t,x)\big)}{t}$$
$$= \liminf_{t \to 0^+} \frac{\langle \varphi'(x), x - H(t,x) \rangle_X}{t} \;\leq\; \|\varphi'(x)\|_X \,,$$

hence we have $|d\varphi|(x) \leq \|\varphi'(x)\|_X$. On the other hand, from Proposition 1.3.23 we have that $\|\varphi'(x)\|_X \leq |d\varphi|(x)$. So finally the required equality must hold.   ⬜

To compare the weak slope with the Clarke subdifferential we need the following auxiliary result.

LEMMA 1.3.2
If X *is a Banach space,* $\psi \in \Gamma_0(X)$ *with* $\psi(0) = 0$ *and*

$$- \|x\|_X \leq \psi(x) \qquad \forall \, x \in X,$$

then there exists $u^* \in X^*$, *such that* $\|u^*\|_{X^*} \leq 1$ *and*

$$\langle u^*, x \rangle_X \leq \psi(x) \qquad \forall \, x \in X.$$

PROOF   Let

$$A \stackrel{df}{=} \big\{ (x, \lambda) \in X \times \mathbb{R} : \|x\|_X < -\lambda \big\}$$
$$C \stackrel{df}{=} \big\{ (y, \eta) \in X \times \mathbb{R} : \psi(y) \leq \eta \big\}.$$

Evidently A and C are nonempty, convex sets, with A open and $A \cap C = \emptyset$ because by hypothesis $- \|x\|_X \leq \psi(x)$. By the Weak Separation Theorem (see Theorem A.3.3), we can find $(x^*, \mu) \in X^* \times \mathbb{R}$, $(x^*, \mu) \neq (0,0)$ and $\gamma \in \mathbb{R}$, such that

$$\langle x^*, y \rangle_X - \mu\eta \;\leq\; \gamma \;\leq\; \langle x^*, x \rangle_X - \mu\lambda \qquad \forall \, (x, \lambda) \in \overline{A}, \; (y, \eta) \in C.$$

Note that $(0,0) \in \overline{A} \cap C$. So it follows that $\gamma = 0$. Also if $\lambda = - \|x\|_X$ we obtain

$$-\mu \|x\|_X \leq \langle x^*, x \rangle_X \qquad \forall \, x \in X.$$

Therefore $\mu \geq 0$ and $\|x^*\|_{X^*} \leq \mu$. If $\mu = 0$, then $x^* = 0$, a contradiction to the fact that $(x^*, \mu) \neq (0, 0)$. So $\mu > 0$ and we define $u^* = \frac{1}{\mu} x^*$. Hence

$$\langle u^*, y \rangle_X \leq \psi(y) \qquad \forall\, y \in X, \ \|u^*\|_{X^*} \leq 1.$$

$\Box$

Now we are ready to compare the weak slope with the Clarke subdifferential.

PROPOSITION 1.3.24

If X is a Banach space and $\varphi \colon X \longrightarrow \mathbb{R}$ is a locally Lipschitz function, then for every $x \in X$ we have

$$|d\varphi|(x) \ \geq \ m^\varphi(x) \ \overset{df}{=} \ \min \{\, \|x^*\|_{X^*} : \ x^* \in \partial\varphi(x) \}.$$

PROOF   According to Proposition 1.3.23, for any $h \in X$ we have

$$\varphi^0(x; h) \ \geq \ \limsup_{y \to x} \overline{D}_+\varphi(y)(h) \ \geq \ -|d\varphi|(x)\, \|h\|_X\, .$$

If $|d\varphi|(x) = 0$, then $\varphi^0(x; h) \geq 0$ for all $h \in X$ and so $0 \in \partial\varphi(x)$ and we have $|d\varphi|(x) = m^\varphi(x)$.

If $|d\varphi|(x) > 0$, then let

$$\psi(h) \ \overset{df}{=} \ \frac{1}{|d\varphi|(x)} \varphi^0(x; h).$$

By virtue of Proposition 1.3.7 we have that the function $X \ni h \longmapsto \psi(h) \in \mathbb{R}$ is continuous and sublinear. So we can apply Lemma 1.3.2 and obtain $u^* \in X^*$, $\|u^*\|_{X^*} \leq 1$, such that

$$\psi(h) \geq \langle u^*, h \rangle_X \qquad \forall\, h \in X.$$

Therefore $|d\varphi|(x)u^* \in \partial\varphi(x)$ and so $m^\varphi(x) \leq |d\varphi|(x)$. $\Box$

1.4   Nonlinear Operators

In this section we discuss some particular aspects of the theory of nonlinear operators between Banach spaces which will be used in the study of boundary value problems. Particular emphasis will be given to operators of monotone type. Also we discuss the Ekeland Variational Principle and its consequences.

1.4.1   Compact Operators

DEFINITION 1.4.1   Let X, Y be two Banach spaces, $D \subseteq X$ nonempty set and $K : D \longrightarrow Y$.

(a) We say that K is **compact** if and only if K is continuous and for every $C \subseteq D$ bounded set we have that $K(C)$ is **relatively compact** (i.e. $\overline{K(C)}$ is compact).

(b) We say that K is **completely continuous** if and only if for any sequence $\{x_n\}_{n \geq 1} \subseteq D$, such that $x_n \xrightarrow{w} x$ for some $x \in D$, we have that $K(x_n) \longrightarrow K(x)$ in Y.

PROPOSITION 1.4.1
If X is reflexive Banach space, Y is a Banach space, $D \subseteq X$ is nonempty closed and convex and $K : D \longrightarrow Y$ is completely continuous,
<u>then</u> K is compact too.

PROOF   Evidently K is continuous. Let $C \subseteq X$ be a nonempty bounded set and let $\{x_n\}_{n \geq 1} \subseteq C$ be a sequence. Because X is reflexive, by passing to a subsequence if necessary, we may assume that $x_n \xrightarrow{w} x \in X$. From the complete continuity of K, we have that $K(x_n) \longrightarrow K(x)$ in Y. Hence $\overline{K(C)}$ is compact and we conclude that K is compact.   ▯

REMARK 1.4.1   The converse of the above proposition is not true. Indeed let $X = Y = L^2(0, 1)$ and let $K : X \longrightarrow Y$ be defined by

$$K(x)(t) \stackrel{df}{=} \int\limits_0^1 x(s)^2 ds \ = \ \|x\|_2^2 \,.$$

Clearly K is compact but it is not completely continuous, since if

$$x_n(s) \stackrel{df}{=} \sin(n\pi s) \quad \forall \, s \in [0, 1], \ n \geq 1,$$

then $x_n \xrightarrow{w} 0$ in $L^2(0, 1)$, but

$$K(x_n) \equiv \frac{1}{2} \quad \forall \, n \geq 1$$

and $K(x_n) \not\longrightarrow K(0) \equiv 0$.

Furthermore, if X is reflexive and $K \in \mathcal{L}(X; Y)$, then it is easy to check that in this case compactness and complete continuity of K are equivalent notions.   ▯

The following proposition gives a useful property of compact operators.

PROPOSITION 1.4.2
If X, Y are two Banach spaces, $D \subseteq X$ is a nonempty set and $\{K_\alpha\}_{\alpha \in J}$ is a net of compact operators from D into Y, such that $K_\alpha \longrightarrow K$ uniformly on bounded subsets of D,
then K: $D \longrightarrow Y$ is also a compact operator.

PROOF   Clearly K is continuous. Let $C \subseteq D$ be nonempty and bounded. For a given $\varepsilon > 0$, we can find $\alpha_0 = \alpha_0(\varepsilon, C) \in J$, such that

$$\|K(x) - K_\alpha(x)\|_Y < \frac{\varepsilon}{2} \qquad \forall \, \alpha \geq \alpha_0, \ x \in C.$$

For a fixed $\alpha \geq \alpha_0$, the set $K_\alpha(C)$ is totally bounded. So we can find $N = N(\varepsilon, C, \alpha)$ and $y_1, \ldots, y_N \in Y$, such that

$$K_\alpha(C) \subseteq \bigcup_{k=1}^{N} B_{\frac{\varepsilon}{2}}(y_k).$$

Let $x_0 \in C$. Then we can find $k_0 \in \{1, \ldots, N\}$, such that

$$\|K_\alpha(x_0) - y_{k_0}\|_Y < \frac{\varepsilon}{2}.$$

We have

$$\|K(x_0) - y_{k_0}\|_Y \ \leq \ \|K(x_0) - K_\alpha(x_0)\|_Y + \|K_\alpha(x_0) - y_{k_0}\|_Y \ < \ \varepsilon,$$

so

$$K(C) \subseteq \bigcup_{k=1}^{N} B_\varepsilon(y_k),$$

which shows that $K(C)$ is totally bounded. Therefore K is compact.   ☐

REMARK 1.4.2   If we define

$$\mathcal{L}_c(X; Y) \stackrel{df}{=} \{K \in \mathcal{L}(X; Y) : \ K \text{ is compact}\},$$

then $\mathcal{L}_c(X; Y)$ with the operator norm is a Banach space.   ☐

Compact operators have a remarkable approximation property. First a definition.

DEFINITION 1.4.2   *Let X, Y be two Banach spaces, $D \subseteq X$ a nonempty set and $K: D \longrightarrow Y$. We say that K is a **finite rank operator** if $R(K)$ lies in a finite dimensional subspace of Y.*

PROPOSITION 1.4.3
If X, Y are Banach spaces and $D \subseteq X$ is nonempty, closed and bounded,
then $K \colon D \longrightarrow Y$ is compact if and only if K is the uniform limit of finite
rank operators.

PROOF $\Longrightarrow$: Note that $\overline{K(D)}$ is compact and so for a given $\varepsilon > 0$, we
can find $N = N(\varepsilon) \in \mathbb{N}$ and $y_1, \ldots, y_N \in K(D)$, such that

$$\overline{K(D)} \subseteq \bigcup_{k=1}^{N} B_{\varepsilon}(y_k).$$

We can find on $\overline{K(D)}$ a continuous partition of unity $\{\varphi_1, \ldots, \varphi_N\}$ subordinate
to the cover $\{B_{\varepsilon}(y_1), \ldots, B_{\varepsilon}(y_N)\}$. Let us set

$$K_{\varepsilon}(x) \overset{df}{=} \sum_{k=1}^{N} \varphi_k\big(K(x)\big) y_k \qquad \forall\, x \in D.$$

If $\varphi_k\big(K(x)\big) > 0$, then $K(x) \in B_{\varepsilon}(y_k)$ and

$$\|K(x) - K_{\varepsilon}(x)\|_Y \;=\; \left\| \sum_{k=1}^{N} \varphi_k\big(K(x)\big)\big(K(x) - y_k\big) \right\|_Y \;<\; \varepsilon \qquad \forall\, x \in D.$$

Evidently K_{ε} is a continuous and finite rank operator.

$\Longleftarrow$: Note that a finite rank operator is compact. So the implication follows
from Proposition 1.4.2 $\Box$

1.4.2   Maximal Monotone Operators

Now we pass to another class of nonlinear operators, which will be a basic
analytical tool in our study of nonlinear boundary value problems. These are
the maximal monotone operators which provide a more general framework
than compact operators, for the study of nonlinear functional equations.

In what follows X is a Banach space and X^* its topological dual. Additional
hypotheses will be introduced as needed. For a given operator $A \colon X \longrightarrow 2^{X^*}$
we define the ***domain*** of A by

$$\text{Dom}\,(A) \overset{df}{=} \{x \in X \colon\; A(x) \neq \emptyset\},$$

the ***range*** of A by

$$R(A) \overset{df}{=} \{x^* \in X^* \colon\; x \in \text{Dom}\,(A),\; x^* \in A(x)\},$$

the **graph** of A by

$$\text{Gr } A \overset{df}{=} \{(x, x^*) \in X \times X^* : x^* \in A(x)\},$$

and the **inverse** $A^{-1}: X^* \longrightarrow 2^X$ of A by

$$A^{-1}(x^*) \overset{df}{=} \{x \in X : x^* \in A(x)\}.$$

DEFINITION 1.4.3   *Let* $A: X \longrightarrow 2^{X^*}$ *be an operator.*

(a) *We say that* A *is* **monotone**, *if*
$$\langle x^* - y^*, x - y \rangle_X \geq 0 \qquad \forall \, (x, x^*), (y, y^*) \in \text{Gr } A;$$

(b) *We say that* A *is* **strictly monotone**, *if*
$$\langle x^* - y^*, x - y \rangle_X > 0 \qquad \forall \, (x, x^*), (y, y^*) \in \text{Gr } A, \; x \neq y;$$

(c) *We say that* A *is* **strongly monotone**, *if there exist* $c > 0$ *and* $p > 1$, *such that*
$$\langle x^* - y^*, x - y \rangle_X \geq c \|x - y\|_X^p \qquad \forall \, (x, x^*), (y, y^*) \in \text{Gr } A;$$

(d) *We say that* A *is* **locally bounded** *at* $x \in \text{Dom}\,(A)$, *if there exist* $M > 0$ *and* $r > 0$, *such that*
$$\|y^*\|_{X^*} \leq M \qquad \forall \, y \in \text{Dom}\,(A) \cap B_r(x), \; y^* \in A(x);$$

(e) *We say that* A *is* **maximal monotone**, *if* A *is monotone and for all* $y \in X$ *and* $y^* \in X^*$ *we have*
$$\left[\langle x^* - y^*, x - y \rangle_X \geq 0 \qquad \forall \, (x, x^*) \in \text{Gr } A \right] \implies (y, y^*) \in \text{Gr } A.$$

REMARK 1.4.3   Definition 1.4.3(e) is equivalent to saying that $\text{Gr } A$ is not properly included in the graph of another monotone map. Evidently if X is reflexive, A is maximal monotone if and only if A^{-1} is. ⬜

PROPOSITION 1.4.4
If $A: X \longrightarrow 2^{X^*}$ *is a monotone operator,*
then A *is locally bounded at every* $x \in \text{int}\,\text{Dom}\,(A)$.

PROOF   First we show that, if $\{x_n^*\}_{n \geq 1} \subseteq X^*$ and $U \subseteq X$ is a nonempty open set, such that
$$\inf_{x \in U} \langle x_n^*, x \rangle_X > -\infty,$$

then

$$\sup_{n \geq 1} \|x_n^*\|_{X^*} < +\infty.$$

Indeed we have

$$\inf_{x \in B_r} \langle x_n^*, x_0 + x \rangle_X > -\infty,$$

with $x_0 \in X$ and $r > 0$ fixed. Therefore

$$\inf \left\{ \langle x_n^*, \pm x \rangle_X : x \in B_r \right\} > -\infty,$$

hence

$$\sup \left\{ | \langle x_n^*, x \rangle_X | : x \in B_r \right\} < +\infty,$$

from which it follows that $\sup_{n \geq 1} \|x_n^*\|_{X^*} < +\infty$.

Next suppose that $x_n \longrightarrow 0$ in X, $\{x_n^*\}_{n \geq 1} \subseteq X^*$ and assume that $\|x^*\|_{X^*} \longrightarrow +\infty$. Then for each $r > 0$ there exists $x \in \overline{B}_r$, such that

$$\liminf_{n \to +\infty} \langle x_n^*, x_n - x \rangle_X = -\infty.$$

Indeed, if there exists an $r > 0$, such that

$$\liminf_{n \to +\infty} \langle x_n^*, x_n - x \rangle_X > -\infty \qquad \forall \, x \in \overline{B}_r,$$

then

$$\overline{B}_r = \bigcup_{k \geq 1} \overline{B}_r^k,$$

with

$$\overline{B}_r^k \stackrel{df}{=} \left\{ x \in \overline{B}_r : \langle x_n^*, x_n - x \rangle_X \geq -k \;\; \forall n \geq 1 \right\}.$$

From the Baire Category Theorem (see Theorem A.1.9) there exists $k \geq 1$, such that $\operatorname{int} \overline{B}_r^k \neq \emptyset$. So

$$\inf \left\{ \langle x_n^*, x \rangle_X : x \in \overline{B}_r \right\} > -\infty$$

and from the first part of the proof, we have that $\{x_n^*\}_{n \geq 1} \subseteq X^*$ is bounded, a contradiction.

Now suppose that the proposition is not true. Then A is not locally bounded at $x_0 \in \operatorname{int} \operatorname{Dom}(A)$. By considering $x \longmapsto A(x + x_0)$ and because monotonicity is invariant under translation, we may assume that $x_0 = 0$. So we can find $(x_n, x_n^*) \in \operatorname{Gr} A$, such that $x_n \longrightarrow 0$ in X and $\|x_n^*\|_{X^*} \longrightarrow +\infty$. Let $r > 0$ be such that $\overline{B}_r \subseteq \operatorname{Dom}(A)$. Then from the previous part of the proof there exists $x \in \overline{B}_r \subseteq \operatorname{Dom}(A)$, such that

$$\liminf_{n \to +\infty} \langle x_n^*, x_n - x \rangle_X = -\infty.$$

For a given $x^* \in A(x)$, from the monotonicity of A we have that

$$\langle x_n^*, x_n - x \rangle_X \geq \langle x^*, x_n - x \rangle_X \qquad \forall \, n \geq 1,$$

hence

$$\liminf_{n \to +\infty} \langle x_n^*, x_n - x \rangle_X \geq \langle x^*, -x \rangle_X > -\infty,$$

a contradiction. $\quad\blacksquare$

PROPOSITION 1.4.5

If $A \colon X \longrightarrow 2^{X^*}$ *is a maximal monotone operator,*
then for all $x \in \mathrm{Dom}\,(A)$, $A(x)$ *is a nonempty, convex and* w^*-*closed set and* $A|_{\mathrm{int}\,\mathrm{Dom}\,(A)}$ *is a norm-to-weak* * *upper semicontinuous multifunction.*

PROOF   By virtue of Definition 1.4.3(e), for all $v \in \mathrm{Dom}\,(A)$ we have

$$A(v) \;=\; \bigcap_{(x,x^*) \in \mathrm{Gr}\, A} \big\{ v^* \in X^* : \langle v^* - x^*, v - x \rangle_X \geq 0 \big\}.$$

In this intersection every set is convex and w^*-closed. Hence so is $A(v)$ and of course it is nonempty since $v \in \mathrm{Dom}\,(A)$.

Next suppose that $\mathrm{int}\,\mathrm{Dom}\,(A) \neq \emptyset$ and $U \subseteq X^*$ is a w^*-open subset. To prove the postulated upper semicontinuity, we need to show that the set

$$A^+(U) \;=\; \big\{ v \in \mathrm{int}\,\mathrm{Dom}\,(A) : A(v) \subseteq U \big\}$$

is strongly open. If this is not the case, then we can find $v \in A^+(U)$ and $\{v_n\}_{n \geq 1} \subseteq \mathrm{int}\,\mathrm{Dom}\,(A)$, such that $v_n \notin A^+(U)$ and $v_n \longrightarrow v$ in X. Because $v_n \notin A^+(U)$, we can find $v_n^* \in A(v_n)$, $v_n^* \notin U$. But from Proposition 1.4.4 we know that A is locally bounded on $\mathrm{int}\,\mathrm{Dom}\,(A)$ and so $\{v_n^*\}_{n \geq 1} \subseteq X^*$ is bounded. Let v^* be a w^*-cluster point of $\{v_n^*\}_{n \geq 1}$. Evidently $v^* \notin U$. Let $\varepsilon > 0$, $y \in \mathrm{Dom}\,(A)$ and $y^* \in A(y)$. We have

$$\langle v_n^* - y^*, v_n - y \rangle_X \geq 0 \qquad \forall \, n \geq 1.$$

Also note that

$$\langle v_n^* - y^*, v - v_n \rangle_X \longrightarrow 0 \qquad \forall \text{ as } n \to +\infty$$

and so there exists $N = N(\varepsilon) \in \mathbb{N}$, such that

$$\langle v_n^* - y^*, v - v_n \rangle_X \geq -\varepsilon \qquad \forall \, n \geq N.$$

So it follows that

$$\langle v_n^* - y^*, v - y \rangle_X \geq -\varepsilon \qquad \forall \, n \geq N.$$

Let $Z \in \mathcal{N}_{w^*}(v)$ be defined by

$$Z \stackrel{df}{=} v^* + \{z^* \in X^* : \langle z^*, v - y \rangle_X < c\}.$$

Since $v^* \in \overline{\{v_1^*, v_2^*, \ldots\}}^{w^*}$ we can find $N_1 \geq N$, such that $v_{N_1}^* \in Z$ and so $\langle v_{N_1}^* - v^*, v - y \rangle_X < \varepsilon$. This yields

$$\begin{aligned}
\langle v^* - y^*, v - y \rangle_X &= \langle v^* - v_{N_1}^*, v - y \rangle_X + \langle v_{N_1}^* - y^*, v - y \rangle_X \\
&> -\varepsilon - \varepsilon = -2\varepsilon.
\end{aligned}$$

Let $\varepsilon \searrow 0$ to obtain

$$\langle v^* - y^*, v - y \rangle_X \geq 0 \qquad \forall y \in \mathrm{Dom}\,(A),\ y^* \in A(y).$$

Since A is maximal monotone it follows that $v^* \in A(v) \subseteq U$, a contradiction. So indeed $A|_{\mathrm{int\,Dom}\,(A)}$ is norm-to-weak* upper semicontinuous. $\quad\square$

In fact there is also a converse of the above proposition.

PROPOSITION 1.4.6

If $A: X \longrightarrow 2^{X^*}$ is a monotone operator, for every $x \in X$, $A(x)$ is a nonempty, convex and w^*-closed set and for all $x, y \in X$, the map

$$\lambda \longmapsto A(\lambda x + (1 - \lambda)y)$$

*has a graph which is closed in $[0,1] \times X_{w^*}^*$,
then A is maximal monotone.*

PROOF   Let $x \in X$ and $x^* \in X^*$ satisfy

$$\langle x^* - y^*, x - y \rangle_X \geq 0 \qquad \forall y \in X,\ y^* \in A(x).$$

We need to show that $x^* \in A(x)$.

If $x^* \notin A(x)$, then by the Strong Separation Theorem (see Theorem A.3.4), we can find $z \in X$, $z \neq 0$, such that

$$\sigma(z, A(x)) < \langle x^*, z \rangle_X.$$

Let $z_\lambda = x + \lambda z$, $\lambda > 0$ and $z_\lambda^* \in A(z_\lambda)$. We have

$$\langle x^* - z_\lambda^*, x - z_\lambda \rangle_X = \langle x^* - z_\lambda^*, -\lambda z \rangle_X \geq 0,$$

hence

$$\langle x^* - z_\lambda^*, z \rangle_X \leq 0.$$

Let $\lambda_n \searrow 0$ and set $z_n = z_{\lambda_n}$, $z_n^* = z_{\lambda_n}^*$. We have that $z_n \longrightarrow z$ in X and because of Proposition 1.4.4, $\{z_n^*\}_{n \geq 1} \subseteq X^*$ is bounded. So we can find a subnet $\{z_\beta^*\}_{\beta \in B}$ of $\{z_n^*\}_{n \geq 1}$, such that $z_\beta^* \xrightarrow{w^*} z^*$ in X^*. We have $z^* \in A(x)$ and $\langle x^* - z^*, z \rangle_X \leq 0$, so

$$\langle x^*, z \rangle_X \leq \langle z^*, z \rangle_X,$$

a contradiction to the choice of z. ▯

Now we shall state some useful consequences of these propositions. But first a definition.

DEFINITION 1.4.4 *Let $A \colon X \longrightarrow X^*$ be a single-valued operator with* $\text{Dom}(A) = X$.

(a) *We say that A is **demicontinuous** if and only if $x_n \longrightarrow x$ in X implies that $A(x_n) \xrightarrow{w^*} A(x)$ in X^*.*

(b) *We say that A is **hemicontinuous** if and only if for all $x, y, z \in X$, the function $[0,1] \ni \lambda \longmapsto \langle A(x + \lambda y), z \rangle_X \in \mathbb{R}$ is continuous.*

REMARK 1.4.4 It is easy to see that for monotone operators $A \colon X \longrightarrow X^*$ with $\text{Dom}(A) = X$, demicontinuity and hemicontinuity are equivalent notions. ▯

COROLLARY 1.4.1
If $A \colon X \longrightarrow 2^{X^*}$ *is a maximal monotone operator,*
then $\text{Gr}\,A$ *is closed in $X \times X_{w^*}^*$ and in $X_w \times X^*$.*

COROLLARY 1.4.2
If $A \colon X \longrightarrow 2^{X^*}$ *is a hemicontinuous and monotone operator,*
then A *is maximal monotone.*

COROLLARY 1.4.3
If $A \colon X \longrightarrow 2^{X^*}$ *is a maximal monotone operator and $D \subseteq \text{Dom}(A)$ is a nonempty and open set,*
then $A|_D$ *is maximal monotone in D.*

The next result determines the structure of $\text{Dom}(A)$ and its proof can be found in Rockafellar (1970).

PROPOSITION 1.4.7
If $A \colon X \longrightarrow 2^{X^*}$ *is a maximal monotone operator and $\text{int}\,\text{Dom}(A) \neq \emptyset$,*

then

$$\begin{aligned} \operatorname{int} \overline{\operatorname{Dom}(A)} &= \operatorname{int} \operatorname{conv} \operatorname{Dom}(A) \text{ (so int Dom}(A) \text{ is convex)} \\ \overline{\operatorname{Dom}(A)} &= \overline{\operatorname{int} \operatorname{Dom}(A)} \qquad \text{(so } \overline{\operatorname{int} \operatorname{Dom}(A)} \text{ is convex).} \end{aligned}$$

An interesting consequence of this proposition is the following result about the range of A.

COROLLARY 1.4.4
If X is a reflexive Banach space, $A \colon X \longrightarrow 2^{X^*}$ is a maximal monotone operator and $\operatorname{int} R(A) \neq \emptyset$,
then $\operatorname{int} R(A)$ *is convex and* $\overline{R(A)}$ *is convex.*

PROOF   Note that $A^{-1} \colon X^* \longrightarrow 2^X$ is maximal monotone. So from Proposition 1.4.7 we have that

$$\operatorname{int} \operatorname{Dom}(A^{-1}) = \operatorname{int} R(A) \quad \text{is convex}$$

and

$$\operatorname{int} \overline{\operatorname{Dom}(A^{-1})} = \overline{\operatorname{Dom}(A^{-1})} = \overline{R(A)} \quad \text{is convex.}$$

$\square$

Before continuing our discussion of maximal monotone operators, we need to make an important observation. Maximal monotonicity is not affected by equivalent renorming of X and X^*. So we often use the existing renorming theorems in the theory of maximal monotone operators. The next Theorem identifies a major class of nonlinear maximal monotone operators. First let us introduce an operator which is an important tool in Banach space theory and nonlinear analysis.

DEFINITION 1.4.5   *The **duality map** of a Banach space X is the map $\mathcal{F} \colon X \longrightarrow 2^{X^*}$ defined by*

$$\mathcal{F}(x) \stackrel{df}{=} \left\{ x^* \in X^* : \langle x^*, x \rangle_X = \|x\|_X^2 = \|x^*\|_{X^*}^2 \right\}.$$

REMARK 1.4.5   From one of the corollaries to the Hahn-Banach theorem we know that for all $x \in X$, $\mathcal{F}(x) \neq \emptyset$. Also it is not difficult to check that

$$\mathcal{F}(x) = \partial \left(\frac{1}{2} \|\cdot\|_X^2 \right)(x) \qquad \forall \, x \in X.$$

The duality map depends on the norm of X in an essential way. So if $\|\cdot\|_1$ and $\|\cdot\|_2$ are two distinct equivalent norms of X and $\mathcal{F}_1$, $\mathcal{F}_2$ are the corresponding duality maps, then $\mathcal{F}_1 \neq \mathcal{F}_2$. If $X = H$ is a Hilbert space identified with its dual, then $\mathcal{F}$ is the identity operator. ▯

THEOREM 1.4.1
If X *is a Banach space and* $\varphi \in \Gamma_0(X)$,
then *the map* $\partial\varphi\colon X \longrightarrow 2^{X^*}$ *is maximal monotone.*

PROOF We do the proof when X is a reflexive Banach space. For the general case we refer to Rockafellar (1970). By the Troyanski Renorming Theorem (see Theorem A.3.9), we may assume that both X and X^* are locally uniformly convex and have Frechet differentiable norms. It is easy to see that $\partial\varphi$ is monotone.

Suppose that $y \in X$, $y^* \in X^*$ and

$$\langle x^* - y^*, x - y \rangle_X \geq 0 \qquad \forall\, x \in \mathrm{Dom}\,(A),\ x^* \in A(x).$$

We have to show that $y^* \in A(y)$. To do this, first we prove that

$$R(\partial\varphi + \mathcal{F}) = X^*.$$

Since X is locally uniformly convex and has Frechet differentiable norm, the map $x \longmapsto \frac{1}{2} \|x\|_X^2$ is strictly convex and the map $x \longmapsto \mathcal{F}(x)$ is single valued and strictly monotone. Let $\psi\colon X \longrightarrow \overline{\mathbb{R}} \overset{df}{=} \mathbb{R} \cup \{+\infty\}$ be defined by

$$\psi(x) \overset{df}{=} \frac{1}{2} \|x\|_X^2 + \varphi(x) - \langle z^*, x \rangle_X\,,$$

for some fixed $z^* \in X^*$. Evidently $\psi \in \Gamma_0(X)$ and

$$\lim_{\|x\|_X \to +\infty} \psi(x) = +\infty.$$

So by Theorem 1.3.4, we can find $z \in X$, such that $\varphi(z) = \inf_X \varphi$. Then from Theorem 1.3.6 we have that

$$0 \in \partial\varphi(z) = \mathcal{F}(z) + \partial\varphi(z) - z^*,$$

and so

$$z^* \in \mathcal{F}(z) + \partial\varphi(z).$$

Because $z^* \in X^*$ was arbitrary, we infer that $R(\partial\varphi + \mathcal{F}) = X^*$.

Using this fact we can find $x \in \mathrm{Dom}\,(\partial\varphi)$ and $x^* \in \partial\varphi(x)$, such that

$$x^* + \mathcal{F}(x) = y^* + \mathcal{F}(y).$$

We have that

$$0 \leq \langle x^* - y^*, x - y \rangle_X = \langle \mathcal{F}(y) - \mathcal{F}(x), x - y \rangle_X \leq 0$$

and the last inequality is strict if $x \neq y$, a contradiction. So

$$y = x \in \mathrm{Dom}\,(\partial \varphi) \qquad \text{and} \qquad y^* = x^* \in \partial \varphi(x) = \partial \varphi(y).$$

$\square$

REMARK 1.4.6   In fact a subdifferential operator is maximal cyclically monotone. An operator $A \colon X \longrightarrow 2^{X^*}$ is **cyclically monotone** if and only if

$$\sum_{k=1}^{n} \langle x_k^*, x_k - x_{k+1} \rangle_X \geq 0,$$

for all $x_k \in \mathrm{Dom}\,(A)$, $x_k^* \in A(x_k)$, $k = 1, \ldots, n$ and all $n \geq 2$, where $x_{n+1} = x_1$. Every maximal cyclically monotone operator (i.e. one whose graph is not properly included in the graph of a cyclically monotone operator) is of the subdifferential type. Every monotone map $a \colon \mathbb{R} \longrightarrow \mathbb{R}$ is cyclically monotone and so if it is maximal monotone, we have $a = \partial \varphi$, with $\varphi \in \Gamma_0(\mathbb{R})$. This fails in higher dimensions. $\square$

The main strength of maximal monotone operators and more generally of operators of monotone type are their surjectivity properties. Central role in this plays the concept of coercivity.

DEFINITION 1.4.6   *Let $A \colon X \longrightarrow 2^{X^*}$ be an operator.*

(a) *We say that A is **coercive** if and only if $\mathrm{Dom}\,(A)$ is bounded or $\mathrm{Dom}\,(A)$ is unbounded and*

$$\frac{inf\,\{\langle x^*, x \rangle_X \colon\; x^* \in A(x)\}}{\|x\|_X} \longrightarrow +\infty \quad \text{as } \|x\|_X \to +\infty,\; x \in \mathrm{Dom}\,(A);$$

(b) *We say that A is **weakly coercive** if and only if $\mathrm{Dom}\,(A)$ is bounded or $\mathrm{Dom}\,(A)$ is unbounded and*

$$\inf\,\{\,\|x^*\|_{X^*} \colon\; x^* \in A(x)\} \longrightarrow +\infty \qquad \text{as } \|x\|_X \to +\infty,\; x \in \mathrm{Dom}\,(A).$$

The next lemma, known as the **Debrunner-Flor Lemma**, is a key tool in proving perturbation and surjectivity results. For a proof we refer to Hu & Papageorgiou (1997, p. 319) or Pascali & Sburlanu (1978, p. 118).

LEMMA 1.4.1 (Debrunner-Flor Lemma)
If X is reflexive Banach space, $C \subseteq X$ is a closed and convex set, $A \colon X \longrightarrow 2^{X^*}$ is a monotone operator with $\mathrm{Dom}\,(A) \subseteq C$ and $B \colon C \longrightarrow X^*$

is a bounded, coercive, demicontinuous and monotone operator,
then there exists $y \in C$, such that

$$\langle x^* + B(y), x - y \rangle_X \ge 0 \qquad \forall\, x \in \text{Dom}\,(A),\ x^* \in A(x).$$

This lemma leads to the first surjectivity result.

THEOREM 1.4.2

If X is a reflexive Banach space, $A: X \longrightarrow 2^{X^*}$ is a maximal monotone operator and $B: X \longrightarrow X^*$ is a bounded, coercive, hemicontinuous monotone operator,
then $R(A + B) = X^*$.

PROOF   Let $y^* \in X^*$ and set $B_1(x) = B(x) - y^*$. It is clear that B_1 has the same properties as B. We use the Debrunner-Flor Lemma (see Lemma 1.4.1) with B_1 instead of B and with $C = X$. So we can find $y \in X$, such that

$$\langle x^* + B(y) - y^*, x - y \rangle_X \ge 0 \qquad \forall\, x \in \text{Dom}\,(A),\ x^* \in A(x).$$

Because A is maximal monotone it follows that $y \in \text{Dom}\,(A)$ and $y^* - B(y) \in A(y)$, hence $y^* \in A(y) + B(y)$. Since $y^* \in X^*$ was arbitrary, we conclude that $R(A + B) = X^*$. $\qquad \Box$

Also we can characterize surjective maximal monotone operators.

THEOREM 1.4.3

If X is a reflexive and $A: X \longrightarrow 2^{X^*}$ is a maximal monotone operator,
then $R(A) = X^*$ if and only if A^{-1} is locally bounded.

PROOF $\Longrightarrow$: Recall that $A^{-1}: X^* \longrightarrow 2^X$ is maximal monotone and $\text{Dom}\,(A^{-1}) = X^*$. Then by Proposition 1.4.4, A^{-1} is locally bounded.

$\Longleftarrow$: We shall show that $R(A)$ is both open and closed in X^*. To show the closedness we consider a sequence $\{x_n^*\}_{n \ge 1} \subseteq R(A)$, such that $x_n^* \longrightarrow x^*$ in X^*. We have $x_n^* \in A(x_n)$ and because A^{-1} is locally bounded it follows that $\{x_n\}_{n \ge 1} \subseteq X$ is bounded. By virtue of the reflexivity of X, we may assume that $x_n \xrightarrow{w} x$ in X. Corollary 1.4.1 implies that $x^* \in A(x)$. So $x^* \in R(A)$ and we have that $R(A)$ is closed.

Next we show that $R(A) \subseteq X^*$ is open. By translation if necessary, we may assume that $0 \in \text{Dom}\,(A)$. Let $x^* \in A(0)$. Because A^{-1} is locally bounded, we can find $r > 0$, such that $A^{-1}(B_r(x^*))$ is bounded in X. Recall that by Troyanski Renorming Theorem (see Theorem A.3.9), we may assume that both X and X^* are locally uniformly convex. Let $y^* \in X^*$ be such that

$$\|y^* - x^*\|_{X^*} < \frac{r}{2}.$$

From Theorem 1.4.2 (with $B = \lambda \mathcal{F}$, $\lambda > 0$) we know that we can find $y_\lambda \in \text{Dom}(A)$ and $y_\lambda^* \in A(y_\lambda)$, such that

$$y_\lambda^* + \lambda \mathcal{F}(y_\lambda) = y^*. \tag{1.6}$$

From the monotonicity of A, we have

$$\langle y^* - \lambda \mathcal{F}(y_\lambda) - x^*, y_\lambda \rangle_X \geq 0$$

(recall that $x = 0$), hence

$$\|y^* - x^*\|_{X^*} \|y_\lambda\|_X - \lambda \|y_\lambda\|_X^2 \geq 0$$

and so

$$\lambda \|y_\lambda\|_X \leq \|y^* - x^*\|_{X^*} < \frac{r}{2} \qquad \forall \lambda > 0.$$

Then from (1.6) we have

$$\|y^* - y_\lambda^*\|_X = \lambda \|\mathcal{F}(y_\lambda)\|_X = \lambda \|y_\lambda\|_X < \frac{r}{2}. \tag{1.7}$$

Therefore

$$\|y_\lambda^* - x^*\|_X \leq \|y_\lambda^* - y^*\|_X + \|y^* - x^*\|_X < \frac{r}{2} + \frac{r}{2} = r \qquad \forall \lambda > 0.$$

Because $A^{-1}(B_r(x^*))$ is bounded, it follows that also $\{y_\lambda\}_{\lambda > 0} \subseteq X$ is bounded. So from (1.7) it follows that $y_\lambda^* \longrightarrow y^*$ in X^* as $\lambda \searrow 0$. Because $y_\lambda^* \in R(A)$ and the latter is closed, we have that $y^* \in R(A)$. Thus $B_{\frac{r}{2}}(x^*) \subseteq R(A)$ and so $R(A)$ is open in X^*. Because $R(A) \subseteq X^*$ is both open and closed, we have that $R(A) = X^*$. ⬜

This theorem has some remarkable consequences.

THEOREM 1.4.4
If X is a reflexive Banach space and $A \colon X \longrightarrow 2^{X^*}$ is maximal monotone and weakly coercive,
<u>*then*</u> *$R(A) = X^*$.*

PROOF   Because A is weakly coercive, then clearly A^{-1} is locally bounded. So by Theorem 1.4.3, $R(A) = X^*$. ⬜

COROLLARY 1.4.5
If X is a reflexive Banach space and $A\colon X \longrightarrow X^*$ is a hemicontinuous, monotone and weakly coercive operator,
then $R(A) = X^*$.

The next theorem gives the main perturbation result for maximal monotone operators. For a proof of it we refer to Barbu (1976, p. 46).

THEOREM 1.4.5
If X is a reflexive Banach space and $A, B\colon X \longrightarrow 2^{X^*}$ are maximal monotone operators, such that
$$\operatorname{int} \operatorname{Dom}(A) \cap \operatorname{Dom}(B) \neq \emptyset,$$

then $A + B$ is maximal monotone too.

REMARK 1.4.7   If $X = H$ is a Hilbert space, then the hypothesis $\operatorname{int} \operatorname{Dom}(A) \cap \operatorname{Dom}(B) \neq \emptyset$ can be replaced by the more general one which says that
$$0 \in \operatorname{int} \big(\operatorname{Dom}(A) \setminus \operatorname{Dom}(B) \big)$$

(note that $\operatorname{int} \operatorname{Dom}(A) \setminus \operatorname{Dom}(B) \subseteq \operatorname{int} \big(\operatorname{Dom}(A) \setminus \operatorname{Dom}(B) \big)$). In fact in this case it can happen that $\operatorname{int} A = \operatorname{int} B = \emptyset$. $\quad\square$

1.4.3   Yosida Approximation

When $X = H$ is a Hilbert space and H is identified with its dual, then the duality map $\mathcal{F}$ is the identity operator. In this case the following operators are very useful in the analysis of problems involving a maximal monotone operator A.

DEFINITION 1.4.7   *Let H be a Hilbert space identified with its dual, $A\colon H \longrightarrow 2^H$ is a maximal monotone operator and $\lambda > 0$.*

(a) *The operator $J_\lambda\colon H \longrightarrow H$ defined by $J_\lambda(x) \stackrel{df}{=} (I + \lambda A)^{-1}(x)$ is called the **resolvent of** A;*

(b) *The operator $A_\lambda\colon H \longrightarrow H$ defined by $A_\lambda(x) \stackrel{df}{=} \frac{1}{\lambda}(I - J_\lambda)(x)$ is called the **Yosida approximation of** A.*

REMARK 1.4.8   Note that $\operatorname{Dom}(J_\lambda) = \operatorname{Dom}(A_\lambda) = H$ and both operators are single-valued. $\quad\square$

In the next proposition we summarize the properties of J_λ and A_λ. For a proof we refer to Brézis (1983, pp. 27–28).

PROPOSITION 1.4.8

If H *is a Hilbert space identified with its dual and* $A: H \longrightarrow 2^H$ *is a maximal monotone operator,*
then for every $\lambda > 0$, *we have*

(a) J_λ *is nonexpansive, i.e.*

$$\|J_\lambda(x) - J_\lambda(y)\|_H \leq \|x - y\|_H \qquad \forall\, x, y \in H;$$

(b) A_λ *is monotone and Lipschitz continuous with constant* $\frac{1}{\lambda}$ *(hence it is maximal monotone);*

(c) $A_\lambda(x) \in A(J_\lambda(x))$ *for all* $x \in H$;

(d) $\|A_\lambda(x)\|_H \leq \|A^0(x)\|_H$ *for all* $x \in \mathrm{Dom}\,(A)$ *with*

$$A^0(x) \stackrel{df}{=} \mathrm{proj}_{A(x)}\{0\}$$

(recall that $A(x) \in P_{fc}(H)$);

(e) $A_\lambda(x) \longrightarrow A^0(x)$ *as* $\lambda \searrow 0$ *for all* $x \in \mathrm{Dom}\,(A)$ *and* $\|A_\lambda(x)\|_H \nearrow +\infty$ *as* $\lambda \searrow 0$ *for all* $x \notin \mathrm{Dom}\,(A)$;

(f) $J_\lambda(x) \longrightarrow \mathrm{proj}_{\overline{\mathrm{Dom}\,(A)}}\{x\}$ *as* $\lambda \searrow 0$ *for all* $x \in H$ *(recall that* $\overline{\mathrm{Dom}\,(A)} \in P_{fc}(H)$; *see Proposition 1.4.7).*

REMARK 1.4.9   If $\overline{\mathrm{Dom}\,(A)} = H$, then $\mathrm{proj}_{\overline{\mathrm{Dom}\,(A)}}\{x\} = x$ and so from Proposition 1.4.8(f) we see that J_λ is an approximation of the identity operator. Moreover, if $\varphi \in \Gamma_0(H)$ and $A = \partial\varphi$, then for $\lambda > 0$ we have

$$A_\lambda = (\partial\varphi)_\lambda = \partial\varphi_\lambda,$$

where $\varphi_\lambda: H \longrightarrow \mathbb{R}$ is the **Moreau-Yosida approximation of** φ, i.e.

$$\varphi_\lambda(x) = \inf_{y \in H}\left[\varphi(y) + \frac{1}{2\lambda}\|x - y\|_H^2\right].$$

The function φ_λ is convex, Frechet differentiable and

$$\varphi_\lambda(x) = \varphi(J_\lambda(x)) + \frac{1}{2\lambda}\|x - J_\lambda(x)\|_H^2 \qquad \forall\, x \in H,\ \lambda > 0.$$

Also for all $x \in H$ and all $\lambda > 0$ we have that

$$\varphi_\lambda(x) \leq \varphi(x)$$

and

$$\varphi_\lambda(x) \nearrow \varphi(x) \qquad \text{as } \lambda \searrow 0.$$

Finally if $\lambda_n \searrow 0$, $x_n \longrightarrow x$ and $\partial \varphi_{\lambda_n}(x_n) \longrightarrow x^*$, then

$$x^* \in \partial \varphi(x) \qquad \text{and} \qquad \varphi(x) \leq \liminf_{n \to +\infty} \varphi_{\lambda_n}(x_n).$$

$\square$

Using the Yosida approximation, we can have a useful perturbation result in Hilbert spaces. For a proof see Barbu (1976, p. 82).

PROPOSITION 1.4.9

If H is a Hilbert space identified with its dual and $A, B \colon H \longrightarrow 2^H$ are two maximal monotone operators, such that $\mathrm{Dom}\,(A) \cap \mathrm{Dom}\,(B) \neq \emptyset$ and

$$(x^*, B_\lambda(x))_H \geq 0 \qquad \forall\, x \in \mathrm{Dom}\,(A),\ x^* \in A(x),\ \lambda > 0,$$

then $A + B$ is maximal monotone.

1.4.4 Pseudomonotone Operators

In the next definition we introduce two useful extensions of the notion of maximal monotonicity.

DEFINITION 1.4.8   *Let X be a reflexive Banach space and $A \colon X \longrightarrow 2^{X^*}$ an operator.*

(a) *We say that A is **pseudomonotone** if*

  (1) *A has nonempty, bounded and convex values;*

  (2) *A is upper semicontinuous from every finite dimensional subspace of X into X_w^*;*

  (3) *if $x_n \xrightarrow{w} x$ in X, $x_n^* \in A(x_n)$ and $\limsup\limits_{n \to +\infty} \langle x_n^*, x_n - x \rangle_X \leq 0$, then for every $y \in X$, there exists $u^*(y) \in A(x)$, such that*

$$\langle u^*(y), x - y \rangle_X \leq \liminf_{n \to +\infty} \langle x_n^*, x_n - y \rangle_X.$$

(b) *We say that A is **generalized pseudomonotone** if*

  "For any sequences $\{x_n\}_{n \geq 1}$ and $\{x_n^*\}_{n \geq 1}$, such that $x_n \xrightarrow{w} x$ in X, $x_n^* \xrightarrow{w} x^*$ in X^*, $x_n^* \in A(x_n)$ and $\limsup\limits_{n \to +\infty} \langle x_n^*, x_n - x \rangle_X \leq 0$, then $\langle x_n^*, x_n \rangle_X \longrightarrow \langle x^*, x \rangle_X$ and $x^* \in A(x)$."

PROPOSITION 1.4.10

If X *is a reflexive Banach space and* $A\colon X \longrightarrow 2^{X^*}$ *is a maximal monotone operator,*

<u>*then*</u> A *is generalized pseudomonotone.*

PROOF   Let $x_n \xrightarrow{w} x$ in X, $x_n^* \xrightarrow{w} x^*$ in X^*, $x_n^* \in A(x_n)$ for $n \geq 1$ and

$$\limsup_{n \to +\infty} \langle x_n^*, x_n - x \rangle_X \leq 0.$$

Let $y \in \mathrm{Dom}\,(A)$, $y^* \in A(y)$. From the monotonicity of A we have that

$$\langle x_n^* - y^*, x_n - y \rangle_X \geq 0 \qquad \forall\, n \geq 1.$$

Also

$$\langle x_n^*, x_n \rangle_X \; = \; \langle x_n^* - y^*, x_n - y \rangle_X + \langle x_n^*, y \rangle_X + \langle y^*, x_n \rangle_X - \langle y^*, y \rangle_X\,.$$

Hence

$$\langle x^*, x \rangle_X \; \geq \; \limsup_{n \to +\infty} \langle x_n^*, x_n \rangle_X \; \geq \; \langle x^*, y \rangle_X + \langle y^*, x \rangle_X + \langle y^*, y \rangle_X$$

and so $\langle x^* - y^*, x - y \rangle_X \geq 0$. Because $y \in \mathrm{Dom}\,(A)$, $y^* \in A(y)$ were arbitrary and A is maximal monotone, it follows that $x^* \in A(x)$. So $\langle x_n^* - x^*, x_n - x \rangle_X \geq 0$ and

$$\begin{aligned} \liminf_{n \to +\infty} \langle x_n^*, x_n \rangle_X \; &\geq \; \lim_{n \to +\infty} \big(\langle x_n^*, x \rangle_X + \langle x^*, x_n \rangle_X - \langle x^*, x \rangle \big) \\ &= \; \langle x^*, x \rangle_X\,. \end{aligned}$$

Therefore finally we have that $\langle x_n^*, x_n \rangle_X \longrightarrow \langle x^*, x \rangle_X$. $\square$

PROPOSITION 1.4.11

If X *is a reflexive Banach space and* $A\colon X \longrightarrow 2^{X^*}$ *is a pseudomonotone operator,*

<u>*then*</u> A *is generalized pseudomonotone.*

PROOF   Let $x_n \xrightarrow{w} x$ in X, $x_n^* \xrightarrow{w} x^*$ in X^*, $x^* \in A(x_n)$ and

$$\limsup_{n \to +\infty} \langle x_n^*, x_n - x \rangle_X \leq 0.$$

In the definition of pseudomonotonicity (Definition 1.4.8(a)) we let $y = x$ and so we have

$$\langle x_n^*, x_n \rangle_X \; \longrightarrow \; \langle x^*, x \rangle_X\,.$$

Also for every $y \in X$, we can find $u^*(y) \in A(x)$, such that

$$\langle u^*(y), x - y \rangle_X \leq \langle x^*, x - y \rangle_X\,. \tag{1.8}$$

We claim that $x^* \in A(x)$. Indeed, if this is not the case, then by the Strong Separation Theorem (see Theorem A.3.4) we can find $v \in X$, such that

$$\langle x^*, v \rangle_X \ < \ \inf \big\{ \langle y^*, v \rangle_X : \ y^* \in A(x) \big\}.$$

Putting $y = x - v$ into (1.8), we obtain

$$\langle u^*(y), v \rangle_X \le \langle x^*, v \rangle_X \qquad \text{with } u^*(y) \in A(x),$$

a contradiction. ∎

We can have the converse of this proposition under some boundedness condition.

PROPOSITION 1.4.12
If X is a reflexive Banach space, $A : X \longrightarrow 2^{X^*}$ is a generalized pseudomonotone and bounded operator (i.e. maps bounded sets into bounded sets) and A has nonempty, closed and convex values,
<u>*then*</u> *A is pseudomonotone.*

PROOF First we show that A is upper semicontinuous from X into $X^*_{w^*}$. By virtue of Proposition 1.2.5 it suffices to show that Gr A is closed in $X \times X^*_{w^*}$. Since the weak closure of a bounded set in a reflexive Banach space is sequentially determined (Kaplansky-Browder Theorem; see e.g. Hu & Papageorgiou (1997, p. 318)), it is enough to check sequential closedness of Gr A in $X \times X^*_w$. So let $x_n^* \in A(x_n)$ and assume that $x_n \longrightarrow x$ in X and $x_n^* \xrightarrow{w} x^*$ in X^*. Then we have

$$\lim_{n \to +\infty} \langle x_n^*, x_n - x \rangle_X = 0$$

and from the generalized pseudomonotonicity of A it follows that $x^* \in A(x)$. Therefore A is upper semicontinuous from X into $X^*_{w^*}$.

Next we show that if $x_n \xrightarrow{w} x$ in X, $x_n^* \in A(x_n)$ and

$$\limsup_{n \to +\infty} \langle x_n^*, x_n - x \rangle_X \le 0,$$

then for every $y \in X$ we can find $u^*(y) \in A(x)$, such that

$$\langle u^*(y), x - y \rangle_X \ \le \ \liminf_{n \to +\infty} \langle x_n^*, x_n - y \rangle_X.$$

Suppose that this is not the case. Then we can find $y \in X$, such that

$$\liminf_{n \to +\infty} \langle x_n^*, x_n - y \rangle_X \ < \ \inf \big\{ \langle v^*, x - y \rangle_X : \ v^* \in A(x) \big\}.$$

Note that $\{ x_n^* \}_{n \ge 1}$ is bounded and so we may assume that $x_n^* \xrightarrow{w} x^*$ in X^*. By virtue of the generalized pseudomonotonicity of A, we have $x^* \in A(x)$ and

$\langle x_n^*, x_n \rangle_X \longrightarrow \langle x^*, x \rangle_X$. So

$$\begin{aligned}
\langle x^*, x - y \rangle_X &= \lim_{n \to +\infty} \langle x_n^*, x_n - y \rangle_X \\
&< \inf \left\{ \langle v^*, x - y \rangle_X : v^* \in A(x) \right\},
\end{aligned}$$

a contradiction. So A is pseudomonotone. □

COROLLARY 1.4.6
If X is a reflexive Banach space and $A \colon X \longrightarrow 2^{X^*}$ is a maximal monotone operator with $\mathrm{Dom}\,(A) = X$,
then A is pseudomonotone.

Pseudomonotonicity is preserved under addition. For the proof see e.g. Hu & Papageorgiou (1997, p. 368).

PROPOSITION 1.4.13
If X is a reflexive Banach space and $A, B \colon X \longrightarrow 2^{X^*}$ are two pseudomonotone operators,
then $A + B$ is pseudomonotone too.

As it was the case with maximal monotone operators, the importance of pseudomonotone operators is due to their remarkable surjectivity properties. More precisely, we have the following fundamental result. Its proof can be found in Hu & Papageorgiou (1997, p. 372) (for multivalued A) or in Zeidler (1990b, p. 875) (for single-valued A).

THEOREM 1.4.6
If X is a reflexive Banach space and $A \colon X \longrightarrow 2^{X^*}$ is pseudomonotone and coercive,
then $R(A) = X^*$.

Finally we mention one more class of nonlinear operators of monotone type.

DEFINITION 1.4.9   *Let X be a reflexive Banach space and suppose that $A \colon X \longrightarrow X^*$ is an operator. We say that A is of **type** $(\mathbf{S})_+$, if for any sequence $\{x_n\}_{n \geq 1} \subseteq X$ and $x \in X$ such that $x_n \xrightarrow{w} x$ in X and $\limsup_{n \to +\infty} \langle A(x_n), x_n - x \rangle_X \leq 0$, we have that $x_n \longrightarrow x$ in X.*

REMARK 1.4.10   It is easy to check that a uniformly monotone operator is of type $(\mathrm{S})_+$. Moreover, if $A \colon X \longrightarrow X^*$ is demicontinuous and of type $(\mathrm{S})_+$, then A is pseudomonotone. □

1.4.5 Nemytskii Operators

An important nonlinear operator which can be found in almost all problems is the Nemytskii (or superposition) operator. So let (Ω, Σ, μ) be a complete σ-finite measure space and $f \colon \Omega \times \mathbb{R}^N \longrightarrow \mathbb{R}$ a Carathéodory function (i.e. for all $x \in \mathbb{R}^N$, the function $\omega \longmapsto f(\omega, x)$ is Σ-measurable and for μ-almost all $\omega \in \Omega$, the function $x \longmapsto f(\omega, x)$ is continuous). It is well known that such functions are jointly measurable, hence superpositionally measurable too (i.e. if $u \colon \Omega \longrightarrow \mathbb{R}^N$ is Σ-measurable, then so is $\omega \longmapsto f(\omega, u(\omega))$). Thus we can define the operator

$$u \longmapsto N_f(u)(\cdot) \stackrel{df}{=} f(\cdot, u(\cdot)),$$

which sends Σ-measurable functions to Σ-measurable functions. This operator is known as the **Nemytskii operator** (or **superposition operator**). The main result concerning this operator is the so-called Krasnoselskii Theorem. The proof can be found in Krasnoselskii (1964, p. 22–29).

THEOREM 1.4.7 (Krasnoselskii Theorem)
Let (Ω, Σ, μ) be a complete σ-finite measure space and $f \colon \Omega \times \mathbb{R}^N \longrightarrow \mathbb{R}$ a Carathéodory function.

(a) *If for μ-almost all $\omega \in \Omega$ and all $(x_1, \ldots, x_N) \in \mathbb{R}^N$ we have*

$$\left| f(\omega, x) \right| \leq a(\omega) + c \sum_{k=1}^{N} |x_k|^{\frac{p_k}{q}},$$

for some $p_1, \ldots, p_N, q \in [1, +\infty)$, $a \in L^q(\Omega)_+$, $c > 0$,
<u>*then*</u>

$$N_f(u) \in L^q(\Omega) \qquad \forall\, u = (u_1, \ldots, u_N) \in \prod_{k=1}^{N} L^{p_k}(\Omega)$$

and $N_f \colon \prod_{k=1}^{N} L^{p_k}(\Omega) \longrightarrow L^q(\Omega)$ is bounded continuous.

(b) *Conversely:* <u>*If*</u> *$N_f \colon \prod_{k=1}^{N} L^{p_k}(\Omega) \longrightarrow L^q(\Omega)$ with $p_1, \ldots, p_N, q \in [1, +\infty)$,*
<u>*then*</u> *there exist $a \in L^q(\Omega)$ and $c > 0$, such that for μ-almost all $\omega \in \Omega$ and all $(x_1, \ldots, x_N) \in \mathbb{R}^N$ we have*

$$\left| f(\omega, x) \right| \leq a(\omega) + c \sum_{k=1}^{N} |x_k|^{\frac{p_k}{q}}.$$

PROPOSITION 1.4.14
Let (Ω, Σ, μ) be a complete σ-finite measure space, $f \colon \Omega \times \mathbb{R} \longrightarrow \mathbb{R}$ a

Carathéodory function and let $F \colon \Omega \times \mathbb{R} \longrightarrow \mathbb{R}$ *be defined by*

$$F(z, \zeta) \overset{df}{=} \int_0^{\zeta} f(z, \xi) d\xi$$

(the **potential function corresponding to** *f). Let* $p \in [1, +\infty)$ *and let* $\psi \colon L^p(\Omega) \longrightarrow \mathbb{R}$ *be defined by*

$$\psi(u) \overset{df}{=} \int_{\Omega} F(z, u(z)) \, dz.$$

If there exist $a \in L^q(\Omega)_+$ *and* $c > 0$*, such that*

$$\left| f(z, \zeta) \right| \leq a(z) + c |\zeta|^{p-1} \qquad \text{for } \mu\text{-a.a. } z \in \Omega \text{ and all } \zeta \in \mathbb{R},$$

then ψ *is continuously differentiable and* $\psi'(u) = N_f(u)$*.*

PROOF   Note that

$$\left| F(z, \zeta) \right| \leq a_1(z) + c_1 |\zeta|^p \quad \text{for a.a. } z \in \Omega \text{ and all } \zeta \in \mathbb{R},$$

for some $a_1 \in L^1(\Omega)$ and $c_1 > 0$. Then $N_f \colon L^p(\Omega) \longrightarrow L^1(\Omega)$ is continuous and from this it follows that ψ is continuous.

Next let

$$r(v) \overset{df}{=} \int_{\Omega} F(z, (u+v)(z)) \, dz - \int_{\Omega} F(z, u(z)) \, dz - \int_{\Omega} f(z, u(z)) v(z) \, dz.$$

From the Mean Value Theorem, we have that

$$F(z, (u+v)(z)) - F(z, u(z)) = \int_0^1 \frac{d}{dt} F(z, u(z) + tv(z)) dt$$

$$= \int_0^1 f(z, u(z) + tv(z)) v(z) \, dt.$$

Therefore we obtain

$$\left| r(v) \right| \leq \int_{\Omega} \int_0^1 \left| f(z, u(z) + tv(z)) - f(z, u(z)) \right| |v(z)| \, dt \, dz.$$

From Fubini's theorem and Hölder's inequality we have

$$\left| r(v) \right| \leq \int_0^1 \| N_f(u + tv) - N_f(u) \|_{p'} \, dt \, \| v \|_p \, .$$

Therefore $\frac{|r(v)|}{\|v\|_p} \longrightarrow 0$ as $\|v\|_p \to 0$, which proves that $\psi'(u) = N_f(u)$, i.e.
$\psi \in C^1(L^p(\Omega))$. □

1.4.6 Ekeland Variational Principle

In the last part of the section we will discuss the Ekeland Variational Principle. This result is one of the major tools of Nonlinear Analysis with a wide range of applications. Roughly speaking, the idea behind the Ekeland Variational Principle is the following. Let (X, d_X) be a complete metric space and $\varphi\colon X \longrightarrow \mathbb{R}$, a lower semicontinuous, bounded below function. Suppose that $\varphi(x_0)$ is nearly a minimum value of φ. Then a small Lipschitz continuous perturbation of φ attains a strict minimum at a point y relatively close to x_0. This fact turned out to be a powerful tool in many different parts of Nonlinear Analysis.

THEOREM 1.4.8 (Ekeland Variational Principle I)
If (X, d_X) is a complete metric space, $\varphi\colon X \longrightarrow \overline{\mathbb{R}}$ is a proper, lower semicontinuous and bounded below function, $x_0 \in \operatorname{dom}\varphi$ and $\delta > 0$ are fixed, then there exists $y \in X$, such that

$$\varphi(y) + \delta d_X(y, x_0) \leq \varphi(x_0)$$
$$\varphi(y) < \varphi(v) + \delta d_X(v, y) \qquad \forall v \neq y.$$

PROOF By changing the metric d to δd, we may assume without loss of generality that $\delta = 1$. Moreover, by considering $\varphi - \inf_X \varphi$ instead of φ, we may also assume without any loss of generality that $\varphi \geq 0$. For each $x \in X$, let

$$S(x) \overset{df}{=} \{z \in X : \varphi(z) + d_X(z, x) \leq \varphi(x)\}.$$

Because φ is lower semicontinuous, the set $S(x) \subseteq X$ is closed and clearly $x \in S(x)$ for every $x \in X$.

First we show that

$$z \in S(x) \implies S(z) \subseteq S(x). \qquad (1.9)$$

If $\varphi(x) = +\infty$, then $S(x) = X$ and so (1.9) holds. So suppose that $x \in \operatorname{dom}\varphi$. Then by definition if $z \in S(x)$ we have

$$\varphi(z) + d_X(z, x) \leq \varphi(x).$$

Also if $h \in S(z)$ then

$$\varphi(h) + d_X(h, z) \leq \varphi(z).$$

Thus we obtain

$$\varphi(h) + d_X(h, x) \leq \varphi(h) + d_X(h, z) + d_X(z, x) \leq \varphi(z) + d_X(z, x) \leq \varphi(x),$$

which proves (1.9).

Next let

$$m_{\inf}(x) \overset{df}{=} \inf\{\varphi(z) : z \in S(x)\} \qquad \forall\, x \in \operatorname{dom}\varphi.$$

For any $z \in S(x)$ we have

$$m_{\inf}(x) \leq \varphi(z) \leq \varphi(x) - d_X(z, x)$$

and so

$$d_X(z, x) \leq \varphi(x) - m_{\inf}(x).$$

Therefore we have

$$\operatorname{diam} S(x) \leq 2\big(\varphi(x) - m_{\inf}(x)\big). \tag{1.10}$$

We define a sequence $\{x_n\}_{n \geq 0} \subseteq X$ in the following way: $x_{n+1} \in S(x_n)$ for $n \geq 0$ is such that

$$\varphi(x_{n+1}) \leq m_{\inf}(x_n) + \frac{1}{2^n}. \tag{1.11}$$

From (1.9) we know that $S(x_{n+1}) \subseteq S(x_n)$ and so

$$m_{\inf}(x_n) \leq m_{\inf}(x_{n+1}) \qquad \forall\, n \geq 0. \tag{1.12}$$

But $m_{\inf}(x) \leq \varphi(x)$ because $x \in S(x)$. This combined with (1.11) and (1.12) gives

$$0 \leq \varphi(x_{n+1}) - m_{\inf}(x_{n+1}) \leq m_{\inf}(x_n) + \frac{1}{2^n} - m_{\inf}(x_{n+1}) \leq \frac{1}{2^n}.$$

Then by (1.10), we see that

$$\operatorname{diam} S(x_n) \longrightarrow 0 \quad \text{as } n \to +\infty.$$

But $\{S(x_n)\}_{n \geq 1}$ is a decreasing sequence of closed sets. Therefore by the Cantor Intersection Theorem (see Theorem A.1.11), we have

$$\bigcap_{n \geq 1} S(x_n) = \{y\}.$$

So $y \in S(x_0)$ which implies that

$$\varphi(y) + d_X(y, x_0) \leq \varphi(x_0)$$

(recall that we have assumed that $\delta = 1$). Moreover, because $y \in S(x_n)$ for all $n \geq 0$, from (1.9) we have

$$S(y) \subseteq \bigcap_{n \geq 0} S(x_n) = \{y\},$$

i.e. $S(y) = \{y\}$. Therefore for $v \in X$, $v \neq y$ we have that $v \notin S(y)$ and so

$$\varphi(y) < \varphi(v) + d_X(y,v)$$

(recall that $\delta = 1$). ☐

COROLLARY 1.4.7 (Ekeland Variational Principle II)
If (X, d_X) *is a complete metric space,* $\varphi \colon X \longrightarrow \overline{\mathbb{R}}$ *is proper, lower semicontinuous and bounded below,* $\varepsilon, \lambda > 0$ *and* $x_0 \in X$ *is such that*

$$\varphi(x_0) \leq \inf_X \varphi + \varepsilon,$$

then there exists $x_\lambda \in X$, *such that*

$$\varphi(x_\lambda) \leq \varphi(x_0), \qquad d_X(x_\lambda, x_0) \leq \lambda,$$
$$\varphi(x_\lambda) \ \leq \ \varphi(x) + \frac{\varepsilon}{\lambda} d_X(x, x_\lambda) \qquad \forall\, x \in X.$$

PROOF Let $\delta = \frac{\varepsilon}{\lambda}$. From Theorem 1.4.8 we know that there exists $x_\lambda \in X$, such that

$$\varphi(x_\lambda) + \frac{\varepsilon}{\lambda} d_X(x_\lambda, x_0) \ \leq \ \varphi(x_0) \ \leq \ \inf_X \varphi + \varepsilon \ \leq \ \varphi(x_\lambda) + \varepsilon,$$

hence $d_X(x_\lambda, x_0) \leq \lambda$ and

$$\varphi(x_\lambda) \ < \ \varphi(x) + \frac{\varepsilon}{\lambda} d_X(x, x_\lambda) \qquad \forall\, x \neq x_\lambda.$$

☐

COROLLARY 1.4.8
If X *is a Banach space and* $\varphi \colon X \longrightarrow \mathbb{R}$ *is a function which is lower semicontinuous, bounded below and Gâteaux differentiable,*
then we can find a minimizing sequence $\{x_n\}_{n \geq 1} \subseteq X$ *(i.e.* $\varphi(x_n) \searrow \inf_X \varphi$*), such that*

$$\|\varphi'_G(x_n)\|_X \ \longrightarrow \ 0 \qquad \text{as } n \to +\infty.$$

PROOF Apply Corollary 1.4.7 with $\lambda = 1$, to obtain $x_\varepsilon \in X$, such that

$$\varphi(x_\varepsilon) \ \leq \ \varphi(x) + \varepsilon \|x_\varepsilon - x\|_X \qquad \forall\, x \in X.$$

Let $h \in X$ and $t > 0$ be arbitrary. Setting $x = x_\varepsilon + th$ we obtain

$$\frac{\varphi(x_\varepsilon) - \varphi(x_\varepsilon + th)}{t} \ \leq \ \varepsilon \|h\|_X.$$

Passing to the limit as $t \to 0$, we obtain

$$-\langle \varphi'_G(x_\varepsilon), h \rangle_X \leq \varepsilon \|h\|_X \qquad \forall\, h \in X$$

and so

$$|\langle \varphi'_G(x_\varepsilon), h \rangle_X| \leq \varepsilon \|h\|_X \qquad \forall\, h \in X.$$

But then $\|\varphi'_G(x_\varepsilon)\|_X \leq \varepsilon$. So if we set $\varepsilon = \frac{1}{n}$, we have the desired sequence. $\Box$

COROLLARY 1.4.9

If X is a Banach space, $\varphi\colon X \longrightarrow \mathbb{R}$ is bounded below and differentiable and $\{x_n\}_{n\geq 1} \subseteq X$ is a minimizing sequence of φ,
then there exists a minimizing sequence $\{y_n\}_{n\geq 1}$ of φ, such that

$$\varphi(y_n) \leq \varphi(x_n), \qquad \|x_n - y_n\|_X \longrightarrow 0, \qquad \text{and} \qquad \|\varphi'(x_n)\|_X \longrightarrow 0.$$

Finally we mention a generalization of the Ekeland Variational Principle (Corollary 1.4.7) which is suitable for variational problems in which the energy functional satisfies a weaker form of compactness condition, the so-called "Cerami condition" (see Section 2.1). The result is due to Zhong (1997), where the interested reader can find the proof.

THEOREM 1.4.9

If $h\colon \mathbb{R}_+ \longrightarrow \mathbb{R}_+$ is a continuous and nondecreasing function, such that $\int\limits_0^{+\infty} \frac{1}{1+h(r)}\, dr = +\infty$, X is a complete metric space, $x_0 \in X$ is fixed, $\varphi\colon X \longrightarrow \overline{\mathbb{R}}$ is a proper, lower semicontinuous and bounded below function, $\varepsilon > 0$, $\varphi(y) \leq \inf\limits_X \varphi + \varepsilon$ and $\lambda > 0$,
then there exists $x_\lambda \in X$, such that $\varphi(x_\lambda) \leq \varphi(y)$, $d_X(x_\lambda, x_0) \leq r_0 + \bar{r}$ and

$$\varphi(x_\lambda) \leq \varphi(x) + \frac{\varepsilon}{\lambda(1 + h\,d_X(x_0, x_\lambda))} d_X(x, x_\lambda) \qquad \forall\, x \in X,$$

where $r_0 = d_X(x_0, y)$ and $\bar{r}$ is such that $\int\limits_{r_0}^{r_0 + \bar{r}} \frac{1}{1+h(r)}\, dr \geq \lambda$.

REMARK 1.4.11   If $h \equiv 0$ and $x_0 = y$, then Theorem 1.4.9 reduces to Corollary 1.4.7 (the Ekeland Variational Principle). $\Box$

We conclude this section with a geometric result, which is helpful in the study of nonlinear problems.

PROPOSITION 1.4.15

If X is a reflexive Banach space, $C \subseteq X$ is nonempty, closed, bounded, convex, $C_1 \subseteq X^*$ is nonempty, convex and for each $x^* \in C_1$ there is an $x \in C$

such that $\langle x^*, x \rangle_X \geq 0$,
then there exists $x_0 \in C$ *such that*

$$\langle x^*, x_0 \rangle_X \geq 0 \qquad \forall\, x^* \in C_1.$$

1.5 Elliptic Differential Equations

In this section we have gathered some fundamental results about elliptic equations. For semilinear problems these results are more or less standard and various versions of them can be found in the basic texts on the subject, such as Gilbarg & Trudinger (2001), Ladyzhenskaya & Uraltseva (1968) and Protter & Weinberger (1967). For quasilinear problems (involving the p-Laplacian) most results were developed in the last decade and are not as well known as for the semilinear problems.

1.5.1 Ordinary Differential Equations

Let $T = [0, b]$ and consider the following problem:

$$\begin{cases} -x''(t) = \lambda x(t) & \text{for a.a. } t \in T, \\ x(0) = x(b) = 0. \end{cases} \tag{1.13}$$

We say that $\lambda \in \mathbb{R}$ is an **eigenvalue** of minus the scalar Laplacian with Dirichlet boundary condition $\left(-x'', W_0^{1,2}(T) \right)$, if (1.13) has a nontrivial solution $x \in W_0^{1,2}(T)$, which is called a corresponding **eigenfunction**. It is well known that (1.13) has a sequence of eigenvalues $0 < \lambda_1 \leq \lambda_2 \leq \ldots \leq \lambda_k \longrightarrow +\infty$ and the corresponding eigenfunctions $\{u_n\}_{n \geq 1} \subseteq L^2(T)$ form an orthonormal basis of $L^2(T)$. In fact we have

$$\lambda_n = \left(\frac{n\pi}{b} \right)^2 \qquad \text{and} \qquad u_n(t) = 2\sin\left(\frac{n\pi t}{b} \right), \ n \geq 1.$$

Similarly, instead of (1.13) we can consider the periodic problem

$$\begin{cases} -x''(t) = \lambda x(t) & \text{for a.a. } t \in T, \\ x(0) = x(b), \ x'(0) = x'(b). \end{cases} \tag{1.14}$$

Again we can say that there is an increasing sequence of eigenvalues $\lambda_0 = 0 < \lambda_1 \leq \lambda_2 \leq \ldots \leq \lambda_k \longrightarrow +\infty$ and a corresponding sequence of eigenfunctions $\{u_n\}_{n \geq 1} \subseteq L^2(T)$ which form an orthonormal basis of $L^2(T)$. More precisely we have

$$\lambda_n = \left(\frac{2n\pi}{b} \right)^2 \qquad \text{and} \qquad u_0(t) = \frac{1}{\sqrt{b}}, \quad u_n(t) = 2\cos\left(\frac{2n\pi t}{b} \right), \ n \geq 1.$$

In general we can state the following result which can be found in Showalter (1977, p. 78). Suppose that X, H are two Hilbert spaces with X densely and compactly embedded in H. We identify H with its dual (pivot space) and have

$$X \subseteq H \subseteq X^*$$

with the embeddings being compact and dense (we say that the spaces (X, H, X^*) form an **evolution triple** or **Gelfand triple**). As usual by $(\cdot, \cdot)_H$ we denote the inner product of H and by $\langle \cdot, \cdot \rangle_X$ the duality brackets for the pair (X, X^*). Let $a \colon X \times X \longrightarrow \mathbb{R}$ be a continuous bilinear form. The form a defines an operator $\mathfrak{a} \in \mathcal{L}(X; X^*)$ by

$$\langle \mathfrak{a}(x), y \rangle_X \overset{df}{=} a(x, y) \qquad \forall\, x, y \in X.$$

We consider the restriction A of $\mathfrak{a}$ on H, i.e. $A \colon H \supseteq D \longrightarrow H$,

$$A(x) \overset{df}{=} \mathfrak{a}(x) \qquad \forall\, x \in D = \{ x \in X : \ \mathfrak{a}(x) \in H \}.$$

We determine the spectrum of A, i.e. all $\lambda \in \mathbb{R}$ for which $A(x) = \lambda x$ has a nontrivial solution, called the eigenfunction corresponding to the eigenvalue λ.

PROPOSITION 1.5.1
If X, H, a *and* A *are as above and there exist* $\mu \in \mathbb{R}$ *and* $c > 0$, *such that*

$$a(x, x) + \mu \|x\|_H^2 \ \geq \ c \|x\|_X^2 \qquad \forall\, x \in X,$$

then *there is an orthonormal sequence of eigenfunctions of* A *which is basis of* H *and the corresponding eigenvalues satisfy*

$$-\mu < \lambda_1 \leq \lambda_2 \leq \ldots \leq \lambda_k \leq +\infty \qquad \text{as } k \to +\infty.$$

REMARK 1.5.1   Of course what was said earlier for the scalar problems (1.13) and (1.14) can be extended to the corresponding vector problems (i.e. with $N > 1$). So for the Dirichlet problem we have

$$\lambda_n = \left(\frac{n\pi}{b} \right)^2 \qquad \text{and} \qquad u_n(t) = a \sin\left(\frac{n\pi t}{b} \right), \ a \in \mathbb{R}^N \ n \geq 1$$

and for the periodic problem

$$\lambda_n = \left(\frac{2n\pi}{b} \right)^2 \qquad \text{and} \qquad u_n(t) = a \cos\left(\frac{2n\pi t}{b} \right), \ a \in \mathbb{R}^N \ n \geq 0.$$

$\square$

Now we turn our attention to the *p-**Laplacian ordinary differential operator*** (scalar and vector). So let $p \in (1, +\infty)$ and let $\psi_p \colon \mathbb{R}^N \longrightarrow \mathbb{R}^N$ be the map

$$\psi_p(\xi) \overset{df}{=} \begin{cases} \|\xi\|_{\mathbb{R}^N}^{p-2} \xi & \text{if } \xi \neq 0, \\ 0 & \text{if } \xi = 0. \end{cases}$$

It is well known that ψ_p is a homeomorphism. We consider the following nonlinear eigenvalue problem:

$$\begin{cases} -\big(\psi_p\big(x'(t)\big)\big)' = \lambda \psi_p\big(x(t)\big) & \text{for a.a. } t \in T, \\ x(0) = x(b) = 0. \end{cases} \tag{1.15}$$

The number $\lambda \in \mathbb{R}$ is an ***eigenvalue*** of minus the p-Laplacian with Dirichlet boundary condition, if problem (1.15) has a nontrivial solution. For this case there is no difference between the scalar case (i.e. with $N = 1$) and the vector case ($N > 1$). The set of eigenvalues is the same. Namely there exist eigenvalues $0 < \lambda_1 \leq \lambda_2 \leq \ldots \lambda_k \longrightarrow +\infty$, and

$$\lambda_n = \left(\frac{n\pi_p}{b}\right)^p \qquad \text{and} \qquad u_n(t) = a \sin_p\left(\frac{n\pi_p t}{b}\right), \ a \in \mathbb{R}^N \ n \geq 1.$$

Here

$$\pi_p \overset{df}{=} 2(p-1)^{\frac{1}{p}} \int\limits_0^1 \frac{dt}{(1-t^p)^{\frac{1}{p}}} = \frac{2\pi(p-1)^{\frac{1}{p}}}{p\sin\left(\frac{\pi}{p}\right)}$$

(observe that $\pi_2 = \pi$) and $\sin_p \colon \mathbb{R} \longrightarrow \mathbb{R}$ is defined by

$$\int\limits_0^{\sin_p t} \frac{ds}{\left(1 - \frac{s^p}{p-1}\right)^{\frac{1}{p}}} = t \qquad \forall \, t \in \left[0, \frac{\pi_p}{2}\right]$$

and then extend the domain of $\sin_p t$ to $\mathbb{R}$ in a similar way as for $\sin t$. This way we obtain a $2\pi_p$-periodic function $\sin_p t$ (for details we refer to works of Ôtani (1984*a*, 1984*b*) and del Pino, Elgueta & Manásevich (1988)).

For the periodic eigenvalue problem, the situation is not as pleasant. The set of eigenvalues of the scalar case is a strict subset of the set of eigenvalues of the vector problem. So we consider the following nonlinear eigenvalue problem:

$$\begin{cases} -\big(\psi_p\big(x'(t)\big)\big)' = \lambda \psi_p\big(x(t)\big) & \text{for a.a. } t \in T, \\ x(0) = x(b), \ x'(0) = x'(b). \end{cases} \tag{1.16}$$

As before every $\lambda \in \mathbb{R}$ for which problem (1.16) has a nontrivial solution is called an ***eigenvalue*** of minus the p-Laplacian with periodic boundary conditions. Let $\varepsilon(p, N)$ be the set of all eigenvalues. Evidently this set is not empty, since clearly $0 \in \varepsilon(p, N)$.

For a given $\mu \in \mathbb{R}$ and $h \in L^1(T; \mathbb{R}^N)$, we also consider the following periodic problem:

$$\begin{cases} -\big(\psi_p(x'(t))\big)' + \mu\psi_p(x(t)) = h(t) & \text{for a.a. } t \in T, \\ x(0) = x(b), \ x'(0) = x'(b). \end{cases} \tag{1.17}$$

In analogy to the linear theory we define the **resolvent set** $\varrho(p, N)$ of minus the p-Laplacian with periodic boundary conditions to be the set of all $\mu \in \mathbb{R}$ for which problem (1.17) has at least one solution for every $h \in L^1(T; \mathbb{R}^N)$. Then we define the **spectrum** $\sigma(p, N)$ of minus the p-Laplacian with periodic boundary conditions to be the set $\sigma(p, N) \overset{df}{=} \mathbb{R} \setminus \varrho(p, N)$. The following result is due to Manásevich & Mawhin (1998).

PROPOSITION 1.5.2
If μ is not an eigenvalue of minus the p-Laplacian with periodic boundary conditions,
<u>*then*</u> *for every $h \in L^1(T; \mathbb{R}^N)$ problem (1.17) has at least one solution.*

So as a consequence of this proposition we have that

$$\sigma(p, N) \subseteq \varepsilon(p, N).$$

In what follows let

$$W_{\text{per}}^{1,p}(T; \mathbb{R}^N) \overset{df}{=} \big\{x \in W^{1,p}(T; \mathbb{R}^N) : \ x(0) = x(b)\big\}.$$

Recall that the embedding $W^{1,p}(T; \mathbb{R}^N) \subseteq C(T; \mathbb{R}^N)$ is continuous and so the evaluations at $t = 0$ and $t = b$ make sense. Let $x \in W_{\text{per}}^{1,p}(T; \mathbb{R}^N)$ be an eigenfunction (i.e. a nontrivial solution of (1.16)) corresponding to some eigenvalue. We take the inner product of (1.16) with $x(t)$. We obtain

$$\lambda = \frac{\|x'\|_p^p}{\|x\|_p^p} \geq 0.$$

So $\varepsilon(p, N) \subseteq \mathbb{R}_+$ and since $0 \in \varepsilon(p, N)$, we have that 0 is the smallest eigenvalue of (1.16). Evidently each eigenvalue for the scalar problem ($N = 1$) is also an eigenvalue for the vector problem ($N > 1$). For the scalar case, by direct integration of (1.16) we obtain all the eigenvalues which are given by

$$\lambda_n = \left(\frac{2n\pi_p}{b}\right)^p = \frac{(p-1)(2n\pi)^p}{\left(p \sin\frac{\pi}{p}\right)^p}, \ n \geq 1.$$

Contrary to the Dirichlet eigenvalue problem, in this case the set $\varepsilon(p, N)$ contains more elements in the vector case ($N > 1$) than in the scalar case ($N = 1$). In fact it can be shown that if $k \in \mathbb{N}$, each nontrivial solution of

$$\begin{cases} y''(t) + \left(\frac{2k\pi}{b}\right)^2 y(t) = 0 & \text{for a.a. } t \in T, \\ y(0) = y(b), \ y'(0) = y'(b), \end{cases}$$

such that

$$(y'(t), y(t))_{\mathbb{R}^N} = 0 \qquad \forall \, t \in T,$$

is an eigenfunction corresponding to the eigenvalue $\left(\frac{2k\pi}{b}\right)^p$ of minus the p-Laplacian with periodic boundary conditions. In general the complete structure of the set $\varepsilon(p, N)$ for $N > 1$ is far from being understood. Nevertheless, there are a few important things that we can say for $\varepsilon(p, N)$ when $N > 1$.

Let $\lambda \in \varepsilon(p, N) \setminus \{0\}$ and let u be a corresponding eigenfunction. Integrating (1.16), we obtain

$$\int_0^b \|u(t)\|_{\mathbb{R}^N}^{p-2} \, u(t) dt = 0.$$

For this reason we introduce the following subset of $W_{\mathrm{per}}^{1,p}(T; \mathbb{R}^N)$:

$$S(p, N) \stackrel{df}{=} \left\{ x \in W_{\mathrm{per}}^{1,p}(T; \mathbb{R}^N) : \|x\|_p = 1 \text{ and } \int_0^b \|x(t)\|_{\mathbb{R}^N}^{p-2} \, x(t) dt = 0 \right\}.$$

Also we introduce the functional $\xi_{p,N} \colon W_{\mathrm{per}}^{1,p}(T; \mathbb{R}^N) \longrightarrow \mathbb{R}_+$ defined by

$$\xi_{p,N}(x) \stackrel{df}{=} \|x'\|_p^p \qquad \forall \, x \in W_{\mathrm{per}}^{1,p}(T; \mathbb{R}^N).$$

Let

$$E(p, N) \stackrel{df}{=} \xi_{p,N}\big(S(p, N)\big).$$

Since $\lambda = \frac{\|x'\|_p^p}{\|x\|_p^p}$, it follows that $\varepsilon(p, N) \setminus \{0\} \subseteq E(p, N)$. We consider the following minimization problem:

$$\inf_{x \in S(p,N)} \xi_{p,N}(x) = m_{\mathrm{inf}}(p, N). \tag{1.18}$$

Exploiting the compactness of the embedding $W_{\mathrm{per}}^{1,p}(T; \mathbb{R}^N) \subseteq L^p(T; \mathbb{R}^N)$ and the weak lower semicontinuity of the norm functional in a Banach space, we have:

PROPOSITION 1.5.3
Problem (1.18) has a unique solution and we have $m_{\mathrm{inf}}(p, N) > 0$.

As a corollary we obtain the following fact known as the **extended Poincaré-Wirtinger inequality** (compare with Theorem 1.1.7).

COROLLARY 1.5.1 (Extended Poincaré-Wirtinger Inequality)
For every $x \in W_{\mathrm{per}}^{1,p}(T; \mathbb{R}^N)$, such that

$$\int_0^b \|x(t)\|_{\mathbb{R}^N}^{p-2} \, x(t) dt = 0,$$

we have

$$m_{\inf}(p, N) \|x\|_p^p \leq \|x'\|_p^p.$$

In fact we can show that for $p \geq 2$, $m_{\inf}(p, N)$ is the smallest positive eigenvalue of minus the p-Laplacian with periodic boundary conditions, i.e. $m_{\inf}(p, N) \in \varepsilon(p, N) \setminus \{0\}$, for $p \geq 2$.

PROPOSITION 1.5.4
For $p \geq 2$, $m_{\inf}(p, N)$ is the smallest positive eigenvalue of (1.16).

PROOF   Let $x \in S(p, N)$ be the solution of (1.18). From the Lagrange multiplier rule we know that there exists $(\alpha, \beta, \gamma_1, \ldots, \gamma_N) \neq 0$, such that for all $y \in W_{\text{per}}^{1,p}(T; \mathbb{R}^N)$,

$$\alpha \int_0^b \left(p \|x'(t)\|_{\mathbb{R}^N}^{p-2} x'(t), y'(t) \right)_{\mathbb{R}^N} dt$$

$$+ \beta \int_0^b \left(p \|x(t)\|_{\mathbb{R}^N}^{p-2} x(t), y(t) \right)_{\mathbb{R}^N} dt$$

$$+ \sum_{k=1}^N \gamma_k \int_0^b \Big[(p-2) \|x(t)\|_{\mathbb{R}^N}^{p-4} \left(x(t), y(t) \right)_{\mathbb{R}^N} x_k(t)$$

$$+ \|x(t)\|_{\mathbb{R}^N}^{p-2} y_k(t) \Big] dt = 0. \qquad (1.19)$$

Let $y = \widehat{\gamma} = (\gamma_1, \ldots, \gamma_N) \in \mathbb{R}^N$. Since $\int_0^b \|x(t)\|_{\mathbb{R}^N}^{p-2} x(t) dt = 0$, we obtain

$$\int_0^b \Big[(p-2) \|x(t)\|_{\mathbb{R}^N}^{p-4} \left(x(t), \widehat{\gamma} \right)_{\mathbb{R}^N}^2 + \|x(t)\|_{\mathbb{R}^N}^{p-2} \|\widehat{\gamma}\|_{\mathbb{R}^N}^2 \Big] dt = 0,$$

hence $\|\widehat{\gamma}\|_{\mathbb{R}^N}^2 \int_0^b \|x(t)\|_{\mathbb{R}^N}^{p-2} dt = 0$ and so $\widehat{\gamma} = 0$. Therefore from (1.19), for all $y \in W_{\text{per}}^{1,p}(T; \mathbb{R}^N)$ we have

$$\alpha \int_0^b \left(p \|x'(t)\|_{\mathbb{R}^N}^{p-2} x'(t), y'(t) \right)_{\mathbb{R}^N} dt$$

$$+ \beta \int_0^b \left(p \|x(t)\|_{\mathbb{R}^N}^{p-2} x(t), y(t) \right)_{\mathbb{R}^N} dt = 0.$$

If $\alpha = 0$, then $\beta \neq 0$ and so

$$\int_0^b \big(\|x(t)\|_{\mathbb{R}^N}^{p-2} x(t), y(t) \big)_{\mathbb{R}^N} dt = 0 \qquad \forall\, y \in W_{\text{per}}^{1,p}(T; \mathbb{R}^N).$$

Let $y = x$. We obtain $\|x'\|_p^p = 0$, a contradiction since $x \in S(p,N)$. So $\alpha \neq 0$ and for all $y \in W_{\text{per}}^{1,p}(T; \mathbb{R}^N)$ we can write that

$$\int_0^b \big(p\,\|x'(t)\|_{\mathbb{R}^N}^{p-2} x'(t), y'(t) \big)_{\mathbb{R}^N} dt$$

$$+ \frac{\beta}{\alpha} \int_0^b \big(p\,\|x(t)\|_{\mathbb{R}^N}^{p-2} x(t), y(t) \big)_{\mathbb{R}^N} dt = 0. \tag{1.20}$$

This implies that

$$\begin{cases} -\big(\|x'(t)\|_{\mathbb{R}^N}^{p-2} x'(t) \big)' = -\frac{\beta}{\alpha}\,\|x(t)\|_{\mathbb{R}^N}^{p-2} x(t) & \text{for a.a. } t \in T, \\ x(0) = x(b), \ x'(0) = x'(b). \end{cases}$$

So $x \in S(p,N)$ is an eigenfunction of (1.16). Moreover, if in (1.20) we set $y = x$, we obtain $\|x'\|_p^p = -\frac{\beta}{\alpha}$ (recall that $\|x\|_p = 1$), i.e. $m_{\text{inf}}(p,N) = -\frac{\beta}{\alpha}$, hence $m_{\text{inf}}(p,N) \in \varepsilon(p,N)\setminus\{0\}$. In fact it is clear from (1.18) that $m_{\text{inf}}(p,N)$ is the smallest positive eigenvalue of (1.16). ∎

REMARK 1.5.2   It is an open problem whether the result is also true for $p \in (1,2)$. If $N = 1$, the answer is affirmative. ∎

In a similar way we can have a corresponding variational characterization of the first eigenvalue λ_1 for the Dirichlet problem (1.15) (Poincaré inequality). Clearly $\lambda_1 > 0$.

PROPOSITION 1.5.5 (Rayleigh quotient)
If $\lambda_1 > 0$ is the first eigenvalue of the minus p-Laplacian with Dirichlet boundary conditions,
then

$$\lambda_1 = \min\left\{ \frac{\|x'\|_p^p}{\|x\|_p^p} : x \in W_0^{1,p}(T; \mathbb{R}^N),\ x \neq 0 \right\}.$$

1.5.2   Partial Differential Equations

Next we pass to partial differential operators. First we examine linear operators and then we consider the p-Laplacian:

$$\Delta_p x \overset{df}{=} \text{div}\,\big(\|\nabla x\|_{\mathbb{R}^N}^{p-2} \nabla x \big).$$

For the linear case, we start with some abstract results about compact symmetric operators in a Hilbert space, which then are used to deal with linear eigenvalue problems with indefinite weights, from which we derive as a special case the spectral properties of minus the Laplacian with Dirichlet boundary conditions.

Let H be a Hilbert space and $A \in \mathcal{L}(H)$ a compact self-adjoint operator.

PROPOSITION 1.5.6
If $\lambda_1 = \sup \left\{ (A(x), x)_H : \|x\|_H = 1 \right\} > 0,$
then *there exists* $u_1 \in H,$ *such that* $\|u_1\|_H = 1,$ $(A(u_1), u_1)_H = \lambda_1$ *and*
$A(u_1) = \lambda_1 u_1.$

PROOF   If $\|x\|_H = 1$, we have

$$(A(x), x)_H \leq \|A\|_{\mathcal{L}(H)}$$

and so $\lambda_1 < +\infty$. Let $\{x_n\}_{n \geq 1} \subseteq H$ be such that $\|x_n\|_H = 1$ and

$$(A(x_n), x_n)_H \nearrow \lambda_1.$$

We may assume that

$$x_n \xrightarrow{w} u_1 \quad \text{in } H.$$

Because A is compact and linear, we have that

$$A(x_n) \longrightarrow A(u_1) \quad \text{in } H$$

(see Remark 1.4.1). Therefore $(A(u_1), u_1)_H = \lambda_1$. Note that $\|u_1\|_H \leq \lambda_1$. If $\|u_1\|_H < \lambda_1$, then we can find $t > 1$, such that $t \|u_1\|_H = 1$ and so we have

$$(A(tu_1), tu_1)_H = t^2 \lambda_1 > \lambda_1,$$

a contradiction to the definition of λ_1. So $\|u_1\|_H = 1$. Set $K_1 = A - \lambda_1 I$. We have $(K_1(u_1), u_1)_H = 0$ and

$$(K_1(x), x)_H \leq 0 \qquad \forall\, x \in H.$$

Take $x = u_1 + th$ with $t \in \mathbb{R}$, $h \in H$. Using the fact that K_1 is self-adjoint, we have

$$2 (K_1(u_1), h)_H + t \|h\|_H^2 \leq 0.$$

Let $t = \eta (K_1(u_1), h)_H$ with $\eta \geq 0$. Then we obtain

$$(K_1(u_1), h)_H \left(2 + \eta \|h\|_H^2 \right) \leq 0$$

and so $(K_1(u_1), h)_H \leq 0$. Since $h \in H$ was arbitrary it follows that $K_1(u_1) = 0$, hence $A(u_1) = \lambda_1 u_1$. □

In a similar fashion, we can also prove the following result.

PROPOSITION 1.5.7
If $\lambda_{-1} \overset{df}{=} \inf \left\{ \left(A(x), x \right)_H : \|x\|_H = 1 \right\} < 0,$
then there exists $u_{-1} \in H$, *such that* $\|u_{-1}\|_H = 1$, $\left(A(u_{-1}), u_{-1} \right)_H = \lambda_{-1}$ *and*
$A(u_{-1}) = \lambda_{-1} u_{-1}.$

REMARK 1.5.3 λ_1 is the largest eigenvalue of A and λ_{-1} is the smallest. In a similar fashion we shall generate the other eigenvalues by considering appropriate restrictions on subsets of $(\mathbb{R}u_1)^{\perp}$ and of $(\mathbb{R}u_{-1})^{\perp}$. The whole process relies on some general fact about self-adjoint operators on H (see the next Proposition). $\qquad\qquad\qquad\qquad\qquad\qquad\qquad\qquad\qquad\qquad\qquad\quad$ □

PROPOSITION 1.5.8
If X *is a subspace of* H *invariant under* A *(i.e.* $A(X) \subseteq X$) *and* $Y = X^{\perp}$,
then Y *is also invariant under* A.

PROOF Let $y \in Y$. Because X is invariant under A, we have

$$\left(A(x), y \right)_H = 0 \qquad \forall\, x \in X.$$

Also since A is self-adjoint we have

$$\left(A(x), y \right)_H = \left(x, A(y) \right)_H \qquad \forall\, (x, y) \in X \times Y.$$

So

$$\left(x, A(y) \right)_H = 0 \qquad \forall\, x \in X,$$

hence $A(y) \in X^{\perp} = Y$, i.e. $A(Y) \subseteq Y$. $\qquad\qquad\qquad\qquad\qquad\qquad\qquad$ □

Using this fact we obtain the following.

PROPOSITION 1.5.9

(a) *If*

$$\lambda_n \overset{df}{=} \sup \left\{ \left(A(x), x \right)_H : \|x\|_H = 1,\ x \perp \operatorname{span}\{u_1, \ldots, u_{n-1}\} \right\} > 0,$$

then there exists $u_n \in H$ *with* $\|u_n\|_H = 1$, *such that*

$$\left(A(u_n), u_n \right)_H = \lambda_n \qquad \text{and} \qquad A(u_n) = \lambda_n u_n;$$

(b) *If*

$$\lambda_{-n} \stackrel{df}{=} \inf \left\{ (A(x), x)_H : \|x\|_H = 1, \, x \perp \text{span}\,\{u_{-1}, \ldots, u_{(n-1)}\} \right\} < 0,$$

then there exists $u_{-n} \in H$ *with* $\|u_{-n}\|_H = 1$, *such that*

$$(A(u_{-n}), u_{-n})_H = \lambda_{-n} \qquad \text{and} \qquad A(u_{-n}) = \lambda_{-n} u_{-n}.$$

This proposition generates all the nonzero eigenvalues of A. To see this, let us recall first the ***Spectral Theorem*** for compact self-adjoint operators on a Hilbert space. For a proof we refer to Brézis (1983, p. 97).

THEOREM 1.5.1 (Spectral Theorem)
If H *is an infinite dimensional Hilbert space and* $A \in \mathcal{L}(H)$ *is a compact self-adjoint operator,*
then

(a) *the spectrum* $\sigma(A)$ *of* A *is a subset of* $\mathbb{R}$;

(b) *every* $\lambda \in \sigma(A) \setminus \{0\}$ *is an eigenvalue of* A;

(c) $\sigma(A)$ *is the closure of the set of all eigenvalues of* A *which is countable with only possible cluster point* 0;

(d) *for every eigenvalue* $\lambda \neq 0$, $N(\lambda I - A) < +\infty$;

(e) *there is an orthonormal basis* $\{u_n\}_{n \geq 1}$ *of* H *formed by eigenvectors of* A, *such that*

$$A(x) = \sum_{n \geq 1} \lambda_n \, (x, u_n)_H \, u_n \qquad \forall \, x \in H,$$

where λ_n *is the eigenvalue corresponding to* u_n.

Using this theorem and recalling that for a self-adjoint operator $A \in \mathcal{L}(H)$, we have

$$\|A\|_{\mathcal{L}(H)} = \sup \left\{ (A(x), x)_H : \|x\|_H = 1 \right\}$$

(see Yosida (1978, p. 201)), we obtain at once the following result.

PROPOSITION 1.5.10
The set $\{\lambda_{-n}, \lambda_n\}_{n \geq 1}$ *generated in Proposition 1.5.9 consists of all the nonzero eigenvalues of* A.

The drawback of Proposition 1.5.9 is that the derivation of λ_n requires the knowledge of $\{\lambda_1, \ldots, \lambda_{n-1}\}$. Similarly for λ_{-n}. The next proposition remedies this.

PROPOSITION 1.5.11

For every $n \geq 1$, we have

(a)

$$\lambda_n = \inf_{\substack{V \subseteq H \\ \dim V = n-1}} \sup_{\substack{x \in V^{\perp} \\ \|x\|_H = 1}} \left(A(x), x\right)_H;$$

(b)

$$\lambda_{-n} = \sup_{\substack{V \subseteq H \\ \dim V = n-1}} \inf_{\substack{x \in V^{\perp} \\ \|x\|_H = 1}} \left(A(x), x\right)_H.$$

PROOF   **(a)** Let

$$m_n = \inf_{\substack{V \subseteq H \\ \dim V = n-1}} \sup_{\substack{x \in V^{\perp} \\ \|x\|_H = 1}} \left(A(x), x\right)_H.$$

Since we can always take $V = \text{span}\{u_1, \ldots, u_{n-1}\}$, we see that $m_n \leq \lambda_n$.

On the other hand let $\{e_1, \ldots, e_{n-1}\} \subseteq H$ be mutually orthogonal vectors and set $V = \text{span}\{e_1, \ldots, e_{n-1}\}$. Let

$$v = \sum_{k=1}^{n} \vartheta_k u_k$$

be such that

$$(v, e_k)_H = 0 \quad \forall\, k \in \{1, \ldots, n-1\} \quad \text{and} \quad \|v\|_H = 1$$

(i.e. $\sum_{k=1}^{n} \vartheta_k^2 = 1$). We have

$$\left(A(v), v\right)_H = \sum_{k=1}^{n} \vartheta_k^2 \lambda_k \geq \lambda_n$$

and so for every $V \subseteq H$ subspace of dimension $n - 1$, we have

$$\sup_{\substack{x \in V^{\perp} \\ \|x\|_H = 1}} \left(A(x), x\right)_H \geq \lambda_n,$$

hence $m_n \geq \lambda_n$. Therefore we conclude that $\lambda_n = m_n$.

(b) Similarly as in **(a)**.   ◻

Still Proposition 1.5.11 has the disadvantage of taking the supremum in the case of λ_n (the infimum in the case of λ_{-n}) over an infinite dimensional subspace of H. The next proposition improves this situation.

PROPOSITION 1.5.12
For every $n \geq 1$, we have

(a)

$$\lambda_n = \sup_{\substack{Y \subseteq H \\ \dim Y = n}} \inf_{\substack{y \in Y \\ \|y\|_H = 1}} \left(A(y), y\right)_H;$$

(b)

$$\lambda_{-n} = \inf_{\substack{Y \subseteq H \\ \dim Y = n}} \sup_{\substack{y \in Y \\ \|y\|_H = 1}} \left(A(y), y\right)_H.$$

PROOF   **(a)** Let

$$M_n = \sup_{\substack{Y \subseteq H \\ \dim Y = n}} \inf_{\substack{y \in Y \\ \|y\|_H = 1}} \left(A(y), y\right)_H.$$

Let $Y = \text{span}\,\{u_1, \ldots, u_n\}$. For all

$$y = \sum_{k=1}^{n} \vartheta_k u_k \in Y \quad \text{with} \quad \sum_{k=1}^{n} \vartheta_k^2 = 1,$$

we have $\|y\|_H^2 = 1$ and

$$\left(A(y), y\right)_H = \sum_{k=1}^{n} \vartheta_k^2 \lambda_k \geq \lambda_n$$

and so $M_n \geq \lambda_n$.

On the other hand for any subspace $Y \subseteq H$ with $\dim Y = n$, we choose $y \perp \text{span}\,\{u_1, \ldots, u_n\}$. By Proposition 1.5.9(a) we have $\left(A(y), y\right)_H \leq \lambda_n$ and so $M_n \leq \lambda_n$. Therefore $\lambda_n = M_n$.

(b) Similarly as in **(a)**.   □

REMARK 1.5.4   In the expressions for λ_n and λ_{-n} in Propositions 1.5.11 and 1.5.12, the infima and suprema are actually attained. So the formulas are min-max expressions.   □

Now we shall apply these abstract results to linear weighted eigenvalue problems. So let $\Omega \subseteq \mathbb{R}^N$ be a bounded domain with a C^1-boundary Γ. Consider the following linear partial differential operator in divergence form:

$$L(x) \overset{df}{=} -\sum_{i,j=1}^{N} D_j\left(a_{ij}(z)D_i x\right) + a_0(z)x$$

where $D_k = \frac{\partial}{\partial z_k}$, $k \in \{1,\dots,N\}$. We consider the following eigenvalue problem:

$$\begin{cases} L(x) = \mu m x & \text{in } \Omega, \\ x|_\Gamma = 0. \end{cases} \tag{1.21}$$

We make the following hypotheses on the coefficient functions a_{ij}, a_0 and on the weight function m.

$\underline{H(a)}$ $a_{ij}, a_0 \in L^\infty(\Omega)$, $a_{ij} = a_{ji}$ for $i \in \{1,\dots,N\}$, $a_0(z) \geq 0$ for almost all $z \in \Omega$ and there exists $c > 0$, such that

$$\sum_{i,j=1}^{N} a_{ij}(z)\xi_i\xi_j \geq c\,\|\xi\|_{\mathbb{R}^N}^2 \quad \text{for a.a. } z \in \Omega \text{ and all } \xi \in \mathbb{R}^N.$$

$\underline{H(m)}$ $m \in L^\infty(\Omega)$.

REMARK 1.5.5 The inequality condition in $H(a)$ is the called **strong ellipticity hypothesis**. We emphasize that the weight function can change sign in Ω (indefinite weight). $\qquad\qquad\qquad\qquad\qquad\qquad\square$

We consider the bilinear form $a: H_0^1(\Omega) \times H_0^1(\Omega) \longrightarrow \mathbb{R}$ defined by

$$a(x,y) \overset{df}{=} \sum_{i,j=1}^{N} \int_\Omega \big(a_{ij}(z)D_ix D_jy + a_0(z)xy\big)dz \quad \forall\, x,y \in H_0^1(\Omega).$$

Evidently a is symmetric (i.e. $a(x,y) = a(y,x)$) and using the Poincaré inequality we can check that

$$|a(x,y)| \leq c_1\,\|x\|_{H^1(\Omega)}\,\|y\|_{H^1(\Omega)} \quad \forall\, x,y \in H_0^1(\Omega)$$

and

$$a(x,x) \geq c_2\,\|x\|_{H^1(\Omega)}^2 \quad \forall\, x \in H_0^1(\Omega),$$

for some $c_1, c_2 > 0$. So we can define $A \in \mathcal{L}\big(H_0^1(\Omega), H^{-1}(\Omega)\big)$ by

$$\langle A(x), y\rangle_{H_0^1(\Omega)} \overset{df}{=} a(x,y) \quad \forall\, x,y \in H_0^1(\Omega).$$

Clearly the operator A is maximal monotone, strongly monotone, self-adjoint and coercive. So according to Theorem 1.4.4, A is surjective. Thus for a given function $f \in L^2(\Omega)$, the equation

$$A(x) = f$$

has a unique solution $x \in H_0^1(\Omega)$ (uniqueness results from the strong monotonicity of A). Let $K: L^2(\Omega) \longrightarrow L^2(\Omega)$ be the linear map which to each

$f \in L^2(\Omega)$ assigns the unique solution x of the operator equation $A(x) = f$. Evidently $K \in \mathcal{L}\left(L^2(\Omega)\right)$ and $R(K) \subseteq H_0^1(\Omega) \subseteq L^2(\Omega)$.

PROPOSITION 1.5.13
If hypotheses $H(a)$ and $H(m)$ hold,
then K is self-adjoint and compact.

PROOF   First we show that K is self-adjoint. For every $g \in L^2(\Omega)$, we have

$$
\begin{aligned}
\left(K(f), g\right)_{L^2(\Omega)} &= \int_\Omega K(f)(z)g(z)dz \\
&= \left\langle A\big(K(g)\big), K(f)\right\rangle_{H_0^1(\Omega)} = \left\langle A\big(K(f)\big), K(g)\right\rangle_{H_0^1(\Omega)} \\
&= \int_\Omega K(g)(z)f(z)dz = \left(K(g), f\right)_{L^2(\Omega)}.
\end{aligned}
$$

Next we show that K is compact. To this end suppose that $f_n \xrightarrow{w} f$ in $L^2(\Omega)$. Set $x_n \overset{df}{=} K(f_n)$. We have

$$
c_2 \left\|x_n\right\|_{H^1(\Omega)}^2 \leq \left\|f_n\right\|_2 \left\|x_n\right\|_{H^1(\Omega)},
$$

hence

$$
c_2 \left\|x_n\right\|_{H^1(\Omega)} \leq \sup_{n \geq 1} \left\|f_n\right\|_2 = M_1 < +\infty,
$$

i.e. the sequence $\{x_n\}_{n \geq 1} \subseteq H_0^1(\Omega)$ is bounded. By passing to a subsequence if necessary, we may assume that $x_n \xrightarrow{w} x$ in $H_0^1(\Omega)$ and $x_n \longrightarrow x$ in $L^2(\Omega)$. We have $A(x_n) \xrightarrow{w} A(x)$ in $H^{-1}(\Omega)$. Since $A(x_n) = f_n$ for all $n \geq 1$, we obtain $A(x) = f$, i.e. $x = K(f)$. This proves the compactness of K. ∎

We introduce the operator $K_m \in \mathcal{L}\left(L^2(\Omega)\right)$ defined by

$$
K_m(f) = K(mf) \qquad \forall f \in L^2(\Omega).
$$

Evidently K_m is self-adjoint and compact.

DEFINITION 1.5.1   *By a **weak solution** of (1.21) we mean a function* $x \in W_0^{1,p}(\Omega)$, *such that*

$$
\sum_{i,j=1}^N \int_\Omega a_{ij}(z)D_i x D_j \vartheta dz = \mu \int_\Omega m(z)x\vartheta dz \qquad \forall \vartheta \in C^\infty\left(\overline{\Omega}\right).
$$

REMARK 1.5.6   A simple integration by parts reveals that we can interpret (1.13) in a pointwise fashion, namely

$$-\sum_{i,j=1}^{N} D_j\big(a_{ij}(z)D_i x(z)\big) = \mu m(z)x(z) \qquad \text{for a.a. } z \in \Omega,$$

i.e. $x \in H_0^1(\Omega)$ is actually a **strong solution**.

From now on when possible we shall be using this pointwise interpretation for our boundary value problems.   ☐

So using K_m and Definition 1.5.1, we can rewrite problem (1.21) as

$$K_m(x) = \frac{1}{\mu}x.$$

We can apply the abstract results in the beginning of the section and obtain the following proposition.

PROPOSITION 1.5.14
The weighted eigenvalue problem (1.21) has a double sequence of eigenvalues μ_n, $n \in \mathbb{Z}$, *such that*

$$\ldots \leq \mu_{-2} \leq \mu_{-1} < 0 < \leq \mu_1 \leq \mu_2 \leq \ldots$$

with variational characterizations given by

$$\mu_n = \sup_{\substack{Y \subseteq H_0^1(\Omega) \\ \dim Y = n}} \inf_{\substack{y \in Y \\ \|y\|_{H^1(\Omega)} = 1}} \int_\Omega my^2 dz$$

and

$$\mu_{-n} = \inf_{\substack{Y \subseteq H_0^1(\Omega) \\ \dim Y = n}} \sup_{\substack{y \in Y \\ \|y\|_{H^1(\Omega)} = 1}} \int_\Omega my^2 dz.$$

Also $\mu_n \longrightarrow +\infty$ *and* $\mu_{-n} \longrightarrow -\infty$ *as* $n \to +\infty$.

The description of the eigenvalues of (1.21) becomes complete with the next proposition (by $|\cdot|_N$ we denote the Lebesgue measure on $\mathbb{R}^N$).

PROPOSITION 1.5.15
If hypotheses $H(a)$ and $H(m)$ hold,

$$\Omega_+ \overset{df}{=} \{m > 0\}, \quad \Omega_- \overset{df}{=} \{m < 0\},$$

then

(a) $|\Omega_+|_N = 0$ implies that there are no positive eigenvalues μ_n;

(b) $|\Omega_-|_N = 0$ implies that there are no negative eigenvalues μ_{-n};

(c) $|\Omega_+|_N > 0$ implies that there is a sequence of positive eigenvalues such that $\frac{1}{\mu_n} \longrightarrow 0$;

(d) $|\Omega_-|_N > 0$ implies that there is a sequence of negative eigenvalues such that $\frac{1}{\mu_{-n}} \longrightarrow 0$.

PROOF   (a) and (b) follow at once from Proposition 1.5.14.

(c) Let $\{B_1, \ldots, B_n\}$ be pairwise disjoint balls in Ω, such that $|B_k \cap \Omega_+| > 0$. Consider $\chi_{B_k \cap \Omega_+} \in L^2(\Omega)$ and choose $\vartheta_k \in C^\infty(\overline{\Omega})$ with supp $\vartheta_k \subseteq B_k$ and $\xi_k = \int_\Omega m\vartheta_k^2 dz > 0$. We can always find such function due to the density of the embedding $C^\infty(\overline{\Omega}) \subseteq L^2(\Omega)$. We show that $\mu_n > 0$. Note that

$$\text{supp}\,\vartheta_k \cap \text{supp}\,\vartheta_i \;=\; \emptyset \qquad \forall\, k \ne i$$

and let

$$Y \stackrel{df}{=} \text{span}\,\{\vartheta_1, \ldots, \vartheta_n\}.$$

For $u = \sum_{k=1}^{n} \gamma_k \vartheta_k \in Y$, $\gamma_k \in \mathbb{R}$, we have

$$\int_\Omega mu^2 dz \;=\; \sum_{k=1}^{n} \gamma_k^2 \int_\Omega m\vartheta_k^2 dz \;=\; \sum_{k=1}^{n} \gamma_k^2 \xi_k \;\ge\; \widehat{\xi} \sum_{k=1}^{n} \gamma_k^2,$$

with $\widehat{\xi} = \min\{\xi_1, \ldots, \xi_n\} > 0$. Also

$$\|u\|_{H^1(\Omega)}^2 \;=\; \sum_{k=1}^{n} \gamma_k^2 \|\vartheta_k\|_{H^1(\Omega)}^2 \;\le\; \eta \sum_{k=1}^{n} \gamma_k^2,$$

with $\eta = \max\{\|\vartheta_1\|_{H^1(\Omega)}^2, \ldots, \|\vartheta_n\|_{H^1(\Omega)}^2\} > 0$. Finally

$$\int_\Omega m \frac{u^2}{\|u\|_{H^1(\Omega)}^2} dz \;\ge\; \int_\Omega m \frac{u^2}{\eta \sum_{k=1}^{n} \gamma_k^2} dz \;\ge\; \frac{\widehat{\xi}}{\eta} \;>\; 0,$$

hence $\mu_n > 0$ (see Proposition 1.5.14).

(d) Similar to the proof of (c).   ▯

Of course μ_n depends on the weight function m. The next proposition explains this dependence. The result follows at once from Proposition 1.5.14.

PROPOSITION 1.5.16
Let hypotheses $H(a)$ and $H(m)$ hold.

(a) *If $m_1(z) \leq m_2(z)$ for almost all $z \in \Omega$ and both $\mu_n(m_1)$ and $\mu_n(m_2)$
exist $(n \in \mathbb{Z})$,*
then $\mu_n(m_1) \geq \mu_n(m_2)$.
*Moreover, if $m_1(z) \leq m_2(z)$ and the inequality is strict on a set of
positive measure, then $\mu_n(m_1) > \mu_n(m_2)$;*

(b) *the map $m \longmapsto \mu_n(m)$ is continuous in the $L^\infty(\Omega)$-norm $(n \in \mathbb{Z})$;*

(c) *If $\Omega' \subseteq \Omega$, then $\mu_n(\Omega') \geq \mu_n(\Omega)$ $(n \in \mathbb{Z})$.*

REMARK 1.5.7 If $\vartheta > 0$, then $\mu_n(\vartheta) = \frac{1}{\vartheta}\mu_n(1)$ and they exist only for
$n \geq 1$. ▯

Next we determine the properties of λ_1 and λ_{-1} and of the corresponding
eigenfunctions. The result can be viewed as a ***Krein-Rutman type Theorem*** for the eigenvalue problem (1.13). For the proof of this result we refer
to Manes & Micheletti (1973) or de Figueiredo (1982).

THEOREM 1.5.2 (Krein-Rutman type Theorem)
If hypotheses $H(a)$ and $H(m)$ hold and $|\Omega_+| > 0$,
*then μ_1 is simple and the corresponding eigenfunction u_1 can be taken to
satisfy*

$$u_1(z) > 0 \qquad \forall\, z \in \Omega.$$

A similar statement holds for μ_{-1} provided that $|\Omega_-| > 0$.

Now assume that $m \equiv 1$, $a_{ij} \equiv 1$ for $1 \leq i,j \leq N$ and $a_0 \equiv 0$. Then we
obtain the negative Laplace operator and the eigenvalue problem is

$$\begin{cases} -\Delta x(z) = \lambda x(z) & \text{for a.a. } z \in \Omega, \\ x|_\Gamma = 0. \end{cases} \tag{1.22}$$

From the previous discussion and in particular from Propositions 1.5.15
and 1.5.16, we can state the following result concerning $\left(-\Delta, H_0^1(\Omega)\right)$.

THEOREM 1.5.3
Consider periodic problem (1.14).

(a) *There exists a sequence $\{\lambda_n\}_{n \geq 1} \subseteq \mathbb{R}_+ \setminus \{0\}$ with $\lambda_n \longrightarrow +\infty$ as
$n \to +\infty$ and orthonormal basis $\{u_n\}_{n \geq 1} \subseteq L^2(\Omega)$, such that the pairs
(λ_n, u_n) solve (1.14) and $u_n \in H_0^1(\Omega) \cap C^\infty(\Omega)$;*

(b) *The eigenelements (λ_n, u_n) have the following variational characterization. If for $x \in H_0^1(\Omega) \setminus \{0\}$,*

$$R(x) \overset{df}{=} \frac{\|\nabla x\|_2^2}{\|x\|_2^2} \qquad \text{(\textbf{the Rayleigh quotient}),}$$

then we have

$$\lambda_1 = \min_{\substack{x \in H_0^1(\Omega) \\ x \neq 0}} R(x) = R(u_1),$$

$$\lambda_n = \max\{R(x) : x \in \text{span}\{u_1,\ldots,u_n\}, \; x \neq 0\} = R(u_n)$$

$$\lambda_n = \min\{R(x) : x \in (\text{span}\{u_1,\ldots,u_{n-1}\})^\perp, \; x \neq 0\}$$

$$= \min_{\substack{Y \subseteq H_0^1(\Omega) \\ \dim Y = n}} \max_{x \in Y} R(x), \qquad n \geq 1.$$

REMARK 1.5.8   The fact that the eigenfunctions u_n belong in $C^\infty(\Omega)$ follows from the standard linear elliptic regularity theory (see e.g. Gilbarg & Trudinger (2001)). If Ω has a C^∞-boundary, then we even have $u_n \in C^\infty(\overline{\Omega})$. Also if $H_0^1(\Omega)$ is equipped with the equivalent norm $\|\nabla x\|_2$ for all $x \in H_0^1(\Omega)$ (by the Poincaré inequality), then

$$\left\{ \frac{u_n}{\sqrt{\lambda}} \right\}_{n \geq 1} \quad \text{is an orthonormal basis for } H_0^1(\Omega).$$

☐

Next we shall formulate the corresponding result to Theorem 1.5.2. First an auxiliary result.

LEMMA 1.5.1
<u>*If*</u> $x \in H_0^1(\Omega) \setminus \{0\}$ *satisfies* $R(x) = \lambda_1$,
<u>*then*</u> x *is an eigenfunction corresponding to* λ_1.

PROOF   Let $y \in H_0^1(\Omega)$ and $t > 0$. From Theorem 1.5.3 we have that

$$\lambda_1 = R(x) \leq R(x + ty).$$

Without any loss of generality we may assume that $\|x\|_2 = 1$. We have

$$\frac{\|\nabla(x + ty)\|_2^2}{\|x + ty\|_2^2} \geq \|\nabla x\|_2^2 = R(x) = \lambda_1$$

and so

$$t^2 \|\nabla y\|_2^2 + 2t \int_\Omega (\nabla x, \nabla y)_{\mathbb{R}^N} \, dz \geq \lambda_1 \left(2t \int_\Omega xy\,dz + t^2 \|y\|_2^2 \right).$$

Dividing by $2t$ and letting $t \to 0^+$, we obtain

$$\int_\Omega (\nabla x, \nabla y)_{\mathbb{R}^N} \, dt \;=\; \lambda_1 \int_\Omega xy \, dz.$$

Since $y \in H_0^1(\Omega)$ was arbitrary, we conclude that x is a solution of the problem (1.14), i.e. x is an eigenfunction. $\qquad\Box$

Using this lemma, we can prove the following specification of Theorem 1.5.2.

THEOREM 1.5.4
The first eigenelement (λ_1, u_1) of $\left(-\Delta, H_0^1(\Omega) \right)$ satisfies $\lambda_1 > 0$ and $u_1(z) > 0$ for all $z \in \Omega$.

PROOF Since $u^+, u^- \in H_0^1(\Omega)$ (see Remark 1.1.10), we have

$$\int_\Omega (\nabla x, \nabla x^+)_{\mathbb{R}^N} \, dz \;=\; \lambda_1 \int_\Omega x x^+ dz$$

and

$$\int_\Omega (\nabla x, \nabla x^-)_{\mathbb{R}^N} \, dz \;=\; \lambda_1 \int_\Omega x x^- \, dz,$$

hence

$$\left\| \nabla x^+ \right\|_2^2 \;=\; \lambda_1 \left\| x^+ \right\|_2^2 \qquad \text{and} \qquad \left\| \nabla x^- \right\|_2^2 \;=\; \lambda_1 \left\| x^- \right\|_2^2.$$

If u_1 changes sign, then $u_1^+, u_1^- \neq 0$ and so $R(u_1^+) = \lambda_1 = R(u_1^-)$. By virtue of Lemma 1.5.1, both u_1^+ and u_1^- are eigenfunctions corresponding to λ_1, i.e.

$$\begin{aligned} -\Delta u_1^+(z) &= \lambda_1 u_1^+(z) \quad \text{for a.a. } z \in \Omega, \\ -\Delta u_1^-(z) &= \lambda_1 u_1^-(z) \quad \text{for a.a. } z \in \Omega. \end{aligned}$$

From the Strong Maximum Principle, we have

$$u_1^+(z) > 0 \qquad \text{and} \qquad u_1^-(z) > 0 \quad \text{for a.a. } z \in \Omega,$$

a contradiction. So u_1 does not change sign in Ω and we can always assume that $u_1(z) > 0$ for all $z \in \Omega$.

Finally let us show that $\lambda_1 > 0$ is simple. If this is not the case, then we can find another eigenfunction v_1 orthogonal to u_1. But as above $v_1(z) > 0$ for all $z \in \Omega$ and so $\int_\Omega u_1 v_1 dz > 0$, a contradiction. $\qquad\Box$

Next we turn our attention to the p-Laplacian partial differential operator $\Delta_p x = \text{div}\left(\|\nabla x\|_{\mathbb{R}^N}^{p-2}\nabla x\right)$ with $p \in (1,+\infty)$. So let $\Omega \subseteq \mathbb{R}^N$ be a bounded domain with $C^{1,\alpha}$-boundary Γ $(0 < \alpha < 1)$ and consider

$$\begin{cases} -\text{div}\left(\|\nabla x(z)\|_{\mathbb{R}^N}^{p-2}\nabla x(z)\right) = \lambda|x(z)|^{p-2}x(z) & \text{for a.a. } z \in \Omega, \\ x|_\Gamma = 0. \end{cases} \tag{1.23}$$

As before $\lambda \in \mathbb{R}$ is an *eigenvalue* of $\left(-\Delta_p, W_0^{1,p}(\Omega)\right)$, if problem (1.23) has a nontrivial solution $x \in W_0^{1,p}(\Omega)$, called the *eigenfunction* corresponding to the eigenvalue λ.

PROPOSITION 1.5.17
There exists a first (principal) eigenvalue $\lambda_1 > 0$ and at least one corresponding eigenfunction $u_1 \neq 0$, $u_1(z) \geq 0$ for almost all $z \in \Omega$.

PROOF   Let

$$\lambda_1 \overset{df}{=} \inf_{\substack{x \in W_0^{1,p}(\Omega) \\ \|x\|_p = 1}} \|\nabla x\|_p^p.$$

Consider a minimizing sequence $\{x_n\}_{n\geq 1}$. We may assume that $x_n \overset{w}{\longrightarrow} u_1$ in $W_0^{1,p}(\Omega)$ and $x_n \longrightarrow u_1$ in $L^p(\Omega)$. We have $\|u_1\|_p = 1$. Also from the weak lower semicontinuity of the norm functional in a Banach space, we have

$$\|\nabla u_1\|_p^p \leq \liminf_{n\to+\infty}\|\nabla x_n\|_p^p = \lambda_1,$$

hence $\|\nabla u_1\|_p^p = \lambda_1$ (since $\|u_1\|_p = 1$). Therefore $\lambda_1 > 0$. Moreover, from the Lagrange multiplier rule we have that (λ_1, u_1) is an eigenelement of (1.23). Finally if u_1 is an eigenfunction corresponding to λ_1, then $|u_1|$ is also an eigenfunction corresponding to λ_1 and so we can say that $u_1 \geq 0$.   ⬚

1.5.3   Regularity Results

To determine further properties of (λ_1, u_1), we need some auxiliary results. We start with a result on Sobolev functions and distributions which is due to Brézis & Browder (1982), where the reader can find its proof.

LEMMA 1.5.2
If $u \in W_0^{m,p}(\Omega)$ with $m \geq 1$ and $p \in (1,+\infty)$, $T \in W^{-m,p'}(\Omega) \cap L_{\text{loc}}^1(\Omega)$ and for some $h \in L^1(\Omega)$ we have $h(z) \leq T(z)u(z)$ for almost all $z \in \Omega$, then $T(u) \in L^1(\Omega)$ and

$$\langle T, u \rangle_{W_0^{m,p}(\Omega)} = \int_\Omega T(z)u(z)dz.$$

Using this general result we shall obtain an $L^\infty(\Omega)$-bound for the first eigen-function u_1. In what follows by p^* we denote the **Sobolev critical exponent** of the Sobolev Embedding Theorem (see Theorem 1.1.5), defined by

$$p^* = \begin{cases} \frac{Np}{N-p} & \text{if } p < N, \\ +\infty & \text{if } p \geq N \end{cases}$$

(see also Theorem 1.1.5).

LEMMA 1.5.3
<u>*If*</u> $u \in W_0^{1,p}(\Omega)$, $p \in (1, +\infty)$, $\Delta_p u \in L^1_{loc}(\Omega)$, *there exist* $c > 0$, $r \in [1, p^*)$, $s \in \left[1, \frac{p^*}{p}\right)$, $a \in L^{s'}(\Omega)$, $\frac{1}{s} + \frac{1}{s'} = 1$, *such that*

$$-u(z)\Delta_p u(z) \leq a(z)|u(z)| + c|u(z)|^r \qquad \text{for a.a. } z \in \Omega,$$

<u>*then*</u> $u \in L^{p_n}(\Omega)$ *for every integer* $n \geq 0$ *where* p_n *is defined by*

$$p_0 = \begin{cases} p^* & \text{if } p^* < +\infty, \\ 2\max[ps, r] & \text{if } p^* = +\infty \end{cases}$$

and

$$p_{n+1} = p_0 + \frac{p_0}{p}\min\left\{p_n - r, \frac{p_n}{s} - 1\right\}.$$

PROOF   Clearly by the Sobolev Embedding Theorem (see Theorem 1.1.5), we have that $u \in L^{p_n}(\Omega)$. Proceeding by induction suppose that for some positive integer n we have $u \in L^{p_n}(\Omega)$. We shall show that $u \in L^{p_{n+1}}(\Omega)$. For each $k \geq 1$ we consider the following truncation of u,

$$v_k(z) = \begin{cases} -k & \text{if } u(z) \leq -k, \\ u(z) & \text{if } -k \leq u(z) \leq k, \\ k & \text{if } k \leq u(z), \end{cases}$$

i.e. $v_k(z) = \min\left\{k, \max\left\{u(z), -k\right\}\right\}$. Also set

$$\vartheta \stackrel{df}{=} \frac{p(p_{n+1} - p_0)}{p_0}.$$

Note that since $p_0 \leq p_n$ for all $n \geq 0$, we have $\vartheta \geq 0$. We have

$$-|v_k(z)|^\vartheta v_k(z)\Delta_p u(z) \leq a(z)|u(z)|^{\vartheta+1} + c|u(z)|^{r+\vartheta} \qquad \text{for a.a. } z \in \Omega. \quad (1.24)$$

Note that

$$\int_\Omega \left(a(z)|u(z)|^{\vartheta+1} + c|u(z)|^{r+\vartheta}\right) dz \leq \|a\|_{s'}\|u\|_{s(\vartheta+1)}^{\vartheta+1} + c\|u\|_{r+\vartheta}^{r+\vartheta}$$

$$\leq \ (\|a\|_{s'} + c)\,(|\Omega| + 1)\left(\|u\|_{p_n}^{p_n} + 1\right)$$

since $p_n = \max\{r + \vartheta, p'(\vartheta + 1)\}$. Because $|v_k|^\vartheta v_k \in W_0^{1,p}(\Omega)$, $\Delta_p u \in W^{-1,p'}(\Omega) \cap L^1_{\mathrm{loc}}(\Omega)$ (see Theorem 1.1.8) and $|v_k|^\vartheta v_k \Delta_p u$ is bounded below by the $L^1(\Omega)$ function $z \longmapsto -a(z)|u(z)|^{\vartheta + 1} - c|u(z)|^{r + \vartheta}$, we can apply Lemma 1.5.3 and Green's identity to obtain

$$\left\langle -\Delta_p u, |v_k|^\vartheta v_k \right\rangle_{W_0^{1,p}(\Omega)} \ = \ -\int_\Omega |v_k|^\vartheta v_k \Delta_p u\, dz$$

$$= \ (\vartheta + 1)\int_\Omega \|\nabla v_k(z)\|_{\mathbb{R}^N}^p\, |v_k(z)|^\vartheta dz$$

$$= \ (\vartheta + 1)\left(\frac{p}{\vartheta + p}\right)^p \int_\Omega \left\|\nabla\left(|v_k(z)|^{\frac{\vartheta}{p}} v_k(z)\right)\right\|_{\mathbb{R}^N}^p dz$$

$$\geq \ \eta^{-p}\frac{p}{\vartheta + p}\,\|v_k\|_{p_0\frac{\vartheta + p}{p}}^{\vartheta + p}\,,$$

where $\eta > 0$ is the constant in the Sobolev inequality, i.e.

$$\|v\|_{p_0} \leq \eta\,\|v\|_{W^{1,p}(\Omega)} \qquad \forall\, v \in W_0^{1,p}(\Omega).$$

So finally we have

$$\|v_k\|_{p_{n+1}}^{\frac{p p_{n+1}}{p_0}} \ \leq \ \eta_1 p_{n+1}^p\left(\|u\|_{p_n}^{p_n} + 1\right),$$

where $\eta_1 = (\|a\|_{s'} + c)\,(|\Omega| + 1)\left(\frac{\eta}{p_0}\right)^p$. Since $v_k(z) \longrightarrow u(z)$ for almost all $z \in \Omega$, we have $v_k \stackrel{w}{\longrightarrow} u$ in $L^{p_{n+1}}(\Omega)$ and so

$$\|u\|_{p_{n+1}}^{\frac{p p_{n+1}}{p_0}} \ \leq \ \liminf_{k \to +\infty} \|v_k\|_{p_{n+1}}^{\frac{p p_{n+1}}{p_0}} \ \leq \ \eta_1 p_{n+1}^p\left(\|u\|_{p_n}^{p_n} + 1\right), \qquad (1.25)$$

from which it follows that $u \in L^{p_{n+1}}(\Omega)$. $\quad\square$

Next we shall derive a better estimation than (1.25). To this end we consider the sequence $\{s_n\}_{n \geq 1}$, defined by

$$\begin{cases} s_0 \stackrel{df}{=} p_0, \\ s_{n+1} \stackrel{df}{=} \big(s_n + s(p - 1)\big)\delta & \forall\, n \geq 1,\ \text{with } \delta = \frac{p_0}{sp}. \end{cases}$$

LEMMA 1.5.4
If the hypotheses of Lemma 1.5.3 hold,
then $u \in L^{s_n}(\Omega)$ for all integers $n \geq 1$ and we have

$$\|u\|_{s_{n+1}}^{\frac{s_{n+1}}{\delta s}} \ \leq \ \gamma s_{n+1}^p\,\|u\|_{s_n}^{\frac{s_n}{s}},$$

with $\gamma > 0$ *being a constant depending only on c, r, s, N, p,* $\|a\|_{s'}$ *and* $\|u\|_{p_0}$.

PROOF We can check easily that for all $n \geq 1$, we have

$$p_n \geq p_0 + \beta_0 \sum_{k=1}^{n} \delta^k = p_0 + \beta_0 \frac{\delta(\delta^n - 1)}{\delta - 1},$$

where $\beta_0 \stackrel{df}{=} \min\{s(p_0 - r), p_0 - s\}$. Since $\beta_0 > 0$ and $\delta > 1$, we see that $p_n \longrightarrow +\infty$ and so by Lemma 1.5.3, we have that $u \in L^r(\Omega)$ for all $r \geq 1$. So in particular we have that $|u|^{r-1} \in L^{s'}(\Omega)$ and because of (1.25), we have that $\left\||u|^{r-1}\right\|_{s'} \leq \gamma_0$ with $\gamma_0 = \gamma_0(c, r, s, N, p, \|a\|_{s'}, \|u\|_{p_0}) > 0$. It follows that

$$\int_{\Omega} \left(a(z)|u|^{\frac{s_n}{s}} + c|u|^{r-1} \right) dz \leq \left(\|a\|_{s'} + c\gamma_0 \right) \|u\|_{s_n}^{\frac{s_n}{s}}.$$

So if in inequality (1.24) we replace ϑ by $\frac{s_n}{s} - 1$ and proceeding as in the proof of the previous lemma, we reach the desired inequality. ⬜

Using the above lemmata we can have an $L^\infty(\Omega)$-estimate for the solutions of quasilinear elliptic problems, which are the prerequisite for regularity results.

THEOREM 1.5.5
If $u \in W_0^{1,p}(\Omega)$, $p \in (1, +\infty)$, $\Delta_p u \in L^1_{\mathrm{loc}}(\Omega)$, *there exist* $c > 0$, $r \in [1, p^*)$, $s \in \left[1, \frac{p^*}{p}\right)$, $a \in L^{s'}(\Omega)$ *with* $\frac{1}{s} + \frac{1}{s'} = 1$, *such that*

$$-u(z)\Delta_p u(z) \leq a(z)|u(z)| + c|u(z)|^r \qquad \text{for a.a. } z \in \Omega,$$

then $u \in L^\infty(\Omega)$ *and* $\|u\|_\infty \leq \gamma$ *with* $\gamma > 0$ *is a constant depending only on c, r, s, N, p,* $\|a\|_{s'}$ *and* $\|u\|_{p_0}$.

PROOF Using the estimation in Lemma 1.5.4, we shall obtain a uniform bound for the sequence $\left\{\|u\|_{s_n}\right\}_{n \geq 1}$. Set

$$\mu_n \stackrel{df}{=} s_n \ln \|u\|_{s_n} \qquad \text{and} \qquad \xi_n \stackrel{df}{=} r \ln\left(\gamma s_{n+1}^p\right).$$

Then we can rewrite the inequality in the conclusion of Lemma 1.5.4 as

$$\mu_{n+1} \leq \delta(\mu_n + \xi_n)$$

and so

$$\mu_n \leq \delta^n \mu_0 + \sum_{k=1}^{n} \delta^k \xi_{n-k} \qquad \forall \, n \geq 1.$$

Since $s_n = \delta^n p_0 + \delta s (p-1) \frac{\delta^n - 1}{\delta - 1}$, we have

$$\delta^n p_0 \leq s_n \leq \delta^n \gamma_1 \quad \text{with} \quad \gamma_1 = p_0 + \delta s \frac{p-1}{\delta - 1}$$

and so

$$\xi_n \leq d_1 + (n+1) d_2,$$

where $d_1 = s \ln(\gamma \gamma_1^p)$ and $d_2 = sp \ln \delta$. With an elementary calculation we obtain

$$\sum_{k=1}^{n} \delta^n \xi_{n-k} \leq \gamma_2 \delta^n,$$

with $\gamma_2 \overset{df}{=} \left(d_1 + d_2 \frac{\delta}{\delta - 1} \right) \frac{\delta}{\delta - 1}$. Finally, we have

$$\|u\|_{s_n} \leq \exp\left(\frac{\mu_n}{\delta^n p_0} \right) \leq \exp\left(\frac{\mu_0 + \gamma_2}{p_0} \right).$$

Therefore it follows that

$$\|u\|_{\infty} \leq \limsup_{n \to +\infty} \|u\|_{s_n} \leq \exp\left(\frac{\mu_0 + \gamma_2}{p_0} \right).$$

$\square$

Using this theorem we can have regularity results. The result that follows is a particular case of a more general result of Lieberman (1988) and we refer to that paper for the proof.

THEOREM 1.5.6
If $u \in W_0^{1,p}(\Omega) \cap L^{\infty}(\Omega)$ *and* $\Delta_p u \in L^{\infty}(\Omega)$,
then $u \in C^{1,\beta}(\overline{\Omega})$ *with* $\beta \in (0,1)$ *and* $\|u\|_{C^{1,\beta}(\overline{\Omega})} \leq \gamma$, *where* $\gamma > 0$ *is a constant depending only on* N, p, $\|u\|_{\infty}$ *and* $\|\Delta_p u\|_{\infty}$.

REMARK 1.5.9   Similar regularity results are valid for the Neumann problem. $\square$

To infer the positivity of the principal eigenfunction u_1, we need the following **Maximum Principle** due to Vázquez (1984), where the reader can find the proof of the result.

THEOREM 1.5.7 (Maximum Principle)
If $\Omega \subseteq \mathbb{R}^N$ *is a bounded domain,* $u \in C^1(\Omega)$, $u(z) \geq 0$ *for all* $z \in \Omega$, $\Delta_p u \in L^2_{\text{loc}}(\Omega)$ *and* $\Delta_p u(z) \leq \beta\big(u(z)\big)$ *for almost all* $z \in \Omega$ *where* $\beta \colon \mathbb{R}_+ \longrightarrow \mathbb{R}$ *is a continuous increasing function, such that* $\beta(0) = 0$ *and either* $\beta(s_0) = 0$

for some $s_0 > 0$ *or* $\int\limits_0^1 \dfrac{ds}{(s\beta(s))^{\frac{1}{p}}} = +\infty,$

<u>*then*</u> *if* $u \neq 0$*, we have* $u(z) > 0$ *for all* $z \in \Omega$*. Moreover, if* $u \in C^1(\Omega \cup \{z_0\})$ *and* $u(z_0) = 0$ *for some* $z_0 \in \Gamma$ *which satisfies the interior ball condition (i.e. there exists a ball* $B \subseteq \Omega$*, such that* $\overline{B} \cap \Gamma = \{z_0\}$*), then* $\frac{\partial u}{\partial n}(z_0) < 0$*, where* $n(z_0)$ *is the exterior unit normal on* Γ*.*

Using Theorems 1.5.5, 1.5.6 and 1.5.7, we can improve the conclusion of Proposition 1.5.17 as follows.

PROPOSITION 1.5.18
There exists a first eigenvalue $\lambda_1 > 0$ *for problem (1.15) and a corresponding eigenfunction* $u_1 \in C^{1,\beta}(\overline{\Omega})$ $(0 < \beta < 1)$ *satisfying* $u_1(z) > 0$ *for all* $z \in \Omega$*. Moreover*

$$\lambda_1 = R_p(u_1) = \inf\left\{ \frac{\|\nabla x\|_p^p}{\|x\|_p^p} : x \in W_0^{1,p}(\Omega),\ x \neq 0 \right\}.$$

In fact we can show that λ_1 is simple and isolated. For a proof of this when Ω is any bounded domain of $\mathbb{R}^N$ (no condition on the boundary Γ), we refer to Lindqvist (1990, 1992).

PROPOSITION 1.5.19
The first eigenvalue λ_1 *of* $\left(-\Delta_p, W_0^{1,p}(\Omega)\right)$ *is positive, simple and isolated (simple means that the corresponding eigenspace is one-dimensional, i.e. if* u *and* v *are two eigenfunctions corresponding to* λ_1*, then* $u = \vartheta v$ *for some* $\vartheta \in \mathbb{R}$*).*

What about higher eigenvalues of $\left(-\Delta_p, W_0^{1,p}(\Omega)\right)$? The ***Lusternik-Schnirelmann theory*** (see e.g. Zeidler (1990b, p. 317)) gives in addition to λ_1 a whole strictly increasing sequence $\{\lambda_n\}_{n\geq 1} \subseteq \mathbb{R}_+$ (not counting multiplicities) for which problem (1.15) has a nontrivial solution. These numbers are defined as follows. Let

$$G \stackrel{df}{=} \left\{ x \in W_0^{1,p}(\Omega) : \|\nabla x\|_p = 1 \right\}$$

and let $\psi \colon G \longrightarrow \mathbb{R}_-$ be given by

$$\psi(x) \stackrel{df}{=} -\|x\|_p^p.$$

We set

$$c_n \stackrel{df}{=} \inf_{K \in \mathcal{A}_n} \sup_{x \in K} \psi(x),$$

where

$$\mathcal{A}_n \stackrel{df}{=} \left\{ K \subseteq G : K \text{ is symmetric, closed and } \gamma(K) \geq n \right\},$$

with γ denoting the **Krasnoselskii $\mathbb{Z}_2$-genus** (see e.g. Struwe (1990, p. 86) or Zeidler (1990b, p. 319)). The sequence $\{\lambda_n\}_{n\geq 1} \subseteq \mathbb{R}_+$, where $\lambda_n \overset{df}{=} -\frac{1}{c_n}$, $n \geq 1$, is strictly increasing and tends to $+\infty$. These numbers are the so-called **Lusternik-Schnirelmann eigenvalues** or **variational eigenvalues** of $\left(-\Delta_p, W_0^{1,p}(\Omega) \right)$. For $p = 2$ (minus the Laplacian), these are the eigenvalues. However, for $p \neq 2$ we cannot say this. Let

$$\lambda_2^* \overset{df}{=} \inf \left\{ \lambda > \lambda_1 : \lambda \text{ is an eigenvalue of } \left(-\Delta_p, W_0^{1,p}(\Omega) \right) \right\}.$$

Since λ_1 is isolated (see Proposition 1.5.19), we have that $\lambda_2^* > \lambda_1$. Anane & Tsouli (1996) proved the following result.

PROPOSITION 1.5.20
$\lambda_2^* = \lambda_2$, *i.e. the second eigenvalue and the second Lusternik-Schnirelmann eigenvalue of* $\left(-\Delta_p, W_0^{1,p}(\Omega) \right)$ *coincide.*

REMARK 1.5.10   For $n \geq 1$, set

$$V_k \overset{df}{=} \left\{ x \in W_0^{1,p}(\Omega) : \ -\mathrm{div}\left(\|\nabla x(z)\|_{\mathbb{R}^N}^{p-2} \nabla x(z) \right) = \lambda_k |x(z)|^{p-2} x(z) \right.$$

$$\left. \text{for a.a. } z \in \Omega \right\}.$$

These are symmetric, closed cones, but in general are not subspaces of $W_0^{1,p}(\Omega)$, unless λ_k is simple. Also if

$$W_n \overset{df}{=} \bigcup_{k=1}^{n} V_k \qquad \text{and} \qquad \widehat{W}_n \overset{df}{=} \bigcup_{k\geq n} V_k,$$

then in contrast to the linear case (i.e. $p = 2$), for $p \neq 2$, in general we do not have the inequalities

$$\|\nabla x\|_p^p \leq \lambda_k \|x\|_p^p \qquad \forall \, x \in W_k$$

and

$$\|\nabla x\|_p^p \geq \lambda_{k+1} \|x\|_p^p \qquad \forall \, x \in \widehat{W}_{k+1}$$

(see Theorem 1.5.3(b)).

This fact is a source of difficulties when dealing with quasilinear problems (in particular in constructing linking sets, see Chapter 4).   ⬚

Let us conclude this section by summarizing the basic regularity results for linear elliptic equations. These results are standard and can be found with the proofs in the books of Gilbarg & Trudinger (2001) and Ladyzhenskaya & Uraltseva (1968).

Let $\Omega \subseteq \mathbb{R}^N$ be a bounded domain with boundary Γ. We consider the boundary value problem

$$\begin{cases} L_1(x) = f & \text{in } \Omega, \\ x|_\Gamma = 0, \end{cases} \qquad (1.26)$$

where L_1 is the linear differential operator in divergence form

$$L_1(x)(z) \overset{df}{=} -\sum_{i,=1}^{N} D_j\big(a_{ij}(z)D_i x(z)\big)$$

and f is a given function (or functional). An important special case is when $a_{ij} = \delta_{ij}$ in which case we obtain minus the Laplacian operator. We impose the following condition on the coefficient functions a_{ij}.

$\underline{H(a)'}$ $a_{ij} \colon \Omega \longrightarrow \mathbb{R}$ are measurable functions $(1 \leq i, j \leq N)$, $a_{ij} = a_{ji}$ and there exists $c > 0$ such that

$$\sum_{i,j=1}^{N} a_{ij}(z)\xi_i \xi_j \geq c \, \|\xi\|_{\mathbb{R}^N}^2 \quad \text{for a.a. } z \in \Omega \text{ and all } \xi \in \mathbb{R}^N.$$

THEOREM 1.5.8
Let hypotheses $H(a)'$ hold.

(a) *If Γ is a C^1-manifold, $a_{ij} \in C(\overline{\Omega})$ and $f \in W^{-1,p'}(\Omega)$ for some $p \in (1, +\infty)$,*
then there exists a unique weak solution $x \in W_0^{1,p}(\Omega)$ of (1.26) and there exists $M > 0$ independent of f, such that

$$\|x\|_{W^{1,p}(\Omega)} \leq M \, \|f\|_{W^{-1,p'}(\Omega)};$$

(b) *If Γ is a $C^{1,1}$-manifold, $a_{ij} \in C^1(\overline{\Omega})$ and $f \in L^p(\Omega)$ for some $p \in (1, +\infty)$,*
then there exists a unique strong solution $x \in W^{2,p}(\Omega) \cap W_0^{1,p}(\Omega)$ of (1.26) (i.e. (1.26) is satisfied pointwise for almost all $z \in \Omega$) and there exists $M > 0$ independent of f, such that

$$\|x\|_{2,p} \leq M \, \|f\|_p;$$

(c) *If Γ is a $C^{2,\alpha}$-manifold, $a_{ij} \in C^{1,\alpha}(\overline{\Omega})$ and $f \in C^{\alpha}(\overline{\Omega})$ for some $\alpha \in (0,1)$,*
then there exists a unique classical solution $x \in C^{2,\alpha}(\overline{\Omega})$ of (1.26) (i.e. (1.26) is satisfied pointwise for all $z \in \Omega$) and there exists $M > 0$ independent of f, such that

$$\|x\|_{C^{2,\alpha}(\overline{\Omega})} \leq M \, \|f\|_{C^{\alpha}(\overline{\Omega})}.$$

REMARK 1.5.11   Parts (a) and (b) are usually referred to as the L^p-*theory* and the L^p-*estimates*. Part (c) is referred to as the **Schauder theory** and the **Schuuder estimates**. If $p > N$, the weak solution is in $C^\alpha(\overline{\Omega})$ with $0 < \alpha < 1 - \frac{N}{p}$ and the strong solution is in $C^{1,\alpha}(\overline{\Omega})$ with $0 < \alpha < 1 - \frac{N}{p}$ (Sobolev Embedding Theorem; see Theorem 1.1.5).   ☐

When the coefficients a_{ij} are discontinuous, there is the following result due to Stampacchia (1965).

THEOREM 1.5.9
If hypotheses $H(a)'$ hold, Γ is a C^2-manifold, $a_{ij} \in L^\infty(\Omega)$ and $f \in L^2(\Omega) \cap \underline{L^p}(\Omega)$ with $p > \frac{N}{2}$,
then problem (1.26) has a unique weak solution $x \in W_0^{1,p}(\Omega)$, such that $x \in C^{0,\alpha}(\overline{\Omega})$ with $0 < \alpha < 1$.

1.6   Remarks

1.1 There are many well-written books on the subject of Sobolev spaces. We have consulted the books of Adams (1975), Brézis (1983), Evans & Gariepy (1992), Gilbarg & Trudinger (2001) and Kufner, John & Fučik (1977). The quasilinear Green formula (Theorem 1.1.9) is due to Kenmochi (1975) and Casas & Fernández (1989). The chain rule presented in Theorem 1.1.13 is due to Marcus & Mizel (1972).

1.2 Set-Valued Analysis (or Multivalued Analysis) is a branch of modern analysis that developed in the last thirty years in synchronization with Nonsmooth Analysis. Set-Valued Analysis appears naturally in many applied areas such as an Optimal Control, Mathematical Economics, Game Theory and Nonsmooth Mechanics. For more details on the theory and the applications of multifunctions we refer to the books of Aubin & Frankowska (1990), Hu & Papageorgiou (1997, 2000), Kisielewicz (1991) and Klein & Thompson (1984). We should also mention the survey paper of Wagner (1977) (see also Wagner (1980)) and its addendum, the paper of Ioffe (1978) with additions from the Soviet literature. These papers provide a very detailed survey of the subject of measurable selectors up until 1980.

1.3 Nonsmooth Analysis appeared in the early sixties to address the needs of modern applications of analytic optimization in fields like Mathematical Economics and Engineering. As we already mentioned it developed in synchronization with Set-Valued Analysis and the two fields provide each other with new tools, ideas and results. First during the sixties and early seventies developed Convex Analysis, initially with the works of Moreau and Rockafellar (building on earlier work of Fenchel) and later with other people also contributing significant results on the subject. The various parts of the theory of Convex Analysis can be found in the books of Aubin & Ekeland (1984), Barbu & Precupanu (1986), Ekeland & Temam (1976), Hiriart Urruty & Lemaréchal (1993), Ioffe & Tihomirov (1979), Rockafellar (1970) and Zeidler (1985). In the early seventies (1973), Clarke, motivated by the fact that a convex function φ is locally Lipschitz in the interior of its effective domain and that a locally Lipschitz function is differentiable almost everywhere (Rademacher's theorem), developed his theory of subdifferentiation for locally Lipschitz functions. This theory and its applications are nicely summarized in the monograph of Clarke (1983). The mean value property (Proposition 1.3.14), which will be a useful tool in what follows, is due to Lebourg (1975). The notion of weak slope (Definition 1.3.10) was introduced by De Giorgi, Marino & Tosques (1980) and was developed further by Canino & Degiovanni (1995), Corvellec & Degiovanni (1997), Degiovanni (1997) and Degiovanni & Marzocchi (1994).

1.4 The first attempts to solve functional equations were based on the notion of compactness. We should mention that unfortunately the terminology on the subject is not uniform. Some authors call compact operators completely continuous and completely continuous operators are called strongly continuous. Other variants are also present in the literature and the reader should be alert and know in what context the term is used. Operators of monotone type have their roots in the calculus of variations and were introduced in order to provide a broader framework of analysis than the one offered by compact operators. The starting point was the gradient of a continuous, convex function. The theory of monotone operators was launched in the early sixties, initially with the works of Kachurowskii (who invented the term monotone operator) and Minty and later by Browder, Brézis, Kenmochi and others who shifted the focus from Hilbert spaces to general dual pairs of Banach spaces and also introduced generalizations of the notion of maximal monotonicity. The theory that was developed can be found in the book of Barbu (1976), Brézis (1983), Browder (1976), Hu & Papageorgiou (1997, 2000), Pascali & Sburlanu (1978), Showalter (1997) and Zeidler (1990b, 1990a). For the Nemytskii operator the main references are the books of Krasnoselskii (1964) and Vaĭnberg (1973). The Ekeland Variational Principle can be viewed as the nonlinear version of the well known Bishop-Phelps Theorem from the theory of Banach spaces. Since its appearance (see Ekeland (1974)), the result has found numerous applications in different fields. For these applications we refer to the papers of Ekeland (1979) and the books of Aubin & Ekeland (1984) and Phelps (1984).

1.5 The eigenvalues for minus the ordinary p-Laplacian with Dirichlet boundary conditions were obtained by Ôtani (1984b) (see also del Pino, Elgueta &

Manásevich (1988)). For the corresponding discussion of minus the ordinary p-Laplacian with periodic boundary conditions we refer to Mawhin (2001). The abstract results about the spectral properties of compact self-adjoint operators can be found in many books on functional analysis. We refer to the books of Akhiezer & Glazman (1961, 1963), Berberian (1976), Gohberg & Goldberg (1981), Halmos (1998) and Kato (1976). The specification of these results on strongly elliptic semilinear partial differential operators follows the work of de Figueiredo (1982). From this analysis follow easily the spectral properties of $\left(-\Delta, H_0^1(\Omega) \right)$. For minus the partial p-Laplacian with Dirichlet boundary conditions (i.e. for $\left(-\Delta_p, W_0^{1,p}(\Omega) \right)$), the first result was obtained by Ôtani (1984a), when $\Omega = B_r(0)$ for some $r > 0$. Later de Thélin (1986) proved that in that setting the first eigenvalue $\lambda_1 > 0$ is simple. The simplicity of λ_1 was proved for more general domains with connected boundary Γ. Anane (1987), assuming that the boundary Γ is a $C^{2,\alpha}$-manifold ($0 < \alpha \leq 1$), proved that $\lambda_1 > 0$ is simple, isolated and that $u_1 > 0$. Anane actually considered a weighted eigenvalue problem with $m \in L^\infty(\Omega)$. Lindqvist (1990, 1992) established the simplicity of λ_1 without any regularity assumptions on the boundary Γ of Ω. In Lindqvist (1990) the author studied the continuity of the dependence of λ_1 on p. In Allegretto & Huang (1995) we find the weighted eigenvalue problem for $\left(-\Delta_p, W^{1,p}(\mathbb{R}^N) \right)$. The result about λ_2 (see Proposition 1.5.20) is due to Anane & Tsouli (1996). Results similar to Theorem 1.5.6 were obtained besides Lieberman (1988) by Di Benedetto (1983) and Tolksdorf (1984). Finally let us mention that if $p = 2$ and u_n is the eigenfunction associated to the eigenvalue λ_n, the number of components of the nodal set $\{u_n \neq 0\}$ is not more than n.

Chapter 2

Critical Point Theory

In this chapter, we develop the basic theories on the existence, characterization and multiplicity of critical points for nondifferentiable (nonsmooth) functionals. Lack of differentiability is encountered in a variety of cases, such as free boundary problems (where we deal with functionals involving a combination of a differentiable and a convex term) and problems with discontinuous nonlinearities (where we deal with locally Lipschitz functionals). The key issue here is to produce a substitute for the usual pseudogradient vector field of the smooth theory and through it obtain a deformation theorem, which is the main tool in the derivation of minimax principles characterizing the critical points. In Section 2.1, we develop a theory for nondifferentiable locally Lipschitz functionals, using the subdifferential theory of Clarke. In Section 2.2 we present a corresponding analysis for locally Lipschitz functionals defined on a closed and convex set. In Section 2.3 we extend the theory to convex perturbations of locally Lipschitz functionals. In Section 2.4 we prove a multiplicity result based on the notion of local linking and we also investigate an alternative formulation of the Palais-Smale condition, suggested by the Ekeland Variational Principle. In Section 2.5 we extend the theory to continuous functionals defined on a metric space. This is done using the notion of weak slope (see Definition 1.3.10). Finally in Section 2.6 we move beyond continuous functions, to multifunctions with closed graph. This chapter provides the abstract results which will be used in the variational analysis of boundary value problems in Chapters 3 and 4.

2.1 Locally Lipschitz Functionals

2.1.1 Compactness-Type Conditions

If Y is a reflexive Banach space and $\varphi\colon Y \longrightarrow \mathbb{R}$ is weakly coercive and it is also weakly sequentially lower semicontinuous (see Definition A.4.2), then we can find $y_0 \in Y$, such that

$$\varphi(y_0) = \inf_{y \in Y} \varphi(y)$$

(i.e. y_0 is a global minimizer of φ). However, many natural functionals φ that we encounter are not bounded at all neither from above nor from below (such functionals are usually called **indefinite**). Therefore we need to be able to identify other types of critical points, which are not of minimum or maximum type. To do this we need a new type of compactness condition.

Let Y be a Banach space and $\varphi \colon Y \longrightarrow \mathbb{R}$ a locally Lipschitz function.

DEFINITION 2.1.1

(a) *Let*

$$m^\varphi(y) \overset{df}{=} \inf_{y^* \in \partial \varphi(y)} \|y^*\|_{Y^*}.$$

*We say that φ satisfies the **nonsmooth Palais-Smale condition at level** $c \in \mathbb{R}$ (**nonsmooth** PS_c-**condition** for short), if the following holds:*

"Every sequence $\{y_n\}_{n \geq 1} \subseteq Y$, such that

$$\varphi(y_n) \longrightarrow c \quad \text{and} \quad m^\varphi(y_n) \longrightarrow 0,$$

has a strongly convergent subsequence."

*If this property holds at every level $c \in \mathbb{R}$, then we simply say that φ satisfies the **nonsmooth Palais-Smale condition** (**nonsmooth** PS-**condition** for short).*

(b) *We say that φ satisfies the **nonsmooth Cerami condition at level** $c \in \mathbb{R}$ (**nonsmooth** C_c-**condition** for short), if the following holds:*

"Every sequence $\{y_n\}_{n \geq 1} \subseteq Y$, such that

$$\varphi(y_n) \longrightarrow c \quad \text{and} \quad \left(1 + \|y_n\|_Y\right) m^\varphi(y_n) \longrightarrow 0,$$

has a strongly convergent subsequence."

*If this property holds at every level $c \in \mathbb{R}$, then we simply say that φ satisfies the **nonsmooth Cerami condition** (**nonsmooth** C-**condition** for short).*

REMARK 2.1.1   The nonsmooth C-condition is slightly weaker than the nonsmooth PS-condition, while it retains the most important implications of the nonsmooth PS-condition. However, in most applications the nonsmooth PS-condition suffices. Nevertheless in order to cover the general case, we shall base our presentation on the nonsmooth C-condition.   ▯

REMARK 2.1.2   The nonsmooth C-condition is a fairly strong condition and it is not satisfied even by "nice" functions. Consider for example the case where $Y = \mathbb{R}$ and $\varphi(y) = c$, $c \in \mathbb{R}$ (constant function). If we consider the sequence $\{n\}_{n\geq 1}$ of natural numbers, then $\varphi(n) = c$ and $(1 + n) m^{\varphi}(n) = 0$ (since $\partial \varphi(n) = \{\varphi'(n)\} = \{0\}$). However, $\{n\}_{n\geq 1}$ has no convergent subsequence. Similarly for the function $\varphi \colon \mathbb{R} \longrightarrow \mathbb{R}$ defined by $\varphi(y) = \cos y$. Consider the sequence $\{n\pi\}_{n\geq 1}$. If Y is finite dimensional, the prototype locally Lipschitz function $\varphi \colon Y \longrightarrow \mathbb{R}$ which satisfies the nonsmooth C-condition is a weakly coercive φ (see Definition A.4.2(a)). Indeed from the boundedness of $\{\varphi(y_n)\}_{n\geq 1}$ it follows that $\{y_n\}_{n\geq 1} \subseteq Y$ is bounded and so it has a convergent subsequence. In fact in what follows we show that more generally weak coercivity and the nonsmooth PS-condition are closely related. □

In the next proposition we make a decisive step in the direction of relating the weak coercivity of φ and the nonsmooth C-condition.

PROPOSITION 2.1.1

If Y *is a Banach space,* $\varphi \colon Y \longrightarrow \mathbb{R}$ *is locally Lipschitz, bounded below and the number*

$$c \overset{df}{=} \liminf_{\|y\|_Y \to +\infty} \varphi(y)$$

is finite,
then there exists a sequence $\{y_n\}_{n\geq 1} \subseteq Y$, *such that* $\|y_n\|_Y \longrightarrow +\infty$, $\varphi(y_n) \longrightarrow c$ *and* $m^{\varphi}(y_n) \longrightarrow 0$.

PROOF For every $n \geq 1$, we can find $M_n \geq 0$, such that for any $y \in Y$ with $\|y\|_Y \geq M_n$, we have $\varphi(y) \geq c - \frac{1}{n}$. We can always choose M_n so that

$$\begin{cases} M_1 \geq 1, \\ M_{n+1} \geq M_n + 2 \text{ for } n \geq 1. \end{cases}$$

We can also find $\{u_n\}_{n\geq 1} \subseteq Y$, such that

$$\varphi(u_n) < c + \frac{1}{n} \qquad \forall\, n \geq 1 \tag{2.1}$$

and

$$\|u_n\|_Y \geq \frac{M_n + 2 + \frac{1}{2\sqrt{n}}}{1 - \frac{1}{2\sqrt{n}}}, \qquad \forall\, n \geq 1. \tag{2.2}$$

Evidently $\|u_n\|_Y > M_n + 2$ for all $n \geq 1$ and so $\|u_n\|_Y \longrightarrow +\infty$.
  Let

$$\eta \overset{df}{=} \inf \{\varphi(y) : \, y \in Y\}.$$

Of course $-\infty < \eta \leq c$. For all $n \geq 1$, let us define $\varphi_n \colon Y \longrightarrow \mathbb{R}$ by

$$\varphi_n(y) \overset{df}{=} \begin{cases} \varphi(y) + c - \eta & \text{if } \|y\|_Y \leq M_n, \\ \varphi(y) + (c - \eta)(M_n + 1 - \|y\|_Y) & \text{if } M_n < \|y\|_Y \leq M_n + 1, \\ \varphi(y) & \text{if } M_n + 1 < \|y\|_Y. \end{cases}$$

Clearly for each $n \geq 1$, φ_n is locally Lipschitz and

$$\eta_n \overset{df}{=} \inf \{\varphi_n(y) : y \in Y\} \geq c - \frac{1}{n} \qquad \forall \, n \geq 1.$$

We apply Theorem 1.4.9 with $h(r) \equiv \|u_n\|_Y$, $x_0 = u_n$, $y = u_n$, $\varepsilon = c + \frac{1}{n} - \eta_n$ and $\lambda = \frac{1}{2\sqrt{n}}$ on the function φ_n. Because of (2.1), we have that

$$\begin{aligned}
\varphi_n(u_n) &= \varphi(u_n) < c + \frac{1}{n} \\
&= \eta_n + \left(c + \frac{1}{n} - \eta_n\right) = \inf_Y \varphi_n + \varepsilon
\end{aligned}$$

and so using Theorem 1.4.9 for all $n \geq 1$, we can find $y_n \in Y$, such that

$$\varphi_n(y_n) \leq \varphi_n(u_n) \qquad \forall \, n \geq 1 \tag{2.3}$$

and

$$\varphi_n(y_n) \leq \varphi_n(y) + \frac{4}{\sqrt{n}\,(1 + \|u_n\|_Y)} \|y_n - y\|_Y \qquad \forall \, n \geq 1, \ y \in Y \tag{2.4}$$

(note that $\varepsilon \leq \frac{2}{n}$) and such that

$$\|y_n - u_n\|_Y \leq \frac{1 + \|u_n\|_Y}{2\sqrt{n}} \tag{2.5}$$

(note that $\bar{r} = \frac{1 + \|u_n\|_Y}{2\sqrt{n}}$ is good for Theorem 1.4.9). In (2.4) we put $y = y_n + tv$ with $t > 0$, $v \in Y$ and obtain

$$-\xi_n \|v\|_Y \leq \frac{\varphi_n(y_n + tv) - \varphi_n(y_n)}{t} \qquad \forall \, n \geq 1, \ t > 0, \ v \in Y, \tag{2.6}$$

with $\xi_n \overset{df}{=} \frac{4}{\sqrt{n}(1 + \|u_n\|_Y)}$ for $n \geq 1$.

Letting $t \searrow 0$, from (2.6) we obtain

$$-\xi_n \|v\|_Y \leq \varphi_n^0(y_n; v) \qquad \forall \, n \geq 1, \ v \in Y.$$

Using (2.5) and (2.2), we see that for all $n \geq 1$, we have

$$\begin{aligned}
\|y_n\|_Y &= \|y_n - u_n + u_n\|_Y \geq \|u_n\|_Y - \|y_n - u_n\|_Y \\
&\geq \|u_n\|_Y \left(1 - \frac{1}{2\sqrt{n}}\right) - \frac{1}{2\sqrt{n}} \geq M_n + 2. \tag{2.7}
\end{aligned}$$

So, from the definition of φ_n, we infer that

$$\varphi_n^0(y_n; v) = \varphi^0(y_n; v) \qquad \forall \, n \geq 1, \ v \in Y.$$

Let us set

$$\psi_n(v) \stackrel{df}{=} \frac{1}{\xi_n}\varphi^0(y_n; v) \qquad \forall\, n \geq 1,\ v \in Y. \tag{2.8}$$

Applying Lemma 1.3.2 (with $\psi = \psi_n$) we obtain $y_n^* \in Y^*$ with $\|y_n^*\|_{Y^{**}} \leq 1$, such that

$$\langle y_n^*, v \rangle_Y \leq \psi_n(v) \qquad \forall\, n \geq 1,\ v \in Y. \tag{2.9}$$

Now from (2.8) and (2.9), we have that

$$\langle \xi_n y_n^*, v \rangle_Y \leq \varphi^0(y_n; v) \qquad \forall\, n \geq 1,\ v \in Y$$

and so

$$\xi_n y_n^* \in \partial\varphi(y_n) \qquad \forall\, n \geq 1,$$

with $\|\xi_n u_n^*\|_{Y^*} = \xi_n$. Hence, using (2.5), for all $n \geq 1$ we have

$$\left(1 + \|y_n\|_Y\right)m^\varphi(y_n) \leq \left(1 + \|y_n\|_Y\right)\xi_n \leq \frac{1 + \|y_n\|_Y}{1 + \|u_n\|_Y} \cdot \frac{4}{\sqrt{n}}$$

$$\leq \frac{1 + \|u_n\|_Y + \|u_n - y_n\|_Y}{1 + \|u_n\|_Y} \cdot \frac{4}{\sqrt{n}} \leq \frac{4}{\sqrt{n}} + \frac{1 + \|u_n\|_Y}{2\sqrt{n}\,(1 + \|u_n\|_Y)} \cdot \frac{4}{\sqrt{n}}$$

$$= \frac{4}{\sqrt{n}} + \frac{2}{n}$$

and so $\left(1 + \|y_n\|_Y\right)m^\varphi(y_n) \longrightarrow 0$ as $n \to +\infty$.

Finally note that since $\|y_n\|_Y \geq M_n + 2$ for $n \geq 1$ (see (2.7)), we have that $\|y_n\|_Y \longrightarrow +\infty$, while from (2.3) and since $\varphi_n(y_n) = \varphi(y_n)$, we infer that $\varphi(y_n) \longrightarrow c$ as $n \to +\infty$. ∎

An immediate consequence of this proposition is the following corollary.

COROLLARY 2.1.1
If Y is a Banach space, $\varphi\colon Y \longrightarrow \mathbb{R}$ is a locally Lipschitz, bounded below function and it satisfies the nonsmooth C-condition,
then φ *is weakly coercive.*

The next proposition shows that in the case of locally Lipschitz functions which are bounded below, the notions of nonsmooth PS-condition and of nonsmooth C-condition are equivalent.

PROPOSITION 2.1.2
If Y is a Banach space, $\varphi\colon Y \longrightarrow \mathbb{R}$ is a locally Lipschitz, bounded below function satisfying the nonsmooth C-condition,
then it satisfies the nonsmooth PS-condition.

PROOF   Let $\{y_n\}_{n\geq 1} \subseteq Y$ be a sequence, such that $\{\varphi(y_n)\}_{n\geq 1}$ is bounded and $m^\varphi(y_n) \longrightarrow 0$. By virtue of Corollary 2.1.1, φ is weakly coercive and so $\{y_n\}_{n\geq 1}$ is bounded. Hence also $\left(1 + \|y_n\|_Y\right)m^\varphi(y_n) \longrightarrow 0$ and because φ satisfies the nonsmooth C-condition, it follows that $\{y_n\}_{n\geq 1}$ has a strongly convergent subsequence.   ⬚

Finally we show some property of the function m^φ needed in what follows.

PROPOSITION 2.1.3
Let X be a reflexive Banach space and $\varphi\colon X \longrightarrow \mathbb{R}$ a locally Lipschitz function. The function

$$X \ni x \longmapsto m^\varphi(x) \in \mathbb{R}$$

is lower semicontinuous.

PROOF   Let

$$L_\lambda \stackrel{df}{=} \left\{x \in X : \; m^\varphi(x) \leq \lambda\right\}.$$

We need to show that for every $\lambda \geq 0$, the set L_λ is closed. So let $\{x_n\}_{n\geq 1} \subseteq L_\lambda$ and suppose that $x_n \longrightarrow x$. Note that in the definition of $m^\varphi(x_n)$, the infimum is actually attained. Indeed recall that the norm in a Banach space is weakly sequentially lower semicontinuous (see Definition A.4.2(b)) and the set $\partial\varphi(x_n) \subseteq X^*$ is nonempty, weakly compact (see Proposition 1.3.8). Hence we can apply the Weierstrass Theorem (see Theorem A.1.5) and obtain

$$x_n^* \in \partial\varphi(x_n), \quad \text{such that} \quad m^\varphi(x_n) = \|x_n^*\|_{X^*} \quad \forall\, n \geq 1.$$

Since $\partial\varphi$ is a bounded multifunction (see Proposition 1.3.8) it follows that the sequence $\{x_n^*\}_{n\geq 1} \subseteq X^*$ is bounded and so, after passing to a subsequence if necessary, we may assume that

$$x_n^* \stackrel{w}{\longrightarrow} x^* \quad \text{in } X^*.$$

Because $x_n^* \in \partial\varphi(x_n)$ for all $n \geq 1$, we have that $x^* \in \partial\varphi(x)$ (see Proposition 1.3.9) and

$$\|x^*\|_{X^*} \leq \liminf_{n\to+\infty} \|x_n^*\|_{X^*} \leq \lambda,$$

hence $m^\varphi(x) \leq \lambda$, i.e. $x \in L_\lambda$ and we have proved that L_λ is closed.   ⬚

2.1.2   Critical Points and Deformation Theorem

Let Y be a Banach space and $\varphi\colon Y \longrightarrow \mathbb{R}$ a locally Lipschitz function.

DEFINITION 2.1.2   *We say that $y \in Y$ is a **critical point of** φ, if $0 \in \partial\varphi(y)$. Here $\partial\varphi(y)$ denotes the generalized subdifferential of φ at $y \in Y$ in the sense of Definition 1.3.7.*

REMARK 2.1.3 From Proposition 1.3.10, we know that if $\varphi \in C^1(Y)$, then $\partial \varphi(y) = \{\varphi'(y)\}$. So in the smooth case (i.e. $\varphi \in C^1(Y)$), the above definition reduces to the classical one. Recall that if $y \in Y$ is a local extremum of φ, then $0 \in \partial \varphi(y)$ (see Proposition 1.3.12(b)). $\quad\Box$

The basic idea in critical point theory is to examine the variations of the topological structure of the level sets of a given function φ. The main technique here is that of "gradient" or "gradient-like" flows, which are used to deform these level sets. In finite dimensions or in the Hilbert space setting with a smooth function φ, the natural candidate is the gradient flow. In the more general setting of a Banach space (where we can no longer identify the space with its dual) still with a smooth function φ, the gradient $\nabla\varphi(x)$ which belongs in the dual space (hence it cannot generate a flow in the original space) is replaced by the so-called "pseudogradient vector field." It is not at all clear how this notion can be generalized to the present nonsmooth setting. In what follows we explain how this can be done and we obtain a deformation theorem, which, next to the compactness condition, is the second main ingredient in the derivation of minimax characterizations of critical points of the locally Lipschitz function φ.

Let X be a reflexive Banach space and $\varphi \colon X \longrightarrow \mathbb{R}$ a locally Lipschitz function. For each $c \in \mathbb{R}$ we introduce the following sets:

$$K^\varphi \stackrel{df}{=} \{x \in X : 0 \in \partial\varphi(x)\},$$

$$K_c^\varphi \stackrel{df}{=} \{x \in X : 0 \in \partial\varphi(x), \ \varphi(x) = c\},$$

$$\varphi^c \stackrel{df}{=} \{x \in X : \varphi(x) \le c\}.$$

So K_c^φ is the set of critical points of φ having "energy" c and φ^c is the closed lower level set of φ at c. Recalling that set

$$\operatorname{Gr}\partial\varphi = \{(x,x^*) \in X \times X^* : \ x^* \in \partial\varphi(x)\}$$

is sequentially closed in $X \times X_w^*$ (here X_w^* denotes the reflexive Banach space X^* furnished with the weak topology; see the proof Proposition 1.3.9), we can easily check that if φ satisfies the nonsmooth C-condition, then the set K_c^φ is compact. In what follows for $\delta > 0$ we set

$$(K_c^\varphi)_\delta \stackrel{df}{=} \{x \in X : \ d_X(x, K_c^\varphi) < \delta\}$$

and by $(K_c^\varphi)_\delta^c$ we denote its complement.

Also for any $c \in \mathbb{R}$ and $\delta, \varepsilon > 0$, we put

$$E_{c,\delta,\varepsilon}^\varphi \stackrel{df}{=} \{x \in X : \ |\varphi(x) - c| \le \varepsilon\} \cap (K_c^\varphi)_\delta^c.$$

LEMMA 2.1.1
If $\varphi\colon X \longrightarrow \mathbb{R}$ *satisfies the nonsmooth* C_c*-condition,*
<u>*then*</u> *for each* $\delta > 0$ *there exist* $\gamma, \varepsilon > 0$, *such that*

$$\gamma \leq \left(1 + \|x\|_X\right) m^\varphi(x) \qquad \forall\, x \in E^\varphi_{c,\delta,\varepsilon}.$$

PROOF   Suppose that the result is not true. Then there exists $\delta > 0$, such that for any $\gamma_n, \varepsilon_n \searrow 0$ we can find $x_n \in (K^\varphi_c)^c_\delta$ with

$$c - \varepsilon_n \leq \varphi(x_n) \leq c + \varepsilon_n \quad \text{and} \quad \left(1 + \|x_n\|_X\right) m^\varphi(x_n) < \gamma_n.$$

So $\left(1 + \|x_n\|_X\right) m^\varphi(x_n) \longrightarrow 0$ and because by hypothesis φ satisfies the nonsmooth C_c-condition, by passing to a subsequence if necessary, we may assume that $x_n \longrightarrow x$ in X. Therefore we have $\varphi(x_n) \longrightarrow \varphi(x) = c$. Moreover, from lower semicontinuity of m^φ (see Proposition 2.1.3), we have that

$$m^\varphi(x) \leq \liminf_{n \to +\infty} m^\varphi(x_n) = 0,$$

hence $m^\varphi(x) = 0$ and so $x \in K^\varphi_c$, a contradiction to the fact that $x_n \in (K^\varphi_c)^c_\delta$ for all $n \geq 1$ (hence also $x \in (K^\varphi_c)^c_\delta$). This proves the lemma. ☐

Using this lemma, we can now obtain a locally Lipschitz vector field which plays the role of the pseudogradient vector field of the smooth theory.

PROPOSITION 2.1.4
If $\varphi\colon X \longrightarrow \mathbb{R}$ *satisfies the nonsmooth* C_c*-condition,* $\delta > 0$ *is given and* $\gamma, \varepsilon > 0$ *are as in Lemma 2.1.1,*
<u>*then*</u> *there exists a locally Lipschitz vector field*

$$v\colon E^\varphi_{c,\delta,\varepsilon} \longrightarrow X,$$

such that

$$\begin{cases} \|v(x)\|_X \leq 1 + \|x\|_X\,, \\ \dfrac{\gamma}{2} \leq \langle x^*, v(x)\rangle_X \end{cases} \qquad \forall\, x \in E^\varphi_{c,\delta,\varepsilon},\ x^* \in \partial\varphi(x).$$

PROOF   Let $x \in E^\varphi_{c,\delta,\varepsilon}$ and take $x^* \in \partial\varphi(x)$, such that $m^\varphi(x) = \|x^*\|_{X^*}$. We have that

$$B_{\|x^*\|_{X^*}} \cap \partial\varphi(x) = \emptyset,$$

where for every $r > 0$, $B_r \overset{df}{=} \{z \in X : \|z\|_X < r\}$. By the Weak Separation Theorem (see Theorem A.3.3) for convex sets, we can find $u(x) \in X$ with $\|u(x)\|_X = 1$, such that

$$\langle z^*, u(x)\rangle_X \leq \langle x^*, u(x)\rangle_X \qquad \forall\, z^* \in B_{\|x^*\|_{X^*}},\ x^* \in \partial\varphi(x).$$

Note that

$$\sup \left\{ \langle z^*, u(x) \rangle_X : z^* \in B_{\|x^*\|_{X^*}} \right\} = \|x^*\|_{X^*} = m^\varphi(x).$$

So by virtue of Lemma 2.1.1, we have

$$\frac{\gamma}{2(1 + \|x\|_X)} < m^\varphi(x) = \|x^*\|_{X^*} \leq \langle x^*, u(x) \rangle_X \qquad \forall \, x^* \in \partial\varphi(x). \tag{2.10}$$

From Proposition 1.3.9 we know that the multifunction $x \longmapsto \partial\varphi(x)$ is upper semicontinuous from X into X^*_w. Hence so is the multifunction $x \longmapsto (1 + \|x\|_X)\partial\varphi(x)$. Let

$$V \stackrel{df}{=} \left\{ y^* \in X^* : \frac{\gamma}{2} < \langle y^*, u(x) \rangle_X \right\}.$$

The set V is weakly open in X^*. Moreover, from (2.10) we see that

$$(1 + \|x\|_X)\partial\varphi(x) \subseteq V.$$

So we can find $r(x) > 0$, such that

$$(1 + \|y\|_X)\partial\varphi(y) \subseteq V \qquad \forall \, y \in B_{r(x)}(x).$$

Hence it follows that

$$\frac{\gamma}{2(1 + \|y\|_X)} < \langle y^*, u(x) \rangle_X \qquad \forall \, y \in B_{r(x)}(x), \ y^* \in \partial\varphi(y). \tag{2.11}$$

The collection $\left\{ B_{r(x)}(x) \right\}_{x \in E^\varphi_{c,\delta,\varepsilon}}$ is an open cover of the set $E^\varphi_{c,\delta,\varepsilon}$. By paracompactness, we can find a locally finite refinement $\{U_i\}_{i \in I}$ of it and a locally Lipschitz partition of unity $\{\xi_i\}_{i \in I}$ subordinate to $\{U_i\}_{i \in I}$. For each $i \in I$, we can find $x_i \in E^\varphi_{c,\delta,\varepsilon}$, such that $U_i \subseteq B_{r(x_i)}(x_i)$. To each such x_i corresponds $u(x_i) \in X$ with $\|u(x_i)\|_X = 1$ for which (2.10) holds with $x = x_i$. We define $v \colon E^\varphi_{c,\delta,\varepsilon} \longrightarrow X$ by

$$v(x) \stackrel{df}{=} (1 + \|x\|_X) \sum_{i \in I} \xi_i(x) u(x_i) \qquad \forall \, x \in E^\varphi_{c,\delta,\varepsilon}.$$

Evidently v is well defined locally Lipschitz and $\|v(x)\|_X \leq 1 + \|x\|_X$ for all $\forall \, x \in E^\varphi_{c,\delta,\varepsilon}$. Moreover, using (2.11), we have that

$$\begin{aligned} \langle x^*, v(x) \rangle_X &= \sum_{i \in I} \xi_i(x) (1 + \|x\|_X) \langle x^*, u(x_i) \rangle_X \\ &> \frac{\gamma}{2} \sum_{i \in I} \xi_i(x) = \frac{\gamma}{2} \qquad \forall \, x \in E^\varphi_{c,\delta,\varepsilon}, \ x^* \in \partial\varphi(x). \end{aligned}$$

$\square$

Using this proposition we can have a deformation theorem, which is the main analytical tool in deriving minimax characterizations of critical points of a locally Lipschitz function φ.

THEOREM 2.1.1 (Deformation Theorem)

If $\varphi \colon X \longrightarrow \mathbb{R}$ satisfies the nonsmooth C_c-condition then for every $\varepsilon_0 > 0$ and for every neighbourhood U of K_c^φ (if $K_c^\varphi = \emptyset$, then $U = \emptyset$) there exist $\varepsilon \in (0, \varepsilon_0)$ and a continuous map $\eta \colon [0,1] \times X \longrightarrow X$, such that for all $(t, x) \in [0,1] \times X$, we have

(a) $\|\eta(t,x) - x\|_X \leq (e+1)(1 + \|x\|_X)\,t;$

(b) *if* $|\varphi(x) - c| \geq \varepsilon_0$ *or* $m^\varphi(x) = 0$, *then* $\eta(t,x) = x;$

(c) $\eta(\{1\} \times \varphi^{c+\varepsilon}) \subseteq \varphi^{c-\varepsilon} \cup U;$

(d) $\varphi(\eta(t,x)) \leq \varphi(x);$

(e) *if* $\eta(t,x) \neq x$, *then* $\varphi(\eta(t,x)) < \varphi(x);$

(f) η *satisfies the semigroup property, i.e.*

$$\eta(s, \cdot) \circ \eta(t, \cdot) = \eta(s+t, \cdot) \qquad \forall\, s, t \in [0,1],\ s + t \leq 1;$$

(g) *for each* $t \in [0,1]$, $\eta(t, \cdot)$ *is a homeomorphism of X;*

(h) *if φ is even, then for every $t \in [0,1]$, $\eta(t, \cdot)$ is odd.*

PROOF   Let $\varepsilon_0 > 0$ and let U be any neighbourhood of K_c^φ ($U = \emptyset$, if $K_c^\varphi = \emptyset$). Recall that the set K_c^φ is compact. So we can find $\delta > 0$, such that $(K_c^\varphi)_{3\delta} \subseteq U$. By virtue of Lemma 2.1.1, we can find $\gamma > 0$ and $\bar{\varepsilon} \in (0, \varepsilon_0)$, such that

$$\gamma \leq \left(1 + \|x\|_X\right) m^\varphi(x) \qquad \forall\, x \in (K_c^\varphi)_\delta^c \cap \left(\varphi^{c+\bar{\varepsilon}} \setminus \varphi^{c-\bar{\varepsilon}}\right).$$

Consider the following two closed subsets of X:

$$C_1 \stackrel{df}{=} \overline{\left(E_{c,\delta,\bar{\varepsilon}}^\varphi\right)}^c = \{x \in X : |\varphi(x) - c| \geq \bar{\varepsilon}\} \cup \overline{(K_c^\varphi)_\delta},$$

$$C_2 \stackrel{df}{=} E_{c,2\delta,\frac{\bar{\varepsilon}}{2}}^\varphi = \{x \in X : |\varphi(x) - c| \leq \frac{\bar{\varepsilon}}{2}\} \cap (K_c^\varphi)_{2\delta}^c.$$

Clearly we have $C_1 \cap C_2 = \emptyset$. So by Urysohn's lemma (see Theorem A.1.12), we can find a locally Lipschitz function $\xi \colon X \longrightarrow [0,1]$, such that $\xi|_{C_1} = 0$ and $\xi|_{C_2} = 1$. Also let $v \colon E_{c,\delta,\bar{\varepsilon}}^\varphi \longrightarrow X$ be a locally Lipschitz vector field obtained in Proposition 2.1.4. Using ξ and v we define $L \colon X \longrightarrow X$ by

$$L(x) \stackrel{df}{=} \begin{cases} -\xi(x)v(x) & \text{if } x \in E_{c,\delta,\bar{\varepsilon}}^\varphi \\ 0 & \text{otherwise} \end{cases} \qquad \forall\, x \in X.$$

Evidently L is locally Lipschitz and

$$\|L(x)\|_X \ = \ \xi(x)\,\|v(x)\|_X \ \leq \ \|v(x)\|_X \ \leq \ 1 + \|x\|_X \qquad \forall\, x \in E^{\varphi}_{c,\delta,\bar{\varepsilon}},$$

so

$$\|L(x)\|_X \ \leq \ 1 + \|x\|_X \qquad \forall\, x \in X \tag{2.12}$$

and from Proposition 2.1.4, we have

$$\langle x^*, L(x)\rangle_X \ = \ -\xi(x)\langle x^*, v(x)\rangle_X \ \leq \ -\xi(x)\frac{\gamma}{2}$$
$$\forall\, x \in E^{\varphi}_{c,\delta,\bar{\varepsilon}},\ x^* \in \partial\varphi(x). \tag{2.13}$$

Fix $x \in X$ and consider the following Cauchy problem in the reflexive Banach space X:

$$\begin{cases} \dfrac{d}{dt}\overline{\eta}(x)(t) \ = \ L\big(\overline{\eta}(x)(t)\big) & \text{for a.a. } t \in [0,1], \\ \overline{\eta}(x)(0) = x. \end{cases} \tag{2.14}$$

The fact that L is locally Lipschitz and has sublinear growth (see estimate (2.12)) implies that problem (2.14) has a unique global Caratheodory solution $\overline{\eta}(x)$, i.e. $\overline{\eta}(x)\colon [0,1] \longrightarrow X$ is an absolutely continuous function which satisfies (2.14) (note that because X is reflexive, $\overline{\eta}(x)$ is differentiable almost everywhere on $[0,1]$). Using (2.12) we have

$$\|\overline{\eta}(x)(t) - x\|_X \ \leq \ \int_0^t \big\|L\big(\overline{\eta}(x)(s)\big)\big\|_X\, ds \tag{2.15}$$

$$\leq \ \int_0^t \big(1 + \|\overline{\eta}(x)(s)\|_X\big)\, ds \ \leq \ \big(1 + \|x\|_X\big)\, t + \int_0^t \|\overline{\eta}(x)(s) - x\|_X\, ds.$$

(a) Using Gronwall's inequality (see Theorem A.4.3) in (2.15) and applying (2.12), we obtain

$$\|\overline{\eta}(x)(t) - x\|_X \ \leq \ \big(1 + \|x\|_X\big)\, t + \int_0^t \big(1 + \|x\|_X\big)\, s e^{t-s}\, ds$$
$$= \ \big(1 + \|x\|_X\big)\, t + \big(1 + \|x\|_X\big)\, e^t\,\big(1 - e^{-t}\big)$$
$$= \ \big(1 + \|x\|_X\big)\, t + \big(1 + \|x\|_X\big)\,\big(e^t - 1\big)$$
$$\leq \ \big(1 + \|x\|_X\big)\, t + \big(1 + \|x\|_X\big)\, e t \ = \ (e+1)\,\big(1 + \|x\|_X\big)\, t.$$

This proves part (a) of the theorem.

(b) If $m^{\varphi}(x) = 0$, then it is clear that $\overline{\eta}(x)(t) = x$ for all $t \in [0,1]$. Let $|\varphi(x) - c| \geq \bar{\varepsilon}$. Then $\xi(x) = 0$ and $L(x) = 0$. So we have

$$\overline{\eta}(x)(t) = x \qquad \forall\, t \in [0,1]$$

(recall that the solution of (2.14) is unique). This proves part (b) of the theorem.

(d) Next let us define $h \colon [0,1] \longrightarrow \mathbb{R}$ by

$$h(t) \stackrel{df}{=} \varphi\big(\overline{\eta}(x)(t)\big).$$

Clearly h is an absolutely continuous function and as such it is differentiable almost everywhere on $[0,1]$. Moreover, from Proposition 1.3.18, we have

$$h'(t) \ \leq \ \max\big\{\langle x^*, \tfrac{d}{dt}\overline{\eta}(x)(t)\rangle_X : \ x^* \in \partial\varphi\big(\overline{\eta}(x)(t)\big)\big\} \quad \text{for a.a. } t \in [0,1],$$

so

$$h'(t) \ \leq \ \max\big\{\langle x^*, L\big(\overline{\eta}(x)(t)\big)\rangle_X : \ x^* \in \partial\varphi\big(\overline{\eta}(x)(t)\big)\big\} \quad \text{for a.a. } t \in [0,1].$$

By Proposition 2.1.4, we have that

$$h'(t) \ \leq \ \begin{cases} -\dfrac{\xi(x)\gamma}{2} & \text{if } x \in E^{\varphi}_{c,\delta,\overline{\varepsilon}} \\[2mm] 0 & \text{otherwise} \end{cases} \qquad \text{for a.a. } t \in [0,1]$$

and so h is nonincreasing. It follows that

$$\varphi\big(\overline{\eta}(x)(t)\big) = h(t) \leq h(0) = \varphi(x) \qquad \text{for a.a. } t \in [0,1], \tag{2.16}$$

which proves part (d) of the theorem.

(e) The inequality (2.16) is strict if $x \in E^{\varphi}_{c,\delta,\overline{\varepsilon}}$. So

$$\text{if } \ \overline{\eta}(x)(t) \neq x \ \text{ then } \ \varphi\big(\overline{\eta}(x)(t)\big) < \varphi(x).$$

This proves part (e) of the theorem.

(c) Let $\varrho > 0$ be such that $\overline{(K^{\varphi}_c)_{2\delta}} \subseteq B_\varrho$. Choose $0 < \varepsilon < \overline{\varepsilon} < \varepsilon_0$, such that

$$4\varepsilon \ \leq \ \gamma \qquad \text{and} \qquad 4\varepsilon(1+\varrho)(e+1) \ \leq \ \delta\gamma. \tag{2.17}$$

We argue by contradiction. Suppose that statement (c) of the theorem is not true. We can find $x \in \varphi^{c+\varepsilon}$, such that

$$\varphi\big(\overline{\eta}(x)(1)\big) > c - \varepsilon \ \text{ and } \ \overline{\eta}(x)(1) \in U^c.$$

From an earlier argument we know that the function

$$t \longmapsto h(t) = \varphi\big(\overline{\eta}(x)(t)\big) \ \text{ is nonincreasing.}$$

So we have

$$c - \varepsilon \ < \ \varphi\big(\overline{\eta}(x)(t)\big) \ \leq \ c + \varepsilon \qquad \forall\, t \in [0,1]. \tag{2.18}$$

Also we have that

$$\overline{\eta}(x)\big([0,1]\big) \cap (K_c^\varphi)_{2\delta} \neq \emptyset. \tag{2.19}$$

Indeed if this intersection is empty, then from (2.13) and the properties of ξ we have that

$$\frac{\gamma}{2} \leq -\int_0^1 h'(s)ds = \varphi(x) - \varphi\big(\overline{\eta}(x)(1)\big). \tag{2.20}$$

But recall that $x \in \varphi^{c+\varepsilon}$. So using this and (2.18) in (2.20), we obtain that $\gamma < 4\varepsilon$, a contradiction to the choice of $\varepsilon > 0$ (see (2.17)). Therefore (2.19) holds.

From (2.19) it follows that we can find $0 \leq t_1 < t_2 \leq 1$, such that

$$\begin{cases} d_X\big(\overline{\eta}(x)(t_1), K_c^\varphi\big) = 2\delta, \\ d_X\big(\overline{\eta}(x)(t_2), K_c^\varphi\big) = 3\delta, \\ 2\delta < d_X\big(\overline{\eta}(x)(t), K_c^\varphi\big) < 3\delta \quad \forall\, t \in (t_1, t_2) \end{cases} \tag{2.21}$$

(recall that $(K_c^\varphi)_{3\delta} \subseteq U$ and $\overline{\eta}(x)(1) \in U^c$). Invoking (2.13) once more, we obtain

$$\frac{\gamma}{2}(t_2 - t_1) \leq -\int_{t_1}^{t_2} h'(s)ds = \varphi\big(\overline{\eta}(x)(t_1)\big) - \varphi\big(\overline{\eta}(x)(t_2)\big) < 2\varepsilon$$

(see (2.18)). Thus

$$t_2 - t_1 < \frac{4\varepsilon}{\gamma}. \tag{2.22}$$

So from (2.21) and using Gronwall's inequality (see Theorem A.4.3) as in part (a), we have

$$\delta \leq \|\overline{\eta}(x)(t_2) - \overline{\eta}(x)(t_1)\|_X \leq \int_{t_1}^{t_2} \big\|L\big(\overline{\eta}(x)(s)\big)\big\|_X ds$$

$$\leq (e+1)\big(1 + \|\overline{\eta}(x)(t_1)\|_X\big)(t_2 - t_1).$$

From (2.21), (2.22) and the choice of $\varrho > 0$ we have

$$\delta < (e+1)(1+\varrho)\frac{4\varepsilon}{\gamma}.$$

But this contradicts the choice of $\varepsilon > 0$ (see (2.17)). This proves part (c) of the theorem.

So if we set

$$\eta(x,t) \overset{df}{=} \overline{\eta}(x)(t) \quad \forall\, (t,x) \in [0,1] \times X,$$

we finish the proof of the theorem. $\quad\square$

2.1.3 Linking Sets

Next we introduce a topological notion, which is crucial in the derivation of the minimax characterizations of the critical points of φ.

DEFINITION 2.1.3 *Let Z be a Hausdorff topological space and let $D \subseteq Z$ be a nonempty subset. We say that the set D is **contractible**, if there exists a continuous function $h\colon [0,1] \times D \longrightarrow Z$ (the so-called **homotopy**) and a point $z_0 \in Z$, such that $h(0,x) = x$ and $h(1,x) = z_0$ for all $x \in D$.*

DEFINITION 2.1.4 *Let Z be a Hausdorff topological space and $A, C \subseteq Z$ two nonempty subsets. We say that the set A **links** C if and only if $A \cap C = \emptyset$ and A is not contractible in $Z \setminus C$.*

REMARK 2.1.4 A contractible set is both simply connected (i.e. every closed path is contractible) and path-connected (i.e. any two points can be connected by a continuous curve). Note that the notions of simple and path-connectedness are independent.

We should point out that in most books on the subject, we find the following definition of the notion of linking:

> "Let Z be a Hausdorff topological space and A and C be two nonempty subsets of Z. We say that A and C **link** if and only if there exists a closed set $B \subseteq Z$ such that $A \subseteq B$, $A \cap C = \emptyset$ and for any map $\vartheta \in C(B; Z)$ with $\vartheta|_A = id_A$, we have $\vartheta(B) \cap C \neq \emptyset$."

Let us denote our definition of linking (see Definition 2.1.4) by L_1 and this last definition by L_2. If Z is a locally convex space, A is the relative boundary of a nonempty bounded convex set $B \subseteq Z$, then we have that $L_1 \Longleftrightarrow L_2$, i.e. the two definitions are equivalent.

First we show that $L_2 \Longrightarrow L_1$. Suppose that the implication is not true. Then we can find $z_0 \in Z$ and a continuous map $h\colon [0,1] \times A \longrightarrow X \setminus C$, such that $h(0,x) = x$ and $h(1,x) = z_0$ for all $x \in A$. Let $v \in \operatorname{rint} B$ and for each $x \in B$, let $r(x) \geq 1$ be such that $(1 - r(x))v + r(x)x \in A$. Evidently the map $x \longmapsto r(x)$ is continuous. Then let us set

$$\gamma(x) \stackrel{df}{=} \left(1 - \frac{1}{r(x)}, (1 - r(x))v + r(x)x\right) \qquad \forall\, x \in B.$$

We see that $\gamma\colon B \longrightarrow [0,1] \times A$ is continuous and

$$\gamma(x) = (0, x) \qquad \forall\, x \in A.$$

Let $\vartheta \stackrel{df}{=} h \circ \gamma\colon B \longrightarrow X \setminus C$. Evidently ϑ is continuous and

$$\begin{cases} \vartheta(x) = h\big(\gamma(x)\big) = h(0,x) = x & \forall\, x \in A, \\ \vartheta(B) \cap C = \emptyset, \end{cases}$$

a contradiction to the fact that L_2 holds.

Next we show that $L_1 \Longrightarrow L_2$. Again suppose that the implication is not true. Then we can find $\vartheta \in C(B; Z)$, such that $\vartheta|_A = id_A$ and $\vartheta(A) \cap C = \emptyset$. For a given $z_0 \in A$ let $h \colon [0, 1] \times A \longrightarrow X$ be defined by

$$h(t, x) = \vartheta\big((1 - t)x + tz_0\big) \qquad \forall \, (t, x) \in [0, 1] \times A.$$

Evidently h is continuous and

$$\begin{cases} h(0, x) = \vartheta(x) = x \\ h(1, x) = \vartheta(z_0) = z_0 \end{cases} \qquad \forall \, x \in A.$$

Moreover, since $\vartheta(A) \cap C = \emptyset$, h has values in $Z \setminus C$, which implies that A is contractible in $X \setminus C$, a contradiction to L_1.

As we shall see in the sequel, the setting used in this equivalence is the one that appears in applications. $\quad\Box$

Next we present some characteristic examples of sets which link. First we prove a simple auxiliary result which we shall need in the discussion of the examples that follow.

LEMMA 2.1.2
If Y is a finite dimensional Banach space, $U \subseteq Y$ is a nonempty, bounded, open set and $y_0 \in U$,
then ∂U is not contractible in $Y \setminus \{y_0\}$.

PROOF   Suppose that the lemma is not true. Then we can find

$$h \colon [0, 1] \times \partial U \longrightarrow Y \setminus \{y_0\},$$

a contraction of ∂U in $Y \setminus \{y_0\}$ to some point $u_0 \in Y$. Let $g \colon \overline{U} \longrightarrow Y$ be the map defined by

$$g(z) = u_0 \qquad \forall \, z \in \overline{U}.$$

Then from the properties of Brouwer's degree, we have

$$\begin{aligned} 1 &= D_B\big(id_{\overline{U}}, U, y_0\big) = D_B\big(h(0, \cdot), U, y_0\big) = D_B\big(h(1, \cdot), U, y_0\big) \\ &= D_B\big(g, U, y_0\big) = 0, \end{aligned}$$

a contradiction. $\quad\Box$

Now we are ready to present examples of sets which link.

EXAMPLE 2.1.1   (a) Let Z be a Banach space, $A = \{x_1, x_2\}$, $C = \partial B_R(x_1) = \{x \in Z : \|x - x_1\|_Z = R\}$ with $R > 0$ and $\|x_1 - x_2\|_Z > R$. It is clear that A is not contractible in $Z \setminus C$. Note that A and C also link in the sense of Definition L$_2$ in Remark 2.1.4.

(b) Let Z be a Banach space, $Z = Y \oplus V$ with $\dim Y < +\infty$, $A = \{x \in Y : \|x\|_Z = R\}$ with $R > 0$ and $C = V$. Then the set A links C. Indeed if this is not the case, then we can find $h \colon [0,1] \times A \longrightarrow Z \setminus C$, a contraction of A in $Z \setminus C$. Let $P_Y \colon Z \longrightarrow Y$ be the projection operator to the finite dimensional subspace Y and let

$$\psi(t,x) = (P_Y \circ h)(t,x) \qquad \forall \, (t,x) \in [0,1] \times A.$$

Then ψ is a contraction of A in $Y \setminus \{0\}$, which contradicts Lemma 2.1.2 (take $U = B_R = \{x \in Y : \|x\|_Z < R\}$). Again the two sets A and C also link in the sense of Definition L$_2$ in Remark 2.1.4.

(c) Let Z be a Banach space, $Z = Y \oplus V$ with $\dim Y < +\infty$, $v_0 \in V$ with $\|v_0\|_Z = 1$, $0 < r < R$. Let

$$B \stackrel{df}{=} \{y + tv_0 : 0 \le t \le R, \ y \in Y, \ \|y\|_Z \le R\}$$

and let A be the boundary of B, hence

$$A = \{y + tv_0 : t \in \{0, R\}, \ \|y\|_Z \le R \text{ or } t \in [0, R], \ \|y\|_Z = R\}$$

and let

$$C \stackrel{df}{=} \{x \in V : \|x\|_Z = r\}.$$

Then the set A links C. As in the previous example, we proceed by contradiction. Suppose that $h \colon [0,1] \times A \longrightarrow Z \setminus C$ is a contraction of A in $Z \setminus C$. Consider the projections $P_Y \colon Z \longrightarrow Y$ and $P_V \colon Z \longrightarrow V$ and set

$$\psi(t,x) \stackrel{df}{=} (P_Y \circ h)(t,x) + \|(P_V \circ h)(t,x)\|_Z v_0.$$

Then ψ is a contraction of A in $(Y \oplus \mathbb{R}v_0) \setminus \{rv_0\}$, which contradicts Lemma 2.1.2. As in the previous examples, the two sets also link in the sense of definition L$_2$ in Remark 2.1.4.   ∎

2.1.4   Minimax Principles

The next abstract minimax principle will generate as byproducts nonsmooth versions of the classical Mountain Pass Theorem, Saddle Point Theorem and Linking Theorem.

Recall that X is a reflexive Banach space and $\varphi \colon X \longrightarrow \mathbb{R}$ is a locally Lipschitz function. For a given subset $A \subseteq X$ by H_A we denote the set of all contractions of A to a point in X.

THEOREM 2.1.2
If $A, C \subseteq X$ are nonempty subsets, A is closed, A links C, φ satisfies the nonsmooth C_c-condition with

$$c \overset{df}{=} \inf_{h \in H_A} \sup_{(t,x) \in [0,1] \times A} (\varphi \circ h)(t,x)$$

and

$$\sup_A \varphi \leq \inf_C \varphi,$$

then $c \geq \inf_C \varphi$ *and c is a critical value of φ.*
Moreover, if $c = \inf_C \varphi$, then there exists a critical point $x_0 \in C$ of φ with $\varphi(x_0) = 0$ (i.e. $K_c^\varphi \cap C \neq \emptyset$).

PROOF Since by hypothesis the set A links the set C, according to Definition 2.1.4, we have

$$h([0,1] \times A) \cap C \neq \emptyset \qquad \forall \, h \in H_A$$

and so it follows that $\inf_C \varphi \leq c$.

Case 1. First assume that $\inf_C \varphi < c$. Suppose that the conclusion of the theorem is not true. Then $K_c^\varphi = \emptyset$. We take $U = \emptyset$, $\varepsilon > 0$ and $\eta \colon [0,1] \times X \longrightarrow X$ as in the Deformation Theorem (see Theorem 2.1.1). From the definition of c we can see that we can find $\widetilde{h} \in H_A$, a contraction of A in X to some point $u_0 \in X$, such that

$$\varphi(\widetilde{h}(t,x)) \leq c + \varepsilon \qquad \forall \, t \in [0,1], \ x \in A. \tag{2.23}$$

Let $\widehat{h} \colon [0,1] \times A \longrightarrow X$ be defined by

$$\widehat{h}(t,x) \overset{df}{=} \begin{cases} \eta(2t,x) & \text{if } 0 \leq t \leq \frac{1}{2} \\ \eta(1, \widetilde{h}(2t-1,x)) & \text{if } \frac{1}{2} < t \leq 1 \end{cases} \qquad \forall \, x \in A. \tag{2.24}$$

Evidently $\widehat{h} \in C([0,1]) \times A; X)$ and

$$\begin{cases} \widehat{h}(0,x) = \eta(0,x) = x \\ \widehat{h}(1,x) = \eta(1,\widetilde{h}(1,x)) = \eta(1,u_0) \end{cases} \qquad \forall \, x \in A.$$

So $\widehat{h} \in H_A$. From Theorem 2.1.1(d), we have

$$\begin{aligned} \varphi(\widehat{h}(t,x)) &= \varphi(\eta(2t,x)) \\ &\leq \varphi(x) \leq \sup_A \varphi < c \qquad \forall \, (t,x) \in [0, \tfrac{1}{2}] \times A. \end{aligned} \tag{2.25}$$

As from (2.23), we have that

$$\widetilde{h}(t,x) \subset \varphi^{c+\varepsilon} \qquad \forall\, (t,x) \in [0,1] \times A,$$

so using Theorem 2.1.1(c), we obtain

$$\begin{aligned} \varphi\big(\widehat{h}(t,x)\big) &= \varphi\big(\eta\big(1,\widetilde{h}(2t-1,x)\big)\big) \\ &\leq\; c-\varepsilon \;<\; c \qquad \forall\, (t,x) \in \big[\tfrac{1}{2},1\big] \times A. \end{aligned} \tag{2.26}$$

But inequalities (2.25) and (2.26) contradict the definition of c. Therefore $K_c^\varphi \neq \emptyset$.

Case 2. Now assume that $c = \inf\limits_C \varphi$. We need to show that $K_c^\varphi \cap C \neq \emptyset$. Proceeding by a contradiction argument, suppose that this intersection is empty and let U be a neighbourhood of K_c^φ, such that $U \cap C = \emptyset$ (recall that K_c^φ is compact). Let $\varepsilon > 0$ and $\eta\colon [0,1] \times X \longrightarrow X$ be as in the Deformation Theorem (see Theorem 2.1.1). As before we choose $\widetilde{h} \in H_A$, such that

$$\varphi\big(\widetilde{h}(t,x)\big) \leq c+\varepsilon \qquad \forall\, (t,x) \in [0,1] \times A.$$

Let $\widehat{h}\colon [0,1] \times A \longrightarrow X$ be as before (see (2.24)). From Theorem 2.1.1(e), we have

$$\eta(2t,x) = x \quad \text{or} \quad \varphi\big(\eta(2t,x)\big) < \varphi(x) \leq \inf_C \varphi = c \qquad \forall\, (t,x) \in \big[0,\tfrac{1}{2}\big] \times A,$$

hence $\eta(2t,x) \in C^c$. But from Theorem 2.1.1(c), we have

$$\eta\big(1,h(2t-1,x)\big) \in \varphi^{c-\varepsilon} \cup U \qquad \forall\, (t,x) \in \big[\tfrac{1}{2},1\big] \times A,$$

while from the choice of U and the definition of c, we have

$$\big(\varphi^{c-\varepsilon} \cup U\big) \cap C = \emptyset.$$

Therefore $\widehat{h}$ is a contraction of A in $X \setminus C$, a contradiction to the fact that A and C link. Thus $K_c^\varphi \cap C \neq \emptyset$ and in particular $K_c^\varphi \neq \emptyset$. This proves the theorem. $\qquad\square$

THEOREM 2.1.3 (Nonsmooth Mountain Pass Theorem)
If there exist $x_1 \in X$ and $r > 0$, such that $\|x_1\|_X > r$,

$$\max\big\{\varphi(0),\varphi(x_1)\big\} \leq \inf_{\|x\|_X = r} \varphi(x) \tag{2.27}$$

and φ satisfies the nonsmooth C_c-condition with

$$c = \inf_{\gamma \in \Gamma} \sup_{t \in [0,1]} \varphi\big(\gamma(t)\big),$$

where

$$\Gamma \overset{df}{=} \{\gamma \in C([0,1]; X) : \gamma(0) = 0, \ \gamma(1) = x_1\},$$

then $c \geq \underset{\|x\|_X = r}{\inf} \varphi(x)$ *and c is a critical value of* φ.

Moreover, if $c = \inf\{\varphi(x) : \|x\|_X = r\}$, *then there exists a critical point* x_0 *of* φ *with* $\varphi(x_0) = c$ *and* $\|x_0\|_X = r$ *(i.e.* $K_c^\varphi \cap \partial B_r \neq \emptyset$).

PROOF   Let

$$A \overset{df}{=} \{0, x_1\},$$

$$C \overset{df}{=} \partial B_r = \{x \in X : \|x\|_X = r\}.$$

Clearly the set A links C (see Example 2.1.1(a)) and $c < +\infty$. Let $\gamma \in \Gamma$ and consider the function $h_\gamma : [0,1] \times A \longrightarrow X$ defined by

$$h_\gamma(t, x) \overset{df}{=} \begin{cases} \gamma(t) & \text{if } x = 0, \\ x_1 & \text{if } x = x_1. \end{cases}$$

Evidently $h_\gamma \in C([0,1] \times A; X)$ and

$$h_\gamma(0,0) = 0, \quad h_\gamma(0, x_1) = x_1,$$
$$h_\gamma(1, x) = x_1, \quad \forall \, x \in A,$$

hence $h_\gamma \in H_A$. So

$$\underset{h \in H_A}{\inf} \underset{(t,x) \in [0,1] \times A}{\sup} \varphi(h(t,x)) \leq \underset{(t,x) \in [0,1] \times A}{\sup} \varphi(h_\gamma(t,x)) = \underset{t \in [0,1]}{\sup} \varphi(\gamma(t)).$$

Since $\gamma \in \Gamma$ was arbitrary, it follows that

$$\underset{h \in H_A}{\inf} \underset{(t,x) \in [0,1] \times A}{\sup} \varphi(h(t,x)) \leq c. \tag{2.28}$$

On the other hand, if $h \in H_A$, then if we define

$$\gamma_h(t) \overset{df}{=} \begin{cases} h(2t, 0) & \text{if } 0 \leq t \leq \frac{1}{2}, \\ h(2 - 2t, x_1) & \text{if } \frac{1}{2} < t \leq 1, \end{cases}$$

we see that $\gamma_h \in C([0,1]; X)$ and

$$\gamma_h(0) = h(0,0) = 0, \quad \gamma_h(1) = h(0, x_1) = x_1,$$

hence $\gamma_h \in \Gamma$. Therefore we have

$$c \leq \underset{t \in [0,1]}{\sup} \varphi(\gamma_h(t)) = \underset{(t,x) \in [0,1] \times A}{\sup} \varphi(h(t,x)).$$

Since $h \in H_A$ was arbitrary, it follows that

$$c \leq \inf_{h \in H_A} \sup_{(t,x) \in [0,1] \times A} \varphi\big(h(t,x)\big). \qquad (2.29)$$

From (2.28) and (2.29) and applying Theorem 2.1.2 we obtain the thesis of the theorem. $\square$

REMARK 2.1.5   In the statement of Theorem 2.1.3, the choice of 0 as the second point in the set A was made only for convenience. In fact we can replace 0 by any other point $x_2 \in X$, such that $\|x_2 - x_1\|_X > r$. Also we emphasize that in Theorem 2.1.3 the inequality (2.27) need not be strict (relaxed boundary conditions). $\square$

THEOREM 2.1.4 (Nonsmooth Saddle Point Theorem)
If $X = Y \oplus V$ *with* $\dim Y < +\infty$, *there exists* $r > 0$, *such that*

$$\max_{\substack{x \in Y \\ \|x\|_X = r}} \varphi(x) \leq \inf_{x \in V} \varphi(x) \qquad (2.30)$$

and φ *satisfies the nonsmooth* C_c*-condition with*

$$c \stackrel{df}{=} \inf_{\gamma \in \Gamma} \sup_{x \in E} \varphi\big(\gamma(x)\big)$$

where

$$\Gamma \stackrel{df}{=} \{\gamma \in C(E; X) : \ \gamma|_{\partial E} = id\},$$
$$E \stackrel{df}{=} \{x \in Y : \ \|x\|_X \leq r\},$$
$$\partial E = \{x \in Y : \ \|x\|_X = r\},$$

then $c \geq \inf_V \varphi$ *and* c *is a critical value of* φ. *Moreover, if* $c = \inf_V \varphi$, *then* $V \cap K_c^\varphi \neq \emptyset$.

PROOF   Let $A \stackrel{df}{=} \partial E$ and $C \stackrel{df}{=} V$. From Example 2.1.1(b) we know that the set A links C. Next let $\gamma \in \Gamma$ and set

$$h_\gamma(t,x) = \gamma\big((1-t)x\big) \qquad \forall \ (t,x) \in [0,1] \times A.$$

Evidently $h_\gamma \in H_A$. So we have

$$\inf_{h \in H_A} \sup_{(t,x) \in [0,1] \times A} \varphi\big(h(t,x)\big) \leq \sup_{(t,x) \in [0,1] \times A} \varphi\big(h_\gamma(t,x)\big)$$
$$= \sup_{(t,x) \in [0,1] \times A} \varphi\big(\gamma((1-t)x)\big) = \sup_{x \in E} \varphi\big(\gamma(x)\big).$$

Since $\gamma \in \Gamma$ was arbitrary, it follows that

$$\inf_{h \in H_A} \sup_{(t,x) \in [0,1] \times A} \varphi\big(h(t,x)\big) \leq c. \tag{2.31}$$

On the other hand, if $h \in H_A$, then for some $z_1 \in X$ we have $h(1,x) = z_1$ for all $x \in A$ and let us define

$$\xi(t,x) \stackrel{df}{=} \begin{cases} h(t,x) & \text{if } (t,x) \in [0,1] \times A, \\ z_1 & \text{if } (t,x) \in \{1\} \times E. \end{cases}$$

Then $\xi \colon ([0,1] \times A) \cup (\{1\} \times E) \longrightarrow X$ is continuous. Let $\psi \colon E \longrightarrow ([0,1] \times A) \cup (\{1\} \times E)$ be a homeomorphism, such that $\psi(A) = \{0\} \times A$. We have that $\xi \circ \psi \in \Gamma$ and so

$$c \leq \sup_{(t,x) \in [0,1] \times A} \varphi\big((\xi \circ \psi)(t,x)\big) = \sup_{(t,x) \in [0,1] \times A} \varphi\big(h(t,x)\big).$$

Since $h \in H_A$ was arbitrary, it follows that

$$c \leq \inf_{h \in H_A} \sup_{(t,x) \in [0,1] \times A} \varphi\big(h(t,x)\big). \tag{2.32}$$

From estimates (2.31) and (2.32) and applying Theorem 2.1.2 we obtain the theorem. ☐

REMARK 2.1.6 Again we emphasize that in the formulation of Theorem 2.1.4, we have relaxed boundary conditions. ☐

THEOREM 2.1.5 (Nonsmooth Linking Theorem)
If $X = Y \oplus V$ *with* $\dim Y < +\infty$, $0 < r < R$ *and* $v_0 \in V$, $\|v_0\|_X = 1$ *are such that*

$$\max_{x \in \partial Q} \varphi(x) \leq \inf_{x \in \partial B_r \cap V} \varphi(x),$$

where

$$Q \stackrel{df}{=} \{x = y + tv_0 : y \in Y, \ 0 \leq t \leq R, \ \|y\|_X \leq R\}$$

and ∂Q *is the boundary of* Q *in* $Y \oplus \mathbb{R}v_0$, *i.e.*

$$\partial Q = \{x = y + tv_0 : y \in Y, \ t \in \{0, R\}, \ \|y\|_X \leq R \text{ or } t \in [0, R], \ \|y\|_X = R\}$$

and φ *satisfies the nonsmooth* C_c-*condition, where*

$$c \stackrel{df}{=} \inf_{\gamma \in \Gamma} \max_{x \in Q} \varphi\big(\gamma(x)\big),$$

with

$$\Gamma \stackrel{df}{=} \{\gamma \in C(Q; X) : \gamma|_{\partial Q} = id\},$$

then $c \geq \inf\limits_{x \in \partial B_r \cap V} \varphi(x)$ *and c is a critical value of* φ.

Moreover, *if* $c = \inf\limits_{x \in \partial B_r \cap V} \varphi(x)$, *then there exists a critical point* $x_0 \in \partial B_r \cap V$

of φ *with* $\varphi(x_0) = c$ *(i.e.* $K_c^\varphi \cap (\partial B_r \cap V) \neq \emptyset)$.

PROOF   Let $A = \partial Q$ and $C = \partial B_r \cap V$. From Example 2.1.1(c) we know that the set A links C. Because Q is compact, we have that $c < +\infty$. As in the proof of Theorem 2.1.4, we can check that

$$c = \inf_{h \in H_A} \sup_{(t,x) \in [0,1] \times A} \varphi\big(h(t,x)\big).$$

So we can apply Theorem 2.1.2 and finish the proof of the theorem.   ⬚

When φ is bounded below (i.e. is not indefinite), then the nonsmooth C_c-condition implies the existence of minimizers which of course are critical points of φ.

THEOREM 2.1.6
If $\varphi \colon X \longrightarrow \mathbb{R}$ *is a bounded below function satisfying the nonsmooth* C_c-*condition with* $c \overset{df}{=} \inf\limits_X \varphi$,
then there exists $x_0 \in X$, *such that* $\varphi(x_0) = \inf\limits_X \varphi$.

PROOF   Applying Theorem 1.4.9 with $\varepsilon = \frac{1}{n}$ for $n \geq 1$, $\lambda = 1$, $h(r) = 1$ and $x_0 = 0$, we generate a sequence $\{x_n\}_{n \geq 1} \subseteq X$, such that $\varphi(x_n) \searrow \inf\limits_X \varphi$ and

$$\varphi(x_n) \leq \varphi(y) + \frac{\frac{1}{n}\|x_n - y\|_X}{1 + \|x_n\|_X} \qquad \forall\, n \geq 1,\ y \in X$$

and so

$$-\frac{\|x_n - y\|_X}{n(1 + \|x_n\|_X)} \leq \varphi(y) - \varphi(x_n) \qquad \forall\, n \geq 1,\ y \in X.$$

For every $n \geq 1$, let $y = x_n + tu$ with $t > 0$ and $u \in X$. We have

$$-\frac{u}{n(1 + \|x_n\|_X)} \leq \frac{\varphi(x_n + tu) - \varphi(x_n)}{t} \qquad \forall\, n \geq 1,\ t > 0,\ u \in X$$

and so

$$-\frac{u}{n(1 + \|x_n\|_X)} \leq \varphi^0(x_n; u) \qquad \forall\, n \geq 1,\ u \in X.$$

Let

$$\psi_n(u) \overset{df}{=} n(1 + \|x_n\|_X)\varphi^0(x_n; u) \qquad \forall\, n \geq 1,\ u \in X.$$

Then ψ_n is sublinear, continuous with $\psi_n(0) = 0$ and for all $u \in X$, we have $-\|u\|_X \leq \psi_n(u)$. So we can apply Lemma 1.3.2 and obtain $y_n^* \in X^*$ with $\|y_n^*\|_{X^*} \leq 1$, such that

$$\langle y_n^*, u \rangle_X \leq \psi_n(u) \qquad \forall n \geq 1, \ u \in X.$$

Let us set

$$x_n^* \stackrel{df}{=} \frac{1}{n\left(1 + \|x_n\|_X\right)} y_n^* \qquad \forall n \geq 1.$$

We have that

$$\langle x_n^*, u \rangle_X \leq \varphi^0(x_n; u) \qquad \forall n \geq 1, \ u \in X.$$

So $x_n^* \in \partial\varphi(x_n)$ for $n \geq 1$ and we have

$$\left(1 + \|x_n\|_X\right) m^\varphi(x_n) \leq \left(1 + \|x_n\|_X\right) \|x_n^*\|_{X^*} = \frac{1}{n}\|y_n^*\|_{X^*} \leq \frac{1}{n}.$$

Since φ satisfies the nonsmooth C_c-condition, by passing to a subsequence if necessary, we may assume that

$$x_n \longrightarrow x_0 \quad \text{in } X.$$

Hence $\varphi(x_n) \searrow \varphi(x_0)$ and so $\varphi(x_0) = \inf_X \varphi$. $\qquad\qquad$ ☐

2.1.5 Existence of Multiple Critical Points

Under a symmetry condition, we can have a multiplicity result about critical points. For this purpose we shall need the notion of Krasnoselskii's genus, first mentioned in Section 1.5. Let us recall the notion.

DEFINITION 2.1.5 *Let Y be a Banach space and let*

$$\mathcal{A}_{cs} \stackrel{df}{=} \left\{ A \subseteq Y : \ A \text{ is closed and } A = -A \right\}$$

*(i.e. $\mathcal{A}_{cs}$ is the family of all closed symmetric subsets of Y). A nonempty subset $A \in \mathcal{A}_{cs}$ is said to have **Krasnoselskii's genus** k (we write $\gamma(A) = k$), if k is the smallest integer with the property that there exists an odd continuous map $h \colon A \longrightarrow \mathbb{R}^k \setminus \{0\}$. If no such k exists we set $\gamma(A) = +\infty$ and if $A = \emptyset$, we set $\gamma(A) = 0$.*

REMARK 2.1.7 *If $A \in \mathcal{A}_{cs}$, by the Tietze's extension theorem (see Theorem A.1.2), any odd $h \in C\left(A; \mathbb{R}^k\right)$ can be extended to $\widehat{h} \in C(Y; \mathbb{R}^k)$. Moreover, setting*

$$\widehat{\widehat{h}}(y) \stackrel{df}{=} \frac{1}{2}\left(\widehat{h}(y) - \widehat{h}(-y)\right),$$

we can choose the extension to be odd. The Krasnoselskii genus is also called the $\mathbb{Z}_2$-**index of Krasnoselskii**, since it is a topological index corresponding to the symmetry group $\mathbb{Z}_2 = \{id_Y, -id_Y\}$. ∏

The next proposition summarizes the basic properties of γ. For a proof of this result we refer to Struwe (1990, p. 87).

PROPOSITION 2.1.5
Let $A, C \in \mathcal{A}_{cs}$ and let $h \in C(Y;Y)$ be an odd map. The following hold:

(a) $\gamma(A) \geq 0$ *and* $\gamma(A) = 0$ *if and only if* $A = \emptyset$;

(b) *if* $h(A) \subseteq C$, *then* $\gamma(A) \leq \gamma(C)$;

(c) *if* $A \subseteq C$, *then* $\gamma(A) \leq \gamma(C)$ *(monotonicity)*;

(d) $\gamma(A \cup C) \leq \gamma(A) + \gamma(C)$ *and if* $\gamma(C) < +\infty$, *then* $\gamma(\overline{A \setminus C}) \geq \gamma(A) - \gamma(C)$
    (subadditivity);

(e) $\gamma(A) \leq \gamma(\overline{h(A)})$ *(supervariance)*;

(f) *if* $A \in \mathcal{A}_{cs}$ *is compact, then* $\gamma(A) < +\infty$ *and there exists* $\delta > 0$, *such that* $\gamma(A) = \gamma(\overline{A_\delta})$, *where*

$$A_\delta \overset{df}{=} \{y \in Y : d_Y(y, A) < \delta\}$$

   (continuity);

(g) *if* U *is a bounded symmetric neighbourhood of the origin in* $\mathbb{R}^k$, *then* $\gamma(\partial U) = k$;

(h) *if* V *is a closed subspace of* Y *with finite codimension and* $A \cap V = \emptyset$, *then* $\gamma(A) \leq \text{codim} \, V$;

(i) *if* $\gamma(A) > 1$, *then* A *contains infinitely many distinct points, while if* $A = \{\pm x_i : i \in I\}$ *with* $0 < \text{card} \, I < +\infty$, *then* $\gamma(A) = 1$.

REMARK 2.1.8   Property (g) illustrates that the notion of genus generalizes the notion of dimension in a linear space. ⏏

Recall that if $c \in \mathbb{R}$, then $\varphi^c \overset{df}{=} \{x \in X : \varphi(x) \leq c\}$ and $K^\varphi \overset{df}{=} \{x \in X : 0 \in \partial\varphi(x)\}$ (the set of critical points of φ). Now we can show the following multiplicity result about critical points of φ.

THEOREM 2.1.7
If X is a reflexive Banach space, $\varphi: X \longrightarrow \mathbb{R}$ is a locally Lipschitz function and

(i) φ *is even and* $\varphi(0) = 0$;

(ii) φ *satisfies the nonsmooth C-condition;*

(iii) *there exist a subspace V of X of finite codimension and numbers $\beta, r > 0$, such that*
$$\varphi(x) \geq \beta \qquad \forall\, x \in \partial B_r \cap V;$$

(iv) *there is a finite dimensional subspace Y of X with* $\dim Y > \operatorname{codim} V$, *such that φ is anticoercive on Y, i.e.*
$$\varphi(x) \longrightarrow -\infty \text{ as } \|x\|_X \to +\infty \text{ with } x \in Y,$$

then $\underline{\text{then}}$ K^φ *has at least* $\dim Y - \operatorname{codim} V$ *pairs of nontrivial elements.*

PROOF Assume that for some $d > 0$, φ^{-d} contains no critical points of φ (otherwise there are infinitely many critical points and so there is nothing to prove). Let $R > r$ be such that $\varphi|_{\partial B_R^Y} \leq -d$ (see hypothesis (vi)). Let

$$m \stackrel{df}{=} \operatorname{codim} V, \quad k \stackrel{df}{=} \dim Y, \quad B \stackrel{df}{=} B_R^Y = \{x \in Y : \|x\|_X \leq R\}.$$

For $j \in \{1, \dots, k\}$, we introduce the following objects:

$$\mathcal{F} \stackrel{df}{=} \{\eta \in C(B; X) : \eta \text{ is odd and } \eta|_{\partial B} \text{ is homotopic to } id|_{\partial B}$$
$$\text{in } \varphi^{-d} \text{ by an odd homotopy}\},$$

$$\Gamma_j \stackrel{df}{=} \{\eta(B \setminus U) : \eta \in \mathcal{F},\ U \text{ is open in } B \text{ and symmetric,}$$
$$U \cap \partial B = \emptyset \text{ and for each } Z \subseteq U \text{ with } Z \in \mathcal{A}_{cs} \text{ we have}$$
$$\gamma(Z) \leq k - j\},$$

$$\Delta_j \stackrel{df}{=} \{A \subseteq X : A \in \mathcal{A}_{cs},\ A \text{ is compact and for each open } W,$$
$$\text{such that } A \subseteq W, \text{ there exists } A_0 \in \Gamma_j \text{ with } A_0 \subseteq W\}.$$

Note that $B \in \Delta_j$ with $A_0 = B$, $U = \emptyset$ and $\eta = id|_B$ so that $\Delta_j \neq \emptyset$ for all $j \in \{1, \dots, k\}$. We introduce the following numbers:

$$c_j \stackrel{df}{=} \inf_{A \in \Delta_j} \sup_{x \in A} \varphi(x) \qquad \forall\, j \in \{1, \dots, k\}.$$

Using standard topological arguments (see Szulkin (1986, Lemmata 4.5 and 4.6)), we obtain:

(a) $c_j \geq \beta$ for all $j \in \{m+1, \dots, k\}$.

(b) $\Delta_{j+1} \subseteq \Delta_j$ for all $j \in \{1, \dots, k-1\}$.

(c) If $A \in \Delta_j$, C is a closed and symmetric set, such that $A \subseteq \text{int}\, C$ and $\psi\colon C \longrightarrow X$ is an odd map, such that $\psi|_{C \cap \varphi^{-d}}$ is homotopic to $id|_{C \cap \varphi^{-d}}$ by an odd homotopy, then $\psi(A) \in \Delta_j$.

(d) If $D \in \mathcal{A}_{cs}$ is compact, $\gamma(D) \leq l$ and $\varphi|_D > -d$, then there exists a number $\delta > 0$, such that for each $A \in \Delta_{j+l}$ we have $A \setminus \text{int}\, D_\delta \in \Delta_j$.

By virtue of (a) and (b), we have $\beta \leq c_{m+1} \leq \ldots \leq c_k$. Suppose that

$$c_{m+1} = \ldots = c_{m+1+l} = c \text{ for some } l \in \{0, \ldots, k - m - 1\}.$$

Because φ is even, we have that K_c^φ is symmetric and of course it is compact (since φ satisfies the nonsmooth C-condition). Note that $c > 0$ (since $c \geq \beta > 0$) and so $0 \notin K_c^\varphi$ (recall that $\varphi(0) = 0$). Hence $K_c^\varphi \in \mathcal{A}_{cs}$.

We shall prove that $l + 1 \leq \gamma(K_c^\varphi)$. Suppose that this is not the case. So $\gamma(K_c^\varphi) \leq l$. By virtue of Proposition 2.1.5(f), we can find $\delta > 0$, such that

$$\gamma(K_c^\varphi) = \gamma\left(\overline{(K_c^\varphi)_\delta}\right).$$

Let $\varepsilon_0 > 0$ and

$$U \overset{df}{=} \text{int}\left(\overline{(K_c^\varphi)_\delta}\right).$$

According to the Deformation Theorem (see Theorem 2.1.1), we can find $\varepsilon \in (0, \varepsilon_0)$ and a continuous map $\eta\colon [0,1] \times X \longrightarrow X$ satisfying statements (a)-(h) of that theorem. In particular, we have that η has the semigroup property

$$\eta(s, \cdot) \circ \eta(t, \cdot) = \eta(s + t, \cdot) \qquad \forall\, s, t \in [0,1],\; s + t \leq 1$$

and for each $t \in [0,1]$, $\eta(t, \cdot)$ is an odd homeomorphism. From Theorem 2.1.1(b) and (c), we have that

$$\eta(1, x) = x \qquad \forall\, x \in \varphi^{c+\varepsilon_0} \setminus \varphi^{c-\varepsilon_0}$$

and

$$\eta(1, \varphi^{c+\varepsilon} \setminus U) \subseteq \varphi^{c-\varepsilon}.$$

We can find $A \in \Delta_{j+l}$, such that

$$\sup_{x \in A} \varphi(x) \leq c + \varepsilon.$$

Because K_c^φ is compact, $\gamma(K_c^\varphi) \leq l$ and $\varphi|_{K_c^\varphi} > -d$, by (d) above and if $\delta > 0$ is small enough, we have that $A \setminus U \in \Delta_j$ (recall the definition of U). Moreover, $A \setminus U \subseteq \varphi^{c+\varepsilon} \setminus U$ and so

$$\eta(1, A \setminus U) \subseteq \varphi^{c-\varepsilon}.$$

Thus if ε_0 is sufficiently small we have $\eta(1, x) = x$ for all $x \in \varphi^{-d}$ and so by (c) above, $\eta(A \setminus U) \in \Delta_j$. We have

$$\sup \left\{ \varphi(x) : \ x \in \eta(1, A \setminus U) \right\} \ \leq \ c - \varepsilon,$$

which contradicts the fact that $\sup \left\{ \varphi(x) : \ x \in \eta(1, A \setminus U) \right\} \geq c$. Therefore $\gamma(K_c^\varphi) \geq l + 1$. In particular then $\gamma(K_{c_j}^\varphi) \geq 1$ and so by Proposition 2.1.5(i), each $K_{c_j}^\varphi$ has at least two antipodal points $\pm x_j$. This produces the claimed number of critical points of φ if all c_j are distinct. If they are not, then $l > 0$ for some j, so $\gamma(K_{c_j}^\varphi) > 1$ and by Proposition 2.1.5(i), φ has infinite number of critical points. ☐

2.2 Constrained Locally Lipschitz Functionals

2.2.1 Critical Points of Constrained Functions

We can also have a corresponding theory for locally Lipschitz functions defined on a closed, convex set. In many applications, we encounter problems with inequality constraints (such as variational inequalities) and we have to deal with energy functionals defined on a closed and convex subset of a Banach space. In order to be able to use variational methods for such problems, we need to have a critical point theory for locally Lipschitz functions defined on a closed, convex set (the constraint set). In the rest of this section we present such a theory.

The mathematical setting is the following. Let X be a reflexive Banach space, $C \subseteq X$ a nonempty, closed and convex set and $\varphi \colon C \longrightarrow \mathbb{R}$ a locally Lipschitz function.

DEFINITION 2.2.1 *For $x \in C$, we define*

$$m_C^\varphi(x) \ \overset{df}{=} \ \inf_{x^* \in \partial\varphi(x)} \ \sup_{\substack{y \in C \\ \|x - y\|_X < 1}} \langle x^*, x - y \rangle_X \,.$$

REMARK 2.2.1 Evidently $m_C^\varphi(x) \geq 0$. This quantity can be viewed as a measure of the generalized slope of φ at $x \in C$. If φ admits an extension $\widehat{\varphi} \in C^1(X)$, then $\partial\varphi(x) = \{\varphi'(x)\}$ (see Proposition 1.3.10) and so we have

$$m_C^\varphi(x) \ = \ \sup_{\substack{y \in C \\ \|x - y\|_X < 1}} \langle \varphi'(x), x - y \rangle_X \,,$$

which is the quantity used in the smooth theory developed by Struwe (1990, p. 147).   ⬚

LEMMA 2.2.1
Function $m_C^\varphi \colon X \longrightarrow \mathbb{R}_+$ is lower semicontinuous.

PROOF   Let X_w^* be the dual space of X, furnished with the weak topology. Let

$$C_1(x) \stackrel{df}{=} (x - C) \cap B_1 \qquad \forall\, x \in C,$$

where $B_1 \stackrel{df}{=} \{z \in X : \|z\|_X < 1\}$. Note that the multifunction $x \longmapsto C_1(x)$ is lower semicontinuous (see Proposition 1.2.9(c)). Consider the map $h \colon X_w^* \times X \longrightarrow \mathbb{R}_+$ defined by

$$h(x^*, x) \stackrel{df}{=} \sup_{y \in C_1(x)} \langle x^*, x - y \rangle_X .$$

We claim that h is lower semicontinuous. It suffices to show that for every $\lambda \in \mathbb{R}_+$, if $(x_n^*, x_n) \longrightarrow (x^*, x)$ in $X_w^* \times X$ and $h(x_n^*, x_n) \leq \lambda$ for all $n \geq 1$, we have $h(x^*, x) \leq \lambda$. Indeed let $y \in C_1(x)$. Then from the lower semicontinuity of C_1 we know that we can find $y_n \in C_1(x_n)$ for $n \geq 1$, such that $y_n \longrightarrow y$ in X (see Proposition 1.2.2(d)). We have

$$\langle x_n^*, x_n - y_n \rangle_X \leq h(x_n^*, x_n) \leq \lambda \qquad \forall\, n \geq 1$$

and so

$$\langle x^*, x - y \rangle_X \leq \lambda.$$

Since $y \in C_1(x)$ was arbitrary it follows that $h(x^*, x) \leq \lambda$, which proves the sequential lower semicontinuity of h on $X_w^* \times X$.

Now we show that m_C^φ is lower semicontinuous. We have

$$m_C^\varphi(x) = \inf_{x^* \in \partial\varphi(x)} h(x^*, x).$$

Let $x_n \longrightarrow x$ in X and suppose that $m_C^\varphi(x_n) \leq \lambda$ for $n \geq 1$. For every $n \geq 1$, we can find $x_n^* \in \partial\varphi(x_n)$, such that $h(x_n^*, x_n) \leq m_C^\varphi(x_n) + \frac{1}{n}$. Evidently $\{x_n^*\}_{n \geq 1} \subseteq X^*$ is bounded and so, by passing to a subsequence if necessary, we may assume that $x_n^* \xrightarrow{\ w\ } x^*$ in X^* (Eberlein-Smulian Theorem; see Theorem A.3.5). We have

$$h(x^*, x) \leq \liminf_{n \to +\infty} h(x_n^*, x_n) \leq \lambda$$

and because $\partial\varphi$ is upper semicontinuous from X into X_w^* (see Proposition 1.3.9) we have that $x^* \in \partial\varphi(x)$. Therefore it follows that $m_C^\varphi(x) \leq \lambda$ from which we conclude the lower semicontinuity of m_C^φ on X.   ⬚

DEFINITION 2.2.2   *A point $x \in C$ is a **critical point of** φ **on** C if $m_C^\varphi(x) = 0$. The value $c = \varphi(x)$ is called **critical value of** φ.*

REMARK 2.2.2   If $C = X$ then

$$m_C^\varphi(x) \;=\; m^\varphi(x) \;=\; \min_{x^* \in \partial\varphi(x)} \|x^*\|_{X^*}\,,$$

which is the quantity used in the unconstrained part of the theory.

In this case $x \in X$ is a critical point of φ if and only if $0 \in \partial\varphi(x)$, which is Definition 2.1.2.   ⬜

We also introduce a version of the nonsmooth C-condition which is suitable for this constrained situation.

DEFINITION 2.2.3   *We say that φ satisfies the **nonsmooth C-condition on the set** C **at the level** $c \in \mathbb{R}$ (the **nonsmooth C_c-condition on** C for short) if and only if any sequence $\{x_n\}_{n \geq 1} \subseteq C$, such that $\varphi(x_n) \longrightarrow c$ and $(1 + \|x_n\|_X)m_C^\varphi(x_n) \longrightarrow 0$ as $n \to +\infty$ has a convergent subsequence. If φ satisfies the nonsmooth C_c-condition on C at every level $c \in \mathbb{R}$, then we simply say that φ satisfies the **nonsmooth C-condition on set** C.*

REMARK 2.2.3   If $C = X$ then the above definition coincides with Definition 2.1.1(b). Similarly we can also define the **nonsmooth PS_c and PS-condition on the set** C.   ⬜

2.2.2   Deformation Theorem

As in the unconstrained case, for any $c \in \mathbb{R}$, $\varepsilon, \delta > 0$, we introduce the following sets:

$$K^\varphi \stackrel{df}{=} \{x \in C : \; m_C^\varphi(x) = 0\}$$

- the set of critical points of φ on C,

$$K_c^\varphi \stackrel{df}{=} \{x \in K^\varphi : \; \varphi(x) = c\}$$

- the set of critical points of φ on C with critical value c,

$$(K_c^\varphi)_\delta \stackrel{df}{=} \{x \in C : \; d_X(x, K_c^\varphi) < \delta\},$$

$$E_{c,\delta,\varepsilon}^\varphi \stackrel{df}{=} \{x \in C : \; |\varphi(x) - c| \leq \varepsilon\} \cap (K_c^\varphi)_\delta^c,$$

$$\varphi^c \stackrel{df}{=} \{x \in C : \; \varphi(x) \leq c\}.$$

LEMMA 2.2.2
<u>If</u> $\varphi\colon C \longrightarrow \mathbb{R}$ *satisfies the nonsmooth* C_c-*condition on* C,
<u>then</u> *for every* $\delta > 0$ *there exist* $\gamma, \varepsilon > 0$, *such that*

$$\gamma \leq \left(1 + \|x\|_X\right) m_C^\varphi(x) \qquad \forall\, x \in E_{c,\delta,\varepsilon}^\varphi.$$

PROOF   Suppose that the result is not true. Then we can find $\delta > 0$ and
a sequence $\{x_n\}_{n \geq 1} \subseteq (K_c^\varphi)_\delta^c$ with $c - \frac{1}{n} \leq \varphi(x_n) \leq c + \frac{1}{n}$, such that

$$\left(1 + \|x_n\|_X\right) m_C^\varphi(x_n) \longrightarrow 0.$$

Since φ satisfies the nonsmooth C_c-condition on C, we may assume, passing
to a subsequence if necessary, that we have

$$x_n \longrightarrow x \quad \text{in } X$$

for some $x \in C$. By lower semicontinuity of m_C^φ (see Lemma 2.2.1), we have
that

$$m_C^\varphi(x) \leq \liminf_{n \to +\infty} m_C^\varphi(x_n)$$

and as $m_C^\varphi(x) \geq 0$, we obtain

$$m_C^\varphi(x) = 0.$$

Thus $x \in K_c^\varphi$, which contradicts the fact that $x \in (K_c^\varphi)_\delta^c$ (since $\{x_n\}_{n \geq 1} \subseteq (K_c^\varphi)_\delta^c$).   ⬚

In analogy to the unconstrained case, we obtain now a locally Lipschitz
vector field.

PROPOSITION 2.2.1
<u>If</u> φ *satisfies the nonsmooth* C_c-*condition on* C, $\delta > 0$ *is given and* $\gamma, \varepsilon > 0$
are as in the Lemma 2.2.2,
<u>then</u> *we can find a locally Lipschitz map*

$$v\colon E_{c,\delta,\varepsilon}^\varphi \longrightarrow X,$$

such that

$$\begin{cases} \|v(x)\|_X \leq 2\left(1 + \|x\|_X\right) \\ \dfrac{\gamma}{4} \leq \langle x^*, v(x)\rangle_X \end{cases} \qquad \forall\, x \in E_{c,\delta,\varepsilon}^\varphi, \ x^* \in \partial\varphi(x).$$

PROOF   Consider the function

$$h(x^*, x) \overset{df}{=} \sup_{\substack{y \in C \\ \|x - y\|_X < 1}} \langle x^*, x - y\rangle_X.$$

Since X is reflexive, by Bauer's maximum principle (see Theorem A.3.11), we can find $u(x) \in X$ with $\|u(x)\|_X = 1$ and $x - u(x) \in C$, such that

$$h(x^*, x) = \langle x^*, u(x) \rangle_X \qquad \forall\, x \in E^{\varphi}_{c,\delta,\varepsilon}, \ x^* \in \partial\varphi(x).$$

Note that

$$m^{\varphi}_C(x) \le h(x^*, x) \qquad \forall\, x \in E^{\varphi}_{c,\delta,\varepsilon}, \ x^* \in \partial\varphi(x).$$

From Lemma 2.2.2, we have

$$\frac{\gamma}{2\,(1 + \|x\|_X)} < m^{\varphi}_C(x) \le \langle x^*, u(x) \rangle_X \qquad \forall\, x \in E^{\varphi}_{c,\delta,\varepsilon}, \ x^* \in \partial\varphi(x). \quad (2.33)$$

The multifunction $x \longmapsto (1 + \|x\|_X)\partial\varphi(x)$ is upper semicontinuous from X into X^*_w.

Let $x \in E^{\varphi}_{c,\delta,\varepsilon}$ and let

$$V \overset{df}{=} \left\{ y^* \in X^* : \frac{\gamma}{2} < \langle y^*, u(x) \rangle_X \right\}.$$

This is a weakly open set in X^* and from (2.33) we can also see that

$$(1 + \|x\|_X)\partial\varphi(x) \subseteq V.$$

Exploiting the upper semicontinuity of the map $x \longmapsto (1 + \|x\|_X)\partial\varphi(x)$ from X into X^*_w, we can find $0 < r(x) < 1$, such that

$$(1 + \|y\|_X)\partial\varphi(y) \subseteq V \qquad \forall\, y \in B_{r(x)}(x) \cap E^{\varphi}_{c,\delta,\varepsilon},$$

hence

$$\frac{\gamma}{2\,(1 + \|y\|_X)} < \langle y^*, u(x) \rangle_X \qquad \forall\, y \in B_{r(x)}(x) \cap E^{\varphi}_{c,\delta,\varepsilon}, \ y^* \in \partial\varphi(y). \quad (2.34)$$

The collection $\left\{ B_{r(x)}(x) \right\}_{x \in E^{\varphi}_{c,\delta,\varepsilon}}$ is an open cover of the set $E^{\varphi}_{c,\delta,\varepsilon}$. So by paracompactness we can find a locally finite refinement $\{U_i\}_{i \in I}$ and a locally Lipschitz partition of unity $\{\xi_i\}_{i \in I}$ subordinate to it. For each $i \in I$, we can find $x_i \in E^{\varphi}_{c,\delta,\varepsilon}$, such that $U_i \subseteq B_{r(x_i)}(x_i)$. To each such $x_i \in E^{\varphi}_{c,\delta,\varepsilon}$ corresponds an element $u(x_i)$ with $\|u(x_i)\|_X = 1$ and $x_i - u(x_i) \in C$, for which (2.33) holds with $x = x_i$.

We define $v \colon E^{\varphi}_{c,\delta,\varepsilon} \longrightarrow X$ by

$$v(x) \overset{df}{=} (1 + \|x\|_X) \sum_{i \in I} \xi_i(x)\bigl(u(x_i) - x_i + x\bigr). \quad (2.35)$$

Clearly v is well defined, locally Lipschitz and $\|v(x)\|_X \le 2\,(1 + \|x\|_X)$. Moreover,

$$
\begin{aligned}
\langle x^*, v(x) \rangle_X ={}& (1 + \|x\|_X) \sum_{i \in I} \xi_i(x)\langle x^*, u(x_i) \rangle_X \\
&+ (1 + \|x\|_X) \sum_{i \in I} \xi_i(x)\,\langle x^*, x - x_i \rangle_X \qquad \forall x^* \in \partial\varphi(x).
\end{aligned}
$$

By using (2.34) and choosing $r(x) \leq \dfrac{\gamma}{4\left(1+\|x\|_X\right)}$, we have that

$$\langle x^*, v(x) \rangle_X \geq \frac{\gamma}{4}.$$

This proves the lemma.   □

Following the steps of the proof of Theorem 2.1.1 with minor modifications, we obtain the following version of the deformation theorem.

THEOREM 2.2.1 (Deformation Theorem for Constrained Functions)
If $\varphi \colon C \longrightarrow \mathbb{R}$ satisfies the nonsmooth C_c-condition on C,
then for every $\varepsilon_0 > 0$ and every neighbourhood U of K_c^φ (if $K_c^\varphi = \emptyset$, then $U = \emptyset$), there exist $\varepsilon \in (0, \varepsilon_0)$ and a continuous map $\eta \colon [0,1] \times C \longrightarrow C$, such that for all $(t, x) \in [0,1] \times C$, we have

(a) $\|\eta(t, x) - x\|_X \leq 2(e+1)\left(1 + \|x\|_X\right)t$;

(b) *if* $|\varphi(x) - c| \geq \varepsilon_0$ *or* $m_C^\varphi(x) = 0$, *then* $\eta(t, x) = x$;

(c) $\eta\left(\{1\} \times \varphi^{c+\varepsilon}\right) \subseteq \varphi^{c-\varepsilon} \cup U$;

(d) $\varphi\big(\eta(t, x)\big) \leq \varphi(x)$;

(e) *if* $\eta(t, x) \neq x$, *then* $\varphi\big(\eta(t, x)\big) < \varphi(x)$.

REMARK 2.2.4   Note that from the proof of Proposition 2.2.1 (see (2.35)), we have that

$$x - \frac{1}{1 + \|x\|_X} v(x) \;=\; \sum_{i \in I} \xi_i(x)\big(x_i - u(x_i)\big) \;\in C.$$

Therefore $-v(x) \in T_C(x)$ (the tangent cone to C at x; see Definition A.4.1), hence $L(x) \in T_C(x)$ (see the proof of Theorem 2.1.1). So by the Nagumo viability theorem (see Theorem A.4.4), problem (2.14) has a flow in C (or equivalently the extension of L on $X \setminus C$, leaves the set C invariant).   □

2.2.3   Minimax Principles

Using the above Deformation Theorem, we can have the first minimax principle.

THEOREM 2.2.2
If M is a compact metric space, $M^* \subseteq M$ is a nonempty, closed set, $\gamma^* \in$

$C(M^*; C)$, $\varphi: C \longrightarrow \mathbb{R}$ *satisfies the nonsmooth* C_c*-condition on* C, *with*

$$c \overset{df}{=} \inf_{\gamma \in \Gamma} \max_{s \in M} \varphi(\gamma(s)),$$

where

$$\Gamma = \{\gamma \in C(M; C): \ \gamma = \gamma^* \text{ on } M^*\}$$

and $c > \beta$ *with*

$$\beta \overset{df}{=} \max_{s \in M^*} \varphi(\gamma^*(s)),$$

then $K_c^\varphi \neq \emptyset$.

PROOF Suppose that the theorem is not true. Then $K_c^\varphi = \emptyset$. Let $\varepsilon_0 = \frac{1}{2}(c - \beta)$. By hypothesis $\varepsilon_0 > 0$. Apply Theorem 2.2.1 to obtain $\varepsilon \in (0, \varepsilon_0)$ and a continuous map $\eta: [0, 1] \times C \longrightarrow C$, which satisfy statements (a)-(e) of that theorem. Take $\widehat{\gamma} \in \Gamma$, such that

$$\max_{s \in M} \varphi(\widehat{\gamma}(s)) \leq c + \varepsilon.$$

Let $\gamma_\eta \in C(M; C)$ be defined by

$$\gamma_\eta(s) = \eta(1, \widehat{\gamma}(s)) \qquad \forall \, s \in M.$$

Note that

$$\gamma_\eta(s) = \eta(1, \gamma^*(s)) \qquad \forall \, s \in M^*.$$

Note that

$$\varepsilon_0 + \varphi(\gamma^*(s)) \ \leq \ \varepsilon_0 + \beta \ = \ \frac{1}{2}(c + \beta) \ < \ c \qquad \forall \, s \in M^*,$$

so

$$\varepsilon_0 \ < \ c - \varphi(\gamma^*(s)) \qquad \forall \, s \in M^*$$

and thus from Theorem 2.2.1(b) we have

$$\eta(t, \gamma^*(s)) \ = \ \gamma^*(s) \qquad \forall \, t \in [0, 1], \ s \in M^*.$$

Therefore $\gamma_\eta|_{M^*} = \gamma^*$ and so $\gamma_\eta \in \Gamma$. From Theorem 2.2.1(c) and recalling that $U = \emptyset$, we have

$$\eta(1, \widehat{\gamma}(s)) \ \in \ \eta(1, \varphi^{c+\varepsilon}) \ \subseteq \ \varphi^{c-\varepsilon} \qquad \forall \, s \in M,$$

so

$$\varphi(\gamma_\eta(s)) \ \leq \ c - \varepsilon \qquad \forall \, s \in M,$$

which contradicts the definition of c. This proves that $K_c^\varphi \neq \emptyset$ and so the theorem holds. $\qquad \square$

We have another minimax principle which is the constrained analog of the nonsmooth Saddle Point Theorem (see Theorem 2.1.4).

THEOREM 2.2.3
If $X = Y \oplus V$ with $\dim Y < +\infty$, $V \cap C \neq \emptyset$, $M = Y \cap C \cap \overline{B}_r$, $M^* = \overline{M} \cap \partial B_r \neq \emptyset$, φ satisfies the nonsmooth C_c-condition on C with

$$c \overset{df}{=} \inf_{\gamma \in \Gamma} \max_{x \in M} \varphi(\gamma(x)),$$

where

$$\Gamma = \{\gamma \in C(M; C): \ \gamma|_{M^*} = id\},$$

and

$$\inf_{V \cap C} \varphi > \max_{M^*} \varphi,$$

then $K_c^\varphi \neq \emptyset$.

PROOF   First note that the set M^* links $V \cap C$ (see Definition 2.1.4 and recall that $\operatorname{rint} C \neq \emptyset$). Let us define

$$\vartheta \overset{df}{=} \inf_{V \cap C} \varphi, \qquad \beta \overset{df}{=} \max_{M^*} \varphi.$$

By hypotheses $\vartheta > \beta$. From Remark 2.1.4, we know that

$$\gamma(M) \cap V \cap C \neq \emptyset \qquad \forall \, \gamma \in \Gamma$$

and so

$$\vartheta \leq \max_{x \in M} \varphi(\gamma(x)) \qquad \forall \, \gamma \in \Gamma. \tag{2.36}$$

We proceed by contradiction. Suppose that $K_c^\varphi = \emptyset$ and let $U \overset{df}{=} \emptyset$ and $\varepsilon_0 \overset{df}{=} \frac{1}{2}(\vartheta - \beta)$. By hypotheses $\varepsilon_0 > 0$. Let $\varepsilon \in (0, \varepsilon_0)$ and $\eta: [0, 1] \times C \longrightarrow C$ be as in the Deformation Theorem (Theorem 2.2.1). Choose $\widehat{\gamma} \in \Gamma$, such that

$$\max_{x \in M} \varphi(\widehat{\gamma}(x)) \leq c + \varepsilon.$$

Let $\gamma_\eta(x) \overset{df}{=} \eta(1, \widehat{\gamma}(x))$. Evidently $\gamma_\eta \in C(M; C)$. Also

$$\gamma_\eta(x) = \eta(1, \widehat{\gamma}(x)) = \eta(1, x) \qquad \forall \, x \in M^*.$$

From the choice of $\varepsilon_0 > 0$ and from (2.36), we have that

$$\varphi(x) \leq \beta < \beta + \varepsilon_0 = \vartheta - \varepsilon_0 < c - \varepsilon_0 \qquad \forall \, x \in M^*.$$

But then from Theorem 2.2.1(b) it follows that $\eta(1, x) = x$ for all $x \in M^*$ and so $\gamma_\eta \in \Gamma$. Then from Theorem 2.2.1(d), we have

$$\varphi(\gamma_\eta(x)) = \varphi(\eta(1, \widehat{\gamma}(x))) \leq \varphi(\widehat{\gamma}(x)) \qquad \forall \, x \in M$$

and so from the choice of $\widehat{\gamma}$, we obtain

$$\varphi(\gamma_\eta(x)) \leq c + \varepsilon \qquad \forall\, x \in M.$$

Therefore $\gamma_\eta(x) \in \varphi^{c+\varepsilon}$ and then from Theorem 2.2.1(c), we have that

$$\varphi(\gamma_\eta(x)) = \varphi(\eta(1,\widehat{\gamma}(x))) \leq c - \varepsilon,$$

a contradiction to the definition of c. This proves the theorem. $\qquad\square$

We conclude with a result on global minima of φ on C.

THEOREM 2.2.4

If $\varphi\colon C \longrightarrow \mathbb{R}$ *is a bounded below function, satisfying the nonsmooth C-condition on* C,
<u>*then*</u> *there exists* $x_0 \in C$, *such that* $\varphi(x_0) = \inf_C \varphi$ *and* x_0 *is a critical point of* φ *on* C.

PROOF Let $\{x_n\}_{n\geq 1} \subseteq C$ be such that $\varphi(x_n) \searrow \inf_C \varphi$. Invoking Theorem 1.4.9 (with $h(r) = r$, $x_0 = 0$, $\varepsilon = \frac{1}{\sqrt{n}}$ and $\lambda = \frac{1}{\varepsilon}$), we obtain a sequence $\{y_n\}_{n\geq 1} \subseteq C$, such that for all $n \geq 1$, we have

$$\varphi(y_n) \leq \varphi(x_n), \qquad \|y_n\|_X \leq \|x_n\|_X + \overline{r}$$

and

$$\varphi(u) \geq \varphi(y_n) - \frac{1}{n\,(1 + \|y_n\|_X)}\,\|u - y_n\|_X \qquad \forall\, u \in C,$$

where $\overline{r} > 0$ is such that

$$\int\limits_{\|y_n\|_X}^{\|y_n\|_X + \overline{r}} \frac{1}{1+r}\,dr \geq \frac{1}{n}.$$

For every $u \in X$ and $t \in (0,1)$, we have

$$-\frac{t\,\|u - y_n\|_X}{n\,(1 + \|y_n\|_X)} \leq \varphi(y_n + t(u - y_n)) - \varphi(y_n) + i_C\,(y_n + t(u - y_n)),$$

where i_C is an indicator function of C, i.e.

$$i_C(v) \stackrel{df}{=} \begin{cases} 0 & \text{if } v \in C, \\ +\infty & \text{if } v \notin C. \end{cases}$$

So

$$-\|u - y_n\|_X \tag{2.37}$$
$$\leq n\,(1 + \|y_n\|_X)\left(\frac{\varphi(y_n + t(u - y_n)) - \varphi(y_n)}{t} + \frac{i_C(y_n + t(u - y_n))}{t}\right).$$

Note that i_C is convex, lower semicontinuous (since $C \in P_{fc}(X)$) and $i_C(y_n) = 0$ for all $n \geq 1$. So

$$\frac{i_C\big(y_n + t(u - y_n)\big)}{t} \leq i_C\big(y_n + u - y_n\big) = i_C(u). \tag{2.38}$$

Also we have

$$\limsup_{t \searrow 0} \frac{\varphi\big(y_n + t(u - y_n)\big) - \varphi(y_n)}{t} \leq \varphi^0(y_n; u - y_n). \tag{2.39}$$

Let us set

$$\begin{aligned} \eta_n(h) &\overset{df}{=} \varphi^0(y_n; h) \\ \vartheta_n(h) &\overset{df}{=} i_C(y_n + h) \end{aligned} \qquad \forall\, n \geq 1,\; h \in X.$$

We see that $\eta_n + \vartheta_n$ is a proper, convex and lower semicontinuous function on X with values in $\overline{R} = \mathbb{R} \cup \{+\infty\}$ (i.e. $\eta_n + \vartheta_n \in \Gamma_0(X)$; see Definition 1.3.1). Also $(\eta_n + \vartheta_n)(0) = 0$. From (2.37), (2.38), (2.39), putting $u = y_n + h$, with $h \in X$, we obtain

$$-\|h\|_X \leq n\,(1 + \|y\|_X)\,(\eta_n + \vartheta_n)\,(h) \qquad \forall\, h \in X,\; n \geq 1.$$

So we can apply Lemma 1.3.2 and obtain $\overline{u}_n^* \in X^*$ with $\|\overline{u}_n^*\|_{X^*} \leq 1$, such that

$$\langle \overline{u}_n^*, h \rangle_X \leq n\,(1 + \|y_n\|_X)\,(\eta_n + \vartheta_n)\,(h) \qquad \forall\, h \in X. \tag{2.40}$$

Since $\eta_n + \vartheta_n \in \Gamma_0(X)$ and $(\eta_n + \vartheta_n)(0) = 0$, from (2.40) it follows that

$$\overline{u}_n^* \in n\,(1 + \|y_n\|_X)\,\partial\,(\eta_n + \vartheta_n)\,(0).$$

But because η_n is continuous, convex (in fact sublinear), from Theorem 1.3.6, we have that

$$\partial(\eta_n + \vartheta_n)(0) = \partial\eta_n(0) + \partial\vartheta_n(0).$$

Note that $\partial\eta_n(0) = \partial\varphi(y_n)$ where the first subdifferential is in the sense of convex analysis (see Definition 1.3.3), while the second is a generalized sub-differential (see Definition 1.3.7). Also $\partial\vartheta_n(0) = \partial i_C(y_n) = N_C(y_n)$ (where $N_C(y_n)$ stands for the normal cone to C at $y_n \in C$; see Definition 1.3.9). Therefore we have $\overline{u}_n^* = \overline{v}_n^* + \overline{w}_n^*$, with $\overline{v}_n^* \in n\,(1 + \|y_n\|_X)\,\partial\varphi(y_n)$ and $\overline{w}_n^* \in n\,(1 + \|y_n\|_X)\,N_C(y_n)$. Set

$$v_n^* \overset{df}{=} \frac{1}{n\,(1 + \|y_n\|_X)}\,\overline{v}_n^* \in \partial\varphi(y_n)$$

$$w_n^* \overset{df}{=} \frac{1}{n\,(1 + \|y_n\|_X)}\,\overline{w}_n^* \in N_C(y_n).$$

From the definition of the normal cone (see Definition 1.3.9), we have

$$\langle w_n^*, y_n - v \rangle_X \geq 0 \qquad \forall\, v \in C. \tag{2.41}$$

By the definition of m_C^φ (Definition 2.2.1), inequality (2.41) and the definition of u_n^*, we have

$$
\begin{aligned}
\big(1 + \|y_n\|_X\big) m_C^\varphi(y_n) &\leq \big(1 + \|y_n\|_X\big) \sigma(v_n^*, C_1(y_n)) \\
&= \big(1 + \|y_n\|_X\big) \sup_{\substack{v \in C \\ \|y_n - v\|_X < 1}} \big\{ \langle u_n^*, y_n - v\rangle_X - \langle w_n^*, y_n - v\rangle_X \big\} \\
&\leq \big(1 + \|y_n\|_X\big) \sigma\big(u_n^*, C_1(y_n)\big) \leq \frac{1}{n}.
\end{aligned}
$$

So

$$
\big(1 + \|y_n\|_X\big) m_C^\varphi(y_n) \longrightarrow 0.
$$

But by hypothesis φ satisfies the nonsmooth C-condition on C. Thus we may assume that $y_n \longrightarrow x_0$ and so $\varphi(x_0) = \inf_C \varphi$. By virtue of Corollary 1.3.8, we have

$$
0 \in \partial\varphi(x_0) + N_C(x_0).
$$

So there exists $x^* \in N_C(x_0)$, such that $-x^* \in \partial\varphi(x_0)$. By the definition of the normal cone, we have that

$$
\langle x^*, x_0 - v\rangle_X \geq 0 \qquad \forall\, v \in C,
$$

hence

$$
\langle -x^*, x_0 - v\rangle_X \leq 0 \qquad \forall\, v \in C
$$

and so $m_C^\varphi(x_0) \leq 0$. But recall that $m_C^\varphi(x_0) \geq 0$. So $m_C^\varphi(x_0) = 0$ and it follows that $x_0 \in C$ is a critical point of φ on C. $\qquad\square$

2.3 Perturbations of Locally Lipschitz Functionals

In this section, we extend the nonsmooth critical point theory to functionals of the form $\varphi = \Phi + \psi$, with Φ locally Lipschitz and ψ a proper, convex and lower semicontinuous functional.

2.3.1 Critical Points of Perturbed Functions

Let X be a reflexive Banach space and let $\varphi\colon X \longrightarrow \overline{\mathbb{R}} = \mathbb{R} \cup \{+\infty\}$ be a functional, such that $\varphi = \Phi + \psi$, with $\Phi\colon X \longrightarrow \mathbb{R}$ locally Lipschitz and $\psi \in \Gamma_0(X)$.

DEFINITION 2.3.1   *We say that $x \in X$ is a **critical point of** φ if and only if*

$$
\Phi^0(x; h) + \psi(x + h) - \psi(x) \geq 0 \qquad \forall\, h \in X.
$$

REMARK 2.3.1   Let $x_0 \in X$ be a critical point of φ. Set

$$
\begin{aligned}
g_1(h) &\overset{df}{=} \Phi^0(x_0; h) \\
g_2(h) &\overset{df}{=} \psi(x_0 + h) - \psi(x_0)
\end{aligned}
\qquad \forall\, h \in X.
$$

Clearly g_1 is continuous and convex (in fact sublinear; see Proposition 1.3.7(a)) and $g_2 \in \Gamma_0(X)$. Moreover, we have that

$$
\partial g_1(0) = \partial \Phi(x_0),
$$

where the first subdifferential is in the sense of convex analysis (see Definition 1.3.3) and the second is a generalized subdifferential (see Definition 1.3.7). Also we have

$$
\partial g_2(0) = \partial \psi(x_0).
$$

From Theorem 1.3.6, we know that

$$
\partial(g_1 + g_2)(0) \; = \; \partial g_1(0) + \partial g_2(0) \; = \; \partial \Phi(x_0) + \partial \psi(x_0).
$$

So $x_0 \in X$ is a critical point of $\varphi = \Phi + \psi$ if and only if $0 \in \partial \Phi(x_0) + \partial \psi(x_0)$. If $\Phi \in C^1(X)$, then $\partial \Phi(x) = \{\Phi'(x)\}$ and so $x_0 \in X$ is a critical point of $\varphi = \Phi + \psi$ if and only if $-\Phi'(x_0) \in \partial \psi(x_0)$. On the other hand, if $\psi = 0$, then $x_0 \in X$ is a critical point of $\varphi = \Phi + \psi$ if and only if $0 \in \partial \Phi(x_0)$, which is the setting of the previous section (see Definition 2.1.2). $\Box$

As before we need a compactness-type condition, which we define next.

DEFINITION 2.3.2   *We say that $\varphi = \Phi + \psi$ satisfies the **generalized nonsmooth Palais-Smale condition at the level** $c \in \mathbb{R}$ (**generalized nonsmooth** PS_c-**condition** for short), if*

"Any sequence $\{x_n\}_{n \geq 1} \subseteq X$, such that $\varphi(x_n) \longrightarrow c$ and

$$
-\varepsilon_n \|y - x_n\|_X \leq \Phi^0(x_n; y - x_n) + \psi(y) - \psi(x_n) \qquad \forall\, y \in X,
$$

with $\varepsilon_n \searrow 0$, has a strongly convergent subsequence."

*If φ satisfies this at every level $c \in \mathbb{R}$ we say that φ satisfies the **generalized nonsmooth Palais-Smale condition** (**generalized nonsmooth** PS-**condition** for short).*

REMARK 2.3.2   Using Lemma 1.3.2, we can equivalently reformulate the above definition as follows:

"Any sequence $\{x_n\}_{n \geq 1} \subseteq X$, such that

$$
\varphi(x_n) \longrightarrow c
$$

and

$$\langle x_n^*, y - x_n \rangle_X \; \leq \; \Phi^0(x_n; y - x_n) + \psi(y) - \psi(x_n) \qquad \forall \, y \in X,$$

for some $x_n^* \longrightarrow 0$ in X^*, has a strongly convergent subsequence."

Using this reformulation we see that when $\psi \equiv 0$ (i.e. $\varphi = \Phi$), we recover Definition 2.1.1(a). $\qquad$ ▯

As in previous sections, for $c \in \mathbb{R}$ we introduce the set of critical points of φ with critical value c:

$$K_c^\varphi \; \overset{df}{=} \; \big\{ x \in X : \; x \text{ is a critical point of } \varphi, \; \varphi(x) = c \big\}.$$

PROPOSITION 2.3.1
If $\{x_n\}_{n \geq 1} \subseteq X$ *is a sequence, such that* $x_n \longrightarrow x_0$, $\varphi(x_n) \longrightarrow c$ *and*

$$-\varepsilon_n \, \|y - x_n\|_X \; \leq \; \Phi^0(x_n; y - x_n) + \psi(y) - \psi(x_n) \qquad \forall \, y \in X, \qquad (2.42)$$

with $\varepsilon_n \searrow 0$, *then* $x_0 \in K_c^\varphi$.

PROOF As $x_n \longrightarrow x_0$, passing to the limit in (2.42), using the upper semicontinuity of Φ^0 (see Proposition 1.3.7(b)) and recalling that $\psi \in \Gamma_0(X)$ we have

$$\begin{aligned} 0 \; &\leq \; \limsup_{n \to +\infty} \Phi^0(x_n; y - x_n) + \psi(y) - \liminf_{n \to +\infty} \psi(x_n) \\ &\leq \; \Phi^0(x_0, y - x_0) + \psi(y) - \psi(x_0) \qquad \forall \, y \in X. \end{aligned}$$

Therefore $x_0 \in X$ is a critical point of $\varphi = \Phi + \psi$. Note that the inequality $\psi(x_0) \leq \liminf\limits_{n \to +\infty} \psi(x_n)$ cannot be strict, because otherwise putting $y = x_0$ we would obtain

$$0 \leq \limsup_{n \to +\infty} \Phi^0(x_n; x_0 - x_n) + \psi(x_0) - \liminf_{n \to +\infty} \psi(x_n) \; < \; 0,$$

a contradiction. So in fact

$$\psi(x_0) \; = \; \liminf_{n \to +\infty} \psi(x_n) \quad \text{and} \quad \varphi(x_0) \; = \; c,$$

i.e. $x_0 \in K_c^\varphi$. $\qquad$ ▯

COROLLARY 2.3.1
The set K_c^φ *is compact in* X.

As a consequence of Proposition 2.3.1, we obtain the first result on the existence of critical points for $\varphi = \Phi + \psi$.

PROPOSITION 2.3.2
If $\varphi = \Phi + \psi$ is bounded below and satisfies the generalized nonsmooth PS_c-condition where $c \stackrel{df}{=} \inf_X \varphi$,
then $K_c^\varphi \neq \emptyset$.

PROOF   We use the Ekeland Variational Principle (see Corollary 1.4.7) with $\varepsilon = \frac{1}{n}$ and $\lambda = 1$ and generate a sequence $\{x_n\}_{n \geq 1} \subseteq X$, such that $\varphi(x_n) \searrow c$ and

$$-\frac{1}{n}\|y - x_n\|_X \leq \varphi(y) - \varphi(x_n) \qquad \forall\, y \in X.$$

Let $y = (1-t)x_n + th$ with $h \in X$ and $t \in [0,1]$. Exploiting the convexity of ψ, we have

$$-\frac{1}{n}\|h - x_n\|_X \leq \frac{1}{t}\big(\Phi\big(x_n + t(h - x_n)\big) - \Phi(x_n)\big) + \psi(h) - \psi(x_n)$$

and so, passing to the limit as $t \searrow 0$, we get

$$-\frac{1}{n}\|h - x_n\|_X \leq \Phi^0(x_n; h - x_n) + \psi(h) - \psi(x_n).$$

Since φ satisfies the generalized nonsmooth PS_c-condition, it follows that by passing to a subsequence if necessary, we may assume that

$$x_n \longrightarrow x_0 \quad \text{in } X,$$

for some $x_0 \in X$. Then from Proposition 2.3.1 we conclude that $x_0 \in K_c^\varphi$. ☐

2.3.2   Generalized Deformation Theorem

The following proposition is a direct consequence of Definition 2.3.2.

PROPOSITION 2.3.3
If $\varphi = \Phi + \psi$ satisfies the generalized nonsmooth PS_c-condition, $\varepsilon_0 > 0$ and U is an open neighbourhood of K_c^φ,
then there exists $\varepsilon \in (0, \varepsilon_0)$, such that for all $x_0 \in \big(\varphi^{c+\varepsilon} \setminus \varphi^{c-\varepsilon}\big) \setminus U$ there exists $y_0 \in X$, such that

$$-3\varepsilon\|y_0 - x_0\|_X > \Phi^0(x_0; y_0 - x_0) + \psi(y_0) - \psi(x_0).$$

Using this proposition, we can prove the last auxiliary result, before formulating the Deformation Theorem for this particular setting.

PROPOSITION 2.3.4
If $\varphi = \Phi + \psi$ satisfies the generalized nonsmooth PS_c-condition, $\varepsilon_0 > 0$, U is a neighbourhood of K_c^φ (with $U = \emptyset$ whenever $K_c^\varphi = \emptyset$) and $\varepsilon > 0$ is such

as in Proposition 2.3.3,
<u>*then*</u> *for every* $x_0 \in \left(\varphi^{c+\varepsilon} \setminus \varphi^{c-\varepsilon}\right) \setminus U$ *there exist* $y_0 \in X$ *and* $r > 0$, *such that*

$$-3\varepsilon \left\| y_0 - z \right\|_X \geq \Phi^0(y; y_0 - y) + \psi(y_0) - \psi(z) \quad \forall y \in B_r(x_0), \ z \in B_{\frac{r}{2}}(x_0).$$

PROOF Let $x_0 \in \left(\varphi^{c+\varepsilon} \setminus \varphi^{c-\varepsilon}\right) \setminus U$ and let y_0 be as claimed in Proposition 2.3.3. Suppose that the proposition is not true. Then we can find two sequences $\{y_n\}_{n \geq 1}$ and $\{z_n\}_{n \geq 1}$ with $y_n \longrightarrow x_0$ and $z_n \longrightarrow x_0$, such that

$$\Phi^0(y_n; y_0 - y_n) + \psi(y_0) - \psi(z_n) \ > \ -3\varepsilon \left\| y_0 - z_n \right\|_X,$$

so

$$\limsup_{n \to +\infty} \Phi^0(y_n; y_0 - y_n) + \psi(y_0) - \liminf_{n \to +\infty} \psi(z_n) \ \geq \ -3\varepsilon \left\| y_0 - x_0 \right\|_X$$

and thus

$$\Phi^0(x_0; y_0 - x_0) + \psi(y_0) - \psi(x_0) \ \geq \ -3\varepsilon \left\| y_0 - x_0 \right\|_X,$$

which contradicts the choice of $y_0 \in X$ (see Proposition 2.3.3). ⬚

Now we are ready for the Deformation Theorem, which will eventually lead to minimax principles.

THEOREM 2.3.1 (Generalized Deformation Theorem)
If $\varphi = \Phi + \psi$ *satisfies the generalized nonsmooth* PS_c-*condition,* U *is an open neighbourhood of* K_c^{φ} *(with* $U = \emptyset$ *if* $K_c^{\varphi} = \emptyset$*) and* $\varepsilon_0 > 0$,
<u>*then*</u> *we can find* $\varepsilon \in (0, \varepsilon_0)$, *such that for each compact* $C \subseteq X \setminus U$ *with* $c - \varepsilon < \inf_C \varphi \leq c + \varepsilon$, *we can find a closed set* D *with* $C \subseteq \mathrm{int}\, D$, $t_0 > 0$ *and a function* $\eta \in C([0, t_0] \times X; X)$, *such that for all* $(t, x) \in [0, t_0] \times X$, *we have*

(a) $\left\| \eta(t, x) - x \right\|_X \leq t;$

(b) $\eta(0, x) = x;$

(c) $\varphi\big(\eta(t, x)\big) \leq \varphi(x);$

(d) *if* $x \in D$, *then* $\varphi\big(\eta(t, x)\big) - \varphi(x) \leq -2\varepsilon t;$

(e) $\sup_{y \in C} \varphi\big(\eta(t, y)\big) - \sup_{y \in C} \varphi(y) \leq -2\varepsilon t.$

PROOF Let $\varepsilon > 0$ be as in Proposition 2.3.3 and let C be any compact subset of $\left(\varphi^{c+\varepsilon} \setminus \varphi^{c-\varepsilon}\right) \setminus U$. Evidently

$$c - \varepsilon < \varphi(x) \qquad \forall\, x \in C.$$

Now for any $x_0 \in C$ we choose $r(x_0) > 0$ as in Proposition 2.3.4 and $y_0 \in X$ as in Proposition 2.3.3. Let $\bar{r}(x_0)$ be such that

$$0 < \bar{r}(x_0) \le \frac{r(x_0)}{4}, \qquad \bar{r}(x_0) < \|x_0 - y_0\|_X$$

(note that from Proposition 2.3.3 it is clear that $x_0 \ne y_0$) and finally such that

$$\varphi|_{B_{\bar{r}(x_0)}(x_0)} > c - \varepsilon.$$

The family $\left\{ B_{\bar{r}(x_0)}(x_0) \right\}_{x_0 \in C}$ is an open cover of C. Since C is compact, we can find a finite set $\{x_1, \ldots, x_N\} \subseteq C$, such that

$$C \subseteq \bigcup_{k=1}^{N} B_{\bar{r}(x_k)}(x_k).$$

Let $\{\xi_k\}_{k=1}^{N}$ be a locally Lipschitz partition of unity subordinate to this cover. Using Tietze's extension theorem (see Theorem A.1.2), we can extend $\sum_{k=1}^{N} \xi_k$ to all of X and have $\sum_{k=1}^{N} \xi_k(x) \le 1$ for all $x \in X$. We introduce the one-parameter family of functions $\eta_t \colon X \longrightarrow X$, $t \ge 0$, defined by

$$\eta_t(x) \stackrel{df}{=} x + t \sum_{k=1}^{N} \xi_k(x) \frac{y_k - x}{\|y_k - x\|_X} \qquad \forall \, x \in X,$$

where $y_k \in X$ is as in Proposition 2.3.3 corresponding to $\varepsilon > 0$ and $x_k \in C$. From the choice of $\bar{r}(x_k) > 0$, we see that η_t is well defined and of course locally Lipschitz. We shall show that $\eta(t, x) \stackrel{df}{=} \eta_t(x)$ is the desired function.

(a) First note that

$$\|\eta(t, x) - x\|_X \le t \sum_{k=1}^{N} \xi_k(x) \le t \qquad \forall \, x \in X,$$

which proves statement (a) of the theorem.

(b) This is an immediate consequence of (a).

(c) Using the Mountain Pass Theorem (see Proposition 1.3.14), for every $u \in X$ with $\|u\|_X \le 1$, we have

$$\Phi(x + tu) = \Phi(x) + \langle x^*, tu \rangle_X,$$

for some $x^* \in \Phi(x + \gamma u)$ and $0 < \gamma < t$ and so

$$\Phi(x + tu) \le \Phi(x) + t\Phi^0(x + \gamma u; u). \tag{2.43}$$

Let us put

$$u \stackrel{df}{=} \sum_{k=1}^{N} \xi_k(x) \frac{y_k - x}{\|y_k - x\|_X} \quad \text{and} \quad v \stackrel{df}{=} x + \gamma u.$$

From (2.43) and the sublinearity of $\Phi^0(x + \gamma u; \cdot)$, we have

$$\Phi(x + tu) \leq \Phi(x) + t\Phi^0 \left(x + \gamma u; \sum_{k+1}^{N} \xi_k(x) \frac{y_k - x}{\|y_k - x\|_X} \right)$$

$$\leq \Phi(x) + t \sum_{k=1}^{N} \frac{\xi_k(x)}{\|y_k - x\|_X} \Phi^0(x + \gamma u; y_k - x)$$

$$\leq \Phi(x) + t \sum_{k=1}^{N} \frac{\xi_k(x)}{\|y_k - x\|_X} \left(\Phi^0(v; y_k - v) + \Phi^0(v; v - x) \right)$$

$$\leq \Phi(x) + t \sum_{k=1}^{N} \frac{\xi_k(x)}{\|y_k - x\|_X} \left(\Phi^0(v; y_k - v) + l_0 \|v - x\|_X \right),$$

with $l_0 > 0$ being the Lipschitz constant of Φ in a neighbourhood of C. Note that since $v = x + \gamma u$ with $0 < \gamma < t$ and $\|u\|_X \leq 1$, we have that $\|v - x\|_X \leq t$. So we obtain

$$\Phi\big(\eta(t, x)\big) = \Phi(x + tu)$$

$$\leq \Phi(x) + t \sum_{k=1}^{N} \frac{\xi_k(x)}{\|y_k - x\|_X} \Phi^0(v; y_k - v) + t^2 \sum_{k=1}^{N} \frac{\xi_k(x)}{\|y_k - x\|_X} l_0$$

$$\leq \Phi(x) + t \sum_{k=1}^{N} \frac{\xi_k(x)}{\|y_k - x\|_X} \Phi^0(v; y_k - v) + \beta t^2 \sum_{k=1}^{N} \xi_k(x), \qquad (2.44)$$

with $\beta \stackrel{df}{=} \max \left\{ \frac{1}{\overline{r}(x_1)}, \ldots, \frac{1}{\overline{r}(x_N)} \right\} > 0$. We have

$$\eta(t, x) = x + tu = x + t \sum_{k=1}^{N} \xi_k(x) \frac{y_k - x}{\|y_k - x\|_X}$$

$$= \left(1 - t \sum_{k=1}^{N} \frac{\xi_k(x)}{\|y_k - x\|_X} \right) x + t \sum_{k=1}^{N} \frac{\xi_k(x)}{\|y_k - x\|_X} y_k.$$

Putting $t_1 \stackrel{df}{=} \min \{ \overline{r}(x_1), \ldots \overline{r}(x_N) \}$, we have that

$$t \sum_{k=1}^{N} \frac{\xi_k(x)}{\|y_k - x\|_X} \leq 1 \qquad \forall \, t \in [0, t_1].$$

Exploiting the convexity of ψ, we obtain

$$\psi\big(\eta(t,x)\big) \leq \left(1 - t\sum_{k=1}^{N}\frac{\xi_k(x)}{\|y_k - x\|_X}\right)\psi(x)$$

$$+ t\sum_{k=1}^{N}\frac{\xi_k(x)}{\|y_k - x\|_X}\psi(y_k) \qquad \forall\, t \in [0, t_1]. \qquad (2.45)$$

Combining (2.44) and (2.45), we have

$$\varphi\big(\eta(t,x)\big) \leq \varphi(x) + t\sum_{k=1}^{N}\frac{\xi_k(x)}{\|y_k - x\|_X}\big(\Phi^0(v; y_k - v) + \psi(y_k) - \psi(x)\big)$$

$$+ \beta t^2 \sum_{k=1}^{N}\xi_k(x).$$

Choose $t_2 \leq t_1$, such that if $t \leq t_2$ we have $v = x + \gamma u \in B_{r(x_i)}(x_i)$ whenever $u \in B_{\frac{r(x_i)}{2}}(x_i)$ (recall that $0 < \gamma < t$). So we can apply Proposition 2.3.4 (with $x_0 = x_k$; note that y_k was chosen for x_k according to this proposition) and obtain

$$\varphi\big(\eta(t,x)\big) \leq \varphi(x) + t\left(\sum_{k=1}^{N}\xi_k(x)\right)(\beta t - 3\varepsilon).$$

If we take $t_0 = \min\left\{t_2, \frac{\varepsilon}{\beta}\right\}$, then we see that

$$\varphi\big(\eta(t,x)\big) \leq \varphi(x) \qquad \forall\, (t,x) \in [0, t_0] \times X.$$

This proves statement (c) of the theorem.

(d) Also if $x \in C \subseteq \bigcup_{k=1}^{N} V_k \subseteq \overline{\bigcup_{k=1}^{N} V_k}$, we have that $\sum_{k=1}^{N}\xi_k(x) = 1$ and so

$$\varphi\big(\eta(t,x)\big) \leq \varphi(x) - 2\varepsilon t, \qquad (2.46)$$

which proves statement (d) of the theorem.

(e) Finally from (2.46) we deduce that

$$\sup_{x\in C}\varphi\big(\eta(t,x)\big) - \sup_{x\in C}\varphi(x) \leq -2\varepsilon t,$$

which proves statement (e) of the theorem.   ⬜

2.3.3   Minimax Principles

As was the case in the previous section, this theorem leads to minimax principles characterizing critical points of $\varphi = \Phi + \psi$. We start with a result which is the counterpart of Theorem 2.1.2 in this new setting. From this we will derive as byproducts versions of the Mountain Pass Theorem, Saddle Point Theorem and Linking Theorem.

THEOREM 2.3.2
If $B \subseteq X$ is nonempty, compact and convex, A is the relative boundary of B, $E \subseteq X$ is nonempty and closed, A and E link, $\varphi = \Phi + \psi$ satisfies the generalized nonsmooth PS_c-condition with

$$c \stackrel{df}{=} \inf_{\gamma \in \Gamma} \max_{x \in B} \varphi(\gamma(x)) < +\infty,$$

where

$$\Gamma \stackrel{df}{=} \{\gamma \in C(B; X) : \gamma|_A = id\}$$

and

$$\sup_A \varphi < \inf_E \varphi$$

then $c \geq \inf_E \varphi$ and $K_c^\varphi \neq \emptyset$.

PROOF   Let us put

$$\mu \stackrel{df}{=} \sup_A \varphi \quad \text{and} \quad \vartheta \stackrel{df}{=} \inf_E \varphi.$$

By hypotheses $\mu < \vartheta$. Consider the set $\Gamma \subseteq C(B; X)$ equipped with the metric

$$d_\infty(\gamma_1, \gamma_2) \stackrel{df}{=} \max_{x \in B} \|\gamma_1(x) - \gamma_2(x)\|_X \qquad \forall \gamma_1, \gamma_2 \in \Gamma.$$

Clearly (Γ, d_∞) is a complete metric space. We consider the function $h \colon \Gamma \longrightarrow \mathbb{R}$ defined by

$$h(\gamma) \stackrel{df}{=} \sup_{x \in B} \varphi(\gamma(x)) \qquad \forall \gamma \in \Gamma.$$

We claim that h is lower semicontinuous. To this end let $\gamma_n \longrightarrow \gamma$ in (Γ, d_∞) and suppose that $h(\gamma_n) \leq \lambda$ for all $n \geq 1$ with $\lambda \in \mathbb{R}$. We need to show that $h(\gamma) \leq \lambda$. For a given $\varepsilon > 0$, we can find $\widehat{x} \in B$, such that $h(\gamma) - \varepsilon \leq \varphi(\gamma(\widehat{x}))$. Note that $\gamma_n(\widehat{x}) \longrightarrow \gamma(\widehat{x})$ and since φ is lower semicontinuous, we have that

$$\varphi(\gamma(\widehat{x})) \leq \liminf_{n \to +\infty} \varphi(\gamma_n(\widehat{x})) \leq \liminf_{n \to +\infty} h(\gamma_n) \leq \lambda,$$

hence $h(\gamma) - \varepsilon \leq \lambda$. Let $\varepsilon \searrow 0$ to obtain $h(\gamma) \leq \lambda$ and conclude that h is lower semicontinuous on (Γ, d_∞).

Because A and E link, we have

$$h(\gamma) \geq \vartheta \qquad \forall \gamma \in \Gamma$$

and so $c \geq \vartheta$.

To prove that $K_c^\varphi \neq \emptyset$, we argue by contradiction. So suppose that $K_c^\varphi = \emptyset$. We apply Theorem 2.3.1 with $U = \emptyset$ and $c_0 - c - \mu > 0$. Let $\varepsilon \in (0, \varepsilon_0)$ be as postulated in Theorem 2.3.1 and additionally such that $\sup\limits_A \varphi < c - \varepsilon$. Let $\bar{\varepsilon} \in (0, \varepsilon)$. Because h is lower semicontinuous and bounded below (by ϑ), we can apply the Ekeland Variational Principle (see Corollary 1.4.7) and obtain $\widehat{\gamma} \in \Gamma$, such that

$$h(\widehat{\gamma}) \leq c + \bar{\varepsilon} \quad \text{and} \quad -\varepsilon d_\infty(\gamma, \widehat{\gamma}) \leq h(\gamma) - h(\widehat{\gamma}) \qquad \forall\, \gamma \in \Gamma. \qquad (2.47)$$

Set

$$C \overset{df}{=} \widehat{\gamma}(B) \cap \{ x \in B : \; c - \bar{\varepsilon} \leq \varphi(\widehat{\gamma}(x)) \leq c + \bar{\varepsilon} \}.$$

Note that $\widehat{\gamma}(B)$ is compact and $\psi|_{\widehat{\gamma}(B)}$ is lower semicontinuous, convex and bounded. Therefore $\psi|_{\widehat{\gamma}(B)}$ is continuous (in fact locally Lipschitz; see Theorem 1.3.2) and so we infer that C is compact. Moreover, from the choice of $\varepsilon, \bar{\varepsilon} > 0$ we see that $C \cap A = \emptyset$ and $C \subseteq \varphi^{c+\varepsilon} \setminus \varphi^{c-\varepsilon}$. We apply Theorem 2.3.1 with U, ε_0, C as above, to obtain $t_0 > 0$ and the continuous deformation $\eta(t, x)$ satisfying all postulates of Theorem 2.3.1.

Let $\widetilde{t} \in (0, t_0)$ and $\widetilde{\gamma} \overset{df}{=} \eta_{\widetilde{t}} \circ \widehat{\gamma}$ (where $\eta_t = \eta(t, \cdot)$). In the proof of Theorem 2.3.1 we can always take the cover $\{V_k\}_{k=1}^N$ of C, to satisfy $A \cap \left(\bigcup\limits_{k=1}^N V_k \right) = \emptyset$. So we have that

$$\eta_t|_A = id_A \qquad \forall\, t \in [0, t_0]$$

(see the definition of $\eta(t, x)$ in the proof of Theorem 2.3.1), hence $\widetilde{\gamma} \in \Gamma$. From Theorem 2.3.1(a), we have that $d_\infty(\widetilde{\gamma}, \widehat{\gamma}) \leq \widetilde{t}$. Note that

$$h(\widetilde{\gamma}) = \sup_{x \in B} \varphi(\widetilde{\gamma}(x)) = \sup_{x \in B} \varphi(\eta_{\widetilde{t}}(\widehat{\gamma}(x))) = \sup_{y \in C} \varphi(\eta_{\widetilde{t}}(y)).$$

Similarly we have $h(\widehat{\gamma}) = \sup\limits_C \varphi$. Then by virtue of Theorem 2.3.1(d), we have that

$$h(\widetilde{\gamma}) - h(\widehat{\gamma}) = \sup_{y \in C} \varphi(\eta_{\widetilde{t}}(y)) - \sup_{y \in C} \varphi(y) \leq -2\varepsilon\widetilde{t} \leq -2\varepsilon d_\infty(\widetilde{\gamma}, \widehat{\gamma}),$$

which contradicts (2.47). So $K_c^\varphi \neq \emptyset$ and this proves the theorem. $\qquad \square$

Now by suitable choice of the sets A, B and E, from the above minimax principle, we shall derive versions of the Mountain Pass Theorem, Saddle Pass Theorem and Linking Theorem.

THEOREM 2.3.3 (Generalized Nonsmooth Mountain Pass Theorem)

If $\varphi = \Phi + \psi$, *there exist* $x_1 \in X$ *and* $r > 0$, *such that* $\|x_1\|_X > r$,

$$\max\{\varphi(0), \varphi(x_1)\} < \vartheta,$$

where

$$\vartheta \stackrel{df}{=} \inf \big\{\varphi(x) : \; \|x\|_X = r\big\},$$

φ *satisfies the generalized nonsmooth* PS_{c_0}*-condition with*

$$c_0 \stackrel{df}{=} \inf_{\xi \in \Gamma_0} \sup_{t \in [0,1]} \varphi\big(\gamma(t)\big) < +\infty,$$

where

$$\Gamma_0 \stackrel{df}{=} \big(\xi \in C\big([0,1]; X\big) : \; \xi(0) = 0, \; \xi(1) = x_1\big),$$

then $c_0 \geq \vartheta$ *and* $K_c^\varphi \neq \emptyset$.

PROOF   Let

$$A \stackrel{df}{=} \{0, x_1\},$$

$$B \stackrel{df}{=} [0, x_1] = \big\{tx_1 : \; t \in [0,1]\big\},$$

$$E \stackrel{df}{=} \partial B_r = \big\{x \in X : \; \|x\|_X = r\big\}.$$

The sets A and E link (see Example 2.1.1(a)), A, B and E are closed and B is compact. Let

$$\Gamma \stackrel{df}{=} \big\{\gamma \in C(B; X) : \; \gamma(0) = 0, \; \gamma(x_1) = x_1\big\}.$$

If $\xi \in \Gamma_0$, we set $\gamma_\xi(tx_1) \stackrel{df}{=} \xi(t)$ and then clearly $\gamma_\xi \in \Gamma$.

On the other hand if $\gamma \in \Gamma$, then if we set $\xi_\gamma(t) \stackrel{df}{=} \gamma(tx_1)$ for $t \in [0,1]$, we have $\xi_\gamma \in \Gamma_0$. Therefore, it follows that $c = c_0$ and by virtue of Theorem 2.3.2 we have that $c_0 \geq \vartheta$ and $K_{c_0}^\varphi \neq \emptyset$.   ▯

REMARK 2.3.3   The point $x_0 = 0$ was chosen for convenience. In fact we can have any other $x_0 \in X$ with $\|x_1 - x_0\|_X > r$ and in this case we put

$$\vartheta \stackrel{df}{=} \inf_{\substack{x \in X \\ \|x - x_0\|_X = r}} \varphi(x).$$

▯

THEOREM 2.3.4 (Generalized Nonsmooth Saddle Point Theorem)

If $\varphi = \Phi + \psi$, $X = Y \oplus V$ *with* $\dim Y < +\infty$, *there exist* $r > 0$, *such that*

$$\sup_{\partial B_r \cap Y} \varphi \; < \; \inf_V \varphi,$$

φ satisfies the generalized nonsmooth PS_c-condition with

$$c \overset{df}{=} \inf_{\gamma \in \Gamma} \sup_{x \in \overline{B}_r \cap Y} \varphi(\gamma(x)) < +\infty,$$

where

$$\Gamma \overset{df}{=} \left\{ \gamma \in C\left(\overline{B}_r \cap Y; X\right) : \ \gamma|_{\partial B_r \cap Y} = id \right\}$$

then $c \geq \inf_V \varphi$ and $K_c^\varphi \neq \emptyset$.

PROOF   Let $A \overset{df}{=} \partial B_r \cap Y$, $B \overset{df}{=} \overline{B}_r \cap Y$ and $E \overset{df}{=} V$. The sets A and E link (see Example 2.1.1(b)). So we can apply Theorem 2.3.2 and finish the proof.   ▯

THEOREM 2.3.5
If $\varphi = \Phi + \psi$, $X = Y \oplus V$ with $\dim Y < +\infty$, there exist $0 < r < R$ and $v_0 \in V$ with $\|v_0\|_X = 1$, such that

$$\sup_{\partial Q} \varphi \ < \ \inf_{\partial B_r \cap V} \varphi,$$

where

$$Q \overset{df}{=} \left\{ y + \lambda v_0 : \ y \in Y, \ \|y\|_X \leq R, \ 0 \leq \lambda \leq R \right\},$$
$$\partial Q \ = \ \left\{ y + \lambda v_0 : \ y \in Y \text{ and } \|y\|_X = R, \ \lambda \in [0, R] \right.$$
$$\left. \text{or } \|y\|_X \leq R, \ \lambda \in \{0, R\} \right\},$$

φ satisfies the generalized nonsmooth PS_c-condition with

$$c \overset{df}{=} \inf_{\gamma \in \Gamma} \sup_{x \in Q} \varphi(\gamma(x)),$$

where

$$\Gamma \overset{df}{=} \left\{ \gamma \in C(Q; X) : \ \gamma|_{\partial Q} = id \right\},$$

then $c \geq \inf_{\partial B_r \cap V} \varphi$ and $K_c^\varphi \neq \emptyset$.

PROOF   Let $A \overset{df}{=} \partial Q$, $B \overset{df}{=} Q$ and $E \overset{df}{=} \partial \overline{B}_r \cap V$. We know that A and E link (see Example 2.1.1(c)). So we can apply Theorem 2.3.2 and set the proof.   ▯

Imposing some symmetry condition on φ, we can have a multiplicity result for functionals $\varphi = \Phi + \psi$. The proof of the result is similar to that of Theorem 2.1.7 and so it is omitted.

THEOREM 2.3.6
<u>If</u> *the functional* $\varphi = \Phi + \psi$ *satisfies the generalized nonsmooth* PS*-condition,* $\overline{\varphi}(0) = 0$, Φ *and* ψ *are even and*

(i) *there exists a subspace* V *of* X *of finite codimension and numbers* $\beta, r \geq 0$, *such that* $\varphi|_{V \cap \partial B_r} \geq \beta$;

(ii) *there exists a finite dimensional subspace* Y *of* X, $\dim Y > \operatorname{codim} V$ *such that* $\varphi(y) \longrightarrow -\infty$ *as* $\|y\|_X \to +\infty$ *with* $y \in Y$,

<u>then</u> φ *has at least* $\dim Y - \operatorname{codim} V$ *distinct pairs of nontrivial critical points.*

An interesting consequence of this Theorem is the following corollary.

COROLLARY 2.3.2
<u>If</u> *the hypotheses of Theorem 2.3.6 hold with (ii) replaced by*

(ii)' *for any positive integer* m *there is an* m-*dimensional subspace* Y *of* X *such that* $\varphi(y) \longrightarrow -\infty$ *as* $\|y\|_X \to +\infty$ *with* $y \in Y$,

<u>then</u> φ *has infinitely many distinct pairs of nontrivial critical points.*

2.4   Local Linking and Extensions

In this section, we derive a multiplicity result based on the notion of local linking at 0 and we also present some extensions of the notions and results of the nonsmooth critical point theory for locally Lipschitz functions.

2.4.1   Local Linking

The mathematical setting is the same as in Section 2.1. Namely X is a reflexive Banach space and $\varphi \colon X \longrightarrow \mathbb{R}$ is a locally Lipschitz function.

DEFINITION 2.4.1   *Suppose that* $X = X_1 \oplus X_2$. *We say that* φ *has a* **local linking at** 0, *if for some* $r > 0$ *we have*

$$\begin{cases} \varphi(x) \geq 0 \ x \in X_1, \ \|x\|_X \leq r, \\ \varphi(x) \leq 0 \ x \in X_2, \ \|x\|_X \leq r. \end{cases}$$

REMARK 2.4.1   If $\varphi \in C^1(X)$, then this condition implies that 0 is a critical point of φ. ▯

If φ is bounded below and satisfies the nonsmooth C-condition, then by virtue of Theorem 2.1.6 we know that there exists a point $y_0 \in X$, such that $\varphi(y_0) = \inf_X \varphi$. If $\inf_X \varphi < \varphi(0)$, then $y_0 \neq 0$. As before, for $c \in \mathbb{R}$, we set

$$\varphi^c \stackrel{df}{=} \{x \in X : \varphi(x) \le c\}.$$

If $c \ge \inf_X \varphi$, then $\varphi^c \neq \emptyset$.

PROPOSITION 2.4.1
If φ is bounded below, $\varphi(0) = 0$, φ satisfies the nonsmooth C-condition, $\inf_X \varphi < \min\{0, c\}$, $y_0 \neq 0$ is a minimizer of φ, $\{y_0, 0\}$ are the only critical points of φ,
then for any neighbourhood U of y_0 and any $\delta > 0$ such that $U \cap B_\delta = \emptyset$, we can find $\gamma > 0$ for which

$$\gamma \le (1 + \|x\|_X) m^\varphi(x) \qquad \forall\, x \in \varphi^c \setminus (U \cup B_\delta).$$

PROOF   Suppose that the proposition is not true. Then we can find $x_n \in \varphi^c \setminus (U \cup B_\delta)$, such that

$$(1 + \|x_n\|_X) m^\varphi(x_n) \longrightarrow 0.$$

Since φ satisfies the nonsmooth C-condition, passing to a subsequence if necessary, we may assume that $x_n \longrightarrow x_0$ in X, for some $x_0 \in \varphi^c \setminus (U \cup B_\delta)$ (note that $\varphi^c \setminus (U \cup B_\delta)$ is closed). From lower semicontinuity of m^φ (see Proposition 2.1.3), we have

$$m^\varphi(x_0) \le \liminf_{n \to +\infty} m^\varphi(x_n)$$

and because $m^\varphi(x_0) \ge 0$ we infer that $m^\varphi(x_0) = 0$. Hence $0 \in \partial\varphi(x_0)$ (i.e. $x_0 \in X$ is a critical point of φ) and since by hypothesis y_0 and 0 are the only critical points of φ, it follows that $x_0 = y_0 \in U$ or $x_0 = 0 \in B_\delta$. But $x_0 \in \varphi^c \setminus (U \cup B_\delta)$, a contradiction. $\quad\Box$

PROPOSITION 2.4.2
If the hypotheses of Proposition 2.4.1 hold,
then there exists a locally Lipschitz map $v \colon \varphi^c \setminus (U \cup B_\delta) \longrightarrow X$, such that

$$\begin{cases} \|v(x)\|_X \le 1 + \|x\|_X \\ \langle x^*, v(x) \rangle_X \ge \frac{\gamma}{2} \end{cases} \qquad \forall\, x \in \varphi^c \setminus (U \cup B_\delta),\ x^* \in \partial\varphi(x),$$

where $\gamma > 0$ is as in Proposition 2.4.1.

PROOF Let $D \stackrel{df}{=} \varphi^c \setminus (U \cup B_\delta)$ and let us fix $x \in D$. Let

$$B_{m^\varphi(x)}^{X^*} \stackrel{df}{=} \{z^* \in X^* : \|z^*\|_{X^*} < m^\varphi(x)\}.$$

We have $B_{m^\varphi(x)}^{X^*} \cap \partial\varphi(x) = \emptyset$. Since both sets are convex and $B_{m^\varphi(x)}^{X^*}$ is open, we can apply the Weak Separation Theorem for convex sets (see Theorem A.3.3) and find $u(x) \in X$ with $\|u(x)\|_X = 1$, such that

$$\langle z^*, u(x) \rangle_X \leq \langle x^*, u(x) \rangle_X \quad \forall z^* \in B_{m^\varphi(x)}^{X^*}, \; x^* \in \partial\varphi(x).$$

Note that

$$\sup_{z^* \in B_{m^\varphi(x)}^{X^*}} \langle z^*, u(x) \rangle_X = m^\varphi(x).$$

Hence by virtue of Proposition 2.4.1, we have that

$$\frac{\gamma}{2(1 + \|x\|_X)} < m^\varphi(x) \leq \langle x^*, u(x) \rangle_X \quad \forall x^* \in \partial\varphi(x). \qquad (2.48)$$

Set

$$V \stackrel{df}{=} \left\{y^* \in X^* : \frac{\gamma}{2} < \langle y^*, u(x) \rangle_X \right\}.$$

Evidently V is weakly open in X^*. We know that the multifunction $y \longmapsto (1 + \|y\|_X)\partial\varphi(y)$ is upper semicontinuous from X into X_w^*. Also from (2.48) we see that

$$(1 + \|x\|_X)\partial\varphi(x) \subseteq V.$$

So we can find $r(x) > 0$, such that

$$(1 + \|y\|_X)\partial\varphi(y) \subseteq V \quad \forall y \in D \cap B_{r(x)}(x)$$

and so

$$\frac{\gamma}{2(1 + \|y\|_X)} < \langle y^*, u(x) \rangle_X \quad \forall y \in D \cap B_{r(x)}(x), \; y^* \in \partial\varphi(y). \quad (2.49)$$

The collection $\{B_{r(x)}(x)\}_{x \in D}$ is an open cover of D. By paracompactness we can find a locally finite refinement $\{U_i\}_{i \in I}$ and a locally Lipschitz partition of unity $\{\xi_i\}_{i \in I}$ subordinate to it. For each $i \in I$, we can find $x_i \in D$, such that $U_i \subseteq B_{r(x_i)}(x_i)$. Moreover, to this $x_i \in D$ corresponds an element $u_i = u(x_i) \in X$ with $\|u_i\|_X = 1$, for which (2.48) holds with $x = x_i$. Now let $v \colon D \longrightarrow X$ be the map defined by

$$v(x) \stackrel{df}{=} (1 + \|x\|_X) \sum_{i \in I} \xi_i(x) u_i \quad \forall x \in D.$$

Evidently this is a well defined locally Lipschitz map and

$$\|v(x)\|_X \leq 1 + \|x\|_X \quad \forall x \in D.$$

In addition, from (2.49) we see that

$$\langle x^*, v(x)\rangle_X = \sum_{i\in I} \xi_i(x)\left(1+\|x\|_X\right)\langle x^*, u_i\rangle_X \geq \frac{\gamma}{2} \qquad \forall\, x \in D,\ x^* \in \partial\varphi(x).$$

$$\square$$

We continue with the hypotheses of Proposition 2.4.1 in effect. Let $\vartheta > 0$ be small enough so that

$$\varphi(x) \leq 0 \qquad \forall\, x \in \overline{B}_\vartheta(y_0).$$

This is possible since $\varphi(y_0) = \inf \varphi < \varphi(0) = 0$. Moreover, we choose $\widetilde{\delta} > 0$ small so that

$$\varphi(y_0) + \widetilde{\delta} < 0, \quad U \subseteq B_\vartheta(y_0) \quad \text{and} \quad U \cap B_{\widetilde{\delta}} = \emptyset,$$

where

$$U \stackrel{df}{=} \left\{x \in X : \varphi(x) < \varphi(y_0) + \widetilde{\delta}\right\}.$$

Indeed, if no such $\widetilde{\delta} > 0$ exists, we can find $\{x_n\}_{n\geq 1} \subseteq X$, such that $\varphi(x_n) \searrow \inf_X \varphi = \varphi(y_0)$ and $\|x_n - y_0\|_X \geq \vartheta$ for all $n \geq 1$. Using Theorem 1.4.9, we can assume without any loss of generality that

$$\left(1 + \|x_n\|_X\right) m^\varphi(x_n) \longrightarrow 0.$$

Because φ satisfies the nonsmooth C-condition, we have that $x_n \longrightarrow x_0$ with $x_0 \neq y_0$ (because $\|x_n - y_0\|_X \geq \vartheta > 0$ for all $n \geq 1$) and $x_0 \neq 0$ (since $U \cap B_\delta = \emptyset$). Then we have

$$m^\varphi(x_0) \leq \liminf_{n\to+\infty} m^\varphi(x_n) = 0,$$

hence $m^\varphi(x_0) = 0$ and so x_0 is a critical point of φ, distinct from y_0 and 0, a contradiction to the hypothesis that y_0 and 0 are the only critical points of φ.

Without any loss of generality we may assume that $1 < \|y_0\|_X$ (recall $y_0 \neq 0$). Also we assume that

$$X = Y \oplus V \quad \text{with} \quad \dim Y < +\infty.$$

Choose $\widetilde{c} > 0$, such that

$$\mathrm{int}\left[\varphi^{\widetilde{c}} \setminus \left(U \cup B_{\widetilde{\delta}}\right)\right] \neq \emptyset \quad \text{and} \quad \widetilde{z} \in \mathrm{int}\left[\varphi^{\widetilde{c}} \setminus \left(U \cup B_{\widetilde{\delta}}\right)\right],$$

with $\varphi(\widetilde{z}) \leq -\widetilde{k}\widetilde{\delta} < 0 = \varphi(0)$, where $\widetilde{k} > 0$ is the Lipschitz constant of φ on $B_{\widetilde{\delta}}$. Let

$$D \stackrel{df}{=} \varphi^{\widetilde{c}} \setminus \left(U \cup B_{\widetilde{\delta}}\right).$$

Now let $v \colon D \longrightarrow X$ be the locally Lipschitz map obtained in Proposition 2.4.2. We consider the following Cauchy problem:

$$
\begin{cases}
\dfrac{d\eta(t)}{dt} = -\dfrac{v\big(\eta(t)\big)}{\big\|v\big(\eta(t)\big)\big\|_X^2} & \text{on } \mathbb{R}_+, \\
\eta(0) = \widetilde{z}.
\end{cases}
\tag{2.50}
$$

Because v is locally Lipschitz, we know that problem (2.50) has a unique local flow.

PROPOSITION 2.4.3

If the hypotheses of Proposition 2.4.1 hold and $\widetilde{c} > 0$, $\widetilde{z} \in X$ and $\widetilde{\delta} > 0$ are chosen as above,
then there exists a finite time $\tau(\widetilde{z}) < +\infty$, such that the flow of (2.50) exists on $[0, \tau(\widetilde{z})]$ and $\varphi\big(\eta(\tau(\widetilde{z}))\big) = \varphi(y_0) + \widetilde{\delta}$.

PROOF Because $\widetilde{z} \in \operatorname{int} D$ (with $D \stackrel{df}{=} \varphi^{\widetilde{c}} \setminus \big(U \cup B_{\widetilde{\delta}}\big)$), the solution of the problem (2.50) exists on a maximal open interval $[0, \tau(\widetilde{z}))$.

Note that for all $t \in [0, \tau(\widetilde{z}))$, $\eta(t) \in D$. By virtue of Corollary 2.1.1, φ is coercive. Hence D is bounded and then so is $\partial\varphi(D) \subseteq X^*$. Therefore we can find $\gamma_1 > 0$, such that

$$
\|x^*\|_{X^*} \le \gamma_1 \qquad \forall\, t \in [0, \tau(\widetilde{z})),\ x^* \in \partial\varphi\big(\eta(t)\big).
\tag{2.51}
$$

From Proposition 2.4.2, we have that

$$
\big\langle x^*, v\big(\eta(t)\big)\big\rangle_X \ge \frac{\gamma}{2} \qquad \forall\, t \in [0, \tau(\widetilde{z})),\ x^* \in \partial\varphi\big(\eta(t)\big)
$$

and so

$$
\|x^*\|_{X^*} \big\|v\big(\eta(t)\big)\big\|_X \ge \frac{\gamma}{2} \qquad \forall\, t \in [0, \tau(\widetilde{z})),\ x^* \in \partial\varphi\big(\eta(t)\big).
$$

From (2.51), we have that

$$
\big\|v\big(\eta(t)\big)\big\|_X \ge \frac{\gamma}{2\gamma_1} \qquad \forall\, t \in [0, \tau(\widetilde{z})).
$$

Note that the function $t \longmapsto \varphi\big(\eta(t)\big)$ is locally Lipschitz and so it is differentiable almost everywhere on $[0, \tau(\widetilde{z}))$. From Proposition 1.3.18 we know that

$$
\begin{aligned}
\frac{d}{dt}\varphi\big(\eta(t)\big) &\le \max_{x^* \in \partial\varphi(\eta(t))} \big\langle x^*, \eta'(t)\big\rangle_X \\
&= \max_{x^* \in \partial\varphi(\eta(t))} \Big\langle x^*, -\frac{v(\eta(t))}{\|v(\eta(t))\|_X^2}\Big\rangle_X \le -\frac{\gamma}{2\gamma_1} \quad \text{for a.a. } t \in \big[0, \tau(\widetilde{z})\big).
\end{aligned}
$$

From this inequality we infer that the function $t \longmapsto \varphi(\eta(t))$ is strictly decreasing on $[0, \tau(\widetilde{z}))$ and

$$\tau(\widetilde{z}) \leq \frac{2\gamma_1}{\gamma}\left(\varphi(z) - \inf_X \varphi\right) < +\infty.$$

We have

$$\varphi(y_0) + \widetilde{\delta} \leq \varphi(\eta(t)) < \varphi(\widetilde{z}) \qquad \forall\, t \in (0, \tau(\widetilde{z})).$$

Therefore $\int_0^{\tau(\widetilde{z})} \eta'(t)\, dt$ exists and this means that $\lim_{t \to \tau(\widetilde{z})} \eta(t) = \eta(\tau(\widetilde{z}))$ exists. Clearly we must have that $\eta(\tau(\widetilde{z})) \in \partial(\varphi^{\widetilde{c}} \cap (U \cup B_{\widetilde{\delta}})^c)$. We know that

$$\partial\left(\varphi^{\widetilde{c}} \cap (U \cup B_{\widetilde{\delta}})^c\right) \subseteq \partial\varphi^{\widetilde{c}} \cup \partial\left((U \cup B_{\widetilde{\delta}})^c\right).$$

Note that $\varphi(\eta(\tau(\widetilde{z}))) < \varphi(\widetilde{z}) < \widetilde{c}$ and so $\eta(\tau(\widetilde{z})) \notin \partial\varphi^{\widetilde{c}}$. So we must have that

$$\eta(\tau(\widetilde{z})) \in \partial(U \cup B_{\widetilde{\delta}})^c = \partial\left(U^c \cap B_{\widetilde{\delta}}^c\right) \subseteq \partial U^c \cup \partial B_{\widetilde{\delta}}^c.$$

If $\eta(\tau(\widetilde{z})) \in \partial B_{\widetilde{\delta}}^c = \partial B_{\widetilde{\delta}}$, then $\|\eta(\tau(\widetilde{z}))\|_X = \widetilde{\delta}$. Because by hypothesis $\varphi(0) = 0$ and $\varphi|_{\overline{B}_{\widetilde{\delta}}}$ is Lipschitz continuous with constant $\widetilde{k} > 0$, we have $|\varphi(\eta(\tau(\widetilde{z})))| \leq \widetilde{k}\widetilde{\delta}$, hence $-\widetilde{k}\widetilde{\delta} \leq \varphi(\eta(\tau(\widetilde{z}))) \leq \widetilde{k}\widetilde{\delta}$. But recall that from the choice of $\widetilde{z}$, we have $\varphi(\eta(\tau(\widetilde{z}))) < \varphi(\widetilde{z}) \leq -\widetilde{k}\widetilde{\delta}$, a contradiction. This shows that $\eta(\tau(\widetilde{z})) \notin \partial B_{\widetilde{\delta}}^c$ and so we must have $\eta(\tau(\widetilde{z})) \in \partial U^c = \partial U$. From this we conclude that $\varphi(\eta(\tau(\widetilde{z}))) = \varphi(y_0) + \widetilde{\delta}$. $\qquad\qquad\square$

To have a multiplicity result based on the notion of local linking (see Definition 2.4.1), we shall need two more auxiliary results. For $a \leq b$, we put

$$R_{a,b} \overset{df}{=} \{x \in X : a \leq \|x\|_X \leq b\}.$$

PROPOSITION 2.4.4
If $0 \leq a \leq b$, $\varphi\colon R_{a,b} \longrightarrow \mathbb{R}$ *is a locally Lipschitz function satisfying the nonsmooth C-condition on every closed subset of* int $R_{a,b}$, φ *does not have any critical points in* int $R_{a,b}$ *and*

$$\vartheta(r) \overset{df}{=} \inf_{\|x\|_X = r} \varphi(x) \qquad \forall\, r \in [a, b],$$

then

$$\vartheta(r) > \min\{\vartheta(r_1), \vartheta(r_2)\} \qquad \forall\, a < r_1 < r < r_2 < b.$$

PROOF   Note that since the set $R_{a,b}$ is bounded, the nonsmooth C-condition and the nonsmooth PS-condition are equivalent on $R_{a,b}$. Suppose that the conclusion of the proposition is not true. So we can find $r \in (r_1, r_2)$, such that $\vartheta(r) \leq \min\{\vartheta(r_1), \vartheta(r_2)\}$. Let $\{x_n\}_{n \geq 1} \subseteq \mathbb{R}$ be a sequence, such that $\varphi(x_n) < \vartheta(r) + \frac{1}{n^2}$. By virtue the Ekeland Variational Principle (see Theorem 1.4.8) applied on the ring R_{r_1, r_2}, we can find $y_n \in R_{r_1, r_2}$, such that

$$\varphi(v) \geq \varphi(y_n) - \frac{1}{n} \|v - y_n\|_X \qquad \forall \, v \in R_{r_1, r_2}$$

and

$$\varphi(y_n) \leq \varphi(x_n) - \frac{1}{n} \|x_n - y_n\|_X \qquad \forall \, n \geq 1.$$

We claim that $y_n \notin \partial R_{r_1, r_2}$. Indeed, if $y_n \in \partial R_{r_1, r_2}$, say $\|y_n\|_X = r_1$, then we have

$$\vartheta(r_1) \leq \varphi(x_n) - \frac{1}{n}(r - r_1) \leq \vartheta(r) + \frac{1}{n^2} - \frac{1}{n}(r - r_1).$$

From this inequality and for $n \geq 1$ large, we have a contradiction to the hypothesis that $\vartheta(r) \leq \min\{\vartheta(r_1), \vartheta(r_2)\}$. So it follows that $y_n \notin \partial R_1$ for at least all large n, let us say for $n \geq n_0$. Then for all $n \geq n_0$, all $h \in X$ and all $t \in (0, 1)$ small, we have $v = y_n + th \in R_{r_1, r_2}$ and so

$$-\frac{t}{n} \|h\|_X \leq \varphi(y_n + th) - \varphi(y_n).$$

Thus

$$-\|h\|_X \leq n \frac{\varphi(y_n + th) - \varphi(y_n)}{t}$$

and so

$$-\|h\|_X \leq n \varphi^0(y_n; h) \qquad \forall \, h \in X, \, n \geq n_0.$$

Invoking Lemma 1.3.2, we obtain $u_n^* \in X^*$ with $\|u_n^*\|_{X^*} \leq 1$, such that

$$\langle u_n^*, h \rangle_X \leq n \varphi^0(y_n; h) \qquad \forall \, h \in X, \, n \geq n_0,$$

so $\frac{1}{n} u_n^* \in \partial \varphi(x_n)$ and $m^\varphi(y_n) \leq \frac{1}{n} \longrightarrow 0$.

So, by passing to a subsequence if necessary, we may assume that $y_n \longrightarrow y \in R_{r_1, r_2}$ and $m^\varphi(y) = 0$, i.e. $0 \in \partial \varphi(y)$, a contradiction to the hypothesis that φ has no critical points in int $R_{a, b}$. $\qquad\square$

PROPOSITION 2.4.5

If $R > 0$, $\varphi \colon \overline{B}_R \longrightarrow \mathbb{R}$ is a locally Lipschitz function satisfying the non-smooth C-condition on every closed subset of B_R, $\varphi(0) = 0$, φ has no critical points on B_R,

$$\varphi(x) > 0 \qquad \forall \, 0 < \|x\|_X < R$$

and

$$\vartheta(r) \overset{df}{=} \inf_{\|x\|_X = r} \varphi(x) \qquad \forall \, r \in [0, R],$$

then *there exists* $r_0 \in (0, R]$, *such that* $\vartheta(r)$ *is strictly increasing on* $[0, r_0)$ *and strictly decreasing on* $[r_0, R)$.

PROOF   It is easy to check that ϑ is upper semicontinuous on $[0, R']$, for any $R' < R$. So we can find $r_0 \in [0, R']$, such that

$$\vartheta(r_0) = \max_{0 \le r \le R'} \vartheta(r).$$

Then the result of this proposition follows from Proposition 2.4.4 by letting $R' \longrightarrow R$.   ⬚

Now we are ready for the multiplicity result when local linking is present.

THEOREM 2.4.1
If $X = Y \oplus V$ *with* $\dim Y < +\infty$, φ *is bounded below, satisfies the nonsmooth* $\bar{C}$*-condition,* $\varphi(0) = 0$, $\inf_X \varphi < 0$ *and there exists* $r > 0$, *such that*

$$\begin{cases} \varphi(x) \le 0 \text{ if } x \in Y, \ \|x\|_X \le r, \\ \varphi(x) \ge 0 \text{ if } x \in V, \ \|x\|_X \le r, \end{cases}$$

then φ *has at least two nontrivial critical points.*

PROOF   From Theorem 2.1.6, we know that there exists $y_0 \in X$, such that $\varphi(y_0) = \inf_X \varphi$. Since $\inf_X \varphi < 0 = \varphi(0)$, we infer that $y_0 \ne 0$. Suppose that y_0 and 0 are the only critical points of φ.

Case 1. $\dim Y > 0$ and $\dim V > 0$.

Without any loss of generality, we may assume that $r = 1 < \|y_0\|_X$. Let $v_0 \in V$ be such that $\|v_0\|_X = 1$. We introduce the set

$$E \stackrel{df}{=} \{ x \in X : x = \lambda v_0 + y, \ y \in Y, \ \lambda \ge 0, \ \|x\|_X \le 1 \}.$$

If $x \in \partial E$, $x \ne v_0$, we have $\|x\|_X = 1$ and we can write this element as $x = \lambda v_0 + \mu y$, with $0 \le \lambda \le 1$, $y \in Y$ with $\|y\|_X = 1$ and $0 < \mu \le 1$. Let $\tilde{c} > 0$, $\tilde{\delta} > 0$ be such as in Proposition 2.4.3. Evidently by choosing $c > \tilde{c}$ large and $\delta \in (0, \tilde{\delta})$ small, we can guarantee that

$$y \in \text{int } D \qquad \forall \, y \in Y, \ \|y\|_X = 1,$$

with

$$D \stackrel{df}{=} \varphi^c \setminus (U \cup B_\delta) \quad \text{and} \quad U \stackrel{df}{=} \{ x \in X : \varphi(x) < \varphi(y_0) + \delta \}.$$

Let η be the flow of (2.50), with $\tilde{z} = y$ and $\tau(y)$ as obtained in Proposition 2.4.3. So we can define the map $p^* \colon \partial E \longrightarrow X$ by

$$p^*(\lambda v_0 + \mu y) = \begin{cases} \eta(2\lambda\tau(y)) & \text{if } \lambda \in \left[0, \tfrac{1}{2}\right], \\ (2\lambda - 1)y_0 + (2 - 2\lambda)\eta(\tau(y)) & \text{if } \lambda \in \left(\tfrac{1}{2}, 1\right]. \end{cases}$$

Clearly p^* is continuous and

$$\begin{aligned} p^*(v_0) &= y_0 \\ p^*(y) &= y & \text{if } y \in Y \text{ with } \|y\|_X \leq 1 \\ \varphi(p^*(x)) &\leq 0 & \forall\, x \in \partial E. \end{aligned}$$

Indeed, if $x = y \in Y$ with $\|y\|_X = 1 = r$, then from the local linking hypothesis, we have that

$$\varphi(p^*(x)) = \varphi(y) \leq 0.$$

If $x = v_0$, then

$$\varphi(p^*(x)) = \varphi(p^*(v_0)) = \varphi(y_0) = \inf_X \varphi < 0.$$

If $x = \lambda v_0 + \mu y$ with $\lambda \in \left[0, \tfrac{1}{2}\right]$, then

$$\varphi(p^*(x)) = \varphi(p^*(\lambda v_0 + \mu y)) = \varphi(\eta(2\lambda\tau(y))) < \varphi(\eta(y)) \leq 0,$$

since $y \in Y$ with $\|y\|_X = 1 = r$ (recall the local linking hypothesis). Finally, if $x = \lambda v_0 + \mu y$ with $\lambda \in \left(\tfrac{1}{2}, 1\right]$, then

$$p^*(x) = (2\lambda - 1)y_0 + (2 - 2\lambda)\eta(\tau(y))$$

and so as λ moves from $\tfrac{1}{2}$ to 1, then $p^*(x)$ covers the segment

$$[\eta(\tau(y)), y_0] = \{x = (1 - \vartheta)\eta(\tau(y)) + \vartheta y_0 : \vartheta \in [0, 1]\}.$$

So

$$\|p^*(x) - y_0\|_X = (2 - 2\lambda)\|\eta(\tau(y)) - y_0\|_X \leq \xi,$$

hence $\varphi(p^*(x)) \leq 0$.

Note that we can find $0 < \gamma_2 \leq 1$, such that

$$\|p^*(x)\|_X \geq \gamma_2 \qquad \forall\, x \in E, \ \|x\|_X = 1.$$

We fix $0 < \varrho < \gamma_2$. Consider the set $p^*(\partial E)$ and the boundary of the ball $\partial B_\varrho^V = \{v \in V : \|v\|_X = \varrho\}$.

Now we will show that for any continuous extension p of p^* on all of E, we have

$$p(E) \cap \partial B_\varrho^V \neq \emptyset. \tag{2.52}$$

Let $P_Y \in \mathcal{L}(X)$ be the projection operator on Y, let $W = V \oplus \mathbb{R}v_0$ and consider the map $G \colon E \longrightarrow W$, defined by

$$G(x) \stackrel{df}{=} P_Y p(x) + \|(id_X - P_Y)p(u)\|_X \, v_0.$$

We need to show that $G(\overline{x}) = \varrho v_0$ for some point $\overline{x} \in E$. Since $\varrho < r = 1$, we have

$$G(x) \neq \varrho x \qquad \forall \, x \in \partial E.$$

So the Brouwer's degree $D_B(G, E, \varrho v_0)$ is defined. Consider the sets

$$C_1 \stackrel{df}{=} \overline{B}_1^V \; = \; \{v \in V : \|v\|_X \leq 1\}$$

$$C_2 \stackrel{df}{=} \partial B_1 \cap E \; = \; \{x \in E : \|x\|_X = 1\}.$$

Clearly $\partial E = C_1 \cup C_2$, $G|_{C_1} = id_{C_1}$ and $G|_{C_2} \geq \gamma_3 > 0$. On ∂E we define

$$\widehat{G}(x) \stackrel{df}{=} \begin{cases} x & \text{if } x \in C_1, \\ \dfrac{G(x)}{\|G(x)\|_X} & \text{if } x \in C_2. \end{cases}$$

If

$$h(t, x) \stackrel{df}{=} t G(x) + (1 - t)\widehat{G}(x) \qquad \forall \, t \in [0, 1], \; x \in E,$$

we see that G and $\widehat{G}$ are homotopic in $W \setminus \{\varrho v_0\}$. Note that $\widehat{G}(C_2) \subseteq C_2$ and that $\widehat{G}|_{\partial C_2} = id$. Since C_2 is homeomorphic to a ball, there is a continuous deformation $\widehat{h}(t, x)$, connecting $\widehat{G}$ and $id|_{C_2}$ and $\widehat{h}(t, \cdot)|_{\partial C_2} = id$ for all $t \in [0, 1]$. Therefore $G|_{\partial E}$ is homotopic to the identity in $W \setminus \{\varrho v_0\}$ and so $D_B(G, E, \varrho v_0) = D_B(I, E, \varrho v_0) = 1$. Thus we can find $\overline{x} \in E$, such that $G(\overline{x}) = \varrho \overline{x}$. This proves (2.52).

Thus $p^*(\partial E)$ and ∂B_ϱ^V link (see Definition 2.1.4 and Remark 2.1.4). Let

$$\Gamma \stackrel{df}{=} \{p \in C(E; X) : \; p|_{\partial E} = p^*\}$$

and set

$$c_0 \stackrel{df}{=} \inf_{p \in \Gamma} \sup_{x \in E} \varphi\big(p(x)\big).$$

Note that $c_0 \geq 0$. Also from Theorem 2.1.2, we know that c_0 is a critical value of φ. If $c_0 > 0$, then the corresponding critical point is the second nontrivial point of φ. If $c_0 = 0$, then again from Theorem 2.1.2, we can produce a critical point of φ located on ∂B_ρ^V with critical value c_0. This is the second nontrivial critical point of φ.

Case 2. $\dim Y = 0$.

If y_0 is the only nonzero critical point of φ, then by the local linking hypothesis we can find $\varrho_1 > 0$, such that

$$\varphi(x) > 0 \qquad \forall \, x \neq 0, \; \|x\|_X < \varrho_1.$$

So by Proposition 2.4.5, we can find $\varrho_2 > 0$ small enough, such that

$$\varphi(x) \geq \gamma_4 > 0 \qquad \forall\, x \in X,\ \|x\|_X = \varrho_2.$$

Since $\varphi(0) = 0 > \varphi(y_0)$, we can apply the nonsmooth Mountain Pass Theorem (see Theorem 2.1.3) and produce a second nontrivial critical point of φ, distinct from y_0.

Case 3. $\dim V = 0$ (in this case we can allow $\dim Y = +\infty$).
  From Proposition 2.4.4 we know that we can find $\varrho_3 > 0$ small, such that

$$\varphi(x) \leq \gamma_5 < 0 \qquad \forall\, x \in X,\ \|x\|_X = \varrho_3.$$

Also recall that φ is coercive (see Corollary 2.1.1). So we can apply Theorem 2.1.3 on the functional $-\varphi$ and for paths joining 0 and u, where $\varphi(u) > 0$ and $y_0 \notin [0, u] \overset{df}{=} \{x \in X : (1 - \lambda)0 + \lambda u,\ \lambda \in [0, 1]\}$. This will produce a second nontrivial critical point of φ distinct from y_0. $\quad\Box$

REMARK 2.4.2 Since φ is bounded below, the nonsmooth PS-condition and the nonsmooth C-condition are equivalent (see Proposition 2.1.2). $\quad\Box$

2.4.2 Minimax Principles

First we prove an extension of the nonsmooth Saddle Point Theorem (see Theorem 2.1.4), which is useful in the analysis of strongly resonant problems (see Chapter 4).
  The mathematical setting remains the same, namely X is a reflexive Banach space and $\varphi \colon X \longrightarrow \mathbb{R}$ is a locally Lipschitz function. As before, we set

$$K_c^\varphi \overset{df}{=} \big\{x \in X : 0 \in \partial\varphi(x),\ \varphi(x) = c\big\}$$

(i.e. the set of critical points of φ with energy level c) and

$$\varphi^c \overset{df}{=} \big\{x \in X : \varphi(x) \leq c\big\}.$$

The next theorem is a minimax principle similar to Theorem 2.2.3.

THEOREM 2.4.2
  If M is a compact metric space, $M^* \subseteq M$ is a closed subspace, $\gamma^* \in \overline{C}(M^*; X)$, φ satisfies the nonsmooth C_c-condition with

$$c \overset{df}{=} \inf_{\gamma \in \Gamma} \max_{s \in M} \varphi\big(\gamma(s)\big),$$

where

$$\Gamma \overset{df}{=} \big\{\gamma \in C(M; X) : \gamma = \gamma^* \text{ on } M^*\big\}$$

and

$$c > \max_{s \in M^*} \varphi\big(\gamma^*(s)\big)$$

<u>*then*</u> $K_c^\varphi \neq \emptyset$.

PROOF   Suppose that the theorem is not true and $K_c^\varphi = \emptyset$. Let

$$a \stackrel{df}{=} \max_{s \in M^*} \varphi\big(\gamma^*(s)\big).$$

By hypothesis $a < c$. Take $\varepsilon_0 \stackrel{df}{=} \frac{1}{2}(c - a)$ and apply Theorem 2.1.1 with $U = \emptyset$. We obtain $\varepsilon \in (0, \varepsilon_0)$ and a continuous homotopy of homeomorphisms $h \colon [0, 1] \times X \longrightarrow X$, such that

$$\begin{align}
h(0, x) &= x \qquad \forall\, x \in X, \tag{2.53}\\
h(t, x) &= x \qquad \forall\, t \in [0, 1],\ x \in X, \tag{2.54}\\
& \qquad \qquad \varphi(x) \notin [c - \varepsilon, c + \varepsilon],\\
h(1, \varphi^{c+\varepsilon}) &\subseteq \varphi^{c-\varepsilon}. \tag{2.55}
\end{align}$$

Note that

$$\varepsilon_0 + \varphi\big(\gamma^*(s)\big) \leq \varepsilon_0 + a = \frac{1}{2}(c + a) < c \qquad \forall\, s \in M^*$$

and so $\big|c - \varphi\big(\gamma^*(s)\big)\big| > \varepsilon_0$. Therefore from (2.54), it follows that

$$h\big(t, \gamma^*(s)\big) = \gamma^*(s) \qquad \forall\, t \in [0, 1],\ s \in M^*. \tag{2.56}$$

From the definition of c, we know that we can find $\overline{\gamma} \in \Gamma$, such that

$$\max_{s \in M} \varphi\big(\overline{\gamma}(s)\big) \leq c + \varepsilon. \tag{2.57}$$

Set

$$\gamma_h \stackrel{df}{=} h\big(1, \overline{\gamma}(s)\big) \qquad \forall\, s \in M.$$

Clearly $\gamma_h \in C(M; X)$. Also since $\overline{\gamma} \in \Gamma$ and from (2.56), we have

$$\gamma_h(s) = h\big(1, \overline{\gamma}(s)\big) = h\big(1, \gamma^*(s)\big) = \gamma^*(s) \qquad \forall\, s \in M^* \tag{2.58}$$

and so $\gamma_h \in \Gamma$. Then from (2.57) and (2.55), we have that

$$\gamma_h(s) = h\big(1, \overline{\gamma}(s)\big) \in h\big(1, \varphi^{c+\varepsilon}\big) \subseteq \varphi^{c-\varepsilon} \qquad \forall\, s \in M$$

and so

$$\varphi\big(\gamma_h(s)\big) \leq c - \varepsilon \qquad \forall\, s \in M,$$

which contradicts the definition of c. This contradiction means that $K_c^\varphi \neq \emptyset$ and so c is a critical value of φ.   □

As a consequence of this general minimax principle, we can have the non-smooth Mountain Pass Theorem (see Theorem 2.1.3) with nonrelaxed boundary condition (i.e. with strict inequality).

COROLLARY 2.4.1 (Nonsmooth Mountain Pass Theorem)
If $\overline{x} \neq 0$ *is such that* $\varphi(\overline{x}) \leq \varphi(0)$, *there exist* $r \in (0, \|\overline{x}\|_X)$ *and* $\mu > \varphi(0)$, *such that*

$$\varphi(x) \geq \mu \qquad \forall\, x \in \partial B_r,$$

φ *satisfies the nonsmooth* C_c-*condition with*

$$c \overset{df}{=} \inf_{\gamma \in \Gamma} \max_{t \in [0,1]} \varphi\big(\gamma(t)\big),$$

where

$$\Gamma \overset{df}{=} \big\{\gamma \in C([0,1];X) : \; \gamma(0) = 0, \; \gamma(1) = \overline{x}\big\},$$

then $K_c^\varphi \neq \emptyset$.

PROOF   Set $M \overset{df}{=} [0,1]$, $M^* \overset{df}{=} \{0,1\}$, $\gamma^*(0) = 0$, $\gamma^*(1) = \overline{x}$ and next apply Theorem 2.4.2. ▯

We can have another corollary, which will lead to a generalization of the nonsmooth Saddle Point Theorem (see Theorem 2.1.4).

COROLLARY 2.4.2
If M, M^*, γ^*, Γ *and* c *are as in Theorem 2.4.2, there exists* $D \subseteq X$, *such that*

$$\gamma(M) \cap D \neq \emptyset \qquad \forall\, \gamma \in \Gamma$$

and

$$\max_{s \in M^*} \varphi\big(\gamma^*(s)\big) < \inf_{x \in D} \varphi(x),$$

then $K_c^\varphi \neq \emptyset$.

PROOF   Note that

$$\inf_{\gamma \in \Gamma} \max_{s \in M} \varphi\big(\gamma(s)\big) \;\geq\; \inf_{x \in D} \varphi(x) \;>\; \max_{s \in M^*} \varphi\big(\gamma^*(s)\big).$$

So we can apply Theorem 2.4.2 and conclude that $K_c^\varphi \neq \emptyset$. ▯

Using this corollary, we can have the following generalization of the nonsmooth Saddle Point Theorem.

THEOREM 2.4.3 (Generalized Nonsmooth Saddle Point Theorem)

If $X = Y \oplus V$ *with* $\dim Y < +\infty$, *there exist* $y_0 \in Y$ *and* $r > \|y_0\|_X$, *such that*

$$\inf_{v \in V} \varphi(y_0 + v) > \max_{y \in Y \cap \partial B_r} \varphi(y),$$

$M = Y \cap \overline{B}_r$, φ *satisfies the nonsmooth* C_c-*condition with*

$$c \overset{df}{=} \inf_{\gamma \in \Gamma} \max_{s \in M} \varphi(\gamma(s)),$$

where

$$\Gamma \overset{df}{=} \{\gamma \in C(M; X) : \gamma|_{Y \cap \partial B_r} = id\},$$

then $K_c^{\varphi} \neq \emptyset$.

PROOF   We want to apply Corollary 2.4.2 with $D = y_0 + V$. To this end, we need to show that

$$\gamma(M) \cap (y_0 + V) \neq \emptyset \qquad \forall \, \gamma \in \Gamma. \tag{2.59}$$

Let $P_Y \in \mathcal{L}(X)$ be the projection on the finite dimensional subspace Y. Clearly (2.59) is equivalent to saying that

$$\forall \gamma \in \Gamma \; \exists y \in M : \; P_Y\big(\gamma(y) - y_0\big) \; = \; P_Y\big(\gamma(y)\big) - y_0 \; = \; 0.$$

To solve this equation for $y \in M$, we proceed as follows. Let $\gamma \in \Gamma$. Note that if $M^* = Y \cap \partial B_r$, then $P_Y \circ \gamma|_{M^*} = id$. From the properties of Brouwer's degree D_B we have

$$D_B(P_Y \circ \gamma - y_0, \operatorname{int} M, 0) \; = \; D_B(P_Y \circ \gamma, \operatorname{int} M, y_0) \; = \; D_B(id_M, \operatorname{int} M, y_0) \; = \; 1,$$

hence we can find $y \in \operatorname{int} M$, such that $(P_Y \circ \gamma)(y) = y_0$. $\Box$

REMARK 2.4.3   If $y_0 = 0$, then Theorem 2.4.3 is just the nonsmooth Saddle Point Theorem. $\Box$

2.4.3   Palais-Smale-Type Conditions

We conclude this section with some additional observations concerning the nonsmooth PS-condition and its role in minimization problems. We start by introducing a type of local PS-condition suggested by the Ekeland Variational Principle, which is quite natural in the present nonsmooth setting.

DEFINITION 2.4.2   *Let* (M, d_M) *be a complete metric space. We say that a function* $\varphi : M \longrightarrow \mathbb{R}$ *satisfies the* $PS_{c,+}^*$-*condition, if*

"Whenever $\{x_n\}_{n\geq 1} \subseteq M$ and $\{\varepsilon_n\}_{n\geq 1}, \{\delta_n\}_{n\geq 1} \subseteq \mathbb{R}_+$ are sequences, such that $\varepsilon_n, \delta_n \searrow 0$, $\varphi(x_n) \longrightarrow c$ and

$$\varphi(x_n) \leq \varphi(y) + \varepsilon_n d_M(x_n, y) \qquad \forall\, y \in M,\ d_M(x_n, y) \leq \delta_n,$$

then $\{x_n\}_{n\geq 1}$ has a convergent subsequence."

*If in this last inequality we interchange x_n and y, we say that φ satisfies the $\mathrm{PS}^*_{c,-}$-**condition**. If φ satisfies both the $\mathrm{PS}^*_{c,+}$-condition and $\mathrm{PS}^*_{c,-}$-condition, then we simply say that it satisfies the PS^*_c-**condition**.*

PROPOSITION 2.4.6

If M is a complete metric space, $\varphi\colon M \longrightarrow \overline{\mathbb{R}}$ is a lower semicontinuous, proper and bounded below function satisfying the $\mathrm{PS}^*_{c,+}$-condition with

$$c \stackrel{df}{=} \inf_M \varphi$$

then there exists $x_0 \in M$, such that $c = \varphi(x_0)$.

PROOF Let $\{y_n\}_{n\geq 1} \subseteq M$ be a minimizing sequence for φ and let $\delta_n = \varphi(y_n) - c > 0$. Of course $\delta_n \searrow 0$. Invoking the Ekeland Variational Principle (see Corollary 1.4.7), with $\varepsilon = \delta_n^2$, $\lambda = \delta_n$, $x_0 = y_n$ for $n \geq 1$, we obtain the sequence $\{x_n\}_{n\geq 1} \subseteq M$, such that

$$\varphi(x_n) \leq \varphi(y_n),$$

$$d_M(x_n, y_n) \leq \delta_n$$

and

$$\varphi(x_n) \leq \varphi(u) + \delta_n d_M(u, x_n) \qquad \forall\, u \in X.$$

Clearly $\varphi(x_n) \longrightarrow c$ and since by hypothesis φ satisfies the $\mathrm{PS}^*_{c,+}$-condition, passing to a subsequence if necessary, we conclude that $x_n \longrightarrow x_0$ in M. Because φ is lower semicontinuous, we have that $\varphi(x_0) = c$. $\quad\Box$

COROLLARY 2.4.3

If X is a reflexive Banach space, $\varphi\colon X \longrightarrow \mathbb{R}$ is a locally Lipschitz and bounded below function satisfying the $\mathrm{PS}^*_{c,+}$-condition, with

$$c \stackrel{df}{=} \inf_X \varphi$$

then $K_c^\varphi \neq \emptyset$.

It is of course natural to ask what is the relation of this new PS-type condition to the nonsmooth PS-condition introduced in Definition 2.1.1(a).

PROPOSITION 2.4.7
_If_ X _is a reflexive Banach space and_ $\varphi\colon X \longrightarrow \mathbb{R}$ _is a locally Lipschitz function,_
then for any $c \in \mathbb{R}$, φ _satisfies the nonsmooth_ PS_c -_condition if and only if it satisfies the_ PS_c^* -_condition._

PROOF   **Nonsmooth** PS_c -**condition** $\Longrightarrow \mathrm{PS}_c^*$ -**condition:**
Let $\{x_n\}_{n\geq 1} \subseteq X$, $\{\varepsilon_n\}_{n\geq 1}, \{\delta_n\}_{n\geq 1} \subseteq \mathbb{R}_+ \setminus \{0\}$ be sequences, such that $\varepsilon_n, \delta_n \searrow 0$, $\varphi(x_n) \longrightarrow c$ and

$$\varphi(x_n) \ \leq \ \varphi(y) + \varepsilon_n \|x_n - y\|_X \qquad \forall \, y \in X : \ d(x_n, y) \ \leq \ \delta_n.$$

Let $y = x_n + tu$, with $u \in X$, $\|u\|_X = 1$ and $0 < t \leq \delta_n$ for $n \geq 1$. We obtain

$$-1 \ \leq \ \frac{1}{\varepsilon_n} \frac{\varphi(x_n + tu) - \varphi(x_n)}{t} \qquad \forall \, 0 < t \leq \delta_n, \ u \in X, \ \|u\|_X = 1,$$

so

$$-1 \ \leq \ \frac{1}{\varepsilon_n} \varphi^0(x_n; u) \qquad \forall \, u \in X, \ \|u\|_X = 1$$

and so

$$-\|u\|_X \ \leq \ \frac{1}{\varepsilon_n} \varphi^0(x_n; u) \qquad \forall \, u \in X.$$

Using Lemma 1.3.2 (note that $\varphi^0(x_n; 0) = 0$ and $\varphi^0(x_n; \cdot) \in \Gamma_0(X)$ by Proposition 1.3.7(a)), we can find $y_n^* \in X^*$, $y_n^* \neq 0$, $\|y_n^*\|_X \leq 1$, such that

$$\varepsilon_n \langle y_n^*, u \rangle_X \ \leq \ \varphi^0(x_n; u) \qquad \forall \, u \in X, \ n \geq 1.$$

If $x_n^* \overset{df}{=} \varepsilon_n y_n^*$, then $x_n^* \in \partial\varphi(x_n)$ and $\|x_n^*\|_{X^*} \longrightarrow 0$. Because φ satisfies the nonsmooth PS_c -condition, we can find a strongly convergent subsequence. Thus the PS_c^* -condition hold.

 PS_c^* -**condition** $\Longrightarrow$ **Nonsmooth** PS_c -**condition:**
Let $\{x_n\}_{n\geq 1} \subseteq X$ be a sequence, such that

$$\varphi(x_n) \ \longrightarrow \ c \quad \text{and} \quad m^\varphi(x_n) \ \longrightarrow \ 0.$$

Let $x_n^* \in \partial\varphi(x_n)$ be such that $m^\varphi(x_n) = \|x_n^*\|_{X^*}$ for $n \geq 1$. Take $\varepsilon_n' \overset{df}{=} \|x_n^*\|_{X^*}$ for $n \geq 1$. By hypothesis, we have that $\varepsilon_n' \searrow 0$. For each $n \geq 1$, let $\{t_m^{(n)}\}_{m\geq 1} \subseteq \mathbb{R}_+ \setminus \{0\}$ be a sequence, such that $t_m^{(n)} \leq \delta_n \searrow 0$ and

$$\frac{\varphi(x_n + t_m^{(n)} h) - \varphi(x_n)}{t_m^{(n)}} \ \longrightarrow \ \varphi^0(x_n; h) \qquad \text{as } m \to +\infty, \qquad \forall \, h \in X.$$

As $x_n^* \in \partial\varphi(x_n)$, for each $n \geq 1$, we can find $m_0 = m_0(n) \geq 1$, such that

$$\langle x_n^*, h \rangle_X - \varepsilon_n' \ \leq \ \frac{1}{t_m^{(n)}} \Big[\varphi(x_n + t_m^{(n)} h) - \varphi(x_n) \Big] \qquad \forall \, m \geq m_0(n)$$

and so

$$\varphi(x_n) \leq \varphi\big(x_n + t_m^{(n)} h\big) + t_m^{(n)} \varepsilon_n' - \Big\langle x_n^*, t_m^{(n)} h \Big\rangle_X$$

$$\leq \varphi\big(x_n + t_m^{(n)} h\big) + 2\varepsilon_n' t_m^{(n)} \qquad \forall\, m \geq m_0(n).$$

If $\varepsilon_n = 2\varepsilon_n'$, then because the PS_c^*-condition holds, we infer that there exists a strongly convergent subsequence. So the nonsmooth PS_c-condition holds. ☐

2.5 Continuous Functionals

In this section, using the notion of weak slope (see Definition 1.3.10), we develop a critical point theory for continuous functionals defined on a metric space.

2.5.1 Compactness-Type Conditions

Let (X, d_X) be a metric space and $\varphi\colon X \longrightarrow \mathbb{R}$ a continuous function. We start by defining what we mean by a critical point of φ in this setting.

DEFINITION 2.5.1 *A point $x_0 \in X$ is said to be a **critical point of** φ, if $|d\varphi|(x_0) = 0$. Then $c = \varphi(x_0)$ is the corresponding **critical value**.*

REMARK 2.5.1 If X is a Banach space and $\varphi \in C^1(X)$, then by virtue of Corollary 1.3.9, the above definition coincides with classical definition of a critical point for a smooth function. If X is a Banach space and φ is locally Lipschitz, then Proposition 1.3.24 implies that the above definition coincides with Definition 2.1.2. ☐

Next we extend the PS-condition to the present general setting.

DEFINITION 2.5.2 *We say that φ satisfies the **extended nonsmooth Palais-Smale condition at level** c (**extended nonsmooth** PS_c-**condition** for short) if the following holds:*

"Every sequence $\{x_n\}_{n \geq 1} \subseteq X$, such that

$$\varphi(x_n) \longrightarrow c \quad \text{and} \quad |d\varphi|(x_n) \longrightarrow 0,$$

has a strongly convergent subsequence."

*If this property holds at every level $c \in \mathbb{R}$, then we simply say that φ satisfies the **extended nonsmooth** PS-**condition**.*

REMARK 2.5.2   The limit x_0 of the convergent subsequence of PS-sequence is necessarily a critical point of φ, since $|d\varphi|(\cdot)$ is lower semicontinuous (see Proposition 1.3.22).   ☐

2.5.2   Deformation Theorem

In order to prove a deformation theorem, we need some auxiliary results. The first one is a general topological result. We omit its proof and refer to Kuratowski (1966, p. 234) for it.

LEMMA 2.5.1
If Y is a metric space and $\{U_\alpha\}_{\alpha \in J}$ is an open cover of Y,
then this open cover admits a locally finite refinement

$$\{V_{k,\beta} :\ k \in \mathbb{N},\ \beta \in S_k\},$$

such that for all $k \in \mathbb{N}$ and all $\beta, \beta' \in S_k$, $\beta \neq \beta'$, we have $V_{k,\beta} \cap V_{k,\beta'} = \emptyset$.

Using this general topological result, we can make the first step towards a deformation theorem.

PROPOSITION 2.5.1
If $\xi \colon X \longrightarrow \mathbb{R}_+$ is a continuous function, such that for all $x \in X$ with $|\overline{d\varphi}|(x) \neq 0$, we have $|d\varphi|(x) > \xi(x)$,
then we can find continuous maps $\eta \colon \mathbb{R}_+ \times X \longrightarrow X$ and $\tau \colon X \longrightarrow \mathbb{R}_+$, such that for every $(t,x) \in \mathbb{R}_+ \times X$, we have:

(a) $d_X\big(x, \eta(t,x)\big) \leq t;$

(b) $\varphi\big(\eta(t,x)\big) \leq \varphi(x);$

(c) *if $t \leq \tau(x)$ then $\varphi\big(\eta(t,x)\big) \leq \varphi(x) - \xi(x)t;$*

(d) *if $|d\varphi|(x) \neq 0$ then $\tau(x) > 0$.*

PROOF   From Proposition 1.3.22 we know that the map $x \longmapsto |d\varphi|(x)$ is lower semicontinuous. So if $x \in X$ is such that $|d\varphi|(x) \neq 0$, we can find $\delta(x) > 0$, such that

$$B_{\delta(x)}(x) \subseteq \{y \in Y :\ |d\varphi|(y) \neq 0\}$$

and

$$\eta(x) \overset{df}{=} \sup_{y \in B_{\delta(x)}(x)} \xi(y) \ < \ |d\varphi|(x).$$

From Proposition 1.3.20, by decreasing $\delta(x) > 0$ if necessary, we can find a continuous map $H_x \colon [0, \delta(x)] \times B_{\delta(x)}(x) \longrightarrow X$, such that

$$\begin{cases} d\big(H_x(t,u),u\big) \leq t \\ \varphi\big(H_x(t,u)\big) \leq \varphi(u) - \eta(x)t \end{cases} \qquad \forall\, (t,u) \in [0,\delta(x)] \times B_{\delta(x)}(x).$$

The collection $\Big\{ B_{\frac{\delta(x)}{2}}(x) \Big\}_{\{x \in X: \, |d\varphi|(x) \neq 0\}}$ is an open cover of the metric space $\{x \in X : |d\varphi|(x) \neq 0\}$. Because of Lemma 2.5.1, we can find a locally Lipschitz refinement $\big\{ V_{k,\beta} : k \in \mathbb{N}, \ \beta \in S_k \big\}$, such that for all $k \in \mathbb{N}$ and all $\beta, \beta' \in S_k$, $\beta \neq \beta'$ we have $V_{k,\beta} \cap V_{k,\beta'} = \emptyset$. Let $\big\{ \vartheta_{k,\beta} : k \in \mathbb{N}, \ \beta \in S_k \big\}$ be a partition of unity subordinate to the refinement $\big\{ V_{k,\beta} : k \in \mathbb{N}, \ \beta \in S_k \big\}$. Since $\overline{V}_{k,\beta} \subseteq \{x \in X : |d\varphi|(x) \neq 0\}$, we can extend each $\vartheta_{k,\beta}$ to all of X by simply setting $\vartheta_{k,\beta}(y) = 0$ for $y \in \{x \in X : |d\varphi|(x) = 0\}$.

For each pair (k, β), let $x_{k,\beta} \in \{x \in X : |d\varphi|(x) \neq 0\}$ be such that $V_{k,\beta} \subseteq B_{\frac{\delta(x_{k,\beta})}{2}}(x_{k,\beta})$. Set $\delta_{k,\beta} \overset{df}{=} \delta(x_{k,\beta})$ and $H_{k,\beta} \overset{df}{=} H_{x_{k,\beta}}$. Let $\widehat{\tau} \colon X \longrightarrow \mathbb{R}_+$ be defined by

$$\widehat{\tau}(x) \overset{df}{=} \begin{cases} \dfrac{1}{4} \min_{x \in \overline{V}_{k,\beta}} \delta_{k,\beta} & \text{if } |d\varphi|(x) \neq 0, \\ 0 & \text{if } |d\varphi|(x) = 0. \end{cases}$$

It is easy to check that $\widehat{\tau}$ is lower semicontinuous. Then we define $\tau \colon X \longrightarrow \mathbb{R}_+$ by

$$\tau(x) \overset{df}{=} \inf_{y \in X} \big(\widehat{\tau}(y) + d(y,x) \big).$$

Clearly τ is continuous and satisfies the following

$$\text{if } |d\varphi|(x) = 0 \text{ then } \tau(x) = 0$$

and

$$\text{if } |d\varphi|(x) \neq 0 \text{ then } 0 < \tau(x) < \frac{1}{2} \min_{x \in \overline{V}_{k,\beta}} \delta_{k,\beta}.$$

Next we shall define a sequence of continuous maps

$$\eta_n \colon \big\{ (t,x) \in \mathbb{R}_+ \times X : t \leq \tau(x) \big\} \longrightarrow X,$$

such that

$$d_X\big(x, \eta_n(t,x)\big) \leq \left(\sum_{k=1}^{n} \sum_{\beta \in S_k} \vartheta_{k,\beta}(x) \right) t \qquad (2.60)$$

and

$$\varphi\big(\eta_n(t,x)\big) \leq \varphi(x) - \xi(x) \left(\sum_{k=1}^{n} \sum_{\beta \in S_k} \vartheta_{k,\beta}(x) \right) t. \qquad (2.61)$$

We do this inductively. First we define

$$\eta_1(t,x) \overset{df}{=} \begin{cases} H_{1,\beta}\big(\vartheta_{1,\beta}(x)t, x\big) & \text{if } x \in \overline{V}_{k,\beta}, \\ x & \text{if } x \notin \bigcup_{\beta \in S_1} V_{1,\beta}. \end{cases}$$

Suppose that we have defined $\{\eta_k\}_{k=1}^{n-1}$ which satisfy (2.60) and (2.61). For every $t \in [0, \tau(x))$ and every $x \in \overline{V}_{n,\beta}$, we have

$$d_X\big(x, \eta_{n-1}(t,x)\big) \leq \left(\sum_{k=1}^{n-1} \sum_{\beta \in S_k} \vartheta_{k,\beta}(x)\right) t \leq \tau(x) < \frac{1}{2}\delta_{n,\beta},$$

hence $\eta_{n-1}(t,x) \in B_{\delta_{n,\beta}}(x_{n,\beta})$. So the map

$$\eta_n(t,x) \overset{df}{=} \begin{cases} H_{n,\beta}\big(\vartheta_{n,\beta}(x)t, \eta_{n-1}(t,x)\big) & \text{if } x \in \overline{V}_{k,\beta}, \\ \eta_{n-1}(t,x) & \text{if } x \notin \bigcup_{\beta \in S_1} V_{1,\beta} \end{cases}$$

is well defined and satisfies (2.60) and (2.61). Therefore by induction, we have generated a sequence of continuous maps η_n, satisfying (2.60) and (2.61).

Let $x \in \{x \in X : |d\varphi|(x) \neq 0\}$. Since $\{V_{k,\beta} : k \in \mathbb{N}, \beta \in S_k\}$ is a locally finite cover of $\{x \in X : |d\varphi|(x) \neq 0\}$, we can find a neighbourhood U of x and $n_0 \in \mathbb{N}$, such that

$$\eta_n(t,x) = \eta_{n_0}(t,x) \qquad \forall\, (t,x) \in [0, \tau(x)] \times U,\ n \geq n_0.$$

Therefore the map $\eta \colon \mathbb{R}_+ \times X \longrightarrow X$ defined by

$$\eta(t,x) \overset{df}{=} \lim_{n \to +\infty} \eta\big(\min\{t, \tau(x)\}, x\big)$$

is continuous on the set $\mathbb{R}_+ \times \{x \in X : |d\varphi|(x) \neq 0\}$. From the estimates (2.60) and (2.61), we have that

$$d_X\big(x, \eta(t,x)\big) \leq \min\{t, \tau(x)\} \leq t,$$

$$\varphi\big(\eta(t,x)\big) \leq \varphi(x)$$

and also

$$\text{if } t \leq \tau(x), \text{ then } \varphi\big(\eta(t,x)\big) \leq \varphi(x) - \xi(x)t$$

and so it follows that η is also continuous on the set

$$\mathbb{R}_+ \times \{x \in X : |d\varphi|(x) = 0\}.$$

$\square$

PROPOSITION 2.5.2
If X is a complete metric space, $C \subseteq X$ is a nonempty and closed set and $\delta, \xi > 0$ are such that

$$\text{if } d(x, C) \leq \delta, \text{ then } |d\varphi|(x) > \xi,$$

<u>*then*</u> *there exists a continuous map $\eta \colon [0, \delta] \times X \longrightarrow X$, such that for every $(t, x) \in [0, \delta] \times X$, we have:*

(a) $d_X\big(x, \eta(t, x)\big) \leq t$;

(b) $\varphi\big(\eta(t, x)\big) \leq \varphi(x)$;

(c) *if $d_X(x, C) \geq \delta$, then $\eta(t, x) = x$;*

(d) *if $x \in C$, then $\varphi\big(\eta(t, x)\big) \leq \varphi(x) - \xi t$.*

PROOF   Consider the function $\psi \colon X \longrightarrow \mathbb{R} \cup \{-\infty\}$ defined by

$$\psi(x) \overset{df}{=} \begin{cases} \xi & \text{if } d_X(x, C) \leq \delta, \\ -\infty & \text{if } d_X(x, C) > \delta. \end{cases}$$

Clearly ψ is upper semicontinuous. On the other hand from Proposition 1.3.22 we know that $|d\psi|(\cdot)$ is lower semicontinuous. Consider the multifunction $G \colon X \longrightarrow 2^{\mathbb{R}} \setminus \{\emptyset\}$ defined by

$$G(x) \overset{df}{=} (\psi(x), |d\psi|(x)).$$

Then G is lower semicontinuous and by virtue of Remark 1.2.7 we can find a continuous map $\widehat{\xi} \colon X \longrightarrow \mathbb{R}$, such that

$$\widehat{\xi}(x) \in G(x) \qquad \forall\, x \in X,$$

hence $\widehat{\xi}(x) < |d\psi|(x)$ and

$$\text{if } d_X(x, C) \leq \delta, \text{ then } \widehat{\xi}(x) > \xi.$$

Set

$$\overline{\xi}(x) \overset{df}{=} \min\{\widehat{\xi}(x), \xi\} \qquad \forall\, x \in X.$$

Clearly $\overline{\xi}$ is continuous and

$$\text{if } d_X(x, C) \leq \delta, \text{ then } \overline{\xi}(x) = \xi$$

and

$$\text{if } |d\varphi|(x) \neq 0, \text{ then } \overline{\xi}(x) < |d\varphi|(x).$$

Apply Proposition 2.5.1 with $\overline{\xi}$, to obtain two continuous maps

$$\tau_1 \colon X \longrightarrow \mathbb{R}_+ \quad \text{and} \quad \eta_1 \colon \mathbb{R}_+ \times X \longrightarrow X,$$

satisfying the postulates of that proposition. Recursively, we define for all $n \geq 2$

$$\tau_n(x) \stackrel{df}{=} \tau_{n-1}(x) + \tau_1\left(\eta_{n-1}\left(x, \tau_{n-1}(x)\right)\right)$$

and

$$\eta_n(t,x) \stackrel{df}{=} \begin{cases} \eta_{n-1}(t,x) & \text{if } 0 \leq t \leq \tau_{n-1}(x), \\ \eta_1\left(t - \tau_{n-1}(x), \eta_{n-1}\left(\tau_{n-1}(x), x\right)\right) & \text{if } \tau_{n-1}(x) \leq t. \end{cases}$$

We claim that for all $(t,x) \in \mathbb{R}_+ \times X$ with $d_X(x,C) + t \leq \delta$, we have $t < \lim_{n \to +\infty} \tau_n(x)$. Suppose that this is not the case and so that $\tau_n(x) \leq t$ for all $n \geq 1$. We have

$$d_X\left(\eta_n(\tau_n(x),x), \eta_{n-1}(\tau_{n-1}(x),x)\right) \leq \tau_n(x) - \tau_{n-1}(x)$$

and so

$$d_X\left(x, \eta_n\left(\tau_n(x),x\right)\right) \leq \tau_n(x),$$

hence

$$d_X\left(\eta_n\left(\tau_n(x),x\right),C\right) \leq d_X\left(\eta_n\left(\tau_n(x),x\right),x\right) + d_X(x,C) \leq \tau_n(x) + d_X(x,C).$$

From this it follows that $\left\{\eta_n\left(\tau_n(x),x\right)\right\}_{n \geq 1}$ is a Cauchy sequence in $\overline{C}_\delta \stackrel{df}{=} \left\{x \in X : d_X(x,C) \leq \delta\right\}$. Let

$$y \stackrel{df}{=} \lim_{n \to +\infty} \eta_n\left(\tau_n(x),x\right).$$

We have $\tau_1(y) = \lim_{n \to +\infty} (\tau_{n+1}(x) - \tau_n(x)) = 0$, a contradiction. So if we define

$$\eta \colon \left\{(t,x) \in \mathbb{R}_+ \times X : d_X(x,C) + t \leq \delta\right\} \longrightarrow X,$$

by

$$\eta(t,x) \stackrel{df}{=} \lim_{n \to +\infty} \eta_n(t,x),$$

then η is continuous and

$$d_X\left(x, \eta(t,x)\right) \leq t \qquad \text{and} \qquad \varphi\left(\eta(t,x)\right) \leq \varphi(x) - \xi t.$$

Moreover, if we set

$$\eta(t,x) \stackrel{df}{=} \eta\left((\delta - d_X(x,C))^+, x\right),$$

whenever $d_X(x,C) + t > \delta$, we see that η is still continuous and so we have defined η on $\mathbb{R}_+ \times X$. $\qquad \blacksquare$

In Proposition 1.3.21 we expressed $|d\varphi|(x)$ in terms of $|dE_\varphi|(x,\lambda)$, where $E_\varphi \colon \text{epi } \varphi \longrightarrow \mathbb{R}$ is the function $E_\varphi(x,\lambda) = \lambda$. On epi φ we consider the metric

$$d_{X \times \mathbb{R}}\left((x,\lambda),(y,\mu)\right) \stackrel{df}{=} \left(d_X(x,y)^2 + (\lambda - \mu)^2\right)^{\frac{1}{2}}.$$

Note that E_φ is Lipschitz continuous with Lipschitz constant 1. By interchanging the roles of φ and E_φ in the proof of Proposition 1.3.21, we can also express $|dE_\varphi|(\cdot,\cdot)$ in terms of $|d\varphi|(\cdot)$. This gives us a device to reduce the study of continuous functions to that of Lipschitz continuous functions.

PROPOSITION 2.5.3
For every $(x,\lambda) \in \text{epi}\,\varphi$, we have

$$|dE_\varphi|(x,\lambda) = \begin{cases} \frac{|d\varphi|(x)}{\sqrt{1+|d\varphi|(x)^2}} & \text{if } \varphi(x) = \lambda \text{ and } |d\varphi|(x) < +\infty, \\ 1 & \text{if } \varphi(x) < \lambda \text{ or } |d\varphi|(x) = +\infty. \end{cases}$$

Now we are ready for a Deformation Theorem for this setting. As before for $c \in \mathbb{R}$ we set

$$K_c^\varphi \stackrel{df}{=} \{x \in X : |d\varphi|(x) = 0, \; \varphi(x) = c\}$$

and

$$\varphi^c \stackrel{df}{=} \{\varphi(x) \leq c\}.$$

THEOREM 2.5.1 (Deformation Theorem)
If X is a complete metric space, $c \in \mathbb{R}$ and φ satisfies the extended nonsmooth $\overline{\text{PS}}_c$-condition,
<u>*then*</u> *for a given $\varepsilon_0 > 0$, a neighbourhood U of K_c^φ (if $K_c^\varphi = \emptyset$, then $U = \emptyset$) and $\vartheta > 0$, there exist $\varepsilon \in (0, \varepsilon_0)$ and a continuous map $\eta \colon [0,1] \times X \longrightarrow X$, such that for every $(t,x) \in [0,1] \times X$, we have:*

(a) $d_X\big(x, \eta(t,x)\big) \leq \vartheta t;$

(b) $\varphi\big(\eta(t,x)\big) \leq \varphi(x);$

(c) *if $\varphi(x) \notin (c - \varepsilon_0, c + \varepsilon_0)$, then $\eta(t,x) = x;$*

(d) $\eta(1, \varphi^{c+\varepsilon} \setminus U) \subseteq \varphi^{c-\varepsilon}.$

PROOF First suppose that φ is Lipschitz continuous with Lipschitz constant 1. Since φ satisfies the extended nonsmooth PS_c-condition and $x \longmapsto |d\varphi|(x)$ is lower semicontinuous (see Proposition 1.3.22), we infer that K_c^φ is compact. Let $\delta_1 > 0$ be such that $(K_c^\varphi)_{2\delta_1} \subseteq U$. Take $\delta, \xi > 0$, such that $2\delta \leq \varepsilon_0$, $\delta \leq \delta_1$ and $|d\varphi|(x) > \xi$ for all $x \in C$, where

$$C \stackrel{df}{=} \{x \in X : \; c - \delta \leq \varphi(x) \leq c + \delta, \; x \notin (K_c^\varphi)_{2\delta_1}\}.$$

Because we have assumed that φ is Lipschitz continuous with constant 1, we have that

$$\text{if } d_X(x, C) \leq \delta, \quad \text{then } |d\varphi|(x) > \xi.$$

Let $\eta_1 \colon [0, \delta] \times X \longrightarrow X$ be a continuous map as in Proposition 2.5.2. Without any loss of generality, we can assume that $\vartheta \leq \delta$ and define $\eta \colon [0, 1] \times X \longrightarrow X$ by

$$\eta(t, x) \stackrel{df}{=} \eta_1(\vartheta t, x).$$

Then statements (a) and (b) follow from the corresponding statements of Proposition 2.5.2 for the function η_1. Since φ is Lipschitz continuous with constant 1, $\varphi(x) \notin (c - \varepsilon_0, c + \varepsilon_0)$ implies that $d_x(x, C) \geq \delta$ and so $\eta(t, x) = x$. Finally, let $\varepsilon \stackrel{df}{=} \min \left\{ \frac{\xi\vartheta}{2}, \delta \right\}$. If $x \in \varphi^{c+\varepsilon} \setminus U$ and $\varphi(x) \geq c - \varepsilon$, we have $x \in C$ and so

$$\varphi\big(\eta(1, x)\big) \;=\; \varphi\big(\eta_1(\vartheta, x)\big) \;\leq\; \varphi(x) - \vartheta\xi \;\leq\; c + \varepsilon - \xi\vartheta \;\leq\; c - \varepsilon$$

(see Proposition 2.5.2). Also if $x \in \varphi^{c+\varepsilon} \setminus U$ and $\varphi(x) \leq c - \varepsilon$, from (b) it follows that $\varphi\big(\eta(1, x)\big) \leq c - \varepsilon$.

Next we drop the hypothesis that φ is Lipschitz continuous with Lipschitz constant 1. Since φ is continuous, epi $\varphi \subseteq X \times \mathbb{R}$ is a closed subset, hence it is complete. Using Proposition 2.5.3, we see that E_φ satisfies the extended nonsmooth $\mathrm{PS_c}$-condition. Moreover, if

$$K_c^{E_\varphi} = \big\{ (x, \lambda) \in X \times \mathbb{R} \colon \; |dE_\varphi|(x, \lambda) = 0, \; E_\varphi(x, \lambda) = c \big\},$$

then $(U \times \mathbb{R}) \cap \mathrm{epi}\, \varphi$ is a neighbourhood of $K_c^{E_\varphi}$ and as we already remarked earlier, the function E_φ is Lipschitz continuous, with Lipschitz constant 1. According to the first part of the proof, we can find $\varepsilon > 0$ and a continuous map $\widehat{\eta} = (\widehat{\eta}_1, \widehat{\eta}_2) \colon [0, 1] \times \mathrm{epi}\, \varphi \longrightarrow \mathrm{epi}\, \varphi$, which satisfies:

(i) $d_{X \times \mathbb{R}}\big((x, \lambda), \widehat{\eta}(t, (x, \lambda))\big) \leq \vartheta t$;

(ii) $\widehat{\eta}_2\big(t, (x, \lambda)\big) \leq \lambda$;

(iii) if $\lambda \notin (c - \varepsilon_0, c + \varepsilon_0)$, then $\widehat{\eta}\big(t, (x, \lambda)\big) = (x, \lambda)$;

(iv) if $\lambda \leq c + \varepsilon$ and $x \notin U$, then $\widehat{\eta}_2\big(1, (x, \lambda)\big) \leq c - \varepsilon$.

We define $\eta \colon [0, 1] \times X \longrightarrow X$ by

$$\eta(t, x) \stackrel{df}{=} \widehat{\eta}_1\big(t, (x, \varphi(x))\big) \qquad \forall\, (t, x) \in [0, 1] \times X.$$

Recall that $\widehat{\eta}$ takes its values in epi φ and so

$$\varphi\big(\widehat{\eta}_1\big(t, (x, \varphi(x))\big)\big) \;\leq\; \widehat{\eta}_2\big(t, (x, \varphi(x))\big).$$

Then statements (a)-(d) follow at once from this definition of η. $\qquad\qquad$ ▯

REMARK 2.5.3   In this case, for $t \in [0, 1]$, the map η is not necessarily a homeomorphism. Consider

$$X \stackrel{df}{=} \big\{ (x, y) \in \mathbb{R}^2 \colon \; y \geq |x| \big\} \cup \big(\{0\} \times \mathbb{R}_-\big)$$

and $\varphi(x,y) \stackrel{df}{=} -y$. Then φ has no critical points and satisfies the extended nonsmooth PS$_c$-condition for every $c \in \mathbb{R}$. However, φ^ε is not homeomorphic to a subset of $\varphi^{-\varepsilon}$, for every $\varepsilon > 0$. ∎

We can have "symmetric" version of Theorem 2.5.1, when X is a Banach space equipped with the group action of the symmetry group $\{id_X, -id_X\} \simeq \mathbb{Z}_2$. Note that the origin is a fixed point of the action. So we have to treat the origin as a critical point, even if we do not know whether $|d\varphi|(0) = 0$ for a continuous and even function $\varphi \colon X \longrightarrow \mathbb{R}$ (it is true if φ is continuous and odd).

THEOREM 2.5.2
If X is a Banach space, $\varphi \colon X \longrightarrow \mathbb{R}$ is continuous and even, $c \in \mathbb{R}$ and φ satisfies the extended nonsmooth PS$_c$-condition,
then for a given $\varepsilon_0 > 0$, a neighbourhood U of $K_c^\varphi \cup \{0\}$ and $\vartheta > 0$, there exist $\varepsilon > 0$ and a continuous map $\eta \colon [0,1] \times X \longrightarrow X$ satisfying (a)-(d) of Theorem 2.5.1 and

(e) *for all $t \in [0,1]$, $\eta(t, \cdot) \colon X \longrightarrow X$ is odd.*

PROOF Because φ is even, it follows that $|d\varphi|(\cdot)$ is even too (hence K_c^φ is a symmetric set). If $x \neq 0$ and $|d\varphi|(x) > \xi > 0$, then by Definition 1.3.10, we can find $\delta > 0$ and a continuous map $H \colon [0, \delta] \times B_\delta(x) \longrightarrow X$, such that

$$\|x - H(t,x)\|_X \leq t \qquad \text{and} \qquad \varphi\big(H(t,x)\big) \leq \varphi(x) - \xi t. \qquad (2.62)$$

We may always suppose that $\delta < \|x\|_X$. We introduce a map

$$\widehat{H} \colon [0, \delta] \times (B_\delta(x) \cup B_\delta(-x)) \longrightarrow X,$$

defined by

$$\widehat{H}(t,y) \stackrel{df}{=} \begin{cases} H(t,y) & \text{if } y \in B_\delta(x), \\ -H(t,-y) & \text{if } y \in B_\delta(-x). \end{cases}$$

Evidently $\widehat{H}$ is continuous, odd with respect to second variable and still satisfies (2.62). So all the constructions of Propositions 2.5.1, 2.5.2 as well as of Theorem 2.5.1 can be repeated in a symmetric fashion and finally give us the result of the theorem. ∎

2.5.3 Minimax Principles

Now let us use the Deformation Theorem (see Theorem 2.5.1) to obtain a minimax principle characterizing critical points of φ.

THEOREM 2.5.3
If X is complete, $A \subseteq B$ nonempty compact subsets of X, $C \subseteq X$ nonempty closed subset, $\gamma^* \colon A \longrightarrow X$ a continuous map,

$$C \cap \psi(A) = \emptyset, \quad C \cap \varphi(B) \neq \emptyset, \quad \max_A(\varphi \circ \gamma^*) \leq \inf_C \varphi,$$

φ satisfies the extended nonsmooth PS_c-condition with

$$c \overset{df}{=} \inf_{\gamma \in \Gamma} \max_B(\varphi \circ \gamma),$$

where

$$\Gamma \overset{df}{=} \left\{ \gamma \in C(B; X) \colon \gamma|_A = \gamma^* \right\} \neq \emptyset,$$

then $K_c^\varphi \neq \emptyset$ and if $c = \inf_C \varphi$, then $K_c^\varphi \cap C \neq \emptyset$.

PROOF   Note that the hypotheses imply that $\inf_C \varphi \leq c$. If $\inf_C \varphi < c$ and by contradiction we assume that $K_c^\varphi = \emptyset$, then arguing as in the first part of the proof of Theorem 2.1.2 (using this time Theorem 2.5.1(c) and (d)) we reach a contradiction.

So suppose that $c = \inf_C \varphi$ and assume by contradiction that $K_c^\varphi \cap C = \emptyset$ and let $\xi > 0$ be such that

$$\begin{aligned} d_X(x, C) > \xi & \quad \forall\, x \in K_c^\varphi \\ d_X(x, y) \geq 2\xi & \quad \forall\, x \in C,\ y \in \psi(C). \end{aligned}$$

Let $\varepsilon > 0$ and $\eta \colon [0,1] \times X \longrightarrow X$ be a continuous map as in Theorem 2.5.1 satisfying

$$\begin{aligned} d_X(x, \eta(t, x)) \leq \xi t \\ \varphi(\eta(1, x)) \leq c - \varepsilon \end{aligned} \quad \forall\, x \in \varphi^{c+\varepsilon} \text{ such that } d_X(x, C) \leq \xi.$$

Let $k \colon X \longrightarrow [0,1]$ be a continuous function, such that

$$k|_{\psi(A)} = 0 \quad \text{and} \quad k|_{\overline{A}_\varepsilon} = 1.$$

Let $\gamma \in \Gamma$ be such that $\max_{\gamma(B)} \leq c + \varepsilon$ and let

$$\widehat{\gamma}(s) \overset{df}{=} \eta(k(\gamma(s)), \gamma(s)) \quad \forall\, s \in B.$$

Evidently $\widehat{\gamma} \in \Gamma$. There exists $s_0 \in B$, such that $\widehat{\gamma}(s_0) \in C$ and so $\varphi(\widehat{\gamma}(s_0)) \geq c$. On the other hand $d(\gamma(s_0), C) \leq \lambda$, so that $k(\gamma(s_0)) = 1$ and

$$\varphi(\widehat{\gamma}(s_0)) = \varphi(\eta(1, \gamma(s_0))) \leq c - \varepsilon,$$

a contradiction.   □

Using Theorem 2.5.2, we can have a symmetric version of the previous theorem.

THEOREM 2.5.4
If X *is a Banach space,* $\varphi\colon X \longrightarrow \mathbb{R}$ *is a continuous and even map and*

(i) *there exist* $r > 0$, $\beta > \varphi(0)$ *and* $V \subseteq X$ *with* $\dim V < +\infty$, *such that* $\varphi|_{V \cap \partial B_r} \geq \beta$;

(ii) *for every* $Y \subseteq X$ *with* $\dim Y < +\infty$, *there exists* $R_Y > 0$, *such that for all* $y \in Y$ *with* $\|y\|_X \geq R_Y$, *we have* $\varphi(y) \leq \varphi(0)$;

(iii) φ *satisfies the extended nonsmooth* PS_c-*condition for every* $c \geq \beta$,

then there exists a sequence $\{x_n\}_{n \geq 1}$ *of critical points of* φ, *such that*

$$\lim_{n \to +\infty} \varphi(x_n) = +\infty.$$

2.6 Multivalued Functionals

In this section, by appropriately extending the notion of weak slope (see Definition 1.3.10), we develop a critical point theory for multivalued functionals which have closed graph.

2.6.1 Compactness-Type Conditions

Let (X, d_X) be a metric space and $F\colon X \longrightarrow 2^{\mathbb{R}} \setminus \{\emptyset\}$, a multifunction with closed graph. The graph of F is the set

$$\mathrm{Gr}\, F \stackrel{df}{=} \{(x, c) \in X \times \mathbb{R}\colon c \in F(x)\}$$

and $X \times \mathbb{R}$ is a metric space with metric

$$d_{X \times \mathbb{R}}\big((x, c), (y, b)\big) \stackrel{df}{=} \big[d_X(x, y)^2 + |c - b|^2\big]^{\frac{1}{2}}.$$

By $\pi_1\colon \mathrm{Gr}\, F \longrightarrow X$ and $\pi_2\colon \mathrm{Gr}\, F \longrightarrow \mathbb{R}$ we denote the projections of $\mathrm{Gr}\, F$ on X and $\mathbb{R}$ respectively, i.e. $\pi_1(x, c) = x$ and $\pi_2(x, c) = c$.

DEFINITION 2.6.1 *Let* $F\colon X \longrightarrow 2^{\mathbb{R}} \setminus \{\emptyset\}$ *be a multifunction with closed graph and let* $(x, c) \in \mathrm{Gr}\, F$. *The* **weak slope** *of* F *at* (x, c), *denoted by* $|dF|(x, c)$, *is the supremum on all* $\xi \geq 0$, *such that there exist* $\delta > 0$ *and a continuous function* $H = (H_1, H_2)\colon [0, \delta] \times B_\delta\big((x, c)\big) \longrightarrow \mathrm{Gr}\, F$, *such that*

(a) $d_{X \times \mathbb{R}}\big((y, b), H(t, (y, b))\big) \leq t\sqrt{1 + \xi^2}$;

(b) $H_2\big(t, (y, b)\big) \leq b - \xi t$.

REMARK 2.6.1   If $F(x) = \{\varphi(x)\}$ with $\varphi \colon X \longrightarrow \mathbb{R}$ a continuous function, then $|dF|\big(x, \varphi(x)\big) = |d\varphi|(x)$. Moreover, if $F \colon X \longrightarrow 2^{\mathbb{R}} \setminus \{\emptyset\}$ is a multifunction with closed graph, we define $G_F \colon \operatorname{Gr} F \longrightarrow \mathbb{R}$ by $G_F(x, c) = c$. Evidently this map is continuous and one can verify that

$$
|dF|(x, c) \;=\; \begin{cases} \dfrac{|dG_F|(x,c)}{\sqrt{1-|dG_F|(x,c)^2}} & \text{if } |dG_F|(x, c) < 1, \\[2mm] +\infty & \text{if } |dG_F|(x, c) = 1 \end{cases}
$$

(see also Proposition 1.3.21). Then using this fact, we deduce at once the lower semicontinuity of $|dF|(\cdot, \cdot)$ (see Proposition 1.3.22). ⬜

PROPOSITION 2.6.1
If $\{(x_n, c_n)\}_{n \geq 1} \subseteq \operatorname{Gr} F$ *and* $(x_n, c_n) \longrightarrow (x, c)$ *in* $X \times \mathbb{R}$,
then
$$
|dF|(x, c) \;\leq\; \liminf_{n \to +\infty} |dF|(x_n, c_n).
$$

Now we introduce the basic notions of critical point and of the Palais-Smale condition.

DEFINITION 2.6.2   *We say that $x \in X$ is a **critical point of F at level** $c \in \mathbb{R}$, if $(x, c) \in \operatorname{Gr} F$ and $|dF|(x, c) = 0$. By K_c^F we denote the set of critical points of F at level c, i.e.*

$$
K_c^F \;\stackrel{df}{=}\; \big\{ x \in X : \; |dF|(x, c) = 0 \big\}.
$$

*Also $c \in \mathbb{R}$ is a **critical value** of F, if $K_c^F \neq \emptyset$.*

REMARK 2.6.2   Note that due to the multivaluedness of F, $x \in X$ can be a critical point at more than one level, i.e. the map associating critical points to critical values is multivalued. If we can find a neighbourhood U of x, such that
$$
-\infty < m^F(x) \leq m^F(y) \qquad \forall\, y \in U,
$$
where
$$
m^F(y) \;\stackrel{df}{=}\; \min \big\{ b \in \mathbb{R} : \; b \in F(y) \big\}
$$
(i.e. x is a local minimum of F), then $x \in K_c^F$ with $c = m^F(x)$. ⬜

DEFINITION 2.6.3   *We say that F satisfies the **multivalued Palais-Smale condition at level** $c \in \mathbb{R}$ (the **multivalued** PS_c **condition** for short), if*

"Every sequence $\{x_n\}_{n \geq 1} \subseteq X$ for which there exist $c_n \in F(x_n)$, such that

$$c_n \longrightarrow c \quad \text{and} \quad |dF|(x_n, c_n) \longrightarrow c,$$

has a convergent subsequence."

2.6.2 Multivalued Deformation Theorem

Using the Lipschitz continuous map $G_F \colon \operatorname{Gr} F \longrightarrow \mathbb{R}$ defined by $G_F(x, c) = c$ and Remark 2.6.1 and following the reasoning of the proof of Theorem 2.5.1, we can have the following Deformation Theorem.

THEOREM 2.6.1 (Multivalued Deformation Theorem)
If $c \in \mathbb{R}$ and F satisfies the multivalued PS_c-condition,
<u>then</u> *for a given $\varepsilon_0 > 0$, a neighbourhood U of $K_c^F \times \{c\}$ (if $K_c^F = \emptyset$, then $U = \emptyset$) and $\xi > 0$, there exist $\varepsilon \in (0, \varepsilon_0)$ and a continuous function $\eta = (\eta_1, \eta_2) \colon [0, 1] \times \operatorname{Gr} F \longrightarrow \operatorname{Gr} F$, such that for every $\big(t, (x, b)\big) \in [0, 1] \times \operatorname{Gr} F$, we have:*

(a) $d_{X \times \mathbb{R}}\big((x, b), \eta\big(t, (x, b)\big)\big) \leq \xi t$;

(b) $\eta_2\big(t, (x, b)\big) \leq b$;

(c) *if* $(x, b) \in \operatorname{Gr} F \setminus \big(X \times (c - \varepsilon_0, c + \varepsilon_0)\big)$, *then* $\eta\big(t, (x, b)\big) = (x, b)$;

(d) $\eta\big(1, \big(\operatorname{Gr} F \cap \big(X \times (-\infty, c + \varepsilon]\big)\big)\big) \setminus U\big) \subseteq X \times (-\infty, c - \varepsilon]$.

COROLLARY 2.6.1
If $c \in \mathbb{R}$, F satisfies the multivalued PS_c-condition, $\xi > 0$, $C \subseteq \operatorname{Gr} F$ is a closed subset, U is a neighbourhood of $K_c^F \times \{c\}$ (if $K_c^F = \emptyset$, then $U = \emptyset$) and V is a neighbourhood of C (if $C = \emptyset$, then $V = \emptyset$)
<u>then</u> *there exist $\varepsilon > 0$ and a continuous function $\eta = (\eta_1, \eta_2) \colon [0, 1] \times \operatorname{Gr} F \longrightarrow \operatorname{Gr} F$, such that for all $\big(t, (x, b)\big) \in [0, 1] \times \operatorname{Gr} F$, we have:*

(a) $d_{X \times \mathbb{R}}\big((x, b), \eta\big(t, (x, b)\big)\big) \leq \xi t$;

(b) $\eta_2\big(t, (x, b)\big) \leq b$;

(c) $\eta = id$ *on* $\big(\{0\} \times \operatorname{Gr} F\big) \cup \big([0, 1] \times C\big)$;

(d) $\eta\big(1, \big(\operatorname{Gr} F \cap \big(X \times (-\infty, c + \varepsilon]\big)\big)\big) \setminus (U \cup V)\big) \subseteq X \times (-\infty, c - \varepsilon]$.

PROOF Let $\overline{\eta} \colon [0, 1] \times \operatorname{Gr} F \longrightarrow \operatorname{Gr} F$ be the continuous deformation postulated by Theorem 2.6.1. Invoking Urysohn's Lemma (see Theorem A.1.12), we can find a continuous map $k \colon X \longrightarrow [0, 1]$, such that

$$k|_C = 0 \quad \text{and} \quad k|_{X \setminus V} = 1.$$

Let $\eta \in C([0,1] \times X; X)$ be defined by

$$\eta(t,x) = \overline{\eta}(k(x)t, x) \qquad \forall \, (t,x) \in [0,1] \times X.$$

Then η is the desired continuous deformation. □

2.6.3 Minimax Principles

Let $E \subseteq \mathrm{Gr}\, F$ and let

$$\Gamma(E) \stackrel{df}{=} \{U \subseteq \mathrm{Gr}\, F : \ E \subseteq U, \ U \neq \emptyset \text{ if } E = \emptyset\}.$$

Also if $S \subseteq D \subseteq \mathrm{Gr}\, F$ are closed subsets and $\psi \colon S \longrightarrow \mathrm{Gr}\, F$ is a continuous function, we set

$$\Gamma(\psi(S), D) \stackrel{df}{=} \{\gamma(D) : \ \gamma = (\gamma_1, \gamma_2) \in C(D; \mathrm{Gr}\, F), \ \gamma|_S = \psi\}.$$

Clearly $\Gamma(\psi(S), D) \subseteq \Gamma(\psi(S))$.

DEFINITION 2.6.4   *Let $E \subseteq \mathrm{Gr}\, F$ and $\widehat{\Gamma} \subseteq \Gamma(E)$. We say that $\widehat{\Gamma}$ is* **invariant with respect to (F, E)-deformations,** *if*

"For every $U \subseteq \widehat{\Gamma}$ and every continuous map $\eta \colon [0,1] \times \mathrm{Gr}\, F \longrightarrow \mathrm{Gr}\, F$, such that

$$\begin{cases} \eta = id & \text{on } (\{0\} \times \mathrm{Gr}\, F) \cup ([0,1] \times E), \\ \eta_2(t, (x,b)) \leq b \ \forall (t, (x,b)) \in [0,1] \times \mathrm{Gr}\, F, \end{cases}$$

we have $\eta(1, U) \in \widehat{\Gamma}$."

Another important notion for what follows is given in the next definition.

DEFINITION 2.6.5   *Let $A, B \subseteq \mathrm{Gr}\, F$ and let $\widehat{\Gamma}$ be a nonempty subset of $\Gamma(E)$. We say that $\widehat{\Gamma}$* **intersects** *A if*

$$U \cap A \neq \emptyset \qquad \forall \, U \in \widehat{\Gamma}.$$

REMARK 2.6.3   In the above definition A and E need not be disjoint and in particular we can have $A = \mathrm{Gr}\, F$. In addition, we can have $E = \emptyset$. □

Now we are ready for the first minimax principle.

THEOREM 2.6.2
If X is a complete metric space, A is a closed subset of $\mathrm{Gr}\, F$, $E \subseteq \mathrm{Gr}\, F$, $\widehat{\Gamma} \subseteq \Gamma(E)$ nonempty and invariant with respect to (F, E)-deformations, $\widehat{\Gamma}$

intersects A,

$$\inf_{U \in \widehat{\Gamma}} \sup \pi_2(U \cap A) \geq \pi_2(E)$$

with strict inequality if $d_X(A, E) = 0$, *and F satisfies the multivalued* PS$_c$-*condition, with*

$$c \stackrel{df}{=} \inf_{U \in \widehat{\Gamma}} \sup \pi_2(U) \in \mathbb{R}$$

<u>*then*</u> $\left(K_c^F \times \{c\}\right) \cap \overline{\overline{\Gamma}} \neq \emptyset$ *where*

$$\overline{\overline{\Gamma}} \stackrel{df}{=} \overline{\bigcup_{U \in \widehat{\Gamma}} U}.$$

Moreover, if $c = \inf_{U \in \widehat{\Gamma}} \sup \pi_2(U \cap A)$, *then* $\left(K_c^F \times \{c\}\right) \cap A \neq \emptyset$.

PROOF Let V be a neighbourhood of $\overline{E}$ and let $\xi > 0$ be fixed in the process of the proof. Let $U = \left(K_c^F \times \{c\}\right)_\xi$. Applying Corollary 2.6.1 (with $C = \overline{E}$) we obtain $\varepsilon > 0$ and a continuous deformation $\eta \colon [0, 1] \times \operatorname{Gr} F \longrightarrow \operatorname{Gr} F$. In particular, we have that

$$\eta = id \qquad \text{on } \left(\{0\} \times \operatorname{Gr} F\right) \cup \left([0, 1] \times \overline{E}\right),$$
$$\eta_2\big(t, (x, b)\big) \leq b \,\forall\big(t, (x, b)\big) \in [0, 1] \times \operatorname{Gr} F.$$

Let $U \in \widehat{\Gamma}$ be such that $\pi_2(U) \leq c + \varepsilon$. By hypothesis we have $\eta(1, U) \in \widehat{\Gamma}$. Let

$$D \stackrel{df}{=} \begin{cases} A \text{ if } c = \inf_{U \in \widehat{\Gamma}} \sup \pi_2(U \cap A), \\ \overline{\overline{\Gamma}} \text{ otherwise.} \end{cases}$$

Suppose that $\left(K_c^F \times \{c\}\right) \cap D = \emptyset$. Since by hypothesis F satisfies the multivalued PS$_c$-condition, K_c^F is compact. So we can take $\xi > 0$, such that

$$d(K_c^F \times \{c\}, D) > 2\xi \quad \text{and} \quad d(A, E) > 2\xi \quad \text{if } c = \sup \pi_2(E).$$

When $c > \sup \pi_2(E)$, we choose $\delta > 0$, such that $c > c - \delta > \sup \pi_2(E)$. Let

$$V \stackrel{df}{=} \begin{cases} E_\xi & \text{if } c = \sup \pi_2(E), \\ \operatorname{Gr} F \cap (X \times (-\infty, c - \delta)) \text{ otherwise.} \end{cases}$$

Let $(\overline{x}, \overline{b}) \in U$ be such that

$$\eta\big(1, (\overline{x}, \overline{b})\big) \in D \quad \text{and} \quad c - \min\{\varepsilon, \delta\} < \eta_2\big(1, (\overline{x}, \overline{b})\big).$$

Because

$$d_{X \times \mathbb{R}}\big((\overline{x}, \overline{b}), \eta(1, (\overline{x}, \overline{b}))\big) \leq \xi \quad \text{and} \quad \eta_2\big(1, (\overline{x}, \overline{b})\big) \leq b,$$

we see that $(\overline{x},\overline{b}) \notin U \cap V$. Hence $\eta\big(1,(\overline{x},\overline{b})\big) \leq c - \varepsilon$, a contradiction. $\qquad\square$

If $A = \operatorname{Gr} F$, $E = \emptyset$ and we use the convention $\sup \pi_2(\emptyset) = -\infty$, then from Theorem 2.6.2, we obtain the following corollary.

COROLLARY 2.6.2
If X is a complete metric space, $\widehat{\Gamma} \subseteq \Gamma(\emptyset)$ is a nonempty subset invariant with respect to $(F,\emptyset)$-deformations and F satisfies the multivalued PS_c-condition, with

$$c \stackrel{df}{=} \inf_{U \in \widehat{\Gamma}} \sup \pi_2(U) \in \mathbb{R},$$

then $\big(K_c^F \times \{c\}\big) \cap \overline{\widehat{\Gamma}} \neq \emptyset.$

The second consequence of Theorem 2.6.2 can be viewed as the multivalued version of the Mountain Pass Theorem.

COROLLARY 2.6.3 (Multivalued Mountain Pass Theorem)
If X is complete, $S \subseteq D \subseteq \operatorname{Gr} F$ are closed subsets, $\psi = (\psi_1,\psi_2)\colon S \longrightarrow \operatorname{Gr} F$ is continuous, $A \subseteq \operatorname{Gr} F$ is closed, $\widehat{\Gamma} \subseteq \Gamma\big(\psi(S),D\big)$ is nonempty and invariant with respect to $\big(F,\psi(S)\big)$-deformations, $\widehat{\Gamma}$ intersects A,

$$\sup \psi_2(S) \leq \inf_{\gamma(D)\in\widehat{\Gamma}} \big(\gamma_2\big(\gamma^{-1}(A)\big)\big),$$

with strict inequality if $d\big(A,\psi(S)\big) = 0$ and F satisfies the multivalued PS_c-condition, with

$$c \stackrel{df}{=} \inf_{\gamma(D)\in\widehat{\Gamma}} \sup \gamma_2(D) \in \mathbb{R}$$

then $\big(K_c^F \times \{c\}\big) \cap \overline{\widehat{\Gamma}} \neq \emptyset.$
Moreover, if $c = \inf_{\gamma(D)\in\widehat{\Gamma}} \sup \gamma_2\big(\gamma^{-1}(A)\big)$, then $\big(K_c^F \times \{c\}\big) \cap A \neq \emptyset.$

REMARK 2.6.4   In the above theorem, if $S \subseteq D \subseteq \operatorname{epi} F$ are compact subsets, then automatically we have $c \in \mathbb{R}$. $\qquad\square$

Let us illustrate these abstract results with concrete example.

EXAMPLE 2.6.1   (a) Let $F\colon [-1,1] \times [0,1] \longrightarrow 2^{\mathbb{R}} \setminus \{\emptyset\}$ be defined by

$$F(x,y) \stackrel{df}{=} \begin{cases} [-2|x|,-3|x|+1] & \text{if } y = 0, \\ \{-3|x|-2y+1\} & \text{otherwise.} \end{cases}$$

Evidently F has closed graph. Fix $S \stackrel{df}{=} \{-1, 1\}$ and $\psi \stackrel{df}{=} (id, 0, -2)$. Using Corollary 2.6.3 with $\Gamma\big(\psi(S), [-1, 1]\big)$, we obtain the critical point $(0, 1)$ with critical level -1. On the other hand, let

$$\widehat{\Gamma} \subseteq \Gamma\big(\psi(S), [-1, 1]\big)$$

be the set invariant with respect to $(F, \psi(S))$-deformations and generated by $\{(x, 0, -2|x|) : x \in [-1, 1]\}$. Again Corollary 2.6.3 gives the critical point $(0, 0)$ with critical level 0. So we see that by considering only deformations η satisfying $\eta\big(t, (x, b)\big) \le b$, we were able to obtain the $(0, 0)$-critical point.

(b) Let $F \colon \left[-\frac{1}{\pi}, \frac{1}{\pi} \right] \longrightarrow 2^{\mathbb{R}} \setminus \{\emptyset\}$ be defined by

$$F(x) \stackrel{df}{=} \begin{cases} \cos\left(\frac{1}{x}\right) - x^2 & \text{if } x \ne 0, \\ [-1, 1] & \text{if } x = 0. \end{cases}$$

We choose

$$D \stackrel{df}{=} \left[-\frac{1}{\pi}, \frac{1}{\pi} \right], \quad S \stackrel{df}{=} \left\{ -\frac{1}{\pi}, \frac{1}{\pi} \right\} \quad \text{and} \quad \psi \stackrel{df}{=} (id, F|_S).$$

Clearly $\Gamma\big(\psi(S), D\big) = \emptyset$. Let

$$\widehat{\Gamma} \stackrel{df}{=} \{\operatorname{Gr} F(D)\} \subseteq \Gamma\big(\psi(S)\big)$$

and

$$A \stackrel{df}{=} \left(\left[-\frac{1}{2\pi}, \frac{1}{\pi} \right] \times \mathbb{R} \right) \cap \operatorname{Gr} F.$$

Theorem 2.6.2 gives the critical point $x = 0$ with critical level 1. Also if $\widehat{\Gamma} \subseteq \Gamma(\emptyset)$ is the subset invariant with respect to $(F, \emptyset)$-deformations and generated by $\operatorname{Gr} F \cap (D \times (-\infty, -1])$, then from Corollary 2.6.2, we obtain the critical point $x = 0$ with critical level -1. $\qquad\square$

2.7 Remarks

2.1 The Palais-Smale condition for smooth functions was introduced by Palais & Smale (1964) in the context of deformation techniques. The Cerami condition for smooth functions was suggested by Cerami (1978) and was used in the context of strongly resonant problems by Bartolo, Benci & Fortunato (1983). The relation between PS-condition and coercivity for smooth functions was first established by Caklović, Li & Willem (1990). Soon thereafter Costa & Silva (1991) proved additional results in this direction. The nonsmooth versions of these results were obtained by Kourogenis & Papageorgiou (2000b).

The nonsmooth critical point theory for locally Lipschitz functions started with the work of Chang (1981). He was able to construct a substitute for the pseudogradient vector field of the smooth theory and use it to obtain non-smooth versions of the Mountain Pass Theorem of Ambrosetti & Rabinowitz (1973) and of the Saddle Point Theorem of Rabinowitz (1978a). Chang (1981) used his theory to study semilinear elliptic boundary value problems with a discontinuous nonlinearity. The more general version of the theory, which we present here and which uses the nonsmooth C-condition and relaxed boundary conditions, is based on the work of Kourogenis & Papageorgiou (2000a). In the smooth case first Pucci & Serrin (1984) proved a version of the Mountain Pass Theorem, where the separating mountain range F has a thickness. Further results in this direction were obtained by Guo, Sun & Qi (1988), Ghoussoub (1993a) and Du (1993). Additional versions of the notion of linking (see Definition 2.1.4 and Remark 2.1.4) can be found in the papers of Silva (1991) and Ding (1994). The smooth version of Theorem 2.1.5 (Linking Theorem) is due to Rabinowitz (1978c). The notion of genus (see Definition 2.1.5) is due to Krasnoselskii (1964), although the definition given here is due to Coffman (1969). The fact that Definition 2.1.5 is equivalent to the original definition of Krasnoselskii was proved by Rabinowitz (1973). Theorem 2.1.7 is essentially due to Szulkin (1986), although Szulkin formulates the result for a different class of nonsmooth functionals ($\varphi = \Phi + \psi$, with $\Phi \in C^1(X)$, $\psi \in \Gamma_0(X)$). The observation that Szulkin's proof, with minor modifications, is also valid in the locally Lipschitz case was made by Goeleven, Motreanu & Panagiotopoulos (1998). Additional multiplicity results for nondifferentiable functionals can be found in the recent work of Marano & Motreanu (2002b).

2.2 The smooth version of the critical point theory for functions defined on closed convex sets was formulated by Struwe (1990). The nonsmooth version is due to Kyritsi & Papageorgiou (to appear a). Critical point theory for smooth functions (using the deformation technique or the Ekeland Variational Principle or both) can be found in the books of Chang (1993), de Figueiredo (1982), Ghoussoub (1993a), Mawhin & Willem (1989), Rabinowitz (1986), Struwe (1990) and Willem (1996).

2.3 The first variation of Chang's theory was formulated by Szulkin (1986), who considered functionals of the form $\varphi = \Phi + \psi$ with $\Phi \in C^1(X)$, $\psi \in \Gamma_0(X)$. Kourogenis, Papadrianos & Papageorgiou (2002) formulated the extension to the case $\varphi = \Phi + \psi$ with Φ locally Lipschitz and $\psi \in \Gamma_0(X)$.

2.4 The notion of local linking was introduced by Liu & Li (1984). Soon thereafter Brézis & Nirenberg (1991) relaxed the assumptions for local linking and proved a theorem on the existence of two nontrivial critical points for C^1-functionals satisfying the Palais-Smale condition. Theorem 2.4.1 extends to a nonsmooth setting the result of Brézis & Nirenberg (1991) and is due to Kandilakis, Kourogenis & Papageorgiou (submitted). Interesting applications of local linking to multiplicity results for semilinear elliptic problems can be found in Li & Willem (1995). Theorem 2.4.2 and its consequence Theorem 2.4.3 (which is useful in the analysis of strongly resonant problems; see Chapter 4) were proved by Gasiński & Papageorgiou (2002*b*). The variants of the Palais-Smale condition given in Definition 2.4.2 and Propositions 2.4.6 and 2.4.7 are due to Costa & Gonçalves (1990).

2.5 The extension of the nonsmooth critical point theory to continuous functions on a metric space was started by Degiovanni & Marzocchi (1994). Additional results and/or applications of this theory can be found in the papers of Canino & Degiovanni (1995), Corvellec, Degiovanni & Marzocchi (1993) and Degiovanni (1997). Another closely related critical point theory for continuous functions on a metric space was developed by Ioffe & Schwartzman (1996).

2.6 The extension of the theory of the previous section to multifunctions with closed graph is due to Frigon (1991).

Chapter 3

Ordinary Differential Equations

This chapter is devoted to the study of nonlinear boundary value problems for ordinary differential equations. We deal with both scalar problems and problems in $\mathbb{R}^N$ ($N > 1$, systems) and in general the right hand side nonlinearity is set-valued (differential inclusions). The formulation of some of the problems is very general and incorporates problems with unilateral constraints (differential variational inequalities). We present a variety of methods, which lead to existence results, multiplicity results and to positive solutions. In Section 3.1, we consider Dirichlet problems in $\mathbb{R}^N$ with a nonhomogeneous, nonlinear differential operator, which contains the ordinary vector p-Laplacian as a special case. The presence in the right hand side of a maximal monotone term (not necessary defined on all of $\mathbb{R}^N$) includes in our framework differential variational inequalities and gradient systems with nonsmooth potential. The solution method that we develop is based on the theory of nonlinear operator of monotone type and on fixed point arguments (a set-valued version of the Leray-Schauder alternative principle). In Section 3.2, we conduct a similar study for periodic systems. Now the right hand side nonlinearity satisfies certain versions of the Hartman and Nagumo-Hartman conditions. Moreover, for the scalar problem we allow the differential operator to depend also on x at the expense of being linear in x'. In Section 3.3, we pass to a more general level and allow the boundary conditions to be nonlinear and multivalued. However, the differential operator is restricted to be the ordinary vector p-Laplacian. The analysis is general enough to achieve a unified treatment of the classical Dirichlet, Neumann and periodic problems. The techniques are analogous to those used in the previous two sections. In Section 3.4 we consider periodic problems (in $\mathbb{R}$ and $\mathbb{R}^N$) in variational form with a nonsmooth potential and we use the nonsmooth critical point theory of Chapter 2 to prove existence and multiplicity theorems. Section 3.5 considers a general scalar nonlinear second order differential inclusion with nonlinear multivalued boundary conditions and uses the methods of "upper-lower" solutions to produce a solution and also "extremal" solutions in the order interval formed by the ordered pair $\underline{\psi} \leq \overline{\psi}$ of the lower and the upper solution respectively. Here the approach is based on truncation and penalization techniques. In the first half of Section 3.6, we deal with a semilinear Sturm-Liouville type system and using a set-valued extension of the "compression-expansion" fixed point theorem, we establish the existence of positive solutions (for the usual partial order on $\mathbb{R}^N$). The hypotheses incorporate a special case, the sublinear and the superlinear problems. The analysis applies also to the Neumann problem. In the second half of Section 3.6, we deal with a scalar Neumann problem driven by a general nonlinear operator, a special case of which is the ordinary scalar p-Laplacian. The solution method is based on a range theorem for nonlinear operators of monotone type and the hypotheses involve a Landesman-Lazer type condition. Finally in Section 3.7 we study Hamiltonian inclusions, i.e. Hamiltonian systems in which the Hamiltonian function is not C^1, but only locally Lipschitz. Without assuming regularity of the locally Lipschitz Hamiltonian H, we obtain conservative solutions. The proof relies on a suitable approximation by smooth functions of the Hamiltonian H.

3.1   Dirichlet Problems

This section is devoted to vector Dirichlet problems driven by nonlinear differential operators which are not necessarily homogeneous and involve a maximal monotone term and a multivalued nonlinearity depending also on the derivative of the unknown function.

3.1.1   Formulation of the Problem

Let $T = [0, b]$. Let us consider the following strongly nonlinear second order differential inclusion

$$\begin{cases} \left(a(x'(t)) \right)' \in A\big(x(t)\big) + F\big(t, x(t), x'(t)\big) & \text{for a.a. } t \in T, \\ x(0) = x(b) = 0. \end{cases} \tag{3.1}$$

Here $a\colon \mathbb{R}^N \longrightarrow \mathbb{R}^N$ is a suitable monotone homeomorphism, which includes the ordinary vector p-Laplacian as a special case, $A\colon \mathbb{R}^N \longrightarrow 2^{\mathbb{R}^N}$ is a maximal monotone map and $F\colon T \times \mathbb{R}^N \times \mathbb{R}^N \longrightarrow 2^{\mathbb{R}^N} \setminus \{\emptyset\}$ is a set-valued nonlinearity. Our approach is primarily based on the theory of nonlinear operators of monotone type (see Section 1.4) and our hypotheses on the multivalued nonlinearity $F(t, \xi, \bar{\xi})$ are minimal, in the sense that besides the usual Carathéodory and p-growth conditions, we only have a general nonresonance condition. Moreover, the presence of the maximal monotone term A incorporates in our framework differential variational inequalities and gradient systems with a nonsmooth potential. The hypotheses on the data of (3.1) are the following:

$\underline{H(A)_1}$ $A\colon \mathbb{R}^N \supseteq \mathrm{Dom}\,(A) \longrightarrow 2^{\mathbb{R}^N}$ is a maximal monotone operator with $0 \in A(0)$.

$\underline{H(a)_1}$ $a\colon \mathbb{R}^N \longrightarrow \mathbb{R}^N$ is a continuous map, such that (*i*) a has one of the following forms:

(*i*)$_1$ $a(\xi) = k_a(\xi)\xi$, where $k_a\colon \mathbb{R}^N \longrightarrow \mathbb{R}$ is a continuous function; or

(*i*)$_2$ $a(\xi) = (k_1^a(\xi_1)\xi_1,\ldots,k_N^a(\xi_N)\xi_N)$, where $k_i^a\colon \mathbb{R} \longrightarrow \mathbb{R}$ are continuous functions for $i \in \{1,\ldots,N\}$;

(*ii*) a is strictly monotone;

(*iii*) for all $\xi \in \mathbb{R}^N$, we have that

$$(a(\xi)\xi,\xi)_{\mathbb{R}^N} \geq \beta \|\xi\|_{\mathbb{R}^N}^p,$$

with $\beta > 0$, $p \in (1,+\infty)$.

REMARK 3.1.1　　If $p \in (1,+\infty)$ and

$$a(\xi) \stackrel{df}{=} \begin{cases} \|\xi\|_{\mathbb{R}^N}^{p-2}\xi & \text{if } \xi \in \mathbb{R}^N \setminus \{0\}, \\ 0 & \text{if } \xi = 0, \end{cases}$$

then hypotheses $H(a)_1$ are satisfied (with $H(a)_1(i)_1$) and we have the ordinary vector p-Laplacian. Also if

$$a(\xi) = \left(|\xi_1|^{p-2}\xi_1,\ldots,|\xi_N|^{p-2}\xi_N\right) \qquad \forall\, \xi \in \mathbb{R}^N,$$

which is a slightly different version of the ordinary vector p-Laplacian, then again hypotheses $H(a)_1$ are satisfied (with $H(a)_1(i)_2$). Note that hypotheses $H(a)_1$ do not impose any growth restrictions on a. So for example, if

$$\vartheta(\xi) \stackrel{df}{=} ce^{\|\xi\|_{\mathbb{R}^N}^p} - \|\xi\|_{\mathbb{R}^N}^p \qquad \forall\, \xi \in \mathbb{R}^N,$$

for some $c > 1$ and

$$a(\xi) \stackrel{df}{=} \vartheta'(x) = \left(ce^{\|\xi\|_{\mathbb{R}^N}^p} - 1\right)\|\xi\|_{\mathbb{R}^N}^p \qquad \forall\, \xi \in \mathbb{R}^N,$$

we see that a satisfies hypotheses $H(a)_1$. Another possibility is to have

$$a(\xi) = \eta\left(\|\xi\|_{\mathbb{R}^N}^p\right)\|\xi\|_{\mathbb{R}^N}^{p-1}\xi \qquad \forall\, \xi \in \mathbb{R}^N,$$

with a continuous map $\eta\colon \mathbb{R}^+ \longrightarrow \mathbb{R}^+$, and $\beta \in (0,\eta(r))$ for all $r \geq 0$ and $r \longmapsto \eta(r^p)r^{p-1}$ strongly increasing. For example, we can have

$$\eta(r) \stackrel{df}{=} 1 + \frac{c}{(1+r)^p} \qquad \forall\, r \geq 0,$$

for some $c > 0$.　　　　　　　　　　　　　　　　　　　　　　　　　　□

$\underline{H(F)_1}$　$F\colon T \times \mathbb{R}^N \times \mathbb{R}^N \longrightarrow P_{kc}(\mathbb{R}^N)$ is a multifunction, such that

(i) for all $\xi, \overline{\xi} \in \mathbb{R}^N$, the multifunction

$$T \ni t \longmapsto F(t, \xi, \overline{\xi}) \in P_{kc}(\mathbb{R}^N)$$

is measurable;

(ii) for almost all $t \in T$, the multifunction

$$\mathbb{R}^N \times \mathbb{R}^N \ni (\xi, \overline{\xi}) \longmapsto F(t, \xi, \overline{\xi}) \in P_{kc}(\mathbb{R}^N)$$

has closed graph;

(iii) for almost all $t \in T$, all $\xi, \overline{\xi} \in \mathbb{R}^N$ and all $u \in F(t, \xi, \overline{\xi})$, we have

$$\|u\|_{\mathbb{R}^N} \leq \gamma_1(t, \|\xi\|_{\mathbb{R}^N}) + \gamma_2(t, \|\xi\|_{\mathbb{R}^N}) \|\overline{\xi}\|_{\mathbb{R}^N}^{p-1},$$

with

$$\sup_{r \in [0,k]} \gamma_1(t, r) \leq \eta_{1,k}(t) \quad \text{and} \quad \sup_{r \in [0,k]} \gamma_2(t, r) \leq \eta_{2,k}(t),$$

where $\eta_{1,k} \in L^2(T)$ and $\eta_{2,k} \in L^\infty(T)$;

(iv) $\displaystyle\liminf_{\|\xi\|_{\mathbb{R}^N} \to +\infty} \left(\inf_{\overline{\xi} \in \mathbb{R}^N} \frac{m(t, \xi, \overline{\xi})}{\|\xi\|_{\mathbb{R}^N}^p} \right) \geq -\vartheta_1(t)$ uniformly for almost all $t \in T$, with $\vartheta_1 \in L^\infty(T)_+$, $\vartheta_1(t) \leq \beta \lambda_1$ for almost all $t \in T$ and the inequality is strict on a set of positive measure. Here

$$m(t, \xi, \overline{\xi}) \stackrel{df}{=} \inf_{u \in F(t, \xi, \overline{\xi})} (u, \xi)_{\mathbb{R}^N} \qquad \forall \, (t, \xi, \overline{\xi}) \in T \times \mathbb{R}^N \times \mathbb{R}^N$$

and $\lambda_1 > 0$ is the first eigenvalue of the ordinary vector p-Laplacian with Dirichlet boundary conditions, i.e. $\lambda_1 = \left(\frac{\pi_p}{b}\right)^p$ (see Section 1.5) and $\beta > 0$ is as in hypotheses $H(a)_1(iii)$.

First we are going to consider a system of nonlinear equations associated to a monotone map such as a. Note that under hypotheses $H(a)_1$, a is a homeomorphism on $\mathbb{R}^N$. Then for $u \in C(T; \mathbb{R}^N)$, we define the **generalized mean value map** $G_u : \mathbb{R}^N \longrightarrow \mathbb{R}^N$, by

$$G_u(\xi) \stackrel{df}{=} \frac{1}{b} \int_0^b a^{-1}(\xi + u(t)) \, dt \qquad \forall \, \xi \in \mathbb{R}^N.$$

PROPOSITION 3.1.1
If hypotheses $H(a)_1$ hold,
then

(a) *for any $u \in C(T; \mathbb{R}^N)$, the system $G_u(\xi) = 0$ has a unique solution $\widehat{\xi}(u) \in \mathbb{R}^N$;*

(b) *the map* $\widehat{\xi}\colon C(T;\mathbb{R}^N) \longrightarrow \mathbb{R}^N$ *defined in* **(a)** *is continuous and bounded (i.e. sends bounded sets to bounded sets).*

PROOF **(a)** By virtue of the strict monotonicity of a, we have that

$$\left(G_u(\xi) - G_u(\overline{\xi}), \xi - \overline{\xi}\right)_{\mathbb{R}^N} > 0 \qquad \forall\, \xi, \overline{\xi} \in \mathbb{R}^N, \ \xi \neq \overline{\xi}$$

and so the solution of $G_u(\xi) = 0$ if it exists, it is unique. Evidently G_u is continuous. Also

$$
\begin{aligned}
\left(G_u(\xi), \xi\right)_{\mathbb{R}^N} &= \frac{1}{b}\int_0^b \left(a^{-1}(\xi + u(t)), \xi + u(t)\right)_{\mathbb{R}^N} dt \\
&\quad - \frac{1}{b}\int_0^b \left(a^{-1}(\xi + u(t)), u(t)\right)_{\mathbb{R}^N} dt \\
&\geq \frac{\beta}{b}\int_0^b \|\xi + u(t)\|_{\mathbb{R}^N}\left(\|\xi + u(t)\|_{\mathbb{R}^N}^{p-1} - \|u\|_\infty\right) dt
\end{aligned}
$$

and so it follows that G_u is weakly coercive (see Definition 1.4.6(b)). From Theorem 1.4.4 it follows that the equation $G_u(\xi) = 0$ has a solution.

(b) Let $B \subseteq C(T;\mathbb{R}^N)$ be a bounded set. We have

$$\int_0^b \left(a^{-1}(\widehat{\xi}(u) + u(t)), \widehat{\xi}(u)\right)_{\mathbb{R}^N} dt = 0$$

and so

$$
\begin{aligned}
&\int_0^b \left(a^{-1}(\widehat{\xi}(u) + u(t)), \widehat{\xi}(u) + u(t)\right)_{\mathbb{R}^N} dt \\
&\qquad = \int_0^b \left(a^{-1}(\widehat{\xi}(u) + u(t)), u(t)\right)_{\mathbb{R}^N} dt \qquad \forall\, u \in B.
\end{aligned}
$$

By Hölder's inequality (see Theorem A.3.12), we have

$$\beta\|\widehat{\xi}(u) + u\|_p^p \leq \|u\|_{p'}\|\widehat{\xi}(u) + u\|_p \qquad \forall\, u \in B,$$

with $\frac{1}{p} + \frac{1}{p'} = 1$, so

$$\beta\|\widehat{\xi}(u) + u\|_p^{p-1} \leq \|u\|_\infty b^{\frac{1}{p'}} \qquad \forall\, u \in B.$$

From this it follows that the set $\widehat{\xi}(B) \subseteq \mathbb{R}^N$ is bounded.

Finally, to show the continuity of $\widehat{\xi}$, let $\{u_n\}_{n \geq 1} \subseteq C(T; \mathbb{R}^N)$ be a sequence, such that

$$u_n \longrightarrow u \quad \text{in } C(T; \mathbb{R}^N),$$

for some $u \in C(T; \mathbb{R}^N)$. Then from the boundedness of $\widehat{\xi}$, it follows that the sequence $\{\widehat{\xi}(u_n)\}_{n \geq 1} \subseteq \mathbb{R}^N$ is bounded. So, we can find a subsequence $\{\widehat{\xi}(u_{n_k})\}_{n_k}$, such that

$$\widehat{\xi}(u_{n_k}) \longrightarrow \widetilde{\xi} \quad \text{in } \mathbb{R}^N \quad \text{as } k \to +\infty$$

for some $\widetilde{\xi} \in \mathbb{R}^N$. Because

$$\int_0^b a^{-1}\big(\widehat{\xi}(u_{n_k}) + u_{n_k}(t)\big)\, dt = 0 \qquad \forall\, k \geq 1,$$

in the limit as $k \longrightarrow +\infty$, we obtain that

$$\int_0^b a^{-1}\big(\widetilde{\xi} + u(t)\big)\, dt = 0$$

and so $\widehat{\xi}(u) = \widetilde{\xi}$. Since every subsequence of $\{\widehat{\xi}(u_n)\}_{n \geq 1}$ has a further subsequence converging to $\widehat{\xi}(u)$, we conclude that $\widehat{\xi}(u_n) \longrightarrow \widehat{\xi}(u)$ in $\mathbb{R}^N$, hence $\widehat{\xi}$ is continuous. ▯

3.1.2 Approximation of the Problem

Now, for every $g \in L^{p'}(T; \mathbb{R}^N)$ (with $\frac{1}{p} + \frac{1}{p'} = 1$), we consider an auxiliary problem. Here for $\lambda > 0$, $J_\lambda \overset{df}{=} (id + \lambda A)^{-1}$ and $A_\lambda \overset{df}{=} \frac{1}{\lambda}(id - J_\lambda)$ is the Yosida approximation of A (see Definition 1.4.7).

$$\begin{cases} -\big(a(x'(t))\big)' + A_\lambda\big(x(t)\big) + \|x(t)\|_{\mathbb{R}^N}^{p-2}\, x(t) = g(t) & \text{for a.a. } t \in T, \\ x(0) = x(b) = 0. \end{cases} \tag{3.2}$$

PROPOSITION 3.1.2

If hypotheses $H(a)_1$ and $H(A)_1$ hold,
then problem (3.2) has a solution $x_0 \in C^1(T; \mathbb{R}^N)$.

PROOF   First let us note that the problem

$$\begin{cases} -\big(a(x'(t))\big)' = g(t) & \text{for a.a. } t \in T, \\ x(0) = x(b) = 0 \end{cases} \tag{3.3}$$

has a unique solution. Indeed, by integrating the equation on $[0, t]$, we obtain

$$a\big(x'(t)\big) \;=\; \xi - H(g)(t) \qquad \forall\, t \in T,$$

with $\xi \in \mathbb{R}^N$ and $H \colon L^{p'}(T; \mathbb{R}^N) \longrightarrow C(T; \mathbb{R}^N)$ being the integral operator, defined by

$$H(h)(t) \;\overset{df}{=}\; \int_0^t h(s)\,ds \qquad \forall\, h \in L^{p'}(T; \mathbb{R}^N),\; t \in T.$$

So, we have that

$$\begin{cases} x'(t) = a^{-1}\big(\xi - H(g)(t)\big) & \text{for a.a. } t \in T, \\ x(0) = x(b) = 0. \end{cases}$$

Integrating on $[0, t]$, we obtain

$$x(t) \;=\; \int_0^t a^{-1}\big(\xi - H(g)(s)\big)\,ds \qquad \forall\, t \in T.$$

Since

$$x(b) \;=\; \int_0^b a^{-1}\big(\xi - H(g)(t)\big)\,dt \;=\; 0,$$

from Proposition 3.1.1, we infer that this last equation has a unique solution $\xi = \widehat{\xi}\big(- H(g)\big)$. Therefore the problem (3.3) has a unique solution $x \in W_0^{1,p}(T; \mathbb{R}^N)$, defined by

$$x(t) \;=\; \int_0^t a^{-1}\big(\widehat{\xi}\big(- H(g)\big) - H(g)(s)\big)\,ds \qquad \forall\, t \in T.$$

Let $K \colon L^{p'}(T; \mathbb{R}^N) \longrightarrow W_0^{1,p}(T; \mathbb{R}^N)$ be the map, which to each forcing term $g \in L^{p'}(T; \mathbb{R}^N)$ assigns the unique solution of (3.3).

Claim 1. The map $K \colon L^{p'}(T; \mathbb{R}^N) \longrightarrow W_0^{1,p}(T; \mathbb{R}^N)$ is completely continuous.

Suppose that $\{g_n\}_{n \geq 1} \subseteq L^{p'}(T; \mathbb{R}^N)$ is a sequence, such that

$$g_n \overset{w}{\longrightarrow} g \quad \text{in } L^{p'}(T; \mathbb{R}^N),$$

for some $g \in L^{p'}(T; \mathbb{R}^N)$ and let $x_n \overset{df}{=} K(g_n)$ for $n \geq 1$. We have that

$$\begin{cases} -\big(a\big(x_n'(t)\big)\big)' = g_n(t) & \text{for a.a. } t \in T, \\ x_n(0) = x_n(b) = 0. \end{cases} \tag{3.4}$$

Taking the inner product with $x_n(t)$, integrating over $T = [0, b]$ and performing integration by parts, we obtain

$$\beta \|x_n'\|_p^p \leq \|g_n\|_{p'} \|x_n\|_p \qquad \forall \, n \geq 1,$$

so by the Poincaré inequality (see Theorem 1.1.6), we have that the sequence $\{x_n\}_{n \geq 1} \subseteq W_0^{1,p}(T; \mathbb{R}^N)$ is bounded.

Because of (3.4), the sequence $\{(a(x_n'(t)))'\}_{n \geq 1} \subseteq L^{p'}(T; \mathbb{R}^N)$ is bounded too. Moreover, recall that

$$x_n'(t) \; = \; a^{-1}\big(\widehat{\xi}\big(-H(g_n)\big) - H(g_n)(t)\big) \qquad \forall \, t \in T, \, n \geq 1.$$

From Proposition 3.1.1(b), we know that $\widehat{\xi} \colon C(T; \mathbb{R}^N) \longrightarrow \mathbb{R}^N$ is continuous and bounded below, while $H \in \mathcal{L}(L^{p'}(T; \mathbb{R}^N), C(T; \mathbb{R}^N))$. In addition, if $N_1 \colon C(T; \mathbb{R}^N) \longrightarrow C(T; \mathbb{R}^N)$ is defined by

$$N_1(y)(\cdot) \; = \; a^{-1}\big(y(\cdot)\big) \qquad \forall \, y \in C(T; \mathbb{R}^N), \tag{3.5}$$

then clearly N_1 is continuous and bounded (i.e. maps bounded sets into bounded sets). So we can find $c_1 > 0$, such that

$$\|x_n'(t)\|_{\mathbb{R}^N} \leq c_1 \qquad \forall \, n \geq 1, \, t \in T,$$

hence for some $c_2 > 0$, we have that

$$\|a(x_n'(t))\| \; \leq \; c_2 \qquad \forall \, n \geq 1, \, t \in T.$$

Therefore, the sequence $\{a(x_n'(\cdot))\}_{n \geq 1} \subseteq W^{1,p'}(T; \mathbb{R}^N)$ is bounded and because the embedding $W^{1,p'}(T; \mathbb{R}^N) \subseteq C(T; \mathbb{R}^N)$ is compact, it follows that the sequence $\{a(x_n'(\cdot))\}_{n \geq 1} \subseteq C(T; \mathbb{R}^N)$ is relatively compact. Using the continuity of the map N_1, it follows that also the sequence $\{x_n'\}_{n \geq 1} \subseteq C(T; \mathbb{R}^N)$ is relatively compact. Therefore, we have proved that in fact the sequence $\{x_n\}_{n \geq 1} \subseteq C^1(T; \mathbb{R}^N)$ is relatively compact and so, after passing to a subsequence if necessary, we may assume that

$$x_n \; \longrightarrow \; x \quad \text{in } C^1(T; \mathbb{R}^N),$$

for some $x \in C^1(T; \mathbb{R}^N)$. Clearly in the limit as $n \to +\infty$, we obtain that x is a solution of (3.3) and so $x = K(g)$, which proves the claim.

Note that in fact we have proved something stronger, namely complete continuity into $C^1(T; \mathbb{R}^N)$.

Next let $N_2 \colon W_0^{1,p}(T; \mathbb{R}^N) \longrightarrow L^{p'}(T; \mathbb{R}^N)$ be defined by

$$N_2(x)(\cdot) \; \stackrel{df}{=} \; -A_\lambda\big(x(\cdot)\big) - \|x(\cdot)\|_{\mathbb{R}^N}^{p-2} x(\cdot) + g(\cdot) \qquad \forall \, x \in W_0^{1,p}(T; \mathbb{R}^N).$$

Clearly this map is continuous and bounded. Note that problem (3.2) is equivalent to the following abstract fixed point problem:

$$x = (K \circ N_2)(x). \tag{3.6}$$

We solve (3.6) by means of the Leray-Schauder Alternative Theorem (see Theorem A.4.1). Let us define

$$S \stackrel{df}{=} \left\{ x \in W_0^{1,p}(T; \mathbb{R}^N) : \ x = \kappa(K \circ N_2)(x), \text{ for some } \kappa \in (0,1) \right\}.$$

Claim 2. The set S is bounded.

To this end let $x \in S$. We have

$$\begin{cases} -\left(a\left(\frac{1}{\kappa}x'(t)\right)\right)' + A_\lambda\big(x(t)\big) + \|x(t)\|_{\mathbb{R}^N}^{p-2} x(t) = g(t) & \text{for a.a. } t \in T, \\ x(0) = x(b) = 0. \end{cases}$$

Taking inner product with $x(t)$, integrating on $T = [0, b]$ and performing integration by parts and since $(A_\lambda(\xi), \xi)_{\mathbb{R}^N} \geq 0$ for all $\xi \in \mathbb{R}^N$, we obtain

$$\frac{\beta}{\kappa^{p-1}} \|x'\|_p^p + \frac{1}{\kappa^{p-1}} \|x\|_p^p \ \leq \ \|g\|_{p'} \|x\|_p.$$

Since $\kappa \in (0,1)$, we have

$$c_3 \|x\|_{W^{1,p}(T;\mathbb{R}^N)}^{p-1} \ \leq \ \|g\|_{p'},$$

for some $c_3 > 0$. Thus, finally we conclude that the set $S \subseteq W_0^{1,p}(T; \mathbb{R}^N)$ is bounded, which proves the claim.

Claims 1 and 2 permit the application of the Leray-Schauder Alternative Principle (see Theorem A.4.1), which gives a solution $x_0 \in W_0^{1,p}(T; \mathbb{R}^N)$ of the fixed point problem (3.6), hence of (3.2) too. Moreover, in the process of the proof, we established that $x_0 \in C^1(T; \mathbb{R}^N)$. ☐

Let

$$D_a \stackrel{df}{=} \left\{ x \in W_0^{1,p}(T; \mathbb{R}^N) : \ a\big(x'(\cdot)\big) \in W^{1,p'}(T; \mathbb{R}^N) \right\}$$

and for $\lambda > 0$, let $U_\lambda \colon L^p(T; \mathbb{R}^N) \supseteq D_a \longrightarrow L^{p'}(T; \mathbb{R}^N)$ be defined by

$$U_\lambda(x)(\cdot) \stackrel{df}{=} -\big(a\big(x'(\cdot)\big)\big)' + \widehat{A}_\lambda(x)(\cdot) \qquad \forall \, x \in D_a,$$

where $\widehat{A}_\lambda(x)(\cdot) \stackrel{df}{=} A_\lambda\big(x(\cdot)\big)$. Note that

$$\widehat{A}_\lambda(x) \in C(T; \mathbb{R}^N) \qquad \forall \, x \in D_a.$$

PROPOSITION 3.1.3

If hypotheses $H(a)_1$ and $H(A)_1$ hold, then U_λ is maximal monotone.

PROOF   Let $J: L^p(T; \mathbb{R}^N) \longrightarrow L^{p'}(T; \mathbb{R}^N)$ be the continuous, strictly monotone (hence maximal monotone too; see Corollary 1.4.2) operator, defined by

$$J(x)(\cdot) \overset{df}{=} \|x(\cdot)\|_{\mathbb{R}^N}^{p-2} x(\cdot) \qquad \forall \, x \in L^p(T; \mathbb{R}^N).$$

From Proposition 3.1.2, it follows that

$$R(U_\lambda + J) = L^{p'}(T; \mathbb{R}^N),$$

i.e. $U_\lambda + J$ is surjective. We claim that this surjectivity property of $U_\lambda + J$ implies the maximal monotonicity of U_λ (see Definition 1.4.3(e)). Indeed, first note that U_λ is monotone (hypothesis $H(a)_1(ii)$). Suppose that for some $y \in L^p(T; \mathbb{R}^N)$ and some $v \in L^{p'}(T; \mathbb{R}^N)$, we have

$$\langle U_\lambda(x) - v, x - y \rangle_p \geq 0 \qquad \forall \, x \in D_a \tag{3.7}$$

(as usual by $\langle \cdot, \cdot \rangle_p$ we denote the duality brackets for the pair of spaces $\left(L^{p'}(T; \mathbb{R}^N), L^p(T; \mathbb{R}^N) \right)$). Since $U_\lambda + J$ is surjective, we can find $x_1 \in D_a$, such that

$$U_\lambda(x_1) + J(x_1) \ = \ v + J(y).$$

We use this in (3.7) with $x = x_1 \in D_a$ and obtain

$$\langle U_\lambda(x_1) - U_\lambda(x_1) - J(x_1) + J(y), x_1 - y \rangle_p \geq 0,$$

and so, from the monotonicity of J, we have

$$0 \geq \langle J(y) - J(x_1), x_1 - y \rangle_p \geq 0,$$

i.e.

$$\langle J(y) - J(x_1), x_1 - y \rangle_p \ = \ 0.$$

From the strict monotonicity of J, we deduce that $y = x_1 \in D_a$ and $v = U_\lambda(x_1)$. This proves the maximality of U_λ.   □

Next, let us consider the following approximation to problem (3.1):

$$\begin{cases} \left(a(x'(t)) \right)' \in A_\lambda(x(t)) + F\left(t, x(t), x'(t) \right) & \text{for a.a. } t \in T, \\ x(0) = x(b) = 0, \end{cases} \tag{3.8}$$

for $\lambda > 0$. To solve this auxiliary problem, we shall use a multivalued version of the Leray-Schauder Alternative Principle, which is due to Bader (2001), where the interested reader can find its proof.

THEOREM 3.1.1 (Multivalued Leray-Schauder Alternative Theorem)
*If X, Y are Banach spaces, $G \colon X \longrightarrow P_{wkc}(Y)$ is upper semicontinuous from X into Y_w (by Y_w we denote the Banach space Y equipped with the weak topology), $K \colon Y \longrightarrow X$ is completely continuous and $\Phi \overset{df}{=} K \circ G$ is compact (i.e. it maps bounded sets of X into relatively compact sets),
<u>then</u> one of the following alternatives holds:*

either
(a) *the set $S \overset{df}{=} \{ x \in X : \ x \in \kappa\Phi(x) \text{ for some } \kappa \in (0,1) \}$ is unbounded;*

or
(b) *Φ has a fixed point.*

REMARK 3.1.2 We emphasize that the composite multifunction Φ need not have convex values. $\Box$

Using this general fixed point principle, we can establish the solvability of (3.8).

PROPOSITION 3.1.4
*If hypotheses $H(a)_1$, $H(A)_1$ and $H(F)_1$ hold,
<u>then</u> problem (3.8) has a solution $x_0 \in C^1(T; \mathbb{R}^N)$.*

PROOF Let $V_\lambda \colon L^p(T; \mathbb{R}^N) \supseteq D_a \longrightarrow L^{p'}(T; \mathbb{R}^N)$ be the nonlinear operator, defined by

$$V_\lambda(x) \overset{df}{=} U_\lambda(x) + J(x) \qquad \forall \, x \in D_a.$$

Because J and U_λ are maximal monotone (see Proposition 3.1.3), V_λ is maximal monotone too as the sum of two maximal monotone operators (see Theorem 1.4.5). Also we have

$$\langle V_\lambda(x), x \rangle_p = \langle U_\lambda(x), x \rangle_p + \langle J(x), x \rangle_p$$
$$\geq \beta \|x'\|_p^p + \|x\|_p^p \qquad \forall \, x \in D_a$$

(note that $\widehat{A}_\lambda(0) = 0$, so by monotonicity, we have $\left\langle \widehat{A}_\lambda(x), x \right\rangle_p \geq 0$). Hence V_λ is coercive (see Definition 1.4.6(a)). So we can apply Theorem 1.4.4 and infer that V_λ is surjective. Moreover, V_λ is strictly monotone since J is. Thus

$$V_\lambda^{-1} \colon L^{p'}(T; \mathbb{R}^N) \longrightarrow D_a \subseteq W_0^{1,p}(T; \mathbb{R}^N)$$

is a well-defined single-valued map.

Claim 1. $V_\lambda^{-1} \colon L^{p'}(T; \mathbb{R}^N) \longrightarrow W_0^{1,p}(T; \mathbb{R}^N)$ is completely continuous.

Let $\{g_n\}_{n\geq 1} \subseteq L^{p'}(T;\mathbb{R}^N)$ be a sequence, such that

$$g_n \xrightarrow{w} g \quad \text{in } L^{p'}(T,\mathbb{R}^N),$$

for some $g \in L^{p'}(T;\mathbb{R}^N)$ and let us set

$$x_n \stackrel{df}{=} V_\lambda^{-1}(g_n) \qquad \forall\, n \geq 1.$$

We have that $V_\lambda(x_n) = g_n$ for $n \geq 1$ and so

$$\begin{cases} -\big(a\big(x_n'(t)\big)\big)' + \|x_n(t)\|_{\mathbb{R}^N}^{p-2}\, x_n(t) + A_\lambda\big(x_n(t)\big) = g_n(t) \\ \qquad\qquad\qquad\qquad\qquad\qquad \text{for a.a. } t \in T, \\ x_n(0) = x_n(b) = 0. \end{cases} \tag{3.9}$$

Taking the inner product with $x_n(t)$, integrating over $T = [0,b]$, performing integration by parts and recalling that $A_\lambda(0) = 0$ (hence also $(A_\lambda(\xi),\xi)_{\mathbb{R}^N} \geq 0$ for all $\xi \in \mathbb{R}^N$), we have

$$\beta \|x_n'\|_p^p + \|x_n\|_p^p \leq \|g_n\|_{p'}\, \|x_n\|_p,$$

so the sequence $\{x_n\}_{n\geq 1} \subseteq W_0^{1,p}(T;\mathbb{R}^N)$ is bounded and it is relatively compact in $C(T;\mathbb{R}^N)$.

Then from (3.9), it follows that the sequence $\big\{\big(a\big(x_n'(\cdot)\big)\big)'\big\}_{n\geq 1} \subseteq L^{p'}(T;\mathbb{R}^N)$ is bounded. Set

$$\widehat{k}(x_n)(t) \stackrel{df}{=} \|x_n(t)\|_{\mathbb{R}^N}^{p-2}\, x_n(t) + A_\lambda\big(x_n(t)\big) - g_n(t) \qquad \forall\, n \geq 1,\ t \in T.$$

Evidently $\widehat{k}(x_n) \in L^{p'}(T;\mathbb{R}^N)$ for $n \geq 1$ and we have that

$$\big(a\big(x_n'(t)\big)\big)' = \widehat{k}(x_n)(t) \quad \text{for a.a. } t \in T \text{ and all } n \geq 1.$$

Therefore

$$a\big(x_n'(t)\big) = a\big(x_n'(0)\big) + \int_0^t \widehat{k}(x_n)(s)\,ds \qquad \forall\, n \geq 1,\ t \in T.$$

Thus, similarly as in the proof of Proposition 3.1.2, we obtain

$$x_n'(t) = a^{-1}\big(a\big(x_n'(0)\big) + H\big(\widehat{k}(x_n)\big)(t)\big) \qquad \forall\, n \geq 1,\ t \in T.$$

Let $\widehat{\xi}\colon C(T;\mathbb{R}^N) \longrightarrow \mathbb{R}^N$ be the continuous and bounded map provided by Proposition 3.1.1. Since $\int_0^b x_n'(t)\,dt = 0$, it follows that

$$a\big(x_n'(t)\big) = \widehat{\xi}\big(H\big(\widehat{k}(x_n)\big)\big) \qquad \forall\, n \geq 1$$

(see Proposition 3.1.1). So, we have

$$x_n'(t) = a^{-1}\big(\widehat{\xi}\big(H\big(\widehat{k}(x_n)\big)\big) + H\big(\widehat{k}(x_n)\big)(t)\big) \qquad \forall\, n \geq 1,\ t \in T. \qquad (3.10)$$

Note that

$$\big\|H\big(\widehat{k}(x_n)\big)\big\|_{C(T;\mathbb{R}^N)} \leq c_4 \qquad \forall\, n \geq 1,$$

for some $c_4 > 0$. Let $N_3\colon C\big(T;\mathbb{R}^N\big) \longrightarrow C\big(T;\mathbb{R}^N\big)$ be the operator, defined by

$$N_3(x)(\cdot) = a^{-1}\big(x(\cdot)\big) \qquad \forall\, x \in C\big(T;\mathbb{R}^N\big)$$

(see also (3.5) in the proof of Proposition 3.1.2). We know that N_3 is also continuous and bounded. Hence from (3.10), it follows that

$$\|x_n'(t)\|_{\mathbb{R}^N} \leq c_5 \qquad \forall\, n \geq 1,\ t \in T,$$

for some $c_5 > 0$ and

$$\big\|a\big(x_n'(t)\big)\big\|_{\mathbb{R}^N} \leq c_6 \qquad \forall\, n \geq 1,\ t \in T,$$

for some $c_6 > 0$. So the sequence $\big\{a\big(x_n'(\cdot)\big)\big\}_{n\geq 1} \subseteq W^{1,p'}\big(T;\mathbb{R}^N\big)$ is bounded, hence relatively compact in $C\big(T;\mathbb{R}^N\big)$. Thus $\{x_n'(\cdot)\}_{n\geq 1} \subseteq C\big(T;\mathbb{R}^N\big)$ is also relatively compact. We have proved that the sequence $\{x_n\}_{n\geq 1} \subseteq C^1\big(T;\mathbb{R}^N\big)$ is relatively compact and so, passing a subsequence if necessary, we may assume that

$$x_n \longrightarrow x \quad \text{in } C^1\big(T;\mathbb{R}^N\big).$$

For every $\psi \in C_c^1\big((0,b);\mathbb{R}^N\big)$, we have

$$\int_0^b \big(a\big(x_n'(t)\big), \psi'(t)\big)_{\mathbb{R}^N}\, dt + \int_0^b \|x_n(t)\|_{\mathbb{R}^N}^{p-2}\, \big(x_n(t), \psi(t)\big)_{\mathbb{R}^N}\, dt$$

$$+ \int_0^b \big(A_\lambda\big(x_n(t)\big), \psi(t)\big)_{\mathbb{R}^N}\, dt \;=\; \int_0^b \big(g_n(t), \psi(t)\big)_{\mathbb{R}^N}\, dt \qquad \forall\, n \geq 1.$$

Passing to the limit as $n \to +\infty$, we obtain

$$\int_0^b \big(a\big(x'(t)\big), \psi'(t)\big)_{\mathbb{R}^N}\, dt + \int_0^b \|x(t)\|_{\mathbb{R}^N}^{p-2}\, \big(x(t), \psi(t)\big)_{\mathbb{R}^N}\, dt$$

$$+ \int_0^b \big(A_\lambda\big(x(t)\big), \psi(t)\big)_{\mathbb{R}^N}\, dt \;=\; \int_0^b \big(g(t), \psi(t)\big)_{\mathbb{R}^N}\, dt.$$

Since $\psi \in C_c^1\big((0,b);\mathbb{R}^N\big)$ was arbitrary, we infer that

$$\begin{cases} -\big(a(x'(t))\big)' + \|x(t)\|_{\mathbb{R}^N}^{p-2}\, x(t) + A_\lambda\big(x(t)\big) = y(t) & \text{for a.a. } t \in T, \\ x(0) = x(b) = 0, \end{cases}$$

so $x = V_\lambda^{-1}(g)$, which proves the claim.

Next let $N_4 \colon W_0^{1,p}\big(T;\mathbb{R}^N\big) \longrightarrow 2^{L^{p'}(T;\mathbb{R}^N)}$ be the multifunction, defined by

$$N_4(x) \overset{df}{=} S_{-F(\cdot,x(\cdot),x'(\cdot))}^{p'} + J(x) \qquad \forall\, x \in W_0^{1,p}\big(T;\mathbb{R}^N\big).$$

Claim 2. N_4 has values in $P_{wkc}\big(L^{p'}\big(T;\mathbb{R}^N\big)\big)$ and it is upper semicontinuous from $W_0^{1,p}\big(T;\mathbb{R}^N\big)$ into $L^{p'}\big(T;\mathbb{R}^N\big)_w$.

First, let us show the nonemptiness of the values of N_4. Indeed, for a given $x \in W_0^{1,p}\big(T;\mathbb{R}^N\big)$, let $\{s_n\}_{n\geq 1}$ and $\{r_n\}_{n\geq 1}$ be two sequences of $\mathbb{R}^N$-valued step functions, such that

$$s_n(t) \longrightarrow x(t) \quad \text{and} \quad r_n(t) \longrightarrow x'(t) \quad \text{for a.a. } t \in T.$$

By virtue of hypothesis $H(F)_1(i)$, for every $n \geq 1$, the function

$$t \longrightarrow F\big(t, s_n(t), r_n(t)\big) \qquad \text{is measurable}$$

and so by the Yankov-von Neumann-Aumann Selection Theorem (see Theorem 1.2.3), we obtain measurable functions $f_n \colon T \longrightarrow \mathbb{R}^N$, such that

$$f_n(t) \in -F\big(t, s_n(t), r_n(t)\big) \quad \text{for a.a. } t \in T \text{ and all } n \geq 1.$$

By virtue of hypothesis $H(F)_1(iii)$, the sequence $\{f_n\}_{n\geq 1} \subseteq L^{p'}\big(T;\mathbb{R}^N\big)$ is bounded and so we may assume that

$$f_n \overset{w}{\longrightarrow} f \quad \text{in } L^{p'}\big(T;\mathbb{R}^N\big).$$

Invoking Proposition 1.2.12, we obtain

$$f(t) \in \text{conv} \limsup_{n\to+\infty} \big(-F\big(t, s_n(t), r_n(t)\big)\big)$$

$$\subseteq -F\big(t, x(t), x'(t)\big) \quad \text{for a.a. } t \in T,$$

where the last inclusion is a consequence of hypothesis $H(F)_1(ii)$. Thus $f \in S_{-F(\cdot,x(\cdot),x'(\cdot))}^{p'}$ and so N_4 has nonempty values. Clearly the values of N_4 are bounded, closed and convex, hence they are elements of $P_{wkc}\big(L^{p'}\big(T;\mathbb{R}^N\big)\big)$.

To prove the upper semicontinuity of the operator N_4 going from the space $W_0^{1,p}\big(T;\mathbb{R}^N\big)$ into $L^{p'}\big(T;\mathbb{R}^N\big)_w$, first note that because of hypothesis $H(F)_1(iii)$, the operator N_4 is bounded. Also the relative weak topology on bounded sets of $L^{p'}(T; \mathbb{R}^N)$ is metrizable. So according to Proposition 1.2.5, to prove the desired upper semicontinuity of N_4, it suffices to show that the graph $\operatorname{Gr} N_4 \subseteq W_0^{1,p}(T; \mathbb{R}^N) \times L^{p'}(T; \mathbb{R}^N)_w$ is sequentially closed. To this end, let us suppose that $\{(x_n, g_n)\}_{n \geq 1} \subseteq \operatorname{Gr} N_4$ is a sequence, such that

$$x_n \longrightarrow x \text{ in } W_0^{1,p}(T; \mathbb{R}^N),$$
$$g_n \overset{w}{\longrightarrow} g \text{ in } L^{p'}(T; \mathbb{R}^N).$$

We may also assume that

$$x_n(t) \longrightarrow x(t) \ \forall t \in T,$$
$$x_n'(t) \longrightarrow x'(t) \text{ for a.a. } t \in T.$$

As before, using Proposition 1.2.12 and hypothesis $H(F)_1(ii)$, we have that

$$g(t) \in \operatorname{conv} \limsup_{n \to +\infty} \left(-F\big(t, x_n(t), x_n'(t)\big) + \|x_n(t)\|_{\mathbb{R}^N}^{p-2} x_n(t) \right)$$

$$\subseteq -F\big(t, x(t), x'(t)\big) + \|x(t)\|_{\mathbb{R}^N}^{p-2} x(t) \quad \text{for a.a. } t \in T,$$

so $g \in N_4(x)$ and thus $(x, g) \in \operatorname{Gr} N_4$. Hence $\operatorname{Gr} N_4$ is sequentially closed in $W_0^{1,p}(T; \mathbb{R}^N) \times L^{p'}(T; \mathbb{R}^N)_w$ and this proves the claim.

We introduce the set

$$S \overset{df}{=} \left\{ x \in W_0^{1,p}(T; \mathbb{R}^N) : \ x \in \kappa\big(V_\lambda^{-1} \circ N_4\big)(x), \text{ for some } \kappa \in (0, 1) \right\}.$$

Claim 3. The set $S \subseteq W_0^{1,p}(T; \mathbb{R}^N)$ is bounded.

Let $x \in S$. For some $\kappa_0 \in (0, 1)$, we have

$$V_\lambda\big(\tfrac{1}{\kappa_0} x\big) \in N_4(x)$$

and so

$$\begin{cases} -\Big(a\big(\tfrac{1}{\kappa_0} x'(t)\big)\Big)' + \tfrac{1}{\kappa_0^{p-1}} \|x(t)\|_{\mathbb{R}^N}^{p-2} x(t) + A_\lambda\big(\tfrac{1}{\kappa_0} x(t)\big) \\ \qquad = f(t) + \|x(t)\|_{\mathbb{R}^N}^{p-2} x(t) \quad \text{for a.a. } t \in T, \\ x(0) = x(b) = 0, \end{cases} \tag{3.11}$$

with $f \in S_{-F(\cdot, x(\cdot), x'(\cdot))}^{p'}$. By virtue of hypothesis $H(F)_1(iv)$, for a given $\varepsilon > 0$, we can find $M_1 = M_1(\varepsilon) > 0$, such that for almost all $t \in T$, all $\xi, \overline{\xi} \in \mathbb{R}^N$ and all $u \in -F(t, \xi, \overline{\xi})$, we have

$$(u, \xi)_{\mathbb{R}^N} \leq \big(\vartheta_1(t) + \varepsilon\big) \|\xi\|_{\mathbb{R}^N}^p.$$

Combining this with hypothesis $H(F)_1(iii)$, we see that for almost all $t \in T$, all $\xi, \overline{\xi} \in \mathbb{R}^N$ and all $u \in -F(t, \xi, \overline{\xi})$, we have

$$(u, \xi)_{\mathbb{R}^N} \leq \big(\vartheta_1(t) + \varepsilon\big) \|\xi\|_{\mathbb{R}^N}^p + \mu_1(t) \|\overline{\xi}\|_{\mathbb{R}^N}^{p-1} + \mu_2(t), \tag{3.12}$$

with $\mu_1 \in L^\infty(T)$, $\mu_2 \in L^1(T)$. Returning to (3.11), taking the inner product with $x(t)$, integrating over T, performing integration by parts on the first integral and using (3.12), we obtain that

$$\frac{\beta}{\kappa_0^{p-1}} \|x'\|_p^p + \frac{1}{\kappa_0^{p-1}} \|x\|_p^p \tag{3.13}$$

$$\leq \int_0^b \left(\vartheta_1(t) + \varepsilon\right) \|x(t)\|_{\mathbb{R}^N}^p \, dt + \|\mu_1\|_\infty \|x'\|_p^{p-1} + \|\mu_2\|_1 .$$

We shall show that there exists $c_7 > 0$, such that

$$\rho(x) \geq c_7 \|x'\|_p^p \qquad \forall\, x \in W_0^{1,p}(T; \mathbb{R}^N), \tag{3.14}$$

where

$$\rho(x) \overset{df}{=} \beta \|x'\|_p^p - \int_0^b \vartheta_1(t) \|x(t)\|_{\mathbb{R}^N}^p \, dt \qquad \forall\, x \in W_0^{1,p}(T; \mathbb{R}^N).$$

Suppose that (3.14) is not true. Exploiting the positive p-homogeneity of ρ, we can find a sequence $\{x_n\}_{n\geq 1} \subseteq W_0^{1,p}(T; \mathbb{R}^N)$ with $\|x_n'\|_p = 1$, such that $\rho(x_n) \searrow 0$ (note that $\rho \geq 0$). By the Poincaré inequality (see Theorem 1.1.6), we may assume that

$$x_n \overset{w}{\longrightarrow} x \text{ in } W_0^{1,p}(T; \mathbb{R}^N),$$
$$x_n \longrightarrow x \text{ in } C(T; \mathbb{R}^N).$$

From the weak lower semicontinuity of the norm functional, we have

$$\beta \|x'\|_p^p - \int_0^b \vartheta_1(t) \|x(t)\|_{\mathbb{R}^N}^p \, dt \leq 0,$$

so

$$\beta \|x'\|_p^p \leq \int_0^b \vartheta_1(t) \|x(t)\|_{\mathbb{R}^N}^p \, dt \leq \beta \lambda_1 \|x\|_p^p$$

and due to the variational characterization of λ_1 (Rayleigh quotient; see Proposition 1.5.5), we obtain

$$\|x'\|_p^p = \lambda_1 \|x\|_p^p .$$

We have

$$\rho(x_n) = \beta - \int_0^b \vartheta_1(t) \|x_n(t)\|_{\mathbb{R}^N}^p \, dt \qquad \forall\, n \geq 1,$$

so

$$\beta = \int_0^b \vartheta_1(t) \|x(t)\|_{\mathbb{R}^N}^p \, dt,$$

i.e. $x \neq 0$.

So it follows that $x = \pm u_1$ (recall that by u_1 we denote the principal eigenfunction of $\left(- \Delta_p, W_0^{1,p}(T; \mathbb{R}^N) \right)$). Therefore $x(t) \neq 0$ for almost all $t \in T$. From hypothesis $H(F)_1(iv)$, we have that

$$\beta \|x'\|_p^p = \int_0^b \vartheta_1(t) \|x(t)\|_{\mathbb{R}^N}^p \, dt < \beta \lambda_1 \|x\|_p^p,$$

so

$$\|x'\|_p^p < \lambda_1 \|x\|_p^p,$$

a contradiction to the variational characterization of $\lambda_1 > 0$. This proves inequality (3.14). Using (3.14) in (3.13) and since $\kappa_0 \in (0,1)$, we have that

$$c_7 \|x'\|_p^p \leq \varepsilon \|x\|_p^p + c_8 \|x'\|_p^{p-1} + c_9,$$

for some $c_8, c_9 > 0$. So by the Poincaré inequality (see Theorem 1.1.6), we obtain

$$c_7 \|x'\|_p^p \leq c_{10} \varepsilon \|x'\|_p^p + c_8 \|x'\|_p^{p-1} + c_9,$$

for some $c_{10} > 0$. Let $\varepsilon \in \left(0, \frac{c_7}{c_{10}} \right)$. We obtain

$$\|x'\|_p^p \leq c_{11} \|x'\|_p^{p-1} + c_{12},$$

for some $c_{11}, c_{12} > 0$. Thus, we conclude that the set $S \subseteq W_0^{1,p}(T; \mathbb{R}^N)$ is bounded. Hence Claim 3 is true.

Claims 1, 2 and 3 permit the application of Theorem 3.1.1, which produces an element $x_0 \in D_a$, such that $x_0 \in V_\lambda^{-1} N_4(x_0)$. Clearly $x_0 \in C^1(T; \mathbb{R}^N)$ and so it solves problem (3.8). $\Box$

Now we shall pass to the limit as $\lambda \searrow 0$ to obtain a solution of the original problem (3.1). To this end, first we prove an auxiliary result. Let $\widehat{A}$ be the **lifting (realization)** of A on the dual pair $\left(L^p(T; \mathbb{R}^N), L^{p'}(T; \mathbb{R}^N) \right)$, namely the operator defined by

$$\widehat{A}(x) \stackrel{df}{=} \left\{ h \in L^{p'}(T; \mathbb{R}^N) : h(t) \in A(x(t)) \text{ for a.a. } t \in T \right\} \qquad \forall \, x \in \widehat{D},$$

where

$$\widehat{D} = \mathrm{Dom}\,(\widehat{A}) \stackrel{df}{=} \left\{ x \in L^p(T; \mathbb{R}^N) : x(t) \in \mathrm{Dom}\,(A) \text{ for a.a. } t \in T \right.$$
$$\left. \text{and } S_{A(x(\cdot))}^{p'} \neq \emptyset \right\}.$$

Clearly $\widehat{D} \supseteq \mathrm{Dom}\,(A)$.

PROPOSITION 3.1.5
If hypotheses $H(A)_1$ *hold,*

then *the multivalued operator* $\widehat{A}\colon L^p\big(T;\mathbb{R}^N\big) \supseteq \widehat{D} \longrightarrow 2^{L^{p'}(T;\mathbb{R}^N)}$ *is maximal monotone.*

PROOF   Clearly $\widehat{A}$ is monotone. Let $J\colon L^p\big(T;\mathbb{R}^N\big) \longrightarrow L^{p'}\big(T;\mathbb{R}^N\big)$ be the operator, defined by

$$J(x)(\cdot) \overset{df}{=} \|x(\cdot)\|_{\mathbb{R}^N}^{p-2}\, x(\cdot) \qquad \forall\, x \in L^p\big(T;\mathbb{R}^N\big)$$

(see the proof of Proposition 3.1.3). We know that J is continuous, strictly monotone, thus maximal monotone too (see Corollary 1.4.2). First we show that

$$R(\widehat{A} + J) = L^{p'}\big(T;\mathbb{R}^N\big). \tag{3.15}$$

For this purpose let $h \in L^{p'}\big(T;\mathbb{R}^N\big)$ and consider the multifunction $\Gamma\colon T \longrightarrow 2^{\mathbb{R}^N}$, defined by

$$\Gamma(t) \overset{df}{=} \big\{\xi \in \mathbb{R}^N \,:\, A(\xi) + \varphi(\xi) \ni h(t)\big\} \qquad \forall\, t \in T,$$

where $\varphi\colon \mathbb{R}^N \longrightarrow \mathbb{R}^N$ is the map, defined by

$$\varphi(\xi) \overset{df}{=} \begin{cases} \|\xi\|_{\mathbb{R}^N}^{p-2}\,\xi & \text{for } \xi \in \mathbb{R}^N \setminus \{0\}, \\ 0 & \text{for } \xi = 0. \end{cases}$$

As φ is continuous and monotone, it is maximal monotone (see Corollary 1.4.2). Thus $A + \varphi$ is maximal monotone as the sum of two maximal monotone maps (see Theorem 1.4.5). Moreover, because $0 \in A(0)$, we have that

$$\big(A(\xi) + \varphi(\xi), \xi\big)_{\mathbb{R}^N} \geq \big(\varphi(\xi), \xi\big)_{\mathbb{R}^N} = \|\xi\|_{\mathbb{R}^N}^p \qquad \forall\, \xi \in \mathbb{R}^N$$

and so $A + \varphi$ is weakly coercive. Therefore from Corollary 1.4.5, we infer that $\Gamma(t) \neq \emptyset$ for all $t \in T$. Note that

$$\mathrm{Gr}\,\Gamma \;=\; \big\{(t,\xi) \in T \times \mathbb{R}^N \,:\, \big(\xi, h(\xi) - \varphi(t)\big) \in \mathrm{Gr}\,A\big\}.$$

Let $k\colon T \times \mathbb{R}^N \longrightarrow \mathbb{R}^N \times \mathbb{R}^N$ be defined by

$$k(t,\xi) \overset{df}{=} \big(\xi, h(\xi) - \varphi(t)\big) \qquad \forall\, (t,\xi) \in T \times \mathbb{R}^N.$$

Evidently k is a Carathéodory function, thus it is jointly measurable. So

$$\mathrm{Gr}\,\Gamma \;=\; k^{-1}(\mathrm{Gr}\,A) \;\in\; \mathcal{L} \times \mathcal{B}(\mathbb{R}^N),$$

with $\mathcal{L}$ being the Lebesgue σ-field of T and $\mathcal{B}(\mathbb{R}^N)$ the Borel σ-field of $\mathbb{R}^N$ (recall that $\operatorname{Gr} A \subseteq \mathbb{R}^N \times \mathbb{R}^N$ is closed, see Corollary 1.4.1). So we can apply the Yankov-von Neumann-Aumann Selection Theorem (see Theorem 1.2.3) and obtain a measurable function $x \colon T \longrightarrow \mathbb{R}^N$, such that

$$x(t) \in \Gamma(t) \quad \text{for a.a. } t \in T.$$

We have that

$$h(t) \in A\big(x(t)\big) + \varphi\big(x(t)\big) \quad \text{for a.a. } t \in T.$$

Taking the inner product with $x(t)$ and recalling that $0 \in A(0)$, we obtain

$$\|x(t)\|_{\mathbb{R}^N}^p \leq \big(h(t), x(t)\big)_{\mathbb{R}^N} \quad \text{for a.a. } t \in T.$$

So

$$\|x(t)\|_{\mathbb{R}^N}^{p-1} \leq \|h(t)\|_{\mathbb{R}^N} \quad \text{for a.a. } t \in T,$$

which implies that $x \in L^p(T; \mathbb{R}^N)$. Therefore $h \in \widehat{A}(x) + J(x)$ and we have proved (3.15).

To show the maximality of $\widehat{A}$, let $y \in L^p(T; \mathbb{R}^N)$ and $v \in L^{p'}(T; \mathbb{R}^N)$ satisfy

$$\langle u - v, x - y \rangle_p \geq 0 \quad \forall \, x \in \widehat{D}, \, u \in \widehat{A}(x).$$

Because of (3.15), we can find $x_1 \in \widehat{D}$ and $u_1 \in \widehat{A}(x_1)$, such that

$$v + J(y) = u_1 + J(x_1).$$

Then, we have

$$0 \leq \big\langle v + J(y) - J(x_1) - v, x_1 - y \big\rangle_p = \big\langle J(y) - J(x_1), x_1 - y \big\rangle_p \leq 0$$

and the last inequality is strict if $x_1 \neq y$ (recall that J is strictly monotone). So we infer that $y = x_1 \in \widehat{D}$ and $v = u_1 \in \widehat{A}(x_1)$, which proves the maximality of $\widehat{A}$. $\qquad\Box$

3.1.3 Existence Results

Using the previous result, we can pass to the limit as $\lambda \searrow 0$ in (3.8) in order to produce a solution of problem (3.1).

THEOREM 3.1.2
If hypotheses $H(A)_1$, $H(a)_1$ and $H(F)_1$ hold,
then problem (3.1) has a solution $x_0 \in C^1(T; \mathbb{R}^N)$.

PROOF   Let $\lambda_n \searrow 0$ and let $x_n \in C^1(T; \mathbb{R}^N)$ for $n \geq 1$ be solutions of the corresponding auxiliary problems (3.8) with $\lambda = \lambda_n$, i.e. for all $n \geq 1$, we have

$$\begin{cases} \left(a(x_n'(t))\right)' = A_{\lambda_n}(x_n(t)) + f_n(t) & \text{for a.a. } t \in T, \\ x_n(0) = x_n(b) = 0, \end{cases} \qquad (3.16)$$

where $f_n \in S_{F(\cdot, x_n(\cdot), x_n'(\cdot))}^{p'}$. Taking inner product with $x_n(t)$, integrating over T, performing integration by parts using hypothesis $H(a)_1(iii)$ and the fact that $A_{\lambda_n}(0) = 0$ for all $n \geq 1$, we obtain

$$\beta \|x_n'\|_p^p \leq \int_0^b \left(-f_n(t), x_n(t)\right)_{\mathbb{R}^N} dt \qquad \forall\, n \geq 1.$$

From (3.12), we have

$$\left(-f_n(t), x_n(t)\right)_{\mathbb{R}^N}$$
$$\leq \left(\vartheta_1(t) + \varepsilon\right) \|x_n(t)\|_{\mathbb{R}^N}^p + \mu_1(t) \|x_n'(t)\|_{\mathbb{R}^N}^{p-1} + \mu_2(t) \quad \text{for a.a. } t \in T,$$

with $\mu_1 \in L^\infty(T)$ and $\mu_2 \in L^1(T)$. So, for all $n \geq 1$, we obtain

$$\beta \|x_n'\|_p^p \leq \int_0^b \vartheta_1(t) \|x_n(t)\|_{\mathbb{R}^N}^p \, dt + \varepsilon \|x_n\|_p^p + c_{13} \|x_n'\|_p^{p-1} + c_{14},$$

for some $c_{13}, c_{14} > 0$. Using the Poincaré inequality (see Theorem 1.1.6), we have

$$c_7 \|x_n'\|_p^p \leq \varepsilon c_{15} \|x_n'\|_p^p + c_{13} \|x_n'\|_p^{p-1} + c_{14} \qquad \forall\, n \geq 1,$$

for some $c_{15} > 0$. Let $\varepsilon \in \left(0, \frac{c_7}{c_{15}}\right)$, to deduce that the sequence $\{x_n'\}_{n \geq 1} \subseteq L^p(T; \mathbb{R}^N)$ is bounded, hence so is the sequence $\{x_n\}_{n \geq 1} \subseteq W_0^{1,p}(T; \mathbb{R}^N)$ (by the Poincaré inequality; see Theorem 1.1.6). Thus, passing to a subsequence if necessary, we may assume that

$$x_n \xrightarrow{w} x \text{ in } W_0^{1,p}(T; \mathbb{R}^N),$$
$$x_n \longrightarrow x \text{ in } C(T; \mathbb{R}^N).$$

Now, taking the inner product of (3.16) with $A_{\lambda_n}(x_n(t))$ and then integrating over T, we obtain

$$\int_0^b \left(-(a(x_n'(t)))', A_{\lambda_n}(x_n(t))\right)_{\mathbb{R}^N} dt + \left\|\widehat{A}_{\lambda_n}(x_n)\right\|_2^2$$
$$\leq \int_0^b \|f_n(t)\|_{\mathbb{R}^N} \left\|A_{\lambda_n}(x_n(t))\right\|_{\mathbb{R}^N} dt \qquad \forall\, n \geq 1. \qquad (3.17)$$

Note that the functions $\widehat{A}_{\lambda_n}(x_n)(\cdot) = A_{\lambda_n}(x_n(\cdot))$ are Lipschitz continuous for $n \geq 1$. By integration by parts and because $x_n(0) = x_n(b) = 0$ and $A_{\lambda_n}(0) = 0$, we have

$$\int_0^b \left(-(a(x_n'(t)))', A_{\lambda_n}(x_n(t))\right)_{\mathbb{R}^N} dt$$

$$= -\left(a(x_n'(b)), A_{\lambda_n}(x_n(b))\right)_{\mathbb{R}^N} + \left(a(x_n'(0)), A_{\lambda_n}(x_n(0))\right)_{\mathbb{R}^N}$$

$$+ \int_0^b \left(a(x_n'(t)), \frac{d}{dt} A_{\lambda_n}(x_n(t))\right)_{\mathbb{R}^N} dt$$

$$= \int_0^b \left(a(x_n'(t)), \frac{d}{dt} A_{\lambda_n}(x_n(t))\right)_{\mathbb{R}^N} dt \qquad \forall \, n \geq 1.$$

Since A_{λ_n} is Lipschitz continuous (see Proposition 1.4.8(b)), by the Rademacher Theorem (see Theorem A.2.4), it is differentiable at every $\xi \in \mathbb{R}^N \setminus S_1$, for some $S_1 \subseteq \mathbb{R}^N$, such that $|S_1|_N = 0$. Due to the monotonicity of A_{λ_n}, we have

$$\left(\overline{\xi}, \frac{A_{\lambda_n}(\xi + s\overline{\xi}) - A_{\lambda_n}(\xi)}{s}\right)_{\mathbb{R}^N} \geq 0 \qquad \forall \, \xi \in \mathbb{R}^N \setminus S_1, \ \overline{\xi} \in \mathbb{R}^N \ s > 0$$

and so

$$\left(\overline{\xi}, A_{\lambda_n}'(\xi)\overline{\xi}\right)_{\mathbb{R}^N} \geq 0 \qquad \forall \, n \geq 1, \ \xi \in \mathbb{R}^N \setminus S_1, \ \overline{\xi} \in \mathbb{R}^N.$$

Moreover, from the Chain Rule for Sobolev Functions (see Proposition 1.1.13(b)), we know that

$$\frac{d}{dt} A_{\lambda_n}(x_n(t)) = A_{\lambda_n}'(x_n(t)) x_n'(t) \quad \text{for a.a. } t \in T.$$

Then, using hypotheses $H(a)_1$, we have

$$\int_0^b \left(-(a(x_n'(t)))', A_{\lambda_n}(x_n(t))\right)_{\mathbb{R}^N} dt$$

$$= \int_0^b \left(a(x_n'(t)), A_{\lambda_n}'(x_n(t)) x_n'(t)\right)_{\mathbb{R}^N} dt \qquad\qquad (3.18)$$

$$= \int_0^b k_a(x_n'(t)) \left(x_n'(t), A_{\lambda_n}'(x_n(t)) x_n'(t)\right)_{\mathbb{R}^N} dt \geq 0 \qquad \forall \, n \geq 1.$$

Since $\sup_{n \geq 1} \|x_n\|_\infty \leq M_2$, for some $M_2 > 0$, from hypothesis $H(F)_1(iii)$, we have

$$\|f_n(t)\|_{\mathbb{R}^N} \leq \gamma_{1,M_2}(t) + \gamma_{2,M_2}(t) \|x_n'(t)\|_{\mathbb{R}^N}^{p-1} \quad \text{for a.a. } t \in T. \qquad (3.19)$$

Because $\gamma_{1,M_2} \in L^2(T)$, $\gamma_{2,M_2} \in L^\infty(T)$ and $x_n' \in C(T;\mathbb{R}^N)$ (since $x_n \in \widehat{D}$), it follows that $f_n \in L^2(T;\mathbb{R}^N)$ for $n \geq 1$. So returning to (3.17) and using (3.18) and the Cauchy-Schwarz inequality (see Theorem A.3.15), we obtain

$$\left\|\widehat{A}_{\lambda_n}(x_n)\right\|_2^2 \leq \|f_n\|_2 \left\|\widehat{A}_{\lambda_n}(x_n)\right\|_2 \qquad \forall\, n \geq 1$$

and so the sequence $\left\{\widehat{A}_{\lambda_n}(x_n)\right\}_{n \geq 1} \subseteq L^2(T;\mathbb{R}^N)$ is bounded. Passing to a subsequence if necessary, we may assume that

$$\widehat{A}_{\lambda_n}(x_n) \xrightarrow{w} \widehat{u} \quad \text{in } L^2(T;\mathbb{R}^N),$$

for some $\widehat{u} \in L^2(T;\mathbb{R}^N)$. Arguing as in Claim 1 in the proof of Proposition 3.1.4, we obtain that the sequence $\{x_n\}_{n \geq 1} \subseteq C^1(T;\mathbb{R}^N)$ is relatively compact. So, we can say that

$$x_n \longrightarrow x_0 \quad \text{in } C^1(T;\mathbb{R}^N),$$

for some $x_0 \in C^1(T;\mathbb{R}^N)$. It follows that

$$a(x_n'(t)) \longrightarrow a(x_0'(t)) \qquad \forall\, t \in T.$$

From (3.19), we see that also the sequence $\{f_n\}_{n \geq 1} \subseteq L^2(T;\mathbb{R}^N)$ is bounded and so, passing to a subsequence if necessary, we have

$$f_n \xrightarrow{w} f \quad \text{in } L^2(T;\mathbb{R}^N).$$

Invoking Proposition 1.2.12, we obtain that $f \in S^2_{F(\cdot,x_0(\cdot),x_0'(\cdot))}$. For all test functions $\psi \in C_c^1((0,b);\mathbb{R}^N)$, we have

$$\int_0^b \left(-a(x_n'(t)), \psi'(t)\right)_{\mathbb{R}^N} dt = \int_0^b \left(A_{\lambda_n}(x_n(t)), \psi(t)\right)_{\mathbb{R}^N} dt$$

$$+ \int_0^b \left(f_n(t), \psi(t)\right)_{\mathbb{R}^N} dt \qquad \forall\, n \geq 1.$$

Passing to the limit as $n \to +\infty$, we obtain

$$\int_0^b \left(-a(x_0'(t)), \psi'(t)\right)_{\mathbb{R}^N} dt = \int_0^b \left(\widehat{u}(t), \psi(t)\right)_{\mathbb{R}^N} dt + \int_0^b \left(f(t), \psi(t)\right)_{\mathbb{R}^N} dt.$$

and

$$\begin{cases} \bigl(a(x_0'(t))\bigr)' = \widehat{u}(t) + f(t) & \text{for a.a. } t \in T, \\ x_0(0) = x_0(b) = 0, \end{cases}$$

with $f \in S^2_{F(\cdot,x_0(\cdot),x_0'(\cdot))}$. We can finish the proof if we show that $\widehat{u} \in S^{p'}_{A(x_0(\cdot))}$. To this end note that $J_{\lambda_n}(x_n(\cdot)) \in W^{1,p}(T;\mathbb{R}^N)$ and

$$\frac{d}{dt} J_{\lambda_n}(x_n(t)) = J'_{\lambda_n}(x_n(t)) x_n'(t) \quad \text{for a.a. } t \in T.$$

Also

$$\bigl\| J'_{\lambda_n}(x_n(t)) \bigr\|_{\mathbb{R}^N} \leq 1 \quad \text{for a.a. } t \in T$$

(see Proposition 1.4.8(a)). Thus putting $\widehat{J}_{\lambda_n}(x_n)(\cdot) \overset{df}{=} J_{\lambda_n}(x_n(\cdot))$, we have that the sequence $\bigl\{ \widehat{J}_{\lambda_n}(x_n) \bigr\}_{n \geq 1} \subseteq W^{1,p}(T;\mathbb{R}^N)$ is bounded and, passing to a subsequence if necessary, we may assume that

$$\begin{aligned} \widehat{J}_{\lambda_n}(x_n) &\overset{w}{\longrightarrow} \widehat{z} \text{ in } W^{1,p}(T;\mathbb{R}^N), \\ \widehat{J}_{\lambda_n}(x_n) &\longrightarrow \widehat{z} \text{ in } C(T;\mathbb{R}^N). \end{aligned}$$

We have

$$J_{\lambda_n}(x_n(t)) + \lambda_n A_{\lambda_n}(x_n(t)) = x_n(t) \quad \forall n \geq 1, \ t \in T$$

and so

$$\widehat{J}_{\lambda_n}(x_n) + \lambda_n \widehat{A}_{\lambda_n}(x_n) = x_n \quad \forall n \geq 1.$$

Because the sequence $\bigl\{ \widehat{A}_{\lambda_n}(x_n) \bigr\}_{n \geq 1} \subseteq L^2(T;\mathbb{R}^N)$ is bounded and $\lambda_n \searrow 0$, we have that

$$\widehat{J}_{\lambda_n}(x_n) \longrightarrow x_0 \quad \text{in } C(T;\mathbb{R}^N).$$

Recall that

$$A_{\lambda_n}(x_n(t)) \in A\bigl(J_{\lambda_n}(x_n(t))\bigr) \quad \forall n \geq 1, \ t \in T,$$

hence

$$\widehat{A}_{\lambda_n}(x_n) \in \widehat{A}\bigl(\widehat{J}_{\lambda_n}(x_n)\bigr) \quad \forall n \geq 1.$$

An appeal to Proposition 3.1.5 and to Corollary 1.4.1 gives that $\widehat{u} \in \widehat{A}(x_0)$, which finishes the proof of the theorem. □

We can relax the hypotheses on a at the expense of strengthening the hypotheses on A. The new conditions on A preclude differential variational inequalities.

<u>$H(A)_2$</u> $A \colon \mathbb{R}^N \supseteq \mathrm{Dom}\,(A) \longrightarrow 2^{\mathbb{R}^N}$ is a maximal monotone operator, such that $\mathrm{Dom}\,(A) = \mathbb{R}^N$.

$\underline{H(a)_2}$ $a: \mathbb{R}^N \longrightarrow \mathbb{R}^N$ is a continuous map, such that

    (*i*) u is strictly monotone;

    (*ii*) we have

$$(a(\xi)\xi, \xi)_{\mathbb{R}^N} \geq \beta \|\xi\|_{\mathbb{R}^N}^p \qquad \forall\, \xi \in \mathbb{R}^N,$$

    with $\beta > 0$, $p \in (1, \infty)$.

$\underline{H(F)_2}$ $F: T \times \mathbb{R}^N \times \mathbb{R}^N \longrightarrow P_{kc}(\mathbb{R}^N)$ is a multifunction satisfying hypotheses $H(F)_1(i) - (iv)$, but with $\eta_{1,k} \in L^{p'}(T)$ (see hypothesis $H(F)_1(iii)$).

THEOREM 3.1.3
If hypotheses $H(A)_2$, $H(a)_2$ and $H(F)_2$ hold,
then problem (3.1) has a solution $x_0 \in C^1(T; \mathbb{R}^N)$.

PROOF   The extra structure for the function a assumed in hypothesis $H(a)_1(i)$, namely that $a(\xi) = k(\xi)\xi$, was only used in the proof of Theorem 3.1.2 to establish the L^2-boundedness of the following sequence $\{\widehat{A}_{\lambda_n}(x_n)\}_{n \geq 1} \subseteq L^2(T; \mathbb{R}^N)$. So all the previous results are still valid in the present setting and we can repeat the first part of the proof of Theorem 3.1.2 to obtain that the sequence $\{x_n\}_{n \geq 1} \subseteq W_0^{1,p}(T; \mathbb{R}^N)$ is bounded, in particular that $\{x_n\}_{n \geq 1}$ is relatively compact in $C(T; \mathbb{R}^N)$. We can find $M_3 > 0$, such that

$$\|x_n(t)\|_{\mathbb{R}^N} \leq M_3 \qquad \forall\, n \geq 1,\ t \in T,$$

for some $M_3 > 0$. Because J_{λ_n} is nonexpansive and $J_{\lambda_n}(0) = 0$, we have

$$J_{\lambda_n}(x_n(t)) \in \overline{B}_{M_3} \qquad \forall\, n \geq 1\ t \in T.$$

where $\overline{B}_{M_3} \overset{df}{=} \{\xi \in \mathbb{R}^N : \|\xi\|_{\mathbb{R}^N} \leq M_3\}$. Recall that

$$A_{\lambda_n}(x_n(t)) \in A(J_{\lambda_n}(x_n(t))) \subseteq A(\overline{B}_{M_3}) \qquad \forall\, n \geq 1,\ t \in T.$$

But A is upper semicontinuous with compact and convex values (see Propositions 1.4.5 and 1.4.6). Therefore $A(\overline{B}_{M_3}) \in P_k(\mathbb{R}^N)$ (see Remark 1.2.1). Hence, we have

$$\|A_{\lambda_n}(x_n(t))\|_{\mathbb{R}^N} \leq M_4 \qquad \forall\, n \geq 1,\ t \in T,$$

for some $M_4 > 0$. The rest of the proof is identical with the corresponding part of the proof of Theorem 3.1.2.   ⬚

3.1.4 Problems with Non-Convex Nonlinearities

We can also prove existence theorems when F is not necessarily convex-valued. In this case our hypotheses on the multivalued nonlinearity F are the following:

$\underline{H(F)_3}$ $F: T \times \mathbb{R}^N \times \mathbb{R}^N \longrightarrow P_k(\mathbb{R}^N)$ is a multifunction, such that

(i) F is graph measurable;

(ii) for almost all $t \in T$, the multifunction

$$\mathbb{R}^N \times \mathbb{R}^N \ni (\xi, \overline{\xi}) \longmapsto F(t, \xi, \overline{\xi}) \in P_k(\mathbb{R}^N)$$

is lower semicontinuous;

(iii) the same as hypothesis $H(F)_1(iii)$;

(iv) the same as hypothesis $H(F)_1(iv)$.

THEOREM 3.1.4
If hypotheses $H(A)_1$, $H(a)_1$ and $H(F)_3$ hold,
then problem (3.1) has a solution $x_0 \in C^1(T; \mathbb{R}^N)$.

PROOF Let $N_5: W_0^{1,p}(\Omega) \longrightarrow P_f(L^{p'}(T; \mathbb{R}^N))$ be the multifunction defined by

$$N_5(x) \overset{df}{=} S^{p'}_{F(\cdot, x(\cdot), x'(\cdot))} \qquad \forall\, x \in W_0^{1,p}(\Omega),$$

i.e. the multivalued Nemytskii operator corresponding to F. We claim that N_5 is lower semicontinuous. According to Proposition 1.2.6(a), it suffices to show that for every $v \in L^{p'}(T; \mathbb{R}^N)$, the function

$$W_0^{1,p}(\Omega) \ni x \longmapsto d_{L^{p'}(T;\mathbb{R}^N)}(v, N_5(x)) \in \mathbb{R}_+$$

is upper semicontinuous. But from Theorem 1.2.8, we have that

$$d_{L^{p'}(T;\mathbb{R}^N)}(v, N_5(x)) = \left(\int_0^b d_{\mathbb{R}^N}(v(t), F(t, x(t), x'(t)))^{p'}\, dt \right)^{\frac{1}{p'}}.$$

Since the map

$$\mathbb{R}^N \times \mathbb{R}^N \ni (\xi, \overline{\xi}) \longmapsto d_{\mathbb{R}^N}(v(t), F(t, \xi, \overline{\xi})) \in \mathbb{R}_+$$

is upper semicontinuous (see hypothesis $H(F)_3(ii)$ and Proposition 1.2.6(a)), from Fatou's lemma (see Proposition A.2.2), we deduce that the function

$$W_0^{1,p}(\Omega) \ni x \longmapsto d_{L^{p'}(T;\mathbb{R}^N)}(v, N_5(x)) \in \mathbb{R}_+$$

is upper semicontinuous, hence the function

$$W_0^{1,p}(\Omega) \ni x \longmapsto N_5(x) \in P_f\big(L^{p'}(T;\mathbb{R}^N)\big)$$

is lower semicontinuous.

Invoking Theorem 1.2.11, we obtain a continuous map

$$\widehat{u} \colon W_0^{1,p}(T;\mathbb{R}^N) \longrightarrow L^{p'}(T;\mathbb{R}^N),$$

such that

$$\widehat{u}(x) \in N_5(x) \qquad \forall\, x \in W_0^{1,p}(T;\mathbb{R}^N).$$

For every $\lambda > 0$, we consider the auxiliary problem

$$\begin{cases} \big(a(x'(t))\big)' = A_\lambda\big(x(t)\big) + \widehat{u}(x)(t) & \text{for a.a. } t \in T, \\ x(0) = x(b) = 0. \end{cases}$$

Reasoning as in the proof of Proposition 3.1.4, we show that this problem has a solution $x_\lambda \in C^1(T;\mathbb{R}^N)$. Letting $\lambda \searrow 0$ as in the proof of Theorem 3.1.2, we obtain a solution $x_0 \in C^1(T;\mathbb{R}^N)$ of problem (3.1). ⬜

As in the "convex" case, we can obtain an existence result with relaxed hypotheses on a at the expense of strengthening the hypotheses on A. In this case we need the following hypotheses on the nonlinearity F.

$\underline{H(F)_4}$ $F \colon T \times \mathbb{R}^N \times \mathbb{R}^N \longrightarrow P_k(\mathbb{R}^N)$ is a multifunction satisfying hypotheses $H(F)_3(i) - (iv)$, but with $\eta_{1,k} \in L^{p'}(T)$ (see hypothesis $H(F)_1(iii)$).

THEOREM 3.1.5
If hypotheses $H(A)_2$, $H(a)_2$ and $H(F)_4$ hold,
then problem (3.1) has a solution $x_0 \in C^1(T;\mathbb{R}^N)$.

EXAMPLE 3.1.1   Let

$$\mathbb{R}_+^N \overset{df}{=} \big\{ \xi = (\xi_1, \ldots, \xi_N) \colon \xi_k \geq 0 \text{ for all } k \in \{1, \ldots, N\} \big\}$$

and

$$i_{\mathbb{R}_+^N}(\xi) \overset{df}{=} \begin{cases} 0 & \text{if } \xi \in \mathbb{R}_+^N, \\ +\infty & \text{if } \xi \notin \mathbb{R}_+^N. \end{cases}$$

Evidently $i_{\mathbb{R}_+^N} \in \Gamma_0(\mathbb{R}^N)$ (see Definition 1.3.1). We consider

$$A(\xi) \overset{df}{=} \partial i_{\mathbb{R}_+^N}(\xi) = N_{\mathbb{R}_+^N}(\xi)$$

$$= \begin{cases} 0 & \text{if } \xi_k > 0 \text{ for all } k \in \{1, \ldots, N\}, \\ -\mathbb{R}_+^N \cap \{x\}^\perp & \text{if } \xi_k = 0 \text{ for at least one } k \in \{1, \ldots, N\} \end{cases}$$

(see Definition 1.3.9). Then problem (3.1) becomes the following multivalued differential variational inequality:

$$\begin{cases} \left(\left(a(x'(t))\right)\right)' \in F\big(t,x(t),x'(t)\big) \\ \qquad \text{a.e. on } \{t \in T : \ x_k(t) > 0 \text{ for all } k \in \{1,\ldots,N\}\}, \\ \left(a\big(x'(t)\big)\right)' \in F\big(t,x(t),x'(t)\big) - u(t) \\ \qquad \text{a.e. on } \{t \in T : \ x_k(t) = 0 \text{ for at least one } k \in \{1,\ldots,N\}\}, \\ u(t) \in \mathbb{R}_+^N, \ \big(u(t),x(t)\big)_{\mathbb{R}^N} = 0 \quad \forall t \in T, \\ x(0) = x(b) = 0, \\ x \in C^1\big(T; \mathbb{R}_+^N\big). \end{cases}$$

If $F = f$ is single-valued, then the last problem takes the following more familiar form (recall that $\xi \le \overline{\xi}$ in $\mathbb{R}^N$ if and only if $\overline{\xi} - \xi \in \mathbb{R}_+^N$).

$$\begin{cases} \left(\left(a(x'(t))\right)\right)' = f\big(t,x(t),x'(t)\big) \\ \qquad \text{a.e. on } \{t \in T : \ x_k(t) > 0 \text{ for all } k \in \{1,\ldots,N\}\}, \\ \left(a\big(x'(t)\big)\right)' \le f\big(t,x(t),x'(t)\big) \\ \qquad \text{a.e. on } \{t \in T : \ x_k(t) = 0 \text{ for at least one } k \in \{1,\ldots,N\}\}, \\ \big(f\big(t,x(t),x'(t)\big) - \big(a\big(x'(t)\big)\big)', x(t)\big)_{\mathbb{R}^N} = 0, \ \text{for a.a. } t \in T, \\ x(0) = x(b) = 0, \\ x \in C^1\big(T; \mathbb{R}_+^N\big). \end{cases}$$

$\Box$

3.2   Periodic Problems

This section deals with periodic problems. Our approach to the resolution of these problems involves tools from the theory of nonlinear operators and from degree theory.

3.2.1   Auxiliary Problems

Let us start with problems that are similar to the ones considered in the previous section, namely they are driven by p-Laplacian like operators. Again our formulation involves a maximal monotone map incorporating this way in our framework second order differential variational inequalities.

Let $T = [0, b]$, with $b > 0$. We start with the investigation of the following auxiliary problem

$$\begin{cases} \big(a\big(x'(t)\big)\big)' = f\big(t,x(t),x'(t)\big) & \text{for a.a. } t \in T, \\ x(0) = x(b), \ x'(0) = x'(b). \end{cases} \tag{3.1}$$

Here $a\colon \mathbb{R}^N \longrightarrow \mathbb{R}^N$ satisfies some monotonicity and coercivity conditions, which implies that a is a homeomorphism. This way we cover a broad class of nonlinear differential operators among which a special case is the ordinary p-Laplacian. More precisely our hypotheses on the data of (3.1) are the following.

$\underline{H(a)_1}$ $a\colon \mathbb{R}^N \longrightarrow \mathbb{R}^N$ is a continuous map, strictly monotone and there exists a function $\eta \in C(\mathbb{R}_+; \mathbb{R}_+)$, such that

$$\lim_{r \to +\infty} \eta(r) \ = \ +\infty$$

and

$$\eta\left(\|\xi\|_{\mathbb{R}^N}\right) \|\xi\|_{\mathbb{R}^N} \ \leq \ \left(a(\xi), \xi\right)_{\mathbb{R}^N} \qquad \forall\, \xi \in \mathbb{R}^N.$$

$\underline{H(f)}$ $f\colon T \times \mathbb{R}^N \times \mathbb{R}^N \longrightarrow \mathbb{R}^N$ is a function, such that

(i) for all $\xi, \overline{\xi} \in \mathbb{R}^N$, the function

$$T \ni t \longmapsto f(t, \xi, \overline{\xi}) \in \mathbb{R}^N$$

is measurable;

(ii) for almost all $t \in T$, the function

$$\mathbb{R}^N \times \mathbb{R}^N \ni (\xi, \overline{\xi}) \longmapsto f(t, \xi, \overline{\xi}) \in \mathbb{R}^N$$

is continuous;

(iii) for every $r > 0$, there exists $\gamma_r \in L^1(T)$, such that

$$\left| f(t, \xi, \overline{\xi}) \right| \ \leq \ \gamma_r(t) \quad \text{for a.a. } t \in T \text{ and all } \xi, \overline{\xi} \in \overline{B}_r \subseteq \mathbb{R}^N.$$

EXAMPLE 3.2.1   (a) Any homeomorphism $a\colon \mathbb{R} \longrightarrow \mathbb{R}$ is either strictly increasing or strictly decreasing. Then in the first case a satisfies hypotheses $H(a)_1$, while in the second case $-a$ satisfies $H(a)_1$.

(b) For $p \in (1, +\infty)$, let $a\colon \mathbb{R}^N \longrightarrow \mathbb{R}^N$ be defined by

$$a(\xi) \ \overset{df}{=} \ \begin{cases} \|\xi\|_{\mathbb{R}^N}^{p-2}\, \xi & \text{if } \xi \in \mathbb{R}^N, \\ 0 & \text{if } \xi = 0. \end{cases}$$

By a well known inequality which, for $p \in (1, 2]$, says that

$$c\left(\|\xi\|_{\mathbb{R}^N} + \|\overline{\xi}\|_{\mathbb{R}^N}\right)^{p-2} \|\xi - \overline{\xi}\|_{\mathbb{R}^N}^p$$
$$\leq \left(\|\xi\|_{\mathbb{R}^N}^{p-2}\, \xi - \|\overline{\xi}\|_{\mathbb{R}^N}^{p-2}\, \overline{\xi}, \xi - \overline{\xi}\right)_{\mathbb{R}^N} \qquad \forall\, \xi, \overline{\xi} \in \mathbb{R}^N,$$

and for $p \geq 2$, says that

$$c \left\| \xi - \overline{\xi} \right\|_{\mathbb{R}^N}^p \leq \left(\|\xi\|_{\mathbb{R}^N}^{p-2} \xi - \|\overline{\xi}\|_{\mathbb{R}^N}^{p-2} \overline{\xi}, \xi - \overline{\xi} \right)_{\mathbb{R}^N} \qquad \forall \, \xi, \overline{\xi} \in \mathbb{R}^N,$$

for some $c > 0$, we see that a satisfies hypotheses $H(a)_1$. This map corresponds to the ordinary p-Laplacian differential operator.

(c) In general, let $a = \nabla \varphi$, where $\varphi \colon \mathbb{R}^N \longrightarrow \mathbb{R}$ is a C^1 and strictly convex function (i.e. a is a potential) and suppose that

$$\eta(\|\xi\|_{\mathbb{R}^N}) \, \|\xi\|_{\mathbb{R}^N} \leq (a(\xi), \xi)_{\mathbb{R}^N} \qquad \forall \, \xi \in \mathbb{R}^N,$$

with η as in hypotheses $H(a)_1$. In particular, let

$$\varphi(\xi) \stackrel{df}{=} e^{\|\xi\|_{\mathbb{R}^N}^2} - \|\xi\|_{\mathbb{R}^N}^2 - 1 \qquad \forall \, \xi \in \mathbb{R}^N.$$

Then $a(\xi) = 2 e^{\|\xi\|_{\mathbb{R}^N}^2} \xi - 2\xi$ and so

$$\left(a(\xi), \xi \right)_{\mathbb{R}^N} = 2 e^{\|\xi\|_{\mathbb{R}^N}^2} \|\xi\|_{\mathbb{R}^N}^2 - 2 \|\xi\|_{\mathbb{R}^N}^2 = 2 \left(e^{\|\xi\|_{\mathbb{R}^N}^2} - 1 \right) \|\xi\|_{\mathbb{R}^N}^2$$

and so hypotheses $H(a)_1$ hold. $\qquad \Box$

The next existence result is due to Manásevich & Mawhin (1998), where the reader can also find its proof, which is based on degree theoretic arguments.

PROPOSITION 3.2.1
If hypotheses $H(a)_1$, $H(f)$ hold and additionally
(i) there exist $\overline{\eta} \in C^1(\mathbb{R}^N; \mathbb{R}^N)$ and $h \in L^1(T)_+$, such that for almost all $t \in T$, we have

$$\begin{cases} 0 \leq \left(a(\overline{\xi}), \overline{\eta}'(\xi)\overline{\xi} \right)_{\mathbb{R}^N} \\ \|f(t, \xi, \overline{\xi})\|_{\mathbb{R}^N} \leq \left(f(t, \xi, \overline{\xi}), \overline{\eta}(\xi) \right)_{\mathbb{R}^N} + h(t) \end{cases} \qquad \forall \, \xi, \overline{\xi} \in \mathbb{R}^N;$$

(ii) there exist functions $\eta_1 \in C(\mathbb{R}_+; \mathbb{R}_+)$ and $h_1 \in L^1(T)_+$, such that

$$\lim_{n \to +\infty} \eta_1(r) = +\infty$$

and

$$\eta_1(\|\xi\|_{\mathbb{R}^N}) - h_1(t) \leq \|f(t, \xi, \overline{\xi})\|_{\mathbb{R}^N} \quad \text{for a.a. } t \in T \text{ and all } \xi, \overline{\xi} \in \mathbb{R}^N;$$

(iii) the Brouwer degree $D(\widehat{f}, B_{R_0}, 0) = 1$, where

$$\widehat{f}(\xi) \stackrel{df}{=} \frac{1}{b} \int_0^b f(t, \xi, 0) \, dt \qquad \forall \, \xi \in \mathbb{R}^N$$

and $B_{R_0} \overset{df}{=} \left\{ \xi \in \mathbb{R}^N : \|\xi\|_{\mathbb{R}^N} < R_0 \right\}$,

then problem (3.1) has a solution

$$x_0 \in C^1(T; \mathbb{R}^N), \text{ such that } a(x_0'(\cdot)) \in W^{1,1}(T; \mathbb{R}^N).$$

Let $A \colon \mathbb{R}^N \supseteq \mathrm{Dom}\,(A) \longrightarrow 2^{\mathbb{R}^N}$ be a maximal monotone map and for $\lambda > 0$, let A_λ be its Yosida approximation (see Definition 1.4.7(b)). Also let $g \in L^{p'}(T; \mathbb{R}^N)$, with $\frac{1}{p} + \frac{1}{p'} = 1$ and consider the following auxiliary periodic problem

$$\begin{cases} -\big(a(x'(t))\big)' + A_\lambda\big(x(t)\big) + \|x(t)\|_{\mathbb{R}^N}^{p-2} x(t) = g(t) & \text{for a.a. } t \in T, \\ x(0) = x(b), \ x'(0) = x'(b). \end{cases} \quad (3.2)$$

Using Proposition 3.2.1 we will solve this problem. By a solution of (3.2), we mean a function $x_0 \in C^1(T; \mathbb{R}^N)$, such that $a(x_0'(\cdot)) \in W^{1,p'}(T; \mathbb{R}^N)$ and it satisfies (3.2).

PROPOSITION 3.2.2
If hypotheses $H(a)_1$ hold, $A \colon \mathbb{R}^N \supseteq \mathrm{Dom}\,(A) \longrightarrow 2^{\mathbb{R}^N}$ is maximal monotone with $0 \in A(0)$ and $g \in L^{p'}(T; \mathbb{R}^N)$,
then problem (3.2) has a unique solution

$$x_0 \in C^1(T; \mathbb{R}^N), \text{ such that } a(x_0'(\cdot)) \in W^{1,p'}(T; \mathbb{R}^N).$$

PROOF   Let $f \colon T \times \mathbb{R}^N \longrightarrow \mathbb{R}^N$ be defined by

$$f(t, \xi) \overset{df}{=} A_\lambda(\xi) + \|\xi\|_{\mathbb{R}^N}^{p-2} \xi - g(t) \qquad \forall\, (t, \xi) \in T \times \mathbb{R}^N$$

and let $\overline{\eta} \in C(\mathbb{R}^N; \mathbb{R}^N)$ be defined by $\overline{\eta}(\xi) \overset{df}{=} \xi$. From hypotheses $H(a)_1$, we have

$$\big(a(\overline{\xi}), \overline{\eta}'(\xi)\overline{\xi}\big)_{\mathbb{R}^N} = \big(a(\overline{\xi}), \overline{\xi}\big)_{\mathbb{R}^N} \geq 0 \qquad \forall\, \xi, \overline{\xi} \in \mathbb{R}^N.$$

Since $\|A_\lambda(\xi)\|_{\mathbb{R}^N} \leq \frac{1}{\lambda} \|\xi\|_{\mathbb{R}^N}$ for $\xi \in \mathbb{R}^N$ and $A_\lambda(0) = 0$, we have that

$$\begin{aligned} \|f(t, \xi)\|_{\mathbb{R}^N} &\leq \|A_\lambda(\xi)\|_{\mathbb{R}^N} + \|\xi\|_{\mathbb{R}^N}^{p-1} + \|g(t)\|_{\mathbb{R}^N} \\ &\leq \frac{1}{\lambda} \|\xi\|_{\mathbb{R}^N} + \|\xi\|_{\mathbb{R}^N}^{p-1} + \|g(t)\|_{\mathbb{R}^N} \quad \text{for a.a. } t \in T \text{ and all } \xi \in \mathbb{R}^N \end{aligned}$$

and

$$\begin{aligned} \big(f(t, \xi), \xi\big)_{\mathbb{R}^N} &= \big(A_\lambda(\xi), \xi\big)_{\mathbb{R}^N} + \|\xi\|_{\mathbb{R}^N}^p - \big(g(t), \xi\big)_{\mathbb{R}^N} \\ &\geq \|\xi\|_{\mathbb{R}^N}^p - \|g(t)\|_{\mathbb{R}^N} \|\xi\|_{\mathbb{R}^N} \quad \text{for a.a. } t \in T \text{ and all } \xi \in \mathbb{R}^N. \end{aligned}$$

Let us set

$$h_1(t) \overset{df}{=} \sup_{r > 0} \left\{ -r^p + r^{p-1} + \frac{1}{\lambda} r + \|g(t)\|_{\mathbb{R}^N}\, r + \|g(t)\|_{\mathbb{R}^N} \right\} \qquad \forall\, t \in T$$

and consider $h \in L^1(T)_+$, defined by $h(t) \overset{df}{=} h_1^+(t)$. Then

$$\|f(t,\xi)\|_{\mathbb{R}^N} \leq (f(t,\xi), \overline{\eta}(\xi))_{\mathbb{R}^N} + h(t) \quad \text{for a.a. } t \in T \text{ and all } \xi \in \mathbb{R}^N.$$

Moreover, note that

$$\|\xi\|_{\mathbb{R}^N}^{p-1} - \|g(t)\|_{\mathbb{R}^N} \leq \|f(t,\xi)\|_{\mathbb{R}^N} \quad \text{for a.a. } t \in T \text{ and all } \xi \in \mathbb{R}^N.$$

Finally, let us define

$$\widehat{f}(\xi) \overset{df}{=} \frac{1}{b} \int_0^b f(t,\xi)\,dt \quad \forall\, \xi \in \mathbb{R}^N$$

$$\overline{g} \overset{df}{=} \frac{1}{b} \int_0^b g(t)\,dt$$

$$H(s,\xi) \overset{df}{=} s\widehat{f}(\xi) + (1-s)\xi \quad \forall\, (s,\xi) \in [0,1] \times \mathbb{R}^N.$$

Let us take $R_0 > \max\left\{1, \|\overline{g}\|_{\mathbb{R}^N}\right\}$. If for some $u \in \partial B_{R_0}$, we have $H(s,u) = 0$, then

$$sA_\lambda(u) + sR_0^{p-2}u + (1-s)u = s\overline{g},$$

so

$$s\left(A_\lambda(u), u\right)_{\mathbb{R}^N} + sR_0^p + (1-s)R_0^2 \leq s\|\overline{g}\|_{\mathbb{R}^N} R_0.$$

Since A_λ is maximal monotone and $A_\lambda(0) = 0$, we have

$$sR_0^{p-1} + (1-s)R_0 \leq s\|\overline{g}\|_{\mathbb{R}^N}$$

and as $\max\left\{1, \|\overline{g}\|_{\mathbb{R}^N}\right\} < R_0$, we have

$$R_0 = sR_0 + (1-s)R_0 \leq s\|\overline{g}\|_{\mathbb{R}^N} < sR_0 \leq R_0,$$

a contradiction. Therefore $0 \notin H(s, \partial B_{R_0})$ and so from the homotopy invariance of the Brouwer degree, we have that

$$D(\widehat{f}, B_{R_0}, 0) = D(id_{\mathbb{R}^N}, B_{R_0}, 0) = 1.$$

So we can apply Proposition 3.2.1 and obtain a solution $x_0 \in C^1(T; \mathbb{R}^N)$ of (3.2). Clearly $a(x_0'(\cdot)) \in W^{1,p'}(T; \mathbb{R}^N)$.

Now, let us show the uniqueness of the solution. For this purpose let $x_0, \overline{x}_0 \in C^1(T; \mathbb{R}^N)$ be two such solutions. Then after integration by parts, we obtain

$$0 = \int_0^b \left(a(x_0'(t)) - a(\overline{x}_0'(t)), x_0'(t) - \overline{x}_0'(t)\right)_{\mathbb{R}^N} dt$$

$$+ \int_0^b \big(A_\lambda(x_0(t)) - A_\lambda(\overline{x}_0(t)), x_0(t) - \overline{x}_0(t) \big)_{\mathbb{R}^N} dt$$

$$+ \int_0^b \big(\|x_0(t)\|_{\mathbb{R}^N}^{p-2} x_0(t) - \|\overline{x}_0(t)\|_{\mathbb{R}^N}^{p-2} \overline{x}_0(t), x_0(t) - \overline{x}_0(t) \big)_{\mathbb{R}^N} dt,$$

so from hypotheses $H(a)_1$ and the fact that A_λ is monotone, we have

$$\int_0^b \big(\|x_0(t)\|_{\mathbb{R}^N}^{p-2} x_0(t) - \|\overline{x}_0(t)\|_{\mathbb{R}^N}^{p-2} \overline{x}_0(t), x_0(t) - \overline{x}_0(t) \big)_{\mathbb{R}^N} dt \ \leq \ 0.$$

Since the map $\mathbb{R}^N \ni \xi \longmapsto \|\xi\|_{\mathbb{R}^N}^{p-2} \xi \in \mathbb{R}^N$ is strictly monotone, from the last inequality, it follows that $x_0(t) = \overline{x}_0(t)$ for all $t \in T$. So the solution of (3.2) is unique. $\qquad \square$

Let

$$\widehat{D} \stackrel{df}{=} \big\{ x \in C^1(T; \mathbb{R}^N) : \ a(x'(\cdot)) \in W^{1,p'}(T; \mathbb{R}^N), \tag{3.3}$$
$$x(0) = x(b), \ x'(0) = x'(b) \big\}.$$

For $\lambda > 0$ consider the operator $S_\lambda : L^p(T; \mathbb{R}^N) \supseteq \widehat{D} \longrightarrow L^{p'}(T; \mathbb{R}^N)$, defined by

$$S_\lambda(x)(\cdot) \stackrel{df}{=} -\big(a(x'(\cdot)) \big)' + \widehat{A}_\lambda(x)(\cdot) \qquad \forall\, x \in \widehat{D}, \tag{3.4}$$

where $\widehat{A}_\lambda(x)(\cdot) \stackrel{df}{=} A_\lambda(x(\cdot))$. Note that

$$A_\lambda(x(\cdot)) \in C(T; \mathbb{R}^N) \qquad \forall\, x \in \widehat{D}.$$

Arguing as in the proof of Proposition 3.1.3, using this time Proposition 3.2.2, we obtain the maximal monotonicity of S_λ.

PROPOSITION 3.2.3
If hypotheses $H(a)_1$ hold and the map

$$A \colon \mathbb{R}^N \supseteq \mathrm{Dom}\, A \longrightarrow 2^{\mathbb{R}^N}$$

is maximal monotone with $0 \in A(0)$,
then the operator S_λ is maximal monotone too.

REMARK 3.2.1   Note that if $p \geq 2$, then

$$A_\lambda(x(\cdot)) \in L^{p'}(T; \mathbb{R}^N) \qquad \forall\, x \in L^p(T; \mathbb{R}^N).$$

Indeed since $A_\lambda(0) = 0$, we have

$$\big\|A_\lambda(x(t))\big\|_{\mathbb{R}^N} \leq \frac{1}{\lambda}\,\|x(t)\|_{\mathbb{R}^N} \quad \text{for a.a. } t \in T$$

and $L^p(T;\mathbb{R}^N) \subseteq L^{p'}(T;\mathbb{R}^N)$ (as $1 < p' \leq 2 \leq p < +\infty$). Moreover, note that

$$A_\lambda(x(\cdot)) \in C(T;\mathbb{R}^N) \quad \forall\, x \in C(T;\mathbb{R}^N).$$

$\square$

In what follows, we shall use the space

$$W^{1,p}_{\mathrm{per}}(T;\mathbb{R}^N) \stackrel{df}{=} \{x \in W^{1,p}(T;\mathbb{R}^N) : x(0) = x(b)\}.$$

Since the embedding $W^{1,p}(T;\mathbb{R}^N) \subseteq C(T;\mathbb{R}^N)$ is continuous, the evaluations at $t = 0$ and $t = b$ make sense.

3.2.2   Formulation of the Problem

Now we are ready to pass to the study of a multivalued problem, which is the periodic counterpart of the problem examined in Section 3.1. So consider the following nonlinear periodic problem:

$$\begin{cases} \big(a(x'(t))\big)' \in A(x(t)) + F(t, x(t), x'(t)) & \text{for a.a. } t \in T, \\ x(0) = x(b), \ x'(0) = x'(b). \end{cases} \tag{3.5}$$

To solve this problem, we need to strengthen a little our hypotheses concerning a. So now our hypotheses on the data of (3.5) are the following:

<u>$H(A)_1$</u> $A\colon \mathbb{R}^N \supseteq \mathrm{Dom}\,(A) \longrightarrow 2^{\mathbb{R}^N}$ is a maximal monotone operator with $0 \in A(0)$.

<u>$H(a)_2$</u> $a\colon \mathbb{R}^N \longrightarrow \mathbb{R}^N$ is a monotone map, such that

(i) a has one of the following forms:

(i)$_1$ $a(\xi) = k_a(\xi)\xi$, where $k_a\colon \mathbb{R}^N \longrightarrow \mathbb{R}_+$ is a continuous function; or

(i)$_2$ $a(\xi) = \big(k_1^a(\xi_1)\xi_1, \ldots, k_N^a(\xi_N)\xi_N\big)$, where $k_i^a\colon \mathbb{R} \longrightarrow \mathbb{R}$ are continuous functions for $i \in \{1, \ldots, N\}$;

(ii) we have that

$$\big(a(\xi), \xi\big)_{\mathbb{R}^N} \geq c_0\,\|\xi\|_{\mathbb{R}^N}^p \quad \forall\, \xi \in \mathbb{R}^N,$$

for some $c_0 > 0$ and $p \in (1, +\infty)$.

<u>$H(F)_1$</u> $F: T \times \mathbb{R}^N \times \mathbb{R}^N \longrightarrow P_{kc}(\mathbb{R}^N)$ is a multifunction, such that

(i) for all $\xi, \overline{\xi} \in \mathbb{R}^N$, the multifunction

$$T \ni t \longmapsto F(t, \xi, \overline{\xi}) \in P_{kc}(\mathbb{R}^N)$$

is graph measurable;

(ii) for almost all $t \in T$, the multifunction

$$\mathbb{R}^N \times \mathbb{R}^N \ni (\xi, \overline{\xi}) \longmapsto F(t, \xi, \overline{\xi}) \in P_{kc}(\mathbb{R}^N)$$

has closed graph;

(iii) for almost all $t \in T$, all $\xi, \overline{\xi} \in \mathbb{R}^N$ and all $v \in F(t, \xi, \overline{\xi})$, we have that

$$(v, \xi)_{\mathbb{R}^N} \geq -c_1 \|\xi\|_{\mathbb{R}^N}^p - c_2 \|\xi\|_{\mathbb{R}^N}^r \|\overline{\xi}\|_{\mathbb{R}^N}^{p-r} - c_3(t) \|\xi\|_{\mathbb{R}^N}^s,$$

with $c_1, c_2 > 0$, $1 \leq r$, $s < p$, $c_3 \in L^1(T)_+$;

(iv) there exists $M_0 > 0$, such that if $\|\xi_0\|_{\mathbb{R}^N} = M_0$ and $(\xi_0, \overline{\xi}_0)_{\mathbb{R}^N} = 0$, we can find $\delta > 0$, such that for almost all $t \in T$, we have

$$\inf \left\{ (v, \xi)_{\mathbb{R}^N} + c_0 \|\overline{\xi}\|_{\mathbb{R}^N}^p : \|\xi - \xi_0\|_{\mathbb{R}^N} + \|\overline{\xi} - \overline{\xi}_0\|_{\mathbb{R}^N} < \delta, \\ v \in F(t, \xi, \overline{\xi}) \right\} \geq 0;$$

(v) for almost all $t \in T$, all $\xi, \overline{\xi} \in \mathbb{R}^N$, $\|\xi\|_{\mathbb{R}^N} \leq M_0$ and all $v \in F(t, \xi, \overline{\xi})$, we have

$$\|v\|_{\mathbb{R}^N} \leq c_4(t) + c_5 \|\overline{\xi}\|_{\mathbb{R}^N}^{p-1},$$

with $c_4 \in L^2(T)_+$, $c_5 > 0$.

REMARK 3.2.2 Hypothesis $H(F)_1(iv)$ is a suitable extension in the present setting of the so-called "Hartman condition." Hartman (1960) was the first to use it in the context of Dirichlet problems for second order systems. He had a single-valued right hand side nonlinearity $F = f$ and he assumed that $f(t, \xi, \overline{\xi})$ is continuous in all three variables. $\qquad \Box$

3.2.3   Approximation of the Problem

As in the Dirichlet case (Section 3.1), the study of problem (3.5) depends on the resolution of the following regular approximation of (3.5):

$$\begin{cases} \left(a(x'(t)) \right)' \in A_\lambda(x(t)) + F(t, x(t), x'(t)) & \text{for a.a. } t \in T, \\ x(0) = x(b), \ x'(0) = x'(b). \end{cases} \tag{3.6}$$

for $\lambda > 0$. As in Section 3.1, the solution of (3.6) will be obtained using fixed point arguments. More precisely, we shall use another multivalued nonlinear version of the Leray-Schauder Alternative Principle due to Bader (2001, Theorem 7), where the interested reader can find its proof.

THEOREM 3.2.1 (Multivalued Nonlinear Leray-Schauder Alternative Principle)
If X, Y are Banach spaces, W is a bounded open subset of X with $0 \in W$, $G \colon \overline{W} \longrightarrow P_{wkc}(Y)$ is upper semicontinuous from $\overline{W}$ into Y_w and bounded and $K \colon Y \longrightarrow X$ is a completely continuous map,
then one of the following alternatives holds:

either
(a) *there exist $x_0 \in \partial W$ and $s \in (0,1)$, such that $x_0 \in s(K \circ G)(x_0)$;*

or
(b) *$K \circ G$ has a fixed point.*

REMARK 3.2.3 Note that the composite multifunction $K \circ G$ need not have convex values. ⬜

Now we can prove the existence of solutions for problem (3.6).

PROPOSITION 3.2.4
If hypothesis $H(a)_2$, $H(A)_1$ and $H(F)_1$ hold,
then problem (3.6) has a solution

$$x_0 \in C^1(T; \mathbb{R}^N), \text{ such that } a\big(x_0'(\cdot)\big) \in W_{\text{per}}^{1,p'}(T; \mathbb{R}^N).$$

PROOF First we do the proof assuming the following stronger version of hypothesis $H(F)_1(iv)$:

$\underline{H(F)_1(iv)'}$ there exists $M_0 > 0$, such that if $\xi_0, \overline{\xi}_0 \in \mathbb{R}^N$, $\|\xi_0\|_{\mathbb{R}^N} = M_0$ and $(\xi_0, \overline{\xi}_0)_{\mathbb{R}^N} = 0$, we can find $\delta, \delta' > 0$, such that for almost all $t \in T$, we have

$$\inf \big\{ (v, \xi)_{\mathbb{R}^N} + c_0 \|\overline{\xi}\|_{\mathbb{R}^N}^p : \|\xi - \xi_0\|_{\mathbb{R}^N} + \|\overline{\xi} - \overline{\xi}_0\|_{\mathbb{R}^N} < \delta, \\ v \in F(t, \xi, \overline{\xi}) \big\} \geq \delta' > 0. \quad (3.7)$$

Let $S_\lambda \colon L^p(T; \mathbb{R}^N) \supseteq \widehat{D} \longrightarrow L^{p'}(T; \mathbb{R}^N)$ be a map defined as in Proposition 3.2.3 (see (3.4) and (3.3)). We know that S_λ is maximal monotone (see Proposition 3.2.3). Also let $J \colon L^p(T; \mathbb{R}^N) \longrightarrow L^{p'}(T; \mathbb{R}^N)$ be defined by

$$J(x)(\cdot) \stackrel{df}{=} \|x(\cdot)\|_{\mathbb{R}^N}^{p-2} x(\cdot) \qquad \forall \, x \in L^p(T; \mathbb{R}^N).$$

This is a continuous, strictly monotone (hence maximal monotone) operator. If we set $V_\lambda \overset{df}{=} S_\lambda + J$, then V_λ is maximal monotone as a sum of two maximal monotone operators (see Theorem 1.4.5). We have

$$\langle V_\lambda(x), x \rangle_p \geq \langle U(x), x \rangle_p + \langle J(x), x \rangle_p \qquad \forall\, x \in \widehat{D},$$

where $U \colon L^p(T; \mathbb{R}^N) \supseteq \widehat{D} \longrightarrow L^{p'}(T; \mathbb{R}^N)$ is the nonlinear differential operator, defined by

$$U(x) \overset{df}{=} -\big(a(x'(\cdot))\big)' \qquad \forall\, x \in \widehat{D}.$$

Performing integration by parts and using hypotheses $H(a)_2$, we obtain

$$\langle V_\lambda(x), x \rangle_p \geq c_0 \|x'\|_p^p + \|x\|_p^p \geq c_6 \|x\|_{W^{1,p}(T;\mathbb{R}^N)}^p,$$

with $c_6 > 0$. From the last inequality, it follows that V_λ is coercive. So, we can apply Theorem 1.4.4 and infer that $R(V_\lambda) = L^{p'}(T; \mathbb{R}^N)$, i.e. V_λ is surjective. Moreover, due to the strict monotonicity of J, V_λ is injective. So we can define

$$K_\lambda \overset{df}{=} V_\lambda^{-1} \colon L^{p'}(T; \mathbb{R}^N) \longrightarrow \widehat{D} \subseteq W_{\mathrm{per}}^{1,p}(T; \mathbb{R}^N).$$

Claim 1. The operator $K_\lambda \colon L^{p'}(T; \mathbb{R}^N) \longrightarrow W_{\mathrm{per}}^{1,p}(T; \mathbb{R}^N)$ is completely continuous.

Suppose that $\{u_n\}_{n \geq 1} \subseteq L^{p'}(T; \mathbb{R}^N)$ is a sequence, such that

$$u_n \overset{w}{\longrightarrow} \overline{u} \quad \text{in } L^{p'}(T; \mathbb{R}^N),$$

for some $\overline{u} \in L^{p'}(T; \mathbb{R}^N)$. We need to show that

$$K_\lambda(u_n) \longrightarrow K_\lambda(\overline{u}) \quad \text{in } W^{1,p}(T; \mathbb{R}^N). \tag{3.8}$$

Let $x_n \overset{df}{=} K_\lambda(u_n)$ for $n \geq 1$. Recall that $\widehat{A}_\lambda(x_n)(\cdot) \overset{df}{=} A_\lambda(x_n(\cdot))$ for $n \geq 1$ and because $x_n \in \widehat{D}$, we have

$$\widehat{A}_\lambda(x_n) \in C(T; \mathbb{R}^N) \qquad \forall\, n \geq 1.$$

We can write that

$$S_\lambda(x_n) + J(x_n) = U(x_n) + \widehat{A}_\lambda(x_n) + J(x_n) = u_n, \tag{3.9}$$

so

$$\langle U(x_n), x_n \rangle_p + \langle \widehat{A}_\lambda(x_n), x_n \rangle_p + \langle J(x_n), x_n \rangle_p = \langle u_n, x_n \rangle_p.$$

As $\widehat{A}_\lambda(0) = 0$, we have

$$c_0 \|x_n'\|_p^p + \|x_n\|_p^p \leq \|u_n\|_{p'} \|x_n\|_p,$$

so

$$c_7 \|x_n\|^p_{W^{1,p}(T;\mathbb{R}^N)} \leq \|u_n\|_{p'} \|x_n\|_p,$$

with $c_7 \overset{df}{=} \min\{c_0, 1\} > 0$. Thus the sequence $\{x_n\}_{n\geq 1} \subseteq W^{1,p}_{\mathrm{per}}(T;\mathbb{R}^N)$ is bounded. Passing to a subsequence if necessary, we may assume that

$$x_n \overset{w}{\longrightarrow} \overline{x} \text{ in } W^{1,p}_{\mathrm{per}}(T;\mathbb{R}^N),$$
$$x_n \longrightarrow \overline{x} \text{ in } L^p(T;\mathbb{R}^N),$$

for some $\overline{x} \in W^{1,p}_{\mathrm{per}}(T;\mathbb{R}^N)$. Recall that V_λ is maximal monotone and so the graph $\mathrm{Gr}\, V_\lambda$ is sequentially closed in $L^p(T;\mathbb{R}^N) \times L^{p'}(T;\mathbb{R}^N)_w$ (see Corollary 1.4.1). Because $u_n = V_\lambda(x_n)$ for $n \geq 1$, we infer that

$$\overline{u} = V_\lambda(\overline{x}) = S_\lambda(\overline{x}) + J(\overline{x}) = U(\overline{x}) + \widehat{A}_\lambda(\overline{x}) + J(\overline{x}).$$

As $x_n \in \widehat{D}$ for $n \geq 1$, we have

$$a\big(x_n'(\cdot)\big) \in W^{1,p'}_{\mathrm{per}}(T;\mathbb{R}^N) \qquad \forall\, n \geq 1.$$

Because

$$W^{1,p'}_{\mathrm{per}}(T;\mathbb{R}^N) = \mathbb{R}^N \oplus V,$$

with

$$V \overset{df}{=} \left\{ x \in W^{1,p'}_{\mathrm{per}}(T;\mathbb{R}^N) : \int_0^b x(t)\, dt = 0 \right\},$$

we have

$$a\big(x_n'(\cdot)\big) = \widehat{a}_n(\cdot) + \overline{a}_n, \quad \text{with } \int_0^b \widehat{a}_n(t)\, dt = 0, \text{ and } \overline{a}_n \in \mathbb{R}^N.$$

From (3.9), it follows that the sequence $\big\{\big(a\big(x_n'(\cdot)\big)\big)'\big\}_{n\geq 1} \subseteq L^{p'}(T;\mathbb{R}^N)$ is bounded, so by the Poincaré-Wirtinger inequality (see Theorem 1.1.7), it follows that the sequence $\{\widehat{a}_n(\cdot)\}_{n\geq 1} \subseteq C(T;\mathbb{R}^N)$ is relatively compact. We have

$$a\big(x_n'(t)\big) = \widehat{a}_n(t) + \overline{a}_n \qquad \forall\, n \geq 1,\ t \in T,$$

so

$$x_n'(t) = a^{-1}\big(\widehat{a}_n(t) + \overline{a}_n\big) \qquad \forall\, n \geq 1,\ t \in T.$$

Integrating over T and using the fact that $x_n(0) = x_n(b)$, we obtain

$$\int_0^b a^{-1}\big(\widehat{a}_n(t) + \overline{a}_n\big)dt = 0.$$

Invoking Proposition 3.1.1(b), we have that the sequence $\{\overline{a}_n\}_{n \geq 1} \subseteq \mathbb{R}^N$ is bounded. Therefore, it follows that the sequence $\{a(x_n'(\cdot))\}_{n \geq 1} \subseteq C(T; \mathbb{R}^N)$ is relatively compact. Hence the sequence $\{a(x_n'(\cdot))\}_{n \geq 1} \subseteq W_{\mathrm{per}}^{1,p'}(T; \mathbb{R}^N)$ is bounded and, passing to a subsequence if necessary, we may assume that

$$a(x_n'(\cdot)) \xrightarrow{w} \beta \quad \text{in } W_{\mathrm{per}}^{1,p'}(T; \mathbb{R}^N)$$

for some $\beta \in W_{\mathrm{per}}^{1,p'}(T; \mathbb{R}^N)$. Because U is maximal monotone, we see that $\beta = U(\overline{x})$. Therefore

$$a(x_n'(\cdot)) \xrightarrow{w} a(\overline{x}'(\cdot)) \text{ in } W_{\mathrm{per}}^{1,p'}(T; \mathbb{R}^N),$$
$$a(x_n'(\cdot)) \longrightarrow a(\overline{x}'(\cdot)) \text{ in } C(T; \mathbb{R}^N).$$

Because $a_0 \colon C(T; \mathbb{R}^N) \longrightarrow C(T; \mathbb{R}^N)$ defined by

$$a_0(x)(\cdot) \overset{df}{=} a^{-1}(x(\cdot)) \qquad \forall x \in C(T; \mathbb{R}^N)$$

is continuous and maps bounded sets to bounded sets, we have that

$$x_n' \longrightarrow \overline{x}' \quad \text{in } C(T; \mathbb{R}^N).$$

Therefore, finally we have that

$$x_n \longrightarrow \overline{x} \quad \text{in } W_{\mathrm{per}}^{1,p}(T; \mathbb{R}^N) \text{ and in } C^1(T; \mathbb{R}^N).$$

Since every subsequence of $\{x_n\}_{n \geq 1}$ has a further subsequence, which strongly converges to $\overline{x} = K_\lambda(\overline{u})$, we conclude that the whole sequence $\{x_n = K_\lambda(u_n)\}_{n \geq 1}$ converges strongly to $\overline{x} = K_\lambda(\overline{u})$ in $W_{\mathrm{per}}^{1,p}(T; \mathbb{R}^N)$. This proves (3.8) and so the claim is true.

Next, let $N_1 \colon \{x \in W_{\mathrm{per}}^{1,p}(T; \mathbb{R}^N) : \|x\|_\infty \leq M_0\} \longrightarrow 2^{L^{p'}(T;\mathbb{R}^N)}$ be the multivalued map, defined by

$$N_1(x) \overset{df}{=} S_{F(\cdot,x(\cdot),x'(\cdot))}^{p'} \qquad \forall x \in W_{\mathrm{per}}^{1,p}(T; \mathbb{R}^N), \|x\|_\infty \leq M_0. \qquad (3.10)$$

Similarly to Claim 2 in the proof of Proposition 3.1.4, we can show that N_1 has values in $P_{wkc}(L^{p'}(T; \mathbb{R}^N))$ and is upper semicontinuous into $L^{p'}(T; \mathbb{R}^N)_w$. Let

$$N_2(x) \overset{df}{=} -N_1(x) + J(x) \qquad \forall x \in W_{\mathrm{per}}^{1,p}(T; \mathbb{R}^N), \|x\|_\infty \leq M_0.$$

Then problem (3.6) is equivalent to the following abstract fixed point problem:

$$x \in (K_\lambda \circ N_2)(x). \qquad (3.11)$$

Let $M_1 > 0$ be such that $M_1^p > \frac{p}{rc_0}\left(c_1 M_0^p b + \frac{rc_2^{\frac{p}{r}} M_0^b b^{\frac{p}{r}}}{c_0 p} + \|c_3\|_1 M_0^s\right)$. Then, we consider the set

$$W \overset{df}{=} \left\{x \in W_{\mathrm{per}}^{1,p}(T;\mathbb{R}^N) : \|x\|_\infty < M_0, \text{ and } \|x'\|_p < M_1\right\}.$$

Note that $W = W_1 \cap W_2$, where

$$W_1 \overset{df}{=} \left\{x \in W_{\mathrm{per}}^{1,p}(T;\mathbb{R}^N) : \|x\|_{C(T;\mathbb{R}^N)} < M_0\right\}$$

$$W_2 \overset{df}{=} \left\{x \in W_{\mathrm{per}}^{1,p}(T;\mathbb{R}^N) : \|x'\|_p < M_1\right\}.$$

Because the embedding $W_{\mathrm{per}}^{1,p}(T;\mathbb{R}^N) \subseteq C(T;\mathbb{R}^N)$ is continuous, we see that W_1 is open and convex in $W_{\mathrm{per}}^{1,p}(T;\mathbb{R}^N)$.

Similarly, since the linear map $L\colon W^{1,p}(T;\mathbb{R}^N) \longrightarrow L^p(T;\mathbb{R}^N)$, defined by

$$L(x) \overset{df}{=} x' \qquad \forall\, x \in W^{1,p}(T;\mathbb{R}^N)$$

is continuous and surjective, we have that the set W_2 is open and convex in $W_{\mathrm{per}}^{1,p}(T;\mathbb{R}^N)$. Therefore W is open and convex and of course bounded in $W_{\mathrm{per}}^{1,p}(T;\mathbb{R}^N)$. Note that

$$\overline{W} = \left\{x \in W_{\mathrm{per}}^{1,p}(T;\mathbb{R}^N) : \|x\|_\infty \leq M_0, \text{ and } \|x'\|_p \leq M_1\right\}.$$

To use Theorem 3.2.1 with $X = W_{\mathrm{per}}^{1,p}(T;\mathbb{R}^N)$, $Y = L^{p'}(T;\mathbb{R}^N)$, $G = N_2$ and $K = K_\lambda$ and solve (3.11), we need to prove the following claim.

Claim 2. For every $x \in \partial W$ and $\kappa \in (0,1)$, we have $x \notin \kappa(K_\lambda \circ N_2)(x)$.

Let $\overline{x} \in \overline{W}$ and suppose that $\overline{x} \in \kappa_0(K_\lambda \circ N_2)(\overline{x})$, for some $\kappa_0 \in (0,1)$. In order to prove the claim, it suffices to show that $\overline{x} \in W$. We have

$$K_\lambda^{-1}\left(\tfrac{1}{\kappa_0}\overline{x}\right) = V_\lambda\left(\tfrac{1}{\kappa_0}\overline{x}\right) \in N_2(\overline{x}),$$

so

$$U\left(\tfrac{1}{\kappa_0}\overline{x}\right) + \widehat{A}_\lambda\left(\tfrac{1}{\kappa_0}\overline{x}\right) + J\left(\tfrac{1}{\kappa_0}\overline{x}\right) = -\overline{f} + J(\overline{x}), \tag{3.12}$$

with $\overline{f} \in N_1(\overline{x})$. Taking the duality brackets with $\overline{x}$, we have

$$\left\langle U\left(\tfrac{1}{\kappa_0}\overline{x}\right), \overline{x}\right\rangle_p + \left\langle \widehat{A}_\lambda\left(\tfrac{1}{\kappa_0}\overline{x}\right), \overline{x}\right\rangle_p + \left\langle J\left(\tfrac{1}{\kappa_0}\overline{x}\right), \overline{x}\right\rangle_p$$
$$= -\langle \overline{f}, \overline{x}\rangle_p + \langle J(\overline{x}), \overline{x}\rangle_p$$

and so

$$\frac{c_0}{\kappa_0^{p-1}}\|\overline{x}'\|_p + \frac{1}{\kappa_0^{p-1}}\|\overline{x}\|_p^p \leq -\langle\overline{f}, \overline{x}\rangle_p + \|\overline{x}\|_p^p.$$

Since $\kappa_0 \in (0,1)$, we have

$$c_0 \|\overline{x}'\|_p^p \le -\kappa_0^{p-1} \langle f, \overline{x} \rangle_p + \left(\kappa_0^{p-1} - 1 \right) \|\overline{x}\|_p^p \le -\kappa_0^{p-1} \langle \overline{f}, \overline{x} \rangle_p. \quad (3.13)$$

Using hypothesis $H(F)_1(iii)$, we have

$$-\kappa_0^{p-1} \langle \overline{f}, \overline{x} \rangle_p \le \kappa_0^{p-1} c_1 \|\overline{x}\|_p^p + \kappa_0^{p-1} c_2 \int_0^b \|\overline{x}(t)\|_{\mathbb{R}^N}^r \|\overline{x}'(t)\|_{\mathbb{R}^N}^{p-r} \, dt$$
$$+ \kappa_0^{p-1} \|c_3\|_1 \|\overline{x}\|_\infty^s. \quad (3.14)$$

Let

$$\vartheta \stackrel{df}{=} \frac{p}{r} \quad \text{and} \quad \vartheta' \stackrel{df}{=} \frac{p}{p-r}.$$

We have that $\frac{1}{\vartheta} + \frac{1}{\vartheta'} = 1$. Using Hölder's inequality (see Theorem A.3.12), we obtain

$$\int_0^b \|\overline{x}(t)\|_{\mathbb{R}^N}^r \|\overline{x}'(t)\|_{\mathbb{R}^N}^{p-r} \, dt$$

$$\le \left(\int_0^b \|\overline{x}(t)\|_{\mathbb{R}^N}^{r\vartheta} \, dt \right)^{\frac{1}{\vartheta}} \left(\int_0^b \|\overline{x}'(t)\|_{\mathbb{R}^N}^{r\vartheta'} \, dt \right)^{\frac{1}{\vartheta'}}$$

$$\le \|\overline{x}\|_p^r \|\overline{x}'\|_p^{p-r}.$$

So, it follows that

$$-\kappa_0^{p-1} \langle \overline{f}, \overline{x} \rangle_p \le \kappa_0^{p-1} c_1 \|\overline{x}\|_p^p + \kappa_0^{p-1} c_2 \|\overline{x}\|_p^r \|\overline{x}'\|_p^{p-r} + \kappa_0^{p-1} \|c_3\|_1 \|\overline{x}\|_\infty^s.$$

Using (3.13), Young's inequality (see Theorem A.4.2) and since $\kappa_0 \in (0,1)$, we have

$$c_0 \|\overline{x}'\|_p^p \le c_1 \|\overline{x}\|_p^p + c_2 \|\overline{x}\|_p^r \|\overline{x}'\|_p^{p-r} + \|c_3\|_1 \|\overline{x}\|_\infty^s$$
$$\le c_1 M_0^p b + c_2 M_0^r b \|\overline{x}'\|_p^{p-r} + \|c_3\|_1 M_0^s$$
$$\le c_1 M_0^p b + \frac{r c_2^{\frac{p}{r}} M_0^p b^{\frac{p}{r}}}{c_0 p} + \frac{c_0 (p-r) \|\overline{x}'\|_p^p}{p} + \|c_3\|_1 M_0^s,$$

so

$$\frac{r c_0}{p} \|\overline{x}'\|_p^p \le c_1 M_0^p b + \frac{r c_2^{\frac{p}{r}} M_0^p b^{\frac{p}{r}}}{c_0 p} + \|c_3\|_1 M_0^s,$$

thus

$$\|\overline{x}'\|_p^p \le \frac{p}{r c_0} \left(c_1 M_0^b b + \frac{r c_2^{\frac{p}{r}} M_0^p b^{\frac{p}{r}}}{c_0 p} + \|c_3\|_1 M_0^s \right) < M_1^p$$

and
$$\|\overline{x}'\|_p < M_1.$$

In order to show that $\overline{x} \in W$, it remains to prove the inequality

$$\|\overline{x}(t)\|_{\mathbb{R}^N} < M_0 \qquad \forall\, t \in T. \tag{3.15}$$

Proceeding by contradiction, suppose that we can find $t_0 \in T$, such that $\|\overline{x}(t_0)\|_{\mathbb{R}^N} = M_0$. Clearly then

$$\|\overline{x}(t_0)\|_{\mathbb{R}^N} = \max_{t \in T} \|\overline{x}(t)\|_{\mathbb{R}^N}.$$

So, if $\overline{\vartheta}(t) \stackrel{df}{=} \frac{1}{p}\|\overline{x}(t)\|_{\mathbb{R}^N}^p$, then $\overline{\vartheta}$ attains its maximum value $\frac{1}{p}M_0^p$ at the point t_0. If $t_0 \in (0, b)$, then $\overline{\vartheta}'(t_0) = 0$ and so

$$\|\overline{x}(t_0)\|_{\mathbb{R}^N}^{p-2}\,(\overline{x}(t_0), \overline{x}'(t_0))_{\mathbb{R}^N} = 0,$$

hence $(\overline{x}(t_0), \overline{x}'(t_0))_{\mathbb{R}^N} = 0$. By virtue of (3.7), for almost all $t \in T$, we have

$$\inf\big\{\, (v, \xi)_{\mathbb{R}^N} + c_0\,\|\overline{\xi}\|_{\mathbb{R}^N}^p \,:\, \|\xi - \overline{x}(t_0)\|_{\mathbb{R}^N} + \|\overline{\xi} - \overline{x}'(t_0)\|_{\mathbb{R}^N} < \delta, $$
$$v \in F(t, \xi, \overline{\xi})\,\big\} \geq \delta' > 0.$$

But $\overline{x} \in \widehat{D}$ and so $\overline{x} \in C^1\big(T; \mathbb{R}^N\big)$. Thus we can find $\delta_1 > 0$, such that

$$\begin{cases} \overline{x}(t) \neq 0 \\ \|\overline{x}(t) - \overline{x}(t_0)\|_{\mathbb{R}^N} + \|\overline{x}'(t) - \overline{x}'(t_0)\|_{\mathbb{R}^N} < \delta \end{cases} \qquad \forall\, t \in (t_0, t_0 + \delta_1].$$

So

$$\big(\overline{f}(t), \overline{x}(t)\big)_{\mathbb{R}^N} + c_0\,\|\overline{x}'(t)\|_{\mathbb{R}^N} \geq \delta' > 0 \quad \text{for a.a. } t \in (t_0, t_0 + \delta_1].$$

From (3.12), we have

$$-\left(a\left(\tfrac{1}{\kappa_0}\overline{x}'(t)\right)\right)' + A_\lambda\left(\tfrac{1}{\kappa_0}\overline{x}(t)\right) + \frac{1}{\kappa_0^{p-1}}\|\overline{x}(t)\|_{\mathbb{R}^N}^{p-2}\,\overline{x}(t)$$
$$= -\overline{f}(t) + \|\overline{x}(t)\|_{\mathbb{R}^N}^{p-2}\,\overline{x}(t) \qquad \text{for a.a. } t \in T,$$

so

$$\overline{f}(t) = \left(a\left(\tfrac{1}{\kappa_0}\overline{x}'(t)\right)\right)' - A_\lambda\left(\tfrac{1}{\kappa_0}\overline{x}(t)\right)$$
$$+ \left(1 - \tfrac{1}{\kappa_0^{p-1}}\right)\|\overline{x}(t)\|_{\mathbb{R}^N}^{p-2}\,\overline{x}(t) \qquad \text{for a.a. } t \in T$$

and

$$\big(\overline{f}(t), \overline{x}(t)\big)_{\mathbb{R}^N} = \left(\left(a\left(\tfrac{1}{\kappa_0}\overline{x}'(t)\right)\right)', \overline{x}(t)\right)_{\mathbb{R}^N}$$
$$- \left(A_\lambda\left(\tfrac{1}{\kappa_0}\overline{x}(t)\right), \overline{x}(t)\right)_{\mathbb{R}^N} + \left(1 - \tfrac{1}{\kappa_0^{p-1}}\right)\|\overline{x}(t)\|_{\mathbb{R}^N}^p \qquad \text{for a.a. } t \in T.$$

Since $\kappa_0 \in (0,1)$ and $\left(A_\lambda \left(\frac{1}{\kappa_0} \overline{x}(t) \right), \overline{x}(t) \right)_{\mathbb{R}^N} \geq 0$, we obtain

$$\left(\left(a \left(\frac{1}{\kappa_0} \overline{x}'(t) \right) \right)', \overline{x}(t) \right)_{\mathbb{R}^N}$$
$$+ c_0 \|\overline{x}'(t)\|_{\mathbb{R}^N} \geq \delta' > 0 \qquad \text{for a.a. } t \in (t_0, t_0 + \delta_1].$$

Integrating this inequality on $[t_0, t]$ with $t \in (t_0, t_0 + \delta_1]$, we have

$$\int_{t_0}^{t} \left(\left(a \left(\frac{1}{\kappa_0} \overline{x}'(s) \right) \right)', \overline{x}(s) \right)_{\mathbb{R}^N} ds + c_0 \int_{t_0}^{t} \|\overline{x}'(s)\|_{\mathbb{R}^N}^p ds$$
$$\geq \delta'(t - t_0) > 0 \qquad \forall t \in (t_0, t_0 + \delta_1].$$

Performing an integration by parts on the first integral, we obtain

$$\left(a \left(\frac{1}{\kappa_0} \overline{x}'(t) \right), \overline{x}(t) \right)_{\mathbb{R}^N} - \left(a \left(\frac{1}{\kappa_0} \overline{x}'(t_0) \right), \overline{x}(t_0) \right)_{\mathbb{R}^N}$$
$$- \int_{t_0}^{t} \left(a \left(\frac{1}{\kappa_0} \overline{x}'(s) \right), \overline{x}'(s) \right)_{\mathbb{R}^N} ds$$
$$+ c_0 \int_{t_0}^{t} \|\overline{x}'(s)\|_{\mathbb{R}^N}^p ds \geq \delta'(t - t_0) > 0 \qquad \forall t \in (t_0, t_0 + \delta_1].$$

Since $a(\xi) = k_a(\xi)\xi$ (see hypothesis $H(a)_2$ and note that the argument is similar if $a(\xi) = \left(k_1^a(\xi_1)\xi_1, \ldots, k_N^a(\xi_N)\xi_N \right)$), we have

$$\left(a \left(\frac{1}{\kappa_0} \overline{x}'(t_0) \right), \overline{x}(t_0) \right)_{\mathbb{R}^N} = k_a \left(\frac{1}{\kappa_0} \overline{x}'(t_0) \right) \frac{1}{\kappa_0} (\overline{x}'(t_0), \overline{x}(t_0))_{\mathbb{R}^N} = 0.$$

So

$$\frac{1}{\kappa_0} k_a \left(\frac{1}{\kappa_0} \overline{x}'(t) \right) (\overline{x}'(t), \overline{x}(t))_{\mathbb{R}^N} - \int_{t_0}^{t} \left(a \left(\frac{1}{\kappa_0} \overline{x}'(s) \right), \overline{x}'(s) \right)_{\mathbb{R}^N} ds$$
$$+ c_0 \int_{t_0}^{t} \|\overline{x}'(s)\|_{\mathbb{R}^N}^p ds \geq \delta'(t - t_0) > 0$$

and

$$\frac{1}{\kappa_0} k_a \left(\frac{1}{\kappa_0} \overline{x}'(t) \right) (\overline{x}'(t), \overline{x}(t))_{\mathbb{R}^N} - \frac{c_0}{\kappa_0^{p-1}} \int_{t_0}^{t} \|\overline{x}'(s)\|_{\mathbb{R}^N}^p ds$$
$$+ c_0 \int_{t_0}^{t} \|\overline{x}'(s)\|_{\mathbb{R}^N}^p ds > 0.$$

Because $\kappa_0 \in (0,1)$, we have

$$(\overline{x}'(t), \overline{x}(t))_{\mathbb{R}^N} > 0$$

and

$$\overline{\vartheta}'(t) = \|\overline{x}(t)\|_{\mathbb{R}^N}^{p-2} (\overline{x}'(t), \overline{x}(t))_{\mathbb{R}^N} > 0,$$

so the function $\overline{\vartheta}$ is increasing on $(t_0, t_0 + \delta_1]$. This contradicts the choice of t_0. Thus we obtain (3.15), which ends the proof of Claim 2.

Therefore, we can apply Theorem 3.2.1 and obtain a solution $x_0 \in \widehat{D}$ of the fixed point problem (3.11). Evidently this is a solution of the auxiliary problem (3.6).

Finally, we remove the stronger hypothesis $H(F)_1(iv)'$ (see (3.7)). To this end let $\varepsilon_n \searrow 0$ and set

$$F_n(t, \xi, \overline{\xi}) \overset{df}{=} F(t, \xi, \overline{\xi}) + \varepsilon_n \xi \qquad \forall\, (t, \xi, \overline{\xi}) \in T \times \mathbb{R}^N \times \mathbb{R}^N.$$

Evidently F_n satisfies hypotheses $H(F)_1$ with $H(F)_1(iv)$ replaced by the stronger condition $H(F)_1(iv)'$. Then, from the previous part of the proof, the auxiliary problem (3.6), with F replaced by F_n, has a solution $x_n \in \widehat{D}$, with

$$\|x_n\|_\infty < M_0 \quad \text{and} \quad \|x'\|_p < M_1 \qquad \forall\, n \geq 1. \tag{3.16}$$

We have

$$U(x_n) + \widehat{A}_\lambda(x_n) + J(x_n) = -f_n + J(x_n) \qquad \forall\, n \geq 1, \tag{3.17}$$

with $f_n \in N_1(x_n)$ for $n \geq 1$. We may assume that

$$x_n \xrightarrow{w} x_0 \text{ in } W_{\text{per}}^{1,p}(T; \mathbb{R}^N),$$
$$f_n \xrightarrow{w} f \text{ in } L^{p'}(T; \mathbb{R}^N).$$

As in the proof of Claim 1, we can show that

$$U(x_0) + \widehat{A}_\lambda(x_0) = f$$

(recall that $\varepsilon_n \searrow 0$) and by hypothesis $H(F)_1(ii)$ and Proposition 1.2.12, we have that $f \in N_1(x)$. Therefore $x_0 \in \widehat{D} \cap \overline{W}$ is a solution of problem (3.6). □

3.2.4 Existence Results

Now that we have solved problem (3.6), by passing to the limit as $\lambda \searrow 0$, we shall obtain our first existence result for problem (3.5).

THEOREM 3.2.2
If hypotheses $H(a)_2$, $H(A)_1$ and $H(F)_1$ hold,
<u>*then*</u> *problem (3.5) has a solution*

$$x_0 \in C^1(T; \mathbb{R}^N), \text{ such that } a(x_0'(\cdot)) \in W_{\mathrm{per}}^{1,2}(T; \mathbb{R}^N).$$

PROOF   Let $\{\lambda_n\}_{n \geq 1} \subseteq (0, +\infty)$ be a sequence, such that $\lambda_n \searrow 0$, and let $\{x_n\}_{n \geq 1} \in C^1(T; \mathbb{R}^N)$ be the sequence of the solutions for the corresponding auxiliary problems (3.6), with $\lambda = \lambda_n$. Evidently the sequence $\{x_n\}_{n \geq 1} \subseteq W_{\mathrm{per}}^{1,p}(T; \mathbb{R}^N)$ is bounded and so, passing to a subsequence if necessary, we may assume that

$$x_n \xrightarrow{w} x_0 \text{ in } W_{\mathrm{per}}^{1,p}(T; \mathbb{R}^N),$$

for some $x_0 \in W_{\mathrm{per}}^{1,p}(T; \mathbb{R}^N)$ We have

$$U(x_n) + \widehat{A}_{\lambda_n}(x_n) = -f_n \qquad \forall\, n \geq 1,$$

with $f_n \in N_1(x_n)$, where N_1 is the operator defined in the proof of Proposition 3.2.4 by (3.10) and

$$\left\langle U(x_n), \widehat{A}_{\lambda_n}(x_n) \right\rangle_p + \left\| \widehat{A}_{\lambda_n}(x_n) \right\|_2^2$$
$$= -\left\langle f_n, \widehat{A}_{\lambda_n}(x_n) \right\rangle_p \qquad \forall\, n \geq 1. \tag{3.18}$$

Let us note that $\widehat{A}_{\lambda_n}(x_n) \in C(T; \mathbb{R}^N)$. From integration by parts and since $x_n(0) = x_n(b)$ and $x_n'(0) = x_n'(b)$, we have

$$\left\langle U(x_n), \widehat{A}_{\lambda_n}(x_n) \right\rangle_p = \int_0^b \left(-\left(a(x_n'(t))\right)', A_{\lambda_n}(x_n(t)) \right)_{\mathbb{R}^N} dt$$

$$= \int_0^b \left(a(x_n'(t)), \tfrac{d}{dt} A_{\lambda_n}(x_n(t)) \right)_{\mathbb{R}^N} dt.$$

Because A_{λ_n} is Lipschitz continuous, by Rademacher's theorem (see Theorem A.2.4), A_{λ_n} is differentiable at every $\xi \in \mathbb{R}^N \setminus E$, for some $E \subseteq \mathbb{R}^N$, such that $|E|_N = 0$. From the monotonicity of A_{λ_n}, we have

$$\left(\overline{\xi}, \tfrac{A_{\lambda_n}(\xi + \lambda \overline{\xi}) - A_{\lambda_n}(\xi)}{\lambda} \right)_{\mathbb{R}^N} \geq 0 \quad \forall \xi \in \mathbb{R}^N \setminus E,\ \overline{\xi} \in \mathbb{R}^N,\ \lambda > 0$$

and so

$$\left(\overline{\xi}, A_{\lambda_n}'(\xi)\overline{\xi} \right)_{\mathbb{R}^N} \geq 0 \qquad \forall\, \xi \in \mathbb{R}^N \setminus E,\ \overline{\xi} \in \mathbb{R}^N. \tag{3.19}$$

Moreover, from the Chain Rule for Sobolev Functions (see Theorem 1.1.13(b)), we have that

$$\frac{d}{dt} A_{\lambda_n}(x_n(t)) = A_{\lambda_n}'(x_n(t))x_n'(t) \quad \text{for a.a. } t \in T.$$

So from (3.19), we have

$$\langle U(x_n), \widehat{A}_{\lambda_n}(x_n) \rangle_p = \int_0^b \left(a(x_n'(t)), A_{\lambda_n}'(x_n(t)) x_n'(t) \right)_{\mathbb{R}^N} dt$$

$$= \int_0^b k_a(x_n'(t)) \left(x_n'(t), A_{\lambda_n}'(x_n(t)) x_n'(t) \right)_{\mathbb{R}^N} dt \geq 0.$$

Using this estimate and hypothesis $H(F)_1(v)$ in (3.18), we obtain

$$\left\| \widehat{A}_{\lambda_n}(x_n) \right\|_2^2 \leq \|f\|_2 \left\| \widehat{A}_{\lambda_n}(x_n) \right\|_2$$

and so the sequence $\{\widehat{A}_{\lambda_n}(x_n)\}_{n \geq 1} \subseteq L^2(T; \mathbb{R}^N)$ is bounded. Passing to a subsequence if necessary, we may assume that

$$\widehat{A}_{\lambda_n}(x_n) \xrightarrow{w} \widehat{u} \quad \text{in } L^2(T; \mathbb{R}^N),$$

for some $\widehat{u} \in L^2(T; \mathbb{R}^N)$. Note that

$$\lambda_n \widehat{A}_{\lambda_n}(x_n) = x_n - \widehat{J}_{\lambda_n}(x_n) \qquad \forall\, n \geq 1,$$

where $\widehat{J}_{\lambda_n}(x_n)(\cdot) \stackrel{df}{=} J_{\lambda_n}(x_n(\cdot))$. So, we infer that

$$\widehat{J}_{\lambda_n}(x_n) \longrightarrow x_0 \quad \text{in } L^2(T; \mathbb{R}^N),$$

for some $x_0 \in L^2(T; \mathbb{R}^N)$. Because

$$A_{\lambda_n}(x_n(t)) \in A(J_{\lambda_n}(x_n(t))) \qquad \forall\, n \geq 1,\ t \in T,$$

from Corollary 1.4.1, we obtain that

$$\widehat{u}(t) \in A(x_0(t)) \quad \text{for a.a. } t \in T.$$

Moreover, because of (3.16) and from hypothesis $H(F)_1(v)$, we have that the sequence $\{f_n\}_{n \geq 1} \subseteq L^2(T; \mathbb{R}^N)$ is bounded. Passing to a subsequence if necessary, we may assume that

$$f_n \xrightarrow{w} f \quad \text{in } L^2(T; \mathbb{R}^N),$$

for some $f \in L^2(T; \mathbb{R}^N)$. Arguing as in the proof of Proposition 3.2.4 (see the proof of Claim 1), we obtain that

$$x_n \longrightarrow x \quad \text{in } C^1(T; \mathbb{R}^N)$$

and so in the limit, we have that $f \in N_1(x_0)$ (see Proposition 1.2.12) and

$$\begin{cases} (a(x_0'(t)))' = \widehat{u}(t) + f(t) \in A(x_0(t)) + F(t, x_0(t), x_0'(t)) \\ \qquad\qquad\qquad\qquad\qquad\qquad\qquad\qquad\qquad \text{for a.a. } t \in T, \\ x_0(0) = x_0(b),\ x_0'(0) = x_0'(b). \end{cases}$$

Therefore $x_0 \in C^1(T; \mathbb{R}^N)$ is a solution of (3.5) and clearly

$$u(x_0'(\cdot)) \subset W_{\text{per}}^{1,2}(T; \mathbb{R}^N).$$

$\square$

If Dom $(A) = \mathbb{R}^N$, we can improve the growth conditions on F and permit multivalued nonlinearities of the Nagumo-Hartman type. The price that we pay for this is that the differential operator is fixed to be the ordinary vector p-Laplacian. So the problem under consideration is the following:

$$\begin{cases} \left(\|x'(t)\|_{\mathbb{R}^N}^{p-2} x'(t)\right)' \in A(x(t)) + F(t, x(t), x'(t)) \\ \qquad\qquad\qquad\qquad\qquad\qquad \text{for a.a. } t \in T, \qquad (3.20) \\ x(0) = x(b), \ x'(0) = x'(b), \end{cases}$$

with $p \in (1, +\infty)$. Our assumptions on the data of (3.20) are the following:

<u>$H(A)_2$</u> $A\colon \mathbb{R}^N \longrightarrow 2^{\mathbb{R}^N}$ is a maximal monotone operator, such that Dom $(A) = \mathbb{R}^N$ and $0 \in A(0)$.

<u>$H(F)_2$</u> $F\colon T \times \mathbb{R}^N \times \mathbb{R}^N \longrightarrow P_{kc}(\mathbb{R}^N)$ is a multifunction, such that

(i) for all $\xi, \bar{\xi} \in \mathbb{R}^N$, the multifunction

$$T \ni t \longmapsto F(t, \xi, \bar{\xi}) \in P_{kc}(\mathbb{R}^N)$$

is graph measurable;

(ii) for almost all $t \in T$, the multifunction

$$\mathbb{R}^N \times \mathbb{R}^N \ni (\xi, \bar{\xi}) \longmapsto F(t, \xi, \bar{\xi}) \in P_{kc}(\mathbb{R}^N)$$

has closed graph;

(iii) for almost all $t \in T$, all $\xi, \bar{\xi} \in \mathbb{R}^N$, such that $\|\xi\|_{\mathbb{R}^N} \leq M_0$ and $\|\bar{\xi}\|_{\mathbb{R}^N}^{p-1} > M_4 > 0$ we have

$$\sup_{v \in F(t, \xi, \bar{\xi})} \|v\|_{\mathbb{R}^N} \leq \eta\left(\|\bar{\xi}\|_{\mathbb{R}^N}^{p-1}\right)$$

with $M_0, M_4 > 0$ and $\eta\colon \mathbb{R}_+ \longrightarrow \mathbb{R}_+ \setminus \{0\}$ is a locally bounded Borel measurable function, such that $\int_{M_4}^{+\infty} \frac{s \, ds}{\eta(s)} = +\infty$;

(iv) if $\|\xi_0\|_{\mathbb{R}^N} = M_0$ (with $M_0 > 0$ as in (iii)) and $(\xi_0, \bar{\xi}_0)_{\mathbb{R}^N} = 0$, we can find $\delta > 0$, such that for almost all $t \in T$, we have

$$\inf \left\{ (v, \xi)_{\mathbb{R}^N} + c_0 \|\bar{\xi}\|_{\mathbb{R}^N}^p : \|\xi - \xi_0\|_{\mathbb{R}^N} + \|\bar{\xi} - \bar{\xi}_0\|_{\mathbb{R}^N} < \delta, \right.$$
$$\left. v \in F(t, \xi, \bar{\xi}) \right\} \geq 0;$$

(v) for all $r > 0$, there exists $\gamma_r \in L^{p'}(T)_+$ (with $\frac{1}{p} + \frac{1}{p'} = 1$), such that for almost all $t \in T$, all $\xi, \overline{\xi} \in \overline{B}_r$ and all $v \in F(t, \xi, \overline{\xi})$, we have $\|v\|_{\mathbb{R}^N} \leq \gamma_r(t)$;

(vi) for almost all $t \in T$, all $\xi, \overline{\xi} \in \mathbb{R}^N$ and all $v \in F(t, \xi, \overline{\xi})$, we have that

$$(v, \xi)_{\mathbb{R}^N} \geq -c_1 \|\xi\|_{\mathbb{R}^N}^p - c_2 \|\xi\|_{\mathbb{R}^N}^r \|\overline{\xi}\|_{\mathbb{R}^N}^{p-r} - c_3(t) \|\xi\|_{\mathbb{R}^N}^s,$$

with $c_1, c_2 > 0$, $1 \leq r, s < p$, $c_3 \in L^1(T)_+$.

REMARK 3.2.4   Hypothesis $H(F)_2(ii)$ is a Bernstein-Nagumo-Hartman growth condition (see Nagumo (1942) and Hartman (2002)) and as we shall see in the sequel, it permits the derivation of pointwise *a priori* bounds for the derivative of the solution. Note that if for almost all $t \in T$, all $\xi, \overline{\xi} \in \mathbb{R}^N$, such that $\|\xi\|_{\mathbb{R}^N} \leq M_0$ and $\|\overline{\xi}\|_{\mathbb{R}^N}^{p-1} \geq M_4$ and all $v \in F(t, \xi, \overline{\xi})$, we have

$$\|v\|_{\mathbb{R}^N} \leq a_1(t) + c_1 \|\xi\|_{\mathbb{R}^N}^{2(p-1)}$$

for some $a_1 \in L^\infty(T)$, $c_1 > 0$, then hypothesis $H(F)_2(iii)$ is satisfied.   ☐

THEOREM 3.2.3
If hypotheses $H(A)_2$ and $H(F)_2$ hold,
<u>*then*</u> *problem (3.20) has a solution*

$$x_0 \in C^1(T; \mathbb{R}^N), \text{ such that } \|x_0'(\cdot)\|_{\mathbb{R}^N}^{p-2} x_0'(\cdot) \in W^{1,p'}(T; \mathbb{R}^N).$$

PROOF   Because of hypotheses $H(A)_2$ and Proposition 1.4.5, the set $\overline{A(\overline{B}_{M_0})} \subseteq \mathbb{R}^N$ is compact and so

$$\vartheta \stackrel{df}{=} \max_{\|\xi\|_{\mathbb{R}^N} \leq M_0} \|A(\xi)\|_{\mathbb{R}^N} < +\infty.$$

Without any loss of generality we may assume that for almost all $r \geq 0$, $0 < \beta \leq \eta(r)$. Set $\eta_1(r) = \vartheta + \eta(r)$. Then, if $\gamma \geq \frac{\vartheta}{\beta} + 1$, we have that $\eta_1(r) \leq \gamma\eta(r)$ for all $r > 0$ and so $\int_{M_0}^\infty \frac{s\,ds}{\eta_1(s)} = +\infty$.

As in the proof of Proposition 3.2.4, we start by assuming first that instead of hypothesis $H(F)_2(iv)$, we have its stronger version (3.7). Let

$$M_4 > \max\left\{ b^{\frac{1}{p}}\left(\frac{p}{rc_0}\left[c_1 M_0^b b + \frac{rc_2^{\frac{p}{r}} M_0^b b^{\frac{p}{r}}}{c_0 p} + \|c_3\|_1 M_0^s\right]\right)^{\frac{p-1}{p}}, M_3 \right\}$$

and take $M_5 > 0$, such that

$$M_5^{p-1} > M_4 \quad \text{and} \quad \int_{M_4}^{M_5^{p-1}} \frac{s\,ds}{\eta_1(s)} ds = M_4.$$

Let

$$C^1_{\mathrm{per}}(T;\mathbb{R}^N) \overset{df}{=} \big\{x \in C^1(T;\mathbb{R}^N) :\ x(0) = x(b),\ x'(0) = x'(b)\big\}.$$

We introduce the following bounded open subset of $C^1_{\mathrm{per}}(T;\mathbb{R}^N)$:

$$W \overset{df}{=} \big\{x \in C^1_{\mathrm{per}}(T;\mathbb{R}^N) :\ \|x\|_\infty < M_0,\ \|x'\|_\infty < M_5\big\}.$$

Evidently

$$\partial W\ =\ \big\{x \in C^1_{\mathrm{per}}(T;\mathbb{R}^N) :\ \|x(t)\|_{\mathbb{R}^N} = M_0,\ \|x'(t)\|_{\mathbb{R}^N} = M_5 \text{ for all } t \in T\big\}.$$

Let $N_F \colon \overline{W} \longrightarrow P_{wkc}\big(L^{p'}(T;\mathbb{R}^N)\big)$ be the multivalued Nemytskii operator, defined by

$$N_F(x) \overset{df}{=} S^{p'}_{F(\cdot,x(\cdot),x'(\cdot))} \qquad \forall\, x \in \overline{W}.$$

As in the proof of Proposition 3.1.4 (see Claim 2), we can check that the multifunction N_F is upper semicontinuous from $\overline{W}$ furnished with the $C^1(T;\mathbb{R}^N)$-norm topology into $L^{p'}(T;\mathbb{R}^N)_w$. For each $g \in L^{p'}(T;\mathbb{R}^N)$, we consider the following periodic problem:

$$\begin{cases} -\big(\|x'(t)\|^{p-2}_{\mathbb{R}^N}\, x'(t)\big)' + \|x(t)\|^{p-2}_{\mathbb{R}^N}\, x(t) = g(t) & \text{for a.a. } t \in T, \\ x(0) = x(b),\ x'(0) = x'(b). \end{cases} \tag{3.21}$$

From Proposition 3.2.2, we know that there exists $x \in C^1(T;\mathbb{R}^N)$, a solution of problem (3.21), which is easily seen to be unique. So we can define the map $K \colon L^{p'}(T;\mathbb{R}^N) \longrightarrow C^1(T;\mathbb{R}^N)$, which to each function $g \in L^{p'}(T;\mathbb{R}^N)$ assigns the unique solution $x \in C^1(T;\mathbb{R}^N)$ of problem (3.21).

Claim 1. The operator $K \colon L^{p'}(T;\mathbb{R}^N) \longrightarrow C^1(T;\mathbb{R}^N)$ is completely continuous.

Let $\{g_n\}_{n\geq 1} \subseteq L^{p'}(T;\mathbb{R}^N)$ be a sequence, such that

$$g_n \overset{w}{\longrightarrow} g \quad \text{in } L^{p'}(T;\mathbb{R}^N),$$

for some $g \in L^{p'}(T;\mathbb{R}^N)$ and set $x_n \overset{df}{=} K(g_n)$ for $n \geq 1$. Taking inner product of (3.21) with $x_n(t)$ and integrating over T, after integration by parts, we obtain

$$\|x'_n\|^p_p + \|x_n\|^p_p \ \leq\ \|g_n\|_{p'}\, \|x_n\|_p,$$

so the sequence $\{x_n\}_{n\geq 1} \subseteq W^{1,p}_{\mathrm{per}}(T;\mathbb{R}^N)$ is bounded. By passing to a subsequence if necessary, we may assume that

$$\begin{aligned} x_n &\overset{w}{\longrightarrow} \overline{x} \quad \text{in } W^{1,p}_{\mathrm{per}}(T;\mathbb{R}^N), \\ x_n &\longrightarrow \overline{x} \quad \text{in } C(T;\mathbb{R}^N). \end{aligned}$$

From the formulation of the problem (3.21), we see that the sequence $\big\{ \|x'_n(\cdot)\|_{\mathbb{R}^N}^{p-2} x'_n(\cdot) \big\}_{n\geq 1} \subseteq L^{p'}\big(T;\mathbb{R}^N\big)$ is bounded and so, after passing to a further subsequence if necessary, we may assume that

$$\big(\|x'_n(\cdot)\|_{\mathbb{R}^N}^{p-2} x'_n(\cdot) \big)' \xrightarrow{w} \widehat{u} \quad \text{in } L^{p'}\big(T;\mathbb{R}^N\big).$$

Since the ordinary vector p-Laplacian with periodic boundary conditions is a maximal monotone map from $L^p\big(T;\mathbb{R}^N\big)$ into $L^{p'}\big(T;\mathbb{R}^N\big)$, it follows that

$$\widehat{u} = \|x'(\cdot)\|_{\mathbb{R}^N}^{p-2} x'(\cdot).$$

Therefore

$$\big(\|x'_n(\cdot)\|_{\mathbb{R}^N}^{p-2} x'_n(\cdot) \big)' \xrightarrow{w} \big(\|\overline{x}'(\cdot)\|_{\mathbb{R}^N}^{p-2} \overline{x}'(\cdot) \big)' \quad \text{in } L^{p'}\big(T;\mathbb{R}^N\big).$$

As the sequence $\big\{ \|x'_n(\cdot)\|_{\mathbb{R}^N}^{p-2} x'_n(\cdot) \big\}_{n\geq 1} \subseteq L^{p'}\big(T;\mathbb{R}^N\big)$ is bounded, we infer that

$$\|x'_n(\cdot)\|_{\mathbb{R}^N}^{p-2} x'_n(\cdot) \xrightarrow{w} \|\overline{x}'(\cdot)\|_{\mathbb{R}^N}^{p-2} \overline{x}'(\cdot) \quad \text{in } L^{p'}\big(T;\mathbb{R}^N\big).$$

So finally, we have

$$\|x'_n(\cdot)\|_{\mathbb{R}^N}^{p-2} x'_n(\cdot) \xrightarrow{w} \|\overline{x}'(\cdot)\|_{\mathbb{R}^N}^{p-2} \overline{x}'(\cdot) \quad \text{in } W^{1,p'}\big(T;\mathbb{R}^N\big).$$

Let $\varphi_p \colon \mathbb{R}^N \longrightarrow \mathbb{R}^N$ be the homeomorphism, defined by

$$\varphi_p(\xi) \stackrel{df}{=} \begin{cases} \|\xi\|_{\mathbb{R}^N}^{p-2} \xi & \text{for } \xi \in \mathbb{R}^N \setminus \{0\}, \\ 0 & \text{for } \xi = 0 \end{cases}$$

and let $\widehat{\psi} \colon C\big(T;\mathbb{R}^N\big) \longrightarrow C\big(T;\mathbb{R}^N\big)$ be defined by

$$\widehat{\psi}(x)(\cdot) \stackrel{df}{=} \varphi_p^{-1}\big(\|x(\cdot)\|_{\mathbb{R}^N}^{p-2} x(\cdot) \big) \qquad \forall\, x \in C\big(T;\mathbb{R}^N\big).$$

Evidently $\widehat{\psi}$ is a bounded continuous map. Therefore, we infer that

$$x'_n \longrightarrow \overline{x}' \quad \text{in } C\big(T;\mathbb{R}^N\big)$$

and so finally

$$x_n \longrightarrow \overline{x} \quad \text{in } C^1\big(T;\mathbb{R}^N\big),$$

which proves the claim.

Let $J \colon C^1\big(T;\mathbb{R}^N\big) \longrightarrow L^{p'}\big(T;\mathbb{R}^N\big)$ be the bounded continuous map, defined by

$$J(x)(\cdot) \stackrel{df}{=} \|x(\cdot)\|_{\mathbb{R}^N}^{p-2} x(\cdot) \qquad \forall\, x \in C^1\big(T;\mathbb{R}^N\big).$$

Also, let $\widehat{A} \colon C^1\big(T;\mathbb{R}^N\big) \longrightarrow 2^{L^{p'}(T;\mathbb{R}^N)}$ be defined by

$$\widehat{A}(x) = S^{p'}_{A(x(\cdot))} \qquad \forall\, x \in C^1\big(T;\mathbb{R}^N\big).$$

Note that since by hypotheses $H(A)_2$, $\mathrm{Dom}\,(A) = \mathbb{R}^N$, we have that

$$\mathrm{Dom}\,(\widehat{A}) = C^1\big(T; \mathbb{R}^N\big).$$

Moreover, from Propositions 1.2.12 and 1.4.5, we see that $\mathrm{Gr}\,\widehat{A}$ is sequentially closed in $C^1\big(T; \mathbb{R}^N\big) \times L^{p'}\big(T; \mathbb{R}^N\big)_w$. Because $\widehat{A}$ is locally bounded (see Proposition 1.4.5) and bounded sets in $L^{p'}\big(T; \mathbb{R}^N\big)$ are relatively weakly compact and the relative weakly topology is metrizable, we conclude that $\widehat{A}$ is upper semicontinuous from $C^1\big(T; \mathbb{R}^N\big)$ into $L^{p'}\big(T; \mathbb{R}^N\big)_w$ (see Proposition 1.2.5). Set

$$N_3(x) \stackrel{df}{=} -N_F(x) + J(x) - \widehat{A}(x) \qquad \forall\, x \in \overline{W}.$$

Clearly $N_3 : \overline{W} \longrightarrow P_{wkc}\big(L^{p'}\big(T; \mathbb{R}^N\big)\big)$ is upper semicontinuous from $\overline{W}$ with the $C^1\big(T; \mathbb{R}^N\big)$-norm topology into $L^{p'}\big(T; \mathbb{R}^N\big)_w$. Problem (3.20) is equivalent to the following abstract fixed point problem:

$$x \in (K \circ N_3)(x). \tag{3.22}$$

Claim 2. For every $x \in \partial W$ and every $\kappa \in (0, 1)$, We have $x \notin \kappa(K \circ N_3)(x)$.

Let $\overline{x} \in \overline{W}$ and suppose that for some $\kappa_0 \in (0, 1)$, we have $\overline{x} \in \kappa_0(K \circ N_3)(\overline{x})$. Then we have

$$K^{-1}\left(\tfrac{1}{\kappa_0}\overline{x}\right) \in N_3(\overline{x}),$$

so

$$\begin{cases} -\frac{1}{\kappa_0^{p-1}}\left(\|\overline{x}'(t)\|_{\mathbb{R}^N}^{p-2}\,\overline{x}'(t)\right)' + A\big(\overline{x}(t)\big) \\ \qquad + \frac{1}{\kappa_0^{p-1}}\|\overline{x}(t)\|_{\mathbb{R}^N}^{p-2}\,\overline{x}(t) = -f(t) + \|\overline{x}(t)\|_{\mathbb{R}^N}^{p-2}\,\overline{x}(t) \\ \hfill \text{for a.a. } t \in T, \\ \overline{x}(0) = \overline{x}(b),\ \overline{x}'(0) = \overline{x}'(b), \end{cases} \tag{3.23}$$

with $f \in S^{p'}_{F(\cdot,\overline{x}(\cdot),\overline{x}'(\cdot))}$. Using (3.23) and arguing as in the proof of Proposition 3.2.4 (Claim 2), we obtain that

$$\|\overline{x}'\|_p^p \le \frac{p}{rc_0}\left(c_1 M_0^b b + \frac{rc_2^{\frac{p}{r}} M_0^p b^{\frac{p}{r}}}{c_0 p} + \|c_3\|_1 M_0^s\right),$$

so

$$\|\overline{x}'\|_p^{p-1} < \frac{1}{b^{\frac{1}{p}}} M_4$$

and by Jensen's inequality (see Theorem A.2.5) and since $\frac{p-1}{p} = \frac{1}{p'}$, we have

$$\int_0^b \|\overline{x}'(t)\|_{\mathbb{R}^N}^{p-1}\, dt < M_4. \tag{3.24}$$

We claim that

$$\|\overline{x}'(t)\|_{\mathbb{R}^N} < M_5 \qquad \forall\, t \in T. \tag{3.25}$$

Suppose that this is not the case. Then we can find $t_0 \in T$, such that $\|\overline{x}'(t_0)\|_{\mathbb{R}^N} = M_5$, hence $\|\overline{x}'(t_0)\|_{\mathbb{R}^N}^{p-1} > M_4$. By virtue of (3.24), we see that there exists $t_1 \in T$, such that $\|\overline{x}'(t_1)\|_{\mathbb{R}^N}^{p-1} = M_4$ (take the $t_1 \in T$ which is closest to t_0). Let $\chi\colon [M_4, +\infty) \longrightarrow \mathbb{R}_+$ be defined by

$$\chi(r) \stackrel{df}{=} \int_{M_4}^{r} \frac{s\,ds}{\eta_1(s)} \qquad \forall\, r \in [M_4, +\infty).$$

Clearly χ is continuous, strictly increasing, $\chi(M_4) = 0$ and $\chi(M_5^{p-1}) = M_4$. By a change of variables (see Theorem 1.1.15), we have

$$M_4 = \chi(M_5^{p-1}) = \left| \chi\big(\|\overline{x}'(t_0)\|_{\mathbb{R}^N}^{p-1} \big) \right|$$

$$= \left| \int_{M_4}^{\|\overline{x}'(t_0)\|_{\mathbb{R}^N}^{p-1}} \frac{s}{\eta_1(s)} ds \right| = \left| \int_{\|\overline{x}'(t_0)\|_{\mathbb{R}^N}^{p-1}}^{\|\overline{x}'(t_1)\|_{\mathbb{R}^N}^{p-1}} \frac{s}{\eta_1(s)} ds \right|$$

$$= \left| \int_{t_0}^{t_1} \frac{\left(\big(\|\overline{x}'(t)\|_{\mathbb{R}^N}^{p-2} \overline{x}'(t)\big)', \|\overline{x}'(t)\|_{\mathbb{R}^N}^{p-2} \overline{x}'(t) \right)_{\mathbb{R}^N}}{\eta_1\big(\|\overline{x}'(t)\|_{\mathbb{R}^N}^{p-1} \big)} dt \right|$$

$$= \left| \int_{t_0}^{t_1} \frac{\|\overline{x}'(t)\|_{\mathbb{R}^N}^{p-1}}{\eta_1\big(\|\overline{x}'(t)\|_{\mathbb{R}^N}^{p-1} \big)} \frac{\left(\big(\|\overline{x}'(t)\|_{\mathbb{R}^N}^{p-2} \overline{x}'(t)\big)', \|\overline{x}'(t)\|_{\mathbb{R}^N}^{p-2} \overline{x}'(t) \right)_{\mathbb{R}^N}}{\|\overline{x}'(t)\|_{\mathbb{R}^N}^{p-1}} dt \right|$$

$$\leq \left| \int_{t_0}^{t_1} \frac{\left\| \big(\|\overline{x}'(t)\|_{\mathbb{R}^N}^{p-2} \overline{x}'(t)\big)' \right\|_{\mathbb{R}^N}}{\eta_1\big(\| \big(\|\overline{x}'(t)\|_{\mathbb{R}^N}^{p-2} \overline{x}'(t)\big)' \|_{\mathbb{R}^N} \big)} \|\overline{x}'(t)\|_{\mathbb{R}^N}^{p-1} \, dt. \right| \tag{3.26}$$

From (3.23) and since $\kappa_0 \in (0,1)$ and $\|A(\overline{x}(t))\|_{\mathbb{R}^N} \leq \vartheta$ for all $t \in T$, we have

$$\left\| \big(\|\overline{x}'(t)\|_{\mathbb{R}^N}^{p-2} \overline{x}'(t)\big)' \right\|_{\mathbb{R}^N} \leq \vartheta + \eta\big(\|\overline{x}'(t)\|_{\mathbb{R}^N}^{p-1} \big) = \eta_1\big(\|\overline{x}'(t)\|_{\mathbb{R}^N}^{p-1} \big)$$

$$= \eta_1\big(\|\overline{x}'(t)\|_{\mathbb{R}^N}^{p-1} \overline{x}(t) \big).$$

Using this and (3.24) in (3.26), we obtain

$$M_4 \leq \left| \int_{t_0}^{t_1} \|\overline{x}'(t)\|_{\mathbb{R}^N}^{p-1} \, dt \right| = \int_{\min\{t_0,t_1\}}^{\max\{t_0,t_1\}} \|\overline{x}'(t)\|_{\mathbb{R}^N}^{p-1} \, dt < M_4,$$

a contradiction. Therefore $\|\overline{x}'(t)\|_{\mathbb{R}^N} < M_5$ for all $t \in T$. To establish the claim we need to show that $\|\overline{x}(t)\|_{\mathbb{R}^N} < M_0$ for all $t \in T$. But this can be shown using the same argument as in the proof of Proposition 3.2.4. So it follows that $\overline{x} \in W$ and this proves the claim.

Claims 1 and 2 permit the use of Theorem 3.2.1, which gives us a solution $x_0 \in C^1(T; \mathbb{R}^N)$ to the fixed point problem (3.22). Clearly such a solution of (3.22) also solves (3.20).

Finally working as in the last part of the proof of Proposition 3.2.4, we can remove the stronger version of hypothesis $H(F)_2(iv)$ and finish the proof of the theorem. $\square$

3.2.5   Problems with Non-Convex Nonlinearities

As for the Dirichlet problem, we can have "nonconvex" versions of the previous results. Our hypotheses on the multivalued nonlinearities F are the following.

$\underline{H(F)_3}$ $F: T \times \mathbb{R}^N \times \mathbb{R}^N \longrightarrow P_k(\mathbb{R}^N)$ is a multifunction, such that

   (i) F is graph measurable;

   (ii) for almost all $t \in T$, the multifunction

$$\mathbb{R}^N \times \mathbb{R}^N \ni (\xi, \overline{\xi}) \longmapsto F(t, \xi, \overline{\xi}) \in P_k(\mathbb{R}^N)$$

   is lower semicontinuous;

$\left.\begin{array}{l}(iii)\\(iv)\\(v)\\(vi)\end{array}\right\}$ are the same as hypotheses $H(F)_1(iii)$, (iv), (v) and (vi).

Using the same proof as in Theorem 3.1.4 (with $W_0^{1,p}(T; \mathbb{R}^N)$ replaced by $W_{\mathrm{per}}^{1,p}(T; \mathbb{R}^N)$), we obtain the following existence theorem.

THEOREM 3.2.4
If hypotheses $H(a)_2$, $H(A)_1$ *and* $H(F)_3$ *hold,*
<u>*then*</u> *problem (3.5) has a solution*

$$x_0 \in C^1(T; \mathbb{R}^N), \text{ such that } a\big(x_0'(\cdot)\big) \in W_{\mathrm{per}}^{1,2}(T; \mathbb{R}^N).$$

Again if $\mathrm{Dom}\,(A) = \mathbb{R}^N$, we can improve the growth conditions on F and permit multivalued nonlinearities of the Nagumo-Hartman type. In this case the hypotheses on the multivalued nonlinearities F are the following:

$\underline{H(F)_4}$ $F: T \times \mathbb{R}^N \times \mathbb{R}^N \longrightarrow P_k(\mathbb{R}^N)$ is a multifunction, such that

   (i) F is graph measurable;

   (ii) for almost all $t \in T$, the multifunction

$$\mathbb{R}^N \times \mathbb{R}^N \ni (\xi, \overline{\xi}) \longrightarrow F(t, \xi, \overline{\xi}) \in P_k(\mathbb{R}^N)$$

   is lower semicontinuous;

$$\left.\begin{array}{c}(iii)\\(iv)\\(v)\\(vi)\end{array}\right\}\ \text{are the same as hypotheses } H(F)_2(iii),\ (iv),\ (v) \text{ and } (vi).$$

THEOREM 3.2.5
*If hypotheses $H(A)_2$ and $H(F)_4$ hold,
then problem (3.20) has a solution*

$$x_0 \in C^1\big(T; \mathbb{R}^N\big),\ \text{ such that }\ \|x_0'(\cdot)\|_{\mathbb{R}^N}^{p-2}\, x_0'(\cdot) \in W^{1,p'}\big(T; \mathbb{R}^N\big).$$

3.2.6 Scalar Problems

Now, let us study a scalar periodic problem (i.e. with $N = 1$), in which the differential operator depends on both x and x' (nonlinearly in x and linearly in x'). More precisely the scalar periodic problem under consideration is the following:

$$\begin{cases} \big(a(x(t))x'(t)\big)' \in A\big(x(t)\big) + F\big(t, x(t), x'(t)\big) & \text{for a.a. } t \in T,\\ x(0) = x(b),\ x'(0) = x'(b). \end{cases} \tag{3.27}$$

Our hypotheses on the data of (3.27) are the following:

$\underline{H(a)_3}$ $a\colon \mathbb{R} \longrightarrow \mathbb{R}$ satisfies

$$a(\zeta)\ \geq\ c_0 \qquad \forall\,\zeta \in \mathbb{R},$$

with $c_0 > 0$ and

$$\big|a(\zeta) - a(\overline{\zeta})\big|\ \leq\ \alpha_a|\zeta - \overline{\zeta}| \qquad \forall\,\zeta, \overline{\zeta} \in \mathbb{R},$$

with $\alpha_a > 0$.

$\underline{H(A)_3}$ $A\colon \mathbb{R} \supseteq \mathrm{Dom}\,(A) \longrightarrow 2^{\mathbb{R}}$ is a maximal monotone map, such that $0 \in A(0)$.

REMARK 3.2.5 From Remark 1.4.6, we know that $A = \partial\varphi$ with $\varphi \in \Gamma_0(\mathbb{R})$. In this case for every $\lambda > 0$, we have $A_\lambda = \partial\varphi_\lambda$, with φ_λ being the Moreau-Yosida regularization of φ (see Remark 1.4.9). $\qquad\Box$

$\underline{H(F)_5}$ $F\colon T \times \mathbb{R} \times \mathbb{R} \longrightarrow P_{fc}(\mathbb{R})$ is a multifunction, such that

(i) for all $\zeta, \overline{\zeta} \in \mathbb{R}$, the multifunction

$$T \ni t \longmapsto F(t, \zeta, \overline{\zeta}) \in P_{fc}(\mathbb{R})$$

is graph measurable;

(*ii*) for almost all $t \in T$, the multifunction

$$\mathbb{R} \times \mathbb{R} \ni (\zeta, \overline{\zeta}) \longmapsto F(t, \zeta, \overline{\zeta}) \in P_{fc}(\mathbb{R})$$

has closed graph;

(*iii*) for almost all $t \in T$, all $\zeta, \overline{\zeta} \in \mathbb{R}$ and all $v \in F(t, \zeta, \overline{\zeta})$, we have

$$v\zeta \geq -c_1|\zeta|^2 - c_2|\zeta||\overline{\zeta}| - c_3(t)|\overline{\zeta}|$$

with $c_1, c_2 \geq 0$, $c_3 \in L^1(T)_+$;

(*iv*) there exists $M_0 > 0$, such that if $|\zeta_0| = M_0$, then we can find $\delta > 0$, such that for almost all $t \in T$, we have

$$\inf\left\{v\zeta + c_0\overline{\zeta}^2 : \ |\zeta - \zeta_0| + c_0|\overline{\zeta}| < \delta, \ v \in F(t, \zeta, \overline{\zeta})\right\} \geq 0;$$

(*v*) for almost all $t \in T$, all $\zeta, \overline{\zeta} \in \mathbb{R}$, $|\zeta| \leq M_0$ and all $v \in F(t, \zeta, \overline{\zeta})$, we have

$$|v| \leq c_4(t) + c_5|\overline{\zeta}|,$$

with $c_4 \in L^2(T)_+$, $c_5 > 0$.

REMARK 3.2.6 From hypotheses $H(F)_5(i)$ and (ii), it follows that

$$F(t, \zeta, \overline{\zeta}) = \left[f_1(t, \zeta, \overline{\zeta}), f_2(t, \zeta, \overline{\zeta})\right] \qquad \forall \ (t, \zeta, \overline{\zeta}) \in T \times \mathbb{R} \times \mathbb{R},$$

with f_1, f_2 two measurable functions, such that for almost all $t \in T$, the functions

$$(\zeta, \overline{\zeta}) \longmapsto -f_1(t, \zeta, \overline{\zeta}) \quad \text{and} \quad (\zeta, \overline{\zeta}) \longmapsto f_2(t, \zeta, \overline{\zeta})$$

are upper semicontinuous. ⬜

As it was the case with problem (3.5), since we do not assume that $\mathrm{Dom}\,(A) = \mathbb{R}$, our analysis of problem (3.27) passes from the study of the following auxiliary periodic problem:

$$\begin{cases} -\big(a\big(x(t)\big)x'(t)\big)' + x(t) + A_\lambda(x(t)) = g(t) & \text{for a.a. } t \in T, \\ x(0) = x(b), \ x'(0) = x'(b), \end{cases} \tag{3.28}$$

with $g \in L^2(T)$ and $\lambda > 0$.

PROPOSITION 3.2.5
If hypotheses $H(a)_3$ and $H(A)_3$ hold,
<u>*then*</u> *for every $g \in L^2(T)$ problem (3.28) has a unique solution*

$$x_0 \in C^1(T) \text{ with } a\big(x_0(\cdot)\big)x_0'(\cdot) \in W^{1,2}_{\mathrm{per}}(T).$$

PROOF   Let $V \colon W^{1,2}_{\mathrm{per}}(T) \longrightarrow \big(W^{1,2}_{\mathrm{per}}(T)\big)^*$ be defined by

$$\big\langle V(x), y \big\rangle_{W^{1,2}_{\mathrm{per}}(T)} = \int_0^b a\big(x(t)\big) x'(t) y'(t) \, dt \qquad \forall \, x, y \in W^{1,2}_{\mathrm{per}}(T).$$

Claim 1. The operator V is pseudomonotone.

Since V is everywhere defined and bounded (see hypotheses $H(a)_3$), it suffices to show that V is generalized pseudomonotone (see Definition 1.4.8(b) and Proposition 1.4.12). To this end assume that

$$x_n \xrightarrow{w} \overline{x} \ \text{ in } W^{1,2}_{\mathrm{per}}(T),$$
$$V(x_n) \xrightarrow{w} \overline{u} \ \text{ in } \big(W^{1,2}_{\mathrm{per}}(T)\big)^* \tag{3.29}$$

and

$$\limsup_{n \to +\infty} \big\langle V(x_n), x_n - \overline{x} \big\rangle_{W^{1,2}_{\mathrm{per}}(T)} \leq 0.$$

We need to show that

$$\overline{u} = V(\overline{x}) \quad \text{and} \quad \big\langle V(x_n), x_n \big\rangle_{W^{1,2}_{\mathrm{per}}(T)} \longrightarrow \big\langle V(\overline{x}), \overline{x} \big\rangle_{W^{1,2}_{\mathrm{per}}(T)}. \tag{3.30}$$

Note that

$$x_n \longrightarrow \overline{x} \ \text{ in } C(T)$$

and so

$$a\big(x_n(\cdot)\big) x'_n(\cdot) \xrightarrow{w} a\big(\overline{x}(\cdot)\big) \overline{x}'(\cdot) \ \text{ in } L^2(T).$$

For every $\vartheta \in C^\infty(0, b)$, we have

$$\int_0^b a\big(x_n(t)\big) x'_n(t) \vartheta'(t) \, dt \longrightarrow \int_0^b a\big(\overline{x}(t)\big) \overline{x}'(t) \vartheta'(t) \, dt,$$

so

$$\big\langle V(x_n), \vartheta \big\rangle_{W^{1,2}_{\mathrm{per}}(T)} \longrightarrow \big\langle \overline{u}, \vartheta \big\rangle_{W^{1,2}_{\mathrm{per}}(T)} \qquad \forall \, \vartheta \in C^\infty(0, b),$$

from which it follows that $\overline{u}(\cdot) = -\big(a\big(\overline{x}(\cdot)\big) \overline{x}'(\cdot)\big)'$, hence $\overline{u} = V(\overline{x})$. Moreover, note that

$$\big\langle V(x_n), x_n - \overline{x} \big\rangle_{W^{1,2}_{\mathrm{per}}(T)} = \int_0^b a\big(x_n(t)\big) \big(x'_n(t) - \overline{x}'(t)\big)^2 dt$$

$$+ \int_0^b a\big(x_n(t)\big) \overline{x}'(t) \big(x'_n(t) - \overline{x}'(t)\big) \, dt.$$

Because

$$\int_0^b a\big(x_n(t)\big)\overline{x}'(t)\big(x_n'(t) - \overline{x}'(t)\big)\,dt \;\longrightarrow\; 0,$$

from the choice of the sequence $\{x_n\}_{n\geq 1} \subseteq W_{\mathrm{per}}^{1,2}(T)$, it follows that $\langle V(x_n), x_n - \overline{x}\rangle_{W_{\mathrm{per}}^{1,2}(T)} \longrightarrow 0$ and because of (3.29), we obtain (3.30). This proves the claim.

Let $\widehat{A}_\lambda\colon L^2(T) \longrightarrow L^2(T)$ be the bounded, continuous and monotone (hence maximal monotone) operator, defined by

$$\widehat{A}_\lambda(u)(\cdot) \overset{df}{=} A_\lambda\big(u(\cdot)\big) \qquad \forall\, u \in L^2(T).$$

Let us set

$$V_1 \overset{df}{=} V + id_{W_{\mathrm{per}}^{1,2}(T)} + \widehat{A}_\lambda.$$

Clearly $V_1\colon W_{\mathrm{per}}^{1,2}(T) \longrightarrow \big(W_{\mathrm{per}}^{1,2}(T)\big)^*$ is a pseudomonotone operator (see Proposition 1.4.13). Moreover, for every $x \in W_{\mathrm{per}}^{1,2}(T)$, we have

$$\langle V_1(x), x\rangle_{W_{\mathrm{per}}^{1,2}(T)} \;\geq\; c_0\,\|x'\|_2^2 + \|x\|_2^2 \;\geq\; c_6\,\|x\|_{W^{1,2}(T)}^2,$$

with $c_6 = \min\{c_0, 1\}$. So the operator V_1 is coercive, hence it is surjective (see Theorem 1.4.6). Therefore, we can find $x_0 \in W_{\mathrm{per}}^{1,2}(T)$, such that $V_1(x_0) = g$. It follows that

$$\begin{cases} -\big(a\big(x_0(t)\big)x_0'(t)\big)' = g(t) - x_0(t) - A_\lambda\big(x_0(t)\big) & \text{for a.a. } t \in T, \\ x_0(0) = x_0(b), \end{cases} \qquad (3.31)$$

so

$$a\big(x_0(\cdot)\big)x_0'(\cdot) \;\in\; W^{1,2}(T)$$

and it follows that the function $t \longmapsto \frac{a(x_0(t))x_0'(t)}{a(x_0(t))} = x_0'(t)$ is continuous, i.e. $x_0 \in C^1(T)$.

For every $y \in W_{\mathrm{per}}^{1,2}(T)$, after integration by parts, we have

$$\int_0^b \big(a\big(x_0(t)\big)x_0'(t)\big)' y(t)\,dt \;=\; a\big(x_0(b)\big)x_0'(b)y(b) - a\big(x_0(0)\big)x_0'(0)y(0)$$

$$-\int_0^b a\big(x_0(t)\big)x_0'(t)y'(t)\,dt.$$

Since $V_1(x_0) = g$ and because of (3.31), we obtain

$$a\big(x_0(0)\big)x_0'(0)y(0) \;=\; a\big(x_0(b)\big)x_0'(b)y(b) \qquad \forall\, y \in W_{\mathrm{per}}^{1,2}(T),$$

so

$$a(x_0(0))x_0'(0) \; = \; a(x_0(b))x_0'(b)$$

and since $x_0(0) = x_0(b)$, we have

$$x_0'(0) \; = \; x_0'(b).$$

This proves that $x_0 \in C^1(T)$ is a solution of (3.28).

Next let us show the uniqueness of this solution. To this end suppose that $x_0, \overline{x}_0 \in C^1(T)$ are two solutions of (3.28). We have

$$\begin{cases} -\big(a(x_0(t))x_0'(t)\big)' + x_0(t) + A_\lambda(x_0(t)) = g(t) & \text{for a.a. } t \in T, \\ x_0(0) = x_0(b), \ x_0'(0) = x_0'(b) \end{cases} \tag{3.32}$$

and

$$\begin{cases} -\big(a(\overline{x}_0(t))\overline{x}_0'(t)\big)' + \overline{x}_0(t) + A_\lambda(\overline{x}_0(t)) = g(t) & \text{for a.a. } t \in T, \\ \overline{x}_0(0) = \overline{x}_0(b), \ \overline{x}_0'(0) = \overline{x}_0'(b). \end{cases} \tag{3.33}$$

Let

$$\eta_\varepsilon(r) \; \stackrel{df}{=} \; \begin{cases} \displaystyle\int_\varepsilon^r \dfrac{ds}{\alpha_a^2 s^2} & \text{if } r \geq \varepsilon, \\ 0 & \text{if } r < \varepsilon, \end{cases}$$

with $\varepsilon > 0$ and with $\alpha_a > 0$ as in hypothesis $H(a)_3$. From Theorem 1.1.13, we know that $\eta_\varepsilon\big((x_0 - \overline{x}_0)(\cdot)\big) \in W^{1,2}(T)$. Subtracting (3.33) from (3.32), multiplying with $\eta_\varepsilon\big((x_0 - \overline{x}_0)(t)\big)$, integrating over T, performing integration by parts on the first integral and using the periodic boundary conditions, we obtain

$$0 \; = \; \int_0^b \big(a(x_0(t))x_0'(t) - a(\overline{x}_0(t))\overline{x}_0'(t)\big)\frac{d}{dt}\eta_\varepsilon\big((x_0 - \overline{x}_0)(t)\big)\, dt$$

$$+ \int_0^b \big(x_0(t) - \overline{x}_0(t)\big)\eta_\varepsilon\big((x_0 - \overline{x}_0)(t)\big)\, dt$$

$$+ \int_0^b \big(A_\lambda(x_0(t)) - A_\lambda(\overline{x}_0(t))\big)\eta_\varepsilon\big((x_0 - \overline{x}_0)(t)\big)\, dt. \tag{3.34}$$

From the monotonicity of the functions η_ε and A_λ, we have

$$\int_0^b \eta_\varepsilon\big((x_0 - \overline{x}_0)(t)\big)\, dt \; = \; \int_{\{x_0 - \overline{x}_0 \geq \varepsilon\}} \eta_\varepsilon\big((x_0 - \overline{x}_0)(t)\big)\, dt \; \geq \; 0$$

and

$$\int_0^b \big(A_\lambda(x_0(t)) - A_\lambda(\overline{x}_0(t))\big)\eta_\varepsilon\big((x_0 - \overline{x}_0)(t)\big)\, dt \; \geq \; 0.$$

Using these inequalities in (3.34), we obtain

$$\int_0^b \left(a\big(x_0(t)\big)x_0'(t) - a\big(\overline{x}_0(t)\big)\overline{x}_0'(t) \right) \frac{d}{dt}\eta_\varepsilon\big((x_0 - \overline{x}_0)(t)\big)\,dt \;\leq\; 0$$

and

$$\int_0^b a\big(x_0(t)\big)\,\big(x_0'(t) - \overline{x}_0'(t)\big)\,\frac{d}{dt}\eta_\varepsilon\big((x_0 - \overline{x}_0)(t)\big)\,dt$$

$$\leq\; -\int_0^b \big(a\big(x_0(t)\big) - a\big(\overline{x}_0(t)\big)\big)\overline{x}_0'(t)\frac{d}{dt}\eta_\varepsilon\big((x_0 - \overline{x}_0)(t)\big)\,dt. \qquad (3.35)$$

First we examine the left hand side of inequality (3.35). From Theorem 1.1.13, we have

$$\int_0^b a\big(x_0(t)\big)\,\big(x_0'(t) - \overline{x}_0'(t)\big)\,\frac{d}{dt}\eta_\varepsilon\big((x_0 - \overline{x}_0)(t)\big)\,dt$$

$$= \int_{T_\varepsilon} a\big(x_0(t)\big)\,\big(x_0'(t) - \overline{x}_0'(t)\big)^2\,\eta_\varepsilon'\big((x_0 - \overline{x}_0)(t)\big)\,dt$$

$$\geq\; c_0 \int_{T_\varepsilon} \frac{\big(x_0'(t) - \overline{x}_0'(t)\big)^2}{\alpha_a^2\big(x_0(t) - \overline{x}_0(t)\big)^2}\,dt, \qquad (3.36)$$

where $T_\varepsilon \stackrel{df}{=} \{t \in T : (x_0 - \overline{x}_0)(t) \geq \varepsilon\}$. Next we examine the right hand side of inequality (3.35). We have

$$-\int_0^b \big(a\big(x_0(t)\big) - a(\overline{x}_0)\big)\,\overline{x}_0'(t)\frac{d}{dt}\eta_\varepsilon\big((x_0 - \overline{x}_0)(t)\big)\,dt$$

$$\leq\; \int_{T_\varepsilon} \alpha_a\big|x_0(t) - \overline{x}_0(t)\big|\overline{x}_0'(t)\frac{x_0'(t) - \overline{x}_0'(t)}{\alpha_a^2\big(x_0(t) - \overline{x}_0(t)\big)^2}\,dt \qquad (3.37)$$

$$=\; \int_{T_\varepsilon} \overline{x}_0'(t)\frac{x_0'(t) - \overline{x}_0'(t)}{\alpha_a\big(x_0(t) - \overline{x}_0(t)\big)}\,dt \;\leq\; \|\overline{x}_0'\|_2 \left(\int_{T_\varepsilon} \frac{\big(x_0'(t) - \overline{x}_0'(t)\big)^2}{\alpha_a^2\big(x_0(t) - \overline{x}_0(t)\big)^2}\,dt \right)^{\frac{1}{2}}.$$

Using (3.36) and (3.37) in (3.35), we obtain

$$c_0 \int_{T_\varepsilon} \frac{\big(x_0'(t) - \overline{x}_0'(t)\big)^2}{\alpha_a^2\big(x_0(t) - \overline{x}_0(t)\big)^2}\,dt \;\leq\; \|\overline{x}_0'\|_2 \left(\int_{T_\varepsilon} \frac{\big(x_0'(t) - \overline{x}_0'(t)\big)^2}{\alpha_a^2\big(x_0(t) - \overline{x}_0(t)\big)^2}\,dt \right)^{\frac{1}{2}},$$

so

$$\int_{T_\varepsilon} \frac{\left(x_0'(t) - \overline{x}_0'(t)\right)^2}{\alpha_a^2\left(x_0(t) - \overline{x}_0(t)\right)^2}\, dt \ \leq\ \frac{1}{c_0^2} \left\|\overline{x}_0'\right\|_2^2$$

and by Jensen's inequality (see Theorem A.2.5), we have

$$\left|\int_{T_\varepsilon} \frac{x_0'(t) - \overline{x}_0'(t)}{\alpha_a\left(x_0(t) - \overline{x}_0(t)\right)}\, dt\right|^{\frac{1}{2}} \ \leq\ \frac{1}{c_0^2}\left\|\overline{x}_0'\right\|_2^2.$$

Performing a change of variables, we obtain

$$\left|\int_\varepsilon^r \frac{du}{u}\right|^{\frac{1}{2}} \ \leq\ c_7 \left\|\overline{x}_0'\right\|_2^2,$$

with $c_7 \overset{df}{=} \frac{1}{c_0^2} > 0$ and so

$$\left(\ln r - \ln \varepsilon\right)^2 \ \leq\ c_7 \left\|\overline{x}_0'\right\|_2^2.$$

Letting $\varepsilon \searrow 0$, we have a contradiction. From this we infer that $|T_\varepsilon|_1 = 0$ for all $\varepsilon > 0$. Hence $x_0 \leq \overline{x}_0$. In a similar fashion by interchanging the roles of x_0 and $\overline{x}_0$, we obtain that $\overline{x}_0 \leq x_0$. Therefore $x_0 = \overline{x}_0$ and this proves the uniqueness of the solution of (3.28).   ☐

Now let

$$\widehat{D} \overset{df}{=} \left\{ x \in C^1(T) : \ a\big(x(\cdot)\big)x'(\cdot) \in W^{1,2}(T), \ x(0) = x(b), \ x'(0) = x'(b)\right\}.$$

Let $U \colon L^2(T) \supseteq \widehat{D} \longrightarrow L^2(T)$ be the nonlinear operator, defined by

$$U(x)(\cdot) \overset{df}{=} -\big(a\big(x(\cdot)\big)x'(\cdot)\big)' \qquad \forall\, x \in \widehat{D}.$$

Set

$$V_\lambda \overset{df}{=} U + id_{\widehat{D}} + \widehat{A}_\lambda \colon L^2(T) \supseteq \widehat{D} \longrightarrow L^2(T).$$

By virtue of Proposition 3.2.5, V_λ is bijective. So the operator

$$K_\lambda = V_\lambda^{-1} \colon L^2(T) \longrightarrow \widehat{D} \subseteq W_{\mathrm{per}}^{1,2}(T)$$

is well defined.

PROPOSITION 3.2.6
If hypotheses $H(A)_3$ and $H(A)_3$ hold
then the operator K_λ is completely continuous.

PROOF   Suppose that $\{u_n\}_{n\geq 1} \subseteq L^2(T)$ is a sequence, such that

$$u_n \xrightarrow{w} \overline{u} \quad \text{in } L^2(T),$$

for some $\overline{u} \in L^2(T)$. We need to show that

$$K_\lambda(u_n) \longrightarrow K_\lambda(\overline{u}) \quad \text{in } W^{1,2}_{per}(T). \tag{3.38}$$

Set $x_n \stackrel{df}{=} K_\lambda(u_n)$ for $n \geq 1$ and $\overline{x} \stackrel{df}{=} K_\lambda(\overline{u})$. We have

$$U(x_n) + x_n + \widehat{A}_\lambda(x_n) = u_n \quad \forall n \geq 1$$

and since $\left(\widehat{A}_\lambda(x_n), x_n\right)_2 \geq 0$, we have

$$(U(x_n), x_n)_2 + \|x_n\|_2^2 \leq (u_n, x_n)_2 \quad \forall\, n \geq 1 \tag{3.39}$$

(here as usual by $(\cdot, \cdot)_2$ we denote the inner product for the Hilbert space $L^2(T)$). Performing an integration by parts and using the periodic boundary conditions (see the definition of $\widehat{D}$), we have

$$(U(x_n), x_n)_2 = -\int_0^b \left(a(x_n(t))x_n'(t)\right)' x_n(t)\, dt = \int_0^b a(x_n(t))\left(x_n'(t)\right)^2 dt$$

$$= (V(x_n), x_n)_2 \geq c_0 \|x_n'\|_2^2 \quad \forall\, n \geq 1, \tag{3.40}$$

where V is as in the proof of Proposition 3.2.5. Using (3.40) in (3.39), we obtain

$$c_0 \|x_n'\|_2^2 + \|x_n\|_2^2 \leq \|u_n\|_2 \|x_n\|_{W^{1,2}(T)} \quad \forall\, n \geq 1,$$

so

$$c_8 \|x_n\|_{W^{1,2}(T)} \leq \|u_n\|_2 \leq \beta \quad \forall\, n \geq 1,$$

for some $\beta > 0$ and $c_8 \stackrel{df}{=} \min\{c_0, 1\}$. Thus the sequence $\{x_n\}_{n\geq 1} \subseteq W^{1,2}_{per}(T)$ is bounded. After passing to a subsequence if necessary, we may assume that

$$x_n \xrightarrow{w} \overline{\overline{x}} \text{ in } W^{1,2}_{per}(T),$$
$$x_n \longrightarrow \overline{\overline{x}} \text{ in } C(T).$$

Clearly we have $(x_n, x_n - \overline{\overline{x}})_2 \longrightarrow 0$ and $\left(\widehat{A}_\lambda(x_n), x_n - \overline{\overline{x}}\right)_2 \longrightarrow 0$ as $n \geq +\infty$. Therefore, we have

$$0 = \lim_{n\to+\infty} (U(x_n), x_n - \overline{\overline{x}})_2 = \left\langle V(x_n), x_n - \overline{\overline{x}}\right\rangle_{W^{1,2}_{per}(T)}.$$

But from the proof of Proposition 3.2.5, we know that V is pseudomonotone. So

$$\left\langle V(x_n), x_n\right\rangle_{W^{1,2}_{per}(T)} \longrightarrow \left\langle V(\overline{\overline{x}}), \overline{\overline{x}}\right\rangle_{W^{1,2}_{per}(T)},$$

hence

$$\left\| \left(a\big(x_n(\cdot) \big) \right)^{\frac{1}{2}} x_n(\cdot) \right\|_2 \longrightarrow \left\| \left(a\big(\overline{\overline{x}}(\cdot) \big) \right)^{\frac{1}{2}} \overline{\overline{x}}(\cdot) \right\|_2.$$

Because

$$\left(a\big(x_n(\cdot) \big) \right)^{\frac{1}{2}} x_n'(\cdot) \xrightarrow{\;w\;} \left(a\big(\overline{\overline{x}}(\cdot) \big) \right)^{\frac{1}{2}} \overline{\overline{x}}'(\cdot) \quad \text{in } L^2(T),$$

it follows that

$$\left(a\big(x_n(\cdot) \big) \right)^{\frac{1}{2}} x_n'(\cdot) \longrightarrow \left(a\big(\overline{\overline{x}}(\cdot) \big) \right)^{\frac{1}{2}} \overline{\overline{x}}'(\cdot) \quad \text{in } L^2(T).$$

Then, we have

$$\int_0^b \left(\big(a(x_n(t)) \big)^{\frac{1}{2}} x_n'(t) - \big(a(\overline{\overline{x}}(t)) \big)^{\frac{1}{2}} \overline{\overline{x}}'(t) \right)^2 dt$$

$$= \int_0^b a\left(\overline{\overline{x}} \right) \left| \left(\frac{a(x_n(t))}{a(\overline{\overline{x}}(t))} \right)^{\frac{1}{2}} \big(x_n'(t) - \overline{\overline{x}}'(t) \big) \right|^2 dt$$

$$\geq c_0 \frac{c_0}{\beta_1} \left\| x_n' - \overline{\overline{x}}' \right\|_2^2,$$

with $\beta_1 = \left\| a\left(\overline{\overline{x}} \right) \right\|_\infty$. So

$$x_n' \longrightarrow \overline{\overline{x}}' \quad \text{in } L^2(T),$$

hence

$$x_n \longrightarrow \overline{\overline{x}} \quad \text{in } W_{\text{per}}^{1,2}(T).$$

We have

$$a\big(x_n(\cdot) \big) x_n'(\cdot) \xrightarrow{\;w\;} a\big(\overline{\overline{x}}(\cdot) \big) \overline{\overline{x}}'(\cdot) \quad \text{in } L^2(T)$$

and directly from the formulation of the problem (3.28), it follows that the sequence $\left\{ \big(a(x_n(\cdot)) x_n'(\cdot) \big)' \right\}_{n \geq 1} \subseteq L^2(T)$ is bounded. Passing to a subsequence if necessary, we can say that

$$\big(a(x_n(\cdot)) x_n'(\cdot) \big)' \xrightarrow{\;w\;} \overline{v} \quad \text{in } L^2(T),$$

for some $\overline{v} \in L^2(T)$ and as before, we can verify that $\overline{v}(\cdot) = \big(a\left(\overline{\overline{x}}(\cdot) \right) \overline{\overline{x}}'(\cdot) \big)'$. Therefore in the limit as $n \to +\infty$, we have

$$\begin{cases} -\big(a(\overline{\overline{x}}(t)) \overline{\overline{x}}'(t) \big)' + \overline{\overline{x}}(t) + A_\lambda \left(\overline{\overline{x}}(t) \right) = \overline{u}(t) & \text{for a.a. } t \in T, \\ \overline{\overline{x}}(0) = \overline{\overline{x}}(b), \; \overline{\overline{x}}'(0) = \overline{\overline{x}}'(b). \end{cases}$$

It follows that $\overline{\overline{x}} = K_\lambda(\overline{u}) = \overline{x}$ and we have (3.38), which completes the proof of the proposition. $\quad\blacksquare$

Now we consider the following regular approximation of problem (3.27):

$$\begin{cases} \big(a(\overline{x}(t))\overline{x}'(t)\big)' \in A_\lambda\big(\overline{x}(t)\big) + F\big(t,\overline{x}(t),\overline{x}'(t)\big) \\ \qquad\qquad\qquad\qquad\qquad\qquad\qquad\text{for a.a. } t \in T, \qquad (3.41) \\ \overline{x}(0) = \overline{x}(b),\ \overline{x}'(0) = \overline{x}'(b), \end{cases}$$

for $\lambda > 0$.

PROPOSITION 3.2.7

If hypotheses $H(a)_3$, $H(A)_3$ and $H(F)_5$ hold,
then problem (3.41) has a solution

$$x_0 \in C^1(T), \text{ such that } a\big(x_0(\cdot)\big)x_0'(\cdot) \in W^{1,2}_{\mathrm{per}}(T).$$

PROOF   The proof of this proposition is similar to the proof of Proposition 3.2.4, if we take $p = 2$, $r = 1$ and $s = 1$. We omit the details. ⧠

Finally, we pass to the limit as $\lambda \searrow 0$ and arguing as in the proof of Theorem 3.2.2, we obtain the following existence result for problem (3.27).

THEOREM 3.2.6

If hypotheses $H(a)_3$, $H(A)_3$ and $H(F)_5$ hold,
then problem (3.27) has a solution

$$x_0 \in C^1(T), \text{such that } a\big(x_0(\cdot)\big)x_0'(\cdot) \in W^{1,2}_{\mathrm{per}}(T).$$

3.3 Nonlinear Boundary Conditions

We study nonlinear second order differential inclusions in $\mathbb{R}^N$ driven by the p-Laplacian, with nonlinear multivalued boundary conditions. This formulation unifies the basic vector boundary value problems (namely Dirichlet, Neumann and periodic) and goes beyond them. As before the presence of the maximal monotone map A, with its domain $\mathrm{Dom}\,(A)$ not necessarily equal to $\mathbb{R}^N$, incorporates in this framework nonlinear differential variational inequalities as well as second order systems with a nonsmooth convex potential. The right hand side nonlinearity is multivalued and of the Nagumo-Hartman type.

The problem under consideration is the following:

$$\begin{cases} \big(\|x'(t)\|_{\mathbb{R}^N}^{p-2}\,x'(t)\big)' \in A\big(x(t)\big) + F\big(t,x(t),x'(t)\big) \\ \qquad\qquad\qquad\qquad\qquad\qquad\qquad\text{for a.a. } t \in T, \qquad (3.42) \\ \big(\varphi_p\big(x'(0)\big),-\varphi_p\big(x'(b)\big)\big) \in \Xi\big(x(0),x(b)\big), \end{cases}$$

where $T = [0, b]$, with $b > 0$ and $p \in (1, +\infty)$.

Here $A \colon \mathbb{R}^N \supseteq \mathrm{Dom}\,(A) \longrightarrow 2^{\mathbb{R}^N}$ is a maximal monotone map, $F \colon T \times \mathbb{R}^N \times \mathbb{R}^N \longrightarrow 2^{\mathbb{R}^N}$ is a multivalued nonlinearity, $\varphi_p \colon \mathbb{R}^N \longrightarrow \mathbb{R}^N$ is the homeomorphism (analytic diffeomorphism on $\mathbb{R}^N \setminus \{0\}$), defined by

$$\varphi_p(\xi) \stackrel{df}{=} \begin{cases} \|\xi\|_{\mathbb{R}^N}^{p-2} \, \xi & \text{if } \xi \in \mathbb{R}^N \setminus \{0\}, \\ 0 & \text{if } \xi = 0 \end{cases}$$

and $\Xi \colon \mathbb{R}^N \times \mathbb{R}^N \supseteq \mathrm{Dom}\,(\Xi) \longrightarrow 2^{\mathbb{R}^N \times \mathbb{R}^N}$ is a maximal monotone map.

The analysis of problem (3.42) starts with the examination of the following auxiliary problem:

$$\begin{cases} -\Big(\|x'(t)\|_{\mathbb{R}^N}^{p-2} \, x'(t)\Big)' + \|x(t)\|_{\mathbb{R}^N}^{p-2} \, x(t) = g(t) & \text{for a.a. } t \in T, \\ \big(\varphi_p\big(x'(0)\big), -\varphi_p\big(x'(b)\big)\big) \in \Xi\big(x(0), x(b)\big). \end{cases} \tag{3.43}$$

Here $g \in L^{p'}\big(T; \mathbb{R}^N\big)$, $p \in (1, +\infty)$ and $\frac{1}{p} + \frac{1}{p'} = 1$. By a solution of (3.43), we mean a function $x \in C^1\big(T; \mathbb{R}^N\big)$, such that $\|x'(\cdot)\|_{\mathbb{R}^N}^{p-2} \, x'(\cdot) \in W^{1,p'}\big(T; \mathbb{R}^N\big)$ and satisfies equation (3.43). In the next proposition, we show that problem (3.43) has a solution.

PROPOSITION 3.3.1

If $\Xi \colon \mathbb{R}^N \times \mathbb{R}^N \supseteq \mathrm{Dom}\,(\Xi) \longrightarrow 2^{\mathbb{R}^N \times \mathbb{R}^N}$ *is a maximal monotone map with* $(0, 0) \in \Xi(0, 0)$,
then *problem (3.43) has a unique solution* $x_0 \in C^1\big(T; \mathbb{R}^N\big)$.

PROOF For given $v, w \in \mathbb{R}^N$, we consider the following nonlinear two-point boundary value problem:

$$\begin{cases} -\big(\|x'(t)\|_{\mathbb{R}^N}^{p-2} \, x'(t)\big)' + \|x(t)\|_{\mathbb{R}^N}^{p-2} \, x(t) = g(t) & \text{for a.a. } t \in T, \\ x(0) = v, \ x(b) = w. \end{cases} \tag{3.44}$$

Let us set $\gamma(t) \stackrel{df}{=} \big(1 - \frac{t}{b}\big) v + \frac{t}{b} w$. Then $\gamma(0) = v$ and $\gamma(b) = w$. We consider the function $y(t) \stackrel{df}{=} x(t) - \gamma(t)$ and rewrite problem (3.44) as a homogeneous Dirichlet problem for y:

$$\begin{cases} -\Big(\|(y + \gamma)'(t)\|_{\mathbb{R}^N}^{p-2} \, (y + \gamma)'(t)\Big)' \\ \qquad + \|(y + \gamma)(t)\|_{\mathbb{R}^N}^{p-2} \, (y + \gamma)(t) = g(t) & \text{for a.a. } t \in T, \\ y(0) = 0, \ y(b) = 0. \end{cases} \tag{3.45}$$

We solve (3.45) for $y \in C^1\big(T; \mathbb{R}^N\big)$ and then $x = y + \gamma$ is a solution of (3.44). To obtain a solution of (3.45), we proceed as follows. Let

$V_1 \colon W_0^{1,p}(T;\mathbb{R}^N) \longrightarrow W^{-1,p'}(T;\mathbb{R}^N)$ be the nonlinear operator, defined by

$$\langle V_1(u), z\rangle_0 \overset{df}{=} \int_0^b \|(u+\gamma)'(t)\|_{\mathbb{R}^N}^{p-2}\big((u+\gamma)'(t), z'(t)\big)_{\mathbb{R}^N}\, dt$$

$$+ \int_0^b \|(u+\gamma)(t)\|_{\mathbb{R}^N}^{p-2}\big((u+\gamma)(t), z(t)\big)_{\mathbb{R}^N}\, dt$$

$$\forall\, u, z \in W_0^{1,p}(T;\mathbb{R}^N).$$

Here by $\langle\cdot,\cdot\rangle_0$ we denote the duality brackets for the following pair of spaces $\big(W_0^{1,p}(T;\mathbb{R}^N), W^{-1,p'}(T;\mathbb{R}^N)\big)$. For $u, z \in W_0^{1,p}(T;\mathbb{R}^N)$, we have

$$\langle V_1(u) - V_1(z), u - z\rangle_0$$

$$= \int_0^b \|(u+\gamma)'(t)\|_{\mathbb{R}^N}^{p-2}\big((u+\gamma)'(t), (u-z)'(t)\big)_{\mathbb{R}^N}\, dt$$

$$+ \int_0^b \|(u+\gamma)(t)\|_{\mathbb{R}^N}^{p-2}\big((u+\gamma)(t), (u-z)(t)\big)_{\mathbb{R}^N}\, dt$$

$$- \int_0^b \|(z+\gamma)'(t)\|_{\mathbb{R}^N}^{p-2}\big((z+\gamma)'(t), (u-z)'(t)\big)_{\mathbb{R}^N}\, dt$$

$$- \int_0^b \|(z+\gamma)(t)\|_{\mathbb{R}^N}^{p-2}\big((z+\gamma)(t), (u-z)(t)\big)_{\mathbb{R}^N}\, dt.$$

Note that

$$\int_0^b \Big(\|(u+\gamma)'(t)\|_{\mathbb{R}^N}^{p-2}\big((u+\gamma)'(t), (u-z)'(t)\big)_{\mathbb{R}^N}$$

$$- \|(z+\gamma)'(t)\|_{\mathbb{R}^N}^{p-2}\big((z+\gamma)'(t), (u-z)'(t)\big)_{\mathbb{R}^N}\Big)\, dt$$

$$\geq \int_0^b \big(\|(u+\gamma)'(t)\|_{\mathbb{R}^N} - \|(z+\gamma)'(t)\|_{\mathbb{R}^N}\big)\times$$

$$\times \Big(\|(u+\gamma)'(t)\|_{\mathbb{R}^N}^{p-1} - \|(z+\gamma)'(t)\|_{\mathbb{R}^N}^{p-1}\Big)\, dt$$

$$\geq \gamma_1 \int_0^b \big|\|(u+\gamma)'(t)\|_{\mathbb{R}^N} - \|(z+\gamma)'(t)\|_{\mathbb{R}^N}\big|^p\, dt, \tag{3.46}$$

for some $\gamma_1 > 0$. Similarly, we obtain

$$
\int_0^b \Big(\big\|(u + \gamma)(t)\big\|_{\mathbb{R}^N}^{p-2} \big((u + \gamma)(t), (u - z)(t)\big)_{\mathbb{R}^N}
$$

$$
- \big\|(z + \gamma)(t)\big\|_{\mathbb{R}^N}^{p-2} \big((z + \gamma)(t), (u - z)(t)\big)_{\mathbb{R}^N} \Big) \, dt
$$

$$
\geq \gamma_2 \int_0^b \big| \|(u + \gamma)(t)\|_{\mathbb{R}^N} - \|(z + \gamma)(t)\|_{\mathbb{R}^N} \big|^p \, dt, \tag{3.47}
$$

with $\gamma_2 > 0$. From (3.46) and (3.47), it follows that

$$
\langle V_1(u) - V_1(z), u - z \rangle_0 \geq 0,
$$

i.e. V_1 is maximal monotone. In fact V_1 is strictly monotone. Indeed, if $\langle V_1(u) - V_1(z), u - z \rangle_0 = 0$, then we have

$$
\int_0^b \big| \|(u + \gamma)'(t)\|_{\mathbb{R}^N} - \|(z + \gamma)'(t)\|_{\mathbb{R}^N} \big|^p \, dt
$$

$$
+ \int_0^b \big| \|(u + \gamma)(t)\|_{\mathbb{R}^N} - \|(z + \gamma)(t)\|_{\mathbb{R}^N} \big|^p \, dt \; = \; 0,
$$

so

$$
\|(u + \gamma)'(t)\|_{\mathbb{R}^N} \; = \; \|(z + \gamma)'(t)\|_{\mathbb{R}^N} \; = \; k_1(t) \quad \text{for a.a. } t \in T
$$

and

$$
\|(u + \gamma)(t)\|_{\mathbb{R}^N} \; = \; \|(z + \gamma)(t)\|_{\mathbb{R}^N} \; = \; k_2(t) \quad \text{for a.a. } t \in T.
$$

So, we obtain

$$
\int_0^b k_1(t)^{p-2} \|(u - z)'(t)\|_{\mathbb{R}^N}^2 \, dt + \int_0^b k_2(t)^{p-2} \|(u - z)(t)\|_{\mathbb{R}^N}^2 \, dt \; = \; 0,
$$

i.e. $u = z$. This proves the strict monotonicity of V_1. Moreover, using the Extended Dominated Convergence Theorem (Vitali's theorem; see Theorem A.2.1), we can check that V_1 is demicontinuous. Finally, for any

$u \in W_0^{1,p}(T; \mathbb{R}^N)$, we have

$$
\begin{aligned}
\langle V_1(u), u \rangle_0 &= \int_0^b \left\| (u + \gamma)'(t) \right\|_{\mathbb{R}^N}^{p-2} \left((u + \gamma)'(t), u'(t) \right)_{\mathbb{R}^N} dt \\
&\quad + \int_0^b \left\| (u + \gamma)(t) \right\|_{\mathbb{R}^N}^{p-2} \left((u + \gamma)(t), u(t) \right)_{\mathbb{R}^N} dt \\
&\geq \left\| u' + \gamma' \right\|_p^p - \left\| u' + \gamma' \right\|_p^{p-1} \left\| \gamma' \right\|_p \\
&\quad + \left\| u + \gamma \right\|_p^p - \left\| u + \gamma \right\|_p^{p-1} \left\| \gamma \right\|_p \\
&\geq \left\| u + \gamma \right\|_{W^{1,p}(T;\mathbb{R}^N)}^p - \gamma_3 \left\| u + \gamma \right\|_{W^{1,p}(T;\mathbb{R}^N)}^{p-1},
\end{aligned}
$$

for some $\gamma_3 > 0$. From the last inequality, we have that V_1 is coercive. Invoking Corollary 1.4.5, we infer that there exists $y \in W_0^{1,p}(T; \mathbb{R}^N)$, such that $V_1(y) = g$. In addition, by virtue of the strict monotonicity of V_1, this y is in fact unique.

Let $\psi \in C_c^\infty((0, b); \mathbb{R}^N)$ and as before by $\langle \cdot, \cdot \rangle_p$ we denote the duality brackets for the pair $(L^p(T; \mathbb{R}^N), L^{p'}(T; \mathbb{R}^N))$. We have

$$
\langle V_1(y), \psi \rangle_0 = \langle g, \psi \rangle_p,
$$

so

$$
\begin{aligned}
&\int_0^b \left\| (y + \gamma)'(t) \right\|_{\mathbb{R}^N}^{p-2} \left((y + \gamma)'(t), \psi'(t) \right)_{\mathbb{R}^N} dt \\
&+ \int_0^b \left\| (y + \gamma)(t) \right\|_{\mathbb{R}^N}^{p-2} \left((y + \gamma)(t), \psi(t) \right)_{\mathbb{R}^N} dt = \int_0^b \left(g(t), \psi(t) \right)_{\mathbb{R}^N} dt
\end{aligned}
$$

and so by Green's identity, we obtain

$$
\begin{aligned}
&- \left\langle \left(\| (y + \gamma)'(\cdot) \|_{\mathbb{R}^N}^{p-2} (y + \gamma)'(\cdot) \right)', \psi \right\rangle_0 \\
&+ \left\langle \| (y + \gamma)(\cdot) \|_{\mathbb{R}^N}^{p-2} (y + \gamma)(\cdot), \psi \right\rangle_0 = \langle g, \psi \rangle_0.
\end{aligned}
$$

Note that

$$
\left(\| (y + \gamma)'(\cdot) \|_{\mathbb{R}^N}^{p-2} (y' + \gamma')(t) \right)' \in W^{-1,p'}(T; \mathbb{R}^N)
$$

(see Theorem 1.1.8). Because of the density of the embedding

$$
C_c^\infty((0, b); \mathbb{R}^N) \subseteq W_0^{1,p}(T; \mathbb{R}^N),
$$

we conclude that

$$- \left(\|(y+\gamma)'(t)\|_{\mathbb{R}^N}^{p-2} (y+\gamma)'(t) \right)'$$
$$+ \|(y+\gamma)(t)\|_{\mathbb{R}^N}^{p-2} (y+\gamma)(t) = g(t) \quad \text{for a.a. } t \in T,$$

hence $\|(y+\gamma)'(\cdot)\|_{\mathbb{R}^N}^{p-2} (y+\gamma)'(\cdot) \in W^{1,p'}(T;\mathbb{R}^N)$. Then, if we set $x = y+\gamma \in C^1(T;\mathbb{R}^N)$, we have that $\|x'(\cdot)\|_{\mathbb{R}^N}^{p-2} x'(\cdot) \in W^{1,p'}(T;\mathbb{R}^N)$ and is the unique solution of (3.44).

Now, let $s\colon \mathbb{R}^N \times \mathbb{R}^N \longrightarrow C^1(T;\mathbb{R}^N)$ be the map, which to each pair $(v,w) \in \mathbb{R}^N \times \mathbb{R}^N$ assigns the unique solution of the two-point boundary value problem (3.44), i.e. $x = s(v,w) \in C^1(T;\mathbb{R}^N)$ is the unique solution of (3.44) established above. Let $\varrho\colon \mathbb{R}^N \times \mathbb{R}^N \longrightarrow \mathbb{R}^N \times \mathbb{R}^N$ be defined by

$$\varrho(v,w) \overset{df}{=} \left(-\|s(v,w)'(0)\|_{\mathbb{R}^N}^{p-2} s(v,w)'(0), \|s(v,w)'(b)\|_{\mathbb{R}^N}^{p-2} s(v,w)'(b) \right).$$

First, we show that ϱ is monotone. Let $x_1 = s(\alpha_1, \beta_1)$ and $x_2 = s(\alpha_2, \beta_2)$. Using Green's identity, we have

$$\left(\varrho(\alpha_1, \beta_1) - \varrho(\alpha_2, \beta_2), \begin{pmatrix} \alpha_1 - \alpha_2 \\ \beta_1 - \beta_2 \end{pmatrix} \right)_{\mathbb{R}^{2N}}$$

$$= \left(\|x_1'(b)\|_{\mathbb{R}^N}^{p-2} x_1'(b) - \|x_2'(b)\|_{\mathbb{R}^N}^{p-2} x_2'(b), \ \beta_1 - \beta_2 \right)_{\mathbb{R}^N}$$

$$- \left(\|x_1'(0)\|_{\mathbb{R}^N}^{p-2} x_1'(0) - \|x_2'(0)\|_{\mathbb{R}^N}^{p-2} x_2'(0), \ \alpha_1 - \alpha_2 \right)_{\mathbb{R}^N}$$

$$= \int_0^b \left(\|x_1'(t)\|_{\mathbb{R}^N}^{p-2} x_1'(t) - \|x_2'(t)\|_{\mathbb{R}^N}^{p-2} x_2'(t), \ x_1'(t) - x_2'(t) \right)_{\mathbb{R}^N}$$

$$+ \int_0^b \left(\left(\|x_1'(t)\|_{\mathbb{R}^N}^{p-2} x_1'(t) - \|x_2'(t)\|_{\mathbb{R}^N}^{p-2} x_2'(t) \right)', \ x_1(t) - x_2(t) \right)_{\mathbb{R}^N}$$

$$\geq \int_0^b \left(\left(\|x_1'(t)\|_{\mathbb{R}^N}^{p-2} x_1'(t) \right)' - \left(\|x_2'(t)\|_{\mathbb{R}^N}^{p-2} x_2'(t) \right)', x_1(t) - x_2(t) \right)_{\mathbb{R}^N}.$$

From (3.44), we obtain

$$\int_0^b \left(\left(\|x_1'(t)\|_{\mathbb{R}^N}^{p-2} x_1'(t) \right)' - \left(\|x_2'(t)\|_{\mathbb{R}^N}^{p-2} x_2'(t) \right)', \ x_1(t) - x_2(t) \right)_{\mathbb{R}^N}$$

$$= \int_0^b \left(\|x_1(t)\|_{\mathbb{R}^N}^{p-2} x_1(t) - \|x_2(t)\|_{\mathbb{R}^N}^{p-2} x_2(t), x_1(t) - x_2(t) \right)_{\mathbb{R}^N} \geq 0,$$

so

$$\left(\varrho(\alpha_1, \beta_1) - \varrho(\alpha_2, \beta_2), \begin{pmatrix} \alpha_1 - \alpha_2 \\ \beta_1 - \beta_2 \end{pmatrix} \right)_{\mathbb{R}^{2N}} \geq 0,$$

i.e. ϱ is monotone.

Next, we show that ϱ is continuous. For this purpose let us assume that $\{\alpha_n\}_{n\geq 1}, \{\beta_n\}_{n\geq 1} \subseteq \mathbb{R}^N$ are two sequences, such that

$$\alpha_n \longrightarrow \alpha \quad \text{and} \quad \beta_n \longrightarrow \beta \quad \text{in } \mathbb{R}^N,$$

for some $\alpha, \beta \in \mathbb{R}^N$ and let $x_n \stackrel{df}{=} s(\alpha_n, \beta_n)$ for $n \geq 1$ and $x \stackrel{df}{=} s(\alpha, \beta)$. As before, we introduce

$$\gamma_n(t) \stackrel{df}{=} \left(1 - \tfrac{t}{b}\right)\alpha_n + \tfrac{t}{b}\beta_n, \qquad \forall\, n \geq 1$$

and

$$\gamma(t) \stackrel{df}{=} \left(1 - \tfrac{t}{b}\right)\alpha + \tfrac{t}{b}\beta$$

and set $y_n = x_n - \gamma_n$ for $n \geq 1$. We have

$$-\int_0^b \left(\left(\|(y_n + \gamma_n)'(t)\|_{\mathbb{R}^N}^{p-2}(y_n + \gamma_n)'(t)\right)', y_n(t)\right)_{\mathbb{R}^N} dt$$

$$+\int_0^b \left(\|(y_n + \gamma_n)(t)\|_{\mathbb{R}^N}^{p-2}(y_n + \gamma_n)(t), y_n(t)\right)_{\mathbb{R}^N} dt$$

$$= \int_0^b (g(t), y_n(t))_{\mathbb{R}^N} dt,$$

so using Green's identity, we have

$$\int_0^b \left(\|(y_n + \gamma_n)'(t)\|_{\mathbb{R}^N}^{p-2}(y_n + \gamma_n)'(t), y_n'(t)\right)_{\mathbb{R}^N} dt$$

$$+\int_0^b \left(\|(y_n + \gamma_n)(t)\|_{\mathbb{R}^N}^{p-2}(y_n + \gamma_n)(t), y_n(t)\right)_{\mathbb{R}^N} dt$$

$$\leq \int_0^b (g(t), y_n(t))_{\mathbb{R}^N} dt,$$

hence

$$\|y_n' + \gamma_n'\|_p^p - \gamma_4 \|y_n' + \gamma_n'\|_p^{p-1} + \|y_n + \gamma_n\|_p^p - \gamma_5 \|y_n + \gamma_n\|_p^{p-1}$$
$$\leq \|g\|_p \|y_n + \gamma_n\|_p + \gamma_6,$$

for some $\gamma_4, \gamma_5, \gamma_6 > 0$ and thus we obtain the boundedness of the sequence $\{x_n = y_n + \gamma_n\}_{n \geq 1} \subseteq W^{1,p}(T; \mathbb{R}^N)$.

It follows that both the sequences $\{\|x_n(\cdot)\|_{\mathbb{R}^N}^{p-2} x_n(\cdot)\}_{n \geq 1} \subseteq L^{p'}(T; \mathbb{R}^N)$ and $\{\|x_n'(\cdot)\|_{\mathbb{R}^N}^{p-2} x_n'(\cdot)\}_{n \geq 1} \subseteq W^{1,p'}(T; \mathbb{R}^N)$ are bounded. So, passing to a subsequence if necessary, we may assume that

$$x_n \xrightarrow{w} u \text{ in } W^{1,p}(T; \mathbb{R}^N),$$
$$\|x_n(\cdot)\|_{\mathbb{R}^N}^{p-2} x_n(\cdot) \xrightarrow{w} w \text{ in } L^{p'}(T; \mathbb{R}^N),$$

for some $u \in W^{1,p}(T; \mathbb{R}^N)$ and $w \in L^{p'}(T; \mathbb{R}^N)$. We have that

$$x_n \longrightarrow u \text{ in } C(T; \mathbb{R}^N)$$

and so $w(\cdot) = \|u(\cdot)\|_{\mathbb{R}^N}^{p-2} u(\cdot)$. Also, we may assume that

$$\|x_n'(\cdot)\|_{\mathbb{R}^N}^{p-2} x_n'(\cdot) \xrightarrow{w} v \text{ in } W^{1,p'}(T; \mathbb{R}^N),$$
$$\|x_n'(\cdot)\|_{\mathbb{R}^N}^{p-2} x_n'(\cdot) \xrightarrow{w} v \text{ in } C(T; \mathbb{R}^N).$$

The map $\widehat{\psi} \colon C(T; \mathbb{R}^N) \longrightarrow C(T; \mathbb{R}^N)$, defined by

$$\widehat{\psi}(y)(\cdot) \stackrel{df}{=} \varphi_p^{-1}(y(\cdot)) = \varphi_{p'}(y(\cdot)) \qquad \forall y \in C(T; \mathbb{R}^N)$$

(since $\varphi_p^{-1} = \varphi_{p'}$, $\frac{1}{p} + \frac{1}{p'} = 1$), is continuous and maps bounded sets into bounded sets. Therefore

$$x_n' \longrightarrow \widehat{\psi}(v) \text{ in } C(T; \mathbb{R}^N)$$

and so $u' = \widehat{\psi}(v)$. We have $v = \varphi_p(u'(\cdot)) = \|u'(\cdot)\|_{\mathbb{R}^N}^{p-2} u'(\cdot)$ and so in the limit as $n \to +\infty$, we obtain

$$\begin{cases} -\left(\|u'(t)\|_{\mathbb{R}^N}^{p-2} u'(t)\right)' + \|u(t)\|_{\mathbb{R}^N}^{p-2} u(t) = g(t) & \text{for a.a. } t \in T, \\ u(0) = \alpha, \ u(b) = \beta, \end{cases}$$

so $u = s(\alpha, \beta) = x$. We infer that $s \colon \mathbb{R}^N \times \mathbb{R}^N \longrightarrow C^1(T; \mathbb{R}^N)$ is continuous. From the continuity of s, we deduce at once the continuity of ϱ.

Finally, we show that ϱ is coercive. Let $x = s(\alpha, \beta)$. Using (3.44), Green's identity and the Hölder's inequality (see Theorem A.3.12), we have

$$\frac{\left(\varrho(\alpha,\beta),\, \binom{\alpha}{\beta}\right)_{\mathbb{R}^{2N}}}{\left\|\binom{\alpha}{\beta}\right\|_{\mathbb{R}^N}}$$

$$= \frac{\|x'(b)\|_{\mathbb{R}^N}^{p-2}\,(x'(b),\beta)_{\mathbb{R}^N} - \|x'(0)\|_{\mathbb{R}^N}^{p-2}\,(x'(0),\alpha)_{\mathbb{R}^N}}{\left\|\binom{\alpha}{\beta}\right\|_{\mathbb{R}^N}}$$

$$= \frac{\int_0^b \left(\left(\|x'(b)\|_{\mathbb{R}^N}^{p-2}\,x'(b)\right)',\, x(t)\right)_{\mathbb{R}^N} dt + \|x'\|_p^p}{\left\|\binom{\alpha}{\beta}\right\|_{\mathbb{R}^N}}$$

$$\geq \frac{\|x\|_p^p + \|x'\|_p^p - \|g\|_q\,\|x\|_p}{\left\|\binom{\alpha}{\beta}\right\|_{\mathbb{R}^N}}.$$

Note that

$$\binom{\alpha}{\beta} \leq 2\,\|x\|_{C\left(T;\mathbb{R}^N\right)} \leq \gamma_7\,\|x\|_{W^{1,p}(T;\mathbb{R}^N)}$$

for some $\gamma_7 > 0$ (here we have used the compactness of the following embedding $W^{1,p}\left(T;\mathbb{R}^N\right) \subseteq C\left(T;\mathbb{R}^N\right)$; see Theorem 1.1.11(b)). Thus, we have

$$\frac{\left(\varrho(\alpha,\beta),\, \binom{\alpha}{\beta}\right)_{\mathbb{R}^{2N}}}{\left\|\binom{\alpha}{\beta}\right\|_{\mathbb{R}^N}} \geq \frac{\|x\|_{W^{1,p}(T;\mathbb{R}^N)}^p - \|g\|_{p'}\,\|x\|_{W^{1,p}(T;\mathbb{R}^N)}}{\gamma_7\,\|x\|_{W^{1,p}(T;\mathbb{R}^N)}},$$

i.e. ϱ is coercive as claimed.

Let $\sigma = \varrho + \Xi\colon \mathbb{R}^N \times \mathbb{R}^N \longrightarrow 2^{\mathbb{R}^N \times \mathbb{R}^N}$. Evidently σ is maximal monotone (see Theorem 1.4.5), while from the coercivity of ϱ and the fact that $\left(\Xi(\alpha,\beta),\binom{\alpha}{\beta}\right)_{\mathbb{R}^{2N}} \geq 0$ (since $(0,0) \in \Xi(0,0)$), we have that σ is coercive. Applying Theorem 1.4.4, we get that σ is surjective and we can find $(\alpha,\beta) \in \mathbb{R}^N \times \mathbb{R}^N$, such that $(0,0) \in \sigma(\alpha,\beta)$. Let $x_0 = s(\alpha,\beta)$. Evidently this is the desired solution of problem (3.43). It is easy to see that this solution is in fact unique. $\quad\square$

Let us still assume that $\Xi\colon \mathbb{R}^N \times \mathbb{R}^N \supseteq \mathrm{Dom}\,(\Xi) \longrightarrow 2^{\mathbb{R}^N \times \mathbb{R}^N}$ is a maximal monotone map with $(0,0) \in \Xi(0,0)$ and $A\colon \mathbb{R}^N \supseteq \mathrm{Dom}\,(A) \longrightarrow 2^{\mathbb{R}^N}$ is a maximal monotone map with $0 \in A(0)$. We define

$$D \stackrel{df}{=} \Big\{ x \in C^1\left(T;\mathbb{R}^N\right): \ \|x'(\cdot)\|_{\mathbb{R}^N}^{p-2}\,x'(\cdot) \in W^{1,p'}\left(T;\mathbb{R}^N\right),$$
$$\left(\varphi_p\big(x'(0)\big),\, -\varphi_p\big(x'(b)\big)\right) \in \Xi\big(x(0),x(b)\big) \Big\}$$

and for every $\lambda > 0$, let $S_\lambda\colon L^p\left(T;\mathbb{R}^N\right) \supseteq D \longrightarrow L^{p'}\left(T;\mathbb{R}^N\right)$ be the operator, defined by

$$S_\lambda(x)(\cdot) \stackrel{df}{=} -\left(\|x'(\cdot)\|_{\mathbb{R}^N}^{p-2}\,x'(\cdot)\right)' + \widehat{A}_\lambda(x)(\cdot) \qquad \forall\, x \in D,$$

where $\widehat{A}_\lambda(x)(\cdot) = A_\lambda\big(x(\cdot)\big)$. We have

$$\widehat{A}_\lambda(x) \in C\big(T;\mathbb{R}^N\big) \qquad \forall\, x \in D.$$

Arguing as in the proof of Proposition 3.1.3, using now Proposition 3.2.2, we obtain the following result.

PROPOSITION 3.3.2
If

$$\Xi\colon \mathbb{R}^N \times \mathbb{R}^N \supseteq \mathrm{Dom}\,(\Xi) \longrightarrow 2^{\mathbb{R}^N \times \mathbb{R}^N}$$

and

$$A\colon \mathbb{R}^N \supseteq \mathrm{Dom}\,(A) \longrightarrow 2^{\mathbb{R}^N}$$

are maximal monotone maps with $(0,0) \in \Xi(0,0)$ and $0 \in A(0)$, then for every $\lambda > 0$, the operator $S_\lambda\colon L^p\big(T;\mathbb{R}^N\big) \supseteq D \longrightarrow L^{p'}\big(T;\mathbb{R}^N\big)$ is maximal monotone.

Now we are ready to introduce the precise hypotheses on the data of (3.42).

<u>$H(A)_1$</u> $A\colon \mathbb{R}^N \supseteq \mathrm{Dom}\,(A) \longrightarrow 2^{\mathbb{R}^N}$ is a maximal monotone map, such that $0 \in A(0)$.

<u>$H(F)_1$</u> $F\colon T \times \mathbb{R}^N \times \mathbb{R}^N \longrightarrow P_{kc}\big(\mathbb{R}^N\big)$ is a multifunction, such that

(i) for all $\xi, \overline{\xi} \in \mathbb{R}^N$, the multifunction

$$T \ni t \longmapsto F(t, \xi, \overline{\xi}) \in \mathbb{R}^N$$

is graph measurable;

(ii) for almost all $t \in T$, the multifunction

$$\mathbb{R}^N \times \mathbb{R}^N \ni (\xi, \overline{\xi}) \longmapsto F(t, \xi, \overline{\xi}) \in \mathbb{R}^N$$

has closed graph;

(iii) for almost all $t \in T$, all $\xi, \overline{\xi} \in \mathbb{R}^N$ and all $v \in F(t, \xi, \overline{\xi})$, we have

$$(v, \xi)_{\mathbb{R}^N} \geq -c_1 \|\xi\|_{\mathbb{R}^N}^p - c_2 \|\xi\|_{\mathbb{R}^N}^r \|\overline{\xi}\|_{\mathbb{R}^N}^{p-r} - c_3(t) \|\xi\|_{\mathbb{R}^N}^s,$$

with $c_1, c_2 > 0$, $r, s \in [1, p)$, $c_3 \in L^1(T)_+$;

(iv) there exists $M_0 > 0$, such that, if $\|\xi_0\|_{\mathbb{R}^N} = M_0$ and $\big(\xi_0, \overline{\xi}_0\big)_{\mathbb{R}^N} = 0$, we can find $\delta > 0$, such that for almost all $t \in T$, we have

$$\inf\Big\{ (v, \xi)_{\mathbb{R}^N} + \|\overline{\xi}\|_{\mathbb{R}^N}^p\,:\, v \in F(t, x, y),$$

$$\|\xi - \xi_0\|_{\mathbb{R}^N} + \|\overline{\xi} - \overline{\xi}_0\|_{\mathbb{R}^N} < \delta \Big\} \geq 0;$$

(v) for almost all $t \in T$, all $\xi, \overline{\xi} \in \mathbb{R}^N$ with $\|\xi\|_{\mathbb{R}^N} \leq M_0$ and all $v \in F(t, \xi, \overline{\xi})$, we have

$$\|v\|_{\mathbb{R}^N} \leq c_4(t) + c_5 \|\overline{\xi}\|_{\mathbb{R}^N}^{p-1},$$

with $c_4 \in L^\eta(T)_+$, where $\eta \overset{df}{=} \max\{2, p'\}$, $c_5 > 0$.

$\underline{H(\Xi)}$ $\Xi \colon \mathbb{R}^N \times \mathbb{R}^N \supseteq \mathrm{Dom}\,(\Xi) \longrightarrow 2^{\mathbb{R}^N \times \mathbb{R}^N}$ is a maximal monotone map, such that $(0,0) \in \Xi(0,0)$ and one of the following statements holds:

(i) for every $(a', d') \in \Xi(a, d)$, we have $(a', a)_{\mathbb{R}^N} \geq 0$ and $(d', d)_{\mathbb{R}^N} \geq 0$; or

(ii) $\mathrm{Dom}\,(\Xi) = \{(a, d) \in \mathbb{R}^N \times \mathbb{R}^N \colon a = d\}$.

We shall also need a "compatibility" condition between the boundary map Ξ and the map A.

$\underline{H_0}$ For all $(a, d) \in \mathrm{Dom}\,(\Xi)$ and all $(a', d') \in \Xi(a, d)$, we have

$$(A_\lambda(a), a')_{\mathbb{R}^N} + (A_\lambda(d), d')_{\mathbb{R}^N} \geq 0 \qquad \forall\, \lambda > 0.$$

REMARK 3.3.1   Let $\psi \colon \mathbb{R}^N \times \mathbb{R}^N \longrightarrow \mathbb{R}$ be convex (thus locally Lipschitz) and $\Xi = \partial \psi$. If by $\partial_i \psi$, for $i = 1, 2$, we denote the partial subdifferential of $\psi(a, d)$ with respect to a and d respectively, then

$$\partial \psi(a, d) \subseteq \partial_1 \psi(a, d) \times \partial_2 \psi(a, d).$$

In this setting the conditions

$$\begin{cases} (A_\lambda(a), a')_{\mathbb{R}^N} \geq 0 \\ (A_\lambda(d), d')_{\mathbb{R}^N} \geq 0 \end{cases} \qquad \forall\, (a, d) \in \mathrm{Dom}\,(\Xi),\; (a', d') \in \Xi(a, d)$$

are equivalent to saying that

$$\psi(J_\lambda(a), d) \leq \psi(a, d) \quad \text{and} \quad \psi(a, J_\lambda(d)) \leq \psi(a, d) \quad \text{respectively.}$$

$\square$

Since we do not assume that $\mathrm{Dom}\,(A) = \mathbb{R}^N$, first we consider the following regular approximation to problem (3.42):

$$\begin{cases} \left(\|x'(t)\|_{\mathbb{R}^N}^{p-2}\, x'(t)\right)' \in A_\lambda\big(x(t)\big) + F\big(t, x(t), x'(t)\big) \\ \hspace{5cm} \text{for a.a. } t \in T, \\ \big(\varphi_p\big(x'(0)\big), -\varphi_p\big(x'(b)\big)\big) \in \Xi\big(x(0), x(b)\big), \end{cases} \qquad (3.48)$$

with $\lambda > 0$ and $p \in (1, +\infty)$. With some minor, obvious modifications in the proof of Proposition 3.2.4, we establish the following existence result for problem (3.48).

PROPOSITION 3.3.3
_If_ hypotheses $H(A)_1$, $H(F)_1$ and $H(\Xi)$ hold,
then problem (3.48) has a solution

$$x_0 \in C^1(T; \mathbb{R}^N) \quad \text{with} \quad \|x_0'(\cdot)\|_{\mathbb{R}^N}^{p-2} x_0'(\cdot) \in W^{1,p'}(T; \mathbb{R}^N).$$

REMARK 3.3.2 Hypothesis $H(\Xi)(i)$ is needed when we examine the case $t_0 = 0$ and $t_0 = b$ (check the proof of Proposition 3.2.4). Hypothesis $H(\Xi)(ii)$ concerns the periodic problem of Section 3.2. $\Box$

As before, passing to the limit as $\lambda \searrow 0$, we obtain a solution for problem (3.42). The proof follows the steps of the Theorem 3.2.2, with minor modifications. We omit the details.

THEOREM 3.3.1
_If_ hypotheses $H(A)_1$, $H(F)_1$, $H(\Xi)$ and H_0 hold,
then problem (3.42) has a solution

$$x_0 \in C^1(T; \mathbb{R}^N), \quad \text{with} \quad \|x_0'(\cdot)\|_{\mathbb{R}^N}^{p-2} x_0'(\cdot) \in W^{1,p'}(T; \mathbb{R}^N).$$

REMARK 3.3.3 If we check the proof of Theorem 3.2.2, we see that in the current setting, we have (keeping the notation of the proof):

$$\left(U(x_n), \widehat{A}_{\lambda_n}(x_n) \right)_p$$

$$= -\int_0^b \left(\left(\|x'(t)\|_{\mathbb{R}^N}^{p-2} x_n'(t) \right)', \widehat{A}_{\lambda_n}(x_n)(t) \right)_{\mathbb{R}^N} dt$$

$$= -\|x_n'(b)\|_{\mathbb{R}^N}^{p-2} \left(x_n'(b), \widehat{A}_{\lambda_n}(x_n)(b) \right)_{\mathbb{R}^N}$$

$$+ \|x_n'(0)\|_{\mathbb{R}^N}^{p-2} \left(x_n'(0), \widehat{A}_{\lambda_n}(x_n)(0) \right)_{\mathbb{R}^N}$$

$$+ \int_0^b \|x_n'(t)\|_{\mathbb{R}^N}^{p-2} \left(x_n'(t), \frac{d}{dt} \widehat{A}_{\lambda_n}(x_n)(t) \right)_{\mathbb{R}^N} dt \geq 0.$$

Here, we have used hypotheses H_0 and the fact that

$$\frac{d}{dt} A_{\lambda_n}(x_n(t)) = A_{\lambda_n}'(x_n(t)) x_n(t) \quad \text{for a.a. } t \in T,$$

with $\left(x'_n(t), A_{\lambda_n}(x_n(t))x'_n(t)\right)_{\mathbb{R}^N} \geq 0$ for almost all $t \in T$. So it is at this point that the compatibility hypothesis H_0 is needed. □

When $\mathrm{Dom}\,(A) = \mathbb{R}^N$, we can improve our hypotheses on the multivalued nonlinearity F and allow for Nagumo-Hartman type conditions. Specifically our hypotheses on the elements of problem (3.42) are the following:

$\underline{H(A)_2}$ $A \colon \mathbb{R}^N \longrightarrow 2^{\mathbb{R}^N}$ is a maximal monotone map, such that

$$\mathrm{Dom}\,(A) = \mathbb{R}^N \quad \text{and} \quad 0 \in A(0).$$

$\underline{H(F)_2}$ $F \colon T \times \mathbb{R}^N \times \mathbb{R}^N \longrightarrow P_{kc}(\mathbb{R}^N)$ is a multifunction, such that

(i) for all $\xi, \overline{\xi} \in \mathbb{R}^N$, the multifunction

$$T \ni t \longmapsto F(t, \xi, \overline{\xi}) \in \mathbb{R}^N$$

is graph measurable;

(ii) for almost all $t \in T$, the multifunction

$$\mathbb{R}^N \times \mathbb{R}^N \ni (\xi, \overline{\xi}) \longmapsto F(t, \xi, \overline{\xi}) \in \mathbb{R}^N$$

has closed graph;

(iii) for almost all $t \in T$, all $\xi, \overline{\xi} \in \mathbb{R}^N$, with $\|\xi\|_{\mathbb{R}^N} \leq M_0$, $\|\overline{\xi}\|_{\mathbb{R}^N}^{p-1} \geq M_1 > 0$, we have

$$\sup_{v \in F(t, \xi, \overline{\xi})} \|v\|_{\mathbb{R}^N} \leq \eta\left(\|\overline{\xi}\|_{\mathbb{R}^N}^{p-1}\right),$$

with $M_0, M_1 > 0$ and a locally bounded Borel measurable function $\eta \colon \mathbb{R}_+ \longrightarrow \mathbb{R}_+ \setminus \{0\}$, such that $\int\limits_{M_1}^{\infty} \frac{s\,ds}{\eta(s)} = +\infty$;

(iv) if $\|\xi_0\|_{\mathbb{R}^N} = M_0$ (with $M_0 > 0$ as in (iii)) and $(\xi_0, \overline{\xi}_0)_{\mathbb{R}^N} = 0$, we can find $\delta > 0$, such that for almost all $t \in T$, we have

$$0 \leq \inf\left\{ (v, \xi)_{\mathbb{R}^N} + \|\overline{\xi}\|_{\mathbb{R}^N}^p \;:\; v \in F(t, x, y), \right.$$
$$\left. \|\xi - \xi_0\|_{\mathbb{R}^N} + \|\overline{\xi} - \overline{\xi}_0\|_{\mathbb{R}^N} < \delta \right\} \geq 0$$

(v) for all $r > 0$, there exists $\gamma_r \in L^{p'}(T)_+$, (with $\frac{1}{p} + \frac{1}{p'} = 1$), such that for almost all $t \in T$, all $\xi, \overline{\xi} \in \mathbb{R}^N$, with $\|\xi\|_{\mathbb{R}^N}, \|\overline{\xi}\|_{\mathbb{R}^N} \leq r$ and all $v \in F(t, \xi, \overline{\xi})$, we have $\|v\|_{\mathbb{R}^N} \leq \gamma_r(t)$.

(*vi*) for almost all $t \in T$, all $\xi, \overline{\xi} \in \mathbb{R}^N$ and all $v \in F(t, \xi, \overline{\xi})$, we have

$$(v, \xi)_{\mathbb{R}^N} \geq -c_1 \|\xi\|_{\mathbb{R}^N}^p - c_2 \|\xi\|_{\mathbb{R}^N}^r \|\overline{\xi}\|_{\mathbb{R}^N}^{p-r} - c_3(t) \|\xi\|_{\mathbb{R}^N}^s ,$$

with $c_1, c_2 > 0$, $r, s \in [1, p)$, $c_3 \in L^1(T)_+$.

Minor modifications in the proof of Theorem 3.2.3 give the following existence theorem for problem (3.42).

THEOREM 3.3.2
If hypotheses $H(A)_2$, $H(F)_2$ and $H(\Xi)$ hold
then problem (3.42) has a solution

$$x_0 \in C^1(T; \mathbb{R}^N) \quad \text{with} \quad \|x_0'(\cdot)\|_{\mathbb{R}^N}^{p-2} x_0'(\cdot) \in W^{1,p'}(T; \mathbb{R}^N).$$

REMARK 3.3.4 Since Dom $(A) = \mathbb{R}^N$, the analysis of the problem (3.42) does not require the study of the auxiliary problem (3.48) (see Section 3.2) and for this reason we do not need the compatibility condition H_0. ∏

Now we show how the general nonlinear multivalued boundary conditions used in this section unify the classical Dirichlet, Neumann and periodic problems.

EXAMPLE 3.3.1 (a) Let $K_1, K_2 \in P_{fc}(\mathbb{R}^N)$ be such that $0 \in K_1 \cap K_2$. Let $\delta_{K_1 \times K_2}$ be the indicator function of the closed, convex set $K_1 \times K_2 \subseteq \mathbb{R}^N \times \mathbb{R}^N$, i.e.

$$\delta_{K_1 \times K_2}(\xi, \overline{\xi}) \stackrel{df}{=} \begin{cases} 0 & \text{if } (\xi, \overline{\xi}) \in K_1 \times K_2, \\ +\infty & \text{otherwise.} \end{cases}$$

We have that $\delta_{K_1 \times K_2} \in \Gamma_0(\mathbb{R}^N \times \mathbb{R}^N)$ (see Definition 1.3.1). Let

$$\Xi \stackrel{df}{=} \partial \delta_{K_1 \times K_2} = N_{K_1 \times K_2} = N_{K_1} \times N_{K_2}$$

(see Definition 1.3.9). Then problem (3.42) becomes:

$$\begin{cases} \left(\|x'(t)\|_{\mathbb{R}^N}^{p-2} x'(t) \right)' \in A(x(t)) + F(t, x(t), x'(t)) \\ \qquad\qquad\qquad\qquad \text{for a.a. } t \in T, \\ x(0) \in K_1, \ x(b) \in K_2, \\ \left(x'(0), x(0) \right)_{\mathbb{R}^N} = \sigma_{\mathbb{R}^N}(x'(0), K_1), \\ \left(-x'(b), x(b) \right)_{\mathbb{R}^N} = \sigma_{\mathbb{R}^N}(-x'(b), K_1) \end{cases} \tag{3.49}$$

(for $\sigma_{\mathbb{R}^N}$ see Definition 1.2.3). Note that since $0 \in K_1 \cap K_2$, we have $(0, 0) \in \Xi(0, 0)$. Moreover, if $(a', d') \in \Xi(a, d) = N_{K_1}(a) \times N_{K_2}(d)$, then from the definition of the normal cone to a convex set (see Remark 1.3.8), we have that $(a', a)_{\mathbb{R}^N} \geq 0$ and $(d', d)_{\mathbb{R}^N} \geq 0$ and so we have satisfied hypothesis $H(\Xi)(i)$.

Assume further that $K_1, K_2 \subseteq \mathbb{R}^N_+$, $\psi = \delta_{\mathbb{R}^N_+}$ and for $x = (x_1, \ldots, x_N)$ define

$$A(x) \overset{df}{=} \partial\psi(x) = N_{\mathbb{R}^N_+}(x)$$
$$= \begin{cases} \{0\} & \text{if } x_k > 0 \text{ for all } k \in \{1, \ldots, N\} \\ -\mathbb{R}^N_+ \cap \{x\}^\perp & \text{if } x_k = 0 \text{ for some } k \in \{1, \ldots, N\}. \end{cases}$$

Evidently A is maximal monotone with $\text{Dom}(A) = \mathbb{R}^N_+$, $0 \in A(0)$ and

$$A_\lambda(x) = \frac{1}{\lambda}\left(x - \text{proj}_{\mathbb{R}^N_+}\{x\}\right) \qquad \forall \lambda > 0$$

and so

$$A_\lambda(x) = 0 \qquad \forall x \in K_1 \cup K_2.$$

Hence the compatibility hypothesis H_0 is satisfied. Then problem (3.49) becomes the following differential variational inclusion:

$$\begin{cases} \left(\|x'(t)\|_{\mathbb{R}^N}^{p-2} x'(t)\right)' \in F\big(t, x(t), x'(t)\big), \\ \qquad \text{for a.a. } t \in \{s \in T : x_k(s) > 0 \text{ for all } k \in \{1, \ldots, N\}\}, \\ \left(\|x'(t)\|_{\mathbb{R}^N}^{p-2} x'(t)\right)' \in F\big(t, x(t), x'(t)\big) - u(t), \\ \qquad \text{for a.a. } t \in \{s \in T : x_k(s) = 0 \text{ for some } k \in \{1, \ldots, N\}\}, \\ x(t) \in \mathbb{R}^N_+ \qquad \forall t \in T, \\ x(0) \in K_1, \ x(b) \in K_2, \\ u \in L^{p'}\big(T; \mathbb{R}^N_+\big), \ \big(x(t), u(t)\big)_{\mathbb{R}^N} = 0 \quad \text{for a.a. } t \in T, \\ \big(x'(0), x(0)\big)_{\mathbb{R}^N} = \sigma_{\mathbb{R}^N}\big(x'(0), K_1\big), \\ \big(-x'(b), x(b)\big)_{\mathbb{R}^N} = \sigma_{\mathbb{R}^N}\big(-x'(b), K_1\big). \end{cases}$$

(b) If in the previous example $K_1 = K_2 = \{0\}$, then $N_{K_1}(0) = N_{K_2}(0) = \mathbb{R}^N$ and so there are no constraints on $x'(0)$ and $x'(b)$. Hence in this case we recover the Dirichlet problem. Moreover, in this case no matter what is A, $A_\lambda(0) = 0$ and so H_0 holds.

(c) If in example (a), $K_1 = K_2 = \mathbb{R}^N$, then for every $\xi \in \mathbb{R}^N$, $N_{K_1}(x) = N_{K_2}(x) = \{0\}$ and so there are no constraints on $x(0)$ and $x(b)$, while $x'(0) = x'(b) = 0$. Therefore the resulting problem is the Neumann problem.

(d) Let

$$K \overset{df}{=} \{(\xi, \overline{\xi}) \in \mathbb{R}^N \times \mathbb{R}^N : \xi = \overline{\xi}\}$$

and set

$$\Xi \overset{df}{=} \partial\delta_K = K^\perp = \{(\xi, \overline{\xi}) \in \mathbb{R}^N \times \mathbb{R}^N : \xi = -\overline{\xi}\}.$$

Note that for this example hypothesis $H(\Xi)(ii)$ holds, while for all $(a', d') \in \Xi(a, d)$, no matter what is the operator A, we have

$$(A_\lambda(a), a')_{\mathbb{R}^N} + (A_\lambda(d), d')_{\mathbb{R}^N} \geq 0.$$

So hypothesis H_0 holds. In this case the resulting problem is the periodic problem.

(e) Let $\Xi \colon \mathbb{R}^N \times \mathbb{R}^N \longrightarrow \mathbb{R}^N \times \mathbb{R}^N$ be defined by

$$\Xi(\xi, \overline{\xi}) \overset{df}{=} \left(\frac{1}{\vartheta^{\frac{1}{p'-1}}} \varphi_p(\xi), \frac{1}{\eta^{\frac{1}{p'-1}}} \varphi_p(\overline{\xi}) \right).$$

Evidently Ξ is monotone, continuous, thus maximal monotone and we have $(0,0) = \Xi(0,0)$. Note that no matter what is A, we have

$$\left(A_\lambda(\xi), \frac{1}{\vartheta^{\frac{1}{p'-1}}} \varphi_p(\xi) \right)_{\mathbb{R}^N} \geq 0 \quad \text{and} \quad \left(A_\lambda(\xi), \frac{1}{\eta^{\frac{1}{p'-1}}} \varphi_p(\xi) \right)_{\mathbb{R}^N} \geq 0$$

and so hypothesis H_0 holds. In this case the resulting problem is a Sturm-Liouville type problem with boundary conditions

$$x(0) - \vartheta x'(0) = 0 \quad \text{and} \quad x(b) + \eta x'(b) = 0.$$

(f) More generally let $\Xi(\xi, \overline{\xi}) \overset{df}{=} \left(\Xi_1(\xi), \Xi_2(\overline{\xi}) \right)$ with $\Xi_1, \Xi_2 \colon \mathbb{R}^N \longrightarrow \mathbb{R}^N$ monotone, continuous maps, such that $\Xi_1(0) = \Xi_2(0) = 0$. Then clearly Ξ is monotone, continuous, hence maximal monotone, $(0,0) = \Xi(0,0)$ and hypothesis $H(\Xi)(i)$ holds. In this case the resulting problem has the nonlinear boundary conditions given by

$$x'(0) = \varphi_{p'} \left(\Xi_1(x(0)) \right) \quad \text{and} \quad -x'(b) = \varphi_{p'} \left(\Xi_2(x(b)) \right).$$

$\square$

3.4 Variational Methods

The purpose of this section is to study nonlinear periodic systems driven by the ordinary vector p-Laplacian and having a nonsmooth potential. The energy functional corresponding to the problem is not C^1 but only locally Lipschitz. So the variational methods cannot be based on the classical smooth critical point theory and instead we use the nonsmooth critical point theory which we developed in Chapter 2. First we prove some existence theorems under different growth conditions on the nonsmooth potential. Then using one of these existence results, we prove the existence of nontrivial homoclinic to zero solutions. Afterwards, we pass to the scalar problem and introduce a Landesman-Lazer type condition, which is more general than the ones used thus far in the literature. We obtain solutions for both the nonlinear and semilinear problems and, moreover, in the semilinear problem we allow resonance at an eigenvalue of arbitrary order. Then we return to nonlinear periodic systems and look for multiple nontrivial periodic solutions. Continuing in this direction, we study a nonlinear periodic eigenvalue problem and we establish the existence of at least three solutions for all values of the parameter near resonance. In the next subsection, moving beyond periodic systems, we examine problems with multivalued nonlinear boundary conditions, similar to the ones used in the previous section. Our formulation is quite general and produces as special cases the Neumann and periodic problems. Finally, we conclude this section with a multiplicity result for scalar and smooth periodic problems, based on the so-called "Second Deformation Theorem" (see Theorem 3.4.15). It is an open problem, whether we can have a nonsmooth version of this result. Such an extension requires a nonsmooth formulation of Second Deformation Theorem, which, to our knowledge, does not exist and appears to be a highly nontrivial task.

3.4.1   Existence Theorems

Let us start the study of the following nonlinear periodic problem:

$$\begin{cases} -\big(\, \|x'(t)\|_{\mathbb{R}^N}^{p-2}\, x'(t)\big)' \in \partial j(t, x(t)) & \text{for a.a. } t \in T, \\ x(0) = x(b), \ x'(0) = x'(b), \end{cases} \tag{3.50}$$

with $T = [0, b]$, $b > 0$ and $p \in (1, +\infty)$. Here $j(t, \xi)$ is a time-dependent potential function which is only locally Lipschitz in the ξ-variable and $\partial j(t, \xi)$ denotes the generalized (Clarke) subdifferential of $j(t, \cdot)$ (see Definition 1.3.7). Our hypotheses on the nonsmooth potential $j(t, \xi)$ are minimal:

$\underline{H(j)_1}$ $j \colon T \times \mathbb{R}^N \longrightarrow \mathbb{R}$ is a function, such that

(i) for all $\xi \in \mathbb{R}^N$, the function

$$T \ni t \longmapsto j(t, \xi) \in \mathbb{R}$$

is measurable and $j(\cdot, 0) \in L^1(T)$;

(ii) for almost all $t \in T$, the function

$$\mathbb{R}^N \ni \xi \longmapsto j(t, \xi) \in \mathbb{R}$$

is locally Lipschitz;

(iii) for almost all $t \in T$, all $\xi \in \mathbb{R}^N$ and all $u \in \partial j(t, \xi)$, we have

$$\|u\|_{\mathbb{R}^N} \ \leq \ a(t) + c(t)\, \|\xi\|_{\mathbb{R}^N}^{\vartheta},$$

with $a, c \in L^\infty(T)_+$, $\vartheta \in [0, p-1)$;

(iv) $\displaystyle \lim_{\|\xi\|_{\mathbb{R}^N} \to +\infty} \frac{1}{\|\xi\|_{\mathbb{R}^N}^{\vartheta p'}} \int_0^b j(t, \xi)\, dt \ = \ +\infty.$

As in Section 3.2, we put

$$W_{\text{per}}^{1,p}(T;\mathbb{R}^N) \overset{df}{=} \left\{ x \in W^{1,p}(T;\mathbb{R}^N) : \ x(0) = x(b) \right\}$$

and consider the energy functional $\varphi \colon W_{\text{per}}^{1,p}(T;\mathbb{R}^N) \longrightarrow \mathbb{R}^N$, defined by

$$\varphi(x) \overset{df}{=} \frac{1}{p} \|x'\|_p^p - \int\limits_0^b j\big(t, x(t)\big)\, dt \qquad \forall\, x \in W_{\text{per}}^{1,p}(T;\mathbb{R}^N).$$

From Theorem 1.3.10, we know that φ is locally Lipschitz.

PROPOSITION 3.4.1
If hypotheses $H(j)_1$ hold,
<u>*then*</u> *φ satisfies the nonsmooth (PS)-condition.*

PROOF According to Definition 2.1.1(a), let $\{x_n\}_{n\geq 1} \subseteq W_{\text{per}}^{1,p}(T;\mathbb{R}^N)$ be a sequence, such that

$$\big|\varphi(x_n)\big| \ \leq\ M_1 \qquad \forall\, n \geq 1 \quad \text{and} \quad m^\varphi(x_n) \longrightarrow 0 \quad \text{as } n \to +\infty,$$

for some $M_1 > 0$, where

$$m^\varphi(x_n) \overset{df}{=} \big\{ \|x^*\|_{(W_{\text{per}}^{1,p}(T;\mathbb{R}^N))^*} : \ x^* \in \partial\varphi(x_n) \big\}.$$

Because $\partial\varphi(x_n) \subseteq \big(W_{\text{per}}^{1,p}(T;\mathbb{R}^N)\big)^*$ is weakly compact for $n \geq 1$ and the norm functional is weakly lower semicontinuous, we can find $x_n^* \in \partial\varphi(x_n)$ for $n \geq 1$, such that

$$m^\varphi(x_n) = \|x_n^*\|_{(W_{\text{per}}^{1,p}(T;\mathbb{R}^N))^*} \qquad \forall\, n \geq 1.$$

Let $A \colon W_{\text{per}}^{1,p}(T;\mathbb{R}^N) \longrightarrow \big(W_{\text{per}}^{1,p}(T;\mathbb{R}^N)\big)^*$ be the nonlinear operator, defined by

$$\langle A(x), y \rangle_{W_{\text{per}}^{1,p}(T;\mathbb{R}^N)} \overset{df}{=} \int\limits_0^b \|x'(t)\|_{\mathbb{R}^N}^{p-2} \big(x'(t), y'(t)\big)_{\mathbb{R}^N} dt$$

$$\forall x, y \in W_{\text{per}}^{1,p}(T;\mathbb{R}^N). \quad (3.51)$$

We know that A is monotone, continuous, hence maximal monotone. Moreover, we have

$$x_n^* \ = \ A(x_n) - u_n \qquad \forall\, n \geq 1,$$

with $u_n \in S_{\partial j(\cdot, x_n(\cdot))}^{p'}$ (see Theorem 1.3.10 and Proposition 1.3.17).

We claim that the sequence $\{x_n\}_{n\geq 1} \subseteq W_{\text{per}}^{1,p}(T;\mathbb{R}^N)$ is bounded. To this end note that

$$W_{\text{per}}^{1,p}(T;\mathbb{R}^N) \ = \ \mathbb{R}^N \oplus V,$$

with

$$V \overset{df}{=} \left\{ v \in W^{1,p}_{\text{per}}(T; \mathbb{R}^N) : \int_0^b v(t) \, dt = 0 \right\}.$$

For every $x \in W^{1,p}_{\text{per}}(T; \mathbb{R}^N)$, we have $x = \overline{x} + \widehat{x}$, with $\overline{x} \in \mathbb{R}^N$ and $\widehat{x} \in V$. From the choice of the sequence $\{x_n\}_{n \geq 1}$, we have

$$\left| \langle A(x_n), \widehat{x}_n \rangle_{W^{1,p}_{\text{per}}(T; \mathbb{R}^N)} - \int_0^b (u_n(t), \widehat{x}_n(t))_{\mathbb{R}^N} \, dt \right| \leq \varepsilon_n \|\widehat{x}_n\|_{\mathbb{R}^N}, \qquad (3.52)$$

with $\varepsilon_n \searrow 0$. From hypothesis $H(j)_1(iii)$, we have that

$$
\begin{aligned}
(u_n(t), \widehat{x}_n(t))_{\mathbb{R}^N} &\leq \left(a(t) + c(t) \|\overline{x}_n + \widehat{x}_n(t)\|_{\mathbb{R}^N}^{\vartheta} \right) \|\widehat{x}_n(t)\|_{\mathbb{R}^N} \\
&\leq a(t) \|\widehat{x}_n(t)\|_{\mathbb{R}^N} + 2^{\vartheta-1} c(t) \|\overline{x}_n(t)\|_{\mathbb{R}^N}^{\vartheta} \|\widehat{x}_n(t)\|_{\mathbb{R}^N} \\
&\quad + 2^{\vartheta-1} c(t) \|\widehat{x}_n(t)\|_{\mathbb{R}^N}^{\vartheta+1} \quad \text{for a.a. } t \in T,
\end{aligned}
$$

so, for any $\varepsilon > 0$, using Young's inequality (see Theorem A.4.2), we have

$$
\left| \int_0^b (u_n(t), \widehat{x}_n(t))_{\mathbb{R}^N} \, dt \right| \leq \|a\|_1 \|\widehat{x}_n\|_{\infty} + 2^{\vartheta-1} \|c\|_1 \|\widehat{x}_n\|_{\infty}^{\vartheta+1} \qquad (3.53)
$$

$$
+ 2^{\vartheta-1} \|c\|_1 \left(\frac{\varepsilon}{p} \|\widehat{x}_n\|_{\infty}^p + \frac{1}{\varepsilon p'} \|\overline{x}_n\|_{\mathbb{R}^N}^{\vartheta p'} \right).
$$

Recall that

$$\|v\|_{\infty} \leq b^{\frac{1}{b}} \|v'\|_p \qquad \forall \, v \in V$$

(see Remark 1.1.11). Using this in (3.53) and applying the Poincaré-Wirtinger inequality (see Theorem 1.1.7), we obtain

$$
\left| \int_0^b (u_n(t), \widehat{x}_n(t))_{\mathbb{R}^N} \, dt \right|
$$

$$
\leq c_1 \|\widehat{x}_n'\|_p + c_2 \|\widehat{x}_n'\|_p^{\vartheta+1} + c_3 \frac{\varepsilon}{p} \|\widehat{x}_n'\|_p^p + c_4 \|\overline{x}_n\|_{\mathbb{R}^N}^{\vartheta p'}, \qquad (3.54)
$$

for some $c_1, c_2, c_3, c_4 > 0$. Using this inequality in (3.52), we have

$$\|\widehat{x}_n'\|_p^p - c_1 \|\widehat{x}_n'\|_p - c_2 \|\widehat{x}_n'\|_p^{\vartheta+1} - c_3 \frac{\varepsilon}{p} \|\widehat{x}_n'\|_p^p - c_4 \|\overline{x}_n\|_{\mathbb{R}^N}^{\vartheta p'} \leq M_2 \|\widehat{x}_n\|_{\mathbb{R}^N},$$

for some $M_2 > 0$ and all $n \geq 1$. So

$$\left(1 - c_3 \frac{\varepsilon}{p} \right) \|\widehat{x}_n'\|_p^p - (c_1 + M_3) \|\widehat{x}_n'\|_p - c_2 \|\widehat{x}_n'\|_p^{\vartheta+1} \leq c_4 \|\overline{x}_n\|_{\mathbb{R}^N}^{\vartheta p'}, \qquad (3.55)$$

for some $M_3 > 0$ and all $n \geq 1$. Choose $\varepsilon > 0$ small so that $c_3 \frac{\varepsilon}{p} < 1$. A new application of Young's inequality with $\varepsilon > 0$ on the second term in the left hand side in (3.55) gives

$$\|\widehat{x}'_n\|_p^p - c_5 \|\widehat{x}'_n\|_p^{\vartheta+1} \leq c_6 \|\overline{x}_n\|_{\mathbb{R}^N}^{\vartheta p'} + c_7 \qquad \forall\, n \geq 1,$$

for some $c_5, c_6, c_7 > 0$. Recall that $\vartheta + 1 < p$. Again Young's inequality (see Theorem A.4.2) on the term $c_5 \|\widehat{x}'_n\|_p^{\vartheta+1}$ with $\varepsilon > 0$ small enough gives

$$\|\widehat{x}'_n\|_p \leq c_8 \|\overline{x}_n\|_{W^{1,p}(T;\mathbb{R}^N)}^{\frac{\vartheta p'}{p}} + c_9 \qquad \forall\, n \geq 1, \tag{3.56}$$

for some $c_8, c_9 > 0$. Via an easy measurable selection argument and using Proposition 1.3.14, we obtain two measurable functions $u_n : T \longrightarrow \mathbb{R}^N$ and $\lambda_n : T \longrightarrow (0,1)$, such that

$$j\big(t, \overline{x}_n + \widehat{x}_n(t)\big) - j\,(t, \overline{x}_n) \;=\; \big(u_n(t), \widehat{x}_n(t)\big)_{\mathbb{R}^N} \quad \text{for a.a. } t \in T$$

and

$$u_n(t) \;\in\; \partial j\big(t, \overline{x}_n + \lambda_n(t)\widehat{x}_n(t)\big) \quad \text{for a.a. } t \in T.$$

So, for all $n \geq 1$, we can write that

$$\varphi(x_n) \;=\; \frac{1}{p}\, \|x'_n\|_p^p - \int_0^b j\big(t, x_n(t)\big)\, dt$$

$$=\; \frac{1}{p}\, \|\widehat{x}'_n\|_p^p - \int_0^b \big(u_n(t), \widehat{x}_n(t)\big)_{\mathbb{R}^N} dt - \int_0^b j(t, \overline{x}_n)\, dt \;\geq\; -M_1.$$

Using (3.54) with $\varepsilon = 1$, we have

$$c_{10} \|\widehat{x}'_n\|_p^p + c_1 \|\widehat{x}'_n\|_p + c_2 \|\widehat{x}'_n\|_p^{\vartheta+1} + c_4 \|\overline{x}_n\|_{W^{1,p}(T;\mathbb{R}^N)}^{\vartheta p'}$$

$$- \int_0^b j(t, \overline{x}_n)\, dt \;\geq\; -M_1 \qquad \forall\, n \geq 1,$$

for some $c_{10} > 0$. Because of (3.56), we can say that

$$c_{11} \|\overline{x}_n\|_{W^{1,p}(T;\mathbb{R}^N)}^{\vartheta p'} + c_{12} \|\overline{x}_n\|_{W^{1,p}(T;\mathbb{R}^N)}^{\frac{\vartheta p'}{p}} + c_{13} \|\overline{x}_n\|_{W^{1,p}(T;\mathbb{R}^N)}^{\vartheta(\vartheta+1)\frac{p'}{p}}$$

$$- \int_0^b j(t, \overline{x}_n)\, dt + c_{14} \;\geq\; -M_1 \qquad \forall\, n \geq 1,$$

for some $c_{11}, c_{12}, c_{13}, c_{14} > 0$. Note that $\vartheta \frac{p'}{p} < \vartheta p'$ and $\vartheta(\vartheta+1)\frac{p'}{p} < \vartheta p'$ (recall that $\vartheta < p - 1$). Suppose that the sequence $\{\overline{x}_n\}_{n \geq 1} \subseteq \mathbb{R}^N$ was unbounded. By passing to a subsequence if necessary, we may assume that $\|\overline{x}_n\|_{\mathbb{R}^N} \longrightarrow +\infty$ and $\|\overline{x}_n\|_{\mathbb{R}^N} \geq 1$ for $n \geq 1$. Thus, for all $n \geq 1$, we have

$$
\|\overline{x}_n\|_{\mathbb{R}^N}^{\vartheta p'} \left(c_{15} - \frac{1}{\|\overline{x}_n\|_{W^{1,p}(T;\mathbb{R}^N)}^{\vartheta p'}} \int_0^b j(t,\overline{x}_n)\, dt + \frac{c_{16}}{\|\overline{x}_n\|_{W^{1,p}(T;\mathbb{R}^N)}^{\vartheta p'}} \right) \geq -M_1,
$$

for some $c_{15}, c_{16} > 0$ and so

$$
c_{15} - \frac{1}{\|\overline{x}_n\|_{W^{1,p}(T;\mathbb{R}^N)}^{\vartheta p'}} \int_0^b j(t,\overline{x}_n)\, dt + \frac{c_{16}}{\|\overline{x}_n\|_{W^{1,p}(T;\mathbb{R}^N)}^{\vartheta p'}} \geq -\frac{M_1}{\|\overline{x}_n\|_{W^{1,p}(T;\mathbb{R}^N)}^{\vartheta p'}}.
$$

Passing to the limit as $n \to +\infty$, we obtain a contradiction to hypothesis $H(j)_1(iv)$. This proves the boundedness of the sequence $\{\overline{x}_n\}_{n\geq 1} \subseteq \mathbb{R}^N$ and then from (3.56), using also the Poincaré-Wirtinger inequality (see Theorem 1.1.7), we infer the boundedness of the sequence $\{\widehat{x}_n\}_{n\geq 1} \subseteq W_{\mathrm{per}}^{1,p}(T;\mathbb{R}^N)$. So, it follows that the sequence $\{x_n\}_{n\geq 1} \subseteq W_{\mathrm{per}}^{1,p}(T;\mathbb{R}^N)$ is bounded and, by passing to a subsequence if necessary, we may assume that

$$
\begin{aligned}
x_n &\xrightarrow{w} x \ \text{ in } W_{\mathrm{per}}^{1,p}(T;\mathbb{R}^N), \\
x_n &\longrightarrow x \ \text{ in } C(T;\mathbb{R}^N),
\end{aligned}
$$

for some $x \in W_{\mathrm{per}}^{1,p}(T;\mathbb{R}^N)$. From the choice of the sequence $\{x_n\}_{n\geq 1} \subseteq W_{\mathrm{per}}^{1,p}(T;\mathbb{R}^N)$, we have

$$
\langle A(x_n), x_n - x \rangle_{W_{\mathrm{per}}^{1,p}(T;\mathbb{R}^N)} - \int_0^b u_n(t)(x_n - x)(t)\, dt \leq \varepsilon_n \|x_n - x\|_{W^{1,p}(T;\mathbb{R}^N)},
$$

with $\varepsilon_n \searrow 0$. By virtue of hypothesis $H(j)_1(iii)$, we have that

$$
\int_0^b u_n(t)(x_n - x)(t)\, dt \longrightarrow 0
$$

and so

$$
\limsup_{n\to+\infty} \langle A(x_n), x_n - x \rangle_{W_{\mathrm{per}}^{1,p}(T;\mathbb{R}^N)} \leq 0.
$$

Because A is maximal monotone, it is also generalized pseudomonotone (see Proposition 1.4.10) and so

$$
\|x_n'\|_p^p = \langle A(x_n), x_n \rangle_{W_{\mathrm{per}}^{1,p}(T;\mathbb{R}^N)} \longrightarrow \langle A(x), x \rangle_{W_{\mathrm{per}}^{1,p}(T;\mathbb{R}^N)} = \|x'\|_p^p.
$$

Recall that $x_n' \xrightarrow{w} x'$ in $L^p(T;\mathbb{R}^N)$. Because the latter space is uniformly convex, from the Kadec-Klee property, we conclude that

$$
x_n' \longrightarrow x' \ \text{ in } L^p(T;\mathbb{R}^N),
$$

hence

$$x_n \longrightarrow x \quad \text{in } W^{1,p}(T; \mathbb{R}^N).$$

$\square$

Using this proposition and the nonsmooth Saddle Point Theorem (see Theorem 2.2.3), we can have the first existence theorem for problem (3.50).

THEOREM 3.4.1
If hypotheses $H(j)_1$ hold,
then problem (3.50) has a solution

$$x_0 \in C^1(T; \mathbb{R}^N) \quad \text{with} \quad \|x_0'(\cdot)\|_{\mathbb{R}^N}^{p-2} x_0'(\cdot) \in W_{\text{per}}^{1,p'}(T; \mathbb{R}^N).$$

PROOF Recall that

$$W_{\text{per}}^{1,p}(T; \mathbb{R}^N) = \mathbb{R}^N \oplus V,$$

with

$$V \overset{df}{=} \left\{ v \in W_{\text{per}}^{1,p}(T; \mathbb{R}^N) : \int_0^b v(t) \, dt = 0 \right\}.$$

Let $v \in V$. We can find $u \in L^{p'}(T; \mathbb{R}^N)$ and a Lebesgue measurable function $\lambda: T \longrightarrow (0,1)$, such that

$$u(t) \in \partial j(t, \lambda(t)v(t)) \quad \text{for a.a. } t \in T$$

and

$$j(t, v(t)) = (u(t), v(t))_{\mathbb{R}^N} + j(t, 0) \quad \text{for a.a. } t \in T.$$

Then, using hypothesis $H(j)_1(iii)$ and the fact that $\|v\|_\infty \leq b^{\frac{1}{p'}} \|v'\|_p$ for all $v \in V$, we have

$$
\begin{aligned}
\varphi(v) &= \frac{1}{p} \|v'\|_p^p - \int_0^b j(t, v(t)) \, dt \\
&= \frac{1}{p} \|v'\|_p^p - \int_0^b (u(t), v(t))_{\mathbb{R}^N} \, dt - \int_0^b j(t, 0) \, dt \\
&\geq \frac{1}{p} \|v'\|_p^p - \|u\|_1 \|v\|_\infty - \int_0^b j(t, 0) \, dt \\
&\geq \frac{1}{p} \|v'\|_p^p - c_{17} \|v'\|_p^{\vartheta+1} - \int_0^b j(t, 0) \, dt.
\end{aligned}
$$

Since $\vartheta < p - 1$, from the last inequality and because $\|v'\|_p$ is a norm equivalent to the Sobolev norm on V (by the Poincaré-Wirtinger inequality; see Theorem 1.1.7), we can conclude that φ is coercive on V.

Let $\xi \in \mathbb{R}^N$. Then

$$\varphi(\xi) = -\int_0^b j(t, \xi)\, dt \longrightarrow -\infty \quad \text{as } \|\xi\|_{\mathbb{R}^N} \to +\infty.$$

So φ is anticoercive on $\mathbb{R}^N$. These two properties of φ and Proposition 3.4.1 permit the use of Theorem 2.2.3, which produces an $x_0 \in W^{1,p}_{\text{per}}(T; \mathbb{R}^N)$, such that $0 \in \partial \varphi(x_0)$. So we have that $A(x_0) = u$, for some $u \in S^{p'}_{\partial j(\cdot, x_0(\cdot))}$. We have

$$\langle A(x_0), h \rangle_{W^{1,p}_{\text{per}}(T; \mathbb{R}^N)} = \int_0^b (u(t), h(t))_{\mathbb{R}^N}\, dt \qquad \forall\, h \in C_c^\infty((0, b); \mathbb{R}^N),$$

so

$$\int_0^b \|x_0'(t)\|_{\mathbb{R}^n}^{p-2} (x_0'(t), h'(t))_{\mathbb{R}^N}\, dt$$

$$= \int_0^b (u(t), h(t))_{\mathbb{R}^N}\, dt \qquad \forall\, h \in C_c^\infty((0, b); \mathbb{R}^N)$$

and noting that $\left(\|x_0'(\cdot)\|_{\mathbb{R}^N}^{p-2} x_0'(\cdot) \right)' \in W^{-1,p'}(T; \mathbb{R}^N)$ (see Theorem 1.1.8), we have

$$\left\langle -\left(\|x_0'(\cdot)\|_{\mathbb{R}^N}^{p-2} x_0'(\cdot) \right)', h \right\rangle_{W_0^{1,p}(T; \mathbb{R}^N)}$$

$$= \int_0^b (u(t), h(t))_{\mathbb{R}^N}\, dt \qquad \forall\, h \in C_c^\infty((0, b); \mathbb{R}^N).$$

Since the embedding $C_c^\infty((0, b); \mathbb{R}^N) \subseteq W_0^{1,p}(T; \mathbb{R}^N)$ is dense (note that $W_0^{1,p}(T; \mathbb{R}^N)$ is predual of $W^{-1,p'}(T; \mathbb{R}^N)$), it follows that

$$\begin{cases} -\left(\|x_0'(t)\|_{\mathbb{R}^N}^{p-2} x_0'(t) \right)' = u(t) & \text{for a.a. } t \in T, \\ x_0(0) = x_0(b), \end{cases}$$

so $\|x_0'(\cdot)\|_{\mathbb{R}^N}^{p-2} x_0'(\cdot) \in W^{1,p'}(T; \mathbb{R}^N)$ and $x_0 \in C^1(T; \mathbb{R}^N)$. For every $v \in$

$W_{\text{per}}^{1,p}(T; \mathbb{R}^N)$, using Green's identity (see Theorem 1.1.9), we have

$$\int_0^b \left(\left(\|x_0'(t)\|_{\mathbb{R}^N}^{p-2} \, x_0'(t) \right)', v(t) \right)_{\mathbb{R}^N}$$

$$= \|x_0'(b)\|_{\mathbb{R}^N}^{p-2} \left(x_0'(b), v(b) \right)_{\mathbb{R}^N} - \|x_0'(0)\|_{\mathbb{R}^N}^{p-2} \left(x_0'(0), v(0) \right)_{\mathbb{R}^N}$$

$$- \left\langle A(x_0), v \right\rangle_{W_{\text{per}}^{1,p}(T; \mathbb{R}^N)},$$

so

$$- \int_0^b \left(u(t), v(t) \right)_{\mathbb{R}^N} + \left\langle A(x_0), v \right\rangle_{W_{\text{per}}^{1,p}(T; \mathbb{R}^N)}$$

$$= \|x_0'(b)\|_{\mathbb{R}^N}^{p-2} \left(x_0'(b), v(b) \right)_{\mathbb{R}^N} - \|x_0'(0)\|_{\mathbb{R}^N}^{p-2} \left(x_0'(0), v(0) \right)_{\mathbb{R}^N}$$

and recalling that $A(x_0) = u$, we have

$$\|x_0'(0)\|_{\mathbb{R}^N}^{p-2} \left(x_0'(0), v(0) \right)_{\mathbb{R}^N} = \|x_0'(b)\|_{\mathbb{R}^N}^{p-2} \left(x_0'(b), v(b) \right)_{\mathbb{R}^N}.$$

Because $v \in W_{\text{per}}^{1,p}(T; \mathbb{R}^N)$ was arbitrary, we conclude that $x_0'(0) = x_0'(b)$, i.e. $x_0 \in C^1(T; \mathbb{R}^N)$ is a solution of problem (3.50) and $\|x_0'(\cdot)\|_{\mathbb{R}^N}^{p-2} \, x_0'(\cdot) \in W^{1,p'}(T; \mathbb{R}^N)$. ⬜

We can have another existence result by changing the asymptotic condition in hypothesis $H(j)_1(iv)$. Namely the new set of hypotheses on the nonsmooth potential j is the following:

<u>$H(j)_2$</u> $j: T \times \mathbb{R}^N \longrightarrow \mathbb{R}$ is a function, such that

$\left. \begin{array}{l} (i) \\ (ii) \\ (iii) \end{array} \right\}$ are the same as hypotheses $H(j)_1(i), (ii)$ and (iii);

(iv) $\displaystyle\lim_{\|\xi\|_{\mathbb{R}^N} \to +\infty} \frac{1}{\|\xi\|_{\mathbb{R}^N}^{\vartheta p'}} \int_0^b j(t, \xi) \, dt = -\infty.$

PROPOSITION 3.4.2
If hypotheses $H(j)_2$ hold, <u>then</u> φ satisfies the nonsmooth PS-condition.

PROOF Let us consider a sequence $\{x_n\}_{n \geq 1} \subseteq W_{\text{per}}^{1,p}(T; \mathbb{R}^N)$, such that

$$|\varphi(x_n)| \leq M_4 \quad \forall\, n \geq 1 \quad \text{and} \quad m^{\varphi}(x_n) \longrightarrow 0 \quad \text{as } n \to +\infty,$$

for some $M_4 > 0$. From the proof of Proposition 3.4.1 (see (3.54)), we have that

$$\frac{1}{p} \left\| x'_n \right\|_p^p - c_1 \left\| \widehat{x}'_n \right\|_p - c_2 \left\| \widehat{x}'_n \right\|_p^{\vartheta+1}$$

$$-c_3 \frac{\varepsilon}{p} \left\| \widehat{x}'_n \right\|_p^p - c_4 \left\| \overline{x}_n \right\|_{\mathbb{R}^N}^{\vartheta p'} - \int_0^b j(t, \overline{x}_n)\, dt \ \leq \ M_4$$

and so

$$\frac{1}{p} \left(1 - c_3 \varepsilon\right) \left\| \widehat{x}'_n \right\|_p^p - c_1 \left\| \widehat{x}'_n \right\|_p - c_2 \left\| \widehat{x}'_n \right\|_p^{\vartheta+1}$$

$$-c_4 \left\| \overline{x}_n \right\|_{\mathbb{R}^N}^{\vartheta p'} - \int_0^b j(t, \overline{x}_n)\, dt \ \leq \ M_4.$$

Choosing $\varepsilon \in \left(0, \frac{1}{c_3}\right)$, we have

$$c_{18} \left\| \widehat{x}'_n \right\|_p^p - c_1 \left\| \widehat{x}'_n \right\|_p - c_2 \left\| \widehat{x}'_n \right\|_p^{\vartheta+1}$$

$$- \left\| \overline{x}_n \right\|_{\mathbb{R}^N}^{\vartheta p'} \left(c_4 + \frac{1}{\left\| \overline{x}_n \right\|_{\mathbb{R}^N}^{\vartheta p'}} \int_0^b j(t, \overline{x}_n)\, dt \right) \ \leq \ M_4,$$

for some $c_{18} > 0$. If the sequence $\{\widehat{x}'_n\}_{n\geq 1} \subseteq L^p(T; \mathbb{R}^N)$ is unbounded, then, passing to a subsequence if necessary, we may assume that $\left\| \widehat{x}'_n \right\|_p \longrightarrow +\infty$ and by virtue of (3.56) this implies that $\left\| \overline{x}_n \right\|_{\mathbb{R}^N} \longrightarrow +\infty$. Since $\vartheta < p - 1$, if we pass to the limit in the last inequality and we use hypothesis $H(j)_2(iv)$, we reach a contradiction. So the sequence $\{\widehat{x}'_n\}_{n\geq 1} \subseteq L^p(T; \mathbb{R}^N)$ is bounded. If the sequence $\{\overline{x}_n\}_{n\geq 1} \subseteq \mathbb{R}^N$ is unbounded, passing to a subsequence if necessary, we may assume that $\left\| \overline{x}_n \right\|_{\mathbb{R}^N} \longrightarrow +\infty$ and then using once again the last inequality and hypothesis $H(j)_2(iv)$, we reach a contradiction. Therefore the sequence $\{\overline{x}_n\}_{n\geq 1} \subseteq \mathbb{R}^N$ is bounded, hence so is the sequence $\{x_n\}_{n\geq 1} \subseteq W^{1,p}(T; \mathbb{R}^N)$.

The rest of the proof goes as the corresponding part of the proof of Proposition 3.4.1.   □

We are ready for the second existence result concerning problem (3.50).

THEOREM 3.4.2
If hypotheses $H(j)_2$ hold,
<u>*then*</u> *problem (3.50) has a solution*

$$x_0 \in C^1(T; \mathbb{R}^N) \quad \text{with} \quad \left\| x'_0(\cdot) \right\|_{\mathbb{R}^N}^{p-2} x'_0(\cdot) \in W_{\text{per}}^{1,p'}(T; \mathbb{R}^N).$$

PROOF   As before (see the proof of Proposition 3.4.1), using (3.54), for every $x \in W_{\mathrm{per}}^{1,p}\big(T; \mathbb{R}^N\big)$, we have

$$
\begin{aligned}
\varphi(x) \;&=\; \frac{1}{p}\,\|x'\|_p^p - \int_0^b j\big(t, x(t)\big)\,dt \\[2mm]
&=\; \frac{1}{p}\,\|\widehat{x}'\|_p^p - \int_0^b \big(u(t), \widehat{x}(t)\big)_{\mathbb{R}^N}\,dt - \int_0^b j(t,\overline{x})\,dt \\[2mm]
&\geq\; \frac{1}{p}\,\|\widehat{x}'\|_p^p - c_1\,\|\widehat{x}'\|_p - c_2\,\|\widehat{x}'\|_p^{\vartheta+1} - c_3\frac{\varepsilon}{p}\,\|\widehat{x}'\|_p^p - c_4\,\|\overline{x}\|_{\mathbb{R}^N}^{\vartheta p'} - \int_0^b j(t,\overline{x})\,dt \\[2mm]
&=\; \frac{1}{p}(1 - c_3\varepsilon)\,\|\widehat{x}'\|_p^p - c_1\,\|\widehat{x}'\|_p^{\vartheta+1} - \|\overline{x}\|_{\mathbb{R}^N}^{\vartheta p'}\left(c_4 + \frac{1}{\|\overline{x}\|_{\mathbb{R}^N}^{\vartheta p'}}\int_0^b j(t,\overline{x}_n)\,dt\right),
\end{aligned}
$$

for some $u \in S_{\partial j(\cdot,\,\overline{x}+\lambda(\cdot)\widehat{x}(\cdot))}^{p'}$. Let $\varepsilon \in \left(0, \frac{1}{c_3}\right)$. Then, from the last inequality, because $\vartheta < p - 1$ and due to hypothesis $H(j)_2(iv)$, we have that φ is coercive (recall the Poincaré–Wirtinger inequality; see Theorem 1.1.7). Then, by virtue of Theorem 2.1.5, we can find $x_0 \in W_{\mathrm{per}}^{1,p}\big(T; \mathbb{R}^N\big)$, such that $0 \in \partial\varphi(x_0)$. As in the proof of Theorem 3.4.1, we can check that

$$
x_0 \in C^1\big(T; \mathbb{R}^N\big) \quad\text{with}\quad \|x_0'(\cdot)\|_{\mathbb{R}^N}^{p-2}\, x_0'(\cdot) \in W_{\mathrm{per}}^{1,p'}\big(T; \mathbb{R}^N\big)
$$

and it solves problem (3.50).   ▯

By strengthening the growth condition on $\partial j(t, \cdot)$, we can weaken the asymptotic condition on the integral of the potential function. More precisely, our hypotheses on j are the following.

<u>$H(j)_3$</u> $j\colon T \times \mathbb{R}^N \longrightarrow \mathbb{R}$ is a function, such that

$\left.\begin{array}{c}(i)\\(ii)\end{array}\right\}$ the same as hypotheses $H(j)_1(i)$ and (ii);

(iii) for almost all $t \in T$, all $\xi \in \mathbb{R}^N$ and all $u \in \partial j(t, \xi)$, we have

$$
\|u\|_{\mathbb{R}^N} \;\leq\; a(t),
$$

with $a \in L^{p'}(T)_+$ (where $\frac{1}{p} + \frac{1}{p'} = 1$);

(iv) $\displaystyle\lim_{\|\xi\|_{\mathbb{R}^N}\to+\infty} \int_0^b j(t, \xi)\,dt \;=\; +\infty.$

REMARK 3.4.1   By Proposition 1.3.14 and hypothesis $H(j)_3(iii)$, we see that

$$\left| j(t,\xi) - j(t,\overline{\xi}) \right| \; \leq \; a(t) \left\| \xi - \overline{\xi} \right\|_{\mathbb{R}^N} \quad \text{for a.a. } t \in T \text{ and all } \xi, \overline{\xi} \in \mathbb{R}^N,$$

hence for almost all $t \in T$, $j(t,\cdot)$ is globally Lipschitz with an $L^{p'}(T)$-Lipschitz constant.   ▯

EXAMPLE 3.4.1   A typical nonsmooth potential satisfying hypotheses $H(j)_3$ is

$$j(t,\xi) \; \overset{df}{=} \; \sin \left\| \xi \right\|_{\mathbb{R}^N} - h(t) \left\| \xi \right\|_{\mathbb{R}^N} \quad \forall \, (t,\xi) \in \times \mathbb{R}^N,$$

with $h \in L^{p'}(T)_+$. Then

$$\partial j(t,\xi) \; = \; \begin{cases} (\cos \left\| \xi \right\|_{\mathbb{R}^N} - h(t)) \dfrac{\xi}{\left\| \xi \right\|_{\mathbb{R}^N}} & \text{if } \xi \neq 0, \\[2mm] (\cos \left\| \xi \right\|_{\mathbb{R}^N} - h(t)) \overline{B}_1 & \text{if } \xi = 0, \end{cases}$$

with $\overline{B}_1 \overset{df}{=} \left(\xi \in \mathbb{R}^N : \left\| \xi \right\|_{\mathbb{R}^N} \leq 1 \right)$. Note that

$$\int_0^b j(t,\xi) \, dt \; = \; b \sin \left\| \xi \right\|_{\mathbb{R}^N} - \left\| \xi \right\|_{\mathbb{R}^N} \left\| h \right\|_1$$

$$\leq \; b - \left\| h \right\|_1 \left\| \xi \right\|_{\mathbb{R}^N} \quad \forall \, (t,\xi) \in T \times \mathbb{R}^N,$$

so

$$\int_0^b j(t,\xi) \, dt \; \longrightarrow \; -\infty \quad \text{as } \left\| \xi \right\|_{\mathbb{R}^N} \to +\infty.$$

On the other hand, a simple nonsmooth potential satisfying $H(j)_1$ is given by

$$j(\xi) \; \overset{df}{=} \; \left\| \xi \right\|_{\mathbb{R}^N}^2 \sin \frac{1}{\left\| \xi \right\|_{\mathbb{R}^N}} \quad \forall \, \xi \in \mathbb{R}^N$$

and one satisfying $H(j)_2$ is given by

$$j(\xi) \; \overset{df}{=} \; - \left\| \xi \right\|_{\mathbb{R}^N}^2 \sin \frac{1}{\left\| \xi \right\|_{\mathbb{R}^N}} \quad \forall \, \xi \in \mathbb{R}^N.$$

Another possibility is the function $j \colon \mathbb{R}^2 \longrightarrow \mathbb{R}$ defined by

$$j(\xi) \; \overset{df}{=} \; \left| \xi_1 - \xi_2 \right| - \frac{1}{2}(\xi_1 + 1)^2 - \frac{1}{2}(\xi_2 + 1)^2 \quad \forall \, \xi = (\xi_1, \xi_2) \in \mathbb{R}^2.$$

In this case $\vartheta = 1$ and if $p > 2$, we see that hypotheses $H(j)_2$ are satisfied.   ▯

PROPOSITION 3.4.3

If hypotheses $H(j)_3$ hold, then φ satisfies the nonsmooth PS-condition.

PROOF We consider a sequence $\{x_n\}_{n \geq 1} \subseteq W^{1,p}_{per}(T; \mathbb{R}^N)$, such that

$$|\varphi(x_n)| \leq M_5 \quad \forall\, n \geq 1 \quad \text{and} \quad m^\varphi(x_n) \longrightarrow 0 \quad \text{as } n \to +\infty,$$

for some $M_5 > 0$. Let $x_n^* \in \partial\varphi(x_n)$ be such that $m^\varphi(x_n) = \|x_n^*\|_{(W^{1,p}_{per}(T;\mathbb{R}^N))^*}$ for $n \geq 1$. We have

$$x_n^* = A(x_n) - u_n \quad \forall\, n \geq 1,$$

with $u_n \in S^{p'}_{j(\cdot, x_n(\cdot))}$. Then, keeping the notation introduced in the proof of Proposition 3.4.1, we have

$$\langle A(x_n), \widehat{x}_n \rangle_{W^{1,p}_{per}(T;\mathbb{R}^N)} - \int_0^b \left(u_n(t), \widehat{x}_n(t) \right)_{\mathbb{R}^N} dt \leq \varepsilon_n \|\widehat{x}_n\|_{W^{1,p}(T;\mathbb{R}^N)},$$

with $\varepsilon_n \searrow 0$. So

$$\|\widehat{x}_n'\|_p^p \leq \varepsilon_n \|\widehat{x}_n\|_{W^{1,p}(T;\mathbb{R}^N)} + \|a\|_1 \|\widehat{x}_n\|_\infty$$

and

$$\|\widehat{x}_n'\|_p^p \leq c_1 \|\widehat{x}_n'\|_p \quad \forall\, n \geq 1,$$

for some $c_1 > 0$ (see Remark 1.1.11). Thus, by the Poincaré-Wirtinger inequality (see Theorem 1.1.7), the sequence $\{\widehat{x}_n\}_{n\geq 1} \subseteq W^{1,p}_{per}(T;\mathbb{R}^N)$ is bounded.

As in the previous proofs, we have

$$-M_5 \leq \varphi(x_n) = \frac{1}{p}\|\widehat{x}_n'\|_p^p - \int_0^b \left(u_n(t), \widehat{x}_n(t) \right)_{\mathbb{R}^N} dt - \int_0^b j(t, \overline{x}_n)\, dt$$

$$\leq \frac{1}{p}\|\widehat{x}_n'\|_p^p + \|a\|_1 \|\widehat{x}_n\|_\infty - \int_0^b j(t, \overline{x}_n)\, dt,$$

with $u_n \in S^{p'}_{\partial j(\cdot, \overline{x}_n + \lambda_n(\cdot)\widehat{x}_n(\cdot))}$. So, by the Poincaré-Wirtinger inequality, we have

$$\int_0^b j(t, \overline{x}_n)\, dt \leq M_6 \quad \forall\, n \geq 1,$$

for some $M_6 > 0$.

If the sequence $\{x_n\}_{n \geq 1} \subseteq W^{1,p}_{\mathrm{per}}(T; \mathbb{R}^N)$ is unbounded, then necessarily $\|\overline{x}_n\|_{W^{1,p}(T;\mathbb{R}^N)} \longrightarrow +\infty$ and so by hypothesis $H(j)_3(iv)$, we obtain

$$\int_0^b j(t, \overline{x}_n)\, dt \longrightarrow +\infty \quad \text{as } n \to +\infty,$$

a contradiction. Therefore the sequence $\{x_n\}_{n \geq 1} \subseteq W^{1,p}_{\mathrm{per}}(T; \mathbb{R}^N)$ is bounded and we continue the proof as before. $\qquad\square$

PROPOSITION 3.4.4
If hypotheses $H(j)_3$ hold,
then $\varphi|_V$ is weakly coercive (see Definition A.4.2(a)).

PROOF   Suppose that the proposition is not true. We can find a sequence $\{v_n\}_{n \geq 1} \subseteq V$, such that

$$\|v_n\|_{W^{1,p}(T;\mathbb{R}^N)} \longrightarrow +\infty \quad \text{and} \quad \varphi(v_n) \leq M_7 \qquad \forall\, n \geq 1,$$

for some $M_7 > 0$. Let us set

$$y_n \overset{df}{=} \frac{v_n}{\|v_n\|_{W^{1,p}(T;\mathbb{R}^N)}} \qquad \forall\, n \geq 1.$$

Passing to a subsequence if necessary, we may assume that

$$y_n \overset{w}{\longrightarrow} y \text{ in } W^{1,p}_{\mathrm{per}}(T; \mathbb{R}^N),$$
$$y_n \longrightarrow y \text{ in } C(T; \mathbb{R}^N).$$

Then

$$\frac{1}{p}\|y_n'\|_p^p - \int_0^b \frac{j(t, v_n(t))}{\|v_n\|_{W^{1,p}(T;\mathbb{R}^N)}^p}\, dt \;\leq\; \frac{M_7}{\|v_n\|_{W^{1,p}(T;\mathbb{R}^N)}^p}. \qquad (3.57)$$

By virtue of hypothesis $H(j)_3(iii)$ and Proposition 1.3.14, we have that

$$\left| j(t, \xi) \right| \;\leq\; a_1(t) + c_2(t)\|\xi\|_{\mathbb{R}^N} \quad \text{for a.a. } t \in T \text{ and all } \xi \in \mathbb{R}^N,$$

with $a_1 \in L^1(T)_+$, $c_2 \in L^{p'}(T)_+$. Using this in (3.57) and passing to the limit as $n \to +\infty$, we obtain that $\|y'\|_p = 0$, hence $y \equiv \overline{\xi} \in \mathbb{R}^N$. Because $y \in V$, it follows that $\overline{\xi} = 0$. Therefore

$$y_n' \longrightarrow 0 \quad \text{in } L^p(T; \mathbb{R}^N),$$

hence

$$y_n \longrightarrow 0 \quad \text{in } W^{1,p}_{\mathrm{per}}(T; \mathbb{R}^N),$$

a contradiction to the fact that $\|y_n\|_{W^{1,p}(T;\mathbb{R}^N)} = 1$ for $n \geq 1$. $\square$

THEOREM 3.4.3
If hypotheses $H(j)_3$ hold,
<u>*then*</u> *problem (3.50) has a solution*

$$x_0 \in C^1(T;\mathbb{R}^N) \quad \text{with} \quad \|x_0'(\cdot)\|_{\mathbb{R}^N}^{p-2}\, x_0'(\cdot) \in W^{1,p'}(T;\mathbb{R}^N).$$

PROOF By virtue of hypothesis $H(j)_3(iv)$, we have that

$$\varphi(\xi) \longrightarrow -\infty \quad \text{as} \quad \|\xi\|_{\mathbb{R}^N} \to +\infty \quad \text{with } \xi \in \mathbb{R}^N.$$

This combined with Propositions 3.4.3 and 3.4.4 permit the use of the non-smooth Saddle Point Theorem (see Theorem 2.1.4), which gives us a solution for problem (3.50). $\square$

In the existence theorems proved thus far, the subdifferential of the non-smooth potential exhibited a growth which is strictly less then $p - 1$ (so in semilinear case, i.e. if $p = 2$, the growth of the superpotential is strictly subquadratic). Now we shall remove this restriction and allow for general growth of the subdifferential. The hypotheses on j are the following.

<u>$H(j)_4$</u> $j\colon T \times \mathbb{R}^N \longrightarrow \mathbb{R}$ is a function, such that

$\left.\begin{array}{l}(i)\\(ii)\end{array}\right\}$ the same as hypotheses $H(j)_1(i)$–(ii) with $j(\cdot,0) \in L^1(T)_+$;

(iii) for almost all $t \in T$, all $\xi \in \mathbb{R}^N$ and all $u \in \partial j(t,\xi)$, we have

$$\|u\|_{\mathbb{R}^N} \leq a(t) + c(t)\,\|\xi\|_{\mathbb{R}^N}^{r-1},$$

with $a, c \in L^\infty(T)$, where $r \in [1, +\infty)$;

(iv) there exist $\mu > p$ and $M_0 > 0$, such that for almost all $t \in T$ and all $\xi \in \mathbb{R}^N$ with $\|\xi\|_{\mathbb{R}^N} \geq M_0$

$$0 < \mu j(t,\xi) \leq -j^0(t,\xi;-\xi);$$

(v) for almost all $t \in T$ and all $\xi \in \mathbb{R}^N$, with $\|\xi\|_{\mathbb{R}^N} \leq 1$, we have

$$j(t,\xi) \leq \frac{1}{pb^p}.$$

REMARK 3.4.2 Suppose that $j(t,\cdot) \in C^1(\mathbb{R}^N)$. Then

$$-j^0(t,\xi;-\xi) = j_\xi'(t,\xi)(\xi) \qquad \forall\, \xi \in \mathbb{R}^N.$$

So hypothesis $H(j)_4(iv)$ is the well known Ambrosetti-Rabinowitz condition (see Ambrosetti & Rabinowitz (1973)), which was introduced in the context of semilinear (i.e. $p = 2$) elliptic problems. It was observed that in the smooth semilinear case this condition implies that the potential is strictly superquadratic. We shall see that the same is true in the present nonsmooth and nonlinear context.   ▯

PROPOSITION 3.4.5

If hypotheses $H(j)_4$ hold, <u>then</u> φ satisfies the nonsmooth PS-condition.

PROOF   First we show that

$$j(t, \xi) \ \geq \ c_3(t) \, \|\xi\|_{\mathbb{R}^N}^{\mu} - a_2(t) \quad \text{for a.a. } t \in T \text{ and all } n \geq 1, \qquad (3.58)$$

with $c_3, a_2 \in L^1(T)_+$. To this end let $L_0 \subseteq \mathbb{R}^N$ be the Lebesgue-null set outside of which hypotheses $H(j)_4(ii) - (v)$ hold. We set

$$\psi(t, \lambda) \ \stackrel{df}{=} \ j(t, \lambda \xi) \qquad \forall \, t \in T \setminus L_0, \ \lambda \geq 0, \ \xi \in \mathbb{R}^N, \ \|\xi\|_{\mathbb{R}^N} \geq M_0.$$

Evidently $\psi(t, \cdot)$ is locally Lipschitz. Moreover, denoting by ∂_λ and ∂_ξ the generalized subdifferential with respect to λ and ξ respectively, from Proposition 1.3.15, we have that

$$\partial \psi(t, \lambda) \ \subseteq \ \left(\partial_\xi j(t, \lambda \xi), \xi \right)_{\mathbb{R}^N},$$

so

$$-\lambda \partial \psi(t, \lambda) \ \subseteq \ \left(\partial_\xi j(t, \lambda \xi), -\lambda \xi \right)_{\mathbb{R}^N}$$

and using hypothesis $H(j)_4(iv)$ and the fact that $\psi(t, \cdot)$ being locally Lipschitz is differentiable almost everywhere and the derivative belongs in the generalized subdifferential, we have

$$\mu \psi(t, \lambda) \ \leq \ \lambda \psi'(t, \lambda).$$

From hypothesis $H(j)_4(iv)$, we know that $\psi(t, \lambda) > 0$, and so

$$\frac{\mu}{\lambda} \ \leq \ \frac{\psi'(t, \lambda)}{\psi(t, \lambda)} \qquad \forall \, \lambda \geq 1.$$

Integrating this inequality from $\lambda = 1$ to $\lambda = \lambda_0 > 1$, we obtain

$$\ln \lambda_0^{\mu} \ \leq \ \ln \frac{\psi(t, \lambda_0)}{\psi(t, 1)}$$

and

$$\lambda_0^{\mu} \psi(t, 1) \ \leq \ \psi(t, \lambda_0).$$

So, we have proved that

$$\lambda^{\mu} j(t,\xi) \leq j(t,\lambda\xi) \qquad \forall\, t \in T \setminus L_0,\ \xi \in \mathbb{R}^N,\ \|\xi\|_{\mathbb{R}^N} \geq M_0,\ \lambda \geq 1.$$

Using this fact, we see that for all $t \in T \setminus L_0$ and all $\xi \in \mathbb{R}^N$, with $\|\xi\|_{\mathbb{R}^N} \geq M_0$, we have

$$j(t,\xi) = j\left(t, \frac{\|\xi\|_{\mathbb{R}^N}}{M_0} \frac{M_0 \xi}{\|\xi\|_{\mathbb{R}^N}}\right) \geq \left(\frac{\|\xi\|_{\mathbb{R}^N}}{M_0}\right)^{\mu} j\left(t, \frac{M_0 \xi}{\|\xi\|_{\mathbb{R}^N}}\right). \tag{3.59}$$

By virtue of Proposition 1.3.14, we have

$$\left| j(t,\xi) - j(t,0) \right| = \left| \left(u(t), \xi \right)_{\mathbb{R}^N} \right| \qquad \forall\, t \in T \setminus L_0,\ \xi \in \mathbb{R}^N,\ \|\xi\|_{\mathbb{R}^N} = M_0,$$

with $u(t) \in \partial j\big(t, \eta(t)\xi\big)$ and $\eta(t) \in (0,1)$, so, from hypothesis $H(j)_4(iii)$, we have

$$j(t,\xi) \leq \beta(t), \tag{3.60}$$

with $\beta \in L^1(T)_+$. So, from (3.59), we obtain

$$j(t,\xi) \geq \left(\frac{\|\xi\|_{\mathbb{R}^N}}{M_0}\right)^{\mu} \beta(t) \qquad \forall\, t \in T \setminus L_0,\ \xi \in \mathbb{R}^N,\ \|\xi\|_{\mathbb{R}^N} \geq M_0.$$

On the other hand for all $t \in T \setminus L_0$, $j(t,\cdot)$ is bounded on B_{M_0}. Therefore, we conclude that (3.58) holds.

To check the nonsmooth PS-condition, let us consider a sequence $\{x_n\}_{n \geq 1} \subseteq W^{1,p}_{\mathrm{per}}(T; \mathbb{R}^N)$, such that

$$\left| \varphi(x_n) \right| \leq M_8 \qquad \forall\, n \geq 1 \quad \text{and} \quad m^{\varphi}(x_n) \longrightarrow 0 \quad \text{as } n \to +\infty,$$

for some $M_8 > 0$. Let $x_n^* \in \partial\varphi(x_n)$ be such that $m^{\varphi}(x_n) = \|x_n^*\|_{(W^{1,p}_{\mathrm{per}}(T;\mathbb{R}^N))^*}$ for $n \geq 1$. We have

$$x_n^* = A(x_n) - u_n \qquad \forall\, n \geq 1,$$

with $u_n \in S^{r'}_{\partial j(\cdot, x_n(\cdot))}$ for $n \geq 1$ and A the same nonlinear operator as in the proof of Proposition 3.4.1 (see (3.51)). We have

$$\frac{\mu}{p} \|x_n'\|_p^p - \int_0^b \mu j\big(t, x_n(t)\big)\, dt \leq \mu M_8 \tag{3.61}$$

and

$$\langle x_n^*, -x_n \rangle_{W^{1,p}_{\mathrm{per}}(T;\mathbb{R}^N)} \leq \varepsilon_n \|x_n\|_{W^{1,p}(T;\mathbb{R}^N)}, \tag{3.62}$$

with $\varepsilon_n \searrow 0$. Adding (3.61) and (3.62), we obtain

$$\left(\frac{\mu}{p} - 1\right) \|x_n'\|_p^p + \int_0^b \left[\big(u_n(t), x_n(t)\big)_{\mathbb{R}^N} - \mu j\big(t, x_n(t)\big) \right] dt$$

$$\leq \mu M_8 + \varepsilon_n \|x_n\|_{W^{1,p}(T;\mathbb{R}^N)},$$

so

$$\left(\frac{\mu}{p} - 1\right) \|x'_n\|_p^p + \int_0^b \left[-j^0\big(t, x_n(t); -x_n(t)\big) - \mu j\big(t, x_n(t)\big) \right] dt$$
$$\leq \ \mu M_8 + \varepsilon_n \|x_n\|_{W^{1,p}(T;\mathbb{R}^N)} \, . \tag{3.63}$$

We analyze the integral in the left hand side of the last inequality. Using hypothesis $H(j)_4(iv)$ and (3.60), we have

$$\int_0^b \left[-j^0\big(t, x_n(t); -x_n(t)\big) - \mu j\big(t, x_n(t)\big) \right] dt \tag{3.64}$$

$$\geq \int_{\{\|x_n\|_{\mathbb{R}^N} < M_0\}} \left[-j^0\big(t, x_n(t); -x_n(t)\big) - \mu j\big(t, x_n(t)\big) \right] dt \ \geq \ -c_4,$$

for some $c_4 > 0$. Returning to (3.63) and using (3.64), we obtain

$$\left(\frac{\mu}{p} - 1\right) \|x'_n\|_p^p \ \leq \ c_5 + \varepsilon_n \|x_n\|_{W^{1,p}(T;\mathbb{R}^N)} \qquad \forall \, n \geq 1, \tag{3.65}$$

for some $c_5 > 0$. Suppose that the sequence $\{x_n\}_{n \geq 1} \subseteq W^{1,p}_{\mathrm{per}}(T;\mathbb{R}^N)$ is unbounded. Passing to a subsequence if necessary, we may assume that $\|x_n\|_{W^{1,p}(T;\mathbb{R}^N)} \longrightarrow +\infty$. Let

$$y_n \stackrel{df}{=} \frac{x_n}{\|x_n\|_{W^{1,p}(T;\mathbb{R}^N)}} \qquad \forall \, n \geq 1.$$

We have

$$y_n \stackrel{w}{\longrightarrow} y \ \text{ in } W^{1,p}_{\mathrm{per}}(T;\mathbb{R}^N) \tag{3.66}$$
$$y_n \longrightarrow y \ \text{ in } C(T;\mathbb{R}^N). \tag{3.67}$$

Dividing (3.65) by $\|x_n\|^p_{W^{1,p}(T;\mathbb{R}^N)}$, we obtain

$$\left(\frac{\mu}{p} - 1\right) \|y'_n\|_p^p \ \leq \ \frac{c_5}{\|x_n\|^p_{W^{1,p}(T;\mathbb{R}^N)}} + \frac{\varepsilon_n}{\|x_n\|^{p-1}_{W^{1,p}(T;\mathbb{R}^N)}} \qquad \forall \, n \geq 1,$$

so passing to the limit as $n \to +\infty$, we get

$$\left(\frac{\mu}{p} - 1\right) \|y'\|_p^p \ \leq \ 0.$$

Thus $y' = 0$ (since $\mu > p$) and so $y \equiv \overline{\xi} \in \mathbb{R}^N$.
    If $\overline{\xi} = 0$, then
$$y_n \longrightarrow 0 \ \text{ in } W^{1,p}_{\mathrm{per}}(T;\mathbb{R}^N),$$

a contradiction to the fact that $\|y_n\|_{W^{1,p}(T;\mathbb{R}^N)} = 1$ for $n \geq 1$. Therefore $\overline{\xi} \neq 0$ and from this we infer that

$$\|x_n(t)\|_{\mathbb{R}^N} \longrightarrow +\infty \quad \text{as } n \to +\infty \qquad \forall\, t \in T.$$

In fact, we shall show that this convergence is uniform in $t \in T$, i.e.

$$\lim_{n\to+\infty} \min_{t\in T} \|x_n(t)\|_{\mathbb{R}^N} = +\infty. \tag{3.68}$$

From (3.66), for a given $\varepsilon \in \left(0, \|\overline{\xi}\|_{\mathbb{R}^N}\right)$, we can find $n_0 \geq 1$, such that

$$\|y_n(t) - \overline{\xi}\|_{\mathbb{R}^N} < \varepsilon \qquad \forall\, n \geq n_0,\ t \in T,$$

hence

$$0 < \|\overline{\xi}\|_{\mathbb{R}^N} - \varepsilon < \|y_n(t)\|_{\mathbb{R}^N} \qquad \forall\, n \geq n_0,\ t \in T.$$

Since $\|x_n\|_{W^{1,p}(T;\mathbb{R}^N)} \longrightarrow +\infty$, for a given $\eta > 0$, we can find $n_1 \geq n_0$, such that

$$0 < \eta \leq \|x_n\|_{W^{1,p}(T;\mathbb{R}^N)} \qquad \forall\, n \geq n_1.$$

Therefore

$$\frac{\|x_n(t)\|_{\mathbb{R}^N}}{\eta} \geq \frac{\|x_n(t)\|_{\mathbb{R}^N}}{\|x_n\|_{W^{1,p}(T;\mathbb{R}^N)}} = \|y_n(t)\|_{\mathbb{R}^N}$$

$$> \|\overline{\xi}\|_{\mathbb{R}^N} - \varepsilon > 0 \qquad \forall\, n \geq n_1,\ t \in T,$$

so

$$\|x_n(t)\|_{\mathbb{R}^N} > \eta\left(\|\overline{\xi}\|_{\mathbb{R}^N} - \varepsilon\right) > 0.$$

Since $\eta > 0$ was arbitrary, it follows that (3.68) holds. So, without any loss of generality, we may assume that

$$\|x_n(t)\|_{\mathbb{R}^N} > 0 \qquad \forall\, n \geq 1,\ t \in T.$$

From the choice of the sequence $\{x_n\}_{n\geq 1} \subseteq W^{1,p}_{\text{per}}(T;\mathbb{R}^N)$, we have

$$-\frac{pM_8}{\|x_n\|^p_{W^{1,p}(T;\mathbb{R}^N)}} \leq \|y_n'\|^p_p - \int_0^b \frac{pj(t, x_n(t))}{\|x_n\|^p_{W^{1,p}(T;\mathbb{R}^N)}}\, dt,$$

so

$$-\frac{pM_8}{\|x_n\|^p_{W^{1,p}(T;\mathbb{R}^N)}} \leq \|y_n'\|^p_p - \int_0^b \frac{pj(t, x_n(t))}{\|x_n(t)\|^p_{\mathbb{R}^N}}\, \|y_n(t)\|^p_{\mathbb{R}^N}\, dt$$

and

$$-\frac{pM_8}{\|x_n\|^p_{W^{1,p}(T;\mathbb{R}^N)}} \leq c_6 - \int_0^b \frac{pc_3(t)\, \|x_n(t)\|^\mu_{\mathbb{R}^N} - pa_2(t)}{\|x_n(t)\|^p_{\mathbb{R}^N}}\, \|y_n(t)\|^p_{\mathbb{R}^N}\, dt.$$

Passing to the limit as $n \to +\infty$ and since $\mu > p$, we obtain a contradiction. This proves the boundedness of the sequence $\{x_n\}_{n\geq 1} \subseteq W^{1,p}_{per}(T;\mathbb{R}^N)$. So, passing to a subsequence if necessary, we may assume that

$$x_n \xrightarrow{w} x \quad \text{in } W^{1,p}_{per}(T;\mathbb{R}^N)$$

and arguing as in the last part of the proof of Proposition 3.4.1, we obtain that

$$x_n \longrightarrow x \quad \text{in } W^{1,p}_{per}(T;\mathbb{R}^N).$$

$\square$

Let $h \in V \setminus \{0\}$ and suppose that $\|h'\|_p = 1$ (recall that by Theorem 1.1.7, $\|h'\|_p$ is equivalent to the Sobolev norm on V). For arbitrary $R > 0$, we define the cylinder set

$$C_R \overset{df}{=} \{x \in W^{1,p}_{per}(T;\mathbb{R}^N) : \ x(t) = \xi + \lambda h(t), \ t \in T,$$
$$\xi \in \mathbb{R}^N, \ \|\xi\|_{\mathbb{R}^N} \leq R, \ \lambda \in [0,R]\}.$$

PROPOSITION 3.4.6
If hypotheses $H(j)_4$ hold and $R > 0$ is large enough, then $\varphi|_{\partial C_R} < 0$.

PROOF   First, we check the lower base of the cylinder C_R (i.e. $\lambda = 0$). In this case $x = \overline{\xi} \in \mathbb{R}^N$ and so, from hypothesis $H(j)_4(iv)$, we get

$$\varphi(x) \ = \ \varphi(\overline{\xi}) \ = \ -\int_0^b j(t,\overline{\xi})\,dt \ < \ 0.$$

Next let

$$V_1 \overset{df}{=} \left\{v \in L^p(T;\mathbb{R}^N) : \ \int_0^b v(t)\,dt = 0\right\}.$$

Evidently $L^p(T;\mathbb{R}^N) = \mathbb{R}^N \oplus V_1$, i.e. for every $x \in L^p(T;\mathbb{R}^N)$, we have

$$x = \overline{x} + v, \quad \text{with} \quad \overline{x} = \int_0^b x(t)\,dt \quad \text{and} \quad v(\cdot) = x(\cdot) - \overline{x}.$$

So, we can find $c_7 > 0$, such that

$$c_7\big(\|\overline{\xi}\|_{\mathbb{R}^N}^\mu + \lambda^\mu \|h\|_p^\mu\big) \ \leq \ \|x\|_p^\mu . \tag{3.69}$$

Because of (3.58), for every $x \in C_R$, we have

$$\varphi(x) \ = \ \frac{1}{p}\|x'\|_p^p - \int_0^b j(t,x(t))\,dt \ \leq \ \frac{\lambda^p}{p} - c_8\|x\|_p^\mu - c_9$$

for some $c_8, c_9 > 0$. So, from (3.69), we get

$$\varphi(x) \leq \frac{\lambda^p}{p} - c_{10} \|c\|_p^\mu - c_{11} \lambda^\mu \|h\|_p^\mu ,$$

for some $c_{10}, c_{11} > 0$. If $x \in C_R$ is on the lateral surface of the cylinder, we have $\|\bar{\xi}\|_{\mathbb{R}^N} = R$ and so

$$\varphi(x) \leq \frac{\lambda^p}{p} - c_{10} R^\mu \leq \frac{R^p}{p} - c_{10} R^\mu$$

and so

$$\varphi(x) < 0 \qquad \forall R \geq R_0$$

(recall that $p < \mu$).

On the other hand, if $x \in C_R$ is on the upper base of the cylinder, then $\lambda = R$ and we have

$$\varphi(x) \leq \frac{R^p}{p} - c_{11} R^\mu \|h\|_p^\mu .$$

Since $\|h\|_p \neq 0$ and $\mu > p$, we can find $R_1 \geq R_0$, such that

$$\varphi(x) < 0 \qquad \forall R \geq R_1.$$

So, we have proved the proposition. ⬛

REMARK 3.4.3 A natural choice for h would be

$$h(t) \stackrel{df}{=} \|u_1(t)\|_{\mathbb{R}^N}^{p-2} u_1(t) \qquad \forall t \in T,$$

with $u_1 \in C^1(T; \mathbb{R}^N)$ being a normalized eigenfunction corresponding to λ_1, the first nonzero eigenvalue of $\left(-\Delta_p, W^{1,p}_{\text{per}}(T; \mathbb{R}^N)\right)$ (see problem (I.1.16)). ⬛

Next let

$$E \stackrel{df}{=} \left\{ v \in V : \|v'\|_p = \frac{1}{b^{p-1}} \right\}.$$

PROPOSITION 3.4.7
If hypotheses $H(j)_4$ hold, <u>then</u> $\inf_E \varphi \geq 0$.

PROOF From Remark 1.1.11, we know that

$$\|v\|_\infty^p \leq b^{p-1} \|v'\|_p^p \qquad \forall v \in V.$$

So

$$\|v\|_\infty^p \leq 1 \qquad \forall v \in E.$$

Note that by hypothesis $H(j)_4(v)$, we have

$$j(t, v(t)) \leq \frac{1}{pb^p} \qquad \forall \, v \in E, \, l \in T$$

and so

$$\varphi(v) \geq \frac{1}{pb^{p-1}} - \frac{1}{pb^{p-1}} = 0.$$

$\square$

These auxiliary results lead to the following existence theorem for problem (3.50).

THEOREM 3.4.4
If hypotheses $H(j)_4$ hold,
then problem (3.50) has a nontrivial solution

$$x_0 \in C^1_{\mathrm{per}}(T; \mathbb{R}^N) \quad \text{with} \quad \|x_0'(\cdot)\|_{\mathbb{R}^N}^{p-2} x_0'(\cdot) \in W^{1,r'}_{\mathrm{per}}(T; \mathbb{R}^N).$$

PROOF Propositions 3.4.5, 3.4.6 and 3.4.7 permit the use of Theorem 2.1.5, which implies the existence of $x_0 \in W^{1,p}_{\mathrm{per}}(T; \mathbb{R}^N)$, such that $0 \in \partial\varphi(x_0)$ and $\varphi(x_0) \geq 0$. Evidently $x_0 \in W^{1,p}_{\mathrm{per}}(T; \mathbb{R}^N)$ is nontrivial since $\varphi(0) \leq 0$ (see Theorem 2.1.5). Since $0 \in \partial\varphi(x_0)$, as in the proof of Theorem 3.4.2, we show that $x_0 \in C^1_{\mathrm{per}}(T; \mathbb{R}^N)$, it solves problem (3.50) and of course $\|x'(\cdot)\|_{\mathbb{R}^N}^{p-2} x'(\cdot) \in W^{1,r'}_{\mathrm{per}}(T; \mathbb{R}^N)$. $\square$

REMARK 3.4.4 Note that if additionally

$$\int_0^b j(t, \xi)\, dt > 0 \qquad \forall \, \xi \in \mathbb{R}^N \setminus \{0\},$$

then the solution x_0 is nonconstant. $\square$

EXAMPLE 3.4.2 A typical function satisfying hypotheses $H(j)_4$ is the function $j: \mathbb{R}^2 \longrightarrow \mathbb{R}$, defined by

$$j(\xi_1, \xi_2) \stackrel{df}{=} \begin{cases} c\left(\xi_1^\mu - \xi_2^2\right) - \sqrt{\xi_1 \xi_2} & \text{if } \xi_1, \xi_2 > 0, \\ c\left(\xi_1^\mu - \xi_2^2\right) & \text{otherwise,} \end{cases}$$

where $0 < c < \frac{1}{pb^p}$ and $\mu > p$. $\square$

The next existence result will be used in the study of homoclinic (to zero) solutions, which is conducted in Subsection 3.4.2. It concerns the following nonlinear and nonsmooth periodic system:

$$\begin{cases} -\big(\|x'(t)\|_{\mathbb{R}^N}^{p-2}\, x'(t) \big)' + g(t)\, \|x(t)\|_{\mathbb{R}^N}^{p-2}\, x(t) \in \partial j\big(t, x(t)\big) \\ \hspace{5cm} \text{for a.a. } t \in T, \quad (3.70) \\ x(0) = x(b),\ x'(0) = x'(b), \end{cases}$$

with $p \in (1, +\infty)$. For this problem our hypotheses are the following:

$\underline{H(g)_1}$ $g \in C(T)$, $g(0) = g(b)$ and there exists $c > 0$, such that

$$g(t) \geq c \qquad \forall\, t \in T.$$

$\underline{H(j)_5}$ $j \colon T \times \mathbb{R}^N \longrightarrow \mathbb{R}$ is a function, such that

(i) for all $\xi \in \mathbb{R}^N$, the function

$$T \ni t \longmapsto j(t, \xi) \in \mathbb{R}$$

is measurable, $j(\cdot, 0) \in L^1(T)$ and $\int_0^b j(t, 0)dt \geq 0$;

(ii) for almost all $t \in T$, the function

$$\mathbb{R}^N \ni \xi \longmapsto j(t, \xi) \in \mathbb{R}$$

is locally Lipschitz;

(iii) for almost all $t \in T$, all $\xi \in \mathbb{R}^N$ and all $u \in \partial j(t, \xi)$, we have

$$\|u\|_{\mathbb{R}^N} \leq a(t) + c(t)\, \|\xi\|_{\mathbb{R}^N}^{r-1},$$

with $a, c \in L^\infty(T)$, $r \in [1, +\infty)$;

(iv) there exist $\mu > p$ and $M > 0$, such that for almost all $t \in T$ and all $\xi \in \mathbb{R}^N$ with $\|\xi\|_{\mathbb{R}^N} \geq M$, we have

$$\mu j(t, \xi) \leq -j^0(t, \xi; -\xi);$$

(v) $\displaystyle \lim_{\|\xi\|_{\mathbb{R}^N} \to 0} \frac{pj(t, \xi)}{\|\xi\|_{\mathbb{R}^N}^p} \leq 0$ uniformly for almost all $t \in T$;

(vi) there exists $\xi_* \in \mathbb{R}^N$, $\|\xi^*\|_{\mathbb{R}^N} \geq M$, such that $\int_0^b j(t, \xi_*)\, dt > 0$.

THEOREM 3.4.5
If hypotheses $H(g)_1$ and $H(j)_5$ hold,
then problem (3.70) has a nontrivial solution

$$x_0 \in C^1\big(T; \mathbb{R}^N\big) \quad \text{with} \quad \|x_0'(\cdot)\|_{\mathbb{R}^N}^{p-2}\, x_0'(\cdot) \in W_{\text{per}}^{1,\infty}\big(T; \mathbb{R}^N\big).$$

PROOF   In this case the energy functional $\varphi \colon W^{1,p}_{\mathrm{per}}(T; \mathbb{R}^N) \longrightarrow \mathbb{R}$ is given by

$$\varphi(x) \stackrel{df}{=} \frac{1}{p} \|x'\|_p^p + \frac{1}{p} \int_0^b g(t) \, \|x(t)\|_{\mathbb{R}^N}^p \, dt - \int_0^b j\big(t, x(\iota)\big) \, dt.$$

Of course φ is locally Lipschitz. We show that φ satisfies the nonsmooth PS-condition. So let $\{x_n\}_{n \geq 1} \subseteq W^{1,p}_{\mathrm{per}}(T; \mathbb{R}^N)$ be such that

$$|\varphi(x_n)| \leq M_9 \quad \forall \, n \geq 1 \quad \text{and} \quad m^\varphi(x_n) \longrightarrow 0 \quad \text{as } n \to +\infty,$$

for some $M_9 > 0$. As in the proof of Proposition 3.4.5, we obtain

$$\left(\frac{\mu}{p} - 1\right) \left(\|x_n'\|_p^p + \int_0^b g(t) \, \|x_n(t)\|_{\mathbb{R}^N}^p \, dt\right) \leq \mu M_9 + \varepsilon_n \, \|x_n\|_{W^{1,p}(T;\mathbb{R}^N)} + c_{12},$$

for some $c_{12} > 0$ and $\varepsilon_n \searrow 0$, so

$$\left(\frac{\mu}{p} - 1\right) \left(\|x_n'\|_p^p + c \, \|x_n\|_p^p\right) \leq \mu M_9 + \varepsilon_n \, \|x_n\|_{W^{1,p}(T;\mathbb{R}^N)} + c_{12}$$

and

$$\|x_n\|_{W^{1,p}(T;\mathbb{R}^N)}^p \leq c_{13} + \varepsilon_n' \, \|x_n\|_{W^{1,p}(T;\mathbb{R}^N)},$$

for some $c_{13} > 0$ and $\varepsilon_n' \searrow 0$. From this inequality, it follows that the sequence $\{x_n\}_{n \geq 1} \subseteq W^{1,p}_{\mathrm{per}}(T; \mathbb{R}^N)$ is bounded and then as before we conclude that φ satisfies the nonsmooth PS-condition.

By virtue of hypothesis $H(j)_5(v)$, for a given $\varepsilon > 0$, we can find $\delta > 0$, such that

$$j(t, \xi) \leq \frac{\varepsilon}{p} \|\xi\|_{\mathbb{R}^N}^p \quad \text{for a.a. } t \in T \text{ and all } \xi \in \mathbb{R}^N, \ \|\xi\|_{\mathbb{R}^N} \leq \delta.$$

On the other hand from hypothesis $H(j)_5(iii)$ and Proposition 1.3.14, we know that

$$j(t, \xi) \leq c_{14} \|\xi\|_{\mathbb{R}^N}^p \quad \text{for a.a. } t \in T \text{ and all } \xi \in \mathbb{R}^N, \ \|\xi\|_{\mathbb{R}^N} \geq \delta,$$

for some $c_{14} > 0$. So finally, we can say that

$$j(t, \xi) \leq \frac{\varepsilon}{p} \|\xi\|_{\mathbb{R}^N}^p + c_{15} \|\xi\|_{\mathbb{R}^N}^s \quad \text{for a.a. } t \in T \text{ and all } \xi \in \mathbb{R}^N,$$

for some $s > \max\{r, p\}$ and some $c_{15} > 0$. Then for all $x \in W^{1,p}_{\mathrm{per}}(T; \mathbb{R}^N)$, we have

$$\varphi(x) = \frac{1}{p} \|x'\|_p^p + \frac{1}{p} \int_0^b g(t) \, \|x(t)\|_{\mathbb{R}^N}^p \, dt - \int_0^b j\big(t, x(t)\big) \, dt$$

$$\geq \frac{1}{p} \|x'\|_p^p + \frac{c}{p} \|x\|_p^p - \frac{\varepsilon}{p} \|x\|_p^p - c_{16} \|x\|_\infty^s,$$

for some $c_{16} > 0$. Recall that the embedding $W^{1,p}_{per}(T; \mathbb{R}^N) \subseteq C(T; \mathbb{R}^N)$ is continuous (in fact compact). So

$$\varphi(x) \geq \frac{1}{p}\left(\|x'\|^p_p + (c - \varepsilon)\|x\|^p_p \right) - c_{17}\|x\|^s_{W^{1,p}(T;\mathbb{R}^N)},$$

for some $c_{17} > 0$. Taking $\varepsilon < c$, we conclude that for all $x \in W^{1,p}_{per}(T; \mathbb{R}^N)$, we have

$$\varphi(x) \geq c_{18}\|x\|^p_{W^{1,p}(T;\mathbb{R}^N)} - c_{17}\|x\|^s_{W^{1,p}(T;\mathbb{R}^N)}, \qquad (3.71)$$

for some $c_{18} > 0$. Since $p < s$, from (3.71), we see that we can find $\varrho > 0$ small enough so that

$$\inf_{\|x\|_{W^{1,p}(T;\mathbb{R}^N)} = \varrho} \varphi(x) = \gamma > 0.$$

On the other hand from the proof of Proposition 3.4.5, because of hypothesis $H(j)_5(iv)$, we have that

$$\lambda^\mu j(t, \xi) \leq j(t, \lambda\xi) \quad \text{for a.a. } t \in T \text{ and all } \xi \in \mathbb{R}^N, \ \|\xi\|_{\mathbb{R}^N} \geq M, \ \lambda \geq 1.$$

Let $\xi_* \in \mathbb{R}^N$ be as postulated by hypothesis $H(j)_5(vi)$. Then for $\lambda \geq 1$, we have

$$\varphi(\lambda\xi_*) = \frac{1}{p}\int_0^b g(t)\|\lambda\xi_*\|^p_{\mathbb{R}^N}\, dt - \int_0^b j(t, \lambda\xi_*)\, dt$$

$$\leq \frac{\lambda^p}{p}\|g\|_\infty \|\xi_*\|^p_{\mathbb{R}^N}\, b - \lambda^\mu \int_0^b j(t, \xi_*)\, dt,$$

so from hypothesis $H(j)_5(vi)$ and recalling that $\mu > p$, we have

$$\varphi(\lambda\xi_*) \longrightarrow -\infty \quad \text{as} \quad \lambda \to +\infty.$$

Hence, we can find $\lambda > 0$ large enough, so that $\|\lambda\xi_*\|_{\mathbb{R}^N} > \varrho$ and $\varphi(\lambda\xi_*) \leq 0$. Moreover, note that

$$\varphi(0) = -\int_0^b j(t, 0)\, dt \leq 0.$$

All these facts permit the use of the nonsmooth Mountain Pass Theorem (see Theorem 2.1.3), which gives $x_0 \in W^{1,p}_{per}(T; \mathbb{R}^N)$, such that

$$\varphi(x_0) \geq \gamma > 0 \geq \varphi(0)$$

(hence $x_0 \neq 0$) and $0 \in \partial\varphi(x_0)$. This is a nontrivial solution of (3.70) belonging in $C^1(T; \mathbb{R}^N)$ with $\|x'_0(\cdot)\|^{p-2}_{\mathbb{R}^N} x'_0(\cdot) \in W^{1,\infty}_{per}(T; \mathbb{R}^N)$. $\qquad\Box$

EXAMPLE 3.4.3　　The nonsmooth potential

$$j(\xi) \stackrel{df}{=} \begin{cases} -\|\xi\|_{\mathbb{R}^N} & \text{if } \|\xi\|_{\mathbb{R}^N} \leq 1, \\ \frac{1}{\mu}\|\xi\|_{\mathbb{R}^N}^{\mu} - \|\xi\|_{\mathbb{R}^N} \ln\|\xi\|_{\mathbb{R}^N} - \frac{\mu+1}{\mu} & \text{if } \|\xi\|_{\mathbb{R}^N} > 1, \end{cases}$$

with $\mu > p$, satisfies the hypotheses $H(j)_5$. Evidently this j does not satisfy hypotheses $H(j)_1$, $H(j)_2$ and $H(j)_3$. ▯

These existence theorems are our main tools in the study of homoclinic solutions which follows.

3.4.2　Homoclinic Solutions

In this subsection we investigate the question of existence of homoclinic (to zero) solutions for the homoclinic problem on $\mathbb{R}^N$ corresponding to (3.70). Namely, we examine the following problem:

$$\begin{cases} -\big(\|x'(t)\|_{\mathbb{R}^N}^{p-2}\, x'(t)\big)' + g(t)\,\|x(t)\|_{\mathbb{R}^N}^{p-2}\, x(t) \in \partial j\big(t, x(t)\big) \\ \qquad\qquad\qquad\qquad\qquad\qquad\qquad\qquad \text{for a.a. } t \in \mathbb{R}, \\ \|x(t)\|_{\mathbb{R}^N} \longrightarrow 0, \ \|x'(t)\|_{\mathbb{R}^N} \longrightarrow 0, \ \text{as } |t| \to +\infty, \end{cases} \qquad (3.72)$$

for $p \in (1, +\infty)$. In this situation our hypotheses on g and j are the following:

$\underline{H(g)_2}$ $g \in C(\mathbb{R}^n)$, g is $2b$-periodic and there exists $c > 0$, such that

$$g(t) \geq c \qquad \forall\, t \in [-b, b].$$

$\underline{H(j)_6}$ $j \colon \mathbb{R} \times \mathbb{R}^N \longrightarrow \mathbb{R}$ is a function, such that

(i) for all $\xi \in \mathbb{R}^N$, the function

$$T \ni t \longmapsto j(t, \xi) \in \mathbb{R}^N$$

is measurable, $2b$-periodic and $j(t, 0) = 0$;

(ii) for almost all $t \in T$, the function

$$\mathbb{R}^N \ni \xi \longmapsto j(t, \xi) \in \mathbb{R}^N$$

is locally Lipschitz;

(iii) for almost all $t \in T$, all $\xi \in \mathbb{R}^N$ and all $u \in \partial j(t, \xi)$, we have

$$\|u\|_{\mathbb{R}^N} \leq a_0(t)\big(1 + \|\xi\|_{\mathbb{R}^N}^{p-1}\big),$$

with $a_0 \in L^\infty(T)_+$;

(iv) there exists $\mu > p$, such that for almost all $t \in T$ and all $\xi \in \mathbb{R}^N$, $\xi \neq 0$, we have
$$\mu j(t,\xi) \leq -j^0(t,\xi;-\xi);$$

(v) we have
$$\limsup_{\|\xi\|_{\mathbb{R}^N} \to 0} \frac{(u(\xi),\xi)_{\mathbb{R}^N}}{\|\xi\|_{\mathbb{R}^N}^p} \leq 0$$
uniformly for almost all $t \in T$ and all $u(\xi) \in \partial j(t,\xi)$;

(vi) there exists $\xi_* \in \mathbb{R}^N$, such that $\int\limits_0^b j(t,\xi_*)\,dt > 0$.

REMARK 3.4.5 Hypothesis $H(j)_6(v)$ implies that
$$\limsup_{\|\xi\|_{\mathbb{R}^N} \to 0} \frac{pj(t,\xi)}{\|\xi\|_{\mathbb{R}^N}^p} \leq 0 \quad \text{uniformly for almost all } t \in T.$$

Indeed from Proposition 1.3.14, we know that
$$j(t,\xi) - j\left(t,\tfrac{\xi}{2}\right) = \left(u(t,\xi),\tfrac{\xi}{2}\right)_{\mathbb{R}^N} \quad \text{for a.a. } t \in T \text{ and all } \xi \in \mathbb{R}^N \setminus \{0\},$$

with $u(t,\xi) \in \partial j\left(t,\lambda\tfrac{\xi}{2}\right)$ and $\lambda \in (1,2)$ (in general λ depends on $t \in T$). So
$$\frac{j(t,\xi)}{\|\xi\|_{\mathbb{R}^N}^p} = \frac{j\left(t,\tfrac{\xi}{2}\right)}{2^p\|\tfrac{\xi}{2}\|_{\mathbb{R}^N}^p} + \frac{\lambda^{p-1}}{2^p}\frac{\left(u,\lambda\tfrac{\xi}{2}\right)_{\mathbb{R}^N}}{\|\lambda\tfrac{\xi}{2}\|_{\mathbb{R}^N}^p}.$$

$\square$

Let us set
$$\gamma(t) \overset{df}{=} \limsup_{\|\xi\|_{\mathbb{R}^N} \to 0} \frac{j(t,\xi)}{\|\xi\|_{\mathbb{R}^N}^p}.$$
Passing to the limit as $\|\xi\|_{\mathbb{R}^N} \to 0$ (note that $\lambda \searrow 1$), we obtain
$$\left(1 - \frac{1}{2^p}\right)\gamma(t) \leq \limsup_{\|\xi\|_{\mathbb{R}^N} \to 0} \frac{\lambda^{p-1}}{2^p}\frac{\left(u,\lambda\tfrac{\xi}{2}\right)_{\mathbb{R}^N}}{\|\lambda\tfrac{\xi}{2}\|_{\mathbb{R}^N}^p},$$
uniformly for almost all $t \in T$.

THEOREM 3.4.6
If hypotheses $H(g)_2$ and $H(j)_6$ hold,
then problem (3.72) has a nontrivial homoclinic solution $x_0 \in C(\mathbb{R};\mathbb{R}^n) \cap W^{1,p}(\mathbb{R};\mathbb{R}^N)$.

PROOF We consider the following auxiliary periodic problem:

$$\begin{cases} -\left(\|x'(t)\|_{\mathbb{R}^N}^{p-2}\, x'(t) \right)' + g(t)\, \|x(t)\|_{\mathbb{R}^N}^{p-2}\, x(t) \in \partial j\big(t, x(t)\big) \\ \qquad\qquad\qquad\qquad\qquad\qquad\qquad \text{for a.a. } t \in T_n, \qquad (3.73) \\ x(-nb) = x(nb),\ x'(-nb) = x'(nb), \end{cases}$$

for $p \in (1, +\infty)$, where $T_n \overset{df}{=} [-nb, nb]$. From Theorem 3.4.5, we know that problem (3.73) has a nontrivial solution $x_n \in C^1_{\mathrm{per}}(T; \mathbb{R}^N)$. Let $\varphi_n \colon W^{1,p}_{\mathrm{per}}(T_n; \mathbb{R}^N) \longrightarrow \mathbb{R}$ be the nonsmooth locally Lipschitz energy functional corresponding to the problem (3.73), i.e.

$$\varphi_n(x) \overset{df}{=} \frac{1}{p}\, \|x'\|_p^p + \frac{1}{p} \int\limits_{-nb}^{nb} g(t)\, \|x(t)\|_{\mathbb{R}^N}^p \ dt - \int\limits_{-nb}^{nb} j\big(t, x(t)\big)\, dt$$

$$\forall x \in W^{1,p}_{\mathrm{per}}(T_n; \mathbb{R}^N).$$

Recall that because of hypotheses $H(j)_6(iv)$, we have

$$\lambda^\mu j(t, \xi) \ \le\ j(t, \lambda\xi) \quad \text{for a.a. } t \in \mathbb{R},\ \text{all } \xi \in \mathbb{R}^N \text{ and all } \lambda \ge 1$$

(see the proof of Proposition 3.4.5).

Consider the integral functional $\psi \colon L^p(T_1; \mathbb{R}^N) \longrightarrow \mathbb{R}$, defined by

$$\psi(x) \overset{df}{=} \int\limits_{-b}^{b} j\big(t, x(t)\big)\, dt \qquad \forall\, x \in L^p(T_1; \mathbb{R}^N).$$

By virtue of hypothesis $H(j)_6(vi)$, we have that $\psi(\xi_*) > 0$. Because ψ is continuous and the embedding $W^{1,p}_0(T_1; \mathbb{R}^N) \subseteq L^p(T_1; \mathbb{R}^N)$ is dense, we can find $\overline{x} \in W^{1,p}_0(T; \mathbb{R}^N)$, such that

$$\psi(\overline{x}) \ =\ \int\limits_{-b}^{b} j(t, \overline{x}(t))\, dt \ >\ 0.$$

Then for $\lambda \ge 1$, we have

$$\varphi_1(\lambda\overline{x}) \ =\ \frac{\lambda^p}{p}\, \|\overline{x}'\|_p^p + \frac{\lambda^p}{p} \int\limits_{-b}^{b} g(t)\, \|\overline{x}(t)\|_{\mathbb{R}^N}^p \ dt - \int\limits_{-b}^{b} j(t, \lambda\overline{x}(t))\, dt$$

$$\le\ \frac{\lambda^p}{p} \left(\|\overline{x}'\|_p^p + 2b\, \|g\|_\infty\, \|\overline{x}\|^p_{L^p(T_1; \mathbb{R}^N)} \right) - \lambda^\mu \int\limits_{-b}^{b} j(t, \overline{x}(t))\, dt.$$

Because $\mu > p$ and $\psi(\overline{x}) = \int\limits_{-b}^{b} j(t, \overline{x}(t))\, dt > 0$, we can find $\lambda_0 \geq 1$, such that for all $\lambda \geq \lambda_0$, we have $\varphi_1(\lambda\overline{x}) < 0$. Let $\widehat{x} \in W_0^{1,p}(T_n; \mathbb{R}^N)$ be defined as follows

$$\widehat{x}(t) \overset{df}{=} \begin{cases} \overline{x}(t) \text{ if } t \in T_1, \\ 0 \quad \text{ if } t \in T_n \setminus T_1. \end{cases}$$

From hypothesis $H(j)_6(i)$, we have that

$$\varphi_n(\lambda\widehat{x}) = \varphi_1(\lambda\overline{x}) \qquad \forall\, \lambda \geq \lambda_0.$$

From the proof of Theorem 3.4.5, we know that the solution $x_n \in C_{\mathrm{per}}^1(T; \mathbb{R}^N)$ of problem (3.72) is obtained via the nonsmooth Mountain Pass Theorem (see Theorem 2.1.3). So

$$\widehat{c}_n \overset{df}{=} \inf_{\gamma \in \Gamma_n} \sup_{t \in [0,1]} \varphi_n\big(\gamma(t)\big) = \varphi_n(x_n) \geq \inf_{\|x\|_{W^{1,p}(T;\mathbb{R}^N)} = \varrho_n} \varphi_n(x) = \xi_n > 0,$$

where

$$\Gamma_n \overset{df}{=} \big\{\gamma \in C\big([0,1]; W^{1,p}(T_n; \mathbb{R}^N)\big) : \gamma(0) = 0,\ \gamma(1) = \lambda\widehat{x}\big\}, \quad \text{with} \quad \lambda \geq \lambda_0$$

and $0 \in \partial\varphi_n(x_n)$ for $n \geq 1$.

Extending by constant, we see that

$$W^{1,p}\big(T_{n_1}; \mathbb{R}^N\big) \subseteq W^{1,p}\big(T_{n_2}; \mathbb{R}^N\big) \quad \text{and} \quad \Gamma_{n_1} \subseteq \Gamma_{n_2} \qquad \forall\, n_1 \leq n_2$$

and consequently

$$\widehat{c}_{n_2} \leq \widehat{c}_{n_1} \qquad \forall\, n_1 \leq n_2.$$

This way we have produced a decreasing sequence $\{\widehat{c}_n\}_{n \geq 1}$ of critical values. For every $n \geq 1$, we have

$$\widehat{c}_n = \varphi_n(x_n) = \frac{1}{p}\|x_n'\|_{L^p(T_n;\mathbb{R}^N)}^p + \frac{1}{p}\int\limits_{-nb}^{nb} g(t)\,\|x_n(t)\|_{\mathbb{R}^N}^p\, dt$$

$$- \int\limits_{-nb}^{nb} j\big(t, x_n(t)\big)\, dt \leq c_1. \tag{3.74}$$

Because $0 \in \partial\varphi_n(x_n)$, we have that

$$A(x_n) + g\,\|x_n\|_{\mathbb{R}^N}^{p-2}\, x_n = u_n,$$

with $u_n \in S^\infty_{\partial j(\cdot, x_n(\cdot))}$. Taking the duality brackets for the pair $\left(W^{1,p}(T_n; \mathbb{R}^N), \left(W^{1,p}(T_n; \mathbb{R}^N)\right)^*\right)$ with $-x_n$, we obtain

$$
- \|x'_n\|^p_{L^p(T_n; \mathbb{R}^N)} - \frac{1}{p} \int_{-nb}^{nb} g(t) \|x_n(t)\|^p_{\mathbb{R}^N} \, dt
$$

$$
= \int_{-nb}^{nb} \left(u_n(t), -x_n(t)\right)_{\mathbb{R}^N} \, dt \leq \int_{-nb}^{nb} j^0\left(t, x_n(t); -x_n(t)\right) dt. \quad (3.75)
$$

Multiplying (3.74) with μ and adding it to (3.75), we obtain

$$
\left(\frac{\mu}{p} - 1\right) \|x'_n\|^p_{L^p(T_n; \mathbb{R}^N)} + \left(\frac{\mu}{p} - 1\right) \int_{-nb}^{nb} g(t) \|x_n(t)\|^p_{\mathbb{R}^N} \, dt
$$

$$
+ \int_{-nb}^{nb} \left(- j^0\left(t, x_n(t); -x_n(t)\right) - \mu j\left(t, x_n(t)\right) \right) dt \leq \mu \widehat{c}_1.
$$

Using hypotheses $H(j)_6(iii)$ and (iv), we see that

$$
\left(\frac{\mu}{p} - 1\right) \|x'_n\|^p_{L^p(T_n; \mathbb{R}^N)} + \left(\frac{\mu}{p} - 1\right) \int_{-nb}^{nb} g(t) \|x_n(t)\|^p_{\mathbb{R}^N} \leq c_{19},
$$

for some $c_{19} > 0$ independent of $n \geq 1$ and so

$$
\|x_n\|_{W^{1,p}(T_n; \mathbb{R}^N)} \leq c_{20} \qquad \forall\, n \geq 1, \quad (3.76)
$$

for some $c_{20} > 0$ independent of $n \geq 1$. Moreover,

$$
\|x_n(t)\|_{\mathbb{R}^N} \leq \|x_n(\tau)\|_{\mathbb{R}^N} + \left\| \int_\tau^t x'_n(s) ds \right\|_{\mathbb{R}^N} \qquad \forall\, t, \tau \in [-nb, nb].
$$

Integrating over $\left[t - \frac{1}{2}, t + \frac{1}{2}\right]$, we obtain

$$
\|x_n(t)\|_{\mathbb{R}^N} \leq \int_{t-\frac{1}{2}}^{t+\frac{1}{2}} \|x_n(\tau)\|_{\mathbb{R}^N} \, d\tau + \int_{t-\frac{1}{2}}^{t+\frac{1}{2}} \left\| \int_\tau^t x'_n(s) ds \right\|_{\mathbb{R}^N} d\tau
$$

$$
\leq 2^{p-1} \left(\int_{t-\frac{1}{2}}^{t+\frac{1}{2}} \left(\|x'_n(s)\|^p_{\mathbb{R}^N} + \|x_n(s)\|^p_{\mathbb{R}^N} \right) ds \right)^{\frac{1}{p}},
$$

from which it follows that

$$\|x_n\|_{L^\infty(T_n;\mathbb{R}^n)} \leq c_{21} \qquad \forall\, n \geq 1, \tag{3.77}$$

for some $c_{21} > 0$ not dependent on $n \geq 1$. We extend by periodicity x_n and u_n to all of $\mathbb{R}$. From (3.76) and since the embedding $W^{1,p}(T_n;\mathbb{R}^N) \subseteq C(T_n;\mathbb{R}^N)$ is compact, we may assume the

$$x_n \longrightarrow x \quad \text{in } C_{\text{loc}}(\mathbb{R};\mathbb{R}^N), \tag{3.78}$$

hence $x \in C(\mathbb{R};\mathbb{R}^n)$. Also because of hypothesis $H(j)_6(iii)$ and (3.77), we have

$$\|u_n(t)\|_{\mathbb{R}^N} \leq \|a_0\|_\infty \left(1 + \|x_n(t)\|_{\mathbb{R}^N}^{r-1}\right) \leq c_{22} \quad \text{for a.a. } t \in \mathbb{R} \text{ and all } n \geq 1,$$

with $c_{22} \overset{df}{=} \|a_0\|_\infty \left(1 + c_{21}^{r-1}\right)$. Note that $c_{22} > 0$ is independent of $n \geq 1$. So by the Alaoglu Theorem (see Theorem A.3.1), we may assume that

$$u_n \overset{w^*}{\longrightarrow} u \text{ in } L^\infty(\mathbb{R};\mathbb{R}^N),$$
$$u_n \longrightarrow u \text{ in } L^{p'}(T_m;\mathbb{R}^N), \quad \forall m \geq 1,$$

where $\frac{1}{p} + \frac{1}{p'} = 1$. Evidently $u \in L^\infty(\mathbb{R};\mathbb{R}^N) \cap L^{p'}_{\text{loc}}(\mathbb{R};\mathbb{R}^N)$, while by using Proposition 1.2.12, we have that

$$u(t) \in \partial j(t, x(t)) \quad \text{for a.a. } t \in T_m \text{ and all } m \geq 1$$

and so

$$u(t) \in \partial j(t, x(t)) \quad \text{for a.a. } t \in \mathbb{R}$$

(recall that $\text{Gr}\,\partial j(t, \cdot)$ is closed; see Proposition 1.3.9). For every $\tau > 0$, we have that

$$\int_{-\tau}^{\tau} \|x_n(t) - x(t)\|_{\mathbb{R}^N}^p \, dt \longrightarrow 0,$$

so

$$\lim_{n \to +\infty} \int_{-\tau}^{\tau} \|x_n(t)\|_{\mathbb{R}^N}^p \, dt = \int_{-\tau}^{\tau} \|x(t)\|_{\mathbb{R}^N}^p \, dt.$$

We can find $n_0 \geq 1$, such that for all $n \geq n_0$, we have $[-\tau, \tau] \subseteq T_{n_0}$ and then using (3.76), we have

$$\int_{-\tau}^{\tau} \|x_n(t)\|_{\mathbb{R}^N}^p \, dt \leq \int_{-n_0 b}^{n_0 b} \|x_n(t)\|_{\mathbb{R}^N}^p \leq c_{20}^p$$

and so

$$\int_{-\tau}^{\tau} \|x(t)\|_{\mathbb{R}^N}^p \, dt \leq c_{20}^p.$$

Because $\tau > 0$ was arbitrary, we infer that $x \in L^p(\mathbb{R}; \mathbb{R}^N)$.

Next let $\vartheta \in C_0^\infty(\mathbb{R}; \mathbb{R}^N)$. Then $\operatorname{supp} \vartheta \subseteq T_n$ for some $n \geq 1$. Performing an integration by parts, we have

$$\left| \int_{\mathbb{R}} \big(x_n(t), \vartheta'(t)\big)_{\mathbb{R}^N} \, dt \right| = \left| \int_{\mathbb{R}} \big(x_n'(t), \vartheta(t)\big)_{\mathbb{R}^N} \, dt \right|,$$

hence we have

$$\left| \int_{\mathbb{R}} \big(x_n(t), \vartheta'(t)\big)_{\mathbb{R}^N} \, dt \right| = \left| \int_{-nb}^{nb} \big(x_n'(t), \vartheta(t)\big)_{\mathbb{R}^N} \, dt \right|$$

$$\leq \|x_n'\|_{L^p(T_n; \mathbb{R}^N)} \|\vartheta\|_{L^{p'}(T_n; \mathbb{R}^N)} \leq c_{20} \|\vartheta\|_{L^{p'}(T_n; \mathbb{R}^N)}.$$

Note that

$$\big(x_n(t), \vartheta'(t)\big)_{\mathbb{R}^N} \longrightarrow \big(x(t), \vartheta'(t)\big)_{\mathbb{R}^N} \quad \text{uniformly on compact sets,}$$

i.e. we have convergence in $C_{\text{loc}}(\mathbb{R}; \mathbb{R}^N)$ and

$$\left| \big(x_n(t), \vartheta'(t)\big)_{\mathbb{R}^N} \right| \leq \|x_n\|_{L^\infty(T_n; \mathbb{R}^n)} \|\vartheta'(t)\|_{\mathbb{R}^N}$$

$$\leq c_{21} \|\vartheta'(t)\|_{\mathbb{R}^N} \quad \text{for a.a. } t \in T_n.$$

Let us set

$$\eta(t) \stackrel{df}{=} \begin{cases} c_{21} \|\vartheta'(t)\|_{\mathbb{R}^N} & \text{for } t \in \operatorname{supp} \vartheta, \\ 0 & \text{otherwise.} \end{cases}$$

Evidently $\eta \in L^1(\mathbb{R})$ and we have

$$\left| \big(x_n(t), \vartheta'(t)\big)_{\mathbb{R}^N} \right| \leq \eta(t) \quad \text{for a.a. } t \in \mathbb{R}.$$

So from the dominated convergence theorem, we have

$$\int_{\mathbb{R}} \big(x_n(t), \vartheta'(t)\big)_{\mathbb{R}^N} \, dt \longrightarrow \int_{\mathbb{R}} \big(x(t), \vartheta'(t)\big)_{\mathbb{R}^N} \, dt,$$

hence

$$\left| \int_{\mathbb{R}} \big(x_n(t), \vartheta'(t)\big)_{\mathbb{R}^N} \, dt \right| \leq c_{21} \|\vartheta\|_{L^{p'}(\mathbb{R}; \mathbb{R}^N)}$$

from which it follows that $x \in W^{1,p}(\mathbb{R}; \mathbb{R}^N)$ (see Proposition 1.1.1).

Also since

$$u_n \xrightarrow{w} u \quad \text{in } L_{\text{loc}}^{p'}(\mathbb{R}; \mathbb{R}^N),$$

we have that

$$\int_{\mathbb{R}} \big(u_n(t), \vartheta(t)\big)_{\mathbb{R}^N} dt \;\longrightarrow\; \int_{\mathbb{R}} \big(u(t), \vartheta(t)\big)_{\mathbb{R}^N} dt,$$

while from (3.78), we infer that

$$\int_{\mathbb{R}} g(t)\, \|x_n(t)\|_{\mathbb{R}^N}^{p-2} \big(x_n(t), \vartheta(t)\big)_{\mathbb{R}^N} dt \;\longrightarrow\; \int_{\mathbb{R}} g(t)\, \|x(t)\|_{\mathbb{R}^N}^{p-2} \big(x(t), \vartheta(t)\big)_{\mathbb{R}^N} dt.$$

Moreover, by integration by parts, we show that

$$\int_{\mathbb{R}} \big(\|x_n'(t)\|_{\mathbb{R}^N}^{p-2}\, x_n'(t), \vartheta'(t)\big)_{\mathbb{R}^N} dt$$
$$= \; -\int_{\mathbb{R}} \Big(\big(\|x_n'(t)\|_{\mathbb{R}^N}^{p-2}\, x_n'(t)\big)', \vartheta(t)\Big)_{\mathbb{R}^N} dt \qquad \forall\, n \geq 1.$$

Because $x_n \in C^1\big(T_n; \mathbb{R}^N\big)$ is a solution of (3.70), we see that

$$\big(\|x_n'(\cdot)\|_{\mathbb{R}^N}^{p-2}\, x_n'(\cdot)\big)' \in L^{p'}\big(T_n; \mathbb{R}^N\big) \qquad \forall\, n \geq 1.$$

So $\|x_n'(\cdot)\|_{\mathbb{R}^N}^{p-2}\, x_n'(\cdot) \in W^{1,p'}\big(T_n; \mathbb{R}^N\big)$ for $n \geq 1$. Thus we may assume that

$$\|x_n'(\cdot)\|_{\mathbb{R}^N}^{p-2}\, x_n'(\cdot) \;\xrightarrow{\;w\;}\; v \quad \text{in } W_{\mathrm{loc}}^{1,p'}\big(\mathbb{R}; \mathbb{R}^N\big),$$

for some $v \in W^{1,p'}\big(\mathbb{R}; \mathbb{R}^N\big)$, hence

$$\|x_n'(\cdot)\|_{\mathbb{R}^N}^{p-2}\, x_n'(\cdot) \;\longrightarrow\; v \text{ in } L_{\mathrm{loc}}^{p'}\big(\mathbb{R}; \mathbb{R}^N\big),$$
$$\|x_n'(\cdot)\|_{\mathbb{R}^N}^{p-2}\, x_n'(\cdot) \;\longrightarrow\; v \text{ in } C_{\mathrm{loc}}\big(\mathbb{R}; \mathbb{R}^N\big).$$

As in previous proofs (see Section 3.2), in the limit as $n \to +\infty$, we obtain

$$-\big(\|x'(t)\|_{\mathbb{R}^N}^{p-2}\, x'(t)\big)' + g(t)\, \|x(t)\|_{\mathbb{R}^N}^{p-2}\, x(t) \;=\; u(t) \quad \text{for a.a. } t \in \mathbb{R},$$

with $u \in L^\infty\big(\mathbb{R}; \mathbb{R}^N\big) \cap L_{\mathrm{loc}}^{p'}\big(\mathbb{R}; \mathbb{R}^N\big)$ and $u(t) \in \partial j\big(t, x(t)\big)$ for almost all $t \in \mathbb{R}$.

Next, we show that $x(\pm\infty) = x'(\pm\infty) = 0$. Recall that $x \in W^{1,p}\big(\mathbb{R}; \mathbb{R}^N\big)$. So from Remark 1.1.8, we have that $\|x(t)\|_{\mathbb{R}^N} \longrightarrow 0$ as $|t| \to +\infty$ and so $x(\pm\infty) = 0$. Since $u(t) \in \partial j\big(t, x(t)\big)$ for almost all $t \in \mathbb{R}$, from hypothesis $H(j)_6(iii)$, we have that

$$\|u(t)\|_{\mathbb{R}^N} \;\leq\; a_0(t)\big(1 + \|x(t)\|_{\mathbb{R}^N}^{p-1}\big) \quad \text{for a.a. } t \in T.$$

Because $x \in W^{1,p}\big(\mathbb{R}; \mathbb{R}^N\big)$, we have that $\|x(\cdot)\|_{\mathbb{R}^N}^{p-2}\, x(\cdot) \in L^{p'}\big(\mathbb{R}; \mathbb{R}^N\big)$ and so we deduce that $u \in L^{p'}\big(\mathbb{R}; \mathbb{R}^N\big)$. Therefore, it follows that $\|x'(\cdot)\|_{\mathbb{R}^N}^{p-2}\, x'(\cdot) \in W^{1,p'}\big(\mathbb{R}; \mathbb{R}^N\big)$ and so

$$\|x'(t)\|_{\mathbb{R}^N}^{p-1} \;\longrightarrow\; 0 \quad \text{as} \quad |t| \to +\infty$$

(see Remark 1.1.8), hence $x'(\pm\infty) = 0$.

It remains to show that x is nontrivial. We have

$$A(x_n) + g\,\|x_n\|_{\mathbb{R}^N}^{p-2}\,x_n \;=\; u_n \qquad \forall\, n \geq 1,$$

hence

$$c\,\|x_n\|_{L^p\left(T_n;\mathbb{R}^N\right)}^p \;\leq\; \int_{-nb}^{nb} \big(u_n(t), x_n(t)\big)_{\mathbb{R}^N} \, dt.$$

Set

$$h_n(t) \;\stackrel{df}{=}\; \begin{cases} \dfrac{\big(u_n(t), x_n(t)\big)_{\mathbb{R}^N}}{\|x_n(t)\|_{\mathbb{R}^N}^p} & \text{if } x_n(t) \neq 0, \\[2mm] 0 & \text{if } x_n(t) = 0. \end{cases}$$

We have

$$c\,\|x_n\|_{L^p\left(T_n;\mathbb{R}^N\right)}^p \;\leq\; \int_{-nb}^{nb} \big(u_n(t), x_n(t)\big)_{\mathbb{R}^N} \, dt$$

$$= \int_{-nb}^{nb} h_n(t)\,\|x_n(t)\|_{\mathbb{R}^N}^p \, dt \;\leq\; \operatorname*{esssup}_{T_n} h_n \, \|x_n\|_{L^p\left(T_n;\mathbb{R}^N\right)}^p.$$

By virtue of hypothesis $H(j)_6(v)$, for a given $\varepsilon > 0$ we can find $\delta > 0$, such that

$$\frac{(u, \xi)_{\mathbb{R}^N}}{\|\xi\|_{\mathbb{R}^N}^p} \;\leq\; \varepsilon \quad \text{for a.a. } t \in \mathbb{R}, \text{ all } \xi \in \mathbb{R}^N,\ \|\xi\|_{\mathbb{R}^N} \leq \delta \text{ and all } u \in \partial j(t,\xi).$$

If $x = 0$ then

$$x_n \;\longrightarrow\; 0 \quad \text{in } C_{\mathrm{loc}}\big(\mathbb{R};\mathbb{R}^N\big)$$

and so we can find $n_0 \geq 1$, such that

$$\|x_n(t)\|_{\mathbb{R}^N} \leq \delta \qquad \forall\, t \in T_n,\ n \geq n_0.$$

Thus for all $n \geq n_0$ and almost all $t \in T_n$, we have $h_n(t) \leq \varepsilon$ and so

$$c \;\leq\; \operatorname*{esssup}_{T_n} h_n \;=\; \operatorname*{esssup}_{\mathbb{R}} h_n \;\leq\; \varepsilon \qquad \forall\, n \geq n_0.$$

(recall that x_n and u_n were extended by periodicity to all of $\mathbb{R}$). Let $\varepsilon \searrow 0$ to reach a contradiction since $0 < c$. This proves that $x \neq 0$.

So $x \in C\big(\mathbb{R};\mathbb{R}^n\big) \cap W^{1,p}\big(\mathbb{R};\mathbb{R}^N\big)$ is a nontrivial, homoclinic (to zero) solution of (3.72). ∎

3.4.3 Scalar Problems

We turn our attention to scalar problems (i.e. $N = 1$). We start with the following problem:

$$\begin{cases} -\big(|x'(t)|^{p-2}x'(t)\big)' \in \partial j\big(t,x(t)\big) - h(t) & \text{for a.a. } t \in T, \\ x(0) = x(b), \ x'(0) = x'(b), \end{cases} \tag{3.79}$$

with $p \in (1,+\infty)$ and $h \in L^1(T)$. We solve (3.79) using a generalized Landesman-Lazer condition, which as we show generalizes all previous such conditions existing in the literature.

The hypotheses on the nonsmooth potential are the following:

<u>$H(j)_7$</u> $j \colon T \times \mathbb{R} \longrightarrow \mathbb{R}$ is a function, such that

(i) for all $\zeta \in \mathbb{R}$, the function

$$T \ni t \longmapsto j(t,\zeta) \in \mathbb{R}$$

is measurable and $j(\cdot,0) \in L^1(T)$;

(ii) for almost all $t \in T$, the function

$$\mathbb{R} \ni \zeta \longmapsto j(t,\zeta) \in \mathbb{R}$$

is locally Lipschitz;

(iii) for almost all $t \in T$, all $\zeta \in \mathbb{R}$ and all $u \in \partial j(t,\zeta)$, we have

$$|u| \ \leq \ a(t) + c(t)|\zeta|^{r-1},$$

with $r \in [1,+\infty)$, $\frac{1}{r} + \frac{1}{r'} = 1$, $a,c \in L^\infty(T)_+$;

(iv) we have

$$\lim_{|\zeta|\to+\infty} \frac{u(\zeta)}{\zeta} = 0$$

uniformly for almost all $t \in T$ and all $u(\zeta) \in \partial j(t,\zeta)$;

(v) there exist functions $j_+, j_- \in L^1(T)$, such that

$$j_+(t) \ = \ \liminf_{\zeta\to+\infty} \frac{j(t,\zeta)}{\zeta} \quad \text{and} \quad j_-(t) \ = \ \limsup_{\zeta\to-\infty} \frac{j(t,\zeta)}{\zeta}$$

uniformly for almost all $t \in T$ and

$$\int_0^b j_-(t)\,dt \ < \ \int_0^b h(t)\,dt \ < \ \int_0^b j_+(t)\,dt.$$

REMARK 3.4.6　Hypothesis $H(j)_7(v)$ is the Landesman-Lazer type condition, which we shall see later it generalizes previous ones of this type.　□

We introduce the nonsmooth, locally Lipschitz functional corresponding to problem (3.79). So let $\varphi\colon W^{1,p}_{\mathrm{per}}(T) \longrightarrow \mathbb{R}$ be defined by

$$\varphi(x) \overset{df}{=} \frac{1}{p}\,\|x'\|^p_p - \int_0^b j\big(t, x(t)\big)\,dt + \int_0^b h(t)x(t)\,dt.$$

PROPOSITION 3.4.8
If hypotheses $H(j)_7$ hold,
then φ satisfies the nonsmooth PS-condition.

PROOF　We consider a Palais-Smale sequence $\{x_n\}_{n\geq 1} \subseteq W^{1,p}_{\mathrm{per}}(T)$. We have

$$\big|\varphi(x_n)\big| \leq M_1 \quad \text{and} \quad m^\varphi(x_n) \longrightarrow 0 \quad \text{as } n \to +\infty,$$

for some $M_1 > 0$.

Let $x_n^* \in \partial\varphi(x_n)$ be such that $m^\varphi(x_n) = \|x_n^*\|_{(W^{1,p}_{\mathrm{per}}(T))^*}$ for $n \geq 1$. We have

$$x_n^* = A(x_n) - u_n + h,$$

with $u_n \in S^{r'}_{\partial j(\cdot, x_n(\cdot))}$.

We shall show that the sequence $\{x_n\}_{n\geq 1} \subseteq W^{1,p}_{\mathrm{per}}(T)$ is bounded. Suppose that this is not the case. We may assume that $\|x_n\|_{W^{1,p}(T;\mathbb{R}^N)} \longrightarrow +\infty$. Let us set

$$y_n \overset{df}{=} \frac{x_n}{\|x_n\|_{W^{1,p}(T;\mathbb{R}^N)}} \qquad \forall\, n \geq 1.$$

We may assume that

$$y_n \overset{w}{\longrightarrow} y \quad \text{in } W^{1,p}_{\mathrm{per}}(T),$$
$$y_n \longrightarrow y \quad \text{in } C_{\mathrm{per}}(T).$$

We have

$$\frac{\big|\varphi(x_n)\big|}{\|x_n\|^p_{W^{1,p}(T;\mathbb{R}^N)}} = \left| \frac{1}{p}\|y_n'\|^p_p - \int_0^b \frac{j\big(t, x_n(t)\big)}{\|x_n\|^p_{W^{1,p}(T;\mathbb{R}^N)}}\,dt \right.$$
$$\left. + \int_0^b \frac{h(t)}{\|x_n\|^{p-1}_{W^{1,p}(T;\mathbb{R}^N)}}\,y_n(t)\,dt \right|. \tag{3.80}$$

Note that

$$\frac{j\big(t, x_n(t)\big)}{\|x_n\|^p_{W^{1,p}(T;\mathbb{R}^N)}} \longrightarrow 0 \quad \text{as } n \to +\infty$$

(see hypothesis $H(j)_7(v)$). So $\frac{1}{p}\|y'\|_p^p \leq 0$, hence $y \equiv \overline{\zeta} \in \mathbb{R}$. If $\overline{\zeta} = 0$, we have $y_n \longrightarrow 0$ in $W_{per}^{1,p}(T)$, a contradiction to the fact that $\|y_n\|_{W^{1,p}(T;\mathbb{R}^N)} = 1$. Therefore $\overline{\zeta} \neq 0$. Without any loss of generality, we may assume that $\overline{\zeta} > 0$ (the analysis is similar if $\overline{\zeta} < 0$). We know that

$$x_n = \overline{x}_n + \widehat{x}_n, \quad \text{with } \overline{x}_n \in \mathbb{R} \quad \text{and} \quad \int_0^b \widehat{x}_n(t)\,dt = 0.$$

Then

$$y_n = \overline{y}_n + \widehat{y}_n \quad \text{with } \overline{y}_n \stackrel{df}{=} \frac{\overline{x}_n}{\|x_n\|_{W^{1,p}(T;\mathbb{R}^N)}} \quad \forall\, n \geq 1$$

and

$$\widehat{y}_n \stackrel{df}{=} \frac{\widehat{x}_n}{\|x_n\|_{W^{1,p}(T;\mathbb{R}^N)}} \quad \forall\, n \geq 1.$$

From the choice of the sequence $\{x_n\}_{n\geq 1} \subseteq W_{per}^{1,p}(T)$, we have

$$-\varepsilon_n \leq \frac{1}{\|x_n\|_{W^{1,p}(T;\mathbb{R}^N)}} \left[\|\widehat{x}_n'\|_p^p - \int_0^b u_n(t)\widehat{x}_n(t)\,dt + \int_0^b h(t)\widehat{x}_n(t)\,dt \right] \leq \varepsilon_n,$$
$$(3.81)$$

with $\varepsilon_n \searrow 0$. Note that

$$x_n(t) \longrightarrow +\infty \quad \forall\, t \in T$$

(since we have assumed that $\overline{\zeta} > 0$). As in the proof of Proposition 3.4.5, we can show that

$$\min_T x_n \longrightarrow +\infty. \qquad (3.82)$$

We have

$$\int_0^b u_n(t)\widehat{x}_n(t)\,dt = \int_{\{x_n \neq 0\}} \frac{u_n(t)}{x_n(t)} x_n(t)\widehat{x}_n(t)\,dt.$$

Evidently $|\{x_n \neq 0\}|_1 \longrightarrow b$ as $n \to +\infty$ (recall that $|\cdot|_1$ denotes the Lebesgue measure on $\mathbb{R}$). Also by virtue of hypothesis $H(j)_7(iv)$, for a given $\varepsilon > 0$ we can find $n_0 \geq 1$, such that

$$\left| \frac{u_n(t)}{x_n(t)} \right| \leq \varepsilon \quad \text{for a.a. } t \in T \text{ and all } n \geq n_0.$$

So

$$\left| \int_{\{x_n \neq 0\}} \frac{u_n(t)}{x_n(t)} x_n(t)\widehat{x}_n(t)\,dt \right| \leq \varepsilon \|\widehat{x}_n\|_2^2 \leq \varepsilon c_{23} \|\widehat{x}_n\|_{W^{1,p}(T;\mathbb{R}^N)}^2 \quad \forall\, n \geq 1,$$

for some $c_{23} > 0$. Using this and the Poincaré-Wirtinger inequality (see Theorem 1.1.7) in (3.81), we obtain

$$\frac{1}{\|x_n\|_{W^{1,p}(T;\mathbb{R}^N)}} \left[c_{24} \|\widehat{x}_n\|^p_{W^{1,p}(T;\mathbb{R}^N)} - \varepsilon c_{23} \|\widehat{x}_n\|^p_{W^{1,p}(T;\mathbb{R}^N)} \right] \leq \varepsilon_n \qquad \forall\, n > 1,$$

for some $c_{24} > 0$. So

$$\left(c_{24} - \varepsilon c_{23} \right) \frac{\|\widehat{x}_n\|^p_{W^{1,p}(T;\mathbb{R}^N)}}{\|x_n\|_{W^{1,p}(T;\mathbb{R}^N)}} \leq \varepsilon_n.$$

We choose $\varepsilon \in \left(0, \frac{c_{24}}{c_{23}} \right)$. We obtain that $\frac{\|\widehat{x}_n\|^p_{W^{1,p}(T;\mathbb{R}^N)}}{\|x_n\|_{W^{1,p}(T;\mathbb{R}^N)}} \longrightarrow 0$ and so again from the Poincaré-Wirtinger inequality (see Theorem 1.1.7), we have

$$\frac{\|x'_n\|^p_p}{\|x_n\|_{W^{1,p}(T;\mathbb{R}^N)}} \longrightarrow 0 \quad \text{as } n \to +\infty. \tag{3.83}$$

For every $n \geq 1$, we have

$$\left| \frac{1}{p} \frac{\|x'_n\|^p_p}{\|x_n\|_{W^{1,p}(T;\mathbb{R}^N)}} - \int_0^b \frac{j\big(t, x_n(t)\big)}{\|x_n\|_{W^{1,p}(T;\mathbb{R}^N)}}\, dt + \int_0^b h(t) y_n(t)\, dt \right|$$

$$\leq \frac{M_1}{\|x_n\|_{W^{1,p}(T;\mathbb{R}^N)}} \tag{3.84}$$

(recall that $\|x_n\|_{W^{1,p}(T;\mathbb{R}^N)} \longrightarrow +\infty$). Note that

$$\int_0^b \frac{j\big(t, x_n(t)\big)}{\|x_n\|_{W^{1,p}(T;\mathbb{R}^N)}}\, dt$$

$$= \int_{\{x_n \neq 0\}} \frac{j\big(t, x_n(t)\big)}{\|x_n\|_{W^{1,p}(T;\mathbb{R}^N)}}\, dt + \int_{\{x_n = 0\}} \frac{j(t, 0)}{\|x_n\|_{W^{1,p}(T;\mathbb{R}^N)}}\, dt. \tag{3.85}$$

Note that

$$\int_{\{x_n = 0\}} \frac{j(t, 0)}{\|x_n\|_{W^{1,p}(T;\mathbb{R}^N)}}\, dt \longrightarrow 0 \quad \text{as } n \to +\infty.$$

Also from hypothesis $H(j)_7(iv)$, we know that for a given $\varepsilon > 0$, we can find $M_2 > 0$, such that

$$\frac{j(t, \zeta)}{\zeta} \geq j_+(t) - \varepsilon \quad \text{for a.a. } t \in T \text{ and all } \zeta \geq M_2.$$

Moreover, because $\min_T x_n \longrightarrow +\infty$, we can find $n_1 \geq 1$, such that

$$j_+(t) - \varepsilon \leq \frac{j\big(t, x_n(t)\big)}{x_n(t)} \quad \text{for a.a. } t \in T \text{ and all } n \geq n_1.$$

We can assume that $y_n(t) > 0$ for all $t \in T$ and so

$$\int\limits_{\{x_n \neq 0\}} (j_+(t) - \varepsilon) y_n(t) \, dt \leq \int\limits_{\{x_n \neq 0\}} \frac{j(t, x_n(t))}{x_n(t)} y_n(t) \, dt.$$

Since $\varepsilon > 0$ was arbitrary, we have

$$\int\limits_0^b j_+(t) \widehat{\zeta} \, dt \leq \liminf_{n \to +\infty} \int\limits_{\{x_n \neq 0\}} \frac{j(t, x_n(t))}{x_n(t)} y_n(t) \, dt.$$

Now from (3.85), we obtain

$$\int\limits_0^b j_+(t) \widehat{\zeta} \, dt \leq \liminf_{n \to +\infty} \int\limits_0^b \frac{j(t, x_n(t))}{x_n(t)} y_n(t) \, dt.$$

Returning to (3.84), using this inequality and (3.83), we obtain

$$\int\limits_0^b j_+(t) \, dt \leq \int\limits_0^b h(t) \, dt,$$

a contradiction to hypothesis $H(j)_7(v)$. So the sequence $\{x_n\}_{n \geq 1} \subseteq W^{1,p}_{\mathrm{per}}(T)$ is bounded and as before we conclude that φ satisfies the nonsmooth Palais-Smale condition. ∎

PROPOSITION 3.4.9
If hypotheses $H(j)_7$ hold,
then $\lim\limits_{\substack{|\zeta| \to +\infty \\ \zeta \in \mathbb{R}}} \varphi(\zeta) = -\infty$, *i.e.* $\varphi|_{\mathbb{R}}$ *is anticoercive.*

PROOF Suppose that the proposition is not true. We can find a sequence $\{\zeta_n\}_{n \geq 1} \subseteq \mathbb{R}$, such that $|\zeta_n| \longrightarrow +\infty$ and $\gamma \leq \varphi(\zeta_n)$ for some $\gamma \in \mathbb{R}$ and all $n \geq 1$. First suppose that $\zeta_n \longrightarrow +\infty$. We may assume that $\zeta_n > 0$ for all $n \geq 1$. We have

$$\frac{\gamma}{\varepsilon_n} \leq -\int\limits_0^b \frac{j(t, \zeta_n)}{\zeta_n} \, dt + \int\limits_0^b h(t) \, dt$$

and so

$$\limsup_{n \to +\infty} \int\limits_0^b \frac{j(t, \zeta_n)}{\zeta_n} \, dt \leq \int\limits_0^b h(t) \, dt,$$

from which, via Fatou's lemma, we obtain that

$$\int_0^b j_+(t)\, dt \;\leq\; \int_0^b h(t)\, dt,$$

a contradiction to hypothesis $H(j)_7(v)$. Similarly, we treat the case $\zeta_n \longrightarrow -\infty$. $\quad\square$

Let us recall that

$$V \;=\; \left\{ v \in W^{1,p}_{\mathrm{per}}(T) : \int_0^b v(t)\, dt = 0 \right\}.$$

PROPOSITION 3.4.10
If hypotheses $H(j)_7$ hold, <u>then</u> $\varphi|_V$ is coercive.

PROOF   We have

$$\varphi(v) \;=\; \frac{1}{p}\, \|v'\|_p^p - \int_0^b j\big(t, v(t)\big)\, dt + \int_0^b h(t) v(t)\, dt \qquad \forall\, v \in V$$

and so, by the Poincaré-Wirtinger inequality (see Theorem 1.1.7), we have

$$
\frac{\varphi(v)}{\|v\|^p_{W^{1,p}(T;\mathbb{R}^N)}}
$$
$$
\geq\; c_{25} - \int_0^b \frac{j\big(t, v(t)\big)}{\|v\|^p_{W^{1,p}(T;\mathbb{R}^N)}}\, dt + \int_0^b \frac{h(t)}{\|v\|^{p-1}_{W^{1,p}(T;\mathbb{R}^N)}} \frac{v(t)}{\|v\|_{W^{1,p}(T;\mathbb{R}^N)}}\, dt,
$$

for some $c_{25} > 0$. If $\|v\|_{W^{1,p}(T;\mathbb{R}^N)} \longrightarrow +\infty$, we obtain that

$$\liminf_{\|v\|_{W^{1,p}(T;\mathbb{R}^N)} \to +\infty} \frac{\varphi(v)}{\|v\|^p_{W^{1,p}(T;\mathbb{R}^N)}} \;\geq\; c_{25} \;>\; 0$$

and so

$$\varphi(v) \;\longrightarrow\; +\infty \quad \text{as } \|v\|_{W^{1,p}(T;\mathbb{R}^N)} \to +\infty,$$

i.e. $\varphi|_V$ is coercive. $\quad\square$

Combining Propositions 3.4.8, 3.4.9 and 3.4.10, we see that we can apply the nonsmooth Saddle Point Theorem (see Theorem 2.1.4) and have the following existence theorem for problem (3.79).

THEOREM 3.4.7
If hypotheses $H(j)_7$ hold,
then problem (3.79) has a solution

$$x_0 \in C_{\text{per}}^1(T) \quad \text{with} \quad \|x_0'(\cdot)\|_{\mathbb{R}^N}^{p-2} x_0'(\cdot) \in W_{\text{per}}^{1,r'}(T).$$

In the literature, we can find the following two versions of the Landesman-Lazer-type condition:

$(LL)_1$ Let

$$g_1(t,\zeta) \stackrel{df}{=} \min\{u : u \in \partial j(t,\zeta)\},$$
$$g_2(t,\zeta) \stackrel{df}{=} \max\{u : u \in \partial j(t,\zeta)\}$$

and let

$$G_1(t,\zeta) \stackrel{df}{=} \begin{cases} \frac{pj(t,\zeta)}{\zeta} - g_1(t,\zeta) & \text{if } \zeta \neq 0, \\ 0 & \text{if } \zeta = 0, \end{cases}$$

$$G_2(t,\zeta) \stackrel{df}{=} \begin{cases} \frac{pj(t,\zeta)}{\zeta} - g_2(t,\zeta) & \text{if } \zeta \neq 0, \\ 0 & \text{if } \zeta = 0. \end{cases}$$

Let also

$$G_1^-(t) \stackrel{df}{=} \limsup_{\zeta \to -\infty} G_1(t,\zeta) \quad \text{and} \quad G_2^+(t) \stackrel{df}{=} \liminf_{\zeta \to +\infty} G_2(t,\zeta)$$

uniformly for almost all $t \in T$, with $G_1^-, G_2^+ \in L^1(T)$. We assume that

$$\int_0^b G_1^-(t)\,dt \; < \; (p-1)\int_0^b h(t)\,dt \; < \; \int_0^b G_2^+(t)\,dt.$$

$(LL)_2$ Let

$$g_-^\infty(t) \stackrel{df}{=} \inf_{\{u_n\}_{n\geq 1}} \left\{ \liminf_{n\to+\infty} u_n : u_n \in \partial j(t,\zeta_n), \; \zeta_n \longrightarrow +\infty \right\}$$

$$g_+^\infty(t) \stackrel{df}{=} \sup_{\{u_n\}_{n\geq 1}} \left\{ \limsup_{n\to+\infty} u_n : u_n \in \partial j(t,\zeta_n), \; \zeta_n \longrightarrow -\infty \right\}$$

uniformly for almost all $t \in T$, with $g_-^\infty, g_+^\infty \in L^1(T)$. We assume that

$$\int_0^b g_-^\infty(t)\,dt \; < \; \int_0^b h(t)\,dt \; < \; \int_0^b g_+^\infty(t)\,dt.$$

Also by (LL) we denote hypothesis $H(j)_6(v)$.

PROPOSITION 3.4.11
If hypotheses $H(F)_7(i), (ii)$ and (v) hold,
then

$$j_-(t) \leq g_-^\infty(t), \quad g_+^\infty(t) \leq j_+(t) \quad \text{for a.a. } t \in T.$$

PROOF   Let $N \subseteq T$ be the Lebesgue-null set, such that $j(t, \cdot)$ is locally Lipschitz for all $T \setminus N$ (see hypothesis $H(j)_7(ii)$). For all $t \in T \setminus N$, the function $j(t, \cdot)$ is differentiable at every $\zeta \in \mathbb{R} \setminus N(t)$, with $|N(t)|_1 = 0$ and we have

$$j_\zeta'(t, \zeta) \in \partial j(t, \zeta) \quad \forall \zeta \in \mathbb{R} \setminus N(t).$$

From the definition of g_+^∞ given in $(LL)_2$, we know that for a given $\varepsilon > 0$, we can find $M_3 > 0$, such that

$$g_+^\infty(t) - \varepsilon \leq j_\zeta'(t, \zeta) \quad \forall \zeta \geq M_3 > 0.$$

We have

$$\frac{j(t, \zeta) - j(t, 0)}{\zeta} = \frac{1}{\zeta} \int_0^\zeta j_\zeta'(t, r)dr = \frac{1}{\zeta} \int_0^{M_3} j_r'(t, r)dr + \frac{1}{\zeta} \int_{M_3}^\zeta j_r'(t, r)dr$$

$$\geq \frac{1}{\zeta} \int_\zeta^{M_3} j_r'(t, r)dr + \frac{\zeta - M_3}{\zeta} \left(g_+^\infty(t) - \varepsilon\right).$$

Passing to the limit as $\zeta \longrightarrow +\infty$, we obtain that

$$g_+^\infty(t) - \varepsilon \leq j_+(t) \quad \text{for a.a. } t \in T \setminus N.$$

Let $\varepsilon \searrow 0$, to obtain

$$g_+^\infty(t) \leq j_+(t) \quad \text{for a.a. } t \in T \setminus N.$$

Similarly we show that

$$j_-(t) \leq g_-^\infty(t) \quad \text{for a.a. } t \in T \setminus N.$$

$\square$

PROPOSITION 3.4.12
If hypotheses $H(j)_7(i), (ii)$ and (v) hold,
then

$$j_-(t) \leq \frac{1}{p-1}G_1^-(t) \quad \text{and} \quad \frac{1}{p-1}G_2^+(t) \leq j_+(t) \quad \text{for a.a. } t \in T.$$

PROOF   Let N be as in the proof of Proposition 3.4.11. Let $t \in T \setminus N$, $\varepsilon > 0$ and let us set

$$k_\varepsilon^+(t) \overset{df}{=} G_2^+(t) - \varepsilon.$$

From the definition of G_2^+ in $(LL)_1$, we know that we can find $M_4 > 0$, such that

$$G_2^+(t) - \varepsilon = k_\varepsilon^+(t) \leq G_2(t,r) \qquad \forall\, r \geq M_4 > 0.$$

So

$$
\begin{aligned}
\frac{k_\varepsilon^+(t)}{r^p} &= \frac{1}{p-1} \frac{d}{dt}\left(-\frac{k_\varepsilon^+(t)}{r^{p-1}}\right) \\
&\leq \frac{G_2(t,r)}{r^p} \leq \frac{pj(t,r)}{r^p} - \frac{u}{r^p} \qquad \forall\, u \in \partial j(t,r) \qquad (3.86)
\end{aligned}
$$

(see $(LL)_1$). From Corollary 1.3.7, we have

$$\partial\left(\frac{j(t,r)}{r^p}\right) \subseteq \frac{r\partial j(t,r) - pj(t,r)}{r^{p+1}} \qquad \forall\, r \geq M_4 > 0. \qquad (3.87)$$

Since the function $r \longmapsto \frac{j(t,r)}{r^p}$ is locally Lipschitz on $(M_4, +\infty)$, it is differentiable for all $\zeta \in \mathbb{R} \setminus N(t)$, with $N(t) \subseteq T$, $|N(t)|_1 = 0$. Let

$$
\vartheta_0(t,r) \overset{df}{=}
\begin{cases}
\dfrac{d}{dr}\left(\dfrac{j(t,r)}{r^p}\right) & \text{if } r \in (M_4, +\infty) \setminus N(t), \\
0 & \text{if } r \in N(t).
\end{cases}
$$

From (3.86) and (3.87), it follows that

$$\frac{1}{p-1} \frac{d}{dr}\left(-\frac{k_\varepsilon^+(t)}{r^{p-1}}\right) \leq -\vartheta_0(t,r) \qquad \forall\, t \in T \setminus N, \; r \in (M_4, +\infty) \setminus N(t).$$

Let $\zeta_1, \zeta_2 \in (M_4, +\infty)$ with $\zeta_1 < \zeta_2$ and integrate the last inequality over $[\zeta_1, \zeta_2]$. We obtain

$$\frac{1}{p-1}\left(-\frac{k_\varepsilon^+(t)}{\zeta_2^{p-1}} + \frac{k_\varepsilon^+(t)}{\zeta_1^{p-1}}\right) \leq -\frac{j(t,\zeta_2)}{\zeta_2^p} + \frac{j(t,\zeta_1)}{\zeta_1^p}.$$

Let $\zeta_2 \longrightarrow +\infty$. Since $\frac{j(t,\zeta_2)}{\zeta_2^p} \longrightarrow 0$ (see hypothesis $H(j)_7(v)$), we obtain

$$\frac{1}{p-1}k_\varepsilon^+(t) \leq \frac{j(t,\zeta_1)}{\zeta_1}.$$

Let $\zeta_1 \longrightarrow +\infty$ and then $\varepsilon \searrow 0$, to conclude that

$$\frac{1}{p-1}G_2^+(t) \leq j_+(t).$$

Similarly we show the second inequality.   ▯

As an immediate consequence of the last two propositions, we have the following result establishing the generality of the (LL)-condition.

PROPOSITION 3.4.13
*If hypotheses $H(j)_7(i), (ii)$ and (v) hold,
then conditions $(LL)_1$ and $(LL)_2$ imply (LL).*

EXAMPLE 3.4.4   The following nonsmooth potential function satisfies the (LL)-condition, but neither of the $(LL)_1$ and $(LL)_2$ (as before for simplicity, we drop the t-dependence):

$$j(\zeta) \overset{df}{=} \max\left\{\zeta^{\frac{1}{3}}, |\zeta|^{\frac{1}{2}}\right\} + \ln(1 + |\zeta|) + \cos\zeta + \zeta.$$

A simple calculation reveals that $j_+ \equiv 1$, $j_- \equiv -1$, but $g_+^\infty = g_-^\infty = G_1^- = G_2^+ \equiv 0$.
    The same can be checked for the potential:

$$j(\zeta) \overset{df}{=} \begin{cases} \ln(1 + |\zeta|) & \text{for } |\zeta| \leq 1, \\ \zeta - 1 + \cos\zeta + \ln 2 - \ln 1 & \text{for } \zeta > 1, \\ -\zeta + 1 + \cos\zeta + \ln 2 - \cos 1 - 2 & \text{for } \zeta < 1. \end{cases}$$

                                                           ▯

In the semilinear case (i.e. $p = 2$), exploiting the Hilbert space features of the problem, we can consider equations resonant at any eigenvalue. So the problem under consideration is the following:

$$\begin{cases} -x''(t) - m^2\omega^2 x(t) \in \partial j\big(t, x(t)\big) - h(t) & \text{for a.a. } t \in T, \\ x(0) = x(b), \ x'(0) = x'(b), \end{cases} \tag{3.88}$$

where $h \in L^1(T)$, $\omega = \frac{2\pi}{b}$, $m \in \mathbb{N}_0 \overset{df}{=} \mathbb{N} \cup \{0\}$. Our hypotheses for j are similar to those in $H(j)_7$.

$\underline{H(j)_8}$ $j: T \times \mathbb{R} \longrightarrow \mathbb{R}$ is a function such that

$\left.\begin{array}{l} (i) \\ (ii) \\ (iii) \\ (iv) \end{array}\right\}$ are the same as $H(F)_7(i), (ii), (iii)$ and (iv);

(v) there exist functions $j_+, j_- \in L^1(T)$, such that

$$j_+(t) = \liminf_{\zeta \to +\infty} \frac{j(t, \zeta)}{\zeta} \quad \text{and} \quad j_-(t) = \limsup_{\zeta \to -\infty} \frac{j(t, \zeta)}{\zeta}$$

uniformly for almost all $t \in T$ and for all $\vartheta \in \mathbb{R}$ we have

$$\int\limits_0^b h(t) \sin(m\omega t + \vartheta) \, dt$$

$$< \int\limits_0^b \left[j_+(t) \sin(m\omega t + \vartheta)^+ - j_-(t) \sin(m\omega t + \vartheta)^- \right] dt.$$

In the variational analysis of problem (3.88), we shall use the following subspace of $W^{1,2}_{\text{per}}(T)$:

$$\overline{H} \stackrel{df}{=} \text{span} \{\sin k\omega t, \, \cos k\omega t : \, k \in \{0, 1, \ldots, m-1\}\}$$

$$N_m \stackrel{df}{=} \text{span} \{\sin m\omega t, \, \cos m\omega t\}$$

and

$$\widehat{H} \stackrel{df}{=} (\overline{H} \oplus N_m)^\perp = \text{span} \{\sin k\omega t, \, \cos k\omega t : \, k \geq m+1\}.$$

REMARK 3.4.7 From Remark 1.5.1, we know that $\left\{ \left(\frac{2n\pi}{b}\right)^2 \right\}_{n \geq 0}$ are the eigenvalues of the negative ordinary Laplacian with periodic boundary conditions and N_m is the eigenspace corresponding to the m-th eigenvalue. Also, we have

$$\overline{H} \stackrel{df}{=} \bigoplus_{i=1}^{m-1} N_i \quad \text{and} \quad \widehat{H} \stackrel{df}{=} \bigoplus_{i=m+1}^{+\infty} N_i.$$

Moreover, we have

$$W^{1,2}_{\text{per}}(T) = \overline{H} \oplus N_m \oplus \widehat{H} \quad \forall \, m \in \mathbb{N}$$

and if $x \in W^{1,2}_{\text{per}}(T)$, we have

$$x = \overline{x} + x^0 + \widehat{x} \quad \text{with } \overline{x} \in \overline{H}, \, x^0 \in N_m \text{ and } \widehat{x} \in \widehat{H}$$

and this decomposition is unique. $\qquad\qquad\square$

LEMMA 3.4.1
There exists $\widehat{c} > 0$, such that

$$\widehat{c} \|x\|^2_{W^{1,2}(T)} \leq \|x'\|^2_2 - \lambda_m \|x\|^2_2 \quad \forall \, x \in \widehat{H}.$$

PROOF   Let

$$\psi(x) \stackrel{df}{=} \|x'\|_2^2 - \lambda_m \|x\|_2^2 \qquad \forall\, x \in \widehat{H}.$$

Suppose that the lemma is not true. We can find a sequence $\{x_n\}_{n \geq 1} \subseteq \widehat{H}$, such that $\psi(x_n) \searrow 0$. Let us set

$$y_n \stackrel{df}{=} \frac{x_n}{\|x_n\|_{W^{1,2}(T)}} \qquad \forall\, n \geq 1.$$

We may assume that

$$y_n \stackrel{w}{\longrightarrow} y \ \text{in } W^{1,2}_{\text{per}}(T),$$
$$y_n \longrightarrow y \ \text{in } L^2(T).$$

So in the limit we obtain

$$\|y'\|_2^2 \ \leq\ \lambda_m \|y\|_2^2.$$

Since $y \in \widehat{H}$, it follows that $y \equiv 0$ and so

$$y_n \ \longrightarrow\ 0 \quad \text{in } W^{1,2}_{\text{per}}(T),$$

a contradiction to the fact that $\|y_n\|_{W^{1,2}(T)} = 1$ for $n \geq 1$.   ☐

In a similar fashion, we can have a corresponding inequality for the space $\overline{H}$.

LEMMA 3.4.2
There exists $\overline{c} > 0$, such that

$$\overline{c} \|x\|_{W^{1,2}(T)}^2 \ \leq\ -\|x'\|_2^2 + \lambda_m \|x\|_2^2 \qquad \forall\, x \in \overline{H}.$$

The nonsmooth, locally Lipschitz energy functional $\varphi \colon W^{1,2}_{\text{per}}(T) \longrightarrow \mathbb{R}$ for problem (3.88) is given by

$$\varphi(x) \stackrel{df}{=} \frac{1}{2}\|x'\|_2^2 - \frac{m^2 w^2}{2}\|x\|_2^2 - \int_0^b j\big(t, x(t)\big)\, dt + \int_0^b h(t)x(t)\, dt \ \ \forall x \in W^{1,2}_{\text{per}}(T).$$

PROPOSITION 3.4.14
If hypotheses $H(j)_8$ hold, then φ satisfies the nonsmooth PS-condition.

PROOF   We consider a Palais-Smale sequence $\{x_n\}_{n \geq 1} \subseteq W^{1,2}_{\text{per}}(T)$ for φ. So we have

$$|\varphi(x_n)| \ \leq\ M_5 \qquad \forall\, n \geq 1 \quad \text{and} \quad m^\varphi(x_n) \ \longrightarrow\ 0 \quad \text{as } n \to +\infty,$$

for some $M_5 > 0$. For each $n \geq 1$, there exists $x_n^* \in \partial\varphi(x_n)$, such that $m^\varphi(x_n) = \|x_n^*\|_{(W_{\mathrm{per}}^{1,p}(T))^*}$. We know that

$$x_n^* = A(x_n) - m^2\omega^2 x_n - u_n + h \qquad \forall\, n \geq 1,$$

with $u_n \in S_{\partial j(\cdot, x_n(\cdot))}^{r'}$ and $A \in \mathcal{L}\big(W_{\mathrm{per}}^{1,2}(T), \big(W_{\mathrm{per}}^{1,2}(T)\big)^*\big)$ being defined by

$$\langle A(x), y \rangle_{W_{\mathrm{per}}^{1,2}(T)} \stackrel{df}{=} \int_0^b x'(t) y'(t)\, dt \qquad \forall\, x, y \in W_{\mathrm{per}}^{1,2}(T).$$

We show that the sequence $\{x_n\}_{n \geq 1} \subseteq W_{\mathrm{per}}^{1,2}(T)$ is bounded. Suppose that this is not true. Passing to a subsequence if necessary, we may assume that $\|x_n\|_{W^{1,2}(T)} \longrightarrow +\infty$. Let us set

$$y_n \stackrel{df}{=} \frac{x_n}{\|x_n\|_{W^{1,2}(T)}} \qquad \forall\, n \geq 1.$$

Passing to a further subsequence if necessary, we have

$$y_n \stackrel{w}{\longrightarrow} y \ \text{ in } W_{\mathrm{per}}^{1,2}(T) \tag{3.89}$$

$$y_n \longrightarrow y \ \text{ in } C_{\mathrm{per}}(T). \tag{3.90}$$

We have

$$\left| \langle x_n^*, y_n - y \rangle_{W_{\mathrm{per}}^{1,2}(T)} \right| \leq \varepsilon_n \|y_2 - y\|_{W^{1,2}(T)},$$

with $\varepsilon_n \searrow 0$ and so

$$\begin{aligned}
\Bigg| \int_0^b & y_n'(t)\big(y_n(t) - y(t)\big)'\, dt - \lambda_m \int_0^b y_n(t)\big(y_n(t) - y(t)\big)\, dt \\
& - \int_0^b \frac{u_n(t)}{\|x_n\|_{W^{1,2}(T)}} \big(y_n(t) - y(t)\big)\, dt + \int_0^b \frac{h(t)}{\|x_n\|_{W^{1,2}(T)}} \big(y_n(t) - y(t)\big)\, dt \Bigg| \\
\leq\ & \frac{\varepsilon_n \|y_n - y\|_{W^{1,2}(T)}}{\|x_n\|_{W^{1,2}(T)}},
\end{aligned} \tag{3.91}$$

with $\lambda_m = m^2\omega^2$. By virtue of hypothesis $H(j)_8(iv)$, for a given $\varepsilon > 0$, we can find $M_6 > 0$, such that

$$\left| \frac{u}{\zeta} \right| \leq \varepsilon \quad \text{ for a.a. } t \in T, \text{ all } \zeta \in \mathbb{R}, \ |\zeta| \geq M_6 \text{ and all } u \in \partial j(t, \zeta).$$

From hypothesis $H(j)_8(iii)$, it follows that

$$\left| \int_0^b \frac{u_n(t)}{\|x_n\|_{W^{1,2}(T)}} \big(y_n(t) - y(t)\big)\, dt \right|$$

$$= \left| \int_{\{|x_n| \geq M_6\}} \frac{u_n(t)}{x_n} y_n(t)\big(y_n(t) - y(t)\big)\, dt \right|$$

$$+ \left| \int_{\{|x_n| < M_6\}} \frac{u_n(t)}{\|x_n\|_{W^{1,2}(T)}} \big(y_n(t) - y(t)\big)\, dt \right|$$

$$\leq \varepsilon \|y_n\|_\infty \|y_n - y\|_1 + \int_{\{|x_n| < M_6\}} \frac{a(t)\big(1 + M^{r-1}\big)}{\|x_n\|_{W^{1,2}(T)}} \big(y_n(t) - y(t)\big)\, dt,$$

so

$$\lim_{n \to +\infty} \int_0^b \frac{u_n(t)}{\|x_n\|_{W^{1,2}(T)}} \big(y_n(t) - y(t)\big)\, dt \; = \; 0.$$

Also, we have that

$$\lim_{n \to +\infty} \int_0^b \frac{h(t)}{\|x_n\|_{W^{1,2}(T)}} \big(y_n(t) - y(t)\big)\, dt \; = \; 0.$$

Thus from (3.91), we obtain

$$\lim_{n \to +\infty} \left[\int_0^b y_n'(t)\big(y_n(t) - y(t)\big)'\, dt - \lambda_m \int_0^b y_n(t)\big(y_n(t) - y(t)\big)\, dt \right] \; = \; 0.$$

Using (3.90), we have

$$\lim_{n \to +\infty} \int_0^b y_n'(t)\big(y_n(t) - y(t)\big)'\, dt \; = \; 0$$

and so

$$\|y_n'\|_2 \; \longrightarrow \; \|y'\|_2 \,.$$

Because $y_n' \xrightarrow{w} y'$ in $L^2(T)$, it follows that $y_n' \longrightarrow y'$ in $L^2(T)$ and so finally

$$y_n \; \longrightarrow \; y \quad \text{in } W^{1,2}(T).$$

Let

$$y_n = \overline{y}_n + y_n^0 + \widehat{y}_n \quad \text{with } \overline{y}_n \in \overline{H}, \; y_n^0 \in N_m, \; \widehat{y}_n \in \widehat{H}, \qquad \forall \, n \geq 1.$$

We use as our test function $v = -\overline{y}_n + y_n^0 + \widehat{y}_n$. We have

$$\left| \langle x_n^*, v \rangle_{W^{1,2}(T)} \right|$$

$$= \left| \int_0^b x_n'(t) \left(-\overline{y}_n + y_n^0 + \widehat{y}_n \right)'(t) \, dt - \lambda_m \int_0^b x_n(t) \left(-\overline{y}_n + y_n^0 + \widehat{y}_n \right)(t) \, dt \right.$$

$$\left. - \int_0^b u_n(t) \left(-\overline{y}_n + y_n^0 + \widehat{y}_n \right)(t) \, dt + \int_0^b h(t) \left(-\overline{y}_n + y_n^0 + \widehat{y}_n \right)(t) \, dt \right|$$

$$\leq \varepsilon_n \left\| -\overline{y}_n + y_n^0 + \widehat{y}_n \right\|_{W^{1,2}(T)} \leq 3\varepsilon_n,$$

with $\varepsilon_n \searrow 0$. So

$$\frac{1}{\|x_n\|_{W^{1,2}(T)}} \left| \int_0^b x_n'(t) \left(-\overline{x}_n' + (x_n^0)' + \widehat{x}_n' \right)(t) \, dt \right.$$

$$- \lambda_m \int_0^b x_n(t) \left(-\overline{x}_n + x_n^0 + \widehat{x}_n \right)(t) \, dt \right|$$

$$- \left| \int_0^b \frac{u_n(t) - h(t)}{\|x_n\|_{W^{1,2}(T)}} \left(-\overline{x}_n + x_n^0 + \widehat{x}_n \right)(t) \, dt \right| \leq 3\varepsilon_n. \quad (3.92)$$

Because $x_n^0 \in N_m$, we have

$$\left\| (x_n^0)' \right\|_2^2 = \lambda_m \left\| x_n^0 \right\|_2^2.$$

Also, using as test function x_n^0 and exploiting the orthogonality relations among the component subspaces, we obtain

$$\left| \frac{1}{\|x_n\|_{W^{1,2}(T)}} \left(\left\| (x_n^0)' \right\|_2^2 - \lambda_m \left\| x_n^0 \right\|_2^2 \right) \right.$$

$$\left. - \int_0^b \frac{u_n(t)}{\|x_n\|_{W^{1,2}(T)}} x_n^0(t) \, dt + \int_0^b \frac{h(t)}{\|x_n\|_{W^{1,2}(T)}} x_n^0(t) \, dt \right| \leq \varepsilon_n,$$

so

$$\int_0^b \frac{u_n(t) - h(t)}{\|x_n\|_{W^{1,2}(T)}} x_n^0(t) \, dt \longrightarrow 0 \quad \text{as } n \to +\infty. \quad (3.93)$$

Because of the orthogonality relations, we have

$$\int_0^b x_n'(t) \left(-\overline{x}_n' + (x_n^0)' + \widehat{x}_n' \right)(t) \, dt = -\left\| \overline{x}_n' \right\|_2^2 + \left\| (x_n^0)' \right\|_2^2 + \left\| \widehat{x}_n' \right\|_2^2$$

and

$$-\lambda_m \int_0^h x_n(t)\left(-\overline{x}_n + x_n^0 + \widehat{x}_n\right)(t)\,dt \;=\; \lambda_m \|\overline{x}_n\|_2^2 - \lambda_m \|x_n^0\|_2^2 - \lambda_m \|\widehat{x}_n\|_2^2.$$

From these equations and Lemmata 3.4.1 and 3.4.2, we obtain

$$\int_0^b x_n'(t)\left(-\overline{x}_n' + (x_n^0)' + \widehat{x}_n'\right)(t)\,dt - \lambda_m \int_0^b x_n(t)\left(-\overline{x}_n + x_n^0 + \widehat{x}_n\right)(t)\,dt$$
$$= \lambda_m \|\overline{x}_n\|_2^2 - \|\overline{x}_n'\|_2^2 - \lambda_m \|\widehat{x}_n\|_2^2 + \|\widehat{x}_n'\|_2^2 \;\geq\; c_{26} \|w_n\|_2^2, \tag{3.94}$$

for some $c_{26} > 0$ and with $w_n = -\overline{x}_n + \widehat{x}_n$ for $n \geq 1$. Moreover, from hypotheses $H(j)_8(iii)$ and (iv) as before, for all $n \geq 1$, we obtain

$$\int_0^b \frac{u_n(t)}{\|x_n\|_{W^{1,2}(T)}}\left(-\overline{x}_n + \widehat{x}_n\right)(t)\,dt \;\leq\; \frac{\varepsilon\|w_n\|_{W^{1,2}(T)}^2}{\|x_n\|_{W^{1,2}(T)}} + \frac{c_{27}\|w_n\|_{W^{1,2}(T)}}{\|x_n\|_{W^{1,2}(T)}}, \tag{3.95}$$

for some $c_{27} > 0$. Finally, we have

$$\int_0^b \frac{h(t)}{\|x_n\|_{W^{1,2}(T)}}\left(-\overline{x}_n + \widehat{x}_n\right)(t)\,dt \;\leq\; \|h\|_1 \|-\overline{y}_n + \widehat{y}_n\|_\infty$$
$$\leq\; c_{28}\frac{\|w_n\|_{W^{1,2}(T)}}{\|x_n\|_{W^{1,2}(T)}} \qquad \forall\, n \geq 1, \tag{3.96}$$

for some $c_{28} > 0$. Using (3.93), (3.94), (3.95) and (3.96) in (3.92), for all $n \geq 1$, we obtain

$$\frac{(c_{26} - \varepsilon)\|w_n\|_{W^{1,2}(T)}^2}{\|x_n\|_{W^{1,2}(T)}} - \frac{c_{29}\|w_n\|_{W^{1,2}(T)}}{\|x_n\|_{W^{1,2}(T)}} \;\leq\; \varepsilon_n',$$

for some $c_{29} > 0$, with $\varepsilon_n \searrow 0$. Let us take $\varepsilon \in (0, c_{26})$. We can write

$$\frac{1}{\|x_n\|_{W^{1,2}(T)}}\left(c_{30}\|w_n\|_{W^{1,2}(T)}^2 - c_{29}\|w_n\|_{W^{1,2}(T)}\right) \;\leq\; \varepsilon_n' \qquad \forall\, n \geq 1,$$

for some $c_{30} > 0$ and so

$$\limsup_{n\to+\infty} \frac{\|w_n\|_{W^{1,2}(T)}^2}{\|x_n\|_{W^{1,2}(T)}}\left(c_{30} - \frac{c_{29}}{\|w_n\|_{W^{1,2}(T)}}\right) \;\leq\; 0.$$

Passing to a subsequence if necessary, we have that

$$\frac{\|w_n\|_{W^{1,2}(T)}^2}{\|x_n\|_{W^{1,2}(T)}} \;\longrightarrow\; 0.$$

Assuming without any loss of generality that $\|x_n\|_{W^{1,2}(T)} \geq 1$ (recall that $\|x_n\|_{W^{1,2}(T)} \longrightarrow +\infty$), we may have that

$$\frac{\|w_n\|^2_{W^{1,2}(T)}}{\|x_n\|_{W^{1,2}(T)}} \geq \frac{\|w_n\|^2_{W^{1,2}(T)}}{\|x_n\|^2_{W^{1,2}(T)}} = \|\overline{y}_n\|^2_{W^{1,2}(T)} + \|\widehat{y}_n\|^2_{W^{1,2}(T)}$$

and so

$$\overline{y}_n \longrightarrow 0 \quad \text{and} \quad \widehat{y}_n \longrightarrow 0 \quad \text{in } W^{1,2}(T)$$

hence $y = y^0 \in N_m$. This means that

$$y(t) = \zeta_1 \sin m\omega t + \zeta_2 \cos \omega t,$$

for some $\zeta_1, \zeta_2 \in \mathbb{R}$. We can write

$$y(t) = r \sin\big(m\omega t + \vartheta\big),$$

with $r = (\zeta_1^2 + \zeta_2^2)^{\frac{1}{2}}$ and $\vartheta = \frac{\zeta_1}{\zeta_2}$. Recall that for all $n \geq 1$, we have

$$|\varphi(x_n)| = \left| \frac{1}{2}\|x_n'\|_2^2 - \frac{\lambda_m}{2}\|x_n\|_2^2 - \int_0^b j(t, x_n(t))\, dt + \int_0^b h(t)x_n(t)\, dt \right| \leq M_6.$$

Since $\|(x_n^0)'\|_2^2 = \lambda_m \|x_n^0\|_2^2$ and exploiting the orthogonality relations among the component spaces, we obtain

$$\frac{|\varphi(x_n)|}{\|x_n\|_{W^{1,2}(T)}}$$

$$= \left| \frac{1}{2}\frac{\|w_n'\|_2^2}{\|x_n\|_{W^{1,2}(T)}} - \frac{\lambda_m}{2}\frac{\|w_n\|_2^2}{\|x_n\|_{W^{1,2}(T)}} - \int_0^b \frac{j(t, x_n(t))}{\|x_n\|_{W^{1,2}(T)}}\, dt + \int_0^b h(t)y_n(t)\, dt \right|$$

$$\leq \frac{M_6}{\|x_n\|_{W^{1,2}(T)}}.$$

Since $\frac{\|w_n\|^2_{W^{1,2}(T)}}{\|x_n\|_{W^{1,2}(T)}} \longrightarrow 0$, it follows that $\frac{\|w_n\|_2^2}{\|x_n\|_{W^{1,2}(T)}}, \frac{\|w_n'\|_2^2}{\|x_n\|_{W^{1,2}(T)}} \longrightarrow 0$ as $n \longrightarrow +\infty$. Moreover, we have

$$\int_0^b \frac{j(t, x_n(t))}{\|x_n\|_{W^{1,2}(T)}}\, dt = \int_{\{x_n \neq 0\}} \frac{j(t, x_n(t))}{x_n(t)} y_n(t)\, dt + \int_{\{x_n = 0\}} \frac{j(t, 0)}{\|x_n\|_{W^{1,2}(T)}}\, dt.$$

Note that on $\{y > 0\}$, we have $x_n(t) \longrightarrow +\infty$ and on $\{y < 0\}$, we have $x_n(t) \longrightarrow -\infty$. In addition

$$\int_{\{x_n = 0\}} \frac{j(t, 0)}{\|x_n\|_{W^{1,2}(T)}}\, dt \longrightarrow 0.$$

Hence via Fatou's lemma (see Proposition A.2.2), we obtain

$$\liminf_{n \to +\infty} \int_0^b \frac{j(t, x_n(t))}{\|x_n\|_{W^{1,2}(T)}} \, dt \geq \int_0^b j_+(t) y^+(t) \, dt - \int_0^b j_-(t) y^-(t) \, dt.$$

Recall that $y \in N_m$, so

$$\liminf_{n \to +\infty} \int_0^b \frac{j(t, x_n(t))}{\|x_n\|_{W^{1,2}(T)}} \, dt$$

$$\geq \int_0^b j_+(t) r \sin(m\omega t + \vartheta)^+ \, dt - \int_0^b j_-(t) r \sin(m\omega t + \vartheta)^- \, dt, \quad (3.97)$$

with $\vartheta \in \mathbb{R}$. Returning to (3.92), passing to the limit as $n \to +\infty$ and using hypothesis $H(j)_8(v)$, we obtain

$$\liminf_{n \to +\infty} \int_0^b \frac{j(t, x_n(t))}{\|x_n\|_{W^{1,2}(T)}} \, dt = \int_0^b h(t) y(t) \, dt = \int_0^b h(t) r \sin(m\omega t + \vartheta) \, dt$$

$$< r \left(\int_0^b j_+(t) \sin(m\omega t + \vartheta)^+ \, dt - \int_0^b j_-(t) \sin(m\omega t + \vartheta)^- \, dt \right). \quad (3.98)$$

Comparing (3.97) and (3.98), we reach a contradiction. So the sequence $\{x_n\}_{n \geq 1} \subseteq W^{1,2}_{\mathrm{per}}(T)$ is bounded and from this it follows that φ satisfies the nonsmooth PS-condition. $\qquad \square$

Let

$$H_1 \stackrel{df}{=} \overline{H} \oplus N_m = \mathrm{span}\left\{ \sin k\omega t, \ \cos k\omega t : \ k \in \{0, 1, \ldots m\} \right\}$$

$$H_2 \stackrel{df}{=} H_1^\perp$$

PROPOSITION 3.4.15
If hypotheses $H(j)_8$ hold,
then $\varphi(x) \longrightarrow -\infty$ as $\|x\|_{W^{1,2}(T)} \to +\infty$, with $x \in H_1$.

PROOF   Suppose that the proposition is not true. Then we can find a sequence $\{x_n\}_{n \geq 1} \subseteq H_1$ and a constant $c_{31} > 0$, such that

$$\|x_n\|_{W^{1,2}(T)} \longrightarrow +\infty \qquad \text{and} \qquad -c_{31} \leq \varphi(x_n) \qquad \forall n \geq 1.$$

We have

$$\frac{1}{2}\left\|x_n'\right\|_2^2 - \frac{\lambda_m}{2}\left\|x_n\right\|_2^2 - \int_0^b j\big(t, x_n(t)\big)\,dt + \int_0^b h(t)x_n(t)\,dt \;\geq\; -c_{31}.$$

Let

$$y_n \;\overset{df}{=}\; \frac{x_n}{\left\|x_n\right\|_{W^{1,2}(T)}} \qquad \forall\, n \geq 1.$$

Passing to a subsequence if necessary, we may assume that

$$\begin{aligned} y_n &\xrightarrow{\;w\;} y \text{ in } W^{1,2}_{\mathrm{per}}(T),\\ y_n &\longrightarrow y \text{ in } C_{\mathrm{per}}(T). \end{aligned}$$

Because $y_n \in H_1$ for $n \geq 1$ and H_1 is finite dimensional, we have

$$y_n \;\longrightarrow\; y \quad \text{in } W^{1,2}_{\mathrm{per}}(T).$$

We also have

$$\frac{1}{2}\left\|y_n'\right\|_2^2 - \frac{\lambda_m}{2}\left\|y_n\right\|_2^2 - \int_0^b \frac{j\big(t, x_n(t)\big)}{\left\|x_n\right\|_{W^{1,2}(T)}^2}\,dt + \int_0^b \frac{h(t)}{\left\|x_n\right\|_{W^{1,2}(T)}}y_n(t)\,dt$$

$$\geq \frac{-c_{31}}{\left\|x_n\right\|_{W^{1,2}(T)}^2} \qquad \forall\, n \geq 1.$$

Note that, from hypothesis $H(j)_8(v)$, we have

$$\int_0^b \frac{j\big(t, x_n(t)\big)}{\left\|x_n\right\|_{W^{1,2}(T)}^2}\,dt \;\longrightarrow\; 0$$

and

$$\int_0^b \frac{h(t)}{\left\|x_n\right\|_{W^{1,2}(T)}}y_n(t)\,dt \;\longrightarrow\; 0.$$

So in the limit as $n \to +\infty$, we obtain

$$\frac{\lambda_m}{2}\left\|y\right\|_2^2 \;\leq\; \frac{1}{2}\left\|y'\right\|_2^2,$$

hence $\left\|y'\right\|_2^2 = \lambda_m\left\|y\right\|_2^2$ (since $y \in H_1$). Therefore $y \in N_m$. From Lemma (3.4.2), we have

$$-\int_0^b j\big(t, x_n(t)\big)\,dt + \int_0^b h(t)x_n(t)\,dt \;\geq\; \varphi(x_n) \;\geq\; -c_{31},$$

so

$$\int_0^b \frac{j(t, x_n(t))}{\|x_n\|_{W^{1,2}(T)}} \, dl \leq \frac{c_{31}}{\|x_n\|_{W^{1,2}(T)}} + \int_0^b h(t) y_n(t) \, dt.$$

As in the proof of Proposition 3.4.14, in the limit as $n \to +\infty$, we have

$$\int_0^b j_+(t) \sin(m\omega t + \vartheta)^+ \, dt - \int_0^b j_-(t) \sin(m\omega t + \vartheta)^- \, dt$$

$$\leq \int_0^b h(t) \sin(m\omega t + \vartheta) \, dt$$

(recall that $y \in N_m$), which contradicts hypothesis $H(j)_8(v)$. Therefore the proposition is true. $\qquad\qquad\qquad\qquad\qquad\qquad\qquad\qquad\qquad\qquad\qquad\qquad\qquad\qquad$ □

PROPOSITION 3.4.16
If hypotheses $H(j)_8$ hold,
then $\varphi(x) \longrightarrow +\infty$ as $\|x\|_{W^{1,2}(T)} \to +\infty$ with $x \in H_2$.

PROOF   Using Lemma 3.4.1, we see that

$$\varphi(x) \geq \widehat{c}\|x\|^2_{W^{1,2}(T)} - \int_0^b j(t, x(t)) \, dt + \int_0^b h(t) x(t) \, dt \qquad \forall\, x \in H_2,$$

so

$$\frac{\varphi(x)}{\|x\|^2_{W^{1,2}(T)}} \geq \widehat{c} - \int_0^b \frac{j(t, x(t))}{\|x\|^2_{W^{1,2}(T)}} \, dt + \int_0^b \frac{h(t)}{\|x\|_{W^{1,2}(T)}} y(t) \, dt \qquad \forall\, x \in H_2$$

and

$$\liminf_{\|x\|_{W^{1,2}(T)} \to +\infty} \frac{\varphi(x)}{\|x\|^2_{W^{1,2}(T)}} \geq \widehat{c} > 0 \qquad \forall\, x \in H_2,$$

i.e. $\varphi|_{H_2}$ is coercive. $\qquad\qquad\qquad\qquad\qquad\qquad\qquad\qquad\qquad\qquad\qquad\qquad\qquad\qquad\qquad\qquad$ □

Propositions 3.4.14, 3.4.15 and 3.4.16 allow the use of the nonsmooth Saddle Point Theorem (see Theorem 2.1.4), which produces a solution for problem (3.88).

THEOREM 3.4.8
If hypotheses $H(j)_8$ hold,
then problem (3.88) has a solution

$$x_0 \in C^1_{\text{per}}(T) \quad \text{with} \quad x'_0(\cdot) \in W^{1,r}_{\text{per}}(T).$$

Next we return to the scalar periodic problems driven by the p-Laplacian. Specifically, we study the following problem:

$$\begin{cases} -\Big(|x'(t)|^{p-2}x'(t)\Big)' \in \partial j\big(t, x(t)\big) & \text{for a.a. } t \in T, \\ x(0) = x(b), \ x'(0) = x'(b), \end{cases} \tag{3.99}$$

with $p \in (1, +\infty)$. Let $\lambda_0 = 0$ and $\lambda_1 > 0$ be the first two eigenvalues of the negative scalar p-Laplacian with periodic boundary conditions (see Section 1.5). Our aim is to achieve the solvability of problem (3.99) with the nonsmooth potential j interacting with λ_0 from the right and staying (in some sense) between $\lambda_0 = 0$ and $\lambda_1 > 0$. We permit full interaction (resonance) with $\lambda_0 = 0$, but only partial interaction with $\lambda_1 > 0$ (see hypotheses $H(j)_9$). In this context the space $W^{1,p}_{per}(T)$ has no orthogonal decomposition using the relevant eigenspaces and the variational characterization of λ_1 is with respect to a cone, not a subspace as in the semilinear case (see Proposition 1.5.4 and Remark 1.5.2). This precludes the use of the nonsmooth Mountain Pass, Saddle Point and Linking Theorems (see Theorems 2.1.3, 2.1.4 and 2.1.5 respectively) and leads to a use of the general minimax principle in Theorem 2.1.2, which requires linking sets. So our task is to produce two suitable linking sets for problem (3.99). For this purpose we introduce the following hypotheses on j. Recall that $\lambda_1 > 0$ denotes the first nonzero eigenvalue of the negative ordinary scalar p-Laplacian with periodic boundary conditions.

$\underline{H(j)_9}$ $j: T \times \mathbb{R} \longrightarrow \mathbb{R}$ is a function, such that

$\left. \begin{matrix} (i) \\ (ii) \end{matrix} \right\}$ the same as hypotheses $H(j)_7(i)$ and (ii);

(iii) we have

$$|u| \leq a_r(t) \quad \text{for a.a. } t \in T, \text{ all } |\zeta| \leq r \text{ and all } u \in \partial j(t, \zeta)$$

with $a_r \in L^\infty(T)_+$;

(iv) we have

$$\lim_{|\zeta| \to +\infty} \big(\zeta u(\zeta) - pj(t, \zeta)\big) = -\infty$$

uniformly for almost all $t \in T$ and all $u(\zeta) \in \partial j(t, \zeta)$;

(v) we have

$$0 \leq \liminf_{|\zeta| \to +\infty} \frac{pj(t, \zeta)}{|\zeta|^p} \leq \limsup_{|\zeta| \to +\infty} \frac{pj(t, \zeta)}{|\zeta|^p} \leq \vartheta(t),$$

uniformly for almost all $t \in T$, with $\vartheta \in L^\infty(T)_+$, $\vartheta(t) \leq \lambda_1$ for almost all $t \in T$ and the inequality is strict on a set of positive measure.

REMARK 3.4.8   Note that hypothesis $H(j)_9(v)$ allows full interaction (resonance) of j with the eigenvalue $\lambda_0 = 0$, but only partial interaction with the eigenvalue $\lambda_1 > 0$ (nonuniform nonresonance).   ▯

The nonsmooth, locally Lipschitz energy functional $\varphi \colon W^{1,p}_{\mathrm{per}}(T) \longrightarrow \mathbb{R}$ of (3.99) is defined by

$$\varphi(x) \stackrel{df}{=} \frac{1}{p}\,\|x'\|^p_p - \int\limits_0^b j\big(t, x(t)\big)\, dt.$$

PROPOSITION 3.4.17
If hypotheses $H(j)_9$ hold, <u>then</u> φ satisfies the nonsmooth C-condition.

PROOF   Let $\{x_n\}_{n \geq 1} \subseteq W^{1,p}_{\mathrm{per}}(T)$ be a Cerami sequence for φ. So, we have

$$\varphi(x_n) \longrightarrow c_{32} \quad \text{and} \quad \big(1 + \|x_n\|_{W^{1,2}(T)}\big) m^\varphi(x_n) \longrightarrow 0 \quad \text{as } n \to +\infty,$$

for some $c_{32} > 0$. Let $x_n^* \in \partial\varphi(x_n)$ be such that $m^\varphi(x_n) = \|x_n^*\|_{(W^{1,p}_{\mathrm{per}}(T))^*}$ for $n \geq 1$. We have

$$x_n^* = A(x_n) - u_n \qquad \forall\, n \geq 1,$$

with $u_n \in S^{p'}_{j(\cdot, x_n(\cdot))}$. Hence

$$\Big|\langle x_n^*, x_n\rangle_{W^{1,p}_{\mathrm{per}}(T)} - p\varphi(x_n) + p c_{32}\Big|$$
$$\leq \|x_n^*\|_{(W^{1,p}_{\mathrm{per}}(T))^*}\, \|x_n\|_{W^{1,2}(T)} + \big|p\varphi(x_n) - p c_{32}\big|$$
$$\leq \big(1 + \|x_n\|_{W^{1,2}(T)}\big) m^\varphi(x_n) + \big|p\varphi(x_n) - p c_{32}\big| \longrightarrow 0,$$

so

$$\int\limits_0^b \big(u_n(t)x_n(t) - p j(t, x_n(t))\big)\, dt \longrightarrow p c_{32}. \qquad (3.100)$$

We shall show that the sequence $\{x_n\}_{n \geq 1} \subseteq W^{1,p}_{\mathrm{per}}(T)$ is bounded. Suppose that this is not the case. Passing to a subsequence if necessary, we may assume that

$$\|x_n\|_{W^{1,p}(T)} \longrightarrow +\infty.$$

Let us set

$$y_n \stackrel{df}{=} \frac{x_n}{\|x_n\|_{W^{1,p}(T)}} \qquad \forall\, n \geq 1.$$

Passing to a subsequence if necessary, we may assume that

$$y_n \stackrel{w}{\longrightarrow} y \text{ in } W^{1,p}_{\mathrm{per}}(T),$$
$$y_n \longrightarrow y \text{ in } C(T).$$

By virtue of hypotheses $H(j)_9(iii)$ and (v) and because of Proposition 1.3.14, we have

$$\left| j(t, \varsigma) \right| \leq a_1(t) + c_{33}|\varsigma|^p \quad \text{for a.a. } t \in T \text{ and all } \varsigma \in \mathbb{R},$$

with $a_1 \in L^\infty(T)_+$ and $c_{33} > 0$. So

$$\frac{\left| j\big(t, x_n(t)\big) \right|}{\|x_n\|_{W^{1,p}(T)}^p} \leq \frac{a_1(t)}{\|x_n\|_{W^{1,p}(T)}^p} + c_{33}|y_n(t)|^p \quad \text{for a.a. } t \in T \text{ and all } n \geq 1.$$

By the Dunford-Pettis Theorem (see Theorem A.3.14), we may assume that

$$\frac{\left| j\big(\cdot, x_n(\cdot)\big) \right|}{\|x_n\|_{W^{1,p}(T)}^p} \xrightarrow{\ w\ } g \quad \text{in } L^1(T),$$

for some $g \in L^1(T)$ and because of hypothesis $H(j)_9(v)$, we have that

$$0 \leq g(t) \leq \vartheta(t)|y(t)|^p \quad \text{for a.a. } t \in T.$$

Recall that

$$\frac{\varphi(x_n)}{\|x_n\|_{W^{1,p}(T)}^p} \leq \frac{M_7}{\|x_n\|_{W^{1,p}(T)}^p} \quad \forall\, n \geq 1,$$

for some $M_7 > 0$. So in the limit as $n \to +\infty$, we have

$$\frac{1}{p} \|y_n'\|_p^p \leq \int_0^b g(t)|y(t)|^p \, dt.$$

Note that $y \neq 0$, or otherwise since

$$y_n \longrightarrow 0 \quad \text{in } W_{\text{per}}^{1,p}(T),$$

we obtain a contradiction to the fact that $\|y_n\|_{W^{1,p}(T)} = 1$ for $n \geq 1$. So if $T_{\neq}^y \overset{df}{=} \{y \neq 0\}$, then $|T_{\neq}^y|_1 > 0$. We have

$$\left| x_n(t) \right| \longrightarrow +\infty \quad \text{as } n \to +\infty \quad \forall\, t \in T_{\neq}^y.$$

Also

$$\int_0^b \big(u_n(t)x_n(t) - pj\big(t, x_n(t)\big)\big)\, dt$$

$$= \int_{T_{\neq}^y} \big(u_n(t)x_n(t) - pj\big(t, x_n(t)\big)\big)\, dt + \int_{T \setminus T_{\neq}^y} \big(u_n(t)x_n(t) - pj\big(t, x_n(t)\big)\big)\, dt.$$

By virtue of hypothesis $H(j)_9(iv)$, we can find $M_8 > 0$, such that

$$\zeta v - pj(t, \zeta) \leq -1 \quad \text{for a.a. } t \in T, \text{ all } \zeta \in \mathbb{R}, \; |\zeta| > M_8 \text{ and all } v \in \partial j(t, \zeta).$$

On the other hand, hypothesis $H(j)_9(iii)$ implies that

$$\zeta v - pj(t, \zeta) \leq M_9 \quad \text{for a.a. } t \in T, \text{ all } \zeta \in \mathbb{R}, \; |\zeta| < M_8 \text{ and all } v \in \partial j(t, \zeta).$$

So finally, we have that

$$\zeta v - pj(t, \zeta) \leq M_{10} \quad \text{for a.a. } t \in T, \text{ all } \zeta \in \mathbb{R} \text{ and all } v \in \partial j(t, \zeta),$$

for some $M_{10} > 0$. Therefore

$$\lim_{n \to +\infty} \int_{T_{\neq}^y} \left(u_n(t) x_n(t) - pj\big(t, x_n(t)\big) \right) dt = -\infty$$

and

$$\lim_{n \to +\infty} \int_{T \backslash T_{\neq}^y} \left(u_n(t) x_n(t) - pj\big(t, x_n(t)\big) \right) dt \leq M_{10} b,$$

so

$$\int_0^b \left(u_n(t) x_n(t) - pj(t, x_n(t)) \right) dt \longrightarrow -\infty \quad \text{as } n \to +\infty. \tag{3.101}$$

Comparing (3.100) and (3.101), we reach a contradiction. This proves that the sequence $\{x_n\}_{n \geq 1} \subseteq W_{\text{per}}^{1,p}(T)$ is bounded and then as before, we can verify the nonsmooth PS-condition. $\qquad \Box$

PROPOSITION 3.4.18
If hypotheses $H(j)_9$ hold, <u>then</u> $\varphi(\zeta) \longrightarrow -\infty$ as $|\zeta| \to +\infty$, with $\zeta \in \mathbb{R}$.

PROOF   From the proof of Proposition 3.4.12, we know that

$$\partial \left(\frac{j(t, \zeta)}{|\zeta|^p} \right) = \frac{\zeta \partial j(t, \zeta) - pj(t, \zeta)}{|\zeta|^{p+1}} \quad \text{for a.a. } t \in T \text{ and all } \zeta > 0.$$

Also, from hypothesis $H(j)_9(iv)$, for a given $\beta > 0$, we can find $M_{11} > 0$, such that

$$\zeta v - pj(t, \zeta) \leq -\beta \quad \text{for a.a. } t \in T, \text{ all } \zeta \geq M_{11} \text{ and all } v \in \partial j(t, \zeta).$$

So it follows that

$$w \leq -\frac{\beta}{\zeta^{p-1}} \quad \text{for a.a. } t \in T, \text{ all } \zeta \geq M_{11} \text{ and all } w \in \partial \left(\frac{j(t, \zeta)}{|\zeta|^p} \right).$$

For $t \in T \setminus N$, with $|N|_1 = 0$, the function $\zeta \longmapsto \frac{j(t,\zeta)}{\zeta^p}$ is locally Lipschitz on $[M_{11}, +\infty)$ and thus differentiable on $[M_{11}, +\infty) \setminus \hat{N}_1(t)$, $|N_1(t)|_1 = 0$. Let us set

$$\vartheta_0(t,r) \stackrel{df}{=} \begin{cases} \dfrac{d}{dr}\left(\dfrac{j(t,r)}{r^p}\right) & \text{if } r \in [M_{11}, +\infty) \setminus N_1(t), \\ 0 & \text{if } r \in N_1(t). \end{cases}$$

We have that

$$\vartheta_0(t,\zeta) \in \partial\left(\frac{j(t,\zeta)}{\zeta^p}\right) \qquad \forall\, t \in T \setminus N,\ \zeta \in [M_{11}, +\infty) \setminus N_1(t).$$

So

$$\vartheta_0(t,\zeta) \leq -\frac{\beta}{\zeta^{p+1}} = \frac{d}{dt}\left(\frac{1}{p}\frac{\beta}{\zeta^p}\right).$$

Integrating this inequality on the interval $[\zeta_1, \zeta_2] \subseteq [M_{11}, +\infty)$ $(\zeta_1 < \zeta_2)$, we obtain

$$\frac{j(t,\zeta_2)}{\zeta_2^p} - \frac{j(t,\zeta_1)}{\zeta_1^p} \leq \frac{\beta}{p}\left(\frac{1}{\zeta_2^p} - \frac{1}{\zeta_1^p}\right).$$

Let $\zeta_2 \longrightarrow +\infty$ and use hypothesis $H(j)_9(v)$ to obtain

$$\frac{j(t,\zeta_1)}{\zeta_1^p} \geq \frac{\beta}{p}\frac{1}{\zeta_1^p}.$$

Hence, we have

$$\frac{\beta}{p} \leq j(t,\zeta_1) \qquad \forall\, t \in T \setminus N,\ \zeta_1 \geq M_{11}$$

and thus

$$\varphi(\zeta) = -\int_0^b j(t,\zeta)\,dt \leq -\frac{\beta}{2}b \qquad \forall\, \zeta \geq M_{11}.$$

Because $\beta > 0$ was arbitrary, we conclude that

$$\varphi(\zeta) \longrightarrow -\infty \quad \text{as } \zeta \to +\infty.$$

$\square$

We introduce the closed cone

$$K \stackrel{df}{=} \left\{ v \in W^{1,p}_{per}(T) : \int_0^b |v(t)|^{p-2}v(t)\,dt = 0 \right\}.$$

PROPOSITION 3.4.19
If hypotheses $H(j)_9$ hold,
then $\varphi(v) \longrightarrow +\infty$ as $\|v\|_{W^{1,p}(T)} \to +\infty$, with $v \in K$.

PROOF   Let us define $\psi \colon K \longrightarrow \mathbb{R}$, by

$$\psi(v) \stackrel{df}{=} \|v'\|_p^p - \int_0^b \vartheta(t)|v(t)|^p \, dt \qquad \forall \, v \in K.$$

We claim that there exists $c_\psi > 0$, such that

$$\psi(v) \geq c_\psi \|v'\|_p^p \qquad \forall \, v \in K.$$

Suppose that this is not true. Since $\psi \geq 0$ (see Corollary 1.5.1), we can find a sequence $\{x_n\}_{n\geq 1} \subseteq W_{\mathrm{per}}^{1,p}(T)$, with $\|x_n'\|_p = 1$, such that $\psi(x_n) \searrow 0$. Note that the sequence $\{x_n\}_{n\geq 1} \subseteq W_{\mathrm{per}}^{1,p}(T)$ is bounded (see Corollary 1.5.1) and so, after passing to a subsequence if necessary, we may assume that

$$x_n \xrightarrow{\ w\ } x \text{ in } W_{\mathrm{per}}^{1,p}(T),$$
$$x_n \longrightarrow x \text{ in } C_{\mathrm{per}}(T).$$

Hence

$$\|v'\|_p^p \leq \int_0^b \vartheta(t)|v(t)|^p \, dt \leq \lambda_1 \|v\|_p^p$$

and so v is a normalized eigenfunction corresponding to the eigenvalue $\lambda_1 > 0$ (see Proposition 1.5.3). Hence $v(t) \neq 0$ for almost all $t \in T$ and so

$$\|v'\|_p^p \leq \int_0^b \vartheta(t)|v(t)|^p \, dt < \lambda_1 \|v\|_p^p,$$

a contradiction to the extended Poincaré-Wirtinger inequality (see Corollary 1.5.1). This proves the claim.

Let $\varepsilon \in (0, c_\psi \lambda_1 p)$. By virtue of hypotheses $H(j)_9(iii)$ and (v), we can find $k_\varepsilon \in L^1(T)_+$, such that

$$j(t, \zeta) \leq \frac{1}{p}\big(\vartheta(t) + \varepsilon\big)|\zeta|^p + k_\varepsilon(t) \quad \text{for a.a. } t \in T \text{ and all } \zeta \in \mathbb{R}.$$

So, from the claim and Corollary 1.5.1, we have

$$\varphi(v) \geq \frac{1}{p}\|v'\|_p^p - \frac{1}{p}\int_0^b \vartheta(t)|v(t)|^p \, dt - \frac{\varepsilon}{p}\|v\|_p^p - c_{34}$$

$$\geq c_\psi \|v'\|_p^p - \frac{\varepsilon}{\lambda_1 p}\|v'\|_p^p - c_{34} \qquad \forall \, v \in K,$$

for some $c_{34} > 0$. From the choice of $\varepsilon > 0$ and from the last inequality, it follows that $\psi|_K$ is coercive.   □

Now we are ready for the existence theorem for problem (3.99).

THEOREM 3.4.9
If hypotheses $H(j)_9$ hold,
then problem (3.99) has a solution

$$x_0 \in C^1_{\text{per}}(T) \quad \text{with} \quad \|x_0'(\cdot)\|^{p-2}_{\mathbb{R}^N} x_0'(\cdot) \in W^{1,p}_{\text{per}}(T).$$

PROOF By virtue of Propositions 3.4.18 and 3.4.19, we can find $\zeta > 0$, such that $\varphi(\pm\zeta) < \inf_K \varphi$. Let

$$E \stackrel{df}{=} [-\zeta, \zeta] = \{x \in W^{1,p}_{\text{per}}(T) : -\zeta \leq x(t) \leq \zeta \text{ for all } t \in T\}$$

$$E_1 \stackrel{df}{=} \{-\zeta, \zeta\}.$$

Evidently $C_1 \cap K = \emptyset$. Moreover, if $\gamma \in C\big(E; W^{1,p}_{\text{per}}(T)\big)$ with $\gamma(\pm\zeta) = \pm\zeta$ and $\chi: W^{1,p}_{\text{per}}(T) \longrightarrow \mathbb{R}$ is the continuous map defined by

$$\chi(x) \stackrel{df}{=} \int_0^b |x(t)|^{p-2} x(t) \, dt,$$

then

$$\chi\big(\gamma(-\zeta)\big) = \chi(-\zeta) < 0 < \chi(\zeta) = \chi(\gamma(\zeta))$$

and so by the intermediate value theorem, we have that

$$\gamma(E) \cap K \neq \emptyset.$$

Hence the sets E_1 and K link (see Definition 2.1.4 and Remark 2.1.4). We can apply Theorem 2.1.2 and obtain a solution of (3.99). $\quad\square$

EXAMPLE 3.4.5 A typical example of a nonsmooth potential satisfying hypotheses $H(j)_9$ is given by the following function

$$j(t, \zeta) \stackrel{df}{=} \frac{\vartheta(t)}{p} |\zeta|^p + \ln(|\zeta| + 1) - \max\{\zeta, 0\} \qquad \forall\, (t, \zeta) \in T \times \mathbb{R},$$

with $\vartheta \in L^\infty(T)_+$ as in hypothesis $H(j)_9(v)$. $\quad\square$

3.4.4 Multiple Periodic Solutions

Thus far we have obtained one solution for the periodic problem. Now our aim is to produce more than one solution (multiple periodic solutions).

The problem under consideration remains the following scalar problem (see also (3.99)):

$$\begin{cases} -\left(|x'(t)|^{p-2} x'(t) \right)' \in \partial j(t, x(t)) & \text{for a.a. } t \in T, \\ x(0) = x(b), \ x'(0) = x'(b), \end{cases} \tag{3.102}$$

for $p \in (1, +\infty)$. In the first multiplicity result we do not allow interaction of the nonsmooth potential j with the beginning $\lambda_0 = 0$ of the spectrum of the negative ordinary scalar p-Laplacian with periodic boundary conditions. More precisely our hypotheses on j are the following:

$\underline{H(j)_{10}}$ $j: T \times \mathbb{R} \longrightarrow \mathbb{R}$ is a function, such that

(i) for all $\zeta \in \mathbb{R}$, the function

$$T \ni t \longmapsto j(t, \zeta) \in \mathbb{R}$$

is measurable;

(ii) for almost all $t \in T$, the function

$$\mathbb{R} \ni \zeta \longmapsto j(t, \zeta) \in \mathbb{R}$$

is locally Lipschitz and $j(t, 0) = 0$;

(iii) for almost all $t \in T$, all $\zeta \in \mathbb{R}$ and all $u \in \partial j(t, \zeta)$, we have

$$|u| \leq a(t) + c(t)|\zeta|^{r-1},$$

where $r \in [1, +\infty)$, $\frac{1}{r} + \frac{1}{r'} = 1$, $a, c \in L^\infty(T)_+$;

(iv) $\lim\limits_{|\zeta| \to +\infty} \dfrac{pj(t, \zeta)}{|\zeta|^p} < 0$ uniformly for almost all $t \in T$;

(v) there exist $\varrho_1 > 0$ and $\mu \in \left(0, \frac{1}{b^p} \right)$, such that

$$0 \leq pj(t, \zeta) \leq \mu |\zeta|^p \quad \text{for a.a. } t \in T \text{ and all } \zeta \in \mathbb{R}, \ |\zeta| \leq \varrho_1.$$

THEOREM 3.4.10
If hypotheses $H(j)_{10}$ hold,
then problem (3.102) has at least two distinct solutions.

PROOF   From hypotheses $H(j)_{10}(iii)$ and (iv), we have

$$j(t, \zeta) \leq -\frac{c_1}{p}|\zeta|^p + a_1(t) \quad \text{for a.a. } t \in T \text{ and all } \zeta \in \mathbb{R},$$

with $c_1 > 0$ and $a_1 \in L^1(T)_+$. Let $\varphi \colon W^{1,p}_{\mathrm{per}}(T) \longrightarrow \mathbb{R}$ be the locally Lipschitz energy functional for problem (3.102), defined by

$$\varphi(x) \overset{df}{=} \frac{1}{p} \|x'\|_p^p - \int_0^b j\big(t, x(t)\big)\, dt \qquad \forall\, x \in W^{1,p}_{\mathrm{per}}(T).$$

We have

$$
\begin{aligned}
\varphi(x) &\geq \frac{1}{p}\|x'\|_p^p + \frac{c_1}{p}\|x\|_p^p - c_2 \\
&\geq \frac{1}{p} c_3 \|x\|_{W^{1,p}(T)}^p - c_2 \qquad \forall\, x \in W^{1,p}_{\mathrm{per}}(T),
\end{aligned}
$$

for some $c_2 > 0$ and $c_3 \overset{df}{=} \{1, c_1\} > 0$. So φ is coercive. Therefore φ is bounded from below and satisfies the nonsmooth PS-condition. Consider the direct sum decomposition

$$W^{1,p}_{\mathrm{per}}(T) \;=\; \mathbb{R} \oplus V,$$

where

$$V \overset{df}{=} \left\{ v \in W^{1,p}_{\mathrm{per}}(T) : \int_0^b v(t)\, dt = 0 \right\}.$$

Note that, from hypothesis $H(j)_{10}(v)$, we have

$$\varphi(\zeta) \;=\; -\int_0^b j(t, \zeta)\, dt \;\leq\; 0 \qquad \forall\, \zeta \in \mathbb{R},\; |\zeta| \leq \varrho_1.$$

From the continuity of the embedding $W^{1,p}_{\mathrm{per}}(T) \subseteq C_{\mathrm{per}}(T)$, we can find $\varrho_2 \in (0, \varrho_1]$, such that

$$|v(t)| \;\leq\; \varrho_1 \qquad \forall\, t \in T,\; v \in V,\; \|v\|_{W^{1,p}(T)} \leq \varrho_2.$$

So, from hypothesis $H(j)_{10}(v)$, we have

$$
\begin{aligned}
\varphi(v) &= \frac{1}{p}\|v'\|_p^p - \int_0^b j\big(t, v(t)\big)\, dt \;=\; \frac{1}{p}\|v'\|_p^p - \int_{\{|v| \leq \varrho_1\}} j\big(t, v(t)\big)\, dt \\
&\geq \frac{1}{p}\|v'\|_p^p - \frac{\mu}{p}\|v\|_p^p \qquad \forall\, v \in V,\; \|v\|_{W^{1,p}(T)} \leq \varrho_2.
\end{aligned}
$$

From Remark 1.1.11, we have that

$$\|v\|_p^p \;\leq\; b\,\|v\|_\infty^p \;\leq\; b^p\,\|v'\|_p^p.$$

So, from hypothesis $H(j)_{10}(v)$, we have

$$\varphi(v) \geq \frac{1}{p}(1 - \mu b^p) \|v'\|_p^p \geq 0 \qquad \forall\, v \in V.$$

Finally note that by virtue of hypothesis $H(j)_{10}(iv)$, we have

$$\inf_{x \in W^{1,p}_{\mathrm{per}}(T)} \varphi(x) \ < \ 0,$$

while $\varphi(0) = 0$. Therefore, we can apply Theorem 2.4.1 and obtain two distinct nontrivial critical points of φ. These are two distinct solutions of problem (3.102). $\square$

We can permit interaction of the nonsmooth potential j with the eigenvalue $\lambda_0 = 0$ at the expense of introducing a new condition for j. Specifically the hypotheses on j are now the following:

$\underline{H(j)_{11}}$ $j \colon T \times \mathbb{R} \longrightarrow \mathbb{R}$ is a function, such that

$\left.\begin{array}{l}(i)\\(ii)\\(iii)\end{array}\right\}$ are the same as $H(j)_{10}(i) - (iii)$;

(iv) $\displaystyle\lim_{|\zeta| \to +\infty} \frac{pj(t,\zeta)}{|\zeta|^p} = 0$ uniformly for almost all $t \in T$;

(v) we have

$$\lim_{|\zeta| \to +\infty} \big(\zeta u(\zeta) - pj(t,\zeta)\big) = +\infty$$

uniformly for almost all $t \in T$ and all $u(\zeta) \in \partial j(t,\zeta)$;

(vi) there exist $\varrho_1 > 0$ and $\mu \in \left(0, \frac{1}{b^p}\right)$, such that

$$0 \ \leq \ pj(t,\zeta) \ \leq \ \mu|\zeta|^p \quad \text{for a.a. } t \in T \text{ and all } \zeta \in \mathbb{R},\ |\zeta| \leq \varrho_1,$$

and for some $\zeta_0 \in \mathbb{R}$, we have $\displaystyle\int_0^b j(t,\zeta_0)\, dt > 0$.

THEOREM 3.4.11
If hypotheses $H(j)_{11}$ hold,
then problem (3.102) has at least two distinct nontrivial solutions.

PROOF   Arguing as in the proof of Proposition 3.4.18, we see that for a given $\beta > 0$, we can find $M_1 > 0$, such that

$$j(t,\zeta) \ \leq \ -\frac{\beta}{p} \quad \text{for a.a. } t \in T \text{ and all } \zeta \geq M_1.$$

So
$$j(t, \varsigma) \longrightarrow -\infty \quad \text{as } \varsigma \to +\infty,$$
uniformly for almost all $t \in T$. In a similar fashion, we show that
$$j(t, \varsigma) \longrightarrow -\infty \quad \text{as } \varsigma \to -\infty,$$
uniformly for almost all $t \in T$. Thus finally
$$j(t, \varsigma) \longrightarrow -\infty \quad \text{as } |\varsigma| \to \infty, \tag{3.103}$$
uniformly for almost all $t \in T$.

We shall show that the locally Lipschitz functional $\varphi \colon W^{1,p}_{\mathrm{per}}(T) \longrightarrow \mathbb{R}$, defined by
$$\varphi(x) \stackrel{df}{=} \frac{1}{p} \|x'\|_p^p - \int_0^b j(t, x(t))\, dt \qquad \forall\, x \in W^{1,p}_{\mathrm{per}}(T),$$

is coercive. Suppose that this is not the case. Then we can find a sequence $\{x_n\}_{n \geq 1} \subseteq W^{1,p}_{\mathrm{per}}(T)$, such that
$$\varphi(x_n) \leq M_2 \quad \forall\, n \geq 1 \qquad \text{and} \qquad \|x_n\|_{W^{1,p}(T)} \longrightarrow +\infty, \tag{3.104}$$
for some $M_2 > 0$. Let us set
$$y_n \stackrel{df}{=} \frac{x_n}{\|x_n\|_{W^{1,p}(T)}} \qquad \forall\, n \geq 1.$$

Passing to a subsequence if necessary, we may assume that
$$y_n \stackrel{w}{\longrightarrow} y \text{ in } W^{1,p}_{\mathrm{per}}(T),$$
$$y_n \longrightarrow y \text{ in } C_{\mathrm{per}}(T).$$

We have
$$\frac{\varphi(x_n)}{\|x_n\|_{W^{1,p}(T)}^p} = \frac{1}{p} \|y_n'\|_p^p - \int_0^b \frac{j(t, x_n(t))}{\|x_n\|_{W^{1,p}(T)}^p}\, dt$$
$$\leq \frac{M_2}{\|x_n\|_{W^{1,p}(T)}^p}. \tag{3.105}$$

From hypothesis $H(j)_{11}(iii)$ and Proposition 1.3.14, we infer that
$$|j(t, \varsigma)| \leq a_2(t) + c_4 |\varsigma|^r \quad \text{for a.a. } t \in T \text{ and all } \varsigma \in \mathbb{R},$$
with $a_2 \in L^1(T)_+$ and $c_4 > 0$. So, if we write
$$\int_0^b \frac{j(t, x_n(t))}{\|x_n\|_{W^{1,p}(T)}^p}\, dt$$
$$= \int_{\{|x_n| < M_2\}} \frac{j(t, x_n(t))}{\|x_n\|_{W^{1,p}(T)}^p}\, dt + \int_{\{|x_n| \geq M_2\}} \frac{j(t, x_n(t))}{\|x_n\|_{W^{1,p}(T)}^p}\, dt,$$

we have

$$\int\limits_{\{|x_n|<M_2\}} \frac{j\big(t,x_n(t)\big)}{\|x_n\|_{W^{1,p}(T)}^p}\, dt \;\leq\; \int\limits_0^b \frac{a_3(t)}{\|x_n\|_{W^{1,p}(T)}^p}\, dt \;\longrightarrow\; 0 \quad \text{as } n \to +\infty,$$

with $a_3 \in L^1(T)_+$ and by virtue of hypothesis $H(j)_{11}(iv)$, we also have

$$\int\limits_{\{|x_n|\geq M_2\}} \frac{j\big(t,x_n(t)\big)}{\|x_n\|_{W^{1,p}(T)}^p}\, dt \;\longrightarrow\; 0 \quad \text{as } n \to +\infty.$$

Thus finally

$$\limsup_{n\to+\infty} \int\limits_0^b \frac{j\big(t,x_n(t)\big)}{\|x_n\|_{W^{1,p}(T)}^p}\, dt \;\leq\; 0.$$

Passing to the limit in (3.105) as $n \to +\infty$, we obtain $\|y'\|_p = 0$, i.e. $y \equiv \overline{\zeta} \in \mathbb{R}$. If $\overline{\zeta} = 0$, then

$$y_n \;\longrightarrow\; 0 \quad \text{in } W^{1,p}_{\mathrm{per}}(T),$$

a contradiction to the fact that $\|y_n\|_{W^{1,p}(T)} = 1$ for all $n \geq 1$. So $\overline{\zeta} \neq 0$ and we have

$$|x_n(t)| \;\longrightarrow\; +\infty \quad \text{as } n \to +\infty \qquad \forall\, t \in T$$

and the convergence is uniform in $t \in T$ (see the proof of Proposition 3.4.5). Then, using (3.103), for a given $\beta_1 > 0$, we can find $n_0 \geq 1$, such that

$$j\big(t,x_n(t)\big) \;\leq\; -\beta_1 \quad \text{for a.a. } t \in T \text{ and all } n \geq n_0.$$

From (3.104), we have

$$-\int\limits_0^b j\big(t,x_n(t)\big)\, dt \;\leq\; M_2 \qquad \forall\, n \geq 1.$$

It follows that $\beta_1 b \leq M_2$ and because $\beta_1 > 0$ was arbitrary, we have a contradiction. Therefore φ is coercive and as such it is bounded below and satisfies the nonsmooth PS-condition. Moreover, $\varphi(0) = 0$ and

$$\inf_{x\in W^{1,p}_{\mathrm{per}}(T)} \varphi(x) \;<\; 0$$

(see hypothesis $H(j)_{11}(vi)$).

Now arguing as in the proof of Theorem 3.4.10, we check that φ satisfies the local linking condition for the direct sum $W^{1,p}_{\mathrm{per}}(T) = \mathbb{R} \oplus V$.

Applying Theorem 2.4.1, we obtain two distinct nontrivial solutions for problem (3.102). $\qquad\qquad\square$

EXAMPLE 3.4.6   The nonsmooth function

$$j(\zeta) \stackrel{df}{=} \begin{cases} \dfrac{\mu}{p}\sin|\zeta|^p & \text{if } |\zeta| \leq \pi, \\ -\dfrac{1}{r}|\zeta|^r & \text{if } |\zeta| > \pi, \end{cases}$$

with $\mu \in \left(0, \frac{1}{bp}\right)$ and $r \in [1,p)$ satisfies hypotheses $H(j)_{11}$. On the other hand the same function but with $r \in [1,+\infty)$ satisfies hypotheses $H(j)_{10}$.   □

In the previous two multiplicity theorems we allowed interaction at zero of the potential with $\lambda_0 = 0$. In the semilinear case (i.e. $p = 2$), this interaction can occur at higher parts of the spectrum. At the same time at infinity, we allow partial interaction of the potential j with $\lambda_0 = 0$ (we even permit crossing of $\lambda_0 = 0$).

The problem that we study is the following:

$$\begin{cases} -x''(t) \in \partial j\big(t, x(t)\big) & \text{for a.a. } t \in T, \\ x(0) = x(b), \ x'(0) = x'(b). \end{cases} \tag{3.106}$$

The precise hypotheses on j are the following:

$\underline{H(j)_{12}}$ $j: T \times \mathbb{R} \longrightarrow \mathbb{R}$ is a function, such that

$\left.\begin{array}{l}(i)\\(ii)\end{array}\right\}$ the same as hypotheses $H(j)_{10}(i)$ and (ii);

(iii) for almost all $t \in T$, all $\zeta \in \mathbb{R}$ and all $u \in \partial j(t, \zeta)$, we have

$$|u| \ \leq \ a(t) + c(t)|\zeta|^{r-1}$$

with $a, c \in L^\infty(T)_+$;

(iv) $\limsup\limits_{|\zeta|\to+\infty} \dfrac{j(t,\zeta)}{\zeta^2} \ \leq \ h(t)$ uniformly for almost all $t \in T$, where

$h \in L^1(T)$ is such that $\int\limits_0^b h(t)\,dt < 0$;

(v) there exist $\varrho_1 > 0$ and $\mu < \lambda_{k+1}$, such that

$$\lambda_k \zeta^2 \ \leq \ 2j(t,\zeta) \ \leq \ \mu\zeta^2 \quad \text{for a.a. } t \in T \text{ and all } \zeta \in \mathbb{R}, \ |\zeta| \leq \varrho_1$$

(recall that $\lambda_k = (k\omega)^2$, where $\omega \stackrel{df}{=} \frac{2\pi}{b}$ and $k \in \mathbb{N}_0$).

THEOREM 3.4.12
If hypotheses $H(j)_{12}$ hold,
then problem (3.106) has at least two distinct nontrivial solutions.

PROOF   Let $\varphi\colon W^{1,2}_{per}(T) \longrightarrow \mathbb{R}$ be the locally Lipschitz functional, defined by

$$\varphi(x) \stackrel{df}{=} \frac{1}{2}\|x'\|_2^2 - \int_0^b j(t, x(t))\, dt \qquad \forall\, x \in W^{1,2}_{per}(T).$$

We claim that φ is coercive. Suppose that this is not the case. We can find a sequence $\{x_n\}_{n\geq 1} \subseteq W^{1,2}_{per}(T)$, such that

$$\|x_n\|_{W^{1,2}(T)} \longrightarrow +\infty \qquad \text{and} \qquad \varphi(x_n) \leq M_3 \qquad \forall\, n \geq 1,$$

for some $M_3 > 0$. Let us set

$$y_n \stackrel{df}{=} \frac{x_n}{\|x_n\|_{W^{1,2}(T)}} \qquad \forall\, n \geq 1.$$

Passing to a subsequence if necessary, we may assume that

$$y_n \stackrel{w}{\longrightarrow} y \text{ in } W^{1,2}_{per}(T),$$
$$y_n \longrightarrow y \text{ in } C_{per}(T).$$

We have

$$\frac{\varphi(x_n)}{\|x_n\|_{W^{1,2}(T)}^2} = \frac{1}{2}\|y_n'\|_2^2 - \int_0^b \frac{j(t, x_n(t))}{\|x_n\|_{W^{1,2}(T)}^2}\, dt \leq \frac{M_3}{\|x_n\|_{W^{1,2}(T)}^2}.$$

As before from hypothesis $H(j)_{12}(iii)$ and Proposition 1.3.14, we see that

$$|j(t, \zeta)| \leq a_4(t) + c_5|\zeta|^2 \quad \text{for a.a. } t \in T \text{ and all } \zeta \in \mathbb{R}, \tag{3.107}$$

with $a_4 \in L^1(T)_+$ and $c_5 > 0$. Hence

$$\frac{|j(t, x_n(t))|}{\|x_n\|_{W^{1,2}(T)}^2} \leq \frac{a_4(t)}{\|x_n\|_{W^{1,2}(T)}^2} + c_5|y_n(t)|^2.$$

So from the Dunford-Pettis Theorem (see Theorem A.3.14), passing to a subsequence if necessary, we may assume that

$$\frac{j(\cdot, x_n(\cdot))}{\|x_n\|_{W^{1,2}(T)}^2} \stackrel{w}{\longrightarrow} \gamma \text{ in } L^1(T),$$

for some $\gamma \in L^1(T)$. Note that

$$|x_n(t)| \longrightarrow +\infty \qquad \forall\, t \in \{y \neq 0\}.$$

Let $\varepsilon > 0$ and define

$$C_n(\varepsilon) \stackrel{df}{=} \left\{ t \in T : \ x_n(t) \neq 0 \text{ and } \frac{j(t, x_n(t))}{x_n(t)^2} \leq h(t) + \varepsilon \right\}.$$

Also let $\chi_n(t) \overset{df}{=} \chi_{C_n}(t)$. By virtue of hypothesis $H(j)_{12}(iv)$, we have

$$\chi_n(t) \longrightarrow 1 \quad \text{for a.a. } t \in \{y \neq 0\}.$$

Moreover

$$\frac{j\big(t,x_n(t)\big)}{\|x_n\|^2_{W^{1,2}(T)}}\chi_n(t) = \frac{j\big(t,x_n(t)\big)}{|x_n(t)|^2}\big|y_n(t)\big|^2\chi_n(t)$$

$$\leq \big(h(t)+\varepsilon\big)\big|y_n(t)\big|^2\chi_n(t) \quad \text{for a.a. } t \in T \text{ and all } n \geq 1.$$

Passing to the weak limit in $L^1(\{y \neq 0\})$ and using Proposition 1.2.12, we obtain

$$\gamma(t) \leq \big(h(t)+\varepsilon\big)y(t)^2 \quad \text{for a.a. } t \in \{y \neq 0\}.$$

Let $\varepsilon \searrow 0$, to obtain

$$\gamma(t) \leq h(t)\big(y(t)\big)^2 \quad \text{for a.a. } t \in \{y \neq 0\}.$$

On the other hand, it is clear that $\gamma(t) = 0$ for almost all $t \in \{y = 0\}$. So finally, we have

$$\gamma(t) \leq h(t)\big(y(t)\big)^2 \quad \text{for a.a. } t \in T$$

and so

$$\gamma(t) = \overline{h}(t)\big(y(t)\big)^2 \quad \text{for a.a. } t \in T,$$

with $\overline{h} \in L^1(T)$, such that $\overline{h}(t) \leq h(t)$ for almost all $t \in T$. In the limit as $n \to +\infty$, we obtain

$$\frac{1}{2}\|y'\|^2_2 \leq \int_0^b \overline{h}(t)\big(y(t)\big)^2\, dt \leq 0,$$

hence $y \equiv \overline{\zeta} \in \mathbb{R}$. If $\overline{\zeta} = 0$, then

$$y_n \longrightarrow 0 \quad \text{in } W^{1,2}_{per}(T),$$

a contradiction since $\|y_n\|_{W^{1,2}(T)} = 1$ for $n \geq 1$. So $\overline{\zeta} \neq 0$ and we have

$$0 \leq \int_0^b \overline{h}(t)\overline{\zeta}^2\, dt < 0,$$

a contradiction. This proves the coercivity of φ. In particular then φ is bounded below and satisfies the nonsmooth PS-condition.

Next let

$$W_k \overset{df}{=} \left\{ x \in W^{1,2}_{per}(T) : x(t) = \sum_{i=0}^{k}\left(a_i \sin\frac{2\pi i}{b}t + b_i \cos\frac{2\pi i}{b}t\right),\ a_i, b_i \in \mathbb{R}\right\}.$$

This is the direct sum of the eigenspaces N_i corresponding to λ_i, $i \in \{0, 1, \dots, k\}$. Let $\varrho_1 > 0$ be as in hypothesis $H(j)_{12}(v)$. From the continuity of the embedding $W^{1,2}_{\mathrm{per}}(T) \subseteq C_{\mathrm{per}}(T)$, we can find $\varrho_2 > 0$, such that

$$|v(t)| \leq \varrho_1 \qquad \forall\, t \in T, \ v \in W_k, \ \|v\|_{W^{1,2}(T)} \leq \varrho_2.$$

Then, from hypothesis $H(j)_{12}(v)$ and since $\dfrac{\|v'\|_2^2}{\|v\|_2^2} \leq \lambda_k$ (by Fourier expansion in $L^2(T)$ and $W^{1,2}_{\mathrm{per}}(T)$), we have

$$\varphi(v) \leq \frac{1}{2} \|v'\|_2^2 - \frac{\lambda_k}{2} \|v\|_2^2 \leq 0 \qquad \forall\, v \in W_k, \ \|v\|_{W^{1,2}(T)} \leq \varrho_2.$$

Next consider $W_k^\perp$. We have

$$
\begin{aligned}
\varphi(w) &= \frac{1}{2} \|w'\|_2^2 - \int_0^b j\big(t, w(t)\big)\, dt \\
&= \frac{1}{2} \|w'\|_2^2 - \int_{\{|w| \leq \varrho_1\}} j\big(t, w(t)\big)\, dt \\
&\qquad - \int_{\{|w| > \varrho_1\}} j\big(t, w(t)\big)\, dt \qquad \forall\, w \in W_k^\perp.
\end{aligned}
$$

Because of (3.107), for a given $s > 2$, we can find $a_5 \in L^1(T)_+$, such that

$$j(t, \zeta) \leq a_5(t)|\zeta|^s \quad \text{for a.a. } t \in T \text{ and all } \zeta \in \mathbb{R} \text{ with } |\zeta| > \varrho_1.$$

Therefore, using hypothesis $H(j)_{12}(v)$, we obtain

$$\varphi(w) \geq \frac{1}{2} \|w'\|_2^2 - \frac{\mu}{2} \|w\|_2^2 - c_6 \|w\|_{W^{1,2}(T)}^s,$$

for some $c_6 > 0$. Since $\dfrac{\|w'\|_2^2}{\|w\|_2^2} \geq \lambda_{k+1} > \mu$ and using the Poincaré-Wirtinger inequality (see Theorem 1.1.7), we have

$$\varphi(w) \geq c_7 \|w\|_{W^{1,2}(T)}^2 - c_6 \|w\|_{W^{1,2}(T)}^s,$$

for some $c_7 > 0$. Because $s > 2$, we can find $\varrho_3 \in \big(0, \min\{\varrho_2, 1\}\big)$, such that

$$\varrho(w) \geq 0 \qquad \forall\, w \in W_k^\perp, \ \|w\|_{W^{1,2}(T)} \leq \varrho_3.$$

Therefore finally we can apply Theorem 2.4.1 and obtain two distinct nontrivial critical points of φ, which are of course two distinct solutions of problem (3.106). $\qquad\square$

EXAMPLE 3.4.7 Consider the locally Lipschitz potential

$$j(t,\zeta) \overset{df}{=} \begin{cases} e^\zeta - x^2 + \mu + 1 - \frac{1}{e} & \text{if } \zeta < -1, \\ \mu\zeta^6 & \text{if } |\zeta| \le 1, \\ h(t)\zeta^2 - \zeta \ln \zeta^2 + \mu - h(t) & \text{if } \zeta > 1. \end{cases}$$

Here $\lambda_l \le \mu < \lambda_{k+1}$ and $h \in L^1(T)_+$ is such that $\int_0^b h(t)\,dt < 0$. Then this function satisfies hypotheses $H(j)_{12}$. □

3.4.5 Nonlinear Eigenvalue Problems

We investigate the following nonlinear eigenvalue problem:

$$\begin{cases} -\left(|x'(t)|^{p-2}x'(t)\right)' - \lambda|x(t)|^{p-2}x(t) \in \partial j\big(t, x(t)\big) \\ \qquad\qquad\qquad\qquad\qquad\qquad\qquad\qquad \text{for a.a. } t \in T, \\ x(0) = x(b), \ x'(0) = x'(b), \end{cases} \quad (3.108)$$

for $p \in (1, +\infty)$ and $\lambda \in \mathbb{R}$. Our aim is to show that this problem has at least three distinct solutions as $\lambda \longrightarrow 0^-$ (problem near resonance).

$\underline{H(j)_{13}}$ $j \colon T \times \mathbb{R} \longrightarrow \mathbb{R}$ is a function, such that

(*i*) for all $\zeta \in \mathbb{R}$, the function

$$T \ni t \longmapsto j(t,\zeta) \in \mathbb{R}$$

is measurable and $j(\cdot, 0) \in L^1(T)$;

(*ii*) for almost all $t \in T$, the function

$$\mathbb{R} \ni \zeta \longmapsto j(t,\zeta) \in \mathbb{R}$$

is locally Lipschitz;

(*iii*) for almost all $t \in T$, all $\zeta \in \mathbb{R}$ and all $u \in \partial j(t,\zeta)$, we have

$$|u| \le a(t) + c(t)|\zeta|^{r-1}$$

with $a, c \in L^\infty(T)_+$, $r \in [1, p)$;

(*iv*) we have

$$\lim_{|\zeta| \to +\infty} \big(\zeta u(\zeta) - pj(t,\zeta)\big) = -\infty$$

uniformly for almost all $t \in T$ and all $u(\zeta) \in \partial j(t,\zeta)$;

(*v*) there exist $M > 0$ and $\zeta_0 > 0$, such that

$$\zeta u \ge \zeta_0 > 0 \quad \text{for a.a. } t \in T \text{ and all } \zeta \in \mathbb{R}, \ |\zeta| \ge M$$
$$\text{and all } u \in \partial j(t,\zeta).$$

THEOREM 3.4.13

If hypotheses $H(j)_{13}$ hold,

then there exists $\widehat{\lambda} < 0$, such that for all $\lambda \in [\widehat{\lambda}, 0)$, problem (3.108) has at least three distinct solutions.

PROOF   For $\lambda < 0$, let $\varphi_\lambda \colon W^{1,p}_{\text{per}}(T) \longrightarrow \mathbb{R}$ be the locally Lipschitz functional, defined by

$$\varphi_\lambda(x) \stackrel{df}{=} \frac{1}{p} \|x'\|_p^p - \frac{\lambda}{p} \|x\|_p^p - \int_0^b j\big(t, x(t)\big)\, dt.$$

We know that

$$\big|j(t, \varsigma)\big| \leq a_1(t) + c_1 |\varsigma|^r \quad \text{for a.a. } t \in T \text{ and all } \varsigma \in \mathbb{R},$$

with $a_1 \in L^1(T)_+$ and $c_1 > 0$. Since $\lambda < 0$, we have that

$$\varphi_\lambda(x) \geq c_2(\lambda) \|x\|^p_{W^{1,p}(T)} - c_3 \|x\|^r_{W^{1,p}(T)} - c_4 \quad \forall\, x \in W^{1,p}_{\text{per}}(T),$$

for some $c_2(\lambda), c_3, c_4 > 0$. But $r < p$, so from the last inequality, we infer that for every $\lambda < 0$, φ_λ is coercive. Hence, it is bounded below and satisfies the nonsmooth PS-condition.

We consider the direct sum decomposition

$$W^{1,p}_{\text{per}}(T) = \mathbb{R} \oplus V,$$

with

$$V \stackrel{df}{=} \left\{ v \in W^{1,p}_{\text{per}}(T) : \int_0^b v(t)\, dt = 0 \right\}.$$

From Remark 1.1.11, we know that

$$\|v\|_\infty \leq b^{\frac{1}{p'}} \|v'\|_p \quad \forall\, v \in V.$$

So, we have

$$\varphi_\lambda(v) \geq \frac{1}{p}\big(1 - \lambda b^p\big) \|v'\|_p^p - c_5 \|v'\|_p^r - c_6 \quad \forall\, v \in V,$$

for some $c_5, c_6 > 0$ and so $\varphi_\lambda|_V$ is coercive uniformly for $\lambda < 0$.

Therefore, we can find $\beta > 0$, such that

$$\varphi_\lambda(v) \geq -\beta \quad \forall\, \lambda < 0,\ v \in V.$$

We consider the following two disjoint open subsets of $W^{1,p}_{\text{per}}(T)$:

$$U^+ = \left\{ x \in W^{1,p}_{\text{per}}(T) : \int_0^b x(t)\, dt > 0 \right\}$$

$$U^- = \left\{ x \in W^{1,p}_{\text{per}}(T) : \int_0^b x(t)\, dt < 0 \right\}.$$

We shall show that there exist $\zeta_1 \in \mathbb{R}$ and $\widehat{\lambda} < 0$, such that

$$\varphi_\lambda(\pm\zeta_1) < -\beta \qquad \forall \, \lambda \in [\widehat{\lambda}, 0). \tag{3.109}$$

First note that for a given $\zeta \in \mathbb{R}$, we have

$$\varphi_\lambda(\zeta) = -\frac{\lambda}{p}|\zeta|^p b - \int_0^b j(t, \zeta) \, dt.$$

By virtue of hypothesis $H(j)_{13}(iv)$, for a given $\eta > \frac{p\beta}{b}$, with $\eta \geq M$, we can find $\zeta_1 > \eta$, such that

$$\zeta u - pj(t, \zeta) \leq -\eta \quad \text{for a.a. } t \in T, \text{ all } |\zeta| \geq \zeta_1 \text{ and all } u \in \partial j(t, \zeta). \tag{3.110}$$

Also by hypothesis $H(j)_{13}(v)$, we can find $\widehat{\lambda} < 0$, such that

$$-\widehat{\lambda}\zeta_1^p \leq \zeta_1 u \quad \text{for a.a. } t \in T \text{ and all } u \in \partial j(t, \zeta_1). \tag{3.111}$$

Using (3.110) and (3.111), we have

$$\varphi_{\widehat{\lambda}}(\zeta_1) = -\frac{\widehat{\lambda}}{p}\zeta_1^p b - \int_0^b j(t, \zeta_1) \, dt \leq \frac{1}{p}\int_0^b (\zeta_1 u - pj(t, \zeta_1)) \, dt$$

$$\leq -\frac{\eta}{p}b < -\beta \qquad \forall \, u \in \partial j(t, \zeta_1).$$

Therefore, we have

$$\varphi_\lambda(\zeta_1) < -\beta \qquad \forall \, \lambda \in [\widehat{\lambda}, 0).$$

Similarly, we show that

$$\varphi_\lambda(-\zeta_1) < -\beta \qquad \forall \, \lambda \in [\widehat{\lambda}, 0).$$

So we have showed (3.109).

Now let

$$m_\pm^\lambda \stackrel{df}{=} \inf_{U^\pm} \varphi_\lambda.$$

By virtue of the coercivity of φ_λ, the numbers $m_\pm^\lambda$ are finite and in addition

$$m_\pm^\lambda < -\beta \qquad \forall \, \lambda \in [\widehat{\lambda}, 0).$$

Moreover, since φ_λ satisfies the nonsmooth PS-condition, we can find $x_\pm^\lambda \in \overline{U^\pm}$, such that

$$\varphi_\lambda\left(x_\pm^\lambda\right) = m_\pm^\lambda \qquad \forall \, \lambda \in [\widehat{\lambda}, 0).$$

If $x_\pm^\lambda \in \partial U^\pm = V$, then $\varphi_\lambda\left(x_\pm^\lambda\right) \geq -\beta$, a contradiction to the fact that $m_\pm^\lambda < -\beta$. So $x_\pm^\lambda \in U^\pm$ and it follows that $x_\pm^\lambda$ are two distinct local minima of φ_λ, thus two distinct critical points of φ_λ. On the other hand, because $\varphi_\lambda|_V > -\beta > \varphi_\lambda(\pm\zeta_1)$, we can apply the nonsmooth Saddle Point Theorem (see Theorem 2.1.4) and obtain $y_\lambda \in W^{1,p}_{per}(T)$, such that

$$\varphi_\lambda(y_\lambda) \geq -\beta > m^\lambda_\pm = \varphi_\lambda(x^\lambda_\pm)$$

and $0 \in \partial\varphi_\lambda(y_\lambda)$. This is the third distinct from the previous two, critical points of φ_λ. Then $x^\lambda_\pm$, y_λ for $\lambda \in [\widehat{\lambda}, 0)$ are the three distinct solutions of (3.108). ▯

EXAMPLE 3.4.8 The potential function

$$j(\zeta) \stackrel{df}{=} \begin{cases} -\zeta & \text{if } \zeta \leq 0, \\ a\zeta^r + \max\{\zeta, \zeta\ln\zeta\} & \text{if } \zeta > 0, \end{cases}$$

with $a > 0$, $r \in [1, p)$, satisfies hypotheses $H(j)_{13}$. ▯

3.4.6 Problems with Nonlinear Boundary Conditions

In this subsection, using a variational approach based on the theory developed in Section 2.3, we study problems with nonlinear multivalued boundary conditions analogous but more restrictive than the ones considered in Section 3.3. The problem under consideration is the following:

$$\begin{cases} -\left(\|x'(t)\|^{p-2}_{\mathbb{R}^N} x'(t)\right)' \in \partial j(t, x(t)) & \text{for a.a. } t \in T, \\ \left(\varphi_p(x'(0)), -\varphi_p(x'(b))\right) \in \Xi(x(0), x(b)), \end{cases} \qquad (3.112)$$

for $p \in (1, +\infty)$. Here $\varphi_p \colon \mathbb{R}^N \longrightarrow \mathbb{R}^N$ is the homeomorphism (analytic diffeomorphism on $\mathbb{R}^N \setminus \{0\}$), defined by

$$\varphi_p(\xi) \stackrel{df}{=} \begin{cases} \|\xi\|^{p-2}_{\mathbb{R}^N} \xi & \text{if } \xi \neq 0, \\ 0 & \text{if } \xi = 0, \end{cases}$$

for $p \in (1, +\infty)$. Our hypotheses on the data of the problem (3.112) are the following:

$\underline{H(j)_{14}}$ $j \colon T \times \mathbb{R}^N \longrightarrow \mathbb{R}$ is a function, such that

(i) for all $\xi \in \mathbb{R}^N$, the function

$$T \ni t \longmapsto j(t, \xi) \in \mathbb{R}$$

is measurable and $j(\cdot, 0) \in L^1(T)$;

(ii) for almost all $t \in T$, the function

$$\mathbb{R}^N \ni \xi \longmapsto j(t, \xi) \in \mathbb{R}$$

is locally Lipschitz;

(iii) for all $r > 0$ there exists $a_r \in L^{p'}(T)_+$, with $\frac{1}{p} + \frac{1}{p'} = 1$, such that for almost all $t \in T$, all $\xi \in \mathbb{R}^N$ with $\|\xi\|_{\mathbb{R}^N} \leq r$ and all $u \in \partial j(t, \xi)$, we have $\|u\|_{\mathbb{R}^N} \leq a_r(t)$;

(iv) there exist $M > 0$ and $\mu > p$, such that for almost all $t \in T$ and all $\xi \in \mathbb{R}^N$ with $\|\xi\|_{\mathbb{R}^N} \geq M$, we have $\mu j(t, \xi) \leq -j^0(t, \xi; -\xi)$;

(v) there exists $h \in L^1(T)$, such that $\|h\|_1 < \frac{1}{b^{p-1}}$, $\int_0^b h(t)\, dt \neq 0$ and

$$\lim_{\|\xi\|_{\mathbb{R}^N} \to +\infty} \frac{pj(t, \xi)}{\|\xi\|_{\mathbb{R}^N}^p} = h(t) \qquad \text{uniformly for a.a. } t \in T.$$

<u>$H(\Xi)$</u> $\Xi \colon \mathbb{R}^N \times \mathbb{R}^N \longrightarrow 2^{\mathbb{R}^N \times \mathbb{R}^N}$ is a multifunction, such that

(i) $\Xi = \partial \vartheta$, with $\vartheta \in \Gamma_0(\mathbb{R}^N \times \mathbb{R}^N)$ such that $0 = \vartheta(0, 0) = \inf \vartheta$;

(ii) Ξ is continuous at some point in $\mathbb{R}^N \times \mathbb{R}^N$;

(iii) we have

$$\vartheta(\lambda \xi, \lambda \xi') \leq \lambda^{c_0} \vartheta(\xi, \xi') \qquad \forall\, \lambda \geq 2,\ (\xi, \xi') \neq (0, 0),$$

with $c_0 \leq \log_2(\mu + 1)$ and

$$\lim_{\substack{\|\xi\|_{\mathbb{R}^N} + \|\xi'\|_{\mathbb{R}^N} \to +\infty \\ (\xi, \xi') \in \mathrm{dom}\, \vartheta}} \frac{\vartheta(\xi, \xi')}{\left(\|\xi\|_{\mathbb{R}^N} + \|\xi'\|_{\mathbb{R}^N}\right)^p} = 0;$$

(iv) there exist $\varrho \geq M$ and $\zeta_0 < 0$, such that

$$-\int_0^b j(t, \xi)\, dt + \vartheta(\xi, \xi) \leq \zeta_0 \qquad \forall\, \xi \in \mathbb{R}^N,\ \|\xi\|_{\mathbb{R}^N} = \varrho.$$

Let $\Phi \colon W^{1,p}(T; \mathbb{R}^N) \longrightarrow \mathbb{R}$ be the locally Lipschitz functional, defined by

$$\Phi(x) \stackrel{df}{=} \frac{1}{p} \|x'\|_p^p - \int_0^b j(t, x(t))\, dt.$$

Also let $\psi \in \Gamma_0\big(W^{1,p}(T; \mathbb{R}^N)\big)$ be the functional, defined by

$$\psi(x) \stackrel{df}{=} \vartheta\big(x(0), x(b)\big) \qquad \forall\, x \in W^{1,p}(T; \mathbb{R}^N).$$

Then the energy functional for problem (3.112) is $\varphi = \Phi + \psi$, which has the form of the functionals considered in Section 2.3.

PROPOSITION 3.4.20
If hypotheses $H(j)_{14}$ and $H(\Xi)$ hold,
<u>*then*</u> *φ satisfies the generalized nonsmooth PS-condition (see Definition 2.3.1).*

PROOF   Let $\{x_n\}_{n\geq 1} \subseteq W^{1,p}(T;\mathbb{R}^N)$ be a sequence, such that

$$|\varphi(x_n)| \leq M_1 \qquad \forall\, n \geq 1 \tag{3.113}$$

and

$$\Phi^0(x_n; y - x_n) + \psi(y) - \psi(x_n)$$
$$\geq -\varepsilon_n \|y - x_n\|_{W^{1,p}(T;\mathbb{R}^N)} \qquad \forall\, n \geq 1,\ y \in W^{1,p}(T;\mathbb{R}^N),$$

for some $M_1 > 0$ and $\varepsilon_n \searrow 0$, where by $\|\cdot\|_{W^{1,p}(T;\mathbb{R}^N)}$ we denote the norm of $W^{1,p}(T;\mathbb{R}^N)$. Let $x_n^* \in \partial\Phi(x_n)$ be such that

$$\langle x_n^*, x_n\rangle_{W^{1,p}(T;\mathbb{R}^N)} = \Phi^0(x_n; x_n)$$

(recall that $\partial\Phi(x_n) \subseteq \left(W^{1,p}(T;\mathbb{R}^N)\right)^*$ is weakly compact and $\Phi^0(x_n; \cdot)$ is the support function of $\partial\Phi(x_n)$). We have

$$x_n^* = A(x_n) - u_n \qquad \forall\, n \geq 1,$$

with $u_n \in S^{p'}_{\partial j(\cdot, x_n(\cdot))}$. Using this in (3.113) and taking as test function $y = 2x_n$, we obtain

$$-\|x_n'\|_p^p - \int_0^b \left(u_n(t), -x_n(t)\right)_{\mathbb{R}^N} dt$$
$$+ \vartheta\big(x_n(0), x_n(b)\big) - \vartheta\big(2x_n(0), 2x_n(b)\big) \leq \varepsilon_n \|x_n\|_{W^{1,p}(T;\mathbb{R}^N)} \qquad \forall\, n \geq 1,$$

so for all $n \geq 1$, we have

$$-\|x_n'\|_p^p - \int_0^b j^0\big(t, x_n(t); -x_n(t)\big)\, dt$$
$$+ \vartheta\big(x_n(0), x_n(b)\big) - \vartheta\big(2x_n(0), 2x_n(b)\big) \leq \varepsilon_n \|x_n\|_{W^{1,p}(T;\mathbb{R}^N)}. \tag{3.114}$$

On the other hand, from the choice of the sequence $\{x_n\}_{n\geq 1} \subseteq W^{1,p}(T;\mathbb{R}^N)$, we have

$$\frac{\mu}{p} \|x_n'\|_p^p - \int_0^b \mu j\big(t, x_n(t)\big)\, dt + \mu\vartheta\big(x_n(0), x_n(b)\big) \leq \mu M_1. \tag{3.115}$$

Adding (3.114) and (3.115), we obtain

$$\left(\frac{\mu}{p} - 1\right) \|x_n'\|_p^p + \int_0^b \left(-j^0\big(t, x_n(t); -x_n(t)\big) - \mu j\big(t, x_n(t)\big)\right) dt$$

$$+ (\mu + 1)\psi(x_n) - \psi(2x_n) \leq \varepsilon_n \|x_n\|_{W^{1,p}(T;\mathbb{R}^N)} + \mu M_1. \qquad (3.116)$$

From hypotheses $H(\Xi)$, we have

$$\psi(2x_n) \leq (\mu + 1)\psi(x_n) \qquad \forall\, n \geq 1.$$

Also note that

$$|j(t, \xi)| \leq a_1(t) + c_1 \|\zeta\|_{\mathbb{R}^N}^p \qquad \text{for a.a. } t \in T \text{ and all } \xi \in \mathbb{R}^N,$$

with $a_1 \in L^1(T)_+$ and $c_1 > 0$. This together with hypothesis $H(j)_{14}(iv)$ imply that

$$\int_0^b \left(-j^0\big(t, x_n(t); -x_n(t)\big) - \mu j\big(t, x_n(t)\big)\right) dt \geq -M_2 \qquad \forall\, n \geq 1,$$

for some $M_2 > 0$. So returning to (3.116), we obtain

$$\left(\frac{\mu}{p} - 1\right) \|x_n'\|_p^p \leq \varepsilon_n \|x_n\|_{W^{1,p}(T;\mathbb{R}^N)} + M_3 \qquad \forall\, n \geq 1, \qquad (3.117)$$

with $M_3 \stackrel{df}{=} (M_2 + \mu M_1)$. Using (3.117), we shall show that the sequence $\{x_n\}_{n \geq 1} \subseteq W^{1,p}(T;\mathbb{R}^N)$ is bounded. Suppose that this is not the case. Passing to a subsequence, we may assume that

$$\|x_n\|_{W^{1,p}(T;\mathbb{R}^N)} \longrightarrow +\infty.$$

Let us set

$$y_n \stackrel{df}{=} \frac{x_n}{\|x_n\|_{W^{1,p}(T;\mathbb{R}^N)}} \qquad \forall\, n \geq 1.$$

At least for a subsequence, we have that

$$y_n \stackrel{w}{\longrightarrow} y \text{ in } W^{1,p}(T;\mathbb{R}^N),$$
$$y_n \longrightarrow y \text{ in } C(T;\mathbb{R}^N).$$

Dividing (3.117) by $\|x_n\|_{W^{1,p}(T;\mathbb{R}^N)}^p$ and passing to the limit as $n \to +\infty$, we obtain

$$y_n \longrightarrow y \text{ in } W^{1,p}(T;\mathbb{R}^N),$$

with $y \equiv \overline{\xi} \in \mathbb{R}^N$ and since $\|y_n\|_{W^{1,p}(T;\mathbb{R}^N)} = 1$, we have that $\overline{\xi} \neq 0$. So

$$\|x_n(t)\|_{\mathbb{R}^N} \longrightarrow +\infty \qquad \text{uniformly in } t \in T$$

(see the proof of Proposition 3.4.5). Then by virtue of hypothesis $H(j)_{14}(v)$, we have that

$$\int\limits_0^b \frac{j\big(t, x_n(t)\big)}{\|x_n\|_{W^{1,p}(T;\mathbb{R}^N)}^p}\, dt \;\longrightarrow\; \frac{1}{p}\int\limits_0^b h(t)\|\widehat{\xi}\|_{\mathbb{R}^N}^p\, dt. \qquad (3.118)$$

Also since $\big(x_n(0), x_n(b)\big) \in \operatorname{dom}\vartheta$ (see (3.113)), from hypotheses $H(\Xi)$, we have that

$$0 \;\leq\; \frac{\vartheta\big(x_n(0), x_n(b)\big)}{\|x_n\|_{W^{1,p}(T;\mathbb{R}^N)}^p} \;\leq\; \frac{\vartheta\big(x_n(0), x_n(b)\big)}{c_2\,\big(\|x_n(0)\|_{\mathbb{R}^N} + \|x_n(b)\|_{\mathbb{R}^N}\big)^p}, \qquad (3.119)$$

for some $c_2 > 0$. Note that the last quantity tends to 0 as $n \to +\infty$. From (3.113), we have

$$-\frac{M_1}{\|x_n\|_{W^{1,p}(T;\mathbb{R}^N)}^p} \;\leq\; \frac{1}{p}\|y_n'\|_p^p - \int\limits_0^b \frac{j\big(t, x_n(t)\big)}{\|x_n\|_{W^{1,p}(T;\mathbb{R}^N)}^p}\, dt + \frac{\vartheta\big(x_n(0), x_n(b)\big)}{\|x_n\|_{W^{1,p}(T;\mathbb{R}^N)}^p}$$

$$\leq\; \frac{M_1}{\|x_n\|_{W^{1,p}(T;\mathbb{R}^N)}^p} \qquad \forall\, n \geq 1.$$

Passing to the limit as $n \to +\infty$, we obtain

$$\frac{1}{p}\int\limits_0^b h(t)\|\widehat{\xi}\|_{\mathbb{R}^N}^p\, dt \;=\; 0,$$

a contradiction to the fact that $\int\limits_0^b h(t)\, dt \neq 0$ (see hypotheses $H(\Xi)$). This proves the boundedness of the sequence $\{x_n\}_{n \geq 1} \subseteq W^{1,p}(T;\mathbb{R}^N)$.

So, passing to a subsequence if necessary, we may assume that

$$x_n \xrightarrow{\;w\;} x \text{ in } W^{1,p}(T;\mathbb{R}^N),$$
$$x_n \longrightarrow x \text{ in } C(T;\mathbb{R}^N).$$

We have

$$-\varepsilon\,\|y - x_n\|_{W^{1,p}(T;\mathbb{R}^N)}$$
$$\leq\; \Phi^0(x_n; y - x_n) + \psi(y) - \psi(x_n) \qquad \forall\, y \in W^{1,p}(T;\mathbb{R}^N),\; n \geq 1.$$

Let us take $y = x$ and $x_n^* \in \partial\Phi(x_n)$, such that

$$\langle x_n^*, x - x_n \rangle_{W^{1,p}(T;\mathbb{R}^N)} \;=\; \Phi^0(x_n; x - x_n) \qquad \forall\, n \geq 1.$$

We have

$$x_n^* \;=\; A(x_n) - v_n \qquad \forall\, n \geq 1,$$

with $v_n \in S^{p'}_{\partial j(\cdot, x_n(\cdot))}$ for $n \geq 1$. So

$$\langle A(x_n), x_n - x \rangle_{W^{1,p}(T;\mathbb{R}^N)} - \int_0^b u_n(t) \big(x_n(t) - x(t)\big)\, dt + \psi(x_n) - \psi(x)$$

$$\leq \varepsilon_n \|x - x_n\|_{W^{1,p}(T;\mathbb{R}^N)}$$

and so

$$\limsup_{n \to +\infty} \langle A(x_n), x_n - x \rangle_{W^{1,p}(T;\mathbb{R}^N)} \leq 0.$$

Since A is maximal monotone, it is generalized pseudomonotone and so

$$\langle A(x_n), x_n \rangle_{W^{1,p}(T;\mathbb{R}^N)} \longrightarrow \langle A(x), x \rangle_{W^{1,p}(T;\mathbb{R}^N)},$$

from which we conclude that

$$x_n \longrightarrow x \quad \text{in } W^{1,p}\big(T; \mathbb{R}^N\big).$$

$\square$

We consider the direct sum decomposition

$$W^{1,p}\big(T; \mathbb{R}^N\big) = \mathbb{R}^N \oplus V,$$

where

$$V \overset{df}{=} \left\{ v \in W^{1,p}\big(T; \mathbb{R}^N\big) : \int_0^b v(t)\, dt = 0 \right\}.$$

PROPOSITION 3.4.21

If hypotheses $H(j)_{14}$ and $H(\Xi)$ hold, <u>then</u> $\vartheta|_V$ is coercive.

PROOF By virtue of hypothesis $H(j)_{14}(v)$, for a given $\varepsilon > 0$, we can find $M_4 = M_4(\varepsilon) > 0$, such that

$$\frac{1}{p}\big(h(t) - \varepsilon\big) \|\xi\|^p_{\mathbb{R}^N} \leq j(t, \xi) \quad \text{for a.a. } t \in T \text{ and all } \xi \in \mathbb{R}^N, \; \|\xi\|_{\mathbb{R}^N} \geq M_4.$$

Also since

$$|j(t, \xi)| \leq a_1(t) + c_1 \|\xi\|^p_{\mathbb{R}^N} \quad \text{for a.a. } t \in T \text{ and all } \xi \in \mathbb{R}^N,$$

with $a_1 \in L^1(T)_+$ and $c_1 > 0$ (see the proof of Proposition 3.4.20), we have

$$|j(t, \xi)| \leq a_2(t) \quad \text{for a.a. } t \in T \text{ and all } \xi \in \mathbb{R}^N, \; \|\xi\|_{\mathbb{R}^N} < M_4,$$

with $a_2 \in L^1(T)_+$. Then, we have

$$\varphi(v) = \frac{1}{p} \|v'\|_p^p - \int_0^b j(t, v(t)) \, dt + \vartheta(v(0), v(b))$$

$$\geq \frac{1}{p} \|v'\|_p^p - \int_{\{\|v(t)\|_{\mathbb{R}^N} \geq M_4\}} \frac{1}{p} (h(t) - \varepsilon) \|v(t)\|_{\mathbb{R}^N}^p \, dt - \int_{\{\|v(t)\|_{\mathbb{R}^N} < M_4\}} a_2(t) \, dt$$

$$\geq \frac{1}{p} \|v'\|_p^p - \frac{1}{p} \|h\|_1 \|v\|_\infty^p - \frac{\varepsilon}{p} \|v\|_\infty^p b - c_3 \qquad \forall \, v \in V, \qquad (3.120)$$

for some $c_3 > 0$. Recall that

$$\|v\|_\infty^p \leq b^{p-1} \|v'\|_p^p \qquad \forall \, v \in V$$

(see Remark 1.1.11). So, we obtain

$$\varphi(v) \geq \frac{1}{p} \left(1 - (\|h\|_1 + \varepsilon b) \, b^{p-1}\right) \|v'\|_p^p - c_3.$$

By hypothesis $H(j)_{14}(v)$, we have $\|h\|_1 < \frac{1}{b^{p-1}}$. So, if $\varepsilon > 0$ is small, we see that

$$1 - (\|h\|_1 + \varepsilon b) \, b^{p-1} > 0$$

and so we conclude that $\varphi|_V$ is coercive.   □

In particular this proposition implies that we can find $\beta_0 > 0$, such that

$$\varphi(v) \geq -\beta_0 \qquad \forall \, v \in V.$$

PROPOSITION 3.4.22
*If hypotheses $H(j)_{14}$ and $H(\Xi)$ hold,
then there exists ϱ_1, such that*

$$\varphi(\xi) < -\beta_0 \qquad \forall \, \xi \in \mathbb{R}^N, \ \|\xi\|_{\mathbb{R}^N} = \varrho_1.$$

PROOF   From the proof of Proposition 3.4.5, we know that due to hypothesis $H(j)_{14}(iv)$, we have

$$\lambda^\mu j(t, \xi) \leq j(t, \lambda \xi) \quad \text{for a.a. } t \in T \text{ and all } \xi \in \mathbb{R}^N, \ \|\xi\|_{\mathbb{R}^N} \geq M, \ \lambda > 1.$$

Let $\varrho \geq M$ as postulated by hypothesis $H(\Xi)(iv)$. Then, for $\xi \in \mathbb{R}^N$ with $\|\xi\|_{\mathbb{R}^N} = \varrho$ and $\lambda > 1$, we have

$$\varphi(\lambda \xi) = -\int_0^b j(t, \lambda \xi) \, dt + \vartheta(\lambda \xi, \lambda \xi)$$

$$\leq -\lambda^\mu \int_0^b j(t, \xi) \, dt + \lambda^{c_0} \vartheta(\xi, \xi) \leq -\lambda^\mu \int_0^b j(t, \xi) \, dt + \lambda^\mu \vartheta(\xi, \xi),$$

since $\lambda > 1$, $\mu \geq c_0$ and $\vartheta \geq 0$. So

$$\frac{\varphi(\lambda\xi)}{\lambda^\mu} \leq -\int_0^b j(t,\xi)\,dt + \vartheta(\xi,\xi) \leq \zeta_0 < 0,$$

with $\zeta_0 < 0$ as in hypothesis $H(\Xi)_{14}(iv)$ and thus

$$\lim_{\lambda\to+\infty} \frac{\varphi(\lambda\xi)}{\lambda^\mu} \leq \zeta_0 < 0.$$

So it follows that

$$\varphi(\lambda\xi) \longrightarrow -\infty \quad \text{as } \lambda \to +\infty.$$

Because $\xi \in \mathbb{R}^N$, $\|\xi\|_{\mathbb{R}^N} = \varrho$ was arbitrary, we see that for some $\varrho_1 > \varrho$, we have

$$\varphi(\xi) < -\beta_0 \qquad \forall\, \xi \in \mathbb{R}^N,\ \|\xi\|_{\mathbb{R}^N} = \varrho_1.$$

$\Box$

Propositions 3.4.20, 3.4.21 and 3.4.22 put us in the framework of the generalized nonsmooth Saddle Point Theorem (see Theorem 2.3.4). So we have the following existence theorem for problem (3.112).

THEOREM 3.4.14
If hypotheses $H(j)_{14}$ and $H(\Xi)$ hold,
<u>*then*</u> *problem (3.112) has a solution*

$$x_0 \in C^1\big(T;\mathbb{R}^N\big) \quad \text{with} \quad \|x_0'(\cdot)\|_{\mathbb{R}^N}^{p-2}\, x_0'(\cdot) \in W^{1,p'}\big(T;\mathbb{R}^N\big).$$

PROOF Invoking Theorem 2.3.4, we obtain $x_0 \in W^{1,p}\big(T;\mathbb{R}^N\big)$, such that

$$0 \leq \Phi^0(x_0;h) + \psi(x_0+h) - \psi(x_0) \qquad \forall\, h \in W^{1,p}\big(T;\mathbb{R}^N\big)$$

(see Definition 2.3.1). Then according to Remark 2.3.1, we can find $x_0^* \in \partial\Phi(x_0)$ and $v_0^* \in \partial\varphi(x_0)$, such that $x_0^* + v_0^* = 0$. We know that

$$x_0^* = A(x_0) - u,$$

with $u \in S^{p'}_{\partial j(\cdot,x_0(\cdot))}$ and $A: W^{1,p}\big(T;\mathbb{R}^N\big) \longrightarrow \big(W^{1,p}(T;\mathbb{R}^N)\big)^*$, the nonlinear operator, defined by

$$\langle A(y),z\rangle_{W^{1,p}(T;\mathbb{R}^N)} \stackrel{df}{=} \int_0^b \|y'(t)\|_{\mathbb{R}^N}^{p-2}\, \big(y'(t),z'(t)\big)_{\mathbb{R}^N}\,dt \quad \forall\, y,z \in W^{1,p}\big(T;\mathbb{R}^N\big).$$

Also $\psi(x_0) = (\vartheta \circ e)(x_0)$, where $e \colon W^{1,p}(T; \mathbb{R}^N) \longrightarrow \mathbb{R}^N \times \mathbb{R}^N$ is the continuous linear operator, defined by

$$e(x) \overset{df}{=} (x(0), x(b)) \qquad \forall\, x \in W^{1,p}(T; \mathbb{R}^N).$$

According to Theorem 1.3.7, we have that

$$\partial\psi(x) \;=\; e^* \partial\vartheta\big(x(0), x(b)\big) \qquad \forall\, x \in W^{1,p}(T; \mathbb{R}^N).$$

Therefore, for some $(v, w) \in \partial\vartheta\big(x_0(0), x_0(b)\big)$, we have

$$A(x_0) - u + e^*(v, w) \;=\; 0.$$

For all $\chi \in C_c^1\big((0, b); \mathbb{R}^N\big)$, we have

$$\langle A(x_0), \chi \rangle_{W^{1,p}(T;\mathbb{R}^N)} - \langle u, \chi \rangle_{W^{1,p}(T;\mathbb{R}^N)} + \langle e^*(v, w), \chi \rangle_{W^{1,p}(T;\mathbb{R}^N)} \;=\; 0.$$

Note that

$$\langle e^*(v, w), \chi \rangle_{W^{1,p}(T;\mathbb{R}^N)} \;=\; (v, \chi(0))_{\mathbb{R}^N} + (w, \chi(0))_{\mathbb{R}^N} \;=\; 0.$$

So we obtain

$$\langle A(x_0), \chi \rangle_{W^{1,p}(T;\mathbb{R}^N)} - \langle u, \chi \rangle_{W^{1,p}(T;\mathbb{R}^N)} \;=\; 0,$$

from which it follows that

$$-\big(\|x_0'(t)\|_{\mathbb{R}^N}^{p-2}\, x_0'(t)\big)' \;=\; u(t) \;\in\; \partial j\big(t, x_0(t)\big) \quad \text{for a.a. } t \in T.$$

Using this and integrating by parts, we have

$$\|x_0'(0)\|_{\mathbb{R}^N}^{p-2} \big(x_0'(0), y'(0)\big)_{\mathbb{R}^N} - \|x_0'(b)\|_{\mathbb{R}^N}^{p-2} \big(x_0'(b), y'(b)\big)_{\mathbb{R}^N}$$
$$= \big(v, y(0)\big)_{\mathbb{R}^N} + \big(w, y(b)\big)_{\mathbb{R}^N}.$$

First assume that $y(b) = 0$. We obtain $\varphi_p\big(x_0'(0)\big) = v$. Then assume that $y(0) = 0$. We obtain $-\varphi_p\big(x_0'(b)\big) = w$. Therefore, we conclude that $x_0 \in C^1(T; \mathbb{R}^N)$ with $\|x_0'(\cdot)\|_{\mathbb{R}^N}^{p-2}\, x_0'(\cdot) \in W^{1,p'}(T; \mathbb{R}^N)$ is a solution of (3.112). $\quad\square$

EXAMPLE 3.4.9

(a) If $\vartheta \equiv 1$ and $\Xi = \partial\vartheta \equiv 0$, then all the hypotheses in $H(\Xi)$ are satisfied and the resulting problem is the Neumann problem. Another situation in the same vein is obtained if we take $K_1 \overset{df}{=} [-1, 1]^N$, $K_2 \overset{df}{=} \mathbb{R}^N$,

$$\vartheta(\xi, \xi') \overset{df}{=} i_{K_1 \times K_2}(\xi, \xi') \;=\; \begin{cases} 0 & \text{if } (\xi, \xi') \in K_1 \times K_2, \\ +\infty & \text{otherwise} \end{cases}$$

and $\Xi(\xi, \xi') \overset{df}{=} \partial\vartheta(\xi, \xi') = N_{K_1}(\xi) \times \{0\}$. Then, if $M \leq 1$, all hypotheses in $H(\Xi)$ are satisfied, provided for some $\varrho \in (0, 1]$, we have

$$\int_0^b j(t, \xi)\, dt \, > \, 0 \qquad \forall\, \xi \in \mathbb{R}^N, \|\xi\|_{\mathbb{R}^N} = \varrho.$$

In this particular case, we have that, if $x = \{x_1, \ldots, x_N\}$ is a solution of problem (3.112), then

(i) $|x_k(0)| \leq 1$ for all $k \in \{1, \ldots, N\}$;

(ii) if $|x_k(0)| < 1$, then $x'_k(0) = 0$;

(iii) if $|x_k(0)| = 1$, then $\|x'(0)\|_{\mathbb{R}^N}^{p-2} x'_k(0) = \lambda x_k(0)$ for some $\lambda > 0$.

On the other hand $x'(b) = 0$ and similarly if we reverse the roles of K_1 and K_2.

Generalizing, suppose that $K_1, K_2 \subseteq \mathbb{R}^N$ are nonempty, closed, convex sets which contain a line and have a nonempty interior containing the origin. Then, if $M > 0$ is small enough (so that $B_M \subseteq K_1 \times K_2$), then with $\vartheta \overset{df}{=} i_{K_1 \times K_2}$ and $\Xi \overset{df}{=} \partial\vartheta$ we satisfy hypotheses $H(\Xi)$.

(b) Let $r \in [1, c_0]$ and set

$$\vartheta(\xi, \xi') \overset{df}{=} \frac{1}{r} \|\xi\|_{\mathbb{R}^N}^r + \frac{1}{r} \|\xi'\|_{\mathbb{R}^N}^r \qquad \forall\, \xi, \xi' \in \mathbb{R}^N.$$

Then

$$\Xi(\xi, \xi') \overset{df}{=} \partial\vartheta(\xi, \xi') \; = \; \left(\|\xi\|_{\mathbb{R}^N}^{r-2}\, \xi,\; \|\xi'\|_{\mathbb{R}^n}^{r-2}\, \xi' \right)$$

and the boundary conditions become

$$\|x'(0)\|_{\mathbb{R}^N}^{p-2}\, x'(0) \; = \; \|x(0)\|_{\mathbb{R}^N}^{r-2}\, x(0) \text{ and } \|x'(b)\|_{\mathbb{R}^N}^{p-2}\, x'(b) \; = \; \|x(b)\|_{\mathbb{R}^N}^{r-2}\, x(b).$$

Generalizing, let $\vartheta_1, \vartheta_2 \colon \mathbb{R}^N \longrightarrow \mathbb{R}_+$ be two monotone, continuous maps, such that $\vartheta_1(0) = \vartheta_2(0)$ and

$$\lim_{\|\xi\|_{\mathbb{R}^N} \to 0} \frac{\vartheta_k(\xi)}{\|\xi\|_{\mathbb{R}^N}^p} = 0 \qquad \forall\, k \in \{1, 2\}.$$

Then for suitable potential $j(t, \xi)$, hypotheses $H(\Xi)$ are satisfied. $\qquad\qquad \Box$

3.4.7 Multiple Solutions for "Smooth" Problems

In this subsection, we deal with a problem in which the potential function is C^1, namely the right hand side nonlinearity is a Carathéodory function. We assume that the equation is strongly resonant at the $\lambda_0 = 0$ eigenvalue of the negative ordinary scalar p-Laplacian with periodic boundary conditions. So if $f(t, \zeta)$ denotes the Carathéodory nonlinearity of the problem, we have that

$$f(t, \zeta) \longrightarrow 0 \quad \text{as } |\zeta| \to +\infty$$

and the C^1-potential function

$$F(t, \zeta) \overset{df}{=} \int\limits_0^\zeta f(t, r)dr$$

has finite limits as $|\zeta| \to +\infty$. Roughly speaking the potential has small rate of increase as $|\zeta| \to +\infty$. Our approach is based on the Second Deformation Theorem. First a definition.

DEFINITION 3.4.1   *Let X be a Hausdorff topological space and $A \subseteq X$ nonempty subset. We say that A is a **strong deformation retract of** X, if there exists $h \in C\big([0, 1] \times X; X\big)$, such that*

(a) *for all $t \in [0, 1]$, we have that $h(t, \cdot)|_A = id_A$;*

(b) *$h(0, \cdot) = id_X$;*

(c) *$h(1, X) \subseteq A$.*

Recall that a continuous map $h \colon [0, 1] \times X \longrightarrow X$, such that $h(0, \cdot) = id_X$, is called a **homotopy**. Often we write $h_t \colon X \longrightarrow X$ for $h(t, \cdot)$. Suppose that U is an open subset of a Banach space X and $\varphi \in C^1(U)$. We say that a homotopy $h \colon [0, 1] \times U \longrightarrow U$ is φ-**decreasing**, if

$$\varphi\big(h(t, x)\big) \leq \varphi\big(h(s, x)\big) \qquad \forall\, s < t,\ x \in U.$$

The next theorem is known as the "Second Deformation Theorem" and for its proof we refer to Chang (1993, p. 23). Recall (see Section 2.1) that for $\varphi \colon X \longrightarrow \mathbb{R}$ and $r \in \mathbb{R}$, we denote

$$\varphi^r \overset{df}{=} \big(x \in X \colon\ \varphi(x) \leq r\big),$$
$$K_r^\varphi \overset{df}{=} \big(x \in X \colon\ \varphi'(x) = 0,\ \varphi(x) = r\big).$$

THEOREM 3.4.15 (Second Deformation Theorem)
<u>*If*</u> *U is an open subset of a Banach space X, $\varphi \in C^1(X)$, $a < b$, φ has no critical values in the interval (a, b), $\varphi^{-1}(\{a\})$ contains at most a finite numbers of critical points of φ and φ satisfies the PS_c-condition for every $c \in [a, b)$,*
<u>*then*</u> *φ^a is a strong deformation retract of $\varphi^b \setminus K_b^\varphi$.*

Moreover, the homotopy $h\colon [0,1] \times \left(\varphi^b \setminus K_b^\varphi\right) \longrightarrow U$ *(see Definition 3.4.1) can be chosen to be* φ*-decreasing.*

REMARK 3.4.9 It is not known to the authors whether this theorem can be extended to nonsmooth locally Lipschitz functions $\varphi\colon X \longrightarrow \mathbb{R}$. It will be very useful to have such an extension. $\square$

The problem under consideration is the following:

$$\begin{cases} -\left(|x'(t)|^{p-2}x'(t)\right)' = f\big(t, x(t)\big) & \text{for a.a. } t \in T, \\ x(0) = x(b), \ x'(0) = x'(b), \end{cases} \tag{3.121}$$

for $p \in (1, +\infty)$. By $F\colon T \times \mathbb{R} \longrightarrow \mathbb{R}$ we denote the potential corresponding to the nonlinearity f, namely

$$F(t, \zeta) \stackrel{df}{=} \int_0^\zeta f(t, r)\,dr \qquad \forall\, (t, \zeta) \in T \times \mathbb{R}.$$

<u>$H(f)$</u> $f\colon T \times \mathbb{R} \longrightarrow \mathbb{R}$ is a function, such that

    (i) for all $\zeta \in \mathbb{R}$, the function

$$T \ni t \longmapsto f(t, \zeta) \in \mathbb{R}$$

    is measurable;

    (ii) for almost all $t \in T$, the function

$$\mathbb{R} \ni \zeta \longmapsto f(t, \zeta) \in \mathbb{R}$$

    is continuous;

    (iii) for almost all $t \in T$ and all $\zeta \in \mathbb{R}$, we have

$$|f(t, \zeta)| \leq a(t) + c(t)|\zeta|^{r-1},$$

    with $a \in L^{r'}(T)_+$, where $r \in [1, +\infty)$, $\frac{1}{r} + \frac{1}{r'} = 1$, $c \in L^\infty(T)_+$;

    (iv) there exist $F_\pm \in L^1(T)$, such that

$$F(t, \zeta) \longrightarrow F_\pm(t) \quad \text{as } \zeta \to \pm\infty \qquad \text{uniformly for a.a. } t \in T$$

    and $\int_0^b F_\pm(t)\,dt \leq 0$;

    (v) we have

$$F(t, \zeta) \leq \frac{1}{pb^p}|\zeta|^p \quad \text{for a.a. } t \in T \text{ and all } \zeta \in \mathbb{R};$$

(vi) there exists $\vartheta \in L^1(T)$ with $\int_0^b \vartheta(t)\, dt > 0$, such that

$$\liminf_{\zeta \longrightarrow 0} \frac{pF(t,\zeta)}{|\zeta|^p} \geq \vartheta(t) \quad \text{uniformly for a.a. } t \in T.$$

Let $\varphi \colon W^{1,p}_{\mathrm{per}}(T) \longrightarrow \mathbb{R}$ be defined by

$$\varphi(x) \overset{df}{=} \frac{1}{p}\|x'\|_p^p - \int_0^b F\big(t, x(t)\big)\, dt.$$

It is well known that $\varphi \in C^1\big(W^{1,p}_{\mathrm{per}}(T)\big)$ and

$$\varphi'(x) \;=\; A(x) - N_f(x) \qquad \forall\, x \in W^{1,p}_{\mathrm{per}}(T),$$

where $A \colon W^{1,p}_{\mathrm{per}}(T) \longrightarrow \big(W^{1,p}_{\mathrm{per}}(T)\big)^*$ is given by

$$\big\langle A(x), y \big\rangle_{W^{1,p}_{\mathrm{per}}(T)} \overset{df}{=} \int_0^b |x'(t)|^{p-2} x'(t) y'(t)\, dt \qquad \forall\, x, y \in W^{1,p}_{\mathrm{per}}(T)$$

and $N_f \colon L^r(T) \longrightarrow L^{r'}(T)$ is defined by

$$N_f(x)(\cdot) \overset{df}{=} f\big(\cdot, x(\cdot)\big) \qquad \forall\, x \in L^r(T),$$

i.e. the Nemytskii operator corresponding to f.

PROPOSITION 3.4.23
If hypotheses $H(f)$ hold,

then φ satisfies the PS_c-condition for every $c < -\int_0^b F_\pm(t)\, dt$.

PROOF Let $\{x_n\}_{n \geq 1} \subseteq W^{1,p}_{\mathrm{per}}(T)$ be a PS_c-sequence for φ, with $c < -\int_0^b F_\pm(t)\, dt$, i.e.

$$\varphi(x_n) \longrightarrow c \quad \text{and} \quad \varphi'(x_n) \longrightarrow 0.$$

We shall show that the sequence $\{x_n\}_{n \geq 1} \subseteq W^{1,p}_{\mathrm{per}}(T)$ is bounded. Suppose that this is not the case. Passing to a subsequence if necessary, we may assume that

$$\|x_n\|_{W^{1,p}(T)} \longrightarrow +\infty$$

and by setting

$$y_n \overset{df}{=} \frac{x_n}{\|x_n\|_{W^{1,p}(T)}} \qquad \forall\, n \geq 1,$$

we can say, at least for a further subsequence, that

$$y_n \overset{w}{\longrightarrow} y \text{ in } W^{1,p}_{\mathrm{per}}(T),$$
$$y_n \longrightarrow y \text{ in } C_{\mathrm{per}}(T).$$

We have

$$\frac{|\varphi(x_n)|}{\|x_n\|^p_{W^{1,p}(T)}} = \left| \frac{1}{p}\|y_n'\|^p_p - \int_0^b \frac{F\big(t,x_n(t)\big)}{\|x_n\|^p_{W^{1,p}(T)}}\, dt \right|$$

$$\leq \frac{M_1}{\|x_n\|^p_{W^{1,p}(T)}} \qquad \forall\, n \geq 1, \qquad (3.122)$$

for some $M_1 > 0$. By virtue of hypotheses $H(f)(iii)$ and (iv), we have that

$$\big|F(t,\varsigma)\big| \leq a(t) \quad \text{for a.a. } t \in T \text{ and all } \varsigma \in \mathbb{R}$$

with $a \in L^1(T)_+$. So

$$\int_0^b \frac{F\big(t,x_n(t)\big)}{\|x_n\|^p_{W^{1,p}(T)}}\, dt \longrightarrow 0 \quad \text{as } n \to +\infty.$$

Thus, by passing to the limit as $n \to +\infty$ in (3.122), we obtain

$$\frac{1}{p}\|y'\|^p_p = 0,$$

i.e. $y \equiv \overline{\varsigma} \in \mathbb{R}$. Note that $\overline{\varsigma} \neq 0$ or otherwise

$$y_n \longrightarrow 0 \quad \text{in } W^{1,p}_{\mathrm{per}}(T),$$

a contradiction to the fact that $\|y_n\|_{W^{1,p}(T)} = 1$ for all $n \geq 1$. Suppose that $\widehat{\varsigma} > 0$ (the analysis is similar if $\widehat{\varsigma} < 0$). We have

$$x_n(t) \longrightarrow +\infty \quad \text{as } n \to +\infty$$

and the convergence is uniform in $t \in T$ (see the proof of Proposition 3.4.5). Recall that $\varphi(x_n) \longrightarrow c$ and so for any $\varepsilon > 0$ there exists $n_0 \geq 1$ large enough, such that

$$-\int_0^b F\big(t,x_n(t)\big)\, dt \leq c + \varepsilon \qquad \forall\, n \geq n_0.$$

Because of hypothesis $H(f)(iv)$, in the limit as $n \to +\infty$ and $\varepsilon \searrow 0$, we obtain

$$- \int_0^b F_+(t)\, dt \ \leq\ c,$$

a contradiction to the choice of level c. So finally the sequence $\{x_n\}_{n \geq 1} \subseteq W^{1,p}_{\mathrm{per}}(T)$ is bounded and thus φ satisfies the PS_c-condition. $\square$

THEOREM 3.4.16

If hypotheses $H(f)$ hold,
<u>*then*</u> *problem (3.119) has at least two nontrivial solutions.*

PROOF   Note that φ is bounded below (hypotheses $H(f)(iii)$ and iv). Also by virtue of hypothesis $H(f)(vi)$, for a given $\varepsilon > 0$, we can find $\delta = \delta(\varepsilon) > 0$, such that

$$\frac{1}{p}\big(\vartheta(t) - \varepsilon\big)|\zeta|^p \ \leq\ F(t, \zeta) \quad \text{for a.a. } t \in T \text{ and all } \zeta \in \mathbb{R},\ |\zeta| < \delta.$$

So if $\zeta \in \mathbb{R}$, we have

$$\varphi(\zeta) \ =\ - \int_0^b F(t, \zeta)\, dt \ \leq\ \frac{|\zeta|^p}{p} \int_0^b \big(\varepsilon - \vartheta(t)\big)\, dt \ =\ \frac{|\zeta|^p \varepsilon b}{p} - \frac{|\zeta|^p}{p} \int_0^b \vartheta(t)\, dt.$$

Because $\int_0^b \vartheta(t)\, dt > 0$ (see hypothesis $H(f)(vi)$), we see that if $\varepsilon > 0$ is small enough, $\varphi(\zeta) < 0$ and so

$$\inf_{x \in W^{1,p}_{\mathrm{per}}(T)} \varphi(x) \ =\ m_0 \ <\ 0 \ \leq\ - \int_0^b F_\pm(t)\, dt$$

(see hypothesis $H(f)(iv)$). So according to Proposition 3.4.23, φ satisfies the PS_{m_0}-condition. Therefore, we can find $x_1 \in W^{1,p}_{\mathrm{per}}(T)$, such that

$$\varphi(x_1) \ =\ m_0 \ <\ \varphi(0)$$

and so $\varphi'(x_1) = 0$ and $x_1 \neq 0$ (see Theorem 2.1.6).

Suppose for the moment that x_1 and 0 are the only critical points of φ. From the previous argument, we know that for a given $\varepsilon > 0$, we can find $\delta > 0$, such that

$$\varphi(\zeta) \ \geq\ \frac{|\zeta|^p}{p}\left(\varepsilon b - \int_0^b \vartheta(t)\, dt\right) \qquad \forall\, \zeta \in \mathbb{R},\ |\zeta| \leq \delta$$

and so for $\varepsilon > 0$ small enough, we have

$$\eta \; = \; \varepsilon b - \int_0^b \vartheta(t)\,dt \; < \; 0.$$

Hence $\varphi(\zeta) \le \frac{|\zeta|^p}{p}\eta < 0$.

We consider the direct sum decomposition

$$W^{1,p}_{\text{per}}(T) \; = \; \mathbb{R} \oplus V,$$

with

$$V \; \stackrel{df}{=} \; \left\{ v \in W^{1,p}_{\text{per}}(T) : \int_0^b v(t)\,dt = 0 \right\}.$$

Because of hypothesis $H(f)(v)$, we have

$$\varphi(v) \; \ge \; \frac{1}{p}\,\|v'\|_p^p - \frac{1}{pb^p}\,\|v\|_p^p \; \ge \; 0$$

(see Remark 1.1.11), so $\inf\limits_V \varphi = 0$.

On the other hand, from the previous considerations, we have

$$\mu \; = \; \sup_{\zeta \in \partial B_\delta \cap \mathbb{R}} \varphi(\zeta) \; < \; 0. \tag{3.123}$$

Let

$$\Gamma \; \stackrel{df}{=} \; \left\{ \gamma \in C(\overline{B}_\delta \cap \mathbb{R}; W^{1,p}_{\text{per}}(T)) : \; \gamma|_{\overline{B}_\delta \cap \mathbb{R}} = id_{\overline{B}_\delta \cap \mathbb{R}} \right\}$$

and h the homotopy postulated by the Second Deformation Theorem (see Theorem 3.4.15). We define $\gamma_0 \colon \overline{B}_\delta \cap \mathbb{R} \longrightarrow W^{1,p}_{\text{per}}(T)$, by

$$\gamma_0(\zeta) \; \stackrel{df}{=} \; \begin{cases} x_1 & \text{if } |\zeta| < \frac{\delta}{2}, \\ h\left(\frac{2(\delta - |\zeta|)}{\delta}, \frac{\delta\zeta}{|\zeta|}\right) & \text{if } |\zeta| \ge \frac{\delta}{2}. \end{cases}$$

Recall that we have assumed that $\{x_1, 0\}$ are the only critical points of φ. Then x_1 is the only minimizer of φ and so from the Second Deformation Theorem (see Theorem 3.4.15), we have that

$$h(1, y) \; = \; x_1 \qquad \forall \, y \in \varphi^0 \setminus \{0\}.$$

So

$$h\left(\tfrac{2(\delta - |\zeta|)}{\delta}, \tfrac{\delta\zeta}{|\zeta|}\right) \; = \; h(1, 2\zeta) \; = \; x_1 \qquad \forall \, \zeta \in \mathbb{R}, \; |\zeta| = \frac{\delta}{2},$$

from which follows the continuity of γ_0. In addition $h(0, \cdot) = id$ (see the Second Deformation Theorem; Theorem 3.4.15) and so $\gamma_0|_{\overline{B}_\delta \cap \mathbb{R}} = id$, i.e. $\gamma_0 \in$

Γ. Moreover, since h is φ-decreasing (see the Second Deformation Theorem; Theorem 3.4.15), we have

$$\varphi\big(h(s,\zeta)\big) \leq \varphi\big(h(t,\zeta)\big) \qquad \forall\, \zeta \in \varphi^0 \setminus \{0\},\ t,s \in [0,1],\ t < s.$$

From this and (3.123), we infer that

$$\varphi\big(\gamma_0(x)\big) < 0 \qquad \forall\, x \in \overline{B}_\delta \cap \mathbb{R}. \tag{3.124}$$

Therefore the sets $\overline{B}_\delta \cap \mathbb{R}$ and V link (see Definition 2.1.4 and Remark 2.1.4). So, we have that

$$\gamma\big(\overline{B}_\delta \cap \mathbb{R}\big) \cap V \neq \emptyset \qquad \forall\, \gamma \in \Gamma,$$

hence

$$\sup_{x \in \overline{B}_\delta \cap \mathbb{R}} \varphi\big(\gamma(x)\big) \geq 0 \qquad \forall\, \gamma \in \Gamma$$

(recall that $\inf\limits_V \varphi = 0$). Hence

$$\sup_{x \in \overline{B}_\delta \cap \mathbb{R}} \varphi\big(\gamma_0(x)\big) = \varphi\big(\gamma_0(x^*)\big) \geq 0, \tag{3.125}$$

for some $x^* \in \overline{B}_\delta \cap \mathbb{R}$. Comparing (3.124) and (3.125), we reach a contradiction. So φ has another critical point x_2, such that $x_2 \neq x_1$ and $x_2 \neq 0$. Then x_1, x_2 are the two nontrivial solutions of (3.121). $\qquad\qquad \Box$

3.5   Method of Upper and Lower Solutions

The method of upper and lower solutions provides a powerful tool to establish the existence of solutions and of multiple solutions for initial and boundary problems of the first and second order. The method generates solutions which are located in the order interval determined by an ordered pair of lower and upper solutions, which serve as lower and upper bounds respectively. In addition, the method coupled with some monotonicity conditions leads to monotone iterative techniques, which generate in a constructive way (amenable to algorithmic analysis and numerical treatment) the extremal solutions (i.e. the smallest and greatest solutions) in the order interval. The method traces its origins in the notions of subharmonic and superharmonic functions of potential theory. For a given problem, one may try several methods to establish a lower and upper solution. There is no methodology (even for single-valued problems), but usually one should try simple functions (such as constants, linear, quadratic, exponential, eigenfunctions of simple operators, solutions of simpler auxiliary equations etc.).

We study the following nonlinear, scalar differential inclusion:

$$\begin{cases} -\left(|x'(t)|^{p-2}x'(t)\right)' \in F\big(t, x(t), x'(t)\big) & \text{for a.a. } t \in T, \\ x'(0) \in \Xi_1\big(x(0)\big), \ -x'(b) \in \Xi_2\big(x(b)\big), \end{cases} \qquad (3.126)$$

for $p \in (1, +\infty)$. Here Ξ_1, Ξ_2 are two maximal monotone operators on $\mathbb{R}$.

We define the notions of solution, lower solution and upper solution for problem (3.126).

DEFINITION 3.5.1

(a) *By a solution of (3.126), we mean a function $x \in C^1(T)$, such that*

(i) $|x'(\cdot)|^{p-2}x'(\cdot) \in W^{1,p'}(T)$, $(\frac{1}{p} + \frac{1}{p'} = 1)$;

(ii) $\left(|x'(t)|^{p-2}x'(t)\right)' = f(t)$ *for almost all $t \in T$, with $f \in S^{p'}_{F(\cdot, x(\cdot), x'(\cdot))}$*;

(iii) $x'(0) \in \Xi_1\big(x(0)\big)$, $-x'(b) \in \Xi_2\big(x(b)\big)$.

(b) *By a lower solution of (3.126), we mean a function $\underline{\psi} \in C^1(T)$, such that*

(i) $|\underline{\psi}'(\cdot)|^{p-2}\underline{\psi}'(\cdot) \in W^{1,p'}(T)$;

(ii) *there exists $\underline{v} \in S^{p'}_{F(\cdot, \underline{\psi}(\cdot), \underline{\psi}'(\cdot))}$ such that $\underline{v}(t) \le \left(|\underline{\psi}'(t)|^{p-2}\underline{\psi}'(t)\right)'$ for almost all $t \in T$;*

(iii) $\underline{\psi}'(0) \in \Xi_1\big(\underline{\psi}(0)\big) + \mathbb{R}_+$, $-\underline{\psi}'(b) \in \Xi_2\big(\underline{\psi}(b)\big) + \mathbb{R}_+$.

(c) *By an upper solution of (3.126), we mean a function $\overline{\psi} \in C^1(T)$, such that*

(i) $|\overline{\psi}'(\cdot)|^{p-2}\overline{\psi}'(\cdot) \in W^{1,p'}(T)$;

(ii) *there exists $\overline{v} \in S^{p'}_{F(\cdot, \overline{\psi}(\cdot), \overline{\psi}'(\cdot))}$ such that $\left(|\overline{\psi}'(t)|^{p-2}\overline{\psi}'(t)\right)' \le \overline{v}(t)$ for almost all $t \in T$;*

(iii) *and* $\overline{\psi}'(0) \in \Xi_1\big(\overline{\psi}(0)\big) - \mathbb{R}_+$, $-\overline{\psi}'(b) \in \Xi_2\big(\overline{\psi}(b)\big) - \mathbb{R}_+$.

We assume that there exists an ordered pair of upper-lower solutions of (3.126), namely:

<u>$H(\psi)$</u> There exist a lower solution $\underline{\psi}$ and upper solution $\overline{\psi}$ of (3.126), such that $\underline{\psi}(t) \le \overline{\psi}(t)$ for all $t \in T$.

Our hypotheses on the multivalued nonlinearity F are the following:

<u>$H(F)$</u> $F: T \times \mathbb{R} \times \mathbb{R} \longrightarrow 2^{\mathbb{R}} \setminus \{\emptyset\}$ is a multifunction, such that

  (i) for all $\zeta, \zeta' \in \mathbb{R}$, the multifunction

$$T \ni t \longmapsto F(t, \zeta, \zeta') \in 2^{\mathbb{R}} \setminus \{\emptyset\}$$

  is graph measurable;

  (ii) for almost all $t \in T$, the multifunction

$$\mathbb{R} \times \mathbb{R} \ni (\zeta, \zeta') \longmapsto F(t, \zeta, \zeta') \in 2^{\mathbb{R}} \setminus \{\emptyset\}$$

  has closed graph;

  (iii) for almost all $t \in T$, all $\zeta \in \left[\underline{\psi}(t), \overline{\psi}(t)\right]$, all $\zeta' \in \mathbb{R}$ and all $u \in F(t, \zeta, \zeta')$ we have

$$|u| \leq \eta\big(|\zeta'|^{p-1}\big)\big(a(t) + c|\zeta'|\big),$$

  with $a \in L^1(T)$, $c > 0$ and where $\eta: \mathbb{R}_+ \longrightarrow \mathbb{R}_+ \setminus \{0\}$ is a Borel measurable function, such that

$$\int\limits_{\vartheta^{p-1}}^{+\infty} \frac{ds}{\eta(s)} > \|a\|_1 + c\left(\max_T \overline{\psi} - \min_T \underline{\psi}\right);$$

  with $\vartheta \stackrel{df}{=} \frac{1}{b} \max\left\{\left|\underline{\psi}(0) - \overline{\psi}(b)\right|, \left|\underline{\psi}(b) - \overline{\psi}(0)\right|\right\}$;

  (iv) for all $r > 0$ there exists $\gamma_r \in L^{p'}(T)_+$, such that for almost all $t \in T$, all $\zeta, \zeta' \in \mathbb{R}$ with $|\zeta|, |\zeta'| \leq r$ and all $u \in F(t, \zeta, \zeta')$, we have that $|u| \leq \gamma_r(t)$.

REMARK 3.5.1   By virtue of hypotheses $H(F)(i)$ and (ii), we have

$$F(t, \zeta, \zeta') = \left[h_1(t, \zeta, \zeta'), \ h_2(t, \zeta, \zeta')\right],$$

where for all $\zeta, \zeta' \in \mathbb{R}$, the functions $t \longmapsto h_i(t, \zeta, \zeta')$ are measurable ($i \in \{1, 2\}$) and for all $t \in T$, the functions $(\zeta, \zeta') \longmapsto -h_1(t, \zeta, \zeta')$ and $(\zeta, \zeta') \longmapsto h_2(t, \zeta, \zeta')$ are upper semicontinuous. Also hypothesis $H(F)(iii)$ is a so-called Nagumo-Wintner growth condition and leads to a pointwise *a priori* bound for the derivative of a solution of problem (3.126). Evidently it is satisfied if for almost all $t \in T$, all $\zeta \in \left[\underline{\psi}(t), \overline{\psi}(t)\right]$, all $\zeta' \in \mathbb{R}$ and all $u \in F(t, \zeta, \zeta')$, we have $|u| \leq a(t) + c|\zeta'|^p$, with $a \in L^1(T)_+$, $c > 0$. $\square$

Our hypotheses on the boundary multifunctions Ξ_1 and Ξ_2 are the following:

$\underline{H(\Xi)}$ $\Xi_1 : \mathbb{R} \supseteq \mathrm{Dom}\,(\Xi_1) \longrightarrow 2^{\mathbb{R}}$ and $\Xi_2 : \mathbb{R} \supseteq \mathrm{Dom}\,(\Xi_2) \longrightarrow 2^{\mathbb{R}}$ are maximal monotone maps and $0 \in \Xi_1(0) \cap \Xi_2(0)$.

REMARK 3.5.2   Hypothesis $H(\Xi)$ implies that there exist functions $k_1, k_2 \in \Gamma_0(\mathbb{R})$, such that $\Xi_1 = \partial k_1$ and $\Xi_2 = \partial k_2$ (see Remark 1.4.6). In fact for $i \in \{1,2\}$, we have

$$k_i(r) \;=\; \int_0^r \Xi_i^0(s)\,ds,$$

with $\Xi_i^0(s) \stackrel{df}{=} \mathrm{proj}_{\Xi_i(s)}(0)$ (the element in $\Xi_i(s)$ with the smallest absolute value). The map $s \longmapsto \Xi_i^0(s)$ is increasing and

$$\Xi_i(s) \;=\; [\Xi_i^0(s^-),\, \Xi_i^0(s^+)] \qquad \forall\, s \in \mathbb{R}.$$

$\square$

Using the Nagumo-Wintner growth condition (see hypothesis $H(F)(iii)$), we derive an *a priori* pointwise bound for the derivative x' of any solution of (3.126) for which we have $\underline{\psi}(t) \le x(t) \le \overline{\psi}(t)$ for all $t \in T$.

PROPOSITION 3.5.1
$\underline{If}$ *hypothesis* $H(F)(iii)$ *holds and* $x \in C^1(T)$ *satisfies*

$$\left(|x'(t)|^{p-2} x'(t)\right)' \;\in\; F\big(t, x(t), x'(t)\big) \qquad \forall\, t \in T$$

and

$$\underline{\psi}(t) \;\le\; x(t) \;\le\; \overline{\psi}(t) \qquad \forall\, t \in T,$$

$\underline{then}$ *there exists* $M_1 = M_1(\overline{\psi}, \underline{\psi}, \eta, a, c) > 0$, *such that*

$$|x'(t)| \;\le\; M_1 \qquad \forall\, t \in T.$$

PROOF   By virtue of hypothesis $H(F)(iii)$, we can find $M_1 > \vartheta$, such that

$$\int_{\vartheta^{p-1}}^{M_1^{p-1}} \frac{ds}{\eta(s)} \;>\; \|a\|_1 + c\left(\max_T \overline{\psi} - \min_T \underline{\psi}\right).$$

We claim that

$$|x'(t)| \;\le\; M_1 \qquad \forall\, t \in T. \tag{3.127}$$

Suppose that this is not the case. Then, we can find $t' \in T$, such that $|x'(t')| > M_1$. From the Mean Value Theorem, we know that there exists $t_0 \in (0, b)$, such that

$$x'(t_0) \;=\; \frac{x(b) - x(0)}{b}.$$

Since

$$\underline{\psi}(b) - \overline{\psi}(0) \leq x(b) - x(0) \leq \overline{\psi}(b) - \underline{\psi}(0),$$

we have

$$|x'(t_0)| = \frac{1}{b}|x(b) - x(0)| \leq \frac{1}{b}\max\left\{|\underline{\psi}(0) - \overline{\psi}(b)|, |\underline{\psi}(b) - \overline{\psi}(0)|\right\}$$
$$= \vartheta < M_1 < |x'(t')|.$$

So $t_0 \neq t'$ and without any loss of generality we may assume that $t_0 < t'$ (the analysis is similar if $t_0 > t'$). Because $x \in C^1(T)$, from the Intermediate Value Theorem, we know that we can find $t_1, t_2 \in [t_0, t')$ with $t_1 < t_2$, such that $|x'(t_1)| = \vartheta$ and $|x'(t_2)| = M_1$. We have the following possibilities:

(a) $x'(t_1) = \vartheta$, $x'(t_2) = M_1$ and $\vartheta < x'(t) < M_1$ for all $t \in (t_1, t_2)$;

(b) $x'(t_1) = -\vartheta$, $x'(t_2) = -M_1$ and $-M_1 < x'(t) < -\vartheta$ for all $t \in (t_1, t_2)$.

We treat possibility (a) (the other case can be analyzed similarly). We have

$$\left(|x'(t)|^{p-2}x'(t)\right)' \in F(t, x(t), x'(t)) \quad \text{for a.a. } t \in T,$$

so

$$\left(|x'(t)|^{p-1}\right)' \leq \left|\left(|x'(t)|^{p-2}x'(t)\right)'\right| \leq |F(t, x(t), x'(t))|$$
$$\leq \eta\left(|x'(t)|^{p-1}\right)\left(a(t) + c|x'(t)|\right) \quad \text{for a.a. } t \in T,$$

thus

$$\frac{\left(|x'(t)|^{p-1}\right)'}{\eta\left(|x'(t)|^{p-1}\right)} \leq a(t) + cx'(t) \quad \text{for a.a. } t \in [t_1, t_2]$$

and since $\underline{\psi}(t) \leq x(t) \leq \overline{\psi}(t)$ for all $t \in T$, we have

$$\int_{t_1}^{t_2} \frac{\left(|x'(t)|^{p-1}\right)'}{\eta\left(|x'(t)|^{p-1}\right)} \, dt \leq \|a\|_1 + c\left(\max_T \overline{\psi} - \min_T \underline{\psi}\right).$$

By a change of variables, we obtain

$$\int_{\vartheta^{p-1}}^{M_1^{p-1}} \frac{ds}{\eta(s)} \leq \|a\|_1 + c\left(\max_T \overline{\psi} - \min_T \underline{\psi}\right),$$

a contradiction to the choice of $M_1 > 0$. $\qquad\qquad\square$

The method of upper and lower solutions involves truncation and penalization techniques, aimed at exploiting the fact that we have a good knowledge of the growth of the multivalued nonlinearity F on the interval $\left[\underline{\psi}(t),\,\overline{\psi}(t)\right]$, $t \in T$. For this reason, we introduce suitable truncation and penalization functions.

Let $M_2 > \max\left\{M_1,\,\|\overline{\psi}'\|_\infty,\,\|\underline{\psi}'\|_\infty\right\}$. The ***truncation function*** $u\colon T \times \mathbb{R} \times \mathbb{R} \longrightarrow \mathbb{R}^2$ is defined by

$$u(t,\zeta,\zeta') \overset{df}{=} \begin{cases} \left(\underline{\psi}(t),\underline{\psi}'(t)\right) & \text{if } \zeta < \underline{\psi}(t), \\ \left(\overline{\psi}(t),\overline{\psi}'(t)\right) & \text{if } \zeta > \overline{\psi}(t), \\ (\zeta, M_2) & \text{if } \underline{\psi}(t) \le \zeta \le \overline{\psi}(t),\ \zeta' > M_2, \\ (\zeta, -M_2) & \text{if } \underline{\psi}(t) \le \zeta \le \overline{\psi}(t),\ \zeta' < -M_2, \\ (\zeta,\zeta') & \text{otherwise.} \end{cases} \quad (3.128)$$

Also we introduce the ***penalty multifunction*** Q and the ***penalty function*** h defined as follows:

$$Q(t,\varsigma) \overset{df}{=} \begin{cases} [\overline{v}(t),+\infty) & \text{if } \zeta > \overline{\psi}(t) \\ \mathbb{R} & \text{if } \underline{\psi}(t) \le \zeta \le \overline{\psi}(t) \\ (-\infty,\underline{v}(t)] & \text{if } \zeta < \underline{\psi}(t) \end{cases} \quad (3.129)$$

and

$$h(t,\varsigma) \overset{df}{=} \begin{cases} |\overline{\psi}(t)|^{p-2}\,\overline{\psi}(t) - |\zeta|^{p-2}\zeta & \text{if } \zeta > \overline{\psi}(t), \\ 0 & \text{if } \underline{\psi}(t) \le \zeta \le \overline{\psi}(t), \\ |\underline{\psi}(t)|^{p-2}\,\underline{\psi}(t) - |\zeta|^{p-2}\zeta & \text{if } \zeta < \underline{\psi}(t). \end{cases} \quad (3.130)$$

Using u, Q and h, we introduce the following modification of F:

$$F_1(t,\varsigma,\varsigma') \overset{df}{=} F\big(t, u(t,\varsigma,\varsigma')\big) \cap Q(t,\varsigma).$$

Clearly $F_1\colon T \times \mathbb{R} \times \mathbb{R} \longrightarrow P_{fc}(\mathbb{R})$ and

$$F_1(t,\varsigma,\varsigma') = F(t,\varsigma,\varsigma') \qquad \text{for a.a. } t \in T \text{ and all } \zeta \in \left[\underline{\psi}(t),\overline{\psi}(t)\right],\ \zeta' \le M_2$$

and

$$|u| \le \gamma_r(t) \qquad \forall\, u \in F_1(t,\varsigma,\varsigma'),$$

with $r \overset{df}{=} \max\left\{M_2,\,\|\overline{\psi}\|_\infty,\,\|\underline{\psi}\|_\infty\right\}$.

Let $G\colon W^{1,p}(T) \longrightarrow 2^{L^{p'}(T)}$ be the multivalued map defined by

$$G(x) \overset{df}{=} \left\{g \in L^{p'}(T)\colon\ -g(t) \in F_1\big(t,x(t),x'(t)\big) - h\big(t,x(t)\big) \text{ for a.a. } t \in T\right\}.$$

PROPOSITION 3.5.2
If hypotheses $H(F)$ hold,
then G has values in $P_{wkc}\big(L^{p'}(T)\big)$ and is upper semicontinuous from $W^{1,p}(T)$

into $L^{p'}(T)_w$ (recall that $L^{p'}(T)_w$ denotes the Lebesgue space $L^{p'}(T)$ equipped with the weak topology).

PROOF   Note that for all $(\zeta, \zeta') \in \mathbb{R}^2$, the multifunction $t \longmapsto F_1(t, \zeta, \zeta')$ is graph measurable, while for almost all $t \in T$ the multifunction $(\zeta, \zeta') \longmapsto F_1(t, \zeta, \zeta')$ has closed graph. So, if $x \in W^{1,p}(T)$, then by approximating x and x' by step functions, we can establish that $S^{p'}_{F_1(\cdot, x(\cdot), x'(\cdot))} \neq \emptyset$ (see Claim 2 in the proof of Proposition 3.1.4). So $G(x) \neq \emptyset$ and it is bounded, closed, convex, hence G has values in $P_{wkc}\big(L^{p'}(T)\big)$.

Next suppose that $C \subseteq L^{p'}(T)$ is a nonempty and weakly closed subset. We need to show that $G^-(C)$ is closed in $W^{1,p}(T)$ (see Definition 1.2.1(a)). To this end let $\{x_n\}_{n \geq 1} \subseteq G^-(C)$ be a sequence, such that

$$x_n \longrightarrow x \quad \text{in } W^{1,p}(T).$$

Let $g_n \in G(x_n) \cap C$ for $n \geq 1$. We have

$$|g_n(t)| \leq \gamma_r(t) + |x_n(t)|^{p-1} + \max\left\{\|\overline{\psi}\|_\infty^{p-1}, \|\underline{\psi}\|_\infty^{p-1}\right\},$$

with $r \overset{df}{=} \max\left\{M_2, \|\overline{\psi}\|_\infty, \|\underline{\psi}\|_\infty\right\}$ and $\gamma_r \in L^{p'}(T)_+$ (see hypothesis $H(F)(iv)$). So by passing to a subsequence if necessary, we may assume that

$$g_n \overset{w}{\longrightarrow} g \quad \text{in } L^{p'}(T).$$

Also because $x_n \longrightarrow x$ in $W^{1,p}(T)$, we have

$$x_n \longrightarrow x \quad \text{in } C(T),$$
$$x'_n \longrightarrow x' \quad \text{in } L^p(T)$$

and at least for a subsequence, we may assume that

$$x'_n(t) \longrightarrow x'(t) \quad \text{for a.a. } t \in T.$$

From the definition of u, we see that

$$u\big(t, x_n(t), x'_n(t)\big) \longrightarrow u\big(t, x(t), x'(t)\big) \quad \text{for a.a. } t \in T,$$

while from the definition of Q, we have

$$\limsup_{n \to +\infty} Q\big(t, x_n(t)\big) \subseteq Q\big(t, x(t)\big) \quad \text{for a.a. } t \in T.$$

Therefore, by Proposition 1.2.9(d), we have

$$\limsup_{n \to +\infty} \big[F\big(t, u(t, x_n(t), x'_n(t))\big) \cap Q\big(t, x_n(t)\big) \big]$$
$$\subseteq F\big(t, u(t, x(t), x'(t))\big) \cap Q\big(t, x(t)\big) \quad \text{for a.a. } t \in T.$$

Invoking Proposition 1.2.12, we obtain that

$$-g(t) \in \text{conv} \limsup_{n\to+\infty} \left[F_1\big(t, x_n(t), x_n'(t)\big) - h\big(t, x_n(t)\big) \right]$$

$$\subseteq F_1\big(t, x(t), x'(t)\big) - h\big(t, x(t)\big) \quad \text{for a.a. } t \in T,$$

thus $g \in G(x) \cap C$ and so $G^-(C)$ is closed, i.e. G is upper semicontinuous into $L^{p'}(T)_w$. □

Let

$$D \stackrel{df}{=} \big\{ x \in C^1(T) : |x'(\cdot)|^{p-2} x'(\cdot) \in W^{1,p'}(T),$$
$$x'(0) \in \Xi_1\big(x(0)\big), \ -x'(b) \in \Xi_2\big(x(b)\big) \big\}$$

and let $S\colon L^p(T) \supseteq D \longrightarrow L^{p'}(T)$ be the differential operator, defined by

$$S(x) \stackrel{df}{=} -\big(|x'(\cdot)|^{p-2} x'(\cdot)\big)' \quad \forall\, x \in D.$$

Then as in Proposition 3.3.2 (see also Propositions 3.3.1 and 3.1.3), we have the following result:

PROPOSITION 3.5.3
If hypotheses $H(\Xi)$ hold,
then $S\colon L^p(T) \supseteq D \longrightarrow L^{p'}(T)$ *is maximal monotone.*

Let $J\colon L^p(T) \longrightarrow L^{p'}(T)$ be the maximal monotone and strictly monotone operator, defined by

$$J(x)(\cdot) \stackrel{df}{=} |x(\cdot)|^{p-2} x(\cdot) \quad \forall\, x \in L^p(T).$$

Then $S + J\colon L^p(T) \supseteq D \longrightarrow L^{p'}(T)$ is maximal monotone and also coercive since

$$\langle S(x), x \rangle_p + \langle J(x), x \rangle_p \geq \langle J(x), x \rangle_p \geq \|x\|_p^p$$

(since $0 \in D$ and $S(0) = 0$). So $S + J$ is surjective and because J is strictly monotone, it is also injective. Therefore

$$K \stackrel{df}{=} (S + J)^{-1}\colon L^{p'}(T) \longrightarrow D \subseteq W^{1,p}(T)$$

is a well-defined, single-valued operator. Arguing as in Claim 1 in the proof of Proposition 3.1.2, we can have the following result.

PROPOSITION 3.5.4
If hypotheses $H(\Xi)$ hold,
then operator $K\colon L^{p'}(T) \longrightarrow D \subseteq W^{1,p}(T)$ *is completely continuous.*

Having these auxiliary results, we can prove an existence theorem for Problem (3.126).

THEOREM 3.5.1
*If hypotheses $H(\psi)$, $H(F)$ and $H(\Xi)$ hold,
then problem (3.126) has a solution*

$$x_0 \in C^1(T) \quad \text{with} \quad |x_0'(\cdot)|^{p-2}x_0'(\cdot) \in W^{1,p'}(T)$$

with $\underline{\psi}(t) \leq x_0(t) \leq \overline{\psi}(t)$ for all $t \in T$.

PROOF   Let $G_1 \colon W^{1,p}(T) \longrightarrow P_{wkc}\big(L^{p'}(T)\big)$ be defined by

$$G_1(x) \stackrel{df}{=} G(x) + J(x) \qquad \forall\, x \in W^{1,p}(T).$$

Note that for every $x \in W^{1,p}(T)$ and every $g_1 \in G_1(x)$, we have

$$\|g_1\|_{p'} \leq \|\gamma_r\|_{p'} + b^{\frac{1}{p'}} \max\left\{\|\overline{\psi}\|_\infty^{p-1}, \|\underline{\psi}\|_\infty^{p-1}\right\} = M_3,$$

with $r \stackrel{df}{=} \max\left\{M_2, \|\overline{\psi}\|_\infty, \|\underline{\psi}\|_\infty\right\}$ (recall the definition of h). Let us set

$$V \stackrel{df}{=} \left\{g \in L^{p'}(T) : \|g\|_{p'} \leq M_3\right\}.$$

Then $K\big(G_1\big(W^{1,p}(T)\big)\big) \subseteq K(V)$ and by virtue of Proposition 3.5.4, $K(V)$ is relatively compact in $W^{1,p}(T)$. So we can apply Theorem 3.2.1 and obtain $x_0 \in D \subseteq W^{1,p}(T)$, such that $x_0 \in K\big(G_1(x_0)\big)$. So $S(x_0) = g$ with $g \in G(x_0)$ and we have

$$\begin{cases} -\big(|x_0'(t)|^{p-2}x_0'(t)\big)' \in F_1\big(t, x_0(t), x_0'(t)\big) - h\big(t, x_0(t)\big) & \text{for a.a. } t \in T, \\ x_0'(0) \in \Xi_1\big(x_0(0)\big), \; -x_0'(b) \in \Xi_2\big(x_0(b)\big). \end{cases}$$

Clearly from the last equation, it follows that

$$x_0 \in C^1(T) \qquad \text{and} \qquad |x_0'(\cdot)|^{p-2}x_0'(\cdot) \in W^{1,p'}(T).$$

We can finish the proof of the theorem if we show that

$$\underline{\psi}(t) \leq x_0(t) \leq \overline{\psi}(t) \qquad \forall\, t \in T \tag{3.131}$$

(because then $u\big(t, x_0(t), x_0'(t)\big) = \big(x_0(t), x_0'(t)\big)$ for all $t \in T$ (see Proposition 3.5.1) and $h\big(t, x_0(t)\big) = 0$ for all $t \in T$).
For some $f \in S^{p'}_{F_1(\cdot, x_0(\cdot), x_0'(\cdot))}$, we have

$$\big(|x_0'(t)|^{p-2}x_0'(t)\big)' = f(t) - h\big(t, x_0(t)\big) \quad \text{for a.a. } t \in T. \tag{3.132}$$

Also, since $\underline{\psi} \in C^1(T)$ is a lower solution for problem (3.126), we have

$$\left(|\underline{\psi}'(t)|^{p-2}\underline{\psi}'(t)\right)' \geq \underline{v}(t) \quad \text{for a.a. } t \in T, \tag{3.133}$$

with $\underline{v} \in S^{p'}_{F(\cdot,\underline{\psi}(\cdot),\underline{\psi}'(\cdot))}$ (see Definition 3.5.1(b)). Moreover, we have

$$\begin{cases} x_0'(0) \in \Xi_1(x_0(0)), & -x_0'(b) \in \Xi_2(x_0(b)), \\ \underline{\psi}'(0) \in \Xi_1(\underline{\psi}(0)) + \mathbb{R}_+, & -\underline{\psi}'(b) \in \Xi_2(\underline{\psi}(b)) + \mathbb{R}_+. \end{cases} \tag{3.134}$$

Subtracting (3.132) from (3.133), we obtain

$$\left(|\underline{\psi}'(t)|^{p-2}\underline{\psi}'(t)\right)' - \left(|x_0'(t)|^{p-2}x_0'(t)\right)'$$
$$\geq \underline{v}(t) - f(t) + h(t,x_0(t)) \quad \text{for a.a. } t \in T,$$

so

$$\int_0^b \left(\left(|\underline{\psi}'(t)|^{p-2}\underline{\psi}'(t)\right)' - \left(|x_0'(t)|^{p-2}x_0'(t)\right)'\right) (\underline{\psi} - x_0)^+(t)\, dt$$

$$\geq \int_0^b \left(\underline{v}(t) - f(t) + h(t,x_0(t))\right)(\underline{\psi} - x_0)^+(t)\, dt.$$

We use Green's identity on the left hand side of the last inequality. So we obtain

$$\left(|\underline{\psi}'(b)|^{p-2}\underline{\psi}'(b) - |x'(b)|^{p-2}x'(b)\right)(\underline{\psi} - x_0)^+(b)$$
$$- \left(|\underline{\psi}'(0)|^{p-2}\underline{\psi}'(0) - |x'(0)|^{p-2}x'(0)\right)(\underline{\psi} - x_0)^+(0)$$
$$- \int_0^b \left(|\underline{\psi}'(t)|^{p-2}\underline{\psi}'(t) - |x_0'(t)|^{p-2}x_0'(t)\right)\left[(\underline{\psi} - x_0)^+\right]'(t)\, dt$$

$$\geq \int_0^b \left(\underline{v}(t) - f(t) + h(t,x_0(t))\right)(\underline{\psi} - x_0)^+(t)\, dt. \tag{3.135}$$

From (3.134), we have

$$-x_0'(b) \in \Xi_2(x_0(b)) \quad \text{and} \quad -\underline{\psi}'(b) \in \Xi_2(\underline{\psi}(b)) + e_b,$$

with $e_b \geq 0$. So, if $x_0(b) \leq \underline{\psi}(b)$ (which is what matters since we are multiplying with $(\underline{\psi}-x_0)^+(b)$), from the monotonicity of Ξ_2, we have $-x_0'(b) \leq -\underline{\psi}'(b)$, hence $\varphi_p(-x_0'(b)) \leq \varphi_p(-\underline{\psi}'(b))$. Recall that $\varphi_p : \mathbb{R} \longrightarrow \mathbb{R}$ is the monotone homeomorphism defined by

$$\varphi_p(\zeta) \stackrel{df}{=} \begin{cases} |\zeta|^{p-2}\zeta & \text{if } \zeta \neq 0, \\ 0 & \text{if } \zeta = 0. \end{cases}$$

So
$$\left|\underline{\psi}'(b)\right|^{p-2}\underline{\psi}'(b) \leq \left|x_0'(b)\right|^{p-2}x_0'(b).$$

Similarly, again from (3.134), we have
$$x_0'(0) \in \Xi_1\big(x_0(0)\big) \quad \text{and} \quad -\underline{\psi}'(0) \in \Xi_1\big(\underline{\psi}(0)\big) + e_0,$$

with $e_0 \geq 0$ and arguing as above, we obtain that
$$\left|x_0'(0)\right|^{p-2}x_0'(0) \leq \left|\underline{\psi}'(0)\right|^{p-2}\underline{\psi}'(0).$$

Using these facts in (3.135), we obtain

$$-\int_0^b \left(\left|\underline{\psi}'(t)\right|^{p-2}\underline{\psi}'(t) - \left|x_0'(t)\right|^{p-2}x_0'(t)\right) \left[\left(\underline{\psi} - x_0\right)^+\right]'(t)\,dt$$

$$\geq \int_0^b \left(\underline{v}(t) - f(t)\right)\left(\underline{\psi} - x_0\right)^+(t)\,dt$$

$$+ \int_0^b h\big(t, x_0(t)\big)\left(\underline{\psi} - x_0\right)^+(t)\,dt. \tag{3.136}$$

Recall that $f \in S^{p'}_{F_1\left(\cdot, x_0(\cdot), x_0'(\cdot)\right)}$. So from the definition of F_1, we see that
$$f(t) \leq \underline{v}(t) \quad \text{for a.a. } t \in \{\underline{\psi} > x_0\}.$$

Hence
$$\int_0^b \left(\underline{v}(t) - f(t)\right)\left[\left(\underline{\psi} - x_0\right)^+\right]'(t)\,dt \geq 0. \tag{3.137}$$

Also from Remark 1.1.10, we know that
$$\left[\left(\underline{\psi} - x_0\right)^+\right]'(t) = \begin{cases} \left(\underline{\psi} - x_0\right)'(t) & \text{if } \underline{\psi}(t) > x_0(t), \\ 0 & \text{if } \underline{\psi}(t) \leq x_0(t), \end{cases} \tag{3.138}$$

while from an elementary inequality (consequence of the convexity of the function $r \longmapsto \frac{1}{p}|r|^p$), we have

$$-\int_0^b \left(\left|\underline{\psi}'(t)\right|^{p-2}\underline{\psi}'(t) - \left|x_0'(t)\right|^{p-2}x_0'(t)\right)\left[\left(\underline{\psi} - x_0\right)^+\right]'(t)\,dt \leq 0. \tag{3.139}$$

Using (3.137), (3.138) and (3.139) in (3.136), we obtain

$$\int_0^b h\big(t, x_0(t)\big)\left(\underline{\psi} - x_0\right)^+(t)\,dt \leq 0,$$

so

$$\int_{\{\underline{\psi}>x_0\}} h\big(t,x_0(t)\big)\big(\underline{\psi}-x_0\big)^+(t)\,dt \;\leq\; 0,$$

and recalling the definition of h, we have

$$\int_{\{\underline{\psi}>x_0\}} \Big(|\underline{\psi}(t)|^{p-2}\underline{\psi}'(t) - |x_0(t)|^{p-2}x_0(t)\Big)\,(\underline{\psi}-x_0)(t)\,dt \;\leq\; 0. \qquad (3.140)$$

But from the strict convexity of the function $r \longmapsto \frac{1}{p}|r|^p$, for every $t \in \{\underline{\psi} > x_0\}$, we have

$$\Big(|\underline{\psi}(t)|^{p-2}\underline{\psi}(t) - |x_0(t)|^{p-2}x_0(t)\Big)\,(\underline{\psi}-x_0)\,(t) \;>\; 0. \qquad (3.141)$$

Comparing (3.140) and (3.141), we infer that

$$\underline{\psi}(t) \;\leq\; x_0(t) \qquad \forall\, t \in T.$$

Similarly, we show that

$$x_0(t) \;\leq\; \overline{\psi}(t) \qquad \forall\, t \in T.$$

This finishes the proof of the theorem.   □

REMARK 3.5.3   Reasoning as in the proof of Theorem 3.1.3, we can have a "nonconvex" version of Theorem 3.5.1. We leave the details to the reader.   □

In fact we can do better than Theorem 3.5.1 and establish the existence of the smallest and of the greatest solutions of (3.126) in the order interval

$$[\underline{\psi},\overline{\psi}] \;\stackrel{df}{=}\; \big\{x \in C^1(T) : \underline{\psi}(t) \leq x(t) \leq \overline{\psi}(t) \text{ for all } t \in T\big\}.$$

In what follows on $L^\infty(T)$ (and on all of its subspaces), we consider the partial order induced by the closed, convex cone

$$L^\infty(T)_+ \;\stackrel{df}{=}\; \Big\{h \in L^\infty(T) : h(t) \geq 0 \text{ for almost all } t \in T\Big\}.$$

So for $h_1, h_2 \in L^\infty(T)$, we say that $h_1 \leq h_2$ if and only if $h_1(t) \leq h_2(t)$ for almost all $t \in T$. Let us set

$$S_1 \;\stackrel{df}{=}\; \big\{x \in C^1(T) : x \text{ is a solution of (3.126) and } x \in [\underline{\psi},\overline{\psi}]\,\big\}.$$

From Theorem 3.5.1, we know that if hypotheses $H(\psi)$, $H(F)$ and $H(\Xi)$ are in effect, then $S_1 \neq \emptyset$. Recall that a set C in a partially ordered space $(X,\leq)$

is a chain (or totally ordered subset), if for every $x, y \in C$, we have that either $x \leq y$ or $y \leq x$.

PROPOSITION 3.5.5

If hypotheses $H(\psi)$, $H(F)$, $H(\Xi)$ hold and $C \subseteq S_1$ is a chain, *then* *C has an upper bound.*

PROOF   Since $C \subseteq L^\infty(T)$ is bounded above and below and $L^\infty(T)$ is a complete lattice, we can find a sequence $\{x_n\}_{n \geq 1} \subseteq C$, such that

$$\sup C \;=\; \sup_{n \geq 1} x_n \;\in\; L^\infty(T)$$

(see e.g. Dunford & Schwartz (1958, p. 336)). Therefore, exploiting the lattice structure of $L^\infty(T)$, we may assume that the sequence $\{x_n\}_{n \geq 1}$ is increasing. Moreover, by the monotone convergence theorem, we have that

$$x_n \;\longrightarrow\; x \quad \text{in } L^p(T).$$

From Proposition 3.5.1, we know that there exists $M_1 = M_1(\overline{\psi}, \underline{\psi}, \eta, a, c) > 0$, such that

$$\left|x'(t)\right| \;\leq\; M_1 \qquad \forall\, t \in T.$$

Let $r \overset{df}{=} \max\left\{M_1, \, \left\|\overline{\psi}\right\|_\infty, \, \left\|\underline{\psi}\right\|_\infty\right\}$. From hypothesis $H(F)(iv)$, we have

$$\left|\left(\left|x_n'(t)\right|^{p-2} x_n'(t)\right)'\right| \;\leq\; \gamma_r(t) \quad \text{for a.a. } t \in T$$

and so the sequence $\left\{\left|x_n'(\cdot)\right|^{p-2} x_n'(\cdot)\right\}_{n \geq 1} \subseteq W^{1,p'}(T)$ is bounded.

Also note that

$$\left\|x_n\right\|_\infty \;\leq\; r, \qquad \left\|x_n'\right\|_\infty \;\leq\; r \qquad \forall\, n \geq 1$$

and so the sequence $\{x_n\}_{n \geq 1} \subseteq W^{1,p}(T)$ is bounded. Thus by passing to a suitable subsequence, we may assume that

$$\left|x_n'(\cdot)\right|^{p-2} x_n'(\cdot) \;\overset{w}{\longrightarrow}\; v_1 \text{ in } W^{1,p'}(T),$$
$$x_n \;\overset{w}{\longrightarrow}\; v_2 \text{ in } W^{1,p}(T).$$

Evidently $v_2 = x$, while by virtue of Proposition 3.5.3, $v_1(\cdot) = \left|x'(\cdot)\right|^{p-2} x'(\cdot)$. Note that

$$x_n \;\overset{w}{\longrightarrow}\; x \qquad\qquad \text{in } C(T),$$
$$\left|x_n'(\cdot)\right|^{p-2} x_n'(\cdot) \;\overset{w}{\longrightarrow}\; \left|x'(\cdot)\right|^{p-2} x'(\cdot) \text{ in } C(T).$$

So invoking Proposition 1.2.12, we obtain

$$\left(\left|x'(t)\right|^{p-2} x'(t)\right)' \;\in\; \text{conv} \limsup_{n \to +\infty} F\big(t, x_n(t), x_n'(t)\big)$$

$$\subseteq\; F\big(t, x(t), x'(t)\big) \quad \text{for a.a. } t \in T.$$

Moreover, since

$$x'_n(0) \in \Xi_1\big(x_n(0)\big) \qquad \text{and} \qquad -x'_n(b) \in \Xi_2\big(x_n(b)\big) \qquad \forall \, n \geq 1,$$

because of hypotheses $H(\Xi)$, we have that

$$x'(0) \in \Xi_1\big(x(0)\big) \qquad \text{and} \qquad -x'(b) \in \Xi_2\big(x(b)\big).$$

Of course

$$\underline{\psi}(t) \leq x(t) \leq \overline{\psi}(t) \qquad \forall \, t \in T.$$

Therefore $x \in S_1$ and so it is an upper bound of C. $\qquad\qquad\qquad \Box$

Recall that if $(Y, \leq)$ is a partially ordered set, we say that Y is directed, if for each pair $y_1, y_2 \in Y$, we can find $y_3 \in Y$, such that $y_1 \leq y_3$ and $y_2 \leq y_3$.

PROPOSITION 3.5.6
If hypotheses $H(\psi)$, $H(F)$ and $H(\Xi)$ hold, <u>then</u> S_1 is directed.

PROOF Let $x_1, x_2 \in S_1$ and set $x_3 \overset{df}{=} \max\{x_1, x_2\}$. We have

$$x_3 \; = \; (x_1 - x_2)^+ + x_2$$

and so $x_3 \in W^{1,p}(T)$ (see Remark 1.1.10). As before we introduce the truncation and penalization functions u, Q and h (see (3.128), (3.129) and (3.130)), but with $\underline{\psi}$ replaced by x, and $\underline{v}$ by $g(t) \overset{df}{=} \min\{f_1(t), \; f_2(t)\}$, where $f_i \in S^{p'}_{F\left(\cdot, x_i(\cdot), x'_i(\cdot)\right)}$ for $i \in \{1, 2\}$, are such that

$$\big(|x'_i(t)|^{p-2} x'_i(t)\big)' \; = \; f_i(t) \quad \text{for a.a. } t \in T.$$

Arguing as in the proof of Theorem 3.5.1, we can find $x \in D$ and $f \in S^{p'}_{F_1(\cdot, x(\cdot), x'(\cdot))}$, such that

$$\begin{cases} -\big(|x'(t)|^{p-2} x'(t)\big)' = f(t) - h\big(t, x(t)\big) & \text{for a.a. } t \in T, \\ x'(0) \in \Xi_1\big(x(0)\big), \; -x'(b) \in \Xi_2\big(x(b)\big). \end{cases}$$

Next we show that

$$x_3(t) \; \leq \; x(t) \; \leq \; \overline{\psi}(t) \qquad \forall \, t \in T. \tag{3.142}$$

We know that for almost all $t \in T$, we have

$$x'_3(t) \; = \; \begin{cases} x'_1(t) & \text{if } x_1(t) \geq x_2(t), \\ x'_2(t) & \text{if } x_1(t) \leq x_2(t) \end{cases}$$

(recall that $(x_1 - x_2)'(t) = 0$ for all $t \in \{x_1 - x_2 = 0\}$; see Remark 1.1.10). Suppose that for some t_0, we have $x(t_0) < x_1(t_0)$. Then, we can find an interval $[l_1, t_2] \subseteq T$, such that

$$x(t) < x_1(t) \qquad \forall\, t \in (t_1, t_2), \tag{3.143}$$

$$x(t_1) = x_1(t_1) \text{ or } t_1 = 0 \text{ with } x(0) < x_1(0), \tag{3.144}$$

$$x(t_2) = x_1(t_2) \text{ or } t_2 = b \text{ with } x(b) < x_1(b). \tag{3.145}$$

We have

$$\left(\left|x'(t)\right|^{p-2} x'(t)\right)' \in F_1\big(t, x(t), x'(t)\big) - h\big(t, x(t)\big) \quad \text{for a.a. } t \in [t_1, t_2],$$

so

$$\left(\left|x'(t)\right|^{p-2} x'(t)\right)' \in F_1\big(t, x_3(t), x_3'(t)\big) \cap \left(-\infty, g(t)\right]$$
$$- \left|x_3(t)\right|^{p-2} x_3(t) + \left|x(t)\right|^{p-2} x(t) \quad \text{for a.a. } t \in [t_1, t_2]$$

and thus

$$\left(\left|x'(t)\right|^{p-2} x'(t)\right)' + \left|x_3(t)\right|^{p-2} x_3(t) - \left|x(t)\right|^{p-2} x(t) \le g(t) \text{ for a.a. } t \in [t_1, t_2].$$

Also since $x_1 \le x_3$ and $\varphi_p \colon \mathbb{R} \longrightarrow \mathbb{R}$ is monotone, it follows that

$$\varphi_p\big(x_1(t)\big) = \left|x_1(t)\right|^{p-2} x_1(t) \le \varphi_p\big(x_3(t)\big) = \left|x_3(t)\right|^{p-2} x_3(t) \qquad \forall\, t \in T.$$

Hence recalling the definition of g, we can write that

$$\left|x_1(t)\right|^{p-2} x_1(t) - \left|x(t)\right|^{p-2} x(t) \le g(t) - \left(\left|x'(t)\right|^{p-2} x'(t)\right)'$$
$$\le \left(\left|x_1'(t)\right|^{p-2} x_1'(t)\right)' - \left(\left|x'(t)\right|^{p-2} x'(t)\right)' \quad \text{for a.a. } t \in [t_1, t_2]. \tag{3.146}$$

We multiply (3.146) with $(x - x_1)(t)$, where $t \in [t_1, t_2]$ and then integrate over $[t_1, t_2]$. Since $(x - x_1)(t) \le 0$ for $t \in [t_1, t_2]$ (see (3.143), (3.144) and (3.145)), integrating by parts, we obtain

$$0 > \int_{t_1}^{t_2} \left(\left|x_1(t)\right|^{p-2} x_1(t) - \left|x(t)\right|^{p-2} x(t)\right)(x - x_1)(t)\, dt$$

$$\ge \int_{t_1}^{t_2} \left(\left(\left|x_1'(t)\right|^{p-2} x_1'(t)\right)' - \left(\left|x'(t)\right|^{p-2} x'(t)\right)'\right)(x - x_1)(t)\, dt$$

$$= \left(\left|x_1'(t_2)\right|^{p-2} x_1'(t_2) - \left|x'(t_2)\right|^{p-2} x'(t_2)\right)(x - x_1)(t_2)$$

$$- \left(\left|x_1'(t_1)\right|^{p-2} x_1'(t_1) - \left|x'(t_1)\right|^{p-2} x'(t_1)\right)(x - x_1)(t_1)$$

$$- \int_{t_1}^{t_2} \left(\left|x_1'(t)\right|^{p-2} x_1'(t) - \left|x'(t)\right|^{p-2} x'(t)\right)(x - x_1)'(t)\, dt \tag{3.147}$$

Using (3.144) and (3.145) and the monotonicity of Ξ_1, Ξ_2 and φ_p, we obtain

$$\left(\left| x_1'(t_1) \right|^{p-2} x_1'(t_1) - \left| x'(t_1) \right|^{p-2} x'(t_1) \right) (x - x_1)(t_1) \leq 0 \qquad (3.148)$$

and

$$\left(\left| x_1'(t_2) \right|^{p-2} x_1'(t_2) - \left| x'(t_2) \right|^{p-2} x'(t_2) \right) (x - x_1)(t_2) \geq 0. \qquad (3.149)$$

Returning to (3.147) and using (3.148) and (3.149), we obtain

$$0 \leq \int_{t_1}^{t_2} \left(\left| x_1'(t) \right|^{p-2} x_1'(t) - \left| x'(t) \right|^{p-2} x'(t) \right) (x - x_1)'(t)\, dt \; < \; 0,$$

a contradiction. So we must have $x_1(t) \leq x(t)$ for all $t \in T$. Similarly, we show that $x_2(t) \leq x(t)$ for all $t \in T$ and so finally

$$x_3(t) \; \leq \; x(t) \qquad \forall\, t \in T.$$

On the other hand as in the proof of Theorem 3.5.1, we can check that

$$x(t) \; \leq \; \overline{\psi}(t) \qquad \forall\, t \in T.$$

Therefore, we have proved that

$$x_3(t) \; \leq \; x(t) \; \leq \; \overline{\psi}(t) \qquad \forall\, t \in T$$

and so

$$F_1\big(t, x(t), x'(t)\big) \; = \; F\big(t, x(t), x'(t)\big)$$

and

$$h\big(t, x(t)\big) \; = \; 0 \qquad \forall\, t \in T.$$

Hence $x \in S_1$ and $x \geq x_3$, which implies that S_1 is directed. $\qquad\square$

Now we are ready to obtain the extremal elements of S_1 in $\left[\underline{\psi}, \overline{\psi} \right]$.

THEOREM 3.5.2
If hypotheses $H(\psi)$, $H(F)$ and $H(\Xi)$ hold,
then there exist the smallest and the greatest solution (extremal solutions) of (3.126) in $\left[\underline{\psi}, \overline{\psi} \right]$.

PROOF   Proposition 3.5.5 and the Kuratowski-Zorn Lemma (see Theorem A.1.1) imply that we can find $x^* \in S_1$, a maximal element of S_1 with respect to the pointwise ordering on $W^{1,p}(T)$ (i.e. the ordering introduced by $L^\infty(T)_+$). Note that, if $x \in S_1$, by virtue of Proposition 3.5.6, we can find

$y \in S_1$, such that $x \leq y$ and $x^* \leq y$. But $x^* \in S_1$ is maximal. So we must have $x^* = y$. Therefore $x \leq x^*$ and because $x \in S_1$ was arbitrary, it follows that $x^* \in S_1$ is the greatest element of S_1.

If on $W^{1,p}(T)$ we consider the reverse partial order $\leq_1$ according to which $x \leq_1 y$ if and only if $y(t) \leq x(t)$ for all $t \in T$, then using the same arguments we can produce $x_* \in S_1$, which is the smallest element of S_1.   ∎

EXAMPLE 3.5.1   **(a)** Suppose that $K_1, K_2 \in P_{fc}(\mathbb{R})$, with $0 \in K_1 \cap K_2$ and set $\Xi_i \stackrel{df}{=} \partial K_i = N_{K_i}$ for $i \in \{1, 2\}$. Then (3.126) becomes

$$\begin{cases} -\left(|x'(t)|^{p-2}x'(t)\right)' \in F\left(t, x(t), x'(t)\right) & \text{for a.a. } t \in T, \\ x(0) \in K_1, \ x(b) \in K_2, \\ x'(0)x(0) = \sigma_{_{\mathbb{R}}}\left(x(0); K_1\right), \ -x'(b)x(b) = \sigma_{_{\mathbb{R}}}\left(x(b); K_2\right). \end{cases} \tag{3.150}$$

Here $\sigma_{_{\mathbb{R}}}(u; K) \stackrel{df}{=} \sup_{k \in K} uk$ is the support function of a set $K \subseteq \mathbb{R}$.
(b) If in (a), $K_1 = K_2 = \{0\}$, then problem (3.150) becomes the Dirichlet problem.
(c) If in (a), $K_1 = K_2 = \mathbb{R}$, then problem (3.150) becomes the Neumann problem.
(d) If

$$\Xi_1(\zeta) \stackrel{df}{=} \frac{1}{\beta}\zeta \qquad \text{and} \qquad \Xi_1(\zeta) \stackrel{df}{=} \frac{1}{\gamma}\zeta,$$

with $\beta, \gamma > 0$, then problem (3.126) becomes the following multivalued Sturm-Liouville problem:

$$\begin{cases} -\left(|x'(t)|^{p-2}x'(t)\right)' \in F\left(t, x(t), x'(t)\right) & \text{for a.a. } t \in T, \\ x(0) - \beta x'(0) = 0, \ x(b) + \gamma x'(b) = 0. \end{cases} \tag{3.151}$$

More generally, we can have

$$\Xi_1(\zeta) \stackrel{df}{=} u_1(\zeta) - \zeta \qquad \text{and} \qquad \Xi_2(\zeta) \stackrel{df}{=} u_2(\zeta) - \zeta,$$

with $u_1, u_2 \colon \mathbb{R} \longrightarrow \mathbb{R}$ two contractions. In this case problem (3.126) becomes

$$\begin{cases} -\left(|x'(t)|^{p-2}x'(t)\right)' \in F\left(t, x(t), x'(t)\right) & \text{for a.a. } t \in T, \\ x(0) + x'(0) = u_1\left(x(0)\right), \ x(b) + x'(b) = u_2\left(x(b)\right). \end{cases} \tag{3.152}$$

∎

The periodic problem is not included in the previous examples. However, a careful reading of the proofs of Propositions 3.5.5 and 3.5.6 and of Theorem 3.5.2 reveals that they are still valid (in fact with simplified arguments) in the context of the periodic problem:

$$\begin{cases} -\left(|x'(t)|^{p-2}x'(t)\right)' \in F\left(t, x(t), x'(t)\right) & \text{for a.a. } t \in T, \\ x(0) = x(b), \ x'(0) = x'(b), \end{cases} \tag{3.153}$$

for $p \in (1, +\infty)$. If $\underline{\psi} \in C^1(T)$ is a lower solution for (3.153), then $\underline{\psi}'(0) \geq \underline{\psi}'(b)$ and if $\overline{\psi} \in C^1(T)$ is an upper solution for (3.153), then $\overline{\psi}'(0) \leq \overline{\psi}'(b)$.

THEOREM 3.5.3
If hypotheses $H(\psi)$, $H(F)$ and $H(\Xi)$ hold,
then problem (3.153) has extremal solutions in the order interval $[\underline{\psi}, \overline{\psi}]$.

3.6 Positive Solutions and Other Methods

In this section, first we investigate the existence of positive solutions for certain second order multivalued boundary value problems and then we deal with single-valued second order boundary value problems driven by p-Laplace like operators and present a new method for the analysis of such problems. This method is based on an abstract result on the range of a sum of nonlinear operators of monotone type. The method is applied on a Neumann problem.

3.6.1 Positive Solutions

We start with the problem of the existence of positive solutions for the following second order multivalued boundary value problem:

$$\begin{cases} -x''(t) \in F\big(t, x(t)\big) & \text{for a.a. } t \in [0, 1], \\ Ax(0) - Bx'(0) = 0, \\ Cx(1) + Dx'(1) = 0, \end{cases} \tag{3.154}$$

where A, B, C, D are diagonal nonnegative definite matrices.

The approach here is based on a fixed point theorem for multifunctions of the compression-expansion type on a cone. More precisely, we shall use the following result which is due to Erbe & Krawcewicz (1991 a), where the interested reader can find its proof. First a definition.

DEFINITION 3.6.1

(a) *Let X be a Banach space and K a nonempty subset of X. We say that K is a **convex cone**, if K is closed, convex, $K \cap (-K) = \{0\}$ and $tK \subseteq K$ for all $t \geq 0$.*

(b) *A multifunction $F: K \longrightarrow P_{kc}(K)$ is said to be **compact**, if it is upper semicontinuous and for every $D \subseteq K$ bounded, the set $\overline{F(D)}$ is compact.*

PROPOSITION 3.6.1

If X *is a Banach space,* U_1, U_2 *are two bounded open sets in* X, $0 \in U_1$, $\overline{U}_1 \subseteq U_2$, K *is a convex cone in* X, $G \colon K \cap (\overline{U}_2 \setminus U_1) \longrightarrow P_{kc}(K)$ *is a compact multifunction and one of the following two conditions is satisfied:*

(i)
$$\sup_{y \in G(x)} \|y\|_X \leq \|x\|_X \text{ for all } x \in K \cap \partial U_1 \text{ and}$$
$$\inf_{y \in G(x)} \|y\|_X \geq \|x\|_X \text{ for all } x \in K \cap \partial U_2;$$

(ii)
$$\inf_{y \in G(x)} \|y\|_X \geq \|x\|_X \text{ for all } x \in K \cap \partial U_1 \text{ and}$$
$$\sup_{y \in G(x)} \|y\|_X \leq \|x\|_X \text{ for all } x \in K \cap \partial U_2,$$

<u>*then*</u> G *has at least one fixed point* $x_0 \in K \cap (\overline{U}_2 \setminus U_1)$, *i.e.* $x_0 \in G(x_0)$.

We impose the following conditions on the matrices of the boundary conditions in problem (3.154):

<u>$H(A\text{--}D)$</u> $A, B, C, D \in \mathbb{R}^{N \times N}$ are matrices, such that

(i) they are diagonal nonnegative definite, i.e.

$$A = \operatorname{diag}(a_k)_{k=1}^N, \qquad B = \operatorname{diag}(b_k)_{k=1}^N,$$
$$C = \operatorname{diag}(c_k)_{k=1}^N, \qquad D = \operatorname{diag}(d_k)_{k=1}^N,$$

with $a_k, b_k, c_k, d_k \geq 0$ for all $k \in \{1, \ldots, N\}$;

(ii) the matrix $AC + AD + BC$ is positive definite, i.e.

$$a_k c_k + a_k d_k + b_k c_k > 0 \qquad \forall\, k \in \{1, \ldots, N\}.$$

By $G \colon [0,1] \times [0,1] \longrightarrow \mathbb{R}^{N \times N}$, we denote the **Green's function** corresponding to the linearization of (3.154):

$$\begin{cases} -x''(t) = 0 & \text{for a.a. } t \in [0,1], \\ Ax(0) - Bx'(0) = 0, \\ Cx(1) + Dx'(1) = 0. \end{cases}$$

Because of hypotheses $H(A\text{--}D)$, it is well known that the function $G(t,s)$ is given by:

$$G(t,s) = \begin{cases} (tA + B)\big((1-s)C + D\big)\big(AC + AD + BC\big)^{-1} & \text{if } t \leq s, \\ (sA + B)\big((1-t)C + D\big)\big(AC + AD + BC\big)^{-1} & \text{if } s < t. \end{cases}$$

Evidently this is a diagonal matrix $G(t,s) = \operatorname{diag}\{G_k(t,s)\}_{k=1}^N$, with diagonal elements, given by

$$G_k(t,s) = \begin{cases} \frac{1}{\beta_k}(ta_k + b_k)\big((1-s)c_k + d_k\big) & \text{if } t \leq s, \\ \frac{1}{\beta_k}(sa_k + b_k)\big((1-t)c_k + d_k\big) & \text{if } t > s, \end{cases}$$

where $\beta_k \overset{df}{=} a_k c_k + a_k d_k + b_k c_k$ for all $k \in \{1, \ldots, N\}$.

Let

$$\gamma \overset{df}{=} \min_{k \in \{1, \ldots, N\}} \left\{ \frac{a_k + 4b_k}{4(a_k + b_k)}, \frac{c_k + 4d_k}{4(c_k + d_k)} \right\}. \tag{3.155}$$

It is easy to check that

$$G_k(t, s) \leq G_k(s, s) \qquad \forall (t, s) \in [0, 1] \times [0, 1] \tag{3.156}$$

and

$$G_k(t, s) \geq \gamma G_k(s, s) \qquad \forall (t, s) \in \left[\frac{1}{4}, \frac{3}{4}\right] \times [0, 1]. \tag{3.157}$$

For every $k \in \{1, \ldots, N\}$, we define $\tau_k \in [0, 1]$ to be such that

$$\int_{\frac{1}{4}}^{\frac{3}{4}} G_k(\tau_k, s) \zeta_0(s) ds = \max_{t \in [0,1]} \int_{\frac{1}{4}}^{\frac{3}{4}} G_k(t, s) \zeta_0(s) ds \qquad \forall k \in \{1, \ldots, N\} \tag{3.158}$$

where $\zeta_0 \colon [0, 1] \longrightarrow \overline{\mathbb{R}}_+$ is a measurable function (see hypothesis $H(F)_1(iv)$ below).

For every $t \in T$, $\xi \in \mathbb{R}^N_+$ and $k \in \{1, \ldots, N\}$, we define

$$m_k^F(t, \xi) \overset{df}{=} \inf\{u_k : u = (u_1, \ldots, u_N), \ u \in F(t, \xi)\}$$

$$M_k^F(t, \xi) \overset{df}{=} \sup\{u_k : u = (u_1, \ldots, u_N), \ u \in F(t, \xi)\}.$$

In what follows on $\mathbb{R}^N$, we consider the l^∞-norm, i.e.

$$\|\xi\|_{l^\infty} \overset{df}{=} \max\{|\xi_k| : k \in \{1, \ldots, N\}\} \qquad \forall \xi = (\zeta_1, \ldots, \xi_N) \in \mathbb{R}^N.$$

Our hypotheses on the multivalued nonlinearity F and the Green's function G are the following:

$\underline{H(F)_1}$ $F : [0, 1] \times \mathbb{R}^N_+ \longrightarrow P_{kc}(\mathbb{R}^N_+)$ is a multifunction, such that

(i) F is graph measurable;

(ii) for almost all $t \in [0, 1]$, the multifunction

$$\mathbb{R}^N_+ \ni \xi \longmapsto F(t, \xi) \in P_{kc}(\mathbb{R}^N_+)$$

has closed graph;

(iii) one of the following two conditions hold:

$(iii)_1$ there exists $g \in L^1(0, 1)_+$, such that for almost all $t \in [0, 1]$, all $\xi \in \mathbb{R}^N_+$ and all $u \in F(t, \xi)$, we have $\|u\|_{l^\infty} \leq g(t)$;

$(iii)_2$ for all $k \in \{1, \ldots, N\}$, we have that

$$\lim_{\substack{\|\xi\|_{\mathbb{R}^N} \to +\infty \\ \xi \in \mathbb{R}_+^N}} M_k^F(t, \xi) = +\infty \quad \text{uniformly for a.a. } t \in [0, 1]$$

and for every $r > 0$ there exists $g_r \in L^1(0, 1)_+$, such that

$$\sup \{ \|u\|_{l^\infty} : u \in F(t, \xi), \ \|\xi\|_{l^\infty} \leq r \} \leq g_r(t)$$

for almost all $t \in [0, 1]$;

(iv) there exist measurable functions $\zeta_0, \zeta_1 : [0, 1] \longrightarrow \overline{\mathbb{R}}_+ \overset{df}{=} \mathbb{R}_+ \cup \{+\infty\}$, such that

$$\liminf_{\substack{\|\xi\|_{l^\infty} \to 0 \\ \xi \in \mathbb{R}_+^N}} \frac{m_k^F(t, \xi)}{\|\xi\|_{l^\infty}} \geq \zeta_0(t) \quad \text{uniformly for a.a. } t \in [0, 1],$$

$$\limsup_{\substack{\|\xi\|_{l^\infty} \to +\infty \\ \xi \in \mathbb{R}_+^N}} \frac{M_k^F(t, \xi)}{\|\xi\|_{l^\infty}} \leq \zeta_1(t) \quad \text{uniformly for a.a. } t \in [0, 1]$$

and for all $k \in \{1, \ldots, N\}$, we have

$$\frac{1}{\gamma} < \int_{\frac{1}{4}}^{\frac{3}{4}} G_k(\tau_k, s) \zeta_0(s) ds \quad \text{and} \quad \int_0^1 G_k(s, s) \zeta_1(s) ds < 1,$$

where γ is defined by (3.155) and $\tau_k \in [0, 1]$ are defined by (3.158).

REMARK 3.6.1   If $\zeta_0 \equiv +\infty$ and $\zeta_1 \equiv 0$, then the problem is "sublinear" (i.e. the growth of the nonlinearity F is sublinear).   ⬚

THEOREM 3.6.1
If hypotheses H(A–D) and H(F)$_1$ hold,
then problem (3.154) has at least one nontrivial solution $x_0 \in W^{1,2}((0, 1); \mathbb{R}^N)$ such that $x_0(t) \in \mathbb{R}_+^N$ for all $t \in [0, 1]$.

PROOF   Let K be the nonempty, closed and convex cone in $C([0, 1]; \mathbb{R}^N)$, defined by

$$K \overset{df}{=} \left\{ x \in C([0, 1]; \mathbb{R}^N) : x(t) = (x_1(t), \ldots, x_N(t)) \in \mathbb{R}_+^N \text{ for all } t \in [0, 1], \right.$$

$$\left. \text{and} \min_{t \in [\frac{1}{4}, \frac{3}{4}]} x_k(t) \geq \gamma \|x_k\|_\infty \text{ for all } k \in \{1, \ldots, N\} \right\}. \quad (3.159)$$

Consider the multivalued operator $V : K \longrightarrow 2^{C([0,1];\mathbb{R}^N)}$, defined for any $x \in K$ by

$$V(x) \overset{df}{=} \left\{ y \in C([0,1]; \mathbb{R}^N) : y(t) = \int_0^1 G(t,s)u(s)ds, \right.$$

$$\left. u \in S^1_{F(\cdot,x(\cdot))} \right\}. \qquad (3.160)$$

Because of hypothesis $H(F)_1(i)$, for all $x \in K$, we have that the multifunction

$$[0,1] \ni t \longmapsto F(t,x(t)) \in 2^{\mathbb{R}^N_+} \qquad \text{is measurable.}$$

This combined with hypothesis $H(F)_1(iii)$ (both alternatives) implies that $S^1_{F(\cdot,x(\cdot))} \neq \emptyset$. Hence V has nonempty values and a straightforward use of the Arzela-Ascoli Theorem (see Theorem A.1.13) implies that

$$V(x) \in P_{kc}\big(C([0,1]; \mathbb{R}^N)\big) \qquad \forall\, x \in K.$$

Next we show that V is compact (i.e. it is upper semicontinuous and maps bounded sets into relatively compact sets). First let $D \subseteq K$ be a nonempty bounded set and let $\tilde{r} \overset{df}{=} \sup_{x \in D} \|x\|_\infty$. Then, we have that

$$\|y(t)\|_{l^\infty} = \left\| \int_0^1 G(t,s)u(s)ds \right\|_{l^\infty} \qquad \forall\, y \in V(D),\ t \in T,$$

for some $u \in S^1_{F(\cdot,x(\cdot))}$, $x \in D$ and so

$$\|y(t)\|_{l^\infty}$$

$$\leq \int_0^1 \|G(t,s)\|_\mathcal{L} \|u(s)\|_{l^\infty}\, ds \ \leq \ M_1 \int_0^1 \tilde{g}(s)ds \ = \ M_1 \|\tilde{g}\|_1 \qquad \forall\, y \in V(D),$$

for $M_1 \overset{df}{=} \max_{k \in \{1,\ldots,N\}} \left[\frac{1}{\beta_k}(a_k + b_k)(c_k + d_k) \right] > 0$ and

$$\tilde{g}(t) \overset{df}{=} \begin{cases} g(t) & \text{if hypothesis } H(F)_1(iii)_1 \text{ holds,} \\ g_{\tilde{r}}(t) & \text{if hypothesis } H(F)_1(iii)_2 \text{ holds.} \end{cases}$$

Moreover, from the definition of G, for a given $\varepsilon > 0$, we can find $\delta > 0$, such that

$$\|G(t',s) - G(t,s)\|_\mathcal{L} \ < \ \varepsilon \qquad \forall\, t,t',s \in [0,1],\ |t - t'| < \delta$$

and so

$$\|y(t') - y(t)\|_{l^\infty}$$

$$\leq \int_0^1 \|G(t',s) - G(t,s)\|_\mathcal{L}\, \tilde{g}(s)ds \ \leq \ \varepsilon \|\tilde{g}\|_1 \qquad \forall\, t,t' \in [0,1],\ |t - t'| < \delta.$$

So by the Arzela-Ascoli Theorem (see Theorem A.1.13), the set $\overline{V(D)} \subseteq C([0,1]; \mathbb{R}^N)$ is compact.

It remains to show that V is upper semicontinuous. Clearly because of hypothesis $H(F)_1(iii)$ (both alternatives), V is locally compact and so by virtue of Proposition 1.2.5, to show the desired upper semicontinuity of V, it suffices to show that $\operatorname{Gr} V$ is closed in $K \times C([0,1]; \mathbb{R}^N)$. To this end let $(x_n, y_n) \in \operatorname{Gr} V$ for $n \geq 1$ and assume that

$$x_n \longrightarrow x \text{ in } C(T; \mathbb{R}^N),$$
$$y_n \longrightarrow y \text{ in } C(T; \mathbb{R}^N).$$

We have

$$y_n(t) = \int_0^1 G(t,s)u_n(s)ds \qquad \forall\, t \in [0,1],\ n \geq 1,$$

for some $u_n \in S^1_{F(\cdot, x_n(\cdot))}$. Let $\hat{r} \stackrel{df}{=} \sup_{n \geq 1} \|x\|_\infty$ and let

$$\hat{g}(t) \stackrel{df}{=} \begin{cases} g(t) & \text{if hypothesis } H(F)_1(iii)_1 \text{ holds} \\ \tilde{g}(t) & \text{if hypothesis } H(F)_1(iii)_2 \text{ holds.} \end{cases}$$

Then

$$\|u_n(t)\|_{l^\infty} \leq \hat{g}(t) \quad \text{for a.a. } t \in [0,1].$$

Since $\hat{g} \in L^1(0,1)_+$, by the Dunford-Pettis Theorem (see Theorem A.3.14), passing to a subsequence if necessary, we may assume that

$$u_n \xrightarrow{w} u \quad \text{in } L^1(0,1; \mathbb{R}^N),$$

for some $u \in L^1(0,1; \mathbb{R}^N)$. Invoking Proposition 1.2.12, we have that

$$u(t) \in \operatorname{conv} \limsup_{n \to +\infty} F(t, x_n(t)) \subseteq F(t, x(t)) \quad \text{for a.a. } t \in [0,1],$$

where the last inclusion is a consequence of the fact that F is convex-valued and for almost all $t \in [0,1]$, $F(t, \cdot)$ has a closed graph (see hypothesis $H(F)_1(ii)$). Hence $u \in S^1_{F(\cdot, x(\cdot))}$.

Since

$$\int_0^1 G(t,s)u_n(s)ds \longrightarrow \int_0^1 G(t,s)u(s)ds \quad \text{as } n \to +\infty,$$

in the limit, we obtain

$$y(t) = \int_0^1 G(t,s)u(s)ds \qquad \forall\, t \in [0,1].$$

As $u \in S^1_{F(\cdot, x(\cdot))}$, we have $y \in V(x)$. This proves the upper semicontinuity and so finally the compactness of V.

Next, we show that

$$V(x) \in K \qquad \forall\, x \in K. \tag{3.161}$$

To this end let $x \in K$ and $y \in V(x)$. Then $y = (y_1, \ldots, y_N) \in C([0,1]; \mathbb{R}^N)$ and

$$y_k(t) = \int_0^1 G_k(t,s) u_k(s) ds \qquad \forall\, t \in [0,1],\ k \in \{1, \ldots, N\},$$

for some $u = (u_1, \ldots, u_N) \in S^1_{F(\cdot, x(\cdot))}$. From (3.157), we have

$$y_k(t) \geq \gamma \int_0^1 G_k(s,s) u_k(s) ds \qquad \forall\, t \in \left[\frac{1}{4}, \frac{3}{4}\right],\ k \in \{1, \ldots, N\}$$

and from (3.156), we get

$$y_k(t) \geq \gamma y_k(s) \qquad \forall\, (t,s) \in \left[\frac{1}{4}, \frac{3}{4}\right] \times [0,1],\ k \in \{1, \ldots, N\}.$$

Thus

$$\min_{t \in \left[\frac{1}{4}, \frac{3}{4}\right]} y_k(t) \geq \gamma \|y_k\|_\infty \qquad \forall\, k \in \{1, \ldots, M\}.$$

So $y \in K$ and (3.161) holds.

By virtue of hypothesis $H(F)_1(iv)$, we have

$$\frac{1}{\gamma} < \int_{\frac{1}{4}}^{\frac{3}{4}} G_k(\tau_k, s) \zeta_0(s) ds \qquad \forall\, k \in \{1, \ldots, N\}.$$

So we can find $\bar{\varepsilon} > 0$ small, so that

$$\frac{1}{\gamma} \leq \int_{\frac{1}{4}}^{\frac{3}{4}} G_k(\tau_k, s) (\zeta_0(s) - \bar{\varepsilon}) ds \qquad \forall\, k \in \{1, \ldots, N\}. \tag{3.162}$$

From hypothesis $H(F)_1(iv)$, for this $\bar{\varepsilon} > 0$, we can find $\bar{\delta} = \bar{\delta}(\bar{\varepsilon}) > 0$, such that for almost all $s \in [0,1]$ and all $\xi \in \mathbb{R}^N$ with $\|\xi\|_{l^\infty} \leq \bar{\delta}$, we have

$$(\zeta_0(s) - \bar{\varepsilon}) \|\xi\|_{l^\infty} \leq m_k^F(s, \xi). \tag{3.163}$$

Suppose that $x \in K$ and $\|x\|_\infty = \bar{\delta}$. Using (3.163), (3.162), the definitions of

K and m_k^F, for every $u = (u_1, \ldots, u_n) \in S_{F(\cdot, x(\cdot))}^1$, we have

$$
\int_0^1 G_k(\tau_k, s) u_k(s) ds \; \geq \; \int_{\frac{1}{4}}^{\frac{3}{4}} G_k(\tau_k, s) u_k(s) ds \; \geq \; \int_{\frac{1}{4}}^{\frac{3}{4}} G_k(\tau_k, s) m_k^F(s, x(s)) ds
$$

$$
\geq \; \int_{\frac{1}{4}}^{\frac{3}{4}} G_k(\tau_k, s) \big(\zeta_0(s) - \overline{\varepsilon} \big) \, \|x(s)\|_{l^\infty} \, ds
$$

$$
\geq \; \gamma \, \|x_k\|_\infty \int_{\frac{1}{4}}^{\frac{3}{4}} G_k(\tau_k, s) \big(\zeta_0(s) - \overline{\varepsilon} \big) ds \; \geq \; \|x_k\|_\infty .
$$

Therefore, if $y \in V(x)$, we have

$$
\|y_k\|_\infty \; \geq \; y_k(\tau_k) \; \geq \; \|x_k\|_\infty \qquad \forall \, k \in \{1, \ldots, N\}
$$

and so

$$
\|y\|_\infty \; \geq \; \overline{\delta} \qquad \forall \, y \in V(x), \; x \in K, \; \|x\|_\infty = \overline{\delta}. \tag{3.164}
$$

Case 1. Assume that hypothesis $H(F)_1(iii)_1$ holds. Set

$$
\overline{\eta} \; \overset{df}{=} \; \max_{k \in \{1, \ldots, N\}} \int_0^1 G_k(s, s) g(s) ds,
$$

with g given in $H(F)_1(iii)_1$. Using (3.156), for all $x \in K$, with $\|x\|_\infty = \overline{\eta}$ and every $u \in S_{F(\cdot, x(\cdot))}^1$, we have

$$
\left\| \int_0^1 G(t, s) u(s) ds \right\|_{l^\infty} \; = \; \max_{k \in \{1, \ldots, N\}} \left| \int_0^1 G_k(t, s) u_k(s) ds \right|
$$

$$
\leq \; \max_{k \in \{1, \ldots, N\}} \int_0^1 G_k(t, s) g(s) ds
$$

$$
\leq \; \overline{\eta} \qquad \forall \, t \in [0, 1].
$$

It follows that

$$
\|y(t)\|_{l^\infty} \; \leq \; \overline{\eta} \qquad \forall \, y \in V(x), \; t \in [0, 1],
$$

so

$$
\|y\|_\infty \; \leq \; \overline{\eta} \qquad \forall \, y \in V(x), \; x \in K, \; \|x\|_\infty = \overline{\eta}. \tag{3.165}
$$

Thus, if $H(F)_1(iii)_1$ is valid and we set

$$
U_1 \; \overset{df}{=} \; \big\{ x \in C\big([0, 1]; \mathbb{R}^N\big) : \; \|x\|_\infty < \overline{\delta} \big\},
$$

$$
U_2 \; \overset{df}{=} \; \big\{ x \in C\big([0, 1]; \mathbb{R}^N\big) : \; \|x\|_\infty < \overline{\eta} \big\},
$$

then by virtue of (3.164) and (3.165), we can apply Proposition 3.6.1 and obtain $x_0 \in K \cap (\overline{U}_2 \setminus U_1)$, such that $x_0 \in V(x_0)$. Evidently this is a nontrivial solution of (3.154), such that $x_0(t) \in \mathbb{R}_+^N$ for all $t \in [0,1]$.

Case 2. Assume that hypothesis $H(F)_1(iii)_2$ holds. By virtue of the last part of hypothesis $H(F)_1(iv)$, we can find $\varepsilon > 0$, such that

$$\int_0^1 G_k(s,s)\big(\varsigma_1(s) + \varepsilon\big)ds + \varepsilon \leq 1 \qquad \forall\, k \in \{1,\dots,N\}. \tag{3.166}$$

Also by virtue of hypothesis $H(F)_1(iv)$, for this $\varepsilon > 0$, we can find $\overline{\overline{r}} \geq 1$, such that for all $k \in \{1,\dots,N\}$, almost all $t \in [0,1]$ and all $\xi \in \mathbb{R}_+^N$ with $\|\xi\|_{l^\infty} \geq \overline{\overline{r}}$, we have

$$M_k^F(t,\xi) \leq \big(\varsigma_1(t) + \varepsilon\big)\|\xi\|_{l^\infty}. \tag{3.167}$$

From $H(F)_1(iii)_2$, we know that for $k \in \{1,\dots,N\}$, we have

$$M_k^F(t,\xi) \leq g_{\overline{\overline{r}}}(t) \quad \text{for a.a. } t \in [0,1] \text{ and all } \xi \in \mathbb{R}_+^N, \; \|\xi\|_{l^\infty} \leq \overline{\overline{r}}. \tag{3.168}$$

By Lusin's theorem (see Theorem A.2.2), for any $\delta > 0$, we can find a closed set $T_\delta \subseteq [0,1]$ with $|T_\delta| > 1 - \delta$, such that $g_{\overline{\overline{r}}}|_{T_\delta}$ is continuous. Let us fix $\delta > 0$, so that

$$\int_{[0,1]\setminus T_\delta} G_k(s,s)g_{\overline{\overline{r}}}(s)ds < \varepsilon \qquad \forall\, k \in \{1,\dots,N\}, \tag{3.169}$$

for some closed set $T_\delta \subseteq [0,1]$ with $|T_\delta| > 1 - \delta$ and $g_{\overline{\overline{r}}}|_{T_\delta}$ is continuous. Let us choose $\overline{\xi} \in \mathbb{R}^N$, such that $\|\overline{\xi}\|_{l^\infty} \geq \overline{\overline{r}}$ and

$$M_k^F(t,\overline{\xi}) > \|g_{\overline{\overline{r}}}\|_{C(T_\delta)} \qquad \forall\, t \in [0,1], \; k \in \{1,\dots,N\} \tag{3.170}$$

(see hypothesis $H(F)_1(iii)_2$). Finally, let $\overline{\overline{\eta}} > \|\overline{\xi}\|_{l^\infty}$.

Note that M_k^F is jointly measurable. Indeed, let $p_k \colon \mathbb{R}^N \longrightarrow \mathbb{R}$ be the projection operator to the k-th component of $\xi = (\varsigma_1,\dots,\xi_N) \in \mathbb{R}^N$. For every $\lambda \geq 0$ and $k \in \{1,\dots,N\}$, let us set

$$L_\lambda^k \overset{df}{=} \big\{(t,\xi) \in [0,1] \times \mathbb{R}^N : p_k(u) \leq \lambda, \; u \in F(t,\xi)\big\}.$$

From Theorem 1.2.14, for every $k \in \{1,\dots,N\}$, we have that

$$L_\lambda^k = \text{proj}_{[0,1]\times\mathbb{R}^N}\big\{[0,1] \times \mathbb{R}^N \times p_k^{-1}((-\infty,\lambda])\big\} \in \mathcal{B}([0,1]) \times \mathcal{B}(\mathbb{R}^N).$$

So

$$\big[(t,\xi) \in [0,1] \times \mathbb{R}^N : M_k^F(t,\xi) \leq \lambda, \; k \in \{1,\dots,N\}\big]$$

$$= \bigcap_{k=1}^N L_\lambda^k \in \mathcal{B}([0,1]) \times \mathcal{B}(\mathbb{R}^N),$$

which proves the joint measurability of M_k^F.

Moreover for almost all $t \in [0,1]$, the function $M_k^F(t, \cdot)$ is upper semicontinuous (see Proposition 1.2.7(b)). So by the Weierstrass Theorem (see Theorem A.1.5) and an easy measurable selection argument, for $k \in \{1, \ldots, N\}$ we can find measurable functions $\widehat{x}^k : [0,1] \longrightarrow \mathbb{R}_+^N$, such that

$$\left\| \widehat{x}^k(t) \right\|_{l^\infty} \leq \overline{\overline{\eta}} \quad \text{and} \quad M_k^F(t, \widehat{x}^k(t)) = \max_{\|\xi\|_{l^\infty} \leq \overline{\overline{\eta}}} M_k^F(t, \xi) \quad \text{for a.a. } t \in [0,1].$$

So, for all $k \in \{1, \ldots, N\}$, almost all $t \in [0,1]$ and all $\xi \in \mathbb{R}_+^N$ with $\|\xi\|_{l^\infty} \leq \overline{\overline{\eta}}$, we have

$$M_k^F(t, \xi) \leq M_k^F(t, \widehat{x}^k(t)). \tag{3.171}$$

Note that

$$\left\| \widehat{x}^k(t) \right\|_{l^\infty} > \overline{\overline{r}} \quad \forall\, t \in T_\delta, \ k \in \{1, \ldots, N\}, \tag{3.172}$$

because otherwise, if $\left\| \widehat{x}^{k_0}(t_0) \right\|_{l^\infty} \leq \overline{\overline{r}}$ for some $t_0 \in T_\delta$ and $k_0 \in \{1, \ldots, N\}$, from (3.168) and (3.170), we would have

$$M_{k_0}^F(t_0, \widehat{x}^{k_0}(t_0)) \leq g_{\overline{\overline{r}}}(t_0) \leq \|g_{\overline{\overline{r}}}\|_{C(T_\delta)} < M_{k_0}^F(t_0, \overline{\xi}),$$

a contradiction to (3.171), as $\left\| \overline{\xi} \right\|_{l^\infty} < \overline{\overline{\eta}}$. Thus, we obtain (3.172).

Let $x \in K$ with $\|x\|_\infty = \overline{\overline{\eta}}$. Using hypothesis $H(F)_1(iii)_2$, (3.156), (3.169), (3.171), (3.167), (3.172), (3.166) and the fact that $\overline{\overline{\eta}} \geq 1$, for all $u = (u_1, \ldots, u_N) \in S_{F(\cdot, x(\cdot))}^1$ and every $k \in \{1, \ldots, N\}$, we have

$$\int_0^1 G_k(t,s) u_k(s)\, ds \leq \int_{T_\delta} G_k(s,s) u_k(s)\, ds + \int_{[0,1] \setminus T_\delta} G_k(s,s) g_{\overline{\overline{r}}}(s)\, ds$$

$$\leq \int_{T_\delta} G_k(s,s) M_k^F(s, x(s))\, ds + \varepsilon \leq \int_{T_\delta} G_k(s,s) M_k^F(s, \widehat{x}^k(s))\, ds + \varepsilon$$

$$\leq \int_{T_\delta} G_k(s,s) \big(\zeta_1(s) + \varepsilon \big) \left\| \widehat{x}^k(s) \right\|_{l^\infty} ds + \varepsilon$$

$$\leq \overline{\overline{\eta}} \int_{T_\delta} G_k(s,s) \big(\zeta_1(s) + \varepsilon \big)\, ds + \varepsilon$$

$$\leq \overline{\overline{\eta}} \left(\int_{T_\delta} G_k(s,s) \big(\zeta_1(s) + \varepsilon \big)\, ds + \varepsilon \right) \leq \overline{\overline{\eta}}.$$

So, we have that

$$\|y\|_\infty \leq \overline{\overline{\eta}} \quad \forall\, y \in V(x), \ x \in K, \ \|x\|_\infty = \overline{\overline{\eta}}. \tag{3.173}$$

Thus, if in this case we let

$$U_1 \overset{df}{=} \left\{ x \in C([0,1]; \mathbb{R}^N) : \ \|x\|_\infty < \overline{\delta} \right\},$$

$$U_2' \overset{df}{=} \left\{ x \in C([0,1]; \mathbb{R}^N) : \ \|x\|_\infty < \overline{\overline{\eta}} \right\}$$

(see (3.164) and (3.173)), we can apply Proposition 3.6.1 and obtain $x_0 \in K \cap (\overline{U}_2 \setminus U_1)$, hence $\overline{\delta} \leq \|x\|_\infty \leq \overline{\overline{\eta}}$ (i.e. $x_0 \neq 0$), such that $x_0 \in V(x_0)$. Evidently this is a nontrivial positive solution of (3.154).   □

As we already mentioned (see Remark 3.6.1), the hypotheses in the previous theorem incorporate as a special case the sublinear problem, i.e. when $\zeta_0 \equiv +\infty$ and $\zeta_1 \equiv 0$. In the next theorem our hypotheses include as a special case the superlinear problem, i.e. when $\zeta_0 \equiv 0$ and $\zeta_1 \equiv +\infty$. Our hypotheses on the multivalued nonlinearity F are the following:

<u>$H(F)_2$</u> $F : [0,1] \times \mathbb{R}_+^N \longrightarrow P_{kc}(\mathbb{R}_+^N)$ is a multifunction, such that

(i) for all $\xi \in \mathbb{R}^N$, the multifunction $[0,1] \ni t \longmapsto F(t,\xi) \in P_{kc}(\mathbb{R}_+^N)$ F is graph measurable;

(ii) for almost all $t \in [0,1]$, the multifunction $\mathbb{R}_+^N \ni \xi \longmapsto F(t,\xi) \in P_{kc}(\mathbb{R}_+^N)$ has a closed graph;

(iii) for all $r > 0$, there exists $g_r \in L^1(0,1)$, such that for all $t \in [0,1]$, all $\xi \in \mathbb{R}_+^N$, with $\|\xi\|_{l^\infty} \leq r$ and all $u \in F(t,\xi)$, we have $\|u\|_{l^\infty} \leq g_r(t)$;

(iv) there exist measurable functions $\zeta_0, \zeta_1 \colon [0,1] \longrightarrow \overline{\mathbb{R}}_+ \overset{df}{=} \mathbb{R}_+ \cup \{+\infty\}$, such that for all $k \in \{1,\ldots,N\}$, we have

$$\liminf_{\substack{\|\xi\|_{l^\infty} \to +\infty \\ \xi \in \mathbb{R}_+^N}} \frac{m_k^F(t,\xi)}{\|\xi\|_{l^\infty}} \geq \zeta_0(t) \quad \text{uniformly for a.a. } t \in [0,1],$$

$$\limsup_{\substack{\|\xi\|_{l^\infty} \to 0 \\ \xi \in \mathbb{R}_+^N}} \frac{M_k^F(t,\xi)}{\|\xi\|_{l^\infty}} \leq \zeta_1(t) \quad \text{uniformly for a.a. } t \in [0,1]$$

and for all $k \in \{1,\ldots,N\}$, we have

$$\frac{1}{\gamma} < \int_{\frac{1}{4}}^{\frac{3}{4}} G_k(\tau_k,s)\zeta_0(s)ds \quad \text{and} \quad \int_0^1 G_k(s,s)\zeta_1(s)ds < 1,$$

where γ is defined by (3.155) and $\tau_k \in [0,1]$ are defined by (3.158).

REMARK 3.6.2   When $\zeta_1 \equiv 0$ and $\zeta_1 \equiv +\infty$, then we have the "super-linear" problem.   □

THEOREM 3.6.2
If hypotheses $H(A\text{--}D)$ and $H(F)_2$ hold, then problem (3.154) has at least

one nontrivial solution $x_0 \in W^{1,2}\big((0,1);\mathbb{R}^N\big)$, *such that* $x_0(t) \in \mathbb{R}_+^N$ *for all* $t \in [0,1]$.

PROOF   Similarly as in the proof of Theorem 3.6.1, let $K \subseteq C\big([0,1];\mathbb{R}^N\big)$ be the nonempty, closed and convex cone in $C\big([0,1];\mathbb{R}^N\big)$, given by (3.159) and $V \colon K \longrightarrow 2^{C([0,1];\mathbb{R}^N)}$ be the multivalued operator, defined by (3.160).

In this case though it is not immediately clear that V has nonempty values. This is so because hypotheses $H(F)_2(i)$ and (ii) in general do not imply joint graph measurability (see Hu & Papageorgiou (1997, Example II.7.2, p. 227)). To establish the nonemptiness of $S^1_{F(\cdot,x(\cdot))}$ for every $x \in K$ (hence of $V(x)$ too), we proceed as follows. Let $x \in K$ and let $\{s_n\}_{n \geq 1}$ be a sequence of simple $\mathbb{R}_+^N$-valued functions, such that

$$\|s_n(t)\|_{l^\infty} \leq \|x\|_\infty \quad \text{for a.a. } t \in [0,1] \text{ and all } n \geq 1$$

and

$$s_n(t) \longrightarrow x(t) \quad \text{for a.a. } t \in [0,1].$$

Then because of hypothesis $H(F)_2(i)$, for every $n \geq 1$, the function $\mathbb{R}^N \ni t \longrightarrow F\big(t,s_n(t)\big) \in 2^{\mathbb{R}^N}$ is graph measurable (hence Lebesgue measurable too; see Theorem 1.2.1). So an easy application of the Yankov-von Neumann-Aumann Selection Theorem (see Theorem 1.2.3) produces measurable functions $u_n \colon [0,1] \longrightarrow \mathbb{R}_+^N$, $n \geq 1$, such that

$$u_n(t) \in F\big(t,s_n(t)\big) \quad \text{for a.a. } t \in [0,1].$$

We have

$$\|u_n\|_{l^\infty} \leq g_{\overline{r}}(t) \quad \text{for a.a. } t \in [0,1],$$

with $\overline{r} \overset{df}{=} \|x\|_\infty$. Hence, invoking the Dunford-Pettis Theorem and passing to a subsequence if necessary, we may assume that

$$u_n \overset{w}{\longrightarrow} u \quad \text{in } L^1\big(0,1;\mathbb{R}^N\big).$$

From Proposition 1.2.12, we have that

$$u(t) \in \operatorname{conv} \limsup_{n \to +\infty} F\big(t,s_n(t)\big) \subseteq F\big(t,x(t)\big) \quad \text{for a.a. } t \in [0,1].$$

Here the last inclusion is a consequence of the fact that F is convex valued and because of hypothesis $H(F)_2(ii)$. So $u \in S^1_{F(\cdot,x(\cdot))}$, which proves that $V(x) \neq \emptyset$. Evidently $V(x)$ is convex and using the Arzela-Ascoli Theorem (see Theorem A.1.13), we have that $V(x) \subseteq C\big([0,1];\mathbb{R}^N\big)$ is also compact. Moreover, arguing as in the proof of Theorem 3.1, we can check that V is compact and $V(K) \subseteq K$.

By virtue of hypothesis $H(F)_2(iv)$, we can find $\bar{\varepsilon} > 0$, so that

$$\int_0^1 G_k(s,s)\big(\zeta_1(s) + \bar{\varepsilon}\big)ds \ \leq \ 1 \qquad \forall\, k \in \{1,\ldots,N\}. \qquad (3.174)$$

Again by hypothesis $H(F)_2(iv)$, for this $\bar{\varepsilon} > 0$, we can find $\bar{\delta} = \bar{\delta}(\bar{\varepsilon}) > 0$, such that for almost all $t \in [0,1]$ and all $\xi \in \mathbb{R}_+^N$ with $0 < \|\xi\|_{l^\infty} \leq \bar{\delta}$, we have

$$M_k^F(t,\xi) \ \leq \ \big(\zeta_1(t) + \bar{\varepsilon}\big)\,\|\xi\|_{l^\infty} \qquad (3.175)$$

and

$$\frac{1}{\gamma} \ \leq \ \int_{\frac{1}{4}}^{\frac{3}{4}} G_k(\tau_k,s)\big(\zeta_0(s) - \bar{\varepsilon}\big)ds. \qquad (3.176)$$

From (3.156), (3.174) and (3.175), for all $x \in K$, with $\|x\|_\infty = \bar{\delta}$, all $u \in S_{F(\cdot,x(\cdot))}^1$ and all $k \in \{1,\ldots,N\}$, we have

$$\int_0^1 G_k(t,s)u_k(s)ds \ \leq \ \int_0^1 G_k(s,s)M_k^F\big(s,x(s)\big)ds$$

$$\leq \ \int_0^1 G_k(s,s)\big(\zeta_1(s) + \bar{\varepsilon}\big)\,\|x(s)\|_{l^\infty}\,ds \ \leq \ \|x\|_\infty.$$

So

$$y_k(t) \ \leq \ \|x\|_\infty \qquad \forall\, t \in [0,1],\ k \in \{1,\ldots,N\}$$

and

$$\|y\|_\infty \ \leq \ \|x\|_\infty \qquad \forall\, x \in K,\ \|x\|_\infty = \bar{\delta},\ y \in V(x). \qquad (3.177)$$

Also again from hypothesis $H(F)_2(iv)$, for our $\bar{\varepsilon} > 0$, we can find $\bar{\eta} = \bar{\eta}(\bar{\varepsilon}) > \bar{\delta} > 0$, such that for almost all $t \in [0,1]$ and all $\xi \in \mathbb{R}_+^N$ with $\|\xi\|_{l^\infty} \geq \bar{\eta}$, we have

$$m_k^F(t,\xi) \ \geq \ \big(\zeta_0(t) - \bar{\varepsilon}\big)\,\|\xi\|_{l^\infty}. \qquad (3.178)$$

Let $x \in K$ be such that $\|x\|_\infty = \bar{\eta}$. Then, from (3.178), the definition of K and the choice of $\bar{\varepsilon} > 0$ (see (3.176)), for all $u \in S_{F(\cdot,x(\cdot))}^1$ and all $k \in \{1,\ldots,N\}$, we have

$$\int_0^1 G_k(\tau_k,s)u_k(s)ds \ \geq \ \int_{\frac{1}{4}}^{\frac{3}{4}} G_k(\tau_k,s)m_k^F\big(s,x(s)\big)ds$$

$$\geq \ \int_{\frac{1}{4}}^{\frac{3}{4}} G_k(\tau_k,s)\big(\zeta_0(s) - \bar{\varepsilon}\big)\,\|x(s)\|_{l^\infty}\,ds$$

$$\geq \ \gamma\|x\|_\infty \int_{\frac{1}{4}}^{\frac{3}{4}} G_k(\tau_k,s)\big(\zeta_0(s) - \bar{\varepsilon}\big)ds \ \geq \ \|x\|_\infty \ = \ \bar{\eta}.$$

So

$$\|y_k\|_\infty \geq y_k(\tau_k) \geq \|x\|_\infty \qquad \forall k \in \{1, \ldots, N\}$$

and

$$\|y\|_\infty \geq \|x\|_\infty \qquad \forall x \in K, \ \|x\|_\infty = \overline{\eta}, \ y \in V(x). \tag{3.179}$$

So let

$$U_1 \overset{df}{=} \{x \in C([0,1]; \mathbb{R}^N) : \|x\|_\infty < \overline{\delta}\},$$

$$U_2 \overset{df}{=} \{x \in C([0,1]; \mathbb{R}^N) : \|x\|_\infty < \overline{\eta}\}$$

(see (3.177) and (3.179)) and apply Theorem 2.1, to obtain $x_0 \in K$, with $\overline{\delta} < \|x_0\|_\infty \leq \overline{\eta}$, such that $x_0 \in V(x_0)$. Evidently this is the desired solution of (3.154), such that $x_0(t) \in \mathbb{R}^N_+$ for all $t \in [0,1]$. $\qquad\square$

It is clear that hypotheses $H(A\text{--}D)$ exclude from the analysis the Neumann problem. Nevertheless using the same reasoning as before (with few minor modifications), we can also treat the Neumann problem:

$$\begin{cases} -x''(t) \in F(t, x(t)) & \text{for a.a. } t \in [0,1], \\ x'(0) = x'(b) = 0. \end{cases} \tag{3.180}$$

To be able to implement the fixed point method used earlier, we pass to the following problem:

$$\begin{cases} -x''(t) + \lambda^2 x(t) \in F(t, x(t)) + \lambda^2 x(t) & \text{for a.a. } t \in [0,1], \\ x'(0) = x'(b) = 0, \end{cases} \tag{3.181}$$

with $\lambda > 0$. The linear differential operator $L(x) = -x'' + \lambda^2 x$, with $\lambda > 0$, with Neumann boundary conditions, has Green's function $\overline{G}(t,s) = \widehat{G}(t,s) id_{\mathbb{R}^N \times \mathbb{R}^N}$, with

$$\widehat{G}(t,s) \overset{df}{=} \begin{cases} \dfrac{\cosh \lambda t \cdot \cosh \lambda(1-s)}{\lambda \sinh \lambda} & \text{if } t \leq s, \\[3mm] \dfrac{\cosh \lambda s \cdot \cosh \lambda(1-t)}{\lambda \sinh \lambda} & \text{if } t > s. \end{cases}$$

Then $\overline{G}(t,s)$ has the same properties as the Green's function G for the Sturm-Liouville problems. In the present case $\gamma = \frac{\cosh \frac{\lambda}{4}}{\cosh \lambda}$.

THEOREM 3.6.3
If hypotheses $H(F)_1$ hold with the Green's function $\overline{G}$ instead of G, then problem (3.180) has at least one nontrivial solution $x_0 \in W^{1,2}((0,1); \mathbb{R}^N)$, such that $x_0(t) \in \mathbb{R}^N_+$ for all $t \in [0,1]$.

REMARK 3.6.3   The above result includes the case of the sublinear Neumann problem ($\zeta_0 \equiv +\infty$, $\zeta_1 \equiv 0$). ▯

THEOREM 3.6.4
If hypotheses $H(F)_2$ hold with the Green's function $\overline{G}$ instead of G,
<u>then</u> *problem (3.180) has at least one nontrivial solution $x_0 \in W^{1,2}\big((0,1); \mathbb{R}^N\big)$, such that $x_0(t) \in \mathbb{R}_+^N$ for all $t \in [0,1]$.*

REMARK 3.6.4   The above result includes the case of the superlinear Neumann problem ($\zeta_0 \equiv 0$, $\zeta_1 \equiv +\infty$). ▯

3.6.2   Method Based on Monotone Operators

Continuing with the Neumann problem, next we present a method based on the theory of operators of monotone type, which we use to study a class of single-valued Neumann problems.

The problem under consideration is the following:

$$\begin{cases} -\big(a\big(|x'(t)|^2\big)x'(t)\big)' + f\big(t, x(t)\big) = v(t) & \text{for a.a. } t \in T, \\ x'(0) = x'(b) = 0, \end{cases} \tag{3.182}$$

where $T = [0, b]$, $b > 0$. Here $a \colon \mathbb{R} \longrightarrow \mathbb{R}$ is a continuous function, which satisfies certain geometric and growth conditions (see hypotheses $H(a)$ below). One possibility is to have $a(r^2) = (r^2)^{\frac{p-2}{2}}$, $p > 1$, in which case the resulting differential operator is the ordinary scalar p-Laplacian considered in previous sections.

We start by introducing a specification of monotone maps (see Definition 1.4.3(a)).

DEFINITION 3.6.2   *Let X be a reflexive Banach space. An operator $A \colon X \supseteq \mathrm{Dom}\,(A) \longrightarrow 2^{X^*}$ is said to be 3-**monotone** if*

$$\langle x^* - y^*, z - x \rangle_X \leq \langle y^* - z^*, y - z \rangle_X \qquad \forall\, (x, x^*), (y, y^*), (z, z^*) \in \mathrm{Gr}\, A.$$

REMARK 3.6.5   It is clear that 3-monotone operator is monotone, but the converse is not true in general. ▯

Another type of operator that we shall need is given in the next definition.

DEFINITION 3.6.3   *Let X be a reflexive Banach space. An operator $A \colon X \supseteq \mathrm{Dom}\,(A) \longrightarrow 2^{X^*}$ is said to be **boundedly inversely compact**, if for any pair of bounded sets $C \subseteq X$ and $C^* \subseteq X^*$, we have that $C \cap A^{-1}(C^*)$ is relatively compact in X.*

REMARK 3.6.6   If $K \colon X^* \longrightarrow X$ maps bounded sets in X^* into relatively compact subsets of X, then the operator $A \overset{df}{=} K^{-1} \colon X \supseteq \mathrm{Dom}\,(A) \longrightarrow 2^{X^*}$ is boundedly inversely compact. So the inverse of a compact operator from X^* into X (see Definition 1.4.1(a)) is boundedly inversely compact.   ⬜

Our approach is based on the following characterization of the range of certain sum of operators, due to Gupta & Hess (1976), where the reader can also find the proof.

PROPOSITION 3.6.2
If X is a reflexive Banach space, A, B_1, B_2 are three operators, such that the following hypotheses holds:

(i) $A \colon X \supseteq \mathrm{Dom}\,(A) \longrightarrow 2^{X^*}$ is monotone;

(ii) $B_1 \colon X \supseteq \mathrm{Dom}\,(B_1) \longrightarrow 2^{X^*}$ is 3-monotone;

(iii) $\mathrm{Dom}\,(A) \subseteq \mathrm{Dom}\,(B_1)$;

(iv) $0 \in (A + B_1)\,(0)$;

(v) $A + B_1 \colon X \supseteq \mathrm{Dom}\,(A) \longrightarrow 2^{X^*}$ is maximal monotone and boundedly inversely compact;

(vi) $B_2 \colon X \longrightarrow X^*$ is demicontinuous;

(vii) for every $k \geq 0$ there exists a constant $c(k) \in \mathbb{R}$, such that

$$\langle B_2(x), x \rangle_X \ \geq \ k\,\|B_2(x)\|_X - c(k) \qquad \forall\, x \in X,$$

then $u \in \mathrm{int}\,\big(R(A) + R(B_1)\big)$ if and only if $u \in R\,(A + B_1 + B_2)$.

REMARK 3.6.7   If $B_2 \equiv 0$, then we can drop the hypothesis that $A + B_1$ is boundedly inversely compact. Moreover, note that by virtue of the condition (vii) imposed on B_2, the operator B_2 is bounded (i.e. maps bounded sets into bounded sets).   ⬜

Our hypotheses on the function a in the differential operator in (3.182) are the following:

$\underline{H(a)}$ $a \colon \mathbb{R} \longrightarrow \mathbb{R}$ is a function, such that

(i) a is continuous and $a(0) = 0$;

(ii) the function $h \colon \mathbb{R} \longrightarrow \mathbb{R}$, defined by

$$h(t) \ \overset{df}{=} \ \frac{1}{2} \int_0^{t^2} a(s)\,ds \qquad \forall\, t \in \mathbb{R},$$

is strictly convex;

(*iii*) there exist constants $c_0^a, c_2^a > 0$, $c_1^a, c_3^a \geq 0$, such that

$$c_0^a |t|^{p-2} - c_1^a \leq a(t^2) \leq c_2^a |t|^{p-2} + c_3^a.$$

REMARK 3.6.8 If a is a polynomial with positive coefficients, it satisfies hypotheses $H(a)$. The same is true if

$$a(t^2) \stackrel{df}{=} 1 + \frac{1}{(1+t^2)^2} \qquad \forall\, t \in \mathbb{R},$$

or

$$a(t^2) \stackrel{df}{=} (t^2)^{\frac{p-2}{2}} \qquad \forall\, t \in \mathbb{R}.$$

The latter case corresponds to the ordinary scalar p-Laplacian. ∎

Note that since by hypothesis h is strictly convex, the function $r \longrightarrow h'(r) = a(r^2)r$ is strictly monotone. Let the operator $A\colon W^{1,p}(T) \longrightarrow (W^{1,p}(T))^*$ be defined by

$$\langle A(x), y\rangle_{W^{1,p}(T)} \stackrel{df}{=} \int_0^b a(|x'(t)|^2)x'(t)y'(t)\,dt \qquad \forall\, x,y \in W^{1,p}(T).$$

It is straightforward to check that A is monotone and demicontinuous. Therefore from Corollary 1.4.2, we infer the following Proposition.

PROPOSITION 3.6.3
If hypotheses $H(a)$ hold,
then the operator $A\colon W^{1,p}(T) \longrightarrow (W^{1,p}(T))^*$ is maximal monotone.

Let A_1 be the restriction of A on $(L^p(T), L^{p'}(T))$, i.e. $A_1\colon L^p(T) \supseteq D_1 \longrightarrow L^{p'}(T)$ is defined by

$$A_1(x) \stackrel{df}{=} A(x) \qquad \forall\, x \in D_1,$$

where

$$D_1 \stackrel{df}{=} (x \in W^{1,p}(T)\colon\ A(x) \in L^{p'}(T)).$$

PROPOSITION 3.6.4
If hypotheses $H(a)$ hold,
then the operator $A_1\colon L^p(T) \supseteq D_1 \longrightarrow L^{p'}(T)$ is maximal monotone.

PROOF   Let $J: L^p(T) \longrightarrow L^{p'}(T)$ be the continuous, strictly monotone (hence maximal monotone too; see Corollary 1.4.2) operator, defined by

$$J(x)(\cdot) \overset{df}{=} |x(\cdot)|^{p-2} x(\cdot) \qquad \forall\, x \in L^p(T).$$

From the proof of Proposition 3.1.3, we know that it is sufficient to show that

$$R(A_1 + J) = L^{p'}(T). \tag{3.183}$$

Let $\widehat{J} = J|_{W^{1,p}(T)}$. Then $\widehat{J}: W^{1,p}(T) \longrightarrow \big(W^{1,p}(T)\big)^*$ is a continuous, monotone, hence maximal monotone (recall that $L^{p'}(T)$ is embedded continuously (in fact compactly) into $\big(W^{1,p}(T)\big)^*$). Then the operator $A + \widehat{J}: W^{1,p}(T) \longrightarrow \big(W^{1,p}(T)\big)^*$ is surjective (see Theorem 1.4.2). So for a given $g \in L^{p'}(T)$, we can find $x \in W^{1,p}(T)$, such that

$$A(x) + \widehat{J}(x) = A(x) + J(x) = g.$$

Hence $A(x) = g - J(x) \in L^{p'}(T)$ and so $x \in D_1$, $A(x) = A_1(x)$. Therefore $g \in R(A_1 + J)$ and we have proved (3.183), which implies the maximality of A_1. ☐

The next two propositions determine precisely the range of A_1.

PROPOSITION 3.6.5
If hypotheses $H(a)$ and $g \in R(A_1)$,
then for some $x \in D_1$, we have

$$\begin{cases} -\big(a\big(|x'(t)|^2\big)x'(t)\big)' = g(t) & \text{for a.a. } t \in T, \\ x'(0) = x'(b) = 0. \end{cases}$$

PROOF   Since $g \in R(A_1)$, then for some $x \in D_1$, we have $A(x) = g$. Let $\vartheta \in C_c^\infty(0,b)$. We have

$$\big\langle A_1(x), \vartheta \big\rangle_{W^{1,p}(T)} = \big(A_1(x), \vartheta\big)_p = (g, \vartheta)_p$$

and so

$$\int_0^b a\big(|x'(t)|^2\big)x'(t)\vartheta'(t)\,dt = \int_0^b g(t)\vartheta(t)\,dt. \tag{3.184}$$

Because of hypotheses $H(a)$ and Theorem 1.1.8 we have that

$$\big(a\big(|x'(t)|^2\big)x'(t)\big)' \in W^{-1,p'}(T) = \big(W_0^{1,p}(T)\big)^*$$

and the embedding $C_c^\infty(0,b) \subseteq W_0^{1,p}(T)$ is dense. Hence

$$\big\langle -\big(a\big(|x'(t)|^2\big)x'(t)\big)', \vartheta \big\rangle_{W_0^{1,p}(T)} = \langle g, \vartheta \rangle_{W_0^{1,p}(T)}$$

and so

$$-\big(a\big(|x'(t)|^2\big)x'(t)\big)' \;=\; g(t) \quad \text{for a.a. } t \in T. \tag{3.185}$$

From this equality, it follows that

$$a\big(|x'(\cdot)|^2\big)x'(\cdot) \;\in\; W^{1,p'}(T) \subseteq C(T)$$

and so $x' \in C(T)$, i.e. $x \in C^1(T)$. Applying Green's identity, from (3.185), we have

$$\int_0^b g(t)y(t)\,dt \;=\; -\int_0^b \big(a\big(|x'(t)|^2\big)x'(t)\big)'y(t)\,dt$$

$$= -a\big(|x'(b)|^2\big)x'(b)y(b) + a\big(|x'(0)|^2\big)x'(0)y(0)$$

$$+ \int_0^b a\big(|x'(t)|^2\big)x'(t)y'(t)\,dt \qquad \forall\, y \in W^{1,p}(T).$$

From (3.184), we obtain

$$a\big(|x'(0)|^2\big)x'(0)y(0) \;=\; a\big(|x'(b)|^2\big)x'(b)y(b).$$

Exploiting the strict monotonicity of $r \longmapsto a(r^2)r$, and since $y \in W^{1,p}(T)$ was arbitrary, we conclude that $x'(0) = x'(b) = 0$. ⬚

REMARK 3.6.9 A byproduct of this proof is that $D_1 \subseteq C^1(T)$. ⬚

Let us set

$$V_1 \;\overset{df}{=}\; \left\{ g \in L^{p'}(T) : \int_0^b g(t)\,dt = 0 \right\}.$$

PROPOSITION 3.6.6
If hypotheses $H(a)$ hold, <u>then</u> $R(A_1) = V_1$.

PROOF   Let $g \in R(A_1)$. By Proposition 3.6.5, we have that

$$\begin{cases} -\big(a\big(|x'(t)|^2\big)x'(t)\big)' = g(t) & \text{for a.a. } t \in T, \\ x'(0) = x'(b) = 0 \end{cases}$$

and so

$$\int_0^b g(t)\,dt \;=\; 0, \qquad \text{i.e. } R(A_1) \subseteq V_1. \tag{3.186}$$

On the other hand, because A is maximal monotone, coercive (see hypothesis $H(a)(ii)$), it is surjective and so for a given $g \in V_1$, we can find $x \in W^{1,p}(T)$, such that $A(x) - g$. Evidently $x \subset D_1$ and $A(x) - A_1(x)$. So it follows that $V_1 \subseteq R(A_1)$. Combining this with (3.186), we get that $R(A_1) = V_1$. $\square$

Now we are ready to introduce our hypotheses on the nonlinearity f and start dealing with the full problem (3.182).

$\underline{H(f)_1}$ $f: T \times \mathbb{R} \longrightarrow \mathbb{R}$ is a function, such that

(i) for all $\zeta \in \mathbb{R}$, the function

$$T \ni t \longmapsto f(t, \zeta) \in \mathbb{R}$$

is measurable;

(ii) for almost all $t \in T$, the function

$$\mathbb{R} \ni \zeta \longmapsto f(t, \zeta) \in \mathbb{R}$$

is continuous;

(iii) for almost all $t \in T$ and all $\zeta \in \mathbb{R}$, we have

$$\left| f(t, \zeta) \right| \leq a(t) + c|\zeta|^{p-1},$$

with $a \in L^{p'}(T)_+$, (where $\frac{1}{p} + \frac{1}{p'} = 1$), $c > 0$;

(iv) there exists $u \in L^p(T)_+$, such that for almost all $t \in T$ and all $\zeta \in \mathbb{R}$ with $|\zeta| \geq u(t)$, we have $f(t, \zeta)\zeta \geq 0$ (generalized sign condition).

Based on hypothesis $H(f)_1(iv)$, we introduce a penalty function $h: T \times \mathbb{R} \longrightarrow \mathbb{R}$, defined by

$$h(t, \zeta) \stackrel{df}{=} \begin{cases} \left(\zeta - u(t) \right)^{p-1} & \text{if } u(t) < \zeta, \\ 0 & \text{if } |\zeta| \leq u(t), \\ -\left(-\zeta - u(t) \right)^{p-1} & \text{if } \zeta < -u(t). \end{cases}$$

It is clear from this definition that h is a Carathéodory function (i.e. measurable in $t \in T$ and continuous in $\zeta \in \mathbb{R}$), hence jointly measurable and we have

$$\left| h(t, \zeta) \right| \leq a_1(t) + c_1|\zeta|^{p-1} \quad \text{for a.a. } t \in T \text{ and all } \zeta \in \mathbb{R},$$

with $a_1 \in L^{p'}(T)_+$, $c_1 > 0$. Using h, we decompose f as $f = f_1 + f_2$, with $f_1, f_2: T \times \mathbb{R} \longrightarrow \mathbb{R}$, defined by

$$f_1(t, \zeta) \stackrel{df}{=} \begin{cases} \min \left\{ \inf_{\zeta' \geq \zeta} f(t, \zeta'), \ h(t, \zeta) \right\} & \text{if } \zeta \geq 0, \\ \max \left\{ \sup_{\zeta' \leq \zeta} f(t, \zeta'), \ h(t, \zeta) \right\} & \text{if } \zeta < 0 \end{cases}$$

and

$$f_2(t, \zeta) \overset{df}{=} f(t, \zeta) - f_1(t, \zeta).$$

Also we introduce

$$f_+(t) \overset{df}{=} \liminf_{\zeta \to +\infty} f(t, \zeta) \qquad \text{and} \qquad f_-(t) \overset{df}{=} \limsup_{\zeta \to -\infty} f(t, \zeta).$$

In the next proposition, we take a closer look of the functions f_1, f_+ and f_-.

PROPOSITION 3.6.7

If hypotheses $H(f)_1$ hold,
then

(a) *for all $\zeta \in \mathbb{R}$, the function $t \longmapsto f_1(t, \zeta)$ is measurable;*

(b) *for almost all $t \in T$, the function $\zeta \longmapsto f_1(t, \zeta)$ is continuous and non-decreasing;*

(c) $\lim_{\zeta \to +\infty} f_1(t, \zeta) = f_+(t);$

(d) $\lim_{\zeta \to -\infty} f_1(t, \zeta) = f_-(t).$

In particular then $f_{\pm\infty}$ are $\mathbb{R}^* \overset{df}{=} \mathbb{R} \cup \{\pm\infty\}$-valued measurable functions.

PROOF   Let us fix $\overline{\zeta} \in \mathbb{R}$ and let $\{\zeta_n\}_{n \geq 1}$ be an enumeration of the rationals in the half-line $\{\zeta \in \mathbb{R} : \overline{\zeta} \leq \zeta\}$. We have

$$\inf_{\zeta \geq \overline{\zeta}} f(t, \zeta) = \inf_{n \geq 1} f(t, \zeta_n)$$

and so it follows that the function $t \longmapsto \inf_{\zeta \geq \overline{\zeta}} f(t, \zeta)$ is measurable on T. Similarly, we have that the function $t \longmapsto \sup_{\zeta \leq \overline{\zeta}} f(t, \zeta)$ is measurable on T. So it follows that for every $\zeta \in \mathbb{R}$ the function $t \longrightarrow f_1(t, \zeta)$ is measurable.

Next let N be the Lebesgue-null subset of T, such that if $t \in T \setminus N$, the function $f(t, \cdot)$ is continuous (see hypothesis $H(f)_1(ii)$). Let us fix $t_0 \in T \setminus N$ and let $\zeta_n \longrightarrow \zeta$. Set

$$m_{t_0}(\zeta) \overset{df}{=} \inf_{\zeta' \geq \zeta} f(t_0, \zeta').$$

From Proposition 1.2.7(a), we know that m_{t_0} is lower semicontinuous. If $\zeta_n \nearrow \zeta$, then from the monotonicity and lower semicontinuity of m_{t_0}, we have that $m_{t_0}(\zeta_n) \longrightarrow m_{t_0}(\zeta)$. So suppose that $\zeta_n \searrow \zeta$. Then for a given $\varepsilon > 0$, we can find $\widehat{\zeta} > \zeta$, such that $f(\widehat{\zeta}) \leq m_{t_0}(\zeta) + \varepsilon$ and because $\zeta_n \searrow \zeta$, we can find $n_0 \geq 1$, such that for all $n \geq n_0$, $\zeta_n \leq \widehat{\zeta}$. So

$$m_{t_0}(\zeta_n) \leq f(\widehat{\zeta}) \leq m_{t_0}(\zeta) + \varepsilon,$$

hence

$$\limsup_{n \to +\infty} m_{t_0}(\zeta_n) \leq m_{t_0}(\zeta) \leq \liminf_{n \to +\infty} m_{t_0}(\zeta_n),$$

i.e. $m_{t_0}(\zeta_n) \longrightarrow m_{t_0}(\zeta)$. This then proves the continuity of the function $\zeta \longmapsto \inf_{\zeta' \geq \zeta} f(t, \zeta')$ on $\mathbb{R}_+$. In a similar fashion, we prove the continuity of $\zeta \longmapsto \sup_{\zeta' \leq \zeta} f(t, \zeta')$ on $\mathbb{R}_-$ and thus conclude the continuity of $f_1(t, \cdot)$. Moreover, note that clearly $f_1(t, \cdot)$ is nondecreasing and

$$f_1(t, \zeta) \longrightarrow f_+(t) \quad \text{as } \zeta \to +\infty,$$

while

$$f_1(t, \zeta) \longrightarrow f_-(t) \quad \text{as } \zeta \to -\infty.$$

$\square$

Evidently $f_2(t, \zeta)$ is a Carathéodory function and

$$\begin{cases} |f_1(t, \zeta)| \leq a_2(t) + c_2 |\zeta|^{p-1}, \\ |f_2(t, \zeta)| \leq a_2(t) + c_2 |\zeta|^{p-1}, \end{cases} \quad \text{for a.a. } t \in T \text{ and all } \zeta \in \mathbb{R},$$

with $a_2 \in L^{p'}(T)_+$, $c_2 > 0$. Let $N_{f_1}, N_{f_2} : L^p(T) \longrightarrow L^{p'}(T)$ be the Nemytskii operators corresponding to the functions f_1 and f_2 respectively, i.e.

$$N_{f_1}(x)(\cdot) \stackrel{df}{=} f_1(\cdot, x(\cdot)) \quad \text{and} \quad N_{f_2}(x)(\cdot) \stackrel{df}{=} f_2(\cdot, x(\cdot)) \qquad \forall\, x \in L^p(T).$$

We know that both N_{f_1} and N_{f_2} are continuous bounded (Krasnoselskii Theorem; see Theorem 1.4.7). Also by virtue of Proposition 3.6.7, N_{f_1} is monotone and so it is easy to check that it is also 3-monotone (see Definition 3.6.2).

PROPOSITION 3.6.8

If hypotheses $H(a)$ and $H(f)$ hold,

then the operator $A_1 + N_{f_1} : L^p(T) \supseteq D_1 \longrightarrow L^{p'}(T)$ is maximal monotone and boundedly inversely compact.

PROOF   The maximal monotonicity of $A_1 + N_{f_1}$ is a consequence of Theorem 1.4.5. To show that it is boundedly inversely compact, let $C \subseteq L^p(T)$ and $C^* \subseteq L^{p'}(T)$ be two nonempty bounded sets. Let us set

$$G \stackrel{df}{=} C \cap (A_1 + N_{f_1})^{-1}(C^*) \subseteq L^p(T).$$

Let $x \in G$. By definition, we have that

$$A_1(x) + N_{f_1}(x) = w \in C^*.$$

Exploiting the monotonicity of N_{f_1} and hypothesis $H(a)(iii)$, we have

$$c_0^a \left\| x' \right\|_p^p - c_3 \leq \left\langle A_1(x), x \right\rangle_p \leq \left\langle A_1(x) + N_{f_1}(x) - N_{f_1}(0), x \right\rangle_p$$
$$= \left\langle w - N_{f_1}(0), x \right\rangle_p \leq \left(\left\| w \right\|_{p'} + \left\| a_2 \right\|_{p'} \right) \left\| x \right\|_p \leq M_2,$$

for some $c_3, M_2 > 0$, since the sets $C \subseteq L^p(T)$ and $C^* \subseteq L^{p'}(T)$ are bounded. Hence G is bounded in $W^{1,p}(T)$ and recall that the embedding $W^{1,p}(T) \subseteq L^p(T)$ is compact. So G is relatively compact in $L^p(T)$ and this proves that $A_1 + N_{f_1}$ is boundedly inversely compact. $\quad\square$

PROPOSITION 3.6.9
If hypotheses $H(f)_1$ hold and $k \geq 0$,
then we can find a constant $c(k) \geq 0$, such that

$$k \left\| N_{f_2}(x) \right\|_{p'} - c(k) \leq \left\langle N_{f_2}(x), x \right\rangle_p \qquad \forall\, x \in L^p(T). \tag{3.187}$$

PROOF   We have

$$\left\langle N_{f_2}(x), x \right\rangle_p = \int_0^b f_2\big(t, x(t)\big) x(t)\, dt$$

$$= \int_{\{|x| \geq u\}} f_2\big(t, x(t)\big) x(t)\, dt + \int_{\{|x| < u\}} f_2\big(t, x(t)\big) x(t)\, dt.$$

Because of hypothesis $H(f)_1(iii)$, we have

$$\int_{\{|x| < u\}} f_2\big(t, x(t)\big) x(t)\, dt \leq M_3,$$

for some $M_3 > 0$. Moreover, we have

$$\left| f_2\big(t, x(t)\big) \right|^{\frac{p'}{p}} \leq \big(2a_2(t)\big)^{\frac{p'}{p}} + (2c_2)^{\frac{p'}{p}} \left| x(t) \right|.$$

By virtue of hypothesis $H(f)_1(iii)$, it follows that

$$\int_{\{|x| \geq u\}} f_2\big(t, x(t)\big) x(t)\, dt = \int_{\{|x| \geq u\}} \left| f_2\big(t, x(t)\big) \right| \left| x(t) \right| dt$$

$$\geq \frac{1}{c_4} \int_{\{|x| \geq u\}} \left(\left| f_2\big(t, x(t)\big) \right|^{1 + \frac{p'}{p}} - \big(2a_2(t)\big)^{\frac{p'}{p}} \left| f_2\big(t, x(t)\big) \right| \right) dt$$

with $c_4 \overset{df}{=} (2c_2)^{-\frac{p'}{p}}$. Using Young's inequality (see Theorem A.4.2) with $\varepsilon > 0$, we obtain

$$\big(2a_2(t)\big)^{\frac{p'}{p}} \left| f_2\big(t, x(t)\big) \right| \leq \frac{1}{\varepsilon^p p}\big(2a_2(t)\big)^{p'} + \frac{\varepsilon^{p'}}{p} \left| f_2\big(t, x(t)\big) \right|^{p'}.$$

So, if we choose $\varepsilon > 0$ such that $1 - \frac{\varepsilon^{p'}}{p'} > 0$, from hypothesis $H(f)_1(iii)$, we have

$$\int\limits_{\{|x| \geq u\}} f_2\big(t, x(t)\big) x(t)\, dt$$

$$\geq \frac{1}{c_4} \int\limits_{\{|x| \geq u\}} \left[\left(1 - \frac{\varepsilon^{p'}}{p'}\right) \big|f_2\big(t, x(t)\big)\big|^{p'} - \frac{1}{\varepsilon^p p} \big(2a_2(t)\big)^{p'} \right] dt$$

$$= c_5(\varepsilon) \int\limits_0^b \big|f_2\big(t, x(t)\big)\big|^{p'} dt - c_5(\varepsilon) \int\limits_{\{|x| < u\}} \big|f_2\big(t, x(t)\big)\big|^{p'} dt - c_6(\varepsilon) \|2a_2\|_{p'}^{p'}$$

$$\geq c_5(\varepsilon) \|N_{f_2}\|_{p'}^{p'} - c_7(\varepsilon,)$$

for some $c_5, c_6, c_7 > 0$. Therefore, we have

$$\big\langle N_{f_2}(x), x \big\rangle_p \geq c_5(\varepsilon) \|N_{f_2}\|_{p'}^{p'} - c_8(\varepsilon),$$

with $c_8(\varepsilon) \overset{df}{=} c_7(\varepsilon) + M_3$. A new use of Young's inequality implies that for a given $k' > 0$, we can find $c_9(k') > 0$, such that

$$\|N_{f_2}(x)\|_{p'} \leq \frac{c_5(\varepsilon)}{k'} \|N_{f_2} x\|_{p'}^{p'} + c_9(k')$$

and so

$$k' \|N_{f_2}(x)\|_{p'} - c_9(k') \leq c_5(\varepsilon) \|N_{f_2}\|_{p'}^{p'}.$$

Therefore, we conclude that for a given $k \geq 0$, we can find $c(k) \geq 0$, such that (3.187) holds. $\square$

Now we are ready for the existence result for problem (3.182).

THEOREM 3.6.5
If hypotheses $H(a)$, $H(f)_1$ hold, $v \in L^{p'}(T)$ and

$$\int\limits_0^b f_-(t)\, dt < \int\limits_0^b v(t)\, dt < \int\limits_0^b f_+(t)\, dt,$$

then problem (3.182) has a solution

$$x_0 \in C^1(T) \quad \text{with} \quad a\big(|x_0'(\cdot)|^2\big) x_0'(\cdot) \in W^{1,p'}(T).$$

PROOF   From Proposition 3.6.7, we know that for almost all $t \in T$, the function $f_1(t, \cdot)$ is nondecreasing and

$$\lim_{\zeta \to +\infty} f_1(t, \zeta) = f_+(t) \quad \text{and} \quad \lim_{\zeta \to -\infty} f_1(t, \zeta) = f_-(t).$$

Invoking the Monotone Convergence Theorem, for a given $\delta > 0$, we can find $n_0 \geq 1$ large enough, such that

$$\int_0^b f_1(t, -n_0)\, dt \; < \; \int_0^b \big(v(t) + u(t)\big)\, dt$$

$$< \int_0^b f_1(t, n_0)\, dt \qquad \forall\, u \in L^{p'}(T),\ \|u\|_{p'} < \delta.$$

By the Intermediate Value Theorem, we can find $\zeta_0 \in (-n_0, n_0)$, such that

$$\int_0^b f_1(t, \zeta_0)\, dt \; = \; \int_0^b \big(v(t) + u(t)\big)\, dt.$$

We write

$$v(t) + u(t) \; = \; v(t) + u(t) - \frac{1}{b}\int_0^b \big(v(s) + u(s)\big)\, ds + \frac{1}{b}\int_0^b \big(v(s) + u(s)\big)\, ds$$

$$= \; v(t) + u(t) - \frac{1}{b}\int_0^b \big(v(s) + u(s)\big)\, ds$$

$$+ \frac{1}{b}\int_0^b f_1(s, \zeta_0)\, ds - f_1(t, \zeta_0) + f_1(t, \zeta_0). \qquad (3.188)$$

Let us set

$$w(t) \;\overset{df}{=}\; v(t) + u(t) - \frac{1}{b}\int_0^b \big(v(s) + u(s)\big)\, ds + \frac{1}{b}\int_0^b f_1(s, \zeta_0)\, ds - f_1(t, \zeta_0).$$

Evidently $w \in L^{p'}(T)$ and $\int_0^b w(t)\, dt = 0$. So from Proposition 3.6.6, we have that $w \in R(A_1)$. Therefore, from (3.188), we have

$$v + u = w + f_1(\cdot, \zeta_0) \in R(A_1) + R(N_{f_1}).$$

Since $u \in L^{p'}(T)$, $\|u\|_{p'} < \delta$ was arbitrary, we see that

$$v \in \text{int}\,\big(R(A_1) + R(N_{f_1})\big).$$

Invoking Proposition 3.6.2, we obtain that $v \in R\big(A_1 + N_{f_1} + N_{f_2}\big)$. Therefore, we can find $x_0 \in D_1 \subseteq C^1(T)$ (see Remark 3.6.9), such that $\big(A_1 + N_{f_1} + N_{f_2}\big)(x_0) = v$. This is the desired solution of (3.182). $\qquad\square$

REMARK 3.6.10   We can have that

$$\int_0^b f_-(t)\, dt \; = \; -\infty \qquad \text{and} \qquad \int_0^b f_+(t)\, dt \; = \; +\infty.$$

The condition on v in the statement of Theorem 3.6.5 is a Landesman-Lazer type condition.   ☐

3.7   Hamiltonian Inclusions

In classical and celestial mechanics the laws governing a system are expressible as a system of Hamiltonian equations

$$\begin{cases} -\dot{p} \; = \; H_x(x,p), \\ \;\; \dot{x} \; = \; H_p(x,p). \end{cases}$$

We write Hamiltonian systems in the form

$$J\dot{y} \; = \; \nabla H(y),$$

where $y = (x,p)$ (x is the state and p is the adjoint variable, the momentum in physics) and J is the standard symplectic matrix on $\mathbb{R}^{2N} = \mathbb{R}^N \times \mathbb{R}^N$, i.e.

$$J \; = \; \begin{bmatrix} 0 & -I \\ I & 0 \end{bmatrix}$$

(here I is the $N \times N$ identity matrix). Note that

$$J^2 = -I, \qquad \text{i.e. } J^{-1} = -J$$

and

$$(Jx,y)_{\mathbb{R}^{2N}} \; = \; -(x, Jy)_{\mathbb{R}^{2N}}, \qquad \text{i.e. } J^* = -J.$$

Exploiting this last property of J (the antisymmetry), we see that, if y is a solution of the Hamiltonian system, then

$$\frac{d}{dt} H\big(y(t)\big) \; = \; \big(\nabla H(y(t)), \dot{y}(t)\big)_{\mathbb{R}^N} \; = \; 0 \qquad \forall\, t \in T = [0,b]$$

and so

$$H\big(y(t)\big) = \text{constant} \qquad \forall\, t \in T,$$

i.e. y is conservative.

Moreover, any "energy surface" $H = c \in \mathbb{R}$ is invariant under the Hamiltonian flow. In this section, we shall study Hamiltonian systems, for which the Hamiltonian function is nonsmooth (i.e. H is not a C^1-function) and it is only locally Lipschitz. In this case the solution of the Hamiltonian system need not be conservative. In this respect, we have the following result. Recall (see Definition 1.3.8) that if X is a Banach space and $\varphi \colon X \longrightarrow \mathbb{R}$ is a locally Lipschitz function, we say that φ is regular at x, if the usual directional derivative

$$\varphi'(x;h) \stackrel{df}{=} \lim_{\lambda \searrow 0} \frac{\varphi(x+\lambda h) - \varphi(x)}{\lambda}$$

exists for every $h \in X$ and is equal to the generalized directional derivative

$$\varphi^0(x;h) \stackrel{df}{=} \limsup_{\substack{y \to x \\ \lambda \searrow 0}} \frac{\varphi(y+\lambda h) - \varphi(y)}{\lambda}.$$

Let us consider the following Hamiltonian inclusion

$$J\dot{y}(t) \in \partial H(y(t)) \quad \text{for a.a. } t \in T, \tag{3.189}$$

where $T = [0,b]$ with $b > 0$.

PROPOSITION 3.7.1
If $H \colon \mathbb{R}^{2N} \longrightarrow \mathbb{R}$ is a locally Lipschitz Hamiltonian, $y \in W^{1,1}(T;\mathbb{R}^{2N})$ is a solution of the Hamiltonian inclusion (3.189) and H is regular at $y(t)$, then $H(y(\cdot))$ is constant on T.

PROOF Let

$$\varphi(t) \stackrel{df}{=} H(y(t)) \qquad \forall \, t \in T.$$

Evidently φ is locally Lipschitz. We need to show that

$$\varphi'(t) = 0 \quad \text{for a.a. } t \in T. \tag{3.190}$$

Let $t \in T$ be a point at which φ' and $\dot{y}$ exist and $J\dot{y}(t) \in \partial H(y(t))$. We know that this set is of full-measure. Recall that

$$(Jv, v)_{\mathbb{R}^{2N}} = 0 \qquad \forall \, v \in \mathbb{R}^{2N}.$$

Hence, using the definition of $\partial H(y(t))$ and the assumptions of the proposition, we get

$$\begin{aligned}
0 &= (J\dot{y}(t), \dot{y}(t))_{\mathbb{R}^{2N}} \leq H^0(y(t); \dot{y}(t)) = H'(y(t); \dot{y}(t)) \\
&= \lim_{\lambda \searrow 0} \frac{H(y(t) + \lambda \dot{y}(t)) - H(y(t))}{\lambda} \\
&= \lim_{\delta \searrow 0} \frac{H(y(t+\delta)) - H(y(t))}{\delta} = \varphi'(t).
\end{aligned}$$

Therefore $0 \leq \varphi'(t)$. On the other hand arguing similarly, we have

$$
\begin{aligned}
\varphi'(t) &= \lim_{\lambda \searrow 0} \frac{H\big(y(t) - \lambda \dot{y}(t)\big) - H\big(y(t)\big)}{-\lambda} \quad - \quad -H'\big(y(t); -\dot{y}(t)\big) \\
&= -H^0\big(y(t); -\dot{y}(t)\big) \leq -\big(-\dot{y}(t), J\dot{y}(t)\big) = 0.
\end{aligned}
$$

Hence $\varphi'(t) \leq 0$ and so we conclude that $\varphi'(t) = 0$ for almost all $t \in T$ as desired. ▯

REMARK 3.7.1   Recall that if H is strictly differentiable at $\xi \in \mathbb{R}^{2N}$ or if H is convex, then H is regular at ξ. ▯

In the sequel we do not assume the regularity of the locally Lipschitz Hamiltonian H and still we are able to obtain conservative solutions of the Hamiltonian inclusion by the use of a C^1-approximation method.

We study the existence of periodic trajectories y of a nonsmooth Hamiltonian system evolving on a given energy surface $S = H^{-1}(c)$, with $c \in \mathbb{R}$. As we already mentioned our approach is based on a C^1-approximation theorem for locally Lipschitz functions, which is actually of independent interest.

THEOREM 3.7.1
If $U \subseteq \mathbb{R}^N$ is a bounded and open set and $V : \mathbb{R}^N \longrightarrow \mathbb{R}$ is a locally Lipschitz function,
<u>*then*</u> *for a given $\varepsilon > 0$, we can find $f \in C^\infty(\mathbb{R}^N; \mathbb{R})$, such that*

(a) $|V(\xi) - f(\xi)| \leq \varepsilon$ *for all $\xi \in \overline{U}$;*

(b) *for every $\xi \in \overline{U}$, we can find $\overline{\xi} \in \overline{U}$ and $\overline{\xi}^* \in \partial V(\overline{\xi})$ such that*

$$
\big\|\xi - \overline{\xi}\big\|_{\mathbb{R}^N} \leq \varepsilon \qquad \text{and} \qquad \big\|\overline{\xi}^* - \nabla f(\xi)\big\|_{\mathbb{R}^N} \leq \varepsilon.
$$

PROOF   **(a)** Let $\vartheta \in C^\infty(\mathbb{R}^N; \mathbb{R})$ be such that $\vartheta \geq 0$,

$$
\operatorname{supp} \vartheta \subseteq \overline{B}_1(0) \qquad \text{and} \qquad \int_{\mathbb{R}^N} \vartheta(\xi)d\xi = 1.
$$

For each $\delta > 0$, we define $V_\delta : \mathbb{R}^N \longrightarrow \mathbb{R}$ by

$$
V_\delta(\xi) \stackrel{df}{=} \int_{\mathbb{R}^N} \vartheta(\overline{\xi})V(\xi - \delta\overline{\xi})d\overline{\xi} \qquad \forall\, \xi \in \mathbb{R}^N. \tag{3.191}
$$

It is well known that $V_\delta \in C^\infty(\mathbb{R}^N; \mathbb{R})$ and

$$
V_\delta \longrightarrow V \qquad \text{uniformly on } \overline{U}, \text{ as } \delta \searrow 0
$$

(see e.g. Denkowski, Migórski & Papageorgiou (2003*a*, p. 342)). So for a given $\varepsilon > 0$, we can find $\delta_1 > 0$ such that

$$|V_\delta(\xi) - V(\xi)| \leq \varepsilon \quad \forall \, \xi \in \overline{U}, \ \delta \in (0, \delta_1].$$

(b) From Theorem 1.3.9, we have

$$\nabla V_\delta(\xi) \in \int_{\mathbb{R}^N} \vartheta(\overline{\xi}) \partial V(\xi - \delta \overline{\xi}) d\overline{\xi},$$

i.e. for each $\xi \in \mathbb{R}^N$, we can find $v_\xi \in S^1_{\partial V(\xi - \delta \cdot)}$, such that

$$\nabla V_\delta(\xi) \ = \ \int_{\mathbb{R}^N} \vartheta(\overline{\xi}) v_\xi(\overline{\xi}) d\overline{\xi}. \tag{3.192}$$

From Proposition 1.3.9, we know that ∂V is an upper semicontinuous multifunction into $P_{kc}(\mathbb{R}^N)$. So for each $\xi \in \overline{U}$, we can find $r_\xi \in (0, \varepsilon)$, such that

$$\partial V\big(B_{r_\xi}(\xi)\big) \subseteq \big[\partial V(\xi)\big]_\varepsilon,$$

where

$$\big[\partial V(\xi)\big]_\varepsilon \ \stackrel{df}{=} \ \big\{ u \in \mathbb{R}^N : \ d_{\mathbb{R}^N}\big(u, \partial V(\xi)\big) < \varepsilon \big\}.$$

The collection $\big\{ B_{r_\xi}(\xi) \big\}_{\xi \in \overline{U}}$ is an open cover of $\overline{U}$ and because $\overline{U}$ is compact, it has a Lebesgue number $\eta > 0$ (i.e. for every $\xi \in \overline{U}$, the open ball $B_\eta(\xi)$ is contained in some $B_{r_z}(z)$, $z \in \overline{U}$). We claim that the function $f = V_\delta$ with $0 < \delta < \min\{\eta, \delta_1\}$ will do the job. For a given $\xi \in \overline{U}$, we know that $B_\eta(\xi) \subseteq B_{r_z}(z)$ for some $z \in \overline{U}$. So $\|\xi - z\|_{\mathbb{R}^N} < r_z < \varepsilon$ and

$$\partial V\big(B_\delta(\xi)\big) \subseteq \partial f\big(B_{r_z}(z)\big) \subseteq \big[\partial V(z)\big]_\varepsilon.$$

Let $v_\xi \in S^1_{\partial V(\xi - \delta \cdot)}$ satisfy (3.192). We have

$$v_\xi(z) \in \partial V(\xi - \delta z) \subseteq \big[\partial V(z)\big]_\varepsilon \quad \text{for a.a. } z \in \overline{B}_1.$$

Since

$$\int_{\{\|z\|_{\mathbb{R}^N} \leq 1\}} \varphi(z) \, dz \ = \ 1$$

and $\big[\partial V(z)\big]_\varepsilon$ is convex, it follows that

$$\nabla V_\delta(\xi) \ = \ \int_{\{\|z\|_{\mathbb{R}^N} \leq 1\}} \varphi(z) v_\xi(z) \, dz \ \in \ \overline{\big[\partial V(z)\big]_\varepsilon}.$$

Therefore, we conclude that $f = V_\delta$ is the desired smooth approximation of V. □

DEFINITION 3.7.1 *A C^1-function f satisfying conditions (a) and (b) of Theorem 3.7.1, will be called an ε-**admissible approximation of** V **on** $\overline{U}$.*

From the theory of smooth Hamiltonian systems (i.e. $H \in C^1$), it is well known that the orbits on an energy surface S are independent of H and depend only on S. More precisely, we have the following lemma.

LEMMA 3.7.1
If $H_1, H_2 \in C^1(\mathbb{R}^{2N}; \mathbb{R})$, $c_1, c_2 \in \mathbb{R}$, $H_1^{-1}(c_1) = H_2^{-1}(c_2) = S$ and c_1, c_2 are regular values of H_1, H_2 respectively (i.e. $\nabla H_1(\xi) \neq 0$ and $\nabla H_2(\xi) \neq 0$ for $\xi \in S$),
then the orbits of the Hamiltonian systems

$$J\dot{x}(t) = \nabla H_1(x(t)) \qquad \text{and} \qquad J\dot{x}(t) = \nabla H_2(x(t))$$

on S are the same.

PROOF   Let $x_1 \in C^1(T; \mathbb{R}^{2N})$ be a solution of $J\dot{x}(t) = \nabla H_1(x(t))$ on S. Because $\nabla H_1(x_1(t))$ and $\nabla H_2(x_1(t))$ are normal to the surface S and continuous, we can find a function $\lambda: T \longrightarrow \mathbb{R}$, such that

$$\nabla H_2(x_1(t)) = \lambda(t)\nabla H_1(x_1(t)) \qquad \forall\, t \in T.$$

Since c_1 and c_2 are regular values of H_1 and H_2 respectively, we have that $\lambda(t) \neq 0$ for all $t \in T$. Moreover, we have

$$\lambda(t) = \frac{\left(\nabla H_2(x_1(t)),\ \nabla H_1(x_1(t))\right)_{\mathbb{R}^{2N}}}{\left\|\nabla H_1(x_1(t))\right\|_{\mathbb{R}^{2N}}^2}.$$

Therefore λ is continuous and so it does not change sign. We define $\psi: T \longrightarrow \mathbb{R}$ by

$$\psi(t) \overset{df}{=} \int_0^t \frac{ds}{\lambda(s)} \qquad \forall\, t \in T.$$

Evidently ψ is continuous and strictly monotone. Let us set $x_2 \overset{df}{=} x_1 \circ \psi^{-1} \in C^1(T; \mathbb{R}^{2N})$. We have

$$\begin{aligned}
J\dot{x}_2(t) &= J(\dot{x}_1 \circ \psi^{-1})(t)\left(\frac{1}{\psi'} \circ \psi^{-1}\right)(t) \\
&= \nabla H_1\left((x_1 \circ \psi^{-1})(t)\right)(\lambda \circ \psi^{-1})(t) \\
&= \nabla H_2((x_1 \circ \psi^{-1})(t)) = \nabla H_2(x_2(t)) \qquad \forall\, t \in \psi(T).
\end{aligned}$$

Hence x_2 is a solution of

$$J\dot{x}(t) \ = \ \nabla H_2\big(x(t)\big) \qquad \forall\, t \in T.$$

$\square$

In the smooth case, using this Lemma and some convexity assumption on the Hamiltonian H, we are able to replace H by another Hamiltonian $\widehat{H}$, which is homogeneous and leads to the same orbits. More precisely, we assume that $H \in C^1\big(\mathbb{R}^{2N};\mathbb{R}\big)$ is convex, $c \in \mathbb{R}$ is a regular value of H, $S = H^{-1}(c)$ and S is the boundary of a compact, convex set C containing the origin as an interior point. We introduce the **gauge function** (**Minkowski functional**) $j_C \colon \mathbb{R}^{2N} \longrightarrow \mathbb{R}_+$ corresponding to C, defined by

$$j_C(\xi) \ = \ \inf\big\{\lambda > 0 : \ \xi \in \lambda C\big\}.$$

Evidently j_C is sublinear. Moreover,

$$\xi \in C \ \text{ if and only if } \ j_C(\xi) \le 1$$

and

$$\xi \in \operatorname{int} C \ \text{ if and only if } \ j_C(\xi) < 1,$$

i.e. j_C characterizes the set C. Then, we set $\widehat{H} \overset{df}{=} j_C^{\frac{3}{2}}$ and we have:

(a) $\widehat{H}^{-1}(1) = S$;

(b) $\widehat{H}$ is homogeneous of degree $\frac{3}{2}$;

(c) there exist $\beta, \gamma > 0$, such that

$$\beta\,\|\xi\|_{\mathbb{R}^{2N}}^{\frac{3}{2}} \ \le \ \widehat{H}(\xi) \ \le \ \gamma\,\|\xi\|_{\mathbb{R}^{2N}}^{\frac{3}{2}} \qquad \forall\, \xi \in \mathbb{R}^{2N};$$

(d) $\widehat{H} \in C^1\big(\mathbb{R}^{2N};\mathbb{R}\big)$;

(e) $\nabla\widehat{H}$ is positively homogeneous of degree $\frac{1}{2}$.

Working with $\widehat{H}$, we can establish the existence of at least one periodic orbit of $J\dot{x}(t) = \nabla H\big(x(t)\big)$, which lies on S. In general a similar approach is not possible in the nonsmooth, locally Lipschitz case, because of the failure of Lemma 3.7.1. In what follows we shall show that under some extra hypotheses, we can still employ the same approach and replace H by a homogeneous $\widehat{H}$. For this purpose let Ω be the $2N \times 2N$ block diagonal matrix, defined by

$$\Omega \ \overset{df}{=} \ \begin{bmatrix} \Omega' & 0 \\ 0 & \Omega' \end{bmatrix},$$

with Ω' being an $N \times N$ diagonal matrix with positive diagonal entries, i.e.

$$\Omega' = \operatorname{diag}\big\{\omega_1,\ldots,\omega_N\big\} \quad \text{with } \omega_k > 0.$$

We set

$$\Gamma \stackrel{df}{=} \left\{ \xi \in \mathbb{R}^{2N} : \frac{1}{2} (\Omega \xi, \xi)_{\mathbb{R}^{2N}} = \sum_{k=1}^{N} \frac{1}{2} \omega_k \left(\xi_k^2 + \xi_{k+N}^2 \right) \leq 1 \right\}$$

(an ellipsoid in $\mathbb{R}^{2N}$)

$$S \stackrel{df}{=} H^{-1}(1)$$
$$\Delta \stackrel{df}{=} \{ \xi \in \mathbb{R}^{2N} : H(\xi) \leq 1 \}.$$

We make the following hypotheses on these items.

$\underline{H(H)}$ $H : \mathbb{R}^{2N} \longrightarrow \mathbb{R}$ is a locally Lipschitz function, such that

(i) S is radially homeomorphic to the unit sphere in $\mathbb{R}^{2N}$, i.e. the function $S \ni \xi \longmapsto \frac{\xi}{\|\xi\|_{\mathbb{R}^{2N}}} \in S^{2N-1}$ is homeomorphism;

(ii) for all $\xi \in S$ and all $u \in \partial H(\xi)$, we have $(u, \xi)_{\mathbb{R}^{2N}} > 0$;

(iii) for some $\beta \in (0, \gamma)$, we have $\beta \Gamma \subseteq \Delta \subseteq \gamma \Gamma$.

We look for periodic solutions on S of the Hamiltonian inclusion

$$J \dot{x}(t) \in \partial H(x(t)) \quad \text{for a.a. } t \in T \tag{3.193}$$

(i.e. periodic solutions with prescribed energy). The difficulty in this problem is that the period and so the underlying function space for the solutions are not known *a priori*. Nevertheless, we shall see that under hypotheses $H(H)$ we can reduce the fixed energy case to the fixed period case.

Because of hypothesis $H(H)(i)$, we can find a unique positive function $\lambda : \mathbb{R}^{2N} \longrightarrow \mathbb{R}_+$, such that

$$\lambda(\xi)\xi \in S \quad \forall \, \xi \in \mathbb{R}^{2N} \setminus \{0\}.$$

This function will play the role of the gauge function of the smooth case, in order to pass to a homogeneous Hamiltonian (see the discussion above).

LEMMA 3.7.2
Let λ be as above.

(a) $\lambda : \mathbb{R}^{2N} \setminus \{0\} \longrightarrow \mathbb{R}_+$ *is locally Lipschitz and positive homogeneous of degree -1;*

(b) $\partial \lambda(\xi) \in -\dfrac{\lambda(\xi)}{\left(\partial H(\lambda(\xi)), \xi \right)_{\mathbb{R}^{2N}}}$ *for all $\xi \in \mathbb{R}^{2N} \setminus \{0\}$.*

PROOF (a) Hypothesis $H(H)(ii)$ and the nonsmooth implicit function theorem (see Theorem 1.3.8) imply that $\lambda: \mathbb{R}^{2N} \setminus \{0\} \longrightarrow \mathbb{R}_+$ is locally Lipschitz. The homogeneity of λ follows from its definition.

(b) Consider the functions $h: \mathbb{R}^{2N} \setminus \{0\} \longrightarrow \mathbb{R}_+$ and $\Lambda: \mathbb{R}^{2N} \setminus \{0\} \longrightarrow \mathbb{R}$, defined by

$$h(\xi) \overset{df}{=} H\big(\lambda(\xi)\xi\big) - 1 \quad \text{and} \quad \Lambda(\xi) \overset{df}{=} \lambda(\xi)\xi \quad \forall\, \xi \in \mathbb{R}^{2N} \setminus \{0\}.$$

Let $\xi \in \mathbb{R}^{2N} \setminus \{0\}$ be a point of differentiability of λ. Note that $h \equiv 0$. So, we have

$$0 = \big(h'(\xi), v\big)_{\mathbb{R}^{2N}} = \lim_{r \to +\infty} \frac{H\big(\lambda(\xi + rv)(\xi + rv)\big) - H\big(\lambda(\xi)\xi\big)}{r} \quad \forall\, v \in \mathbb{R}^{2N}.$$

Invoking Proposition 1.3.15, we can find $\xi_v^* \in \partial H\big(\lambda(\xi)\xi\big)$, such that

$$0 = \big(h'(\xi), v\big)_{\mathbb{R}^{2N}} = \big(\xi_v^*, \Lambda'(\xi)v\big)_{\mathbb{R}^{2N}}.$$

Because the set $\partial H\big(\lambda(\xi)\xi\big)$ is bounded and convex, we can apply Proposition 1.4.15, and obtain $\widehat{\xi} \in \partial H\big(\lambda(\xi)\xi\big)$, such that

$$\big(\widehat{\xi}, \Lambda'(\xi)v\big)_{\mathbb{R}^N} = 0 \quad \forall\, v \in \mathbb{R}^{2N}.$$

So, we have

$$\big(\widehat{\xi}, (\lambda'(\xi), v)_{\mathbb{R}^{2N}}\, \xi + \lambda(\xi)v\big)_{\mathbb{R}^{2N}}$$
$$= \big((\widehat{\xi}, \xi)_{\mathbb{R}^{2N}}\lambda'(\xi) + \lambda(\xi)\widehat{\xi},\ v\big)_{\mathbb{R}^{2N}} = 0 \quad \forall\, v \in \mathbb{R}^{2N}.$$

Thus, for each differentiability point $\xi \in \mathbb{R}^{2N} \setminus \{0\}$ of λ, we have

$$\lambda'(\xi) = -\frac{\lambda(\xi)}{(\widehat{\xi}, \xi)_{\mathbb{R}^{2N}}}\widehat{\xi} \in -\frac{\lambda(\xi)}{\big(\partial H(\lambda(\xi)), \xi\big)_{\mathbb{R}^{2N}}}\partial H\big(\lambda(\xi)\xi\big). \tag{3.194}$$

Recall that

$$\partial\lambda(\xi) = \operatorname{conv}\Big\{\xi^* = \lim_{n \to +\infty} \lambda'(\xi_n) : \xi_n \longrightarrow \xi,\ \text{and}$$

$$\lambda \text{ is differentiable at } \xi_n,\ n \geq 1\Big\}$$

(see Remark 1.3.5). From this and inclusion (3.194), we deduce part (b) of the lemma. $\qquad\square$

Now we are ready to introduce the locally Lipschitz homogeneous Hamiltonian function, which will replace H in our analysis. So let $\widehat{H}: \mathbb{R}^{2N} \longrightarrow \mathbb{R}$

be defined by

$$\widehat{H}(0) \overset{df}{=} 0$$

$$\widehat{H}(\xi) \overset{df}{=} \left(\frac{1}{\lambda(\xi)}\right)^2 \qquad \forall\, \xi \in \mathbb{R}^{2N} \setminus \{0\}.$$

PROPOSITION 3.7.2
We have

(a) $\widehat{H}$ *is locally Lipschitz, positively homogeneous of degree 2 and* $H^{-1}(1) = \widehat{H}^{-1}(1) = S;$

(b) $\partial\widehat{H}$ *is positively homogeneous of degree 1,* $\partial\widehat{H}(0) = \{0\}$ *and*

$$\partial\widehat{H}(\xi) \subseteq \frac{2\widehat{H}(\xi)}{\big(\partial H(\lambda(\xi)\xi),\xi\big)_{\mathbb{R}^{2N}}} \partial H\big(\lambda(\xi)\xi\big) \qquad \forall\, \xi \in \mathbb{R}^{2N}\setminus\{0\};$$

(c) *we have*

$$(\xi^*,\xi)_{\mathbb{R}^{2N}} = 2\widehat{H}(\xi) \qquad \forall\, \xi \in \mathbb{R}^{2N},\ \xi^* \in \partial\widehat{H}(\xi);$$

(d) *the distinct periodic orbits of*

$$J\dot{x}(t) = \partial\widehat{H}\big(x(t)\big) \qquad \text{on } S$$

coincide with the distinct periodic orbits of (3.193) on S.

PROOF   (a–b) These properties follow from the definition of $\widehat{H}$ and Lemma 3.7.2.

(c) Because $\widehat{H}$ is positively homogeneous of degree 2, at any differentiable point $\xi \in \mathbb{R}^{2N}$ of $\widehat{H}$, we have

$$\big(\nabla\widehat{H}(\xi),\xi\big)_{\mathbb{R}^{2N}} = 2\widehat{H}(\xi).$$

Using as before the definition of $\partial\widehat{H}(\xi)$ for functions defined on $\mathbb{R}^{2N}$ (see Remark 1.3.5), we conclude that statement (c) of the Proposition holds.

(d) Let $x_0 \in W^{1,1}_{\text{per}}(T;\mathbb{R}^{2N})$ be a solution of the Hamiltonian inclusion

$$J\dot{x}(t) \in \partial\widehat{H}\big(x(t)\big) \qquad \forall\, t \in T.$$

Then we can find $\widehat{x}_0 \in S^\infty_{\partial\widehat{H}(x_0(\cdot))}$, such that

$$J\dot{x}_0(t) = \widehat{x}_0(t) \quad \text{for a.a. } t \in T.$$

According to part (c), we can find $\mu \in L^\infty(T)$, $\mu > 0$ and $\eta \in S^\infty_{\partial H(x_0(\cdot))}$, such that

$$\widehat{x}_0(t) \;=\; \mu(t)\eta(t) \quad \text{for a.a. } t \in T.$$

Let us set

$$k(t) \;\overset{df}{=}\; \int_0^t \mu(s)ds \qquad \forall\, t \in T,$$

which is strictly increasing and let

$$w(\tau) \;\overset{df}{=}\; x_0\big(k^{-1}(\tau)\big) \qquad \forall\, \tau \in [0, \widehat{T}],$$

with $\widehat{T} \overset{df}{=} \int_0^T \mu(s)ds$. Then w is a periodic orbit (with period $\widehat{T}$) on S of the Hamiltonian inclusion (3.193). $\qquad\square$

REMARK 3.7.2 Let

$$\varrho \;\overset{df}{=}\; \inf_{\substack{\xi \in S \\ \xi^*,\,\overline{\xi}^* \in \partial H(\xi)}} \frac{(\xi^*, \xi)_{\mathbb{R}^{2N}}}{\big\|\overline{\xi}^*\big\|_{\mathbb{R}^{2N}}}.$$

Then from Proposition 3.7.2(c), we have

$$\partial\widehat{H}(\xi) \;\subseteq\; \frac{2}{(\partial H(\xi), \xi)_{\mathbb{R}^{2N}}} \partial H(\xi) \qquad \forall\, \xi \in S$$

(recall that $\widehat{H}|_S = 1$; see Proposition 3.7.2(a)). It follows that

$$\big\|\overline{\xi}^*\big\|_{\mathbb{R}^{2N}} \;\leq\; \frac{2}{\varrho} \;=\; \frac{(\xi^*, \xi)_{\mathbb{R}^{2N}}}{\varrho} \qquad \forall\, \xi \in S,\ \xi^*,\overline{\xi}^* \in \partial\widehat{H}(\xi)$$

(see Proposition 3.7.2(c) and recall that $\widehat{H}(\xi) = 1$ for $\xi \in S$). Therefore, we have

$$(\xi^*, \xi)_{\mathbb{R}^{2N}} \;\geq\; \varrho\big\|\overline{\xi}^*\big\|_{\mathbb{R}^{2N}} \qquad \forall\, \xi \in S,\ \xi^*,\overline{\xi}^* \in \partial\widehat{H}(\xi).$$

$\qquad\square$

In light of Proposition 3.7.2(d), to obtain periodic orbits for the Hamiltonian inclusion (3.193), it suffices to produce periodic solutions for the Hamiltonian inclusion

$$J\dot{x}(t) \;\in\; \partial\widehat{H}\big(x(t)\big) \quad \text{for a.a. } t \in T. \tag{3.195}$$

To solve this inclusion, we shall use the approximation result given in Theorem 3.7.1. To this end let

$$M \;\overset{df}{=}\; \sup_{\xi \in S} \|\xi\|_{\mathbb{R}^{2N}}$$

and consider a sequence $\{\varepsilon_n\}_{n\geq 1}$ of positive numbers, such that $\varepsilon_n \searrow 0$. Using Theorem 3.7.1, for every $n \geq 1$, we can find $H_n \in C^\infty(\mathbb{R}^{2N};\mathbb{R})$ an ε_n-admissible approximation of $\widehat{H}$ on $B_{M+1} \stackrel{df}{=} \{\xi \in \mathbb{R}^{2N} : \|\xi\|_{\mathbb{R}^{2N}} < M+1\}$. Also let us set

$$S_n \stackrel{df}{=} H_n^{-1}(1) \qquad \forall\, n \geq 1.$$

Directly from the properties of $\widehat{H}$ and H_n, we obtain the following result.

PROPOSITION 3.7.3
For $n \geq 1$ sufficiently large, we have

(a) S_n *is radially diffeomorphic to the unit sphere in* $\mathbb{R}^{2N}$ *and* $S_n \subseteq B_{M+1}$;

(b) *there are positive sequences* $\{c_n\}_{n\geq 1}$ *and* $\{d_n\}_{n\geq 1}$, *such that* $c_n \searrow 0$, $d_n \searrow 0$ *and*

$$\|\xi - \lambda(\xi)\xi\|_{\mathbb{R}^{2N}} \leq c_n \quad \text{and} \quad \left|(\nabla H_n(\xi),\xi)_{\mathbb{R}^{2N}} - 2\right| \leq d_n \qquad \forall\, \xi \in S_n.$$

By virtue of Proposition 3.7.3, for $n \geq 1$ sufficiently large, we can find a unique function $\lambda_n \colon \mathbb{R}^{2N} \setminus \{0\} \longrightarrow \mathbb{R}$, $\lambda_n > 0$, such that

$$\lambda_n(\xi)\xi \in S_n \qquad \forall\, \xi \in \mathbb{R}^{2N} \setminus \{0\}.$$

Then as before, we introduce

$$\widehat{H}_n(0) \stackrel{df}{=} 0$$

$$\widehat{H}_n(\xi) \stackrel{df}{=} \left(\frac{1}{\lambda_n(\xi)}\right)^2 \qquad \forall\, \xi \in \mathbb{R}^{2N} \setminus \{0\}.$$

PROPOSITION 3.7.4
For $n \geq 1$ sufficiently large, we have

(a) $\widehat{H}_n \in C^1(\mathbb{R}^{2N};\mathbb{R})$, $\nabla\widehat{H}_n \colon \mathbb{R}^{2N} \longrightarrow \mathbb{R}$ *is locally Lipschitz*, $\widehat{H}_n$ *is positively homogeneous of degree 2*, $\widehat{H}_n^{-1}(1) = H_n^{-1}(1) = S_n$, $\nabla\widehat{H}_n$ *is positively homogeneous of degree 1 and*

$$(\nabla\widehat{H}_n(\xi),\xi)_{\mathbb{R}^{2N}} = 2\widehat{H}_n(\xi) \qquad \forall\, \xi \in \mathbb{R}^{2N};$$

(b) *there is a positive sequence* $\{e_n\}_{n\geq 1}$, *such that* $e_n \searrow 0$ *and*

$$\left\|\nabla\widehat{H}_n(\xi) - \nabla H_n(\xi)\right\|_{\mathbb{R}^{2N}} \leq e_n \qquad \forall\, \xi \in S_n.$$

PROOF   (a) This follows from the definition of $\widehat{H}_n$.

(b) As in the proof of Lemma 3.7.1, we can find a function $\lambda_n \colon S_n \longrightarrow \mathbb{R}$, $\lambda_n > 0$, such that

$$\nabla \widehat{H}_n(\xi) = \lambda_n(\xi) \nabla H_n(\xi) \qquad \forall \, \xi \in S_n$$

and so

$$2 = \left(\nabla \widehat{H}_n(\xi), \xi \right)_{\mathbb{R}^{2N}} = \lambda_n(\xi) \left(\nabla H_n(\xi), \xi \right)_{\mathbb{R}^{2N}} \qquad \forall \, \xi \in S_n$$

(see part (a)). Finally use Proposition 3.7.3(b) to finish the proof. $\qquad\square$

In what follows, without any loss of generality, we will assume that Propositions 3.7.3 and 3.7.4 hold for all $n \geq 1$.

PROPOSITION 3.7.5

There is a positive sequence $\{\widehat{\varepsilon}_n\}_{n \geq 1}$, such that $\widehat{\varepsilon}_n \searrow 0$ and for all $n \geq 1$, we have

(a) $\left| \widehat{H}_n(\xi) - \widehat{H}(\xi) \right| \leq \widehat{\varepsilon}_n \|\xi\|_{\mathbb{R}^{2N}}^2$ *for all $\xi \in \mathbb{R}^{2N}$;*

(b) *for every $\xi \in \mathbb{R}^{2N}$ we can find $\overline{\xi} \in \mathbb{R}^{2N}$ and $\xi^* \in \partial \widehat{H}(\xi)$, such that*

$$\left\| \xi - \overline{\xi} \right\|_{\mathbb{R}^{2N}} \leq \widehat{\varepsilon}_n \|\xi\|_{\mathbb{R}^{2N}} \qquad \text{and} \qquad \left\| \nabla \widehat{H}_n(\xi) - \xi^* \right\|_{\mathbb{R}^{2N}} \leq \widehat{\varepsilon}_n \|\xi\|_{\mathbb{R}^{2N}} .$$

PROOF   **(a)** From Theorem 3.7.1, we have

$$\left| \widehat{H}_n(\xi) - \widehat{H}(\xi) \right| = \left| H_n(\xi) - \widehat{H}(\xi) \right| \leq \varepsilon_n \qquad \forall \, \xi \in S_n.$$

Because of the compactness of S_n and the homogeneity of $\widehat{H}_n$ and $\widehat{H}$, there is a positive sequence $\{\varepsilon_n'\}_{n \geq 1}$, such that $\varepsilon_n' \searrow 0$ and

$$\left| \widehat{H}_n(\xi') - \widehat{H}(\xi') \right| \leq \varepsilon_n' \qquad \forall \, \xi' \in S^{2N-1}.$$

(b) Using Proposition 3.7.4 and the definition of H_n (see also Theorem 3.7.1), we can find a positive sequence $\{a_n\}_{n \geq 1}$, such that $a_n \searrow 0$ and for every $n \geq 1$ and every $\xi \in S_n$, we can find $\overline{\xi} \in \mathbb{R}^{2N}$ and $\xi^* \in \partial \widehat{H}(\overline{\xi})$, such that

$$\left\| \xi - \overline{\xi} \right\|_{\mathbb{R}^{2N}} \leq a_n \qquad \text{and} \qquad \left\| \nabla \widehat{H}_n(\xi) - \xi^* \right\|_{\mathbb{R}^{2N}} \leq a_n.$$

As before because S_n is compact and $\nabla \widehat{H}_n$ and $\partial \widehat{H}$ are both positive homogeneous, we can find a positive sequence $\{b_n\}_{n \geq 1}$, such that $b_n \searrow 0$ and for every $n \geq 1$ and every $\xi \in S$, we can find $\overline{\xi} \in \mathbb{R}^{2N}$ and $\xi^* \in \partial \widehat{H}(\overline{\xi})$, such that

$$\left\| \xi - \overline{\xi} \right\|_{\mathbb{R}^{2N}} \leq b_n \qquad \text{and} \qquad \left\| \nabla \widehat{H}_n(\xi) - \xi^* \right\|_{\mathbb{R}^{2N}} \leq b_n.$$

Let us set $\widehat{\varepsilon}_n \overset{df}{=} \min\{\varepsilon'_n, b_n\}$. Evidently $\widehat{\varepsilon}'_n \searrow 0$ and from the positive homogeneity of $\widehat{H}_n$, $\widehat{H}$, $\nabla\widehat{H}_n$, $\partial\widehat{H}$, we see that the conclusions of the Proposition hold. $\Box$

PROPOSITION 3.7.6
There exist constants $a, b, \widehat{a}, \widehat{a} > 0$, such that

(a) $a\,\|\xi\|^2_{\mathbb{R}^{2N}} \leq \widehat{H}(\xi) \leq b\,\|\xi\|^2_{\mathbb{R}^{2N}}$, $a\,\|\xi\|^2_{\mathbb{R}^{2N}} \leq \widehat{H}_n(\xi) \leq b\,\|\xi\|^2_{\mathbb{R}^{2N}}$ *for all $\xi \in \mathbb{R}^{2N}$ and all $n \geq 1$;*

(b) $\widehat{a}\,\|\xi\|_{\mathbb{R}^{2N}} \leq \|\xi^*\|_{\mathbb{R}^{2N}} \leq \widehat{b}\,\|\xi\|_{\mathbb{R}^{2N}}$, $\widehat{a}\,\|\xi\|_{\mathbb{R}^{2N}} \leq \|\nabla\widehat{H}_n(\xi)\| \leq \widehat{b}\,\|\xi\|_{\mathbb{R}^{2N}}$ *for all $\xi \in \mathbb{R}^{2N}$, all $\xi^* \in \partial\widehat{H}(\xi)$ and all $n \geq 1$.*

PROOF   We prove part (b) only. The proof of part (a) is similar (and in fact simpler).

Let $\delta \in (0, 1)$. We can find $\widehat{a}_1, \widehat{b}_1 > 0$, such that

$$\widehat{a}_1 \leq \|\xi^*\|_{\mathbb{R}^{2N}} \leq \widehat{b}_1 \qquad \forall\, z \in \overline{R}_{1-\delta, 1+\delta},\ \xi^* \in \partial\widehat{H}(\xi),$$

where $R_{a,b} \overset{df}{=} \{\xi \in \mathbb{R}^{2N} : a < \|\xi\|_{\mathbb{R}^{2N}} < b\}$. Since $\widehat{\varepsilon}_n \searrow 0$ (see Proposition 3.7.5), we may assume that $\widehat{\varepsilon}_n < \min\{\delta, \widehat{a}_1\}$ for $n \geq 1$. Let us set $\widehat{a} \overset{df}{=} \widehat{a}_1 - \widehat{\varepsilon}_1$ and $\widehat{b} \overset{df}{=} \widehat{b}_1 + \widehat{\varepsilon}_1$. Then, by virtue of Proposition 3.7.5, we have

$$\widehat{a} \leq \|\nabla\widehat{H}_n(\xi)\|_{\mathbb{R}^{2N}} \leq \widehat{b} \qquad \forall\, \xi \in S^{2N-1},\ n \geq 1.$$

Because of the homogeneity of $\nabla\widehat{H}_n$ and $\partial\widehat{H}$, we obtain (b). $\Box$

Now in addition to Hamiltonian inclusion (3.195), we consider the problem:

$$J\dot{x}(t) = \nabla\widehat{H}_n\big(x(t)\big) \qquad \forall\, t \in T,\ n \geq 1. \tag{3.196}$$

We shall obtain the conservative solutions of problem (3.195) from the "almost" solutions of the smooth Hamiltonian systems (3.196). The approach is variational. For this purpose let

$$S^1 \overset{df}{=} \{e^{i\vartheta} : \vartheta \in [0, 2\pi]\} = \mathbb{R}/2\pi\mathbb{Z}.$$

We introduce the potential functions $F_n, F \colon L^2(S^1; \mathbb{R}^{2N}) \longrightarrow \mathbb{R}$, defined by

$$F_n(x) \overset{df}{=} \frac{1}{2\pi} \int_0^{2\pi} \widehat{H}_n\big(x(t)\big)\, dt \qquad \forall\, x \in L^2(S^1; \mathbb{R}^{2N}),\ n \geq 1$$

$$F(x) \overset{df}{=} \frac{1}{2\pi} \int_0^{2\pi} \widehat{H}\big(x(t)\big)\, dt \qquad \forall\, x \in L^2(S^1; \mathbb{R}^{2N}).$$

We note that $F_n \in C\big(L^2(S^1;\mathbb{R}^{2N})\big)$, $F_n' : L^2(S^1;\mathbb{R}^{2N}) \longrightarrow L^2(S^1;\mathbb{R}^{2N})$ are locally Lipschitz and

$$\langle F_n'(x), y \rangle_2 \;=\; \frac{1}{2\pi} \int_0^{2\pi} \big(\nabla \widehat{H}_n(x(t)),\; y(t)\big)_{\mathbb{R}^{2N}} dt \qquad \forall\, x, y \in L^2(S^1;\mathbb{R}^{2N}).$$

Hereafter, by $\langle \cdot, \cdot \rangle_2$ we denote the inner product of the Hilbert space $L^2(S^1;\mathbb{R}^{2N})$ and by $\|\cdot\|_2$ the norm of $L^2(S^1;\mathbb{R}^{2N})$.

Also by virtue of Theorem 1.3.10, F is locally Lipschitz on $L^2(S^1;\mathbb{R}^{2N})$ (Lipschitz on every bounded set in $L^2(S^1;\mathbb{R}^{2N})$) and

$$\langle \partial F(x), y \rangle_2 \;=\; \frac{1}{2\pi} \int_0^{2\pi} \big(\partial \widehat{H}(x(t)),\; y(t)\big)_{\mathbb{R}^{2N}} dt \qquad \forall\, x, y \in L^2(S^1;\mathbb{R}^{2N})$$

(see Theorem 1.3.10). Note that F_n and F are positively homogeneous of degree 2. By Proposition 3.7.6, we have

$$\begin{cases} a\,\|x\|_2^2 \le F(x) \le b\,\|x\|_2^2 \\ a\,\|x\|_2^2 \le F_n(x) \le b\,\|x\|_2^2 \end{cases} \qquad \forall x \in L^2(S^1;\mathbb{R}^{2N}),\; n \ge 1.$$

Moreover, from Proposition 3.7.5, we have

$$\big|F_n(x) - F(x)\big| \;\le\; \varepsilon_n \,\|x\|_2^2 \qquad \forall\, x \in L^2(S^1;\mathbb{R}^{2N}),\; n \ge 1,$$

with $\varepsilon_n \searrow 0$.

We introduce the multivalued Nemytskii operator $N_{\partial \widehat{H}} : L^2(S^1;\mathbb{R}^{2N}) \longrightarrow P_{wkc}\big(L^2(S^1;\mathbb{R}^{2N})\big)$ corresponding to $\partial \widehat{H}$, defined by

$$N_{\partial \widehat{H}}(x) \;\overset{df}{=}\; S^2_{\partial \widehat{H}(x(\cdot))} \qquad \forall\, x \in L^2(S^1;\mathbb{R}^{2N}).$$

Recalling that $\partial \widehat{H}$ is upper semicontinuous, we can easily check that $N_{\partial \widehat{H}}$ is upper semicontinuous from $L^2(S^1;\mathbb{R}^{2N})$ with strong topology, into $L^2(S^1;\mathbb{R}^{2N})$ with weak topology. Note that F_n' and $N_{\partial \widehat{H}}$ are positively homogeneous of degree 1 and

$$\begin{cases} \langle F_n'(x), x \rangle_2 = 2F_n(x) \\ \langle u, x \rangle_2 = 2F(x) \end{cases} \qquad \forall\, x \in L^2(S^1;\mathbb{R}^{2N}),\; u \in N_{\partial \widehat{H}}(x)$$

(see Proposition 3.7.4(a)). Also by virtue of Proposition 3.7.6(b), we have

$$\begin{cases} \widehat{a}\,\|x\|_2 \le \|F_n'(x)\|_2 \le \widehat{b}\,\|x\|_2 \\ \widehat{a}\,\|x\|_2 \le \|u\|_2 \le \widehat{b}\,\|x\|_2 \end{cases} \qquad \forall\, x \in L^2(S^1;\mathbb{R}^{2N}),\; u \in N_{\partial \widehat{H}}(x),\; n \ge 1.$$

So, it follows that the sequence $\{F_n'(x)\}_{n \ge 1}$ is uniformly bounded on bounded subsets of $L^2(S^1;\mathbb{R}^{2N})$.

Now we introduce the multifunctions $K_n \colon \mathbb{R}^{2N} \longrightarrow 2^{\mathbb{R}^{2N} \times \mathbb{R}^{2N}}$, defined by

$$K_n(\xi) \stackrel{df}{=} \left\{ (\bar{\xi}, \bar{\xi}^*) \in \mathbb{R}^{2N} \times \mathbb{R}^{2N} \colon \; \left\| \zeta \quad \bar{\xi} \right\|_{\mathbb{R}^{2N}} \leq \varepsilon_n \|\xi\|_{\mathbb{R}^{2N}}, \bar{\xi}^* \in \partial \widehat{H}(\bar{\xi}), \right.$$

$$\left. \left\| \nabla \widehat{H}_n(\xi) - \bar{\xi}^* \right\|_{\mathbb{R}^{2N}} \leq \varepsilon_n \|\xi\|_{\mathbb{R}^{2N}} \right\} \qquad \forall \, \xi \in \mathbb{R}^{2N}, \; n \geq 1.$$

PROPOSITION 3.7.7
For each $n \geq 1$, K_n is $P_k\big(\mathbb{R}^{2N} \times \mathbb{R}^{2N}\big)$-valued and upper semicontinuous.

PROOF   From Proposition 3.7.5, we see that

$$K_n(\xi) \neq \emptyset \qquad \forall \, \xi \in \mathbb{R}^{2N}, \; n \geq 1.$$

The compactness of the values of K_n and its upper semicontinuity follow easily from the fact that ∂H is upper semicontinuous from $\mathbb{R}^{2N}$ into $P_{kc}\big(\mathbb{R}^{2N}\big)$ (see Proposition 1.3.9). $\qquad\qquad \Box$

PROPOSITION 3.7.8
For every $x \in L^2\big(S^1; \mathbb{R}^{2N}\big)$ and every $n \geq 1$, we can find $y \in L^2\big(S^1; \mathbb{R}^{2N}\big)$ and $u \in N_{\partial \widehat{H}}(y)$, such that

$$\|x - y\|_2 \leq \varepsilon_n \|x\|_2 \qquad \text{and} \qquad \left\| N_{\partial \widehat{H}}(x) - u \right\|_2 \leq \varepsilon_n \|x\|_2.$$

PROOF   Invoking Proposition 3.7.7 and the Yankov-von Neumann-Aumann Selection Theorem (see Theorem 1.2.3), we obtain $(y, u) \in S^2_{K_n(x(\cdot))}$. So we have

$$\|x - y\|_2 \leq \varepsilon_n \|x\|_2 \qquad \text{and} \qquad \left\| \nabla \widehat{H}_n(x) - u \right\|_2 \leq \varepsilon_n \|x\|_2,$$

with $u \in N_{\partial \widehat{H}}(y)$. $\qquad\qquad \Box$

PROPOSITION 3.7.9
<u>*If*</u> *$\{x_n\}_{n \geq 1} \subseteq L^2\big(S^1; \mathbb{R}^{2N}\big)$ is a sequence, such that*

$$x_n \longrightarrow x \quad \text{in } L^2\big(S^1; \mathbb{R}^{2N}\big)$$

for some $x \in L^2\big(S^1; \mathbb{R}^{2N}\big)$,
<u>*then*</u> *we can find a subsequence $\{x_{n_k}\}_{k \geq 1}$ of $\{x_n\}_{n \geq 1}$ and $u \in N_{\partial \widehat{H}}$, such that*

$$N_{\nabla \widehat{H}_{n_k}}\big(x_{n_k}\big) \xrightarrow{\; w \;} u \quad \text{in } L^2\big(S^1; \mathbb{R}^{2N}\big),$$

for some $u \in L^2\big(S^1; \mathbb{R}^{2N}\big)$ i.e. $w\text{-}\limsup\limits_{n \to +\infty} \big\{ N_{\nabla \widehat{H}_{n_k}}(x_n) \big\} \subseteq N_{\partial \widehat{H}}(x).$

PROOF From Proposition 3.7.8, we know that for every $n \geq 1$, we can find $y_n \in L^2(S^1; \mathbb{R}^{2N})$ and $u_n \in N_{\partial \widehat{H}}(y_n)$, such that

$$\|x_n - y_n\|_2 \leq \varepsilon_n \|x_n\|_2 \qquad \text{and} \qquad \|\nabla \widehat{H}_n(x) - u_n\|_2 \leq \varepsilon_n \|x_n\|_2 .$$

It follows that

$$y_n \longrightarrow x \quad \text{in } L^2(S^1; \mathbb{R}^{2N})$$

and

$$\|\nabla \widehat{H}_n(x_n) - u_n\|_2 \longrightarrow 0.$$

Note that $N_{\partial \widehat{H}}$ is a **bounded multifunction** (i.e. maps bounded sets into bounded sets). So it follows that the sequence $\{u_n\}_{n \geq 1} \subseteq L^2(S^1; \mathbb{R}^{2N})$ is bounded and so we can find a subsequence $\{u_{n_k}\}_{k \geq 1}$, such that

$$u_{n_k} \overset{w}{\longrightarrow} u \quad \text{in } L^2(S^1; \mathbb{R}^{2N}).$$

Recall that $\operatorname{Gr} N_{\partial \widehat{H}}$ is sequentially closed in $L^2(S^1; \mathbb{R}^{2N}) \times L^2(S^1; \mathbb{R}^{2N})_w$. Therefore $u \in N_{\partial \widehat{H}}(x)$ and this finishes the proof. $\quad\square$

Consider the densely defined self-adjoint linear differential operator $A \colon L^2(S^1; \mathbb{R}^{2N}) \supseteq \operatorname{Dom}(A) \longrightarrow L^2(S^1; \mathbb{R}^{2N})$, defined by

$$Ax \overset{df}{=} J\dot{x} \qquad \forall\, x \in \operatorname{Dom}(A),$$

where

$$\operatorname{Dom}(A) \overset{df}{=} W^{1,2}_{\mathrm{per}}\big((0, 2\pi); \mathbb{R}^{2N}\big) = W^{1,2}(S^1; \mathbb{R}^{2N}).$$

Note that $\sigma(A) = \mathbb{Z}$, $\ker A = \mathbb{R}^{2N}$ and the normalized eigenvectors corresponding to the eigenvalue $k \in \mathbb{Z}$ are $\varphi_{k_j}(t) = e^{ikt}\varepsilon_j$, where $\{\varepsilon_1, \dots, \varepsilon_{2N}\}$ is a canonical basis of $\mathbb{R}^{2N}$. Any $x \in L^2(S^1; \mathbb{R}^{2N})$ has the Fourier expansion

$$x(t) = \sum_{k \in \mathbb{Z}} x_k e^{ikt} = \sum_{\substack{k \in \mathbb{Z} \\ j \in \{1, \dots, 2N\}}} x_{kj} \varphi_{kj},$$

with $x_k \in \mathbb{R}^{2N}$, $x_{kj} \in \mathbb{R}$. Then, we set

$$Z \overset{df}{=} H^{\frac{1}{2}}(S^1; \mathbb{R}^{2N}) = \left\{ x \in L^2(S^1; \mathbb{R}^{2N}) : \sum_{k \in \mathbb{Z}} (1 + |k|)|x_k|^2 < +\infty \right\},$$

equipped with the inner product, given by

$$(x, y)_Z \overset{df}{=} \sum_{k \in \mathbb{Z}} (1 + |k|)\, (x_k, y_k)_{\mathbb{R}^{2N}} .$$

Let $i: Z \longrightarrow L^2(S^1; \mathbb{R}^{2N})$ be the embedding map. We know that it is compact and then so is its adjoint $i^*: L^2(S^1; \mathbb{R}^{2N}) \longrightarrow Z^*$. Let us set

$$G_n \overset{df}{=} F_n \circ i, \quad G \overset{df}{=} F \circ i, \quad E_n \overset{df}{=} i^* \circ F_n' \circ i, \quad \widehat{E} \overset{df}{=} i^* \circ N_{\partial \widehat{H}} \circ i.$$

We have

$$G_n(x) = \frac{1}{2\pi} \int_0^{2\pi} \widehat{H}_n(x(t)) \, dt \qquad \forall \, x \in Z, \ n \geq 1$$

$$G(x) = \frac{1}{2\pi} \int_0^{2\pi} \widehat{H}(x(t)) \, dt \qquad \forall \, x \in Z, \ n \geq 1$$

$$\langle E_n(x), v \rangle_Z = \frac{1}{2\pi} \int_0^{2\pi} \left(\nabla \widehat{H}_n(x(t)), v(t) \right)_{\mathbb{R}^{2N}} dt \qquad \forall \, x, v \in Z, \ n \geq 1$$

$$\langle u, v \rangle_Z = \frac{1}{2\pi} \int_0^{2\pi} \left(u(t), v(t) \right)_{\mathbb{R}^{2N}} dt \qquad \forall \, x, v \in Z, \ u \in \widehat{E}(x).$$

Exploiting the compactness of the embedding $Z \subseteq L^2(S^1; \mathbb{R}^{2N})$ and the previous propositions, we can state the following result.

PROPOSITION 3.7.10
We have

(a) $G_n, G: Z \longrightarrow \mathbb{R}$ *(for* $n \geq 1$*) are positively homogeneous of degree 2,* $G_n \in C^1(Z; \mathbb{R})$, $G_n': Z \longrightarrow Z^*$ *and* $G: Z \longrightarrow \mathbb{R}$ *are both locally Lipschitz maps and*

$$|G_n(x) - G(x)| \leq \varepsilon_n \|x\|_2^2 \qquad \forall \, x \in Z, \ n \geq 1;$$

(b) $E_n: Z \longrightarrow Z^*$ *for* $n \geq 1$ *are positively homogeneous of degree 1, compact and*

$$\langle E_n(x), x \rangle_Z = 2G_n(x) \qquad \forall \, x \in Z;$$

(c) $E: Z \longrightarrow P_{wkc}(Z)$ *is positively homogeneous of degree 1, it is upper semicontinuous from* Z *with the norm topology into* Z *with the weak topology and*

$$\langle u, x \rangle_Z = 2G(x) \qquad \forall \, x \in Z, \ u \in E(x);$$

(d) *for any sequence* $\{x_n\}_{n \geq 1} \subseteq Z$*, such that* $x_n \overset{w}{\longrightarrow} x$ *in* Z*, we can find a subsequence* $\{x_{n_k}\}_{k \geq 1}$ *and* $u \in \widehat{E}(x)$*, such that* $E_{n_k}(x_{n_k}) \longrightarrow u$ *in* Z^**;*

(e) *we have*

$$
\begin{aligned}
\widehat{a}\,\|x\|_2^2 &\leq G_n(x) \leq \widehat{b}\,\|x\|_2^2 && \forall\, x \in Z,\ n \geq 1, \\
\widehat{a}\,\|x\|_2^2 &\leq G(x) \leq \widehat{b}\,\|x\|_2^2 && \forall\, x \in Z, \\
\|E_n(x)\|_{Z^*} &\leq \widehat{b}\,\|x\|_2 \leq \widehat{b}\,\|x\|_Z && \forall\, x \in Z,\ n \geq 1, \\
\|u\|_{Z^*} &\leq \widehat{b}\,\|x\|_2 \leq \widehat{b}\,\|x\|_Z && \forall\, x \in Z,\ u \in \widehat{E}(x).
\end{aligned}
$$

Consider the continuous integral operator $Q\colon C^1\big(S^1;\mathbb{R}^{2N}\big) \longrightarrow \mathbb{R}$, defined by

$$
Q(x) \overset{df}{=} \frac{1}{2\pi} \int_0^{2\pi} \big(x_1(t), \dot{x}_2(t)\big)_{\mathbb{R}^N}\, dt \qquad \forall\, x = (x_1, x_2) \in C^1\big(S^1;\mathbb{R}^{2N}\big).
$$

We extend Q to Z (still denoted by Q). We have

$$
Q(x) = \frac{1}{2}\,\langle Lx, x\rangle_Z \qquad \forall\, x \in X,
$$

where $L \in \mathcal{L}(Z; Z^*)$ is bounded and self-adjoint. We decompose the space Z using L. So

$$
Z = Z^+ \oplus Z^- \oplus Z^0,
$$

where Z^+, Z^- and Z^0 are the positive, negative and null eigenspaces of L, respectively. Let us also set

$$
C_n \overset{df}{=} \{x \in Z : G_n(x) = 1\} \qquad \forall\, n \geq 1,
$$

$$
C \overset{df}{=} \{x \in Z : G(x) = 1\}.
$$

By virtue of Proposition 3.7.10(e), we see that

$$
\frac{1}{\widehat{b}} \leq \|x\|_2^2 \leq \frac{1}{\widehat{a}} \qquad \forall\, x \in C \cup \left(\bigcup_{n \geq 1} C_n\right). \tag{3.197}
$$

Because of the homogeneity of G_n and G, the sets C_n (for $n \geq 1$) and C are radially homeomorphic to the unit sphere in Z. Let $P_n\colon C_n \longrightarrow C$ for $n \geq 1$ be the radial projection.

PROPOSITION 3.7.11
For every $n \geq 1$, we have

(a) $\left|\dfrac{\|x\|_Z^2}{\|P_n(x)\|_Z^2} - 1\right| \leq \dfrac{\varepsilon_n}{\widehat{a}}$ *for all $x \in C_n$;*

(b) $\big|Q(x) - Q\big(P_n(x)\big)\big| \leq \dfrac{\varepsilon_n}{\widehat{a}}\,\big|Q\big(P_n(x)\big)\big|$ *for all $x \in C_n$.*

PROOF   (a) Let $x \in C_n$ and set $v \stackrel{df}{=} P_n(x)$ and $\lambda \stackrel{df}{=} \frac{\|x\|_Z}{\|v\|_Z}$. We have $x = \lambda v$ (recall that P_n is the radial projection of C_n on C). Then

$$0 = G_n(x) - G(v) = G_n(x) - G(x) + G(x) - G(v)$$

and so

$$\left| G_n(x) - G(x) \right| = \left| G(x) - G(v) \right|. \tag{3.198}$$

Note that

$$\left| G(x) - G(v) \right| = \left| G(\lambda v) - G(v) \right| = \left| \lambda^2 - 1 \right| G(v) = \left| \lambda^2 - 1 \right|, \tag{3.199}$$

while from Proposition 3.7.10(a) and (3.199), we have

$$\left| G_n(x) - G(x) \right| \leq \varepsilon_n \|x\|_2^2 \leq \frac{\varepsilon_n}{\widehat{a}}. \tag{3.200}$$

Combining (3.198), (3.199) and (3.200), we obtain that

$$\left| \lambda^2 - 1 \right| = \left| \frac{\|x\|^2}{\|P_n(x)\|_Z^2} - 1 \right| \leq \frac{\varepsilon_n}{\widehat{a}}.$$

(b) We have

$$\begin{aligned} \left| Q(x) - Q(P_n(x)) \right| &= \left| Q(x) - Q(v) \right| = \left| Q(\lambda v) - Q(v) \right| \\ &= \left| \lambda^2 - 1 \right| \left| Q(x) \right| \leq \frac{\varepsilon_n}{\widehat{a}} \left| Q(P_n(x)) \right|. \end{aligned}$$

$\square$

Consider the restriction of Q on the set S_n, denoted by $Q|_{S_n}$. We know that

$$\nabla \left(Q|_{C_n} \right)(x) = Lx - \frac{\left\langle Lx, \mathcal{F}^{-1}E_n(x) \right\rangle_Z}{\|E_n(x)\|_Z^2} E_n(x) \qquad \forall \, x \in C_n.$$

Here $\mathcal{F} \colon Z \longrightarrow Z^*$ is the duality map of Z^* (the canonical isomorphism since Z is a Hilbert Space). Similarly, for every $x \in C$, we have

$$\partial \left(Q|_C \right)(x) = \left\{ y \in Z^* : \; y = Lx - \frac{\left\langle Lx, \mathcal{F}^{-1}(u) \right\rangle_Z}{\|u\|_Z^2} u, \; u \in \widehat{E}(x) \right\}.$$

The next proposition establishes a kind of Palais-Smale condition for $Q|_{C_n}$ for $n \geq 1$.

PROPOSITION 3.7.12
If $x_n \in C_n$ for $n \geq 1$, $\{Q(x_n)\}_{n \geq 1}$ is bounded and

$$\nabla \left(Q|_{C_n} \right)(x_n) \longrightarrow 0 \quad \text{as } n \to +\infty,$$

then there is a subsequence $\{x_{n_k}\}_{k \geq 1}$ of $\{x_n\}_{n \geq 1}$, such that

$$x_{n_k} \longrightarrow x \quad \text{in } Z$$

for some $x \in C$ and $0 \in \partial(Q|_C)(x)$.

PROOF Let

$$x_n = x_n^+ + x_n^- + x_n^0 \qquad \forall\, n \geq 1,$$

with

$$x_n^+ \in Z^+,\ x_n^- \in Z^-, x_n^0 \in Z^0 \qquad \forall\, n \geq 1.$$

From (3.197), we have that the sequence $\{x_n\}_{n \geq 1} \subseteq L^2(S^1; \mathbb{R}^{2N})$ is bounded. Hence the sequence $\{x_n^0\}_{n \geq 1} \subseteq Z^0$ is L^2-bounded and because $\dim Z^0 < +\infty$, we may assume that the sequence $\{x_n^0\}_{n \geq 1}$ has a convergent subsequence. Also since by hypothesis the sequence $\{Q(x_n)\}_{n \geq 1}$ is bounded, we have

$$-c_1 + c_2 \left\|x_n^+\right\|_Z \ \leq\ \left\|x_n^-\right\|_Z \ \leq\ c_3 \left\|x_n^+\right\|_Z + c_4 \qquad \forall\, n \geq 1,$$

for some $c_1, c_2, c_3, c_4 > 0$. We introduce

$$y_n \overset{df}{=} \nabla(Q|_{C_n})(x_n) \ =\ Lx_n - \lambda_n E_n(x_n) \qquad \forall\, n \geq 1, \qquad (3.201)$$

where

$$\lambda_n \overset{df}{=} \frac{\langle Lx_n, \mathcal{F}^{-1}\big(E_n(x_n)\big)\rangle_Z}{\|E_n(x_n)\|_Z^2} \qquad \forall\, n \geq 1.$$

By hypothesis, we know that $y_n \longrightarrow 0$ in Z^*. Taking duality brackets with x_n of y_n (see (3.201)) and recalling that

$$\langle E_n(x_n), x_n \rangle_Z \ =\ 2G_n(x_n) \ =\ 2 \qquad \forall\, n \geq 1$$

(see Proposition 3.7.10(b) and recall that $G_n(z) = 1$ for all $z \in C_n$), we obtain

$$\langle y_n, x_n \rangle_Z \ =\ \langle Lx_n, x_n \rangle_Z - 2\lambda_n \qquad \forall\, n \geq 1$$

and so

$$|\lambda_n| \ \leq\ \frac{1}{2}|\langle Lx_n, x_n\rangle_Z| + \frac{1}{2}\|y_n\|_{Z^*}\|x_n\|_Z$$
$$\leq\ c_5 + c_6\|x_n\|_Z \qquad \forall\, n \geq 1, \quad (3.202)$$

for some $c_5, c_6 > 0$. Also taking duality brackets of y_n (see (3.201)) with x_n^+, we obtain

$$c_7\left\|x_n^+\right\|_Z^2 \ \leq\ \langle Lx_n^+, x_n^+\rangle_Z$$
$$\leq\ \|y_n\|_{Z^*}\left\|x_n^+\right\|_Z + |\lambda_n|\,\|F_n'(x_n)\|_2\left\|x_n^+\right\|_2 \qquad \forall\, n \geq 1,$$

for some $c_7 > 0$. Recall that since the sequence $\{x_n\}_{n \geq 1} \subseteq L^2\big(S^1; \mathbb{R}^{2N}\big)$ is bounded, so the sequence $\big\{F_n'(x_n)\big\}_{n \geq 1} \subseteq L^2\big(S^1; \mathbb{R}^{2N}\big)$ is bounded too. Using this and (3.202), we obtain that

$$\left\| x_n^+ \right\|_Z^2 \ \leq \ c_8 + c_9 \left\| x_n \right\|_Z \qquad \forall\, n \geq 1, \tag{3.203}$$

for some $c_8, c_9 > 0$. Similarly, we show that

$$\left\| x_n^- \right\|_Z^2 \ \leq \ c_{10} + c_{11} \left\| x_n \right\|_Z \qquad \forall\, n \geq 1, \tag{3.204}$$

for some $c_{10}, c_{11} > 0$. From (3.203) and (3.204) and the boundedness of the sequence $\big\{x_n^0\big\}_{n \geq 1} \subseteq Z^0$, it follows that

$$\left\| x_n \right\|_Z^2 \ \leq \ c_{12} + c_{13} \left\| x_n \right\|_Z \qquad \forall\, n \geq 1,$$

for some $c_{12}, c_{13} > 0$. So the sequence $\{x_n\}_{n \geq 1} \subseteq Z$ is bounded.

Hence, passing to a subsequence if necessary, we may assume that

$$x_n \ \xrightarrow{w} \ x \quad \text{in } Z.$$

By virtue of Proposition 3.7.10(d), we can find a subsequence $\{x_{n_k}\}_{k \geq 1}$ of $\{x_n\}_{n \geq 1}$ and $u \in \widehat{E}(x)$, such that

$$E_{n_k}(x_{n_k}) \ \longrightarrow \ u \quad \text{in } Z^*.$$

Note that

$$Lx_{n_k} \ \xrightarrow{w} \ Lx \quad \text{in } Z^*.$$

Since $\mathcal{F}^{-1}$ is continuous linear from Z^* into Z, it follows that

$$\lambda_{n_k} \ \longrightarrow \ \lambda \ \overset{df}{=} \ \frac{\big\langle Lx, \mathcal{F}^{-1}(u) \big\rangle_Z}{\|u\|_Z^2}.$$

From (3.201) and since

$$y_{n_k} \ \longrightarrow \ 0 \quad \text{in } Z^*,$$

we obtain

$$Lx_{n_k} \ \longrightarrow \ Lx \quad \text{in } Z.$$

Since $L^{-1}|_{Z^+}$ and $L^{-1}|_{Z^-}$ are well-defined compact linear operators, we infer that

$$x_{n_k} \ \longrightarrow \ x \quad \text{in } Z.$$

So in the limit, we obtain

$$Lx - u \ = \ 0.$$

Because $G(x) = \lim_{k \to +\infty} G_{n_k}(x_{n_k}) = 1$, we see that $x \in C$ and also $0 \in \partial(Q|_C)(x)$. $\square$

PROPOSITION 3.7.13

If $x_n \in C_n$ *for* $n \geq 1$,

$$x_n \longrightarrow x \quad \text{in } Z \qquad \text{and} \qquad \nabla(Q|_{C_n}) \longrightarrow 0 \quad \text{in } Z^*,$$

then

(a) $x \in H^1(S^1; \mathbb{R}^{2N})$;

(b) $x(t) \in S$ *for all* $t \in [0, 2\pi]$;

(c) $J\dot{x}(t) \in \lambda \partial H(x(t))$ *for almost all* $t \in T$, *with* $\lambda = Q(x) = \lim\limits_{n \to +\infty} Q(x_n)$.

PROOF By Proposition 3.7.12, $x \in C$ and $0 \in \partial(Q|_C)(x) = 0$. So we have

$$Lx \in \lambda \widehat{E}(x). \tag{3.205}$$

Taking duality brackets with $x \in C$ and using the fact that $\langle u, x \rangle_Z = 2$ for all $u \in \widehat{E}(x)$ (see Proposition 3.7.10(c)) and recall that $G(x) = 1$ since $x \in C$), we infer that

$$\lambda = \frac{1}{2} \langle Lx, x \rangle_Z = Q(x) = \lim_{n \to +\infty} Q(x_n).$$

From (3.205), it follows that $x \in W^{1,2}(S^1; \mathbb{R}^{2N})$ and satisfies $J\dot{x}(t) \in \lambda \partial H(x(t))$ for almost all $t \in [0, 2\pi]$.

We need to show that

$$x(t) \in S \qquad \forall\, t \in [0, 2\pi]. \tag{3.206}$$

If $\lambda = 0$, the conclusion is clear. So suppose that $\lambda \neq 0$. Let $z_n \overset{df}{=} Lx - \lambda E_n(x)$ for $n \geq 1$ and note that

$$z_n \longrightarrow 0 \quad \text{in } Z^*,$$

so

$$\langle z_n, v \rangle_Z = \frac{1}{2\pi} \int\limits_0^{2\pi} \left(J\dot{x}(t) - \lambda \nabla \widehat{H}_n(x(t)),\, v(t) \right)_{\mathbb{R}^{2N}} dt \qquad \forall\, v \in Z.$$

Thus

$$J\dot{x} - \lambda \nabla \widehat{H}_n(x) \overset{w}{\longrightarrow} 0 \quad \text{in } L^2(S^1; \mathbb{R}^{2N})$$

and

$$\int\limits_0^t \left(J\dot{x}(s) - \lambda \nabla \widehat{H}_n(x(s)),\, \dot{x}(s) \right)_{\mathbb{R}^{2N}} ds \longrightarrow 0 \qquad \forall\, t \in [0, 2\pi],$$

so

$$\int\limits_0^t \left(\nabla \widehat{H}_n\big(x(s)\big), \dot{x}(s) \right)_{\mathbb{R}^{2N}} ds \;\longrightarrow\; 0 \quad \text{as } n \to +\infty \qquad \forall\, t \in [0, 2\pi].$$

It follows that for any $t \in [0, 2\pi]$, we have

$$\widehat{H}_n\big(x(t)\big) - \widehat{H}_n\big(x(0)\big) \;=\; \int\limits_0^t \frac{d}{ds}\widehat{H}_n\big(x(s)\big)ds \;=\; \int\limits_0^t \left(\nabla \widehat{H}_n\big(x(s)\big), \dot{x}(s) \right)_{\mathbb{R}^{2N}} ds,$$

with the last term tending to zero as $n \to +\infty$, so

$$\widehat{H}\big(x(t)\big) \;=\; \widehat{H}\big(x(0)\big) \;=\; c \qquad \forall\, t \in [0, 2\pi].$$

Since $x \in C$, we have

$$1 \;=\; \frac{1}{2\pi}\int\limits_0^{2\pi} \widehat{H}\big(x(t)\big)\, dt \;=\; c$$

and so we obtain (3.206). $\square$

Now exploiting the homogeneity of $\partial\widehat{H}$ and Propositions 3.7.13 and 3.7.2(d), we have an existence theorem for periodic orbits of the nonsmooth Hamiltonian systems (3.193).

PROPOSITION 3.7.14
If $x \in W^{1,2}\big(S^1; \mathbb{R}^{2N}\big) = W^{1,2}_{\mathrm{per}}\big((0, 2\pi); \mathbb{R}^{2N}\big)$ *is a solution of*

$$J\dot{x}(t) \;\in\; \lambda\partial\widehat{H}\big(x(t)\big) \quad \text{for a.a. } t \in [0, 2\pi]$$

and $\lambda = Q(x) > 0$,
then $z(t) = x\left(\frac{t}{\lambda}\right)$ *is a* $2\pi\lambda$-*periodic solution of (3.193).*

To establish a general existence and multiplicity theorem for the nonsmooth Hamiltonian system (3.193), we need to develop an "index theory" with respect to an S^1-action, which has a nontrivial fixed point space. Then, we shall construct critical values using a minimax principal of the Lusternik-Schnirelmann type.

DEFINITION 3.7.2   *Elements of* S^1 *will be denoted either by* $\vartheta \in \mathbb{R}/2\pi\mathbb{Z}$ *or by* $e^{i\vartheta}$.

(a) *A **representation of the group** S^1 **over** Z is a family $\{T_\vartheta\}_{\vartheta \in S^1} \subseteq \mathcal{L}(Z)$, such that*

(i) $T_0 = \mathrm{id}_Z$;

(ii) $T_{\vartheta_1 + \vartheta_2} = T_{\vartheta_1} \circ T_{\vartheta_2}$;

(iii) *the function* $(\vartheta, z) \longmapsto T_\vartheta(z)$ *is continuous.*

(b) *A representation* $\{T_\vartheta\}_{\vartheta \in S^1} \subseteq \mathcal{L}(Z)$ *is* **isometric** *or* **unitary***, if*

$$\|T_\vartheta(z)\|_Z = \|z\|_Z \qquad \forall\, \vartheta \in S^1,\ z \in Z.$$

(c) *A set* $D \subseteq Z$ *is said to be* **invariant under the representation** $\{T_\vartheta\}_{\vartheta \in S^1} \subseteq \mathcal{L}(Z)$*, if*

$$T_\vartheta(D) = D \qquad \forall\, \vartheta \in S^1.$$

(d) *A function* $\varphi \colon Z \longrightarrow \mathbb{R}$ *is said to be* **invariant under the representation** $\{T_\vartheta\}_{\vartheta \in S^1} \subseteq \mathcal{L}(Z)$*, if*

$$\varphi(T_\vartheta z) = \varphi(z) \qquad \forall\, \vartheta \in S^1,\ z \in Z.$$

(e) *A representation* $\{R_\vartheta\}_{\vartheta \in S^1} \subseteq \mathcal{L}(Z)$ *of* S^1 *over* $\mathbb{R}^{2N}$ *(or an* S^1-*action on* $\mathbb{R}^{2N}$*) is said to be* **regular***, if it has a trivial fixed point space, i.e.*

$$\left[R_\vartheta x = x \text{ for all } \vartheta \in S^1\right] \implies x = 0.$$

(f) *A map* $\psi \colon Z \longrightarrow \mathbb{R}^{2N}$ *is said to bc* **equivariant** *with respect to the pair* (T, R)*, if*

$$\psi \circ T_\vartheta = R_\vartheta \circ \psi \qquad \forall\, \vartheta \in S^1.$$

(g) *By* $M_N(A; \mathbb{R})$ *we denote the space of all continuous maps* $\eta \colon A \longrightarrow \mathbb{R}^{2N} \setminus \{0\}$*, which are equivariant with respect to* (T, R)*.*

EXAMPLE 3.7.1   For a given $\alpha = \{\alpha_1, \ldots, \alpha_N\} \in \mathbb{Z}^N$ an example of a S^1-action on $\mathbb{R}^{2N}$ ($\simeq \mathbb{C}^N$) is given by

$$R_\vartheta^\alpha(x) \overset{df}{=} \left(e^{i\alpha_1 \vartheta} x_1, \ldots, e^{i\alpha_N \vartheta} x_N\right).$$

In this example R^α is regular if and only if $\alpha_1, \ldots, \alpha_N$ are all nonzero. Any S^1-action on $\mathbb{R}^{2N}$ is of the form R^α for some $\alpha = (\alpha_1, \ldots, \alpha_N)$ in some orthonormal basis (Peter-Weyl theorem). The set S^{2N-1} is invariant under any S^1-action. Also if R is an S^1-action on $\mathbb{R}^{2N}$, then we can find an invariant matrix $B \colon \mathbb{R}^{2N} \longrightarrow \mathbb{R}^{2N}$, such that $BR_\vartheta B^{-1}$ is an isometric S^1-action on $\mathbb{R}^{2N}$. Moreover, any isometric S^1-action on $\mathbb{R}^{2N}$, $\{R_\vartheta\}_{\vartheta \in S^1}$ has the matrix representation $R_\vartheta = \text{diag}\{M_1, \ldots, M_k\}$, where M_j is either of order 1 and $M_j = 1$ or is of order 2 and for some $n \geq 1$, we have

$$M_j = \begin{bmatrix} \cos n\vartheta & -\sin n\vartheta \\ \sin n\vartheta & \cos n\vartheta \end{bmatrix}.$$

Finally, if R is an isometric action on $\mathbb{R}^k$ and

$$\text{Fix}\,(R) \overset{df}{=} \{\xi \in \mathbb{R}^k : R_\vartheta \xi = \xi \text{ for all } \vartheta \in S^1\} = \{0\},$$

then k is even.   □

In our framework, there is a natural S^1-action on Z, namely the time shifts. So, we set

$$T_\vartheta(z) \overset{df}{=} z(\cdot + \vartheta) \qquad \forall\, \vartheta \in S^1,\ z \in Z.$$

Clearly C and Q introduced earlier are invariant under this action. By Z^0, we denote the space of fixed points of T, i.e.

$$Z^0 \overset{df}{=} \{z \in Z : T_\vartheta z = z \text{ for all } \vartheta \in S^1\}.$$

We define two classes of subsets of Z:

$$\mathcal{T} \overset{df}{=} \{A \subseteq Z : A \cap Z^0 = \emptyset,\ A \text{ is closed and } T\text{-invariant}\}$$

$$\mathcal{T}_c \overset{df}{=} \{A \in \mathcal{T} : A \text{ is compact}\}.$$

For these classes of subsets, we can have a ***geometrical S^1-index theory***.

DEFINITION 3.7.3   *For $A \in \mathcal{T}$, we define*

$$\gamma_b(A) \overset{df}{=} \inf\{k \in \mathbb{N} : \text{there exists a regular } S^1\text{-action } R \text{ on } \mathbb{R}^{2k}$$
$$\text{with } M_k(A; R) \neq \emptyset\}.$$

As usual $\gamma_b(A) = +\infty$ if $A \neq \emptyset$ and no such $k \in \mathbb{N}$ can be found and $\gamma_b(\emptyset) = 0$.

REMARK 3.7.3   First note the similarity with the Krasnoselskii genus (index) introduced in Section 1.5, where the compact group is $\{id, -id\}$. The quantity γ_b is a ***topological index*** and so it satisfies the following properties:

(a) $\gamma_b(A) \geq 0$ and $\gamma_b(A) = 0$ if and only if $A = \emptyset$ (definiteness);

(b) if $A \subseteq B$ then $\gamma_b(A) \leq \gamma_b(B)$ (monotonicity);

(c) $\gamma_b(A \cap B) \leq \gamma_b(A) + \gamma_b(B)$ (subadditivity);

(d) if $h \in C(Z; Z)$ and $h \circ T_\vartheta = T_\vartheta \circ h$ for all $\vartheta \in S^1$ (i.e. h is an equivariant map), then $\gamma_b(A) \leq \gamma_b\big(\overline{h(A)}\big)$ (supervariance);

(e) if $A \in \mathcal{T}_c$, then $\gamma_b(A) < +\infty$ and there exists a T-invariant neighbourhood U of A, such that $\gamma_b(\overline{U}) = \gamma_b(A)$ (continuity);

(f) if $z \notin Z^0$, then $\gamma_b\left(\bigcup_{\vartheta \in S^1} T_\vartheta z\right) = 1$ (normalization).

For details we refer to Benci (1981). ⬚

In addition to the index γ_b, we define a relative index. For this purpose we introduce the following class of subsets of Z:

$$\mathcal{F} \overset{df}{=} \{A \subseteq Z \setminus \{0\} : \ A \text{ is closed and } T\text{-invariant}\}.$$

Let $X \subseteq Z$ be a closed linear subspace invariant under the action, such that $Z^0 \subseteq X^\perp$. We set $Y = \big(Z^0 \oplus X\big)^\perp$ and denote by P_Y and P_{Z^0} the orthogonal projections onto Y and Z^0 respectively. Also $P_1 \overset{df}{=} \mathrm{P}_Y + \mathrm{P}_{Z^0}$.

DEFINITION 3.7.4 *Suppose that $A \in \mathcal{T}$ and let R be a regular S^1-action on $\mathbb{R}^{2N}$. By $D_N(A; R)$ we denote the set of all continuous maps $h \colon A \longrightarrow X^\perp \times \mathbb{R}^{2N}$, $h = (h_1, h_2)$ having the properties:*

(a) $(0,0) \notin h(A)$;

(b) *h is equivariant with respect to (T, R) in the following sense:*

$$h(T_\vartheta z) \ = \ \big(T_\vartheta h_1(z), \ R_\vartheta h_2(z)\big) \qquad \forall \, \vartheta \in S^1, \ z \in Z;$$

(c) *$h(z) = (z,0)$ for all $z \in A \cap Z^0$;*

(d) *$\mathrm{P}_Y h_1 = \mathrm{P}_Y + K$ with $K \colon A \longrightarrow Y$ compact.*

REMARK 3.7.4 If $X^\perp = Z^0$, then condition (d) of the above definition is automatically satisfied (with $K \equiv 0$). ⬚

DEFINITION 3.7.5 *Let $A \in \mathcal{F}$. The relative index of A with respect to X, denoted by $\gamma_{br}(A|X) = \gamma_{br}(A)$, is defined by*

$$\gamma_{br}(A) \overset{df}{=} \inf \big\{k \in \mathbb{N} : \text{ there exists a regular } S^1\text{-action on } \mathbb{R}^{2k}$$
$$\text{with } D_k(A; R) \neq \emptyset\big\} \qquad \forall \, A \in \mathcal{F}.$$

As usual $\gamma_{br}(A) = +\infty$, if $A \neq \emptyset$ and no such k can be found and $\gamma_{br}(\emptyset) = 0$.

Now returning to our problem and denoting by $P \colon Z \setminus \{0\} \longrightarrow \partial B_1$ the radial projection, we introduce the following items for $k, n \in \mathbb{N}$:

$$\Gamma_k(C_n) \stackrel{df}{=} \{B \in \mathcal{F} \colon B \subseteq C_n, \ \gamma_{br}(PB) \geq k\},$$

$$\Gamma_k(C) \stackrel{df}{=} \{B \in \mathcal{F} \colon B \subseteq C, \ \gamma_{br}(PB) \geq k\},$$

$$c_{k,n} \stackrel{df}{=} \inf_{B \in \Gamma_k(C_n)} \sup_{z \in B} Q(z),$$

$$c_k \stackrel{df}{=} \inf_{B \in \Gamma_k(C)} \sup_{z \in B} Q(z).$$

Since C and C_n for $n \geq 1$ are radially homeomorphic to the unit sphere in Z, we see that $B \in \Gamma_k(C_n)$ if and only if $P_n B \in \Gamma_k(C)$. Proposition 3.7.11 implies that

$$\lim_{n \to +\infty} c_{k,n} = c_k \qquad \forall \, k \geq 1.$$

Moreover, note that

$$\sup_{z \in B} Q(z) \ \geq \ \inf_{C \cap Z^+} Q(z) \ \geq \ \frac{\mu_1}{2(\mu_1 + 1)} \inf_{z \in C \cap Z^+} \|z\|_Z^2 \ > \ 0$$

(μ_1 being the first positive eigenvalue of A) and so $0 < c_1 \leq c_2 \leq \ldots$.

DEFINITION 3.7.6   *For any $c \in \mathbb{R}$, we define*

$$K_c \stackrel{df}{=} \Big\{x \in C \colon Q(x) = x, \ \text{there exist } x_{n_k} \in C_{n_k},$$

$$\text{such that } x_{n_k} \longrightarrow x \text{ and } \nabla\big(Q|_{C_{n_k}}\big)(x_{n_k}) \longrightarrow 0\Big\}.$$

From this definition and Proposition 3.7.11, we have:

PROPOSITION 3.7.15
$K_c \subseteq C$ *is compact.*

PROPOSITION 3.7.16
If U is an open neighbourhood of K_c,
then there exist $\delta, \varepsilon > 0$ and $n_0 \geq 1$, such that

$$\big\|\nabla(Q|_{C_n})(x)\big\|_{Z^*} \geq 0 \qquad \forall \, x \in (Q|_{C_n})^{-1}\big([c - \varepsilon, c + \varepsilon]\big) \setminus U, \ n \geq n_0.$$

PROOF   Suppose that the conclusion of the proposition is not true. Then, we can find $x_{n_k} \in C_{n_k}$, $x_{n_k} \notin U$, such that

$$Q(x_{n_k}) \longrightarrow c \qquad \text{and} \qquad \nabla(Q|_{C_{n_k}})(x_{n_k}) \longrightarrow 0.$$

By virtue of Proposition 3.7.11, the sequence $\{x_{n_k}\}_{k \geq 1}$ has a subsequence which converges to $x \in K_c \subseteq U$, a contradiction. $\square$

Using this proposition and arguing similarly as in the proof of Theorem 2.1.1, we obtain the following ***deformation-type result***. For detailed proof see Berestycki, Lasry, Mancini & Ruf (1985, p. 27).

PROPOSITION 3.7.17

If U is a S^1-invariant neighbourhood of K_c,
<u>*then*</u> *there exist $\varepsilon \in (0, \varepsilon_0)$ and $n_0 \geq 1$, such that for every $n \geq n_0$, there is an S^1-equivariant continuous map $\eta_n \colon Z \longrightarrow Z$ (i.e. $T_\vartheta \circ \eta_n = \eta_n \circ T_\vartheta$ for all $\vartheta \in S^1$), such that*

(a) *if $A \in \Gamma_k(C_n)$, then $\eta_n(A) \in \Gamma_k(C_n)$;*

(b) $\eta_n(x) = x$ *for all* $x \in \left(Q|_{C_n}\right)^{-1}\left((-\infty, c - \varepsilon_0] \cup [c + \varepsilon_0, +\infty)\right)$;

(c) $\eta_n\left(\left(Q|_{C_n}\right)^{-1}\left(-\infty, c + \varepsilon\right]\right) \setminus U\right) \subseteq \left(Q|_{C_n}\right)^{-1}\left((-\infty, c - \varepsilon]\right)$.

Using this deformation result, we can establish that $\{c_k\}_{k \geq 1}$ is a sequence of critical values of $Q|_C$ and estimate the γ_b-index of the set K_{c_k}.

PROPOSITION 3.7.18

We have

(a) $K_{c_k} \neq \emptyset$ *for each $k \in \mathbb{N}$;*

(b) <u>*If for some $k, m \geq 1$, we have $c_k = c_{k+1} = \ldots = c_{k+m-1} = c$,*</u> <u>*then*</u> $\gamma_b(K_c) \geq m$.

PROOF **(a)** Let

$$K_{c_{k,n}} \overset{df}{=} \{x \in C_n : Q(x) = c_{k,n} \text{ and } \nabla\left(Q|_{C_n}\right)(x) = 0\}.$$

Using standard arguments we show that $K_{c_{k,n}} \neq \emptyset$ for all $k, n \geq 1$ (see Berestycki, Lasry, Mancini & Ruf (1985, Theorem 8.8, p. 269)). Since $c_{k,n} \longrightarrow c_k$ as $n \to +\infty$, it follows that $K_{c_k} \neq \emptyset$ (see Proposition 3.7.12).

(b) Now suppose that for some $n, m \geq 1$, we have $c_n = c_{n_1} = \ldots = c_{n+m-1} = c$. From the compactness of K_c (see Proposition 3.7.15) and property (e) of the index γ_b (see Remark 3.7.3), we can find a S^1-invariant neighbourhood U of K_c, such that

$$\gamma_b\left(\overline{U}\right) = \gamma_b(K_c).$$

Invoking Proposition 3.7.17, we can find $\varepsilon \in (0, \varepsilon_0)$ and $n_0 \geq 1$ as postulated by that result. Because $c_{k,n} \longrightarrow c_k$ as $n \to +\infty$, we can take $n \geq n_0$ large such that

$$|c_{k+i,n} - c| < \varepsilon \qquad \forall\, i \in \{0, 1, \ldots, m-1\}.$$

Let η_n be the S^1-equivariant deformation postulated by Proposition 3.7.17 and let $A \subseteq \Gamma_{k+m-1}(C_n)$, such that $\sup Q(A) \leq c + \varepsilon$ (recall the definition of $c_{k+m-1} = c$). Then

$$\eta_n\left(\overline{A \setminus \overline{U}}\right) \subseteq \left(Q|_{C_n}\right)^{-1}\left((-\infty, c - \varepsilon_0]\right).$$

It follows that $\gamma_{br}\left(\eta_n\left(A \setminus \overline{U}\right)\right) < k$ and so $\gamma_{br}\left(\overline{A \setminus \overline{U}}\right) < k$ (see Remark 3.7.3). But we know that

$$\gamma_{br}(A) - \gamma_b\left(\overline{U}\right) \leq \gamma_{br}\left(\overline{A \setminus \overline{U}}\right)$$

(see Berestycki, Lasry, Mancini & Ruf (1985, p. 260)). So we obtain

$$\gamma_b\left(\overline{U}\right) \;\geq\; \gamma_{br}(A) - \gamma_{br}\left(\overline{A \setminus \overline{U}}\right) \;>\; k + m - 1 - k \;=\; m - 1$$

and so

$$\gamma_b(K_c) \;\geq\; m.$$

$\square$

Now let $\widehat{A} \colon L^2\left(S^1; \mathbb{R}^{2N}\right) \longrightarrow L^2\left(S^1; \mathbb{R}^{2N}\right)$ be defined by

$$\widehat{A}(x)(\cdot) \;=\; \Omega\left(x(\cdot)\right) \qquad \forall\, x \in L^2\left(S^1; \mathbb{R}^{2N}\right).$$

Also let $A \colon Z \longrightarrow Z^*$ be defined by $A \stackrel{df}{=} i^* \circ \widehat{A} \circ i$ and let $\ldots \leq \mu_{-2} \leq \mu_{-1} \leq \mu_0 = 0 \leq \mu_1 \leq \mu_2 \leq \ldots$ be the eigenvalues of $Lz = \mu Az$. In obtaining the multiplicity result for the periodic orbits of our nonsmooth Hamiltonian system, the relation between c_k and μ_k is crucial.

PROPOSITION 3.7.19
We have

$$\beta^2 \mu_k \;\leq\; c_k \;\leq\; \gamma^2 \mu_k \qquad \forall\, k \geq 1.$$

PROOF   Let

$$\widehat{C} \stackrel{df}{=} \left\{z \in Z \colon \frac{1}{2}\langle Az, z\rangle_Z = 1\right\}$$

and let $\widehat{P} \colon C \longrightarrow \widehat{C}$ be the radial projection onto $\widehat{C}$. Let $z \in C$ and $\lambda z \in \widehat{C}$ with $\lambda > 0$. We have

$$\frac{1}{\gamma} \;\leq\; \lambda \;\leq\; \frac{1}{\beta}.$$

Let $A \in \mathcal{F}$, $A \subseteq C$. If $z \in A$ and $\lambda z \in \widehat{P}(A)$, we have

$$Q(z) \;=\; \frac{1}{\lambda^2} Q(\lambda z) \;\leq\; \gamma^2 Q(\lambda z) \;\leq\; \gamma^2 \sup_{\widehat{P}(A)} Q$$

and so

$$\sup_A Q \le \gamma^2 \sup_{\widehat{P}(A)} Q. \tag{3.207}$$

Similarly, we also have that

$$\beta^2 \sup_{\widehat{P}(A)} Q \le \sup_A Q. \tag{3.208}$$

Now note that $A \in \Gamma_k(C)$ if and only if $\widehat{P}(A) \in \Gamma_k(\widehat{C})$. So from (3.207) and (3.208), it follows that

$$\beta^2 \mu_k \le c_k \le \gamma^2 \mu_k \qquad \forall\, k \ge 1.$$

$$\square$$

We shall also need some lower bounds for the periods of the solutions of some relevant differential inclusions. So let $b > 0$ and set $S_b = \mathbb{R}/_{b\mathbb{Z}}$. We consider $T = [0, b]$ and $L^2(T; \mathbb{R}^k)$ with the inner product $\langle x, y \rangle_2 \overset{df}{=} \int_0^b \big(x(t), y(t)\big)_{\mathbb{R}^k} dt$.

LEMMA 3.7.3

If $x \in H^1(S_b; \mathbb{R}^k)$ *and* $y \in L^2(T; \mathbb{R}^k)$ *with* $\int_0^b y(t)\, dt = 0$,

then $b\, \|\dot{x}\|_2 \, \|y\|_2 \ge 2\pi \big| \langle x, y \rangle_2 \big|$.

PROOF   From the Wirtinger inequality (see Remark 1.1.11), we have

$$2\pi\, \|y\|_2 \le b\, \|\dot{x}\|_2 .$$

Let $v \in \mathbb{R}^k$ be such that $\int_0^b \big(x(t) - v\big)\, dt = 0$. Then, we have

$$\big| \langle x, y \rangle_2 \big| \;=\; \big| \langle x - v, y \rangle_2 \big| \;\le\; \|x - v\|_2 \, \|y\|_2 \;\le\; \frac{b}{2\pi}\, \|\dot{x}\|_2 \, \|y\|_2 .$$

$$\square$$

PROPOSITION 3.7.20

If $V \colon \mathbb{R}^k \longrightarrow 2^{\mathbb{R}^k} \setminus \{\emptyset\}$ *is a multifunction,* $A \in \mathcal{L}(\mathbb{R}^k)$ *and*

$$(u, A\xi)_{\mathbb{R}^k} \;\ge\; \|u\|^2 \qquad \forall\, \xi \in \mathbb{R}^k,\; u \in V(\xi),$$

then if $x \in H^1(S_b; \mathbb{R}^k)$ *is a solution of*

$$\dot{x}(t) \in V\big(x(t)\big) \quad \text{for a.a. } t \in [0, b],$$

we have that $\|A - A^*\|_{\mathcal{L}} \geq \frac{4\pi}{b}$.

PROOF   Note that

$$\int_0^b \left(A^* x(t), \dot{x}(t)\right)_{\mathbb{R}^k} dt \; = \; \int_0^b \left(x(t), A\dot{x}(t)\right)_{\mathbb{R}^k} dt \; = \; -\int_0^b \left(\dot{x}(t), Ax(t)\right)_{\mathbb{R}^k} dt$$

and so

$$\left\langle \dot{x}, (A - A^*)x \right\rangle_2 \; = \; 2 \left\langle \dot{x}, Ax \right\rangle_2$$

Let $u \in S^2_{V(x(\cdot))}$ be such that

$$\dot{x}(t) \; = \; u(t) \; \in \; V\big(x(t)\big) \quad \text{for a.a. } t \in [0, b].$$

Using our hypothesis on A and V, we have

$$\left\langle \dot{x}, (A - A^*)x \right\rangle_2 \; = \; 2 \int_0^b \left(u(t), Ax(t)\right)_{\mathbb{R}^k} dt \; \geq \; 2 \|u\|_2^2 \; = \; 2 \|\dot{x}\|_2^2. \quad (3.209)$$

Because of Lemma 3.7.3, we obtain

$$b \|\dot{x}\|_2 \left\|(A - A^*)\dot{x}\right\|_2 \; \geq \; 2\pi \left|\left\langle \dot{x}, (A - A^*)x \right\rangle_2\right|,$$

so from (3.209), we have

$$b \|\dot{x}\|_2 \|A - A^*\|_{\mathcal{L}} \|\dot{x}\|_2 \; \geq \; 4\pi \|\dot{x}\|_2^2$$

and

$$b \|A - A^*\|_{\mathcal{L}} \; \geq \; 4\pi.$$

$\square$

PROPOSITION 3.7.21
If $H_1 : \mathbb{R}^{2N} \longrightarrow \mathbb{R}$ *is a locally Lipschitz function for which there exists* $\eta > 0$, *such that*

$$(u, \xi)_{\mathbb{R}^{2N}} \; \geq \; \eta \|\xi\|_{\mathbb{R}^{2N}}^2 \qquad \forall \, \xi \in \mathbb{R}^{2N}, \; u \in \partial H_1(\xi)$$

and $x \in H^1\big(S_b; \mathbb{R}^{2N}\big) = H^1_{\text{per}}\big(T; \mathbb{R}^{2N}\big)$ *is a nontrivial b-periodic solution of* $J\dot{x} \in \partial H_1(x)$,
then $b \geq 2\pi\eta$.

PROOF   Apply Proposition 3.7.20 with $V \stackrel{df}{=} -J\partial H_1$ and $A \stackrel{df}{=} -\frac{1}{\eta} J$. $\square$

PROPOSITION 3.7.22

If $H_1 \colon \mathbb{R}^{2N} \longrightarrow \mathbb{R}$ is a locally Lipschitz function, $H_1(x) > 0$ for all $x \neq 0$, $\overline{H}_1$ is positively homogeneous of degree 2 and there exists $\varrho > 0$, such that

$$(u, \xi)_{\mathbb{R}^{2N}} \geq \varrho \|v\|_{\mathbb{R}^{2N}} \qquad \forall\, \xi \in S_1 = H_1^{-1}(1),\ u, v \in \partial H(\xi),$$

then for every $x \in H^1\big(S_b; \mathbb{R}^{2N}\big) = H^1_{\mathrm{per}}\big(T; \mathbb{R}^{2N}\big)$ nontrivial b-periodic solution of $J\ddot{x} \in \partial H_1(x)$, we have $b \geq \pi \varrho^2$.

PROOF   By virtue of the homogeneity of H_1, we have

$$(u, \xi)_{\mathbb{R}^{2N}} = 2 \qquad \forall\, \xi \in S^1,\ u \in \partial H_1(\xi).$$

So, we have

$$\frac{\varrho}{2} \|u\|_{\mathbb{R}^{2N}} \leq 1 \qquad \forall\, \xi \in S^1,\ u \in \partial H_1(\xi).$$

Note that ∂H_1 is positively homogeneous of degree 1. So

$$(u, \xi)_{\mathbb{R}^{2N}} \geq \frac{\varrho^2}{2} \|v\|^2_{\mathbb{R}^{2N}} \qquad \forall\, \xi \in \mathbb{R}^{2N},\ u, v \in \partial H_1(\xi).$$

Hence, we can apply Proposition 3.7.21, with $\eta \overset{df}{=} \frac{\varrho^2}{2}$ and conclude that $b \geq \pi \varrho^2$. $\square$

We note that if $z \in Z$ is a critical point of $Q|_C$, then $z_k(t) \overset{df}{=} z(kt)$ for $k \geq 1$ are critical points with $Q(z_k) = kQ(z)$. But all these points z_k give rise to the same orbit on the prescribed energy surface $S = \widehat{H}^{-1}(1)$. Moreover, if $\widehat{z}$ is a critical point with minimal period $\frac{2\pi}{k}$, then $z(t) = \widehat{z}\big(\frac{t}{k}\big)$ is a critical point having minimal period 2π and $\widehat{z} = z_k$. We call z the ***primitive critical point*** corresponding to $\widehat{z}$. Let $l \in \mathbb{N}$ be the number of equivalence classes of the set $\{\omega_1, \ldots, \omega_N\}$ in $\mathbb{R}/_{\mathbb{Q}}$. Thus, relabelling ω_k, we assume that elements $\omega_1, \ldots, \omega_N$ are ordered as follows

$$\omega_1^1, \ldots, \omega_{p_1}^1, \omega_1^2, \ldots, \omega_{p_2}^2,\ \ldots\ , \omega_1^l, \ldots, \omega_{p_l}^l,$$

with $\omega_j^i \in \mathbb{R}^+$, $p_1 + \ldots + p_l = N$ and $\omega_j^i = n_j^i \omega^i$, $n_j^i \in \mathbb{N}$ for $j \in \{1, \ldots, p_l\}$, where ω^i is defined as the largest positive real number satisfying the last equality (i.e. ω^i is the largest common integral divider of the $\{\omega_1^i, \ldots, \omega_{p_i}^i\}$). Note that $\frac{\omega^i}{\omega^j} \notin \mathbb{Q}$ for all $i \neq j$. Also, let

$$\delta_1 \overset{df}{=} \min_{i \in \{1, \ldots, l\}} \frac{\omega^i \varrho^2}{2\beta^2},$$

$$\delta_2 \overset{df}{=} \min \left\{ \frac{n\omega^i}{m\omega^j} - 1 :\ 1 \leq n, m < 1 + \frac{1}{\delta_1},\ 1 \leq i \neq j \leq l,\ n\omega^i > m\omega^j \right\}.$$

Because $\frac{\omega^i}{\omega^j} \notin \mathbb{Q}$ for all $i \neq j$, we see that $\delta_2 > 0$. Then, we define

$$\delta \overset{df}{=} \min\{\delta_1, \delta_2\} > 0.$$

Consider the critical levels of $Q|_{\widehat{C}}$ given by $\left\{\frac{\beta^2}{\omega^1}, \ldots, \frac{\beta^2}{\omega^l}\right\}$. For the positive eigenvalues $\{\mu_k\}_{k \geq 1}$, we have that

$$\mu_k = \inf_{B \in \Gamma_k(\widehat{C})} \sup_{z \in B} Q(z) \qquad \forall\, k \geq 1$$

(see Berestycki, Lasry, Mancini & Ruf (1985, p. 267)). So m_i critical levels of minimax type coincide with $\frac{\beta^2}{\omega^i}$ for $i \in \{1, \ldots, l\}$. Then by virtue of Proposition 3.7.19, there are m_1 critical levels of $Q|_C$ in $I_i \overset{df}{=} \left[\frac{\beta^2}{\omega^i}, \frac{\gamma^2}{\omega^i}\right]$ for $i \in \{1, \ldots, l\}$. Let

$$U_i \overset{df}{=} \big\{z \in C : z \text{ is a primitive critical point}$$
$$\text{of } Q|_C, \ z(jt) \in I_i \text{ for some } j \in \mathbb{N}.\big\}$$

LEMMA 3.7.4
If $\left(\frac{\gamma}{\beta}\right)^2 < 1 + \delta$, *then* $\operatorname{card} U_i \geq m_i$ *for* $i \in \{1, \ldots, l\}$.

PROOF   By virtue of Proposition 3.7.18(b), if two or more critical levels coincide, then the corresponding critical set has γ_b-index greater or equal 2. So there are infinitely many distinct closed (i.e. periodic) orbits on S. So, we may assume that there are m_i distinct critical levels in I_i. Let $z_1, \ldots, z_{m_i}$ be the corresponding primitive critical points. We have

$$z_j(t) = \widehat{z}_j\left(\frac{t}{h_j}\right) \qquad \forall\, j \in \{1, \ldots, m_i\}.$$

Recall that to z_j corresponds a $\frac{2\pi}{h_j} Q(\widehat{z}_j)$-periodic solution of (3.193) (see Proposition 3.7.14) and so by Proposition 3.7.22, we have

$$\frac{1}{h_j} Q(\widehat{z}_j) \geq \frac{\varrho^2}{2} \qquad \forall\, j \in \{1, \ldots, m_j\}.$$

Let $c \overset{df}{=} Q(\widehat{z}_j)$. Since $\frac{2\beta^2}{\omega_j \varrho^2} \leq \frac{1}{\delta_1}$, we have

$$\frac{\beta^2}{2} \varrho^2 h_j \leq c \leq \frac{\gamma^2}{\omega_j} < \frac{\beta^2}{2} \varrho^2 \left(1 + \frac{1}{\delta_1}\right),$$

hence $h_j < \frac{1}{\delta_1} + 1$. Now, we show that the primitive critical points $\{z_1, \ldots, z_{m_i}\}$ are all distinct. Indeed, if this is not the case, we would have $w = z_j = z_r$ for some $1 \leq j < r \leq m_i$, i.e.

$$w(h_j t) = \widehat{z}_j(t) \qquad \text{and} \qquad w(h_r t) = \widehat{z}_r(t),$$

for some $h_j < h_r < \frac{1}{\delta_1} + 1$. So

$$\frac{1}{\omega_j} \leq Q(\widehat{z}_j) = h_j Q(w) = \frac{h_j}{h_r} Q(\widehat{z}_r) \leq \frac{h_j \gamma^2}{h_r \beta^2 \omega_j}$$

and so

$$\frac{\gamma^2}{\beta^2} \geq \frac{h_r}{h_j} \geq \frac{h_r}{h_{r-1}} > \frac{\frac{1}{\delta_1} + 1}{\frac{1}{\delta_1}} = 1 + \delta_1,$$

a contradiction to our assumption. ☐

We have found l sets $U_1, \ldots, U_l$ of primitive critical points corresponding to the l families $\omega_1, \ldots, \omega_l$. Because $\sum\limits_{i=1}^{l} m_i = N$, the next lemma will lead us to the multiplicity theorem.

LEMMA 3.7.5
We have
$$U_i \cap U_j = \emptyset \qquad \forall\, i \neq j.$$

PROOF Suppose that the lemma is not true and let $w \in U_i \cap U_j$ for some $i \neq j$. There exist $\widehat{z} \in Q^{-1}(I_i)$ and $\widehat{y} \in Q^{-1}(I_j)$, such that

$$w(t) = \widehat{z}\left(\frac{t}{h}\right) = \widehat{y}\left(\frac{t}{k}\right),$$

for some $h, k < \frac{1}{\delta_1} + 1$ (see the proof of Lemma 3.7.4). Recall that

$$Q(w) = \frac{1}{h} Q(\widehat{z}) \in \frac{1}{h} I_i \qquad \text{and} \qquad Q(w) = \frac{1}{h} Q(\widehat{y}) \in \frac{1}{k} I_j.$$

So, we have $\frac{1}{h} I_i \cap \frac{1}{k} I_j \neq \emptyset$. On the other hand, assuming $\frac{k}{h} \frac{\omega^j}{\omega^i} > 1$ and using the inequality $\frac{\gamma^2}{\beta^2} < 1 + \delta$, we obtain

$$\frac{k\omega^j}{h\omega^i} \geq 1 + \delta_2 \geq \frac{\gamma^2}{\beta^2}.$$

From this it follows that

$$\frac{1}{k}\left[\frac{\beta^2}{\omega^j}, \frac{\gamma^2}{\omega^j}\right] \cap \frac{1}{h}\left[\frac{\beta^2}{\omega^i}, \frac{\gamma^2}{\omega^i}\right] = \emptyset,$$

a contradiction. ☐

So now we are ready for the multiplicity result for periodic orbits of the nonsmooth Hamiltonian system (3.193).

THEOREM 3.7.2
_If_ hypotheses $H(H)$ hold,
then there is a constant $\delta > 0$, such that $\frac{\gamma^2}{\beta^2} < 1 + \delta$ and (3.193) has at least N periodic orbits on $S = H^{-1}(1)$.

3.8   Remarks

3.1 During the last decade a great deal of attention was given to various separated two-point boundary value problems involving the so-called ordinary p-Laplacian. The Dirichlet problem, which is of course easier to deal with, was examined by many authors. Indicatively, we mention the works of Addou (1999), Boccardo, Drábek, Giachetti & Kučera (1986), De Coster (1995), del Pino, Elgueta & Manásevich (1988, 1991), del Pino & Manásevich (1991), Drábek & Robinson (1999), Manásevich & Mawhin (2000), Mawhin (2000), Mawhin & Ureña (2002), Ôtani (1984a), Wang, Gao & Lin (1995), Yang & Guo (1995), and Zhong (1997). Besides existence theorems, in the aforementioned works, we find multiplicity results (see De Coster (1995)), results on the existence of positive solutions (see De Coster (1995), Yang & Guo (1995)), eigenvalue problems (see Addou (1999), Boccardo, Drábek, Giachetti & Kučera (1986), del Pino, Elgueta & Manásevich (1988), Ôtani (1984a)), comparison results and Sturmian theory (see del Pino, Elgueta & Manásevich (1991), del Pino & Manásevich (1991)), problems with right-hand side nonlinearities depending also on x' (see Manásevich & Mawhin (2000), Mawhin & Ureña (2002)). Soon thereafter the p-Laplacian map $\varphi_p(x) = \left(\|x'(\cdot)\|_{\mathbb{R}^N}^p x'(\cdot)\right)'$ was replaced by a one-dimensional possibly nonhomogeneous operator φ. We refer to the works of Dang & Oppenheimer (1996), Guo (1993), García Huidobro, Manásevich & Zanolin (1993), Manásevich & Mawhin (2000), Mawhin (2000), O'Regan (1993) and Ubilla (1995a). Second order multivalued Dirichlet problems were first studied by Erbe & Krawcewicz (1991a, 1991b), Frigon (1990, 1991, 1995), Halidias & Papageorgiou (1998a, 2000a), Kandilakis & Papageorgiou (1996), Pruszko (1984) (semilinear problems, i.e. $p = 2$, with no maximal monotone term A) and Matzakos & Papageorgiou (2001) (nonlinear problems with an operator A of the subdifferential type, F independent of x' but a depending also on x).

3.2 The periodic problem presents more difficulties. The kernel of the differential operator is nontrivial and the Poincaré inequality is no longer true. We only have the Poincaré-Wirtinger inequality valid in the topological complement of the kernel of the p-Laplacian. The main contributions in this direction were made by Mawhin. More precisely, the following papers deal with the periodic problem using a variety of tools: Dang & Oppenheimer (1996), del Pino, Manásevich & Murúa (1992), Fabry & Fayyad (1992), Guo (1993), Gasiński & Papageorgiou (2002*a*, 2003*b*), Kyritsi, Matzakos & Papageorgiou (2002), Manásevich & Mawhin (1998), Mawhin (1979, 1993, 1995, 2000, 2001), Mawhin & Ureña (2002) and Papageorgiou & Papageorgiou (to appear). In these works the approach is degree theoretical and only in Gasiński & Papageorgiou (2002*a*, 2003*b*) and Papageorgiou & Papageorgiou (to appear), the solution method is variational based on the nonsmooth critical point theory since the potential function is only locally Lipschitz in $\xi \in \mathbb{R}^N$ and not C^1. Multiplicity results can be found in the works of del Pino, Manásevich & Murúa (1992), Gasiński & Papageorgiou (2002*a*, 2003*b*). Eigenvalue problems and a comprehensive study of the spectral properties of the ordinary p-Laplacian with periodic boundary conditions can be found in the paper of Mawhin (2001). Periodic problems with the nonlinearity satisfying Hartman and Nagumo-Hartman conditions were investigated by Mawhin (2000) and Mawhin & Ureña (2002) respectively. Extensions to multivalued problems with a maximal monotone term A were obtained by Gasiński & Papageorgiou (2003*a*, 2003*b*), Papageorgiou & Papageorgiou (2004*a*, submitted). Our presentation in this section is primarily based on the paper of Papageorgiou & Papageorgiou (submitted).

3.3 Nonlinear boundary conditions were first considered in the context of semilinear (i.e. $p = 2$) problems. We refer to the works of Thompson (1996) (single-valued problems) and Erbe & Krawcewicz (1991*b*), Halidias & Papageorgiou (1998*a*, 2000*a*), Kandilakis & Papageorgiou (1996), Palmucci & Papalini (2002) (multivalued problems). Nonlinear problems involving the p-Laplacian or generalizations of it can be found in the works of Gasiński & Papageorgiou (2003*c*), Halidias & Papageorgiou (2000*a*), Kourogenis (2002), Matzakos & Papageorgiou (2001), Papageorgiou & Papageorgiou (2004*a*, submitted), Papageorgiou & Yannakakis (1999).

3.4 The periodic problem has been approached using primarily degree theoretic methods (see e.g. the works of Manásevich & Mawhin (1998, 2000), Mawhin (2000, 2001) and the references therein). Variational techniques can be found in the works of Ahmad & Lazer (1984), Fonda & Lupo (1989), Gasiński & Papageorgiou (2002*a*, 2003*b*), Ma & Tang (2002), Mawhin (2000), Mawhin & Willem (1989), Papageorgiou & Papageorgiou (to appear), Papageorgiou & Yannakakis (1999), Tang (1998*a*, 1998*b*, 1999), Tang & Wu (2001*b*), Wu & Tang (1999). From these works, only Gasiński & Papageorgiou (2002*a*, 2003*b*) and Papageorgiou & Papageorgiou (to appear) deal with problems driven by the ordinary p-Laplacian $(p > 1)$ and having a nonsmooth potential. The rest consider the semilinear problem (i.e. $p = 2$) and assume that the potential is a C^1-function. Subsection 3.4.1 is primary based on the work of Papageorgiou & Papageorgiou (to appear). The results of Subsection 3.4.2 on the existence of homoclinic to zero solutions appear to be new. In the past the issue of homoclinic solutions was addressed in the context of semilinear problems. The only nonlinear work in this direction is that of Ubilla (1995b). Our proof is modelled after the approach of Rabinowitz (1990) who was the first to use variational methods for homoclinic solutions. Recently Hu & Papageorgiou (submitted) obtained positive homoclinic solutions for a nonlinear periodic problem with nonsmooth potential, extending and improving an earlier work of Korman & Lazer (1994). Similarly the results of Subsection 3.4.3 are also new. The Landesman-Lazer type condition $(LL)_1$ was used by Tang (1998c) (in a smooth context, i.e. $G_1 = G_2$), while condition $(LL)_2$ can be found in Goeleven, Motreanu & Panagiotopoulos (1998) and is a natural extension to the present nonsmooth (hence multivalued) setting of the classical Landesman-Lazer condition (see Landesman & Lazer (1969/1970), Lazer & Leach (1969) and Mawhin & Willem (1989)). The results of Subsection 3.4.4 on the existence of multiple periodic solutions can be found in Gasiński & Papageorgiou (2003c). Additional multiplicity results for nonlinear nonsmooth problems can be found in Gasiński & Papageorgiou (2002a), Papageorgiou & Papageorgiou (to appear), Papageorgiou & Yannakakis (2004). Another work on multiplicity periodic solutions for problems involving the ordinary p-Laplacian but with a smooth potential is that of del Pino, Manásevich & Murúa (1992). Their approach uses degree theory and the right hand side nonlinearity $f(t, \xi)$ is jointly continuous. Their conditions on f also require that asymptotically there is no interaction between the nonlinearity and the Fučik spectrum of the ordinary scalar p-Laplacian. Subsection 3.4.5 extends a corresponding semilinear, smooth result of Mawhin & Schmitt (1990). Subsection 3.4.6 is an application of the abstract theory developed by Kourogenis, Papadrianos & Papageorgiou (2002). Finally Subsection 3.4.7 is based on the paper of Papageorgiou & Papageorgiou (submitted) and presents a new use of the "Second Deformation Theorem". It remains an interesting open question whether this result can be extended to "nonsmooth" problems.

3.5 The method of upper-lower solutions was first introduced by Perron (1923), who used the so-called "sub-harmonic functions" in potential theory. Later Nagumo (1942) used upper and lower solutions to study second order differential equations with Dirichlet boundary conditions. Since then many authors have used this method primarily in the context of single-valued differential equations with linear boundary conditions (Dirichlet, Neumann, Sturm-Liouville or periodic). We refer to the books of Gaines & Mawhin (1977), Heikkilä & Lakshmikantham (1994), Kuzin & Pohozaev (1997) and the papers of Cabada & Nieto (1990), Erbe (1982), Fabry & Habets (1986), Frigon (1990), Gao & Wang (1995), Granas, Guenther & Lee (1985), Nieto (1989), Omari (1986), Papageorgiou & Papalini (2001), Rachunková (1999) and Wang, Cabada & Nieto (1993). The method was extended to multivalued problems by Frigon (1991, 1995), Frigon & Granas (1991), Halidias & Papageorgiou (1998c), Palmucci & Papalini (2002) (semilinear problems) and Bader & Papageorgiou (2002) and Douka & Papageorgiou (submitted). Our presentation here is based on the work of Douka & Papageorgiou (submitted).

3.6 The use of fixed point theorems on cones (such as Proposition 3.6.1) in order to obtain solutions of constant sign can be traced in the study of radial solutions for semilinear elliptic equations on an annulus. We refer to the papers of Garaizar (1987), Wang (1994) and the references therein. Earlier approaches to the subject involved the shooting method, see Bandle, Coffman & Marcus (1987), Coffman & Marcus (1989) and the book of Kuzin & Pohozaev (1997). The fixed point method was then adopted to deal with the problem of existence of multiple positive solutions for two-point boundary value problems. We mention the works of Erbe, Hu & Wang (1994), Erbe & Wang (1994), Fink & Gatica (1993), Ma (2000) (semilinear problems) and Agarwal, Lü & O'Regan (2002), Wang (2003) (nonlinear problems). The method was extended to differential inclusions by Erbe & Krawcewicz (1991b), who for this purpose established the compression-expansion type fixed point result for multifunctions on cone, stated in Proposition 3.6.1. Analogous single-valued results can be found in the books of Denkowski, Migórski & Papageorgiou (2003a) and Guo & Lakshmikantham (1988). Our presentation here is based on the paper of Filippakis, Gasiński & Papageorgiou (to appear b), which improves the work of Erbe & Krawcewicz (1991b). The second method presented in this section is based on Proposition 3.6.2, which is a range result for sums of nonlinear operators of monotone type and is due to Gupta & Hess (1976). An analogous method based on accretive operators can be found in Calvert & Gupta (1978) and Li & Zhen (1995). Our presentation here is based on the work of Kourogenis & Papageorgiou (1998b).

3.7 The problem of existence and multiplicity of periodic trajectories of a Hamiltonian system on a given energy surface has attracted the interest of many mathematicians. Rabinowitz (1978a) and Weinstein (1978) proved the existence of at least one periodic trajectory under certain smoothness and convexity conditions on the Hamiltonian function H. Ekeland & Lasry (1980) proved a global multiplicity result under the assumption that $\Omega' = I$ and the sublevel sets of H are convex. Their result was improved by Ambrosetti & Mancini (1982). These works were eventually extended and unified by Berestycki, Lasry, Mancini & Ruf (1985), who assumed more general geometric conditions on the energy surface $\Sigma = \{\xi \in \mathbb{R}^{2N} : H(\xi) = c\}$. In all these works H is at least C^2. Fan (1992) extended the work of Berestycki, Lasry, Mancini & Ruf (1985), by assuming that H is only locally Lipschitz (hence the Hamiltonian equation becomes a Hamiltonian inclusion). The approach of Fan is based on the approximation result presented in Theorem 3.7.1. This result was actually used earlier by Barbu (1984) to obtain necessary conditions for nonsmooth optimal control problems. The result proved also to be useful in other contexts such as obtaining solutions for first order systems using nonsmooth guiding functions (see Filippakis, Gasiński & Papageorgiou (to appear a)). Our presentation is based on the work of Fan (1992).

Chapter 4

Elliptic Equations

This chapter is devoted to the study of nonlinear elliptic equations. We concentrate on problems in which the potential function is not necessarily smooth. This introduces a multivalued feature in the problem and requires a different approach based on new analytical tools. Prominent among them is the nonsmooth critical point theory (see Chapter 2) and the methods and techniques of multivalued analysis, nonsmooth analysis and nonlinear operator theory (see Chapter 1). We examine different methods for the analysis of nonlinear elliptic equations, such as the variational method (i.e. the energy method within Sobolev spaces), the method of upper and lower solutions, the method of nonlinear operators of monotone type and the maximum principle method to produce solutions of constant sign. In Section 4.1 we consider problems at resonance. Such problems were first studied by Landesman-Lazer for equations driven by the Laplace differential operator and with a smooth potential. They introduced a "classification" of resonant problems depending on the rate of growth of the right hand side nonlinearity and a sufficient (and sometimes also necessary) condition involving the asymptotic values at $\pm\infty$ of the nonlinearity and which since then is known as "(LL)-condition". Here we introduce a new (LL)-type condition which we show generalizes all previous ones existing in the literature. Using a variational approach and the generalized (LL)-condition, we solve a resonant elliptic equation involving the Laplace and a nonsmooth potential. Then we pass to resonant nonsmooth problems driven by the p-Laplacian and allow the nonsmooth potential asymptotically at $\pm\infty$ to cross in a certain sense $\lambda_1 > 0$, the principal eigenvalue of $\left(-\Delta_p, W_0^{1,p}(\Omega) \right)$. This, combined with the lack of full knowledge of the spectrum of this operator, creates serious difficulties, which require new tools based on the Lusternik-Schnirelmann theory. In the last part of this section we examine a variational-hemivariational inequality, in particular an obstacle problem with a nonsmooth potential. Using the nonsmooth critical point theory for convex perturbations of locally Lipschitz functions and (LL)-conditions, we prove the existence of solutions. In Section 4.2 we study Neumann problems for equations driven by the p-Laplacian. In the first part of the section we discuss the spectrum of $\left(-\Delta_p, W^{1,p}(\Omega) \right)$, while in the second part we study homogeneous and nonhomogeneous Neumann problems with nonsmooth potential. Our analysis is based on variational methods which employ the nonsmooth critical point theory. In Section 4.3 we examine a

generalized Dirichlet problem with an area-type term. The appropriate space
for the analysis of this problem is the space of functions of bounded variation
on Ω, which we examine in the first part of the section. In addition for the
needs of the variational problem, we developed an equivariant version of the
critical point theory in Section 2.5. In Section 4.4, we study nonlinear elliptic
equations which are not in variational form. So our approach changes and
uses the theory of nonlinear operators of monotone type and the multivalued
Leray-Schauder alternative principle. In Section 4.5 we continue with nonlinear problems of nonvariational nature. Now the problems are approached
using the method of upper and lower solutions. The presence of multivalued terms in the equations complicates things and requires a more involved
argument. We also prove a theorem establishing the existence of extremal solutions in the order interval determined by the ordered pair of upper and lower
solutions. In Section 4.6 we look for multiple solutions for elliptic problems.
We start with semilinear equations with a nonsmooth potential and resonant
at higher eigenvalues. Our hypotheses allow for "double-double resonance"
(at 0 and at $\pm\infty$) and our approach is based on a nonsmooth variant of the
"reduction method", which we develop. Then we consider nonlinear equations
involving the p-Laplacian. To deal with this case, we prove a result relating
local minimizers in $C_0^1(\overline{\Omega})$ and in $W_0^{1,p}(\Omega)$ for a nonsmooth functional. In our
analysis we also use maximum principle techniques. In Section 4.7 we look
for solutions of constant sign (in particular positive solutions) of equations
driven by the p-Laplacian and having a nonsmooth potential. Using a combination of the method of upper-lower solutions with variational techniques,
we prove multiplicity results. Finally, in Section 4.8 we examine nonlinear
equations with a discontinuous nonlinearity. We insist on the single-valued
interpretation of the problem and prove existence and multiplicity results.

4.1   Problems at Resonance

In this section we study semilinear and nonlinear elliptic problems at resonance and with a nonsmooth potential. The study of resonant problems
started with the seminal work of Landesman & Lazer (1969/1970), who produced sufficient conditions (which in certain circumstances are also necessary)
for the existence of solutions for some "smooth" semilinear Dirichlet problems.
These conditions involve the asymptotic values at $\pm\infty$ of the right hand side
nonlinearity (i.e. of the derivative of the potential) and are since known as
"Landesman-Lazer conditions." Let us try to explain further the situation and
motivate the analysis that follows, using the classical semilinear C^1-framework
of Landesman & Lazer (1969/1970).

4.1.1 Semilinear Problems at Resonance

So let $\Omega \subseteq \mathbb{R}^N$ be a bounded domain and $f \colon \mathbb{R} \longrightarrow \mathbb{R}$ a continuous function. Consider the following semilinear Dirichlet problem:

$$\begin{cases} -\Delta x(z) = f(x(z)) & \text{for a.a. } z \in \Omega, \\ x|_{\partial\Omega} = 0. \end{cases} \tag{4.1}$$

In order to produce results concerning the existence and multiplicity of solutions for problem (4.1), we need to know the asymptotic behaviour of the nonlinearity f. First Landesman & Lazer (1969/1970) and since then many works have assumed that the nonlinearity f is asymptotically linear at infinity. This means that

$$\lim_{|\zeta| \to +\infty} \frac{f(\zeta)}{\zeta} = \lambda \in \mathbb{R}.$$

So we can write

$$f(x) = \lambda\zeta - h(\zeta), \quad \text{with} \quad \lim_{|\zeta| \to +\infty} \frac{h(\zeta)}{\zeta} = 0.$$

If $\lambda = \lambda_k$ (an eigenvalue of $(-\Delta, H_0^1(\Omega))$), we say that problem (4.1) is resonant at infinity. In fact there are several degrees of resonance at infinity depending on the rate of growth of h. More precisely, the smaller the rate of growth of h, the "stronger" the resonance. Actually, we can distinguish the following three cases:

(a) $h(\zeta) \longrightarrow h_{\pm}$ as $\zeta \to \pm\infty$, with $(h_+, h_-) \neq (0,0)$.

(b) $\displaystyle\lim_{|\zeta| \to +\infty} h(\zeta) = 0$ and $\displaystyle\lim_{|\zeta| \to +\infty} \int_0^{\zeta} h(s)\,ds = \pm\infty$.

(c) $\displaystyle\lim_{|\zeta| \to +\infty} h(\zeta) = 0$ and $\displaystyle\lim_{|\zeta| \to +\infty} \int_0^{\zeta} h(s)\,ds = \xi \in \mathbb{R}$.

The last situation is usually called "strong resonance" and is of special interest because it exhibits a certain lack of compactness. The Landesman-Lazer type conditions are a crucial tool in the analysis of resonant problems.

In this section we study resonant (including strongly resonant) elliptic equations, both semilinear and quasilinear (driven by the p-Laplacian) with a non-smooth potential function (hemivariational inequalities).

Let $\Omega \subseteq \mathbb{R}^N$ be a bounded domain with a C^1-boundary $\partial\Omega$, let λ_k be the k-th eigenvalue of $(-\Delta, H_0^1(\Omega))$ and let $j(z, \cdot)$ be a locally Lipschitz potential function with $\partial j(z, \cdot)$ being the generalized subdifferential (see Definition 1.3.7). We start with the following problem:

$$\begin{cases} -\Delta x(z) - \lambda_k x(z) \in \partial j(z, x(z)) & \text{for a.a. } z \in \Omega, \\ x|_{\partial\Omega} = 0. \end{cases} \tag{4.2}$$

In what follows by $\{\lambda_n\}_{n\geq 1}$ we denote the sequence of distinct eigenvalues of the operator $\big(-\Delta, H_0^1(\Omega)\big)$. From Theorem 1.5.3, we know that $\lambda_n \longrightarrow +\infty$ as $n \to +\infty$. Also there is an orthonormal basis $\{u_n\}_{n\geq 1} \subseteq L^2(\Omega)$ with $u_n \in H_0^1(\Omega) \cap C^\infty(\Omega) \cap C^1(\overline{\Omega})$ which are eigenfunctions corresponding to the eigenvalues $\{\lambda_n\}_{n\geq 1}$. Moreover, the sequence $\big\{\frac{1}{\sqrt{\lambda_n}}u_n\big\}_{n\geq 1}$ is an orthonormal sequence of $H_0^1(\Omega)$. For every integer $m \geq 1$, let $E(\lambda_m)$ be the eigenspace corresponding to the eigenvalue λ_m. We set

$$\overline{X}_{m-1} \stackrel{df}{=} \bigoplus_{i=1}^{m-1} E(\lambda_i) \qquad \forall\, m \geq 1,$$

$$X_m^0 \stackrel{df}{=} E(\lambda_m),$$

$$\widehat{X}_{m+1} \stackrel{df}{=} \bigoplus_{i=m+1}^{\infty} E(\lambda_i) \qquad \forall\, m \geq 1.$$

We have the following orthogonal direct sum decomposition:

$$H_0^1(\Omega) \;=\; \overline{X}_{m-1} \oplus X_m^0 \oplus \widehat{X}_{m+1}.$$

Also we set

$$W_m \stackrel{df}{=} \overline{X}_{m-1} \oplus X_m^0 \quad \text{and} \quad V_m \stackrel{df}{=} X_m^0 \oplus \widehat{X}_{m+1} \qquad \forall\, m \geq 1.$$

Our hypotheses on the nonsmooth potential function $j(z, \zeta)$ are the following:

$\underline{H(j)_1}$ $j \colon \Omega \times \mathbb{R} \longrightarrow \mathbb{R}$ is a function, such that

    (i) for every $\zeta \in \mathbb{R}$, the function

$$\Omega \ni z \longrightarrow j(z, \zeta) \in \mathbb{R}$$

        is measurable and $j(\cdot, 0) \in L^1(\Omega)$;

    (ii) for almost all $z \in \Omega$, the function

$$\mathbb{R} \ni \zeta \longrightarrow j(z, \zeta) \in \mathbb{R}$$

        is locally Lipschitz;

    (iii) for all $M > 0$, there exists $a_M \in L^\infty(\Omega)$, such that

$$|u| \leq a_M(z) \quad \text{for a.a. } z \in \Omega, \text{ all } |\zeta| \leq M \text{ and all } u \in \partial j(z, \zeta);$$

    (iv) $\displaystyle\lim_{|\zeta| \to +\infty} \frac{u}{\zeta} = 0$ uniformly for almost all $z \in \Omega$ and all $u \in \partial j(z, \zeta)$;

(v) there exist two functions $j_\pm \in L^1(\Omega)$, such that for almost all $z \in \Omega$, we have

$$j_+(z) = \liminf_{\zeta \to +\infty} \frac{j(z,\zeta)}{\zeta} \quad \text{and} \quad j_-(z) = \limsup_{\zeta \to +\infty} \frac{j(z,\zeta)}{\zeta}$$

and

$$\int_\Omega \left(j_+(z) x^+(z) - j_-(z) x^-(z) \right) dz > 0 \qquad \forall\, x \in X_k^0.$$

REMARK 4.1.1 Hypothesis $H(j)_1(v)$ is the generalized Landesman-Lazer condition. Later in this section we shall compare it with other conditions existing in the literature. Recall that

$$x^+ = \max\{x,0\} \quad \text{and} \quad x^- = \max\{-x,0\}.$$

$\square$

We introduce the energy functional $\varphi: H_0^1(\Omega) \longrightarrow \mathbb{R}$, defined by

$$\varphi(x) \overset{df}{=} \frac{1}{2} \|\nabla x\|_2^2 - \frac{\lambda_k}{2} \|x\|_2^2 - \int_\Omega j(z, x(z))\, dz.$$

From Theorem 1.3.10 and Proposition 1.3.17, we know that φ is locally Lipschitz.

PROPOSITION 4.1.1
If hypotheses $H(j)_1$ hold, then φ satisfies the nonsmooth PS-condition.

PROOF We consider a sequence $\{x_n\}_{n\geq 1} \subseteq H_0^1(\Omega)$, such that

$$|\varphi(x_n)| \leq M_1 \qquad \forall\, n \geq 1,$$

for some $M_1 > 0$ and

$$m^\varphi(x_n) \longrightarrow 0.$$

Let $x_n^* \in \partial\varphi(x_n)$ be such that $m^\varphi(x_n) = \|x_n^*\|_{H^{-1}(\Omega)}$, $n \geq 1$. This is possible because the set $\partial\varphi(x_n) \subseteq H^{-1}(\Omega)$ is weakly compact and the norm functional is weakly lower semicontinuous. Let $A: H_0^1(\Omega) \longrightarrow H^{-1}(\Omega)$ be defined by

$$\langle A(x), y \rangle_{H_0^1(\Omega)} \overset{df}{=} \int_\Omega \left(\nabla x(z), \nabla y(z) \right)_{\mathbb{R}^N} dz \qquad \forall\, x, y \in H_0^1(\Omega).$$

The operator A is monotone, hence maximal monotone (see Corollary 1.4.2). For every $n \geq 1$, we have

$$x_n^* = A(x_n) - \lambda_k x_n - u_n^* \qquad \forall\, n \geq 1,$$

with $u_n^* \in L^2(\Omega)$ and $u_n^*(z) \in \partial j(z, x_n(z))$ for almost all $z \in \Omega$ (see Theorem 1.3.10).

We claim that the sequence $\{x_n\}_{n \geq 1} \subseteq H_0^1(\Omega)$ is bounded. Suppose that this is not true. Then, passing to a subsequence if necessary, we may assume that

$$\|x_n\|_{H^1(\Omega)} \longrightarrow +\infty.$$

Let us set

$$y_n \overset{df}{=} \frac{x_n}{\|x_n\|_{H^1(\Omega)}} \qquad \forall\, n \geq 1.$$

Evidently $\|y_n\|_{H^1(\Omega)} = 1$ and since the embedding $H_0^1(\Omega) \subseteq L^2(\Omega)$ is compact, by passing to a subsequence if necessary, we may assume that

$$\begin{aligned}
y_n &\overset{w}{\longrightarrow} y \quad \text{in } H_0^1(\Omega), \\
y_n &\longrightarrow y \quad \text{in } L^2(\Omega), \\
y_n(z) &\longrightarrow y(z) \text{ for a.a. } z \in \Omega, \\
\big|y_n(z)\big| &\leq k(z) \text{ for a.a. } z \in \Omega \text{ and all } n \geq 1,
\end{aligned}$$

with $k \in L^2(\Omega)_+$. From the choice of the sequence $\{x_n\}_{n \geq 1} \subseteq H_0^1(\Omega)$, we have that

$$\big|\langle x_n^*, v \rangle_{H_0^1(\Omega)}\big| \leq \varepsilon_n \|v\|_{H^1(\Omega)} \qquad \forall\, v \in H_0^1(\Omega), \tag{4.3}$$

with $\varepsilon_n \searrow 0$, so

$$\left| \int_\Omega (\nabla y_n, \nabla v)_{\mathbb{R}^N}\, dz - \lambda_k \int_\Omega y_n v\, dz - \int_\Omega \frac{u_n}{\|x_n\|_{H^1(\Omega)}} v\, dz \right|$$
$$\leq \varepsilon_n \frac{\|v\|_{H^1(\Omega)}}{\|x_n\|_{H^1(\Omega)}} \qquad \forall\, v \in H_0^1(\Omega). \tag{4.4}$$

Note that hypotheses $H(j)_1(iii)$ and (iv) imply that for almost all $z \in \Omega$, all $\zeta \in \mathbb{R}$ and all $u \in \partial j(z, \zeta)$, we have that

$$|u| \leq \widehat{a}(z) + \widehat{c}|\zeta|,$$

with $\widehat{a} \in L^{r'}(\Omega)$, $\widehat{c} > 0$. So we have

$$\int_\Omega \frac{u_n(z)}{\|x_n\|_{H^1(\Omega)}} (y_n - y)(z)\, dz \leq \int_\Omega \left(\frac{\widehat{a}(z)}{\|x_n\|_{H^1(\Omega)}} + \widehat{c}|y_n(z)| \right) (y_n - y)(z)\, dz,$$

with the right hand side tending to zero as $n \to +\infty$. If in (4.4), we set $v = y_n - y \in H_0^1(\Omega)$ and use the above convergence, we obtain

$$\lim_{n \to +\infty} \left[\langle A(y_n), y_n - y \rangle_{H_0^1(\Omega)} - \lambda_k \int_\Omega y_n(z)(y_n - y)(z)\, dz \right] = 0.$$

Evidently

$$\int_\Omega y_n(z)(y_n - y)(z)\, dz \longrightarrow 0.$$

Hence

$$\lim_{n \to +\infty} \langle A(y_n), y_n - y \rangle_{H_0^1(\Omega)} = 0.$$

Because

$$A(y_n) \xrightarrow{w} A(y) \quad \text{in } H^{-1}(\Omega),$$

we have that

$$\langle A(y_n), y_n \rangle_{H_0^1(\Omega)} \longrightarrow \langle A(y), y \rangle_{H_0^1(\Omega)},$$

hence

$$\|\nabla y_n\|_2 \longrightarrow \|\nabla y\|_2.$$

Since

$$\nabla y_n \xrightarrow{w} \nabla y \quad \text{in } L^2(\Omega; \mathbb{R}^N),$$

it follows that

$$\nabla y_n \longrightarrow \nabla y \quad \text{in } L^2(\Omega; \mathbb{R}^N)$$

(Kadec-Klee property) and so we conclude that

$$y_n \longrightarrow y \quad \text{in } H_0^1(\Omega).$$

We consider the decomposition

$$y_n = \overline{y}_n + y_n^0 + \widehat{y}_n \qquad \forall\, n \geq 1,$$

with

$$\overline{y}_n \in \overline{X}_{k-1}, \quad y_n^0 \in X_k^0 \quad \text{and} \quad \widehat{y}_n \in \widehat{X}_{k+1} \qquad \forall\, n \geq 1.$$

Using in (4.3), as a test function $v = -\overline{y}_n + y_n^0 + \widehat{y}_n \in H_0^1(\Omega)$, we obtain

$$\left| \int_\Omega \left(\nabla x_n(z), \left(-\nabla \overline{y}_n + \nabla y_n^0 + \nabla \widehat{y}_n \right)(z) \right)_{\mathbb{R}^N} \right.$$

$$-\lambda_k \int_\Omega x_n(z) \left(-\overline{y}_n + y_n^0 + \widehat{y}_n \right)(z)\, dz$$

$$\left. - \int_\Omega u_n(z) \left(-\overline{y}_n + y_n^0 + \widehat{y}_n \right)(z)\, dz \right|$$

$$\leq \varepsilon_n \left\| -\overline{y}_n + y_n^0 + \widehat{y}_n \right\|_{H^1(\Omega)} \leq 3\varepsilon_n,$$

so

$$\frac{1}{\|x_n\|_{H^1(\Omega)}} \left| \int_\Omega \left(\nabla x_n(z), \left(-\nabla \overline{x}_n \mid \nabla x_n^0 + \nabla \widehat{x}_n \right)(z) \right)_{\mathbb{R}^N} \right.$$

$$\left. -\lambda_k \int_\Omega x_n(z) \left(-\overline{x}_n + x_n^0 + \widehat{x}_n \right)(z)\, dz \right|$$

$$-\left| \int_\Omega \frac{u_n(z)}{\|x_n\|_{H^1(\Omega)}} \left(-\overline{x}_n + x_n^0 + \widehat{x}_n \right)(z)\, dz \right| \leq 3\varepsilon_n. \quad (4.5)$$

Also, if in (4.3), we use as a test function $v = y_n^0$, we obtain

$$\left| \frac{1}{\|x_n\|_{H^1(\Omega)}} \left(\left\|\nabla x_n^0\right\|_2^2 - \lambda_k \left\|x_n^0\right\|_2^2 \right) - \int_\Omega \frac{u_n(z)}{\|x_n\|_{H^1(\Omega)}} x_n^0(z)\, dz \right| \leq \varepsilon_n.$$

But recall that

$$\left\|\nabla x_n^0\right\|_2^2 = \lambda_k \left\|x_n^0\right\|_2^2 \qquad \forall\, n \geq 1.$$

So we infer that

$$\lim_{n \to +\infty} \int_\Omega \frac{u_n(z)}{\|x_n\|_{H^1(\Omega)}} x_n^0(z)\, dz = 0. \qquad (4.6)$$

From the variational characterization of the eigenvalues of $\left(-\Delta, H_0^1(\Omega) \right)$ (see Theorem 1.5.3), we have

$$\|\nabla \overline{x}_n\|_{H^1(\Omega)} \leq \lambda_{k-1} \|\overline{x}_n\|_2^2 \quad \text{and} \quad \|\nabla \widehat{x}_n\|_2^2 \geq \lambda_{k+1} \|\widehat{x}_n\|_2^2 \qquad \forall\, n \geq 1.$$

Using these inequalities and the orthogonality relations among the subspaces $\widehat{X}_{k-1}$, X_k^0 and $\widehat{X}_{k+1}$, we obtain

$$\int_\Omega \left(\nabla x_n(z), \left(-\nabla \widehat{x}_n + \nabla x_n^0 + \nabla \widehat{x}_n \right)(z) \right)_{\mathbb{R}^N} dz$$

$$= -\|\nabla \widehat{x}_n\|_2^2 + \left\|\nabla x_n^0\right\|_2^2 + \|\nabla \widehat{x}_n\|_2^2$$

and

$$-\lambda_k \int_\Omega x_n(z) \left(-\overline{x}_n + x_n^0 + \widehat{x}_n \right)(z)\, dz = \lambda_n \|\overline{x}_n\|_2^2 - \lambda_k \left\|x_n^0\right\|_2^2 - \lambda_k \|\widehat{x}_n\|_2^2$$

$$\geq \frac{\lambda_k}{\lambda_{k-1}} \|\nabla \overline{x}_n\|_2^2 - \left\|\nabla x_n^0\right\|_2^2 - \frac{\lambda_k}{\lambda_{k+1}} \|\nabla \widehat{x}_n\|_2^2.$$

So we can write that

$$\int_\Omega \left(\nabla x_n(z), \left(-\nabla \overline{x}_n + \nabla x_n^0 + \nabla \widehat{x}_n \right)(z) \right)_{\mathbb{R}^N} dz$$

$$-\lambda_k \int_\Omega x_n(z)\big(-\overline{x}_n + x_n^0 + \widehat{x}_n\big)(z)\,dz$$

$$\geq \left(\frac{\lambda_k}{\lambda_{k-1}} - 1\right)\|\nabla \overline{x}_n\|_2^2 + \left(1 - \frac{\lambda_k}{\lambda_{k+1}}\right)\|\nabla \widehat{x}_n\|_2^2$$

$$= \frac{\lambda_k - \lambda_{k-1}}{\lambda_{k-1}}\|\nabla \overline{x}_n\|_2^2 + \frac{\lambda_{k+1} - \lambda_k}{\lambda_{k+1}}\|\nabla \widehat{x}_n\|_2^2. \tag{4.7}$$

By virtue of hypothesis $H(j)_1(iv)$, for a given $\varepsilon > 0$, we can find $M_2 = M_2(\varepsilon) > 0$, such that

$$\frac{|u|}{|\zeta|} \leq \varepsilon \quad \text{for a.a. } z \in \Omega, \text{ all } |\zeta| \geq M_2 \text{ and all } u \in \partial j(z, \zeta).$$

In addition, from hypothesis $H(j)_1(iii)$, we have that

$$|u| \leq a_{M_2}(z) \quad \text{for a.a. } z \in \Omega, \text{ all } |\zeta| < M_2 \text{ and all } u \in \partial j(z, \zeta),$$

with $a_{M_2} \in L^{r'}(\Omega)$. Thus, we can write that

$$\left| \int_\Omega \frac{u_n(z)}{\|x_n\|_{H^1(\Omega)}} \big(-\overline{x}_n + \widehat{x}_n\big)(z)\,dz \right|$$

$$= \Big| \frac{1}{\|x_n\|_{H^1(\Omega)}} \int_{\{|x_n| \geq M_2\}} \frac{u_n(z)}{x_n(z)} x_n(z)\big(-\overline{x}_n + \widehat{x}_n\big)(z)\,dz$$

$$+ \int_{\{|x_n| < M_2\}} \frac{|u_n(z)|}{\|x_n\|_{H^1(\Omega)}} \big(-\overline{x}_n + \widehat{x}_n\big)(z)\,dz\Big|$$

$$\leq \frac{\varepsilon}{\|x_n\|_{H^1(\Omega)}} \beta_1 \left(\|\nabla \overline{x}_n\|_2^2 + \|\nabla \widehat{x}_n\|_2^2\right)$$

$$+ \frac{\beta_2}{\|x_n\|_{H^1(\Omega)}} \left(\|\nabla \overline{x}_n\|_2 + \|\nabla \widehat{x}_n\|_2\right), \tag{4.8}$$

for some $\beta_1, \beta_2 > 0$. Using (4.6), (4.7) and (4.8) in (4.5) and choosing $\varepsilon > 0$ small enough, we have

$$\limsup_{n \to +\infty} \frac{1}{\|x_n\|_{H^1(\Omega)}} \left(\beta_3 \left(\|\nabla \overline{x}_n\|_2^2 + \|\nabla \widehat{x}_n\|_2^2\right) - \beta_2 \left(\|\nabla \overline{x}_n\|_2 + \|\nabla \widehat{x}_n\|_2\right)\right) \leq 0,$$

for some $\beta_3 > 0$. We set $w_n = \overline{x}_n + \widehat{x}_n$ for $n \geq 1$. We have

$$\|\nabla w_n\|_2^2 = \|\nabla \overline{x}_n\|_2^2 + \|\nabla \widehat{x}_n\|_2^2$$

and

$$\tfrac{1}{2}\left(\|\nabla \overline{x}_n\|_2 + \|\nabla \widehat{x}_n\|_2\right) \leq \|\nabla w_n\|_2.$$

So we can write that

$$\limsup_{n \to +\infty} \frac{1}{\|x_n\|_{H^1(\Omega)}} \left(\beta_3 \|\nabla w_n\|_2^2 - 2\beta_2 \|\nabla w_n\|_2\right) \leq 0$$

and thus

$$\limsup_{n \to +\infty} \frac{\|\nabla w_n\|_2^2}{\|x_n\|_{H^1(\Omega)}} \left(\beta_3 - \frac{2\beta_2}{\|\nabla w_n\|_2}\right) \leq 0. \tag{4.9}$$

If the sequence $\{\nabla w_n\}_{n \geq 1} \subseteq L^2(\Omega; \mathbb{R}^N)$ is bounded, then

$$\frac{\|\nabla w_n\|_2^2}{\|x_n\|_{H^1(\Omega)}} \longrightarrow 0.$$

If the sequence $\{\nabla w_n\}_{n \geq 1} \subseteq L^2(\Omega; \mathbb{R}^N)$ is unbounded, then by passing to a subsequence if necessary, we may assume that

$$\|\nabla w_n\|_2 \longrightarrow +\infty.$$

So we can find $n_0 \geq 1$, such that

$$\frac{2\beta_2}{\|\nabla w_n\|_2} < \beta_3 \qquad \forall \, n \geq n_0.$$

Hence from (4.9), it follows that

$$\limsup_{n \to +\infty} \frac{\|\nabla w_n\|_2^2}{\|x_n\|_{H^1(\Omega)}} \leq 0.$$

Therefore, in both cases, we have that

$$\lim_{n \to +\infty} \frac{\|\nabla w_n\|_2^2}{\|x_n\|_{H^1(\Omega)}} = 0. \tag{4.10}$$

Recall that

$$\|\nabla w_n\|_2^2 = \|\nabla \overline{x}_n\|_2^2 + \|\nabla \widehat{x}_n\|_2^2,$$

hence

$$\frac{\|\nabla w_n\|_2^2}{\|x_n\|_{H^1(\Omega)}} = \frac{\|\nabla \overline{x}_n\|_2^2}{\|x_n\|_{H^1(\Omega)}} + \frac{\|\nabla \widehat{x}_n\|_2^2}{\|x_n\|_{H^1(\Omega)}}.$$

Assuming without any loss of generality that $\|x_n\|_{H^1(\Omega)} \geq 1$ for $n \geq 1$ (recall that $\|x_n\|_{H^1(\Omega)} \longrightarrow +\infty$), we have

$$\frac{\|\nabla w_n\|_2^2}{\|x_n\|_{H^1(\Omega)}} \geq \frac{\|\nabla w_n\|_2^2}{\|x_n\|_{H^1(\Omega)}^2} = \|\nabla \overline{y}_n\|_2^2 + \|\nabla \widehat{y}_n\|_2^2,$$

so

$$\|\nabla \overline{y}_n\|_2 \longrightarrow 0, \qquad \|\nabla \widehat{y}_n\|_2 \longrightarrow 0$$

and so

$$\overline{y}_n \longrightarrow 0, \quad \widehat{y}_n \longrightarrow 0 \quad \text{in } H_0^1(\Omega),$$

hence $y = y^0 \in E(\lambda_k)$.

From the choice of the sequence $\{x_n\}_{n\geq 1} \subseteq H_0^1(\Omega)$, we have

$$\left| \frac{1}{2} \|\nabla x_n\|_2^2 - \frac{\lambda_k}{2} \|x_n\|_2^2 - \int_\Omega j(z, x_n(z)) \, dz \right| \leq M_1 \qquad \forall \, n \geq 1.$$

Dividing by $\|x_n\|_{H^1(\Omega)}$ and exploiting the orthogonality relations among the components $\overline{x}_n$, x_n^0 and $\widehat{x}_n$, for $n \geq 1$ and the equality $\|\nabla x_n^0\|_2^2 = \lambda_k \|x_n^0\|_2^2$, we obtain

$$\left| \frac{1}{2} \frac{\|\nabla w_n\|_2^2}{\|x_n\|_{H^1(\Omega)}} - \frac{\lambda_k}{2} \frac{\|w_n\|_2^2}{\|x_n\|_{H^1(\Omega)}} - \int_\Omega \frac{j(z, x_n(z))}{\|x_n\|_{H^1(\Omega)}} \, dz \right|$$

$$\leq \frac{M_1}{\|x_n\|_{H^1(\Omega)}} \qquad \forall \, n \geq 1. \tag{4.11}$$

From (4.10), we know that

$$\frac{1}{2} \frac{\|\nabla w_n\|_2^2}{\|x_n\|_{H^1(\Omega)}} \longrightarrow 0.$$

Also from the Poincaré inequality, we have

$$\frac{\lambda_k}{2} \frac{\|w_n\|_2^2}{\|x_n\|_2} \leq \frac{\lambda_k}{2} \beta_4 \frac{\|\nabla w_n\|_2^2}{\|x_n\|_{H^1(\Omega)}} \qquad \forall \, n \geq 1,$$

for some $\beta_4 > 0$, so

$$\frac{\lambda_k}{2} \frac{\|w_n\|_2^2}{\|x_n\|_2} \longrightarrow 0.$$

Recall that

$$y_n \longrightarrow y = y^0 \quad \text{in } H_0^1(\Omega)$$

and $y^0 \in E(\lambda_k)$. By the unique continuation property of the eigenspace $E(\lambda_k)$ and since $y \neq 0$, it follows that $y(z) \neq 0$ for almost all $z \in \Omega$. Note that

$$x_n(z) \longrightarrow +\infty \text{ on } \{y > 0\},$$
$$x_n(z) \longrightarrow -\infty \text{ on } \{y < 0\}.$$

For a given $\varepsilon > 0$ and any $n \geq 1$, we define the following sets:

$$\Omega_{n,\varepsilon}^+ \stackrel{df}{=} \left\{ z \in \Omega : x_n(z) > 0, \ \frac{j(z, x_n(z))}{x_n(z)} \geq j_+(z) - \varepsilon \right\},$$

$$\Omega_{n,\varepsilon}^- \stackrel{df}{=} \left\{ z \in \Omega : x_n(z) < 0, \ \frac{j(z, x_n(z))}{x_n(z)} \leq j_-(z) + \varepsilon \right\},$$

$$\Omega^+ \stackrel{df}{=} \{z \in \Omega : y(z) > 0\},$$

$$\Omega^- \stackrel{df}{=} \{z \in \Omega : y(z) < 0\},$$

$$\Omega_{n,\varepsilon}^0 \stackrel{df}{=} \left(\Omega_{n,\varepsilon}^+ \cup \Omega_{n,\varepsilon}^- \right)^c.$$

Then

$$\chi_{\Omega_{n,\varepsilon}^{+}}(z) \; \to \; \chi_{\Omega^{+}}(z) \qquad \text{for a.a. } z \in \Omega,$$
$$\chi_{\Omega_{n,\varepsilon}^{-}}(z) \; \longrightarrow \; \chi_{\Omega^{-}}(z) \qquad \text{for a.a. } z \in \Omega$$

and

$$\left|\Omega_{n,\varepsilon}^{0}\right|_{N} \; \longrightarrow \; 0.$$

Then, we have

$$\int_{\Omega} \frac{j(z, x_n(z))}{\|x_n\|_{H^1(\Omega)}} \, dz$$

$$= \int_{\Omega_{n,\varepsilon}^{+}} \frac{j(z, x_n(z))}{x_n(z)} y_n(z) \, dz + \int_{\Omega_{n,\varepsilon}^{-}} \frac{j(z, x_n(z))}{x_n(z)} y_n(z) \, dz + \int_{\Omega_{n,\varepsilon}^{0}} \frac{j(z, x_n(z))}{\|x_n\|_{H^1(\Omega)}} \, dz$$

$$\geq \int_{\Omega_{n,\varepsilon}^{+}} \left(j_{+}(z) - \varepsilon\right) y_n(z) \, dz + \int_{\Omega_{n,\varepsilon}^{-}} \left(j_{-}(z) + \varepsilon\right) y_n(z) \, dz + \int_{\Omega_{n,\varepsilon}^{0}} \frac{j(z, x_n(z))}{\|x_n\|_{H^1(\Omega)}} \, dz.$$

Note that

$$\int_{\Omega_{n,\varepsilon}^{0}} \frac{j(z, x_n(z))}{\|x_n\|_{H^1(\Omega)}} \, dz \; \longrightarrow \; 0.$$

If we pass to the limit as $n \to +\infty$ in the last inequality and letting $\varepsilon \searrow 0$, we obtain

$$\liminf_{n \to +\infty} \int_{\Omega} \frac{j(z, x_n(z))}{\|x_n\|_{H^1(\Omega)}} \, dz \; \geq \; \int_{\Omega} j_{+}(z) y^{+}(z) \, dz - \int_{\Omega} j_{-}(z) y^{-}(z) \, dz.$$

Returning to (4.11), passing to the limit as $n \to +\infty$ and using the last inequality, we obtain

$$\int_{\Omega} j_{+}(z) y^{+}(z) \, dz - \int_{\Omega} j_{-}(z) y^{-}(z) \, dz \; \leq \; 0,$$

a contradiction to the hypothesis $H(j)_1(v)$. This proves that the sequence $\{x_n\}_{n \geq 1} \subseteq H_0^1(\Omega)$ is bounded and so we may assume that

$$x_n \xrightarrow{w} x \text{ in } H_0^1(\Omega),$$
$$x_n \longrightarrow x \text{ in } L^2(\Omega).$$

So we have

$$\int_{\Omega} x_n(z)\big(x_n - x\big)(z) \, dz \; \longrightarrow \; 0 \quad \text{and} \quad \int_{\Omega} u_n(z)\big(x_n - x\big)(z) \, dz.$$

It follows that

$$\langle A(x_n), x_n - x \rangle_{H_0^1(\Omega)} \longrightarrow 0$$

and so

$$\|\nabla x_n\|_2 \longrightarrow \|\nabla x\|_2.$$

Since

$$\nabla x_n \xrightarrow{w} \nabla x \quad \text{in } L^2(\Omega; \mathbb{R}^N),$$

we infer that

$$\nabla x_n \longrightarrow \nabla x \quad \text{in } L^2(\Omega; \mathbb{R}^N)$$

and so finally

$$x_n \longrightarrow x \quad \text{in } H_0^1(\Omega).$$

$\square$

Recall that

$$W_k \stackrel{df}{=} \overline{X}_k \oplus X_k^0 = \bigoplus_{i=1}^{k} E(\lambda_i).$$

Evidently

$$W_k^\perp = \widehat{X}_{k+1}.$$

PROPOSITION 4.1.2
If hypotheses $H(j)_1$ hold,
then $\varphi|_{W_k}$ is anticoercive, i.e.

$$\varphi(x) \longrightarrow -\infty \quad \text{as } \|x\|_{H^1(\Omega)} \to +\infty, \ x \in W_k.$$

PROOF Suppose that the proposition is not true. Then we can find a sequence $\{x_n\}_{n\geq 1} \subseteq W_k$ and a constant $\beta_5 \in \mathbb{R}$, such that

$$\varphi(x_n) \geq \beta_5 \quad \forall n \geq 1.$$

Putting

$$y_n \stackrel{df}{=} \frac{x_n}{\|x_n\|_{H^1(\Omega)}} \quad \forall n \geq 1,$$

we have

$$\frac{1}{2}\|\nabla y_n\|_2^2 - \frac{\lambda_k}{2}\|y_n\|_2^2 - \int_\Omega \frac{j(z, x_n(z))}{\|x_n\|_{H^1(\Omega)}^2} \, dz \geq \frac{\beta_5}{\|x_n\|_{H^1(\Omega)}^2}. \tag{4.12}$$

Passing to a subsequence if necessary, we may assume that

$$y_n \xrightarrow{w} y \quad \text{in } H^1(\Omega),$$

with $y \in W_k$. Since W_k is finite dimensional, we have that

$$y_n \longrightarrow y \quad \text{in } H^1(\Omega).$$

From hypotheses $H(j)_1(iii)$ and (iv) and the mean value theorem for locally Lipschitz functions (see Proposition 1.3.14), for a given $\varepsilon > 0$, we can find $\widehat{a}_\varepsilon \in L^\infty(\Omega)$, such that

$$|j(z, \varsigma)| \leq |j(z, 0)| + \widehat{a}_\varepsilon(z)|\varsigma| + \varepsilon\varsigma^2 \quad \text{for a.a. } z \in \Omega \text{ and all } \varsigma \in \mathbb{R}.$$

So we have

$$\left| \int_\Omega \frac{j(z, x_n(z))}{\|x_n\|^2_{H^1(\Omega)}} \right| \leq \int_\Omega \frac{|j(z, 0)|}{\|x_n\|^2_{H^1(\Omega)}} + \int_\Omega \frac{\widehat{a}_\varepsilon(z)}{\|x_n\|_{H^1(\Omega)}} |y_n(z)| \, dz + \varepsilon \|y_n\|^2_2.$$

As $\varepsilon > 0$ was arbitrary, we have

$$\int_\Omega \frac{j(z, x_n(z))}{\|x_n\|^2_{H^1(\Omega)}} \longrightarrow 0.$$

Passing to the limit in (4.12), we obtain

$$\|\nabla y\|^2_2 \leq \lambda_k \|y\|^2_2,$$

so from Theorem 1.5.3(b), we have

$$\|\nabla y\|^2_2 = \lambda_k \|y\|^2_2$$

and thus $y \in X_k^0 \setminus \{0\}$.

From the choice of the sequence $\{x_n\}_{n \geq 1} \subseteq W_k$ and since

$$\frac{1}{2} \|\nabla x_n\|^2_2 \leq \frac{\lambda_k}{2} \|x_n\|^2_2$$

(see Theorem 1.5.3(b)), we have

$$\frac{\beta_5}{\|x_n\|_{H^1(\Omega)}} \leq -\int_\Omega \frac{j(z, x_n(z))}{\|x_n\|_{H^1(\Omega)}} \, dz,$$

so

$$\limsup_{n \to +\infty} \int_\Omega \frac{j(z, x_n(z))}{\|x_n\|_{H^1(\Omega)}} \, dz \leq 0.$$

Arguing as in the proof of Proposition 4.1.1, since $y(z) \neq 0$ for almost all $z \in \Omega$ (unique continuation property), we obtain that

$$\int_\Omega \left(j_+(z)y^+(z) - j_-(z)y^-(z) \right) dz \leq 0,$$

which contradicts hypothesis $H(j)_1(v)$. Therefore $\varphi|_{W_k}$ is anticoercive. □

PROPOSITION 4.1.3

If hypotheses $h(j)_1$ hold, <u>then</u> φ is weakly coercive on $W_k^\perp = \widehat{X}_{k+1}$.

PROOF If $x \in W_k^\perp = \widehat{X}_{k+1}$ and $x \neq 0$, we know that

$$\lambda_{k+1} \|x\|_2^2 \leq \|\nabla x\|_2^2$$

(see Theorem 1.5.3(b)). So, we have

$$\varphi(x) = \frac{1}{2}\|\nabla x\|_2^2 - \frac{\lambda_k}{2}\|x\|_2^2 - \int_\Omega j(z, x(z))\, dz$$

$$\geq \frac{1}{2}\left(1 - \frac{\lambda_k}{\lambda_{k+1}}\right)\|\nabla x\|_2^2 - \int_\Omega j(z, x(z))\, dz$$

and putting $y \stackrel{df}{=} \frac{x}{\|x\|_{H^1(\Omega)}}$, we have

$$\frac{\varphi(x)}{\|x\|_{H^1(\Omega)}^2} \geq \frac{1}{2}\left(1 - \frac{\lambda_k}{\lambda_{k+1}}\right)\|\nabla y\|_2^2 - \int_\Omega \frac{j(z, x(z))}{\|x\|_{H^1(\Omega)}^2}\, dz. \qquad (4.13)$$

From the proof of Proposition 4.1.2, we know that

$$\int_\Omega \frac{j(z, x(z))}{\|x\|_{H^1(\Omega)}^2}\, dz \longrightarrow 0 \quad \text{as } \|x\|_{H^1(\Omega)} \to +\infty.$$

So, if we pass to the limit as $\|x\|_{H^1(\Omega)} \to +\infty$ in (4.13), we obtain

$$\liminf_{\|x\|_{H^1(\Omega)} \to +\infty} \frac{\varphi(x)}{\|x\|_{H^1(\Omega)}^2} \geq \frac{1}{2}\left(1 - \frac{\lambda_k}{\lambda_{k+1}}\right)\|\nabla y\|_2^2 > 0,$$

thus

$$\varphi(x) \longrightarrow +\infty \quad \text{as } \|x\|_{H^1(\Omega)} \to +\infty, \text{ with } x \in W_k^\perp = \widehat{X}_{k+1}.$$

□

The three propositions produce the geometry of the nonsmooth Saddle Point Theorem (see Theorem 2.1.4) and give the following existence theorem for problem (4.2).

THEOREM 4.1.1

If hypotheses $H(j)_1$ hold, <u>then</u> problem (4.2) has a solution $x_0 \in H_0^1(\Omega)$.

PROOF   Proposition 4.1.3 implies the existence of $\beta_6 \in \mathbb{R}$, such that

$$\varphi(x) \geq \beta_6 \quad \forall\, x \in W_k^\perp = \widehat{X}_{k+1}.$$

On the other hand Proposition 4.1.2 implies the existence of $\varrho > 0$ large enough, such that

$$\varphi(x) < \beta_6 \quad \forall\, x \in W_k,\ \|x\|_{H^1(\Omega)} = \varrho.$$

These facts together with Proposition 4.1.1 permit the use of the nonsmooth Saddle Point Theorem (see Theorem 2.1.4), which gives $x_0 \in H_0^1(\Omega)$, such that $0 \in \partial\varphi(x_0)$. Hence

$$A(x_0) - \lambda_k x_0 - u_0^* = 0,$$

with $u_0^* \in L^2(\Omega)$, $u_0^*(z) \in \partial j\big(z, x_0(z)\big)$ for almost all $z \in \Omega$. Let $\vartheta \in C_c^\infty(\Omega)$. We have

$$\langle A(x_0), \vartheta \rangle_{H_0^1(\Omega)} - \lambda_k \int_\Omega x_0(z)\vartheta(z)\,dz = \int_\Omega u_0^*(z)\vartheta(z)\,dz,$$

so

$$\int_\Omega \big(\nabla x_0(z), \nabla\vartheta(z)\big)_{\mathbb{R}^N} dz - \lambda_k \int_\Omega x_0(z)\vartheta(z)\,dz = \int_\Omega u_0^*(z)\vartheta(z)\,dz.$$

Because

$$\Delta x_0 = \operatorname{div}\nabla x_0 \in H^{-1}(\Omega) = \big(H_0^1(\Omega)\big)^*$$

(see Theorem 1.1.8), by Green's identity (see Theorem 1.1.9), we obtain

$$\langle -\Delta x_0, \vartheta \rangle_{H_0^1(\Omega)} - \lambda_k \langle x_0, \vartheta \rangle_{H_0^1(\Omega)} = \langle u_0^*, \vartheta \rangle_{H_0^1(\Omega)} \quad \forall\, \vartheta \in C_c^\infty(\Omega).$$

Since the embedding $C_c^\infty(\Omega) \subseteq H_0^1(\Omega)$ is dense, we infer that

$$\begin{cases} -\Delta x_0(z) - \lambda_k x_0(z) = u_0^*(z) \in \partial j\big(z, x(z)\big) & \text{for a.a. } z \in \Omega, \\ x|_{\partial\Omega} = 0 \end{cases}$$

and so $x_0 \in H_0^1(\Omega)$ solves (4.2).   □

A careful reading of the proof of the previous proposition reveals that we can modify the generalized Landesman-Lazer type condition (see hypothesis $H(j)_1(v)$) and still have an existence theorem. The modified hypotheses on the nonsmooth potential are the following:

$\underline{H(j)_2}$ $j: \Omega \times \mathbb{R} \longrightarrow \mathbb{R}$ is a function, such that

$$\left.\begin{array}{l}(i)\\(ii)\\(iii)\\(iv)\end{array}\right\} \text{ are the same as hypotheses } H(j)_1(i), (ii), (iii) \text{ and } (iv);$$

(v) there exist two functions $j_\pm \in L^1(\Omega)$, such that for almost all $z \in \Omega$, we have

$$j_+(z) = \limsup_{\zeta \to +\infty} \frac{j(z, \zeta)}{\zeta} \quad \text{and} \quad j_-(z) = \liminf_{\zeta \to +\infty} \frac{j(z, \zeta)}{\zeta}$$

and

$$\int_\Omega (j_+(z)x^+(z) - j_-(z)x^-(z))dz < 0 \qquad \forall\, x \in X_k^0.$$

THEOREM 4.1.2
If hypotheses $H(j)_2$ hold, <u>then</u> problem (4.2) has a solution $x_0 \in H_0^1(\Omega)$.

PROOF   As before, we consider the locally Lipschitz energy functional φ. Again φ satisfies the nonsmooth PS-condition. Indeed the proof of Proposition 4.1.1 remains the same and only near the end we modify it as follows. We define

$$\Omega_{n,\varepsilon}^+ \overset{df}{=} \left\{ z \in \Omega : x_n(z) > 0, \; \frac{j(z, x_n(z))}{x_n(z)} \le j_+(z) + \varepsilon \right\},$$

$$\Omega_{n,\varepsilon}^- \overset{df}{=} \left\{ z \in \Omega : x_n(z) < 0, \; \frac{j(z, x_n(z))}{x_n(z)} \ge j_-(z) - \varepsilon \right\},$$

$$\Omega^+ \overset{df}{=} \{z \in \Omega : y(z) > 0\},$$

$$\Omega^- \overset{df}{=} \{z \in \Omega : y(z) < 0\},$$

$$\Omega_{n,\varepsilon}^0 \overset{df}{=} \left(\Omega_{n,\varepsilon}^+ \cup \Omega_{n,\varepsilon}^-\right)^c.$$

Again

$$\chi_{\Omega_{n,\varepsilon}^+}(z) \longrightarrow \chi_{\Omega^+}(z) \quad \text{for a.a. } z \in \Omega,$$

$$\chi_{\Omega_{n,\varepsilon}^-}(z) \longrightarrow \chi_{\Omega^-}(z) \quad \text{for a.a. } z \in \Omega$$

and

$$\left|\Omega_{n,\varepsilon}^0\right|_N \longrightarrow 0.$$

Then, we have

$$\int\limits_{\Omega} \frac{j(z, x_n(z))}{\|x_n\|_{H^1(\Omega)}}\, dz$$

$$= \int\limits_{\Omega_{n,\varepsilon}^+} \frac{j(z, x_n(z))}{x_n(z)} y_n(z)\, dz + \int\limits_{\Omega_{n,\varepsilon}^-} \frac{j(z, x_n(z))}{x_n(z)} y_n(z)\, dz + \int\limits_{\Omega_{n,\varepsilon}^0} \frac{j(z, x_n(z))}{\|x_n\|_{H^1(\Omega)}}\, dz$$

$$\leq \int\limits_{\Omega_{n,\varepsilon}^+} (j_+(z) + \varepsilon) y_n(z)\, dz + \int\limits_{\Omega_{n,\varepsilon}^-} (j_-(z) - \varepsilon) y_n(z)\, dz + \int\limits_{\Omega_{n,\varepsilon}^0} \frac{j(z, x_n(z))}{\|x_n\|_{H^1(\Omega)}}\, dz.$$

Passing to the limit as $n \to +\infty$ and letting $\varepsilon \searrow 0$, we obtain

$$\limsup_{n \to +\infty} \int\limits_{\Omega} \frac{j(z, x_n(z))}{\|x_n\|_{H^1(\Omega)}}\, dz \leq \int\limits_{\Omega} \left(j_+(z) y^+(z) - j_-(z) y^-(z) \right) dz.$$

But from (4.11), we have that

$$\liminf_{n \to +\infty} \int\limits_{\Omega} \frac{j(z, x_n(z))}{\|x_n\|_{H^1(\Omega)}}\, dz \geq 0$$

and so we contradict hypothesis $H(j)_2(v)$. With a similar modification, we establish the validity of Proposition 4.1.2, namely that $\varphi|_{W_k}$ is anticoercive. Finally, it is clear that the proof of Proposition 4.1.3 is not affected by the modification in the generalized Landesman-Lazer type condition. So finally we can apply the nonsmooth Saddle Point Theorem (see Theorem 2.1.4) and produce a solution of (4.2). □

Next we show that the Landesman-Lazer type conditions used in Theorems 4.1.1 and 4.1.2 are more general than the ones existing in the literature. For this purpose we introduce the following functions:

$$g_+^\infty(z) \overset{df}{=} \inf_{\{u_n\}_{n \geq 1}} \{\liminf u_n^*, \ u_n^* \in \partial j(z, x_n), \ x_n \to +\infty\},$$

$$g_-^\infty(z) \overset{df}{=} \sup_{\{u_n\}_{n \geq 1}} \{\limsup u_n^*, \ u_n^* \in \partial j(z, x_n), \ x_n \to -\infty\}.$$

In the smooth case and if we set

$$f(z, \zeta) = \partial j(z, \zeta),$$

then

$$g_+^\infty(z) = \liminf_{\zeta \to +\infty} f(z, \zeta) \quad \text{and} \quad g_-^\infty(z) = \limsup_{\zeta \to -\infty} f(z, \zeta)$$

and so we recover the classical Landesman-Lazer conditions. In fact originally it was assumed by Landesman & Lazer (1969/1970) that the limits at $+\infty$ and at $-\infty$ actually existed.

Also we introduce the functions

$$g_{\min}(z, \zeta) \stackrel{df}{=} \min_{u^* \in \partial j(z,\zeta)} u^*$$

$$g_{\max}(z, \zeta) \stackrel{df}{=} \max_{u^* \in \partial j(z,\zeta)} u^*$$

$$G_{\min}(z, \zeta) \stackrel{df}{=} \begin{cases} \frac{2j(z,\zeta)}{\zeta} - g_{\min}(z, \zeta) & \text{if } \zeta \neq 0, \\ 0 & \text{if } \zeta = 0, \end{cases}$$

$$G_{\max}(z, \zeta) \stackrel{df}{=} \begin{cases} \frac{2j(z,\zeta)}{\zeta} - g_{\max}(z, \zeta) & \text{if } \zeta \neq 0, \\ 0 & \text{if } \zeta = 0, \end{cases}$$

$$G_{\min}^-(z) \stackrel{df}{=} \limsup_{\zeta \to -\infty} G_{\min}(z, \zeta),$$

$$G_{\max}^-(z) \stackrel{df}{=} \liminf_{\zeta \to +\infty} G_{\max}(z, \zeta).$$

Note that in the smooth case and if $f(z, \zeta) = \partial j(z, \zeta)$, then

$$g_{\min}(z, \zeta) = g_{\max}(z, \zeta) = f(z, \zeta),$$
$$G_{\min}(z, \zeta) = G_{\max}(z, \zeta).$$

Using these items, a Landesman-Lazer type condition was introduced recently in the context of ordinary differential equations and of elliptic problems (semilinear and nonlinear) with smooth potential. For the relevant literature see Section 4.9.

The two propositions that follow show that the Landesman-Lazer type condition employed here (see hypothesis $H(j)_1(v)$) is more general. In what follows $j(z, \zeta)$ satisfies hypotheses $H(j)_1$.

PROPOSITION 4.1.4
If hypotheses $H(j)_1$ hold,
then
$$g_+^\infty(z) \leq j_+(z) \quad \text{and} \quad g_-^\infty(z) \geq j_-(z) \quad \text{for a.a. } z \in \Omega.$$

PROOF By virtue of hypothesis $H(j)_1(ii)$, there exists a set $D \subseteq \Omega$ with $|D|_N = 0$, such that the function $j(z, \cdot)$ is locally Lipschitz for every $z \in \Omega \setminus D$. So for fixed $z \in \Omega \setminus D$, the function $j(z, \cdot)$ is differentiable at every $\zeta \in \mathbb{R} \setminus N(z)$, with $|N(z)|_N = 0$. We have

$$j_r'(z, r) \in \partial j(z, r) \quad \forall z \in \Omega \setminus D, \ r \in \mathbb{R} \setminus N(z).$$

From the definition of $g_+^\infty(z)$, we know that for a given $\varepsilon > 0$, we can find $M_3 = M_3(\varepsilon) > 0$, such that

$$g_+^\infty(z) - \varepsilon(z) \leq j_r'(z, r) \qquad \forall\, r \geq M_3 > 0.$$

For $\zeta \geq M_3 > 0$, we have

$$\frac{j(z, \zeta) - j(z, 0)}{\zeta}$$

$$= \frac{1}{\zeta} \int_0^\zeta j_r'(z, r)\, dr \ = \ \frac{1}{\zeta} \int_0^{M_3} j_r'(z, r)\, dr + \frac{1}{\zeta} \int_{M_3}^\zeta j_r'(z, r)\, dr$$

$$\geq \frac{1}{\zeta} \int_0^{M_3} j_r'(z, r)\, dr + \frac{\zeta - M_3}{\zeta} \big(g_+^\infty(z) - \varepsilon\big).$$

Passing to the limit as $\zeta \to +\infty$, we obtain

$$j_+(z) \ \geq \ g_+^\infty(z) - \varepsilon.$$

Let $\varepsilon \searrow 0$ to conclude that

$$g_+^\infty(z) \ \leq \ j_+(z) \quad \text{for a.a. } z \in \Omega.$$

Similarly, we show that

$$g_-^\infty(z) \ \geq \ j_-(z) \quad \text{for a.a. } z \in \Omega.$$

$$\square$$

REMARK 4.1.2   This proposition implies that the Landesman-Lazer type conditions employed here (see hypothesis $H(j)_1(v)$ or $H(j)_2(v)$) are more general than the original Landesman-Lazer condition. $\square$

PROPOSITION 4.1.5
If hypotheses $H(j)_1$ hold,
then

$$G_{\max}^+(z) \ \leq \ j_+(z) \quad \text{and} \quad G_{\max}^-(z) \ \geq \ j_-(z) \quad \text{for a.a. } z \in \Omega.$$

PROOF   Let $D \subseteq \Omega$ be the Lebesgue-null set outside of which $j(z, \cdot)$ is locally Lipschitz and the asymptotic limits $j_\pm(z)$ exist. Let $z \in \Omega \setminus D$, $\varepsilon > 0$ and $k_\varepsilon^+(z) \overset{df}{=} G_{\max}^+(z) - \varepsilon$. From the definition of $G_{\max}^+$, we know that we can find $M_4 = M_4(\varepsilon) > 0$, such that for all $r \geq M_4 > 0$, we have

$$G_{\max}^+(z) - \varepsilon \ = \ k_\varepsilon^+(z) \ \leq \ G_{\max}(z, r),$$

so

$$\frac{k_\varepsilon^+(z)}{r^2} = \frac{d}{dr}\left(-\frac{k_\varepsilon^+(z)}{r}\right) \le \frac{G_{\max}(z,r)}{r^2}.$$

From the definition of $G_{\max}$, we have

$$\frac{G_{\max}(z,r)}{r^2} = \frac{2j(z,r)}{r^3} - \frac{g_{\max}(z,r)}{r^2} \le \frac{2j(z,r)}{r^3} - \frac{u^*}{r^2} \qquad \forall\, u^* \in \partial j(z,r).$$

The function $r \longmapsto \frac{j(z,r)}{r}$ is locally Lipschitz on $[M_4,+\infty)$ and from Corollary 1.3.7, we have

$$\partial\left(\frac{j(z,r)}{r^2}\right) \subseteq \frac{\partial j(z,r)}{r^2} - \frac{2j(z,r)}{r^3}$$

and thus

$$\frac{d}{dr}\left(-\frac{k_\varepsilon^+(z)}{r}\right) \le -\vartheta(z,r) \qquad \forall\, \vartheta(z,r) \in \partial j\left(\frac{j(z,r)}{r^2}\right).$$

The function $r \longmapsto \frac{j(z,r)}{r^2}$ is locally Lipschitz on $[M_4,+\infty)$ and it is differentiable at every $r \in [M_4,+\infty) \setminus N(z)$, with $|N(z)|_1 = 0$. We set

$$\vartheta_0(z,r) \stackrel{df}{=} \begin{cases} \frac{d}{dr}\left(\frac{j(z,r)}{r^2}\right) & \text{if } r \in [M_4,+\infty) \setminus N(z), \\ 0 & \text{otherwise.} \end{cases}$$

Since

$$\vartheta_0(z,r) \in \partial j(z,r) \qquad \forall\, z \in \Omega \setminus D,\ r \in [M_4,+\infty) \setminus N(z),$$

we can write that

$$\tfrac{d}{dr}\left(-\tfrac{k_\varepsilon^+(z)}{r}\right) \le -\vartheta_0(z,r) \qquad \forall\, r \in [M_4,+\infty) \setminus N(z).$$

Integrating on $[\zeta,\overline{\zeta}]$ with $\zeta,\overline{\zeta} \in [M_4,+\infty)$ and $\zeta < \overline{\zeta}$, we obtain

$$-\frac{k_\varepsilon^+(z)}{\overline{\zeta}} + \frac{k_\varepsilon^+(z)}{\zeta} \le \int_\zeta^{\overline{\zeta}} -\vartheta_0(z,r)\,dr = \int_\zeta^{\overline{\zeta}} -\frac{d}{dr}\left(\frac{j(z,r)}{r^2}\right)\,dr$$

$$= -\frac{j(z,\overline{\zeta})}{\overline{\zeta}^2} + \frac{j(z,\zeta)}{\zeta^2}. \tag{4.14}$$

By virtue of hypotheses $H(j)_1(iii)$ and (iv), we have

$$-\beta_7 r - \varepsilon r \overline{r} \le u^*(z,r)\overline{r} \qquad \forall\, z \in \Omega \setminus D,\ r,\overline{r} \ge 0,\ u^* \in \partial j(z,r),$$

for some $\beta_7 > 0$. Applying the mean value theorem for locally Lipschitz functions (see Proposition 1.3.14), we obtain

$$j(z,r) - j(z,0) = u^* r,$$

with $u^* \in \partial j(z, t_z r)$, $t_z \in (0, 1)$, so

$$j(z, r) \geq -\varepsilon t_z r^2 - \beta_7 r + j(z, 0) \geq -\varepsilon r^2 - \beta_7 r + j(z, 0)$$

and since $\varepsilon > 0$ was arbitrary, we get

$$\liminf_{r \to +\infty} \frac{j(z, r)}{r^2} \geq 0. \tag{4.15}$$

So, if in (4.14) we pass to the limit as $\overline{\zeta} \to +\infty$, using also (4.15), we obtain

$$k_{\varepsilon}^+(z) \leq \frac{j(z, \zeta)}{\zeta}$$

and since $\varepsilon > 0$ was arbitrary, we get

$$G_{\max}^+(z) \leq \liminf_{\zeta \to +\infty} \frac{j(z, \zeta)}{\zeta} = j_+(z).$$

Similarly, we establish that

$$G_{\min}^-(z) \geq j_-(z) \quad \text{for a.a. } z \in \Omega.$$

$\Box$

EXAMPLE 4.1.1   Consider the nonsmooth locally Lipschitz potential function $j(\zeta)$, defined by

$$j(\zeta) \overset{df}{=} \max\left\{ \zeta^{\frac{1}{3}}, |\zeta|^{\frac{1}{2}} \right\} + \ln\left(1 + |\zeta|\right) + \cos\zeta + \zeta.$$

Then we can check that $j_- = -1$, $j_+ = 1$, but $g_-^\infty = g_+^\infty = G_{\min}^- = G_{\max}^+ = 0$. So the Landesman-Lazer type condition introduced above is more general than the other two.

Similarly the function

$$j(\zeta) \overset{df}{=} \begin{cases} \ln(1 + |\zeta|) & \text{if } |\zeta| \leq 1, \\ \zeta - 1 + \cos\zeta + \ln 2 - \cos 1 & \text{if } \zeta > 1, \\ -\zeta + 1 + \cos\zeta + \ln 2 - \cos 1 & \text{if } \zeta < -1. \end{cases}$$

Again we have $j_- = -1$, $j_+ = 1$, $g_-^\infty = g_+^\infty = G_{\min}^- = G_{\max}^+ = 0$. $\Box$

4.1.2   Nonlinear Problems at Resonance

Next we turn our attention to nonlinear problems driven by the p-Laplacian. Let $\Omega \subseteq \mathbb{R}^N$ be a bounded domain with a $C^{1,\alpha}$ boundary $(0 < \alpha < 1)$. The problem under consideration is the following:

$$\begin{cases} -\text{div}\left(\|\nabla x(z)\|_{\mathbb{R}^N}^{p-2} \nabla x(z) \right) \in \partial j\left(z, x(z)\right) & \text{for a.a. } z \in \Omega, \\ x|_{\partial\Omega} = 0, \end{cases} \tag{4.16}$$

with $p \in (1, +\infty)$. As we already saw in the context of semilinear problems the resolution of inclusions like (4.16) involves the asymptotic behaviour of the nonsmooth potential $j(z, \zeta)$ at $\pm\infty$. When these asymptotic values interfere with the spectrum of the differential operator, then we have a resonant problem. In the present nonlinear setting, we encounter serious difficulties which are due to the lack of full knowledge of the spectrum of $\left(-\Delta_p, W_0^{1,p}(\Omega)\right)$ and the lack of variational expressions for the higher eigenvalues (analogous to the ones produced in Theorem 1.5.3(b) for the semilinear case $(p = 2)$). We do not even have variational expressions with respect to certain suitable cones in $W_0^{1,p}(\Omega)$ (see Remark 1.5.10). This fact makes it difficult to produce appropriate linking sets. Recall (see Section 1.5) that the first eigenvalue λ_1 of $\left(-\Delta_p, W_0^{1,p}(\Omega)\right)$ is positive, isolated and simple (i.e. the corresponding eigenspace is one dimensional). Moreover, the corresponding normalized eigenfunction u_1 satisfies $u_1 \in C^{1,\beta}\left(\overline{\Omega}\right)$ $(\beta \in (0,1))$ (see Theorem 1.5.6) and $u_1(z) > 0$ for all $z \in \Omega$ (see Theorem 1.5.7). The Lusternik-Schnirelmann theory gives, in addition to λ_1, a whole strictly increasing sequence $\{\lambda_n\}_{n \geq 1} \subseteq \mathbb{R}_+$ of eigenvalues, known as "variational eigenvalues" (or "Lusternik-Schnirelmann eigenvalues"). These numbers are defined as follows. Let

$$G \stackrel{df}{=} \left\{x \in W_0^{1,p}(\Omega) : \|\nabla x\|_p = 1\right\}$$

and let $\psi \colon G \longrightarrow \mathbb{R}_-$ be given by

$$\psi(x) \stackrel{df}{=} -\|x\|_p^p.$$

We set

$$\widetilde{c}_n \stackrel{df}{=} \inf_{K \in \mathcal{A}_n} \sup_{x \in K} \psi(x), \tag{4.17}$$

where

$$\mathcal{A}_n \stackrel{df}{=} \left\{K \subseteq G : K \text{ is symmetric, closed and } \gamma(K) \geq n\right\},$$

with γ being the Krasnoselskii $\mathbb{Z}_2$-genus. Then

$$\lambda_n \stackrel{df}{=} -\frac{1}{\widetilde{c}_n} \qquad \forall \, n \geq 1.$$

If $p = 2$ (semilinear case), then these are all the eigenvalues of $\left(-\Delta, H_0^1(\Omega)\right)$. For $p \neq 2$, we cannot say this. We only know that

$$\lambda_2 = \inf\left\{\lambda > \lambda_1 : \lambda \text{ is an eigenvalue of } \left(-\Delta_p, W_0^{1,p}(\Omega)\right)\right\}.$$

Also, for any $k \geq 1$, let

$$V_k \stackrel{df}{=} \left\{x \in W_0^{1,p}(\Omega) : -\mathrm{div}\left(\|\nabla x(z)\|_{\mathbb{R}^N}^{p-2} \nabla x(z)\right) = \lambda_k |x(z)|^{p-2} x(z)\right.$$

$$\left. \text{for a.a. } z \in \Omega\right\}.$$

These are symmetric closed cones, but in general not subspaces of $W_0^{1,p}(\Omega)$, unless λ_k is simple. Let us set

$$W_n \stackrel{df}{=} \bigcup_{k=1}^{n} V_k \quad \text{and} \quad \widehat{W}_n \stackrel{df}{=} \bigcup_{k=n+1}^{\infty} V_k.$$

We cannot say (as in the semilinear case) that

$$\|\nabla x\|_p^p \leq \lambda_k \|x\|_p^p \qquad \forall\, x \in W_k$$

and

$$\|\nabla x\|_p^p \geq \lambda_{k+1} \|x\|_p^p \qquad \forall\, x \in \widehat{W}_k.$$

This negative fact is the source of difficulties when we deal with nonlinear problems, in which the nonsmooth potential asymptotically at $\pm\infty$ goes beyond the first eigenvalue $\lambda_1 > 0$. For this reason almost all works involving the p-Laplacian use asymptotic conditions at $\pm\infty$ restricted from above by $\lambda_1 > 0$. In that situation the Mountain Pass geometry is satisfied and we obtain critical points. When we cross $\lambda_1 > 0$, we have to abandon the Mountain Pass Theorem and employ general minimax principles involving linking sets (see Theorem 2.1.3). In this context we face the difficulty of constructing links between the sets.

Our hypotheses for the nonsmooth potential function $j(z,\zeta)$ of (4.16) are the following:

$\underline{H(j)_3}$ $j\colon \Omega \times \mathbb{R} \longrightarrow \mathbb{R}$ is a function, such that

(i) for every $\zeta \in \mathbb{R}$, the function

$$\Omega \ni z \longrightarrow j(z,\zeta) \in \mathbb{R}$$

is measurable and $j(\cdot,0) \in L^1(\Omega)$;

(ii) for almost all $z \in \Omega$, the function

$$\mathbb{R} \ni \zeta \longrightarrow j(z,\zeta) \in \mathbb{R}$$

is locally Lipschitz;

(iii) there exist $a \in L^\infty(\Omega)$, $c > 0$ and $r \in [1, p^*)$, such that

$$|u| \leq a(z) + c|\zeta|^{r-1} \quad \text{for a.a. } z \in \Omega,\ \text{all } \zeta \in \mathbb{R} \text{ and all } u \in \partial j(z,\zeta);$$

(iv) $\displaystyle\lim_{|\zeta| \to +\infty} \big(u\zeta - pj(z,\zeta)\big) = -\infty$ uniformly for almost all $z \in \Omega$ and all $u \in \partial j(z,\zeta)$;

(v) $\displaystyle \lambda_1 \leq \liminf_{|\zeta| \to +\infty} \frac{pj(z,\zeta)}{|\zeta|^p} \leq \limsup_{|\zeta| \to +\infty} \frac{pj(z,\zeta)}{|\zeta|^p} < \lambda_2$ uniformly for almost all $z \in \Omega$.

We introduce the locally Lipschitz energy functional $\varphi\colon W_0^{1,p}(\Omega) \longrightarrow \mathbb{R}$, defined by

$$\varphi(x) \stackrel{df}{=} \frac{1}{p} \|\nabla x\|_p^p - \int_\Omega j\big(z, x(z)\big)\, dz \qquad \forall\, x \in W_0^{1,p}(\Omega).$$

PROPOSITION 4.1.6
If hypotheses $H(j)_3$ hold, <u>then</u> φ satisfies the nonsmooth C-condition.

PROOF Let $\{x_n\}_{n\geq 1} \subseteq W_0^{1,p}(\Omega)$ be a sequence, such that

$$\varphi(x_n) \longrightarrow c_1 \quad \text{and} \quad \big(1 + \|x_n\|_{W^{1,p}(\Omega)}\big) m^\varphi(x_n) \longrightarrow 0.$$

Let $x_n^* \in \partial\varphi(x_n)$ be such that $\|x_n^*\|_{W^{1,p}(\Omega)} = m^\varphi(x_n)$ for $n \geq 1$. We have

$$x_n^* = A(x_n) - u_n^* \qquad \forall\, n \geq 1,$$

where $A\colon W_0^{1,p}(\Omega) \longrightarrow W^{-1,p'}(\Omega)$ is defined by

$$\langle A(x), y\rangle_{W_0^{1,p}(\Omega)} = \int_\Omega \|\nabla x(z)\|_{\mathbb{R}^N}^{p-2} \big(\nabla x(z), \nabla y(z)\big)_{\mathbb{R}^N}\, dz \qquad \forall\, x, y \in W_0^{1,p}(\Omega)$$

and $u_n^* \in L^{p'}(\Omega)$, $u_n^*(z) \in \partial j(z, x_n(z))$ for almost all $z \in \Omega$ and all $n \geq 1$. Note that A is monotone, demicontinuous, thus maximal monotone (see Corollary 1.4.2). From the choice of the sequence $\{x_n\}_{n\geq 1} \subseteq W_0^{1,p}(\Omega)$, we have

$$\left| \langle x_n^*, x_n\rangle_{W_0^{1,p}(\Omega)} - p\varphi(x_n) + pc_1 \right|$$
$$\leq \|x_n^*\|_{W^{-1,p'}(\Omega)} \|x_n\|_{W^{1,p}(\Omega)} + |p\varphi(x_n) - pc_1|$$
$$\leq m^\varphi(x_n)\big(1 + \|x_n\|_{W^{1,p}(\Omega)}\big) + p|\varphi(x_n) - c_1|, \qquad (4.18)$$

with the last term tending to zero as $n \to +\infty$. Note that

$$\langle x_n^*, x_n\rangle_{W_0^{1,p}(\Omega)} = \langle A(x_n), x_n\rangle_{W_0^{1,p}(\Omega)} - \int_\Omega u_n^*(z) x_n(z)\, dz$$

$$= \|\nabla x_n\|_p^p - \int_\Omega u_n^*(z) x_n(z)\, dz.$$

So from (4.18) above, we obtain

$$p\varphi(x_n) - \langle x_n^*, x_n\rangle_{W_0^{1,p}(\Omega)}$$
$$= \int_\Omega \big(u_n^*(z) x_n(z) - pj(z, x_n(z))\big)\, dz \longrightarrow pc_1. \qquad (4.19)$$

We claim that the sequence $\{x_n\}_{n\geq 1} \subseteq W_0^{1,p}(\Omega)$ is bounded. Suppose that this is not the case. Then by passing to a subsequence if necessary, we may assume that

$$\|x_n\|_{W^{1,p}(\Omega)} \longrightarrow +\infty.$$

Let

$$y_n \overset{df}{=} \frac{x_n}{\|x_n\|_{W^{1,p}(\Omega)}} \qquad \forall\, n \geq 1.$$

We may assume that

$$\begin{aligned}
y_n &\overset{w}{\longrightarrow} y && \text{in } W_0^{1,p}(\Omega), \\
y_n &\longrightarrow y && \text{in } L^p(\Omega) \text{ and in } L^r(\Omega), \\
y_n(z) &\longrightarrow y(z) && \text{for a.a. } z \in \Omega, \\
|y_n(z)| &\leq k(z) && \text{for a.a. } z \in \Omega \text{ and all } n \geq 1,
\end{aligned}$$

with $k \in L^m(\Omega)$ and $m = \max\{p, r\}$.

Because of hypothesis $H(j)_3(v)$, we can find $\eta \in [\lambda_1, \lambda_2)$ and $M_5 > 0$, such that

$$j(z, \zeta) \leq \frac{\eta}{p}|\zeta|^p \quad \text{for a.a. } z \in \Omega \text{ and all } |\zeta| \geq M_5.$$

Also note that from hypothesis $H(j)_3(iii)$ and the mean value theorem for locally Lipschitz functions (see Proposition 1.3.14), we have

$$|j(z, \zeta)| \leq a_1(z) + c_2|\zeta|^r \quad \text{for a.a. } z \in \Omega \text{ and all } \zeta \in \mathbb{R},$$

with $a_1 \in L^1(\Omega)$ and $c_2 > 0$ (recall that $j(\cdot, 0) \in L^1(\Omega)$). So

$$|j(z, \zeta)| \leq a_2(z) \quad \text{for a.a. } z \in \Omega \text{ and all } |\zeta| < M_5,$$

with $a_2 \in L^1(\Omega)$. Thus we can say that

$$j(z, \zeta) \leq \frac{\eta}{p}|\zeta|^p + a_2(z). \tag{4.20}$$

For every $n \geq 1$, we have

$$\begin{aligned}
\frac{\varphi(x_n)}{\|x_n\|_{W^{1,p}(\Omega)}^p} &= \frac{1}{p}\|\nabla y_n\|_p^p - \int_\Omega \frac{j(z, x_n(z))}{\|x_n\|_{W^{1,p}(\Omega)}^p}\, dz \\
&\geq \frac{1}{p}\|\nabla y_n\|_p^p - \frac{\eta}{p}\int_\Omega |y_n(z)|^p\, dz - \int_\Omega \frac{a_2(z)}{\|x_n\|_{W^{1,p}(\Omega)}^p}\, dz.
\end{aligned} \tag{4.21}$$

If $y = 0$, then from (4.21), we infer that

$$\|\nabla y_n\|_p \longrightarrow 0,$$

hence

$$y_n \longrightarrow 0 \quad \text{in } W_0^{1,p}(\Omega),$$

a contradiction to the fact that $\|y_n\|_{W^{1,p}(\Omega)} = 1$ for all $n \geq 1$. So $y \neq 0$.

Because of hypothesis $H(j)_3(iv)$, we can find $M_6 > 0$, such that

$$u^* \zeta - pj(z, \zeta) \leq -1 \quad \text{for a.a. } z \in \Omega, \text{ all } |\zeta| \geq M_6 \text{ and all } u^* \in \partial j(z, \zeta).$$

On the other hand, from hypothesis $H(j)_3(iii)$ and (4.20), we see that

$$u^* \zeta - pj(z, \zeta) \leq c_3 \quad \text{for a.a. } z \in \Omega, \text{ all } |\zeta| \leq M_6 \text{ and all } u^* \in \partial j(z, \zeta),$$

for some $c_3 > 0$. Therefore

$$u^* \zeta - pj(z, \zeta) \leq c_3 \quad \text{for a.a. } z \in \Omega, \text{ all } \zeta \in \mathbb{R} \text{ and all } u^* \in \partial j(z, \zeta). \quad (4.22)$$

Let us set

$$C \stackrel{df}{=} \{z \in \Omega : y(z) \neq 0\}.$$

Evidently $|C|_N > 0$ and for almost all $z \in C$, we have that

$$|x_n(z)| \longrightarrow +\infty.$$

So, using (4.22), we have

$$\int_\Omega \left(u_n^*(z) x_n(z) - pj\big(z, x_n(z)\big) \right) dz$$

$$= \int_C \left(u_n^*(z) x_n(z) - pj\big(z, x_n(z)\big) \right) dz + \int_{C^c} \left(u_n^*(z) x_n(z) - pj\big(z, x_n(z)\big) \right) dz$$

$$\leq \int_C \left(u_n^*(z) x_n(z) - pj\big(z, x_n(z)\big) \right) dz + c_3 |C^c|_N.$$

So by Fatou's lemma and hypothesis $H(j)_3(iv)$, we have

$$\lim_{n \to +\infty} \int_\Omega \left(u_n^*(z) x_n(z) - pj\big(z, x_n(z)\big) \right) dz = -\infty,$$

which contradicts (4.19). This proves that the sequence $\{x_n\}_{n \geq 1} \subseteq W_0^{1,p}(\Omega)$ is bounded and so, passing to a subsequence if necessary, we may assume that

$$x_n \xrightarrow{w} x \text{ in } W_0^{1,p}(\Omega),$$
$$x_n \longrightarrow x \text{ in } L^r(\Omega)$$

(since $r < p^*$). We have

$$\left| \langle x_n^*, x_n - x \rangle_{W^{1,p}(\Omega)} \right|$$

$$= \left| \langle A(x_n), x_n - x \rangle_{W^{1,p}(\Omega)} - \int_\Omega u_n^*(z)(x_n - x)(z) \, dz \right| \leq \varepsilon_n,$$

with $\varepsilon_n \searrow 0$. By hypothesis $H(j)_3(iii)$, the sequence $\{u_n^*\}_{n \geq 1} \subseteq L^{r'}(\Omega)$ is bounded and so

$$\int_\Omega u_n^*(z)(x_n - x)(z)\, dz \longrightarrow 0.$$

Hence

$$\lim_{n \to +\infty} \langle A(x_n), x_n - x \rangle_{W_0^{1,p}(\Omega)} = 0.$$

Because A is maximal monotone, it is generalized pseudomonotone (see Proposition 1.4.11) and so

$$\langle A(x_n), x_n \rangle_{W_0^{1,p}(\Omega)} \longrightarrow \langle A(x), x \rangle_{W_0^{1,p}(\Omega)},$$

hence

$$\|\nabla x_n\|_p \longrightarrow \|\nabla x\|_p.$$

Since

$$\nabla x_n \xrightarrow{w} \nabla x \quad \text{in } L^p(\Omega; \mathbb{R}^N)$$

and the latter space is uniformly convex, we have that

$$\nabla x_n \longrightarrow \nabla x \quad \text{in } L^p(\Omega; \mathbb{R}^N)$$

(see Kadec-Klee property) and so

$$x_n \longrightarrow x \quad \text{in } W_0^{1,p}(\Omega).$$

$$\square$$

In the next proposition, we show the anticoercivity of φ on $\mathbb{R}u_1$ (the eigenspace of λ_1).

PROPOSITION 4.1.7
If hypotheses $H(j)_3$ *hold,* <u>*then*</u> $\varphi(tu_1) \longrightarrow -\infty$ *as* $|t| \to +\infty$.

PROOF   Let us set

$$\vartheta(z, \varsigma) = j(z, \varsigma) - \frac{\lambda_1}{p}|\varsigma|^p.$$

Clearly for every $\varsigma \in \mathbb{R}$, the function $z \longmapsto \vartheta(z, \varsigma)$ is measurable and for almost all $z \in \Omega$, the function $\varsigma \longmapsto \vartheta(z, \varsigma)$ is locally Lipschitz. From Corollary 1.3.7, we know that the function $\varsigma \longmapsto \frac{\vartheta(z,\varsigma)}{|\varsigma|^p}$ is locally Lipschitz on $(0, +\infty)$ and for almost all $z \in \Omega$ and all $\varsigma > 0$, we have

$$\partial \left(\frac{\vartheta(z,\varsigma)}{|\varsigma|^p} \right) = \frac{|\varsigma|^p \partial \vartheta(z, \varsigma) - |\varsigma|^{p-2} \varsigma \vartheta(z, \varsigma)}{|\varsigma|^{2p}}$$

$$= |\zeta|^{p-1} \frac{\zeta \partial \vartheta(z,\zeta) - p\vartheta(z,\zeta)}{|\zeta|^{2p}}$$

$$= \frac{\zeta \partial j(z,\zeta) - \lambda_1 |\zeta|^p - pj(z,\zeta) + \lambda_1 |\zeta|^p}{|\zeta|^{p+1}}$$

$$= \frac{\zeta \partial j(z,\zeta) - pj(z,\zeta)}{|\zeta|^{p+1}}. \tag{4.23}$$

By virtue of hypothesis $H(j)_3(iv)$, for a given $\mu > 0$, we can find $M_7 > 0$, such that

$$u^* \zeta - pj(z,\zeta) \leq -\mu \quad \text{for a.a. } z \in \Omega, \text{ all } \zeta > M_7 \text{ and all } u^* \in \partial j(z,\zeta),$$

so, using also (4.23), we have

$$v \leq -\frac{\mu}{|\zeta|^{p+1}} \qquad \forall \, v \in \partial \left(\frac{\vartheta(z,\zeta)}{|\zeta|^p} \right). \tag{4.24}$$

For all $z \in \Omega \setminus D$, where $D \subseteq \Omega$, $|D|_N = 0$, the function $\zeta \longmapsto \frac{\vartheta(z,\zeta)}{|\zeta|^p}$ is locally Lipschitz on $(M_7, +\infty)$ and so it is differentiable at every $\zeta \in (M_7, +\infty) \setminus N(z)$, with $|N(z)|_1 = 0$. Let us set

$$\gamma_0(z,\zeta) \overset{df}{=} \begin{cases} \frac{d}{d\zeta} \left(\frac{\vartheta(z,\zeta)}{\zeta^p} \right) & \text{if } \zeta \in (M_7, +\infty) \setminus N(z), \\ 0 & \text{if } \zeta \in N(z). \end{cases}$$

Then from (4.24), we have

$$\gamma_0(z,\zeta) \leq -\frac{\mu}{\zeta^{p+1}} = \frac{d}{d\zeta} \left(\frac{\mu}{p\zeta^p} \right) \qquad \forall \, z \in \Omega \setminus D, \, \zeta \in (M_7, +\infty) \setminus E(z).$$

Integrating this inequality on $[\zeta_1, \zeta_2]$, with $M_7 < \zeta_1 < \zeta_2 < +\infty$, we obtain

$$\int_{\zeta_1}^{\zeta_2} \gamma_0(z,\zeta) \, d\zeta = \int_{\zeta_1}^{\zeta_2} \frac{d}{ds} \left(\frac{\vartheta(z,\zeta)}{\zeta^p} \right) d\zeta \leq \frac{\mu}{p} \int_{\zeta_1}^{\zeta_2} \left(\frac{d}{d\zeta} \frac{1}{\zeta^p} \right) d\zeta,$$

so

$$\frac{\vartheta(z,\zeta_2)}{\zeta_2^p} - \frac{\vartheta(z,\zeta_1)}{\zeta_1^p} \leq \frac{\mu}{p} \left(\frac{1}{\zeta_2^p} - \frac{1}{\zeta_1^p} \right). \tag{4.25}$$

Let $\zeta_2 \longrightarrow +\infty$. By virtue of hypothesis $H(j)_3(v)$, we have that

$$\liminf_{\zeta_2 \to +\infty} \frac{\vartheta(z,\zeta_2)}{\zeta_2^p} \geq 0 \quad \text{uniformly for a.a. } z \in \Omega.$$

So from (4.25), we obtain

$$\frac{\mu}{p} \leq \vartheta(z,\zeta_1) \qquad \forall \, \zeta_1 \in (M_7, +\infty). \tag{4.26}$$

For $t > 0$, let

$$L_t \stackrel{df}{=} \big\{ z \in \Omega : \; t u_1(z) > M_7 \big\}.$$

Recall that $u_1(z) > 0$ for all $z \in \Omega$. So it follows that $\big| L_t^c \big|_N \longrightarrow 0$ as $t \to +\infty$. For $t > 0$, we have

$$
\begin{aligned}
\varphi(tu_1) \; &= \; \frac{t^p}{p} \|\nabla u_1\|_p^p - \int_\Omega j\big(z, tu_1(z)\big)\, dz \\[2mm]
&= \; \frac{t^p}{p} \|\nabla u_1\|_p^p - \int_{L_t} j\big(z, tu_1(z)\big)\, dz - \int_{L_t^c} j\big(z, tu_1(z)\big)\, dz \\[2mm]
&\leq \; \frac{t^p}{p} \|\nabla u_1\|_p^p - \frac{\mu}{p} |L_t|_N - \frac{t^p \lambda_1}{p} \|u_1\|_p^p + \frac{\lambda_1}{p} \int_{L_t^c} |tu_1(z)|^p\, dz \\[2mm]
&\quad + \int_{L_t^c} a_1(z)\, dz + c_2 \int_{L_t^c} |tu_1(z)|^r\, dz \\[2mm]
&= \; -\frac{\mu}{p} |L_t|_N + \frac{\lambda_1}{p} M_7^p \big| L_t^c \big|_N + \int_{L_t^c} a_1(z)\, dz + c_2 M_7^r \big| L_t^c \big|_N,
\end{aligned}
$$

so

$$\limsup_{t \to +\infty} \varphi(tu_1) \; \leq \; -\frac{\mu}{p} |\Omega|_N.$$

But $\mu > 0$ was arbitrary. So we conclude that

$$\varphi(tu_1) \; \longrightarrow \; -\infty \quad \text{as } t \to +\infty.$$

Similarly we show that

$$\varphi(tu_1) \; \longrightarrow \; -\infty \quad \text{as } t \to -\infty.$$

This proves that $\varphi|_{\mathbb{R}u_1}$ is anticoercive.   ☐

We introduce the symmetric closed cone

$$K \stackrel{df}{=} \big\{ x \in W_0^{1,p}(\Omega) : \; \|\nabla x\|_p^p = \lambda_2 \|x\|_p^p \big\}.$$

Next we show that $\varphi|_K$ is weakly coercive.

PROPOSITION 4.1.8
If hypotheses $H(j)_3$ hold, then

$$\varphi(v) \; \longrightarrow \; +\infty \quad \text{as } \|v\|_{W^{1,p}(\Omega)} \to +\infty, \; v \in K. \tag{4.27}$$

PROOF Recall that we can find $\eta \in [\lambda_1, \lambda_2)$ and $M_5 > 0$, such that

$$j(z, \zeta) \leq \frac{\eta}{p} |\zeta|^p \quad \text{for a.a. } z \in \Omega \text{ and all } |\zeta| \geq M_5$$

(see the proof of Proposition 4.1.6). So, if $v \in K$, we have

$$
\begin{aligned}
\varphi(v) &= \frac{1}{p} \|\nabla v\|_p^p - \int_\Omega j\big(z, v(z)\big)\, dz \\
&= \frac{1}{p} \|\nabla v\|_p^p - \int_{\{|v| \geq M_5\}} j\big(z, v(z)\big)\, dz - \int_{\{|v| < M_5\}} j\big(z, v(z)\big)\, dz \\
&\geq \frac{1}{p} \|\nabla v\|_p^p - \frac{\eta}{p} \|v\|_p^p - c_4 = \frac{1}{p}\left(1 - \frac{\eta}{\lambda_2}\right) \|\nabla v\|_p^p - c_4,
\end{aligned}
$$

for some $c_4 > 0$. Because $\eta < \lambda_2$, we obtain (4.27). $\quad\square$

Using the above auxiliary results concerning φ and the general minimax principle of Theorem 2.1.3, we obtain an existence theorem of problem (4.16).

THEOREM 4.1.3
If hypotheses $h(j)_3$ hold, <u>then</u> problem (4.16) has a solution $x_0 \in W_0^{1,p}(\Omega)$.

PROOF By virtue of Proposition 4.1.8, we can find $k_1 > -\infty$, such that

$$k_1 = \inf_{v \in K} \varphi(v).$$

Also because of Proposition 4.1.7, we can find $t^* > 0$ large enough, such that

$$\varphi(t^* u_1) < k_1.$$

Let

$$
\begin{aligned}
G &\overset{df}{=} \left\{ x \in W_0^{1,p}(\Omega) : \|\nabla x\|_p = 1 \right\} \\
U &\overset{df}{=} \left\{ x \in G : -\psi(x) = \|x\|_p^p > -\tilde{c}_2 \right\},
\end{aligned}
$$

where $\tilde{c}_2 < 0$ is defined in (4.17). Note that the set U is open in G. Moreover, because

$$\|\pm u_1\|_p^p = \frac{1}{\lambda_1} \|\pm \nabla u_1\|_p^p = \frac{1}{\lambda_1} = -\tilde{c}_2 > -\tilde{c}_1,$$

we see that $\pm u_1 \in U$. We claim that u_1 and $-u_1$ belong to different path-connected components of the set U. Suppose that this is not true. This means that we can find $\xi \in C\big([0,1]; U\big)$, such that $\xi(0) = u_1$ and $\xi(1) = -u_1$ (i.e. a continuous curve in U joining u_1 and $-u_1$). Let us set

$$H \overset{df}{=} \xi\big([0,1]\big) \cup \big(-\xi\big([0,1]\big)\big).$$

Evidently it is symmetric, compact and $\gamma(H) > 1$, i.e. $H \in \mathcal{A}_2$. Moreover, from the definition of U and since $H \subseteq U$, we have that

$$\sup_{x \in H} \psi(x) \ < \ \tilde{c}_2,$$

which contradicts (4.17). So indeed u_1 and $-u_1$ belong in different path-connected components of the set U.

Let W be the path-connected component of U, which contains u_1. Then $-W$ is the path-connected component of U, which contains $-u_1$. We set

$$E \overset{df}{=} t^*W \quad \text{and} \quad S \overset{df}{=} E \cup (-E).$$

Because $\lambda_1 = -\frac{1}{\tilde{c}_2}$, we have

$$\|\nabla w\|_p^p \ < \ \lambda_2 \|w\|_p^p \qquad \forall\, w \in S$$

and

$$\|\nabla w\|_p^p \ = \ \lambda_2 \|w\|_p^p \qquad \forall\, w \in \partial S.$$

Therefore $\partial S \subseteq K$. Let us set

$$
\begin{aligned}
C &\overset{df}{=} [-t^*u_1, t^*u_1] \\
&= \{ h \in W_0^{1,p}(\Omega) : \ h = \lambda(-t^*u_1) + (1-\lambda)t^*u_1 \text{ for some } \lambda \in [0,1] \} \\
C_1 &\overset{df}{=} \partial C \ = \ \{ -t^*u_1, t^*u_1 \} \\
D &\overset{df}{=} V.
\end{aligned}
$$

We claim that C_1 and D link in $W_0^{1,p}(\Omega)$ (see Definition 2.1.4 and Remark 2.1.4). To this end first recall that

$$\varphi(\pm t^* u_1) \ < \ k_1 \ = \ \inf_D \varphi,$$

hence $C_1 \cap V = \emptyset$. Now let $\vartheta \in C\big(C; W_0^{1,p}(\Omega)\big)$ be such that

$$\vartheta(\pm t^* u_1) \ = \ \pm t^* u_1.$$

We have that $\vartheta(C) \cap V \supseteq \vartheta(C) \cap \partial S \neq \emptyset$, which proves that the sets C_1 and D link in $W_0^{1,p}(\Omega)$. So we can apply Theorem 2.1.3 and produce $x_0 \in W_0^{1,p}(\Omega)$, such that $0 \in \partial\varphi(x_0)$. As in the proof of Theorem 4.1.1, we can check that this inclusion implies that $x_0 \in W_0^{1,p}(\Omega)$ is a solution of (4.16). $\quad\square$

REMARK 4.1.3   Let

$$j(z, \zeta) \ = \ j(\zeta) \overset{df}{=} \frac{\lambda_1}{p}|\zeta|^p + k(\zeta),$$

with

$$k(\zeta) \overset{df}{=} \begin{cases} \pm \zeta \ln \zeta & \text{if } \zeta > 0, \\ |\zeta| & \text{if } \zeta \le 0. \end{cases}$$

It is easy to check that this locally Lipschitz function satisfies hypotheses $H(j)_3$ with resonance at λ_1 (see hypothesis $H(j)_3(v)$). $\quad\square$

4.1.3   Variational-Hemivariational Inequality at Resonance

Let $\Omega \subseteq \mathbb{R}^N$ be a bounded domain with $C^{1,\alpha}$ boundary ($\alpha \in (0,1)$). The problem under consideration is the following:

$$
\begin{cases}
\displaystyle \int_\Omega \|\nabla x(z)\|_{\mathbb{R}^N}^{p-2} \left(\nabla x(z), \nabla y(z) - \nabla x(z)\right)_{\mathbb{R}^N} dz \\[2mm]
\quad -\lambda_1 \int_\Omega |x(z)|^{p-2} x(z)\big(y(z) - x(z)\big) dz, \\[2mm]
\quad \geq \int_\Omega u^*(z)\big(y(z) - x(z)\big) dz \\[2mm]
\text{for all } y \in C_g, \text{ with } g \in W^{1,p}(\Omega),\ g(z) \leq 0 \text{ for a.a. } z \in \Omega \\[1mm]
\text{and for some } u^* \in L^{p'}(\Omega), \text{ with} \\[1mm]
\quad u^*(z) \in \partial j\big(z, x(z)\big) \text{ for a.a. } z \in \Omega,
\end{cases}
\tag{4.28}
$$

where $p \in (1, +\infty)$ and

$$
C_g \stackrel{df}{=} \big\{ x \in W_0^{1,p}(\Omega) :\ x(z) \geq g(z) \text{ for a.a. } z \in \Omega \big\}.
$$

Problem (4.28) is the weak form of an obstacle problem, with the obstacle being the function $g \in W^{1,p}(\Omega)$ and with a nonsmooth potential (hence a discontinuous right hand side nonlinearity). We approach the problem using variational methods and in particular the critical point theory for convex perturbations of locally Lipschitz functions (see Section 2.3).

In our hypotheses as well as in the analysis of problem (4.28), we shall need the following quantities, which we already encountered in a semilinear context (i.e. $p = 2$) in the comparison of the different Landesman-Lazer type conditions. We define:

$$
g_{\min}(z, \zeta) \stackrel{df}{=} \min_{u^* \in \partial j(z,\zeta)} u^*,
$$

$$
g_{\max}(z, \zeta) \stackrel{df}{=} \max_{u^* \in \partial j(z,\zeta)} u^*,
$$

$$
G_{\min}(z, \zeta) \stackrel{df}{=}
\begin{cases}
\dfrac{pj(z,\zeta)}{\zeta} - g_{\min}(z, \zeta) & \text{if } \zeta \neq 0, \\[2mm]
0 & \text{if } \zeta = 0,
\end{cases}
$$

$$
G_{\max}(z, \zeta) \stackrel{df}{=}
\begin{cases}
\dfrac{pj(z,\zeta)}{\zeta} - g_{\max}(z, \zeta) & \text{if } \zeta \neq 0, \\[2mm]
0 & \text{if } \zeta = 0.
\end{cases}
$$

Our hypotheses for the nonsmooth potential $j(z, \zeta)$ are the following:

$\underline{H(j)_4}$ $j \colon \Omega \times \mathbb{R} \longrightarrow \mathbb{R}$ is a function, such that

$\quad (i)$ for every $\zeta \in \mathbb{R}$, the function

$$
\Omega \ni z \longrightarrow j(z, \zeta) \in \mathbb{R}
$$

is measurable and $j(z, 0) = 0$ for almost all $z \in \Omega$;

(ii) for almost all $z \in \Omega$, the function

$$\mathbb{R} \ni \zeta \longrightarrow j(z, \zeta) \in \mathbb{R}$$

is locally Lipschitz;

(iii) for all $M > 0$, there exists $a_M \in L^\infty(\Omega)$, such that

$$|u| \leq a_M(z) \quad \text{for a.a. } z \in \Omega, \text{ all } |\zeta| \leq M \text{ and all } u \in \partial j(z, \zeta);$$

(iv) $\lim\limits_{|\zeta| \to +\infty} \dfrac{u}{|\zeta|^{p-1}} = 0$ uniformly for almost all $z \in \Omega$ and all $u \in \partial j(z, \zeta)$;

(v) there exist functions $G_+, \vartheta \in L^{p'}(\Omega)$ ($\frac{1}{p} + \frac{1}{p'} = 1$), such that

$$G_+(z) = \liminf_{\zeta \to +\infty} G(z, \zeta)$$

uniformly for almost all $z \in \Omega$,

$$\int_\Omega G_+(z) u_1(z)\, dz > 0$$

and in addition

$$G_{\min}(z, \zeta) \leq \vartheta(z) \quad \text{for a.a. } z \in \Omega;$$

(vi) $\limsup\limits_{\zeta \to 0} \dfrac{pj(z, \zeta)}{|\zeta|^p} < 0$ uniformly for almost all $z \in \Omega$.

Let $\varphi_1 \colon W_0^{1,p}(\Omega) \longrightarrow \mathbb{R}$ be the locally Lipschitz function, defined by

$$\varphi_1(x) \overset{df}{=} \frac{1}{p} \|\nabla x\|_p^p - \frac{\lambda_1}{p} \|x\|_p^p - \int_\Omega j(z, x(z))\, dz \qquad \forall\, x \in W_0^{1,p}(\Omega).$$

Also let $\varphi_2 \colon W_0^{1,p}(\Omega) \longrightarrow \overline{\mathbb{R}}$ be the function defined by

$$\varphi_2(x) \overset{df}{=} i_{C_g}(x) = \begin{cases} 0 & \text{if } x \in C_g, \\ +\infty & \text{if } x \notin C_g, \end{cases}$$

i.e. the indicator function of the set C_g. Since C_g is closed and convex in $W_0^{1,p}(\Omega)$, we have that $i_{C_g} \in \Gamma_0\big(W_0^{1,p}(\Omega)\big)$ (see Definition 1.3.1). We set

$$\varphi \overset{df}{=} \varphi_1 + \varphi_2.$$

PROPOSITION 4.1.9
If hypotheses $H(j)_4$ hold,
then φ satisfies the generalized nonsmooth PS-condition (see Definition 2.3.2).

PROOF   Let $\{x_n\}_{n\geq 1} \subseteq W_0^{1,p}(\Omega)$ be a sequence, such that

$$\left|\varphi(x_n)\right| \leq M_8 \qquad \forall\, n \geq 1$$

for some $M_8 > 0$ and

$$\varphi_1^0\big(x_n; v - x_n\big) + \varphi_2(v) - \varphi_2(x_n) \geq -\varepsilon \left\|v - x_n\right\|_{W^{1,p}(\Omega)} \qquad \forall\, v \in W_0^{1,p}(\Omega).$$

Evidently $\{x_n\}_{n\geq 1} \subseteq C_g$. We can find $x_n^* \in \partial\varphi_1(x_n)$, such that

$$\varphi_1^0\big(x_n; v - x_n\big) = \langle x_n^*, v - x_n\rangle_{W_0^{1,p}(\Omega)}.$$

This is a consequence of the fact that $\varphi_1^0(x_n; \cdot)$ is the support function of the set $\partial\varphi_1(x_n) \subseteq W^{-1,p'}(\Omega)$ and the latter is weakly compact. We have

$$x_n^* = A(x_n) - \lambda_1 |x_n|^{p-2} x_n - u_n^* \qquad \forall\, n \geq 1,$$

with $A \colon W_0^{1,p}(\Omega) \longrightarrow W^{-1,p'}(\Omega)$ being the maximal monotone operator, defined by

$$\langle A(x), y\rangle_{W_0^{1,p}(\Omega)} \stackrel{df}{=} \int_\Omega \|\nabla x(z)\|_{\mathbb{R}^N}^{p-2} \big(\nabla x(z), \nabla y(z)\big)_{\mathbb{R}^N} \, dz \qquad \forall\, x, y \in W_0^{1,p}(\Omega)$$

and $u_n^* \in L^{p'}(\Omega)$, $u_n^*(z) \in \partial j\big(z, x_n(z)\big)$ for almost all $z \in \Omega$ and all $n \geq 1$.

We claim that the sequence $\{x_n\}_{n\geq 1} \subseteq W_0^{1,p}(\Omega)$ is bounded. Suppose that this is not the case. By passing to a subsequence if necessary, we may assume that

$$\|x_n\|_{W^{1,p}(\Omega)} \longrightarrow +\infty.$$

Let us set

$$y_n \stackrel{df}{=} \frac{x_n}{\|x_n\|_{W^{1,p}(\Omega)}} \qquad \forall\, n \geq 1.$$

Passing to a subsequence if necessary, we may assume that

$$\begin{aligned} y_n &\xrightarrow{w} y \quad \text{in } W_0^{1,p}(\Omega), \\ y_n &\longrightarrow y \quad \text{in } L^p(\Omega), \\ y_n(z) &\longrightarrow y(z) \text{ for a.a. } z \in \Omega, \\ |y_n(z)| &\leq k(z) \text{ for a.a. } z \in \Omega \text{ and all } n \geq 1, \end{aligned}$$

with $k \in L^p(\Omega)$. By virtue of hypotheses $H(j)_4(iii)$ and (iv), for a given $\varepsilon > 0$, we can find $\hat{a}_\varepsilon \in L^\infty(\Omega)$, such that

$$|u| \leq \hat{a}_\varepsilon(z) + \varepsilon|\zeta|^{p-1} \quad \text{for a.a. } z \in \Omega, \text{ all } \zeta \in \mathbb{R} \text{ and all } u \in \partial j(z, \zeta).$$

Using the mean value theorem for locally Lipschitz functions (see Proposition 1.3.14), we see that for almost all $z \in \Omega$ and all $\zeta \in \mathbb{R}$, we have

$$|j(z, \zeta)| \leq |j(z, 0)| + \hat{a}_\varepsilon(z)|\zeta| + \varepsilon|\zeta|^p \leq \hat{\beta}_\varepsilon(z) + 2\varepsilon|\zeta|^p,$$

with $\widehat{\beta}_\varepsilon \in L^\infty(\Omega)_+$. So we have

$$\limsup_{n\to+\infty} \left| \int_\Omega \frac{j(z, x_n(z))}{\|x_n\|_{W^{1,p}(\Omega)}^p} \, dz \right| \leq 2\varepsilon.$$

Since $\varepsilon > 0$ was arbitrary, we infer that

$$\int_\Omega \frac{j(z, x_n(z))}{\|x_n\|_{W^{1,p}(\Omega)}^p} \, dz \longrightarrow 0.$$

From the choice of the sequence $\{x_n\}_{n\geq 1} \subseteq W_0^{1,p}(\Omega)$, we have

$$\frac{\varphi(x_n)}{\|x_n\|_{W^{1,p}(\Omega)}^p} = \frac{1}{p}\|\nabla y_n\|_p^p - \frac{\lambda_1}{p}\|y_n\|_p^p - \int_\Omega \frac{j(z, x_n(z))}{\|x_n\|_{W^{1,p}(\Omega)}^p} \, dz \leq \frac{M_8}{\|x_n\|_{W^{1,p}(\Omega)}^p}$$

(note that $\varphi_2(x_n) = 0$ because $x_n \in C_g$ for $n \geq 1$). Passing to the limit as $n \to +\infty$, we obtain

$$\frac{1}{p}\|\nabla y\|_p^p \leq \frac{\lambda_1}{p}\|y\|_p^p,$$

so from Proposition 1.5.5, we have

$$\|\nabla y\|_p^p = \lambda_1 \|y\|_p^p,$$

hence $y = \pm u_1$ or $y = 0$.

If $y = 0$, then

$$\|\nabla y_n\|_{W^{1,p}(\Omega)} \longrightarrow 0$$

and so by the Poincaré inequality (see Theorem 1.1.6), we have that

$$y_n \longrightarrow 0 \quad \text{in } W_0^{1,p}(\Omega),$$

which contradicts the fact that $\|y_n\|_{W^{1,p}(\Omega)} = 1$ for all $n \geq 1$. So $y = \pm u_1$. Note that $x_n(z) \geq g(z)$ for almost all $z \in \Omega$ and so

$$y_n(z) \geq \frac{g(z)}{\|x_n(z)\|_{W^{1,p}(\Omega)}} \quad \text{for a.a. } z \in \Omega$$

and so passing to the limit as $n \to +\infty$, we obtain

$$y(z) \geq 0 \quad \text{for a.a. } z \in \Omega.$$

Therefore $y = u_1$.

Recall that for all $n \geq 1$ and all $v \in W_0^{1,p}(\Omega)$, we have

$$\langle A(x_n), v - x_n \rangle_{W^{1,p}(\Omega)} - \lambda_1 \int_\Omega |x_n(z)|^{p-2} x_n(z)(v - x_n)(z) \, dz$$

$$- \int_\Omega u_n^*(z)(v - x_n)(z) \, dz + \varphi_2(v) - \varphi_2(x_n) \geq -\varepsilon_n \|v - x_n\|_{W^{1,p}(\Omega)},$$

with $\varepsilon_n \searrow 0$.

Putting $v \equiv 0 \in C_g$, we have

$$\|\nabla x_n\|_p^p - \lambda_1 \|x_n\|_p^p - \int_\Omega u_n^*(z)x_n(z)\,dz \;\leq\; \varepsilon_n \|x_n\|_{W^{1,p}(\Omega)}. \qquad (4.29)$$

Also since $|\varphi(x_n)| \leq M_8$ for all $n \geq 1$, we have

$$-\|\nabla x_n\|_p^p + \lambda_1 \|x_n\|_p^p + \int_\Omega pj\big(z,x_n(z)\big)\,dz \;\leq\; pM_8. \qquad (4.30)$$

Adding (4.29) and (4.30), we obtain

$$\int_\Omega \big(pj\big(z,x_n(z)\big) - u_n^*(z)x_n(z)\big)\,dz \;\leq\; \varepsilon_n \|x_n\|_{W^{1,p}(\Omega)} + pM_8,$$

so

$$\int_\Omega \left(\frac{pj(z,x_n(z))}{\|x_n\|_{W^{1,p}(\Omega)}} - u_n^*(z)y_n(z) \right) dz \;\leq\; \varepsilon_n + \frac{pM_8}{\|x_n\|_{W^{1,p}(\Omega)}}.$$

Let us set

$$h_n(z) \;\overset{df}{=}\; \begin{cases} \dfrac{j(z,x_n(z))}{x_n(z)} & \text{if } x_n(z) \neq 0, \\[2mm] 0 & \text{if } x_n(z) = 0. \end{cases}$$

Using also hypothesis $H(j)_4(v)$, we can write that

$$\varepsilon_n + \frac{pM_8}{\|x_n\|_{W^{1,p}(\Omega)}}$$

$$\geq \int_\Omega \frac{pj(z,x_n(z))}{\|x_n\|_{W^{1,p}(\Omega)}}\,dz - \int_\Omega u_n^*(z)y_n(z)\,dz$$

$$\geq \int_\Omega ph_n(z)\,dz - \int_{\{y_n<0\}} g_{\min}\big(z,x_n(z)\big)y_n(z)\,dz$$

$$\qquad - \int_{\{y_n>0\}} g_{\max}\big(z,x_n(z)\big)y_n(z)\,dz$$

$$= \int_{\{y_n>0\}} G_{\max}\big(z,x_n(z)\big)y_n(z)\,dz - \int_{\{y_n<0\}} G_{\min}\big(z,x_n(z)\big)y_n(z)\,dz$$

$$\geq \int_{\{y_n>0\}} G_{\max}\big(z,x_n(z)\big)y_n(z)\,dz - \int_{\{y_n<0\}} \vartheta(z)y_n(z)\,dz. \qquad (4.31)$$

Recall that $y = u_1 > 0$. So

$$x_n(z) \;\longrightarrow\; +\infty \quad \text{for a.a. } z \in \Omega.$$

Therefore, we have

$$\big|\{y_n > 0\}\big|_N \;\longrightarrow\; |\Omega|_N \quad \text{and} \quad \big|\{y_n < 0\}\big|_N \;\longrightarrow\; 0.$$

Also from the definition of G_+, for a given $\varepsilon > 0$, we can find $M_9 > 0 = M_9(\varepsilon) > 0$, such that

$$G_{\max}(z,\zeta) \;\geq\; G_+(z) - \varepsilon \quad \text{for a.a. } z \in \Omega \text{ and all } \zeta > M_9.$$

On the other hand, from hypothesis $H(j)_4(iii)$, we have

$$\big|G_{\max}(z,\zeta)\big| \;\leq\; a_{M_9}(z) \quad \text{for a.a. } z \in \Omega \text{ and all } \zeta \in [-M_9, M_9].$$

So, we have

$$G_{\max}(z,\zeta) \;\geq\; G_+(z) - \varepsilon - \bar{a}(z) \quad \text{for a.a. } z \in \Omega \text{ and all } \zeta \geq 0, \qquad (4.32)$$

with $\bar{a} \in L^{p'}(\Omega)_+$. Passing to the limit as $n \to +\infty$ in (4.31) and using Fatou's lemma (note that (4.32) permits the use of Fatou's lemma), we obtain

$$\int_\Omega G_+(z) u_1(z)\, dz \;\leq\; 0,$$

which contradicts hypothesis $H(j)_4(v)$. From this contradiction, it follows that the sequence $\{x_n\}_{n\geq 1} \subseteq C_g$ is bounded and passing to a subsequence if necessary, we may assume that

$$\begin{aligned}
x_n &\xrightarrow{w} x \quad \text{in } W_0^{1,p}(\Omega), \\
x_n &\longrightarrow x \quad \text{in } L^p(\Omega),
\end{aligned} \qquad (4.33)$$

with $x \in C_g$. From the choice of the sequence $\{x_n\}_{n\geq 1} \subseteq C_g$ we have

$$\begin{aligned}
&\langle A(x_n), x_n - v\rangle_{W^{1,p}(\Omega)} - \lambda_1 \int_\Omega |x_n(z)|^{p-2} x_n(z)(x_n - v)(z)\, dz \\
&- \int_\Omega u_n^*(z)(x_n - v)(z)\, dz \;\leq\; \varepsilon_n \|x_n - v\|_{W^{1,p}(\Omega)} \qquad \forall\, v \in C_g,\ n \geq 1.
\end{aligned}$$

Putting $v = x \in C_g$, we have

$$\begin{aligned}
&\langle A(x_n), x_n - x\rangle_{W^{1,p}(\Omega)} - \lambda_1 \int_\Omega |x_n(z)|^{p-2} x_n(z)(x_n - x)(z)\, dz \\
&- \int_\Omega u_n^*(z)(x_n - x)(z)\, dz \;\leq\; \varepsilon_n \|x_n - x\|_{W^{1,p}(\Omega)} \qquad \forall\, n \geq 1.
\end{aligned}$$

From (4.33), we see that

$$\int_\Omega |x_n(z)|^{p-2} x_n(z)(x_n - x)(z)\, dz \;\longrightarrow\; 0$$

and

$$\int_{\Omega} u_n^*(z)\big(x_n - x\big)(z)\,dz \;\longrightarrow\; 0.$$

So

$$\limsup_{n \to +\infty} \langle A(x_n), x_n - x \rangle_{W^{1,p}(\Omega)} \;\leq\; 0.$$

But A is maximal monotone, thus generalized pseudomonotone. Therefore, we have

$$\langle A(x_n), x_n \rangle_{W^{1,p}(\Omega)} \;\longrightarrow\; \langle A(x), x \rangle_{W^{1,p}(\Omega)}$$

and so

$$\|\nabla x_n\|_{W^{1,p}(\Omega)} \;\longrightarrow\; \|\nabla\|_{W^{1,p}(\Omega)}.$$

Since

$$\nabla x_n \;\xrightarrow{w}\; \nabla x \quad \text{in } L^p\big(\Omega; \mathbb{R}^N\big),$$

as before exploiting the Kadec-Klee property of the uniformly convex space $L^p\big(\Omega; \mathbb{R}^N\big)$, we have that

$$\nabla x_n \;\longrightarrow\; \nabla x \quad \text{in } L^p\big(\Omega; \mathbb{R}^N\big),$$

hence

$$x_n \;\longrightarrow\; x \quad \text{in } W_0^{1,p}(\Omega).$$

$\square$

We consider the direct sum decomposition

$$W_0^{1,p}(\Omega) \;=\; \mathbb{R}u_1 \oplus V_{u_1},$$

with

$$V_{u_1} \;\overset{df}{=}\; \left\{ v \in W_0^{1,p}(\Omega) : \int_{\Omega} u_1(z)^{p-1} v(z)\,dz = 0 \right\}.$$

Next we show that $\varphi|_{R_+ u_1}$ is anticoercive.

PROPOSITION 4.1.10

If hypothesis $H(j)_4$ hold, <u>then</u>

$$\varphi(tu_1) \;\longrightarrow\; -\infty \quad \text{as } t \to +\infty.$$

PROOF For every $t > 0$, we have that $tu_1 \in C_g$ and so

$$\varphi(tu_1) \;=\; -\int_{\Omega} j\big(z, tu_1(z)\big)\,dz.$$

From the definition of G_+, we know that for a given $\varepsilon > 0$, we can find $M_{10} = M_{10}(\varepsilon) > 0$, such that

$$k_\varepsilon^+(z) = G_+(z) - \varepsilon \leq G_{\max}(z, \zeta) \qquad \forall\, z \in \Omega \setminus D, \; \zeta > M_{10},$$

for some $D \subseteq \Omega$, such that $|D|_N = 0$. Thus

$$\frac{G_{\max}(z, \zeta)}{\zeta^p} \geq \frac{k_\varepsilon^+(z)}{\zeta^p} = \frac{d}{d\zeta}\left(-\frac{1}{p-1}\frac{k_\varepsilon^+(z)}{\zeta^{p-1}}\right). \qquad (4.34)$$

For all $z \in \Omega \setminus D$, all $\zeta \geq M_{10}$ and all $u \in \partial j(z, \zeta)$, we have

$$\frac{G_{\max}(z, \zeta)}{\zeta^p} = \frac{pj(z, \zeta)}{\zeta^{p+1}} - \frac{g_{\max}(z, \zeta)}{\zeta^p} \leq \frac{pj(z, \zeta)}{\zeta^{p+1}} - \frac{u}{\zeta^p}.$$

From Corollary 1.3.7, we know that for all $z \in \Omega \setminus D$ and all $\zeta > M_{10}$, the function $\zeta \longmapsto \frac{pj(z,\zeta)}{\zeta^p}$ is locally Lipschitz and

$$\partial\left(\frac{j(z,\zeta)}{\zeta^p}\right) \subseteq \frac{\zeta^p \partial j(z, \zeta) - pj(z, \zeta)\zeta^{p-1}}{\zeta^{2p}} = \frac{\partial j(z, \zeta)}{\zeta^p} - \frac{pj(z, \zeta)}{\zeta^{p+1}}. \qquad (4.35)$$

Therefore using (4.34) and (4.35), for all $z \in \Omega \setminus D$, all $\zeta > M_{10}$ and all $v \in \partial\left(\frac{j(z,\zeta)}{\zeta^p}\right)$, we have

$$v \leq -\frac{1}{\zeta^p}G_{\max}(z, \zeta) \leq \frac{d}{d\zeta}\left(\frac{1}{p-1}\frac{k_\varepsilon^+(z)}{\zeta^{p-1}}\right).$$

For every $z \in \Omega \setminus D$, the function $\zeta \longmapsto \frac{j(z,\zeta)}{\zeta^p}$ is locally Lipschitz on $(M_{10}, +\infty)$ and so it is differentiable at all $\zeta \in (M_{10}, +\infty) \setminus E(z)$, with $|E(z)|_1 = 0$. We define

$$v_0(z, \zeta) \stackrel{df}{=} \begin{cases} \frac{d}{d\zeta}\left(\frac{j(z,\zeta)}{\zeta^p}\right) & \text{if } \zeta \in (M_{10}, +\infty) \setminus E(z), \\ 0 & \text{otherwise.} \end{cases}$$

For fixed $z \in \Omega \setminus D$ and $\zeta \in (M_{10}, +\infty) \setminus E(z)$, we have

$$v_0(z, \zeta) = \frac{d}{d\zeta}\left(\frac{j(z, \zeta)}{\zeta^p}\right) \in \partial\left(\frac{j(z, \zeta)}{\zeta^p}\right) \leq \frac{d}{d\zeta}\left(\frac{1}{p-1}\frac{k_\varepsilon^+(z)}{\zeta^{p-1}}\right).$$

If $M_{10} < \zeta_1 < \zeta_2 < +\infty$ and we integrate the above inequality on $[\zeta_1, \zeta_2]$, we obtain

$$\frac{j(z, \zeta_2)}{\zeta_2^p} - \frac{j(z, \zeta_1)}{\zeta_1^p} \leq \frac{k_\varepsilon^+(z)}{p-1}\left(\frac{1}{\zeta_2^{p-1}} - \frac{1}{\zeta_1^{p-1}}\right). \qquad (4.36)$$

From the proof of Proposition 4.1.9, we know that

$$|j(z, \zeta_2)| \leq \widehat{\beta}_\varepsilon(z) + 2\varepsilon|\zeta_2|^p \qquad \forall\, z \in \Omega \setminus D, \; \zeta_2 \in \mathbb{R},$$

with $\widehat{\beta}_{\varepsilon} \in L^{\infty}(\Omega)_{+}$. As $\varepsilon > 0$ was arbitrary, we have

$$\lim_{\zeta_2 \to +\infty} \frac{j(z, \zeta_2)}{\zeta_2^p} = 0.$$

Therefore, if in (4.36) above we let $\zeta_2 \to +\infty$, we obtain

$$\frac{j(z, \zeta_1)}{\zeta_1^p} \geq \frac{k_{\varepsilon}^{+}(z)}{p-1} \frac{1}{\zeta_1^{p-1}},$$

so

$$\frac{j(z, \zeta_1)}{\zeta_1} \geq \frac{k_{\varepsilon}^{+}(z)}{p-1}$$

and since $\varepsilon > 0$ was arbitrary, we obtain

$$\liminf_{\zeta_1 \to +\infty} \frac{j(z, \zeta_1)}{\zeta_1} \geq \frac{1}{p-1} G_{+}(z). \tag{4.37}$$

If the proposition was not true, we could find $M_{11} > 0$ and a sequence $\{t_n\}_{n \geq 1} \subseteq \mathbb{R}$, such that $t_n \to +\infty$ and

$$-M_{11} \leq \varphi(t_n u_1) = -\int_{\Omega} j(z, t_n u_1(z)) \, dz,$$

so

$$-\frac{M_{11}}{t_n} \leq \frac{\varphi(t_n u_1)}{t_n} = -\int_{\Omega} \frac{j(z, t_n u_1(z))}{t_n u_1(z)} u_1(z) \, dz.$$

Because of (4.37), we can use Fatou's lemma and obtain

$$\int_{\Omega} G_{+}(z) u_1(z) \, dz \leq 0,$$

which contradicts hypothesis $H(j)_4(v)$. This proves the proposition. □

With the next proposition, we shall establish that φ satisfies a Mountain Pass-type geometry.

PROPOSITION 4.1.11
If hypotheses $H(j)_4$ hold,
<u>*then*</u>

$$\varphi(x) \geq \beta_1 \|x\|_{W^{1,p}(\Omega)}^p - \beta_2 \|x\|_{W^{1,p}(\Omega)}^{\eta} \qquad \forall \, x \in W_0^{1,p}(\Omega),$$

for some $\beta_1, \beta_2 > 0$, $\eta \in (p, p^*]$.

PROOF   By virtue of hypothesis $H(j)_4(vi)$, we can find $\varepsilon > 0$ and $\delta > 0$, such that

$$j(z, \zeta) \leq \frac{\varepsilon}{p}|\zeta|^p \qquad \forall\, \zeta \in [-\delta, \delta]. \tag{4.38}$$

From the proof of Proposition 4.1.9, we know that

$$\left| j(z, \zeta) \right| \leq \widehat{\beta}(z) + \widehat{c}|\zeta|^p \quad \text{for a.a. } z \in \Omega \text{ and all } \zeta \in \mathbb{R},$$

with $\widehat{\beta} \in L^{\infty}(\Omega)_+$, $\widehat{c} > 0$. Combining this with (4.38), we obtain that

$$j(z, \zeta) \leq -\frac{\varepsilon}{p}|\zeta|^p + c_5|\zeta|^{\eta} \quad \text{for a.a. } z \in \Omega \text{ and all } \zeta \in \mathbb{R},$$

with $c_5 > 0$, $\eta \in (p, p^*]$. So for all $x \in W_0^{1,p}(\Omega)$, we have

$$\varphi(x) = \frac{1}{p}\|\nabla x\|_p^p - \frac{\lambda_1}{p}\|x\|_p^p - \int_{\Omega} j\big(z, x(z)\big)\, dz$$

$$\geq \frac{1}{p}\|\nabla x\|_p^p - \frac{\lambda_1}{p}\|x\|_p^p + \frac{\varepsilon}{p}\|x\|_p^p - c_5\|x\|_{\eta}^{\eta}.$$

Using the Poincaré inequality (see Theorem 1.1.6) and the fact that the embedding $W_0^{1,p}(\Omega) \subseteq L^{\eta}(\Omega)$ is continuous (since $\eta \in (p, p^*]$), we infer that

$$\varphi(x) \geq \beta_1 \|x\|_{W^{1,p}(\Omega)}^p - \beta_2 \|x\|_{W^{1,p}(\Omega)}^{\eta} \qquad \forall\, x \in W_0^{1,p}(\Omega),$$

with $\beta_1, \beta_2 > 0$. $\qquad\qquad\qquad\qquad\qquad\qquad\qquad\qquad\qquad\qquad$ ⬚

Now we are ready for an existence theorem concerning problem (4.28).

THEOREM 4.1.4
If hypotheses $H(j)_4$ hold,
then problem (4.28) has a nontrivial solution $x_0 \in W_0^{1,p}(\Omega)$.

PROOF   Note that because of Proposition 4.1.11, if $r > 0$ is small, we have

$$\varphi(x) \geq c_6 \qquad \forall\, x \in W_0^{1,p}(\Omega),\ \|x\|_{W^{1,p}(\Omega)} = r,$$

for some $c_6 > 0$. Also Proposition 4.1.10 implies that there exists $t > 0$, such that

$$\varphi(t u_1) \leq \varphi(0) \leq 0.$$

These facts combined with Proposition 4.1.9 permit the application of Theorem 2.3.3, which gives $x_0 \in C_g$, such that

$$\varphi(x_0) \geq c_6 > \varphi(0)$$

(hence $x_0 \neq 0$) and

$$\varphi_1^0(x_0; h) + \varphi_2(x_0 + h) - \varphi_2(x_0) \geq 0 \qquad \forall\, h \in W_0^{1,p}(\Omega).$$

From the second inequality, it follows that we can find $x_0^* \in \partial\varphi_1(x_0)$ and $v_0^* \in \partial\varphi_2(x_0)$, such that $x_0^* + v_0^* = 0$ (the subdifferential of φ_1 is the generalized subdifferential (see Definition 1.3.7), while the subdifferential of φ_2 is the convex subdifferential (see Definition 1.3.3)). We know that

$$x_0^* = A(x_0) - \lambda_1 |x_0|^{p-2} x_0 - u_0^*,$$

with $u_0^* \in L^{p'}(\Omega)$, $u_0^*(z) \in \partial j(z, x_0(z))$ for almost all $z \in \Omega$ and $v_0^* \in \partial\varphi_2(x_0) = N_{C_g}(x_0)$ (the normal cone to C_g at x_0), hence

$$\langle x_0^*, y - x_0 \rangle_{W_0^{1,p}(\Omega)} \leq 0 \qquad \forall\, y \in C_g$$

(see Remark 1.3.8). So, we have

$$\left\langle A(x_0) - \lambda_1 |x_0|^{p-2} x_0 - u_0^*, \left(y - x_0\right) \right\rangle_{W_0^{1,p}(\Omega)} \geq 0 \qquad \forall\, y \in C_g,$$

so

$$\int_\Omega \|\nabla x_0(z)\|_{\mathbb{R}^N}^{p-2} \left(\nabla x_0(z), \nabla y(z) - \nabla x_0(z)\right)_{\mathbb{R}^N} dz$$

$$- \lambda_1 \int_\Omega |x_0(z)|^{p-2} x_0(z) \big(y - x_0\big)(z)\, dz$$

$$\geq \int_\Omega u_0^*(z) \big(y - x_0\big)(z)\, dz \qquad \forall\, y \in C_g$$

and thus $x_0 \in W_0^{1,p}(\Omega)$ is a solution of problem (4.28). $\qquad\square$

EXAMPLE 4.1.2 Let $r \in (1, p)$. The nonsmooth, locally Lipschitz function

$$j(\zeta) \stackrel{df}{=} \begin{cases} e^\zeta + \zeta - \frac{1}{e} & \text{if } \zeta < -1, \\ -|\zeta|^r & \text{if } \zeta \in [-1, 1], \\ \zeta + \sin\zeta - \ln(2\zeta - 1) - 2 & \text{if } \zeta > 1 \end{cases}$$

satisfies hypotheses $H(j)_4$. $\qquad\square$

4.1.4 Strongly Resonant Problems

We conclude this section with the analysis of a problem which is strongly resonant in the sense described in the beginning of this Section. However, the problem that we study is nonlinear driven by the p-Laplacian. Namely let

$\Omega \subseteq \mathbb{R}^N$ be a bounded domain with $C^{1,\alpha}$ boundary ($\alpha \in (0,1)$) and consider the following nonlinear elliptic inclusion:

$$\begin{cases} -\text{div}\left(\|\nabla x(z)\|_{\mathbb{R}^N}^{p-2}\nabla x(z)\right) - \lambda_1|x(z)|^{p-2}x(z) \in \partial j\left(z, x(z)\right) \\ \qquad\qquad\qquad\qquad\qquad\qquad\qquad \text{for a.a. } z \in \Omega, \qquad (4.39) \\ x|_{\partial\Omega} = 0. \end{cases}$$

Our hypotheses for the nonsmooth potential $j(z,\zeta)$ are the following:

$\underline{H(j)_5}$ $j: \Omega \times \mathbb{R} \longrightarrow \mathbb{R}$ is a function, such that

(*i*) for every $\zeta \in \mathbb{R}$, the function

$$\Omega \ni z \longrightarrow j(z,\zeta) \in \mathbb{R}$$

is measurable and $j(z,0) = 0$ for almost all $z \in \Omega$;

(*ii*) for almost all $z \in \Omega$, the function

$$\mathbb{R} \ni \zeta \longrightarrow j(z,\zeta) \in \mathbb{R}$$

is locally Lipschitz;

(*iii*) there exists $a \in L^\infty(\Omega)$, $c > 0$ and $r \in [1, p^*)$, such that

$$|u| \leq a(z) + c|\zeta|^{r-1} \quad \text{for a.a. } z \in \Omega, \text{ all } \zeta \in \mathbb{R} \text{ and all } u \in \partial j(z,\zeta);$$

(*iv*) there exist functions $j_\pm \in L^1(\Omega)$, such that

$$\int_\Omega j_\pm(z)\,dz \ \leq \ 0$$

and

$$\limsup_{\zeta \to \pm\infty} j(z,\zeta) \ \leq \ j_\pm(z)$$

uniformly for almost all $z \in \Omega$;

(*v*) there exists $\delta > 0$, such that

$$j(z,\zeta) \ \geq \ 0 \quad \text{for a.a. } z \in \Omega \text{ and all } \zeta \in [0, \delta]$$

or

$$j(z,\zeta) \ \geq \ 0 \quad \text{for a.a. } z \in \Omega \text{ and all } \zeta \in [-\delta, 0]$$

(local sign condition);

REMARK 4.1.4   Hypothesis $H(j)_5(iv)$ classifies the problem as strongly resonant. ⬚

We consider the locally Lipschitz energy functional $\varphi\colon W_0^{1,p}(\Omega) \longrightarrow \mathbb{R}$, defined by

$$\varphi(x) \overset{df}{=} \frac{1}{p}\|\nabla x\|_p^p - \frac{\lambda_1}{p}\|x\|_p^p - \int_\Omega j\big(z, x(z)\big)\,dz \qquad \forall\, x \in W_0^{1,p}(\Omega).$$

The next proposition reveals the main difficulty that we encounter with strongly resonant problems, which is a lack of compactness, namely the non-smooth PS-condition is not satisfied at all levels.

PROPOSITION 4.1.12

If hypotheses $H(j)_5$ *hold,*
then φ *satisfies the nonsmooth* PS_c*-condition for all* $c \in \mathbb{R}$*, such that* $c < \min\big\{-\int_\Omega j_-(z)\,dz, -\int_\Omega j_+(z)\,dz\big\}$.

PROOF Let $c < \min\big\{-\int_\Omega j_-(z)\,dz, -\int_\Omega j_+(z)\,dz\big\}$ and consider a sequence $\{x_n\}_{n\geq 1} \subseteq W_0^{1,p}(\Omega)$, such that

$$\varphi(x_n) \longrightarrow c \quad \text{and} \quad m^\varphi(x_n) \longrightarrow 0.$$

Let $x_n^* \in \partial\varphi(x_n)$ be such that $m^\varphi(x_n) = \|x_n^*\|_{W^{-1,p'}(\Omega)}$ for $n \geq 1$. We have

$$x_n^* = A(x_n) - \lambda_1|x_n|^{p-2}x_n - u_n^*,$$

with $A\colon W_0^{1,p}(\Omega) \longrightarrow W^{-1,p'}(\Omega)$ as in the proof of Proposition 4.1.9 and $u_n^* \in L^{r'}(\Omega)$, $u_n^*(z) \in \partial j\big(z, x_n(z)\big)$ for almost all $z \in \Omega$ and all $n \geq 1$. We claim that the sequence $\{x_n\}_{n\geq 1} \subseteq W_0^{1,p}(\Omega)$ is bounded. If this is not true, then, passing to a subsequence if necessary, we may assume that

$$\|x_n\|_{W^{1,p}(\Omega)} \longrightarrow +\infty.$$

We set

$$y_n \overset{df}{=} \frac{x_n}{\|x_n\|_{W^{1,p}(\Omega)}} \qquad \forall\, n \geq 1$$

and passing to a further subsequence if necessary, we may assume that

$$y_n \overset{w}{\longrightarrow} y \text{ in } W_0^{1,p}(\Omega),$$
$$y_n \longrightarrow y \text{ in } L^s(\Omega),$$

for any $s \in [1, p^*)$. By virtue of hypothesis $H(j)_5(iv)$ (the strong resonance hypothesis), we can find $M_{12} > 0$, such that

$$\begin{cases} \big|j(z,\zeta) - j_+(z)\big| \leq 1 & \text{for a.a. } z \in \Omega \text{ and all } \zeta \geq M_{12}, \\ \big|j(z,\zeta) - j_-(z)\big| \leq 1 & \text{for a.a. } z \in \Omega \text{ and all } \zeta \leq -M_{12}, \end{cases}$$

so

$$\begin{cases} \left| j(z,\varsigma) \right| \leq \left| j_+(z) \right| + 1 & \text{for a.a. } z \in \Omega \text{ and all } \varsigma \geq M_{12}, \\ \left| j(z,\varsigma) \right| \leq \left| j_-(z) \right| + 1 & \text{for a.a. } z \in \Omega \text{ and all } \varsigma \leq -M_{12}. \end{cases}$$

On the other hand from the mean value theorem for locally Lipschitz functions (see Proposition 1.3.14), we know that

$$\left| j(z,\varsigma) - j(z,0) \right| = \left| j(z,\varsigma) \right| = \left| u^*\varsigma \right| \quad \text{for a.a. } z \in \Omega \text{ and all } \varsigma \in [-M_{12}, M_{12}],$$

for some $u^* \in \partial j\big(z, \lambda(z)\varsigma\big)$, $\lambda(z) \in (0,1)$, so

$$\left| j(z,\varsigma) \right| \leq a_{M_{12}}(z) M_{12}.$$

Therefore, we can say that

$$\left| j(z,\varsigma) \right| \leq \widehat{a}(z) \quad \text{for a.a. } z \in \Omega \text{ and all } \varsigma \in \mathbb{R}, \tag{4.40}$$

with $\widehat{a} \in L^1(\Omega)_+$. From the choice of the sequence $\{x_n\}_{n\geq 1} \subseteq W_0^{1,p}(\Omega)$, we have

$$\frac{\varphi(x_n)}{\|x_n\|_{W^{1,p}(\Omega)}^p} \leq \frac{M_{13}}{\|x_n\|_{W^{1,p}(\Omega)}^p} \quad \forall\, n \geq 1,$$

for some $M_{13} > 0$, so

$$\frac{1}{p} \|\nabla y_n\|_p^p - \frac{\lambda_1}{p} \|y_n\|_p^p - \int_\Omega \frac{j(z, x_n(z))}{\|x_n\|_p^p}\, dz \leq \frac{M_{13}}{\|x_n\|_p^p}.$$

Note that

$$\left| \int_\Omega \frac{j(z, x_n(z))}{\|x_n\|_p^p}\, dz \right| \leq \frac{\|\widehat{a}\|_1}{\|x_n\|_p^p} \longrightarrow 0.$$

So exploiting the weak lower semicontinuity of the norm functional in a Banach space, in the limit as $n \to +\infty$, we obtain

$$\frac{1}{p} \|\nabla y\|_p^p \leq \frac{\lambda_1}{p} \|y\|_p^p.$$

Using also Proposition 1.5.18, we have

$$\|\nabla y\|_p^p = \lambda_1 \|y\|_p^p$$

and so $y = \pm u_1$ or $y = 0$.

If $y = 0$, then we have

$$\|\nabla y_n\|_p \longrightarrow 0$$

and so

$$y_n \longrightarrow 0 \quad \text{in } W_0^{1,p}(\Omega),$$

a contradiction to the fact that $\|y_n\|_{W^{1,p}(\Omega)} = 1$ for all $n \geq 1$. So $y = \pm u_1$. Suppose that $y = u_1$. Because $u_1(z) > 0$ for all $z \in \Omega$ (see Proposition 1.5.18), we must have that

$$x_n(z) \longrightarrow +\infty \quad \text{for a.a. } z \in \Omega.$$

From the choice of the sequence $\{x_n\}_{n \geq 1} \subseteq W_0^{1,p}(\Omega)$, for a given $\varepsilon > 0$, we can find $n_0 = n_0(\varepsilon) \geq 1$, such that

$$\varphi(x_n) \leq c + \varepsilon \quad \forall n \geq n_0,$$

so

$$\frac{1}{p}\|\nabla x_n\|_p^p - \frac{\lambda_1}{p}\|x_n\|_p^p - \int_\Omega j(z, x_n(z))\, dz \leq c + \varepsilon.$$

As $\lambda_1 \|x_n\|_p^p \leq \|\nabla x_n\|_p^p$, we have

$$-\int_\Omega j(z, x_n(z))\, dz \leq c + \varepsilon.$$

By Fatou's lemma and hypothesis $H(j)_5(iv)$, we have that

$$-\int_\Omega j_+(z)\, dz \leq -\int_\Omega \limsup_{n \to +\infty} j(z, x_n(z))\, dz$$

$$\leq -\limsup_{n \to +\infty} \int_\Omega j(z, x_n(z)) \leq c + \varepsilon.$$

Since $\varepsilon > 0$ was arbitrary, we obtain

$$-\int_\Omega j_+(z)\, dz \leq c,$$

a contradiction. If we suppose that $y = -u_1$, then in a similar fashion we obtain

$$-\int_\Omega j_-(z)\, dz \leq c,$$

a contradiction. Therefore the sequence $\{x_n\}_{n \geq 1} \subseteq W_0^{1,p}(\Omega)$ is bounded and then arguing as in the last part of the proof of Proposition 4.1.9, we conclude that φ satisfies the nonsmooth PS_c-condition for $c < \min\{-\int_\Omega j_+(z)\, dz, -\int_\Omega j_-(z)\, dz\}$. $\square$

Using this proposition, we can have the following existence theorem for problem (4.39).

THEOREM 4.1.5
If hypotheses $H(j)_5$ hold, <u>then</u> problem (4.39) has a solution $x_0 \in W_0^{1,p}(\Omega)$.

PROOF   Without any loss of generality we assume that the first option in the local sign condition (i.e. hypothesis $H(j)_5(v)$ holds), namely that

$$j(z, \zeta) \geq 0 \quad \text{for a.a. } z \in \Omega \text{ and all } \zeta \in (0, \delta].$$

The analysis is similar if the other option is in effect. From Proposition 1.5.18, we know that $u_1 \in C^{1,\beta}(\overline{\Omega})$ ($\beta \in (0,1)$). So we can find $\delta_1 > 0$, such that

$$0 < tu_1(z) \leq \delta \quad \forall z \in \Omega, \ t \in (0, \delta_1].$$

Then hypothesis $H(j)_5(v)$ (the local sign condition) implies that

$$j\big(z, tu_1(z)\big) \geq 0 \quad \text{for a.a. } z \in \Omega$$

and so

$$\varphi(tu_1) \leq 0$$

(recall that $\|\nabla u_1\|_p^p = \lambda_1 \|u_1\|_p^p$). Therefore

$$m_0 \overset{df}{=} \inf_{W_0^{1,p}(\Omega)} \varphi \leq 0.$$

Note that $-\infty \leq m_0$, because φ is bounded below (see (4.40) in the proof of Proposition 4.1.12 and recall that $\lambda_1 \|x\|_p^p \leq \|\nabla x\|_p^p$ for all $x \in W_0^{1,p}(\Omega)$).

If $m_0 < 0$, then by virtue of Proposition 4.1.12, φ satisfies the nonsmooth PS_{m_0}-condition and so we can find $x_0 \in W_0^{1,p}(\Omega)$, such that

$$\varphi(x_0) = m_0 < 0 = \varphi(0)$$

(i.e. $x_0 \neq 0$) and $0 \in \partial\varphi(x_0)$. This last inclusion implies that $x_0 \in W_0^{1,p}(\Omega)$ is a nontrivial solution of (4.39).

If $m_0 = 0$, then

$$\varphi(tu_1) = 0 = m_0 \quad \forall t \in (0, \delta_1]$$

and so φ has a continuum of minima, thus (4.39) has a nontrivial solution $x_0 \in W_0^{1,p}(\Omega)$. ▯

4.2   Neumann Problems

In this section we study nonlinear Neumann problems driven by the p-Laplacian with a nonsmooth potential. The natural space for the analysis of Neumann problems is the Sobolev space $W^{1,p}(\Omega)$, where the Poincaré inequality is no longer true (see Theorem 1.1.6). This takes away a very convenient analytical tool and makes the study of the Neumann problem more difficult. We shall examine two problems. The first with a homogeneous Neumann boundary condition and the second with an inhomogeneous multivalued Neumann boundary condition. Both problems are approached using variational methods based on the nonsmooth critical point theory and employing Landesman-Lazer type conditions, like the ones considered in the previous section for Dirichlet problems.

4.2.1 Spectrum of $\left(-\Delta_p, W^{1,p}(\Omega)\right)$

We shall start with some useful observations about the spectrum of $\left(-\Delta_p, W^{1,p}(\Omega)\right)$. Let $\Omega \subseteq \mathbb{R}^N$ be a bounded domain with $C^{1,\alpha}$ boundary ($\alpha \in (0,1)$). We consider the following nonlinear eigenvalue problem:

$$\begin{cases} -\operatorname{div}\left(\|\nabla x(z)\|_{\mathbb{R}^N}^{p-2} \nabla x(z)\right) = \lambda |x(z)|^{p-2} x(z) & \text{for a.a. } z \in \Omega, \\ \frac{\partial x}{\partial n} = 0 & \text{on } \partial\Omega, \end{cases} \tag{4.41}$$

with $\lambda > 0$, $p \in (1, \infty)$. Here $\frac{\partial x}{\partial n}$ stands for the directional derivative in the direction of the outward unit normal on $\partial\Omega$. The boundary condition is understood in the sense of traces. We say that $\lambda \in \mathbb{R}$ is an eigenvalue of $\left(-\Delta_p, W^{1,p}(\Omega)\right)$, if problem (4.41) has a nontrivial solution, known as an eigenfunction corresponding to the eigenvalue λ. Note that by nonlinear regularity theory (see Remark 1.5.9), every eigenfunction of $\left(-\Delta_p, W^{1,p}(\Omega)\right)$ belongs in $C^{1,\beta}(\overline{\Omega})$ ($\beta \in (0,1)$). Moreover, $\lambda = 0$ is an eigenvalue with eigenfunctions, the constant functions. More precisely we have:

PROPOSITION 4.2.1
The eigenvalue λ_0 of (4.41) is the first eigenvalue and is isolated and simple.

PROOF First we note that problem (4.41) cannot have negative eigenvalues. Indeed, if $\lambda < 0$ is an eigenvalue with a corresponding eigenfunction x, if we multiply with $x(z)$ and integrate on Ω, via Green's identity (see Theorem 1.1.9), we obtain

$$\|\nabla x\|_p^p = \lambda \|x\|_p^p,$$

which cannot be true for $\lambda < 0$.

The simplicity of $\lambda_0 = 0$ is a direct consequence of the fact that

$$0 = \inf_{\substack{x \in W^{1,p}(\Omega) \\ x \neq 0}} \frac{\|\nabla x\|_p^p}{\|x\|_p^p}.$$

Finally suppose that $\lambda_0 = 0$ is not isolated. So we can find a sequence of nonzero eigenvalues $\{\lambda_n\}_{n\geq 1}$ of $(-\Delta_p, W^{1,p}(\Omega))$, such that $\lambda_n \searrow 0$ as $n \to +\infty$. Consider a sequence of associated eigenfunctions $\{x_n\}_{n\geq 1} \subseteq C^1(\overline{\Omega})$ with $\|x_n\|_p = 1$ for $n \geq 1$. We have

$$\lambda_n = \frac{\|\nabla x_n\|_p^p}{\|x_n\|_p^p} = \|\nabla x_n\|_p^p \searrow 0 \qquad \forall\, n \to +\infty$$

and so the sequence $\{x_n\}_{n\geq 1} \subseteq W^{1,p}(\Omega)$ is bounded. By passing to a subsequence if necessary, we may assume that

$$x_n \xrightarrow{w} x \text{ in } W^{1,p}(\Omega),$$
$$x_n \longrightarrow x \text{ in } L^p(\Omega),$$

for some $x \in W^{1,p}(\Omega)$. We have that $\|x\|_p = 1$ and $\|\nabla x\|_p = 0$, so

$$x = \frac{\pm 1}{|\Omega|_N^{\frac{1}{p}}}.$$

Recall that

$$\int_\Omega \|\nabla x_n(z)\|_{\mathbb{R}^N}^{p-2} \left(\nabla x_n(z), \nabla u(z)\right)_{\mathbb{R}^N} dz$$

$$= \lambda_n \int_\Omega |x_n(z)|^{p-2} x_n(z) u(z)\, dz \qquad \forall\, u \in W^{1,p}(\Omega).$$

So if $u \equiv 1$, we obtain

$$\int_\Omega |x_n(z)|^{p-2} x_n(z)\, dz = 0$$

and so by passing to the limit as $n \to +\infty$, we obtain

$$\int_\Omega |x(z)|^{p-2} x(z)\, dz = 0,$$

a contradiction.   ∎

If $\lambda > 0$ is a nonzero eigenvalue and u is an associated eigenfunction, then integrating the partial differential equation in (4.41) and using Green's identity (see Theorem 1.1.9), we obtain

$$\int_\Omega |u(z)|^{p-2} u(z)\, dz = 0.$$

So naturally, we are led to the consideration of the following set

$$C(p) \stackrel{df}{=} \left\{ x \in W^{1,p}(\Omega) : \; \|x\|_p = 1, \; \int_\Omega |x(z)|^{p-2} x(z) \, dz = 0 \right\}$$

and of the function $\psi_p \colon W^{1,p}(\Omega) \longrightarrow \mathbb{R}$, defined by

$$\psi_p(x) \stackrel{df}{=} \|\nabla x\|_p^p \qquad \forall \, x \in W^{1,p}(\Omega).$$

For these items, we consider the following minimization problem:

$$\inf_{x \in C(p)} \psi_p(x) \; = \; \lambda_1(p). \tag{4.42}$$

PROPOSITION 4.2.2
Problem (4.42) has a solution $\lambda_1 = \lambda_1(p) > 0$ which is attained in $C(p)$.

PROOF   Consider a minimizing sequence $\{x_n\}_{n \geq 1} \subseteq C(p)$, i.e. $\psi_p(x_n) \searrow \lambda_1$. Evidently the sequence $\{x_n\}_{n \geq 1} \subseteq W^{1,p}(\Omega)$ is bounded and so, passing to a subsequence if necessary, we may assume that

$$\begin{aligned}
x_n &\xrightarrow{w} x && \text{in } W^{1,p}(\Omega), \\
x_n &\longrightarrow x && \text{in } L^p(\Omega), \\
x_n(z) &\longrightarrow x(z) && \text{for a.a. } z \in \Omega, \\
|x_n(z)| &\leq k(z) && \text{for a.a. } z \in \Omega \text{ and all } n \geq 1,
\end{aligned}$$

with $k \in L^p(\Omega)$. Note that the sequence $\left\{ |x_n(\cdot)|^{p-2} x_n(\cdot) \right\}_{n \geq 1} \subseteq L^{p'}(\Omega)$ (with $\frac{1}{p} + \frac{1}{p'} = 1$) is bounded and

$$|x_n(z)|^{p-2} x_n(z) \; \longrightarrow \; |x(z)|^{p-2} x(z) \quad \text{for a.a. } z \in \Omega.$$

So it follows that

$$|x_n(\cdot)|^{p-2} x_n(\cdot) \; \longrightarrow \; |x(\cdot)|^{p-2} x(\cdot) \quad \text{in } L^{p'}(\Omega),$$

hence

$$\int_\Omega |x(z)|^{p-2} x(z) \, dz \; = \; 0 \quad \text{and} \quad \|x\|_p = 1,$$

i.e. $x \in C(p)$. Also from the weak lower semicontinuity of the norm functional in a Banach space, we have that $\|\nabla x\|_p^p \leq \lambda_1$, hence

$$\|\nabla x\|_p^p \; = \; \lambda_1.$$

Since $x \in C(p)$, then x is a nonconstant element in $W^{1,p}(\Omega)$ and so $\lambda_1 > 0$. $\square$

An immediate consequence of Proposition 4.2.2, is the following Poincaré-Wirtinger type inequality.

COROLLARY 4.2.1

If $x \in W^{1,p}(\Omega)$ *is such that*

$$\int_\Omega |x(z)|^{p-2} x(z)\,dz \;=\; 0,$$

then

$$\lambda_1 \|x\|_p^p \;\leq\; \|\nabla x\|_p^p\,.$$

In fact for $p \geq 2$, we can show that $\lambda_1 > 0$ is the first nonzero eigenvalue of $\big(-\Delta_p, W^{1,p}(\Omega)\big)$.

PROPOSITION 4.2.3

If $p \geq 2$,
then the number λ_1 *is the first nonzero eigenvalue of* $\big(-\Delta_p, W^{1,p}(\Omega)\big)$.

PROOF   Let $x \in C(p)$ be a solution of problem (4.42). Then by virtue of the Lagrange multiplier rule, we can find $a, b, c \in \mathbb{R}$, not all of them equal to zero, such that for all $v \in W^{1,p}(\Omega)$, we have

$$ap \int_\Omega \|\nabla x(z)\|_{\mathbb{R}^N}^{p-2} \big(\nabla x(z), \nabla v(z)\big)_{\mathbb{R}^N}\,dz + bp \int_\Omega |x(z)|^{p-2} x(z) v(z)\,dz$$

$$+ c(p-2) \int_\Omega |x(z)|^{p-2} v(z)\,dz + c \int_\Omega |x(z)|^{p-2} v(z)\,dz \;=\; 0. \qquad (4.43)$$

Taking $v \equiv c$ and recalling that $x \in C(p)$, we obtain

$$\big(c^2(p-2) + c^2\big) \int_\Omega |x(z)|^{p-2}\,dz \;=\; 0,$$

so $c = 0$. Thus (4.43) becomes

$$a \int_\Omega \|\nabla x(z)\|_{\mathbb{R}^N}^{p-2} \big(\nabla x(z), \nabla v(z)\big)_{\mathbb{R}^N}$$

$$+ b \int_\Omega |x(z)|^{p-2} x(z) v(z)\,dz \;=\; 0 \qquad \forall\, v \in W^{1,p}(\Omega).$$

Suppose that $a = 0$. Then we have

$$b \int_\Omega |x(z)|^{p-2} x(z) v(z)\,dz \;=\; 0 \qquad \forall\, v \in W^{1,p}(\Omega).$$

Taking $v = x$, we obtain

$$b \, \|x\|_p^p \; = \; 0,$$

hence $b = 0$, a contradiction to the fact that the Lagrange multipliers cannot be all equal to zero. So $a \neq 0$ and without any loss of generality, we may assume that $a = 1$. So we have

$$\int_\Omega \|\nabla x(z)\|_{\mathbb{R}^N}^{p-2} \left(\nabla x(z), \nabla v(z)\right)_{\mathbb{R}^N}$$

$$+ \, b \int_\Omega |x(z)|^{p-2} x(z) v(z) \, dz \; = \; 0 \qquad \forall \, v \in W^{1,p}(\Omega).$$

Using as a test function $v = x$, we infer that $b = -\lambda_1$.

Then via Green's identity (see Theorem 1.1.9), we conclude that x is an eigenfunction of $\left(-\Delta_p, W^{1,p}(\Omega)\right)$ for the eigenvalue $\lambda_1 > 0$, which is clearly the first eigenvalue. $\qquad\qquad\square$

From this analysis we deduce that $\lambda_0 = 0$ is the only principal eigenvalue of $\left(-\Delta_p, W^{1,p}(\Omega)\right)$. When a weight $m(z)$ is introduced in the eigenvalue problem (4.41), then the situation is more involved, in particular if the weight m changes sign in Ω. We continue with a bounded domain $\Omega \subseteq \mathbb{R}^N$ which has a $C^{1,\alpha}$ boundary $\partial\Omega$ ($\alpha \in (0,1)$). We consider the following weighted version of problem (4.41):

$$\begin{cases} -\mathrm{div}\left(\|\nabla x(z)\|_{\mathbb{R}^N}^{p-2}\nabla x(z)\right) = \lambda m(z)|x(z)|^{p-2}x(z) + g(z) \\ \qquad\qquad\qquad\qquad\qquad\qquad \text{for a.a. } z \in \Omega, \qquad (4.44) \\ \frac{\partial x}{\partial n} = 0 \quad \text{on } \partial\Omega, \end{cases}$$

with $p \in (1,\infty)$ and $m, g \in L^\infty(\Omega)$. We shall proceed in our study of problem (4.44), with (unless otherwise stated) the assumption that m changes sign in Ω, i.e. $\left|\{z \in \Omega : \; m(z) > 0\}\right|_N > 0$ and $\left|\{z \in \Omega : \; m(z) < 0\}\right|_N > 0$. Also without any loss of generality, we may assume that $\|m\|_\infty < 1$. From non-linear regularity theory (see Theorem 1.5.6 and Remark 1.5.9), any solution $x \in W^{1,p}(\Omega)$ of problem (4.44) actually belongs in $C^{1,\beta}(\overline{\Omega})$ ($\beta \in (0,1)$) and so the boundary condition "$\frac{\partial x}{\partial n} = 0$ on $\partial\Omega$" is understood in the usual pointwise sense.

We start with a form of the maximum principle, which will be helpful in what follows.

PROPOSITION 4.2.4
<u>*If*</u> $x \in W^{1,p}(\Omega)$ *is a solution of the Neumann problem*

$$\begin{cases} -\mathrm{div}\left(\|\nabla x(z)\|_{\mathbb{R}^N}^{p-2}\nabla x(z)\right) + a(z)|x(z)|^{p-2}x(z) = g(z) \\ \qquad\qquad\qquad\qquad\qquad\qquad \text{for a.a. } z \in \Omega, \qquad (4.45) \\ \frac{\partial x}{\partial n} = 0 \quad \text{on } \partial\Omega, \end{cases}$$

with $p \in (1, \infty)$ and $a, g \in L^\infty(\Omega)_+$, $g \neq 0$,
then $x(z) > 0$ for all $z \in \overline{\Omega}$.

PROOF   As we observed for the solutions of problem (4.44), the solution $x \in W^{1,p}(\Omega)$ of problem (4.45) actually belongs in $C^{1,\beta}(\overline{\Omega})$ ($\beta \in (0,1)$) and so the boundary condition is interpreted in the usual pointwise sense. We use as a test function $-x^- \in W^{1,p}(\Omega)$ (see Remark 1.1.10) and we obtain

$$\|\nabla x^-\|_p^p + \int_\Omega a(z)|x^-(z)|^p \, dz \; = \; -\int_\Omega g(z)x^-(z) \, dz \; \leq \; 0,$$

so $x^- = 0$ and thus $x(z) \geq 0$ for all $z \in \Omega$. Invoking Theorem 1.5.7, we conclude that $x(z) > 0$ for all $z \in \overline{\Omega}$ (recall that $\frac{\partial x}{\partial n}|_{\partial \Omega} = 0$). $\qquad \blacksquare$

This Proposition suggests that for the problem

$$\begin{cases} -\mathrm{div}\left(\|\nabla x(z)\|_{\mathbb{R}^N}^{p-2} \nabla x(z)\right) = \lambda m(z)|x(z)|^{p-2}x(z) \\ \qquad\qquad\qquad\qquad\qquad\qquad\qquad\qquad \text{for a.a. } z \in \Omega, \\ \frac{\partial x}{\partial n} = 0 \quad \text{on } \partial\Omega \end{cases} \qquad (4.46)$$

(with $p \in (1, \infty)$) we should look for the principal eigenvalues (i.e. for the eigenvalues with associated eigenfunctions which do not change sign). Evidently, as was the case with problem (4.41), $\lambda_0 = 0$ is a principal eigenvalue, with nonzero constants as eigenfunctions. Moreover, if we rewrite the equation in (4.44) (with g possibly identically zero) as

$$-\mathrm{div}\left(\|\nabla x(z)\|_{\mathbb{R}^N}^{p-2} \nabla x(z)\right) \pm \lambda|x(z)|^{p-2}x(z)$$
$$= \lambda\big(m(z) \pm 1\big)|x(z)|^{p-2}x(z) + g(z) \quad \text{for a.a. } z \in \Omega$$

(here "$+$" is used if $\lambda > 0$ and "$-$" if $\lambda < 0$) and recalling that we have assumed that $|m(z)| < 1$ for almost all $z \in \Omega$, from Proposition 4.2.4, we infer the following corollary.

COROLLARY 4.2.2
If $x \in C^1(\overline{\Omega})$ *is a nontrivial, nonnegative solution of (4.44) or of (4.46), then* $x(z) > 0$ *for all* $z \in \overline{\Omega}$.

Let

$$E(p) \stackrel{df}{=} \left\{ x \in W^{1,p}(\Omega) : \int_\Omega m(z)|x(z)|^p \, dz = 1 \right\}.$$

LEMMA 4.2.1
<u>If</u> $\int_\Omega m(z)\, dz < 0$,
<u>then</u> there exists $c > 0$, such that

$$\|\nabla x\|_p^p \geq c\,\|x\|_p^p \qquad \forall\, x \in E(p)$$

PROOF Suppose that the lemma is not true. Then we can find a sequence $\{x_n\}_{n\geq 1} \in E(p)$, such that

$$\|\nabla x_n\|_p^p \leq \frac{1}{n}\,\|x_n\|_p^p.$$

First assume that the sequence $\{x_n\}_{n\geq 1} \in L^p(\Omega)$ is unbounded. Passing to a subsequence if necessary, we may assume that

$$\|x_n\|_p \longrightarrow +\infty.$$

Let

$$y_n \overset{df}{=} \frac{x_n}{\|x_n\|_p}.$$

Then

$$\|\nabla y_n\|_p^p \leq \frac{1}{n} \qquad \forall\, n \geq 1$$

and so the sequence $\{y_n\}_{n\geq 1} \subseteq W^{1,p}(\Omega)$ is bounded. Passing to a further subsequence if necessary, we may assume that

$$y_n \overset{w}{\longrightarrow} y \text{ in } W^{1,p}(\Omega),$$
$$y_n \longrightarrow y \text{ in } L^p(\Omega),$$

for some $y \in E(p)$, such that $\|\nabla y\|_p = 0$, hence $y \equiv constant$. But these facts contradict the hypothesis that $\int_\Omega m(z)\, dz < 0$. So we must have that the sequence $\{x_n\}_{n\geq 1} \subseteq L^p(\Omega)$ is bounded. But then so is the sequence $\{x_n\}_{n\geq 1} \subseteq W^{1,p}(\Omega)$ and repeating the above argument we reach again a contradiction. Therefore the lemma is true. ☐

We introduce the following quantity

$$\lambda_m^* \overset{df}{=} \inf\left\{\|\nabla x\|_p^p : x \in W^{1,p}(\Omega), \int_\Omega m(z)|x(z)|^p\, dz = 1\right\}. \qquad (4.47)$$

PROPOSITION 4.2.5
Let λ_m^* be as above.

(a) <u>If</u> $\int\limits_\Omega m(z)\, dz < 0$,

   <u>then</u> $\lambda_m^* > 0$, λ_m^* is a principal eigenvalue and $(0, \lambda_m^*)$ does not contain any eigenvalues of (4.46);

(b) *If* $\int\limits_{\Omega} m(z)\,dz \geq 0$, *then* $\lambda_m^* = 0$.

PROOF   **(a)** From Lemma 4.2.1 and its proof, we see that the infimum in (4.47) is obtained and so it follows that $\lambda_m^* > 0$. Moreover, if $x \in E(p)$ is the function realizing this infimum, by replacing it with $|x|$ if necessary, we can assume without any loss of generality that $x \geq 0$, $x \neq 0$. From the Lagrange multiplier rule, we can find $a, b \in \mathbb{R}$ not both equal to zero, such that

$$a \int\limits_{\Omega} \|\nabla x(z)\|_{\mathbb{R}^N}^{p-2} \left(\nabla x(z), \nabla v(z)\right)_{\mathbb{R}^N} dz$$

$$+ b \int\limits_{\Omega} m(z)|x(z)|^{p-2} x(z) v(z)\, dz \;=\; 0 \qquad \forall\, v \in W^{1,p}(\Omega).$$

If $a = 0$, then by using as a test function x itself, we obtain

$$b \int\limits_{\Omega} m(z)|x(z)|^p\, dz \;=\; 0.$$

Because $x \in E(p)$, it follows that $b = 0$, a contradiction to the fact that $(a,b) \neq (0,0)$. So $a \neq 0$ and we have

$$\int\limits_{\Omega} \|\nabla x(z)\|_{\mathbb{R}^N}^{p-2} \left(\nabla x(z), \nabla v(z)\right)_{\mathbb{R}^N} dz$$

$$+ \lambda \int\limits_{\Omega} m(z)|x(z)|^{p-2} x(z) v(z)\, dz \;=\; 0 \qquad \forall\, v \in W^{1,p}(\Omega),$$

with $\lambda \overset{df}{=} \frac{b}{a}$. From Green's identity (see Theorem 1.1.9), we see that x solves problem (4.46). Moreover, using once more as a test function $v = x$, we obtain that $\lambda = \lambda_m^*$. This proves that λ_m^* is a principal eigenvalue. In addition, it is clear that no $\lambda \in \left(0, \lambda_m^*\right)$ is an eigenvalue of (4.46) (see (4.47)).

(b) If $\int\limits_{\Omega} m(z)\,dz > 0$, then $E(p)$ contains constants, namely $\pm\dfrac{1}{\left(\int_{\Omega} m(z)\,dz\right)^{\frac{1}{p}}}$, and so $\lambda_m^* = 0$. The case $\int\limits_{\Omega} m(z)\,dz = 0$ is a little more involved. Indeed, let $v \in C_c^{\infty}\left(\overline{\Omega}\right)_+$ be such that $\int\limits_{\Omega} m(z)v(z)\,dz > 0$. For every $\varepsilon > 0$, we consider

$$x_\varepsilon \overset{df}{=} \frac{1 + \varepsilon v}{\left(\int_{\Omega} m(z)(1 + \varepsilon v(z))^p\, dz\right)^{\frac{1}{p}}}.$$

Note that using the binomial expansion, we can check that for $\varepsilon > 0$ sufficiently small the denominator in the definition of x_ε does not vanish. Note that for all such small $\varepsilon > 0$, $x_\varepsilon \in E(p)$ and

$$\|\nabla x_\varepsilon\|_p^p \longrightarrow 0 \quad \text{as } \varepsilon \searrow 0.$$

So $\lambda_m^* = 0$. $\qquad\qquad\qquad\qquad\qquad\qquad\qquad\qquad\qquad\qquad\qquad\qquad$ □

For $C^1(\Omega)$-functions $x > 0$ and $y \geq 0$, Picone's identity for $p = 2$ says the following:

$$\|\nabla y(z)\|_{\mathbb{R}^N}^2 + \left(\frac{y(z)}{x(z)}\right)^2 \|\nabla x(z)\|_{\mathbb{R}^N}^2 - 2\frac{y(z)}{x(z)} (\nabla x(z), \nabla y(z))_{\mathbb{R}^N}$$

$$= \|\nabla y(z)\|_{\mathbb{R}^N}^2 - \nabla\left(\frac{y^2}{x}\right)(z)\nabla x(z) \geq 0 \quad \forall z \in \Omega.$$

This identity can be extended for any $p > 1$ (problems with the p-Laplacian).

PROPOSITION 4.2.6
If $x, y \in C^1(\Omega)$, $x > 0$, $y \geq 0$ *and we set*

$$L(y,x)(z) \stackrel{df}{=} \|\nabla y(z)\|_{\mathbb{R}^N}^p + (p-1)\left(\frac{y(z)}{x(z)}\right)^p \|\nabla x(z)\|_{\mathbb{R}^N}^p$$

$$- p\left(\frac{y(z)}{x(z)}\right)^{p-1} \|\nabla x(z)\|_{\mathbb{R}^N}^{p-2} (\nabla x(z), \nabla y(z))_{\mathbb{R}^N},$$

$$R(y,x)(z) \stackrel{df}{=} \|\nabla y(z)\|_{\mathbb{R}^N}^p - \|\nabla x(z)\|_{\mathbb{R}^N}^{p-2} \left(\nabla x(z), \nabla\left(\frac{y^p}{x^{p-1}}\right)(z)\right)_{\mathbb{R}^N},$$

then $R(y,x) = L(y,x) \geq 0$ *and* $L(y,x) = 0$ *if and only if* $\nabla\left(\frac{y}{x}\right)(z) = 0$ *on* Ω, *i.e.* $y = kx$ *for some* $k \in \mathbb{R}$.

PROOF By expanding $R(y,x)$, we can easily check that $R(y,x) = L(y,x)$. Rewrite $L(y,x)(z)$ as follows:

$$L(y,x)(z) \stackrel{df}{=} \|\nabla y(z)\|_{\mathbb{R}^N}^p - p\left(\frac{y(z)}{x(z)}\right)^{p-1} \|\nabla x(z)\|_{\mathbb{R}^N}^{p-2} \|\nabla x(z)\|_{\mathbb{R}^N} \|\nabla y(z)\|_{\mathbb{R}^N}$$

$$+ (p-1)\left(\frac{y(z)}{x(z)}\right)^p \|\nabla x(z)\|_{\mathbb{R}^N}^p$$

$$+ p\left(\frac{y(z)}{x(z)}\right)^{p-1} \|\nabla x(z)\|_{\mathbb{R}^N}^{p-2} \big[\|\nabla x(z)\|_{\mathbb{R}^N} \|\nabla y(z)\|_{\mathbb{R}^N}$$

$$- (\nabla x(z), \nabla y(z))_{\mathbb{R}^N}\big]. \quad (4.48)$$

Note that if $\vartheta = \|\nabla y(z)\|_p$ and $\eta = \left(\frac{y(z)}{x(z)} \|\nabla x(z)\|_{\mathbb{R}^N}\right)^{p-1}$ and we apply Young's inequality with data ϑ, η (see Theorem A.4.2), we infer that

$$\|\nabla y(z)\|_p^p + (p-1)\left(\frac{y(z)}{x(z)}\right)^p \|\nabla x(z)\|_{\mathbb{R}^N}^p$$

$$\geq p\left(\frac{y(z)}{x(z)}\right)^{p-1} \|\nabla x(z)\|_{\mathbb{R}^N}^{p-2} \|\nabla x(z)\|_{\mathbb{R}^N} \|\nabla y(z)\|_{\mathbb{R}^N} \qquad (4.49)$$

(recall that $p - 1 = \frac{p}{p'}$). Also from the Cauchy-Schwarz inequality (see Theorem A.3.15), we have

$$p\left(\tfrac{y(z)}{x(z)}\right)^{p-1} \|\nabla x(z)\|_{\mathbb{R}^N}^{p-2} \left[\|\nabla x(z)\|_{\mathbb{R}^N} \|\nabla y(z)\|_{\mathbb{R}^N} - (\nabla x(z), \nabla y(z))_{\mathbb{R}^N}\right] \geq 0.$$

Using (4.49) and the last inequality in (4.48), we conclude that

$$L(y, x)(z) \geq 0 \qquad \forall\, z \in \Omega.$$

If $L(y, x)(z_0) = 0$ and $y(z_0) \neq 0$, we must have

$$\|\nabla y(z_0)\|_{\mathbb{R}^N} \;=\; \left(\tfrac{y(z_0)}{x(z_0)}\right) \|\nabla x(z_0)\|_{\mathbb{R}^N}$$

(condition for equality in Young's inequality) and

$$\|\nabla x(z_0)\|_{\mathbb{R}^N} \|\nabla y(z_0)\|_{\mathbb{R}^N} \;=\; (\nabla x(z_0), \nabla y(z_0))_{\mathbb{R}^N}.$$

So

$$\nabla y(z_0) \;=\; \left(\tfrac{y(z_0)}{x(z_0)}\right)\nabla x(z_0),$$

hence

$$\nabla\left(\tfrac{y}{x}\right)(z_0) \;=\; 0.$$

On the other hand, if

$$S \overset{df}{=} \{z \in \Omega : y(z) = 0\},$$

then, from Remark 1.1.10, we have that

$$\nabla y(z) \;=\; 0 \qquad \forall\, z \in S$$

and so

$$\nabla\left(\tfrac{y}{x}\right)(z) \;=\; 0 \qquad \forall\, z \in S.$$

Therefore finally

$$\nabla\left(\tfrac{y}{x}\right)(z) \;=\; 0 \qquad \forall\, z \in \Omega,$$

hence $y = kx$ for some $k \in \mathbb{R}$. $\qquad\qquad\qquad\qquad\qquad\qquad\quad$ ▯

Using this identity, we can prove the following inequality.

LEMMA 4.2.2
If $x \in C^1(\overline{\Omega})$ is a solution of (4.44) or of (4.46) and $x(z) > 0$ for all $z \in \overline{\Omega}$, then

$$\frac{gu^p}{x^{p-1}} \in L^1(\Omega) \qquad \forall\, u \in W^{1,p}(\Omega) \cap L^\infty(\Omega) \cap C^1(\Omega)_+$$

and

$$\lambda \int_{\Omega} m(z)u(z)^p \, dz + \int_{\Omega} g(z) \frac{u(z)^p}{x(z)^{p-1}} \, dz \leq \|\nabla u\|_p^p.$$

Moreover, equality holds if and only if $u = kx$ for some $k \in \mathbb{R}$.

PROOF We use Proposition 4.2.6 with data x and u. Then for any subdomain $\Omega_0 \subseteq \Omega$ with compact closure in Ω (i.e. $\Omega_0 \subseteq\subseteq \Omega$), we have

$$0 \leq \int_{\Omega_0} L(u,x)(z) \, dz \leq \int_{\Omega} L(u,x)(z) \, dz = \int_{\Omega} R(u,x)(z) \, dz$$

$$= \|\nabla u\|_p^p - \int_{\Omega} \|\nabla x\|_{\mathbb{R}^N}^p \left(\nabla x(z), \nabla \left(\tfrac{u^p}{x^{p-1}} \right)(z) \right)_{\mathbb{R}^N} dz. \qquad (4.50)$$

Since $x(z) > 0$ for all $z \in \overline{\Omega}$, we have that $\frac{u^p}{x^{p-1}} \in W^{1,p}(\Omega)$. So it can be used as a test function in equation (4.44) or in equation (4.46) to obtain

$$\int_{\Omega} \|\nabla x(z)\|_{\mathbb{R}^N}^{p-2} \left(\nabla x(z), \nabla \left(\tfrac{u^p}{x^{p-1}} \right)(z) \right)_{\mathbb{R}^N} dz$$

$$= \lambda \int_{\Omega} |x(z)|^{p-2} x(z) \frac{u(z)^p}{x(z)^{p-1}} \, dz + \int_{\Omega} g(z) \frac{u(z)^p}{x(z)^{p-1}} \, dz.$$

Using this equality in (4.50), we have

$$0 \leq \int_{\Omega_0} L(u,x)(z) \, dz$$

$$\leq \|\nabla u\|_p^p - \lambda \int_{\Omega} m(z)u(z)^p \, dz - \int_{\Omega} g(z) \frac{u(z)^p}{x(z)^{p-1}} \, dz. \qquad (4.51)$$

From this, we have the inequality of the Lemma. Moreover, if equality holds, then from (4.51), we have that $L(u,x)(z) = 0$ for all $z \in \Omega_0$ and since $\Omega_0 \subseteq\subseteq \Omega$ was arbitrary, we conclude that $L(u,x)(z) = 0$ for all $z \in \Omega$. Invoking Proposition 4.2.6, we have that $u = kx$ for some $k \in \mathbb{R}$. ∎

Using this lemma we can show that problem (4.44) cannot have a nontrivial nonnegative solution if $\int_{\Omega} m(z) \, dz \leq 0$ and $\lambda \notin [0, \lambda_m^*]$.

PROPOSITION 4.2.7
If $\int_{\Omega} m(z) \, dz \leq 0$ and $\lambda \notin [0, \lambda_m^*]$,
then problems (4.44) and (4.46) have no solution $x \neq 0$, $x \geq 0$.

PROOF   Suppose that $x_0 \neq 0$, $x_0 \geq 0$ is a solution of (4.44) or of (4.46) for some $\lambda \in \mathbb{R}$. From Corollary 4.2.2, we have that $x_0(z) > 0$ for all $z \in \overline{\Omega}$. So we can apply Lemma 4.2.2 and obtain

$$\lambda \int_{\Omega} m(z)|u(z)|^p \, dz \ \leq \ \|\nabla u\|_p^p \qquad \forall \, u \in W^{1,p}(\Omega) \cap L^{\infty}(\Omega) \cap C^1(\Omega)_+.$$

By virtue of Theorem 1.1.1 and Remark 1.1.10, the above inequality holds for any $u \in W^{1,p}(\Omega)$. So from (4.47), we infer that $\lambda \leq \lambda_m^*$ and $-\lambda \leq -\lambda_{-m}^*$. Since $\int_{\Omega} -m(z) \, dz \geq 0$, from Proposition 2.5.3(b), we have $\lambda_{-m}^* = 0$ and so $\lambda \geq 0$. Therefore $\lambda \in \left[0, \lambda_m^*\right]$ and this completes the proof.   ☐

This Proposition allows us to complete Proposition 4.2.5 as follows.

COROLLARY 4.2.3
Let $\int_{\Omega} m(z) \, dz \leq 0$.

(a) *If $\int_{\Omega} m(z) \, dz < 0$,*
   then $\lambda_m^* > 0$ is the unique nonzero principal eigenvalue.

(b) *If $\int_{\Omega} m(z) \, dz = 0$, then $\lambda_m^* = 0$ is the unique principal eigenvalue.*

Let

$$F(p) \ \overset{df}{=} \ \left\{ x \in W^{1,p}(\Omega) : \int_{\Omega} m(z)|x(z)|^p \, dz \leq 0 \right\}.$$

LEMMA 4.2.3
If $\int_{\Omega} m(z) \, dz \neq 0$ and $\lambda > 0$,
then

$$\|\nabla x\|_p^p - \lambda \int_{\Omega} m(z) \, |x(z)|^p \, dz \ \geq \ c \, \|x\|_p^p \qquad \forall \, x \in F(p),$$

for some $c > 0$.

PROOF   Suppose that the Lemma is not true. Then we can find a sequence $\{x_n\}_{n \geq 1} \subseteq F(p)$, such that

$$\|\nabla x_n\|_p^p - \lambda \int_{\Omega} m(z) \, |x_n(z)|^p \, dz \ \leq \ \frac{1}{n} \, \|x_n\|_p^p.$$

Let

$$y_n \ \overset{df}{=} \ \frac{x_n}{\|x_n\|_p}.$$

We have

$$0 \leq \|\nabla y_n\|_p^p \leq \|\nabla y_n\|_p^p - \lambda \int_\Omega m(z)|y_n(z)|^p \, dz \leq \frac{1}{n} \qquad (4.52)$$

(since $y_n \in F(p)$ for $n \geq 1$). So

$$\|\nabla y_n\|_p \longrightarrow 0,$$

hence

$$y_n \longrightarrow c \in \mathbb{R} \quad \text{in } W^{1,p}(\Omega),$$

for some $c \neq 0$. Then from (4.52), we have

$$\lambda \int_\Omega m(z)|c|^p \, dz = 0,$$

hence

$$\int_\Omega m(z) \, dz = 0,$$

a contradiction. $\qquad\qquad$ □

Using this lemma, we can have the complete picture about problem (4.44), when $\int_\Omega m(z) \, dz \leq 0$ and $\lambda \in [0, \lambda_m^*]$.

PROPOSITION 4.2.8
If $\int_\Omega m(z) \, dz \leq 0$ and $g \in L^\infty(\Omega)_+ \setminus \{0\}$,
then problem (4.44) has no solution when $\lambda = 0$ or $\lambda = \lambda_m^*$ and has a unique solution which is strictly positive on $\overline{\Omega}$ when $\lambda \in (0, \lambda_m^*)$.

PROOF Recall that if $\int_\Omega m(z) \, dz = 0$, then $\lambda_m^* = 0$ (see Proposition 4.2.5(b)).

If $\lambda = 0$, then we use as a test function $v \equiv 1$ and obtain

$$0 = \int_\Omega g(z) \, dz > 0,$$

a contradiction. So problem (4.44) has no solution in this case.

If $\lambda = \lambda_m^*$, first we show that if a solution x exists, then $x \geq 0$. Suppose that this is not the case. Then $x^- \neq 0$ and we can use $-x^- \in W^{1,p}(\Omega)$ as a test function. We obtain

$$\|\nabla x^-\|_p^p = \lambda_m^* \int_\Omega m(z)|x^-(z)|^p \, dz - \int_\Omega g(z)x^-(z) \, dz.$$

Since $g \geq 0$, we have that

$$\int_{\Omega} g(z)x^-(z)\, dz \;\geq\; 0$$

and so recalling the definition of λ_m^* (see (4.47)), we have that

$$\int_{\Omega} g(z)x^-(z)\, dz \;=\; 0$$

and x^- is a solution of (4.47). Then as before from the Lagrange multiplier rule, we have

$$\begin{cases} -\mathrm{div}\left(\|\nabla x^-(z)\|_{\mathbb{R}^N}^{p-2}\nabla x^-(z)\right) = \lambda_m^* m(z)|x^-(z)|^{p-2}x^-(z) & \text{for a.a. } z \in \Omega, \\ \frac{\partial x^-}{\partial n} = 0 & \text{on } \partial\Omega. \end{cases}$$

Then from Corollary 4.2.2, we have that $x^-(z) > 0$ for all $z \in \overline{\Omega}$, a contradiction to the fact that $\int_{\Omega} g(z)x^-(z)\, dz = 0$. Therefore $x \geq 0$ and invoking once more Corollary 4.2.2, we have that $x(z) > 0$ for all $z \in \overline{\Omega}$. Applying Lemma 4.2.2, we have

$$\lambda_m^* \int_{\Omega} m(z)u(z)^p\, dz + \int_{\Omega} g(z)\frac{u(z)^p}{x(z)^{p-1}}\, dz$$
$$\leq \;\|\nabla u\|_p^p \qquad \forall\, u \in W^{1,p}(\Omega) \cap L^\infty(\Omega) \cap C^1(\Omega)_+.$$

If we take as u the positive eigenfunction of (4.46) associated to $\lambda_m^* \leq 0$, we have

$$\int_{\Omega} g(z)\frac{u(z)^p}{x(z)^{p-1}}\, dz \;\leq\; 0,$$

which is impossible since $u > 0$, $g \geq 0$, $g \neq 0$ and $x > 0$.

Now we examine the case when $\lambda \in \left(0, \lambda_m^*\right)$. First we establish the existence of a solution. To this end, we introduce the energy functional

$$\varphi(x) \;\stackrel{df}{=}\; \frac{1}{p}\|\nabla x\|_p^p - \frac{\lambda}{p}\int_{\Omega} m(z)|x(z)|^p\, dz - \int_{\Omega} g(z)x(z)\, dz \qquad \forall\, x \in W^{1,p}(\Omega).$$

Exploiting the compactness of the embedding $W^{1,p}(\Omega) \subseteq L^p(\Omega)$ and the weak lower semicontinuity of the norm functional in a Banach space, we see that φ is weakly lower semicontinuous on $W^{1,p}(\Omega)$.

First suppose that $x \in W^{1,p}(\Omega)$ is such that $\int_{\Omega} m(z)|x(z)|^p\, dz > 0$. From (4.47) and Lemma 4.2.1, we have

$$\varphi(x) \;\geq\; \frac{1}{p}\left(1 - \frac{\lambda}{\lambda_m^*}\right)\|\nabla x\|_p^p - \int_{\Omega} g(z)x(z)\, dz$$

$$\geq c_1 \|\nabla x\|_p^p + c_2 \|x\|_p^p - \int_\Omega g(z)x(z)\,dz, \qquad (4.53)$$

for some $c_1, c_2 > 0$. Next suppose that $x \in W^{1,p}(\Omega)$ is such that $\int_\Omega m(z)|x(z)|^p\,dz \leq 0$. Because $\lambda > 0$, from Lemma 4.2.3, we have

$$\varphi(x) \geq \frac{c}{p} \|x\|_p^p - \int_\Omega g(z)x(z)\,dz, \qquad (4.54)$$

while because $\lambda \int_\Omega m(z)|x(z)|^p\,dz \leq 0$, we have

$$\varphi(x) \geq \frac{1}{p} \|\nabla x\|_p^p - \int_\Omega g(z)x(z)\,dz. \qquad (4.55)$$

Adding (4.54) and (4.55), we have

$$\varphi(x) \geq \frac{1}{2p} \|\nabla x\|_p^p + \frac{c}{2p} \|x\|_p^p - \int_\Omega g(z)x(z)\,dz. \qquad (4.56)$$

From (4.53) and (4.56), we conclude that φ is coercive. So by the Weierstrass Theorem (see Theorem A.1.5), we can find $x \in W^{1,p}(\Omega)$, such that $\varphi(x) = \inf \varphi$. We have $\varphi'(x) = 0$ and this implies that x is a solution of (4.44). As before using $-x^-$ as a test function, we obtain that $x \geq 0$, $x \neq 0$ and then Corollary 4.2.2 implies that $x > 0$ on $\overline{\Omega}$.

Finally let us prove the uniqueness of this solution. Suppose that y is another solution of (4.44). We have $y > 0$ and using Lemma 4.2.2 with $u = y$, we obtain

$$\lambda \int_\Omega m(z)y(z)^p\,dz + \int_\Omega g(z)\frac{y(z)^p}{x(z)^{p-1}}\,dz$$

$$\leq \|\nabla y\|_p^p = \lambda \int_\Omega m(z)|y(z)|^p\,dz + \int_\Omega g(z)y(z)\,dz \qquad (4.57)$$

and so

$$\int_\Omega g(z)y(z)\left(1 - \frac{y(z)^{p-1}}{x(z)^{p-1}}\right)\,dz \geq 0. \qquad (4.58)$$

Interchanging the roles of x and y in the above arguments, we obtain

$$\int_\Omega g(z)x(z)\left(1 - \frac{x(z)^{p-1}}{y(z)^{p-1}}\right)\,dz \geq 0. \qquad (4.59)$$

Adding (4.58) and (4.59) we get

$$\int_\Omega g(z)\left[y(z)\left(1 - \frac{y(z)^{p-1}}{x(z)^{p-1}}\right) + x(z)\left(1 - \frac{x(z)^{p-1}}{y(z)^{p-1}}\right)\right]\,dz \geq 0.$$

But because

$$x(z)^{p-1}y(z)^{p-1} \leq \frac{x(z)^{2p-1} + y(z)^{2p-1}}{x(z) + y(z)},$$

we see that

$$y(z)\left(1 - \frac{y(z)^{p-1}}{x(z)^{p-1}}\right) + x(z)\left(1 - \frac{x(z)^{p-1}}{y(z)^{p-1}}\right) \leq 0.$$

So we infer that

$$\int_\Omega g(z)\left[y(z)\left(1 - \frac{y(z)^{p-1}}{x(z)^{p-1}}\right) + x(z)\left(1 - \frac{x(z)^{p-1}}{y(z)^{p-1}}\right)\right] dz = 0.$$

From this and (4.57), it follows that

$$\lambda \int_\Omega m(z)y(z)^p \, dz + \int_\Omega g(z)\frac{y(z)^p}{x(z)^{p-1}} \, dz = \|\nabla y\|_p^p$$

and this by Lemma 4.2.2 implies that $y = kx$ for some $k \in \mathbb{R}$. Returning to (4.44) and since $g \neq 0$, we conclude that $k = 1$, i.e. $x = y$. ∎

PROPOSITION 4.2.9

If $\int_\Omega m(z)\,dz \leq 0$,

then the principal eigenvalues 0 and λ_m^* are simple.

PROOF   This is certainly clear for $\lambda = 0$. Now let x be an eigenfunction associated to λ_m^*. As before, we can assume that $x > 0$. Similarly for any other eigenfunction y, we can say that $y > 0$. Using Lemma 4.2.2 with $u = y$, we have

$$\lambda_m^* \int_\Omega m(z)y(z)^p \, dz \leq \|\nabla y\|_p^p$$

and in fact we must have equality (see (4.46)). So once again Lemma 4.2.2 tells us that $x = ky$ with $k \in \mathbb{R}$. ∎

4.2.2 Homogeneous Neumann Problems

Now we pass to the study of the nonlinear elliptic problems with nonsmooth potential and Neumann boundary condition. First we investigate the problem with homogeneous Neumann boundary condition. More precisely on a bounded domain $\Omega \subseteq \mathbb{R}^N$ with a $C^{1,\alpha}$ boundary $\partial\Omega$ ($\alpha \in (0,1)$), we consider the following problem:

$$\begin{cases} -\operatorname{div}\left(\|\nabla x(z)\|_{\mathbb{R}^N}^{p-2}\nabla x(z)\right) \in \partial j(z, x(z)) & \text{for a.a. } z \in \Omega, \\ \frac{\partial x}{\partial n} = 0 & \text{on } \partial\Omega. \end{cases} \qquad (4.60)$$

The boundary condition is understood in the sense of traces. We shall study problem (4.60) using a general Landesman-Lazer type condition, which was already used in the previous section. First we introduce the functions:

$$g_{\min}(z, \zeta) \stackrel{df}{=} \min_{u^* \in \partial j(z, \zeta)} u^*,$$

$$g_{\max}(z, \zeta) \stackrel{df}{=} \max_{u^* \in \partial j(z, \zeta)} u^*,$$

$$G_{\min}(z, \zeta) \stackrel{df}{=} \begin{cases} \frac{pj(z, \zeta)}{\zeta} - g_{\min}(z, \zeta) & \text{if } \zeta \neq 0, \\ 0 & \text{if } \zeta = 0, \end{cases}$$

$$G_{\max}(z, \zeta) \stackrel{df}{=} \begin{cases} \frac{pj(z, \zeta)}{\zeta} - g_{\max}(z, \zeta) & \text{if } \zeta \neq 0, \\ 0 & \text{if } \zeta = 0, \end{cases}$$

$$G_{\min}^-(z, \zeta) \stackrel{df}{=} \limsup_{\zeta \to -\infty} G_{\min}(z, \zeta),$$

$$G_{\max}^+(z, \zeta) \stackrel{df}{=} \liminf_{\zeta \to +\infty} G_{\max}(z, \zeta).$$

Our hypotheses on the nonsmooth potential function $j(z, \zeta)$ are the following:

$\underline{H(j)_1}$ $j \colon \Omega \times \mathbb{R} \longrightarrow \mathbb{R}$ is a function, such that

   (*i*) for every $\zeta \in \mathbb{R}$, the function

$$\Omega \ni z \longrightarrow j(z, \zeta) \in \mathbb{R}$$

   is measurable;

   (*ii*) for almost all $z \in \Omega$, the function

$$\mathbb{R} \ni \zeta \longrightarrow j(z, \zeta) \in \mathbb{R}$$

   is locally Lipschitz;

   (*iii*) for every $M > 0$, there exists $a_M \in L^\infty(\Omega)$, such that

$$|u| \leq a_M(z) \quad \text{for a.a. } z \in \Omega, \text{ all } |\zeta| \leq M \text{ and all } u \in \partial j(z, \zeta);$$

   (*iv*) $\displaystyle \lim_{|\zeta| \to +\infty} \frac{u}{\zeta^{p-1}} = 0$ uniformly for almost all $z \in \Omega$ and all $u \in \partial j(z, \zeta)$;

   (*v*) $G_{\min}^-, G_{\max}^+ \in L^1(\Omega)$ and

$$\int_\Omega G_{\min}^-(z)\, dz \; < \; 0 \; < \; \int_\Omega G_{\max}^+(z)\, dz.$$

We consider the locally Lipschitz energy functional $\varphi\colon W^{1,p}(\Omega) \longrightarrow \mathbb{R}$, defined by

$$\varphi(x) \stackrel{df}{=} \frac{1}{p}\|\nabla x\|_p^p - \int_\Omega j\big(z, x(z)\big)\, dz.$$

PROPOSITION 4.2.10
If hypotheses $H(j)_1$ hold, then φ satisfies the nonsmooth PS-condition.

PROOF   By considering the potential function $j_1(z, \zeta) = j(z, \zeta) - j(z, 0)$ if necessary, we may assume that $j(z, 0) = 0$ for almost all $z \in \Omega$.
Let $\{x_n\}_{n \geq 1} \subseteq W^{1,p}(\Omega)$ be a sequence, such that

$$|\varphi(x_n)| \leq M_1 \text{ for all } n \geq 1 \quad \text{and} \quad m^\varphi(x_n) \longrightarrow 0,$$

for some $M_1 > 0$. We consider $x_n^* \in \partial\varphi(x_n)$ for $n \geq 1$, such that $m^\varphi(x_n) = \|x_n^*\|_{(W^{1,p}(\Omega))^*}$. We have

$$x_n^* = A(x_n) - u_n^* \qquad \forall\, n \geq 1,$$

with $A\colon W^{1,p}(\Omega) \longrightarrow \big(W^{1,p}(\Omega)\big)^*$ the nonlinear operator, defined by

$$\langle A(x), y\rangle_{W^{1,p}(\Omega)} \stackrel{df}{=} \int_\Omega \|\nabla x(z)\|_{\mathbb{R}^N}^{p-2} \big(\nabla x(z), \nabla y(z)\big)_{\mathbb{R}^N}\, dz \qquad \forall\, x, y \in W^{1,p}(\Omega)$$

and $u_n^* \in L^{p'}(\Omega)$, with $u_n^*(z) \in \partial j\big(z, x_n(z)\big)$ for almost all $z \in \Omega$ and all $n \geq 1$. We claim that the sequence $\{x_n\}_{n \geq 1} \subseteq W^{1,p}(\Omega)$ is bounded. Suppose that this is not the case. Passing to a subsequence if necessary, we may assume that

$$\|x_n\|_{W^{1,p}(\Omega)} \longrightarrow +\infty.$$

Let

$$y_n \stackrel{df}{=} \frac{x_n}{\|x_n\|_{W^{1,p}(\Omega)}} \qquad \forall\, n \geq 1.$$

Passing to a further subsequence if necessary, we can say that

$$\begin{aligned}
y_n &\xrightarrow{w} y \quad \text{in } W^{1,p}(\Omega), \\
y_n &\longrightarrow y \quad \text{in } L^p(\Omega), \\
y_n(z) &\longrightarrow y(z) \text{ for a.a. } z \in \Omega, \\
|y_n(z)| &\leq k(z) \text{ for a.a. } z \in \Omega \text{ and all } n \geq 1,
\end{aligned}$$

with $k \in L^p(\Omega)_+$. By virtue of hypothesis $H(j)_1(iv)$, for a given $\varepsilon > 0$, we can find $M_2 = M_2(\varepsilon) > 0$, such that

$$|u| \leq \varepsilon |\zeta|^{p-1} \quad \text{for a.a. } z \in \Omega, \text{ all } |\zeta| \geq M_2 \text{ and all } u \in \partial j(z, \zeta).$$

On the other hand from hypothesis $H(j)_1(iii)$, we know that

$$|u| \leq a_{M_2(\varepsilon)}(z) \quad \text{for a.a. } z \in \Omega, \text{ all } |\zeta| < M_2 \text{ and all } u \in \partial j(z,\zeta),$$

with $a_{M_2(\varepsilon)} \in L^{p'}(\Omega)$. So we can say that

$$|u| \leq a_{M_2(\varepsilon)}(z) + \varepsilon|\zeta|^{p-1} \quad \text{for a.a. } z \in \Omega, \text{ all } \zeta \in \mathbb{R} \text{ and all } u \in \partial j(z,\zeta).$$

Using the mean value theorem for locally Lipschitz functions (see Proposition 1.3.15), we obtain that for almost all $z \in \Omega$ and all $\zeta \in \mathbb{R}$, we have

$$\big|j(z,\zeta)\big| \leq \big|j(z,0)\big| + |u||\zeta| \leq \beta_\varepsilon(z) + 2\varepsilon|\zeta|^p,$$

with $\beta_\varepsilon \in L^1(\Omega)_+$. So for all $n \geq 1$, we have

$$\left| \int_\Omega \frac{j(z,x_n(z))}{\|x_n\|^p_{W^{1,p}(\Omega)}} \, dz \right| \leq \frac{\|\beta_\varepsilon\|_1}{\|x_n\|^p_{W^{1,p}(\Omega)}} + 2\varepsilon \|y_n\|^p_p \leq \frac{\|\beta_\varepsilon\|_1}{\|x_n\|^p_{W^{1,p}(\Omega)}} + 2\varepsilon.$$

Letting $n \to +\infty$ and recalling that $\varepsilon > 0$ was arbitrary, we infer that

$$\lim_{n \to +\infty} \int_\Omega \frac{j(z,x_n(z))}{\|x_n\|^p_{W^{1,p}(\Omega)}} \, dz = 0.$$

So, we have

$$\frac{1}{p} \|\nabla y\|^p_p \leq \liminf_{n \to +\infty} \frac{\varphi(x_n)}{\|x_n\|^p_{W^{1,p}(\Omega)}} \leq \lim_{n \to +\infty} \frac{M_1}{\|x_n\|^p_{W^{1,p}(\Omega)}} = 0,$$

hence $y \equiv \xi \in \mathbb{R}$ (i.e. y is a constant function).

If $\xi = 0$, then

$$y_n \longrightarrow 0 \quad \text{in } W^{1,p}(\Omega),$$

a contradiction to the fact that $\|y_n\|_{W^{1,p}(\Omega)} = 1$ for all $n \geq 1$. So $\xi \neq 0$ and we may assume that $\xi > 0$ (the analysis is similar if $\xi < 0$). From the choice of the sequence $\{x_n\}_{n \geq 1} \subseteq W^{1,p}(\Omega)$, we have

$$\langle x_n^*, y_n \rangle_{W^{1,p}(\Omega)} - \frac{p\varphi(x_n)}{\|x_n\|_{W^{1,p}(\Omega)}} \leq \varepsilon_n + \frac{pM_1}{\|x_n\|_{W^{1,p}(\Omega)}} \qquad \forall\, n \geq 1,$$

with $\varepsilon_n \searrow 0$, so

$$\int_\Omega \frac{pj(z,x_n(z))}{\|x_n\|_{W^{1,p}(\Omega)}} \, dz - \int_\Omega u_n^*(z)y_n(z) \, dz \leq \varepsilon_n + \frac{pM_1}{\|x_n\|_{W^{1,p}(\Omega)}} \qquad \forall\, n \geq 1.$$

Let us set

$$\vartheta_n(z) \stackrel{df}{=} \begin{cases} \frac{j(z,x_n(z))}{x_n(z)} & \text{if } x_n(z) \neq 0 \\ 0 & \text{if } x_n(z) = 0 \end{cases} \qquad \forall\, n \geq 1.$$

We have

$$\frac{pM_1}{\|x_n\|_{W^{1,p}(\Omega)}} + \varepsilon_n \; > \; \int_\Omega \frac{pj(z, x_n(z))}{x_n(z)}\, dz - \int_\Omega u_n^*(z) y_n(z)\, dz$$

$$\geq \int_{\{x_n \neq 0\}} p\vartheta_n(z) y_n(z)\, dz - \int_\Omega u_n^*(z) y_n(z)\, dz - \int_\Omega \frac{|j(z,0)|}{\|x_n\|_{W^{1,p}(\Omega)}}\, dz$$

$$\geq \int_{\{y_n > 0\}} p\vartheta_n(z) y_n(z)\, dz - \int_{\{y_n > 0\}} g_{\max}\big(z, x_n(z)\big) y_n(z)\, dz$$

$$+ \int_{\{y_n < 0\}} p\vartheta_n(z) y_n(z)\, dz - \int_{\{y_n < 0\}} g_{\min}\big(z, x_n(z)\big) y_n(z)\, dz$$

$$= \int_{\{y_n > 0\}} G_{\max}\big(z, x_n(z)\big) y_n(z)\, dz + \int_{\{y_n < 0\}} G_{\min}\big(z, x_n(z)\big) y_n(z)\, dz.$$

Let $0 < \varepsilon < \xi$. Then

$$\{y_n \leq 0\} \; \subseteq \; \{y_n \leq \xi - \varepsilon\} \cup \{y_n \geq \xi + \varepsilon\} \; = \; \{\,|y_n - \xi| \geq \varepsilon\}.$$

But since

$$y_n(z) \; \longrightarrow \; \xi \quad \text{for a.a. } z \in \Omega,$$

we also have convergence in measure. Therefore

$$\big|\{|y_n - \xi| \geq \varepsilon\}\big|_N \; \longrightarrow \; 0,$$

hence

$$\big|\{y_n \leq 0\}\big|_N \; \longrightarrow \; 0.$$

It follows that

$$\big|\{y_n > 0\}\big|_N \; \longrightarrow \; |\Omega|_N.$$

So using Fatou's lemma (its use is permissible by virtue of hypotheses $H(j)_1(iii)$ and (v)), we obtain

$$\xi \int_\Omega G_{\max}^+(z)\, dz \; \leq \; 0,$$

a contradiction to hypothesis $H(j)_1(v)$. This proves that the sequence $\{x_n\}_{n \geq 1} \subseteq W^{1,p}(\Omega)$ is bounded. Passing to a subsequence if necessary, we may assume that

$$x_n \; \xrightarrow{w} \; x \quad \text{in } W^{1,p}(\Omega)$$

and we finish the proof as in Proposition 4.2.7.   ∎

Next we show that $\varphi|_{\mathbb{R}}$ is anticoercive.

PROPOSITION 4.2.11

If hypotheses $H(j)_1$ *hold,*
<u>*then*</u>

$$\varphi(c) \longrightarrow -\infty \quad \text{as } |c| \to +\infty, \ c \in \mathbb{R}.$$

PROOF   Arguing as in the proof of Proposition 4.2.11, we can show that for a given $\varepsilon > 0$, we have

$$\liminf_{c \to +\infty} \frac{j(z,c)}{c} \geq \frac{\vartheta_\varepsilon^+(z)}{p-1} \quad \text{for a.a. } z \in \Omega,$$

and

$$\limsup_{c \to -\infty} \frac{j(z,c)}{c} \leq \frac{\vartheta_\varepsilon^-(z)}{p-1} \quad \text{for a.a. } z \in \Omega,$$

where

$$\vartheta_\varepsilon^+(z) \stackrel{df}{=} G_{\max}^+(z) - \varepsilon \quad \text{and} \quad \vartheta_\varepsilon^-(z) \stackrel{df}{=} G_{\min}^-(z) + \varepsilon.$$

Now suppose that the Proposition was not true. Then we can find a sequence $\{c_n\}_{n \geq 1} \subseteq \mathbb{R}$, such that

$$|c_n| \longrightarrow +\infty \quad \text{and} \quad \varphi(c_n) \geq -M_3 \quad \forall \, n \geq 1,$$

for some $M_3 > 0$. Suppose that $c_n \longrightarrow +\infty$. We have

$$\varphi(c_n) = -\int_\Omega j(z, c_n) \, dz \geq -M_3$$

and so

$$\frac{\varphi(c_n)}{c_n} = -\int_\Omega \frac{j(z, c_n)}{c_n} \, dz \geq -\frac{M_3}{c_n}.$$

Via Fatou's lemma, we have

$$\int_\Omega \liminf_{n \to +\infty} \frac{j(z, c_n)}{c_n} \, dz \leq 0,$$

so

$$\int_\Omega \frac{\vartheta_\varepsilon^+(z)}{p-1} \, dz \leq 0.$$

Let $\varepsilon \searrow 0$, to obtain that

$$\int_\Omega G_{\max}^+(z) \, dz \leq 0,$$

a contradiction to hypothesis $H(j)_1(v)$. Similarly, if we suppose that $c_n \longrightarrow -\infty$, we obtain

$$\int_\Omega G_{\min}^-(z)\,dz \geq 0,$$

again a contradiction. $\square$

Let

$$C \stackrel{df}{=} \left\{ x \in W^{1,p}(\Omega) : \int_\Omega |x(z)|^{p-1} x(z)\,dz = 0 \right\}.$$

Evidently this is a closed cone in $W^{1,p}(\Omega)$.

PROPOSITION 4.2.12
If hypotheses $H(j)_1$ hold, <u>then</u> $\varphi|_C$ is weakly coercive.

PROOF   From the proof of Proposition 4.2.10, we know that for a given $\varepsilon > 0$, we can find $\beta_\varepsilon \in L^1(\Omega)_+$, such that

$$j(z,\zeta) \leq \beta_\varepsilon(z) + 2\varepsilon|\zeta|^p \quad \text{for a.a. } z \in \Omega \text{ and all } \zeta \in \mathbb{R}.$$

So, from Corollary 4.2.1, if $x \in C$, we can write that

$$\varphi(x) \geq \frac{1}{p}\|\nabla x\|_p^p - 2\varepsilon\|x\|_p^p - \|\beta_\varepsilon\|_1$$

$$\geq \left(\frac{1}{p} - \frac{2\varepsilon}{\lambda_1}\right)\|\nabla x\|_p^p - \|\beta_\varepsilon\|_1.$$

We choose $\varepsilon > 0$ so that $\varepsilon < \frac{\lambda_1}{2p}$. Then from the last inequality and Corollary 4.2.1, we conclude that $\varphi|_C$ is weakly coercive. $\square$

Having these auxiliary results, we can prove an existence theorem for problem (4.60).

THEOREM 4.2.1
If hypotheses $H(j)_1$ hold, <u>then</u> problem (4.60) has a solution $x_0 \in W^{1,p}(\Omega)$.

PROOF   By virtue of Proposition 4.2.12, we can find $M_4 > 0$, such that

$$\varphi(x) \geq -M_4 \quad \forall\, x \in C.$$

Also because of Proposition 4.2.11, we can find $c > 0$, such that $\varphi(\pm c) < -M_4$. Consider the sets

$$E \stackrel{df}{=} \{y \in W^{1,p}(\Omega) : -c \leq y(z) \leq c \text{ for a.a. } z \in \Omega\},$$

$$E_1 \stackrel{df}{=} \{\pm c\}.$$

We claim that E_1 and C link in $W^{1,p}(\Omega)$. To this end note that $E_1 \cap C = \emptyset$ and let $\vartheta \in C\big(E; W^{1,p}(\Omega)\big)$, such that $\vartheta|_{E_1} = id_{E_1}$. Introduce the map $\psi \colon W^{1,p}(\Omega) \longrightarrow \mathbb{R}$, defined by

$$\psi(u) \stackrel{df}{=} \int_{\Omega} |u(z)|^{p-2} u(z)\, dz$$

and let $\psi_1 = \psi \circ \vartheta$. Evidently $\psi_1 \in C(E; \mathbb{R})$ and $\psi_1(-c) < 0 < \psi_1(c)$. From the intermediate value theorem, we can find $x_0 \in E$, such that $\psi_1(x_0) = \psi\big(\vartheta(x_0)\big) = 0$, hence $\vartheta(x_0) \in C$ and so $\vartheta(E) \cap C \neq \emptyset$, which establishes the linking of E_1 and C in $W^{1,p}(\Omega)$ (see Remark 2.1.4). Apply Theorem 2.1.3 to obtain $x_0 \in W^{1,p}(\Omega)$, such that $0 \in \partial\varphi(x_0)$. We have

$$A(x_0) = u_0^*, \tag{4.61}$$

with $u_0^* \in L^{p'}(\Omega)$, $u_0^*(z) \in \partial j\big(z, x_0(z)\big)$ for a.a. $z \in \Omega$. Let $\eta \in C_c^1(\Omega)$. Since

$$-\mathrm{div}\,\big(\, \|\nabla x_0\|_{\mathbb{R}^N}^{p-2}\, \nabla x_0 \big) \;\in\; W^{-1,p'}(\Omega) = \big(W_0^{1,p}(\Omega)\big)^*$$

(see Theorem 1.1.8), after integration by parts, we have

$$\int_{\Omega} \|\nabla x_0(z)\|_{\mathbb{R}^N}^{p-2}\, \big(\nabla x_0(z), \nabla \eta(z)\big)_{\mathbb{R}^N}\, dz$$

$$= \Big\langle -\mathrm{div}\,\big(\, \|\nabla x_0(\cdot)\|_{\mathbb{R}^N}^{p-2}\big), \eta \Big\rangle_{W^{1,p}(\Omega)} = \int_{\Omega} u_0^*(z)\eta(z)\, dz.$$

Exploiting the fact that the embedding $C_c^{\infty}(\Omega) \subseteq W^{1,p}(\Omega)$ is dense, from the above equality, we infer that

$$-\mathrm{div}\,\big(\, \|\nabla x_0(z)\|_{\mathbb{R}^N}^{p-2}\, \nabla x_0(z)\big)$$
$$= u_0^*(z) \in \partial j\big(z, x_0(z)\big) \quad \text{for a.a. } z \in \Omega. \tag{4.62}$$

Also from Green's identity (see Theorem 1.1.9), for every $v \in W^{1,p}(\Omega)$, we have

$$\int_{\Omega} \big(\mathrm{div}\, \|\nabla x_0(z)\|_{\mathbb{R}^N}^{p-2}\, \nabla x_0(z)\big) v(z)\, dz$$

$$+ \int_{\Omega} \|\nabla x_0(z)\|_{\mathbb{R}^N}^{p-2}\, \big(\nabla x_0(z), \nabla v(z)\big)_{\mathbb{R}^N}\, dz = \Big\langle \frac{\partial x_0}{\partial n_p}, \gamma_0(x) \Big\rangle_{\partial\Omega},$$

where

$$\frac{\partial x_0}{\partial n_p}(z) \stackrel{df}{=} \|\nabla x_0(z)\|_{\mathbb{R}^N}^{p-2}\, \big(\nabla x_0(z), n(z)\big)_{\mathbb{R}^N},$$

with $n(z)$ being the outward unit normal at $z \in \partial\Omega$ and $\langle \cdot, \cdot \rangle_{\partial\Omega}$ denotes the duality brackets for the pair $\left(W^{\frac{1}{p'},p}(\partial\Omega), W^{-\frac{1}{p'},p'}(\partial\Omega) \right)$ ($\frac{1}{p} + \frac{1}{p'} = 1$). From (4.61) and (4.62), we have that

$$\left\langle \frac{\partial x_0}{\partial n_p}, \gamma_0(x) \right\rangle_{\partial\Omega} = 0.$$

Since $\gamma_0 \left(W^{1,p}(\Omega) \right) = W^{-\frac{1}{p'},p'}(\partial\Omega)$, we conclude that

$$\frac{\partial x_0}{\partial n_p} = 0 \quad \text{in } W^{-\frac{1}{p'},p'}(\partial\Omega).$$

So $x_0 \in W^{1,p}(\Omega)$ is a solution of (4.60). $\Box$

REMARK 4.2.1   Since in hypothesis $H(j)_1(iii)$, $a_M \in L^\infty(\Omega)$, from the nonlinear regularity theory (see Remark 1.5.9), we have $x_0 \in C^{1,\beta}(\overline{\Omega})$ ($\beta \in (0,1)$) and so the boundary condition is interpreted pointwise on $\partial\Omega$. $\Box$

We can reverse the inequality in hypothesis $H(j)_1(v)$ (Landesman-Lazer type condition) and still have a solution of problem (4.60). More precisely, our hypotheses on $j(z, \zeta)$ are the following:

$\underline{H(j)_2}$ $j \colon \Omega \times \mathbb{R} \longrightarrow \mathbb{R}$ is a function, such that

$$\left.\begin{array}{l} (i) \\ (ii) \\ (iii) \\ (iv) \end{array}\right\} \text{are the same as hypotheses } H(j)_1(i), (ii), (iii) \text{ and } (iv);$$

(v) there exist two functions $\widehat{G}^-_{min}, \widehat{G}^+_{max} \in L^1(\Omega)$, such that

$$\widehat{G}^-_{min}(z) = \liminf_{\zeta \to -\infty} G_{min}(z, \zeta),$$

$$\widehat{G}^+_{max}(z) = \limsup_{\zeta \to +\infty} G_{max}(z, \zeta)$$

uniformly for almost all $z \in \Omega$ and

$$\int_\Omega \widehat{G}^+_{max}(z)\,dz < 0 < \int_\Omega \widehat{G}^-_{min}(z)\,dz.$$

A slight modification of the proof of Proposition 4.2.10 gives the following result.

PROPOSITION 4.2.13
If hypotheses $H(j)_2$ hold, then φ satisfies the nonsmooth PS-condition.

In this case we can show global weak coercivity of φ.

PROPOSITION 4.2.14

If hypotheses $H(j)_2$ hold,
then

$$\varphi(x) \longrightarrow +\infty \quad \text{as} \quad \|x\|_{W^{1,p}(\Omega)} \to +\infty.$$

PROOF Suppose that the result of the Proposition is not true. Then we can find a sequence $\{x_n\}_{n\geq 1} \subseteq W^{1,p}(\Omega)$ and $M_5 > 0$, such that

$$\|x_n\|_{W^{1,p}(\Omega)} \longrightarrow +\infty \quad \text{and} \quad \varphi(x_n) \leq M_5 \qquad \forall\, n \geq 1.$$

Let

$$y_n \overset{df}{=} \frac{x_n}{\|x_n\|_{W^{1,p}(\Omega)}} \qquad \forall\, n \geq 1.$$

Passing to a subsequence if necessary, we may assume that

$$y_n \overset{w}{\longrightarrow} y \text{ in } W^{1,p}(\Omega),$$
$$y_n \longrightarrow y \text{ in } L^p(\Omega)$$

and as before we can check that $y \equiv \xi \in \mathbb{R}$, $\xi \neq 0$. Assume that $\xi > 0$. We have

$$-\int_\Omega \frac{j(z, x_n(z))}{\|x_n\|_{W^{1,p}(\Omega)}}\, dz \;\leq\; \frac{\varphi(x_n)}{\|x_n\|_{W^{1,p}(\Omega)}} \;\leq\; \frac{M_5}{\|x_n\|_{W^{1,p}(\Omega)}}. \tag{4.63}$$

A slight modification of the reasoning in the proof of Proposition 4.1.10 gives that

$$\limsup_{y \to +\infty} \frac{j(z, y)}{y} \;\leq\; \frac{\widehat{G}^+_{max}(z)}{p-1} \quad \text{for a.a. } z \in \Omega$$

and

$$\liminf_{y \to -\infty} \frac{j(z, y)}{y} \;\geq\; \frac{\widehat{G}^-_{min}(z)}{p-1} \quad \text{for a.a. } z \in \Omega.$$

Via Fatou's lemma, from the first inequality we have

$$\xi \int_\Omega \widehat{G}^+_{max}(z)\, dz \;\geq\; 0,$$

a contradiction to hypothesis $H(j)_2(v)$. Similarly, if $\xi < 0$, we obtain

$$\xi \int_\Omega \widehat{G}^-_{min}(z)\, dz \;\geq\; 0,$$

again a contradiction to hypothesis $H(j)_2(v)$. $\qquad\qquad\square$

The two Propositions above lead to the following existence result.

THEOREM 4.2.2
If hypotheses $H(j)_2$ hold, <u>then</u> problem (4.60) has a solution $x_0 \in W^{1,p}(\Omega)$.

PROOF   From Proposition 2.1.2, we know that there exists $x_0 \in W^{1,p}(\Omega)$, such that $0 \in \partial\varphi(x_0)$. We finish as in the proof of Theorem 4.2.1.   ▯

REMARK 4.2.2   Consider the locally Lipschitz function

$$j(\zeta) \overset{df}{=} \begin{cases} e^\zeta - \zeta - 1 & \text{if } \zeta < 0, \\ 0 & \text{if } \zeta = 0, \\ 2\zeta + \sin\zeta - \ln(1+\zeta^2) & \text{if } \zeta > 0. \end{cases}$$

Then

$$\partial j(\zeta) = \begin{cases} e^\zeta - 1 & \text{if } \zeta > 0, \\ [0,3] & \text{if } \zeta = 0, \\ 2 + 2\cos\zeta - \frac{2\zeta}{1+\zeta^2} & \text{if } \zeta < 0. \end{cases}$$

A simple calculation reveals that $G_{\min}^- = -(p-1)$ and $G_{\max}^+ = 2p-4$. So if $p > 2$, then $G_{\min}^- < 0 < G_{\max}^+$ and the Landesman-Lazer type hypothesis $H(j)_1(v)$ is satisfied. On the other hand the usual Landesman-Lazer condition, which requires that

$$\limsup_{\zeta\to-\infty} \max_{u^* \in \partial j(\zeta)} u^* < 0 < \liminf_{\zeta\to+\infty} \min_{u^* \in \partial j(\zeta)} u^*,$$

is not satisfied since

$$\liminf_{\zeta\to+\infty} \min_{u^* \in \partial j(\zeta)} u^* = 2 - 2 = 0.$$

▯

4.2.3   Nonhomogeneous Neumann Problem

Next we pass to the nonhomogeneous Neumann problem. So we shall study the following equation

$$\begin{cases} -\text{div}\left(\|\nabla x(z)\|_{\mathbb{R}^N}^{p-2}\nabla x(z)\right) \in \partial j(z, x(z)) & \text{for a.a. } z \in \Omega, \\ \frac{\partial x}{\partial n} \in \partial k(z, \gamma_0(x)(z)) & \text{for a.a. } z \in \partial\Omega, \end{cases} \quad (4.64)$$

where $p \in (0,1)$ and γ_0 is the trace operator. In this case we shall employ the usual Landesman-Lazer condition for the nonsmooth potential $j(z, \zeta)$. The precise hypotheses on the data of (4.64) are the following:

$\underline{H(j)_3}$ $j: \Omega \times \mathbb{R} \longrightarrow \mathbb{R}$ is a function, such that

(i) for every $\zeta \in \mathbb{R}$, the function

$$\Omega \ni z \longrightarrow j(z, \zeta) \in \mathbb{R}$$

is measurable;

(ii) for almost all $z \in \Omega$, the function

$$\mathbb{R} \ni \zeta \longrightarrow j(z, \zeta) \in \mathbb{R}$$

is locally Lipschitz;

(iii) there exists a function $a \in L^\infty(\Omega)_+$, such that

$$|u| \leq a(z) \quad \text{for a.a. } z \in \Omega, \text{ all } \zeta \in \mathbb{R} \text{ and all } u \in \partial j(z, \zeta);$$

(iv) there exist functions $f_+, f_- \in L^{p'}(\Omega)_+$, such that

$$g_{\max}(z, \zeta) \longrightarrow f_-(z) \quad \text{as } \zeta \to -\infty \quad \text{for a.a. } z \in \Omega$$

and

$$g_{\min}(z, \zeta) \longrightarrow f_+(z) \quad \text{as } \zeta \to +\infty \quad \text{for a.a. } z \in \Omega,$$

where

$$g_{\max}(z, \zeta) \overset{df}{=} \max_{u^* \in \partial j(z, \zeta)} u^* \quad \text{and} \quad g_{\min}(z, \zeta) \overset{df}{=} \min_{u^* \in \partial j(z, \zeta)} u^*$$

and

$$\int_\Omega f_-(z)\, dz \; < \; 0 \; < \; \int_\Omega f_+(z)\, dz.$$

$\underline{H(k)}$ $k: \partial\Omega \times \mathbb{R} \longrightarrow \mathbb{R}_+$ is a function, such that

(i) for every $\zeta \in \mathbb{R}$, the function

$$\partial\Omega \ni z \longrightarrow k(z, \zeta) \in \mathbb{R}_+$$

is measurable and $k(\cdot, \zeta) \in L^1(\partial\Omega)$;

(ii) for almost all $z \in \partial\Omega$, the function

$$\mathbb{R} \ni \zeta \longrightarrow k(z, \zeta) \in \mathbb{R}_+$$

is locally Lipschitz (on $\partial\Omega$ we consider the $(N-1)$-dimensional Hausdorff (surface) measure σ);

(*iii*) for almost all $z \in \partial\Omega$, all $\zeta \in \mathbb{R}$ and all $v \in \partial k(z, \zeta)$, we have

$$|v| \leq \widehat{a}(z) + c|\zeta|^{r-1},$$

with $\widehat{a} \in L^{\infty}(\Omega)$, $\widehat{c} > 0$, $r \in [1, p)$;

(*iv*) there exists $M > 0$, such that for almost all $z \in \partial\Omega$, all $|\zeta| \geq M$ and all $v \in \partial k(z, \zeta)$, we have

$$pk(z, \zeta) - v\zeta \geq 0.$$

In what follows $\gamma_0 \colon W^{1,p}(\Omega) \longrightarrow L^p(\partial\Omega)$ stands for the trace operator (see Theorem 1.1.3). The energy functional $\varphi \colon W^{1,p}(\Omega) \longrightarrow \mathbb{R}$ corresponding to problem (4.64) is given by

$$\varphi(x) \stackrel{df}{=} \frac{1}{p} \|\nabla x\|_p^p - \int_{\Omega} j\big(z, x(z)\big) \, dz - \int_{\partial\Omega} k\big(z, \gamma_0(x)(z)\big) \, d\sigma.$$

The function φ is locally Lipschitz (see Theorem 1.3.9).

PROPOSITION 4.2.15
If hypotheses $H(j)_3$ and $H(k)$ hold,
then φ *satisfies the nonsmooth PS-condition.*

PROOF   Let $\{x_n\}_{n\geq 1} \subseteq W^{1,p}(\Omega)$ be a sequence, such that

$$|\varphi(x_n)| \leq M_6 \quad \forall \, n \geq 1 \quad \text{and} \quad m^{\varphi}(x_n) \longrightarrow 0,$$

for some $M_6 > 0$. We consider $x_n^* \in \partial\varphi(x_n)$, such that $\|x_n^*\|_{(W^{1,p}(\Omega))^*} = m^{\varphi}(x_n)$ for $n \geq 1$. Let $\widehat{J}_k \colon L^p(\partial\Omega) \longrightarrow \mathbb{R}$ be the integral functional, defined by

$$\widehat{J}_k(h) \stackrel{df}{=} \int_{\partial\Omega} k\big(z, h(z)\big) \, d\sigma \quad \forall \, h \in L^p(\partial\Omega).$$

We know that $\widehat{J}_k$ is locally Lipschitz (see Theorem 1.3.9). Set

$$J_k \stackrel{df}{=} \widehat{J}_k \circ \gamma_0 \colon W^{1,p}(\Omega) \longrightarrow \mathbb{R}.$$

Then J_k is locally Lipschitz too and from the chain rule for locally Lipschitz functions (see Proposition 1.3.15), we have that

$$\partial J_k(x) \subseteq \gamma_0^* \partial\widehat{J}_k\big(\gamma_0(x)\big) \quad \forall \, x \in W^{1,p}(\Omega).$$

So, we have

$$x_n^* = A(x_n) - u_n^* - \gamma_0^*(v_n) \quad \forall \, n \geq 1,$$

with $A\colon W^{1,p}(\Omega) \longrightarrow \left(W^{1,p}(\Omega)\right)^*$ the nonlinear operator defined by

$$\langle A(x), y\rangle_{W^{1,p}(\Omega)} \;=\; \int_{\Omega} \|\nabla x(z)\|_{\mathbb{R}^N}^{p-2} \left(\nabla x(z), \nabla y(z)\right)_{\mathbb{R}^N} dz \quad \forall x, y \in W^{1,p}(\Omega),$$

$u_n^* \in L^{p'}(\Omega)$, $u_n^*(z) \in \partial j(z, x_n(z))$ for almost all $z \in \Omega$, $v_n \in L^{r'}(\partial\Omega)$, $v_n(z) \in \partial k(z, \gamma_0(x_n)(z))$ for almost all $z \in \partial\Omega$.

We claim that the sequence $\{x_n\}_{n \geq 1} \subseteq W^{1,p}(\Omega)$ is bounded. Suppose for the moment that this is not the case. Passing to a subsequence if necessary, we may assume that

$$\|x_n\|_{W^{1,p}(\Omega)} \longrightarrow +\infty.$$

We set

$$y_n \stackrel{df}{=} \frac{x_n}{\|x_n\|_{W^{1,p}(\Omega)}} \qquad \forall\, n \geq 1.$$

Evidently, passing to a subsequence if necessary, we may assume that

$$y_n \stackrel{w}{\longrightarrow} y \text{ in } W^{1,p}(\Omega),$$
$$y_n \longrightarrow y \text{ in } L^p(\Omega),$$

for some $y \in W^{1,p}(\Omega)$. From the choice of the sequence $\{x_n\}_{n \geq 1} W^{1,p}(\Omega)$, we have

$$\frac{1}{p}\|\nabla y_n\|_p^p - \int_{\Omega} \frac{j(z, x_n(z))}{\|x_n\|_{W^{1,p}(\Omega)}^p} \, dz - \int_{\partial\Omega} \frac{k(z, \gamma_0(x_n)(z))}{\|x_n\|_{W^{1,p}(\Omega)}^p} \, d\sigma \;\leq\; \frac{M_6}{\|x_n\|_{W^{1,p}(\Omega)}^p}.$$
$$(4.65)$$

Note that without any loss of generality, we may assume that $j(z, 0) = 0$ for almost all $z \in \Omega$. Then by virtue of the mean value theorem for locally Lipschitz functions (see Proposition 1.3.14), we have

$$\left|j(z, \varsigma)\right| \;\leq\; a_1(z) + c_3|\varsigma| \quad \text{for a.a. } z \in \Omega \text{ and all } \varsigma \in \mathbb{R},$$

with $a_1 \in L^1(\Omega)_+$, $c_3 > 0$ (see hypothesis $H(j)_3(iii)$). Similarly, we have

$$\left|k(z, \varsigma)\right| \;\leq\; a_2(z) + c_4|\varsigma|^r \quad \text{for a.a. } z \in \partial\Omega \text{ and all } \varsigma \in \mathbb{R},$$

with $a_2 \in L^1(\partial\Omega)_+$, $c_4 > 0$ (see hypotheses $H(k)(i)$ and (iii)). It follows that

$$\int_{\Omega} \frac{j(z, x_n(z))}{\|x_n\|_{W^{1,p}(\Omega)}^p} \, dz \longrightarrow 0 \quad \text{and} \quad \int_{\partial\Omega} \frac{k(z, \gamma_0(x_n)(z))}{\|x_n\|_{W^{1,p}(\Omega)}^p} \, d\sigma \longrightarrow 0.$$

So, passing to the limit as $n \to +\infty$ in (4.65), we obtain $\|\nabla y\|_p = 0$, i.e. $y \equiv \xi \in \mathbb{R}$. It cannot happen that $\xi = 0$, because then

$$y_n \longrightarrow 0 \quad \text{in } W^{1,p}(\Omega),$$

a contradiction to the fact that $\|y_n\|_{W^{1,p}(\Omega)} = 1$. So $\xi \neq 0$ and we assume that $\xi > 0$ (the analysis is similar if $\xi < 0$). Recall that

$$|p\varphi(x_n)| \leq pM_6 \qquad \forall\, n \geq 1$$

and

$$\left|\langle x_n^*, x_n\rangle_{W^{1,p}(\Omega)}\right| \leq \varepsilon_n \|x_n\|_{W^{1,p}(\Omega)} \qquad \forall\, n \geq 1,$$

with $\varepsilon_n \searrow 0$, so

$$-\|\nabla x_n\|_p^p + \int_\Omega pj\big(z, x_n(z)\big)\, dz + \int_{\partial\Omega} pk\big(z, \gamma_0(x_n)(z)\big)\, d\sigma \leq pM_6 \qquad (4.66)$$

and

$$\|\nabla x_n\|_p^p - \int_\Omega u_n^*(z)x_n(z)\, dz - \int_{\partial\Omega} v_n(z)\gamma_0(x_n)(z)\, d\sigma \leq \varepsilon_n \|x_n\|_{W^{1,p}(\Omega)}. \qquad (4.67)$$

Adding (4.66) and (4.67), we obtain

$$\int_\Omega \big(pj\big(z, x_n(z)\big) - u_n^*(z)x_n(z)\big)\, dz$$
$$+ \int_{\partial\Omega} \big(pk\big(z, \gamma_0(x_n)(z)\big) - v_n(z)\gamma_0(x_n)(z)\big)\, d\sigma$$
$$\leq pM_6 + \varepsilon_n \|x_n\|_{W^{1,p}(\Omega)}. \qquad (4.68)$$

We examine the first integral in the left hand side of (4.68). Since $\xi > 0$, we have

$$x_n(z) \longrightarrow +\infty \quad \text{for a.a. } z \in \Omega.$$

For a given $\varepsilon \in (0,1)$, from the mean value theorem for locally Lipschitz functions (see Proposition 1.3.15), we have

$$j\big(z, x_n(z)\big) = j\big(z, \varepsilon x_n(z)\big) + w_n^*(z)(1 - \varepsilon)x_n(z) \quad \text{for a.a. } z \in \Omega \setminus D,$$

for some $D \subseteq \Omega$, $|D|_N = 0$, $w_n^*(z) \in \partial j\big(z, w_n(z)\big)$, $w_n(z) = \big(1 - t_n(z)\big)x_n(z) + t_n(z)\varepsilon x_n(z)$, with $t_n(z) \in (0,1)$, $n \geq 1$.

We may assume that $x_n(z) > 0$ for all $n \geq 1$ (recall that $x_n(z) \to +\infty$ for all $z \in \Omega \setminus D$). So we have

$$w_n(z) = x_n(z) - t_n(z)(1 - \varepsilon)x_n(z) \geq x_n(z) - (1 - \varepsilon)x_n(z) = \varepsilon x_n(z),$$

hence

$$w_n(z) \longrightarrow +\infty.$$

Then by virtue of hypothesis $H(j)_3(iv)$, we have that

$$w_n^*(z) \longrightarrow f_+(z).$$

We can find $n_0 = n_0(\varepsilon, z) \geq 1$, such that, if $n \geq n_0$, we have

$$\left| w_n^*(z) - f_+(z) \right| < \varepsilon.$$

So for $n \geq n_0$, we have

$$\frac{pj(z, x_n(z))}{x_n(z)} = \frac{pj(z, \varepsilon x_n(z))}{x_n(z)} + p(1 - \varepsilon) w_n^*(z).$$

Recall that

$$\left| j(z, \zeta) \right| \leq a_1(z) + c_3 |\zeta| \quad \text{for a.a. } z \in \Omega \text{ and all } \zeta \in \mathbb{R}.$$

Hence, we can write that

$$\frac{-pa_1(z) - pc_3 \varepsilon x_n(z)}{x_n(z)} + p(1 - \varepsilon)\left(-\varepsilon + f_+(z) \right)$$

$$\leq \frac{pj(z, x_n(z))}{x_n(z)} \leq \frac{pa_1(z) + pc_3 \varepsilon x_n(z)}{x_n(z)} + p(1 - \varepsilon)\left(\varepsilon + f_+(z) \right).$$

Since $x_n(z) \longrightarrow +\infty$ and $\varepsilon > 0$ was arbitrary, we infer that

$$\frac{pj(z, x_n(z))}{x_n(z)} \longrightarrow pf_+(z) \quad \text{for a.a. } z \in \Omega.$$

Then, we have

$$\int_\Omega \frac{pj(z, x_n(z))}{\|x_n\|_{W^{1,p}(\Omega)}} \, dz = \int_{\{x_n \neq 0\}} \frac{pj(z, x_n(z))}{x_n(z)} y_n(z) \, dz$$

$$= \int_{\{x_n > 0\}} \frac{pj(z, x_n(z))}{x_n(z)} y_n(z) \, dz + \int_{\{x_n < 0\}} \frac{pj(z, x_n(z))}{x_n(z)} y_n(z) \, dz$$

(recall that $j(z, 0) = 0$ for almost all $z \in \Omega$). Because

$$\left| \frac{pj(z, x_n(z))}{x_n(z)} \right| \leq a_3(z) \quad \text{for a.a. } z \in \Omega \text{ and all } n \geq 1$$

(by virtue of the mean value theorem and hypothesis $H(j)_3(iii)$) and since $\left| \{x_n > 0\} \right|_N \longrightarrow |\Omega|_N$ (hence $\left| \{x_n < 0\} \right|_N \longrightarrow 0$), passing to the limit, we obtain

$$\int_\Omega \frac{pj(z, x_n(z))}{\|x_n\|_{W^{1,p}(\Omega)}} \, dz \longrightarrow p\xi \int_\Omega f_+(z) \, dz.$$

Also, from hypothesis $H(j)_3(iv)$, we have

$$\int_\Omega u_n^*(z) y_n(z) \, dz \longrightarrow \xi \int_\Omega f_+(z) \, dz.$$

So finally

$$\frac{1}{\|x_n\|_{W^{1,p}(\Omega)}} \int_\Omega \left(pj\big(z, x_n(z)\big) - u_n^*(z) x_n(z) \right) dz$$

$$\longrightarrow \xi(p-1) \int_\Omega f_+(z) \, dz. \tag{4.69}$$

Next, we examine the second integral in the left hand side of (4.68). We have

$$\int_{\partial\Omega} \left(pk\big(z, \gamma_0(x_n)(z)\big) - v_n(z)\gamma_0(x_n)(z) \right) d\sigma$$

$$= \int_{\{|\gamma_0(x_n)| \geq M\}} \left(pk\big(z, \gamma_0(x_n)(z)\big) - v_n(z)\gamma_0(x_n)(z) \right) d\sigma$$

$$+ \int_{\{|\gamma_0(x_n)| < M\}} \left(pk\big(z, \gamma_0(x_n)(z)\big) - v_n(z)\gamma_0(x_n)(z) \right) d\sigma$$

$$\geq \int_{\{|\gamma_0(x_n)| < M\}} \left(pk\big(z, \gamma_0(x_n)(z)\big) - v_n(z)\gamma_0(x_n)(z) \right) d\sigma$$

(see hypothesis $H(k)(iv)$), so

$$\frac{1}{\|x_n\|_{W^{1,p}(\Omega)}} \int_{\partial\Omega} \left(pk\big(z, \gamma_0(x_n)(z)\big) - v_n(z)\gamma_0(x_n)(z) \right) d\sigma \geq \frac{\beta}{\|x_n\|_{W^{1,p}(\Omega)}},$$

for some $\beta \in \mathbb{R}$ and

$$\liminf_{n \to +\infty} \frac{1}{\|x_n\|_{W^{1,p}(\Omega)}} \int_{\partial\Omega} \left(pk\big(z, \gamma_0(x_n)(z)\big) - v_n(z)\gamma_0(x_n)(z) \right) d\sigma \geq 0. \tag{4.70}$$

So, if we return to (4.68), divide with $\|x_n\|_{W^{1,p}(\Omega)}$, pass the limit as $n \to +\infty$ and use (4.69) and (4.70), we obtain

$$\xi(p-1) \int_\Omega f_+(z) \, dz \leq 0$$

and so

$$\int_\Omega f_+(z) \, dz \leq 0,$$

a contradiction to hypothesis $H(j)_3(iv)$.

Similarly, if we assume that $\xi < 0$, we reach the contradiction $0 \leq \int_\Omega f_-(z) \, dz$.

Therefore the sequence $\{x_n\}_{n\geq 1} \subseteq W^{1,p}(\Omega)$ is bounded and passing to a subsequence if necessary, we may assume that

$$x_n \xrightarrow{w} x_0 \text{ in } W^{1,p}(\Omega),$$
$$x_n \longrightarrow x_0 \text{ in } L^p(\Omega),$$

for some $x_0 \in W^{1,p}(\Omega)$. From these facts as in previous proofs, exploiting the maximal monotonicity of the operator A and the Kadec-Klee property of $L^p(\Omega; \mathbb{R}^N)$, we conclude that

$$x_n \longrightarrow x_0 \quad \text{in } W^{1,p}(\Omega).$$

$\square$

We consider the direct sum decomposition

$$W^{1,p}(\Omega) = \mathbb{R} \oplus V,$$

with

$$V \stackrel{df}{=} \left\{ x \in W^{1,p}(\Omega) : \int_\Omega x(z)\,dz = 0 \right\}.$$

We examine the behaviour of φ on each one of the components of this direct sum.

PROPOSITION 4.2.16
If hypotheses $h(j)_3$ and $H(k)$ hold, <u>then</u> $\varphi|_V$ is weakly coercive.

PROOF For every $v \in V$, we have

$$\varphi(v) \geq \frac{1}{p} \|\nabla v\|_p^p - c_3 \|v\|_p - \|a_1\|_1 - c_2 \|\gamma_0(v)\|_{L^{r'}(\Omega)}^r - \|a_2\|_1$$

$$\geq \frac{1}{p} \|\nabla v\|_p^p - c_3 \|v\|_p - c_5 \|v\|_p^r - c_6,$$

for some $c_5, c_6 > 0$. Invoking the Poincaré-Wirtinger inequality (see Theorem 1.1.7), we obtain

$$\varphi(v) \geq \frac{1}{p} \|\nabla v\|_p^p - c_7 \|\nabla v\|_p^r - c_8,$$

for some $c_7, c_8 > 0$. Because $r < p$ (see hypothesis $H(k)(iii)$), we conclude that $\varphi|_V$ is indeed weakly coercive. $\square$

PROPOSITION 4.2.17
If hypotheses $H(j)_3$ and $H(k)$ hold,
<u>*then*</u>

$$\varphi(\xi) \longrightarrow -\infty \quad \text{as } |\xi| \to +\infty, \ \xi \in \mathbb{R}.$$

PROOF   Since $k \geq 1$, for all $\xi \in \mathbb{R}$, we have

$$\varphi(\xi) \; = \; -\int_{\Omega} j(z,\xi)\,dz - \int_{\partial\Omega} k(z,\xi)\,d\sigma \; \leq \; -\int_{\Omega} j(z,\xi)\,dz,$$

so

$$\varphi(\xi) \; \leq \; -\xi \int_{\Omega} \frac{j(z,\xi)}{\xi}\,dz \qquad \forall\, \xi \neq 0. \tag{4.71}$$

But from the proof of Proposition 4.2.15, we know that for almost all $z \in \Omega$, we have

$$\frac{j(z,\zeta)}{\zeta} \; \longrightarrow \; f_{\pm}(z) \quad \text{as } \zeta \to \pm\infty.$$

Since

$$\left| \frac{j(z,\zeta)}{\zeta} \right| \; \leq \; a_3(z) \quad \text{for a.a. } z \in \Omega,$$

from the dominated convergence theorem, we have that

$$\int_{\Omega} \frac{j(z,\zeta)}{\zeta}\,dz \; \longrightarrow \; \int_{\Omega} f_{\pm}(z)\,dz \quad \text{as } \zeta \to \pm\infty,$$

so

$$\varphi(\zeta) \; \longrightarrow \; -\infty \quad \text{as } |\zeta| \to +\infty$$

(see (4.71) and hypothesis $H(j)_3(iv)$).   ⬜

These Propositions lead to the following existence result for problem (4.64).

THEOREM 4.2.3
If hypotheses $H(j)_3$ and $H(k)$ hold,
then problem (4.64) has a solution $x_0 \in W^{1,p}(\Omega)$.

PROOF   Propositions 4.2.15, 4.2.16 and 4.2.17 permit the use of the nonsmooth Saddle Point Theorem (see Theorem 2.1.4), which furnishes an $x_0 \in W^{1,p}(\Omega)$, such that $0 \in \partial\varphi(x_0)$. So, we have

$$A(x_0) - u_0^* \; = \; \gamma_0^*(v), \tag{4.72}$$

where $u_0^* \in L^{p'}(\Omega)$, $u_0^*(z) \in \partial j\big(z, x_0(z)\big)$ for almost all $z \in \Omega$ and $v \in L^{r'}(\partial\Omega)$, $v(z) \in \partial k\big(z, \gamma_0(x_0)(z)\big)$ for almost all $z \in \partial\Omega$. For every $\vartheta \in C_c^{\infty}(\Omega)$, we have

$$\langle A(x_0), \vartheta \rangle_{W^{1,p}(\Omega)} - \int_{\Omega} u_0^*(z)\vartheta(z)\,dz \; = \; \int_{\partial\Omega} v(z)\gamma_0(\vartheta)(z)\,d\sigma \; = \; 0$$

(since $\gamma_0(\vartheta) = 0$). We know that

$$-\mathrm{div}\left(\|\nabla x_0(\cdot)\|_{\mathbb{R}^N}^{p-2}\nabla x_0(\cdot)\right) \in W^{-1,p'}(\Omega) = \left(W_0^{1,p}(\Omega)\right)^*$$

(see Theorem 1.1.8). So, after integrating by parts (Green's identity), we have

$$\langle A(x_0), \vartheta\rangle_{W^{1,p}(\Omega)} = \int_\Omega \|\nabla x_0(z)\|_{\mathbb{R}^N}^{p-2}\left(\nabla x(z), \nabla\vartheta(z)\right)_{\mathbb{R}^N}dz$$

$$= \left\langle -\mathrm{div}\left(\|\nabla x_0(\cdot)\|_{\mathbb{R}^N}^{p-2}\nabla x_0(\cdot)\right), \vartheta\right\rangle_{W^{1,p}(\Omega)} = \langle u_0^*, \vartheta\rangle_{W^{1,p}(\Omega)}.$$

Recalling that the embedding $C_c^\infty(\Omega) \subseteq W^{1,p}(\Omega)$ is dense, we infer that

$$-\mathrm{div}\left(\|\nabla x_0(z)\|_{\mathbb{R}^N}^{p-2}\nabla x_0(z)\right)$$
$$= u_0^*(z) \in \partial j\big(z, x_0(z)\big) \quad \text{for a.a. } z \in \Omega. \tag{4.73}$$

From Green's identity (see Theorem 1.1.9), for all $y \in W^{1,p}(\Omega)$, we have

$$\int_\Omega \|\nabla x_0(z)\|_{\mathbb{R}^N}^{p-2}\left(\nabla x_0(z), \nabla y(z)\right)_{\mathbb{R}^N}dz$$

$$+ \int_\Omega \mathrm{div}\left(\|\nabla x_0(z)\|_{\mathbb{R}^N}^{p-2}\nabla x_0(z)\right)y(z)\,dz = \left(\frac{\partial x_0}{\partial n_p}, \gamma_0(y)\right)_{\partial\Omega}, \tag{4.74}$$

where as before $\langle\cdot,\cdot\rangle_{\partial\Omega}$ denotes the duality brackets for the pair $\left(W^{\frac{1}{p'},p}(\partial\Omega), W^{-\frac{1}{p'},p'}(\partial\Omega)\right)$. Because of (4.72), we have

$$\int_\Omega \|\nabla x_0(z)\|_{\mathbb{R}^N}^{p-2}\left(\nabla x_0(z), \nabla y(z)\right)_{\mathbb{R}^N}dz$$

$$= \int_\Omega u_n^*(z)y(z)\,dz + \langle v, \gamma_0(x)\rangle_{\partial\Omega} \tag{4.75}$$

(recall that $R(\gamma_0) = W^{\frac{1}{p'},p}(\partial\Omega)$; see Section 1.1). Using (4.75) and (4.74) and taking into account (4.73), we obtain

$$\langle v, \gamma_0(y)\rangle_{\partial\Omega} = \left\langle\frac{\partial x_0}{\partial n_p}, \gamma_0(y)\right\rangle_{\partial\Omega} \quad \forall\, y \in W^{1,p}(\Omega). \tag{4.76}$$

We know that $\gamma_0\big(W^{1,p}(\Omega)\big) = W^{\frac{1}{p'},p}(\partial\Omega)$. So from (4.76), it follows that

$$\frac{\partial x_0}{\partial n_p}(z) = v(z) \in \partial k\big(z, \gamma_0(x_0)(z)\big) \quad \text{for a.a. } z \in \partial\Omega.$$

Therefore $x_0 \in W^{1,p}(\Omega)$ is a solution of (4.64). $\qquad\square$

4.3   Problems with an Area-Type Term

In this section we shall use the methods of the nonsmooth critical point theory (in particular the material from Section 2.5) to study a class of functionals which includes as a particular case the area integral of the nonparametric minimal hypersurface problem. In that problem, we seek to minimize the integral

$$\int_\Omega \sqrt{1 + \|\nabla x(z)\|_{\mathbb{R}^N}^2}\, dz$$

among all the functions x, such that $x|_{\partial\Omega} = g$. This is the problem of minimizing the area of a hypersurface that can be represented as a graph over a domain in $\Omega \subseteq \mathbb{R}^N$ with a prescribed boundary (the points $(z, g(z))$, $z \in \partial\Omega$). The Euler-Lagrange equation of this calculus of variations problem is the following Dirichlet problem (minimal hypersurface equation):

$$\begin{cases} -\mathrm{div}\left(\dfrac{\nabla x(z)}{\sqrt{1 + \|\nabla x(z)\|_{\mathbb{R}^N}^2}}\right) = 0 & \text{in } \Omega, \\ x|_{\partial\Omega} = g & \text{on } \partial\Omega. \end{cases}$$

The special feature of this problem is that it is only coercive on a nonreflexive space built on $L^1(\Omega)$, which we shall introduce in the sequel. This fact makes the variational methods for the minimal hypersurface problem really challenging.

As we already mentioned, in this section we shall be concerned with a more general problem which incorporates the minimal hypersurface problem as a special case. So, let $\Omega \subseteq \mathbb{R}^N$ be a bounded domain with a Lipschitz boundary $\partial\Omega$. We are given two functions $\vartheta\colon \mathbb{R}^N \longrightarrow \mathbb{R}$ and $f\colon \Omega \times \mathbb{R} \longrightarrow \mathbb{R}$ satisfying the following properties:

$\underline{H(\vartheta)}$ $\vartheta\colon \mathbb{R}^N \longrightarrow \mathbb{R}$ is a convex, even function and there exist $c_1, c_2, c_3 > 0$, such that

$$c_1 \|u\|_{\mathbb{R}^N} - c_2 \leq \vartheta(u) \leq c_3\left(1 + \|u\|_{\mathbb{R}^N}\right) \qquad \forall\, u \in \mathbb{R}^N.$$

$\underline{H(f)}$ $f\colon \Omega \times \mathbb{R} \longrightarrow \mathbb{R}$ is a function, such that

(i) for every $\zeta \in \mathbb{R}$, the function

$$\Omega \ni z \longrightarrow f(z, \zeta) \in \mathbb{R}$$

is measurable;

(ii) for almost all $z \in \Omega$, the function

$$\mathbb{R} \ni \zeta \longrightarrow f(z,\zeta) \in \mathbb{R}$$

is continuous and odd;

(iii) there exist $p \in \left(1, \frac{N}{N-1}\right)$, $a \in L^{p'}(\Omega)$ $(\frac{1}{p} + \frac{1}{p'} = 1)$ and $c > 0$, such that

$$\left|f(z,\zeta)\right| \leq a(z) + c|\zeta|^{p-1} \quad \text{for a.a. } z \in \Omega \text{ and all } \zeta \in \mathbb{R};$$

(iv) there exist $\mu > 1$ and $M > 0$, such that

$$0 < \mu F(z,\zeta) \leq \zeta f(z,\zeta) \quad \text{for a.a. } z \in \Omega \text{ and all } |\zeta| \geq M,$$

where

$$F(z,\zeta) \stackrel{df}{=} \int_0^\zeta f(z,r)\,dr$$

(the potential function corresponding to f).

We introduce the energy functional φ, defined by

$$\varphi(x) \stackrel{df}{=} \int_\Omega \vartheta\big(\nabla x(z)\big)\,dz - \int_\Omega F\big(z, x(z)\big)\,dz$$

and we assume homogeneous Dirichlet boundary condition, i.e. $x \in W_0^{1,1}(\Omega)$. To have coercivity of φ, the natural space to consider is the space $BV(\Omega)$, which is the space of functions of bounded variation on Ω. So, let us begin by introducing this space. A function $x \in L^1(\Omega)$ is of bounded variation, if its partial derivatives in the sense of distributions are signed measures of bounded variation. More precisely, we make the following definition.

DEFINITION 4.3.1   *We say that a function $x\colon \Omega \longrightarrow \mathbb{R}$ is of **bounded variation** (denoted by $x \in BV(\Omega)$) if and only if $x \in L^1(\Omega)$ and there are signed measures $\mu_k\colon \Omega \longrightarrow \mathbb{R}$, $k \in \{1,\dots,N\}$ of bounded variation, such that*

$$\int_\Omega x(z) D_k\varphi(z)\,dz = -\int_\Omega \varphi(z)\,d\mu_k \qquad \forall\, \varphi \in C_c^\infty(\Omega), \ k \in \{1,\dots,N\}.$$

Therefore ∇x is a vector measure with finite total variation, i.e.

$$\|\nabla x\|_{TV(\Omega)} = \sup_{\substack{u \in C_c^\infty(\Omega;\mathbb{R}^N) \\ \|u\|_\infty \leq 1}} \int_\Omega x(z)\mathrm{div}\,u(z)\,dz < +\infty.$$

As usual, we set

$$\text{div } u \stackrel{df}{=} \sum_{i=1}^{N} D_k u_k, \qquad \forall \; u = (u_1, \ldots, u_N) \in C_c^1(\Omega; \mathbb{R}^N).$$

The space $BV(\Omega)$ endowed with the norm

$$\|x\|_{BV(\Omega)} \stackrel{df}{=} \|x\|_1 + \|\nabla x\|_{TV(\Omega)}$$

becomes a Banach space.

EXAMPLE 4.3.1   (a) If $x \in C^1(\overline{\Omega})$, then $x \in BV(\Omega)$ and $\|\nabla x\|_{TV(\Omega)} = \|\nabla x\|_1$. Indeed, let $u = (u_1, \ldots, u_N) \in C_c^\infty(\Omega; \mathbb{R}^N)$. Performing an integration by parts, we obtain

$$\left| \int_\Omega x(z) \text{div } u(z) \, dz \right| = \left| \int_\Omega \big(\nabla x(z), u(z) \big)_{\mathbb{R}^N} \, dz \right|$$

$$\leq \int_\Omega \|\nabla x(z)\|_{\mathbb{R}^N} \|u(z)\|_{\mathbb{R}^N} \, dz \leq \|u\|_\infty \int_\Omega \|\nabla x(z)\|_{\mathbb{R}^N} \, dz,$$

so

$$\|\nabla x\|_{TV(\Omega)} \leq \|\nabla x\|_1.$$

On the other hand by a simple measurable selection argument, we can produce a measurable function $u: \Omega \longrightarrow \mathbb{R}^N$, such that $\|u(z)\|_{\mathbb{R}^N} = 1$ and

$$\big(\nabla x(z), u(z) \big)_{\mathbb{R}^N} = \|\nabla x(z)\|_{\mathbb{R}^N} \quad \text{for a.a. } z \in \Omega.$$

Let $\{u_n\}_{n \geq 1} \subseteq C_c^\infty(\Omega; \mathbb{R}^N)$ be a sequence, such that

$$u_n \longrightarrow u \quad \text{in } L^1(\Omega; \mathbb{R}^N)$$

(recall that embedding $C_c^\infty(\Omega; \mathbb{R}^N) \subseteq L^p(\Omega; \mathbb{R}^N)$ is dense for all $p \in [1, +\infty)$). Then, we have

$$\int_\Omega \big(\nabla x(z), u_n(z) \big)_{\mathbb{R}^N} \, dz \longrightarrow \int_\Omega \big(\nabla x(z), u(z) \big)_{\mathbb{R}^N} \, dz = \|\nabla x\|_1,$$

hence

$$\|\nabla x\|_1 \leq \|\nabla x\|_{TV(\Omega)}.$$

Therefore, we conclude that

$$\|\nabla x\|_1 = \|\nabla x\|_{TV(\Omega)}.$$

More precisely, we have that $W^{1,1}(\Omega) \subseteq BV(\Omega)$ and

$$\|\nabla x\|_1 = \|\nabla x\|_{TV(\Omega)} \qquad \forall\, x \in W^{1,1}(\Omega).$$

This is a multivariate analog of the well known fact that an absolutely continuous function $h\colon [0,b] \longrightarrow \mathbb{R}$ is of bounded variation and $\|h\|_{TV(\Omega)} = \|h'\|_1$

(b) Let $C \subseteq \mathbb{R}^N$ be a bounded open set with a C^2-boundary ∂C. An application of the divergence theorem reveals that χ_C is of bounded variation in Ω. Indeed, if $u \in C_c^\infty(\Omega; \mathbb{R}^N)$ with $\|u\|_\infty \leq 1$, then

$$\int_C \operatorname{div} u(z)\, dz = \int_{\partial C} \big(u(z), n(z)\big)_{\mathbb{R}^N}\, d\sigma_{N-1}$$
$$\leq \sigma_{N-1}(\Omega \cap \partial C) < +\infty, \tag{4.77}$$

with $n(z)$ being the outward unit normal at $z \in \partial C$ and σ_{N-1} is the $(N-1)$-dimensional Hausdorff measure. Therefore $\|\nabla \chi_C\|_{TV(\Omega)} < +\infty$. In fact, we have

$$\|\nabla \chi_C\|_{TV(\Omega)} = \sigma_{N-1}(\Omega \cap \partial C).$$

To see this note that since by hypothesis ∂C is a C^2-manifold, we can find an open set $U \supseteq \partial C$, such that $\widehat{d}(z) \stackrel{df}{=} d(z, C)$ is C^1 on $U \setminus \partial C$ and

$$\nabla \widehat{d}(z) = \frac{z - \operatorname{proj}_{\partial C}(z)}{\widehat{d}(z)},$$

with $\operatorname{proj}_{\partial C}(z)$ being the unique point in $\partial \Omega$, which is nearest to z. So the outward unit normal n to ∂C has an extension $\widehat{n} \in C_c^1(\mathbb{R}^N)$, such that $|\widehat{n}(z)| \leq 1$ for all $z \in \mathbb{R}^N$. So, if $u \in C_c^\infty(\Omega)$ and $v = u\widehat{n}$, by the divergence theorem, we have

$$\int_C \operatorname{div} v(z)\, dz = \int_C \operatorname{div}(u\widehat{n})(z)\, dz = \int_{\partial C} u(z)\, d\sigma_{N-1},$$

so

$$\|\nabla \chi_C\|_{TV(\Omega)} \geq \sup_{\substack{u \in C_c^\infty(\Omega) \\ \|u\|_\infty \leq 1}} \int_{\partial C} u(z)\, d\sigma_{N-1} = \sigma_{N-1}(\Omega \cap \partial C)$$

and from (4.77), we get

$$\|\nabla \chi_C\|_{TV(\Omega)} = \sigma_{N-1}(\Omega \cap \partial C).$$

$\square$

REMARK 4.3.1 The Banach space $BV(\Omega)$ is embedded compactly in $L^p(\Omega)$ ($p \in [1, +\infty)$) (see Giusti (1984)). $\square$

In this section we are concerned with the following minimization problem:

$$\inf \big\{ \varphi(x) : \; x \in BV(\Omega), \text{ with homogeneous Dirichlet boundary condition} \big\}. \tag{4.78}$$

Problem (4.78) can be viewed as the generalized Dirichlet problem for the functional φ. Since the functional φ need not be weakly lower semicontinuous on $BV(\Omega)$, to deal with (4.78) in an effective way, we need to pass to the Γ_w-regularization of φ. To describe this new functional we need to introduce the following notion from Convex Analysis.

DEFINITION 4.3.2   *Let* $\psi \colon \mathbb{R}^N \longrightarrow \overline{\mathbb{R}}$ *be proper, convex and lower semicontinuous (i.e.* $\psi \in \Gamma_0(\mathbb{R}^N)$; *see Definition 1.3.1). The* ***recession function*** ψ^∞ *of* ψ *is defined by*

$$\psi^\infty(y) \;\overset{df}{=}\; \sup_{v \in \mathrm{dom}\,\psi} \big(\psi(y+v) - \psi(v) \big)$$

$$= \sup_{t>0} \frac{\psi(v+ty) - \psi(v)}{t}$$

$$= \lim_{t \to +\infty} \frac{\psi(v+ty) - \psi(v)}{t} \;=\; \lim_{t \to +\infty} \frac{\psi(v+ty)}{t}.$$

In fact the second, third and fourth equalities above are actually independent of $v \in \mathrm{dom}\,\psi = \{ v \in \mathbb{R}^N : \; \psi(v) < +\infty \}$.

Also we need to recall a definition and a classical result from Measure Theory.

DEFINITION 4.3.3   *Let* $\mathcal{M}(\Omega; \mathbb{R}^N)$ *be the space of all vector measures on* Ω *with values in* $\mathbb{R}^N$ *which are of bounded variation. Let* $m \in \mathcal{M}(\Omega; \mathbb{R}^N)$ *and let* μ *be a* σ-*finite Borel measure on* Ω.

(a) *We say that* m *is an* ***absolutely continuous measure with respect to*** μ *(denoted by* $m \prec\prec \mu$) *if and only if* $|m|(C) = 0$ *whenever* $C \in \mathcal{B}(\Omega)$ *and* $\mu(C) = 0$.

(b) *We say that* m *is* ***singular with respect to*** μ *(denoted by* $m \perp \mu$ *if and only if* $|m|(\Omega \setminus C) = 0$ *for some* $C \in \mathcal{B}(\Omega)$ *with* $\mu(C) = 0$.

In what follows, for a given $u \in L^1(\Omega; \mathbb{R}^N)$, we shall denote by $u.\mu$ (or simply by u when no confusion is possible) the element of $\mathcal{M}(\Omega; \mathbb{R}^N)$, defined by

$$(u.\mu)(C) \;\overset{df}{=}\; \int_C u(z)\, d\mu(z) \qquad \forall\, C \in \mathcal{B}(\Omega).$$

From the Radon-Nikodym Theorem, we know that if $m \prec\prec \mu$, then $m = u.\mu$ for some $u \in L^1(\Omega; \mathbb{R}^N)$. The function u is often denoted by $\frac{dm}{d\mu}$ and it is called the **Radon-Nikodym derivative of** m **with respect to** μ.

The elements of $\mathcal{M}(\Omega; \mathbb{R}^N)$ admit the following decomposition, known as the **Lebesgue decomposition**.

PROPOSITION 4.3.1

If $m \in \mathcal{M}(\Omega; \mathbb{R}^N)$,

then *there exists a unique function* $u \in L^1(\Omega; \mathbb{R}^N)$ *and a unique measure* $m^s \in \mathcal{M}(\Omega; \mathbb{R}^N)$, *such that*

(i) $m = u.\mu + m^s$;

(ii) $m^s \perp \mu$ *(clearly* $u.\mu \prec\prec \mu$*)*.

Using these notions from Convex Analysis and from Measure Theory, the Γ-relaxation in $BV(\Omega)$ with the weak topology was calculated by Anzellotti, Buttazzo & Dal Maso (1986), who obtained the following functional

$$\overline{\varphi}(x) \stackrel{df}{=} \int_{\Omega} \vartheta\big(\nabla x^a(z)\big)\, dz + \int_{\Omega} \vartheta^{\infty}\left(\frac{\nabla x^s}{|\nabla x^s|}(z)\right) d\big(|\nabla x^s|\big)(z)$$

$$- \int_{\Omega} F\big(z, x(z)\big)\, dz \qquad \forall\, x \in BV(\Omega),$$

where $\nabla x = \nabla x^a + \nabla x^s$ is the Lebesgue decomposition of the vector measure ∇x (see Proposition 4.3.1), $|\nabla x^s|$ is the total variation of the singular part, $\frac{\nabla x^s}{|\nabla x^s|}$ is the Radon-Nikodym derivative of ∇x^s with respect to $|\nabla x^s|$ and ϑ^{∞} is the recession functional of ϑ.

In what follows, for convenience, we set

$$\overline{\varphi}_0(x) \stackrel{df}{=} \int_{\Omega} \vartheta\big(\nabla x^a(z)\big)\, dz + \int_{\Omega} \vartheta^{\infty}\left(\frac{\nabla x^s}{|\nabla x^s|}(z)\right) d\big(|\nabla x^s|\big)(z)$$

$$\overline{\varphi}_1(x) \stackrel{df}{=} -\int_{\Omega} F\big(z, x(z)\big)\, dz \qquad \forall\, x \in BV(\Omega).$$

Evidently $\overline{\varphi}_0 \in \Gamma_0\big(BV(\Omega)\big)$ and $\overline{\varphi}_1 \in C^1\big(BV(\Omega)\big)$. So $\overline{\varphi} = \overline{\varphi}_0 + \overline{\varphi}_1$ fits in the framework considered in Section 2.3. Then by virtue of Definition 2.3.1, we have:

DEFINITION 4.3.4 *We say that* $x \in BV(\Omega)$ *is a **critical point of** $\overline{\varphi}$* *if and only if*

$$\overline{\varphi}_0(y) \geq \overline{\varphi}_0(x) + \int_{\Omega} f\big(z, x(z)\big)\big(y - x\big)(z)\, dz \qquad \forall\, y \in BV(\Omega).$$

We shall show that $\overline{\varphi}$ has a sequence of critical points $\{x_n\}_{n\geq 1} \subseteq BV(\Omega)$ with corresponding critical values increasing to $+\infty$. The presence of area-type term $\overline{\varphi}_0$ causes some difficulties. First of all it makes the energy functional nonsmooth and so we need to appeal to the nonsmooth critical point theory in order to use variational methods in the analysis of the generalized Dirichlet problem. Second, by having $\overline{\varphi}$ defined on $BV(\Omega)$, we are unable to satisfy the generalized nonsmooth PS-condition (see Definition 2.3.2).

EXAMPLE 4.3.2   Let $\Omega = (0,1) \subseteq \mathbb{R}$, $\vartheta(\zeta) = |\zeta|$, $f(z,\zeta) = \zeta$. We can check that for each $n \geq 1$,

$$x_n(z) \stackrel{df}{=} \begin{cases} 4 \text{ if } \frac{1}{n+2} < z < \frac{1}{2} + \frac{1}{n+2}, \\ 0 \text{ otherwise} \end{cases}$$

is a critical point of φ and $\varphi(x_n) = 4$. However, the sequence $\{x_n\}_{n\geq 1}$ has no subsequence which strongly converges in $BV(\Omega)$. Therefore the generalized PS-condition is not satisfied.   ⬜

To overcome this difficulty, we extend the energy functional φ on all of $L^p(\Omega)$ by setting $\varphi(x) = +\infty$ if $x \in L^p(\Omega) \setminus BV(\Omega)$. In the space $L^p(\Omega)$ we can recover the necessary compactness. To prove a multiplicity result for our generalized Dirichlet problem, we need some abstract results, which essentially are the adaptation to an equivariant setting of the material from Section 2.5.

DEFINITION 4.3.5   *Let G be a topological group and X a complete space.*

(a) *A **representation** of G over the metric space X will be a continuous map $u\colon G \times X \longrightarrow X$, such that if $L(g)$ denotes the map $x \longmapsto u(g,x)$, then $L(g_1 g_2) = L(g_1) \circ L(g_2)$ and $L(e) = id$, where $e \in G$ is the group unit.*

(b) *A function $\varphi\colon X \longrightarrow \mathbb{R}$ is said to be **G-invariant**, if*
$$\varphi\big(u(g,x)\big) \ = \ \varphi(x) \qquad \forall \, g \in G.$$

(c) *A set $C \subseteq X$ is **G-invariant**, if its characteristic function is G-invariant, i.e. if*
$$u(g,C) \ = \ C \qquad \forall \, g \in G.$$

(d) *A function $f\colon X \longrightarrow X$ is said to be **G-equivariant**, if*
$$f\big(u(g,x)\big) \ = \ u\,(g,f(x)) \qquad \forall \, (g,x) \in G \times X.$$

(e) *A homotopy $h\colon [0,1] \times X \longrightarrow X$ will be called **G-invariant**, if for each $t \in [0,1]$, the function $h(t,\cdot)$ is G-equivariant from X into itself.*

(f) *The representation u is said to be* **isometric***, if*

$$|u(g,x) - u(g,y)| = d_X(x,y) \qquad \forall\, g \in G,\ x,y \in X.$$

In what follows let G be a compact topological group which acts by isometric transformations on a complete metric space X. We often say that X is a **complete metric G-space**. In the next definition, we adapt to the equivariant setting a notion introduced in Definition 1.3.10.

DEFINITION 4.3.6 *Let $\varphi \colon X \longrightarrow \mathbb{R}$ be a continuous G-invariant function and let $x \in X$. We introduce $|d_G\varphi|(x)$ to be the supremum of all $\xi \geq 0$, such that there exist an invariant neighbourhood U of x and a continuous map $H \colon [0,\delta] \times U \longrightarrow X$, such that*

$$d_X\big(H(t,u),u\big) \leq t \quad \text{and} \quad \varphi\big(H(t,u),u\big) \leq \varphi(u) - \xi t \qquad \forall\, (t,u) \in [0,\delta] \times U$$

and for all $t \in T$, $H(t,\cdot)$ is G-invariant (i.e. for all $g \in G$ and all $u \in U$, we have $H(t,gu) = gH(t,u)$; see Definition 4.3.5(d)). The extended real number $|d_G\varphi|(x)$ is called the **G-invariant weak slope** *of φ at x.*

REMARK 4.3.2 The function $|d_G\varphi| \colon X \longrightarrow \mathbb{R}_+$ is invariant and lower semicontinuous. Note that if $G = \{e\}$ (the trivial group), then the above definition reduces to Definition 1.3.10 (see also Proposition 1.3.20). $\quad\square$

Consider the space $X \times \mathbb{R}$, which becomes a complete metric space when endowed with the metric

$$d_{X \times \mathbb{R}}\big((x,\lambda),(y,\mu)\big) \overset{df}{=} \big(d_X(x,y)^2 + (\lambda - \mu)^2\big)^{\frac{1}{2}}.$$

Also for a given function $\varphi \colon X \longrightarrow \overline{\mathbb{R}}$, we introduce the function $E_\varphi \colon \mathrm{epi}\,\varphi \longrightarrow \mathbb{R}$, defined by

$$E_\varphi(x,\lambda) \overset{df}{=} \lambda \qquad \forall\, (x,\lambda) \in X \times \mathbb{R},$$

which is Lipschitz continuous. Note that $X \times \mathbb{R}$ becomes a complete metric G-space through the action $\widehat{u} \colon G \times (X \times \mathbb{R}) \longrightarrow X \times \mathbb{R}$, defined by

$$\widehat{u}\big(g,(x,\lambda)\big) \overset{df}{=} \big(u(g,x),\lambda\big).$$

Clearly, if φ is G-invariant, then so is E_φ. Using E_φ and Definition 4.3.6, we can make the following definition.

DEFINITION 4.3.7 *Let $\varphi \colon X \longrightarrow \overline{\mathbb{R}}$ be a lower semicontinuous, G-invariant function and let $x \in \mathrm{dom}\,\varphi$. We set*

$$|d_G\varphi|(x) \overset{df}{=} \begin{cases} \dfrac{|d_G E_\varphi|(x,\varphi(x))}{\sqrt{1 - |d_G E_\varphi|(x,\varphi(x))^2}} & \text{if } |d_G E_\varphi|(x,\varphi(x)) < 1, \\[2ex] +\infty & \text{if } |d_G E_\varphi|(x,\varphi(x)) = 1. \end{cases}$$

REMARK 4.3.3   An immediate question is whether this definition is consistent with Definition 4.3.6. The answer is yes and this can be established as in Proposition 1.3.22. So now we have the notion of the G-invariant weak slope for lower semicontinuous functions $\varphi \colon X \longrightarrow \overline{\mathbb{R}}$.   ⊔

Now that we have the G-invariant weak slope, we shall consider critical orbits instead of critical points (see Definition 2.5.1).

DEFINITION 4.3.8   *Let* $\varphi \colon X \longrightarrow \overline{\mathbb{R}}$ *be a lower semicontinuous, G-invariant function. An orbit $C \subseteq \operatorname{dom} \varphi$ is said to be* **critical**, *if* $|d_G\varphi|(x) = 0$ *for every (equivalently for some) $x \in C$.*

We can also extend to the present G-invariant setting, the PS-condition introduced in Definition 2.5.2.

DEFINITION 4.3.9   *Let* $\varphi \colon X \longrightarrow \overline{\mathbb{R}}$ *be a lower semicontinuous, G-invariant function and let $c \in \mathbb{R}$. We say that φ satisfies the* **extended nonsmooth G-Palais-Smale condition at level c** *(extended nonsmooth* $G - \mathrm{PS}_c$*-condition for short), if the following holds:*

"Every sequence $\{x_n\}_{n \geq 1} \subseteq \operatorname{dom}\varphi$, such that

$$\varphi(x_n) \longrightarrow c \quad \text{and} \quad |d_G\varphi|(x_n) \longrightarrow 0,$$

has a strongly convergent subsequence."

If this property holds at every level $c \in \mathbb{R}$, then we say that φ satisfies the **extended nonsmooth G-Palais-Smale condition** *(extended nonsmooth* $G - \mathrm{PS}$*-condition for short).*

For a lower semicontinuous, G-invariant function $\varphi \colon X \longrightarrow \overline{\mathbb{R}}$ and for every $c \in \mathbb{R}$, we introduce the following two sets:

$$\varphi^c \stackrel{df}{=} \{x \in X \colon \varphi(x) \leq c\},$$
$$K_c^\varphi \stackrel{df}{=} \{x \in X \colon |d_G\varphi|(x) = 0, \ \varphi(x) = c\}.$$

A careful reading of the proof of Theorem 2.5.1, reveals that we can restate that deformation result in a G-invariant form.

THEOREM 4.3.1 (Deformation Theorem)
If X is a complete metric G-space, $\varphi \colon X \longrightarrow \mathbb{R}$ is a continuous G-invariant function, $c \in \mathbb{R}$ and φ satisfies the extended nonsmooth $G - \mathrm{PS}_c$-condition, then for a given $\varepsilon_0 > 0$, a G-invariant neighbourhood U of K_c^φ (if $K_c^\varphi = \emptyset$, then $U = \emptyset$) and $\vartheta > 0$, there exist $\varepsilon \in (0, \varepsilon_0)$ and a continuous map $\eta \colon [0,1] \times X \longrightarrow X$, such that for every $(t, x) \in [0,1] \times X$, we have:

(a) $d_X\big(x, \eta(t, x)\big) \leq \vartheta t$;

(b) $\varphi\big(\eta(t, x)\big) \leq \varphi(x)$;

(c) *if* $\varphi(x) \notin (c - \varepsilon_0, c + \varepsilon_0)$, *then* $\eta(t, x) = x$;

(d) $\eta(1, \varphi^{c+\varepsilon} \setminus U) \subseteq \varphi^{c-\varepsilon}$

(e) *for every* $t \in [0, 1]$, $\eta(t, \cdot)$ *is G-invariant.*

Using this deformation result, we can have an abstract multiplicity result for lower semicontinuous functions. For this purpose, we consider a Banach space X and the symmetry group $G = \mathbb{Z}_2$. We consider the representation of G over X consisting of $\{id, -id\}$. This way X becomes a G-space.

THEOREM 4.3.2
If $\varphi\colon X \longrightarrow \overline{\mathbb{R}}$ *is a lower semicontinuous and even function, there exists a strictly increasing sequence* $\{V_n\}_{n \geq 1}$ *of finite dimensional subspaces of* X, *such that*

(i) *there exist a closed subspace* Y *of* X, $r > 0$ *and* $\beta > \varphi(0)$, *such that*
$$X = V_0 \oplus Y \quad \text{and} \quad \varphi|_{\partial B_r \cap Y} \geq \beta;$$

(ii) *there exists a sequence* $\{R_n\}_{n \geq 1} \subseteq (r, +\infty)$, *such that*
$$\varphi(v) \geq \varphi(0) \quad \forall\, v \in V_n,\ \|v\|_X \geq R_n;$$

(iii) E_φ *satisfies the extended nonsmooth* PS_c-*condition for every* $c \geq \beta$;

(iv) $|d_G E_\varphi|(0, \lambda) \neq 0$ *whenever* $\lambda \geq \beta$

then there exists a sequence $\{(x_n, \lambda_n)\}_{n \geq 1} \subseteq \operatorname{epi} \varphi$, *such that* $|dE_\varphi|(x_n, \lambda_n) = 0$ *for all* $n \geq 1$ *and* $E_\varphi(x_n, \lambda_n) = \lambda_n \longrightarrow +\infty$.

PROOF Without any loss of generality, we can assume that $\varphi(0) = 0$.
The function $E_\varphi\colon \operatorname{epi} \varphi \longrightarrow \mathbb{R}$ is Lipschitz continuous and G-invariant (since by hypothesis φ is even). Moreover, it is easy to see that
$$|d_G E_\varphi|(x, \lambda) = |dE_\varphi|(x, \lambda) \quad \forall\, x \neq 0.$$
Hence the function E_φ actually satisfies the extended nonsmooth $G - \mathrm{PS}_c$-condition for every $c \geq \beta$. Also $K_c^\varphi \subseteq (X \setminus \{0\}) \times \{c\}$.
Let
$$\mathcal{A} \stackrel{df}{=} \big(C \subseteq X \setminus \{0\} : C \text{ is closed},\ C = -C\big)$$
(i.e. $\mathcal{A}$ is the class of closed, symmetric (with respect to the origin) subsets of $X \setminus \{0\}$). We consider the Krasnoselskii genus $\gamma\colon \mathcal{A} \longrightarrow \mathbb{R}_+$, defined by
$$\gamma(C) \stackrel{df}{=} \begin{cases} \inf\{m \in \mathbb{N} : \text{ there exists } h \in C(C; \mathbb{R}^m \setminus \{0\}),\ h \text{ is odd}\}, \\ +\infty \text{ if no such } m \in \mathbb{N} \text{ exists, in particular if } 0 \in C, \end{cases}$$

for $C \neq \emptyset$ and $\gamma(\emptyset) \stackrel{df}{=} 0$. Note that for any $C \in \mathcal{A}$, by the Tietze's extension theorem (see Theorem A.1.2) an odd map $h \in C(C; \mathbb{R}^m)$ can be extended to a map $\overline{h} \in C(X; \mathbb{R}^m)$ and in fact by considering $\widehat{h}(x) \stackrel{df}{=} \frac{1}{2}\big(\overline{h}(x) - \overline{h}(-x)\big)$, we can assume that this extension is odd too. The Krasnoselskii genus generalizes the notion of dimension of a linear space and its properties can be found in Proposition 2.1.5.

Let $k \stackrel{df}{=} \dim V_0$ and without any loss of generality assume that $\dim V_n = k + n$ for $n \geq 1$. Let

$$D_n \stackrel{df}{=} \overline{B}_{R_n} \cap V_n.$$

We introduce

$$\Psi \stackrel{df}{=} \big\{\psi \in C(D_n; \mathrm{epi}\, \varphi) : \ \psi \text{ is } G - \text{equivariant}$$
$$\text{and } \psi(u) = (u, 0) \text{ for all } u \in \partial B_{R_n} \cap V_n \big\}.$$

$$\Gamma_i \stackrel{df}{=} \big\{\psi\big(\overline{D_n \setminus C}\big) : \ \psi \in \Psi_n, \ n \geq i, \ C \in \mathcal{A} \text{ and } \gamma(C) \leq n - i \big\}.$$

The sets Γ_i possess the following properties:

(a) $\Gamma_i \neq \emptyset$ for all $i \in \mathbb{N}$. Indeed note that $id \in \Psi_n$.

(b) $\Gamma_{i+1} \subseteq \Gamma_i$ (monotonicity). Let $A = \psi\big(\overline{D_n \setminus C}\big) \in \Gamma_{i+1}$. Then $\psi \in \Psi_n$, $n \geq i+1 > i$, $C \in \mathcal{A}$ and $\gamma(C) \leq n - (i+1) \leq n - i$ and so $A \in \Gamma_i$.

(c) If $\vartheta \in C(\mathrm{epi}\, \varphi, \mathrm{epi}\, \varphi)$ is G-equivariant and $\vartheta(x, 0) = (x, 0)$ for all $x \in \partial B_{R_n} \cap V_n$ and all $n \geq 1$, then $\vartheta(A) \in \Gamma_i$ for all $A \in \Gamma_i$ (invariance). Let ϑ be as above and set $A = \psi\big(\overline{D_n \setminus C}\big) \in \Gamma_i$. Then $\vartheta \circ \psi$ is G-equivariant, belongs in $C(D_n; \mathrm{epi}\, \varphi)$ and $\vartheta \circ \psi|_{\partial B_{R_n} \cap V_n} = (id, 0)$. Therefore $\vartheta \circ \psi \in \Psi_n$ and $(\vartheta \circ \psi)\big(\overline{D_n \setminus C}\big) = \vartheta(A) \in \Gamma_i$.

(d) If $A \in \Gamma_i$, $E \in \mathcal{A}$ and $\gamma(E) \leq k < i$, then $\overline{A \setminus (E \times \mathbb{R})} \in \Gamma_{i-k}$ (excision). Let $A = \psi\big(\overline{D_n \setminus C}\big) \in \Gamma_i$ and $C \in \mathcal{A}$, with $\gamma(C) \leq k < i$. We claim that

$$\overline{A \setminus (E \times \mathbb{R})} \ = \ \psi\big(\overline{D_n \setminus \big(C \cup \psi^{-1}(E \times \mathbb{R})\big)}\big). \tag{4.79}$$

Suppose that $(c, \lambda) \in \psi\big(D_n \setminus \big(C \cup \psi^{-1}(E \times \mathbb{R})\big)\big)$. Then

$$(c, \lambda) \ \in \ \psi(D_n \setminus C) \setminus (E \times \mathbb{R}) \ = \ A \setminus (E \times \mathbb{R}) \ \subseteq \ \overline{A \setminus (E \times \mathbb{R})}.$$

Hence

$$\psi\big(D_n \setminus \big(C \cup \psi^{-1}(E \times \mathbb{R})\big)\big) \ \subseteq \ \overline{A \setminus (E \times \mathbb{R})}. \tag{4.80}$$

On the other hand, let $(c, \lambda) \in A \setminus (E \times \mathbb{R})$. Then $(c, \lambda) = \psi(w)$, with

$$w \ \in \ \big(\overline{D_n \setminus C}\big) \setminus \psi^{-1}(E \times \mathbb{R}) \ \subseteq \ \overline{D_n \setminus \big(C \cup \psi^{-1}(E \times \mathbb{R})\big)}.$$

So we infer that

$$A \setminus (E \times \mathbb{R}) \ \subseteq \ \psi\big(\overline{D_n \setminus \big(C \cup \psi^{-1}(E \times \mathbb{R})\big)}\big). \tag{4.81}$$

From (4.80) and (4.81), we conclude that (4.79) holds.

(e) If $A \in \Gamma_i$ and $L \stackrel{df}{=} \big\{(y,\lambda) \in \operatorname{epi}\varphi : \ y \in \partial B_r \cap Y\big\}$, then $A \cap L \neq \emptyset$. Let $p \colon X \times \mathbb{R} \longrightarrow \mathbb{R}$ be the linear continuous map defined by $p(x,\mu) = x$. Then we can easily check that

$$A \cap L \ \neq \emptyset \text{ if and only if } p(C) \cap \partial B_r \cap Y \neq \emptyset. \qquad (4.82)$$

Let $A = \psi(D_n \setminus C)$. Then $p \circ \psi \in C(D_n; X)$ is odd and $p \circ \psi|_{\partial B_{R_n} \cap V_n}$. Then from Rabinowitz (1986, Proposition 9.23, p. 57), we have that

$$p(A) \cap \partial B_r \cap Y \ \neq \emptyset$$

and this proves that $A \cap L \neq \emptyset$ (see (4.82)).

We introduce the following minimax values of E_φ:

$$c_i \ \stackrel{df}{=} \ \inf_{A \in \Gamma_i} \ \max_{(x,\mu) \in C} E_\varphi(x,\mu) \qquad \forall \, i \geq 1.$$

Claim 1. $c_i \geq \beta > 0$ for all $i \geq 1$.

If $A \in \Gamma_i$, then from (4.82) and hypothesis (ii), we have that

$$\max_{x \in A} \varphi(x) \ \geq \ \beta,$$

hence $c_i \geq \beta$ for all $i \geq 1$.

In the next claim we show that the numbers c_i are critical values of E_φ and also make an observation about their multiplicity.

Claim 2. If $c_i = \cdots = c_{i+m} = c$, then $\gamma(K_c^\varphi) \geq m + 1$.

We have $\varphi(0) = 0$ and $c \geq \beta > 0$ (see Claim 1). Therefore $0 \notin K_c^\varphi$. Because $K_c^\varphi \subseteq X$ is compact (due to the extended nonsmooth PS-condition), we have $K_c^\varphi \in \mathcal{A}$. If $\gamma(K_c^\varphi) \leq m$, then from the continuity properties of the Krasnoselskii genus (see Struwe (1990, Proposition 5.4(5), p. 87)), we can find $\delta > 0$ such that $\gamma\big(\overline{(K_c^\varphi)_\delta}\big) \leq m$ (here $(K_c^\varphi)_\delta \stackrel{df}{=} \{x \in X : \ d_x(x, K_c^\varphi) < \delta\}$). We apply Theorem 4.3.1 for the functional E_φ with $U \stackrel{df}{=} (K_c^\varphi)_\delta \times \mathbb{R}$ and $\varepsilon_0 \stackrel{df}{=} \frac{\beta}{2}$. We can find $\eta \in C\big([0,1] \times (X \times \mathbb{R}), X \times \mathbb{R}\big)$ and $\varepsilon \in (0, \varepsilon_0)$ such that $\eta(t, \cdot)$ is odd for every $t \in [0,1]$ and

$$\eta\big(1, E_\varphi^{c+\varepsilon} \setminus U\big) \ \subseteq \ E_\varphi^{c-\varepsilon}. \qquad (4.83)$$

We choose $C \in \Gamma_{i+m}$ such that

$$\max_{(x,\mu) \in C} E_\varphi(x,\mu) \ \leq \ c + \varepsilon.$$

From (d) (the excision property of the family $\{\Gamma_i\}_{i\geq 1}$), we have that $\overline{C \setminus U} \in \Gamma_i$. By virtue of hypothesis (ii) and Theorem 4.3.1(c), we have that

$$\eta(1,\cdot)|_{\partial B_{R_n} \cap V_n} = id_{\partial B_{R_n} \cap V_n} \qquad \forall\, n \geq 1.$$

So from (c) (the invariance property of the family $\{\Gamma_i\}_{i\geq 1}$), we have that

$$\eta\big(1, \overline{C \setminus U}\big) \in \Gamma_i$$

and thus from (4.83), we infer that

$$\max_{\eta(1,\overline{C\setminus U})} E_\varphi \leq c - \varepsilon,$$

a contradiction to the definition of c.

The next claim completes the proof of the Theorem.

Claim 3. $c_i \longrightarrow +\infty$ as $u \to +\infty$.

Because of (b) (the monotonicity property of the family $\{\Gamma_i\}_{i\geq 1}$), we have that the sequence $\{c_i\}_{i\geq 1}$ is increasing. Suppose that the claim is not true and the sequence $\{c_i\}_{i\geq 1}$ is bounded. We have that

$$c_i \longrightarrow \bar{c} \quad \text{as } i \to +\infty.$$

If $c_i = \bar{c}$ for all $i \geq i_0$, then by virtue of Claim 2 we have that $\gamma(K_{\bar{c}}^\varphi) = +\infty$, which is not possible since $K_{\bar{c}}^\varphi$ is compact. Therefore $\bar{c} > c_i$ for all $i \geq 1$. Let us set

$$S \overset{df}{=} \big\{(x,\mu) \in X \times \mathbb{R} : c_1 \leq E_\varphi(x,\mu) \leq \bar{c},\ |dE_\varphi|(x,\mu) = 0\big\}.$$

This set is compact too (since E_φ satisfies the extended nonsmooth PS$_c$-condition for $c \geq \beta$) and so $\gamma(S) < +\infty$. We can find $\delta > 0$ such that

$$\gamma\big(\overline{S}_\delta\big) = \gamma(S) = k,$$

where

$$S_\delta \overset{df}{=} \big\{(x,\mu) \in X \times \mathbb{R} : d\big((x,\mu),S\big) < \delta\big\}.$$

Theorem 4.3.1 with $c \overset{df}{=} \bar{c}$, $\varepsilon_0 \overset{df}{=} \bar{c} - c_k$ and $U \overset{df}{=} S_\delta$ yields an $\varepsilon > 0$ and a continuous odd deformation map such that

$$\eta\big(1, E_\varphi^{\bar{c}+\varepsilon} \setminus U\big) \subseteq E_\varphi^{\bar{c}-\varepsilon}.$$

Choose $i \in \mathbb{N}$ such that $c_i > \bar{c} - \varepsilon$ and $A \in \Gamma_{i+k}$ such that

$$\max_A E_\varphi \leq \bar{c} + \varepsilon.$$

Arguing as in the proof of Claim 3, we obtain $\overline{A \setminus U} \in \Gamma_i$ and the same holds for $\eta(1, \overline{A \setminus U})$ provided that $\eta(1, \cdot)|_{\partial B_{R_n} \cap V_n} = id$ for all $n \geq i$. But by hypothesis $E_\varphi|_{\partial B_{R_n} \cap V_n} \leq 0$ for all $n \geq 1$ and $\overline{c} - \varepsilon_0 = c_k \geq c_0 \geq \beta > 0$. Therefore $\eta(1, \overline{A \setminus U}) \in \Gamma_i$ and so

$$c_i \leq \max_{\eta(1, \overline{A \setminus U})} E_\varphi \leq \overline{c} - \varepsilon < c_i,$$

a contradiction. $\qquad\qquad\qquad\qquad\qquad\qquad\qquad\qquad\qquad\qquad\qquad\qquad\quad$ ⬜

REMARK 4.3.4 The hypothesis that $|d_G E_\varphi|(0, \lambda) \neq 0$ for all $\lambda \geq \beta$ cannot be removed. Indeed let X be an infinite dimensional Banach space and let $\varphi \colon X \longrightarrow \mathbb{R}$ be defined by

$$\varphi(x) \stackrel{df}{=} \begin{cases} 0 & \text{if } x = 0, \\ 1 - \|x\|_X & \text{if } x \neq 0. \end{cases}$$

Then all the other hypotheses of Theorem 4.3.2 are satisfied. However, the only critical points of E_φ are $(0, 0)$ and $(0, 1)$. $\qquad\qquad\qquad\qquad\quad$ ⬜

In the three propositions that follow, we examine the weak slope of functions $\varphi = \varphi_0 + \varphi_1$ with φ_0 lower semicontinuous and $\varphi_1 \in C^1(X)$ or φ_1 locally Lipschitz.

PROPOSITION 4.3.2
If $\varphi = \varphi_0 + \varphi_1$ *with* φ_0 *proper, lower semicontinuous,* φ_1 *locally Lipschitz,* $x \in \operatorname{dom} \varphi_0$ *and*

$$\lim_{r \to 0^+} \sup_{\substack{y, z \in B_r(x) \\ y \neq z}} \frac{|\varphi_1(y) - \varphi_1(z)|}{\|y - z\|} = 0,$$

then for every $\lambda \geq \varphi(x)$, *we have*

$$|dE_\varphi|(x, \lambda) = |dE_{\varphi_0}|(x, \lambda)$$

In particular $|d\varphi|(x) = |d\varphi_0|(x)$.

PROOF For $\lambda \geq \varphi(x)$, first we show that

$$|dE_{\varphi_0}|(x, \lambda - \varphi_1(x)) \leq |dE_\varphi|(x, \lambda). \tag{4.84}$$

If $|dE_{\varphi_0}|(x, \lambda - \varphi_1(x)) = 0$, then (4.84) is immediate. So suppose that $0 < \tau < |dE_{\varphi_0}|(x, \lambda - \varphi_1(x))$.

Let $\varepsilon > 0$ be given and let $\delta > 0$ be small enough so that φ_1 is Lipschitz continuous with constant $\varepsilon > 0$ in $B_{2\delta}(x)$ and there exists $H \colon [0, \delta] \times$

$\left(B_\delta(x, \varphi(x)) \cap \operatorname{epi} \varphi\right) \longrightarrow X$ as in the definition of weak slope (see Definition 1.3.10). Let $\delta' \in (0, \delta]$ be such that

$$\left(y, \mu - \varphi_1(y)\right) \in B_\delta\left((x, \lambda - \varphi_1(x))\right) \qquad \forall \, (y, \mu) \in B_{\delta'}\left((x, \lambda)\right).$$

Then let $K \colon [0, \delta'] \times \left(B_{\delta'}\left((x, \lambda)\right) \cap \operatorname{epi} \varphi\right) \longrightarrow \operatorname{epi} \varphi$ be defined by

$$K\left(t, (y, \mu)\right) \overset{df}{=} \left(K_1\left(\tfrac{t}{1+\varepsilon}, (y, \mu - \varphi_1(y))\right),\right.$$
$$\left. K_2\left(\tfrac{t}{1+\varepsilon}, (y, \mu - \varphi_1(y))\right) + \varphi_1\left(\tfrac{t}{1+\varepsilon}, K_1\left(\tfrac{t}{1+\varepsilon}, (y, \mu - \varphi_1(y))\right)\right)\right).$$

From the triangle inequality, we have

$$d\left(K\left((1+\varepsilon)s, (y, \mu + \varphi_1(y))\right), (y, \mu + \varphi_1(y))\right)$$
$$= d\left(K\left(s, (y, \mu)\right), (y, \mu + \varphi_1(y) - \varphi_1(K_2(s, (y, \mu))))\right)$$
$$\leq d\left(K\left(s, (y, \mu)\right), (y, \mu)\right) + \left|\varphi_1\left(K_1(s, (y, \mu))\right) - \varphi_1(y)\right|$$
$$\leq s + \varepsilon s = (1+\varepsilon)s.$$

Moreover, we have

$$E_\varphi\left(K\left(t, (y, \mu)\right)\right) = K_2\left(\tfrac{t}{1+\varepsilon}, (y, \mu - \varphi_1(y))\right) + \varphi_1\left(\tfrac{t}{1+\varepsilon}, (y, \mu - \varphi_1(y))\right)$$
$$\leq \mu - \varphi_1(y) - \tau\tfrac{t}{1+\varepsilon} + \varphi_1\left(K_1\left(\tfrac{t}{1+\varepsilon}, (y, \mu - \varphi_1(y))\right)\right)$$
$$\leq E_\varphi(y, \mu) - \left(\tfrac{\tau}{1+\varepsilon} - \varepsilon\right) t,$$

so

$$|dE_\varphi|(x, \lambda) \geq \frac{\tau}{1+\varepsilon} - \varepsilon.$$

Let $\varepsilon \searrow 0$ to obtain that $|dE_\varphi|(x, \lambda) \geq \tau$, which implies that

$$|dE_\varphi|(x, \lambda) \geq |dE_{\varphi_0}|\left(x, \lambda - \varphi_1(x)\right).$$

Therefore (4.84) is true.

On the other hand by replacing φ_0 with φ and φ_1 with $-\varphi_1$, we obtain

$$|dE_\varphi|(x, \lambda) \leq |dE_{\varphi_0}|\left(x, \lambda - \varphi_1(x)\right). \tag{4.85}$$

From (4.84) and (4.85), we conclude that the equalities of the Proposition hold. $\square$

Next we shall consider C^1 perturbations of convex functions. But first we need to introduce and briefly comment on two notions related to the concept of weak slope introduced in Definition 1.3.10.

DEFINITION 4.3.10   Let $\varphi\colon X \longrightarrow \overline{\mathbb{R}}$ be a function.

(a) If $\varphi\colon X \longrightarrow \overline{\mathbb{R}}$ is proper, lower semicontinuous, $x \in \operatorname{dom}\varphi$, the **strong slope** $|D\varphi|(x)$ is defined by

$$|D\varphi|(x) \overset{df}{=} \begin{cases} \underset{y \to x}{\lim\sup} \dfrac{\varphi(x) - \varphi(y)}{\|x - y\|_X} & \text{if } x \text{ is not a local minimum,} \\ 0 & \text{if } x \text{ is a local minimum.} \end{cases}$$

(b) If $\varphi \in \Gamma_0(X)$ and $x \in \operatorname{dom}\varphi$, the **extended subdifferential** of φ at x is the set

$$\partial^-\varphi(x) \overset{df}{=} \left\{ x^* \in X^* : \ \underset{y \to x}{\lim\inf} \frac{\varphi(y) - \varphi(x) - \langle x^*, y - x\rangle_X}{\|x - y\|_X} \geq 0 \right\}.$$

We can easily verify the following properties.

PROPOSITION 4.3.3
If $\varphi\colon X \longrightarrow \overline{\mathbb{R}}$ is proper, lower semicontinuous and $x \in \operatorname{dom}\varphi$,
then the following are true:

(1) If $\psi\colon X \longrightarrow \mathbb{R}$ is Frechet differentiable at x, then

$$\partial^-(\varphi + \psi)(x) \ = \ \{x^* + \psi'(x) : \ x^* \in \partial^-\varphi(x)\}.$$

(2) If $x^* \in \partial^-\varphi(x)$, then

$$\langle x^*, h\rangle_X \ \leq \ \underset{\lambda \searrow 0}{\lim\inf} \frac{\varphi(x + \lambda h) - \varphi(x)}{\lambda} \qquad \forall\, h \in X.$$

(3) If φ is convex, then $\partial^-\varphi(x)$ coincides with the convex subdifferential (see Definition 1.3.3).

(4) If $x^* \in \partial^-\varphi(x)$, then

$$|d\varphi|(x) \ \leq \ |D\varphi|(x) \ \leq \ \|x^*\|_{X^*}.$$

(5) $\partial^-\varphi(x) \subseteq X^*$ is strongly closed and convex.

Now we are ready to consider C^1 perturbations of convex functions.

PROPOSITION 4.3.4
If $\varphi = \varphi_0 + \varphi_1$ with $\varphi_0 \in \Gamma_0(X)$ and $\varphi_1 \in C^1(X)$,
then the following are true:

(1) $|d\varphi|(x) = |D\varphi|(x)$ for all $x \in \operatorname{dom}\varphi$;

(2) *for every $x \in \mathrm{dom}\, \varphi$, we have $|d\varphi|(x) < +\infty$ if and only if $\partial^{-}\varphi(x) \neq \emptyset$ and in this case we have*

$$|d\varphi|(x) \;=\; \min_{x^{*} \in \partial^{-}\varphi(x)} \|x^{*}\|_{X^{*}} .$$

PROOF   From Proposition 4.3.3 we know that $\partial^{-}\varphi(x)$ is w^{*}-closed and

$$\partial^{-}\varphi(x) \;=\; \{x^{*} + \varphi_{1}'(x) : \; x \in \partial^{-}\varphi_{0}(x) = \partial\varphi_{0}(x)\}.$$

It follows that $\partial^{-}\varphi(x)$ is w^{*}-closed too. Hence, if $\partial^{-}\varphi(x) \neq \emptyset$, then we can find $x_{0}^{*} \in \partial^{-}\varphi(x)$ such that $\|x_{0}^{*}\|_{X^{*}} = \inf_{x^{*} \in \partial^{-}\varphi(x)} \|x^{*}\|_{X^{*}}$.

Without any loss of generality we can say that $\varphi_{1}'(x) = 0$. Then by virtue of Proposition 4.3.2 we can assume that $\varphi_{1} \equiv 0$.

By Proposition 4.3.3(4), we have

$$\left[\partial^{-}\varphi(x) \neq \emptyset\right] \;\Longrightarrow\; \left[|d\varphi|(x) \leq |D\varphi|(x) \leq \min_{x^{*} \in \partial^{-}\varphi(x)} \|x^{*}\|_{X^{*}}\right]. \qquad (4.86)$$

Suppose that $0 \notin \partial^{-}\varphi(x)$ and let $\tau > 0$ be such that

$$\|x^{*}\|_{X^{*}} \geq \tau \qquad \forall\, x^{*} \in \partial^{-}\varphi(x).$$

By virtue of Lemma 1.3.2, we can find $y \in X$ such that

$$\varphi(y) \;<\; \varphi(x) - \tau\, \|y - x\|_{X} .$$

Because φ is lower semicontinuous, we can find $\delta > 0$ such that

$$\varphi(y) \;<\; \varphi(z) - \tau\, \|y - z\|_{X} \qquad \forall\, z \in B_{\delta}(x).$$

By decreasing $\delta > 0$ if necessary, we may assume that $y \notin B_{2\delta}(x)$. We define $H \colon [0, \delta] \times B_{\delta}(x) \longrightarrow X$ by

$$H(t, u) \;\overset{df}{=}\; u + t\frac{y - u}{\|y - u\|_{X}} \qquad \forall\, (t, u) \in [0, \delta] \times B_{\delta}(x).$$

Clearly H is continuous and $\|H(t, u) - u\|_{X} = t$. Moreover, since $0 \leq \frac{t}{\|y - u\|_{X}} \leq 1$, we have

$$\varphi\big(H(t, u)\big) \;\leq\; \varphi(u) + \frac{t}{\|y - u\|_{X}} \big(\varphi(y) - \varphi(u)\big) \;\leq\; \varphi(u) - \tau t,$$

so

$$|d\varphi|(x) \;\geq\; \tau.$$

Therefore $\partial^{-}\varphi(x) = \emptyset$ implies that $|d\varphi|(x) = +\infty$, while $\partial^{-}\varphi(x) \neq \emptyset$ implies that

$$|d\varphi|(x) \;\geq\; \min_{x^{*} \in \partial^{-}\varphi(x)} \|x^{*}\|_{X^{*}} . \qquad (4.87)$$

From (4.86) and (4.87), we conclude that

$$|d\varphi|(x) = \min_{x^* \in \partial^- \varphi(x)} \|x^*\|_{X^*} .$$

$\square$

COROLLARY 4.3.1
If $\varphi \in C^1(X)$,
then $|d\varphi|(x) = |D\varphi|(x) = \|\varphi'(x)\|_X$ *for all* $x \in X$.

COROLLARY 4.3.2
If $\varphi = \varphi_0 + \varphi_1$ *with* $\varphi_0 \in \Gamma_0(X)$ *and* $\varphi_1 \in C^1(X)$,
then for every $x \in \operatorname{dom} \varphi$ *with* $|d\varphi|(x) < +\infty$, *there exists* $u^* \in X^*$ *such that*
$\|u^*\|_{X^*} = |d\varphi|(x)$ *and*

$$\varphi_0(y) \geq \varphi_0(x) - \langle \varphi_1'(x), y - x \rangle_X + \langle u^*, y - x \rangle_X \qquad \forall \, y \in X.$$

The weak slope exhibits a better behaviour when we deal with a continuous function. When $\varphi : X \longrightarrow \overline{\mathbb{R}}$ is proper and lower semicontinuous, we can still transfer the analysis to the continuous case by considering the Lipschitz continuous function $E_\varphi : \operatorname{epi} \varphi \longrightarrow \mathbb{R}$ defined by $E_\varphi(x, \lambda) \overset{df}{=} \lambda$. Then by virtue of Definition 4.3.7, $|d\varphi|(x) = 0$ if and only if $|dE_\varphi|(x, \varphi(x)) = 0$ and $\{x_n\}_{n \geq 1} \subseteq X$ is a Palais-Smale sequence for φ if and only if $\{(x_n, \varphi(x_n))\}_{n \geq 1} \subseteq \operatorname{epi} \varphi$ is a Palais-Smale sequence for E_φ. The main difficulty in dealing with E_φ is that in general we do not know the behaviour of $|dE_\varphi|(x, \lambda)$ when $\varphi(x) < \lambda$. However, this can be remedied when we deal with a C^1 perturbation of a convex function, as is the case in the problem of this section (see also Section 2.3).

PROPOSITION 4.3.5
If $\varphi = \varphi_0 + \varphi_1$, $\varphi_0 \in \Gamma_0(X)$ *and* $\varphi_1 \in C^1(X)$,
then for every $x \in \operatorname{dom} \varphi$ *and* $\lambda > \varphi(x)$, *we have* $E_\varphi(x, \lambda) = 1$.

PROOF Clearly we may assume that $\varphi_1'(x) = 0$. This by virtue of Proposition 4.3.2 permits us to assume that $\varphi_1 \equiv 0$.
Let $H : [0, \delta] \times B_\delta((x, \lambda)) \longrightarrow \operatorname{epi} \varphi$ be defined by

$$H(t, (y, \mu)) \overset{df}{=} \left(y + \frac{t(x-y)}{\sqrt{\|y-x\|_X^2 + |\mu - \varphi(x)|^2}}, \mu - \frac{t(\mu - \varphi(x))}{\sqrt{\|y-x\|_X^2 + |\mu - \varphi(x)|^2}} \right),$$

with $\delta > 0$ such that $\varphi(x) < \lambda - 2\delta$. Since

$$\varphi\left(y + \frac{s(x-y)}{\sqrt{\|y-x\|_X^2 + |\mu - \varphi(x)|^2}} \right) \leq \varphi(y) + \frac{s}{\sqrt{\|y-x\|_X^2 + |\mu - \varphi(x)|^2}} (\varphi(x) - \varphi(y))$$

$$\leq \mu - (\mu - \varphi(x)) \frac{s}{\sqrt{\|y-x\|_X^2 + |\mu - \varphi(x)|^2}},$$

we infer that $H\big(t,(y,\mu)\big) \in$ epi φ.   Clearly H is continuous and $d\big(H\big(t,(y,\mu)\big),(y,\mu)\big) = t$.

On the other hand we have

$$E_\varphi\big(H\big(s,(y,\mu)\big)\big) \;=\; \mu - \big(\mu - \varphi(x)\big)\,\frac{s}{\sqrt{\|y-x\|_X^2 + |\mu - \varphi(x)|^2}}$$

$$=\; E_\varphi(y,\mu) - \frac{\mu - \varphi(x)}{\sqrt{\|y-x\|_X^2 + |\mu - \varphi(x)|^2}}\,s$$

$$\leq\; E_\varphi(y,\mu) - \frac{\lambda - \delta - \varphi(x)}{\sqrt{\delta^2 + (\lambda + \delta - \varphi(x))^2}}\,s,$$

so

$$|dE_\varphi|(x,\lambda) \;\geq\; \frac{\lambda - \delta - \varphi(x)}{\sqrt{\delta^2 + (\lambda + \delta - \varphi(x))^2}}.$$

Let $\delta \searrow 0$, to conclude that $|dE_\varphi|(x,\lambda) \geq 1$, hence $|dE_\varphi|(x,\lambda) = 1$.   □

As a consequence of this Proposition, we have:

PROPOSITION 4.3.6

If $\varphi = \varphi_0 + \varphi_1$, $\varphi_0 \in \Gamma_0(X)$, $\varphi_1 \in C^1(X)$ *and* φ *is even,*
then for every $(x,\lambda) \in$ epi φ *we have*

(1) *if* $\varphi(x) < \lambda$, *then* $|d_G E_\varphi|(x,\lambda) = |dE_\varphi|(x,\lambda) = 1$;

(2) $|d_G E_\varphi|\big(x,\varphi(x)\big) = |dE_\varphi|\big(x,\varphi(x)\big)$;

(3) $|dE_\varphi|\big(0,\varphi(0)\big) = 0$.

PROOF   The only thing that needs to be proved is that $|dE_\varphi|\big(0,\varphi(0)\big) = 0$. Without any loss of generality we assume that both φ_0 and φ_1 are even. So we have $\varphi_1'(0) = 0$ and by virtue of Proposition 4.3.2, we may assume that $\varphi_1 \equiv 0$. Then φ is convex, even and so 0 is a minimizer of φ. Hence $\big(0,\varphi(0)\big)$ is a minimizer of E_φ and so $|dE_\varphi|\big(0,\varphi(0)\big) = 0$.   □

Now we are ready for an abstract multiplicity result for the generalized Dirichlet problem under consideration.

THEOREM 4.3.3

If $\varphi = \varphi_0 + \varphi_1$, $\varphi_0 \in \Gamma_0(X)$, $\varphi_1 \in C^1(X)$ *and* φ *is even,*
then there exists a strictly increasing sequence $\{V_n\}_{n\geq 1}$ *of finite dimensional subspaces of* X *such that*

(1) *there exist a closed subspace* Y *of* X, $r > 0$ *and* $\beta > \varphi(0)$ *such that*

$$X \;=\; V_0 \oplus Y \quad \text{and} \quad \varphi|_{\partial B_r \cap Y} \;\geq\; \beta;$$

(2) *there exists a sequence* $\{R_n\}_{n \geq 1} \subseteq (r, +\infty)$ *such that*

$$\varphi(v) \leq \varphi(0) \qquad \forall\, v \in V_n, \; \|v\|_X \geq R_n;$$

(3) *(a generalized nonsmooth PS-condition) if* $c \geq \beta$, $\{x_n^*\}_{n \geq 1} \subseteq X^*$, $x_n^* \longrightarrow$ 0 *in* X^*, $\{x_n\}_{n \geq 1} \subseteq X$, $\varphi(x_n) \longrightarrow c$ *and*

$$\varphi_0(u) \geq \varphi_0(x_n) - \langle \varphi_1'(x_n), u - x_n \rangle_X + \langle x_n^*, u - x_n \rangle_X \;\; \forall u \in X, \; n \geq 1,$$

then the sequence $\{x_n\}_{n \geq 1}$ *admits a subsequence which is strongly convergent in* X,

<u>*then*</u> *there exists a sequence* $\{x_n\}_{n \geq 1} \subseteq \operatorname{dom} \varphi$ *such that* $\varphi(x_n) \longrightarrow +\infty$ *and*

$$\varphi_0(u) \geq \varphi_0(x_n) - \langle \varphi_1'(x_n), u - x_n \rangle_X \qquad \forall\, y \in X$$

(i.e. $-\varphi_1'(x_n) \in \partial \varphi_0(x_n)$ *for* $n \geq 1$).

PROOF By virtue of Corollary 4.3.2 and Proposition 4.3.6, we see that E_φ satisfies the extended nonsmooth PS-condition for $c \geq \beta$. Moreover,

$$|dE_\varphi|(0, \lambda) \neq 0 \qquad \forall\, \lambda \geq \beta$$

(see Proposition 4.3.6). So we can apply Theorem 4.3.3 and obtain a sequence $\{(x_n, \lambda_n)\}_{n \geq 1} \subseteq \operatorname{epi} \varphi$ such that

$$|dE_\varphi|(x_n, \lambda_n) = 0 \quad \text{and} \quad E_\varphi(x_n, \lambda_n) = \lambda_n \longrightarrow +\infty.$$

Then according to Proposition 4.3.6, we have

$$\varphi(x_n) = \lambda_n \qquad \forall\, n \geq 1,$$

while

$$|d\varphi|(x_n) = 0 \qquad \forall\, n \geq 1.$$

This completes the proof of the theorem. $\qquad\qquad\qquad\qquad\qquad$ ☐

Now we can start dealing with our generalized Dirichlet problem. Our aim is to eventually apply the abstract multiplicity result in Theorem 4.3.3. To this end we need some auxiliary material. Recall that $\varphi \colon L^p(\Omega) \longrightarrow \overline{\mathbb{R}}$ is defined by

$$\varphi(x) \overset{df}{=} \begin{cases} \varphi_0(x) + \varphi_1(x) & \text{if } x \in BV(\Omega), \\ +\infty & \text{if } x \in L^p(\Omega) \setminus BV(\Omega), \end{cases}$$

with

$$\varphi_0(x) \overset{df}{=} \int_\Omega \vartheta(\nabla x^a(z))\, dz + \int_\Omega \vartheta^\infty \left(\frac{\nabla x^s}{|\nabla x^s|}(z) \right) d|\nabla x^s|(z)$$

and

$$\varphi_1(x) \stackrel{df}{=} -\int_\Omega F\big(z, x(z)\big)\, dz.$$

The lemmata that follow will pave the way for a multiplicity theorem based on Theorem 4.3.3.

LEMMA 4.3.1
If $\{x_n\}_{n\geq 1} \subseteq BV(\Omega)$ is a sequence such that $\varphi(x_n) \longrightarrow c \in \mathbb{R}$, $u_n^* \longrightarrow 0$ in $L^{p'}(\Omega)$ and

$$\varphi_0(y) \geq \varphi_0(x_n) + \int_\Omega f\big(z, x_n(z)\big)\big(y - x_n\big)(z)\, dz$$

$$+ \int_\Omega u_n^*(z)\big(y - x_n\big)(z)\, dz \qquad \forall\, y \in L^p(\Omega),$$

then the sequence $\{x_n\}_{n\geq 1} \subseteq BV(\Omega)$ admits a subsequence which converges strongly in $L^p(\Omega)$.

PROOF   Choosing as a test function $y = 2x_n$ and using hypotheses $H(j)(iii)$ and (iv), for all $n \geq 1$ we have

$$\varphi_0(2x_n) \geq \varphi_0(x_n) + \int_\Omega f\big(z, x_n(z)\big)x_n(z)\, dz + \int_\Omega u_n^*(z)x_n(z)\, dz$$

$$\geq \varphi_0(x_n) + \int_\Omega \mu F\big(z, x_n(z)\big)\, dz + \int_\Omega u_n^*(z)x_n(z)\, dz - c_4,$$

for some $c_4 > 0$, so

$$\mu\varphi(x_n) - \int_\Omega u_n^*(z)x_n(z)\, dz + c_4 \geq (\mu + 1)\varphi_0(x_n) - \varphi_0(2x_n).$$

By virtue of hypothesis $H(\vartheta)$, we have that ϑ is Lipschitz continuous and

$$\lim_{\|u\|\to+\infty} \frac{2\vartheta(u) - \vartheta(2u)}{\|u\|} = \lim_{\|u\|\to+\infty} \frac{\vartheta(0)}{\|u\|} = 0$$

and so

$$(\mu + 1)\vartheta(u) - \vartheta(2u) \geq \frac{\mu - 1}{2}\vartheta(u) - M_1, \tag{4.88}$$

for some $M_1 > 0$ and

$$(\mu + 1)\vartheta^\infty(u) - \vartheta^\infty(2u) \geq \frac{\mu - 1}{2}\vartheta^\infty(u). \tag{4.89}$$

Also we have

$$c_5 \|x\|_{BV(\Omega)} - c_6 \;\leq\; \varphi_0(u) \;\leq\; c_7 \left(1 + \|x\|_{BV(\Omega)}\right) \qquad \forall\, x \in BV(\Omega), \quad (4.90)$$

for some $c_5, c_6, c_7 > 0$. Then using (4.88), (4.89) and (4.90), we obtain that

$$\mu\varphi(x_n) + \|u_n^*\|_{p'} \|x_n\|_p + c_4 \;\geq\; \frac{\mu - 1}{2}\varphi_0(x_n) - M_1 |\Omega|_N$$

$$\geq\; \frac{\mu - 1}{2}\left(c_5 \|x_n\|_{BV(\Omega)} - c_6\right) - M_1 |\Omega|_N$$

and thus the sequence $\{x_n\}_{n\geq 1} \subseteq BV(\Omega)$ is bounded.

Because the embedding $BV(\Omega) \subseteq L^p(\Omega)$ is compact, we have the result of the Lemma.   ☐

LEMMA 4.3.2

There exists a strictly increasing sequence $\{\overline{V}_n\}_{n\geq 1}$ of finite dimensional subspaces of $BV(\Omega) \cap L^\infty(\Omega)$ and a strictly decreasing sequence $\{Y_n\}_{n\geq 1}$ of closed subspaces of $L^p(\Omega)$ such that

$$L^p(\Omega) \;=\; \overline{V}_n \oplus Y_n \quad\text{and}\quad \bigcap_{n=1}^{\infty} Y_n \;=\; \{0\}.$$

PROOF   Let $m > \frac{N}{2}$ and let $\{e_n\}_{n\geq 1}$ be an orthogonal basis of the Hilbert space $H^{-m}(\Omega) = \left(H_0^m(\Omega)\right)^*$ consisting of elements of $H_0^1(\Omega)$. Let

$$\overline{V}_n \;\stackrel{df}{=}\; \mathrm{span}\{e_k\}_{k=0}^{n} \quad\text{and}\quad \widehat{Y}_n \;\stackrel{df}{=}\; \overline{\mathrm{span}}\{e_k\}_{k=n+1}^{\infty}.$$

Then

$$H^{-1}(\Omega) \;=\; \overline{V}_n \oplus \widehat{Y}_n.$$

Evidently

$$\bigcap_{n=1}^{\infty} \widehat{Y}_n \;=\; \{0\}.$$

Let us set

$$Y_n \;\stackrel{df}{=}\; \widehat{Y}_n \cap L^p(\Omega) \qquad \forall\, n \geq 1.$$

Then the sequences $\{\overline{V}_n\}_{n\geq 1}$ and $\{Y_n\}_{n\geq 1}$ have the desired properties.   ☐

LEMMA 4.3.3

If the sequence $\{Y_n\}_{n\geq 1}$ is as in the previous lemma, <u>*then*</u>

$$\sup_{n\geq 1} \inf_{\substack{x \in Y_n \\ \|x\|_p = 1}} \varphi_0(x) \;=\; +\infty.$$

PROOF   Suppose that the result of the Lemma is not true. Then we can find a sequence $\{x_n\}_{n\geq 1} \subseteq BV(\Omega)$, with $x_n \in Y_n$, $\|x_n\|_p = 1$ and $\{\varphi(x_n)\}_{n\geq 1}$ is bounded. By passing to a subsequence if necessary, we may assume that

$$x_n \longrightarrow x \quad \text{in } L^p(\Omega)$$

for some $x \in L^p(\Omega)$ with $\|x\|_p = 1$. Also

$$x \in \bigcap_{n=1}^{\infty} Y_n = \{0\},$$

a contradiction.   ☐

Now we are ready for the multiplicity result for the generalized Dirichlet problem.

THEOREM 4.3.4
If hypotheses $H(\vartheta)$ and $H(f)$ hold,
then there exists a sequence $\{x_n\}_{n\geq 1} \subseteq BV(\Omega)$ of critical points of the functional φ such that $\varphi(x_n) \longrightarrow +\infty$.

PROOF   Let $\{\overline{V}_n\}_{n\geq 1}$ and $\{Y_n\}_{n\geq 1}$ be sequences of subspaces as in Lemma 4.3.2. Note that φ_1 is bounded on bounded subsets of $L^p(\Omega)$ (see hypothesis $H(j)(iii)$). So by virtue of Lemma 4.3.3 we can find $n_0 \geq 1$ such that

$$\varphi(x) \geq \varphi(0) + 1 \quad \forall\, x \in Y_{n_0}, \|x\|_p = 1.$$

Let us set $Y \overset{df}{=} Y_{n_0}$ and $V_n \overset{df}{=} \overline{V}_{n_0+n}$. Because of hypothesis $H(j)(iv)$, we have

$$F(z,\zeta) \geq \frac{1}{M^\mu} F(z,M)|\zeta|^\mu - \xi(z) \quad \text{for a.a. } z \in \Omega \text{ and all } \zeta \in \mathbb{R}, \quad (4.91)$$

for some $\xi \in L^1(\Omega)$. From (4.90) and (4.91) we deduce that for some $R_n > 1$ we have

$$\varphi(x_n) \leq \varphi(0) \quad \forall\, x \in V_n, \|x\|_p \geq R_n.$$

This combined with Lemma 4.3.1 permits the use of Theorem 4.3.3, which gives the desired sequence of critical points of φ.   ☐

4.4   Strongly Nonlinear Problems

In this section we study boundary value problems driven by general nonlinear differential operators (which include as a special case the p-Laplacian) and have unilateral constraints (hemivariational inequalities and variational inequalities). We consider three different boundary value problems, none of which is in variational form. So now our approach is different and it is based on nonlinear operator theory (in particular operators of monotone type) and on degree theoretic arguments.

We start with hemivariational inequalities. Let $\Omega \subseteq \mathbb{R}^N$ be a bounded domain with a C^1-boundary $\partial\Omega$. The first problem that we shall study is the following:

$$\begin{cases} -\operatorname{div} a\big(z, x(z), \nabla x(z)\big) - \partial j\big(z, x(z)\big) \ni f\big(z, x(z), \nabla x(z)\big) \\ \qquad\qquad\qquad\qquad\qquad\qquad \text{for a.a. } z \in \Omega, \qquad (4.92) \\ x|_{\partial\Omega} = 0. \end{cases}$$

Here $a: \Omega \times \mathbb{R} \times \mathbb{R}^N \longrightarrow 2^{\mathbb{R}^N} \setminus \{\emptyset\}$ and for every $x \in W_0^{1,p}(\Omega)$, the first multivalued term in (4.92) is interpreted as

$$\big\{\operatorname{div} v: \ v \in L^{p'}\big(\Omega; \mathbb{R}^N\big), \ v(z) \in a\big(z, x(z), \nabla x(z)\big) \text{ for a.a. } z \in \Omega\big\},$$

with $1 < p' < +\infty$ (i.e. $v \in S_{a(\cdot, x(\cdot), \nabla x(\cdot))}^{p'}$). Then by a solution of problem (4.92), we mean a function $x \in W_0^{1,p}(\Omega)$ ($\frac{1}{p} + \frac{1}{p'} = 1$) for which there exist $v \in S_{a(\cdot, x(\cdot), \nabla x(\cdot))}^{p'}$ and $u \in S_{\partial j(\cdot, x(\cdot))}^{p'}$, such that

$$-\operatorname{div} v(z) - u(z) \ = \ f\big(z, x(z), \nabla x(z)\big) \quad \text{for a.a. } z \in \Omega.$$

Our hypotheses on the data of (4.92) are the following:

$\underline{H(a)_1}$ $a: \Omega \times \mathbb{R} \times \mathbb{R}^N \longrightarrow P_{kc}\big(\mathbb{R}^N\big)$ is a multifunction, such that

  (*i*) a is graph measurable;

  (*ii*) for almost all $z \in \Omega$ and all $\zeta \in \mathbb{R}$, the function

  $$\mathbb{R}^N \ni \xi \longmapsto a(z, \zeta, \xi) \in P_{kc}\big(\mathbb{R}^N\big)$$

  is strictly monotone;

  (*iii*) for almost all $z \in \Omega$, the function

  $$\mathbb{R} \times \mathbb{R}^N \ni (\zeta, \xi) \longmapsto a(z, \zeta, \xi) \in P_{kc}\big(\mathbb{R}^N\big)$$

  has closed graph and for almost all $z \in \Omega$ and all $\xi \in \mathbb{R}^N$, the function

  $$\mathbb{R} \ni \zeta \longmapsto a(z, \zeta, \xi) \in P_{kc}\big(\mathbb{R}^N\big)$$

  is lower semicontinuous;

  (*iv*) for almost all $z \in \Omega$, all $\zeta \in \mathbb{R}$, all $\xi \in \mathbb{R}^N$ and all $v \in a(z, \zeta, \xi)$, we have that

  $$\|v\|_{\mathbb{R}^N} \ \leq \ b_1(z) + c_1\big(|\zeta|^{p-1} + \|\xi\|_{\mathbb{R}^N}^{p-1}\big),$$

  with $b_1 \in L^{p'}(\Omega)_+$, $c_1 > 0$, $p \in (1, +\infty)$, $\frac{1}{p} + \frac{1}{p'} = 1$;

(v) for almost all $z \in \Omega$, all $\zeta \in \mathbb{R}$, all $\xi \in \mathbb{R}^N$ and all $v \in a(z,\zeta,\xi)$, we have that

$$\langle v, \xi \rangle_{\mathbb{R}^N} \geq \eta_1 \|\xi\|_{\mathbb{R}^N}^p - \eta_2,$$

with $\eta_1, \eta_2 > 0$.

REMARK 4.4.1   A possibility for the multifunction $a(z,\zeta,\xi)$, is when $a(z,\zeta,\xi) = \beta(z,\zeta)\partial\phi(z,\xi)$, β a Carathéodory function (i.e. $z \longmapsto \beta(z,\zeta)$ measurable and $\zeta \longmapsto \beta(z,\zeta)$ continuous), $\beta \geq 0$ and $\phi(z,\zeta)$ is Carathéodory too and for almost all $z \in \Omega$, the function $\xi \longmapsto \phi(z,\xi)$ is strictly convex, not necessarily differentiable. By $\partial\phi(x,\xi)$ we denote the subdifferential in the sense of convex analysis of the function $\phi(z,\cdot)$ (see Definition 1.3.3). If $\beta \equiv 1$ and $\phi(z,\xi) = \phi(\xi) = \frac{1}{p}\|\xi\|_{\mathbb{R}^N}^p$, then the resulting differential operator is the p-Laplacian.   ⬛

$\underline{H(j)_1}$ $j \colon \Omega \times \mathbb{R} \longrightarrow \mathbb{R}$ is a function, such that

(i) for all $\zeta \in \mathbb{R}$, the function

$$\Omega \ni z \longmapsto j(z,\zeta) \in \mathbb{R}$$

is measurable;

(ii) for almost all $z \in \Omega$, the function

$$\mathbb{R} \ni \zeta \longmapsto j(z,\zeta) \in \mathbb{R}$$

is locally Lipschitz;

(iii) for almost all $z \in \Omega$, all $\zeta \in \mathbb{R}$ and all $u \in \partial j(z,\zeta)$, we have that

$$|u| \leq b_2(z) + c_2|\zeta|^{r-1},$$

with $b_2 \in L^\infty(\Omega)$, $c_2 > 0$, $1 < r < p^*$;

(iv) there exists $\vartheta \in L^\infty(\Omega)_+$, such that

$$\limsup_{|\zeta| \to +\infty} \frac{u}{|\zeta|^{p-2}\zeta} \leq \vartheta(z),$$

uniformly for almost all $z \in \Omega$ and all $u \in \partial j(z,\zeta)$, with $\vartheta(z) \leq \lambda_1\eta_1$ for almost all $z \in \Omega$ and the inequality is strict on a set of positive measure (as before λ_1 is the first eigenvalue of $\left(-\Delta_p, W_0^{1,p}(\Omega)\right)$ and $\eta_1 > 0$ is as in hypothesis $H(a)_1(v)$).

$\underline{H(f)_1}$ $f \colon \Omega \times \mathbb{R} \times \mathbb{R}^N \longrightarrow \mathbb{R}$ is a function, such that

(i) for all $(\zeta, \xi) \in \mathbb{R} \times \mathbb{R}^N$, the function

$$\Omega \ni z \longmapsto f(z, \zeta, \xi) \in \mathbb{R}$$

is measurable;

(ii) for almost all $z \in \Omega$, the function

$$\mathbb{R} \times \mathbb{R}^N \ni (\zeta, \xi) \longmapsto f(z, \zeta, \xi) \in \mathbb{R}$$

is continuous;

(iii) for almost all $z \in \Omega$, all $\zeta \in \mathbb{R}$ and all $\xi \in \mathbb{R}^N$, we have that

$$|f(z, \zeta, \xi)| \leq b_3(z) + c_3 \left(|\zeta|^{\vartheta-1} + \|\xi\|_{\mathbb{R}^N}^{\vartheta-1} \right),$$

with $b_3 \in L^{p'}(\Omega)$, $c_3 > 0$, $1 \leq \vartheta < p$.

We consider the multivalued operator $V: W_0^{1,p}(\Omega) \longrightarrow P_{wkc}\left(W^{-1,p'}(\Omega)\right)$, defined by

$$V(x) \stackrel{df}{=} \left\{ -\mathrm{div}\, v : \ v \in S^{p'}_{a(\cdot, x(\cdot), \nabla x(\cdot))} \right\} \qquad \forall\, x \in W_0^{1,p}(\Omega).$$

Moreover, for fixed $x \in W_0^{1,p}(\Omega)$, we consider the auxiliary operator $K_x: W_0^{1,p}(\Omega) \longrightarrow 2^{W^{-1,p'}(\Omega)}$, defined by

$$K_x(y) \stackrel{df}{=} \left\{ -\mathrm{div}\, u : \ u \in S^{p'}_{a(\cdot, x(\cdot), \nabla y(\cdot))} \right\} \qquad \forall\, y \in W_0^{1,p}(\Omega).$$

LEMMA 4.4.1
If hypothesis $H(a)_1$ hold and $x \in W_0^{1,p}(\Omega)$,
then the operator K_x is maximal monotone.

PROOF By virtue of hypothesis $H(a)_1(i)$, for every $x, y \in W_0^{1,p}(\Omega)$, the multifunction

$$\Omega \ni z \longmapsto a(z, x(z), \nabla y(z)) \in P_{kc}\left(\mathbb{R}^N\right)$$

is measurable and so via the Yankov-von Neumann-Aumann Selection Theorem (see Theorem 1.2.4) and hypothesis $H(a)_1(iv)$, we infer that

$$S^{p'}_{a(\cdot, x(\cdot), \nabla y(\cdot))} \neq \emptyset.$$

Therefore K_x has nonempty values which clearly are convex, closed, bounded (thus convex, w-compact). Because of hypothesis $H(a)_1(ii)$, K_x is monotone. So according to Proposition 1.4.6, in order to establish the desired maximality of K_x, it suffices to show that for every $y, h \in W_0^{1,p}(\Omega)$, the multifunction $[0, 1] \ni t \longmapsto K_x(y + th) \in 2^{W^{-1,p'}(\Omega)}$ is upper semicontinuous from $[0, 1]$ into

$W^{-1,p'}(\Omega)_w$. To this end let $C \subseteq W^{-1,p'}(\Omega)$ be a nonempty, weakly closed subset and let us set

$$R_{y,h}^x(t) \overset{df}{=} K_x(y + th) \qquad \forall \ y, h \in W_0^{1,p}(\Omega).$$

We need to show that

$$\left(R_{y,h}^x\right)^- (C) \ = \ \left\{ t \in [0,1] : \ R_{y,h}^x(t) \cap C \neq \emptyset \right\}$$

is closed in $[0, 1]$ (see Proposition 1.2.1). Let $\{t_n\}_{n \geq 1} \subseteq \left(R_{y,h}^x\right)^- (C)$ and assume that $t_n \longrightarrow t$. Let us take $v_n \in R_{y,h}^x(t_n) \cap C$ for $n \geq 1$. We have that

$$v_n \ = \ -\operatorname{div} u_n, \quad \text{with } u_n \in S_{a(\cdot,x(\cdot),\nabla(y+t_nh)(\cdot))}^{p'}.$$

Because of hypothesis $H(a)_1(iv)$, we see that the sequence $\{u_n\}_{n \geq 1} \subseteq L^{p'}(\Omega; \mathbb{R}^N)$ is bounded and so, by passing to a subsequence if necessary, we may assume that

$$u_n \ \overset{w}{\longrightarrow} \ u \quad \text{in } L^{p'}(\Omega; \mathbb{R}^N).$$

Invoking Proposition 1.2.12 and exploiting hypothesis $H(a)_1(iii)$, we infer that $u \in S_{a(\cdot,x(\cdot),\nabla(y+th)(\cdot))}^{p'}$. Note that

$$-\operatorname{div} u_n \ \overset{w}{\longrightarrow} \ -\operatorname{div} u \quad \text{in } W^{-1,p'}(\Omega)$$

and so

$$v \ = \ -\operatorname{div} u \ \in \ R_{y,h}^x(t) \cap C.$$

Hence $t \in \left(R_{y,h}^x\right)^- (C)$ and we have proved the desired upper semicontinuity of $t \longmapsto K_x(y + th)$. $\square$

Using this Lemma, we can prove the following result concerning the multifunction V.

PROPOSITION 4.4.1
If hypotheses $H(a)_1$ hold,
<u>*then*</u> *V is an operator of type* (S)$_+$ *(see Definition 1.4.9).*

PROOF   Let $\{x_n\}_{n \geq 1} \subseteq W_0^{1,p}(\Omega)$ be a sequence, such that

$$x_n \ \overset{w}{\longrightarrow} \ x \quad \text{in } W_0^{1,p}(\Omega)$$

and let $v_n^* \in V(x_n)$ for $n \geq 1$ be such that

$$\limsup_{n \to +\infty} \langle v_n^*, x_n - x \rangle_{W_0^{1,p}(\Omega)} \ \leq \ 0.$$

We need to show that

$$x_n \longrightarrow x \quad \text{in } W_0^{1,p}(\Omega)$$

(note that because of hypothesis $H(a)_1(iv)$, V is bounded). We have that

$$v_n^* = -\operatorname{div} v_n \quad \text{with } v_n \in S_{a(\cdot, x_n(\cdot), \nabla x_n(\cdot))}^{p'}.$$

We may assume that

$$v_n \xrightarrow{w} v \quad \text{in } L^{p'}(\Omega; \mathbb{R}^N)$$

(see hypothesis $H(a)_1(iv)$). Hence

$$v_n^* \longrightarrow v^* = -\operatorname{div} v \quad \text{in } W^{-1,p'}(\Omega).$$

Let $y \in W_0^{1,p}(\Omega)$ and let $w \in S_{a(\cdot, x(\cdot), \nabla y(\cdot))}^{p'}$. For each $n \geq 1$, we introduce the multifunction $L_n \colon \Omega \longrightarrow 2^{\mathbb{R}^N}$, defined by

$$L_n(z) \overset{df}{=} \{\xi \in a(z, x_n(z), \nabla y(z)) : \\ \|w(z) - \xi\|_{\mathbb{R}^N} = d_X(w(z), a(z, x_n(z), \nabla y(z))).\}$$

Evidently $L_n(z) \neq \emptyset$ for almost all $z \in \Omega$ and $\operatorname{Gr} L_n \in \mathcal{L} \times \mathcal{B}(\mathbb{R}^N)$, with $\mathcal{L}$ being the Lebesgue σ-field of Ω and $\mathcal{B}(\mathbb{R}^N)$ the Borel σ-field of $\mathbb{R}^N$. So via the Yankov-von Neumann-Aumann Selection Theorem (see Theorem 1.2.4), we produce $w_n \in S_{a(\cdot, x_n(\cdot), \nabla y(\cdot))}^{p'}$, such that

$$w_n(z) \in L_n(z) \quad \text{for a.a. } z \in \Omega.$$

We have

$$\|w(z) - w_n(z)\|_{\mathbb{R}^N} = d_{\mathbb{R}^N}(w(z), a(z, x_n(z), \nabla y(z))) \\ \leq h_{\mathbb{R}^N}^*(a(z, x(z), \nabla y(z)), a(z, x_n(z), \nabla y(z))) \quad \text{for a.a. } z \in \Omega. \quad (4.93)$$

Since the embedding $W_0^{1,p}(\Omega) \subseteq L^p(\Omega)$ is compact, we may assume that

$$x_n \longrightarrow x \quad \text{in } L^p(\Omega), \\ x_n(z) \longrightarrow x(z) \text{ for a.a. } z \in \Omega.$$

From (4.93) and hypothesis $H(a)_1(iii)$, we have that

$$w_n(z) \longrightarrow w(z) \quad \text{for a.a. } z \in \Omega$$

(see Definition 1.2.5(b)). Then from the Extended Dominated Convergence Theorem (see Theorem A.2.1), we infer that

$$w_n \longrightarrow w \quad \text{in } L^{p'}(\Omega; \mathbb{R}^N).$$

Exploiting the monotonicity of $a(z, x_n(z), \cdot)$ and Green's identity (see Theorem 1.1.9), we have that

$$
\begin{aligned}
0 \;\le\; & \int_\Omega \big(v_n(z) - w_n(z), \nabla x_n(z) - \nabla y(z)\big)_{\mathbb{R}^N} \, dz \\
= \;& \int_\Omega \big(v_n(z), \nabla x_n(z) - \nabla x(z)\big)_{\mathbb{R}^N} \, dz + \int_\Omega \big(v_n(z), \nabla x(z) - \nabla y(z)\big)_{\mathbb{R}^N} \, dz \\
& + \int_\Omega \big(w_n(z), \nabla y(z) - \nabla x_n(z)\big)_{\mathbb{R}^N} \, dz \\
= \;& \langle v_n^*, x_n - x\rangle_{W_0^{1,p}(\Omega)} + \int_\Omega \big(v_n(z), \nabla x(z) - \nabla y(z)\big)_{\mathbb{R}^N} \, dz \\
& + \int_\Omega \big(w_n(z), \nabla y(z) - \nabla x_n(z)\big)_{\mathbb{R}^N} \, dz.
\end{aligned}
\tag{4.94}
$$

By hypothesis, we have that

$$
\limsup_{n \to +\infty} \langle v_n^*, x_n - x\rangle_{W_0^{1,p}(\Omega)} \;\le\; 0.
$$

Also, recall that $v_n \xrightarrow{w} v$ and $w_n \longrightarrow w$ in $L^{p'}(\Omega; \mathbb{R}^N)$. So passing to the limit as $n \to +\infty$ in (4.94), we obtain

$$
\begin{aligned}
0 \;\le\; & \int_\Omega \big(v(z), \nabla x(z) - \nabla y(z)\big)_{\mathbb{R}^N} \, dz + \int_\Omega \big(w(z), \nabla y(z) - \nabla x(z)\big)_{\mathbb{R}^N} \, dz \\
= \;& \langle -\operatorname{div} v - (-\operatorname{div} w), x - y\rangle_{W_0^{1,p}(\Omega)}.
\end{aligned}
$$

But $(y, -\operatorname{div} w) \in \operatorname{Gr} K_x$ was arbitrary. So, by virtue of Lemma 4.4.1, it follows that $-\operatorname{div} v \in K_x(z)$, hence $v \in S^{p'}_{a(\cdot, x(\cdot), \nabla x(\cdot))}$.

As above through a measurable selection argument, we can generate $u_n \in S^{p'}_{a(\cdot, x_n(\cdot), \nabla x(\cdot))}$ for $n \ge 1$, such that

$$
u_n \longrightarrow v \quad \text{in } L^{p'}(\Omega; \mathbb{R}^N).
$$

Let

$$
u_n^* \stackrel{df}{=} -\operatorname{div} u_n \quad \text{and} \quad v^* \stackrel{df}{=} -\operatorname{div} v.
$$

We have that

$$
\limsup_{n \to +\infty} \langle v_n^* - v^*, x_n - x\rangle_{W_0^{1,p}(\Omega)} \;\le\; 0,
$$

so

$$
\limsup_{n \to +\infty} \Big[\langle v_n^* - u_n^*, x_n - x\rangle_{W_0^{1,p}(\Omega)} + \langle u_n^* - v^*, x_n - x\rangle_{W_0^{1,p}(\Omega)} \Big] \;\le\; 0.
$$

Thus

$$\limsup_{n \to +\infty} \langle v_n^*, x_n - x \rangle_{W_0^{1,p}(\Omega)} + \liminf_{n \to +\infty} \langle u_n^* - v^*, x_n - x \rangle_{W_0^{1,p}(\Omega)} \leq 0$$

and since $u_n^* \longrightarrow v^*$ in $W^{-1,p'}(\Omega)$, we have

$$\limsup_{n \to +\infty} \langle v_n^* - u_n^*, x_n - x \rangle_{W_0^{1,p}(\Omega)} \leq 0. \tag{4.95}$$

On the other hand, from the monotonicity of $a(z, x_n(z), \cdot)$, we have

$$\langle v_n^* - u_n^*, x_n - x \rangle_{W_0^{1,p}(\Omega)} = \int_\Omega \left(v_n(z) - u_n(z), \nabla x_n(z) - \nabla x(z) \right)_{\mathbb{R}^N} dz \geq 0,$$

so

$$\liminf_{n \to +\infty} \langle v_n^* - u_n^*, x_n - x \rangle_{W_0^{1,p}(\Omega)} \geq 0. \tag{4.96}$$

From (4.95) and (4.96), it follows that

$$\langle v_n^* - u_n^*, x_n - x \rangle_{W_0^{1,p}(\Omega)} = \int_\Omega \left(v_n(z) - u_n(z), \nabla x_n(z) - \nabla x(z) \right)_{\mathbb{R}^N} dz \longrightarrow 0.$$

Note that the integrand

$$\beta_n(z) \stackrel{df}{=} \left(v_n(z) - u_n(z), \nabla x_n(z) - \nabla x(z) \right)_{\mathbb{R}^N}$$

is nonnegative and so we may assume that

$$\beta_n(z) \longrightarrow 0 \quad \text{for a.a. } z \in \Omega$$

and

$$\left| \beta_n(z) \right| \leq k_1(z) \quad \text{for a.a. } z \in \Omega \text{ and all } n \geq 1,$$

with $k_1 \in L^1(\Omega)$.

Because of hypotheses $H(a)_1(iv)$ and (v), for all $z \in \Omega \setminus D$, $|D|_N = 0$, we have that

$$k_1(z) \geq \left(v_n(z) - u_n(z), \nabla x_n(z) - \nabla x(z) \right)_{\mathbb{R}^N}$$
$$\geq \eta_1 \left(\|\nabla x_n(z)\|_{\mathbb{R}^N}^p + \|\nabla x(z)\|_{\mathbb{R}^N}^p \right) - 2\eta_2$$
$$- \|\nabla x_n(z)\|_{\mathbb{R}^N} \left(b_1(z) + c_1 |x_n(z)|^{p-1} + c_1 \|\nabla x(z)\|_{\mathbb{R}^N}^{p-1} \right)$$
$$- \|\nabla x(z)\|_{\mathbb{R}^N} \left(b_1(z) + c_1 |x_n(z)|^{p-1} + c_1 \|\nabla x_n(z)\|_{\mathbb{R}^N}^{p-1} \right). \tag{4.97}$$

Since $x_n \xrightarrow{w} x$ in $W_0^{1,p}(\Omega)$, by passing to a subsequence if necessary, we may assume that

$$x_n \longrightarrow x \qquad \text{in } L^p(\Omega),$$
$$x_n(z) \longrightarrow x(z) \quad \text{for a.a. } z \in \Omega,$$
$$|x_n(z)| \leq k_2(z) \text{ for a.a. } z \in \Omega \text{ and all } n \geq 1,$$

with $k_2 \in L^p(\Omega)$. So from (4.97), it follows that for all $z \in \Omega \setminus D$, the sequence $\{\nabla x_n(z)\}_{n \geq 1} \subseteq \mathbb{R}^N$ is bounded. Thus by passing to a subsequence if necessary (the subsequence in general depends on $z \in \Omega \setminus D$), we may assume that

$$\nabla x_n(z) \longrightarrow \widehat{\xi}(z).$$

Fix $z \in \Omega \setminus D$. We can find

$$g_n(z) \in a(z, x(z), \widehat{\xi}(z)),$$

such that

$$
\begin{aligned}
\|v_n(z) - g_n(z)\|_{\mathbb{R}^N} &= d_{\mathbb{R}^N}\left(v_n(z), a(z, x(z), \widehat{\xi}(z))\right) \\
&\leq h^*_{\mathbb{R}^N}\left(a(z, x_n(z), \nabla x_n(z)), a(z, x(z), \widehat{\xi}(z))\right) \qquad \forall\, n \geq 1. \quad (4.98)
\end{aligned}
$$

From the definition of the excess function $h^*_{\mathbb{R}^N}$ (see Definition 1.2.4), we see that we can find $s_n(z) \in a(z, x_n(z), \nabla x_n(z))$, for $n \geq 1$, such that

$$d_{\mathbb{R}^N}\left(s_n(z), a(z, x_n(z), \widehat{\xi}(z))\right) = h^*_{\mathbb{R}^N}\left(a(z, x_n(z), \nabla x_n(z)), a(z, x(z), \widehat{\xi}(z))\right).$$

Evidently the sequence $\{s_n(z)\}_{n \geq 1} \subseteq \mathbb{R}^N$ is bounded (see hypothesis $H(a)_1(\mathrm{iv})$) and so, by passing to a subsequence if necessary, we may assume that

$$s_n(z) \longrightarrow s(z) \quad \text{in } \mathbb{R}^N.$$

We have

$$\left(x_n(z), \nabla x_n(z), s_n(z)\right) \in \operatorname{Gr} a(z, \cdot, \cdot) \qquad \forall\, n \geq 1$$

and so from hypothesis $H(a)_1(\mathrm{iii})$, we have that

$$\left(x(z), \widehat{\xi}(z), s(z)\right) \in \operatorname{Gr} a(z, \cdot, \cdot),$$

hence

$$h^*_{\mathbb{R}^N}\left(a(z, x_n(z), \nabla x_n(z)), a(z, x(z), \widehat{\xi}(z))\right) \longrightarrow 0.$$

By virtue of (4.98), it follows that

$$\|v_n(z) - g_n(z)\|_{\mathbb{R}^N} \longrightarrow 0.$$

Note that $\{g_n(z)\}_{n \geq 1} \subseteq a(z, x(z), \widehat{\xi}(z)) \in P_{kc}(\mathbb{R}^N)$. So, we may assume that

$$g_n(z) \longrightarrow \widehat{g}(z) \in a(z, x(z), \widehat{\xi}(z)) \qquad \forall\, z \in \Omega \setminus D.$$

Recall that for all $z \in \Omega \setminus D$, we have

$$\beta_n(z) \longrightarrow 0.$$

Therefore, in the limit as $n \to +\infty$, we have that

$$\left(\widehat{g}(z) - v(z), \widehat{\xi}(z) - \nabla x(z)\right)_{\mathbb{R}^N} = 0,$$

with $\widehat{g}(z) \in a\left(z, x(z), \widehat{\xi}(z)\right)$ and $v(z) \in a\left(z, x(z), \nabla x(z)\right)$ for $z \in \Omega \setminus D$. But $a\left(z, x(z), \cdot\right)$ is strictly monotone. So it follows that

$$\widehat{\xi}(z) = \nabla x(z) \qquad \forall\, z \in \Omega \setminus D.$$

Thus, for the original sequence $\{\nabla x_n(z)\}_{n \geq 1}$, we have that

$$\nabla x_n(z) \longrightarrow \nabla x(z) \qquad \forall\, z \in \Omega \setminus D.$$

Also from (4.97), for all $z \in \Omega \setminus D$, we have

$$\begin{aligned}
& \eta_1 \|\nabla x_n(z)\|_{\mathbb{R}^N}^p \\
& \leq\ k_1(z) + \eta_1 \|\nabla x(z)\|_{\mathbb{R}^N}^p + 2\eta_2 \\
& \quad + \|\nabla x_n(z)\|_{\mathbb{R}^N} \left(b_1(z) + c_1 |x_n(z)|^{p-1} + c_1 \|\nabla x(z)\|_{\mathbb{R}^N}^{p-1}\right) \\
& \quad + \|\nabla x(z)\|_{\mathbb{R}^N} \left(b_1(z) + c_1 |x_n(z)|^{p-1} + c_1 \|\nabla x_n(z)\|_{\mathbb{R}^N}^{p-1}\right).
\end{aligned}$$

Using Young's inequality (see Theorem A.4.2) with $\varepsilon > 0$ small, we obtain

$$\begin{aligned}
\eta_3(\varepsilon) \|\nabla x_n(z)\|_{\mathbb{R}^N}^p &\leq\ k_1(z) + \eta_1 \|\nabla x(z)\|_{\mathbb{R}^N}^p + 2\eta_2 \\
& \quad + \eta_4(\varepsilon) \left(b_1(z)^{p'} + c_1^{p'} |x_n(z)|^p + c_1^{p'} \|\nabla x(z)\|_{\mathbb{R}^N}^p\right) \\
& \quad + b_1(z) \|\nabla x(z)\|_{\mathbb{R}^N}^p + c_1 |x_n(z)|^{p-1} \|\nabla x(z)\|_{\mathbb{R}^N}^p \\
& \quad + \eta_5(\varepsilon) \|\nabla x(z)\|_{\mathbb{R}^N}^p \qquad \forall\, z \in \Omega \setminus D,\ n \geq 1,
\end{aligned}$$

with $\eta_3(\varepsilon), \eta_4(\varepsilon), \eta_5(\varepsilon) > 0$. Thus the sequence $\left\{\|\nabla x_n(\cdot)\|_{\mathbb{R}^N}^p\right\}_{n \geq 1}$ is uniformly integrable.

Applying the Extended Dominated Convergence Theorem (see Theorem A.2.1), we obtain

$$\nabla x_n \longrightarrow \nabla x \quad \text{in } L^p\left(\Omega; \mathbb{R}^N\right)$$

and so

$$x_n \longrightarrow x \quad \text{in } W_0^{1,p}(\Omega).$$

This proves that V is of type $(S)_+$. $\qquad \square$

Let $G \colon W_0^{1,p}(\Omega) \longrightarrow P_{wkc}\left(L^{r'}(\Omega)\right)$ be the multifunction, defined by

$$G(x) \stackrel{df}{=} S_{\partial j(\cdot, x(\cdot))}^{r'} \qquad \forall\, x \in W_0^{1,p}(\Omega)$$

and let $N_f\colon W_0^{1,p}(\Omega) \longrightarrow L^{p'}(\Omega)$ be the Nemytskii operator corresponding to the function f, i.e.

$$N_f(x)(\cdot) \stackrel{df}{=} f\big(\cdot, x(\cdot), \nabla x(\cdot)\big) \qquad \forall\, x \in W_0^{1,p}(\Omega).$$

We introduce the multivalued operator $R\colon W_0^{1,p}(\Omega) \longrightarrow P_{wkc}\big(W^{-1,p'}(\Omega)\big)$, defined by

$$R(x) \stackrel{df}{=} V(x) - G(x) - N_f(x) \qquad \forall\, x \in W_0^{1,p}(\Omega).$$

PROPOSITION 4.4.2
If hypotheses $H(a)_1$, $H(j)_1$ and $H(f)_1$ hold, then R is pseudomonotone.

PROOF   Observe that R is defined on all of $W_0^{1,p}(\Omega)$, has convex, closed values and it is bounded. So according to Proposition 1.4.11 in order to show that R is pseudomonotone, it suffices to show that R is generalized pseudomonotone. So suppose that

$$x_n \xrightarrow{w} x \ \text{in } W_0^{1,p}(\Omega), \tag{4.99}$$
$$x_n^* \xrightarrow{w} x^* \ \text{in } W^{-1,p'}(\Omega),$$
$$x_n^* \in R(x_n) \ \text{for } n \geq 1$$

and

$$\limsup_{n\to+\infty} \langle x_n^*, x_n - x\rangle_{W_0^{1,p}(\Omega)} \leq 0.$$

We have to show that $x^* \in R(x)$ and $\langle x_n^*, x_n\rangle_{W_0^{1,p}(\Omega)} \longrightarrow \langle x^*, x\rangle_{W_0^{1,p}(\Omega)}$ (see Definition 1.4.8(b)). From the definition of R, we have

$$x_n^* = v_n^* - g_n - N_f(x_n) \qquad \forall\, n \geq 1,$$

with $v_n^* \in V(x_n)$, $g_n \in G(x_n)$ for $n \geq 1$.
   Note that because of hypothesis $H(j)_1(iii)$, we have that the sequence $\{g_n\}_{n\geq 1} \subseteq L^{r'}(\Omega)$ is bounded and so, by passing to a subsequence if necessary, we may assume that

$$g_n \xrightarrow{w} g \ \text{in } L^{r'}(\Omega). \tag{4.100}$$

From (4.99) and the compactness of the embedding $W_0^{1,p}(\Omega) \subseteq L^p(\Omega)$, passing to a further subsequence if necessary, we have that

$$x_n \longrightarrow x \qquad \text{in } L^p(\Omega),$$
$$x_n(z) \longrightarrow x(z) \ \text{for a.a. } z \in \Omega,$$
$$|x_n(z)| \leq k_1(z) \ \text{for a.a. } z \in \Omega \text{ and all } n \geq 1,$$

for some $k_1 \in L^p(\Omega)$. Then using Propositions 1.2.12 and 1.3.11, we obtain that $g \in G(x)$.

Moreover, since $r < p^*$, we have that

$$x_n \longrightarrow x \quad \text{in } L^r(\Omega)$$

and

$$\langle g_n, x_n - x \rangle_{W_0^{1,p}(\Omega)} = \int_\Omega g_n(z)(x_n - x)(z)\, dz \longrightarrow 0.$$

Also, from hypothesis $H(j)_1(iii)$, we have that the sequence $\{N_f(x_n)\}_{n \geq 1} \subseteq L^{p'}(\Omega)$ is bounded and so

$$\langle N_f(x_n), x_n - x \rangle_{W_0^{1,p}(\Omega)} = \int_\Omega f(z, x_n(z), \nabla x_n(z))(x_n - x)(z)\, dz \longrightarrow 0.$$

Thus, finally

$$\limsup_{n \to +\infty} \langle v_n^*, x_n - x \rangle_{W_0^{1,p}(\Omega)} = \limsup_{n \to +\infty} \langle x_n^*, x_n - x \rangle \leq 0. \qquad (4.101)$$

But from Proposition 4.4.1, we know that the operator V is of type $(S)_+$. So (4.101) implies that

$$x_n \longrightarrow x \quad \text{in } W_0^{1,p}(\Omega).$$

Then

$$N_f(x_n) \longrightarrow N_f(x) \quad \text{in } L^{p'}(\Omega) \text{ and in } W^{-1,p'}(\Omega).$$

Similarly, from (4.100) and the continuity of the embedding $L^{r'}(\Omega) \subseteq W^{-1,p'}(\Omega)$ (since $1 < r < p^*$), we have that

$$g_n \xrightarrow{w} g \quad \text{in } W^{-1,p'}(\Omega).$$

Also

$$v_n^* = -\operatorname{div} v_n \qquad \forall\, n \geq 1,$$

with $v_n \in S^{p'}_{a(\cdot, x_n(\cdot), \nabla x_n(\cdot))}$ for $n \geq 1$. Passing to a subsequence, we may assume that

$$v_n \xrightarrow{w} v \quad \text{in } L^{p'}(\Omega; \mathbb{R}^N),$$

with $v \in L^{p'}(\Omega; \mathbb{R}^N)$ and so

$$v_n^* \xrightarrow{w} v^* = -\operatorname{div} v \quad \text{in } W^{-1,p'}(\Omega).$$

Moreover, $v \in S^{p'}_{a(\cdot, x(\cdot), \nabla x(\cdot))}$ and so $v^* \in V(x)$. In the limit as $n \to +\infty$, we have

$$x^* = v^* - g - N_f(x),$$

with $v^* \in V(x)$, $g \in G(x)$ and so

$$x^* \in R(x).$$

Also, from the fact that $x_n \longrightarrow x$ in $W_0^{1,p}(\Omega)$, we have that

$$\langle x_n^*, x_n \rangle_{W_0^{1,p}(\Omega)} \longrightarrow \langle x^*, x \rangle_{W_0^{1,p}(\Omega)}$$

and this finishes the proof.   ☐

PROPOSITION 4.4.3
If hypotheses $H(a)_1$, $H(j)_1$ and $H(f)_1$ hold, then R is coercive.

PROOF   First we establish the following claim.

Claim 1. There exists $\beta > 0$, such that

$$\psi(x) \overset{df}{=} \eta_1 \|\nabla x\|_p^p - \int_\Omega \vartheta(z)|x(z)|^p \, dz$$

$$\geq \beta \|\nabla x\|_p^p \quad \forall \, x \in W_0^{1,p}(\Omega). \qquad (4.102)$$

Note that $\psi \geq 0$ (see Proposition 1.5.17 and hypothesis $H(j)_1(iv)$). Suppose that the claim is not true. Then we can find a sequence $\{x_n\}_{n \geq 1} \subseteq W_0^{1,p}(\Omega)$ with $\|\nabla x_n\|_p = 1$, such that $\psi(x_n) \searrow 0$. Using the Poincaré inequality (see Theorem 1.1.6), we may assume that

$$x_n \overset{w}{\longrightarrow} x \quad \text{in } W_0^{1,p}(\Omega)$$

and by the Sobolev embedding theorem (see Theorem 1.1.5), passing to a subsequence if necessary, we may assume that

$$x_n \longrightarrow x \quad \text{in } L^p(\Omega).$$

Because of the weak lower semicontinuity of the norm functional, in the limit as $n \to +\infty$, we have

$$0 \leq \psi(x) = \eta_1 \|\nabla x\|_p^p - \int_\Omega \vartheta(z)|x(z)|^p \, dz \leq 0,$$

i.e. $\psi(x) = 0$ and so

$$\eta_1 \|\nabla x\|_p^p = \int_\Omega \vartheta(z)|x(z)|^p \, dz \leq \eta_1 \lambda_1 \|x\|_p^p, \qquad (4.103)$$

thus $x = 0$ or $x = \pm u_1$, where u_1 is the normalized eigenfunction of $\left(-\Delta_p, W_0^{1,p}(\Omega) \right)$ corresponding to $\lambda_1 > 0$. If $x = 0$, then $\|\nabla x_n\|_p \longrightarrow 0$, a contradiction to the fact that $\|\nabla x_n\|_p = 1$ for all $n \geq 1$. So $x = \pm u_1$. Then from (4.103), hypothesis $H(j)_1(iv)$ and since $u_1(z) > 0$ for all $z \in \Omega$, we have that

$$\|\nabla x\|_p^p < \lambda_1 \|x\|_p^p,$$

a contradiction. This proves the claim.

By virtue of hypotheses $H(j)_1(iii)$ and $H(j)_1(iv)$, for almost all $z \in \Omega$ and all $u \in \partial j(z, \zeta)$, we have

$$u \leq \left(\vartheta(z) + \frac{\lambda_1 \beta}{2} \right) |\zeta|^{p-2} \zeta + \gamma_1(z) \quad \text{if } \zeta \geq 0$$

$$u \geq \left(\vartheta(z) + \frac{\lambda_1 \beta}{2} \right) |\zeta|^{p-2} \zeta - \gamma_1(z) \quad \text{if } \zeta \leq 0,$$

with $\gamma_1 \in L^{r'}(\Omega)_+$ For $x^* \in R(x)$, we have

$$x^* = v^* - u^* - N_f(x),$$

with $v^* \in V(x)$ and $u^* \in G(x)$. As $v^* = -\operatorname{div} v$, $v \in S^{p'}_{a(\cdot, x(\cdot), \nabla x(\cdot))}$ and using hypothesis $H(a)_1(v)$, we have

$$\langle x^*, x \rangle_{W_0^{1,p}(\Omega)}$$

$$= \langle v^*, x \rangle_{W_0^{1,p}(\Omega)} - \int_\Omega u^*(z) x(z) \, dz - \int_\Omega f\big(z, x(z), \nabla x(z)\big) x(z) \, dz$$

$$= \int_\Omega \big(v(z), \nabla x(z)\big)_{\mathbb{R}^N} \, dz - \int_\Omega u^*(z) x(z) \, dz - \int_\Omega f\big(z, x(z), \nabla x(z)\big) x(z) \, dz$$

$$\geq \eta_1 \|\nabla x\|_p^p - \eta_2 |\Omega|_N - \int_\Omega u^*(z) x(z) \, dz$$

$$- \int_\Omega f\big(z, x(z), \nabla x(z)\big) x(z) \, dz. \tag{4.104}$$

Also, we have

$$\int_\Omega u^*(z) x(z) \, dz \leq \int_\Omega \left(\vartheta(z) + \frac{\lambda_1 \beta}{2} \right) |x(z)|^p \, dz + \|\gamma_1\|_{r'} \|x\|_r$$

and

$$\int_\Omega f\big(z, x(z), \nabla x(z)\big) x(z) \, dz \leq \|b_3\|_{p'} \|x\|_p + c_4 \|x\|_{W^{1,p}(\Omega)}^\vartheta + c_5 \|\nabla x\|_p^\vartheta,$$

for some $c_4, c_5 > 0$ (see hypothesis $H(j)_1(iii)$).
Using these inequalities in (4.104), we obtain

$$\langle x^*, x \rangle_{W_0^{1,p}(\Omega)} \geq \eta_1 \|\nabla x\|_p^p - \eta_2 |\Omega|_N - \int_\Omega \vartheta |x(z)|^p \, dz - \frac{\lambda_1 \beta}{2} \|x\|_p^p$$

$$- \|\gamma_1\|_{r'} \|x\|_r - \|b_3\|_{p'} \|x\|_p - c_4 \|x\|_{W^{1,p}(\Omega)}^\vartheta - c_5 \|\nabla x\|_p^\vartheta$$

$$\geq \frac{\beta}{2} \|\nabla x\|_p^p - c_6 \|\nabla x\|_p^\vartheta - c_7,$$

for some $c_6, c_7 > 0$ (see (4.102)). This implies that R is coercive (recall that $\vartheta < p$). □

Now we are ready for the existence result concerning problem (4.92).

THEOREM 4.4.1
If hypotheses $H(a)_1$, $H(j)_1$ and $H(f)_1$ hold,
<u>*then*</u> *problem (4.92) has a solution $x_0 \in W_0^{1,p}(\Omega)$.*

PROOF   Propositions 4.4.1 and 4.4.2 permit the application of Theorem 1.4.6, which gives an element $x_0 \in W_0^{1,p}(\Omega)$, such that $0 \in R(x_0)$. So we can find $v \in S^{p'}_{a(\cdot,x_0(\cdot),\nabla x_0(\cdot))}$ and $u^* \in G(x_0)$, such that

$$-\operatorname{div} v \;=\; u^* + N_f(x_0) \;\in\; L^s(\Omega),$$

with $s = \min\{r', p'\}$, thus

$$\begin{cases} -\operatorname{div} v(z) = u^*(z) + f\big(z, x_0(z), \nabla x_0(z)\big) & \text{for a.a. } z \in \Omega, \\ x_0|_{\partial\Omega} = 0, \end{cases}$$

i.e. $x_0 \in W_0^{1,p}(\Omega)$ solves (4.92).   □

The second problem that we study in this section is again a hemivariational inequality, not in variational form and at resonance at infinity. More precisely, the problem under consideration is the following:

$$\begin{cases} -\operatorname{div} a\big(z, \nabla x(z)\big) - \lambda^* |x(z)|^{p-2} x(z) \in \partial j\big(z, x(z)\big) \\ \hspace{6cm} \text{for a.a. } z \in \Omega, \qquad (4.105) \\ x|_{\partial\Omega} = 0, \end{cases}$$

where $\lambda^* = \lambda_1 c_1$, $c_1 > 0$, $p \in (1, +\infty)$.
Our hypotheses on the functions $a(z, \xi)$ and $j(z, \zeta)$ are the following:

<u>$H(a)_2$</u> $a: \Omega \times \mathbb{R}^N \longrightarrow \mathbb{R}^N$ is a function, such that

(*i*) for every $\xi \in \mathbb{R}^N$, the function

$$\Omega \ni z \longmapsto a(z, \xi) \in \mathbb{R}^N$$

is measurable;

(*ii*) for almost all $z \in \Omega$, the function

$$\mathbb{R}^N \ni \xi \longmapsto a(z, \xi) \in \mathbb{R}^N$$

is continuous and strictly monotone and $a(z, 0) = 0$;

(*iii*) for almost all $z \in \Omega$ and all $\xi \in \mathbb{R}^N$, we have that

$$\left(a(z,\xi),\xi\right)_{\mathbb{R}^N} \geq c_1 \|\xi\|_{\mathbb{R}^N}^p .$$

REMARK 4.4.2   If $\vartheta \in L^\infty(\Omega)_+$ and $a(z,\xi) = \vartheta(z) \|\xi\|_{\mathbb{R}^N}^{p-2} \xi$, then we have a generalized p-Laplacian operator.   ▯

$\underline{H(j)_2}$ $j \colon \Omega \times \mathbb{R} \longrightarrow \mathbb{R}$ is a function, such that

(*i*) for all $\zeta \in \mathbb{R}$, the function

$$\Omega \ni z \longmapsto j(z,\zeta) \in \mathbb{R}$$

is measurable;

(*ii*) for almost all $z \in \Omega$, the function

$$\mathbb{R} \ni \zeta \longmapsto j(z,\zeta) \in \mathbb{R}$$

is locally Lipschitz;

(*iii*) we have that

$$|u| \leq \beta(z) \quad \text{for a.a. } z \in \Omega, \text{ all } \zeta \in \mathbb{R} \text{ and all } u \in \partial j(z,\zeta),$$

with $\beta \in L^{p'}(\Omega)$ (where $\frac{1}{p} + \frac{1}{p'} = 1$);

(*iv*) if

$$g_1(z,\zeta) \overset{df}{=} \min_{u \in \partial j(z,\zeta)} u \quad \text{and} \quad g_2(z,\zeta) \overset{df}{=} \max_{u \in \partial j(z,\zeta)} u,$$

there exist functions $g_-, g_+ \in L^1(\Omega)$, such that

$$g_-(z) = \liminf_{\zeta \to -\infty} g_1(z,\zeta) \quad \text{for a.a. } z \in \Omega,$$
$$g_+(z) = \limsup_{\zeta \to +\infty} g_2(z,\zeta) \quad \text{for a.a. } z \in \Omega$$

and

$$\int_\Omega g_+(z) u_1(z)\, dz < 0 < \int_\Omega g_-(z) u_1(z)\, dz.$$

REMARK 4.4.3   Hypothesis $H(j)_2(iv)$ is a Landesman-Lazer type condition. Assuming without any loss of generality that $j(\cdot,\cdot)$ is Borel measurable, we can easily check that the function

$$\Omega \times \mathbb{R} \times \mathbb{R} \ni (z,\zeta;h) \longrightarrow j^0(z,\zeta;h) \in \mathbb{R}$$

is Borel measurable and so

$$\text{Gr}\,\partial j \;\in\; \mathcal{B}\big(\Omega \times \mathbb{R} \times \mathbb{R}\big) \;=\; \mathcal{B}(\Omega) \times \mathcal{B}(\mathbb{R}) \times \mathcal{B}(\mathbb{R}),$$

where

$$\text{Gr}\,\partial j \;=\; \{(z,\zeta,u) \in \Omega \times \mathbb{R} \times \mathbb{R} : \; u \in \partial j(z,\zeta)\}$$

and with $\mathcal{B}(\Omega)$, $\mathcal{B}(\mathbb{R})$ being the Borel σ-fields of Ω and $\mathbb{R}$ respectively. For every $\mu \in \mathbb{R}$, we have

$$\{(z,\zeta) \in \Omega \times \mathbb{R} : \; g_1(z,\zeta) < \mu\} \;=\; \text{proj}_{\Omega \times \mathbb{R}}\big(\text{Gr}\,\partial j \cap (\Omega \times \mathbb{R} \times (-\infty,\mu))\big).$$

Since the subdifferential multifunction is compact-valued, from the second projection theorem (see the Theorem 1.2.14), we have that

$$\{(z,\zeta) \in \Omega \times \mathbb{R} : \; g_1(z,\zeta) < \mu\} \;\in\; \mathcal{B}(\Omega \times \mathbb{R}) \;=\; \mathcal{B}(\Omega) \times \mathcal{B}(\mathbb{R}),$$

i.e. g_1 is a Borel measurable function. Similarly, we show that g_2 is a Borel measurable function. $\qquad\square$

Our approach will be based on the multivalued Leray-Schauder alternative theorem (see Theorem 3.1.1).

THEOREM 4.4.2
If *hypotheses* $H(a)_2$ *and* $H(j)_2$ *hold,*
then *problem (4.105) has a solution* $x_0 \in W_0^{1,p}(\Omega)$.

PROOF   Let $A\colon W_0^{1,p}(\Omega) \longrightarrow W^{-1,p'}(\Omega)$ be the nonlinear operator, defined by

$$\langle A(x), y\rangle_{W_0^{1,p}(\Omega)} \;\overset{df}{=}\; \int_\Omega \big(a(z,\nabla x(z)), \nabla y(z)\big)_{\mathbb{R}^N} dz \qquad \forall\, x,y \in W_0^{1,p}(\Omega).$$

It is easy to check that A is strictly monotone, demicontinuous, thus maximal monotone. Let $\widehat{A}$ be the restriction of A in $L^{p'}(\Omega)$, i.e. $\widehat{A}\colon L^p(\Omega) \supseteq D(A) \longrightarrow L^{p'}(\Omega)$ is defined by

$$\widehat{A}(x) \;\overset{df}{=}\; A(x) \qquad \forall\, x \in D(A),$$

with

$$D(A) \;\overset{df}{=}\; \{x \in W_0^{1,p}(\Omega) : \; A(x) \in L^{p'}(\Omega)\}$$

(recall that $L^{p'}(\Omega) \subseteq W^{-1,p'}(\Omega)$). We claim that $\widehat{A}$ is maximal monotone. Evidently $\widehat{A}$ is monotone. We also show that if $J\colon L^p(\Omega) \longrightarrow L^{p'}(\Omega)$ is defined by

$$J(x)(\cdot) \;\overset{df}{=}\; |x(\cdot)|^{p-2} x(\cdot)$$

(note that J is continuous, strictly monotone, hence maximal monotone), then $\widehat{A} + J$ is surjective, i.e. $R(\widehat{A} + J) = L^{p'}(\Omega)$.

Note that the operator $A + J : W_0^{1,p}(\Omega) \longrightarrow W^{-1,p'}(\Omega)$ is maximal monotone (see Theorem 1.4.5) and coercive. Thus it is surjective (see Theorem 1.4.4). So for a given $h \in L^{p'}(\Omega)$, we can find $x \in W_0^{1,p}(\Omega)$, such that $A(x) + J(x) = h$, hence $A(x) = h - J(x) \in L^{p'}(\Omega)$, i.e. $x \in D(A)$ and so $A(x) = \widehat{A}(x)$. Since $h \in L^{p'}(\Omega)$ was arbitrary, we infer that $R(\widehat{A} + J) = L^{p'}(\Omega)$.

Using this surjectivity, we can show that $\widehat{A}$ is maximal monotone. Indeed, let $y \in L^p(\Omega)$ and $v \in L^{p'}(\Omega)$ be such that

$$\left\langle \widehat{A}(x) - v, x - y \right\rangle_{L^p(\Omega)} \geq 0 \qquad \forall \, x \in D(A). \tag{4.106}$$

Since $\widehat{A} + J$ is surjective, we can find $x_1 \in D$, such that

$$\widehat{A}(x_1) + J(x_1) \; = \; v + J(y).$$

So, if in (4.106), we set $x = x_1 \in D(A)$, we obtain

$$\left\langle J(y) - J(x_1), x_1 - y \right\rangle_{L^p(\Omega)} \geq 0 \qquad \forall \, x \in D(A).$$

Recalling that J is strictly monotone, we conclude that $y = x_1 \in D(A)$ and $v = \widehat{A}(x_1)$, which shows that $\widehat{A}$ is maximal monotone.

Next, let $V \overset{df}{=} \widehat{A} + J : L^p(\Omega) \supseteq D(A) \longrightarrow L^{p'}(\Omega)$. This map is maximal monotone, strictly monotone and coercive. So $V^{-1} : L^{p'}(\Omega) \longrightarrow D(A) \subseteq W_0^{1,p}(\Omega)$ is well defined.

Claim 1. V^{-1} is completely continuous.

Let $\{v_n\}_{n \geq 1} \in L^{p'}(\Omega)$ and $\{x_n\}_{n \geq 1} \subseteq D(A)$ be sequences, such that

$$v_n \overset{w}{\longrightarrow} v \quad \text{in } L^{p'}(\Omega),$$

for some $v \in L^{p'}(\Omega)$ and $x_n = V^{-1}(v_n)$ for $n \geq 1$. We have

$$\widehat{A}(x_n) + J(x_n) \; = \; v_n,$$

thus

$$\left\langle \widehat{A}(x_n), x_n \right\rangle_{L^p(\Omega)} + \left\langle J(x_n), x_n \right\rangle_{L^p(\Omega)} \; = \; \left\langle v_n, x_n \right\rangle_{L^p(\Omega)}.$$

Using hypothesis $H(j)_2(iii)$, we get

$$c_1 \|\nabla x_n\|_p^p + \|x_n\|_p^p \; \leq \; \|v_n\|_{p'} \|x_n\|_p$$

and so the sequence $\{x_n\}_{n \geq 1} \subseteq W_0^{1,p}(\Omega)$ is bounded. By passing to a subsequence if necessary, we may assume that

$$x_n \overset{w}{\longrightarrow} x \text{ in } W_0^{1,p}(\Omega),$$
$$x_n \longrightarrow x \text{ in } L^p(\Omega),$$

for some $x \in W_0^{1,p}(\Omega)$. Since $(x_n, v_n) \in \operatorname{Gr} V$ and V is maximal monotone, we have that $(x, v) \in \operatorname{Gr} V$ (see Corollary 1.4.1), hence $x = V^{-1}(v)$. Also, for all $n \geq 1$, we have

$$\left\langle \widehat{A}(x_n), x_n - x \right\rangle_{L^p(\Omega)} + \left\langle J(x_n), x_n - x \right\rangle_{L^p(\Omega)} = \left\langle v_n, x_n - x \right\rangle_{L^p(\Omega)},$$

so

$$\left\langle A(x_n), x_n - x \right\rangle_{W_0^{1,p}(\Omega)} + \left\langle J(x_n), x_n - x \right\rangle_{L^p(\Omega)} = \left\langle v_n, x_n - x \right\rangle_{L^p(\Omega)}.$$

Because the sequences $\{J(x_n)\}_{n\geq 1}, \{x_n\}_{n\geq 1} \subseteq L^{p'}(\Omega)$ are bounded and $x_n \longrightarrow x$ in $L^p(\Omega)$, we have that

$$\left\langle J(x_n), x_n - x \right\rangle_{L^p(\Omega)} \longrightarrow 0$$

and

$$\left\langle v_n, x_n - x \right\rangle_{L^p(\Omega)} \longrightarrow 0$$

and so

$$\lim_{n \to +\infty} \left\langle A(x_n), x_n - x \right\rangle_{W_0^{1,p}(\Omega)} = 0.$$

But from Proposition 4.4.1, we know that V is of type $(S)_+$. So, it follows that

$$x_n \longrightarrow x \quad \text{in } W_0^{1,p}(\Omega),$$

which proves the complete continuity of V.

Let $G \colon L^p(\Omega) \longrightarrow P_{wkc}\big(L^{p'}(\Omega)\big)$ be the multifunction, defined by

$$G(x) \overset{df}{=} S_{\partial j(\cdot, x(\cdot))}^{p'}.$$

Claim 2. G is upper semicontinuous from $L^p(\Omega)$ into $L^{p'}(\Omega)_w$.

Since G is bounded (see hypothesis $H(j)_2(iii)$), it is locally compact into $L^{p'}(\Omega)_w$. So according to Proposition 1.2.5 and because bounded sets in $L^{p'}(\Omega)_w$ are metrizable, it suffices to show that $\operatorname{Gr} G$ is sequentially closed in $L^p(\Omega) \times L^{p'}(\Omega)_w$. So let $\big\{(x_n, u_n)\big\}_{n\geq 1} \subseteq \operatorname{Gr} G$ be a sequence such that

$$x_n \longrightarrow x \text{ in } L^p(\Omega),$$
$$u_n \overset{w}{\longrightarrow} u \text{ in } L^{p'}(\Omega),$$

for some $x \in L^p(\Omega)$ and $u \in L^{p'}(\Omega)$. We may assume that

$$x_n(z) \longrightarrow x(z) \quad \text{for a.a. } z \in \Omega$$

and using Proposition 1.2.12, we have

$$u(z) \in \operatorname{conv} \limsup_{n \to +\infty} \partial j\big(z, x_n(z)\big) \subseteq \partial j\big(z, x(z)\big) \quad \text{for a.a. } z \in \Omega,$$

where the last inclusion is a consequence of the closedness of the graph of $\zeta \longmapsto \partial j(z, \zeta)$ and of the convexity of the value of ∂j. Therefore $u \in G(x)$ and this proves the claim.

Let $G_1 = G + (\lambda^* + 1)J|_{W_0^{1,p}(\Omega)}$. Clearly G_1 is upper semicontinuous from $W_0^{1,p}(\Omega)$ into $L^{p'}(\Omega)_w$ with nonempty, weakly compact and convex values. We consider the following multivalued fixed point problem:

$$x \in V^{-1}G_1(x). \tag{4.107}$$

Because of Claims 1 and 2, to solve (4.107), we can use the multivalued Leray-Schauder Alternative Theorem (see Theorem 3.1.1). So we need to show that the set

$$S \overset{df}{=} \left\{ x \in W_0^{1,p}(\Omega) : x \in tV^{-1}G_1(x), \ t \in (0,1) \right\}$$

is bounded uniformly in $t \in (0,1)$. Suppose that this is not true. Then, we can find $x_n \in D(A)$ and $t_n \in (0,1)$, $n \geq 1$, such that $\|x_n\|_{W^{1,p}(\Omega)} \longrightarrow +\infty$, $t_n \longrightarrow t$ and for all $n \geq 1$, we have

$$V\left(\tfrac{1}{t_n}x_n\right) = w_n,$$

with $w_n \in G_1(x_n)$, $w_n = u_n^* + (\lambda^* + 1)J(x_n)$ and $u_n^* \in G(x_n)$. Thus

$$A\left(\tfrac{1}{t_n}x_n\right) + J\left(\tfrac{1}{t_n}x_n\right) = u_n^* + (\lambda^* + 1)J(x_n) \qquad \forall\, n \geq 1. \tag{4.108}$$

From hypothesis $H(j)_2(iii)$, we have that the sequence $\{u_n^*\}_{n \geq 1} \subseteq L^{p'}(\Omega)$ is bounded and so we may assume that

$$u_n^* \overset{w}{\longrightarrow} u^* \quad \text{in } L^{p'}(\Omega).$$

Let us set

$$y_n \overset{df}{=} \frac{x_n}{\|x_n\|_{W^{1,p}(\Omega)}} \qquad \forall\, n \geq 1.$$

Passing to a subsequence if necessary, we may assume that

$$y_n \overset{w}{\longrightarrow} y \quad \text{in } W_0^{1,p}(\Omega),$$
$$y_n \longrightarrow y \quad \text{in } L^p(\Omega),$$

for some $y \in W_0^{1,p}(\Omega)$. We act on (4.108) with $\frac{1}{\|x_n\|_{W^{1,p}(\Omega)}^{p-1}} y_n$. We obtain

$$\frac{1}{t_n^{p-1}}\big\langle A(y_n), y_n \big\rangle_{W_0^{1,p}(\Omega)} + \frac{1}{t_n^{p-1}}\big\langle J(y_n), y_n \big\rangle_{L^p(\Omega)}$$

$$= \left\langle \frac{u_n^*}{\|x_n\|_{W^{1,p}(\Omega)}^{p-1}}, y_n \right\rangle_{L^p(\Omega)} + (\lambda^* + 1)\big\langle J(y_n), y_n \big\rangle_{L^p(\Omega)},$$

so

$$c_1 \|\nabla y_n\|_p^p + \|y_n\|_p^p \leq t_n^{p-1} \left\langle \frac{u_n^*}{\|x_n\|_{W^{1,p}(\Omega)}^{p-1}}, y_n \right\rangle_{L^p(\Omega)} + t_n^{p-1}(\lambda^* + 1) \|y_n\|_p^p$$

and recalling that $0 < t_n < 1$ for $n \geq 1$, we have

$$c_1 \|\nabla y_n\|_p^p \leq t_n^{p-1} \left\langle \frac{u_n^*}{\|x_n\|_{W^{1,p}(\Omega)}^{p-1}}, y_n \right\rangle_{L^p(\Omega)} + t_n^{p-1}\lambda^* \|y_n\|_p^p. \qquad (4.109)$$

Note that

$$\frac{u_n^*}{\|x_n\|_{W^{1,p}(\Omega)}^{p-1}} \longrightarrow 0 \quad \text{in } L^{p'}(\Omega)$$

(see hypothesis $H(j)_2(iii)$ and recall that $\|x_n\|_{W^{1,p}(\Omega)} \longrightarrow +\infty$). So by passing to the limit as $n \to +\infty$, we obtain

$$c_1 \|\nabla y\|_p^p \leq t^{p-1}\lambda^* \|y\|_p^p \leq c_1\lambda_1 \|y\|_p^p$$

(since $0 \leq t \leq 1$). Thus

$$\|\nabla y\|_p^p = \lambda_1 \|y\|_p^p$$

and so $y = \pm u_1$ or $y = 0$ and $t_n \longrightarrow 1^-$.

If $y = 0$, then

$$y_n \longrightarrow 0 \quad \text{in } W_0^{1,p}(\Omega),$$

a contradiction to the fact that $\|y_n\|_{W^{1,p}(\Omega)} = 1$ for all $n \geq 1$. So $y = \pm u_1$.

First assume that $y = u_1$. We have

$$c_1 \|\nabla y\|_p^p + \left(1 - t_n^{p-1}(\lambda^* + 1)\right) \|y_n\|_p^p$$

$$\leq \frac{t_n^{p-1}}{\|x_n\|_{W^{1,p}(\Omega)}^{p-1}} \langle u_n^*, y_n \rangle_{L^p(\Omega)} \qquad \forall\, n \geq 1.$$

Since $t_n \in (0,1)$ and $\lambda^* = c_1\lambda_1$, we see that

$$c_1 \|\nabla y\|_p^p + \left(1 - t_n^{p-1}(\lambda^* + 1)\right) \|y_n\|_p^p > 0 \qquad \forall\, n \geq 1,$$

so

$$\frac{t_n^{p-1}}{\|x_n\|_{W^{1,p}(\Omega)}^{p-1}} \langle u_n^*, y_n \rangle_{L^p(\Omega)} > 0 \qquad \forall\, n \geq 1$$

and thus

$$\langle u_n^*, y_n \rangle_{L^p(\Omega)} = \int_\Omega u_n^*(z) y_n(z)\, dz > 0 \qquad \forall\, n \geq 1.$$

Since $y = u_1$, we see that

$$x_n(z) \longrightarrow +\infty \quad \text{for a.a. } z \in \Omega \text{ as } n \to +\infty.$$

We have $u_n^*(z) \leq g_2\big(z, x_n(z)\big)$ for almost all $z \in \Omega$, hence

$$u_n^*(z)u_1(z) \leq g_2\big(z, x_n(z)\big)u_1(z) \quad \text{for a.a. } z \in \Omega.$$

From Fatou's lemma (see Theorem A.2.2) and since

$$u_n^* \xrightarrow{w} u^* \in L^{p'}(\Omega),$$

we have that

$$0 \leq \int_\Omega u^*(z)u_1(z)\,dz \leq \int_\Omega g_+(z)u_1(z)\,dz,$$

a contradiction to hypothesis $H(j)_1(iv)$.
Similarly, if $y = -u_1$, we obtain that

$$\int_\Omega g_-(z)u_1(z)\,dz \leq \int_\Omega u^*(z)u_1(z)\,dz \leq 0,$$

again a contradiction to hypothesis $H(j)_1(iv)$. This proves that S is bounded uniformly in $t \in (0, 1)$ and so by Theorem 3.1.1, we can find $x_0 \in W_0^{1,p}(\Omega)$, such that

$$V(x_0) \in G_1(x_0),$$

so

$$A(x_0) + J(x_0) \in G(x_0) + (\lambda^* + 1)J(x_0).$$

Thus

$$A(x_0) - \lambda^* J(x_0) \in G(x_0)$$

and so

$$\begin{cases} -\operatorname{div} a\big(z, \nabla x_0(z)\big) - \lambda^* |x_0(z)|^{p-2} x_0(z) \in \partial j\big(z, x_0(z)\big) & \text{for a.a. } z \in \Omega, \\ x_0|_{\partial\Omega} = 0. \end{cases}$$

Therefore $x_0 \in W_0^{1,p}(\Omega)$ is a solution of (4.105). $\quad\square$

The third problem that we shall examine is a nonlinear elliptic variational inequality:

$$\begin{cases} -\operatorname{div} a\big(z, x(z), \nabla x(z)\big) + \beta\big(x(z)\big) \ni f\big(z, x(z)\big) & \text{for a.a. } z \in \Omega, \\ x|_{\partial\Omega} = 0. \end{cases} \qquad (4.110)$$

As for problem (4.92), $a \colon \Omega \times \mathbb{R} \times \mathbb{R}^N \longrightarrow 2^{\mathbb{R}^N} \setminus \{\emptyset\}$, while $\beta \colon \mathbb{R} \longrightarrow 2^{\mathbb{R}}$ is a maximal monotone graph in $\mathbb{R}^2$ and so $\beta = \partial j$, with $j \in \Gamma_0(\mathbb{R})$ (see Remark 1.4.6). As before

$$\operatorname{div} a\big(z, x(z), \nabla x(z)\big) \stackrel{df}{=} \Big\{\operatorname{div} v(z) : v \in S_{a(\cdot, x(\cdot), \nabla x(\cdot))}^q\Big\},$$

with $q \in (1, +\infty)$.

The precise hypotheses on the data of (4.110) are the following:

$\underline{H(a)_3}$ $a \colon \Omega \times \mathbb{R} \times \mathbb{R}^N \longrightarrow P_{kc}(\mathbb{R}^N)$ is a multifunction, such that

   (i) a is graph measurable;

   (ii) for almost all $z \in \Omega$ and all $\zeta \in \mathbb{R}$, the multivalued operator

$$\mathbb{R}^N \ni \xi \longmapsto a(z, \zeta, \xi) \in P_{kc}(\mathbb{R}^N)$$

   is maximal monotone with $0 \in a(z, \zeta, 0)$;

   (iii) for almost all $z \in \Omega$ and all $\xi \in \mathbb{R}^N$, the multivalued operator

$$\mathbb{R} \ni \zeta \longmapsto a(z, \zeta, \xi) \in P_{kc}(\mathbb{R}^N)$$

   is lower semicontinuous;

   (iv) for almost all $z \in \Omega$, the multivalued operator

$$\mathbb{R} \times \mathbb{R}^N \ni (\zeta, \xi) \longmapsto a(z, \zeta, \xi) \in P_{kc}(\mathbb{R}^N)$$

   has closed graph;

   (v) for almost all $z \in \Omega$, all $\zeta \in \mathbb{R}$, all $\xi \in \mathbb{R}^N$ and all $v \in a(z, \zeta, \xi)$, we have

$$\|v\|_{\mathbb{R}^N} \leq b_1(z) + c_1\big(|\zeta|^{p-1} + \|\xi\|_{\mathbb{R}^N}^{p-1}\big),$$

   with $b_1 \in L^{p'}(\Omega)$, $c_1 > 0$, $p \in [2, +\infty)$, $\frac{1}{p} + \frac{1}{p'} = 1$;

   (vi) for almost all $z \in \Omega$, all $\zeta \in \mathbb{R}$, all $\xi \in \mathbb{R}^N$ and all $v \in a(z, \zeta, \xi)$, we have

$$(v, \xi)_{\mathbb{R}^N} \geq \eta_1 \|\xi\|_{\mathbb{R}^N}^p - \eta_2,$$

   with $\eta_1, \eta_2 > 0$.

$\underline{H(\beta)_1}$ $\beta = \partial j$ with $j \in \Gamma_0(\mathbb{R})$, $j \geq 0$ and $j(0) = \min\limits_{\mathbb{R}} j$.

REMARK 4.4.4   Hypothesis $H(\beta)_1$ implies that $0 \in \partial j(0)$. Moreover, it is easy to check that for all $\lambda > 0$, we have that

$$j_\lambda(0) = 0 \quad \text{and} \quad 0 = j_\lambda'(0) = \partial j_\lambda(0)$$

(see Remark 1.4.6). $\Box$

$\underline{H(f)_2}$ $f \colon \Omega \times \mathbb{R} \times \longrightarrow \mathbb{R}$ is a function, such that (*i*) for all $\zeta \in \mathbb{R}$, the function

$$\Omega \ni z \longmapsto f(z, \zeta) \in \mathbb{R}$$

is measurable;

(*ii*) for almost all $z \in \Omega$, the function

$$\mathbb{R} \ni \zeta \longmapsto f(z, \zeta) \in \mathbb{R}$$

is continuous;

(*iii*) for almost all $z \in \Omega$ and all $\zeta \in \mathbb{R}$, we have

$$|f(z, \zeta)| \leq b_2(z) + c_2|\zeta|^{\vartheta-1},$$

with $b_2 \in L^{\vartheta'}(\Omega)$, $c_2 > 0$, $\vartheta > 1$ is such that

$$\begin{cases} \vartheta \leq p' \text{ if } 2 < p, \\ \vartheta < 2 \text{ if } 2 = p, \end{cases} \quad \text{and} \quad \frac{1}{\vartheta} + \frac{1}{\vartheta'} = 1.$$

As we did for problem (4.92), we consider the multivalued operator $V \colon W_0^{1,p}(\Omega) \longrightarrow P_{wkc}\big(W^{-1,p'}(\Omega)\big)$, defined by

$$V(x) \stackrel{df}{=} \left\{ -\operatorname{div} v : v \in S^{p'}_{a(\cdot, x(\cdot), \nabla x(\cdot))} \right\} \qquad \forall\, x \in W_0^{1,p}(\Omega).$$

A careful reading of the proof of Proposition 4.4.1 reveals that the following is true about the operator V (in fact the proof is now simplified).

PROPOSITION 4.4.4
If hypotheses $H(a)_3$ hold, <u>then</u> V is pseudomonotone.

Now let $j_\lambda \colon \mathbb{R} \longrightarrow \mathbb{R}$ be the Moreau-Yosida approximation of j, i.e.

$$j_\lambda(\zeta) \stackrel{df}{=} \inf_{\zeta' \in \mathbb{R}} \left[j(\zeta') + \frac{1}{2\lambda}|\zeta - \zeta'|^2 \right]$$

(see Remark 1.4.9) and let $G_\lambda \colon L^p(\Omega) \longrightarrow \mathbb{R}$ be the integral operator, defined by

$$G_\lambda(x) \stackrel{df}{=} \int_\Omega j_\lambda\big(x(z)\big)\,dz \qquad \forall\, x \in L^p(\Omega).$$

Then $G_\lambda(\cdot)$ is convex and

$$\partial G_\lambda(x) = G'_\lambda(x) = N_{j'_\lambda}(x),$$

with $N_{j'_\lambda}$ being the Nemytskii operator corresponding to j'_λ, i.e.

$$N_{j'_\lambda}(x)(\cdot) = j'_\lambda(x(\cdot)) \qquad \forall\, x \in L^p(\Omega).$$

Also let $N_f \colon L^{\vartheta}(\Omega) \longrightarrow L^{\vartheta'}(\Omega)$ be the Nemytskii operator corresponding to f, i.e.

$$N_f(x)(\cdot) \stackrel{df}{=} f\big(\cdot, x(\cdot)\big) \qquad \forall\, x \in L^{\vartheta}(\Omega).$$

By virtue of the Krasnoselskii Theorem (see Theorem 1.4.6), this operator is continuous and bounded. We consider the following operator inclusion:

$$0 \in V(x) + G'_{\lambda}(x) - N_f(x). \tag{4.111}$$

PROPOSITION 4.4.5

If hypotheses $H(a)_3$ and $H(f)_2$ hold and $\lambda > 0$,
then problem (4.111) has at least one solution $x_0 \in W_0^{1,p}(\Omega)$.

PROOF   From the compactness of the embeddings $W_0^{1,p}(\Omega) \subseteq L^p(\Omega)$ and $W_0^{1,p}(\Omega) \subseteq L^{\vartheta}(\Omega)$, we infer that the maps

$$W_0^{1,p}(\Omega) \ni x \longmapsto G'_{\lambda}(x) = N_{j'_{\lambda}}(x) \in W^{-1,p'}(\Omega)$$

and

$$W_0^{1,p}(\Omega) \ni x \longmapsto N_f(x) \in W^{-1,p'}(\Omega)$$

are completely continuous. From this and Propositions 4.4.4 and 1.4.13, it follows that the map

$$W_0^{1,p}(\Omega) \ni x \longmapsto V(x) + G'_{\lambda}(x) + N_f(x) \in W^{-1,p'}(\Omega)$$

is pseudomonotone.

If $x \in W_0^{1,p}(\Omega)$, for every $v^* \in V(x)$, we have

$$\langle v^*, x \rangle_{W_0^{1,p}(\Omega)} + \langle G'_{\lambda}(x), x \rangle_{W_0^{1,p}(\Omega)} - \langle N_f(x), x \rangle_{W_0^{1,p}(\Omega)}$$

$$= \langle v^*, x \rangle_{W_0^{1,p}(\Omega)} + \int_{\Omega} j'_{\lambda}\big(x(z)\big) x(z)\, dz - \int_{\Omega} f\big(z, x(z)\big) x(z)\, dz$$

$$\geq \eta_1 \|\nabla x\|_p^p - \eta_2 |\Omega|_N + \int_{\Omega} j'_{\lambda}\big(x(z)\big) x(z)\, dz - \int_{\Omega} f\big(z, x(z)\big) x(z)\, dz$$

(see hypothesis $H(a)_3(vi)$).

Since j'_{λ} is monotone and $j'_{\lambda}(0) = 0$, we have that

$$j'_{\lambda}\big(x(z)\big) x(z) \geq 0 \quad \text{for a.a. } z \in \Omega.$$

Also from the Hölder's inequality (see Theorem A.3.12) and hypothesis $H(f)_2(iii)$, we have that

$$\int_{\Omega} f\big(z, x(z)\big) x(z)\, dz \leq c_3 \|x\|_{\vartheta}^{\vartheta} + c_4 \leq c_5 \|\nabla x\|_p^{\vartheta} + c_6$$

with $c_3, c_4, c_5, c_6 > 0$. In the last inequality, we have used the fact that the embedding $W_0^{1,p}(\Omega) \subseteq L^\vartheta(\Omega)$ is continuous (in fact compact) since $\vartheta < p$ and the Poincaré inequality (see Theorem 1.1.6). Therefore, we have

$$\langle v^*, x\rangle_{W_0^{1,p}(\Omega)} + \langle G_\lambda'(x), x\rangle_{W_0^{1,p}(\Omega)} - \langle N_f(x), x\rangle_{W_0^{1,p}(\Omega)}$$
$$\geq \ \eta_1 \|\nabla x\|_p^p - \eta_2 |\Omega|_N - c_5 \|\nabla x\|_p^\vartheta - c_6.$$

Because $\vartheta < p$, it follows that the multivalued operator

$$x \ \longmapsto \ V(x) + G_\lambda'(x) - N_f(x)$$

is coercive. But a pseudomonotone coercive operator is surjective (see Theorem 1.4.6). So problem (4.111) has a solution $x_0 \in W_0^{1,p}(\Omega)$. □

Next we shall pass to the limit as $\lambda \longrightarrow 0^+$ to obtain a solution for the original problem (4.110).

THEOREM 4.4.3
If hypotheses $H(a)_3$, $H(\beta)_1$ and $H(f)_2$ hold,
<u>*then*</u> *problem (4.110) has a solution $x_0 \in W_0^{1,p}(\Omega)$.*

PROOF Let $\{\lambda_n\}_{n\geq 1} \subseteq \mathbb{R}$ be a sequence, such that $\lambda_n \searrow 0$ and let $x_n \in W_0^{1,p}(\Omega)$ be solutions of problem (4.111) when $\lambda = \lambda_n$ (see Proposition 4.4.5). Set

$$u_n \ \overset{df}{=} \ N_{j_{\lambda_n}'}(x_n) \ = \ G_{\lambda_n}'(x_n) \ = \ \partial G_{\lambda_n}(x_n) \qquad \forall \, n \geq 1.$$

For some $v_n \in S_{a(\cdot, x_n(\cdot), \nabla x_n(\cdot))}^{p'}$, we have

$$\langle -\mathrm{div}\, v_n, x_n\rangle_{W_0^{1,p}(\Omega)} + \langle u_n, x_n\rangle_{W_0^{1,p}(\Omega)} \ = \ \langle N_f(x_n), x_n\rangle_{W_0^{1,p}(\Omega)},$$

thus

$$\int_\Omega (v_n(z), \nabla x_n(z))_{\mathbb{R}^N} \, dz + \int_\Omega u_n(z) x_n(z)\, dz \ = \ \int_\Omega f(z, x_n(z)) x_n(z)\, dz$$

and so

$$\eta_1 \|\nabla x_n\|_p^p - \eta_2 |\Omega|_N \ \leq \ \|N_f(x_n)\|_{\vartheta'} \|x_n\|_\vartheta \ \leq \ c_5 \|\nabla x_n\|_p^\vartheta + c_6,$$

with $c_5, c_6 > 0$ (see the proof of Proposition 4.4.5). Since $\vartheta < p$, it follows that the sequence $\{x_n\}_{n\geq 1} \subseteq W_0^{1,p}(\Omega)$ is bounded.

For every $n \geq 1$ let $\eta_n : \mathbb{R} \longrightarrow \mathbb{R}$ be the Lipschitz continuous function, defined by

$$\eta_n(r) \ \overset{df}{=} \ |\beta_{\lambda_n}(r)|^{p-2} \beta_{\lambda_n}(r) \qquad \forall \, r \in \mathbb{R},$$

with $\beta_{\lambda_n} : \mathbb{R} \longrightarrow \mathbb{R}$ being the Yosida approximation of the maximal monotone map β. We know that

$$\beta_{\lambda_n} \ = \ \partial j_{\lambda_n} \ = \ j'_{\lambda_n} \qquad \forall\, n \geq 1$$

(see Remark 1.4.9). From Theorem 1.1.13(a), we know that $\eta_n\big(x_n(\cdot)\big) \in W_0^{1,p}(\Omega)$. Using $\eta_n\big(x_n(\cdot)\big)$ as a test function, we obtain

$$\int_\Omega \big(v_n(z), \nabla \eta_n\big(x_n(z)\big)\big)_{\mathbb{R}^N}\, dz + \int_\Omega \big|\beta_{\lambda_n}\big(x_n(z)\big)\big|^p\, dz$$
$$= \int_\Omega f\big(z, x_n(z)\big) \eta_n\big(x_n(z)\big)\, dz. \qquad (4.112)$$

From the Chain Rule for Sobolev functions (see Theorem 1.1.12(b)), for every $k \in \{1, 2, \ldots, N\}$, we have

$$D_k \eta_n\big(x_n(z)\big) \ = \ (p-1)\big|\beta_{\lambda_n}\big(x_n(z)\big)\big|^{p-2} \beta'_{\lambda_n}\big(x_n(z)\big) D_k x_n(z) \quad \text{for a.a. } z \in \Omega,$$

so

$$\int_\Omega \big(v_n(z), \nabla \eta_n\big(x_n(z)\big)\big)_{\mathbb{R}^N}\, dz \qquad\qquad\qquad (4.113)$$
$$= \ (p-1) \int_\Omega \big|\beta_{\lambda_n}\big(x_n(z)\big)\big|^{p-2} \beta'_{\lambda_n}\big(x_n(z)\big) \big(v_n(z), \nabla x_n(z)\big)_{\mathbb{R}^N}\, dz.$$

Note that $\beta'_{\lambda_n}\big(x_n(z)\big) \geq 0$ for almost all $z \in \Omega$, while from the monotonicity of $a\big(z, x_n(z), \cdot\big)$ and the fact that $0 \in a\big(z, x_n(z), 0\big)$ for almost all $z \in \Omega$ (see hypothesis $H(a)_3(ii)$), we have that

$$\big(v_n(z), \nabla x_n(z)\big)_{\mathbb{R}^N} \ \geq \ 0 \quad \text{for a.a. } z \in \Omega.$$

Therefore, from (4.113), it follows that

$$0 \ \leq \ \int_\Omega \big(v_n(z), \nabla \eta_n\big(x_n(z)\big)\big)_{\mathbb{R}^N}\, dz.$$

From hypothesis $H(f)_2(iii)$, we have that $N_f(x_n) \in L^{\vartheta'}(\Omega) \subseteq L^p(\Omega)$ for all $n \geq 1$. So, we have

$$\int_\Omega f\big(z, x_n(z)\big) \eta_n\big(x_n(z)\big)\, dz \ \leq \ \|N_f(x_n)\|_p \, \|\eta_n(x_n)\|_{p'}$$
$$= \ \|N_f(x_n)\|_p \, \|\beta_{\lambda_n}(x_n)\|_p^{p-1}.$$

Using this in (4.112), we obtain

$$\|\beta_{\lambda_n}(x_n)\|_p^p \leq \|N_f(x_n)\|_p \|\beta_{\lambda_n}(x_n)\|_p^{p-1} \qquad \forall\, n \geq 1.$$

But the sequence $\{N_f(x_n)\}_{n\geq 1} \subseteq L^p(\Omega)$ is bounded, hence the sequence $\{\beta_{\lambda_n}(x_n)\}_{n\geq 1} \subseteq L^p(\Omega)$ is bounded. By passing to a subsequence if necessary, we may assume that

$$x_n \xrightarrow{w} x_0 \text{ in } W_0^{1,p}(\Omega),$$
$$x_n \longrightarrow x_0 \text{ in } L^p(\Omega),$$
$$\beta_{\lambda_n}(x_n) \xrightarrow{w} w_0 \text{ in } L^p(\Omega),$$

for some $x_0 \in W_0^{1,p}(\Omega)$ and $w_0 \in L^p(\Omega)$.

Let $G\colon W_0^{1,p}(\Omega) \longrightarrow \overline{\mathbb{R}} \stackrel{df}{=} \mathbb{R} \cup \{+\infty\}$ be the integral operator, defined by

$$G(x) \stackrel{df}{=} \int_{\Omega} j(x(z))\, dz.$$

Evidently $G \in \Gamma_0(W_0^{1,p}(\Omega))$ and its Moreau-Yosida approximation is given by G_λ. Note that

$$\partial G_\lambda(y) = G'_\lambda(y) = \beta_\lambda(y(\cdot)) \qquad \forall\, \lambda > 0,\ y \in W_0^{1,p}(\Omega).$$

From Remark 1.4.9, we have that $w \in \partial G(x)$ and so $w(z) \in \beta(x(z))$ for almost all $z \in \Omega$.

For every $n \geq 1$, we have

$$-\mathrm{div}\, v_n + G'_{\lambda_n}(x_n) = N_f(x_n).$$

By virtue of hypothesis $H(a)_3(iii)$, we may assume that

$$v_n \xrightarrow{w} v_0 \quad \text{in } L^{p'}(\Omega; \mathbb{R}^N)$$

for some $v_0 \in L^{p'}(\Omega; \mathbb{R}^N)$ and so

$$\mathrm{div}\, v_n \xrightarrow{w} \mathrm{div}\, v_0 \quad \text{in } W^{-1,p'}(\Omega).$$

Taking duality brackets with $x_n - x_0$, we obtain

$$\langle -\mathrm{div}\, v_n, x_n - x_0 \rangle_{W_0^{1,p}(\Omega)} - \int_{\Omega} \beta_{\lambda_n}(x_n(z))(x_n - x_0)(z)\, dz$$
$$= \int_{\Omega} f(z, x_n(z))(x_n - x_0)(z)\, dz.$$

Because $p \geq 2$ (hence $p' \leq 2$), we have that

$$x_n \longrightarrow x_0 \quad \text{in } L^{p'}(\Omega)$$

and so

$$\int_\Omega \beta_{\lambda_n}(x_n(z))(x_n - x_0)(z)\, dz \longrightarrow 0$$

and

$$\int_\Omega f(z, x_n(z))(x_n - x_0)(z)\, dz \longrightarrow 0.$$

It follows that

$$\lim_{n \to +\infty} \langle -\operatorname{div} v_n, x_n - x_0 \rangle_{W_0^{1,p}(\Omega)} = 0.$$

Because V is pseudomonotone (see Proposition 4.4.4), we infer that $-\operatorname{div} v_0 \in V(x)$ and so $v_0 \in S_{a(\cdot,x_0(\cdot),\nabla x_0(\cdot))}$. We have

$$-\operatorname{div} v_0 + w_0 = N_f(x_0),$$

thus

$$-\operatorname{div} v_0 = N_f(x_0) - w_0 \in L^p(\Omega),$$

with $w_0(z) \in \beta(x_0(z))$ for almost all $z \in \Omega$. So

$$\begin{cases} -\operatorname{div} v_0(z) + \beta(x_0(z)) \ni f(z, x_0(z)) & \text{for a.a. } z \in \Omega, \\ x_0|_{\partial\Omega} = 0. \end{cases}$$

This proves that $x_0 \in W_0^{1,p}(\Omega)$ is a solution of problem (4.110). $\quad\square$

4.5   Method of Upper and Lower Solutions

In Section 3.3 we have seen the method of upper lower solutions in the context of second ordinary differential inclusions. In this section we return to this method and apply it to second order elliptic inclusions similar to the ones considered in the previous section. The method of upper and lower solutions provides a powerful and flexible mechanism to prove existence and comparison theorems for a broad class of nonlinear elliptic partial differential equations. The presence of the ordered pair of upper and lower solutions permits the relaxation of the growth condition on the nonlinearity. However, in our case the presence of the multivalued subdifferential term complicates matters and requires more delicate reasoning. First, by coupling the method of upper and lower solutions with appropriate truncation and penalization techniques, we establish the existence of at least one solution in the ordered interval $[\underline{\psi}, \overline{\psi}]$ formed by an ordered pair $\{\underline{\psi}, \overline{\psi}\}$ of a lower and of an upper solution. Subsequently for a subclass of problems, in which the right hand nonlinearity is independent of the gradient of the unknown function, we show that in the order interval $[\underline{\psi}, \overline{\psi}]$ there is a maximum solution and a minimum solution (for the usual pointwise ordering, which the Sobolev space $W_0^{1,p}(\Omega)$ inherits form the Banach space $L^p(\Omega)$). Such solutions are known as *extremal solutions*. In the context of semilinear problems, such solutions usually can be obtained using some monotone iterative scheme which suggests ways for the numerical treatment of the equation.

4.5.1 Existence of Solutions

Let $\Omega \in \mathbb{R}^N$ be a bounded domain with a C^1 boundary $\partial\Omega$. The problem under consideration is the following:

$$\begin{cases} -\operatorname{div} a\big(z, x(z), \nabla x(z)\big) + f\big(z, x(z), \nabla x(z)\big) + \partial j\big(z, x(z)\big) \\ \qquad\qquad\qquad\qquad\qquad \ni \vartheta\big(x(z)\big) \quad \text{for a.a. } z \in \Omega, \qquad (4.114) \\ x|_{\partial\Omega} = 0. \end{cases}$$

As in the previous section, $a\colon \Omega \times \mathbb{R} \times \mathbb{R}^N \longrightarrow 2^{\mathbb{R}^N} \setminus \{\emptyset\}$ and for every $x \in W_0^{1,p}(\Omega)$, the differential operator $\operatorname{div} a\big(z, x(z), \nabla x(z)\big)$ has the following interpretation:

$$\operatorname{div} a\big(z, x(z), \nabla x(z)\big) \stackrel{df}{=} \Big\{ \operatorname{div} v(z) \colon \ v \in L^{p'}\big(\Omega; \mathbb{R}^N\big),$$

$$v(z) \in a\big(z, x(z), \nabla x(z)\big) \text{ for a.a. } z \in \Omega \Big\},$$

with $\frac{1}{p} + \frac{1}{p'} = 1$ (i.e. $v \in S_{a(\cdot, x(\cdot), \nabla x(\cdot))}^{p'}$). Then by a solution of problem (4.114), we mean a function $x \in W_0^{1,p}(\Omega)$ for which there exist $v \in S_{a(\cdot, x(\cdot), \nabla x(\cdot))}^{p'}$ and $u \in S_{\partial j(\cdot, x(\cdot))}^{p'}$, such that

$$-\operatorname{div} v(z) - u(z) \ = \ f\big(z, x(z), \nabla x(z)\big) \quad \text{for a.a. } z \in \Omega.$$

Again we introduce the multifunction $V\colon W_0^{1,p}(\Omega) \longrightarrow P_{wkc}\big(W^{-1,p'}(\Omega)\big)$ defined by

$$V(x) \stackrel{df}{=} \Big\{ -\operatorname{div} v \colon \ v \in S_{a(\cdot, x(\cdot), \nabla x(\cdot))}^{p'} \Big\} \qquad \forall\, x \in W_0^{1,p}(\Omega).$$

We start the analysis of problem (4.114) with the definition of upper and lower solutions for the problem.

DEFINITION 4.5.1

(a) *A function* $\overline{\psi} \in W^{1,p}(\Omega)$, $\overline{\psi}|_{\partial\Omega} \geq 0$ *is an* **upper solution** *for problem (4.114), if there exist* $v_+^* \in V(\overline{\psi})$ *and* $u_+ \in S_{\partial j(\cdot, \overline{\psi}(\cdot))}^{r'}$ $\big(\frac{1}{r} + \frac{1}{r'} = 1,$

$1 \leq r < p^*$) *such that*

$$\langle v_+^*, y \rangle_{W_0^{1,p}(\Omega)} + \int_\Omega f\big(z, \overline{\psi}(z), \nabla\overline{\psi}(z)\big) y(z) \, dz + \int_\Omega u_+(z) y(z) \, dz$$

$$\geq \int_\Omega \vartheta\big(\overline{\psi}(z)\big) y(z) \, dz \qquad \forall \, y \in W_0^{1,p}(\Omega)_+.$$

(b) *A function* $\underline{\psi} \in W^{1,p}(\Omega)$, $\underline{\psi}|_{\partial\Omega} \leq 0$ *is a **lower solution** for problem (4.114), if there exist* $v_-^* \in V(\underline{\psi})$ *and* $u_- \in S_{\partial j(\cdot, \underline{\psi}(\cdot))}^{r'}$ $(\frac{1}{r} + \frac{1}{r'} = 1,$ $1 \leq r < p^*$) *such that*

$$\langle v_-^*, y \rangle_{W_0^{1,p}(\Omega)} + \int_\Omega f\big(z, \underline{\psi}(z), \nabla\underline{\psi}(z)\big) y(z) \, dz + \int_\Omega u_-(z) y(z) \, dz$$

$$\leq \int_\Omega \vartheta\big(\underline{\psi}(z)\big) y(z) \, dz \qquad \forall \, y \in W_0^{1,p}(\Omega)_+.$$

We assume that there exists an ordered pair of upper and lower solutions.

$\underline{H_0}$ There exist an upper solution $\overline{\psi}$ and a lower solution $\underline{\psi}$ such that

$$\underline{\psi}(z) \leq \overline{\psi}(z) \quad \text{for a.a. } z \in \Omega.$$

Now we are ready to introduce the hypotheses on the data of problem (4.114).

$\underline{H(a)_1}$ $a: \Omega \times \mathbb{R} \times \mathbb{R}^N \longrightarrow P_{kc}(\mathbb{R}^N)$ is a multifunction, such that

    (i) a is graph measurable;

    (ii) for almost all $z \in \Omega$ and all $\zeta \in \mathbb{R}$, the multifunction

$$\mathbb{R}^N \ni \xi \longmapsto a(z, \zeta, \xi) \in P_{kc}(\mathbb{R}^N)$$

    is strictly monotone;

    (iii) for almost all $z \in \Omega$, the multifunction

$$\mathbb{R} \times \mathbb{R}^N \ni (\zeta, \xi) \longmapsto a(z, \zeta, \xi) \in P_{kc}(\mathbb{R}^N)$$

    has closed graph and for almost all $z \in \Omega$ and all $\xi \in \mathbb{R}^N$, the multifunction

$$\mathbb{R} \ni \zeta \longmapsto a(z, \zeta, \xi) \in P_{kc}(\mathbb{R}^N)$$

    is lower semicontinuous;

(*iv*) for almost all $z \in \Omega$, all $\zeta \in \mathbb{R}$, all $\xi \in \mathbb{R}^N$ and all $v \in a(z, \zeta, \xi)$, we have that

$$\|v\|_{\mathbb{R}^N} \leq b_1(z) + c_1 \left(|\zeta|^{p-1} + \|\xi\|_{\mathbb{R}^N}^{p-1} \right),$$

with $b_1 \in L^{p'}(\Omega)_+$, $c_1 > 0$, $p \in [2, +\infty)$, $\frac{1}{p} + \frac{1}{p'} = 1$;

(*v*) for almost all $z \in \Omega$, all $\zeta \in \mathbb{R}$, all $\xi \in \mathbb{R}^N$ and all $v \in a(z, \zeta, \xi)$, we have that

$$\langle v, \xi \rangle_{\mathbb{R}^N} \geq \eta_1 \|\xi\|_{\mathbb{R}^N}^p - \eta_2,$$

with $\eta_1, \eta_2 > 0$.

REMARK 4.5.1   These hypotheses are the same as hypotheses $H(a)_1$ in Section 4.4. ▯

$\underline{H(j)_1}$ $j \colon \Omega \times \mathbb{R} \longrightarrow \mathbb{R}$ is a function, such that

(*i*) for all $\zeta \in \mathbb{R}$, the function

$$\Omega \ni z \longmapsto j(z, \zeta) \in \mathbb{R}$$

is measurable;

(*ii*) for almost all $z \in \Omega$, the function

$$\mathbb{R} \ni \zeta \longmapsto j(z, \zeta) \in \mathbb{R}$$

is locally Lipschitz;

(*iii*) for almost all $z \in \Omega$, all $\zeta \in \left[\underline{\psi}(z), \overline{\psi}(z) \right]$ and all $u \in \partial j(z, \zeta)$, we have that

$$|u| \leq b_2(z),$$

with $b_2 \in L^{r'}(\Omega)$ $1 \leq r < p^*$, $\frac{1}{r} + \frac{1}{r'} = 1$.

$\underline{H(f)_1}$ $f \colon \Omega \times \mathbb{R} \times \mathbb{R}^N \longrightarrow \mathbb{R}$ is a function such that

(*i*) for all $(\zeta, \xi) \in \mathbb{R} \times \mathbb{R}^N$, the function

$$\Omega \ni z \longmapsto f(z, \zeta, \xi) \in \mathbb{R}$$

is measurable;

(*ii*) for almost all $z \in \Omega$, the function

$$\mathbb{R} \times \mathbb{R}^N \ni (\zeta, \xi) \longmapsto f(z, \zeta, \xi) \in \mathbb{R}$$

is continuous;

(*iii*) for almost all $z \in \Omega$, all $\zeta \in \left[\underline{\psi}(z), \overline{\psi}(z)\right]$ and all $\xi \in \mathbb{R}^N$, we have that

$$\left|f(z, \zeta, \xi)\right| \leq h_3(z) + c_3 \left\|\xi\right\|_{\mathbb{R}^N},$$

with $b_3 \in L^{p'}(\Omega)$ $(\frac{1}{p} + \frac{1}{p'} = 1)$, $c_3 > 0$.

$\underline{H(\vartheta)}$ $\vartheta: \mathbb{R} \longrightarrow \mathbb{R}$ is a function such that

(*i*) we have $\vartheta\big(\overline{\psi}(\cdot)\big), \vartheta\big(\underline{\psi}(\cdot)\big) \in L^{p'}(\Omega)$

(*ii*) for some $M > 0$, the function

$$\mathbb{R} \ni \zeta \longmapsto \vartheta(\zeta) + M\zeta \in \mathbb{R}$$

is nondecreasing.

As we already saw in Section 3.5, the method of upper and lower solutions involves truncation and penalization techniques, which aim at exploiting the fact that we control the data of the problem in the interval $\left[\underline{\psi}(z), \overline{\psi}(z)\right]$ (see hypotheses $H(j)_1(iii)$ and $H(f)_1(iii)$). So we introduce the following items.

First the ***truncation map*** $\tau: W^{1,p}(\Omega) \longrightarrow W^{1,p}(\Omega)$ is defined by

$$\tau(x)(z) \stackrel{df}{=} \begin{cases} \overline{\psi}(z) & \text{if } \overline{\psi}(z) < x(z), \\ x(z) & \text{if } \underline{\psi}(z) \leq x(z) \leq \overline{\psi}(z), \\ \underline{\psi}(z) & \text{if } x(z) < \underline{\psi}(z). \end{cases}$$

It is easy to see that τ is continuous. Second, we introduce the ***penalty function*** $\beta: \Omega \times \mathbb{R} \longrightarrow \mathbb{R}$ defined by

$$\beta(z, \zeta) \stackrel{df}{=} \begin{cases} |\zeta|^{p-2}\zeta - |\overline{\psi}(z)|^{p-2}\overline{\psi}(z) & \text{if } \overline{\psi}(z) < \zeta, \\ 0 & \text{if } \underline{\psi}(z) \leq \zeta \leq \overline{\psi}(z), \\ |\zeta|^{p-2}\zeta - |\underline{\psi}(z)|^{p-2}\underline{\psi}(z) & \text{if } \zeta < \underline{\psi}(z). \end{cases}$$

Evidently $\beta(z, \zeta)$ is a Carathéodory function (i.e. measurable in $z \in \Omega$, continuous in $\zeta \in \mathbb{R}$). Also we have

$$\left|\beta(z, \zeta)\right| \leq b_4(z) + c_4 |\zeta|^{p-1} \quad \text{for a.a. } z \in \Omega \text{ and all } \zeta \in \mathbb{R},$$

with $b_4 \in L^{p'}(\Omega)$, $c_4 > 0$ and

$$\int_\Omega \beta\big(z, x(z)\big)x(z)\,dz \geq c_5 \left\|x\right\|_p^p - c_6 \quad \forall\, x \in L^p(\Omega),$$

with $c_5, c_6 > 0$. Third we introduce a ***penalty multifunction*** $Q: \Omega \times \mathbb{R} \longrightarrow P_{fc}(\mathbb{R})$ defined by

$$Q(z, \zeta) \stackrel{df}{=} \begin{cases} [u_+(z), +\infty) & \text{if } \overline{\psi}(z) < \zeta, \\ \mathbb{R} & \text{if } \underline{\psi}(z) \leq \zeta \leq \overline{\psi}(z), \\ (-\infty, u_-(z)] & \text{if } \zeta < \underline{\psi}(z). \end{cases}$$

Finally, we set

$$E\big(z, x(z)\big) \stackrel{df}{=} \partial j\big(z, \tau(x)(z)\big) \cap Q\big(z, x(z)\big) \qquad \forall\, x \in W_0^{1,p}(\Omega).$$

In what follows by K we denote the following order interval in $W^{1,p}(\Omega)$:

$$K \stackrel{df}{=} [\underline{\psi}, \overline{\psi}] = \Big\{ x \in W^{1,p}(\Omega) : \underline{\psi}(z) \leq x(z) \leq \overline{\psi}(z) \text{ for a.a. } z \in \Omega \Big\}.$$

Now for fixed $w \in K$ we consider the following auxiliary problem:

$$\begin{cases} -\mathrm{div}\, a\big(z, \tau(x)(z), \nabla x(z)\big) + f\big(z, \tau(x)(z), \nabla \tau(x)(z)\big) \\ \quad + M\tau(x)(z) + \varrho\beta\big(z, x(z)\big) + E\big(z, x(z)\big) \\ \quad \ni \vartheta\big(w(z)\big) + Mw(z) \quad \text{for a.a. } z \in \Omega, \\ x|_{\partial\Omega} = 0, \quad \varrho > 0. \end{cases} \qquad (4.115)$$

First we solve this auxiliary problem. For this purpose we introduce the multifunction $V_1 \colon W_0^{1,p}(\Omega) \longrightarrow P_{wkc}\big(W^{-1,p'}(\Omega)\big)$ defined by

$$V_1(x) \stackrel{df}{=} \Big\{ -\mathrm{div}\, v : v \in S_{a(\cdot, \tau(x)(\cdot), \nabla x(\cdot))}^{p'} \Big\} \qquad \forall\, x \in W_0^{1,p}(\Omega).$$

From Proposition 4.4.1 we know the following (the presence of the truncation map does not affect the argument, only simplifies it):

PROPOSITION 4.5.1
If hypotheses $H(a)_1$ hold,
then V_1 is an operator of type $(S)_+$ (see Definition 1.4.9).

Next we introduce some more maps. So let $N_{f,\tau} \colon W_0^{1,p}(\Omega) \longrightarrow L^{p'}(\Omega)$ be defined by

$$N_{f,\tau}(x)(\cdot) \stackrel{df}{=} f\big(\cdot, \tau(x)(\cdot), \nabla\tau(x)(\cdot)\big) \qquad \forall\, x \in W_0^{1,p}(\Omega)$$

and let $N_\beta \colon L^p(\Omega) \longrightarrow L^{p'}(\Omega)$ be defined by

$$N_\beta(x)(\cdot) \stackrel{df}{=} \beta\big(\cdot, x(\cdot)\big) \qquad \forall\, x \in L^p(\Omega).$$

Also let $N_E \colon W_0^{1,p}(\Omega) \longrightarrow P_{fc}\big(L^{r'}(\Omega)\big)$ be defined by

$$N_E(x) \stackrel{df}{=} S_{E(\cdot, x(\cdot))}^{r'} \qquad \forall\, x \in W_0^{1,p}(\Omega).$$

Recall that the embeddings $L^{p'}(\Omega) \subseteq W^{-1,p'}(\Omega)$ and $L^{r'}(\Omega) \subseteq W^{-1,p'}(\Omega)$ are compact (we have assumed that $r < p*$). So we can define the multifunction $U_\varrho \colon W_0^{1,p}(\Omega) \longrightarrow P_{wkc}\big(W^{-1,p'}(\Omega)\big)$ by

$$U_\varrho(x) \stackrel{df}{=} V_1(x) + N_{f,\tau}(x) + M\tau(x) + \varrho N_\beta(x) + N_E(x) \qquad \forall\, x \in W_0^{1,p}(\Omega).$$

PROPOSITION 4.5.2
*If hypotheses H_0, $H(a)_1$, $H(j)_1$, $H(f)_1$ and $H(\vartheta)$ hold,
then U_ϱ is pseudomonotone and for $\varrho > 0$ large enough, U_ϱ is coercive.*

PROOF   Because U_ϱ has values in $P_{wkc}(W^{-1,p'}(\Omega))$ and it is bounded,
it suffices to show that U_ϱ is generalized pseudomonotone (see Proposition 1.4.11). To this end, let

$$x_n \xrightarrow{w} x \text{ in } W_0^{1,p}(\Omega),$$
$$x_n^* \xrightarrow{w} x \text{ in } W^{-1,p'}(\Omega),$$

with $x_n^* \in U_\varrho(x_n)$ for $n \geq 1$ and suppose that

$$\limsup_{n \to +\infty} \langle x_n^*, x_n - x \rangle_{W_0^{1,p}(\Omega)} \leq 0.$$

We need to show that $x^* \in U_\varrho(x)$ and

$$\langle x_n^*, x_n \rangle_{W_0^{1,p}(\Omega)} \longrightarrow \langle x^*, x \rangle_{W_0^{1,p}(\Omega)}.$$

We have

$$x_n^* = v_n^* + N_{f,\tau}(x_n) + M\tau(x_n) + \varrho N_\beta(x_n) + g_n \qquad \forall\, n \geq 1,$$

with $v_n^* \in V_1(x_n)$, $g_n \in N_E(x_n)$ for $n \geq 1$. Because of hypothesis $H(j)_1(iii)$,
we see that the sequence $\{g_n\}_{n \geq 1} \subseteq L^{r'}(\Omega)$ is bounded and so passing to a
subsequence if necessary, we may assume that

$$g_n \xrightarrow{w} g \text{ in } L^{r'}(\Omega).$$

From the compactness of the embedding $W_0^{1,p}(\Omega) \subseteq L^p(\Omega)$ and the continuity
of τ, we have that

$$x_n \longrightarrow x \quad \text{in } L^p(\Omega),$$
$$\tau(x_n) \longrightarrow \tau(x) \text{ in } L^p(\Omega).$$

Moreover, by passing to a further subsequence if necessary, we may assume
that

$$x_n(z) \longrightarrow x(z) \qquad \text{for a.a. } z \in \Omega,$$
$$\tau(x_n)(z) \longrightarrow \tau(x)(z) \text{ for a.a. } z \in \Omega.$$

Using Proposition 1.2.12, we have

$$\begin{aligned}
g(z) &\in \text{conv} \limsup_{n \to +\infty} \left[\partial j\big(z, \tau(x_n)(z)\big) \cap Q\big(z, x_n(z)\big) \right] \\
&\subseteq \text{conv} \left[\limsup_{n \to +\infty} \partial j\big(z, \tau(x_n)(z)\big) \cap \limsup_{n \to +\infty} Q\big(z, x_n(z)\big) \right] \\
&\subseteq \text{conv} \left[\partial j\big(z, \tau(x)(z)\big) \cap Q\big(z, x(z)\big) \right] \quad \text{for a.a. } z \in \Omega.
\end{aligned}$$

The last inclusion is a consequence of the fact that both the multifunctions $\zeta \longmapsto \partial j(z, \zeta)$ and $\zeta \longrightarrow Q(z, \zeta)$ have a closed graph. So we infer that $g \in G(x)$.

Also, because of the compactness of the embedding $W_0^{1,p}(\Omega) \subseteq L^r(\Omega)$ (recall that $r < p^*$) we have

$$x_n \longrightarrow x \quad \text{in } L^r(\Omega).$$

Therefore

$$\langle g_n, x_n - x \rangle_{W_0^{1,p}(\Omega)} = \int_\Omega g_n(z)(x_n - x)(z)\, dz \longrightarrow 0.$$

Similarly since

$$x_n \longrightarrow x \quad \text{in } L^p(\Omega)$$

and the sequences $\{N_\beta(x_n)\}_{n\geq 1}, \{N_{f,\tau}(x_n)\}_{n\geq 1} \subseteq L^{p'}(\Omega)$ are bounded, we have

$$\langle N_\beta(x_n), x_n - x \rangle_{W_0^{1,p}(\Omega)} \longrightarrow 0,$$
$$\langle N_{f,\tau}(x_n), x_n - x \rangle_{W_0^{1,p}(\Omega)} \longrightarrow 0.$$

So it follows that

$$\limsup_{n\to+\infty} \langle v_n^*, x_n - x \rangle_{W_0^{1,p}(\Omega)} \leq 0.$$

But from Proposition 4.5.1 we know that V_1 is of type $(S)_+$. Hence we deduce that

$$x_n \longrightarrow x \quad \text{in } W_0^{1,p}(\Omega)$$

and so

$$N_{f,\tau}(x_n) \longrightarrow N_{f,\tau}(x) \quad \text{in } L^{p'}(\Omega) \text{ (and in } W^{-1,p'}(\Omega) \text{ too)}.$$

Also

$$N_\beta(x_n) \longrightarrow N_\beta(x) \quad \text{in } W^{-1,p'}(\Omega).$$

By definition we have

$$v_n^* = -\operatorname{div} v_n \quad \forall\, n \geq 1,$$

with $v_n \in S_{a(\cdot,\tau(x_n)(\cdot),\nabla x_n(\cdot))}^{p'}$ and

$$v_n \longrightarrow v \quad \text{in } L^{p'}(\Omega; \mathbb{R}^N).$$

Hence

$$v_n^* = -\operatorname{div} v_n \xrightarrow{\ w\ } -\operatorname{div} v = v^* \quad \text{in } W^{-1,p'}(\Omega).$$

Because

$$\tau(x_n) \longrightarrow \tau(x) \quad \text{in } W_0^{1,p}(\Omega)$$

(continuity of τ) and since $\mathrm{Gr}\, a(z, \cdot, \cdot)$ is closed in $\mathbb{R} \times \mathbb{R}^N \times \mathbb{R}^N$ (see hypothesis $H(a)_1(iii)$), as before with the help of Proposition 1.2.12, we obtain

$$v(z) \in \mathrm{conv} \limsup_{n \to +\infty} a\big(z, \tau(x_n)(z), \nabla x(z)\big)$$

$$\subseteq a\big(z, \tau(x)(z), \nabla x(z)\big) \quad \text{for a.a. } z \in \Omega,$$

i.e. $v^* \in V_1(x)$. Thus finally in the limit as $n \to +\infty$, we have

$$x^* = v^* + N_{f,\tau}(x) + M\tau(x) + \varrho N_\beta(x) + g,$$

with $v^* \in V_1(x)$, i.e. $x^* \in U_\varrho(x)$.

Also it is clear from the above arguments that

$$\langle x_n^*, x_n \rangle_{W_0^{1,p}(\Omega)} \longrightarrow \langle x^*, x \rangle_{W_0^{1,p}(\Omega)}.$$

This proves the pseudomonotonicity of U_ϱ.

Next we show that for $\varrho > 0$ large enough, the operator U_ϱ is coercive. Let $x \in W_0^{1,p}(\Omega)$ and $x^* \in U_\varrho$. We have

$$\langle x^*, x \rangle_{W_0^{1,p}(\Omega)} = \langle v^*, x \rangle_{W_0^{1,p}(\Omega)} + \langle N_{f,\tau}(x), x \rangle_{W_0^{1,p}(\Omega)} + \langle M\tau(x), x \rangle_{W_0^{1,p}(\Omega)}$$
$$+ \varrho \langle N_\beta(x), x \rangle_{W_0^{1,p}(\Omega)} + \langle g, x \rangle_{W_0^{1,p}(\Omega)}. \tag{4.116}$$

From hypothesis $H(a)_1(v)$ we know that

$$\langle v^*, x \rangle_{W_0^{1,p}(\Omega)} \geq \eta_1 \|\nabla x\|_p^p - \eta_2 |\Omega|_N. \tag{4.117}$$

Also from hypothesis $H(f)_1(iii)$ we have that

$$\langle N_{f,\tau}(x), x \rangle_{W_0^{1,p}(\Omega)} \geq -c_7 \|x\|_{W^{1,p}(\Omega)}^{p-1} \|x\|_p - c_8 \|x\|_p - c_9,$$

for some $c_7, c_8, c_9 > 0$. Using Young's inequality with $\varepsilon > 0$ we obtain

$$\|x\|_{W^{1,p}(\Omega)}^{p-1} \|x\|_p \leq \frac{1}{\varepsilon^p p} \|x\|_p^p + \frac{\varepsilon^{p'}}{p'} \|x\|_{W^{1,p}(\Omega)}^p$$

(recall that $p - 1 = \frac{p}{p'}$). So it follows that

$$\langle N_{f,\tau}(x), x \rangle_{W_0^{1,p}(\Omega)} \geq -c_7 \frac{1}{\varepsilon^p p} \|x\|_{W^{1,p}(\Omega)}^{p-1} - c_7 \frac{\varepsilon^{p'}}{p'} \|x\|_p - c_8 \|x\|_p - c_9. \tag{4.118}$$

From the properties of the penalty function $\beta(z, \zeta)$ we have

$$\varrho \langle N_\beta(x), x \rangle_{W_0^{1,p}(\Omega)} \geq \varrho c_5 \|x\|_p^p - \varrho c_6. \tag{4.119}$$

Finally because of hypothesis $H(j)_1(iii)$ we have

$$\langle g, x \rangle_{W_0^{1,p}(\Omega)} \geq -c_{10} \|x\|_{W^{1,p}(\Omega)}, \tag{4.120}$$

for some $c_{10} > 0$. Using (4.117), (4.118), (4.119) and (4.120) in (4.116) and exploiting the Poincaré inequality, we have

$$\langle x^*, x \rangle_{W_0^{1,p}(\Omega)} \geq \left(c_{11} - c_7 \frac{\varepsilon^{p'}}{p'} \right) \|x\|_{W^{1,p}(\Omega)}^p + \left(\varrho c_5 - \frac{c_7}{\varepsilon^p p} \right) \|x\|_p^p$$
$$- c_{10} \|x\|_{W^{1,p}(\Omega)} - c_{12}(\varrho), \qquad (4.121)$$

for some $c_{11}, c_{12}(\varrho) > 0$. First we choose $\varepsilon > 0$ small so that $c_{11} > c_7 \frac{\varepsilon^{p'}}{p'}$. Then with this choice of $\varepsilon > 0$, we choose $\varrho > 0$ large enough so that $\varrho c_5 > \frac{c_7}{\varepsilon^p p}$. With these choices, from (4.120) we infer that U_ϱ is coercive. ☐

Having this Proposition, we can now solve the auxiliary problem (4.115).

PROPOSITION 4.5.3
If hypotheses H_0, $H(a)_1$, $H(j)_1$, $H(f)_1$ and $H(\vartheta)$ hold,
then for $\varrho > 0$ large enough, problem (4.115) has a solution $x_0 \in W_0^{1,p}(\Omega)$.

PROOF From Proposition 4.5.2 we know that for $\varrho > 0$ large enough, U_ϱ is pseudomonotone and coercive. Hence it is surjective (see Theorem 1.4.6). Therefore we can find $x_0 \in W_0^{1,p}(\Omega)$ such that

$$U_\varrho(x_0) = \vartheta(w) + Mw \in L^{p'}(\Omega) \subseteq W^{-1,p'}(\Omega)$$

(recall that $2 \leq p$). Clearly this is a solution of (4.115). ☐

With the help of the auxiliary problem (4.115), now we shall solve the original problem (4.114). This will be done with the use of the following fixed point theorem of Heikkilä & Hu (1993), where the interested reader can find its proof.

THEOREM 4.5.1
If X is a separable, reflexive ordered Banach space, $K \subseteq X$ is a nonempty and weakly closed subset and $S \colon K \longrightarrow 2^K \setminus \{\emptyset\}$ is a multifunction with weakly closed values, $S(K)$ is bounded and

(i) _the set_
$$M \overset{df}{=} \left\{ x \in K : \ x \leq y \text{ for some } y \in S(x) \right\}$$

is nonempty and

(ii) _if $x_1 \leq y_1$, $y_1 \in S(x_1)$ and $x_1 \leq x_2$, then we can find $y_2 \in S(x_2)$ such that $y_1 \leq y_2$,_

then S has a fixed point, i.e. there exists $x \in K$ such that $x \in S(x)$.

Using this fixed point theorem and the auxiliary problem (4.115), we can prove the following existence theorem for problem (4.114).

THEOREM 4.5.2
If hypotheses H_0, $H(a)_1$, $H(j)_1$, $H(f)_1$ *and* $H(\vartheta)$ *hold,*
<u>then</u> *problem (4.114) has a solution* $x_0 \in W_0^{1,p}(\Omega)$.

PROOF   Let

$$K \overset{df}{=} [\underline{\psi}, \overline{\psi}] = \left\{ x \in W^{1,p}(\Omega) : \underline{\psi}(z) \leq x(z) \leq \overline{\psi}(z) \text{ for a.a. } z \in \Omega \right\}.$$

Clearly K is weakly closed in $W^{1,p}(\Omega)$. We consider the multifunction $S \colon K \longrightarrow 2^{W_0^{1,p}(\Omega)} \setminus \{\emptyset\}$ which to each $w \in K$ assigns the set of solutions of the auxiliary problem (4.115). From Proposition 4.5.3 we know that

$$S(w) \neq \emptyset \qquad \forall\, w \in K.$$

Moreover, arguing as in the proof of Proposition 4.5.2, we can check that for each $w \in K$, the set $S(w) \subseteq W^{1,p}(\Omega)$ is weakly closed.

Claim 1. $S(K) \subseteq K$.

Let $w \in K$ and $x \in S(w)$. We have

$$v^* + N_{f,\tau}(x) + M\tau(x) + \varrho N_\beta(x) + g = \vartheta(w) + Mw, \tag{4.122}$$

with $v^* \in V_1(x)$ (hence $v^* = -\operatorname{div} v$ with $v \in S^{p'}_{a(\cdot,\tau(x)(\cdot),\nabla x(\cdot))}$) and $g \in N_E(x)$. Since $\underline{\psi} \in W^{1,p}(\Omega)$ is a lower solution for problem (4.114), using as a test function $(\underline{\psi} - x)^+ \in W_0^{1,p}(\Omega)$ (recall that $\underline{\psi}|_{\partial\Omega} \leq 0$), we have

$$\int_\Omega \left(v_-(z), \nabla(\underline{\psi} - x)^+(z) \right)_{\mathbb{R}^N} dz + \int_\Omega f(z, \underline{\psi}(z), \nabla\underline{\psi}(z))(\underline{\psi} - x)^+(z)\, dz$$
$$+ \int_\Omega u_-(z)(\underline{\psi} - x)^+(z)\, dz \leq \int_\Omega \vartheta(\underline{\psi})(z)(\underline{\psi} - x)^+(z)\, dz, \tag{4.123}$$

where $v_-^* = -\operatorname{div} v_-$ with $v_- \in S^{p'}_{a(\cdot,\underline{\psi}(\cdot),\nabla\underline{\psi}(\cdot))}$.
   Also if on (4.122) we act with $(\underline{\psi} - x)^+$, we obtain

$$\int_\Omega \left(v(z), \nabla(\underline{\psi} - x)^+(z) \right)_{\mathbb{R}^N} dz + \int_\Omega f(z, \tau(x)(z), \nabla\tau(x)(z))(\underline{\psi} - x)^+(z)\, dz$$
$$+ \int_\Omega \varrho\beta(z, x(z))(\underline{\psi} - x)^+(z)\, dz + \int_\Omega g(z)(\underline{\psi} - x)^+(z)\, dz$$

$$= \int_{\Omega} \vartheta(w)(z)(\underline{\psi} - x)^+(z)\,dz + \int_{\Omega} M\big(w - \tau(x)\big)(z)(\underline{\psi} - x)^+(z)\,dz. \quad (4.124)$$

Subtracting (4.124) from (4.123) and using the definitions of τ, β, N_E and the monotonicity of $a\big(z, \underline{\psi}(z), \cdot\big)$ (see hypothesis $H(a)_1(ii)$) and hypotheses $H(\vartheta)$, we obtain

$$-\varrho \int_{\Omega} \beta\big(z, x(z)\big)(\underline{\psi} - x)^+(z)\,dz$$

$$\leq \int_{\{\underline{\psi} > x\}} \big(\vartheta(\underline{\psi}) + M\underline{\psi} - \vartheta(w) - Mw\big)(z)(\underline{\psi} - x)(z)\,dz \; \leq \; 0$$

and so

$$- \int_{\{\underline{\psi} > x\}} \big(|x(z)|^{p-2}x(z) - |\underline{\psi}(z)|^{p-2}\underline{\psi}(z)\big)(\underline{\psi} - x)(z)\,dz \; \leq \; 0,$$

a contradiction unless $\underline{\psi}(z) \leq x(z)$ for almost all $z \in \Omega$. Similarly we can show that $x(z) \leq \overline{\psi}(z)$ for almost all $z \in \Omega$. This proves Claim 1.

Claim 2. If $w_1 \in K$, $w_1 \leq x_1$, $x_1 \in S(w_1)$ and $w_1 \leq w_2$, then we can find $x_2 \in S(w_2)$ such that $x_1 \leq x_2$ (recall that in $W^{1,p}(\Omega)$ we consider the partial order induced by the positive cone $L^p(\Omega)_+$, i.e. the pointwise partial order).

Since $x_1 \in S(w_1) \subseteq K$ (see Claim 1), we have

$$N_\beta(x_1) = 0, \quad \tau(x_1) = x_1, \quad \nabla\tau(x_1) = \nabla x_1 \quad \text{and} \quad Q\big(z, x_1(z)\big) = \mathbb{R}.$$

So we can write that

$$v_1^* + N_f(x_1) + Mx_1 + g = \vartheta(w_1) + Mw_1,$$

with $v_1^* = -\mathrm{div}\,v_1$, $v_1 \in S^{p'}_{a(\cdot, x_1(\cdot), \nabla x_1(\cdot))}$, $N_f(x_1)(\cdot) \overset{df}{=} f\big(\cdot, x_1(\cdot), \nabla x_1(\cdot)\big)$ and $g \in S^{r'}_{\partial j(\cdot, x_1(\cdot))}$. Since $w_1 \leq w_2$, from hypotheses $H(\vartheta)$ we have

$$\vartheta\big(w_1(z)\big) + Mw_1(z) \; \leq \; \vartheta\big(w_2(z)\big) + Mw_2(z) \quad \text{for a.a. } z \in \Omega.$$

So for all $y \in W_0^{1,p}(\Omega)_+$ we have

$$\langle v_1^*, y \rangle_{W_0^{1,p}(\Omega)} + \int_{\Omega} f\big(z, x_1(z), \nabla x_1(z)\big)y(z)\,dz$$

$$+ M \int_{\Omega} x_1(z)y(z)\,dz + \int_{\Omega} g(z)y(z)\,dz$$

$$\leq \int_{\Omega} \big(\vartheta(w_2) + Mw_2\big)(z)y(z)\,dz,$$

from which we infer that $x_1 \in W_0^{1,p}(\Omega)$ is a lower solution for problem

$$\begin{cases} -\operatorname{div} a\big(z, x(z), \nabla x(z)\big) + f\big(z, x(z), \nabla x(z)\big) \\ \quad + Mx(z) + \partial j\big(z, x(z)\big) \\ \quad \ni \vartheta\big(w_2(z)\big) + Mw_2(z) \quad \text{for a.a. } z \in \Omega, \\ x|_{\partial\Omega} = 0 \end{cases} \tag{4.125}$$

(see Definition 4.5.1(b)). Note that $\overline{\psi} \in W^{1,p}(\Omega)$ is an upper solution of (4.125) too (see Definition 4.5.1(a) and hypotheses $H(\vartheta)$). So arguing as in the proof of Proposition 4.5.3 and the proof of Claim 1, we can produce a solution $x_2 \in W_0^{1,p}(\Omega)$ of problem (4.125) such that

$$x_1(z) \leq x_2(z) \leq \overline{\psi}(z) \quad \text{for a.a. } z \in \Omega.$$

Therefore $x_2 \in S(w_2)$ and $x_1 \leq x_2$. This proves Claim 2.

Because of Claims 1 and 2, we can apply Theorem 4.5.1 with data $X \overset{df}{=} W^{1,p}(\Omega)$, $K \overset{df}{=} [\underline{\psi}, \overline{\psi}]$ and S, the solution multifunction defined in the beginning of the proof. Note that by virtue of the coercivity of U_ϱ, the set $S(K) \subseteq W_0^{1,p}(\Omega)$ is bounded. We obtain $x_0 \in W_0^{1,p}(\Omega) \cap K$ such that $x_0 \in S(x_0)$. Evidently this is a solution of (4.114). $\qquad\Box$

4.5.2   Existence of Extremal Solutions

Now that we have established the existence of at least one solution for problem (4.114) in the order interval $[\underline{\psi}, \overline{\psi}]$, we ask the question of whether among all these solutions, there is the greatest and the smallest solution for the pointwise ordering on $W_0^{1,p}(\Omega)$. Such solutions, if they exist, are known as ***extremal solutions*** of problem (4.114) in the order interval $[\underline{\psi}, \overline{\psi}]$. We shall produce extremal solutions for a particular case of problem (4.114) with suitable monotonicity structure (variational inequality). More precisely, the problem under consideration is the following:

$$\begin{cases} -\operatorname{div} a\big(z, x(z), \nabla x(z)\big) \in \partial j\big(z, x(z)\big) + f\big(z, x(z)\big) \\ \qquad\qquad\qquad\qquad\qquad\qquad\qquad \text{for a.a. } z \in \Omega, \\ x|_{\partial\Omega} = 0. \end{cases} \tag{4.126}$$

Our hypotheses on the data of (4.126) are the following:

$\underline{H(a)_2}$ $a\colon \Omega \times \mathbb{R} \times \mathbb{R}^N \longrightarrow \mathbb{R}^N$ is a function, such that

       (i) for every $(\zeta, \xi) \in \mathbb{R} \times \mathbb{R}^N$, the function

$$\Omega \ni z \longmapsto a(z, \zeta, \xi) \in \mathbb{R}^N$$

       is measurable;

(ii) for almost all $z \in \Omega$, the function

$$\mathbb{R} \times \mathbb{R}^N \ni (\zeta, \xi) \longmapsto a(z, \zeta, \xi) \in \mathbb{R}^N$$

is continuous;

(iii) for almost all $z \in \Omega$ and all $\zeta \in \mathbb{R}$, the function

$$\mathbb{R}^N \ni \xi \longmapsto a(z, \zeta, \xi) \in \mathbb{R}^N$$

is monotone;

(iv) for almost all $z \in \Omega$, all $\zeta, \zeta' \in \mathbb{R}$ and all $\xi \in \mathbb{R}^N$ we have

$$\|a(z, \zeta, \xi)\|_{\mathbb{R}^N} \ \leq \ b_1(z) + c_1 \left(|\zeta|^{p-1} + \|\xi\|_{\mathbb{R}^N}^{p-1} \right),$$

with $b_1 \in L^{p'}(\Omega)_+$, $c_1 > 0$, $p \in [2, +\infty)$, $\frac{1}{p} + \frac{1}{p'} = 1$ and

$$\|a(z, \zeta, \xi) - a(z, \zeta', \xi)\|_{\mathbb{R}^N}$$
$$\leq \ \left[\widehat{c}(|\zeta| + |\zeta'| + \|\xi\|_{\mathbb{R}^N})^{p-1} + k(z) \right] |\zeta - \zeta'|,$$

with $\widehat{c} > 0$, $k \in L^{p'}(\Omega)$;

(v) for almost all $z \in \Omega$, all $\zeta \in \mathbb{R}$ and all $\xi \in \mathbb{R}^N$ we have

$$\left(a(z, \zeta, \xi), \xi \right)_{\mathbb{R}^N} \ \geq \ \eta_1 \|\xi\|_{\mathbb{R}^N}^p - \eta_2,$$

with $\eta_1, \eta_2 > 0$.

$\underline{H(j)_2}$ $j \colon \Omega \times \mathbb{R} \longrightarrow \mathbb{R}$ is a function such that

(i) for all $\zeta \in \mathbb{R}$, the function

$$\Omega \ni z \longmapsto j(z, \zeta) \in \mathbb{R}$$

is measurable;

(ii) for almost all $z \in \Omega$, the function

$$\mathbb{R} \ni \zeta \longmapsto j(z, \zeta) \in \mathbb{R}$$

is concave;

(iii) for almost all $z \in \Omega$, all $\zeta \in \left[\underline{\psi}(z), \overline{\psi}(z) \right]$ and all $u \in \partial j(z, \zeta)$, we have that

$$|u| \ \leq \ \beta(z),$$

with $\beta \in L^{r'}(\Omega)$.

REMARK 4.5.2   By virtue of hypothesis $H(j)_2(ii)$ for almost all $z \in \Omega$, the function $j(z, \cdot)$ is locally Lipschitz and so the generalized subdifferential in (4.126) makes sense and can also be interpreted as

$$\partial j(z, \zeta) = \left\{ u \in \mathbb{R} : \ j(z, \zeta') - j(z, \zeta) \leq u(\zeta - \zeta') \text{ for all } \zeta' \in \mathbb{R} \right\}$$

(concave subdifferential). In particular therefore, for almost all $z \in \Omega$, the multifunction $\mathbb{R} \ni x \longmapsto -\partial j(z, \zeta)$ is maximal monotone.   ⬚

$\underline{H(f)_2}$ $f : \Omega \times \mathbb{R} \longrightarrow \mathbb{R}$ is a function such that

(i) for all $\zeta \in \mathbb{R}$, the function

$$\Omega \ni z \longmapsto f(z, \zeta) \in \mathbb{R}$$

is measurable;

(ii) for almost all $z \in \Omega$, the function

$$\mathbb{R} \ni \zeta \longmapsto f(z, \zeta) \in \mathbb{R}$$

is continuous and nonincreasing;

(iii) we have $f\big(\cdot, \underline{\psi}(\cdot)\big), f\big(\cdot, \overline{\psi}(\cdot)\big) \in L^{p'}(\Omega)$.

REMARK 4.5.3   We have

$$f\big(\cdot, x(\cdot)\big) \in L^{p'}(\Omega) \qquad \forall\, x \in K,$$

where

$$K \overset{df}{=} [\underline{\psi}, \overline{\psi}] = \left\{ x \in W^{1,p}(\Omega) : \ \underline{\psi}(z) \leq x(z) \leq \overline{\psi}(z) \text{ for a.a. } z \in \Omega \right\}.$$

⬚

Our analysis of problem (4.126) begins with a result which shows that the set of solutions of (4.126) exhibits a lattice structure.

PROPOSITION 4.5.4
If hypotheses $H(a)_2$, $H(j)_2$ and $H(f)_2$ hold and $x, y \in W_0^{1,p}(\Omega)$ are solutions of (4.126),
then $v = \min\{x, y\} \in W_0^{1,p}(\Omega)$ and $w = \max\{x, y\} \in W_0^{1,p}(\Omega)$ are both solutions of (4.126).

PROOF   Since $x, y \in W_0^{1,p}(\Omega)$ are solutions of (4.126), we can find $u_1, u_2 \in L^{p'}(\Omega)$ such that

$$-\text{div } a\big(z, x(z), \nabla x(z)\big) = u_1(z) + f\big(z, x(z)\big) \quad \text{for a.a. } z \in \Omega, \qquad (4.127)$$

with $u_1(z) \in \partial j\big(z, x(z)\big)$ for almost all $z \in \Omega$,

$$-\text{div } a\big(z, y(z), \nabla y(z)\big) = u_2(z) + f\big(z, y(z)\big) \quad \text{for a.a. } z \in \Omega, \qquad (4.128)$$

with $u_2(z) \in \partial j\big(z, y(z)\big)$ for almost all $z \in \Omega$.

First we show that for every $\vartheta \in C_0^1(\Omega)$ we have

$$\int\limits_{\{y < x\}} \big(a\big(z, x(z), \nabla x(z)\big), \nabla \vartheta(z)\big)_{\mathbb{R}^N} dz$$

$$- \int\limits_{\{y < x\}} u_1(z)\vartheta(z)\, dz - \int\limits_{\{y < x\}} f\big(z, x(z)\big)\vartheta(z)\, dz$$

$$= \int\limits_{\{y < x\}} \big(a\big(z, y(z), \nabla y(z)\big), \nabla \vartheta(z)\big)_{\mathbb{R}^N} dz$$

$$- \int\limits_{\{y < x\}} u_2(z)\vartheta(z)\, dz - \int\limits_{\{y < x\}} f\big(z, y(z)\big)\vartheta(z)\, dz. \qquad (4.129)$$

To show this, for every $\varepsilon > 0$ we introduce the truncation function $\sigma_\varepsilon : \mathbb{R} \longrightarrow \mathbb{R}$ defined by

$$\sigma_\varepsilon(t) \overset{df}{=} \begin{cases} \varepsilon & \text{if } \varepsilon < t, \\ t & \text{if } |t| \leq \varepsilon, \\ -\varepsilon & \text{if } t < -\varepsilon. \end{cases}$$

Clearly σ_ε is Lipschitz continuous and $\sigma_\varepsilon\big((x - y)^+(\cdot)\big)\vartheta(\cdot) \in W_0^{1,p}(\Omega)$ (see Theorem 1.1.12 and Remark 1.1.10). So using this as a test function, we obtain

$$0 = \int\limits_{\Omega} \big(a\big(z, x(z), \nabla x(z)\big) - a\big(z, y(z), \nabla y(z)\big), \nabla\big(\sigma_\varepsilon\big((x-y)^+(z)\big)\vartheta(z)\big)\big)_{\mathbb{R}^N} dz$$

$$- \int\limits_{\Omega} (u_1 - u_2)(z)\sigma_\varepsilon\big((x - y)^+(z)\big)\vartheta(z)\, dz$$

$$- \int\limits_{\Omega} \big(f\big(z, x(z)\big) - f\big(z, y(z)\big)\big)\sigma_\varepsilon\big((x - y)^+(z)\big)\vartheta(z)\, dz.$$

Recall that

$$\nabla\big(\sigma_\varepsilon\big((x - y)^+(\cdot)\big)\vartheta(\cdot)\big) = \nabla\big(\sigma_\varepsilon\big((x - y)^+(\cdot)\big)\big)\vartheta(\cdot) + \sigma_\varepsilon\big((x - y)^+(\cdot)\big)\nabla\vartheta(\cdot)$$

$$= \begin{cases} \vartheta\nabla(x - y) + (x - y)\nabla\vartheta & \text{on } \{0 < x - y < \varepsilon\}, \\ \varepsilon\nabla\vartheta & \text{on } \{\varepsilon \leq x - y\}, \\ 0 & \text{on } \{x - y \leq 0\} \end{cases}$$

(see Remark 1.1.10). So we obtain

$$0 = \int_{\Omega} \left(a\big(z, x(z), \nabla x(z)\big) - a\big(z, y(z), \nabla y(z)\big), \nabla \big(x - y\big)(z) \right)_{\mathbb{R}^N} \vartheta(z)\, dz$$

$$+ \int_{\Omega} \left(a\big(z, x(z), \nabla x(z)\big) - a\big(z, y(z), \nabla y(z)\big), \nabla \vartheta(z) \right)_{\mathbb{R}^N} \sigma_\varepsilon \big((x - y)^+(z)\big)\, dz$$

$$- \int_{\Omega} (u_1 - u_2)(z)\sigma_\varepsilon \big((x - y)^+(z)\big)\vartheta(z)\, dz$$

$$- \int_{\Omega} \big(f(z, x(z)) - f(z, y(z))\big)\sigma_\varepsilon \big((x - y)^+(z)\big)\vartheta(z)\, dz.$$

Dividing with $\varepsilon > 0$, we get

$$\frac{1}{\varepsilon} \int_{\{0 < x - y < \varepsilon\}} \left(a\big(z, x(z), \nabla x(z)\big) - a\big(z, y(z), \nabla y(z)\big), \nabla \big(x - y\big)(z) \right)_{\mathbb{R}^N} \vartheta(z)\, dz$$

$$= \int_{\Omega} \left(a\big(z, y(z), \nabla y(z)\big) - a\big(z, x(z), \nabla x(z)\big), \nabla \vartheta(z) \right)_{\mathbb{R}^N} \frac{\sigma_\varepsilon \big((x - y)^+(z)\big)}{\varepsilon}\, dz$$

$$+ \int_{\Omega} (u_1 - u_2)(z)\frac{\sigma_\varepsilon \big((x - y)^+(z)\big)\vartheta(z)}{\varepsilon}\, dz$$

$$+ \int_{\Omega} \big(f(z, x(z)) - f(z, y(z))\big)\frac{\sigma_\varepsilon \big((x - y)^+(z)\big)\vartheta(z)}{\varepsilon}\, dz. \qquad (4.130)$$

We estimate the left hand side of (4.130). Acting on (4.127) and (4.128) with the test function $\sigma_\varepsilon \big((x - y)^+(\cdot)\big) \in W_0^{1,p}(\Omega)$ and then subtracting, we obtain

$$\int_{\Omega} \left(a\big(z, x(z), \nabla x(z)\big) - a\big(z, y(z), \nabla y(z)\big), \nabla \sigma_\varepsilon \big((x - y)^+(z)\big) \right)_{\mathbb{R}^N}\, dz$$

$$= \int_{\{0 < x - y < \varepsilon\}} \left(a\big(z, x(z), \nabla x(z)\big) - a\big(z, y(z), \nabla y(z)\big), \nabla \big(x - y\big)(z) \right)_{\mathbb{R}^N}\, dz$$

$$= \int_{\Omega} (u_1 - u_2)(z)\sigma_\varepsilon \big((x - y)^+(z)\big)\, dz$$

$$+ \int_{\Omega} \big(f(z, x(z)) - f(z, y(z))\big)\sigma_\varepsilon \big((x - y)^+(z)\big)\, dz.$$

Note that $-\partial j(z, \varsigma) = \partial(-j)(z, \varsigma)$ (see Proposition 1.3.13(a)). So the function $x \longmapsto -\partial j(z, \varsigma)$ is maximal monotone (see hypothesis $H(j)_2(ii)$). Therefore we have

$$\int_{\Omega} (u_1 - u_2)(z)\sigma_\varepsilon \big((x - y)^+(z)\big)\, dz \leq 0.$$

Similarly, by virtue of hypothesis $H(j)_2(ii)$ we have

$$\int_\Omega (f(z, x(z)) - f(z, y(z)))\sigma_\varepsilon((x - y)^+(z))\, dz \ \leq\ 0.$$

Thus we can write that

$$\int_{\{0 < x - y < \varepsilon\}} (a(z, x(z), \nabla x(z)) - a(z, y(z), \nabla y(z)), \nabla(x - y)(z))_{\mathbb{R}^N}\, dz \ \leq\ 0$$

and so

$$\int_{\{0 < x - y < \varepsilon\}} (a(z, x(z), \nabla x(z)) - a(z, x(z), \nabla y(z)), \nabla(x - y)(z))_{\mathbb{R}^N}\, dz$$

$$\leq \int_{\{0 < x - y < \varepsilon\}} (a(z, y(z), \nabla y(z)) - a(z, x(z), \nabla y(z)), \nabla(x - y)(z))_{\mathbb{R}^N}\, dz.$$

$$(4.131)$$

Using hypothesis $H(a)_2(iv)$ we have

$$\left| \int_{\{0 < x - y < \varepsilon\}} (a(z, y(z), \nabla y(z)) - a(z, x(z), \nabla y(z)), \nabla(x - y)(z))_{\mathbb{R}^N}\, dz \right|$$

$$\leq \int_{\{0 < x - y < \varepsilon\}} \left[c_2(|x(z)| + |y(z)| + \|\nabla y(z)\|_{\mathbb{R}^N})^{p-2} + k(z) \right] \times$$

$$\times |x(z) - y(z)|\, \|\nabla(x - y)(z)\|_{\mathbb{R}^N}\, dz$$

$$\leq \varepsilon \int_{\{0 < x - y < \varepsilon\}} \mu(z)\, \|\nabla(x - y)(z)\|_{\mathbb{R}^N}\, dz, \qquad (4.132)$$

with

$$\mu(\cdot) \overset{df}{=} c_2(|x(\cdot)| + |y(\cdot)| + \|\nabla y(\cdot)\|_{\mathbb{R}^N})^{p-2} + k(\cdot) \in L^{p'}(\Omega).$$

Use this inequality in (4.131), note that the left hand side of (4.131) is non-negative (see hypothesis $H(a)_2(iii)$) and divide by $\varepsilon > 0$, to obtain

$$0 \leq \frac{1}{\varepsilon} \int_{\{0 < x - y < \varepsilon\}} (a(z, x(z), \nabla x(z)) - a(z, x(z), \nabla y(z)), \nabla(x - y)(z))_{\mathbb{R}^N}\, dz$$

$$\leq \int_{\{0 < x - y < \varepsilon\}} \mu(z)\, \|\nabla(x - y)\|_{\mathbb{R}^N}\, dz.$$

Note that as $\varepsilon \searrow 0$,

$$\{0 < x - y < \varepsilon\} \ \longrightarrow\ \{x = y\}$$

and

$$\nabla(x - y)(z) = 0 \quad \text{for a.a. } z \in \{x = y\}$$

(see Remark 1.1.10). Therefore, it follows that

$$\lim_{\varepsilon \searrow 0} \frac{1}{\varepsilon} \int\limits_{\{0 < x - y < \varepsilon\}} \left(a(z, x(z), \nabla x(z)) - a(z, x(z), \nabla y(z)), \nabla(x - y)(z) \right)_{\mathbb{R}^N} dz$$
$$= 0. \quad (4.133)$$

Also note that

$$\frac{1}{\varepsilon} \sigma_\varepsilon \left((x - y)^+(z) \right) \longrightarrow \chi_{\{y < x\}}(z) \quad \text{for a.a. } z \in \Omega \quad \text{as } \varepsilon \searrow 0$$

and for all $\varepsilon > 0$ we have

$$0 \leq \frac{1}{\varepsilon} \sigma_\varepsilon \left((x - y)^+(z) \right) \leq 1 \quad \text{for a.a. } z \in \Omega.$$

So from the Lebesgue dominated convergence theorem we have

$$\int\limits_\Omega \left(a(z, y(z), \nabla y(z)) - a(z, x(z), \nabla x(z)), \nabla \vartheta(z) \right)_{\mathbb{R}^N} \frac{\sigma_\varepsilon \left((x - y)^+(z) \right)}{\varepsilon} dz$$
$$\longrightarrow \int\limits_{\{y < x\}} \left(a(z, y(z), \nabla y(z)) - a(z, x(z), \nabla x(z)), \nabla \vartheta(z) \right)_{\mathbb{R}^N} dz, \quad (4.134)$$

$$\int\limits_\Omega (u_1 - u_2)(z) \frac{\sigma_\varepsilon \left((x - y)^+(z) \right)}{\varepsilon} \vartheta(z) dz$$
$$\longrightarrow \int\limits_{\{y < x\}} (u_1 - u_2)(z) \vartheta(z) dz \quad (4.135)$$

and

$$\int\limits_\Omega \left(f(z, x(z)) - f(z, y(z)) \right) \frac{\sigma_\varepsilon \left((x - y)^+(z) \right)}{\varepsilon} \vartheta(z) dz$$
$$\longrightarrow \int\limits_{\{y < x\}} \left(f(z, x(z)) - f(z, y(z)) \right) \vartheta(z) dz \quad (4.136)$$

as $\varepsilon \searrow 0$. Returning to (4.130), passing to the limit as $\varepsilon \searrow 0$ and using (4.133),

(4.134), (4.135) and (4.136) we obtain

$$
0 = \int_{\{y<x\}} \big(a(z,y(z),\nabla y(z)) - a(z,x(z),\nabla x(z)), \nabla\vartheta(z)\big)_{\mathbb{R}^N}
$$
$$
+ \int_{\{y<x\}} (u_1 - u_2)(z)\vartheta(z)\,dz
$$
$$
+ \int_{\{y<x\}} \big(f(z,x(z)) - f(z,y(z))\big)\vartheta(z)\,dz.
$$

From this we obtain (4.129). In fact since the embedding $C_c^1(\Omega) \subseteq W_0^{1,p}(\Omega)$ is dense, it follows that (4.129) holds for all $\vartheta \in W_0^{1,p}(\Omega)$.

Let $v \stackrel{df}{=} \min\{x,y\} \in W_0^{1,p}(\Omega)$ and let us set

$$
\widehat{u} \stackrel{df}{=} \chi_{\{x\leq y\}} u_1 + \chi_{\{y<x\}} u_2 \in S^{p'}_{\partial j(\cdot,v(\cdot))}.
$$

Using (4.129), for all $\vartheta \in W_0^{1,p}(\Omega)$ we have

$$
\int_\Omega \big(a(z,v(z),\nabla v(z)),\nabla\vartheta(z)\big)_{\mathbb{R}^N} - \int_\Omega \widehat{u}(z)\vartheta(z)\,dz - \int_\Omega f(z,v(z))\vartheta(z)\,dz
$$
$$
= \int_{\{x\leq y\}} \big(a(z,x(z),\nabla x(z)),\nabla\vartheta(z)\big)_{\mathbb{R}^N} + \int_{\{y<x\}} \big(a(z,y(z),\nabla y(z)),\nabla\vartheta(z)\big)_{\mathbb{R}^N}
$$
$$
- \int_{\{x\leq y\}} u_1(z)\vartheta(z)\,dz - \int_{\{y<x\}} u_2(z)\vartheta(z)\,dz
$$
$$
- \int_{\{x\leq y\}} f(z,x(z))\vartheta(z)\,dz - \int_{\{y<x\}} f(z,y(z))\vartheta(z)\,dz
$$
$$
= \int_{\{x\leq y\}} \big(a(z,x(z),\nabla x(z)),\nabla\vartheta(z)\big)_{\mathbb{R}^N} + \int_{\{y<x\}} \big(a(z,x(z),\nabla x(z)),\nabla\vartheta(z)\big)_{\mathbb{R}^N}
$$
$$
- \int_{\{x\leq y\}} u_1(z)\vartheta(z)\,dz - \int_{\{y<x\}} u_1(z)\vartheta(z)\,dz
$$
$$
- \int_{\{x\leq y\}} f(z,x(z))\vartheta(z)\,dz - \int_{\{y<x\}} f(z,x(z))\vartheta(z)\,dz
$$
$$
= \int_\Omega \big(a(z,x(z),\nabla x(z)),\nabla\vartheta(z)\big)_{\mathbb{R}^N} - \int_\Omega u_1(z)\vartheta(z)\,dz
$$
$$
- \int_\Omega f(z,x(z))\vartheta(z)\,dz = 0
$$

so

$$\int_\Omega \big(a(z, v(z), \nabla v(z)), \nabla \vartheta(z)\big)_{\mathbb{R}^N} - \int_\Omega \widehat{u}(z)\vartheta(z)\,dz$$

$$- \int_\Omega f(z, v(z))\vartheta(z)\,dz \;=\; 0 \qquad \forall\, \vartheta \in W_0^{1,p}(\Omega)$$

and we infer that

$$\begin{cases} -\operatorname{div} a\big(z, v(z), \nabla v(z)\big) \;=\; \widehat{u}(z) + f\big(z, v(z)\big) \in \partial j\big(z, v(z)\big) + f\big(z, v(z)\big) \\ \hspace{7cm} \text{for a.a. } z \in \Omega, \\ v|_{\partial\Omega} = 0. \end{cases}$$

This proves that $v = \min\{x, y\} \in W_0^{1,p}(\Omega)$ is a solution of (4.126). In a similar fashion we show that $w = \max\{x, y\} \in W_0^{1,p}(\Omega)$ is a solution of (4.126).   ☐

Exploiting this lattice structure of the solution set of (4.126) we can establish the existence of extremal solutions on the order interval $K = [\underline{\psi}, \overline{\psi}]$.

THEOREM 4.5.3
If hypotheses H_0, $H(a)_2$, $H(j)_2$ and $H(f)_2$ hold,
<u>*then*</u> *problem (4.126) has extremal solutions in the order interval $K = [\underline{\psi}, \overline{\psi}]$.*

PROOF   Let

$$\mathcal{Y} \stackrel{df}{=} \left\{ x \in K = [\underline{\psi}, \overline{\psi}] : \; x \in W_0^{1,p}(\Omega) \text{ is a solution of (4.126)} \right\}.$$

From Theorem 4.5.2 we know that $\mathcal{Y} \neq \emptyset$ (note that in the present setting V is only pseudomonotone, but this is compensated by the fact that f is independent of the $\xi \in \mathbb{R}^N$-variable). Let $\mathcal{T}$ be a chain of the set in $\mathcal{Y}$ (i.e. $\mathcal{T}$ is a linearly (totally) ordered subset of $\mathcal{Y}$; recall that on $W_0^{1,p}(\Omega)$ we consider the pointwise ordering induced by $L^p(\Omega)_+$, i.e. $x \leq y$ if and only if $x(z) \leq y(z)$ for almost all $z \in \Omega$). Note that $\mathcal{Y}$ is order bounded in $L^p(\Omega)$ and so we can define $w = \sup \mathcal{T}$ in $L^p(\Omega)$. In fact we can find an increasing sequence $\{x_n\}_{n \geq 1}$ such that $w = \sup_{n \geq 1} x_n$, hence

$$x_n \longrightarrow w \quad \text{in } L^p(\Omega)$$

(monotone convergence theorem). By definition we have

$$-\operatorname{div} a\big(z, x_n(z), \nabla x_n(z)\big) \;=\; u_n(z) + f\big(z, x_n(z)\big) \quad \text{for a.a. } z \in \Omega,$$

with $u_n(z) \in \partial j\big(z, x_n(z)\big)$ for almost all $z \in \Omega$ and all $n \geq 1$.

By hypothesis $H(j)_2(iii)$ we have that

$$|u_n(z)| \leq \beta(z) \quad \text{for a.a. } z \in \Omega$$

and by hypotheses $H(f)_2(ii)$ and (iii) we have

$$|f(z, x_n(z))| \leq \max\left\{-f(z, \overline{\psi}(z)), f(z, \underline{\psi}(z))\right\} \quad \text{for a.a. } z \in \Omega \text{ and all } n \geq 1.$$

So multiplying with $x_n(z)$, integrating over Ω, using Green's identity, exploiting the above bounds and the fact that

$$|x_n(z)| \leq \max\left\{\overline{\psi}(z), -\underline{\psi}(z)\right\} \quad \text{for a.a. } z \in \Omega \text{ and all } n \geq 1$$

and using hypothesis $H(a)_2(v)$ we infer that the sequence $\{\nabla x_n\}_{n \geq 1} \subseteq L^p(\Omega; \mathbb{R}^N)$ is bounded and so the sequence $\{x_n\}_{n \geq 1} \subseteq W_0^{1,p}(\Omega)$ is bounded. Therefore it follows that

$$x_n \xrightarrow{w} w \quad \text{in } W_0^{1,p}(\Omega).$$

If as before $A: W_0^{1,p}(\Omega) \longrightarrow W^{-1,p'}(\Omega)$ is the nonlinear operator defined by

$$\langle A(x), y \rangle_{W_0^{1,p}(\Omega)} \stackrel{df}{=} \int_\Omega \left(a(z, x(z), \nabla x(z)), \nabla y(z)\right)_{\mathbb{R}^N} \quad \forall\, x, y \in W_0^{1,p}(\Omega),$$

we know that A is a monotone, demicontinuous (thus maximal monotone) and bounded operator. We have

$$A(x_n) = u_n + N_f(x_n) \quad \forall\, n \geq 1$$

(recall that $N_f(y)(\cdot) = f(\cdot, y(\cdot))$). Note that, by passing to a subsequence if necessary, we may assume that

$$u_n \xrightarrow{w} u \quad \text{in } L^{p'}(\Omega)$$

and because $\operatorname{Gr} \partial j(z, \cdot)$ is closed for almost all $z \in \Omega$, we have that $u \in S_{\partial j(\cdot, w(\cdot))}^{p'}$. Also

$$N_f(x_n) \longrightarrow N_f(w) \quad \text{in } L^{p'}(\Omega)$$

(see hypotheses $H(f)_2(ii)$ and (iii)) and as in previous proofs exploiting the generalized pseudomonotonicity of A we have

$$A(x_n) \xrightarrow{w} A(w) \quad \text{in } W^{-1,p'}(\Omega).$$

Thus in the limit as $n \to +\infty$ we have

$$A(w) = u + N_f(w),$$

with $u \in S^{p'}_{\partial j(\cdot, w(\cdot))}$. Thus

$$\begin{cases} -\operatorname{div} a\big(z, w(z), \nabla w(z)\big) \in \partial j\big(z, w(z)\big) + f\big(z, w(z)\big) & \text{for a.a. } z \in \Omega, \\ w|_{\partial \Omega} = 0 \end{cases}$$

and so $w \in \mathcal{Y}$. Applying the Kuratowski-Zorn Lemma (see Theorem A.1.1) we infer that $\mathcal{Y}$ has a maximal element $x_{max} \in \mathcal{Y}$. Because of Proposition 4.5.4 we have that x_{max} is the greatest element of $\mathcal{Y}$. Similarly we produce $x_{min} \in \mathcal{Y}$ the smallest element. Then $\{x_{min}, x_{max}\} \subseteq W_0^{1,p}(\Omega)$ are the extremal solutions of (4.126) in $K = [\underline{\psi}, \overline{\psi}]$ $\Box$

4.6   Multiplicity Results

In this section we study resonant elliptic equations, semilinear (i.e. $p = 2$) and nonlinear (i.e. $p \neq 2$) with a nonsmooth potential. We go beyond the analysis conducted in Section 4.1 and look for conditions guaranteeing the existence of multiple solutions. We start with semilinear problems and eventually move to nonlinear ones. Our hypotheses allow the nonsmooth potential to interact asymptotically at $\pm\infty$ with two consecutive eigenvalues of higher order of $\big(-\Delta, H_0^1(\Omega)\big)$. Such problems are often called *double resonance problems*. Note that, while the principal eigenfunction u_1 is strictly positive and $\frac{\partial u_1}{\partial n} < 0$ on $\partial \Omega$ (by the maximum principle), this is no longer true in higher parts of the spectrum, where the corresponding eigenfunctions change sign. This is a source of difficulties and requires a more delicate analysis for problems resonant at higher parts of the spectrum. In addition our hypotheses may also permit for double resonance at the origin too, in which case we can speak of a *double-double resonance problem*.

Our approach is variational and a basic tool in our arguments is Theorem 2.4.1.

4.6.1   Semilinear Problems

The first problem that we study is the following semilinear elliptic Dirichlet equation. Throughout this section $\Omega \subseteq \mathbb{R}^N$ is a bounded domain with a C^2-boundary $\partial \Omega$.

$$\begin{cases} -\Delta x(z) - \lambda_k x(z) \in \partial j\big(z, x(z)\big) & \text{for a.a. } z \in \Omega, \\ x|_{\partial \Omega} = 0. \end{cases} \tag{4.137}$$

Here $k \geq 1$ is a fixed integer, $\{\lambda_n\}_{n \geq 1}$ is the increasing sequence of distinct eigenvalues of $\big(-\Delta, H_0^1(\Omega)\big)$ (see Theorem 1.5.3), $j(z, \zeta)$ is an integrand locally Lipschitz in the $\zeta \in \mathbb{R}$-variable and $\partial j(z, \zeta)$ denotes the generalized subdifferential of $\zeta \longmapsto j(z, \zeta)$. To obtain a multiplicity result for problem (4.137) we shall develop and use a nonsmooth version of the so-called *reduction method*. As the name suggests, this method reduces the problem to the search of critical points of a functional defined on a finite dimensional Banach space. Dealing with a functional defined on a finite dimensional vector space of course has obvious advantages.

Our hypotheses on the nonsmooth potential $j(z, \zeta)$ are the following:

$\underline{H(j)_1}$ $j \colon \Omega \times \mathbb{R} \longrightarrow \mathbb{R}$ is a function, such that

(i) for every $\zeta \in \mathbb{R}$, the function

$$\Omega \ni z \longrightarrow j(z, \zeta) \in \mathbb{R}$$

is measurable and $j(z, 0) = 0$ for almost all $z \in \Omega$;

(ii) for almost all $z \in \Omega$, the function

$$\mathbb{R} \ni \zeta \longrightarrow j(z, \zeta) \in \mathbb{R}$$

is locally Lipschitz;

(iii) for almost all $z \in \Omega$, all $\zeta \in \mathbb{R}$ and all $u \in \partial j(z, \zeta)$, we have

$$|u| \leq a_1(z) + c_1 |\zeta|^{r-1},$$

with $a_1 \in L^\infty(\Omega)_+$, $c_1 > 0$ and $r \in [1, 2^*)$ where as usual

$$2^* \overset{df}{=} \begin{cases} \frac{2N}{N-2} & \text{if } N > 2, \\ +\infty & \text{if } N \leq 2; \end{cases}$$

(iv) $\displaystyle\lim_{|\zeta| \to +\infty} \left[\zeta u - 2j(z, \zeta) \right] = -\infty$ uniformly for almost all $z \in \Omega$ and all $u \in \partial j(z, \zeta)$;

(v) there exists $l \in L^\infty(\Omega)$ such that $l(z) \leq \lambda_{k+1} - \lambda_k$ for almost all $z \in \Omega$, the inequality is strict on a set of positive measure and for almost all $z \in \Omega$, all $\zeta_1, \zeta_2 \in \mathbb{R}$ with $\zeta_1 \neq \zeta_2$ and all $v_1 \in \partial j(z, \zeta_1)$, $v_2 \in \partial j(z, \zeta_2)$ we have

$$\frac{v_1 - v_2}{x_1 - x_2} \leq l(z);$$

(vi) there exist $\beta \in L^\infty(\Omega)_-$ and $\delta_0 > 0$ such that for some integer $1 \leq m \leq k$ we have

$$\beta(z) \leq \lambda_m - \lambda_k$$

with strict inequality on a set of positive measure and we have

$$\lambda_{m-1} - \lambda_k \leq \frac{2j(z, \zeta)}{\zeta^2} \leq \beta(z) \quad \text{for a.a. } z \in \Omega \text{ and all } |\zeta| \leq \delta_0;$$

(vii) there exists $\gamma \in L^\infty(\Omega)_+$ with $\gamma(z) \leq \lambda_{k+1} - \lambda_k$ for almost all $z \in \Omega$ with strict inequality on a set of positive measure and

$$0 \leq \liminf_{|\zeta| \to +\infty} \frac{2j(z,\zeta)}{\zeta^2} \leq \limsup_{|\zeta| \to +\infty} \leq \frac{2j(z,\zeta)}{\zeta^2} \leq \gamma(z)$$

uniformly for almost all $z \in \Omega$.

REMARK 4.6.1   Hypothesis $H(j)_1(vi)$ implies that we have a double resonance at the origin. This resonance is complete from below and incomplete from above. A similar double resonance situation occurs asymptotically at $\pm\infty$, by virtue of hypothesis $H(j)_1(vii)$. So these two hypotheses $H(j)_1(vi)$ and (vii) classify problem (4.137) as a *double-double resonance problem*. Recall that due to hypothesis $H(j)_1(ii)$ for almost all $z \in \Omega$, the function $\zeta \longmapsto j(z,\zeta)$ is almost everywhere differentiable and hypothesis $H(j)_1(iv)$ implies that this derivative can only have downward discontinuities. In the smooth case (i.e. $j(z,\cdot) \in C^1$), hypothesis $H(j)_1(iv)$ is a unilateral Lipschitz condition on $j'_\zeta(z,\cdot)$. $\square$

EXAMPLE 4.6.1   The following function satisfies hypotheses $H(j)_1$. For simplicity we drop the z-dependence.

$$j(\zeta) \overset{df}{=} \begin{cases} -\zeta_1 \zeta - 2\mu - 4\zeta_1 & \text{if } \zeta \in (-\infty, -4), \\ \frac{\mu}{2}\zeta^2 + 3\mu\zeta + 2\mu & \text{if } \zeta \in [-4, -1), \\ -\frac{\mu}{2}\zeta^2 & \text{if } \zeta \in [-1, 1), \\ \frac{\mu}{2}\zeta^2 - 3\mu\zeta + 2\mu & \text{if } \zeta \in [1, 4), \\ -\zeta_2 \zeta - 2\mu - 4\zeta_2 & \text{if } \zeta \in [4, +\infty). \end{cases}$$

Here $\lambda_k - \lambda_m < \mu < \min\{\lambda_k - \lambda_{m-1}, \lambda_{k+1} - \lambda_k\}$, $\zeta_1, \zeta_2 \in (0, \mu)$. We can take $l(z) \equiv \mu$ and note that

$$\frac{j(\zeta)}{\zeta^2} \longrightarrow 0 \quad \text{as } |\zeta| \to +\infty$$

so we have complete resonance at $\pm\infty$. $\square$

For each integer $n \geq 1$, let $E(\lambda_n)$ be the eigenspace corresponding to the eigenvalue λ_n. Recall that the eigenspace $E(\lambda_n) \subseteq H_0^1(\Omega) \subseteq C^\infty(\Omega)$ has the unique continuation property, namely if $u \in E(\lambda_n)$ is such that it vanishes on a set of positive measure, then $u(z) = 0$ for all $z \in \Omega$. We set

$$\overline{X} \overset{df}{=} \overline{X}_k = \bigoplus_{i=1}^{k-1} E(\lambda_i) \quad \text{and} \quad \widehat{X} \overset{df}{=} \widehat{X}_k = \overline{\bigoplus_{i=k+1}^{\infty} E(\lambda_i)}.$$

We know that

$$H_0^1(\Omega) = \overline{X} \oplus E(\lambda_k) \oplus \widehat{X}.$$

Also we set

$$\overline{X}_0 \overset{df}{=} \overline{X}_k \oplus E(\lambda_k) = \bigoplus_{i=1}^{k} E(\lambda_i).$$

Let $\varphi \colon H_0^1(\Omega) \longrightarrow \mathbb{R}$ be the energy functional defined by

$$\varphi(x) \overset{df}{=} \frac{1}{2} \|\nabla x\|_2^2 - \frac{\lambda_k}{2} \|x\|_2^2 - \int_\Omega j(z, x(z)) \, dz \qquad \forall \, x \in H_0^1(\Omega).$$

We know that φ is locally Lipschitz (see hypotheses $H(j)_1(ii)$ and (iii)). Let $u \in \overline{X}_0$ and consider the following minimization problem:

$$\inf_{v \in \widehat{X}} \varphi(u + v). \tag{4.138}$$

Because we do not identify $H_0^1(\Omega)$ with its dual, we have that

$$\left(H_0^1(\Omega) \right)^* = H^{-1}(\Omega) = \overline{X}_0^* \oplus \widehat{X}^*.$$

We start with a straightforward lemma, analogous to the claim in the proof of Proposition 4.4.3.

LEMMA 4.6.1
If $n \geq 1$ *and* $\beta \in L^\infty(\Omega)_+$ *with* $\beta(z) \leq \lambda_{n+1}$ *for almost all* $z \in \Omega$ *and the inequality is strict on a set of positive measure,*
<u>*then*</u> *there exists* $\xi_1 > 0$ *such that*

$$\|\nabla x\|_2^2 - \int_\Omega \beta(z) |x(z)|^2 \, dz \geq \xi_1 \|\nabla x\|_2^2 \qquad \forall \, x \in \widehat{X}_n.$$

PROOF   Consider the function $\vartheta \colon H_0^1(\Omega) \longrightarrow \mathbb{R}$ defined by

$$\vartheta(x) \overset{df}{=} \|\nabla x\|_2^2 - \int_\Omega \beta(z) |x(z)|^2 \, dz.$$

From Theorem 1.5.3(b), we have that $\vartheta \geq 0$. Suppose that the conclusion of the Lemma is not true. Then we can find a sequence $\{x_m\}_{m \geq 1} \subseteq \widehat{X}_n$ such that $\|\nabla x_m\|_2 = 1$ for $m \geq 1$ and $\vartheta(x_m) \searrow 0$. By passing to a subsequence if necessary, we may assume that

$$x_m \overset{w}{\longrightarrow} x \text{ in } H_0^1(\Omega),$$
$$x_m \longrightarrow x \text{ in } L^2(\Omega),$$

with $x \in \widehat{X}_n$. Exploiting the weak lower semicontinuity of the norm functional, in the limit as $m \to +\infty$ we obtain

$$\vartheta(x) = \|\nabla x\|_2^2 - \int_\Omega \beta(z) |x(z)|^2 \, dz \leq 0,$$

so

$$\|\nabla x\|_2^2 \leq \int_\Omega \beta(z) |x(z)|^2 \, dz \leq \lambda_{n+1} \|x\|_2^2$$

and thus

$$\|\nabla x\|_2^2 = \lambda_{n+1} \|x\|_2^2 \qquad\qquad (4.139)$$

(since $x \in \widehat{X}_n$; see Theorem 1.5.3(b)).

If $x \equiv 0$, then

$$\|\nabla x_m\|_2 \longrightarrow 0,$$

a contradiction to the fact that $\|\nabla x_m\|_2 = 1$ for $m \geq 1$. Hence $x \neq 0$ and so from (4.139) it follows that $x \in E(\lambda_{n+1})$. By hypothesis $\beta(z) < \lambda_{n+1}$ on a set of positive measure and so by virtue of the unique continuation property of $E(\lambda_{n+1})$ we have

$$\|\nabla x\|_2^2 = \int_\Omega \beta(z) |x(z)|^2 \, dz < \lambda_{n+1} \|x\|_2^2,$$

which contradicts (4.139).   □

The next proposition is the basic step in the implementation of the reduction method to the present nonsmooth setting.

PROPOSITION 4.6.1

If hypotheses $H(j)_1$ hold,

<u>*then*</u> *there exists a continuous map $\vartheta \colon \overline{X}_0 \longrightarrow \widehat{X}$ such that for every $u \in \overline{X}_0$ we have*

$$\inf_{v \in \widehat{X}} \varphi(u + v) = \varphi(u + \vartheta(u))$$

and $\vartheta(u) \in \widehat{X}$ is the unique solution of the operator inclusion

$$0 \in \mathrm{P}_{\widehat{X}^*} \partial\varphi(u + v),$$

with $u \in \overline{X}_0$ fixed and $\mathrm{P}_{\widehat{X}^*}$ being the orthogonal projection on $\widehat{X}^* = \left[\overline{X}_0^*\right]^\perp$.

PROOF   For a fixed $u \in \overline{X}_0$ we introduce the map $\varphi_u \colon H_0^1(\Omega) \longrightarrow \mathbb{R}$ defined by

$$\varphi_u(w) \stackrel{df}{=} \varphi(u + w) \qquad \forall \, w \in H_0^1(\Omega).$$

For every $w, h \in H^1_0(\Omega)$, we have that

$$\varphi^0_u(w; h) = \limsup_{\substack{w' \to w \\ t \searrow 0}} \frac{\varphi_u(w' + th) - \varphi_u(w')}{t}$$

$$= \limsup_{\substack{w' \to w \\ t \searrow 0}} \frac{\varphi(u + w' + th) - \varphi(u + w')}{t} = \varphi^0(u + w; h),$$

so

$$\partial \varphi_u(w) = \partial \varphi(u + w) \qquad \forall\, w \in H^1_0(\Omega). \tag{4.140}$$

Let $\widehat{i} \colon \widehat{X} \longrightarrow H^1_0(\Omega)$ be the inclusion map and let $\widehat{\varphi}_u \colon \widehat{X} \longrightarrow \mathbb{R}$ be defined by

$$\widehat{\varphi}_u(v) = \varphi(u + v) \qquad \forall\, v \in \widehat{X}.$$

We have that $\varphi_u \circ \widehat{i} = \widehat{\varphi}_u$ and so

$$\partial\big(\varphi_u \circ \widehat{i}\big)(v) = \partial \widehat{\varphi}_u(v) \qquad \forall\, v \in \widehat{X}. \tag{4.141}$$

But from Proposition 1.3.15(b) (see also Remark 1.3.6), we have

$$\partial\big(\varphi_u \circ \widehat{i}\big)(v) \subseteq \widehat{i}^* \partial \varphi\big(\widehat{i}(v)\big) = \mathrm{P}_{\widehat{X}*} \partial \varphi_u\big(\widehat{i}(v)\big) \qquad \forall\, v \in \widehat{X},$$

since $\widehat{i}^* = \mathrm{P}_{\widehat{X}*}$. Then from (4.140) and (4.141) it follows that

$$\partial \widehat{\varphi}_u(v) \subseteq \mathrm{P}_{\widehat{X}*} \partial \varphi_u\big(\widehat{i}(v)\big) = \mathrm{P}_{\widehat{X}*} \partial \varphi(u + v) \qquad \forall\, v \in \widehat{X}. \tag{4.142}$$

Note that we have

$$x^* = A(x) - \lambda_k x - h \qquad \forall\, x \in H^1_0(\Omega),\ x^* \in \partial \varphi(x), \tag{4.143}$$

with $A \in \mathcal{L}\big(H^1_0(\Omega), H^{-1}(\Omega)\big)$ being the monotone operator defined by

$$\langle A(x), y \rangle_{H^1_0(\Omega)} \stackrel{df}{=} \int_\Omega \big(\nabla x(z), \nabla y(z)\big)_{\mathbb{R}^N} dz \qquad \forall\, x, y \in H^1_0(\Omega)$$

and $h \in L^{r'}(\Omega)$ (where $\frac{1}{r} + \frac{1}{r'} = 1$) such that $h(z) \in \partial j\big(z, x(z)\big)$ for almost all $z \in \Omega$. So if $v_1, v_2 \in \widehat{X}$ and $x^*_1 \in \partial \widehat{\varphi}_u(v_1)$, $x^*_2 \in \partial \widehat{\varphi}_u(v_2)$, using (4.142) and (4.143) we have

$$x^*_i = \mathrm{P}_{\widehat{X}*} A(u + v_i) - \lambda_k v_i - \mathrm{P}_{\widehat{X}*} h_i \qquad \text{for } i \in \{1, 2\}, \tag{4.144}$$

where $h_i \in L^{r'}(\Omega) \subseteq H^{-1}(\Omega)$ (recall $r < 2^*$) and $h_i(z) \in \partial j\big(z, (u + v_i)(z)\big)$ for almost all $z \in \Omega$ and $i \in \{1, 2\}$. Since $\mathrm{P}^*_{\widehat{X}*} = \widehat{i}$, we have that

$$\big\langle \mathrm{P}_{\widehat{X}*}\big(A(u + v_1) - A(u + v_2)\big), v_1 - v_2 \big\rangle_{\widehat{X}}$$

$$= \big\langle A(u + v_1) - A(u + v_2), \widehat{i}(v_1) - \widehat{i}(v_2) \big\rangle_{H^1_0(\Omega)}$$

$$= \big\langle A(u + v_1) - A(u + v_2), v_1 - v_2 \big\rangle_{H^1_0(\Omega)} = \|\nabla v_1 - \nabla v_2\|^2_2.$$

By hypothesis $H(j)_1(v)$ we have that

$$\frac{h_1(z) - h_2(z)}{v_1(z) - v_2(z)} \leq l(z) \quad \text{for a.a. } z \in \{v_1 \neq v_2\}. \tag{4.145}$$

Using (4.144) and (4.145), we obtain that

$$\langle x_1^* - x_2^*, v_1 - v_2 \rangle_{\widehat{X}}$$
$$= \|\nabla v_1 - \nabla v_2\|_2^2 - \lambda_k \|v_1 - v_2\|_2^2 - \int_{\Omega} \big(h_1(z) - h_2(z)\big)\big(v_1(z) - v_2(z)\big) \, dz$$
$$\geq \|\nabla v_1 - \nabla v_2\|_2^2 - \lambda_k \|v_1 - v_2\|_2^2 - \int_{\Omega} l(z)|v_1(z) - v_2(z)|^2 \, dz.$$

By hypothesis $H(j)_1(v)$, we know that

$$l(z) \leq \lambda_{k+1} - \lambda_k \quad \text{for a.a. } z \in \Omega,$$

with strict inequality on a set of positive measure. So we can apply Lemma 4.6.1 (with $\beta(z) \overset{df}{=} l(z) + \lambda_k$) and obtain $\xi_1 > 0$, such that

$$\langle x_1^* - x_2^*, v_1 - v_2 \rangle_{\widehat{X}} \geq \xi_1 \|\nabla v_1 - \nabla v_2\|_2^2.$$

This implies that the multifunction $v \longmapsto \partial \widehat{\varphi}_u(v)$ is strongly monotone in the dual pair $(\widehat{X}, \widehat{X}^*)$. Hence the function $\widehat{X} \ni v \longmapsto \widehat{\varphi}_u(v) \in \mathbb{R}$ is strongly convex, i.e. the function $\widehat{X} \ni v \longmapsto \widehat{\varphi}_u(v) - \frac{\xi_1}{2} \|v\|_{H^1(\Omega)}^2 \in \mathbb{R}$ is convex. This means that $v \longmapsto \partial \widehat{\varphi}_u(v)$ is maximal monotone and strongly monotone in the dual pair $(\widehat{X}, \widehat{X}^*)$.

Let $v \in \widehat{X}$, $x^* \in \partial \widehat{\varphi}_u(v)$ and $y^* \in \partial \widehat{\varphi}_u(0)$. Then we have

$$\langle x^*, v \rangle_{\widehat{X}} = \langle x^* - y^*, v \rangle_{\widehat{X}} + \langle y^*, v \rangle_{\widehat{X}}$$
$$\geq \xi_1 \|\nabla v\|_2^2 - \xi_2 \|y^*\|_{H^{-1}(\Omega)} \|\nabla v\|_2,$$

for some $\xi_2 > 0$, so the multifunction $v \longmapsto \partial \varphi_u(v)$ is coercive.

Therefore we have shown that the multifunction $v \longmapsto \partial \widehat{\varphi}_u(v)$ is maximal monotone, coercive. So it is surjective (see Theorem 1.4.4). Thus, we can find $v_0 \in \widehat{X}$, such that

$$0 \in \partial \widehat{\varphi}_u(v_0) \quad \text{and} \quad \inf_{v \in \widehat{X}} \varphi(u+v) = \varphi(u+v_0).$$

This minimizer $v_0 \in \widehat{X}$ is unique, due to the strong convexity of $\widehat{\varphi}_u$. So we can define a map $\vartheta \colon \overline{X}_0 \longrightarrow \widehat{X}$ which to each fixed $u \in \overline{X}_0$ assigns the unique solution $v_0 \in \widehat{X}$ of the minimization problem (4.138). Then, from (4.142) we have

$$0 \in \partial \widehat{\varphi}_u(\vartheta(u)) \subseteq P_{\widehat{X}^*} \partial \varphi(u + \vartheta(u))$$

and
$$\inf_{v \in \widehat{X}} \varphi(u + v) = \varphi(u + \vartheta(u)).$$

It remains to show that ϑ is continuous. So let
$$u_n \longrightarrow u \quad \text{in } \overline{X}_0.$$

If $v_n \longrightarrow v$ in $\widehat{X}$, then we have $\widehat{\varphi}_{u_n}(v_n) \longrightarrow \widehat{\varphi}_u(v)$ (in fact it is easy to see that actually $\widehat{\varphi}_{u_n} \longrightarrow \widehat{\varphi}_u$ in $C(\widehat{X})$). On the other hand, if
$$v_n \xrightarrow{\; w \;} v \quad \text{in } \widehat{X},$$

then because of the weak lower semicontinuity of the norm functional in a Banach space and the compactness of the embedding $H_0^1(\Omega) \subseteq L^2(\Omega)$ we have
$$\widehat{\varphi}_u(v) \leq \liminf_{n \to +\infty} \widehat{\varphi}_{u_n}(v_n).$$

It follows that
$$\widehat{\varphi}_{u_n} \xrightarrow{\; M \;} \widehat{\varphi}_u$$

and so from Remark 1.2.10, we have
$$\operatorname{Gr} \partial \widehat{\varphi}_{u_n} \xrightarrow{\; K \;} \operatorname{Gr} \partial \widehat{\varphi}_u \quad \text{as } n \to +\infty.$$

Because $0 \in \partial \widehat{\varphi}_u(\vartheta(u))$, we can find $v_n^* \in \partial \widehat{\varphi}_{u_n}(v_n)$, for $n \geq 1$ with $v_n \longrightarrow \vartheta(u)$ in $\widehat{X}$ such that $v_n^* \longrightarrow 0$ in $\widehat{X}^*$. Recall that $0 \in \partial \widehat{\varphi}_{u_n}(\vartheta(u_n))$, for $n \geq 1$ and from the strong monotonicity of $\partial \widehat{\varphi}_{u_n}$ we have that
$$\xi_1 \big\| \nabla v_n - \nabla \vartheta(u_n) \big\|_2^2 \leq \langle v_n^*, v_n - \vartheta(u_n) \rangle_{\widehat{X}},$$

for some $\xi_1 > 0$, so
$$\| \nabla v_n - \nabla \vartheta(u_n) \|_2 \leq \frac{1}{\xi_1} \| v_n^* \|_{H^{-1}(\Omega)} \longrightarrow 0 \quad \text{as } n \to +\infty.$$

Thus
$$\| \nabla \vartheta(u) - \nabla \vartheta(u_n) \|_2 \leq \| \nabla \vartheta(u) - \nabla v_n \|_2 + \| \nabla v_n - \nabla \vartheta(u) \|_2 \longrightarrow 0$$

and then from the Poincaré inequality (see Theorem 1.1.6), we have
$$\vartheta(u_n) \longrightarrow \vartheta(u) \quad \text{in } \widehat{X}.$$

This proves the continuity of ϑ and finishes the proof of the Proposition. $\quad \Box$

Using Proposition 4.6.1, we can define the map $\overline{\varphi} \colon \overline{X}_0 \longrightarrow \mathbb{R}$, by
$$\overline{\varphi}(u) \overset{df}{=} \varphi(u + \vartheta(u)) \qquad \forall\, u \in \overline{X}_0.$$

From the definition of ϑ, for all $u, h \in \overline{X}_0$ we have that

$$\begin{aligned}
\overline{\varphi}(u+h) - \overline{\varphi}(u) &= \varphi\big(u+h+\vartheta(u+h)\big) - \varphi\big(u+\vartheta(u)\big) \\
&\leq \varphi\big(u+h+\vartheta(u)\big) - \varphi\big(u+\vartheta(u)\big).
\end{aligned} \qquad (4.146)$$

Similarly from the definition of ϑ, for all $u, h \in \overline{X}_0$ we obtain

$$\begin{aligned}
\overline{\varphi}(u) - \overline{\varphi}(u+h) &\leq \varphi\big(u+\vartheta(u)\big) - \varphi\big(u+h+\vartheta(u+h)\big) \\
&\leq \varphi\big(u+\vartheta(u+h)\big) - \varphi\big(u+h+\vartheta(u+h)\big).
\end{aligned} \qquad (4.147)$$

From (4.146), (4.147) and the fact that φ is locally Lipschitz, we conclude that $\overline{\varphi}$ is locally Lipschitz.

Now we show that

$$\partial\overline{\varphi}(u) \subseteq \mathrm{P}_{\overline{X}_0^*}\partial\varphi(u+\vartheta(u)) \qquad \forall\, u \in \overline{X}_0. \qquad (4.148)$$

Note that for all $u, h \in \overline{X}_0$ we have

$$\begin{aligned}
\overline{\varphi}^0(u; h) &= \limsup_{\substack{u' \to u \\ t \searrow 0}} \frac{\overline{\varphi}(u'+th) - \overline{\varphi}(u')}{t} \\
&= \limsup_{\substack{u' \to u \\ t \searrow 0}} \frac{\varphi\big(u'+th+\vartheta(u'+th)\big) - \varphi\big(u'+\vartheta(u')\big)}{t} \\
&\leq \limsup_{\substack{u' \to u \\ t \searrow 0}} \frac{\varphi\big(u'+th+\vartheta(u')\big) - \varphi\big(u'+\vartheta(u')\big)}{t} \\
&\leq \varphi^0\big(u+\vartheta(u); h\big).
\end{aligned}$$

Let $\bar{i}_0 \colon \overline{X}_0 \longrightarrow H_0^1(\Omega)$ be the inclusion map. We have $\bar{i}_0^* = \mathrm{P}_{\overline{X}_0^*}$. So for all $u, h \in \overline{X}_0$ we have that

$$\begin{aligned}
\overline{\varphi}^0(u; h) &\leq \varphi^0\big(u+\vartheta(u); \bar{i}_0(h)\big) \\
&= \sup_{u^* \in \partial\varphi(u+\vartheta(u))} \big\langle u^*, \bar{i}_0(h) \big\rangle_{H_0^1(\Omega)} = \sup_{u^* \in \partial\varphi(u+\vartheta(u))} \big\langle \mathrm{P}_{\overline{X}_0^*}(u^*), h \big\rangle_{\overline{X}_0}.
\end{aligned}$$

Suppose that $u_0^* \in \partial\overline{\varphi}(u)$. From the definition of the Clarke directional derivative (see Definition 1.3.6) we have

$$\big\langle u_0^*, h \big\rangle_{\overline{X}_0} \leq \overline{\varphi}^0(u; h) \qquad \forall\, h \in \overline{X}_0,$$

so

$$\big\langle u_0^*, h \big\rangle_{\overline{X}_0} \leq \sup_{u^* \in \partial\varphi(u+\vartheta(u))} \big\langle \mathrm{P}_{\overline{X}_0^*}(u^*), h \big\rangle_{\overline{X}_0} \qquad \forall\, h \in \overline{X}_0$$

and thus

$$u_0^* \in \mathrm{P}_{\overline{X}_0^*}\partial\varphi\big(u+\vartheta(u)\big).$$

This establishes inclusion (4.148).

In what follows $\psi = -\overline{\varphi}$. Evidently ψ is locally Lipschitz on the finite dimensional space $\overline{X}_0$. We introduce two more subspaces of $H_0^1(\Omega)$.

$$Y \stackrel{df}{=} \bigoplus_{i=1}^{m-1} E(\lambda_i) \quad \text{and} \quad V \stackrel{df}{=} \bigoplus_{i=m}^{k} E(\lambda_i).$$

Here $1 \leq m \leq k$ is as in hypothesis $H(j)_1(vi)$. We have

$$\overline{X}_0 = \overline{X} \oplus E(\lambda_k) = Y \oplus V.$$

In the next proposition we show that $\psi = -\overline{\varphi}$ satisfies the local linking condition (see Definition 2.4.1).

PROPOSITION 4.6.2
If hypotheses $H(j)_1$ hold,
<u>*then*</u> *there exists $\delta > 0$ such that*

$$\begin{cases} \psi(u) \leq 0 \text{ if } u \in V, \|u\|_{H^1(\Omega)} \leq \delta, \\ \psi(u) \geq 0 \text{ if } u \in Y, \|u\|_{H^1(\Omega)} \leq \delta. \end{cases}$$

PROOF Because $Y \subseteq C(\overline{\Omega})$ is finite dimensional, we know that on it all norms are equivalent and so we can find $M_1 > 0$ such that

$$\sup_{z \in \Omega} |u(z)| \leq M_1 \|u\|_{H^1(\Omega)} \quad \forall u \in Y.$$

We consider $\beta \in L^\infty(\Omega)$ and $\delta_0 > 0$ as in hypothesis $H(j)_1(vi)$. Thus if we choose $\delta' \stackrel{df}{=} \frac{\delta_0}{M_1}$, we have

$$\sup_{z \in \Omega} |u(z)| \leq \delta_0 \quad \forall u \in Y, \|u\|_{H^1(\Omega)} \leq \delta'.$$

By virtue of hypothesis $H(j)_1(vi)$, the definition of ϑ and Theorem 1.5.3(b), for all $u \in Y$ with $\|u\|_{H^1(\Omega)} \leq \delta'$ we have

$$\psi(u) = -\overline{\varphi}(u) \geq -\frac{1}{2}\|\nabla u\|_2^2 + \frac{\lambda_k}{2}\|u\|_2^2 + \int_\Omega j(z, u(z))\, dz$$

$$\geq -\frac{1}{2}\|\nabla u\|_2^2 + \frac{\lambda_k}{2}\|u\|_2^2 + \frac{\lambda_{m-1} - \lambda_k}{2}\|u\|_2^2$$

$$= -\frac{1}{2}\|\nabla u\|_2^2 + \frac{\lambda_{m-1}}{2}\|u\|_2^2 \geq 0.$$

Also for all $u \in V$, we have

$$\psi(u) = -\varphi(u + \vartheta(u)) \tag{4.149}$$

$$= -\frac{1}{2}\|\nabla(u + \vartheta(u))\|_2^2 + \frac{\lambda_k}{2}\|u + \vartheta(u)\|_2^2 + \int_\Omega j(z, (u + \vartheta(u))(z))\, dz.$$

From hypothesis $H(j)_1(vi)$ we know that

$$j(z, \zeta) \leq \frac{1}{2}\beta(z)\zeta^2 \quad \text{for a.a. } z \in \Omega \text{ and all } |\zeta| \leq \delta_0.$$

On the other hand, by the mean value theorem for locally Lipschitz functions (see Proposition 1.3.15) and hypothesis $H(j)_1(iii)$, we see that

$$j(z, \zeta) \leq c_2|\zeta|^\eta \quad \text{for a.a. } z \in \Omega \text{ and all } |\zeta| > \delta_0,$$

for some $2 < \eta \leq 2^*$ and $c_2 > 0$. Therefore finally we can say that

$$j(z, \zeta) \leq \frac{1}{2}\beta(z)\zeta^2 + c_3|\zeta|^\eta \quad \text{for a.a. } z \in \Omega \text{ and all } \zeta \in \mathbb{R},$$

with $c_3 \stackrel{df}{=} c_2 + \frac{1}{2}\|\beta\|_\infty > 0$. Using this in (4.149), we obtain

$$\psi(u) \leq -\frac{1}{2}\left\|\nabla\big(u + \vartheta(u)\big)\right\|_2^2 + \frac{\lambda_k}{2}\left\|u + \vartheta(u)\right\|_2^2 \qquad (4.150)$$

$$+\frac{1}{2}\int_\Omega \beta(z)\big|(u + \vartheta(u))(z)\big|^2 \, dz + c_3\left\|u + \vartheta(u)\right\|_\eta^\eta \qquad \forall\, u \in V.$$

Since $\beta(z) + \lambda_k \leq \lambda_m$ for almost all $z \in \Omega$ with strict inequality on a set of positive measure (see hypothesis $H(j)_1(vi)$) and because

$$u + \vartheta(u) \in \widehat{X}_{m-1} = \overline{\bigoplus_{i=m}^{+\infty} E(\lambda_1)},$$

we can apply Lemma 4.6.1 and obtain $\xi_1 > 0$ such that

$$\left\|\nabla\big(u + \vartheta(u)\big)\right\|_2^2 - \int_\Omega \big(\beta(z) + \lambda_k\big)\big|(u + \vartheta(u))(z)\big|^2 \, dz$$

$$\geq \xi_1\left\|\nabla\big(u + \vartheta(u)\big)\right\|_2^2 \qquad \forall\, u \in V.$$

Using this inequality in (4.150) together with the Sobolev Embedding Theorem (see Theorem 1.1.5) and the Poincaré inequality (see Theorem 1.1.6), we have

$$\psi(u) \leq -\frac{\xi_1}{2}\left\|\nabla\big(u + \vartheta(u)\big)\right\|_2^2 + c_4\left\|\nabla\big(u + \vartheta(u)\big)\right\|_2^\eta \qquad \forall\, v \in V, \quad (4.151)$$

for some $c_4 > 0$. Recall that $2 < \eta$. So from (4.151) and the Poincaré inequality (see Theorem 1.1.6), we see that we can find $\delta'' > 0$ such that

$$\psi(u) \leq 0 \qquad \forall\, u \in V, \ \|u\|_{H^1(\Omega)} \leq \delta''.$$

Choosing $\delta \stackrel{df}{=} \min\{\delta', \delta''\}$, we conclude the proof of the Proposition. $\quad\square$

Our aim is to use Theorem 2.4.1. For this purpose we need to show that ψ satisfies the nonsmooth C-condition. First we show this for the functional φ. To do this we need the following addendum to the well known Egorov's theorem.

LEMMA 4.6.2
If $C \subseteq \mathbb{R}^N$ is a measurable set with $|C|_N < +\infty$, $u_n \colon C \longrightarrow \mathbb{R}$ for $n \geq 1$ is a sequence of Lebesgue measurable functions such that

$$u_n(z) \longrightarrow +\infty \quad \text{for a.a. } z \in C,$$

<u>*then*</u> *for every $\delta > 0$, we can find $C_\delta \subseteq C$ measurable with $|C \setminus C_\delta|_N < \delta$ such that*

$$u_n(z) \longrightarrow +\infty \quad \text{uniformly on } C_\delta.$$

PROOF Without any loss of generality we may assume that

$$u_n(z) \longrightarrow +\infty \quad \forall\, z \in C.$$

For every $n \geq 1$ and $M > 0$ we define

$$C_{n,M} \stackrel{df}{=} \bigcap_{k=n+1}^{+\infty} \left\{ z \in C : \left| u_k(z) \right| > M \right\}.$$

Evidently $C_{n,M}$ is measurable, $C_{n,M} \subseteq C_{m,M}$ if $n < m$ and

$$C = \bigcup_{n=1}^{\infty} C_{n,M}.$$

So

$$|C|_N = \lim_{n \to +\infty} |C_{n,M}|_N,$$

from which it follows that

$$\lim_{n \to +\infty} |C \setminus C_{n,M}|_N = 0.$$

So for every $m \geq 1$ we can find $n_m \geq 1$ such that

$$|C \setminus C_{n,M}|_N < \frac{\delta}{2^m} \quad \forall\, n \geq n_m.$$

Let us set

$$C_\delta \stackrel{df}{=} \bigcap_{m=1}^{\infty} C_{n_m, M}.$$

We have

$$
\begin{aligned}
|C \setminus C_\delta|_N &= \left| \bigcup_{m=1}^{\infty} (C \setminus C_{n_m, M}) \right|_N \\
&\leq \sum_{m=1}^{\infty} |(C \setminus C_{n_m, M})|_N < \sum_{m=1}^{\infty} \frac{\delta}{2^m} = \delta.
\end{aligned}
$$

For every $M > 0$ let us choose $m_0 \geq M$. We have $C_\delta \subseteq C_{n_{m_0}, m_0}$ and so

$$
u_n(z) \geq m_0 \geq M \qquad \forall \, n \geq n_{m_0}, \, z \in C_\delta.
$$

Therefore we have that $u_n(z) \longrightarrow +\infty$ uniformly on C_δ. ⬚

PROPOSITION 4.6.3

If hypotheses $H(j)_1$ hold,
then φ *satisfies the nonsmooth C-condition.*

PROOF   Let $\{x_n\}_{n \geq 1} \subseteq H_0^1(\Omega)$ be a sequence, such that

$$
\varphi(x_n) \longrightarrow c \quad \text{and} \quad (1 + \|x_n\|_{H^1(\Omega)}) m^\varphi(x_n) \longrightarrow 0.
$$

Since the norm functional is weakly lower semicontinuous and $\partial\varphi(x_n) \subseteq H^{-1}(\Omega)$ is weakly compact, from the Weierstrass theorem (see Theorem A.1.5), we know that there exists $x_n^* \in \partial\varphi(x_n)$, such that $\|x_n^*\|_{H^{-1}(\Omega)} = m^\varphi(x_n)$ for all $n \geq 1$. We know that

$$
x_n^* = A(x_n) - \lambda_k x_n - h_n \qquad \forall \, n \geq 1,
$$

with $A \in \mathcal{L}(H_0^1(\Omega), H^{-1}(\Omega))$ being the maximal monotone operator defined by

$$
\langle A(x), y \rangle_{H_0^1(\Omega)} \stackrel{df}{=} \int_\Omega (\nabla x(z), \nabla y(z))_{\mathbb{R}^N} \, dz \qquad \forall \, x, y \in H_0^1(\Omega)
$$

and $h_n \in L^{r'}(\Omega)$ (with $\frac{1}{r} + \frac{1}{r'} = 1$) is such that $h_n(z) \in \partial j(z, x_n(z))$ for almost all $z \in \Omega$ (see Theorem 1.3.10). From the choice of the sequence $\{x_n\}_{n \geq 1} \subseteq H_0^1(\Omega)$ we have

$$
\begin{aligned}
\left| \langle x_n^*, x_n \rangle_{H_0^1(\Omega)} - 2\varphi(x_n) + 2c \right| \\
\leq \|x_n\|_{H^1(\Omega)} \|x_n^*\|_{H^{-1}(\Omega)} + 2|\varphi(x_n) - c| \\
\leq (1 + \|x_n\|_{H^1(\Omega)}) m^\varphi(x_n) + 2|\varphi(x_n) - c| \longrightarrow 0. \qquad (4.152)
\end{aligned}
$$

Note that

$$
\langle x_n^*, x_n \rangle_{H_0^1(\Omega)} = \langle A(x_n), x_n \rangle_{H_0^1(\Omega)} - \lambda_k \|x_n\|_2^2 - \int_\Omega h_n(z) x_n(z) \, dz.
$$

Using this in (4.152), we obtain

$$\int_{\Omega}\left(h_n(z)x_n(z) - 2j\big(z, x_n(z)\big)\right)dz$$
$$= 2\varphi(x_n) - \langle x_n^*, x_n\rangle_{H_0^1(\Omega)} \longrightarrow 2c. \tag{4.153}$$

We claim that the sequence $\{x_n\}_{n\geq 1} \subseteq H_0^1(\Omega)$ is bounded. Suppose that this is not the case. By passing to a subsequence if necessary, we may assume that

$$\|x_n\|_{H^1(\Omega)} \longrightarrow +\infty.$$

Let

$$y_n \stackrel{df}{=} \frac{x_n}{\|x_n\|_{H^1(\Omega)}} \qquad \forall\, n \geq 1.$$

Since $\|y_n\|_{H^1(\Omega)} = 1$, passing to a subsequence if necessary, we may assume that

$$\begin{aligned}
y_n \xrightarrow{w} y \quad &\text{in } H_0^1(\Omega),\\
y_n \longrightarrow y \quad &\text{in } L^2(\Omega),\\
y_n(z) \longrightarrow y(z) \quad &\text{for a.a. } z \in \Omega,\\
|y_n(z)| \leq k(z) \quad &\text{for a.a. } z \in \Omega \text{ and all } n \geq 1,
\end{aligned}$$

for some $k \in L^2(\Omega)$. By virtue of hypothesis $H(j)_1(vii)$, for a given $\varepsilon > 0$ we can find $M_2 = M_2(\varepsilon) > 0$ such that

$$j(z, \zeta) \leq \frac{1}{2}\big(\gamma(z) + \varepsilon\big)\zeta^2 \quad \text{for a.a. } z \in \Omega \text{ and all } |\zeta| \geq M_2.$$

On the other hand, as before via the mean value theorem for locally Lipschitz functions (see Proposition 1.3.14) and hypothesis $H(j)_1(iii)$ we have

$$\big|j(z, \zeta)\big| \leq \xi_3 \quad \text{for a.a. } z \in \Omega \text{ and all } |\zeta| < M_2,$$

for some $\xi_3 > 0$. So we can say that

$$j(z, \zeta) \leq \frac{1}{2}\big(\gamma(z) + \varepsilon\big)\zeta^2 + \xi_3 \quad \text{for a.a. } z \in \Omega \text{ and all } \zeta \in \mathbb{R}.$$

Then for every $n \geq 1$, we have

$$\begin{aligned}
\frac{\varphi(x_n)}{\|x_n\|_{H^1(\Omega)}^2} &= \frac{1}{2}\|\nabla y_n\|_2^2 - \frac{\lambda_k}{2}\|y_n\|_2^2 - \int_{\Omega}\frac{j(z, x_n(z))}{\|x_n\|_{H^1(\Omega)}^2}\,dz\\
&\geq \frac{c_5}{2} - \frac{\lambda_k}{2}\|y_n\|_2^2 - \frac{1}{2}\int_{\Omega}\gamma(z)y_n(z)^2\,dz - \frac{\varepsilon}{2}\|y_n\|_2^2 - \frac{c_6}{\|x_n\|_{H^1(\Omega)}^2},
\end{aligned}$$

for some $c_5, c_6 > 0$ (by the Poincaré inequality; see Theorem 1.1.6). Passing to the limit as $n \to +\infty$, we obtain

$$0 \geq \frac{1}{2}\big(c_5 - (\lambda_k + \|\gamma\|_\infty + \varepsilon)\,\|y\|_2^2\big)$$

and thus $y \neq 0$.

By virtue of hypothesis $H(j)_1(iv)$, we can find $M_3 > 0$ such that

$$u\zeta - 2j(z, \zeta) \leq -1 \quad \text{for a.a. } z \in \Omega, \text{ all } |\zeta| \geq M_3 \text{ and all } u \in \partial j(z, \zeta).$$

On the other hand, as above, we have

$$|j(z, \zeta)| \leq \xi_4 \quad \text{for a.a. } z \in \Omega \text{ and all } |\zeta| < M_3,$$

for some $\xi_4 > 0$. Therefore using hypothesis $H(j)_1(iii)$, we see that

$$|u\zeta - 2j(z, \zeta)| \leq \xi_5 \quad \text{for a.a. } z \in \Omega, \text{ all } |\zeta| < M_3 \text{ and all } u \in \partial j(z, \zeta),$$

for some $\xi_5 > 0$. Thus finally we can say that

$$u\zeta - 2j(z, \zeta) \leq \xi_5 \quad \text{for a.a. } z \in \Omega, \text{ all } \zeta \in \mathbb{R} \text{ and all } u \in \partial j(z, \zeta). \quad (4.154)$$

Let

$$C \overset{df}{=} \left\{ z \in \Omega : y(z) \neq 0 \right\}.$$

Since $y \neq 0$ we have $|C|_N > 0$ and for all $z \in C$, we have that

$$|x_n(z)| \longrightarrow +\infty.$$

Using Lemma 4.6.2, for a given $\delta \in (0, |C|_N)$, we can find a measurable subset $C_1 \subseteq C$ such that $|C \setminus C_1|_N < \delta$ and

$$|x_n(z)| \longrightarrow +\infty \quad \text{uniformly for all } z \in C_1.$$

Then hypothesis $H(j)_1(iv)$ implies that

$$\int_{C_1} \left(h_n(z)x_n(z) - 2j(z, x_n(z)) \right) dz \longrightarrow -\infty.$$

From (4.154) we know that

$$h_n(z)x_n(z) - 2j(z, x_n(z)) \leq \xi_5 \quad \text{for a.a. } z \in \Omega \setminus C_1$$

and so

$$\int_{\Omega} \left(h_n(z)x_n(z) - 2j(z, x_n(z)) \right) dz$$

$$\leq \int_{C_1} \left(h_n(z)x_n(z) - 2j(z, x_n(z)) \right) dz + \xi_5 |(\Omega \setminus C_1)_c|_N \longrightarrow -\infty,$$

which contradicts (4.153). This proves the boundedness of $\{x_n\}_{n \geq 1} \subseteq H_0^1(\Omega)$ and so passing to a subsequence if necessary, we may assume that

$$x_n \overset{w}{\longrightarrow} x_0 \text{ in } H_0^1(\Omega),$$
$$x_n \longrightarrow x_0 \text{ in } L^p(\Omega),$$

for some $x_0 \in H_0^1(\Omega)$. From the choice of the sequence $\{x_n\}_{n \geq 1} \subseteq H_0^1(\Omega)$ we have

$$
\left| \langle x_n^*, x_n - x_0 \rangle_{H_0^1(\Omega)} \right|
$$

$$
= \left| \langle A(x_n), x_n - x_0 \rangle_{H_0^1(\Omega)} - \lambda_k \int_\Omega x_n(z)(x_n - x_0)(z)\, dz \right.
$$

$$
\left. - \int_\Omega h_n(z)(x_n - x_0)(z)\, dz \right|
$$

$$
\leq \varepsilon_n \| x_n - x_0 \|_{H^1(\Omega)} ,
$$

with $\varepsilon_n \searrow 0$. Note that

$$
\int_\Omega x_n(z)(x_n - x_0)(z)\, dz \longrightarrow 0
$$

and

$$
\int_\Omega h_n(z)(x_n - x_0)(z)\, dz \longrightarrow 0.
$$

So it follows that

$$
\lim_{n \to +\infty} \langle A(x_n), x_n - x_0 \rangle_{H_0^1(\Omega)} = 0.
$$

From the maximal monotonicity of A we have

$$
\langle A(x_n), x_n \rangle_{H_0^1(\Omega)} \longrightarrow \langle A(x_0), x_0 \rangle_{H_0^1(\Omega)}
$$

and so

$$
\| \nabla x_n \|_2 \longrightarrow \| \nabla x_0 \|_2 .
$$

Because $\nabla x_n \xrightarrow{w} \nabla x_0$ in $L^2(\Omega; \mathbb{R}^N)$, from the Kadec-Klee property of Hilbert spaces, we conclude that

$$
\nabla x_n \longrightarrow \nabla x_0 \quad \text{in } L^2(\Omega; \mathbb{R}^N)
$$

and so finally

$$
x_n \longrightarrow x_0 \quad \text{in } H_0^1(\Omega).
$$

□

Now we can show that $\overline{\varphi}$ (hence ψ too) satisfies the nonsmooth C-condition.

PROPOSITION 4.6.4
If hypotheses $H(j)_1$ hold,
<u>*then*</u> *$\overline{\varphi}$ satisfies the nonsmooth C-condition.*

PROOF   Let $c \in \mathbb{R}$ and let $\{u_n\}_{n \geq 1} \subseteq \overline{X}_0$ be a sequence, such that

$$\overline{\varphi}(u_n) \longrightarrow c \quad \text{and} \quad \left(1 + \|u_n\|_{H^1(\Omega)}\right) m^{\overline{\varphi}}(u_n) \longrightarrow 0.$$

As before we can find $\overline{v}_n^* \in \partial \overline{\varphi}(u_n)$ such that $m^{\overline{\varphi}}(u_n) = \|\overline{v}_n^*\|_{H^{-1}(\Omega)}$. From (4.148) we know that we can find $v_n^* \in \partial \varphi\left(u_n + \vartheta(u_n)\right)$ such that

$$\overline{v}_n^* = \mathrm{P}_{\overline{X}_0^*} v_n^* \qquad \forall\, n \geq 1.$$

Recall that for $n \geq 1$, $0 \in \mathrm{P}_{\widehat{X}^*} \partial \varphi\left(u_n + \vartheta(u_n)\right)$ (see Proposition 4.6.1). Therefore we may assume that $v_n^* = \overline{v}_n^* + 0$ for $n \geq 1$ with $\overline{v}_n^* \in \overline{X}_0^*$, $0 \in \widehat{X}^*$. We have

$$\left(1 + \|x_n\|_{H^1(\Omega)}\right) \|v_n^*\|_{H^1(\Omega)} \longrightarrow 0 \quad \text{with } v_n^* \in \partial \varphi(x_n + \vartheta(x_n)).$$

Using Proposition 4.6.3, we can extract a subsequence of $\{x_n\}_{n \geq 1}$, which is strongly convergent. This proves the proposition.   ☐

Next we shall show that $\psi = -\overline{\varphi}$ is bounded below. To do this we shall need the following Lemma.

LEMMA 4.6.3
If $V \subseteq H_0^1(\Omega)$ is a finite dimensional subspace with unique continuation property and $V \subseteq L^\infty(\Omega)$,
then

(a) *for every $\varepsilon_1 > 0$ we can find $\delta(\varepsilon_1) > 0$ such that*

$$\left|\left\{z \in \Omega: \ |x(z)| < \delta(\varepsilon_1)\,\|x\|_{H^1(\Omega)}\right\}\right|_N < \varepsilon_1 \qquad \forall\, x \in V \setminus \{0\};$$

(b) *for every $\varepsilon_2 > 0$ we can find $\delta(\varepsilon_2) > 0$ such that*

$$\left|\left\{z \in \Omega: \ |y(z)| > \delta(\varepsilon_2)\,\|y\|_{H^1(\Omega)}\right\}\right|_N < \varepsilon_2 \qquad \forall\, y \in V^\perp.$$

PROOF   **(a)** Since all norms on V are equivalent, we can consider the $\|\cdot\|_\infty$-norm on V. Evidently it is enough to show that for a given $\varepsilon_1 > 0$ we can find $\delta(\varepsilon_1) > 0$ such that

$$\left|\left\{z \in \Omega: \ |x(z)| < \delta(\varepsilon_1)\right\}\right|_N < \varepsilon_1 \qquad \forall\, x \in \partial B_r.$$

To this end we show that for every $x \in \partial B_r$ there exists $\delta(x, \varepsilon_1) > 0$ such that

$$\left|\left\{z \in \Omega: \ |v(z)| < \delta(x, \varepsilon_1)\right\}\right|_N < \varepsilon_1 \qquad \forall\, v \in V, \ \|v - x\|_{H^1(\Omega)} < \delta(x, \varepsilon_1).$$
$$(4.155)$$

We argue by contradiction. So suppose that there exists $x_0 \in \partial B_r$ such that for every $n \geq 1$ we can find $v_n \in V$ such that

$$\|v_n - x_0\|_{H^1(\Omega)} < \frac{1}{2n} \quad \text{and} \quad \left|\left\{z \in \Omega : |v_n(z)| < \tfrac{1}{2n}\right\}\right|_N \geq \varepsilon_1.$$

We introduce the set

$$\Omega_n \overset{df}{=} \left\{z \in \Omega : |v_n(z)| < \tfrac{1}{2n}\right\} \qquad \forall\, n \geq 1.$$

Then

$$\Omega_n \subseteq \left\{z \in \Omega : |x_0(z)| < \tfrac{1}{n}\right\} \qquad \forall\, n \geq 1$$

and so

$$\left|\left\{z \in \Omega : |x_0(z)| < \tfrac{1}{n}\right\}\right|_N \geq \varepsilon_1 \qquad \forall\, n \geq 1.$$

By the unique continuation property of V, it follows that $x_0 \equiv 0$, a contradiction to the fact that $x_0 \in \partial B_r$. This proves (4.155).

Now consider the family $\left\{B_{\delta(x,\varepsilon_1)}(x)\right\}_{x \in \partial B_1}$. This is an open cover of the compact set ∂B_1 (recall that V is finite dimensional). So we can find a finite subcover $\left\{B_{\delta(x_k,\varepsilon_1)}(x_k)\right\}_{k=1}^m$. Let

$$\delta(\varepsilon_1) \overset{df}{=} \min\left\{\delta(x_k, \varepsilon_1)\right\}_{k=1}^m.$$

Then this $\delta(\varepsilon_1) > 0$ does the job.

(b) Suppose that the result is not true. Then there exists $\varepsilon_0 > 0$ such that for every $n \geq 1$ we can find $y_n \in V^{\perp}$ with $\|y_n\|_{H^1(\Omega)} = 1$ satisfying

$$\left|\left\{z \in \Omega : |y_n(z)| > n\right\}\right|_N \geq \varepsilon_0.$$

By passing to a suitable subsequence if necessary, we may assume that

$$y_n \overset{w}{\longrightarrow} y \quad \text{in } H_0^1(\Omega).$$

Because of the compactness of the embedding $H_0^1(\Omega) \subseteq L^2(\Omega)$, we have that

$$y_n \longrightarrow y \quad \text{in } L^2(\Omega).$$

For every $n \geq 1$ we have

$$\int_{\Omega} |y_n(z)|^2\, dz \geq \int_{\{z \in \Omega : |y_n(z)| \geq n\}} |y_n(z)|^2\, dz \geq \varepsilon_0 n^2$$

and so

$$\lim_{n \to +\infty} \int_{\Omega} |y_n(z)|^2\, dz = +\infty,$$

a contradiction to the fact that $y_n \longrightarrow y$ in $L^2(\Omega)$. $\square$

Using this lemma we can show that $\psi = -\overline{\varphi}$ is bounded below.

PROPOSITION 4.6.5
If *hypotheses* $H(j)_1$ *hold,*
then ψ *is bounded below.*

PROOF   We show that $-\left(\varphi|_{\overline{X}_0}\right)$ is bounded below. Then because $-\psi = \overline{\varphi} \leq \left(\varphi|_{\overline{X}_0}\right)$, we can conclude that ψ is bounded below. To this end we proceed by contradiction. Suppose that $-\left(\varphi|_{\overline{X}_0}\right)$ is not bounded below. Then we can find a sequence $\{x_n\}_{n\geq 1} \subseteq \overline{X}_0$, such that

$$\varphi(x_n) \geq n \qquad \forall\, n \geq 1 \tag{4.156}$$

and

$$\|x_n\|_{H^1(\Omega)} \longrightarrow +\infty.$$

By virtue of hypothesis $H(j)_1(vii)$, for a given $\varepsilon > 0$, we can find $M_4 = M_4(\varepsilon) > 0$ such that

$$-\frac{\varepsilon}{2}\zeta^2 \leq j(x,\zeta) \quad \text{for a.a. } z \in \Omega \text{ and all } |\zeta| \geq M_4.$$

On the other hand, as before via the mean value theorem for locally Lipschitz functions, we can find $\xi_6 > 0$ such that

$$\left|j(z,\zeta)\right| \leq \xi_6 \quad \text{for a.a. } z \in \Omega \text{ and all } |\zeta| \leq M_4.$$

So finally we see that

$$-\frac{\varepsilon}{2}\zeta^2 - \xi_6 \leq j(z,\zeta) \quad \text{for a.a. } z \in \Omega \text{ and all } \zeta \in \mathbb{R}. \tag{4.157}$$

We write

$$x_n = \overline{x}_n + \overline{\overline{x}}_n, \quad \text{with } \overline{x}_n \in \overline{X}, \ \overline{\overline{x}}_n \in E(\lambda_k) \quad n \geq 1.$$

First assume that

$$\frac{\|\nabla \overline{x}_n\|_2}{\|\nabla x_n\|_2} \longrightarrow \mu \neq 0. \tag{4.158}$$

Exploiting the orthogonality among $\overline{X}$ and $E(\lambda_k)$, the fact that $\left\|\nabla \overline{\overline{x}}_n\right\|_2^2 = \lambda_k \left\|\overline{\overline{x}}_n\right\|_2^2$ and (4.157) we have

$$
\begin{aligned}
\varphi(x_n) &= \frac{1}{2}\left\|\nabla \overline{x}_n\right\|_2^2 - \frac{\lambda_k}{2}\left\|\overline{\overline{x}}_n\right\|_2^2 - \int_\Omega j\big(z, x_n(z)\big)\, dz \\
&\leq \frac{1}{2}\left\|\nabla \overline{x}_n\right\|_2^2 - \frac{\lambda_k}{2}\left\|\overline{\overline{x}}_n\right\|_2^2 + \frac{\varepsilon}{2}\left\|x_n\right\|_2^2 + \xi_6|\Omega|_N.
\end{aligned}
$$

Using the variational characterization of λ_k given in Theorem 1.5.3(b), we obtain

$$\varphi(x_n) \leq \frac{1}{2}\left(1 - \frac{\lambda_k}{\lambda_{k-1}}\right)\|\nabla\overline{x}_n\|_2^2 + \frac{\varepsilon}{2}\|x_n\|_2^2 + \xi_6|\Omega|_N$$

$$\leq \frac{1}{2}\|\nabla x_n\|_2^2\left(\left(1 - \frac{\lambda_k}{\lambda_{k-1}}\right)\frac{\|\nabla\overline{x}_n\|_2^2}{\|\nabla x_n\|_2^2} + \frac{\varepsilon}{\lambda_k}\right) + \xi_6|\Omega|_N. \quad (4.159)$$

Recall that by hypothesis

$$\|\nabla x_n\|_2 \longrightarrow +\infty,$$

so from (4.158) and recalling that $\lambda_{k-1} < \lambda_k$, by passing to the limit as $n \to +\infty$ in (4.159), we see that $\varphi(x_n) \longrightarrow -\infty$, a contradiction to (4.156).

Next assume that

$$\frac{\|\nabla\overline{x}_n\|_2}{\|\nabla x_n\|_2} \longrightarrow 0.$$

Arguing as in the proof of Proposition 4.1.7, via hypotheses $H(j)_1(iv)$ and (vii), for a given $\eta > 0$ we can find $M_5 = M_5(\eta) > 0$ such that

$$j(z,\zeta) \geq \frac{\eta}{2} \quad \text{for a.a. } z \in \Omega \text{ and all } |\zeta| \geq M_5.$$

Since $\eta > 0$ was arbitrary, it follows that

$$j(z,\zeta) \longrightarrow +\infty \quad \text{as } \zeta \to +\infty \text{ uniformly for a.a. } z \in \Omega,$$
$$j(z,\zeta) \longrightarrow +\infty \quad \text{as } \zeta \to -\infty \text{ uniformly for a.a. } z \in \Omega.$$

Therefore

$$j(z,\zeta) \longrightarrow +\infty \quad \text{as } |\zeta| \to +\infty \text{ uniformly for a.a. } z \in \Omega. \quad (4.160)$$

Using Lemma 4.6.3(a) we have that for a given $\varepsilon > 0$ we can find $\gamma_\varepsilon > 0$ such that

$$\left|\left\{z \in \Omega : |v(z)| < \gamma_\varepsilon \|v\|_{H^1(\Omega)}\right\}\right|_N < \varepsilon \quad \forall\, v \in E(\lambda_k)$$

(recall that $E(\lambda_k)$ has the unique continuation property). Let us set

$$C_n \stackrel{df}{=} \left\{z \in \Omega : |\overline{\overline{x}}_n(z)| \geq \gamma_\varepsilon \|\overline{\overline{x}}_n\|_{H^1(\Omega)}\right\} \quad \forall\, n \geq 1.$$

We have that $|\Omega \setminus C_n|_N < \varepsilon$ for $n \geq 1$.

Now let us show that the sequence $\{\overline{x}_n\}_{n\geq 1}$ is bounded in $H_0^1(\Omega)$. Suppose that this is not true. We may assume that

$$\|\overline{x}_n\|_{H^1(\Omega)} \longrightarrow +\infty.$$

Recall that

$$\varphi(x_n) \leq \frac{1}{2}\left(1 - \frac{\lambda_k}{\lambda_{k-1}}\right)\|\nabla \overline{x}_n\|_2^2 - \int_\Omega j\big(z, x_n(z)\big)\, dz \qquad \forall\, n \geq 1.$$

Because of (4.160) we can find a constant $\sigma_0 > 0$ such that

$$j(z, \varsigma) \geq 0 \quad \text{for a.a. } z \in \Omega \text{ and all } |\varsigma| > \sigma_0.$$

Hence it follows that

$$
\begin{aligned}
-\int_\Omega j\big(z, x_n(z)\big)\, dz &= - \int_{\{z \in \Omega:\ |x_n(z)| > \sigma_0\}} j\big(z, x_n(z)\big)\, dz \\
&\quad - \int_{\{z \in \Omega:\ |x_n(z)| \leq \sigma_0\}} j\big(z, x_n(z)\big)\, dz \\
&\leq - \int_{\{z \in \Omega:\ |x_n(z)| \leq \sigma_0\}} j\big(z, x_n(z)\big)\, dz \ \leq\ \xi_7 |\Omega|_N,
\end{aligned}
$$

for some $\xi_7 > 0$ (see hypothesis $H(j)_1(iii)$). So we obtain that

$$\varphi(x_n) \leq \frac{1}{2}\left(1 - \frac{\lambda_k}{\lambda_{k-1}}\right)\|\nabla \overline{x}_n\|_2^2 + \xi_7 |\Omega|_N.$$

Since $\lambda_{k-1} < \lambda_k$ and we assumed that $\|\overline{x}_n\|_{H^1(\Omega)} \to +\infty$, we arrive to the conclusion that

$$\varphi(x_n) \longrightarrow -\infty,$$

a contradiction to (4.156). This proves that the sequence $\{\overline{x}_n\}_{n \geq 1}$ is bounded in $H_0^1(\Omega)$.

Because $\overline{x}_n \in \overline{X}$ and the space $\overline{X} \subseteq C(\overline{\Omega})$ is finite dimensional, we can find $\xi_8 > 0$ such that

$$\big|\overline{x}_n(z)\big| \leq \xi_8 \qquad \forall\, z \in \Omega,\ n \geq 1.$$

By virtue of (4.160), for a given $\eta_1 > 0$ we can find $M_6 = M_6(\eta_1) > 0$ such that

$$j(z, \varsigma) \geq \eta_1 \quad \text{for a.a. } z \in \Omega \text{ and all } |\varsigma| \geq M_6.$$

Let

$$D_n \overset{df}{=} \big\{z \in \Omega : |x_n(z)| \geq M_6\big\} \qquad \forall\, n \geq 1.$$

If $z_0 \in C_n$, then

$$\big|x_n(z_0)\big| \geq \big|\overline{\overline{x}}_n(z_0)\big| - \big|\overline{x}_n(z_0)\big| \geq \gamma_\varepsilon \big\|\overline{\overline{x}}_n\big\|_{H^1(\Omega)} - \xi_8.$$

Note that $\left\|\overline{\overline{x}}_n\right\|_{H^1(\Omega)} \longrightarrow +\infty$. So there exists $n_0 \geq 1$ large enough such that

$$\gamma_\varepsilon \left\|\overline{\overline{x}}_n\right\|_{H^1(\Omega)} - \xi_8 \geq M_6 \qquad \forall\, n \geq n_0$$

and so $z_0 \in D_n$ for $n \geq n_0$, i.e.

$$C_n \subseteq D_n \qquad \forall\, n \geq n_0.$$

Using (4.157) and the fact that $|\Omega \setminus D_n|_N < \varepsilon$ (as $\Omega \setminus D_n \subseteq \Omega \setminus C_n$), for $n \geq n_0$ we have that

$$\int_Z j(z, x_n(z))\, dz = \int_{D_n} j(z, x_n(z))\, dz + \int_{\Omega \setminus D_n} j(z, x_n(z))\, dz$$

$$\geq \eta_1 |D_n|_N - \left(\frac{\varepsilon}{2} M_6^2 + \xi_6\right) |\Omega \setminus D_n|_N \geq \eta_1 |D_n|_N - \left(\frac{\varepsilon}{2} M_6^2 + \xi_6\right)\varepsilon$$

$$\geq \eta_1\big(|\Omega|_N - \varepsilon\big) - \left(\frac{\varepsilon}{2} M_6^2 + \xi_6\right)\varepsilon,$$

so

$$\liminf_{n \to +\infty} \int_Z j(z, x_n(z))\, dz \geq \eta_1\big(|\Omega|_N - \varepsilon\big) - \left(\frac{\varepsilon}{2} M_6^2 + \xi_6\right)\varepsilon.$$

Since $\varepsilon > 0$ was arbitrary, we let $\varepsilon \searrow 0$ and obtain

$$\liminf_{n \to +\infty} \int_Z j(z, x_n(z))\, dz \geq \eta_1 |\Omega|_N.$$

Because $\eta_1 > 0$ was arbitrary, we conclude that

$$\int_Z j(z, x_n(z))\, dz \longrightarrow +\infty.$$

From the choice of the sequence $\{x_n\}_{n \geq 1} \subseteq \overline{X}_0$, we have

$$n \leq \varphi(x_n) \leq \frac{1}{2}\left(1 - \frac{\lambda_k}{\lambda_{k-1}}\right) \|\nabla \overline{x}_n\|_2^2 - \int_Z j(z, x_n(z))\, dz \longrightarrow -\infty,$$

a contradiction. Therefore $-\left(\varphi|_{\overline{X}_0}\right)$ is bounded below and so ψ is bounded below too. $\qquad\square$

Now we are ready for our multiplicity result concerning problem (4.137).

THEOREM 4.6.1
If hypotheses $H(j)_1$ hold,
then problem (4.137) has at least two nontrivial solutions.

PROOF   Note that $\psi(0) = 0$. If $\inf\limits_{\overline{X}_0} \psi = 0$, then by virtue of Proposition 4.6.2, all $x \in V$ with $\|x\|_V \leq \delta$ are critical points of ψ.

So assume that $\inf\limits_{\overline{X}_0} \psi < 0$. Then Propositions 4.6.2, 4.6.4 and 4.6.5 permit the use of Theorem 2.4.1, which gives two nontrivial critical points $x_1, x_2 \in \overline{X}_0$ of ψ, i.e.

$$0 \in \partial\psi(x_i) \quad \text{for } i = 1, 2.$$

Then $0 \in \partial\overline{\varphi}(x_i)$ for $i = 1, 2$ and finally from (4.148) we have

$$0 \in \mathrm{P}_{\overline{X}_0^*}\partial\varphi\big(x_i + \vartheta(x_i)\big) \quad \text{for } i = 1, 2.$$

Recall that $0 \in \mathrm{P}_{\widehat{X}^*}\partial\varphi\big(x_i + \vartheta(x_i)\big)$ for $i = 1, 2$, and $\widehat{X} = \overline{X}_0^{\perp}$. So we infer that

$$0 \in \partial\varphi(x_i + \vartheta(x_i)) \quad \text{for } i = 1, 2.$$

Let us set $u_i = x_i + \vartheta(x_i)$ for $i = 1, 2$. Then u_1 and u_2 are two nontrivial critical points of φ, thus they are two nontrivial solutions of (4.137). $\quad\square$

Next we turn our attention to the following problem:

$$\begin{cases} -\Delta x(z) - \lambda_1 x(z) \in \partial j\big(z, x(z)\big) & \text{for a.a. } z \in \Omega, \\ x|_{\partial\Omega} = 0. \end{cases} \tag{4.161}$$

Now we will have a different set of conditions for the nonsmooth potential, which will allow for partial resonance at infinity and for full resonance from above and below at the origin between successive eigenvalues of $\big(-\Delta, H_0^1(\Omega)\big)$. As before $\{\lambda_n\}_{n\geq 1}$ is the sequence of distinct eigenvalues of $\big(-\Delta, H_0^1(\Omega)\big)$. In this case our hypotheses on the nonsmooth potential function j are the following:

$\underline{H(j)_2}$ $j \colon Z \times \mathbb{R} \longrightarrow \mathbb{R}$ is a function, such that

(i) for every $\zeta \in \mathbb{R}$, the function

$$\Omega \ni z \longrightarrow j(z, \zeta) \in \mathbb{R}$$

is measurable and $j(z, 0) = 0$ for almost all $z \in \Omega$;

(ii) for almost all $z \in \Omega$ the function

$$\mathbb{R} \ni \zeta \longrightarrow j(z, \zeta) \in \mathbb{R}$$

is locally Lipschitz;

(iii) we have

$$|u| \leq a_1(z) + c_1|\zeta| \quad \text{for a.a. } z \in \Omega, \text{ all } \zeta \in \mathbb{R} \text{ and all } u \in \partial j(z, \zeta),$$

with $a_1 \in L^{\infty}(\Omega)_+$, $c_1 > 0$;

(iv) there exists $\vartheta \in L^\infty(\Omega)$ such that $\vartheta(z) \leq 0$ for almost all $z \in \Omega$ with strict inequality on a set of positive measure and

$$\limsup_{|\zeta| \to +\infty} \frac{2j(z,\zeta)}{\zeta^2} \leq \vartheta(z)$$

uniformly for almost all $z \in \Omega$;

(v) there exists $\delta > 0$ such that for almost all $z \in \Omega$ and for all $\zeta \in \mathbb{R}$ with $|\zeta| \leq \delta$ we have

$$(\lambda_k - \lambda_1)\zeta^2 \leq 2j(z,\zeta) \leq (\lambda_{k+1} - \lambda_1)\zeta^2 \qquad \forall\, k \geq 1.$$

REMARK 4.6.2 Hypothesis $H(j)_2(v)$ is satisfied if for example we have

$$\lambda_k - \lambda_1 < \liminf_{\zeta \to 0} \frac{2j(z,\zeta)}{\zeta^2} \leq \limsup_{\zeta \to 0} \frac{2j(z,\zeta)}{\zeta^2} < \lambda_{k+1} - \lambda_1. \qquad (4.162)$$

Note that our requirement is more general than the above inequalities since we allow interaction (resonance) of $\frac{2j(z,\zeta)+\lambda_1\zeta^2}{\zeta^2}$ near the origin with the successive eigenvalues λ_k and λ_{k+1}, while condition (4.162) does not permit this. So with $H(j)_2(v)$ we can have double resonance between λ_k and λ_{k+1} at the origin. $\Box$

We consider the energy functional $\varphi : H_0^1(\Omega) \longrightarrow \mathbb{R}$, defined by

$$\varphi(x) \overset{df}{=} \frac{1}{2}\|\nabla x\|_2^2 - \frac{\lambda_k}{2}\|x\|_2^2 - \int_\Omega j(z, x(z))\, dz \qquad \forall\, x \in H_0^1(\Omega).$$

We know that φ is locally Lipschitz.

PROPOSITION 4.6.6
If hypotheses $H(j)_2$ hold, <u>then</u> φ is coercive.

PROOF By virtue of hypothesis $H(j)_2(iv)$ for a given $\varepsilon > 0$ we can find $M_1 = M_1(\varepsilon) > 0$, such that

$$2j(z,\zeta) \leq \varepsilon\zeta^2 \quad \text{for a.a. } z \in \Omega \text{ and all } |\zeta| \geq M_1. \qquad (4.163)$$

Then for every $x \in H_0^1(\Omega)$ we have

$$\varphi(x) = \frac{1}{2}\|\nabla x\|_2^2 - \frac{\lambda_1}{2}\|x\|_2^2 - \int_\Omega j(z, x(z))\, dz$$

$$= \frac{1}{2}\|\nabla x\|_2^2 - \frac{\lambda_1}{2}\|x\|_2^2$$

$$- \int\limits_{\{|\zeta| \geq M_1\}} j\big(z, x(z)\big)\, dz - \int\limits_{\{|\zeta| < M_1\}} j\big(z, x(z)\big)\, dz. \quad (4.164)$$

Because of (4.163), we have

$$\int\limits_{\{|\zeta| \geq M_1\}} j\big(z, x(z)\big)\, dz \ \leq\ \frac{\varepsilon}{2}\, \|x\|_2^2 \qquad \forall\, x \in H_0^1(\Omega). \quad (4.165)$$

Also from hypothesis $H(j)_2(iii)$ and the mean value theorem for locally Lipschitz functions (see Proposition 1.3.15) we have

$$\big|j(z, \zeta)\big| \ \leq\ a_2(z) + c_2 |\zeta|^2 \quad \text{for a.a. } z \in \Omega \text{ and all } \zeta \in \mathbb{R},$$

$a_2 \in L^\infty(\Omega)_+$ $c_2 > 0$. Hence

$$\int\limits_{\{|\zeta| < M_1\}} j\big(z, x(z)\big)\, dz \ \leq\ c_3 \qquad \forall\, x \in H_0^1(\Omega), \quad (4.166)$$

for some $c_3 > 0$. Using (4.165) and (4.166) in (4.164) we obtain

$$\varphi(x) \ \geq\ \frac{1}{2}\, \|\nabla x\|_2^2 - \frac{\lambda_1}{2}\, \|x\|_2^2 - \frac{\varepsilon}{2}\, \|x\|_2^2 - c_3 \qquad \forall\, x \in H_0^1(\Omega). \quad (4.167)$$

For $x \in H_0^1(\Omega)$ we write

$$x \ = \ x^1 + x^\perp, \quad \text{with } x^1 \in E(\lambda_1),\ x^\perp \in E(\lambda_1)^\perp,$$

i.e. $x^1 = t u_1$ for some $t \in \mathbb{R}$, where u_1 is the normalized principal eigenfunction of $\big(-\Delta, H_0^1(\Omega)\big)$. Exploiting the orthogonality between $E(\lambda_1)$ and $E(\lambda_1)^\perp$ and the fact that $\|\nabla u_1\|_2^2 = \lambda_1 \|u_1\|_2^2$ (see Theorem 1.5.3(b)) from (4.167) we have

$$\varphi(x) \ \geq\ \frac{1}{2}\, \big\|\nabla x^\perp\big\|_2^2 - \frac{\lambda_1}{2}\, \big\|x^\perp\big\|_2^2 - \frac{\varepsilon}{2}\, \|x\|_2^2 - c_3 \qquad \forall\, x \in H_0^1(\Omega). \quad (4.168)$$

Recall that

$$\begin{aligned}
\|\nabla v\|_2^2 &\geq \lambda_2 \|v\|_2^2 &\quad \forall\, v \in E(\lambda_1)^\perp, \\
\|\nabla y\|_2^2 &\geq \lambda_1 \|y\|_2^2 &\quad \forall\, y \in H_0^1(\Omega)
\end{aligned}$$

(see Theorem 1.5.3(b)). So for all $x \in H_0^1(\Omega)$, we have

$$\varphi(x) \ \geq\ \frac{1}{2}\Big(1 - \frac{\lambda_1}{\lambda_2} - \frac{\varepsilon}{\lambda_1}\Big)\, \big\|\nabla x^\perp\big\|_2^2 - \frac{\varepsilon}{2\lambda_1}\, \big\|x^1\big\|_2^2 - c_3 \qquad \forall\, x \in H_0^1(\Omega). \quad (4.169)$$

With the help of this estimate, we shall show that φ is coercive. To this end consider a sequence $\{x_n\}_{n \geq 1} \subseteq H_0^1(\Omega)$ such that

$$\|x_n\|_{H^1(\Omega)} \ \longrightarrow\ +\infty.$$

We have

$$x_n = x_n^1 + x_n^\perp, \quad \text{with} \quad x_n^1 = t_n u_1, \ t_n \in \mathbb{R} \quad \text{and} \quad x_n^\perp \in E(\lambda_1)^\perp \qquad \forall\, n \geq 1.$$

If the sequence $\{x_n^1\}_{n \geq 1} \subseteq H_0^1(\Omega)$ is bounded, then we must have

$$\left\| \nabla x_n^\perp \right\|_2 \longrightarrow +\infty$$

and so from (4.169) it follows that

$$\varphi(x_n) \longrightarrow +\infty,$$

i.e. φ is coercive.

So assume that the sequence $\{x_n^1\}_{n \geq 1} \subseteq H_0^1(\Omega)$ is unbounded. By passing to a suitable subsequence if necessary, we may assume that

$$\left\| x_n^1 \right\|_{H^1(\Omega)} = |t_n| \longrightarrow +\infty.$$

The sequence $\left\{ \dfrac{t_n^2}{\|x_n\|_{H^1(\Omega)}^2} \right\}_{n \geq 1}$ is in the interval $[0, 1]$ and so passing to a further subsequence if necessary, we may assume that

$$\frac{t_n^2}{\|x_n\|_{H^1(\Omega)}^2} \longrightarrow \eta,$$

with $\eta \in [0, 1]$.

Case 1. $0 \leq \eta < 1$.

From (4.169) and since $\|\nabla u_1\|_2 = 1$ we have

$$\varphi(x_n) \geq \frac{1}{2}\left(1 - \frac{\lambda_1}{\lambda_2} - \frac{\varepsilon}{\lambda_1} \right) \left\| \nabla x_n^\perp \right\|_2^2 - \frac{\varepsilon t_n^2}{2\lambda_1} - c_3 \qquad \forall\, n \geq 1,$$

so

$$\frac{\varphi(x_n)}{\|\nabla x_n\|_2^2} \geq \frac{1}{2}\left(1 - \frac{\lambda_1}{\lambda_2} - \frac{\varepsilon}{\lambda_1} \right) \frac{\left\| \nabla x_n^\perp \right\|_2^2}{\|\nabla x_n\|_2^2} - \frac{\varepsilon}{2\lambda_1} \frac{t_n^2}{\|\nabla x_n\|_2^2} - \frac{c_3}{\|\nabla x_n\|_2^2}. \qquad (4.170)$$

Since $H_0^1(\Omega) = E(\lambda_1) \oplus E(\lambda_1)^\perp$, we have

$$1 = \frac{\|\nabla x_n\|_2^2}{\|\nabla x_n\|_2^2} = \frac{\left\| \nabla x_n^\perp \right\|_2^2 + t_n^2 \|\nabla u_1\|_2^2}{\|\nabla x_n\|_2^2} = \frac{\left\| \nabla x_n^\perp \right\|_2^2 + t_n^2}{\|\nabla x_n\|_2^2}.$$

So, we obtain that

$$\frac{\left\| \nabla x_n^\perp \right\|_2^2}{\|x_n\|_{H^1(\Omega)}^2} \longrightarrow 1 - \eta.$$

By passing to the limit as $n \to +\infty$ in (4.170), we obtain

$$\liminf_{n \to +\infty} \frac{\varphi(x_n)}{\|x_n\|_2^2} \geq \frac{1}{2} \left(1 - \frac{\lambda_1}{\lambda_2} - \frac{\varepsilon}{\lambda_1} \right) (1 - \eta) - \frac{\varepsilon \eta}{2\lambda_1}.$$

If we choose $\varepsilon > 0$ small enough so that $\frac{\lambda_1}{\lambda_2} + \frac{\varepsilon}{\lambda_1} < 1$ (recall that $\lambda_1 < \lambda_2$) we infer that

$$\liminf_{n \to +\infty} \frac{\varphi(x_n)}{\|x_n\|_{H^1(\Omega)}^2} > 0,$$

hence

$$\varphi(x_n) \longrightarrow +\infty,$$

which proves the coercivity of φ.

Case 2. Suppose that $\eta = 1$, i.e.

$$\frac{t_n^2}{\|x_n\|_{H^1(\Omega)}^2} \longrightarrow 1.$$

We have

$$\frac{\left\| \nabla x_n^{\perp} \right\|_2^2}{\|\nabla x_n\|_2^2} \longrightarrow 0.$$

First, we will show that

$$\limsup_{n \to +\infty} \frac{\int_\Omega j\big(z, x_n(z)\big)\, dz}{\|x_n\|_{H^1(\Omega)}^2} < 0. \tag{4.171}$$

From Lemma 4.6.3 we know for given $\varepsilon_1, \varepsilon_2 > 0$, we can find $\delta_1 = \delta_1(\varepsilon_1) \in (0, \varepsilon_1)$ and $\delta_2 = \delta_2(\varepsilon_2) \in (0, \varepsilon_2)$ such that

$$\left| \left\{ z \in \Omega : \ |v(z)| < \delta_1 \|v\|_2 \right\} \right|_N < \varepsilon_1 \qquad \forall\, v \in E(\lambda_1)$$

and

$$\left| \left\{ z \in \Omega : \ |y(z)| > \delta_2 \|\nabla y\|_2 \right\} \right|_N < \varepsilon_2 \qquad \forall\, y \in E(\lambda_1)^{\perp}.$$

We introduce the following two Lebesgue measurable subsets of Ω:

$$A_n \overset{df}{=} \left\{ z \in \Omega : \ |x_n^1(z)| \geq \delta_1 \left\| \nabla x_n^1 \right\|_2 \right\}$$
$$C_n \overset{df}{=} \left\{ z \in \Omega : \ |x_n^{\perp}(z)| \leq \delta_2 \left\| \nabla x_n^{\perp} \right\|_2 \right\},$$

for $n \geq 1$. Evidently $|\Omega \setminus A_n|_N < \varepsilon_1$ and $|\Omega \setminus C_n|_N < \varepsilon_2$. So if we choose $\varepsilon_1, \varepsilon_2 > 0$ small enough, we may have $|A_n \cap C_n|_N \neq \emptyset$ for all $n \geq 1$. Because $E(\lambda_1) \subseteq C\left(\overline{\Omega}\right)$ is finite dimensional we can find $c_4 > 0$, such that

$$|v(z)| \leq c_4 \|\nabla v\|_2 \qquad \forall\, y \in E(\lambda_1),\ z \in \Omega.$$

Then for $z \in A_n \cap C_n$ and $n \geq 1$, we have

$$\frac{|x_n(z)|}{\|\nabla x_n\|_2} \leq \frac{|x_n^1(z)|}{\|\nabla x_n\|_2} + \frac{|x_n^\perp(z)|}{\|\nabla x_n\|_2} \leq c_4 \frac{\|\nabla x_0^1\|_2}{\|\nabla x_n\|_2} + \delta_2 \frac{\|\nabla x_n^\perp\|_2}{\|\nabla x_n\|_2}$$

and so

$$\limsup_{n \to +\infty} \frac{|x_n(z)|}{\|\nabla x_n\|_2} \chi_{A_n \cap C_n}(z) \leq c_4 \tag{4.172}$$

uniformly for almost all $z \in \Omega$.

Also for $z \in C_n \cap A_n$ and $n \geq 1$, we have

$$\frac{|x_n(z)|}{\|\nabla x_n\|_2} \geq \frac{|x_n^1(z)|}{\|\nabla x_n\|_2} - \frac{|x_n^\perp(z)|}{\|\nabla x_n\|_2} \geq \delta_1 \frac{\|\nabla x_n^1\|_2}{\|\nabla x_n\|_2} - \delta_2 \frac{\|\nabla x_n^\perp\|_2}{\|\nabla x_n\|_2}$$

and so

$$\liminf_{n \to +\infty} \frac{|x_n(z)|}{\|\nabla x_n\|_2} \chi_{A_n \cap C_n}(z) \geq \delta_1 \tag{4.173}$$

uniformly for almost all $z \in \Omega$.

From hypotheses $H(j)_2(iii)$ and (iv) and the mean value theorem for locally Lipschitz functions (see Proposition 1.3.15) we infer that for a given $\varepsilon > 0$, we can find $c_5 = c_5(\varepsilon) > 0$ such that

$$j(z, \zeta) \leq \frac{1}{2}(\vartheta(z) + \varepsilon)\zeta^2 + c_5 \quad \text{for a.a. } z \in \Omega \text{ and all } \zeta \in \mathbb{R}.$$

So using also (4.172) and (4.173), for a given $\varepsilon' > 0$, we can find $n_0 = n_0(\varepsilon') \geq 1$ such that for all $n \geq n_0$ we have

$$\int_{A_n \cap C_n} \frac{j(z, x_n(z))}{\|\nabla x_n\|_2^2} \, dz$$

$$\leq \frac{1}{2} \int_{A_n \cap C_n} (\vartheta(z) + \varepsilon) \frac{x_n(z)^2}{\|\nabla x_n\|_2^2} \, dz + c_5 \frac{|A_n \cap C_n|_N}{\|\nabla x_n\|_2^2}$$

$$\leq \frac{1}{2}(c_4 + \varepsilon')^2 \int_{\{\vartheta > -\varepsilon\}} (\vartheta(z) + \varepsilon) \, dz$$

$$+ \frac{1}{2}(\delta_1 - \varepsilon')^2 \int_{\{\vartheta < -\varepsilon\}} (\vartheta(z) + \varepsilon) \, dz + \varepsilon'$$

$$\leq \frac{1}{2}(c_4 + \varepsilon')^2 \int_\Omega (\vartheta(z) + \varepsilon)^+ \, dz$$

$$- \frac{1}{2}(\delta_1 - \varepsilon')^2 \int_\Omega (\vartheta(z) + \varepsilon)^- \, dz + \varepsilon'. \tag{4.174}$$

On the other hand for $z \in C_n \setminus A_n$ we have

$$\frac{|x_n(z)|}{\|\nabla x_n\|_2} \leq \frac{|x_n^1(z)|}{\|\nabla x_n\|_2} + \frac{|x_n^\perp(z)|}{\|\nabla x_n\|_2} \leq \delta_1 \frac{\|\nabla x_n^1\|_2}{\|\nabla x_n\|_2} + \delta_2 \frac{\|\nabla x_n^\perp\|_2}{\|\nabla x_n\|_2}$$

$$= \delta_1 \frac{t_n}{\|\nabla x_n\|_2} + \delta_2 \frac{\|\nabla x_n^\perp\|_2}{\|\nabla x_n\|_2}$$

and so

$$\limsup_{n\to+\infty} \frac{|x_n(z)|}{\|\nabla x_n\|_2} \chi_{A_n \setminus C_n}(z) \leq \delta_1 \tag{4.175}$$

uniformly for almost all $z \in \Omega$.

Recall that

$$|j(z,\zeta)| \leq a_2(z) + c_2|\zeta|^2$$

$a_2 \in L^\infty(\Omega)_+$, $c_2 > 0$. Combining this with (4.175) we see that for all $n \geq n_0$ we have

$$\int_{C_n \setminus A_n} \frac{|j(z,x_n(z))|}{\|\nabla x_n\|_2^2}\,dz \leq \int_{C_n \setminus A_n} \frac{a_2(z) + c_2|x_n(z)|^2}{\|\nabla x_n\|_2^2}\,dz$$
$$\leq \varepsilon' + c_2(\delta_1 + \varepsilon')^2 |\Omega \setminus A_n|_N \leq \varepsilon' + c_2(\delta_1 + \varepsilon')^2 \varepsilon_1. \tag{4.176}$$

Finally, for all $n \geq n_0$, we have

$$\int_{\Omega \setminus C_n} \frac{|j(z,x_n(z))|}{\|\nabla x_n\|_2^2}\,dz \leq \int_{\Omega \setminus C_n} \frac{a_2(z) + c_2|x_n(z)|^2}{\|\nabla x_n\|_2^2}\,dz$$
$$\leq \varepsilon' + c_6|\Omega \setminus C_n|_N \leq \varepsilon' + c_6\varepsilon_2. \tag{4.177}$$

From (4.174), (4.176) and (4.177) it follows that

$$\limsup_{n\to+\infty} \int_\Omega \frac{j(z,x_n(z))}{\|\nabla x_n\|_2^2}\,dz$$
$$\leq \frac{1}{2}(c_4 + \varepsilon')^2 \int_\Omega \big(\vartheta(z) + \varepsilon\big)^+\,dz - \frac{(\delta_1 - \varepsilon')^2}{2} \int_\Omega \big(\vartheta(z) + \varepsilon\big)^-\,dz$$
$$+3\varepsilon' + c_2(\delta_1 + \varepsilon')^2 \varepsilon_1 + c_6\varepsilon_2.$$

Recall that $\varepsilon, \varepsilon', \varepsilon_1, \varepsilon_2 > 0$ are arbitrary. Let $\varepsilon, \varepsilon', \varepsilon_2 \searrow 0$ and note that $\big(\vartheta(z) + \varepsilon\big)^+ \to 0$ and $-\big(\vartheta(z) + \varepsilon\big)^- \to \vartheta(z)$ for almost all $z \in \Omega$ (recall that $\vartheta(z) \leq 0$ for almost all $z \in \Omega$). So in the limit we obtain

$$\limsup_{n\to+\infty} \int_\Omega \frac{j(z,x_n(z))}{\|\nabla x_n\|_2^2}\,dz \leq \frac{1}{2}\delta_1^2 \int_\Omega \vartheta(z)\,dz + c_2\delta_1^2\varepsilon_1 \leq \frac{1}{2}\varepsilon_1^2 \int_\Omega \vartheta(z)\,dz + c_2\varepsilon_1^3$$

(recall that $\delta_1 \in (0, \varepsilon_1)$). Hence

$$\limsup_{n\to+\infty} \int_\Omega \frac{j(z,x_n(z))}{\|\nabla x_n\|_2^2}\,dz \leq \varepsilon_1^2 \left(\frac{1}{2}\int_\Omega \vartheta(z)\,dz + c_2\varepsilon_1\right).$$

Since $\int_\Omega \vartheta(z)\,dz < 0$ (see hypothesis $H(j)_2(iv)$) and $\varepsilon_1 > 0$ was arbitrary, we conclude that (4.171) is true.

Now because $\lambda_1 \|x_n\|_2^2 \leq \|\nabla x_n\|_2^2$ for all $n \geq 1$, we have

$$\varphi(x_n) = \frac{1}{2}\|\nabla x_n\|_2^2 - \frac{\lambda_1}{2}\|\nabla x_n\|_2^2 - \int_\Omega j(z, x_n(z))\, dz \geq -\int_\Omega j(z, x_n(z))\, dz,$$

so

$$\frac{\varphi(x_n)}{\|\nabla x_n\|_2^2} \geq -\int_\Omega \frac{j(z, x_n(z))}{\|\nabla x_n\|_2^2}\, dz \qquad \forall\, n \geq 1.$$

Using also (4.171), we get

$$\liminf_{n \to +\infty} \frac{\varphi(x_n)}{\|\nabla x_n\|_2^2} > 0.$$

From this it follows that at least for a subsequence we have that $\varphi(x_n) \longrightarrow +\infty$. Because every subsequence of $\{\varphi(x_n)\}_{n \geq 1}$ has a further subsequence such that $\varphi(x_n) \longrightarrow +\infty$, we conclude that for the original sequence we also have $\varphi(x_n) \longrightarrow +\infty$, hence φ is coercive in this case too. ∎

REMARK 4.6.3 In particular the coercivity of φ implies that φ is bounded below and satisfies the nonsmooth PS-condition. ∎

Let

$$Y_k \overset{df}{=} \bigoplus_{i=1}^k E(\lambda_i) \quad \text{and} \quad V_k \overset{df}{=} \bigoplus_{i=k+1}^\infty E(\lambda_i).$$

We have that

$$H_0^1(\Omega) = Y_k \oplus V_k, \quad V_k = Y_k^\perp \quad \text{and} \quad \dim Y_k < \infty.$$

In the next proposition we show that φ has the local linking geometry for the direct sum decomposition $Y_k \oplus V_k$ (see Definition 2.4.1).

PROPOSITION 4.6.7
If hypotheses $H(j)_2$ hold,
then there exists $r > 0$ such that

$$\begin{cases} \varphi(x) \leq 0 \text{ if } x \in Y, \ \|x\|_{H^1(\Omega)} \leq r, \\ \varphi(x) \geq 0 \text{ if } x \in V, \ \|x\|_{H^1(\Omega)} \leq r. \end{cases}$$

PROOF Note that $Y_k \subseteq C(\overline{\Omega})$ and it is finite dimensional. So we can find $r_1 > 0$ such that

$$|x(z)| \leq \frac{\delta}{3} < \delta \qquad \forall\, z \in \Omega, \ x \in Y, \ \|x\|_{H^1(\Omega)} \leq r_1,$$

with δ as in hypothesis $H(j)_2(v)$. Therefore, for $x \in Y_k$ with $\|x\|_{H^1(\Omega)} \leq r_1$ we have

$$\varphi(x) = \frac{1}{2}\|\nabla x\|_2^2 - \frac{\lambda_1}{2}\|x\|_2^2 - \int_\Omega j(z, x(z))\, dz$$

$$\leq \frac{1}{2}\|\nabla x\|_2^2 - \frac{\lambda_1}{2}\|x\|_2^2 - \frac{\lambda_k - \lambda_1}{2}\|x\|_2^2 = \frac{1}{2}\|\nabla x\|_2^2 - \frac{\lambda_k}{2}\|x\|_2^2 \leq 0.$$

Next, let $x \in V_k$. We decompose x as follows:

$$x = \overline{x} + \widehat{x}, \quad \text{with } \overline{x} \in E(\lambda_{k+1}) \text{ and } \widehat{x} \in \bigoplus_{i=k+2}^{\infty} E(\lambda_i).$$

Using hypothesis $H(j)_2(v)$ and the fact that $\|\nabla \overline{x}\|_2^2 = \lambda_{k+1}\|\overline{x}\|_2^2$ we have

$$\varphi(x) = \frac{1}{2}\|\nabla x\|_2^2 - \frac{\lambda_1}{2}\|x\|_2^2 - \int_\Omega j(z, x(z))\, dz$$

$$= \frac{1}{2}\|\nabla x\|_2^2 - \frac{\lambda_1}{2}\|x\|_2^2 - \int_{\{|x| \leq \delta\}} j(z, x(z))\, dz - \int_{\{|x| > \delta\}} j(z, x(z))\, dz$$

$$\geq \frac{1}{2}\|\nabla \widehat{x}\|_2^2 - \frac{\lambda_1}{2}\|\widehat{x}\|_2^2 + \int_{\{|x| \leq \delta\}} \left[\frac{\lambda_{k+1} - \lambda_1}{2}x(z)^2 - j(z, x(z))\right] dz$$

$$+ \int_{\{|x| > \delta\}} \left[\frac{\lambda_{k+1} - \lambda_1}{2}x(z)^2 - j(z, x(z))\right] dz - \frac{\lambda_{k+1} - \lambda_1}{2}\|\widehat{x}\|_2^2$$

$$\geq \frac{1}{2}\|\nabla \widehat{x}\|_2^2 - \frac{\lambda_{k+1}}{2}\|\widehat{x}\|_2^2$$

$$+ \int_{\{|x| > \delta\}} \left[\frac{\lambda_{k+1} - \lambda_1}{2}x(z)^2 - j(z, x(z))\right] dz. \tag{4.178}$$

Moreover, we claim that we can find $r_2 > 0$, such that for almost all $z \in \{|x| \geq \delta\}$ and all $x \in V$ with $\|x\|_{H^1(\Omega)} \leq r_2$, we have

$$\frac{1}{2}|x(z)| \leq |\widehat{x}(z)|. \tag{4.179}$$

Indeed, by virtue of the finite dimensionality of $E(\lambda_{k+1}) \subseteq C(\overline{\Omega})$, we can find $\xi_1 > 0$ such that

$$|u(z)| \leq \xi_1 \|u\|_{H^1(\Omega)} \quad \forall z \in \Omega, \ u \in E(\lambda_{k+1}).$$

Let $r_2 = \frac{\delta}{2\xi_1}$ and suppose that $\|x\|_{H^1(\Omega)} \leq r_2$. For almost all $z \in \{|x| \geq \delta\}$, we have

$$|x(z)| \leq |\overline{x}(z)| + |\widehat{x}(z)| \leq \xi_1 \|\overline{x}\|_{H^1(\Omega)} + |\widehat{x}(z)| \leq \xi_1 \|x\|_{H^1(\Omega)} + |\widehat{x}(z)|$$

$$\leq \frac{\delta}{2} + |\widehat{x}(z)| \leq \frac{1}{2}|x(z)| + |\widehat{x}(z)|,$$

thus we have obtained (4.179).

In addition, by virtue of the fact that

$$\left|j(z,\zeta)\right| \;\leq\; a_2(z) + c_2|\zeta| \quad \text{for a.a. } z \in \Omega \text{ and all } \zeta \in \mathbb{R},$$

with $a_2 \in L^\infty(\Omega)_+$, $c_2 > 0$ (see the proof of Proposition 4.6.6), we can say that

$$\left|j\big(z, x(z)\big)\right| \;\leq\; c_7 \big|x(z)\big|^\gamma \quad \text{for a.a. } z \in \{|x| > \delta\}, \tag{4.180}$$

for some $c_7 > 0$ and with $\gamma \in (2, 2^*]$, where 2^* is the critical Sobolev exponent, i.e.

$$2^* \stackrel{df}{=} \begin{cases} \dfrac{2N}{N-2} & \text{if } N > 2, \\[2mm] +\infty & \text{if } N \leq 2. \end{cases}$$

From (4.179), (4.180) and the continuity of the embedding $H_0^1(\Omega) \subseteq L^\gamma(\Omega)$ (as $\gamma \leq 2^*$) we can write that

$$\left| \int_{\{|x|>\delta\}} \left[\frac{\lambda_{k+1} - \lambda_1}{2} x(z)^2 - j\big(z, x(z)\big) \right] dz \right|$$
$$\leq\; c_8 \int_{\{|x|>\delta\}} \big|\widehat{x}(z)\big|^\gamma \, dz \;\leq\; c_9 \, \|\nabla\widehat{x}\|_2^\gamma, \tag{4.181}$$

for some $c_8, c_9 > 0$. Returning to (4.178) and using (4.181) we obtain

$$\begin{aligned} \varphi(x) \;&\geq\; \frac{1}{2}\|\nabla\widehat{x}\|_2\, 2 - \frac{\lambda_{k+1}}{2}\|\widehat{x}\|_2^2 - c_9 \|\nabla\widehat{x}\|_2^\gamma \\[1mm] &\geq\; \frac{1}{2}\left(1 - \frac{\lambda_{k+1}}{\lambda_{k+2}}\right)\|\nabla\widehat{x}\|_2^2 - c_9 \|\nabla\widehat{x}\|_2^\gamma \end{aligned}$$

(see Theorem 1.5.3(b)). Since $\gamma > 2$, we can find $r_3 > 0$ small enough so that

$$\varphi(x) \;\geq\; 0 \qquad \forall\, x \in V,\ \|x\|_{H^1(\Omega)} \leq r_3.$$

Finally, choosing $r \stackrel{df}{=} \min\{r_1, r_2, r_3\}$, we achieve the local linking geometry. $\square$

Now we are ready for the multiplicity theorem.

THEOREM 4.6.2
If hypotheses $H(j)_2$ hold,
<u>*then*</u> *problem (4.161) has at least two nontrivial solutions.*

PROOF First note that $\varphi(0) = 0$ (recall that $j(z, 0) = 0$ for almost all $z \in \Omega$). Also if $v \in E(\lambda_{k-1}) \subseteq C(\overline{\Omega})$ is such that $|v(z)| \leq \delta$ for all $z \in \overline{\Omega}$, then using hypothesis $H(j)_2(v)$ we have

$$
\begin{aligned}
\varphi(v) &= \frac{1}{2}\|\nabla v\|_2^2 - \frac{\lambda_1}{2}\|v\|_2^2 - \int_\Omega j\big(z, v(z)\big)\, dz \\[2mm]
&\leq \frac{1}{2}\|\nabla v\|_2^2 - \frac{\lambda_1}{2}\|v\|_2^2 - \frac{\lambda_k - \lambda_1}{2}\|v\|_2^2 \\[2mm]
&= \frac{1}{2}\|\nabla v\|_2^2 - \frac{\lambda_k}{2}\|v\|_2^2 \leq \frac{\lambda_{k-1}}{2}\|v\|_2^2 - \frac{\lambda_k}{2}\|v\|_2^2 < 0,
\end{aligned}
$$

so $\inf\limits_{H_0^1(\Omega)} \varphi < 0$.

This fact combined with Propositions 4.6.6 and 4.6.7 permit the use of Theorem 2.4.1 which produces two nontrivial critical points of φ. These are two nontrivial solutions of (4.161). $\qquad\square$

4.6.2   Nonlinear Problems

Next we consider nonlinear problems driven by the p-Laplacian differential operator. So we consider the following nonlinear elliptic inclusion:

$$
\begin{cases}
-\operatorname{div}\big(\|\nabla x(z)\|_{\mathbb{R}^N}^{p-2}\,\nabla x(z)\big) - \lambda_1 |x(z)|^{p-2} x(z) \in \partial j\big(z, x(z)\big) \\
\qquad\qquad\qquad\qquad\qquad\qquad\qquad\qquad\qquad \text{for a.a. } z \in \Omega, \qquad (4.182) \\
x\big|_{\partial\Omega} = 0,
\end{cases}
$$

with $p \in [2, +\infty)$. Recall that $\Omega \subseteq \mathbb{R}^N$ is a bounded domain with a C^2-boundary $\partial\Omega$ and $\lambda_1 > 0$ is the principal eigenvalue of $\big(-\Delta_p, W_0^{1,p}(\Omega)\big)$ (see Proposition 1.5.17). As we did for the semilinear problems considered earlier, we shall establish the existence of at least two distinct solutions for problem (4.182). Our hypotheses allow for different asymptotic behaviour of $\partial j(z, \cdot)$ at $\pm\infty$. More specifically, we may have complete resonance at $-\infty$ (see hypotheses $H(j)_1$).

We begin our analysis of problem (4.182), by establishing the existence of a smooth strictly positive solution. For this purpose, our hypotheses on the nonsmooth potential $j(z, \zeta)$ are the following:

$\underline{H(j)_3}$ $j: \Omega \times \mathbb{R} \longrightarrow \mathbb{R}$ is a function, such that

    (*i*) for every $\zeta \in \mathbb{R}$, the function

$$
\Omega \ni z \longrightarrow j(z, \zeta) \in \mathbb{R}
$$

    is measurable, $j(\cdot, 0) \in L^\infty(\Omega)$, $\int_\Omega j(z, 0)\, dz \leq 0$ and $\partial j(z, 0) \subseteq \mathbb{R}_+$ for almost all $z \in \Omega$;

    (*ii*) for almost all $z \in \Omega$, the function

$$
\mathbb{R} \ni \zeta \longrightarrow j(z, \zeta) \in \mathbb{R}
$$

    is locally Lipschitz;

(*iii*) for almost all $z \in \Omega$, all $\zeta \geq 0$ and all $u \in \partial j(z, \zeta)$, we have

$$u \geq -c_0 \zeta^{p-1}$$

and for almost all $z \in \Omega$, all $\zeta \in \mathbb{R}$ and all $u \in \partial j(z, \zeta)$, we have

$$|u| \leq a_1(z) + c_1 |\zeta|^{p-1},$$

with $a_1 \in L^\infty(\Omega)_+$, $c_0, c_1 > 0$;

(*iv*) there exists $\vartheta \in L^\infty(\Omega)$ such that $\vartheta(z) \leq 0$ for almost all $z \in \Omega$, the inequality is strict on a set of positive measure and

$$\limsup_{\zeta \to +\infty} \frac{u}{\zeta^{p-1}} \leq \vartheta(z)$$

uniformly for almost all $z \in \Omega$ and all $u \in \partial j(z, \zeta)$;

(*v*) there exists $\eta \in L^\infty(\Omega)$ such that $\eta(z) \geq 0$ for almost all $z \in \Omega$, the inequality is strict on a set of positive measure and

$$\liminf_{\zeta \to 0^+} \frac{pj(z, \zeta)}{\zeta^p} \geq \eta(z)$$

uniformly for almost all $z \in \Omega$;

(*vi*) there exists $M > 0$ such that

$$u \geq 0 \quad \text{for a.a. } z \in \Omega, \text{ all } \zeta \geq M \text{ and all } u \in \partial j(z, \zeta)$$

or

$$u \leq 0 \quad \text{for a.a. } z \in \Omega, \text{ all } \zeta \geq M \text{ and all } u \in \partial j(z, \zeta).$$

REMARK 4.6.4 If $\operatorname*{esssup}_{\Omega} \vartheta < 0$, then hypothesis $H(j)_3(vi)$ follows from hypothesis $H(j)_3(iv)$. If $p = 2$ (semilinear case) and the potential is smooth (i.e. $j(z, \cdot) \in C^1(\Omega)$), then hypotheses $H(j)_3(iv)$ and (v) incorporate problems with right hand side nonlinearity which is sublinear at 0^+ and superlinear at $+\infty$. Finally note that hypotheses $H(j)_3(iv)$ and (v) imply that we may have incomplete resonance at 0^+ and $+\infty$ (nonuniform nonresonance). $\square$

Let

$$C \overset{df}{=} \left\{ x \in W_0^{1,p}(\Omega) : x(z) \geq 0 \text{ for a.a. } z \in \Omega \right\}$$

(i.e. C is the positive cone for $W_0^{1,p}(\Omega)$ equipped with the usual pointwise ordering). We introduce two functionals $\varphi_1 : W_0^{1,p}(\Omega) \longrightarrow \mathbb{R}$ and

$\varphi_2 \colon W_0^{1,p}(\Omega) \longrightarrow \overline{\mathbb{R}}$ defined by

$$\varphi_1(x) \overset{df}{=} \frac{1}{p} \|\nabla x\|_p^p - \lambda_1 \|x\|_p^p - \int_\Omega j\big(z, x(z)\big) \, dz$$

$$\varphi_2(x) \overset{df}{=} i_C(x) \; = \; \begin{cases} 0 & \text{if } x \in C, \\ +\infty & \text{if } x \notin C \end{cases}$$

(indicator function of $C \in P_{fc}\big(W_0^{1,p}(\Omega)\big)$).

We know that φ_1 is locally Lipschitz and $\varphi_2 \in \Gamma_0\big(W_0^{1,p}(\Omega)\big)$ (see Definition 1.3.1). We set

$$\varphi \overset{df}{=} \varphi_1 + \varphi_2.$$

Note that φ fits in the setting of Section 2.3 and so we can check whether it satisfies the generalized nonsmooth PS-condition (see Definition 2.3.2).

PROPOSITION 4.6.8
If hypotheses $H(j)_3$ hold,
<u>then</u> φ satisfies the generalized nonsmooth PS-condition.

PROOF   Let $\{x_n\}_{n \geq 1} \subseteq W_0^{1,p}(\Omega)$ be a sequence such that

$$\big|\varphi(x_n)\big| \; \leq \; M_1 \qquad \forall \, n \geq 1,$$

for some $M_1 > 0$ and

$$\varphi_1^0(x_n; y - x_n) + \varphi_2(y) - \varphi_2(x_n)$$
$$\geq \; -\varepsilon_n \|y - x_n\|_{W^{1,p}(\Omega)} \qquad \forall \, y \in W_0^{1,p}(\Omega), \; n \geq 1,$$

with $\varepsilon_n \searrow 0$.

Note that

$$x_n \in C \qquad \forall \, n \geq 1.$$

Let $y = 0$. We obtain

$$\varphi_1^0(x_n; -x_n) \; \geq \; -\varepsilon_n \|x_n\|_{W^{1,p}(\Omega)} \qquad \forall \, n \geq 1.$$

We know that $\varphi_1^0(x_n; \cdot)$ is the support function of the subdifferential set $\partial\varphi_1(x_n)$, which is nonempty, convex and weakly compact. So we can find $x_n^* \in \partial\varphi_1(x_n)$ such that

$$\varphi_1^0(x_n; -x_n) \; = \; \langle x_n^*, -x_n \rangle_{W_0^{1,p}(\Omega)}.$$

So we have

$$\langle x_n^*, x_n \rangle_{W_0^{1,p}(\Omega)} \; \leq \; \varepsilon_n \|x_n\|_{W^{1,p}(\Omega)} \qquad \forall \, n \geq 1.$$

We know that

$$x_n^* = A(x_n) - \lambda_1 |x_n|^{p-2} x_n - u_n^* \qquad \forall\, n \geq 1,$$

with $A\colon W_0^{1,p}(\Omega) \longrightarrow W^{-1,p'}(\Omega)$ being as before the maximal monotone operator defined by

$$\langle A(x), y\rangle_{W_0^{1,p}(\Omega)} \stackrel{df}{=} \int_\Omega \|\nabla x(z)\|_{\mathbb{R}^N}^{p-2} \left(\nabla x(z), \nabla y(z)\right)_{\mathbb{R}^N} dz \qquad \forall\, x,y \in W_0^{1,p}(\Omega)$$

and $u_n^* \in L^{p'}(\Omega)$ with $u_n^*(z) \in \partial j\big(z, x_n(z)\big)$ for almost all $z \in \Omega$. So, for all $n \geq 1$, we have

$$\langle x_n^*, x_n\rangle_{W_0^{1,p}(\Omega)} = \langle A(x_n), x_n\rangle_{W_0^{1,p}(\Omega)} - \lambda_1 \|x_n\|_p^p - \int_\Omega u_n^*(z) x_n(z)\, dz$$

$$= \|\nabla x_n\|_p^p - \lambda_1 \|x_n\|_p^p - \int_\Omega u_n^*(z) x_n(z)\, dz \leq \varepsilon_n \|x_n\|_{W^{1,p}(\Omega)}. \qquad (4.183)$$

We claim that the sequence $\{x_n\}_{n\geq 1} \subseteq W_0^{1,p}(\Omega)$ is bounded. Suppose that this is not true. Then by passing to a suitable subsequence, we may assume that

$$\|x_n\|_{W^{1,p}(\Omega)} \longrightarrow +\infty.$$

Let us set

$$y_n \stackrel{df}{=} \frac{x_n}{\|x_n\|_{W^{1,p}(\Omega)}} \qquad \forall\, n \geq 1.$$

Passing to a further subsequence if necessary, we may assume that

$$\begin{aligned} y_n &\stackrel{w}{\longrightarrow} y \quad \text{in } W_0^{1,p}(\Omega),\\ y_n &\longrightarrow y \quad \text{in } L^p(\Omega),\\ y_n(z) &\longrightarrow y(z) \text{ for a.a. } z \in \Omega,\\ |y_n(z)| &\leq k(z) \text{ for a.a. } z \in \Omega \text{ and all } n \geq 1, \end{aligned}$$

with $k \in L^p(\Omega)$ and $y \in W_0^{1,p}(\Omega)$. From hypothesis $H(j)_3(iii)$ we have

$$\frac{|u_n(z)|}{\|x_n\|_{W^{1,p}(\Omega)}^{p-1}} \leq \frac{a_1(z)}{\|x_n\|_{W^{1,p}(\Omega)}^{p-1}} + c_1 |y_n(z)|^{p-1} \quad \text{for a.a. } z \in \Omega \qquad (4.184)$$

and so the sequence $\left\{ \frac{u_n(\cdot)}{\|x_n\|_{W^{1,p}(\Omega)}^{p-1}} \right\}_{n\geq 1} \subseteq L^{p'}(\Omega)$ is bounded.

Thus, by passing to a subsequence if necessary, we may assume that

$$\frac{u_n(z)}{\|x_n\|_{W^{1,p}(\Omega)}^{p-1}} \stackrel{w}{\longrightarrow} h \quad \text{in } L^{p'}(\Omega),$$

for some $h \in L^{p'}(\Omega)$. Since $x_n \in C$ we have

$$y_n \in C \qquad \forall\, n \geq 1$$

and so $y \in C$. This implies that

$$x_n(z) \longrightarrow +\infty \quad \text{for a.a. } z \in \{y \neq 0\} = \{y > 0\}.$$

For a given $\varepsilon > 0$ and any $n \geq 1$, we introduce the set

$$\Omega_{\varepsilon,n}^+ \stackrel{df}{=} \left\{ z \in \Omega : x_n(z) > 0, \ \frac{u_n(z)}{\|x_n\|_{W^{1,p}(\Omega)}^{p-1}} \leq \vartheta(z) + \varepsilon \right\}.$$

Because of hypothesis $H(j)_3(iv)$, we know that

$$\chi_{\Omega_{\varepsilon,n}^+}(z) \longrightarrow 1 \quad \text{for a.a. } z \in \{y > 0\}.$$

Note that

$$\chi_{\Omega_{\varepsilon,n}^+}(z) \frac{u_n(z)}{\|x_n\|_{W^{1,p}(\Omega)}^{p-1}} = \chi_{\Omega_{\varepsilon,n}^+}(z) \frac{u_n(z)}{x_n(z)^{p-1}} y_n(z)^{p-1}$$

$$\leq \chi_{\Omega_{\varepsilon,n}^+}(z)\big(\vartheta(z) + \varepsilon\big) y_n(z)^{p-1}. \qquad (4.185)$$

Passing to the weak limits in $L^1(\{y \neq 0\})$ and using Proposition 1.2.12, from (4.185), we obtain

$$h(z) \leq \big(\vartheta(z) + \varepsilon\big) y(z)^{p-1} \quad \text{for a.a. } z \in \{y > 0\}.$$

Since $\varepsilon > 0$ was arbitrary, it follows that

$$h(z) \leq \vartheta(z) y(z)^{p-1} \quad \text{for a.a. } z \in \{y > 0\} = \{y \neq 0\}.$$

Moreover, from (4.184), it is clear that

$$h(z) = 0 \quad \text{for a.a. } z \in \{y = 0\}.$$

Therefore, we conclude that

$$h(z) \leq \vartheta(z) y(z)^{p-1} \quad \text{for a.a. } z \in \Omega. \qquad (4.186)$$

Now we return to (4.183) and divide with $\|x_n\|_{W^{1,p}(\Omega)}^p$. We obtain

$$\|\nabla y_n\|_p^p - \lambda_1 \|y_n\|_p^p - \int_\Omega \frac{u_n(z)}{\|x_n\|_{W^{1,p}(\Omega)}^{p-1}} y_n(z)\, dz \leq \frac{\varepsilon_n}{\|x_n\|_{W^{1,p}(\Omega)}^{p-1}} \qquad \forall\, n \geq 1.$$

We pass to the limit as $n \to +\infty$. Using (4.186) and hypothesis $H(j)_3(iv)$, we have

$$\|\nabla y\|_p^p = \lambda_1 \|y\|_p^p$$

and so, from Propositions 1.5.18 and 1.5.19 and recalling that $y \in C$, we have

$$y = u_1 \quad \text{or} \quad y = 0.$$

If $y = 0$, then

$$\nabla y_n \longrightarrow 0 \quad \text{in } L^p(\Omega; \mathbb{R}^N),$$

hence

$$y_n \longrightarrow 0 \quad \text{in } W_0^{1,p}(\Omega),$$

a contradiction to the fact that $\|y_n\|_{W^{1,p}(\Omega)} = 1$ for $n \geq 1$. So $y = u_1$. We know that

$$y(z) = u_1(z) > 0 \qquad \forall z \in \Omega$$

(see Proposition 1.5.18). So

$$\int_\Omega \vartheta(z) y(z)^p \, dz < 0$$

and thus

$$\|\nabla y\|_p^p < \lambda_1 \|y\|_p^p,$$

a contradiction.

This proves that the sequence $\{x_n\}_{n \geq 1} \subseteq C$ is bounded in $W_0^{1,p}(\Omega)$ and so by passing to a subsequence if necessary, we may assume that

$$x_n \xrightarrow{w} x \text{ in } W_0^{1,p}(\Omega),$$
$$x_n \longrightarrow x \text{ in } L^p(\Omega),$$

with $x \in W_0^{1,p}(\Omega)$. From the choice of the sequence $\{x_n\}_{n \geq 1} \subseteq C$ we have that

$$\langle A(x_n), y - x_n \rangle_{W_0^{1,p}(\Omega)} - \lambda_1 \int_\Omega |x_n(z)|^{p-2} x_n(z) (y - x_n)(z) \, dz$$

$$- \int_\Omega u_n^*(z) (y - x_n)(z) \, dz + \varphi_2(y) - \varphi_2(x_n)$$

$$\geq -\varepsilon_n \|y - x_n\|_{W^{1,p}(\Omega)} \qquad \forall y \in C.$$

Let $y = x \in C$. We obtain

$$\langle A(x_n), x_n - x \rangle_{W_0^{1,p}(\Omega)} - \lambda_1 \int_\Omega |x_n(z)|^{p-2} x_n(z) (x_n - x)(z) \, dz$$

$$- \int_\Omega u_n^*(z) (x_n - x)(z) \, dz \leq \varepsilon_n \|x_n - x\|_{W^{1,p}(\Omega)}.$$

Clearly

$$\int_\Omega |x_n(z)|^{p-2} x_n(z) (x_n - x)(z)\, dz \longrightarrow 0,$$

$$\int_\Omega u_n^*(z)(x_n - x)(z)\, dz \longrightarrow 0$$

and

$$\varepsilon_n \|x_n - x\|_{W^{1,p}(\Omega)} \longrightarrow 0.$$

So

$$\limsup_{n\to+\infty} \langle A(x_n), x_n - x\rangle_{W_0^{1,p}(\Omega)} \leq 0.$$

As before, exploiting the maximal monotonicity of A, we get

$$\|\nabla x_n\|_p \longrightarrow \|\nabla x\|_p$$

and this, via the Kadec-Klee property, implies that

$$\nabla x_n \longrightarrow \nabla x \quad \text{in } L^p(\Omega; \mathbb{R}^N),$$

hence

$$x_n \longrightarrow x \quad \text{in } W_0^{1,p}(\Omega).$$

$\square$

Using this proposition, we can prove the existence of a smooth strictly positive solution for problem (4.182).

PROPOSITION 4.6.9

If hypotheses $H(j)_3$ hold,
then problem (4.182) has a solution $x_0 \in C_0^1(\overline{\Omega})$ such that $x_0(z) > 0$ for all $z \in \Omega$ and $\frac{\partial x_0}{\partial n}(z) < 0$ for all $z \in \partial\Omega$.

PROOF   We claim that φ is bounded below. To this end, note that

$$\varphi(x) = \begin{cases} \varphi_1(x) & \text{if } x \in C, \\ +\infty & \text{otherwise,} \end{cases} \quad \forall\, x \in W_0^{1,p}(\Omega).$$

From hypothesis $H(j)_3(iv)$, we know that for a given $\varepsilon > 0$, we can find $M_2 = M_2(\varepsilon) > 0$ such that

$$u \leq (\vartheta(z) + \varepsilon)\zeta^{p-1} \quad \text{for a.a. } z \in \Omega, \text{ all } \zeta \geq M_2 \text{ and all } u \in \partial j(z, \zeta).$$

For a given $x \in C$ and every $z \in \Omega$ we define the set

$$E(z) \stackrel{df}{=} \Big\{ (\lambda, v) \in (0,1) \times \mathbb{R} : v \in \partial j\big(z, \lambda x(z) + (1-\lambda) M_2\big),$$

$$j\big(z, x(z)\big) - j(z, M_2) = v\big(x(z) - M_2\big) \Big\}.$$

From the mean value theorem for the locally Lipschitz functions (see Proposition 1.3.15), we know that

$$E(z) \neq \emptyset \quad \text{for a.a. } z \in \Omega.$$

By redefining E on a Lebesgue-null set, we may assume that

$$E(z) \neq \emptyset \quad \forall z \in \Omega.$$

Let us set

$$\widehat{x}(z, \lambda) \overset{df}{=} \lambda x(z) + (1 - \lambda)M_2.$$

Then for every $h \in \mathbb{R}$ consider the function

$$\gamma_h(z, \lambda) \overset{df}{=} j^0\big(z, \widehat{x}(z, h); h\big).$$

We claim that the function $(z, \lambda) \longmapsto \gamma_h(z, \lambda)$ is measurable. From Definition 1.3.6, we have

$$\gamma_h(z, \lambda) \tag{4.187}$$
$$= \inf_{m \geq 1} \sup_{r, s \in \mathbb{Q} \cap \left(-\frac{1}{m}, \frac{1}{m}\right)} \frac{j(z, \widehat{x}(z, \lambda) + r + sh) - j(z, \widehat{x}(z, \lambda) + r)}{s}.$$

Note that the function $(z, \zeta) \longmapsto j(z, \zeta)$ is $\mathcal{L} \times \mathcal{B}(\mathbb{R})$-measurable, with $\mathcal{L}$ being the Lebesgue σ-field of Ω and $\mathcal{B}(\mathbb{R})$ is the Borel σ-field of $\mathbb{R}$. So from (4.187) it follows that the function $(z, \lambda) \longmapsto \gamma_h(z, \lambda)$ is $\mathcal{L} \times \mathcal{B}(0, 1)$-measurable. Let us set

$$S(z, \lambda) \overset{df}{=} \partial j\big(z, \widehat{x}(z, \lambda)\big)$$

and let $\{h_m\}_{m \geq 1}$ be an enumeration of the rational numbers. Exploiting the continuity of $j^0\big(z, \widehat{x}(z, \lambda); \cdot\big)$, we have

$$\mathrm{Gr}\, S = \left\{(z, \lambda, u) \in \Omega \times (0, 1) \times \mathbb{R} : u \in S(z, \lambda)\right\}$$

$$= \bigcap_{m \geq 1} \Big\{(z, \lambda, u) \in \Omega \times (0, 1) \times \mathbb{R} :$$

$$uh_m \leq j^0\big(z, \widehat{x}(z, \lambda); h_m\big) = \gamma_{h_m}(z, \lambda)\Big\}$$

$$\in \mathcal{L} \times \mathcal{B}(0, 1) \times \mathcal{B}(\mathbb{R}).$$

Then we have

$$\mathrm{Gr}\, E = \left\{(z, \lambda, u) \in \Omega \times (0, 1) \times \mathbb{R} : (\lambda, v) \in E(z)\right\}$$

$$= \mathrm{Gr}\, S \cap \Big\{(z, \lambda, u) \in \Omega \times (0, 1) \times \mathbb{R} :$$

$$j\big(z, x(z)\big) - j(z, M_2) = v\big(x(z) - M_2\big)\Big\}$$

$$\in \mathcal{L} \times \mathcal{B}(0, 1) \times \mathcal{B}(\mathbb{R}).$$

So invoking the Yankov-von Neumann-Aumann selection theorem (see Theorem 1.2.4), we obtain measurable maps $\lambda \colon \Omega \longrightarrow (0,1)$ and $v \colon \Omega \longrightarrow \mathbb{R}$ such that

$$\big(\lambda(z), v(z)\big) \in E(z) \quad \text{for a.a. } z \in \Omega.$$

Hence we have

$$j\big(z, x(z)\big) - j\big(z, M_2\big) = v(z)\big(x(z) - M_2\big) \quad \text{for a.a. } z \in \Omega$$

and

$$v(z) \in \partial j\big(z, \lambda(z)x(z) + \big(1 - \lambda(z)\big)M_2\big) \quad \text{for a.a. } z \in \Omega.$$

Evidently $v \in L^{p'}(\Omega)$ (see hypothesis $H(j)_3(iii)$). We have

$$
\begin{aligned}
\int_\Omega j\big(z, x(z)\big)\, dz &= \int_\Omega v(z)\big(x(z) - M_2\big)\, dz + \int_\Omega j(z, M_2)\, dz \\
&= \int_{\{x \geq M_2\}} v(z)\big(x(z) - M_2\big)\, dz + \int_{\{x < M_2\}} v(z)\big(x(z) - M_2\big)\, dz \\
&\quad + \int_\Omega j(z, M_2)\, dz.
\end{aligned}
\tag{4.188}
$$

Since $v(z) \in \partial j\big(z, \lambda(z)x(z) + \big(1 - \lambda(z)\big)M_2\big)$ for almost all $z \in \Omega$, from the choice of $M_2 > 0$ in the beginning of the proof, we have

$$
\begin{aligned}
v(z) &\leq \big(\vartheta(z) + \varepsilon\big)\big(\lambda(z)x(z) + \big(1 - \lambda(z)\big)M_2\big)^{p-1} \\
&\leq \vartheta(z)M_2^{p-1} + \varepsilon x(z)^{p-1} \quad \text{for a.a. } z \in \{x \geq M_2\},
\end{aligned}
$$

so

$$
\begin{aligned}
v(z)\big(x(z) - M_2\big) &\leq \vartheta(z)M_2^{p-1}\big(x(z) - M_2\big) + \varepsilon x(z)^{p-1}\big(x(z) - M_2\big) \\
&\leq \varepsilon x(z)^p \quad \text{for a.a. } z \in \{x \geq M_2\}
\end{aligned}
$$

(recall that $\vartheta(z) \leq 0$ for almost all $z \in \Omega$; see hypothesis $H(j)_3(iv)$). So we can write that

$$\int_{\{x \geq M_2\}} v(z)\big(x(z) - M_2\big)\, dz \leq \varepsilon \|x\|_p^p. \tag{4.189}$$

Also from hypothesis $H(j)_3(iii)$, we have

$$\big|v(z)\big| \leq a_1(z) + c_1 M_2^{p-1} \quad \text{for a.a. } z \in \{x < M_2\}.$$

It follows that

$$\left| \int_{\{0 \leq x < M_2\}} v(z)\big(x(z) - M_2\big)\, dz \right| \leq \int_{\{0 \leq x < M_2\}} \big|v(z)\big|\big(\big|x(z)\big| + M_2\big)\, dz$$

$$\leq 2M_2 \int_{\{0 \leq x < M_2\}} |v(z)| \, dz \ \leq \ c_2, \tag{4.190}$$

for some $c_2 > 0$. Using (4.189) and (4.190) in (4.188), we obtain

$$\int_\Omega j\big(z, x(z)\big) \, dz \ \leq \ \varepsilon \|x\|_p^p + c_3,$$

with $c_3 \overset{df}{=} c_2 + \int_\Omega |j(z, M_2)| \, dz$. As $\varepsilon > 0$ was arbitrary, we have

$$\int_\Omega j\big(z, x(z)\big) \, dz \ \leq \ c_3.$$

Therefore for all $x \in C$, we have

$$\varphi_1(x) \ \geq \ \frac{1}{p} \|\nabla x\|_p^p - \frac{\lambda_1}{p} \|x\|_p^p - c_3 \ \geq \ -c_3$$

(see Proposition 1.5.18) and

$$\varphi(x) \ \geq \ -c_3 \qquad \forall \, x \in W_0^{1,p}(\Omega).$$

This proves that indeed φ is bounded below.

Now let

$$m_0 \overset{df}{=} \inf_{W_0^{1,p}(\Omega)} \varphi.$$

Because φ is bounded below and clearly it is lower semicontinuous, we can apply the Ekeland variational principle (see Corollary 1.4.7) and obtain a sequence $\{x_n\}_{n \geq 1} \subseteq C$ such that

$$\varphi(x_n) \searrow m_0 \quad \text{and} \quad -\frac{1}{n} \|y - x_n\|_{W^{1,p}(\Omega)} \leq \varphi(y) - \varphi(x_n) \qquad \forall \, y \in W_0^{1,p}(\Omega).$$

Let us set

$$y \overset{df}{=} (1 - \lambda)x_n + \lambda w,$$

with $w \in W_0^{1,p}(\Omega)$, $\lambda \in (0, 1)$. Exploiting the convexity of φ_2, we have

$$-\frac{\lambda}{n} \|w - x_n\|_{W^{1,p}(\Omega)} \ \leq \ \varphi(y) - \varphi(x_n) \ \leq \ \varphi_1\big(x_n + \lambda(w - x_n)\big) + \lambda \varphi_2(w).$$

We divide by $\lambda > 0$ and let $\lambda \to 0^+$. We obtain

$$-\frac{1}{n} \|w - x_n\|_{W^{1,p}(\Omega)} \ \leq \ \varphi_1^0(x_n; w - x_n) + \varphi_2(w) \qquad \forall \, w \in W_0^{1,p}(\Omega). \tag{4.191}$$

Because of Proposition 4.6.8, we may assume that

$$x_n \longrightarrow x_0 \quad \text{in } W_0^{1,p}(\Omega).$$

So $x_0 \in C$ and $m_0 = \varphi(x_0)$. Also from (4.191) and since the function $(x, h) \longmapsto \varphi_1^0(x; h)$ is upper semicontinuous (see Proposition 1.3.7(b)), in the limit as $n \to +\infty$, we obtain

$$0 \leq \varphi_1^0(x_0; w - x_0) + \varphi_2(w) \qquad \forall \, w \in W_0^{1,p}(\Omega). \qquad (4.192)$$

From (4.192) it follows that there exist $x^* \in \partial\varphi_1(x_0)$ and $v^* \in \partial\varphi_2(x_0) = N_C(x_0)$ (the normal cone to C at x_0) such that $x^* + v^* = 0$ (see Remark 2.3.1). Recall that

$$x^* = A(x_0) - \lambda_1 |x_0|^{p-2} x_0 - u^*,$$

with $u^* \in L^{p'}(\Omega)$, $u^*(z) \in \partial j(z, x_0(z))$ for almost all $z \in \Omega$. So we have

$$\begin{aligned}
0 &\leq \langle x^*, w - x_0 \rangle_{W_0^{1,p}(\Omega)} \\
&= \langle A(x_0) - \lambda_1 |x_0|^{p-2} x_0 - u^*, w - x_0 \rangle_{W_0^{1,p}(\Omega)} \qquad \forall \, w \in C.
\end{aligned}$$

Let $h \in W_0^{1,p}(\Omega)$ and $\varepsilon > 0$. We use as a test function

$$w \overset{df}{=} (x_0 + \varepsilon h)^+ = (x_0 + \varepsilon h) + (x_0 + \varepsilon h)^-.$$

It follows that

$$0 \leq \langle x^*, \varepsilon h \rangle_{W_0^{1,p}(\Omega)} + \langle x^*, (x_0 + \varepsilon h)^- \rangle_{W_0^{1,p}(\Omega)}$$

and so

$$\begin{aligned}
&- \langle A(x_0), (x_0 + \varepsilon h)^- \rangle_{W_0^{1,p}(\Omega)} + \lambda_1 \int_\Omega |x_0(z)|^{p-2} x_0(z)(x_0 + \varepsilon h)^-(z) \, dz \\
&+ \int_\Omega u^*(z)(x_0 + \varepsilon h)^-(z) \, dz \ \leq \ \varepsilon \langle x^*, h \rangle_{W_0^{1,p}(\Omega)}. \qquad (4.193)
\end{aligned}$$

Let us set

$$\Omega_\varepsilon^- \overset{df}{=} \{z \in \Omega : (x_0 + \varepsilon h)(z) < 0\}$$

and recall that

$$\nabla(x_0 + \varepsilon h)^-(z) = \begin{cases} -\nabla(x_0 + \varepsilon h)(z) & \text{for a.a. } z \in \Omega_\varepsilon^-, \\ 0 & \text{otherwise} \end{cases}$$

(see Remark 1.1.10). So we can write that

$$- \langle A(x_0), (x_0 + \varepsilon h)^- \rangle_{W_0^{1,p}(\Omega)}$$

$$= - \int_\Omega \|\nabla x_0(z)\|_{\mathbb{R}^N} \left(\nabla x_0(z), \nabla(x_0 + \varepsilon h)^-(z)\right)_{\mathbb{R}^N} dz$$

$$= \int_{\Omega_\varepsilon^-} \|\nabla x_0(z)\|_{\mathbb{R}^N}^p \left(\nabla x_0(z), \nabla(x_0 + \varepsilon h)(z)\right)_{\mathbb{R}^N} dz$$

$$\geq \varepsilon \int_{\Omega_\varepsilon^-} \|\nabla x_0(z)\|_{\mathbb{R}^N}^{p-2} \left(\nabla x_0(z), \nabla h(z)\right)_{\mathbb{R}^N} dz. \geq 0. \qquad (4.194)$$

Also we have

$$\lambda_1 \int_\Omega |x_0(z)|^{p-2} x_0(z)(x_0 + \varepsilon h)^-(z)\, dz$$

$$= -\lambda_1 \int_{\Omega_\varepsilon^-} |x_0(z)|^{p-2} x_0(z)(x_0 + \varepsilon h)(z)\, dz \geq 0. \qquad (4.195)$$

In addition, we have

$$\int_\Omega u^*(z)(x_0 + \varepsilon h)^-(z)\, dz = -\int_{\Omega_\varepsilon^-} u^*(z)(x_0 + \varepsilon h)(z)\, dz$$

$$= -\int_{\Omega_\varepsilon^- \cap \{x_0 < M\}} u^*(z)(x_0 + \varepsilon h)(z)\, dz$$

$$- \int_{\Omega_\varepsilon^- \cap \{x_0 \geq M\}} u^*(z)(x_0 + \varepsilon h)(z)\, dz, \qquad (4.196)$$

with $M > 0$ as in hypothesis $H(j)_3(vi)$. First we assume that

$$u \geq 0 \quad \text{for a.a. } z \in \Omega, \text{ all } \zeta \geq M \text{ and all } u \in \partial j(z, \zeta)$$

(first option in hypothesis $H(j)_3(vi)$). Then

$$- \int_{\Omega_\varepsilon^- \cap \{x_0 \geq M\}} u(z)(x_0 + \varepsilon h)(z)\, dz \geq 0. \qquad (4.197)$$

Also since by hypothesis $\partial j(z, 0) \subseteq \mathbb{R}_+$ for almost all $z \in \Omega$, we have that

$$u^*(z) \geq 0 \quad \text{for a.a. } z \in \Omega_\varepsilon^- \cap \{x_0 = 0\}.$$

Moreover, note that since $x_0(z) \geq 0$ for almost all $z \in \Omega$, we have that

$$h(z) < 0 \quad \text{for a.a. } z \in \Omega_\varepsilon^-.$$

So we obtain

$$- \int_{\Omega_\varepsilon^- \cap \{x_0 = 0\}} u^*(z)\varepsilon h(z)\, dz \geq 0,$$

and using also hypothesis $H(j)_3(iii)$ and the fact that $x(z) \geq 0$ for almost all $z \in \Omega$, we have

$$- \int_{\Omega_\varepsilon^- \cap \{x_0 < M\}} u^*(z)(x_0 + \varepsilon h)(z)\, dz \; \geq \; - \int_{\Omega_\varepsilon^- \cap \{0 < x_0 < M\}} u^*(z)(x_0 + \varepsilon h)(z)\, dz$$

$$\geq \; c_4 \int_{\Omega_\varepsilon^- \cap \{0 < x_0 < M\}} (x_0 + \varepsilon h)(z)\, dz \; \geq \; \varepsilon c_4 \int_{\Omega_\varepsilon^- \cap \{0 < x_0 < M\}} h(z)\, dz, \qquad (4.198)$$

for some $c_4 > 0$. We use (4.197) and (4.198) in (4.196) to obtain

$$\int_\Omega u^*(z)(x_0 + \varepsilon h)^-(z)\, dz \; \geq \; \varepsilon c_4 \int_{\Omega_\varepsilon^- \cap \{0 < x_0 < M\}} h(z)\, dz. \qquad (4.199)$$

We return to (4.193) and use (4.194), (4.195) and (4.199). We have

$$\langle x^*, h \rangle_{W_0^{1,p}(\Omega)} \; \geq \; \int_{\Omega_\varepsilon^-} \|\nabla x_0(z)\|_{\mathbb{R}^N}^{p-2} \left(\nabla x_0(z), \nabla h(z) \right)_{\mathbb{R}^N} dz$$

$$+ \; c_4 \int_{\Omega_\varepsilon^- \cap \{0 < x_0 < M\}} h(z)\, dz. \qquad (4.200)$$

We know that
$$\nabla x_0(z) \; = \; 0 \quad \text{for a.a. } z \in \{x_0 = 0\}$$

(see Remark 1.1.10). Also note that

$$\left| \Omega_\varepsilon^- \cap \{0 < x_0 < M\} \right|_N \; \longrightarrow \; 0 \quad \text{as } \varepsilon \searrow 0.$$

So if in (4.200) we let $\varepsilon \searrow 0$, we obtain

$$\langle x^*, h \rangle_{W_0^{1,p}(\Omega)} \; \geq \; 0 \qquad \forall\, h \in W_0^{1,p}(\Omega),$$

so

$$x^* \; = \; A(x_0) - \lambda_1 |x_0|^{p-2} x_0 - u^* \; = \; 0$$

and

$$\begin{cases} -\text{div}\left(\|\nabla x_0(z)\|_{\mathbb{R}^N}^{p-2} \nabla x_0(z) \right) - \lambda_1 |x_0(z)|^{p-2} x_0(z) = u^*(z) & \text{for a.a. } z \in \Omega, \\ x_0|_{\partial\Omega} = 0. \end{cases}$$

Next assume that

$$u \; \leq \; 0 \quad \text{for a.a. } z \in \Omega, \text{ all } \zeta \geq M \text{ and all } u \in \partial j(z, \zeta)$$

(second option in hypothesis $H(j)_3(vi)$). Then, since $-u^*(z)x_0(z) \geq 0$ for almost all $z \in \{x_0 \geq M\}$, we have

$$- \int_{\Omega_\varepsilon^- \cap \{x_0 < M\}} u^*(z)(x_0 + \varepsilon h)(z)\, dz \; \geq \; c_4 \int_{\Omega_\varepsilon^- \cap \{x_0 < M\}} (x_0 + \varepsilon h)(z)\, dz$$

$$\geq \varepsilon c_4 \int_{\Omega_\varepsilon^- \cap \{x_0 < M\}} h(z)\, dz \qquad (4.201)$$

and

$$-\int_{\Omega_\varepsilon^- \cap \{x_0 \geq M\}} u^*(z)(x_0 + \varepsilon h)(z)\, dz \geq -\varepsilon \int_{\Omega_\varepsilon^- \cap \{x_0 \geq M\}} u^*(z) h(z)\, dz. \qquad (4.202)$$

Using (4.194), (4.195), (4.202) in (4.193), we obtain

$$\langle x^*, h \rangle_{W_0^{1,p}(\Omega)} \geq \int_{\Omega_\varepsilon^-} \|\nabla x_0(z)\|_{\mathbb{R}^N}^{p-2} \left(\nabla x_0(z), \nabla h(z) \right)_{\mathbb{R}^N} dz$$

$$+ c_4 \int_{\Omega_\varepsilon^- \cap \{0 < x_0 < M\}} h(z)\, dz - \int_{\Omega_\varepsilon^- \cap \{x_0 \geq M\}} u^*(z) h(z)\, dz.$$

Note that

$$\left| \Omega_\varepsilon^- \cap \{x_0 \geq M\} \right|_N \longrightarrow 0 \quad \text{as } \varepsilon \searrow 0.$$

So if we pass to the limit as $\varepsilon \searrow 0$ in the last inequality, as before we obtain

$$\langle x^*, h \rangle_{W_0^{1,p}(\Omega)} \geq 0 \qquad \forall\, h \in W_0^{1,p}(\Omega),$$

so

$$x^* = A(x_0) - \lambda_1 |x_0|^{p-2} x_0 - u^* = 0$$

and

$$\begin{cases} -\text{div}\left(\|\nabla x_0(z)\|_{\mathbb{R}^N}^{p-2} \nabla x_0(z) \right) - \lambda_1 |x_0(z)|^{p-2} x_0(z) = u^*(z) & \text{for a.a. } z \in \Omega, \\ x_0|_{\partial\Omega} = 0. \end{cases}$$

Since $x_0 \in C$, we have that $x_0(z) \geq 0$ for almost all $z \in \Omega$, while $u^*(z) \in \partial j(z, x_0(z))$ for almost all $z \in \Omega$. So by virtue of hypothesis $H(j)_3(iii)$, we have

$$-c_0 x_0(z)^{p-1} \leq u^*(z) \quad \text{for a.a. } z \in \Omega$$

and so

$$\Delta_p x_0(z) = \text{div}\left(\|\nabla x_0(z)\|_{\mathbb{R}^N}^{p-2} \nabla x_0(z) \right)$$
$$\leq c_0 x_0(z)^{p-1} - \lambda_1 |x_0(z)|^{p-2} x_0(z) \quad \text{for a.a. } z \in \Omega. \qquad (4.203)$$

First we show that x_0 is nontrivial. By virtue of hypothesis $H(j)_3(v)$, for a given $\xi > 0$, we can find $\delta = \delta(\xi) > 0$, such that

$$\frac{1}{p}(\eta(z) - \xi)\zeta^p \leq j(z, \zeta) \quad \text{for a.a. } z \in \Omega \text{ and all } \zeta \in [0, \delta]. \qquad (4.204)$$

Let u_1 be the principal eigenfunction of $\left(-\Delta_p, W_0^{1,p}(\Omega)\right)$. We know that $u_1 \in C^1(\overline{\Omega})$ and $u_1(z) > 0$ for all $z \in \Omega$ (see Proposition 1.5.18). So we can find $\varepsilon_0 > 0$ small, such that $0 < \varepsilon_0 u_1(z) < \delta$ for all $z \in \Omega$. Then, using also (4.204), for $\varepsilon \in (0, \varepsilon_0)$, we can write that

$$\varphi(\varepsilon u_1) = \varphi_1(\varepsilon u_1) = \frac{\varepsilon^p}{p}\|\nabla u_1\|_p^p - \frac{\lambda_1 \varepsilon^p}{p}\|u_1\|_p^p - \int_{\Omega} j\big(z, \varepsilon u_1(z)\big)\,dz$$

$$= -\int_{\Omega} j\big(z, \varepsilon u_1(z)\big)\,dz \leq \frac{\xi \varepsilon^p}{p}\|u_1\|_p^p - \frac{\varepsilon^p}{p}\int_{\Omega}\eta(z) u_1(z)^p\,dz.$$

Because $\xi > 0$ was arbitrary, letting $\xi \searrow 0$ and using hypothesis $H(j)_3(v)$, we obtain

$$\varphi(\varepsilon u_1) \leq -\frac{\varepsilon^p}{p}\int_{\Omega}\eta(z) u_1(z)^p\,dz < 0$$

and so

$$\inf_{W_0^{1,p}(\Omega)} \varphi = \varphi(x_0) < 0 \leq \varphi(0),$$

i.e. $x_0 \neq 0$.

From (4.203) and using Theorem 1.5.5, we have that $x_0 \in L^{\infty}(\Omega)_+$ and then by virtue of Theorem 1.5.6, we infer that $x_0 \in C^{1,\beta}(\overline{\Omega})$ for some $\beta \in (0,1)$. Because

$$\Delta_p x_0(z) \leq (c_1 + \lambda_1) x_0(z)^{p-1} \quad \text{for a.a. } z \in \Omega,$$

we can apply Theorem 1.5.7 and conclude that

$$x_0(z) > 0 \quad \forall z \in \Omega$$

and

$$\frac{\partial x_0(z)}{\partial n} < 0 \quad \forall z \in \partial\Omega.$$

This completes the proof of the proposition. $\square$

REMARK 4.6.5   We know that if

$$C_0^1(\overline{\Omega}) \overset{df}{=} \{x \in C^1(\overline{\Omega}) : x(z) = 0 \text{ for all } z \in \partial\Omega\}$$

and $C_0^1(\overline{\Omega})_+$ denotes the positive cone of $C_0^1(\overline{\Omega})$ for the pointwise partial ordering, i.e.

$$C_0^1(\overline{\Omega})_+ \overset{df}{=} \{x \in C_0^1(\overline{\Omega}) : x(z) \geq 0 \text{ for all } z \in \overline{\Omega}\},$$

then

$$\text{int } C_0^1(\overline{\Omega})_+ = \left\{ x \in C_0^1(\overline{\Omega}) : x(z) > 0 \text{ for all } z \in \Omega \right.$$

$$\left. \text{and } \frac{\partial x(z)}{\partial n} < 0 \text{ for all } z \in \partial\Omega \right\}.$$

Therefore the conclusion of Proposition 4.6.9 implies that the solution x_0 belongs in int $C_0^1(\overline{\Omega})_+$. $\Box$

In order to produce a second solution for problem (4.182), distinct from x_0, we shall need an auxiliary result, which is actually of independent interest and relates local minimizers in $C_0^1(\overline{\Omega})$ and in $W_0^{1,p}(\Omega)$ respectively, for a broad class of nonsmooth functionals.

To this end, let $\psi \colon W_0^{1,p}(\Omega) \longrightarrow \mathbb{R}$ be defined by

$$\psi(x) \overset{df}{=} \frac{1}{p} \|\nabla x\|_p^p - \int_\Omega j_0\big(z, x(z)\big)\, dz.$$

Our hypotheses on the integrand (nonsmooth potential) $j_0(z, \zeta)$ are the following:

$\underline{H(j_0)}$ $j_0 \colon \Omega \times \mathbb{R} \longrightarrow \mathbb{R}$ is a function, such that

(i) for every $\zeta \in \mathbb{R}$, the function

$$\Omega \ni z \longrightarrow j_0(z, \zeta) \in \mathbb{R}$$

is measurable and $j_0(\cdot, 0) \in L^\infty(\Omega)$;

(ii) for almost all $z \in \Omega$, the function

$$\mathbb{R} \ni \zeta \longrightarrow j_0(z, \zeta) \in \mathbb{R}$$

is locally Lipschitz;

(iii) for almost all $z \in \Omega$, all $\zeta \in \mathbb{R}$ and all $u \in \partial j_0(z, \zeta)$, we have that

$$|u| \leq a_0 + c_0|\zeta|^{s-1},$$

with $a_0 \in L^\infty(\Omega)_+$, $c_0 > 0$ and $s \in [1, p^*)$.

These hypotheses imply that ψ is locally Lipschitz on $W_0^{1,p}(\Omega)$.

PROPOSITION 4.6.10
If hypotheses $H(j_0)$ hold and $x_0 \in W_0^{1,p}(\Omega)$ is a local $C_0^1(\overline{\Omega})$-minimizer of ψ, i.e. there exists $r > 0$ such that

$$\psi(x_0) \leq \psi(x_0 + y) \qquad \forall\, y \in C_0^1(\overline{\Omega}), \ \|y\|_{C_0^1(\overline{\Omega})} \leq r,$$

then x_0 is a local minimizer of ψ in $W_0^{1,p}(\Omega)$, i.e. there exists $r_0 > 0$ such that

$$\psi(x_0) \leq \psi(x_0 + u) \qquad \forall\, u \in W_0^{1,p}(\Omega), \|u\|_{W^{1,p}(\Omega)} \leq r_0.$$

PROOF   Let $h \in C_0^1(\overline{\Omega})$. For $\lambda > 0$ small enough, we have

$$0 \leq \frac{\psi(x_0 + \lambda h) - \psi(x_0)}{\lambda},$$

so

$$0 \leq \psi^0(x_0; h) \qquad \forall\, h \in C_0^1(\overline{\Omega}).$$

Since $\psi^0(x_0; \cdot)$ is continuous (see Proposition 1.3.7(a)) and the embedding $C_0^1(\overline{\Omega}) \subseteq W_0^{1,p}(\Omega)$ is dense, it follows that

$$0 \leq \psi^0(x_0; h) \qquad \forall\, h \in W_0^{1,p}(\Omega)$$

and so $0 \in \partial\psi(x_0)$. This inclusion means that

$$A(x_0) = u^*,$$

for some $u^* \in L^{s'}(\Omega)$ ($\frac{1}{s} + \frac{1}{s'} = 1$), $u^*(z) \in \partial j_0\big(z, x_0(z)\big)$ for almost all $z \in \Omega$. Hence

$$-\operatorname{div}\big(\,\|\nabla x_0(z)\|_{\mathbb{R}^N}^{p-2}\,\nabla x_0(z)\big) = u^*(z) \quad \text{for a.a. } z \in \Omega.$$

As before, via Theorem 1.5.6, we have that $x_0 \in C_0^{1,\beta}(\overline{\Omega})$ for some $\beta \in (0,1)$.

Now suppose that the Proposition is not true. So for a given $\varepsilon > 0$, we can find

$$x_\varepsilon \in \overline{B_\varepsilon(x_0)},$$

such that

$$\psi(x_\varepsilon) = \inf_{x \in \overline{B_\varepsilon(x_0)}} \psi(x) < \psi(x_0).$$

Invoking the nonsmooth Lagrange multiplier rule (see Proposition 1.3.17), we can find constants $\lambda_\varepsilon \geq 0 \geq \mu_\varepsilon$, not both equal to zero, such that

$$\lambda_\varepsilon x_\varepsilon^* = \mu_\varepsilon A(x_\varepsilon - x_0),$$

with $x_\varepsilon^* \in \partial\psi(x_\varepsilon)$. If $\lambda_\varepsilon = 0$, then $\mu_\varepsilon A(x_\varepsilon - x_0) = 0$ and taking duality brackets with $x_\varepsilon - x_0$, we obtain

$$\mu_\varepsilon \|\nabla(x_\varepsilon - x_0)\|_p^p = 0,$$

i.e. $\mu_\varepsilon = 0$, a contradiction. So $\lambda_\varepsilon > 0$ and without any loss of generality we can assume that $\lambda_\varepsilon = 1$. Then we have

$$A(x_\varepsilon) - \mu_\varepsilon A(x_\varepsilon - x_0) = u_\varepsilon^*,$$

with $u_\varepsilon^* \in L^{s'}(\Omega)$, $u_\varepsilon^*(z) \in \partial j_0(z, x_\varepsilon(z))$ for almost all $z \in \Omega$. From the sensitivity analysis of optimization problems (see Dontchev & Zolezzi (1994, p. 361)), we know that the map $(0,1] \ni \varepsilon \longmapsto \mu_\varepsilon$ is continuous. Using standard nonlinear regularity theory (see Ladyzhenskaya & Uraltseva (1968, Theorem 7.1, p. 286)), we obtain a constant $M_3 > 0$ independent on $\varepsilon \in (0,1]$ such that

$$\|x_\varepsilon\|_\infty \leq M_3 \qquad \forall \varepsilon \in (0,1].$$

Moreover, by considering the vector field

$$V_\varepsilon(z,\xi) \overset{df}{=} \|\xi\|_{\mathbb{R}^N}^{p-2} \xi - \mu_\varepsilon \|\xi - \sigma(z)\|_{\mathbb{R}^N}^{p-2} (\xi - \sigma(z)) - \mu_\varepsilon \|\sigma(z)\|_{\mathbb{R}^N}^{p-2} \sigma(z),$$

with $\sigma = \nabla x_0 \in C^\beta(\overline{\Omega})$, we see that

$$\begin{cases} -\operatorname{div} V_\varepsilon(z, \nabla x_\varepsilon(z)) = u_\varepsilon^*(z) - \mu_\varepsilon u(z) & \text{for a.a. } z \in \Omega, \\ x_\varepsilon|_{\partial\Omega} = 0. \end{cases} \qquad (4.205)$$

A simple calculation reveals that we can find $c_0 = c_0(p) > 1$ (depending only on p) such that for all $z \in \Omega$ and all $\xi \in \mathbb{R}^N$ we have

$$\frac{1}{c_0} \left((1 + |\mu_\varepsilon|) \|\xi\|_{\mathbb{R}^N}^{p-2} + |\mu_\varepsilon| \|\sigma(z)\|_{\mathbb{R}^N}^{p-2} \right)$$
$$\leq \left(\|\xi\|_{\mathbb{R}^N}^{p-2} + |\mu_\varepsilon| \|\sigma(z) - \xi\|_{\mathbb{R}^N}^{p-2} \right)$$
$$\leq c_0 \left((1 + |\mu_\varepsilon|) \|\xi\|_{\mathbb{R}^N}^{p-2} + |\mu_\varepsilon| \|\sigma(z)\|_{\mathbb{R}^N}^{p-2} \right). \qquad (4.206)$$

From (4.206) it follows that if $\mu_\varepsilon \in [-1, 0]$, then

$$\left(\nabla_\xi V_\varepsilon(z,\xi)\eta, \eta \right)_{\mathbb{R}^N} \geq \vartheta \left(\gamma(z) + \|\xi\|_{\mathbb{R}^N}^{p-2} \right) \|\eta\|_{\mathbb{R}^N}^2$$

and

$$\left\| \nabla_\xi V_\varepsilon(z,\xi) \right\|_{\mathbb{R}^N} \leq \varrho \left(\gamma(z) + \|\xi\|_{\mathbb{R}^N}^{p-2} \right),$$

with $\gamma \in L^\infty(\Omega)_+$ and $\vartheta, \varrho > 0$.

So from (4.205) and the nonlinear regularity theory up to the boundary (see Di Benedetto (1983, 1995, Chapter IX) and Lieberman (1988, Theorem 1)), we can find $\beta \in (0,1)$ and $M_4 > 0$ both independent of $\varepsilon \in (0,1]$ such that

$$\|x_\varepsilon\|_{C_0^{1,\beta}(\overline{\Omega})} \leq M_4.$$

If $\mu_\varepsilon < -1$, then by setting $y_\varepsilon = x_\varepsilon - x_0$ we have

$$A(y_\varepsilon) = \frac{1}{\mu_\varepsilon} A(y_\varepsilon + x_0) - \frac{1}{\mu_\varepsilon} u_\varepsilon^*,$$

with $u_\varepsilon^* \in L^{s'}(\Omega)$, $u_\varepsilon^*(z) \in \partial j(z, y_\varepsilon(z) + x_0(z))$ for almost all $z \in \Omega$.

Also we have

$$A(x_0) = u^*.$$

Therefore we deduce that

$$-\operatorname{div}\left(\|\nabla y_\varepsilon(z)\|_{\mathbb{R}^N}^{p-2}\nabla y_\varepsilon(z)\right) - \frac{1}{\mu_\varepsilon}\|\nabla(y_\varepsilon+x_0)(z)\|_{\mathbb{R}^N}^{p-2}\nabla(y_\varepsilon+x_0)(z)$$

$$+ \frac{1}{\mu_\varepsilon}\|\nabla y_\varepsilon(z)\|_{\mathbb{R}^N}^{p-2}\nabla y_\varepsilon(z)$$

$$= \frac{1}{\mu_\varepsilon}\left(u^*(z) - u_\varepsilon^*(z)\right) \quad \text{for a.a. } z \in \Omega, \tag{4.207}$$

which is an equation similar to (4.205). Note that the vector field

$$W_\varepsilon(z,\xi) \stackrel{df}{=} \|\xi\|_{\mathbb{R}^N}^{p-2}\xi - \frac{1}{\mu_\varepsilon}\|\xi+\sigma(z)\|_{\mathbb{R}^N}^{p-2}\left(\xi+\sigma(z)\right) + \frac{1}{\mu_\varepsilon}\|\sigma(z)\|_{\mathbb{R}^N}^{p-2}\sigma(z)$$

satisfies the same estimates as $V_\varepsilon(z,\xi)$. Therefore from (4.207), as before, we obtain $\beta \in (0,1)$ and $M_5 > 0$ both independent of $\varepsilon \in (0,1]$ such that

$$\|x_\varepsilon\|_{C_0^{1,\beta}(\overline{\Omega})} \leq M_5.$$

But we know that the embedding $C_0^{1,\beta}(\overline{\Omega}) \subseteq C_0^1(\overline{\Omega})$ is compact (see Kufner, John & Fučik (1977, p. 38)). Therefore, we can find $\varepsilon_n \in (0,1]$ with $\varepsilon_n \searrow 0$ and $\widehat{x} \in C_0^1(\overline{\Omega})$ such that

$$x_{\varepsilon_n} \longrightarrow \widehat{x} \quad \text{in } C_0^1(\overline{\Omega}).$$

Note that by the construction of the sequence $\{x_{\varepsilon_n}\}_{n\geq 1}$, we have

$$x_{\varepsilon_n} \longrightarrow x_0 \quad \text{in } W_0^{1,p}(\Omega).$$

It follows that $\widehat{x} = x_0$. Thus we can find $n_0 \geq 1$ such that

$$\|x_{\varepsilon_n} - x_0\|_{C_0^1(\overline{\Omega})} \leq r \quad \forall\, n \geq n_0$$

so

$$\psi(x_0) \leq \psi(x_{\varepsilon_n}) < \psi(x_0),$$

a contradiction. This completes the proof of the Proposition. $\quad\square$

If we look back at hypotheses $H(j)_3$, we see that only an asymptotic condition at $+\infty$ was imposed (see hypothesis $H(j)_3(iv)$). Now by imposing additional asymptotic conditions at $-\infty$ and with the help of Proposition 4.6.10, we obtain a second solution for problem (4.182), distinct from x_0. Note that the new asymptotic condition permits complete resonance at $-\infty$.

$\underline{H(j)_4}$ $j \colon \Omega \times \mathbb{R} \longrightarrow \mathbb{R}$ is a function such that

   (i) for every $\zeta \in \mathbb{R}$, the function

$$\Omega \ni z \longrightarrow j(z,\zeta) \in \mathbb{R}$$

   is measurable, $j(\cdot,0) \in L^\infty(\Omega)$, $\int_\Omega j(z,0)\,dz \leq 0$;

(*ii*) for almost all $z \in \Omega$, the function

$$\mathbb{R} \ni \zeta \longrightarrow j(z, \zeta) \in \mathbb{R}$$

is locally Lipschitz;

(*iii*) for almost all $z \in \Omega$, all $\zeta \in \mathbb{R}$, and all $u \in \partial j(z, \zeta)$, we have

$$|u| \le \widetilde{a}_1(z) + \widetilde{c}_1 |\zeta|^{p-1},$$

with $\widetilde{a}_1 \in L^\infty(\Omega)_+$, $\widetilde{c}_1 > 0$ and for almost all $z \in \Omega$, all $\zeta \ge 0$ and all $u \in \partial j(z, \zeta)$, we have

$$-\widetilde{c}_2 \zeta^{p-1} \le u,$$

with $\widetilde{c}_2 > 0$;

(*iv*) there exist $\vartheta \in L^\infty(\Omega)$ and $j_- \in L^1(\Omega)$ such that $\vartheta(z) \le 0$ for almost all $z \in \Omega$ with strict inequality on a set of positive measure,

$$\int_\Omega j_-(z) u_1(z)^2 \, dz < 0,$$

$$\limsup_{\zeta \to +\infty} \frac{u}{\zeta^{p-1}} \le \vartheta(z),$$

$$\limsup_{\zeta \to -\infty} \frac{u}{|\zeta|^{p-2}\zeta} = 0$$

and

$$\limsup_{\zeta \to -\infty} \frac{j(z, \zeta)}{\zeta^2} \le j_-(z)$$

uniformly for almost all $z \in \Omega$ and all $u \in \partial j(z, \zeta)$;

(*v*) there exists $\eta \in L^\infty(\Omega)$ such that $\eta(z) \ge 0$ for almost all $z \in \Omega$ with strict inequality on a set of positive measure and

$$\liminf_{\zeta \to 0^+} \frac{p j(z, \zeta)}{\zeta^p} \ge \eta(z)$$

uniformly for almost all $z \in \Omega$;

(*vi*) there exists $M > 0$ such that

$$u \le 0 \quad \text{for a.a. } z \in \Omega, \text{ all } \zeta \ge M \text{ and all } u \in \partial j(z, \zeta).$$

REMARK 4.6.6 The function

$$j(\zeta) \overset{df}{=} \begin{cases} -|\zeta|^r & \text{if } \zeta < 0, \\ -\zeta^p + \zeta & \text{if } \zeta \ge 0, \end{cases}$$

with $2 \leq r < p$, satisfies hypotheses $H(j)_4$. Here for simplicity we have dropped the z-dependence. ▯

THEOREM 4.6.3
If hypotheses $H(j)_4$ hold,
then problem (4.182) has at least two distinct solutions $x_0, x_1 \in W_0^{1,p}(\Omega)$, with $x_0 \in C_0^1(\overline{\Omega})$, $x_0(z) > 0$ for all $z \in \Omega$ and $\frac{\partial x_0(z)}{\partial n} < 0$ for all $z \in \partial\Omega$.

PROOF   One solution $x_0 \in C_0^1(\overline{\Omega})$ was obtained in Proposition 4.6.9. For this solution we know that $x_0(z) > 0$ for all $z \in \Omega$ and $\frac{\partial x_0(z)}{\partial n} < 0$ for all $z \in \partial\Omega$.

Next we show that

$$\liminf_{t \to -\infty} \varphi_1(tu_1) = -\infty.$$

To this end, we have

$$\varphi_1(tu_1) = -\int_\Omega j\big(z, tu_1(z)\big)\, dz$$

(since $\|\nabla u_1\|_p^p = \lambda_1 \|u_1\|_p^p$). Suppose for the moment that the convergence we want to show is not true. Then we can find $M_6 > 0$ such that

$$\varphi(tu_1) = -\int_\Omega j\big(z, tu_1(z)\big)\, dz \geq -M_6 \qquad \forall\, t < 0$$

and so

$$\int_\Omega \frac{j(z, tu_1(z))}{t^2 u_1(z)^2} u_1(z)^2\, dz \geq -\frac{M_6}{t^2} \qquad \forall\, t < 0.$$

Let $t_n \to -\infty$. From Fatou's lemma and hypothesis $H(j)_4(iv)$, we have that

$$0 \leq \limsup_{n \to +\infty} \int_\Omega \frac{j(z, t_n u_1(z))}{t_n^2 u_1^2(z)} u_1^2(z)\, dz \leq \int_\Omega j_-(z) u_1(z)^2\, dz < 0,$$

a contradiction. This proves that $\liminf\limits_{t \to -\infty} \varphi(tu_1) = -\infty$.

Recall that $x_0 \in \mathrm{int}\, C_0^1(\Omega)_+$ (see Remark 4.6.5). So we can find $\varepsilon_1 > 0$ such that

$$\overline{B_{\varepsilon_1}^{C_0^1(\Omega)}}(x_0) \overset{df}{=} \left\{ x \in C_0^1(\Omega) : \|x - x_0\|_{C_0^1(\overline{\Omega})} \leq \varepsilon_1 \right\} \subseteq C_0^1(\overline{\Omega})_+.$$

Since $\varphi(x_0) = \inf\limits_{W_0^{1,p}(\Omega)} \varphi$ (see the proof of Proposition 4.6.9), we have

$$\varphi_1(x_0) = \varphi(x_0) \leq \varphi(v) = \varphi_1(v) \qquad \forall\, v \in \overline{B_{\varepsilon_1}^{C_0^1(\Omega)}}(x_0).$$

Therefore x_0 is a local $C_0^1(\overline{\Omega})$-minimizer for the functional φ_1. We can apply Proposition 4.6.10 to deduce that x_0 is also a local $W_0^{1,p}(\Omega)$-minimizer for φ_1. This means that we can find $\varepsilon_2 > 0$ such that

$$\varphi_1(x_0) \leq \varphi_1(y) \qquad \forall\, y \in \overline{B_{\varepsilon_2}^{W_0^{1,p}(\Omega)}}(x_0).$$

Evidently, we may assume that x_0 is a strict minimizer or otherwise we are done. So we have

$$\varphi_1(x_0) < \varphi_1(y) \qquad \forall\, y \in \partial B_{\varepsilon_2}^{W_0^{1,p}(\Omega)}(x_0).$$

A careful reading of the proof of Proposition 4.6.8 reveals that as a byproduct, we have that the locally Lipschitz function φ_1 satisfies the nonsmooth PS-condition. Choose $t < 0$ such that

$$\|t u_1 - x_0\|_{W^{1,p}(\Omega)} > \varepsilon_2 \quad \text{and} \quad \varphi_1(t u_1) \leq \varphi_1(x_0)$$

(recall that as we proved earlier $\varphi_1|_{\mathbb{R} u_1}$ is anticoercive). All these facts imply that the nonsmooth Mountain Pass geometry is in place and so we can apply Theorem 2.1.3 and obtain $x_1 \in W_0^{1,p}(\Omega)$ such that

$$0 \in \partial\varphi_1(x_1) \quad \text{and} \quad \varphi_1(x_0) < \inf_{y \in \partial B_{\varepsilon_2}^{W_0^{1,p}(\Omega)}} \varphi_1(y) \leq \varphi_1(x_1),$$

so $x_1 \neq x_0$ and x_1 is a solution of problem (4.182). ▯

REMARK 4.6.7 If there exists $\Omega_1 \subseteq \Omega$, with $|\Omega_1|_N > 0$ such that $0 \notin \partial j(z, 0)$ for all $z \in \Omega_1$, then x_1 is nontrivial too. ▯

4.7 Positive Solutions

We continue our investigations initiated in the previous section and we look for multiple solutions of constant sign. We do this in the context of the following nonlinear elliptic eigenvalue problem with a nonsmooth potential (hemivariational inequality). So let $\Omega \subseteq \mathbb{R}^N$ be a bounded domain with a C^2-boundary $\partial\Omega$. The problem under consideration is the following:

$$\begin{cases} -\mathrm{div}\,\big(\|\nabla x(z)\|_{\mathbb{R}^N}^{p-2}\,\nabla x(z)\big) - \lambda |x(z)|^{p-2} x(z) \in \partial j\big(z, x(z)\big) \\ \qquad\qquad\qquad\qquad\qquad\qquad\qquad \text{for a.a. } z \in \Omega, \qquad (4.208) \\ x|_{\partial\Omega} = 0, \end{cases}$$

where $p \in [2, +\infty)$, $\lambda > 0$. As before, $j(z, \zeta)$ is a potential function which is only locally Lipschitz and $\partial j(z, \zeta)$ stands for Clarke's subdifferential of

$\zeta \longmapsto j(z,\zeta)$. We are concerned with the case when $\lambda > 0$ approaches $\lambda_1 > 0$, the principal eigenvalue of $\left(-\Delta_p, W_0^{1,p}(\Omega)\right)$, from above. Under certain hypotheses regulating the behaviour of the nonsmooth potential $j(z,\zeta)$ and of its subdifferential $\partial j(z,\zeta)$ near zero and near $\pm\infty$, we show that as $\lambda \to \lambda_1^+$, problem (4.208) has at least three solutions of constant sign, one negative and two positive solutions. Moreover, one of the solutions is smooth. Our approach is variational based on the nonsmooth critical point theory for locally Lipschitz functions, coupled with method of upper and lower solutions.

Our hypotheses on the nonsmooth potential function $j(z,\zeta)$ are the following:

<u>$H(j)_1$</u> $j: \Omega \times \mathbb{R} \longrightarrow \mathbb{R}$ is a function, such that

    (i) for every $\zeta \in \mathbb{R}$, the function

$$\Omega \ni z \longrightarrow j(z,\zeta) \in \mathbb{R}$$

    is measurable and $j(z,0) = 0$ for almost all $z \in \Omega$;

    (ii) for almost all $z \in \Omega$, the function

$$\mathbb{R} \ni \zeta \longrightarrow j(z,\zeta) \in \mathbb{R}$$

    is locally Lipschitz;

    (iii) for almost all $z \in \Omega$, all $\zeta \in \mathbb{R}$ and all $u \in \partial j(z,\zeta)$, we have

$$|u| \leq \tilde{a}_1(z) + \tilde{c}_1 |\zeta|^{r-1},$$

    with $\tilde{a}_1 \in L^\infty(\Omega)_+$, $\tilde{c}_1 > 0$ and $r \in [1, p^*)$ where as usual

$$p^* \stackrel{df}{=} \begin{cases} \frac{pN}{N-p} & \text{if } N > p, \\ +\infty & \text{if } N \leq p; \end{cases}$$

    (iv) there exists a function $\vartheta \in L^\infty(\Omega)$ such that $\vartheta(z) \leq 0$ for almost all $z \in \Omega$, the inequality is strict on a set of positive measure and

$$\limsup_{\zeta \to -\infty} \frac{u}{|\zeta|^{p-2}\zeta} \leq \vartheta(z) \quad \text{and} \quad \limsup_{\zeta \to +\infty} \frac{pj(z,\zeta)}{\zeta^p} < 0$$

    uniformly for almost all $z \in \Omega$ and all $u \in \partial j(z,\zeta)$;

    (v) we have

$$\liminf_{\zeta \to 0^-} \frac{u}{|\zeta|^{p-2}\zeta} \geq 0 \quad \text{and} \quad \liminf_{\zeta \to 0^+} \frac{pj(z,\zeta)}{\zeta^p} \geq 0$$

    uniformly for almost all $z \in \Omega$ and all $u \in \partial j(z,\zeta)$;

(*vi*) we have

$$-\widehat{c}_1 \zeta^{p-1} \;\leq\; u \quad \text{for a.a. } z \in \Omega, \text{ all } \zeta \geq 0 \text{ and all } u \in \partial j(z, \zeta)$$

with $0 < \widehat{c}_1 \leq \lambda_1$.

For every $\lambda > 0$ we introduce the locally Lipschitz functional $\varphi_\lambda \colon W_0^{1,p}(\Omega) \longrightarrow \mathbb{R}$, defined by

$$\varphi_\lambda(x) \;\overset{df}{=}\; \frac{1}{p} \|\nabla x\|_p^p - \frac{\lambda}{p} \|x\|_p^p - \int_\Omega j\big(z, x(z)\big)\, dz \qquad \forall\, x \in W_0^{1,p}(\Omega).$$

First we shall establish the existence of a strictly negative solution. This will be done by employing the method of upper and lower solutions. Using our conditions on the asymptotic behaviour of the subdifferential $\partial j(z, \zeta)$ at $-\infty$ and at 0^-, we shall produce an ordered pair of a lower and of an upper solution respectively and then show that there is a solution of (4.208) in the ordered interval determined by the lower and the upper solutions. In fact the lower and the upper solutions that we generate are stronger then the corresponding ones used in Section 4.5, since they satisfy the following more restrictive definition (compare with Definition 4.5.1).

DEFINITION 4.7.1

(a) *A function $\overline{x} \in W^{1,p}(\Omega)$ is an **upper solution** of problem (4.208), if $\overline{x}|_{\partial\Omega} \geq 0$ and*

$$\int_\Omega \|\nabla\overline{x}(z)\|_{\mathbb{R}^N}^{p-2}\, \big(\nabla\overline{x}(z), \nabla v(z)\big)_{\mathbb{R}^N}\, dz - \lambda \int_\Omega |\overline{x}(z)|^{p-2}\overline{x}(z)v(z)\, dz$$

$$\geq \int_\Omega u(z)v(z)\, dz,$$

for all $v \in W_0^{1,p}(\Omega)$, $v \geq 0$ and all $u \in L^{r'}(\Omega)$ $(\frac{1}{r} + \frac{1}{r'} = 1)$ with $u(z) \in \partial j\big(z, x(z)\big)$ for almost all $z \in \Omega$.

(b) *A function $\underline{x} \in W^{1,p}(\Omega)$ is a **lower solution** of problem (4.208), if $\underline{x}|_{\partial\Omega} \leq 0$ and*

$$\int_\Omega \|\nabla\underline{x}(z)\|_{\mathbb{R}^N}^{p-2}\, \big(\nabla\underline{x}(z), \nabla v(z)\big)_{\mathbb{R}^N}\, dz - \lambda \int_\Omega |\underline{x}(z)|^{p-2}\underline{x}(z)v(z)\, dz$$

$$\leq \int_\Omega u(z)v(z)\, dz,$$

for all $v \in W_0^{1,p}(\Omega)$, $v \geq 0$ and all $u \in L^{r'}(\Omega)$ $(\frac{1}{r} + \frac{1}{r'} = 1)$ with $u(z) \in \partial j\big(z, x(z)\big)$ for almost all $z \in \Omega$.

Let $\xi\colon W_0^{1,p}(\Omega) \longrightarrow \mathbb{R}$ be defined by

$$\xi(x) \stackrel{df}{=} \|\nabla x(z)\|_{\mathbb{R}^N}^{p-2} - \int_\Omega \big(\lambda_1 + \vartheta(z)\big)|x(z)|^p \, dz.$$

Arguing as in the proof of Proposition 4.4.3 (see the Claim), we can have the following lemma.

LEMMA 4.7.1
If $\vartheta \in L^\infty(\Omega)$, $\vartheta(z) \le 0$ *for almost all* $z \in \Omega$ *and the inequality is strict on a set of positive measure,*
<u>*then*</u> *there exists* $\beta > 0$ *such that*

$$\xi(x) \ge \beta\,\|\nabla x\|_p^p \qquad \forall\, x \in W_0^{1,p}(\Omega).$$

First we prove the existence of a strictly negative solution.

PROPOSITION 4.7.1
If *hypotheses* $H(j)_1(i) - (v)$ *hold,*
<u>*then*</u> *there exists* $\varepsilon_0 > 0$ *such that for all* $\lambda \in (\lambda_1, \lambda_1 + \varepsilon_0)$ *problem (4.208) has a solution* $y_1 \in W_0^{1,p}(\Omega)$ *such that* $y_1(z) < 0$ *for almost all* $z \in \Omega$.

PROOF   From hypothesis $H(j)_1(iii)$ and the first limit (at $-\infty$) of hypothesis $H(j)_1(iv)$, for a given $\varepsilon > 0$, we can find $\gamma_\varepsilon \in L^\infty(\Omega)_+$, $\gamma_\varepsilon \ne 0$, such that for almost all $z \in \Omega$, all $\zeta \in \mathbb{R}$ and all $u \in \partial j(z, \zeta)$, we have

$$\big(\vartheta(z) + \varepsilon\big)|\zeta|^{p-2}\zeta - \gamma_\varepsilon(z) \le u. \qquad (4.209)$$

Then we consider the following auxiliary boundary value problem:

$$\begin{cases} -\mathrm{div}\,\big(\,\|\nabla x(z)\|_{\mathbb{R}^N}^{p-2}\,\nabla x(z)\big) - \lambda|x(z)|^{p-2}x(z) \\ \qquad = \big(\vartheta(z) + \varepsilon\big)|x(z)|^{p-2}x(z) - \gamma_\varepsilon(z) \quad \text{for a.a. } z \in \Omega, \\ x|_{\partial\Omega} = 0. \end{cases} \qquad (4.210)$$

As in previous proofs, we introduce the maximal monotone operator $A\colon W_0^{1,p}(\Omega) \longrightarrow W^{-1,p'}(\Omega)$, defined by

$$\langle A(x), y\rangle_{W_0^{1,p}(\Omega)} \stackrel{df}{=} \int_\Omega \|\nabla x(z)\|_{\mathbb{R}^N}^p \big(\nabla x(z), \nabla y(z)\big)_{\mathbb{R}^N} \qquad \forall\, x, y \in W_0^{1,p}(\Omega).$$

Also let $S\colon W_0^{1,p}(\Omega) \longrightarrow L^{p'}(\Omega)$ $(\frac{1}{p} + \frac{1}{p} = 1)$ be the nonlinear operator defined by

$$S(x)(\cdot) \stackrel{df}{=} \big(\lambda + \vartheta(\cdot) + \varepsilon\big)|x(\cdot)|^{p-2}x(\cdot) \qquad \forall\, x \in W_0^{1,p}(\Omega).$$

Since the embedding $W_0^{1,p}(\Omega) \subseteq L^p(\Omega)$ is compact, it follows that S is completely continuous (see Theorem 1.4.6). Let us set

$$V \stackrel{df}{=} A - S \colon W_0^{1,p}(\Omega) \longrightarrow W^{-1,p'}(\Omega)$$

(recall that the embedding $L^{p'}(\Omega) \subseteq W^{-1,p'}(\Omega)$ is compact). Evidently V is pseudomonotone. Also if $\varepsilon = \varepsilon_0 < \frac{\lambda_1 \beta}{2}$ with $\beta > 0$ as in Lemma 4.7.1, then for $\lambda \in (\lambda_1, \lambda_1 + \varepsilon_0)$ and all $x \in W_0^{1,p}(\Omega)$, we have

$$\langle V(x), x \rangle_{W_0^{1,p}(\Omega)} = \|\nabla x\|_p^p - \int_\Omega \left(\lambda_1 + \vartheta(z) + \varepsilon\right)|x(z)|^p \, dz$$

$$\geq \beta \|\nabla x\|_p^p - \frac{2\varepsilon_0}{\lambda_1} \|\nabla x\|_p^p = \beta_1 \|\nabla x\|_p^p,$$

with $\beta_1 \stackrel{df}{=} \beta - \frac{2\varepsilon_0}{\lambda_1} > 0$. So V is coercive.

A pseudomonotone, coercive operator is surjective (see Theorem 1.4.6). Thus we can find $v_1 \in W_0^{1,p}(\Omega)$ such that

$$V(v_1) = -\gamma_{\varepsilon_0},$$

with $\gamma_{\varepsilon_0} \in L^\infty(\Omega)_+$. So we have

$$A(v_1) - S(v_1) = -\gamma_{\varepsilon_0}. \tag{4.211}$$

Acting with the test function $v_1^+ \in W_0^{1,p}(\Omega)$ and using the fact that $\gamma_{\varepsilon_0} \geq 0$, we obtain

$$\|\nabla v_1^+\|_p^p - \int_\Omega \left(\lambda + \vartheta(z) + \varepsilon_0\right)|v_1^+(z)|^p \, dz = -\int_\Omega \gamma_{\varepsilon_0}(z) v_1^+(z) \, dz \leq 0$$

so

$$\frac{\beta}{2} \|\nabla v_1^+\|_p^p \leq 0,$$

i.e. $v_1^+ = 0$ and so $v_1 \leq 0$.

Because $\gamma_{\varepsilon_0} \neq 0$, we can see that $v_1 \neq 0$. From (4.211) it follows that $v_1 \in W_0^{1,p}(\Omega)$ is a solution of the auxiliary problem (4.210), with $\varepsilon = \varepsilon_0$, $\gamma_\varepsilon = \gamma_{\varepsilon_0}$ and $\lambda \in (\lambda_1, \lambda_1 + \varepsilon_0)$. We show that v_1 is a strictly negative, smooth lower solution of problem (4.208).

By virtue of Theorem 1.5.5, we have that $v_1 \in L^\infty(\Omega)_-$ and then invoking Theorem 1.5.6, we obtain $v_1 \in C_0^{1,\eta}(\overline{\Omega})$ for some $\eta \in (0,1)$. Moreover, from (4.210) and since $\gamma_{\varepsilon_0} \geq 0$, we have

$$\Delta_p(-v_1)(z) \leq \left|\lambda + \vartheta(z) + \varepsilon\right|\left(-v_1(z)\right)^{p-1} \quad \text{for a.a. } z \in \Omega.$$

Thus by virtue of Theorem 1.5.7, we have that

$$v_1(z) < 0 \quad \text{for a.a. } z \in \Omega \quad \text{and} \quad \frac{\partial v_1(z)}{\partial n} > 0 \quad \text{for a.a. } z \in \partial\Omega$$

(n is the outer unit normal on $\partial\Omega$). This means that $v_1 \in -\mathrm{int}\,C_0^1(\overline{\Omega})_+$ (see Remark 4.6.5).

For every $y \in W_0^{1,p}(\Omega)$, $y \geq 0$, we have

$$
\int_\Omega \|\nabla v_1(z)\|_{\mathbb{R}^N}^{p-2} \left(\nabla v_1(z), \nabla y(z)\right)_{\mathbb{R}^N} dz - \lambda \int_\Omega |v_1(z)|^{p-2} v_1(z) y(z)\, dz
$$

$$
= \int_\Omega \left(\vartheta(z) + \varepsilon_0\right) |v_1(z)|^{p-2} v_1(z) y(z)\, dz - \int_\Omega \gamma_{\varepsilon_0}(z) y(z)\, dz
$$

$$
\leq \int_\Omega u(z) y(z)\, dz,
$$

for all $u \in L^{r'}(\Omega)$ with $u(z) \in \partial j(z, v_1(z))$ for almost all $z \in \Omega$ (see (4.209)). This proves that v_1 is a strictly negative, smooth lower solution of problem (4.208) (see Definition 4.7.1(b)).

By virtue of the first limit in hypothesis $H(j)_1(v)$ (asymptotic condition at 0^-), we can find $\delta > 0$ such that

$$
\frac{u}{|\zeta|^{p-2}\zeta} \geq \lambda_1 - \lambda \quad \text{for a.a. } z \in \Omega,\ \text{all } \zeta \in (-\delta, 0) \text{ and all } u \in \partial j(z, \zeta)
$$

(recall that $\lambda \in (\lambda_1, \lambda_1 + \varepsilon_0)$) and so

$$
u \leq (\lambda_1 - \lambda)|\zeta|^{p-2}\zeta. \tag{4.212}
$$

Let $u_1 \in C_0^1(\overline{\Omega})$ be the principal eigenfunction of $\left(-\Delta_p, W_0^{1,p}(\Omega)\right)$. We know that $u_1(z) > 0$ for all $z \in \Omega$ (see Proposition 1.5.18). Therefore we can find $\xi > 0$ small enough such that

$$
-\xi u_1(z) \in (-\delta, 0) \qquad \forall\, z \in \Omega
$$

and

$$
v_1(z) < -\xi u_1(z) \qquad \forall\, z \in \Omega
$$

(recall that $v_1 \in -\mathrm{int}\,C_0^1(\overline{\Omega})_+$). For all $u \in \partial j\big(z, -\xi u_1(z)\big)$, we have

$$
\begin{aligned}
-\Delta_p(-\xi u_1) &= \xi^{p-1}\Delta_p u_1 = -\lambda_1(\xi u_1)^{p-1} = \lambda_1|-\xi u_1|^{p-2}(-\xi u_1) \\
&= (\lambda_1 - \lambda)|-\xi u_1|^{p-2}(-\xi u_1) + \lambda|-\xi u_1|^{p-2}(-\xi u_1) \\
&\geq u + \lambda|-\xi u_1|^{p-2}(-\xi u_1)
\end{aligned}
$$

(see (4.212)). From this last inequality, it follows that $v_2 = -\xi u_1$ is a strictly negative, smooth upper solution of problem (4.208) (see Definition 4.7.1(a)). Note that

$$
\frac{\partial v_2(z)}{\partial n} > 0 \qquad \forall\, z \in \partial\Omega.
$$

So we have produced an ordered pair (v_1, v_2) of a lower and of an upper solution for problem (4.208) (we have $v_1(z) < v_2(z) = -\xi u_1(z)$ for all $z \in \Omega$).

Following the techniques employed in Section 4.5, we shall produce a solution of problem (4.208) in the order interval

$$[v_1, v_2] \overset{df}{=} \{x \in W_0^{1,p}(\Omega) : \ v_1(z) \le x(z) \le v_2(z) \quad \text{for a.a. } z \in \Omega\}.$$

To this end, we introduce the truncation map $\tau \colon W^{1,p}(\Omega) \longrightarrow W^{1,p}(\Omega)$, defined by

$$\tau(x)(z) \overset{df}{=} \begin{cases} v_2(z) \text{ if } v_2(z) < x(z), \\ x(z) \ \text{ if } v_1(z) \le x(z) \le v_2(z), \\ v_1(z) \text{ if } x(z) < v_1(z). \end{cases}$$

Clearly τ is continuous and bounded and it has the same properties as a map from $L^p(\Omega)$ into itself. In addition, note that

$$\|\tau(x)\|_p^p \ \le \ \|x\|_p^p + c_0,$$

for some $c_0 > 0$. Also we introduce the penalty function $\beta \colon \Omega \times \mathbb{R} \longrightarrow \mathbb{R}$, defined by

$$\beta(z, \zeta) \overset{df}{=} \begin{cases} |\zeta|^{p-2}\zeta - |v_2(z)|^{p-2} v_2(z) \text{ if } v_2(z) < \zeta, \\ 0 \qquad\qquad\qquad\qquad\quad \text{ if } v_1(z) \le \zeta \le v_2(z), \\ |\zeta|^{p-2}\zeta - |v_1(z)|^{p-2} v_1(z) \text{ if } \zeta < v_1(z). \end{cases}$$

Evidently $\beta(z, \zeta)$ is a Carathéodory function and we have

$$|\beta(z, \zeta)| \ \le \ a_1(z) + c_1|\zeta|^{p-1} \quad \text{for a.a. } z \in \Omega \text{ and all } \zeta \in \mathbb{R},$$

with $a_1 \in L^{p'}(\Omega)$ ($\frac{1}{p} + \frac{1}{p'} = 1$), $c_1 > 0$ and

$$\int_\Omega \beta\big(z, x(z)\big) x(z) \, dz \ \ge \ c_2 \|x\|_p^p - c_3,$$

for some $c_2, c_3 > 0$. We consider the following auxiliary boundary value problem:

$$\begin{cases} -\operatorname{div}\big(\|\nabla x(z)\|_{\mathbb{R}^N}^{p-2} \nabla x(z)\big) + \varrho\beta\big(z, x(z)\big) - \lambda\big|\tau(x)(z)\big|^{p-2} \tau(x)(z) \\ \qquad\qquad\qquad\qquad\qquad\qquad \in \partial j\big(z, \tau(x)(z)\big) \quad \text{for a.a. } z \in \Omega, \\ x|_{\partial\Omega} = 0, \end{cases}$$

$$(4.213)$$

with $\varrho > 0$. Arguing as in the proof of Proposition 4.5.3, we can show that for $\varrho > 0$ large enough, we have that problem (4.213) has a solution $y_1 \in W_0^{1,p}(\Omega)$ and so as in the proof of Theorem 4.5.2, we can show that

$$v_1(z) \ \le \ y_1(z) \ \le \ v_2(z) \quad \text{for a.a. } z \in \Omega.$$

Therefore

$$\beta\big(z, y_1(z)\big) \ = \ 0 \quad \text{for a.a. } z \in \Omega \quad \text{and} \quad \tau(y_1) = y_1,$$

which imply that $y_1 \in W_0^{1,p}(\Omega)$ is a solution of problem (4.208) and $y_1(z) < 0$ for almost all $z \in \Omega$. ∎

Next we prove a general result, which establishes the existence of a second nontrivial critical point of a locally Lipschitz functional in the presence of a local minimizer of constant sign. More precisely, consider the functional $\psi_0 \colon W_0^{1,p}(\Omega) \longrightarrow \mathbb{R}$, defined by

$$\psi_0(x) \stackrel{df}{=} \frac{1}{p}\|\nabla x\|_p^p - \int_\Omega j_0\big(z, x(z)\big)\,dz.$$

Our hypotheses on the nonsmooth potential $j_0(z, \zeta)$ are the following:

$\underline{H(j_0)}$ $j_0 \colon \Omega \times \mathbb{R} \longrightarrow \mathbb{R}$ is a function, such that

(i) for every $\zeta \in \mathbb{R}$, the function

$$\Omega \ni z \longrightarrow j_0(z, \zeta) \in \mathbb{R}$$

is measurable and $j_0(\cdot, 0) \in L^\infty(\Omega)$;

(ii) for almost all $z \in \Omega$, the function

$$\mathbb{R} \ni \zeta \longrightarrow j_0(z, \zeta) \in \mathbb{R}$$

is locally Lipschitz;

(iii) for almost all $z \in \Omega$, all $\zeta \geq 0$ and all $u \in \partial j_0(z, \zeta)$, we have

$$|u| \leq \tilde{a}_0(z) + \tilde{c}_0|\zeta|^{r-1},$$

with $\tilde{a}_0 \in L^\infty(\Omega)_+$, $\tilde{c}_0 > 0$ and $r \in [1, p^*)$ where as usual

$$p^* \stackrel{df}{=} \begin{cases} \frac{pN}{N-p} & \text{if } N > p, \\ +\infty & \text{if } N \leq p; \end{cases}$$

(iv) there exists a function $\vartheta_0 \in L^\infty(\Omega)_+$ such that $\vartheta_0(z) \leq \lambda_1$ for almost all $z \in \Omega$, the inequality is strict on a set of positive measure and

$$\limsup_{\zeta \to +\infty} \frac{p j_0(z, \zeta)}{\zeta^p} \leq \vartheta_0(z)$$

uniformly for almost all $z \in \Omega$.

Suppose that $x_0 \in W_0^{1,p}(\Omega)$, $x_0 \neq 0$, $x_0(z) \geq 0$ for almost all $z \in \Omega$ is a local minimizer of the functional ψ_0. Then we know that $0 \in \partial\psi_0(x_0)$ and so we have

$$A(x_0) = u_0^*,$$

with $A\colon W_0^{1,p}(\Omega) \longrightarrow W^{-1,p'}(\Omega)$ as in the proof of Proposition 4.7.1 and $u_0^* \in L^{r'}(\Omega)$ $(\frac{1}{r} + \frac{1}{r'} = 1)$, $u_0^*(z) \in \partial j_0(z, x_0(z))$ for almost all $z \in \Omega$. Hence we have

$$\begin{cases} -\operatorname{div}\left(\|\nabla x_0(z)\|_{\mathbb{R}^N}^{p-2}\,\nabla x_0(z)\right) = u_0^*(z) & \text{for a.a. } z \in \Omega, \\ x|_{\partial\Omega} = 0. \end{cases}$$

From Theorem 1.5.5, it follows that $x_0 \in L^\infty(\Omega)$. Also let $\tau\colon \mathbb{R} \longrightarrow \mathbb{R}$ be the rump function defined by

$$\tau_0(\zeta) \overset{df}{=} \begin{cases} \zeta & \text{if } \zeta > 0, \\ 0 & \text{if } \zeta \leq 0. \end{cases}$$

Clearly τ_0 is Lipschitz continuous and so if we set

$$\widehat{j}_0(z, \zeta) \overset{df}{=} j_0\big(z, \tau_0(\zeta) + x_0(z)\big),$$

we see that $\widehat{j}_0(z, \cdot)$ is locally Lipschitz too. From Proposition 1.3.16, we know that

$$\partial\widehat{j}_0(z, \zeta) \subseteq \begin{cases} \{0\} & \text{if } \zeta < 0, \\ \operatorname{conv}\big\{\eta\partial j_0\big(z, x_0(z)\big) : \eta \in [0, 1]\big\} & \text{if } \zeta = 0, \\ \partial j_0\big(z, x + x_0(z)\big) & \text{if } \zeta > 0. \end{cases} \quad (4.214)$$

Let $\widehat{\psi}_0\colon W_0^{1,p}(\Omega) \longrightarrow \mathbb{R}$ be the locally Lipschitz functional defined by

$$\widehat{\psi}_0(x) \overset{df}{=} \frac{1}{p}\left[\|\nabla(x + x_0)\|_p^p - \|\nabla x_0\|_p^p\right] - \int_\Omega \widehat{j}_0\big(z, x(z)\big)\,dz$$

$$+ \int_\Omega u_0^*(z)x^-(z)\,dz + \xi,$$

with $\xi \overset{df}{=} \int_\Omega j_0\big(z, x_0(z)\big)\,dz$.

PROPOSITION 4.7.2
If hypotheses $H(j_0)$ hold,
then $\widehat{\psi}_0$ *satisfies the nonsmooth PS-condition.*

PROOF Let $\{x_n\}_{n\geq 1} \subseteq W_0^{1,p}(\Omega)$ be a sequence such that

$$m^{\widehat{\psi}_0}(x_n) \longrightarrow 0 \quad \text{and} \quad \big|\widehat{\psi}_0(x_n)\big| \leq M_1,$$

for some $M_1 > 0$. Let $x_n^* \in \partial\widehat{\psi}_0(x_n)$ be such that $m^{\widehat{\psi}_0}(x_n) = \|x_n^*\|_{W^{-1,p'}(\Omega)}$ for $n \geq 1$. We have

$$x_n^* = A(x_n + x_0) - u_n^* - u_0^*\chi_{\Omega_n^-},$$

with $u_n^* \in L^{r'}(\Omega)$, $u_n^*(z) \in \partial \widehat{j_0}\big(z, x_n(z)\big)$ for almost all $z \in \Omega$ and

$$\Omega_n^- \overset{df}{=} \big\{ z \in \Omega : x_n(z) < 0 \big\}.$$

From the choice of the sequence $\{x_n\}_{n \geq 1} \subseteq W_0^{1,p}(\Omega)$, we have

$$\left| \langle x_n^*, v \rangle_{W_0^{1,p}(\Omega)} \right| \leq \varepsilon_n \|v\|_{W^{1,p}(\Omega)} \qquad \forall\, v \in W_0^{1,p}(\Omega),$$

with $\varepsilon_n \searrow 0$. Choose as a test function $v = -x_n^- \in W_0^{1,p}(\Omega)$. We obtain

$$\langle A(x_n + x_0), -x_n^- \rangle_{W_0^{1,p}(\Omega)} + \int_\Omega u_n^*(z) x_n^-(z)\, dz + \int_\Omega u_0^*(z) x_n^-(z)\, dz$$

$$\leq \varepsilon_n \|x_n^-\|_{W^{1,p}(\Omega)} . \tag{4.215}$$

Note that

$$\langle A(x_n + x_0), -x_n^- \rangle_{W_0^{1,p}(\Omega)}$$

$$= \int_\Omega \|\nabla(x_n + x_0)(z)\|_{\mathbb{R}^N}^{p-2} \big(\nabla(x_n + x_0)(z), -\nabla x_n^-(z) \big)_{\mathbb{R}^N} dz$$

$$= \int_\Omega \|\nabla(x_0 - x_n^-)(z)\|_{\mathbb{R}^N}^{p-2} \big(\nabla(x_0 - x_n^-)(z), \nabla x_n^-(z) \big)_{\mathbb{R}^N} dz$$

$$= \int_\Omega \|\nabla(x_0 - x_n^-)(z)\|_{\mathbb{R}^N}^{p-2} \big(\nabla(x_0 - x_n^-)(z), \nabla(x_0 - x_n^-)(z) \big)_{\mathbb{R}^N} dz$$

$$- \int_\Omega \|\nabla(x_0 - x_n^-)(z)\|_{\mathbb{R}^N}^{p-2} \big(\nabla(x_0 - x_n^-)(z), \nabla x_0(z) \big)_{\mathbb{R}^N} dz$$

$$\geq \|\nabla(x_0 - x_n^-)\|_p^p - \|\nabla(x_0 - x_n^-)\|_p^{p-1} \|\nabla x_0\|_p . \tag{4.216}$$

Also because $u_n^*(z) \in \partial \widehat{j_0}\big(z, x_n(z)\big)$ for almost all $z \in \Omega$, we have $u_n^*(z) = 0$ for almost all $z \in \Omega_n^-$ (see (4.214)). So

$$\int_\Omega u_n^*(z) x_n^-(z)\, dz = 0. \tag{4.217}$$

Returning to (4.215) and using (4.216) and (4.217), we obtain

$$\|\nabla(x_0 - x_n^-)\|_p^p \leq \|\nabla(x_0 - x_n^-)\|_p^{p-1} \|\nabla x_0\|_p$$
$$+ c_4 \|x_n^-\|_{W^{1,p}(\Omega)} \qquad \forall\, n \geq 1, \tag{4.218}$$

for some $c_4 > 0$. By virtue of the Poincaré inequality (see Theorem 1.1.6), (4.218) implies that the sequence $\{x_n^-\}_{n \geq 1} \subseteq W_0^{1,p}(\Omega)$ is bounded.

From the choice of the sequence $\{x_n\}_{n \geq 1} \subseteq W_0^{1,p}(\Omega)$, we have

$$\left| \frac{1}{p} \left[\|\nabla(x_n + x_0)\|_p^p - \|\nabla x_0\|_p^p \right] - \int_\Omega \widehat{j}_0\big(z, x_n(z)\big) \, dz \right.$$

$$\left. + \int_\Omega u_0^*(z) x_n^-(z) \, dz + \xi \right| \leq M_1 \quad \forall n \geq 1. \quad (4.219)$$

Suppose that the sequence $\{x_n^+ + x_0\}_{n \geq 1} \subseteq W_0^{1,p}(\Omega)$ was unbounded. By selecting a suitable subsequence, we may assume that

$$\|x_n^+ + x_0\|_{W^{1,p}(\Omega)} \longrightarrow +\infty.$$

Let us set

$$y_n \overset{df}{=} \frac{x_n^+ + x_0}{\|x_n^+ + x_0\|_{W^{1,p}(\Omega)}} \qquad \forall\, n \geq 1.$$

We may assume that

$$
\begin{aligned}
y_n &\overset{w}{\longrightarrow} y &&\text{in } W_0^{1,p}(\Omega), \\
y_n &\longrightarrow y &&\text{in } L^p(\Omega), \\
y_n(z) &\longrightarrow y(z) &&\text{for a.a. } z \in \Omega, \\
|y_n(z)| &\leq k(z) &&\text{for a.a. } z \in \Omega \text{ and all } n \geq 1,
\end{aligned}
$$

with $k \in L^p(\Omega)$. Note that

$$\|\nabla(x_n + x_0)\|_p^p - \|\nabla x_0\|_p^p = \left[\|\nabla(x_n^+ + x_0)\|_p^p - \|\nabla x_0\|_p^p \right]$$

$$+ \left[\|\nabla(x_0 - x_n^-)\|_p^p - \|\nabla x_0\|_p^p \right]. \quad (4.220)$$

Using (4.220) in (4.219) and recalling that the sequence $\{x_n^-\}_{n \geq 1} \subseteq W_0^{1,p}(\Omega)$ is bounded, we obtain

$$\frac{1}{p} \|\nabla(x_n^+ + x_0)\|_p^p - \int_\Omega \widehat{j}_0\big(z, x_n(z)\big) \, dz \leq M_2 \qquad \forall\, n \geq 1,$$

for some $M_2 > 0$. Dividing by $\|x_n^+ + x_0\|_{W^{1,p}(\Omega)}^p$, we have

$$\frac{1}{p} \|\nabla y_n\|_p^p - \int_\Omega \frac{\widehat{j}_0(z, x_n(z))}{\|x_n^+ + x_0\|_{W^{1,p}(\Omega)}^p} \, dz$$

$$\leq \frac{M_2}{\|x_n^+ + x_0\|_{W^{1,p}(\Omega)}^p} \qquad \forall\, n \geq 1. \quad (4.221)$$

Using hypotheses $H(j_0)(iii)$ and (iv), we see that for a given $\varepsilon > 0$, we can find $a_\varepsilon \in L^\infty(\Omega)_+$ such that

$$j_0(z, \varsigma) \leq a_\varepsilon(z) + \frac{1}{p}(\vartheta_0(z) + \varepsilon)|\varsigma|^p \quad \text{for a.a. } z \in \Omega \text{ and all } \varsigma \geq 0. \quad (4.222)$$

Let us set

$$\widehat{\Omega}_{\varepsilon,n}^+ \stackrel{df}{=} \left\{ z \in \Omega : \ x_n^+(z) + x_0(z) > 0 \right.$$

$$\left. \text{and } \frac{j_0(z, x_n^+(z) + x_0(z))}{|x_n^+(z) + x_0(z)|^p} \leq \frac{1}{p}\big(\vartheta_0(z) + \varepsilon\big) \right\}$$

(recall that $\widehat{j}_0|_{\Omega \times \mathbb{R}_+} = j_0$). From hypothesis $H(j_0)(iv)$, we have that

$$\chi_{\widehat{\Omega}_{\varepsilon,n}^+}(z) \longrightarrow 1 \quad \text{for a.a. } z \in \{y > 0\}$$

(since $x_n^+(z) + x_0(z) \longrightarrow +\infty$ for almost all $z \in \{y > 0\}$). We have

$$\frac{j_0(z, x_n^+(z) + x_0(z))}{\|x_n^+ + x_0\|_{W^{1,p}(\Omega)}^p} \chi_{\widehat{\Omega}_{\varepsilon,n}^+}(z) = \frac{j_0(z, x_n^+(z) + x_0(z))}{|x_n^+(z) + x_0(z)|^p} |y_n(z)|^p \chi_{\widehat{\Omega}_{\varepsilon,n}^+}(z)$$

$$\leq \frac{1}{p}\big(\vartheta_0(z) + \varepsilon\big)|y_n(z)|^p \chi_{\widehat{\Omega}_{\varepsilon,n}^+}(z) \quad \text{for a.a. } z \in \Omega.$$

Because $y \geq 0$, we infer that

$$\limsup_{n \to +\infty} \frac{j_0(z, x_n^+(z) + x_0(z))}{\|x_n^+ + x_0\|_{W^{1,p}(\Omega)}^p} \leq \frac{1}{p}\big(\vartheta_0(z) + \varepsilon\big)|y(z)|^p \quad \text{for a.a. } z \in \Omega.$$

Since $\varepsilon > 0$ was arbitrary, we let $\varepsilon \searrow 0$ and conclude that

$$\limsup_{n \to +\infty} \frac{j_0(z, x_n^+(z) + x_0(z))}{\|x_n^+ + x_0\|_{W^{1,p}(\Omega)}^p} \leq \frac{1}{p}\vartheta_0(z)|y(z)|^p \quad \text{for a.a. } z \in \Omega. \qquad (4.223)$$

Because of (4.222), we can use Fatou's lemma and obtain that

$$\limsup_{n \to +\infty} \int_\Omega \frac{j_0(z, x_n^+(z) + x_0(z))}{\|x_n^+ + x_0\|_{W^{1,p}(\Omega)}^p}\, dz$$

$$\leq \frac{1}{p} \int_\Omega \vartheta_0(z)|y(z)|^p\, dz \quad \text{for a.a. } z \in \Omega. \qquad (4.224)$$

So passing to the limit in (4.221) and using the weak lower semicontinuity of the norm functional, we obtain

$$\frac{1}{p}\|\nabla y\|_p^p \leq \frac{1}{p} \int_\Omega \vartheta(z)|y(z)|^p\, dz \leq \frac{\lambda_1}{p}\|y\|_p^p$$

(see (4.224) and hypothesis $H(j_0)(iv)$). Hence $y = 0$ or $y = u_1$ (since $y \geq 0$). If $y = 0$, then

$$\|\nabla y_n\|_p \longrightarrow 0$$

and so
$$y_n \longrightarrow 0 \quad \text{in } W_0^{1,p}(\Omega),$$

a contradiction to the fact that $\|y_n\|_{W^{1,p}(\Omega)} = 1$ for all $n \geq 1$. So $y = u_1$. But recall that $u_1(z) > 0$ for all $z \in \Omega$ (see Proposition 1.5.18). So we obtain

$$\frac{1}{p}\|\nabla y\|_p^p \leq \frac{1}{p}\int_\Omega \vartheta(z)|y(z)|^p\,dz < \frac{\lambda_1}{p}\|y\|_p^p,$$

a contradiction to the variational characterization of $\lambda_1 > 0$ (see Proposition 1.5.18). This proves the boundedness of the sequence $\{x_n^+ + x_0\}_{n \geq 1} \subseteq W_0^{1,p}(\Omega)$, hence also of the sequence $\{x_n^+\}_{n \geq 1} \subseteq W_0^{1,p}(\Omega)$. Thus we may assume that
$$x_n \xrightarrow{w} x \quad \text{in } W_0^{1,p}(\Omega),$$
$$x_n \longrightarrow x \quad \text{in } L^p(\Omega),$$

for some $x \in W_0^{1,p}(\Omega)$. We have

$$\langle A(x_n + x_0), x_n - x \rangle_{W_0^{1,p}(\Omega)} - \int_\Omega u_n^*(z)(x_n - x)(z)\,dz$$

$$- \int_{\Omega_n^-} u_0^*(z)(x_n - x)(z)\,dz \leq \varepsilon_n \|x_n - x\|_{W^{1,p}(\Omega)}.$$

Note that

$$\int_\Omega u_n^*(z)(x_n - x)(z)\,dz \longrightarrow 0 \quad \text{and} \quad \int_{\Omega_n^-} u_0^*(z)(x_n - x)(z)\,dz \longrightarrow 0.$$

Hence
$$\limsup_{n \to +\infty} \langle A(x_n + x_0), x_n - x \rangle_{W_0^{1,p}(\Omega)} \leq 0.$$

Because A is maximal monotone, it is generalized pseudomonotone and so
$$\|\nabla(x_n + x_0)\|_p \longrightarrow \|\nabla(x + x_0)\|_p,$$

thus by the Kadec-Klee property, we have
$$x_n + x_0 \longrightarrow x + x_0 \quad \text{in } W_0^{1,p}(\Omega)$$

and
$$x_n \longrightarrow x \quad \text{in } W_0^{1,p}(\Omega).$$

$\square$

Using this proposition and the nonsmooth Mountain Pass Theorem (see Theorem 2.1.3), we can prove an intermediate multiplicity result, which we shall need in the sequel and which is also of independent interest.

PROPOSITION 4.7.3
<u>*If*</u>

$$j_0(z, \zeta) \overset{df}{=} \tilde{j}(z, \zeta) + \frac{\lambda}{p} |\zeta|^p,$$

with $\lambda > \lambda_1$, *hypotheses* $H(j_0)$ *hold and* $x_0 \in W_0^{1,p}(\Omega)$, $x_0 \neq 0$, $x_0 \geq 0$ *is a local minimizer of* ψ_0 *on* $W_0^{1,p}(\Omega)$,
<u>*then*</u> ψ_0 *has at least one more nontrivial critical point* $v \in W_0^{1,p}(\Omega)$ *with* $v \geq x_0$, $v \neq x_0$.

PROOF   We assume that x_0 is an isolated local minimum, or otherwise we are done. So we can find $0 < \varrho < \|x_0\|_{W^{1,p}(\Omega)}$ such that

$$\psi_0(x_0) < \psi_0(y) \qquad \forall \, y \in \overline{B_\varrho(x_0)}. \tag{4.225}$$

We claim that we can find $\gamma > 0$ such that

$$\gamma \leq \psi_0(y) - \psi_0(x_0) \qquad \forall \, y \in \overline{B_\varrho(x_0)} \setminus B_{\frac{\varrho}{2}}(x_0). \tag{4.226}$$

Suppose that (4.226) is not true. The we can find a sequence $\{y_n\}_{n \geq 1} \subseteq W_0^{1,p}(\Omega)$ such that

$$\frac{\varrho}{2} \leq \|y_n - x_0\|_{W^{1,p}(\Omega)} \leq \varrho \quad \text{and} \quad \psi_0(y_n) \searrow \psi_0(x_0).$$

By passing to a subsequence if necessary, we may assume that

$$\begin{aligned}
y_n &\overset{w}{\longrightarrow} y \quad \text{in } W_0^{1,p}(\Omega), \\
y_n &\longrightarrow y \quad \text{in } L^\eta(\Omega), \\
y_n(z) &\longrightarrow y(z) \text{ for a.a. } z \in \Omega, \\
|y_n(z)| &\leq k(z) \text{ for a.a. } z \in \Omega \text{ and all } n \geq 1,
\end{aligned}$$

with $k \in L^\eta(\Omega)$ and $\eta = \max\{r, p\} < p^*$.
Note that from hypothesis $H(j_0)(iii)$ and the mean value theorem for locally Lipschitz functionals (see Proposition 1.3.15), we have

$$|j_0(z, \zeta)| \leq \tilde{a}_0(z) + \tilde{c}_0 |\zeta|^r$$

with $\tilde{a}_0 \in L^\infty(\Omega)_+$, $\tilde{c}_0 > 0$. Then via the Lebesgue dominated convergence theorem and the weak lower semicontinuity of the norm functional, we have

$$\psi_0(y) \leq \liminf_{n \to +\infty} \psi_0(y_n)$$

and so

$$\psi_0(y) = \psi_0(x_0),$$

with $\frac{\varrho}{2} \leq \|y - x_0\|_{W^{1,p}(\Omega)} \leq \varrho$. This contradicts (4.225) and so (4.226) is true.

Next for all $y \in W_0^{1,p}(\Omega)$, we have

$$
\begin{aligned}
\widehat{\psi}_0(y) &= \frac{1}{p}\left[\|\nabla(y + x_0)\|_p^p - \|\nabla x_0\|_p^p\right] \\
&\quad - \int_\Omega \widehat{j}_0\big(z, y(z)\big)\, dz + \int_\Omega u_0^*(z) y^-(z)\, dz + \xi \\
&= \frac{1}{p}\left[\|\nabla(y^+ + x_0)\|_p^p - \|\nabla x_0\|_p^p\right] + \frac{1}{p}\left[\|\nabla(x_0 - y^-)\|_p^p - \|\nabla x_0\|_p^p\right] \\
&\quad - \int_\Omega \widehat{j}_0\big(z, y(z)\big)\, dz + \int_\Omega u_0^*(z) y^-(z)\, dz + \xi.
\end{aligned}
\tag{4.227}
$$

Recall that $A(x_0) = u_0^*$. So using as a test function $y^- \in W_0^{1,p}(\Omega)$, we obtain

$$
\int_\Omega \|\nabla x_0(z)\|_{\mathbb{R}^N}^{p-2}\big(\nabla x_0(z), \nabla y^-(z)\big)_{\mathbb{R}^N}\, dz = \int_\Omega u_0^*(z) y^-(z)\, dz.
\tag{4.228}
$$

Note that the function $x \longmapsto \frac{1}{p}\|\nabla x\|_p^p$ is strongly convex. Therefore we have

$$
\begin{aligned}
&\frac{1}{p}\big[\|\nabla(x_0 - y^-)\|_p^p - \|\nabla x_0\|_p^p\big] \\
&+ \int_\Omega \|\nabla x_0(z)\|_{\mathbb{R}^N}^{p-2}\big(\nabla x_0(z), \nabla y^-(z)\big)_{\mathbb{R}^N}\, dz \geq c_5 \|\nabla y^-\|_p^p,
\end{aligned}
\tag{4.229}
$$

for some $c_5 > 0$. Using (4.228) and (4.229) in (4.227), we obtain

$$
\begin{aligned}
\widehat{\psi}_0(y) &\geq \frac{1}{p}\big|\|\nabla(y^+ + x_0)\|_p^p - \|\nabla x_0\|_p^p\big| + c_5 \|\nabla y^-\|_p^p - \int_\Omega \widehat{j}_0\big(z, y(z)\big)\, dz + \xi \\
&= \psi_0(y^+ + x_0) - \psi_0(x_0) + c_5 \|\nabla y^-\|_p^p.
\end{aligned}
\tag{4.230}
$$

If $\|y + x_0 - x_0\|_{W^{1,p}(\Omega)} = \|y\|_{W^{1,p}(\Omega)} = \|y^+\|_{W^{1,p}(\Omega)} + \|y^-\|_{W^{1,p}(\Omega)} = \varrho$, we must have

$$
\|y^+\|_{W^{1,p}(\Omega)} \geq \frac{\varrho}{2} \quad \text{or} \quad \|y^-\|_{W^{1,p}(\Omega)} \geq \frac{\varrho}{2}.
$$

First suppose that $\|y\|_{W^{1,p}(\Omega)} = \varrho$ and $\|y^+\|_{W^{1,p}(\Omega)} \geq \frac{\varrho}{2}$. Then from (4.230) and (4.226), we have

$$
0 < \gamma \leq \widehat{\psi}_0(y).
$$

On the other hand, if $\|y\|_{W^{1,p}(\Omega)} = \varrho$ and $\|y^-\|_{W^{1,p}(\Omega)} \geq \frac{\varrho}{2}$, then by the Poincaré inequality (see Theorem 1.1.6), we have

$$
\|\nabla y^-\|_p \geq \gamma_1,
$$

for some $\gamma_1 > 0$ and $\psi_0(y^+ + x_0) - \psi(x_0) \geq 0$. So we obtain that

$$
0 < c_5 \gamma_1 \leq \widehat{\psi}_0(y).
$$

Therefore, if $\|y\|_{W^{1,p}(\Omega)} = \varrho$, we have

$$0 \; < \; \min\{\gamma, c_5\gamma_1\} \; \leq \; \widehat{\psi}_0(y). \tag{4.231}$$

Note that $\widehat{\psi}_0(0) = 0$ (recall the choice of ξ). Since $u_1 \in C_0^1(\overline{\Omega})$ is an order unit of the ordered Banach space $C_0^1(\overline{\Omega})$ (recall that $u_1 \in \mathrm{int}\, C_0^1(\overline{\Omega})$), then we can find $t > 0$ large enough, such that

$$tu_1 - x_0 \geq 0 \quad \text{and} \quad \|tu_1 - x_0\|_{W^{1,p}(\Omega)} > \varrho.$$

Then exploiting the form of $j_0(z, \zeta)$, we have

$$\begin{aligned}
\psi_0(tu_1 - x_0) \; = \; & \frac{t^p}{p}\left(1 - \frac{\lambda}{\lambda_1}\right)\|\nabla u_1\|_p^p \\
& - \frac{1}{p}\|\nabla x_0\|_p^p - \int_\Omega \left(\widetilde{j}\big(z, tu_1(z)\big) - \widetilde{j}\big(t, x_0(z)\big)\right) dz,
\end{aligned}$$

hence for $t > 0$ large enough, we will have $\psi_0(tu_1 - x_0) \leq 0$.

These facts in conjunction with Proposition 4.7.2 permit the use of the nonsmooth Mountain Pass Theorem (see Theorem 2.1.3), which gives $y \in W_0^{1,p}(\Omega)$ such that $0 < \gamma_2 \leq \widehat{\psi}_0(y)$, hence $y \neq 0$ and $0 \in \partial\widehat{\psi}_0(y)$. So we have

$$A(y + x_0) - u^* - u_0^*\chi_{\Omega^-} \; = \; 0,$$

where $u \in L^{r'}(\Omega)$, $u^*(z) \in \partial j_0\big(z, y(z)\big)$ for almost all $z \in \Omega$ and

$$\Omega^- \overset{df}{=} \big\{z \in \Omega : \; y(z) < 0\big\}.$$

Using as a test function $y^- \in W_0^{1,p}(\Omega)$, we obtain

$$\big\langle A(y + x_0), y^- \big\rangle_{W_0^{1,p}(\Omega)} - \int_\Omega u^*(z)y^-(z)\, dz - \int_\Omega u_0^*(z)y^-(z)\, dz \; = \; 0.$$

From (4.214), we know that $u^*(z) = 0$ for almost all $z \in \Omega^-$. So using (4.228), we have

$$\int_\Omega \|\nabla(y + x_0)(z)\|_{\mathbb{R}^N}^{p-2}\big(\nabla(y + x_0)(z), \nabla y^-(z)\big)_{\mathbb{R}^N}\, dz$$

$$- \int_\Omega \|\nabla x_0(z)\|_{\mathbb{R}^N}^{p-2}\big(\nabla x_0(z), \nabla y^-(z)\big)_{\mathbb{R}^N}\, dz \; = \; 0.$$

Thus

$$\int_\Omega \|\nabla(x_0 - y^-)(x)\|_{\mathbb{R}^N}^{p-2}\big(\nabla(x_0 - y^-)(z), \nabla y^-(z)\big)_{\mathbb{R}^N}\, dz$$

$$- \int_{\Omega} \|\nabla x_0(z)\|_{\mathbb{R}^N}^{p-2} \left(\nabla x_0(z), \nabla y^-(z)\right)_{\mathbb{R}^N} dz = 0. \quad (4.232)$$

But the function $u \longmapsto \frac{1}{p}\|u\|_{\mathbb{R}^N}^p$ is strictly monotone on $\mathbb{R}^N$. So from (4.232) it follows that

$$\nabla y^-(z) = 0 \quad \text{for a.a. } z \in \Omega,$$

hence $y^-(z) = 0$ for almost all $z \in \Omega$ and so $y \geq 0$, $y \neq 0$.

Let us set $v = x_0 + y$. We have that $0 \in \partial \psi_0(v)$, i.e. $v \geq x_0 \geq 0$, $v \neq x_0 \neq 0$ is another critical point of ψ_0 distinct from x_0. $\qquad\square$

Now we return to problem (4.208) and produce for it two distinct positive solutions, provided that λ approaches λ_1 from above.

PROPOSITION 4.7.4

If hypotheses $H(j)_1$ hold,
then there exists ε_1 such that for all $\lambda \in (\lambda_1, \lambda_1 + \varepsilon_1)$, problem (4.208) has at least two solutions $y_2, y_3 \in W_0^{1,p}(\Omega)$ such that $y_2 \in C_0^1(\overline{\Omega})$, $0 < y_2(z) \leq y_3(z)$ for almost all $z \in \Omega$ and $y_2 \neq y_3$.

PROOF   As before let $\tau_0 \colon \mathbb{R} \longrightarrow \mathbb{R}$ be the Lipschitz continuous function defined by

$$\tau_0(\zeta) \stackrel{df}{=} \begin{cases} \zeta \text{ if } \zeta > 0, \\ 0 \text{ if } \zeta \leq 0 \end{cases}$$

(the rump function). Let us set

$$j_1(z, \zeta) \stackrel{df}{=} j\big(z, \tau_0(\zeta)\big).$$

Clearly for almost all $z \in \Omega$, $j_1(z, \cdot)$ is locally Lipschitz. For $\lambda > 0$, we introduce the locally Lipschitz functional $\psi_{1,\lambda} \colon W_0^{1,p}(\Omega) \longrightarrow \mathbb{R}$, defined by

$$\psi_{1,\lambda}(x) \stackrel{df}{=} \frac{1}{p}\|\nabla x\|_p^p - \frac{\lambda}{p}\|x^+\|_p^p - \int_{\Omega} j_1\big(z, x(z)\big) dz \qquad \forall\, x \in W_0^{1,p}(\Omega).$$

From the second limit in hypothesis $H(j)_1(iv)$ (the asymptotic behaviour at $+\infty$ of $j(z, \cdot)$) we know that we can find $\varepsilon_1 > 0$ and $M_3 > 0$ such that

$$j(z, \zeta) \leq -\frac{\varepsilon_1}{p}\zeta^p \quad \text{for a.a. } z \in \Omega \text{ and all } \zeta \geq M_3.$$

On the other hand from the mean value theorem for locally Lipschitz functions (see Proposition 1.3.15) and hypothesis $H(j)_1(iii)$, we have

$$|j(z, \zeta)| \leq \beta_2(z) \quad \text{for a.a. } z \in \Omega \text{ and all } |\zeta| < M_3,$$

with $\beta_2 \in L^\infty(\Omega)_+$. So finally we can say that

$$j(z, \zeta) \leq -\frac{\varepsilon_1}{p}\zeta^p + \beta_3(z) \quad \text{for a.a. } z \in \Omega \text{ and all } \zeta \geq 0, \qquad (4.233)$$

with $\beta_3(z) \stackrel{df}{=} \beta_2 + \frac{\varepsilon_1}{p}M_3^p$.

Then, since $\tau_0(x)(z) = x^+(z)$, using (4.233) and Lemma 4.7.1 and noting that $\|x\|_p \geq \|x^+\|_p$, for all $x \in W_0^{1,p}(\Omega)$, we have

$$
\begin{aligned}
\psi_{1,\lambda}(x) &= \frac{1}{p}\|\nabla x\|_p^p - \frac{\lambda}{p}\|x^+\|_p^p - \int_\Omega j_1(z, x(z))\, dz \\[2mm]
&\geq \frac{1}{p}\|\nabla x\|_p^p - \frac{\lambda}{p}\|x^+\|_p^p - \int_\Omega j_1(z, x^+(z))\, dz \\[2mm]
&\geq \frac{1}{p}\|\nabla x\|_p^p - \frac{\lambda}{p}\|x^+\|_p^p + \frac{\varepsilon_1}{p}\|x^+\|_p^p - c_6 \\[2mm]
&\geq \xi\|\nabla x\|_p^p - c_6,
\end{aligned}
$$

for some $c_6 > 0$, $\xi = \xi(\lambda) > 0$ and for $\lambda \in (\lambda_1, \lambda_1 + \varepsilon_1)$. Thus by the Poincaré inequality (see Theorem 1.1.6), $\psi_{1,\lambda}$ is coercive if $\lambda \in (\lambda_1, \lambda_1 + \varepsilon)$.

Because the embedding $W_0^{1,p}(\Omega) \subseteq L^p(\Omega)$ is compact, we can easily verify that $\psi_{1,\lambda}$ is weakly lower semicontinuous. So we can use the Weierstrass theorem (see Theorem A.1.5) and obtain $x_0 \in W_0^{1,p}(\Omega)$ such that

$$\psi_{1,\lambda}(x_0) = \inf_{x \in W_0^{1,p}(\Omega)} \psi_{1,\lambda}(x).$$

We claim that $\psi_{1,\lambda}(x_0) < 0$. Indeed, from the second limit in hypothesis $H(j)_1(v)$ (the asymptotic behaviour at 0^+ of $j(z, \cdot)$), for a given $\varepsilon > 0$, we can find $\delta = \delta(\varepsilon) > 0$ such that

$$j(z, \zeta) = j_1(z, \zeta) \geq -\frac{\varepsilon}{p}\zeta^p \quad \text{for a.a. } z \in \Omega \text{ and all } \zeta \in (0, \delta).$$

Choose $\xi_1 > 0$ small enough so that $\xi_1 u_1(z) < \delta$ for all $z \in \overline{\Omega}$. We have

$$j(z, \xi_1 u_1(z)) = j_1(z, \xi_1 u_1(z)) \geq -\frac{\varepsilon}{p}(\xi_1 u_1(z))^p$$

and so

$$
\begin{aligned}
\psi_{1,\lambda}(\xi_1 u_1) &\leq \frac{\xi_1^p}{p}\|\nabla u_1\|_p^p - \frac{\lambda \xi_1^p}{p}\|u_1\|_p^p + \frac{\varepsilon \xi_1^p}{p}\|u_1\|_p^p \\[2mm]
&= \frac{\xi_1^p}{p}\left(1 - \frac{\lambda - \varepsilon}{\lambda_1}\right)\|\nabla u_1\|_p^p \qquad (4.234)
\end{aligned}
$$

(recall that $\|\nabla u_1\|_p^p = \lambda_1\|u_1\|_p^p$). But $\lambda > \lambda_1$. So if we choose $\varepsilon > 0$ small enough, we have $\lambda - \varepsilon > \lambda_1$ and so from (4.234) it follows that $\psi_{1,\lambda}(\xi_1 u_1) < 0$ and thus

$$\psi_{1,\lambda}(x_0) = \inf_{W_0^{1,p}(\Omega)} \psi_{1,\lambda} < 0.$$

Moreover, $\psi_{1,\lambda}(x_0) < 0 = \psi_{1,\lambda}(0)$ implies that $x_0 \neq 0$. We have $0 \in \partial\psi_{1,\lambda}(x_0)$ and so

$$A(x_0) - \lambda|x_0^+|^{p-2}x_0^+ = u_0^*, \tag{4.235}$$

with $u_0^* \in L^{r'}(\Omega)$, $u_0^*(z) \in \partial j_1(x, x_0(z))$ for almost all $z \in \Omega$. Use as a test function $x_0^- \in W_0^{1,p}(\Omega)$. From (4.214), we know that $u_0^*(z) = 0$ for almost all $z \in \{x_0 < 0\}$. So we obtain

$$\|\nabla x_0^-\|_p^p = 0,$$

i.e. $x_0^- = 0$ and so $x_0 \geq 0$, $x_0 \neq 0$. Because of hypothesis $H(j)_1(vi)$, we have

$$-\mathrm{div}\left(\|\nabla x_0(z)\|_{\mathbb{R}^N}^{p-2}\nabla x_0(z)\right) \leq \lambda x_0(z)^{p-2} + u_0^*(z) \quad \text{for a.a. } z \in \Omega,$$

and so from Theorem 1.5.5, it follows that $x_0 \in L^\infty(\Omega)$. Then Theorem 1.5.6 implies that $x_0 \in C_0^{1,\alpha}(\overline{\Omega})$ for some $\alpha \in (0,1)$. Also because of hypothesis $H(j)_1(vi)$, we have

$$\Delta_p x_0(z) = -\left|\lambda x_0(z)^{p-2} + u_0^*(z)\right| \leq \lambda x_0(z)^{p-2} - u_0^*(z)$$
$$\leq \lambda x_0(z)^{p-1} + \hat{c}_1 x_0(z)^{p-1} = c_7 x_0(z)^{p-1} \quad \text{for a.a. } z \in \Omega,$$

for some $c_7 > 0$. Invoking Theorem 1.5.7, we obtain

$$x_0(z) > 0 \quad \forall z \in \Omega \quad \text{and} \quad \frac{\partial x_0(z)}{\partial n} < 0 \quad \forall z \in \partial\Omega. \tag{4.236}$$

From (4.236) it follows that $x_0 \in \mathrm{int}\, C_0^1(\overline{\Omega})_+$. So we can find $\varepsilon > 0$ such that

$$\overline{B_\varepsilon(x_0)} = \left\{y \in C_0^1(\overline{\Omega}) : \|y - x_0\|_{C_0^1(\overline{\Omega})} \leq \varepsilon\right\} \subseteq C_0^1(\overline{\Omega})_+.$$

Then

$$\varphi_\lambda(x_0) = \psi_{1,\lambda}(x_0) \leq \psi_{1,\lambda}(y) = \varphi_\lambda(y) \quad \forall y \in \overline{B_\varepsilon(x_0)}$$

and so $x_0 \in C_0^1(\overline{\Omega})_+$ is a local $C_0^1(\Omega)$-minimizer of φ.

By virtue of Proposition 4.6.10, x_0 is also a local $W_0^{1,p}(\Omega)$-minimizer of φ_λ. Moreover, if

$$j_0(z,\zeta) \stackrel{df}{=} \frac{\lambda}{p}|\zeta|^p + j(z,\zeta),$$

then it satisfies hypotheses $H(j_0)$. Indeed, hypotheses $H(j_0)(i) - (iii)$ are clear. Also we have

$$\limsup_{\zeta \to +\infty} \frac{pj_0(z,\zeta)}{\zeta^p} = \limsup_{\zeta \to +\infty}\left[\lambda - \frac{pj(z,\zeta)}{\zeta^p}\right] < \lambda - \varepsilon_1 < \lambda_1$$

and note that $H(j_0)(v)$ is valid by virtue of hypothesis $H(j)(vi)$. So we can apply Proposition 4.7.3 and produce a second critical point $v \in W_0^{1,p}(\Omega)$ of

φ such that $x_0 \leq v$, $x_0 \neq v$. Let us set $y_2 \stackrel{df}{=} x_0$, $y_3 \stackrel{df}{=} v$. Then y_2, y_3 are the desired solutions of (4.208) with $0 < y_2(z) \leq y_3(z)$ for almost all $z \in \Omega$, $y_2 \in C_0^1(\overline{\Omega})$ and $y_2 \neq y_3$. $\quad\square$

Combining Proposition 4.7.4 with Proposition 4.7.1 we obtain the following "three solutions theorem" concerning problem (4.208).

THEOREM 4.7.1

If hypotheses $H(j)_1$ hold,
<u>*then*</u> *we can find $\varepsilon_2 > 0$ such that for all $\lambda \in (\lambda_1, \lambda_1 + \varepsilon_2)$ problem (4.208) has at least three solutions $y_1, y_2, y_3 \in W_0^{1,p}(\Omega)$, with $y_2 \in C_0^1(\overline{\Omega})$ and*

$$y_1(z) < 0 < y_2(z) \leq y_3(z) \quad \text{for a.a. } z \in \Omega$$

and $y_2 \neq y_3$.

We can have another multiplicity result for problem (4.208), which is valid for all $\lambda > \lambda_1$. In this case the hypotheses on the nonsmooth potential $j(z, \zeta)$ are the following:

<u>$H(j)_2$</u> $\ j: \Omega \times \mathbb{R} \longrightarrow \mathbb{R}$ is a function, such that

(i) for every $\zeta \in \mathbb{R}$, the function

$$\Omega \ni z \longrightarrow j(z, \zeta) \in \mathbb{R}$$

is measurable and $j(z, 0) = 0$ for almost all $z \in \Omega$;

(ii) for almost all $z \in \Omega$, the function

$$\mathbb{R} \ni \zeta \longrightarrow j(z, \zeta) \in \mathbb{R}$$

is locally Lipschitz;

(iii) for almost all $z \in \Omega$, all $\zeta \in \mathbb{R}$ and all $u \in \partial j(z, \zeta)$, we have

$$|u| \leq \tilde{a}_2(z) + \hat{c}_2 |\zeta|^{r-1},$$

with $\tilde{a}_2 \in L^\infty(\Omega)_+$, $\hat{c}_2 > 0$ and $r \in [1, p^*)$ where as usual

$$p^* \stackrel{df}{=} \begin{cases} \frac{pN}{N-p} & \text{if } N > p, \\ +\infty & \text{if } N \leq p; \end{cases}$$

(iv) we have

$$\lim_{\zeta \to -\infty} \frac{u}{|\zeta|^{p-2}\zeta} = -\infty \quad \text{and} \quad \lim_{\zeta \to +\infty} \frac{pj(z, \zeta)}{\zeta^p} = +\infty$$

uniformly for almost all $z \in \Omega$ and all $u \in \partial j(z, \zeta)$;

(v) we have

$$\lim_{\zeta \to 0^-} \frac{u}{|\zeta|^{p-2}\zeta} \geq 0 \quad \text{and} \quad \lim_{\zeta \to 0^+} \frac{pj(z,\zeta)}{\zeta^p} \geq 0$$

uniformly for almost all $z \in \Omega$ and all $u \in \partial j(z,\zeta)$.

As before, we consider the locally Lipschitz functional $\varphi_\lambda \colon W_0^{1,p}(\Omega) \longrightarrow \mathbb{R}$ defined by

$$\varphi_\lambda(x) \stackrel{df}{=} \frac{1}{p}\|\nabla x\|_p^p - \frac{\lambda}{p}\|x\|_p^p - \int_\Omega j\big(z,x(z)\big)\,dz \qquad \forall\, x \in W_0^{1,p}(\Omega).$$

THEOREM 4.7.2
If hypotheses $H(j)_2$ hold and $\lambda > \lambda_1$,
then problem (4.208) has two solutions $y_1, y_2 \in W_0^{1,p}(\Omega)$ such that

$$y_1(z) \;<\; 0 \;<\; y_2(z) \quad \text{for a.a. } z \in \Omega$$

and $y_2 \in C_0^1(\overline{\Omega})$.

PROOF From the first limit in hypothesis $H(j)_2(iv)$ (the asymptotic behaviour at $-\infty$ of $\partial j(z,\cdot)$), we know that we can find $\eta < 0$ such that

$$\frac{u}{|\eta|^{p-2}\eta} \;<\; -\lambda \quad \text{for a.a. } z \in \Omega \text{ and all } u \in \partial j(z,\eta)$$

and so

$$u \;>\; -\lambda|\eta|^{p-2}\eta.$$

So if we set $\underline{u}(z) \stackrel{df}{=} \eta$, then from the above inequality it follows that $\underline{u}$ is a lower solution of problem (4.208) (see Definition 4.7.1(a)).

Also, as in the proof of Proposition 4.7.1, let $\overline{u}(z) \stackrel{df}{=} -\xi u_1(z)$, with $\xi > 0$ small enough so that

$$-\xi\overline{u}(z) \;\in\; (-\delta,0) \qquad \forall\, z \in \Omega,$$

with $\eta < -\delta$. Then we know that $\overline{u}$ is an upper solution of problem (4.208) (see the proof of Proposition 4.7.1). The argument in the proof of Proposition 4.7.1 produces a solution $y_1 \in W_0^{1,p}(\Omega)$ of problem (4.208) such that

$$\underline{u}(z) \;\leq\; y_1(z) \;\leq\; \overline{u}(z) \;=\; -\xi u_1(z) \quad \text{for a.a. } z \in \Omega.$$

Now, as in the proof of Proposition 4.7.4, let $\psi_{1,\lambda} \colon W_0^{1,p}(\Omega) \longrightarrow \mathbb{R}$ be the locally Lipschitz functional, defined by

$$\psi_{1,\lambda}(x) \stackrel{df}{=} \frac{1}{p}\|\nabla x\|_p^p - \frac{1}{p}\big\|x^+\big\|_p^p - \int_\Omega j_1\big(z,x(z)\big)\,dz,$$

where $j_1(z, \zeta) \overset{df}{=} j(z, \tau_0(\zeta))$. By virtue of the second limit in hypothesis $H(j)_2(iv)$ (the asymptotic behaviour at $+\infty$ of $j(z, \cdot)$), we can find $M_4 > 0$ such that

$$j(z, \zeta) \leq -\frac{\lambda}{p}\zeta^p \quad \text{for a.a. } z \in \Omega \text{ and all } \zeta \geq M_4.$$

Also from the mean value theorem for locally Lipschitz functions (see Proposition 1.3.15) and hypothesis $H(j)_2(iii)$, we have

$$\left| j(z, \zeta) \right| \leq \beta_4(z) \quad \text{for a.a. } z \in \Omega \text{ and all } |\zeta| < M_4,$$

with $\beta_4 \in L^\infty(\Omega)_+$. So finally we can say that

$$j(z, \zeta) \leq -\frac{\lambda}{p}\zeta^p + \beta_5(z) \quad \text{for a.a. } z \in \Omega \text{ and all } \zeta \geq 0,$$

with $\beta_5(z) \overset{df}{=} \beta_4(z) + \frac{\lambda}{p}M_4^p$. Then for all $x \in W_0^{1,p}(\Omega)$, we have

$$\psi_{1,\lambda}(x) = \frac{1}{p}\|\nabla x\|_p^p - \frac{\lambda}{p}\|x^+\|_p^p - \int_\Omega j_1(z, x(z)) \, dz$$

$$\geq \frac{1}{p}\|\nabla x\|_p^p - \frac{\lambda}{p}\|x^+\|_p^p + \frac{\lambda}{p}\|x^+\|_p^p - c_8 = \frac{1}{p}\|\nabla x\|_p^p - c_8,$$

for some $c_8 > 0$. Thus $\psi_{1,\lambda}$ is coercive.

By the Weierstrass theorem, we find $x_0 \in W_0^{1,p}(\Omega)$ such that

$$\psi_{1,\lambda}(x_0) = \inf_{W_0^{1,p}(\Omega)} \psi_{1,\lambda}.$$

As in the proof of Proposition 4.7.4, we can check that

$$\psi_{1,\lambda}(x_0) < 0 = \psi_{1,\lambda}(0),$$

(so $x_0 \neq 0$), $x_0 \in C_0^1(\overline{\Omega})$, $x_0(z) > 0$ for all $z \in \Omega$ and x_0 solves problem (4.208). Let us set $y_2 \overset{df}{=} x_0$ and we have finished the proof of the theorem. ☐

EXAMPLE 4.7.1

(a) For simplicity we drop the z-dependence. Consider the following non-smooth, locally Lipschitz potential:

$$j_1(\zeta) \overset{df}{=} \begin{cases} -\frac{|\zeta|^p}{p}\ln|\zeta| + \frac{\lambda_1 - 1}{p} & \text{if } \zeta \in (-\infty, -1), \\ -\frac{1}{p}|\zeta|^p + \frac{\lambda_1}{p} & \text{if } \zeta \in [-1, 0], \\ \frac{\lambda_1}{p}e^{-|\zeta|^p} & \text{if } \zeta \in (0, 1], \\ -\frac{\overline{\beta}}{p}\zeta^p + \frac{1}{p}\left(\frac{\lambda_1}{e} + \overline{\beta}\right) & \text{if } \zeta \in (1, +\infty), \end{cases}$$

where $\overline{\beta} \in (0, \lambda_1]$. We can check that j_1 satisfies hypotheses $H(j)_1$.

(b) Consider the nonsmooth locally Lipschitz function

$$j_2(\zeta) \stackrel{df}{=} \frac{1}{r}|\zeta|^r + \bar{c} \sin |\zeta|,$$

with $r \in (p, p^*)$, $\bar{c} > 0$. We can check that j_2 satisfies hypotheses $H(j)_2$.

$\square$

4.8 Problems with Discontinuous Nonlinearities

The nonsmooth critical point theory provides the right tools to treat elliptic problems with a discontinuous right hand side nonlinearity. Let $\Omega \subseteq \mathbb{R}^N$ be a bounded domain with a C^2-boundary $\partial\Omega$. We consider the following nonlinear elliptic problem:

$$\begin{cases} -\mathrm{div}\left(\|\nabla x(z)\|_{\mathbb{R}^N}^{p-2} \nabla x(z)\right) = f\big(z, x(z)\big) & \text{for a.a. } z \in \Omega, \\ x|_{\partial\Omega} = 0, \end{cases} \tag{4.237}$$

with $p \in (1, +\infty)$. We do not require that the right hand side nonlinearity $f(z, \zeta)$ is continuous in the $\zeta \in \mathbb{R}$ variable. The possible presence of discontinuities precludes the use of the classical (smooth) critical point theory and demands a different approach. The nonsmooth critical point theory for locally Lipschitz functions provides a suitable such analytical framework.

There are two possible ways to interpret a solution for problem (4.237). The first has a multivalued character. Roughly speaking, at every discontinuity point we fill in the gap and this way we obtain a multifunction which replaces the right hand side function, transforming problem (4.237) to an elliptic inclusion. More precisely, we introduce:

$$\underline{f}(z, \zeta) \stackrel{df}{=} \liminf_{\zeta' \to \zeta} f(z, \zeta') = \lim_{\varepsilon \searrow 0} \inf_{|\zeta' - \zeta| \le \varepsilon} f(z, \zeta')$$

$$\overline{f}(z, \zeta) \stackrel{df}{=} \limsup_{\zeta' \to \zeta} f(z, \zeta') = \lim_{\varepsilon \searrow 0} \sup_{|\zeta' - \zeta| \le \varepsilon} f(z, \zeta').$$

Evidently $\underline{f} \le \overline{f}$ and if $\zeta \in \mathbb{R}$ is a continuity point for $f(z, \cdot)$, then $\underline{f}(z, \zeta) = \overline{f}(z, \zeta)$. We introduce the multifunction

$$\widehat{f}(z, \zeta) \stackrel{df}{=} \left[\underline{f}(z, \zeta), \overline{f}(z, \zeta)\right]$$

and instead of (4.237), we consider the following elliptic inclusion:

$$\begin{cases} -\mathrm{div}\left(\|\nabla x(z)\|_{\mathbb{R}^N}^{p-2} \nabla x(z)\right) \in \widehat{f}\big(z, x(z)\big) & \text{for a.a. } z \in \Omega, \\ x|_{\partial\Omega} = 0, \end{cases} \tag{4.238}$$

with $p \in (1, +\infty)$.

Suppose that

$$|f(z, \zeta)| \leq a_0(z) + c_0 |\zeta|^{r-1} \quad \text{for a.a. } z \in \Omega \text{ and all } \zeta \in \mathbb{R},$$

with $r \in [1, p^*)$, $a_0 \in L^{r'}(\Omega)_+$ $(\frac{1}{r} + \frac{1}{r'} = 1)$, $c_0 > 0$.

Let f be superpositionally measurable (i.e. for all measurable functions $x \colon \Omega \longrightarrow \mathbb{R}$, the function $z \longmapsto f(z, x(z))$ is measurable too; note that if f is jointly measurable, it is also superpositionally measurable). We introduce

$$F(z, \zeta) \overset{df}{=} \int_0^\zeta f(z, s)\, ds$$

(the potential function corresponding to f) and consider the integral functional $I_F \colon L^r(\Omega) \longrightarrow \mathbb{R}$ defined by

$$I_F(x) \overset{df}{=} \int_\Omega F(z, x(z))\, dz \qquad \forall\, x \in L^r(\Omega).$$

We know that I_F is locally Lipschitz (see Theorem 1.3.9). The next proposition explains why problem (4.238) can be viewed as a hemivariational inequality.

PROPOSITION 4.8.1

If functions f, $\underline{f}$ and $\overline{f}$ defined above are superpositionally measurable, then

$$\partial I_F(x) \subseteq \left\{ v \in L^{r'}(\Omega) : \underline{f}(z, x(z)) \leq v(z) \leq \overline{f}(z, x(z)) \text{ for a.a. } z \in \Omega \right\}.$$

PROOF Let $h \in L^r(\Omega)$. By definition, we have

$$
\begin{aligned}
I_F^0(x; h) &= \limsup_{\substack{u \to 0 \\ t \searrow 0}} \frac{I_F(x + u + th) - I_F(x + u)}{t} \\
&= \limsup_{\substack{u \to 0 \\ t \searrow 0}} \frac{1}{t} \int_\Omega \int_{(x+u)(z)}^{(x+u+th)(z)} f(z, s)\, ds\, dz.
\end{aligned}
$$

Performing a change of variables and using Fatou's lemma, we obtain

$$
\begin{aligned}
I_F^0(x;h) &\leq \int_{\Omega} \limsup_{\substack{u \to 0 \\ t \searrow 0}} \int_0^1 f\big(z, x(z) + u(z) + \xi th(z)\big) h(z)\, d\xi\, dz \\
&= \int_{\{h>0\}} \overline{f}\big(z, x(z)\big) h(z)\, dz + \int_{\{h<0\}} \underline{f}\big(z, x(z)\big) h(z)\, dz.
\end{aligned}
$$

Recall that $v \in \partial I_F(x)$ if and only if $(v, h)_{L^r(\Omega)} \leq I_F^0(x;h)$. So we have

$$
\begin{aligned}
\int_{\Omega} v(z) h(z)\, dz &\leq \int_{\{h>0\}} \overline{f}\big(z, x(z)\big) h(z)\, dz \\
&\quad + \int_{\{h<0\}} \underline{f}\big(z, x(z)\big) h(z)\, dz \qquad \forall\, h \in L^r(\Omega)
\end{aligned}
$$

and so

$$
v(z) \in \big[\underline{f}\big(z, x(z)\big), \overline{f}\big(z, x(z)\big)\big] \quad \text{for a.a. } z \in \Omega.
$$

$\square$

REMARK 4.8.1 If $f(z, \cdot)$ is monotone, say increasing, then

$$
\widehat{f}(z, \varsigma) = \big[f(z, \varsigma^-), f(z, \varsigma^+)\big]
$$

and so $v(z) \in \big[f(z, x^-(z)), f(z, x^+(z))\big]$. $\square$

DEFINITION 4.8.1 *A function* $x \in W_0^{1,p}(\Omega)$ *for which*

$$
-\Delta_p x(z) = u(z) \quad \text{for a.a. } z \in \Omega
$$

with $u(z) \in \big[\underline{f}(z, x(z)), \overline{f}(z, x(z))\big]$ *for almost all* $z \in \Omega$ *is said to be an* **M-solution** *of problem (4.237).*

There is another solution concept, more restrictive, but more interesting, since it preserves the single-valued character of the problem.

DEFINITION 4.8.2 *A function* $x \in W_0^{1,p}(\Omega)$ *for which*

$$
-\Delta_p x(z) = f(z, x(z)) \quad \text{for a.a. } z \in \Omega
$$

is said to be an **S-solution** *of problem (4.237).*

REMARK 4.8.2   Clearly an S-solution is automatically an M-solution too, but the converse is not in general true.   ▯

In this section we are interested in the existence of multiple S-solutions for problem (4.237). For this purpose, we introduce the following hypotheses on the discontinuous nonlinearity $f(z, \zeta)$:

$\underline{H(f)}$ $f \colon \Omega \times \mathbb{R} \longrightarrow \mathbb{R}$ is a function, such that

(i) f is Borel measurable and $f(z, 0) = 0$ for almost all $z \in \Omega$;

(ii) for almost all $z \in \Omega$, the function

$$\mathbb{R} \ni \zeta \longrightarrow f(z, \zeta) \in \mathbb{R}$$

is nondecreasing and the jump discontinuity points $\{r_n\}_{n \geq 1}$ are independent of z;

(iii) we have

$$|f(z, \zeta)| \ \leq \ a_f(z) + c_f |\zeta|^{p-1} \quad \text{for a.a. } z \in \Omega \text{ and all } \zeta \in \mathbb{R},$$

with $a_f \in L^\infty(\Omega)_+$, $c_f > 0$;

(iv) there exist functions $\vartheta_1, \vartheta_2, \vartheta_3 \in L^\infty(\Omega)_+$ such that

$$\vartheta_1(z) \ \leq \ \lambda_1 \ \leq \ \vartheta_2(z) \quad \text{for a.a. } z \in \Omega,$$

with strict inequalities on sets (in general different) of positive measure and

$$\limsup_{\zeta \to +\infty} \frac{f(z, \zeta)}{\zeta^{p-1}} \ \leq \ \vartheta_1(z) \quad \text{uniformly for almost all } z \in \Omega,$$

$$\vartheta_2(z) \ \leq \ \liminf_{\zeta \to -\infty} \frac{f(z, \zeta)}{|\zeta|^{p-2}\zeta} \quad \text{uniformly for almost all } z \in \Omega,$$

$$\limsup_{\zeta \to -\infty} \frac{f(z, \zeta)}{|\zeta|^{p-2}\zeta} \ \leq \ \vartheta_3(z) \quad \text{uniformly for almost all } z \in \Omega$$

and

$$\vartheta_2(z) \ \leq \ \liminf_{\zeta \to -\infty} \frac{p F(z, \zeta)}{|\zeta|^p} \quad \text{uniformly for almost all } z \in \Omega;$$

(v) we have

$$\limsup_{\zeta \to 0^+} \frac{f(z, \zeta)}{\zeta^{p-1}} \ > \ \lambda_1$$

uniformly for almost all $z \in \Omega$.

REMARK 4.8.3   Hypothesis $H(f)(iv)$ implies that the nonlinearity $f(z, \zeta)$ exhibits asymmetric asymptotic behaviour at $\pm\infty$, crossing the principal eigenvalue $\lambda_1 > 0$ (jumping nonlinearity). ▯

By virtue of the first limit in hypothesis $H(f)(iv)$ (the asymptotic behaviour at $+\infty$ of $f(z, \cdot)$), for a given $\varepsilon > 0$, we can find $M_1 = M_1(\varepsilon) > 0$ such that

$$f(z, \zeta) \leq (\vartheta_1(z) + \varepsilon)\zeta^{p-1} \quad \text{for a.a. } z \in \Omega \text{ and all } \zeta > M_1.$$

On the other hand, because of hypothesis $H(f)(iii)$, we can find $a_\varepsilon \in L^\infty(\Omega)_+$ such that

$$|f(z, \zeta)| \leq a_\varepsilon(z) \quad \text{for a.a. } z \in \Omega \text{ and all } |\zeta| \leq M_1.$$

Thus finally, we can write

$$f(z, \zeta) \leq (\vartheta_1(z) + \varepsilon)\zeta^{p-1} + a_\varepsilon(z) \quad \text{for a.a. } z \in \Omega \text{ and all } \zeta \geq 0. \quad (4.239)$$

We consider the following auxiliary problem:

$$\begin{cases} -\operatorname{div}\left(\|\nabla x(z)\|_{\mathbb{R}^N}^{p-2} \nabla x(z) \right) = (\vartheta_1(z) + \varepsilon)|x(z)|^{p-2} x(z) + a_\varepsilon(z) \\ \qquad\qquad\qquad\qquad\qquad\qquad\qquad\qquad \text{for a.a. } z \in \Omega, \quad (4.240) \\ x|_{\partial\Omega} = 0. \end{cases}$$

Arguing as in the proof of Proposition 4.7.1, we can have the following existence result concerning the auxiliary result (4.240).

PROPOSITION 4.8.2
If hypotheses $H(f)$ hold,
then for all $\varepsilon > 0$ small enough, problem (4.240) has a solution $x_\varepsilon \in C_0^1(\overline{\Omega})$ with

$$x_\varepsilon(z) > 0 \quad \forall z \in \Omega \quad \text{and} \quad \frac{\partial x_\varepsilon}{\partial n}(z) \leq 0 \quad \forall z \in \partial\Omega.$$

REMARK 4.8.4   By virtue of (4.239) any such solution x_ε is an upper solution of problem (4.237). ▯

We fix $\varepsilon > 0$ small enough and let $\overline{u} \stackrel{df}{=} x_\varepsilon \in \operatorname{int} C_0^1(\overline{\Omega})_+$. Because of hypothesis $H(f)(v)$, we can find $\delta > 0$ and $\mu > \lambda_1$ such that

$$\mu\zeta^{p-1} \leq f(z, \zeta) \quad \text{for a.a. } z \in \Omega \text{ and all } \zeta \in (0, \delta).$$

Choose $\xi > 0$ small enough, so that $\xi u_1(z) \in (0, \delta)$ for all $z \in \Omega$ and $\xi u_1(z) \leq x_\varepsilon(z)$ for all $z \in \Omega$. (as always u_1 is the principal eigenvalue of $\left(-\Delta_p, W_0^{1,p}(\Omega) \right)$). Let us set $\underline{u} \stackrel{df}{=} \xi u_1 \in C_0^1(\overline{\Omega})$. Evidently $\underline{u}$ is a lower solution for problem (4.237).

We introduce the truncation map $\tau \colon \Omega \times \mathbb{R} \longrightarrow \mathbb{R}$ defined by

$$\tau(z, \zeta) \stackrel{df}{=} \begin{cases} \overline{u}(z) & \text{if } \overline{u}(z) < \zeta, \\ \zeta & \text{if } \underline{u}(z) \leq \zeta \leq \overline{u}, \\ \underline{u}(z) & \text{if } \zeta < \underline{u}(z). \end{cases}$$

Clearly τ is a Carathéodory function, thus it is jointly measurable. Let us set

$$\widehat{f}(z, \zeta) \stackrel{df}{=} f\bigl(z, \tau(z, \zeta)\bigr) \qquad \forall \, (z, \zeta) \in \Omega \times \mathbb{R}.$$

Then $\widehat{f}$ is Borel measurable and for almost all $z \in \Omega$, $\widehat{f}(z, \cdot)$ is nondecreasing on $\mathbb{R}$. Moreover, we have

$$\bigl|\widehat{f}(z, \zeta)\bigr| \leq \eta_1(z) \quad \text{for a.a. } z \in \Omega \text{ and all } \zeta \in \mathbb{R}, \tag{4.241}$$

with $\eta_1 \in L^{\infty}(\Omega)_+$. We consider the following nonlinear elliptic problem:

$$\begin{cases} -\operatorname{div}\bigl(\|\nabla x(z)\|_{\mathbb{R}^N}^{p-2} \nabla x(z)\bigr) = \widehat{f}\bigl(z, x(z)\bigr) & \text{for a.a. } z \in \Omega, \\ x|_{\partial \Omega} = 0. \end{cases} \tag{4.242}$$

PROPOSITION 4.8.3
If hypotheses $H(f)$ hold,
<u>*then*</u> *problem (4.242) has a solution $x_0 \in C_0^1(\overline{\Omega})$.*

PROOF   Let

$$\widehat{F}(z, \zeta) \stackrel{df}{=} \int\limits_0^\zeta \widehat{f}(z, s) \, ds$$

and consider the energy functional $\widehat{\varphi} \colon W_0^{1,p}(\Omega) \longrightarrow \mathbb{R}$ defined by

$$\widehat{\varphi}(x) \stackrel{df}{=} \frac{1}{p} \|\nabla x\|_p^p - \int\limits_\Omega \widehat{F}(z, x(z)) \, dz \qquad \forall \, x \in W_0^{1,p}(\Omega).$$

Note that

$$\bigl|\widehat{F}(z, \zeta)\bigr| \leq \int\limits_0^{|\zeta|} \bigl|\widehat{f}(z, s)\bigr| \, ds \leq \|\eta_1\|_\infty \, |\zeta|$$

(see (4.241)). So for every $x \in W_0^{1,p}(\Omega)$, we have

$$\widehat{\varphi}(x) \geq \frac{1}{p} \|\nabla x\|_p^p - c_1 \|\nabla x\|_p,$$

for some $c_1 > 0$, so $\widehat{\varphi}$ is coercive.

Also it is easy to see that $\widehat{\varphi}$ is weakly lower semicontinuous. Hence by the Weierstrass theorem, we can find $x_0 \in W_0^{1,p}(\Omega)$ such that

$$\widehat{\varphi}(x_0) = \inf_{x \in W_0^{1,p}(\Omega)} \widehat{\varphi}(x). \tag{4.243}$$

Since $\widehat{\varphi}$ is locally Lipschitz, we have that

$$0 \in \partial\widehat{\varphi}(x_0)$$

and so

$$A(x_0) = v, \tag{4.244}$$

where $A \colon W_0^{1,p}(\Omega) \longrightarrow W^{-1,p'}(\Omega)$ $(\frac{1}{p} + \frac{1}{p'} = 1)$ is the nonlinear monotone operator defined by

$$\langle A(x), y \rangle_{W_0^{1,p}(\Omega)} \overset{df}{=} \int_\Omega \|\nabla x(z)\|_{\mathbb{R}^N}^{p-2} \left(\nabla x(z), \nabla y(z)\right)_{\mathbb{R}^N} dz \qquad \forall\, x, y \in W_0^{1,p}(\Omega)$$

and $v \in L^{p'}(\Omega)$ with $v(z) \in \partial\widehat{f}(z, x_0(z))$ for almost all $z \in \Omega \setminus D$ where

$$D \overset{df}{=} \bigcup_{n=1}^{\infty} \left\{ z \in \Omega \colon x_0(z) = r_n \right\}$$

and $v(z) \in \left[\widehat{f}(z, x_0(z)^-), \widehat{f}(z, x_0(z)^+)\right]$ for almost all $z \in D$. For all $\varepsilon > 0$, we have

$$\begin{aligned}
0 &\leq \frac{\widehat{\varphi}(x_0 + \varepsilon u_1) - \widehat{\varphi}(x_0)}{\varepsilon} \\
&= \frac{1}{p} \left[\frac{\|\nabla(x_0 + \varepsilon u_1)\|_p^p - \|\nabla x_0\|_p^p}{\varepsilon} \right] \\
&\quad - \int_\Omega \frac{\widehat{F}(z, x_0(z) + \varepsilon u_1(z)) - \widehat{F}(z, x_0(z))}{\varepsilon}\, dz.
\end{aligned}$$

Passing to the limit as $\varepsilon \searrow 0$ and using (4.244), we obtain

$$\begin{aligned}
0 &\leq \langle A(x_0), u_1 \rangle_{W_0^{1,p}(\Omega)} - \int_\Omega \widehat{f}(z, x_0(z)^+) u_1(z)\, dz \\
&= \int_\Omega \left(v(z) - \widehat{f}(z, x_0(z)^+)\right) u_1(z)\, dz \\
&= \int_D \left(v(z) - \widehat{f}(z, x_0(z)^+)\right) u_1(z)\, dz. \tag{4.245}
\end{aligned}$$

Because $v(z) \leq \widehat{f}(z, x_0(z)^+)$ for almost all $z \in \Omega$ and $u_1(z) > 0$ for all $z \in \Omega$, if $|D|_N > 0$, from (4.245) we have that

$$v(z) = \widehat{f}(z, x_0(z)) \quad \text{for a.a. } z \in D. \qquad (4.246)$$

Similarly, for $\varepsilon > 0$, we have

$$0 \leq \frac{\varphi(x_0 - \varepsilon u_1) - \varphi(x_0)}{\varepsilon}$$

and so, using also (4.244), we have

$$
\begin{aligned}
0 &\leq \langle A(x_0), -u_1 \rangle_{W_0^{1,p}(\Omega)} + \int_\Omega \widehat{f}(z, x_0(z)^-) u_1(z)\, dz \\
&= \int_\Omega \big(-v(z) + \widehat{f}(z, x_0(z)^-) \big) u_1(z)\, dz \\
&= \int_D \big(-v(z) + \widehat{f}(z, x_0(z)^-) \big) u_1(z)\, dz.
\end{aligned}
\qquad (4.247)
$$

Recall that $\widehat{f}(z, x_0(z)^-) \leq v(z)$ for almost all $z \in \Omega$. So from (4.247) (since we have assumed that $|D|_N > 0$), we have

$$v(z) = \widehat{f}(z, x_0(z)) \quad \text{for a.a. } z \in D. \qquad (4.248)$$

But

$$\widehat{f}(z, x_0(z)^-) < \widehat{f}(z, x_0(z)^+) \quad \text{for a.a. } z \in D.$$

This combined with (4.246) and (4.248) leads to a contradiction. So $|D|_N = 0$ and we have

$$v(z) = \widehat{f}(z, x_0(z)) \quad \text{for a.a. } z \in \Omega.$$

Therefore $x \in W_0^{1,p}(\Omega)$ solves problem (4.242). Moreover, because

$$-x_0(z)\Delta_p x_0(z) = -x_0(z)\widehat{f}(z, x_0(z)) \leq \eta_1(z)|x_0(z)| \quad \text{for a.a. } z \in \Omega$$

(see (4.241)), we can apply Theorem 1.5.5 and have that $x_0 \in L^\infty(\Omega)$. Since $x_0 \in W_0^{1,p}(\Omega) \cap L^\infty(\Omega)$ and $\Delta_p x_0 \in L^\infty(\Omega)$, via Theorem 1.5.6, we have that $x_0 \in C_0^1(\overline{\Omega})$. ∎

To prove multiplicity results for the S-solutions of problem (4.237), we need to have a strong comparison principle. To this end, first we prove the following result:

LEMMA 4.8.1
If $x, y \in W^{1,p}(\Omega)$, $u, v \in L^\infty(\Omega)$, $x(z) \geq y(z)$ and $u(z) \geq v(z)$ for almost all $z \in \Omega$,

$$-\mathrm{div}\left(\|\nabla x(z)\|_{\mathbb{R}^N}^{p-2} \nabla x(z) \right) = u(z) \quad \text{for a.a. } z \in \Omega,$$
$$-\mathrm{div}\left(\|\nabla y(z)\|_{\mathbb{R}^N}^{p-2} \nabla y(z) \right) = v(z) \quad \text{for a.a. } z \in \Omega$$

and the set $C \overset{df}{=} \{z \in \Omega : \ x(z) = y(z)\}$ *is compact,*
then $C = \emptyset$.

PROOF   From Theorems 1.5.5 and 1.5.6, we obtain that $x, y \in C_0^{1,\alpha}(\overline{\Omega})$
for some $\alpha \in (0,1)$. Suppose that $C \neq \emptyset$ and let Ω_1 be a relatively compact
set with C^2-boundary such that

$$C \subseteq \Omega_1 \subseteq \overline{\Omega}_1 \subseteq \Omega$$

and

$$y(z) \ < \ x(z) \qquad \forall \, z \in \Omega_1 \setminus C.$$

Let $\varepsilon > 0$ and consider the following two auxiliary problems

$$\begin{cases} -\mathrm{div}\left(\left(\varepsilon + \|\nabla\sigma(z)\|_{\mathbb{R}^N}^2\right)^{\frac{p-2}{2}} \nabla\sigma(z)\right) = u(z) & \text{for a.a. } z \in \Omega_1, \\ \sigma|_{\partial\Omega_1} = x|_{\partial\Omega_1}, \end{cases} \qquad (4.249)$$

and

$$\begin{cases} -\mathrm{div}\left(\left(\varepsilon + \|\nabla\tau(z)\|_{\mathbb{R}^N}^2\right)^{\frac{p-2}{2}} \nabla\tau(z)\right) = v(z) & \text{for a.a. } z \in \Omega_1, \\ \tau|_{\partial\Omega_1} = x|_{\partial\Omega_1}. \end{cases} \qquad (4.250)$$

Exploiting the maximal monotonicity of the differential operator, we infer that
the two problems (4.249) and (4.250) have solutions $x_\varepsilon, y_\varepsilon \in W^{1,p}(\Omega_1)$ respectively (these solutions are unique due to strict monotonicity of the operator).
Note that

$$\|\nabla x_\varepsilon\|_{L^p(\Omega_1)} \ \leqslant \ M_2 \quad \text{and} \quad \|\nabla y_\varepsilon\|_{L^p(\Omega_1)} \ \leqslant \ M_2 \qquad \forall \, \varepsilon \in (0,1),$$

for some $M_2 > 0$. Moreover, from nonlinear regularity theory (see Theo-
rem 1.5.6) we have that $x_\varepsilon, y_\varepsilon \in C_0^{1,\alpha}(\overline{\Omega}_1)$. From the Poincaré inequality (see
Theorem 1.1.6), we have that

$$\begin{aligned} x_\varepsilon &\xrightarrow{w} x \text{ in } W_0^{1,p}(\Omega) & \text{as } \varepsilon \searrow 0, \\ y_\varepsilon &\xrightarrow{w} y \text{ in } W_0^{1,p}(\Omega) & \text{as } \varepsilon \searrow 0, \\ x_\varepsilon &\longrightarrow x \text{ in } C_{\mathrm{loc}}^{1,\alpha'}(\Omega_1) & \text{as } \varepsilon \searrow 0, \\ y_\varepsilon &\longrightarrow y \text{ in } C_{\mathrm{loc}}^{1,\alpha'}(\Omega_1) & \text{as } \varepsilon \searrow 0, \end{aligned}$$

for all $\alpha' \in (0,\alpha)$ (recall that the embedding $C_{\mathrm{loc}}^{1,\alpha}(\Omega_1) \subseteq C_{\mathrm{loc}}^{1,\alpha'}(\Omega_1)$ is
compact). Let Ω_2 be a relatively compact open subset of Ω such that
$C \subseteq \Omega_2 \subseteq \overline{\Omega}_2 \subseteq \Omega_1$ and let

$$\vartheta \overset{df}{=} \min_{\partial\Omega_2}(x - y) \ > \ 0$$

and $w_\varepsilon \overset{df}{=} x_\varepsilon - y_\varepsilon$ for $\varepsilon > 0$. Let us take $\varepsilon > 0$ small enough so that

$$\|x - x_\varepsilon\|_{L^\infty(\Omega_2)} < \frac{\vartheta}{4} \quad \text{and} \quad \|y - y_\varepsilon\|_{L^\infty(\Omega_2)} < \frac{\vartheta}{4}.$$

It follows that

$$x_\varepsilon - y_\varepsilon > \frac{\vartheta}{2} \quad \text{on } \partial\Omega_2 \quad \text{and} \quad x_\varepsilon - y_\varepsilon < \frac{\vartheta}{2} \quad \text{on } C.$$

Invoking the mean value theorem, we have

$$-\sum_{i,j=1}^{N} \frac{\partial}{\partial z_i}\left(a_{ij}^\varepsilon(z)\frac{\partial w_\varepsilon}{\partial z_j}(z)\right) = u(z) - v(z) \quad \text{for a.a. } z \in \Omega_2, \qquad (4.251)$$

with

$$\begin{aligned}
a_{ij}^\varepsilon(z) \overset{df}{=} \ & \left(\varepsilon + \|t_i\nabla x_\varepsilon(z) + (1 - t_i)\nabla y_\varepsilon(z)\|_{\mathbb{R}^N}\right)^{\frac{p-4}{2}} \\
& + \left[\delta_{ij}\left(\varepsilon + \|t_i\nabla x_\varepsilon(z) + (1 - t_i)\nabla y_\varepsilon(z)\|_{\mathbb{R}^N}\right)\right. \\
& + (p-2)\left(t_i\frac{\partial x_\varepsilon(z)}{\partial z_i} + (1 - t_i)\frac{\partial y_\varepsilon(z)}{\partial z_i}\right)\left(t_i\frac{\partial x_\varepsilon(z)}{\partial z_j} + (1 - t_i)\frac{\partial y_\varepsilon(z)}{\partial z_j}\right)\Bigg],
\end{aligned}$$

with $t_i \in (0, 1)$. Let us set

$$\varrho \overset{df}{=} \inf_{z \in \Omega_2} (x_\varepsilon - y_\varepsilon)(z) \quad \text{and} \quad C_\varrho \overset{df}{=} \{z \in \Omega_2 : (x_\varepsilon - y_\varepsilon)(z) = \varrho\}.$$

Then $C_\varrho \subseteq \Omega_2$ is nonempty, compact and

$$\nabla x_\varepsilon(z) = \nabla y_\varepsilon(z) \quad \text{for a.a. } z \in C_\varrho$$

(see Remark 1.1.10). So we have

$$a_{ij}^\varepsilon(z) = \left(\varepsilon + \|\nabla x_\varepsilon(z)\|_{\mathbb{R}^N}^2\right)^{\frac{p-4}{2}}\left(\delta_{ij}\left(\varepsilon + \|\nabla x_\varepsilon(z)\|_{\mathbb{R}^N}^2\right) + (p-2)\frac{\partial x_\varepsilon(z)}{\partial z_i}\frac{\partial x_\varepsilon(z)}{\partial z_j}\right)$$

and

$$\sum_{i,j=1}^{N} a_{ij}^\varepsilon(z)\xi_i\xi_j \geq \eta\|\xi\|_{\mathbb{R}^N}^2 \qquad \forall\, \xi \in \mathbb{R}^N,\ z \in C_\varrho, \qquad (4.252)$$

for some $\eta > 0$. The last inequality is a consequence of the fact that $\left(a_{ij}^\varepsilon(z)\right)_{i,j=1}^{N}$ is the Hessian matrix of the convex function $\mathbb{R} \ni \xi \longmapsto \frac{1}{p}\left(\varepsilon + \|\xi\|_{\mathbb{R}^N}^2\right)^{\frac{p}{2}}$ at the point $\xi = \nabla x_\varepsilon$. We consider an open neighbourhood Ω_ϱ of C_ϱ such that $\overline{\Omega}_\varrho \subseteq \Omega_2$, $x_\varepsilon - y_\varepsilon > \varrho$ on $\partial\Omega_\varrho$ and inequality (4.252) is valid on $\partial\Omega_\varrho$ with $\eta > 0$ replaced by $\frac{1}{2}\eta$. It is possible to find such a neighbourhood because ∇x_ε and ∇y_ε are both continuous on Ω_2. Finally invoking the strong maximum principle, we infer that $(x_\varepsilon - y_\varepsilon)|_{\Omega_\varrho} = w_\varepsilon|_{\Omega_\varrho}$ is constant, a contradiction. This proves that $C \neq \emptyset$. $\quad\blacksquare$

This lemma leads to the following extension of the Hopf boundary result.

PROPOSITION 4.8.4
If $x, y \in C_0^1(\overline{\Omega})$, $u, v \in L^\infty(\Omega)$ *and* $u(z) \geq v(z) \geq 0$ *for almost all* $z \in \Omega$,

$$-\operatorname{div}\left(\|\nabla x(z)\|_{\mathbb{R}^N}^{p-2} \nabla x(z)\right) = u(z) \quad \text{for a.a. } z \in \Omega,$$

$$-\operatorname{div}\left(\|\nabla y(z)\|_{\mathbb{R}^N}^{p-2} \nabla y(z)\right) = v(z) \quad \text{for a.a. } z \in \Omega$$

and the set $C_0 \stackrel{df}{=} \{z \in \Omega : u(z) = v(z)\}$ *has an empty interior,* *then* $x(z) > y(z)$ *for all* $z \in \Omega$ *and* $\frac{\partial(x-y)}{\partial n}(z) < 0$ *for all* $z \in \partial\Omega$.

PROOF From our hypothesis about the set C_0, it follows that

$$\left|\{z \in \Omega : u(z) > 0\}\right|_N > 0$$

and so x is not the zero function. So we can apply Theorem 1.5.7 and obtain that

$$x(z) > 0 \quad \forall z \in \Omega$$

and

$$\frac{\partial x}{\partial n}(z) < 0 \quad \forall z \in \partial\Omega.$$

Moreover, since $-\Delta_p x = u$, $-\Delta_p y = v$ and $u \geq v$, by acting with the test function $(y - x)^+$, we can easily infer that $x \geq y \geq 0$. Therefore Lemma 4.8.1 implies that the coincidence set $C \stackrel{df}{=} \{z \in \Omega : x(z) = y(z)\}$ cannot be nonempty compact. Suppose that $C \neq \emptyset$. Then we can find a sequence $\{z_n\}_{n \geq 1} \subseteq C$ and $z_0 \in \partial\Omega$ such that

$$z_n \longrightarrow z_0 \quad \text{in } \mathbb{R}^N.$$

Since by hypothesis $x, y \in C_0^1(\overline{\Omega})$ and as we saw in the beginning of the proof $x \in \operatorname{int} C_0^1(\overline{\Omega})_+$, we have

$$\frac{\partial x}{\partial n}(z_0) = \frac{\partial y}{\partial n}(z_0) < 0. \tag{4.253}$$

The function $w = x - y$ satisfies

$$-\sum_{i,j=1}^{N} \frac{\partial}{\partial z_j}\left(a_{ij}(z)\frac{\partial}{\partial z_i}w(z)\right) = u(z) - v(z) \geq 0 \quad \text{for a.a. } z \in \Omega,$$

with $a_{ij} = a_{ij}^0$ (see (4.251) with $\varepsilon = 0$). In particular then

$$a_{ij}(z_0) = \left|\frac{\partial x}{\partial n}(z_0)\right|^{p-4}\left(\delta_{ij}\left|\frac{\partial x}{\partial n}(z_0)\right|^2 + (p-2)\frac{\partial}{\partial z_i}x(z_0)\frac{\partial}{\partial z_j}x(z_0)\right).$$

Because $\nabla x, \nabla y \in C\left(\overline{\Omega}\right)$, we can find all ball $B \subseteq \Omega$ such that $z_0 \in \partial B$ and the elliptic operator with coefficients a_{ij} is strictly elliptic on B. Hence either $w = 0$ on B, which is impossible because int $C_0 = \emptyset$ or $w > 0$ on B and $\frac{\partial w}{\partial n}(z_0) < 0$, which contradicts (4.253). This proves that $C = \emptyset$ and so $x(z) > y(z)$ for all $z \in \Omega$. The inequality

$$\frac{\partial x}{\partial n} < \frac{\partial y}{\partial n} \quad \text{on } \partial\Omega$$

is established similarly since

$$\frac{\partial x}{\partial n} \leq \frac{\partial y}{\partial n} \leq 0 \quad \text{on } \partial\Omega$$

and

$$\frac{\partial x}{\partial n}(z_0) < 0.$$

$\square$

Having this strong comparison principle, we can now show that problem (4.237) has at least two smooth S-solutions, one of which is strictly positive.

THEOREM 4.8.1
If hypotheses $H(f)$ hold and

$$0 \notin \left[f(z, r_n^-), f(z, r_n^+)\right] \quad \text{for a.a. } z \in \Omega \text{ and all } n \geq 1$$

then problem (4.237) has at least two S-solutions $x_0, u_0 \in C_0^{1,\alpha}(\overline{\Omega})$ with $\alpha \in (0,1)$ and

$$x_0(z) > 0 \quad \forall z \in \Omega, \qquad \frac{\partial x_0}{\partial n}(z) < 0 \quad \forall z \in \partial\Omega.$$

PROOF   Let $x_0 \in C_0^{1,\alpha}(\overline{\Omega})$ be the solution of problem (4.242) established in Proposition 4.8.3. Recall that $\underline{u} = \xi u_1$. So we have

$$A(\underline{u}) = \lambda_1 |\underline{u}|^{p-1} < \mu|\underline{u}|^{p-1} \leq f(\cdot, \underline{u}(\cdot)) = \widehat{f}(\cdot, \underline{u}(\cdot))$$

(recall the definition of $\widehat{f}$). Using as a test function $(\underline{u} - x_0) \in W_0^{1,p}(\Omega)$, we have

$$\langle A(\underline{u}) - A(x_0), (\underline{u} - x_0)^+ \rangle_{W_0^{1,p}(\Omega)}$$

$$< \int_\Omega \left(\widehat{f}(z, \underline{u}(z)) - \widehat{f}(z, x_0(z)) \right) (\underline{u} - x_0)^+(z) \, dz = 0.$$

Also from the strong monotonicity of A, we have

$$\langle A(\underline{u}) - A(x_0), (\underline{u} - x_0)^+ \rangle_{W_0^{1,p}(\Omega)} \geq c_2 \left\| \nabla(\underline{u} - x_0)^+ \right\|_p^p,$$

for some $c_2 > 0$ and so

$$\left\| \nabla(\underline{u} - x_0)^+ \right\|_p = 0,$$

i.e. $\underline{u} \leq x_0$. In a similar fashion, using this time (4.239), we obtain that $x_0 \leq \overline{u}$. Therefore

$$\underline{u}(z) \leq x_0(z) \leq \overline{u}(z) \quad \text{for a.a. } z \in \Omega$$

and so from the definition of $\widehat{f}$, it follows that

$$\widehat{f}(z, x_0(z)) = f(z, x_0(z)) \quad \text{for a.a. } z \in \Omega.$$

Exploiting the monotonicity of $f(z, \cdot)$ and the definition of $\underline{u}$, we have

$$-\Delta_p \underline{u}(z) = \lambda_1 \underline{u}(z)^{p-1} < f(z, \underline{u}(z))$$
$$\leq f(z, x_0(z)) = -\Delta_p x_0(z) \quad \forall z \in \Omega.$$

From Proposition 4.8.4, it follows that $x_0 - \underline{u} \in \operatorname{int} C_0^1(\Omega)_+$. Also we have

$$-\Delta_p x_0(z) = f(z, x_0(z)) \leq f(z, \overline{u}(z))$$
$$< (\vartheta_1(z) + \varepsilon)\overline{u}(z)^{p-1} + a_\varepsilon(z) = -\Delta_p \overline{u}(z) \quad \text{for a.a. } z \in \Omega$$

(see (4.239)). Again Proposition 4.8.4 implies that $\overline{u} - x_0 \in \operatorname{int} C_0^1(\Omega)_+$. Now let $\varphi \colon W_0^{1,p}(\Omega) \longrightarrow \mathbb{R}$ be the locally Lipschitz functional defined by

$$\varphi(x) \stackrel{df}{=} \frac{1}{p} \|\nabla x\|_p^p - \int_\Omega F(z, x(z)) \, dz.$$

Since $x_0 - \underline{u}, \overline{u} - x_0 \in \operatorname{int} C_0^1(\overline{\Omega})_+$, we can find $\delta > 0$ small such that

$$\varphi(x) = \widehat{\varphi}(x) \quad \forall x \in C_0^1(\overline{\Omega}), \ \|x - x_0\|_{C_0^1(\overline{\Omega})} \leq \delta$$

and so x_0 is a local $C_0^1(\overline{\Omega})$-minimizer of φ. Invoking Proposition 4.6.10, we obtain that x_0 is a local $W_0^{1,p}(\Omega)$-minimizer of φ. Using this fact and arguing as in the proof of Proposition 4.8.3, we deduce that $|D|_N = 0$ and so we have

$$\begin{cases} -\operatorname{div}\left(\|\nabla x_0(z)\|_{\mathbb{R}^N}^{p-2} \nabla x_0(z) \right) = f(z, x_0(z)) & \text{for a.a. } z \in \Omega, \\ x_0|_{\partial\Omega} = 0, \end{cases}$$

so $x_0 \in C_0^{1,\alpha}(\overline{\Omega})$ is an S-solution of (4.237).

Moreover, because $x_0 \geq \xi u_1 = \underline{u}$, we have that $x_0(z) > 0$ for all $z \in \Omega$. Also

$$-\Delta_p x_0(z) = f(z, x_0(z)) \geq f(z, 0) = 0 \quad \text{for a.a. } z \in \Omega,$$

hence

$$\Delta_p x_0(z) \leq 0 \quad \text{for a.a. } z \in \Omega$$

and by virtue of Theorem 1.5.7, we conclude that

$$\frac{\partial x_0}{\partial n}(z) < 0 \quad \text{for a.a. } z \in \partial\Omega.$$

Next with the help of the nonsmooth Mountain Pass Theorem (see Theorem 2.1.3), we shall produce a second smooth S-solution for problem (4.237).

Claim: φ satisfies the nonsmooth PS-condition.

Let $\{x_n\}_{n \geq 1} \subseteq W_0^{1,p}(\Omega)$ be a sequence such that

$$|\varphi(x_n)| \leq M_3 \quad \forall\, n \geq 1 \quad \text{and} \quad m^\varphi(x_n) \longrightarrow 0,$$

for some $M_3 > 0$. We choose $x_n^* \in \partial\varphi(x_n)$ such that $m^\varphi(x_n) = \|x_n^*\|_{W^{-1,p'}(\Omega)}$ for $n \geq 1$. We know that

$$x_n^* = A(x_n) - v_n,$$

with $v_n \in L^{p'}(\Omega)$, $f(z, x_n(z)^-) \leq v_n(z) \leq f(z, x_n(z)^+)$ for almost all $z \in \Omega$. Using as a test function $x_n^+ \in W_0^{1,p}(\Omega)$ we have

$$\|\nabla x_N^+\|_p^p \leq \int_\Omega v_n(z) x_n^+(z)\, dz + \varepsilon_n \|x_n^+\|_{W^{1,p}(\Omega)},$$

with $\varepsilon_n \searrow 0$. Note that

$$v_n(z) \leq 0 \quad \text{for a.a. } z \in \{x_n \leq 0\}$$

and so

$$v_n(z) \leq (\vartheta_1(z) + \varepsilon)|x_n^+(z)|^{p-1} + a_\varepsilon(z) \quad \text{for a.a. } z \in \Omega.$$

Thus we can write that

$$\|\nabla x_n^+\|_p^p - \int_\Omega \vartheta_1(z)|x_n^+(z)|^p\, dz \leq \varepsilon\|x_n^+\|_p + \varepsilon_n\|x_n^+\|_{W^{1,p}(\Omega)} + c_3 \quad \forall\, n \geq 1,$$

for some $c_3 > 0$, thus

$$\gamma\|\nabla x_n^+\|_p^p - \frac{\varepsilon}{\lambda_1}\|\nabla x_n^+\|_p^p \leq \varepsilon_n\|x_n^+\|_{W^{1,p}(\Omega)} + c_3 \quad \forall\, n \geq 1,$$

for some $\gamma > 0$ (see Lemma 4.7.1) and so the sequence $\{x_n^+\}_{n \geq 1} \subseteq W_0^{1,p}(\Omega)$ is bounded.

Suppose that the sequence $\{x_n\}_{n \geq 1} \subseteq W_0^{1,p}(\Omega)$ is not bounded. Since $\|x_n\|_{W^{1,p}(\Omega)} = \|x_n^+\|_{W^{1,p}(\Omega)} + \|x_n^-\|_{W^{1,p}(\Omega)}$, by selecting a suitable subsequence if necessary, we may assume that

$$\|x_n^-\|_{W^{1,p}(\Omega)} \longrightarrow +\infty.$$

Let us set

$$y_n \stackrel{df}{=} \frac{x_n^-}{\|x_n\|_{W^{1,p}(\Omega)}} \qquad \forall \, n \geq 1.$$

We can see that

$$\begin{aligned}
y_n &\stackrel{w}{\longrightarrow} y && \text{in } W_0^{1,p}(\Omega), \\
y_n &\longrightarrow y && \text{in } L^p(\Omega), \\
y_n(z) &\longrightarrow y(z) && \text{for a.a. } z \in \Omega, \\
|y_n(z)| &\leq k(z) && \text{for a.a. } z \in \Omega \text{ and all } n \geq 1,
\end{aligned}$$

with $k \in L^p(\Omega)$. An argument similar to that in the proof of Proposition 4.6.8 shows that

$$\frac{f(\cdot, x_n(\cdot))}{\|x_n^-\|_{W^{1,p}(\Omega)}^{p-1}} \stackrel{w}{\longrightarrow} h \quad \text{in } L^{p'}(\Omega),$$

with $h(z) = g(z)y(z)^{p-1}$, where $g \in L^\infty(\Omega)$ is such that $\vartheta_2(z) \leq g(z) \leq \vartheta_3(z)$ for almost all $z \in \Omega$. For every $w \in W_0^{1,p}(\Omega)$ we have

$$\left| \langle A(x_n), w \rangle_{W_0^{1,p}(\Omega)} - \int_\Omega f(z, x_n(z)) w(z) \, dz \right| \leq \varepsilon_n \|w\|_{W^{1,p}(\Omega)},$$

so

$$\left| \int_\Omega \frac{\|\nabla x_n(z)\|_{\mathbb{R}^N}^{p-2}}{\|x_n^-\|_{W^{1,p}(\Omega)}^{p-1}} \left(\nabla x_n^+(z), \nabla w(z) \right)_{\mathbb{R}^N} dz \right.$$
$$- \int_\Omega \frac{\|\nabla x_n(z)\|_{\mathbb{R}^N}^{p-2}}{\|x_n^-\|_{W^{1,p}(\Omega)}^{p-1}} \left(\nabla x_n^-(z), \nabla w(z) \right)_{\mathbb{R}^N} dz$$
$$\left. - \int_\Omega \frac{f(z, x_n(z))}{\|x_n^-\|_{W^{1,p}(\Omega)}^{p-1}} w(z) \, dz \right| \leq \varepsilon_n \|w\|_{W^{1,p}(\Omega)}. \quad (4.254)$$

Since the sequence $\{x_n^+\}_{n \geq 1} \subseteq W_0^{1,p}(\Omega)$ is bounded, we have

$$\int_\Omega \frac{\|\nabla x_n(z)\|_{\mathbb{R}^N}^{p-2}}{\|x_n^-\|_{W^{1,p}(\Omega)}^{p-1}} \left(\nabla x_n^+(z), \nabla w(z) \right)_{\mathbb{R}^N} dz \longrightarrow 0.$$

So if we pass to the limit as $n \to +\infty$ in (4.254), we obtain

$$\int_{\Omega} \|\nabla y(z)\|_{\mathbb{R}^N}^{p-2} \left(\nabla y(z), \nabla w(z) \right)_{\mathbb{R}^N} dz$$

$$= \int_{\Omega} g(z) y(z)^{p-1} dz \qquad \forall \, w \in W_0^{1,p}(\Omega)$$

and so

$$\begin{cases} -\text{div}\left(\|\nabla y(z)\|_{\mathbb{R}^N}^{p-2} \nabla y(z) \right) = g(z) y(z)^{p-1} & \text{for a.a. } z \in \Omega, \\ y|_{\partial\Omega} = 0. \end{cases}$$

From Theorems 1.5.5 and 1.5.6, we have that $y \in C_0^{1,\alpha}(\overline{\Omega})$ (with $\alpha \in (0,1)$) and $y \geq 0$. So, if $y \neq 0$, invoking Theorem 1.5.7 (nonlinear strong maximum principle), we have $y(z) > 0$ for all $z \in \Omega$. Therefore y is an eigenfunction corresponding to the eigenvalue $\lambda = 1$, of the weighted nonlinear eigenvalue problem:

$$\begin{cases} -\text{div}\left(\|\nabla u(z)\|_{\mathbb{R}^N}^{p-2} \nabla u(z) \right) = \lambda g(z) |u(z)|^{p-2} u(z) & \text{for a.a. } z \in \Omega, \\ u|_{\partial\Omega} = 0. \end{cases}$$

Note that since $g \geq 0$, $\lambda = 1$ is the principal eigenvalue, since otherwise the corresponding eigenfunction must change sign (see Section 1.5, where $g \equiv 1$). We have

$$\lambda_1 = \frac{\|\nabla y\|_p^p}{\int_{\Omega} g(z)|y(z)|^p dz} = \inf_{\substack{u \in W_0^{1,p}(\Omega) \\ u \neq 0}} \frac{\|\nabla y\|_p^p}{\int_{\Omega} g(z)|y(z)|^p dz}$$

$$\leq \frac{\|\nabla u_1\|_p^p}{\int_{\Omega} g(z)|u_1(z)|^p dz} < \frac{\|\nabla y\|_p^p}{\lambda_1 \|u_1\|_p^p} = 1,$$

a contradiction. Hence $y \equiv 0$. But then

$$\nabla y_n \longrightarrow 0 \quad \text{in } L^p(\Omega; \mathbb{R}^N)$$

and so

$$y_n \longrightarrow 0 \quad \text{in } W_0^{1,p}(\Omega),$$

a contradiction to the fact that $\|y_n\|_{W^{1,p}(\Omega)} = 1$ for $n \geq 1$. This proves that the sequence $\{x_n\}_{n \geq 1} \subseteq W_0^{1,p}(\Omega)$ is bounded. Then as in previous cases, exploiting the maximal monotonicity of A, we conclude that the claim is true.

Now for $t > 0$, we have

$$\varphi(-tu_1) = \frac{t^p}{p} \|\nabla u_1\|_p^p - \int_{\Omega} F(z, -tu_1(z)) \, dz,$$

so

$$\frac{\varphi(-tu_1)}{t^p} = \frac{1}{p}\|\nabla u_1\|_p^p - \frac{1}{p}\int_\Omega \frac{pF(z, -tu_1(z))}{t^p u_1(z)} u_1(z)\, dz.$$

Using Fatou's lemma and the last inequality in hypothesis $H(f)(iv)$ (the asymptotic behaviour at $-\infty$ of $F(z, \cdot)$), we obtain

$$\limsup_{n\to+\infty}\frac{\varphi(-tu_1)}{t^p} \le \frac{1}{p}\|\nabla u_1\|_p^p - \frac{1}{p}\int_\Omega \vartheta_2(z)u_1(z)^p\, dz$$

$$< \frac{1}{p}\|\nabla u_1\|_p^p - \frac{\lambda_1}{p}\|u_1\|_p^p = 0,$$

so

$$\varphi(tu_1) \longrightarrow -\infty \quad \text{as } t\to+\infty.$$

Therefore we can find $t_0 > 0$ large enough, such that

$$\varphi(-tu_1) \le \varphi(x_0) \qquad \forall\, t \ge t_0. \tag{4.255}$$

Recall that x_0 is a local $W_0^{1,p}(\Omega)$-minimizer of φ. We may assume that it is a strict local minimizer (or otherwise we are done). So we can find $\delta > 0$ such that

$$\varphi(x_0) < \varphi(v) \qquad \forall\, v \in \overline{B}_\delta(x_0)\setminus\{x_0\}. \tag{4.256}$$

Let $t \ge t_0 \ge \delta$ in (4.255). Then (4.255), (4.256) and the claim permit the use of the nonsmooth Mountain Pass Theorem (see Theorem 2.1.3), which gives us $u_0 \in W_0^{1,p}(\Omega)$ such that

$$0 \in \partial\varphi(u_0) \quad \text{and} \quad \varphi(u_0) = c \overset{df}{=} \inf_{\gamma\in\Gamma}\max_{t\in[0,1]}\varphi\big(\gamma(t)\big) \ge \varphi(x_0),$$

where

$$\Gamma \overset{df}{=} \big\{\gamma \in C\big([0,1]; W_0^{1,p}(\Omega)\big) : \gamma(0) = x_0,\ \gamma(1) = -tu_1\big\}.$$

If $\varphi(x_0) < c$, then $u_0 \ne x_0$ and we have

$$A(u_0) = v_0,$$

where $v_0 \in L^{p'}(\Omega)$, $f\big(z, u_0(z)^-\big) \le v_0(z) \le f\big(z, u_0(z)^+\big)$ for almost all $z \in \Omega$. From Remark 1.1.10, we have that

$$\nabla u_0(z) = 0 \quad \text{for a.a. } z \in D.$$

But recall that

$$0 \notin \left[f(z, r_n^-), f(z, r_n^+)\right] \quad \text{for a.a. } z \in \Omega \text{ and all } n \ge 1.$$

Therefore $|D|_N = 0$ and we conclude that u_0 is a second S-solution of problem (4.237).

On the other hand, if $\varphi(x_0) = c$, then from nonsmooth Mountain Pass Theorem (see Theorem 2.1.3), we know that we can find $u_0 \in \partial B_\delta(x_0)$ (hence $u_0 \neq x_0$) such that $0 \in \partial\varphi(u_0)$, $\varphi(u_0) = c$. Again we check that u_0 is a second S-solution of problem (4.237).

Finally from the nonlinear regularity theory (see Theorems 1.5.5 and 1.5.6, we conclude that $u_0 \in C_0^{1,\alpha}(\overline{\Omega})$ (with $\alpha \in (0,1)$). $\square$

4.9   Remarks

4.1 Landesman & Lazer (1969/1970) were the first to consider "resonant problems." They produced sufficient conditions for the existence of solutions for semilinear problems (i.e. $p = 2$) with a continuous right hand side nonlinearity, which reaches nonzero finite asymptotic limits at $\pm\infty$ (case (a) in the classification described in the beginning of the section). Extensions of their work can be found in Brézis & Nirenberg (1991) and the references therein. Ahmad, Lazer & Paul (1976) were the first to consider problems in which the nonlinearity vanishes asymptotically at $\pm\infty$, while the potential function asymptotically goes to $\pm\infty$ as $|\zeta| \to +\infty$ (case (b) in the classification). The first multiplicity results for these two cases can be found in Ambrosetti & Mancini (1978a, 1978b) and Hess (1978). Strongly resonant problems were first investigated by Thews (1980), Bartolo, Benci & Fortunato (1983) (who coined the term strong resonance) and Ward (1984). The Landesman-Lazer condition ((LL)-condition for short) using the quantities g_-^∞ and G_+^∞ essentially goes back to the work of Landesman-Lazer. The present form, suitable for hemivariational inequalities, was used by Goeleven, Motreanu & Panagiotopoulos (1998). The (LL)-condition using G_1^- and G_2^+ is an extension to the present multivalued setting, of a condition introduced in the context of ordinary differential equations by Tang (1998c) (see also Wu & Tang (2001) where this condition is used in the context of semilinear elliptic problems with smooth potential). Still another (LL)-condition can be found in Landesman, Robinson & Rumbos (1995). For nonlinear problems driven by the p-Laplacian differential operator, (LL)-type conditions were used by Arcoya & Orsina (1997) (the original (LL)-condition) and by Bouchala & Drábek (2000) (the condition of Tang mentioned earlier). When dealing with problems involving the p-Laplacian, the lack of full knowledge of the spectrum of $\left(-\Delta_p, W_0^{1,p}(\Omega)\right)$ and the loss of linearity and of the variational expressions for the higher eigenvalues, forces the use of asymptotic conditions at $\pm\infty$, which stay below the first eigenvalue $\lambda_1 > 0$. We refer for example to the works of Anane & Gossez (1990), Costa & Magalhães (1995) and El Hachimi & Gossez

(1994). In Costa & Magalhães (1995), the authors permit interaction with $\lambda_1 > 0$, but do not allow that the quotient $\frac{pj(z,\zeta)}{|\zeta|^p}$ goes beyond $\lambda_1 > 0$. When this happens we encounter serious difficulties in producing linking sets, which are necessary in order to exploit the existing minimax principles.

Various forms of obstacle problems and of general variational inequalities with smooth potential were studied in the last decade, using a variety of methods. We mention the work of Quittner (1989), Le (1997) based on bifurcation methods, the work of Szulkin (1986), Le & Schmitt (1995) who follow a variational approach, the work of Szulkin (1985) who uses degree theory, the work of Ang, Schmitt & Le (1990) based on recession arguments and the recent works of Le (2000, 2001) where the method of upper and lower solution is employed.

In addition to the "strong resonant" works mentioned earlier, more recently results concerning such problems were obtained by Costa & Silva (1993) and Gonçalves & Miyagaki (1992, 1995). However these works, as well as the earlier ones, deal with semilinear problems with smooth potential.

On the subject of hemivariational inequalities, we have the works of Bocea, Motreanu & Panagiotopoulos (2000), Bocea, Panagiotopoulos & Rădulescu (1999), Cîrstea & Rădulescu (2000), Gasiński & Papageorgiou (2001c, 2001d, 2002b), Gazzola & Rădulescu (2000), Marano & Motreanu (2002b), Motreanu (1995, 2001), Motreanu & Panagiotopoulos (1995a, 1995b, 1995c, 1996, 1997), Naniewicz (1994, 1997), Rădulescu & Panagiotopoulos (1998), Rădulescu (1993) (semilinear scalar or vector problems) and Gasiński & Papageorgiou (1999, 2000, 2001b, 2001e, 2002b), Kourogenis, Papadrianos & Papageorgiou (2002), Kyritsi & Papageorgiou (2001, 2004, to appeara, to appearb), Marano & Motreanu (2002a), Motreanu & Rădulescu (2000), Panagiotopoulos, Fundo & Rădulescu (1999), Papageorgiou & Smyrlis (2003) (nonlinear problems driven by the p-Laplacian differential operator or other nonlinear operators). We also mention the books of Naniewicz & Panagiotopoulos (1995) and Panagiotopoulos (1988, 1993), where the interested reader can find concrete applications in mechanics, engineering and economic problems.

4.2 The spectrum of $\left(-\Delta_p, W^{1,p}(\Omega) \right)$ (Neumann problem) was investigated by Huang (1990) and Godoy, Gossez & Paczka (2002). The extension of Picone's identity in Proposition 4.2.6 was proved by Allegretto & Huang (1998) and was then used by the same authors (see Allegretto & Huang (1999)) to obtain comparison principles and positive solutions, for elliptic problems driven by the p-Laplacian. In contrast to the Dirichlet problem, the Neumann problem with the p-Laplacian has not been studied much. Only recently there have been the works of Binding, Drábek & Huang (1997), Faraci (2003), Hu & Papageorgiou (2001) (problems with a continuous right hand side nonlinearity), Hu, Matzakos & Papageorgiou (2001) and Papalini (2002, 2003) (problems with a discontinuous nonlinearity). General hemivariational inequalities, with Neumann boundary condition, were studied by Marano & Motreanu (2002a) (homogeneous boundary condition) and Papageorgiou & Smyrlis (2003) (nonhomogeneous, multivalued boundary condition). The work of Papageorgiou & Smyrlis (2003) extends earlier ones by Szulkin (1986) and Halidias & Papageorgiou (2000b), where the potential function is smooth. Our presentation here of the nonlinear, nonhomogeneous Neumann problem is based on the aforementioned work of Papageorgiou & Smyrlis (2003).

4.3 A detailed study of functions of boundary variation (i.e. of the space $BV(\Omega)$) can be found in the books of Giusti (1984) and Ziemer (1989). The equivariant reformulation of the nonsmooth critical point theory for continuous functionals can be found in Marzocchi (1995). Analogous equivariant extensions of the smooth theory can be found in the books of Ghoussoub (1993a), Mawhin & Willem (1989), Struwe (1990) and Willem (1996). Our analysis of the problems with an area-type term is based on the paper of Marzocchi (1995). Related results can be found in Degiovanni, Marzocchi & Rădulescu (2000) and Marino (1989).

4.4 Problems driven by a nonlinear differential operator exhibiting a different dependence on its higher order part and its lower order terms were already considered in the sixties and led to the weakening of the concept of monotonicity and the introduction of new broader classes of nonlinear operators, such as pseudomonotone operators. The first works in this direction were produced by Browder (1965, 1977), Ton (1971) and Hess (1973a, 1973b). In Brézis & Browder (1978) and Webb (1980), there are no growth restrictions on the nonlinearities and instead a sign condition is used. Related are also the works of Hess (1976) and Rudd & Schmitt (2002). We mention also the work of Landes (1980), who employed the method of Galerkin approximations. More recently, such fully nonlinear problems were studied in the presence of unilateral constraints (such as problems (4.92), (4.105) and (4.110) in Section 4.4). We refer to the works of Gasiński & Papageorgiou (to appear), Papageorgiou & Smyrlis (2003), Naniewicz (1994, 1995), Panagiotopoulos (1991), Simon (1988, 1990) and Liu (1999). Our analysis here follows the methods developed by Gasiński & Papageorgiou (to appear) and by Papageorgiou & Smyrlis (2003).

4.5 The first ones to employ the method of upper and lower solutions to nonlinear elliptic problems were Schoenenberger-Deuel & Hess (1974/75). Since then there have been several other works in this direction. We mention those by Carl (1992, 1997, 2001), Carl & Dietrich (1995), Carl & Heikkilä (1992, 1998), Dancer & Sweers (1989), Delgado & Suárez (2000), Kourogenis & Papageorgiou (2003), Papageorgiou & Smyrlis (2003) and the references therein. Our approach here is based on the last two of the above mentioned works.

4.6 The reduction method was first developed for smooth problems by Castro & Lazer (1979) and Thews (1980). Double resonance problems were first investigated by Berestycki & de Figueiredo (1981), who also suggested the term "double resonance." In their analysis, they use the interval $[\lambda_1, \lambda_2]$

and this is crucial in their approach, because it depends on the fact that the principal eigenfunction u_1 is strictly positive and $\frac{\partial u_1}{\partial n} < 0$ for $z \in \partial\Omega$. As we know, this is no longer true in higher parts of the spectrum. The principal eigenfunction is the only one with constant sign. So for higher parts of the spectrum of $\left(-\Delta, H_0^1(\Omega)\right)$, the analysis of double resonance is more delicate. For smooth problems this was done in the papers of Các (1988), Hirano & Nishimura (1993), Robinson (1993), Costa & Silva (1993), Landesman, Robinson & Rumbos (1995), Iannacci & Nkashama (1995), Tang & Wu (2001 *a*), Su & Tang (2001), Zou & Liu (2001) and Su (2002). In Các (1988), the derivative of the smooth potential is unbounded with sublinear growth and he employs certain asymptotic conditions along the eigenspaces. Hirano & Nishimura (1993) assume that the potential function is C^2 and subquadratic and their method of proof uses the reduction technique as well as some multiplicity results, which the authors prove using minimax methods. In Robinson (1993), the derivative of the potential is unbounded and his method of proof uses degree theory. However, he obtains only one solution. Costa & Silva (1993) have a potential with sublinear derivative and produce only one solution via degree theoretic arguments. Landesman, Robinson & Rumbos (1995) have $k = 1$ (resonance at the first eigenvalue) and using degree theory in conjunction with a convenient (LL)-condition, they prove the existence of at least two nontrivial solutions. In Iannacci & Nkashama (1995), the derivative of the potential is bounded and they make use of a certain kind of (LL)-type condition. Tang & Wu (2001 *a*) use the reduction method and minimax techniques to improve the results of Các (1988) mentioned earlier. Finally Su & Tang (2001), Zou & Liu (2001) and Su (2002) have a C^2-potential and use Morse theory and critical groups in conjunction with an (LL)-condition to deal with the problem.

In our study of problem (4.161), we allow for complete resonance at the origin, which is not the case in the above mentioned works.

There are very few multiplicity results for equations involving the p-Laplacian. We mention the papers of Ambrosetti, Garcia Azorero & Peral (1996), Chen & Li (2002), Guo (1996) and Wei & Wu (1992). In all these works the potential is smooth and has a specific expression. Proposition 4.6.10 was first proved for semilinear (i.e. $p = 2$) problems with C^1-potential, by Brézis & Nirenberg (1991). The extension presented here is twofold. First we do not require that $p = 2$ (so we move beyond the semilinear setting) and second the potential function need not be smooth (nonsmooth case). This extension is due to Kyritsi & Papageorgiou (2004). Recently the result of Brézis & Nirenberg (1991) was extended to functionals defined on smooth manifolds by Tehrani (1996). The result of Tehrani was extended in the nonlinear case (i.e. $p \neq 2$) by Kyritsi & Papageorgiou (preprint). Other multiplicity results for hemivariational inequalities can be found in Bocea, Motreanu & Panagiotopoulos (2000), Cîrstea & Rădulescu (2000), Degiovanni, Marzocchi & Rădulescu (2000), Gasiński & Papageorgiou (2001 *c*), Goeleven, Motreanu & Panagiotopoulos (1997, 1998), Marano & Motreanu (2002 *b*), Motreanu (2001),

Motreanu & Panagiotopoulos (1996) (semilinear problems) and Gasiński & Papageorgiou (2000, 2001a, 2001c, 2001e, 2003d), Kyritsi & Papageorgiou (2004, to appear a) and Marano & Motreanu (2002a) (for problems driven by p-Laplacian).

4.7 Problem (4.208) with $\lambda \to \lambda_1^+$ was first considered by Mawhin & Schmitt (1988, 1990), when $N = 1$ (ordinary differential equation) and $p = 2$ (semilinear equation), with periodic boundary condition. Using degree theoretic arguments and bifurcation theory, coupled with some (LL)-conditions, they proved that the problem has two solutions as $\lambda \to 0^+$ (recall that 0 is the first eigenvalue of the periodic problem; see Mawhin & Schmitt (1988)) and they proved the existence of three solutions as $\lambda \to 0^-$ (see Mawhin & Schmitt (1990)). In Mawhin & Schmitt (1990), the authors examine also the Dirichlet problem and prove a corresponding result for it. The work of Mawhin-Schmitt was extended to semilinear partial differential equations by Chiappinelli, Mawhin & Nugari (1992), who, for the Dirichlet problem, proved that as $\lambda \to \lambda_1^-$, the problem has two solutions. Soon thereafter Chiappinelli & de Figueiredo (1993) obtained a result for systems. In these works, the method of proof is based on bifurcation theory combined with degree theory. More recently, Ramos & Sanchez (1997) examined the semilinear smooth Dirichlet partial differential equation and employing variational methods, they proved a "three solutions theorem" for the cases $\lambda \to \lambda_1^-$ and $\lambda \to \lambda_1^+$. The work of Ramos-Sanchez was extended to nonlinear (driven by the p-Laplacian), nonsmooth Dirichlet problems as $\lambda \to \lambda_1^-$, by Gasiński & Papageorgiou (2000). Similar problems for $p = 2$ (semilinear equations) and with a smooth potential independent on $z \in \Omega$ were investigated by Struwe (1990, 1982), and Ambrosetti & Lupo (1984). Assuming a superlinear behaviour of infinity and at zero of the derivative of the potential, they proved that for $\lambda > \lambda_2$ (with λ_2, the second eigenvalue of $\left(-\Delta, H_0^1(\Omega) \right)$), the problem has three solutions. Their approach is variational and Ambrosetti-Lupo used Morse theory (Morse inequalities) and for this reason assumed that the potential is a C^2-function. We should also mention the work of Ambrosetti, Brézis & Cerami (1994), where the interplay of convex and concave nonlinearities is studied for semilinear equations. Finally we mention that several papers studied nonlinear eigenvalue problems of the form

$$\begin{cases} -\Delta x(z) = \lambda f\big(x(z)\big) & \text{for a.a. } z \in \Omega, \\ x|_{\partial\Omega} = 0, \ x \geq 0, \end{cases}$$

for $\lambda > 0$ under the assumption that $f: \mathbb{R} \longrightarrow \mathbb{R}$ is continuous, positive, monotone. For this reason such problems were named ***positone***. For a well-written survey of such problems we refer to Lions (1982). If the nonlinearity $f: \mathbb{R} \longrightarrow \mathbb{R}$ is continuous, monotone and $f(0) < 0$ (for example $f(\zeta) = \zeta^p - \varepsilon$, $\varepsilon > 0$, $p > 1$), then the eigenvalue problem is called ***semipositone*** and for a survey of the literature for such problems we refer to Castro, Maya & Shivaji (2003).

Our presentation here uses ideas from Kyritsi & Papageorgiou (2004).

4.8 The two different interpretations of problems with discontinuous non-linearities and the corresponding solution notions can be found in Stuart (1976/1977, 1978). In fact Stuart calls the S-solutions, solutions of type I and the M-solutions, solutions of type II. In Stuart (1976, 1976/1977) we find examples of ordinary and partial differential equations, illustrating that these two notions are in general distinct. There have been several works concerning problems with discontinuous nonlinearities. The solution methods they employ include upper-lower solutions, variational techniques (based on non-smooth critical point theory), degree theory, bifurcation techniques, the dual action principle of Clarke and fixed point methods. We mention the works of Ambrosetti & Badiale (1989), Ambrosetti & Turner (1988), Badiale (1995), Badiale & Tarantello (1997), Bertsch & Klaver (1991), Bouguima (1995), Carl (1992, 1997, 2001), Carl & Heikkilä (1992, 1998), Cerami (1983), Chang (1978, 1980, 1981), Halidias & Papageorgiou (1997/1998b, 2000b), Heikkilä (1990), Hu, Kourogenis & Papageorgiou (1999), Kourogenis & Papageorgiou (1998a, 2000a, 2001), Marano (1995), Massabó (1980), Massabó & Stuart (1978), Mizoguchi (1991), Stuart (1976/1977, 1978) and Stuart & Toland (1980). The comparison principle in Proposition 4.8.4 is due to Guedda & Véron (1989). Similar comparison principles can be found in Allegretto & Huang (1999), Damascelli (1998) and García-Melián & Sabina de Lis (1998).

Chapter A

Appendix

A.1   Set Theory and Topology

DEFINITION A.1.1

(a) *If X is a set, then a **partial order** on X is a reflexive, transitive and antisymmetric binary relation on X. So we denote the partial order by $\leq$ (as it is customary). For all $x, y, z \in X$, we have $x \leq x$ (reflexivity), $x \leq y$ and $y \leq z$ imply $x \leq z$ (transitivity) and $x \leq y$, $y \leq x$ imply $x = y$ (antisymmetry).*

(b) *A **total order** (or **linear order**) is a partial order $\leq$ with the property that if $x \neq y$, then $x \leq y$ or $y \leq x$.*

(c) *A **chain** in a partially ordered set X is a subset on which the order is total.*

(d) *Let X be a partially ordered set. An **upper bound for a set** $A \subseteq X$ is an element $x \in X$, such that $a \leq x$ for all $a \in A$. An element $x \in A$ is a **maximal element** of A, if there is no $y \neq x$ in A for which $x \leq y$. A **greatest element** of A is an element $x \in A$ satisfying $y \leq x$ for all $y \in A$. Evidently every greatest element is maximal.*

THEOREM A.1.1 (Kuratowski-Zorn Lemma)
If in a partially ordered set X, every chain has an upper bound, then X has a maximal element.

DEFINITION A.1.2   *Let X be a Hausdorff topological space.*

(a) *A **neighbourhood** of a point $x \in X$ is any open set U, such that $x \in U$. The collection of all neighbourhoods of $x \in X$ is called the **filter of neighbourhoods** of x and is denoted by $\mathcal{N}(x)$.*

(b) *We say that X is **regular**, if for a given nonempty closed set $C \subseteq X$ and $x \notin C$, we can find open sets U, V, such that $C \subseteq U$, $x \in V$ and $U \cap V = \emptyset$.*

(c) *We say that X is **normal**, if for two given disjoint nonempty closed sets C and D, we can find open sets U, V, such that $C \subseteq U$, $D \subseteq V$ and $U \cap V = \emptyset$.*

THEOREM A.1.2 (Tietze Extension Theorem)

A Hausdorff topological space is normal if and only if every continuous function defined on a closed subset of X with values in $[a, b] \subseteq \mathbb{R}$ has a continuous extension on all of X with values in the same interval $[a, b]$.

DEFINITION A.1.3   *A topological space X is an **absolute retract** whenever*

(a) *X is metrizable (see Definition A.1.11) and*

(b) *for any metrizable space V and nonempty closed set $A \subseteq V$, every continuous function $\varphi \colon A \longrightarrow X$ admits a continuous extension on all of V.*

THEOREM A.1.3 (Dugundji Extension Theorem)

If V is a metrizable space, $A \subseteq V$ is a nonempty and closed set, X is a locally convex space and $\varphi \colon A \longrightarrow X$ is a continuous function,
then there exists a continuous function $\widehat{\varphi} \colon V \longrightarrow X$, such that $\widehat{\varphi}|_A = \varphi$ and $\widehat{\varphi}(V) \subseteq \operatorname{conv} \varphi(A)$.

REMARK A.1.1   The theorem is valid if we replace X by any convex set $C \subseteq X$. By virtue of this, every nonempty convex and metrizable subset of a locally convex space is an absolute retract.   ⬜

DEFINITION A.1.4   *Let (X, τ) be a Hausdorff topological space.*

(a) *A **base** for the topology τ is a subfamily $\mathcal{B}$ of τ, such that for every $x \in X$ and every open set U containing x, we can find $V \in \mathcal{B}$, such that $x \in V \subseteq U$. Evidently, each $U \in \tau$ is a union of elements in $\mathcal{B}$.*

(b) *A **local base** at $x \in X$ is a subfamily $\mathcal{B}(x)$ of $\mathcal{N}(x)$, such that for every $U \in \mathcal{N}(x)$, we can find $V \in \mathcal{B}(x)$ with $V \subseteq U$.*

(c) *We say that X is **first countable**, if every point $x \in X$ has a countable local base.*

(d) *We say that X is **second countable**, if it has a countable base.*

(e) *We say that X is **separable**, if there exists a countable set $D \subseteq X$, such that $\overline{D} = X$.*

REMARK A.1.2 Evidently a second countable space is first countable, but the converse is not in general true. Every second countable space is separable, but the converse is not in general true. However, for metric spaces second countability and separability are equivalent notions. □

DEFINITION A.1.5

(a) *A **direction** $\prec$ on a set is a reflexive and transitive binary relation with the property that for any $\alpha, \beta \in D$, we can find $\gamma \in D$, such that $\alpha \prec \gamma$ and $\beta \prec \gamma$ (i.e. each pair of elements in D has an upper bound). A **directed set** is any set D equipped with a direction $\prec$.*

(b) *A **net** in a set X is a function $x \colon D \longrightarrow X$, where D is a directed set. The directed set D is called the index set of the net. It is customary to denote the function x by $\{x_\alpha\}_{\alpha \in D}$.*

(c) *A net $\{x_\alpha\}_{\alpha \in D}$ in a Hausdorff topological space X converges to $x \in X$, if for a given $U \in \mathcal{N}(x)$, we can find $\alpha_0 \in D$, such that for all $\alpha \in D$, $\alpha_0 \prec \alpha$, we have that $x_\alpha \in U$. The limit $x \in X$ is unique (Hausdorff property).*

REMARK A.1.3 Sequences are a particular case of nets. Nets are used to describe topological notions and properties, when sequences do not suffice (i.e. the space is not first countable). For example if X is a Hausdorff topological space and $A \subseteq X$ is nonempty, $x \in \overline{A}$ if and only if there exists a net $\{x_\alpha\}_{\alpha \in D} \subseteq A$, such that $x_\alpha \longrightarrow x$. The notion of subnet, which is defined next, generalizes the notion of a subsequence. □

DEFINITION A.1.6 *A net $\{y_\beta\}_{\beta \in B}$ is a **subnet** of $\{x_\alpha\}_{\alpha \in D}$, if there exists a function $\varphi \colon B \longrightarrow D$, such that:*

(a) *$y_\beta = x_{\varphi(\beta)}$ for every $\beta \in B$;*

(b) *for each $\alpha_0 \in D$, we can find $\beta_0 \in B$, such that $\beta_0 \prec \beta$ in B implies that $\alpha_0 \prec \varphi(\beta)$ in D.*

DEFINITION A.1.7 *Let X be a Hausdorff topological space and $K \subseteq X$. We say that K is a **compact set**, if every open cover of K has a finite subcover, i.e. if every family $\{U_i\}_{i \in I}$ of open sets satisfying $K \subseteq \bigcup_{i \in I} U_i$ has a finite subfamily $\{U_{i_1}, \ldots, U_{i_N}\}$, such that $K \subseteq \bigcup_{k=1}^{N} U_{i_k}$. We say that K is a **relatively compact set**, if $\overline{K}$ is compact.*

REMARK A.1.4   In $\mathbb{R}^N$ compact sets have a very convenient description, namely $C \subseteq \mathbb{R}^N$ is compact if and only if C is closed and bounded. The next theorem gives equivalent characterization of compact sets. $\square$

THEOREM A.1.4

For a Hausdorff topological space X, the following statements are equivalent:

(1) *X is compact;*

(2) *Every family of closed subsets of X with the finite intersection property (i.e. every finite subfamily has a nonempty intersection) has a nonempty intersection;*

(3) *Every net has a convergent subnet.*

REMARK A.1.5   In metric spaces compactness, countable compactness and sequentially compactness are all equivalent. We say that a Hausdorff topological space is **countably compact**, if every countable open cover has a finite subcover. Also we say that it is **sequentially compact**, if every sequence has a convergent subsequence. Compact subsets of a Hausdorff topological space are closed. Every continuous function between Hausdorff topological spaces carries compact sets to compact sets. Moreover, a bijective continuous function from a compact space onto a Hausdorff topological space is a homeomorphism. $\square$

DEFINITION A.1.8   *Let X be a Hausdorff topological space and let $\varphi \colon X \longrightarrow \mathbb{R}^* \overset{df}{=} \mathbb{R} \cup \{\pm\infty\}$ be a function.*

(a) *We say that φ is a **lower semicontinuous function**, if for all $\lambda \in \mathbb{R}$, the set $\{x \in X : \varphi(x) \leq \lambda\}$ is closed.*

(b) *We say that φ is a **upper semicontinuous function**, if $-\varphi$ is lower semicontinuous.*

PROPOSITION A.1.1

If X is a Hausdorff topological space and $\varphi \colon X \longrightarrow \mathbb{R}^*$ is a function, then the following statements are equivalent:

(1) *φ is lower semicontinuous;*

(2) *$\operatorname{epi}\varphi \overset{df}{=} \{(x,\lambda) \in X \times \mathbb{R} : \varphi(x) \leq \lambda\}$ is closed in $X \times \mathbb{R}$;*

(3) *$x_\alpha \longrightarrow x$ in X implies that $\varphi(x) \leq \liminf \varphi(x_\alpha)$.*

REMARK A.1.6 A similar result holds for upper semicontinuous functions, with the obvious modifications. In statement (3), the nets can be replaced by sequences if X is a first countable space. ▯

THEOREM A.1.5 (Weierstrass Theorem)
If X *is a compact Hausdorff topological space and* $\varphi \colon X \longrightarrow \overline{\mathbb{R}} \overset{df}{=} \mathbb{R} \cup \{+\infty\}$ *is a lower semicontinuous function,*
then the set $M(\varphi) \overset{df}{=} \left\{ x \in X : \varphi(x) = \inf\limits_X \varphi \right\}$ *is nonempty and compact.*

REMARK A.1.7 An analogous result holds for the maximization of upper semicontinuous functions on compact spaces. ▯

DEFINITION A.1.9 *Let* X *be a Hausdorff topological space.*

(a) *We say that* X *is a **locally compact space**, if every point* $x \in X$ *has a relatively compact neighbourhood.*

(b) *We say that* X *is a* σ***-compact space**, if* $X = \bigcup\limits_{n \geq 1} K_n$ *with* K_n *compact.*

REMARK A.1.8 Every finite dimensional vector space is locally compact and σ-compact. In a locally compact space every point has a local base consisting of relatively compact sets. Also if X is locally compact and σ-compact, then $X = \bigcup\limits_{n=1}^{\infty} \overline{K_n}$, with K_n compact and $\overline{K_n} \subseteq \operatorname{int} K_{n+1}$ for all $n \geq 1$ (hence $X = \bigcup\limits_{n=1}^{\infty} K_n$). ▯

THEOREM A.1.6
If (X, τ) *is a noncompact, locally compact space,*

$$X_\infty \overset{df}{=} X \cup \{\infty\}, \quad \text{with } \infty \notin X,$$
$$\tau_\infty \overset{df}{=} \tau \cup \{X_\infty \setminus K : K \subseteq X \text{ is compact}\},$$

then τ_∞ *is a Hausdorff topology on* X_∞, (X_∞, τ_∞) *is a compact space and* X *is an open dense set in* X_∞.

REMARK A.1.9 The space (X_∞, τ_∞) is called the **Alexandrov one-point compactification of** X. As an example, the one-point compactification $\mathbb{R}_\infty$ of $\mathbb{R}$ is homeomorphic to a circle. ▯

DEFINITION A.1.10   *Let X be a set.*

(a) *Let $\mathcal{V} = \{V_i\}_{i \in I}$ and $\mathcal{U} = \{U_j\}_{j \in J}$ be two covers of X. We say that $\mathcal{U}$ is a **refinement** of $\mathcal{V}$, if for each $j \in J$, we can find $i \in I$ with $U_j \subseteq V_i$. A collection of subsets $\{W_\alpha\}_{\alpha \in A}$ of a Hausdorff topological space X is **locally finite**, if every $x \in X$ has a neighbourhood U, which meets at most finitely many W_α.*

(b) *A Hausdorff topological space X is said to be **paracompact**, if every open cover of X has a locally finite refinement.*

(c) *A **partition of unity** on a set X is a family $\{\varphi_i\}_{i \in I}$ of functions $\varphi_i \colon X \longrightarrow [0,1]$, such that at each $x \in X$, only finitely many functions are nonzero and $\sum_{i \in I} \varphi_i(x) = 1$ (we use the convention that the sum of an arbitrary collection of zeros is zero). A partition of unity is **subordinate** to a cover $\mathcal{Y}$ of X, if each function vanishes outside some element of $\mathcal{Y}$. If X is a Hausdorff topological space, a partition of unity $\{\varphi_i\}_{i \in I}$ is said to be **continuous**, if each function $\varphi_i \colon X \longrightarrow [0,1]$ for $i \in I$ is continuous. Finally, we say that a partition of unity $\{\varphi_i\}_{i \in I}$ is **locally finite**, if every point $x \in X$ has a neighbourhood on which all but finitely many of the functions φ_i vanish.*

REMARK A.1.10   Compact and metrizable topological spaces are paracompact. A paracompact space is normal. ▯

THEOREM A.1.7
A Hausdorff topological space X is paracompact if and only if every open cover of X has a continuous locally finite partition of unity subordinated to it.

DEFINITION A.1.11   *A topological space (X, τ) is **metrizable**, if the topology τ is generated by some metric. It is **separable** if it has a countable dense subset.*

REMARK A.1.11   The metric generating the topology of a metrizable space is not unique. If we fix such a metric d_X, then (X, d_X) is referred to as a metric space. For metrizable spaces separability and second countability are equivalent notions. Separable metrizable spaces cannot be too large. They have at most the cardinality of the continuum. In general Hausdorff topological spaces, separability may not be inherited by its subspaces. However, for metrizable spaces, separability is an hereditary property. ▯

DEFINITION A.1.12

(a) *A Hausdorff topological space X is* **completely metrizable** *(or* **topologically complete***), if there exists a metric d_X generating the topology of X and (X, d_X) is complete.*

(b) *A* **Polish space** *is a Hausdorff topological space which is separable and completely metrizable.*

THEOREM A.1.8
A metric space (X, d_X) is completely metrizable if and only if it is a G_δ in its completion for d_X.

THEOREM A.1.9 (Baire Category Theorem)
If X is a completely metrizable space and $X = \bigcup\limits_{n=1}^{\infty} C_n$ with C_n closed for all $n \geq 1$,
then there exists at least one $n_0 \geq 1$, such that $\operatorname{int} C_{n_0} \neq \emptyset$.

DEFINITION A.1.13   *Let (X, d_X) be a metric space and let $\varphi \colon X \longrightarrow X$ be a function. We say that φ is a d_X-contraction, if*

$$d_X\big(\varphi(x), \varphi(y)\big) \;\leq\; k d_X(x, y) \qquad \forall\, x, y \in X,$$

with $k \in [0, 1)$.

DEFINITION A.1.14   *Let (X, d_X) be a metric space and let $\varphi \colon X \longrightarrow X$ be a function. We say that φ is d_X-nonexpansive if*

$$d_X\big(\varphi(x), \varphi(y)\big) \;\leq\; d_X(x, y) \qquad \forall\, x, y \in X.$$

THEOREM A.1.10 (Banach Fixed Point Theorem)
If (X, d_X) is a complete metric space and $\varphi \colon X \longrightarrow X$ is a d_X-contraction, then φ has a unique fixed point $\widehat{x} \in X$ (i.e. $\varphi(\widehat{x}) = \widehat{x}$) and for any $x_0 \in X$, the sequence $\{x_n\}_{n \geq 1}$ with $x_n \stackrel{df}{=} \varphi(x_{n-1})$ for $n \geq 1$ converges to $\widehat{x}$.

THEOREM A.1.11 (Cantor Intersection Theorem)
A metric space (X, d_X) is complete if and only if for every decreasing sequence $\{C_n\}_{n \geq 1}$ of closed sets with $\operatorname{diam} C_n \longrightarrow 0$, we have that $\bigcap\limits_{n=1}^{\infty} C_n$ is a singleton.

THEOREM A.1.12 (Urysohn Lemma)
If (X, d_X) *is a metric space and* $A, C \subseteq X$ *are two nonempty disjoint closed sets,*
then there exists a locally Lipschitz function $\varphi \colon X \longrightarrow [0, 1]$, *such that* $A = \varphi^{-1}(0)$ *and* $C = \varphi^{-1}(1)$.

REMARK A.1.12   This functional separation property is in fact equivalent to saying that every closed set is a G_δ-set.   ⬚

DEFINITION A.1.15   *Let* (X, d_X), (Y, d_Y) *be two metric spaces and* $\mathcal{F}$ *is a family of continuous functions* $f \colon X \longrightarrow Y$. *We say that family* $\mathcal{F}$ *is* **equicontinuous**, *if for every* $x \in X$ *and* $\varepsilon > 0$, *we can find* $\delta = \delta(\varepsilon, x)$, *such that for all* $y \in X$ *with* $d_X(x, y) < \delta$, *we have that* $d_Y\big(f(x), f(y)\big) < \varepsilon$ *for all* $f \in \mathcal{F}$.

If for every $\varepsilon > 0$, *we can find* $\delta = \delta(\varepsilon) > 0$, *such that when* $d_X(x, y) < \delta$, *we have that* $d_Y\big(f(x), f(y)\big) < \varepsilon$ *for all* $f \in \mathcal{F}$, *then we say that* $\mathcal{F}$ *is* **uniformly equicontinuous**.

PROPOSITION A.1.2
If (X, d_X) *is a compact metric space and* (Y, d_Y) *is a metric space,*
then any equicontinuous family of functions from X *into* Y *is uniformly equicontinuous.*

THEOREM A.1.13 (Arzela-Ascoli Theorem)
If (X, d_X) *is a compact metric space and* $K \subseteq C(X)$,
then K *is relatively compact for the* $d_{\sup}$-*metric on* $C(K)$ *(recall that* $d_{\sup}(f, g) = \max_{x \in X} |f(x) - g(x)|$ *for all* $f, g \in C(X)$*) if and only if it is uniformly bounded and equicontinuous (thus uniformly equicontinuous).*

A.2   Measure Theory

DEFINITION A.2.1   *Let* (Ω, Σ, μ) *be a finite measure space and let* C *be a family of integrable functions on* Ω *(i.e.* $C \subseteq L^1(\Omega)$*). We say that* C *is a* **uniformly integrable set**, *if*

$$\sup_{f \in C} \int_{\{|f| \geq c\}} |f(\omega)| \, d\mu \longrightarrow 0 \qquad \text{as } c \to +\infty.$$

REMARK A.2.1   If there exists an integrable function $h \colon \Omega \longrightarrow \mathbb{R}_+^* \overset{df}{=} \mathbb{R}_+ \cup \{+\infty\}$, such that $\big|f(\omega)\big| \leq h(\omega)$ μ-almost everywhere for $f \in C$, then C is uniformly integrable.   ⬚

PROPOSITION A.2.1

If (Ω, Σ, μ) *is a finite measure space and* C *is a family of integrable functions on* Ω *(i.e.* $C \subseteq L^1(\Omega)$*),*
then C *is uniformly integrable if and only if the following conditions hold:*

(a) $\displaystyle \sup_{f \in C} \int_\Omega |f| \, d\mu < +\infty,$

(b) *for every* $\varepsilon > 0$*, there exists* $\delta > 0$*, such that, if* $A \in \Sigma$*, with* $\mu(A) \leq \delta$*,*
then

$$\sup_{f \in C} \int_\Omega |f| \, d\mu \ \leq \ \varepsilon.$$

REMARK A.2.2   We can show that condition (a) is a consequence of condition (b), if the measure μ is nonatomic. Recall that an **atom** of μ is a set $A \in \Sigma$, with $0 < \mu(A)$ and such that for every $C \subseteq A$ either $\mu(C) = 0$ or $\mu(C) = \mu(A)$. A measure without atoms is called **nonatomic**. The main example of atoms are singletons $\{\omega\}$ that have positive measure. The Lebesgue measure is nonatomic.   ⬚

PROPOSITION A.2.2 (Fatou's Lemma)

If (Ω, Σ, μ) *is a finite measure space and the sequence* $\{f_n\}_{n \geq 1} \subseteq L^1(\Omega)$ *is uniformly integrable, then*

$$\int_\Omega \left(\liminf_{n \to +\infty} f_n \right) d\mu \ \leq \ \liminf_{n \to +\infty} \int_\Omega f_n \, d\mu$$

$$\leq \ \limsup_{n \to +\infty} \int_\Omega f_n \, d\mu \ \leq \ \int_\Omega \left(\limsup_{n \to +\infty} f_n \right) d\mu.$$

THEOREM A.2.1 (Vitali's Theorem)

If (Ω, Σ, μ) *is a finite measure space, the sequence* $\{f_n\}_{n \geq 1} \subseteq L^1(\Omega)$ *is uniformly integrable and* $f_n \xrightarrow{\mu} f$*, then*

$$\int_\Omega f_n \, d\mu \ \longrightarrow \ \int_\Omega f \, d\mu.$$

REMARK A.2.3   The above convergence theorem is also known as the "Extended Dominated Convergence Theorem."   ⬚

THEOREM A.2.2 (Lusin Theorem)
If X is a Polish space, Y is a separable metric space, $f: X \longrightarrow Y$ is a Borel measurable function and μ is a finite Borel measure on X,
then for a given $\varepsilon > 0$, we can find a compact set $K_\varepsilon \subseteq X$, such that $\mu\left(K_\varepsilon^c\right) \leq \varepsilon$ and $f|_{K_\varepsilon}$ is continuous.

DEFINITION A.2.2   *Let X be a Banach space. A function $f: [a, b] \longrightarrow X$ is said to be **absolutely continuous**, if for every $\varepsilon > 0$, we can find $\delta > 0$, such that for any disjoint subintervals $[a_k, b_k)$ of $[a, b]$ for $k \geq 1$, with $a_k < b_k$ and $\sum_{k \geq 1} (b_k - a_k) < \delta$, we have*

$$\sum_{k \geq 1} |f(b_k) - f(a_k)| \; < \; \varepsilon.$$

REMARK A.2.4   An absolutely continuous function is of bounded variation. The converse is not true.   ⬚

THEOREM A.2.3 (Lebesgue Theorem)
If $f: [a, b] \longrightarrow \mathbb{R}^N$ is absolutely continuous,
then f' exists almost everywhere, $f \in L^1\left(a, b; \mathbb{R}^N\right)$ and

$$f(x) \; = \; f(a) + \int_a^x f'(s)ds \quad \forall x \in [a, b].$$

REMARK A.2.5   The above theorem remains true if $\mathbb{R}^N$ is replaced by a Banach space X provided that X has the Radon-Nikodym property. Reflexive spaces and separable dual Banach spaces have the Radon-Nikodym property.
⬚

THEOREM A.2.4 (Rademacher Theorem)
If $f: \mathbb{R}^N \longrightarrow \mathbb{R}^m$ is locally Lipschitz,
then f is almost everywhere differentiable (on $\mathbb{R}^N$ we use the Lebesgue measure).

REMARK A.2.6   There is a generalization of this theorem to functions $f: X \longrightarrow Y$, where X is a separable Banach space and Y is a Banach space with the Radon-Nikodym property, provided that we use the notion of Haar-null set. $\square$

THEOREM A.2.5 (Jensen Inequality)

If (Ω, Σ, μ) is a finite measure space, I is an open interval in $\mathbb{R}$, $\varphi\colon I \longrightarrow \mathbb{R}$ is a convex function, $f \in L^1(\Omega)$ with $f(\Omega) \subseteq I$ and $\varphi \circ f \in L^1(\Omega)$, then

$$\varphi\left(\frac{1}{\mu(\Omega)}\int_\Omega f\,d\mu\right) \leq \frac{1}{\mu(\Omega)}\int_\Omega (\varphi \circ f)\,d\mu.$$

THEOREM A.2.6 (Egorov Theorem)

If (Ω, Σ, μ) is a finite measure space, (X, d_X) is a metric space, $f_n, f\colon \Omega \longrightarrow X$ are Σ-measurable functions and $f_n(\omega) \longrightarrow f(\omega)$ for μ-almost all $\omega \in \Omega$, then for any $\varepsilon > 0$, we can find $A \in \Sigma$ with $\mu(A^c) < \varepsilon$, such that $f_n \longrightarrow f$ uniformly on A, i.e.

$$\lim_{n \to +\infty} \sup_{\omega \in A} d_X\big(f_n(\omega), f(\omega)\big) = 0.$$

DEFINITION A.2.3

*Let X be a metric space (for example X is a subset of $\mathbb{R}^N$). Let $0 \leq m < +\infty$ and $\delta > 0$. The m-**dimensional Hausdorff measure** of the nonempty set X is given by*

$$H^m(X) \stackrel{df}{=} \lim_{\delta \to 0} H_\delta^m(X),$$

where

$$H_\delta^m(X) \stackrel{df}{=} \inf \sum_k \left(\frac{1}{2^k}\operatorname{diam} B_k\right),$$

the infimum being taken over all at most countable coverings $\{B_k\}_{k \geq 1}$ of the set X by subsets B_k with $\operatorname{diam} B_k \leq \delta$ for all $k \geq 1$. If no such covering exists, then we set $H_\delta^m(X) = +\infty$.

REMARK A.2.7

The number $H_\delta^m(X)$ is called the outer δ-Hausdorff measure of X. The limit $\lim_{\delta \to 0} H_\delta^m(X)$ is well defined, since the sequence $\{H_\delta^m(X)\}_{\delta > 0}$ is increasing as $\delta \to 0^+$. Therefore

$$H^m(X) = \sup_{\delta > 0} H_\delta^m(X).$$

Note that

$$0 \leq H^m(X) \leq +\infty.$$

If $m = 0$, then we use the convention "$0^m = 1$." Thus, if X is a singleton, we have that

$$H^0(X) = 1 \quad \text{and} \quad H^m(X) = 0 \qquad \forall\, m > 0.$$

For all $m \geq 0$ and all $\delta > 0$, we set

$$H^m(\emptyset) \stackrel{df}{=} 0 \quad \text{and} \quad H^m_\delta(\emptyset) \stackrel{df}{=} 0.$$

Also, if $V_m = \Gamma\left(\frac{1}{2}\right)^m \Gamma\left(\frac{m}{2} + 1\right)$, $m \geq 0$ (V_m is the volume of the m-dimensional unit ball in $\mathbb{R}^m$), then $\sigma^m(X) = V_m H^m(X)$ is the normalized Hausdorff measure of X. If X is a sufficiently smooth m-dimensional surface in $\mathbb{R}^N$, $N > m$, then $\sigma^m(X)$ equals the classical surface measure. More precisely, this is the case if X is an m-dimensional C^1-submanifold of $\mathbb{R}^N$, $N > m$. ▯

A.3   Functional Analysis

DEFINITION A.3.1   *Let X be a vector space with a Hausdorff topology τ.*

(a) *We say that (X, τ) is a **topological vector space**, if the operations of vector addition and scalar multiplication are continuous.*

(b) *If the topological vector space (X, τ) has a base consisting of convex sets, then it is called a **locally convex space**.*

(c) *A **norm** on X is a function $\|\cdot\|_X : X \longrightarrow \mathbb{R}$, such that*

    (1) $\|x\|_X \geq 0$ *for all $x \in X$ and $\|x\|_X = 0$ if and only if $x = 0$;*

    (2) $\|\lambda x\|_X = |\lambda|\, \|x\|_X$ *for all $(\lambda, x) \in \mathbb{R} \times X$;*

    (3) $\|x + y\|_X \leq \|x\|_X + \|y\|_X$ *for all $x, y \in X$ (triangle inequality).*

    *The pair $(X, \|\cdot\|_X)$ is called a **normed space** or a **normed vector space**.*

(d) *Let X be a normed space. The **metric induced by the norm** of X is the metric d_X on X, defined by the formula*

$$d_X(x, y) \stackrel{df}{=} \|x - y\|_X \qquad \forall\, x, y \in X.$$

    *The **norm topology** of X is the topology obtained from this metric.*

(e) *A **Banach norm** is a norm that induces a complete metric. A normed space is a **Banach space**, if its norm is a Banach norm.*

REMARK A.3.1   A closed subspace of a Banach space is also a Banach space with the inherited norm.   ⊓

DEFINITION A.3.2   *Let X be a normed space. The (**topological**) **dual** X^* of X is the vector space of all continuous linear functionals on X. This is a Banach space with norm given by*

$$\|x^*\|_{X^*} \stackrel{df}{=} \sup_{x \in \overline{B}_1} |x^*(x)|.$$

REMARK A.3.2   Instead of $x^*(x)$, we write $\langle x^*, x \rangle_X$, because we may wish to think as x acting on x^* (if we consider x as an element of X^{**}). Then $\langle \cdot, \cdot \rangle_X$ are the **duality brackets** for the pair (X, X^*).   ⊓

PROPOSITION A.3.1
If X is a normed space and $x \in X$, <u>then</u>

$$\|x\|_X = \sup_{x^* \in \overline{B}_1^{X^*}} |\langle x^*, x \rangle_X|$$

and the supremum is attained.

PROPOSITION A.3.2
If X is a normed space and $i_0 \colon X \longrightarrow X^{**}$ is defined by

$$i_0(x)(x^*) \stackrel{df}{=} \langle x^*, x \rangle_X \qquad \forall\, x \in X,\ x^* \in X^*,$$

*<u>then</u> i_0 is an isometric isomorphism into X^{**} and $i_0(X)$ is closed subspace of X^{**} if and only if X is a Banach space. This map i_0 is called the **canonical embedding** of X into X^{**}.*

DEFINITION A.3.3   *A normed space X is **reflexive**, if $i_0(X) = X^{**}$, with i_0 being the canonical embedding of X into X^{**}.*

REMARK A.3.3   Every reflexive normed space is in fact a Banach space. All finite dimensional normed spaces are reflexive. The same can be said about a normed space isomorphic to a reflexive normed space.   ⊓

DEFINITION A.3.4   *Let X be a normed space.*

(a) *The smallest topology on X making all elements of X^* continuous is called the **weak topology** of X and it is denoted by $w(X, X^*)$ or simply* by w. Also X_w denotes the normed space X furnished with the weak topology.

(b) *The smallest topology on X^* making all elements of X continuous is called the **weak* topology** of X^* and it is denoted by $w(X^*, X)$ or simply by w^*. Also $X^*_{w^*}$ denotes the Banach space X^* furnished with the weak*-topology.*

REMARK A.3.4   For infinite dimensional normed spaces the weak topology is always coarser than the norm topology and it is never metrizable. If X is finite dimensional, then the weak topology and the norm topology coincide. Also in X^*, we always have $w^* \subseteq w$ and the two coincide if and only if X is reflexive. Note that

$$(X_w)^* = X^*, \quad (X^*_{w^*})^* = X \quad \text{and} \quad (X^*_w)^* = X^{**}.$$

The space X_w is a locally convex topological vector space which is regular (in fact completely regular). ▯

THEOREM A.3.1 (Alaoglu Theorem)
If X is a normed space and $\overline{B}_1^{X^*} \stackrel{df}{=} \{x^* \in X^* : \ \|x^*\|_{X^*} \leq 1\}$,
then $\overline{B}_1^{X^*}$ is w^*-compact.

REMARK A.3.5   So every bounded subset of X^* is relatively w^*-compact. Hence a bounded and w^*-closed subset of X^* is w^*-compact. ▯

THEOREM A.3.2 (Goldstine Theorem)
If X is a normed space and $i_0 \colon X \longrightarrow X^{**}$ is the canonical embedding,
then $i_0\left(\overline{B}_1^X\right)$ is w^*-dense in $\overline{B}_1^{X^{**}}$.

THEOREM A.3.3 (Weak Separation Theorem)
If X is a locally convex vector space, $A, C \subseteq X$ are two nonempty, disjoint convex sets and $\mathrm{int}\, A \neq \emptyset$,
then there exists $x^* \in X^* \setminus \{0\}$, such that

$$\langle x^*, a \rangle_X \ \leq \ \langle x^*, c \rangle_X \qquad \forall\, a \in A, \ c \in C.$$

THEOREM A.3.4 (Strong Separation Theorem)
If X is a locally convex vector space, $A, C \subseteq X$ are two nonempty, disjoint convex sets and A is compact,
then there exist $x^* \in X^* \setminus \{0\}$ and $\varepsilon > 0$, such that

$$\sup_{a \in A} \langle x^*, a \rangle_X \ \leq \ \inf_{c \in C} \langle x^*, c \rangle_X - \varepsilon.$$

REMARK A.3.6   Often in applications X is a Banach space endowed with the weak topology. ▯

THEOREM A.3.5 (Eberlein-Smulian Theorem)
If X is a normed space and $A \subseteq X$,
then the following statements are equivalent:

(a) *A is (relatively) weakly compact;*

(b) *A is (relatively) weakly countably compact;*

(c) *A is (relatively) weakly sequentially compact;*

REMARK A.3.7   So if A is a relatively weakly compact set in a normed space X and $x_0 \in \overline{A}^w$, then there is a sequence $\{x_n\}_{n \geq 1} \subseteq A$, such that $x_n \xrightarrow{w} x_0$. So weak compactness and relative weak compactness are sequentially determined, although the relative weak topology need not be metrizable. Next, we provide conditions under which metrizability of the weak topology occurs. ▯

THEOREM A.3.6
Let X be a Banach space.

(a) *If X is a separable Banach space and $A \subseteq X^*$ is bounded, then the relative weak* topology on A is metrizable.*

(b) *If X is a Banach space, X^* is separable and $A \subseteq X$ is bounded, then the relative weak topology on A is metrizable.*

(c) *If X is a separable Banach space and $A \subseteq X$ is weakly compact, then the relative weak topology on A is metrizable.*

REMARK A.3.8   Note that, if the dual of a Banach space is separable, then so is the space. However, separability of X does not necessarily imply separability of X^* (e.g. l^1 and $l^\infty = (l^1)^*$). But, if X is reflexive, then X is separable if and only if X^* is separable. ▯

PROPOSITION A.3.3
If $x_\alpha \xrightarrow{w} x$ in a normed space X and $x_\alpha^* \xrightarrow{w^*} x^*$ in X^*, then

$$\|x\|_X \leq \liminf \|x_\alpha\|_X \quad \text{and} \quad \|x^*\|_X \leq \liminf \|x_\alpha^*\|_{X^*} .$$

REMARK A.3.9   The above Proposition says that the norm functional on X is weakly lower semicontinuous and the norm functional on X^* is weakly* lower semicontinuous. ▯

THEOREM A.3.7 (Mazur Lemma)
If X is a normed space and $A \subseteq X$ is convex, <u>then</u> $\overline{A} = \overline{A}^w$.

REMARK A.3.10   In particular then in a normed space, a convex set is closed if and only if it is weakly closed.   ☐

THEOREM A.3.8
If X is a reflexive Banach space,
<u>then</u> a convex subset of X^* is weakly closed if and only if it is weakly sequentially closed.

DEFINITION A.3.5   *A Banach space X is said to be **locally uniformly convex**, if for any $\varepsilon > 0$ and $x \in X$ with $\|x\|_X = 1$, we can find $\delta = \delta(\varepsilon, x) > 0$, such that if $\|x - y\|_X \geq \varepsilon$, we have*

$$\left\| \tfrac{x+y}{2} \right\|_X \leq 1 - \delta \qquad \forall\, y \in X,\ \|y\|_X = 1.$$

*If $\delta > 0$ can be chosen independent of x , then we say that the space X is **uniformly convex**.*

REMARK A.3.11   Uniformly convex Banach spaces are reflexive (Milman-Pettis Theorem). Locally uniformly convex spaces have the Kadec-Klee property, i.e. if $x_n \overset{w}{\longrightarrow} x$ in X and $\|x_n\|_X \longrightarrow \|x\|_X$, then $x_n \longrightarrow x$ in X .   ☐

THEOREM A.3.9 (Troyanski Renorming Theorem)
Every reflexive Banach space can be given an equivalent norm so that both the space and its dual are locally uniformly convex and both have Frechet differentiable norm.

DEFINITION A.3.6   *Let X be a locally convex space and let $C \subseteq X$ be a nonempty set. An **extreme subset** of C is a nonempty set $D \subseteq X$, such that, if $x = \lambda y + (1 - \lambda)z \in D$ where $\lambda \in (0,1)$, then $y, z \in D$. A singleton extreme set of C is an **extreme point** of C . The set of extreme points of C is denoted by $\operatorname{ext} C$.*

THEOREM A.3.10 (Krein-Milman Theorem)
If X is a locally convex space and $C \subseteq X$ is a nonempty, convex and compact set,
<u>then</u> $C = \overline{\operatorname{conv}} \operatorname{ext} C$.

THEOREM A.3.11 (Bauer Maximum Principle)

If X is a locally convex space and $C \subseteq X$ is a nonempty, compact and convex set,
then every upper semicontinuous convex function on C achieves its maximum at an extreme point of C.

THEOREM A.3.12 (Hölder Inequality)

If (Ω, Σ, μ) is a measure space, $p \in [1, +\infty)$, $1 < p' \le +\infty$, $\frac{1}{p} + \frac{1}{p'} = 1$ (if $p = 1$, then $p' = +\infty$) and $f \in L^p(\Omega)$, $g \in L^{p'}(\Omega)$,
then

(a) $fg \in L^1(\Omega)$;
(b) $\|fg\|_1 \le \|f\|_p \|g\|_{p'}$.

THEOREM A.3.13 (Riesz Representation Theorem)

If (Ω, Σ, μ) is a σ-finite measure space and $p \in [1, +\infty)$, $1 < p' \le +\infty$, $\frac{1}{p} + \frac{1}{p'} = 1$ (if $p = 1$, then $p' = +\infty$),
then $u\colon L^{p'}(\Omega) \longrightarrow \left(L^p(\Omega) \right)^*$, defined by

$$u(g)(f) \stackrel{df}{=} \int_\Omega fg\,d\mu \qquad \forall\, f \in L^p(\Omega)$$

produces a linear isometry of $L^{p'}(\Omega)$ onto $\left(L^p(\Omega) \right)^*$. So we can write that

$$\left(L^p(\Omega) \right)^* = L^{p'}(\Omega).$$

THEOREM A.3.14 (Dunford-Pettis Theorem)

If (Ω, Σ, μ) is a finite measure space and $C \subseteq L^1(\Omega)$,
then C is relatively weakly compact in $L^1(\Omega)$ if and only if it is uniformly integrable.

THEOREM A.3.15 (Cauchy-Schwarz Inequality)

If H is an inner product space with $(\cdot, \cdot)_H$ its inner product,
then

$$\left| (x, y)_H \right| \le \|x\|_H \|y\|_H \qquad \forall\, x, y \in H.$$

A.4   Nonlinear Analysis

THEOREM A.4.1 (Leray-Schauder Alternative Theorem)
If X is a Banach space and $K \colon X \longrightarrow X$ is compact,
<u>*then*</u> *one of the following two statements holds:*

(a) *the set $F \stackrel{df}{=} \{x = \lambda K(x) \colon 0 < \lambda < 1\}$ is unbounded, or*

(b) K *has a fixed point (i.e. there exists $x \in X$, such that $x = K(x)$).*

REMARK A.4.1   A multivalued generalization of this principle can be found in Theorem 3.2.1.   ▯

THEOREM A.4.2 (Young Inequality)
If $a, b \geq 0$, $1 < p, p' < +\infty$, $\frac{1}{p} + \frac{1}{p'} = 1$ and $\varepsilon > 0$,
<u>*then*</u> $ab \leq \frac{\varepsilon}{p} a^p + \frac{1}{\varepsilon p'} b^{p'}$.

THEOREM A.4.3 (Gronwall Inequality)
If $T = [0, b]$, $x \in C(T)$, $k \in L^1(T)_+$, $h \in L^1(T)$ and

$$x(t) \leq h(t) + \int_0^t k(s)x(s)ds \qquad \forall\, t \in T,$$

<u>*then*</u>

$$x(t) \leq h(t) + \int_0^t \exp\left(\int_s^t k(r)dr\right) k(s)h(s)ds \qquad \forall\, s \in T.$$

REMARK A.4.2   If $h(t) = h_0$ for all $t \in T$, then we obtain

$$x(t) \leq h_0 \int_0^t k(s)ds \qquad \forall\, t \in T$$

and this inequality is also known as the **Bellman inequality**.   ▯

DEFINITION A.4.1   *Let X be a Banach space and let $C \subseteq X$ be a nonempty subset. The **tangent cone** to C at x is defined by*

$$T_C(x) \stackrel{df}{=} \bigcap_{\varepsilon > 0} \bigcap_{\eta > 0} \bigcup_{0 < \lambda \leq \eta} \left(\frac{1}{\lambda}(C - x) + \varepsilon B_1^X\right).$$

REMARK A.4.3   It follows from the above definition that

$$T_C(x) \;=\; \limsup_{\lambda \to 0^+} \frac{C - x}{\lambda}$$

and so $v \in T_C(x)$ if and only if there exist sequences $\{\lambda_n\}_{n\geq 1} \subseteq \mathbb{R}$ of strictly positive numbers and $\{u_n\}_{n\geq 1} \subseteq X$, such that

(a) $u_n \longrightarrow v$ in X;

(b) $\lambda_n \searrow 0$;

(c) $x + \lambda_n u_n \in C$ for all $n \geq 1$.

Also

$$v \in T_C(x) \quad \text{if and only if} \quad \lim_{\lambda \to 0^+} \frac{d_X(x + \lambda h, C)}{\lambda} = 0.$$

If C is convex, then

$$T_C(x) \;=\; \overline{\bigcup_{\lambda > 0} \frac{1}{\lambda}(C - x)}$$

and the tangent cone is convex (and of course closed). $\quad\Box$

THEOREM A.4.4 (Nagumo Viability Theorem)
If X is a Banach space, $C \subseteq X$ is a nonempty, closed and convex set, $f \colon C \longrightarrow X$ is locally Lipschitz and for all $x \in C$, $f(x) \in T_C(x)$, then for every $x_0 \in C$, the Cauchy problem

$$\begin{cases} \dot{x}(t) = f\big(x(t)\big) & \text{for a.a. } t \in T = [0, b], \\ x(0) = x_0, \end{cases}$$

has a unique solution $x \in C(T; X)$ with $x(t) \in C$ for all $t \in T$.

DEFINITION A.4.2   *Let X be a Banach space.*

(a) *If $\varphi \colon X \longrightarrow \mathbb{R}$ is a function, we say that φ is **weakly coercive**, if*

$$\varphi(x) \longrightarrow +\infty \quad \text{as } \|x\|_X \to +\infty.$$

(b) *A function $\varphi \colon X \longrightarrow \overline{\mathbb{R}}$ is **weakly sequentially lower semicontinuous**, if $x_n \xrightarrow{w} x$ in X implies that $\varphi(x) \leq \liminf\limits_{n \to +\infty} \varphi(x_n)$.*

DEFINITION A.4.3   *We say that multifunction $F \colon X \longrightarrow P_f(Y)$ is **locally compact** if for every $x \in X$, we can find $U \in \mathcal{N}(x)$, such that $\overline{F(U)}$ is compact in Y.*

List of Symbols

$B_r(x)$	p. 15	open r-ball centered at x
$B_r^X(x)$	p. 15	open r-ball in a metric space X centered at x
B_r	p. 15	open r-ball centered at the origin
$F^+(A)$	p. 15	strong inverse image of A under the multifunction F
$F^-(A)$	p. 15	weak inverse image of A under the multifunction F
$\{x_\alpha\}_{\alpha \in J}$	p. 16	net
$\mathrm{Gr}\,F$	p. 16	graph of a multifunction F
$d_X(\cdot, A)$	p. 17	distance function from a set $A \subseteq X$
$\sigma_X(\cdot, A)$	p. 17	support function of a set $A \subseteq X$
$\mathbb{R}^*$	p. 17	$\mathbb{R}^* \overset{df}{=} \mathbb{R} \cup \{\pm\infty\}$
$\overline{\mathbb{R}}$	p. 18	$\overline{\mathbb{R}} \overset{df}{=} \mathbb{R} \cup \{+\infty\}$
$h_X^*(A, C)$	p. 18	excess of A over C where $A, C \subseteq X$
$h_X(A, C)$	p. 18	Hausdorff distance of A and C where $A, C \subseteq X$
$\mathbb{R}_+$	p. 19	interval $[0, +\infty)$
(Ω, Σ)	p. 21	measurable space
$\mathcal{B}(X)$	p. 21	Borel σ-field of X
$\widehat{\Sigma}$	p. 21	universal σ-field
$X_{w^*}^*$	p. 23	dual Banach space X^* furnished with the w^*-topology
$L^0(\Omega; X)$	p. 25	space of Lebesgue measurable functions
S_F	p. 25	space of measurable selections of a multifunction F
S_F^p	p. 25	space of $L^p(\Omega; X)$-selections of a multifunction F
I_u	p. 25	integral operator for u
$\mathrm{ext}\,C$	p. 27	set of extremal points of a set C
$\mathrm{ext}\,F$	p. 27	set of extremal points of a multifunction F
$\|\cdot\|_w$	p. 27	weak norm in $L^1(T; X)$
$L_w^1(T; X)$	p. 27	$L^1(T; X)$ space equipped with weak norm
CS_F^w	p. 28	selections of F, continuous from K into $L_w^1(T; X)$
$\tau\text{-}\liminf\limits_{n \to +\infty} A_n$	p. 29	τ-Kuratowski limit inferior of a sequence of sets $\{A_n\}_{n \geq 1}$

$\overline{\mathrm{conv}}\,^{\tau}C$	p. 55	the closure of the convex hull of C in τ-topology
$T_C(x)$	p. 60	tangent cone to C at x
$N_C(x)$	p. 60	normal cone to C at x
$|d\varphi|(x)$	p. 62	weak slope of φ at x
m^{φ}	p. 67	$m^{\varphi}(x) \overset{df}{=} \min\left\{\, \|x^*\|_{X^*} :\; x^* \in \partial\varphi(x) \right\}$
$\mathcal{L}_c(X;Y)$	p. 69	space of $L \in \mathcal{L}(X;Y)$ such that L is compact
$\mathrm{Dom}\,(A)$	p. 70	domain of the multivalued operator $A\colon X \longrightarrow 2^{X^*}$
$R(A)$	p. 70	range of the multivalued operator $A\colon X \longrightarrow 2^{X^*}$
$\mathrm{Gr}\,A$	p. 70	graph of the multivalued operator $A\colon X \longrightarrow 2^{X^*}$
A^{-1}	p. 70	inverse of the multivalued operator $A\colon X \longrightarrow 2^{X^*}$
$\mathcal{F}$	p. 76	duality map of a Banach space X
J_{λ}	p. 81	resolvent of a maximal monotone operator $A\colon H \longrightarrow 2^H$
A_{λ}	p. 81	Yosida approximation of a maximal monotone operator $A\colon H \longrightarrow 2^H$
A^0	p. 82	$A^0(x) \overset{df}{=} \mathrm{proj}_{A(x)}\{0\}$
φ_{λ}	p. 82	Moreau-Yosida approximation of a function $\varphi \in \Gamma_0(H)$
$(\mathrm{S})_+$	p. 86	type of an operator $A\colon X \to X^*$
$\varrho(p, N)$	p. 96	resolvent set
$\sigma(p, N)$	p. 96	spectrum, i.e. $\sigma(p, N) \overset{df}{=} \mathbb{R} \setminus \varrho(p, N)$
$W^{1,p}_{\mathrm{per}}(T; \mathbb{R}^N)$	p. 96	Sobolev space of periodic functions
$C(T; \mathbb{R}^N)$	p. 96	space of continuous functions from T into $\mathbb{R}^N$
λ_1	p. 99	first eigenvalue
Δ_p	p. 99	p-Laplacian differential operator, i.e. $\Delta_p x \overset{df}{=} \mathrm{div}\left(\, \|\nabla x\|^{p-2}_{\mathbb{R}^N}\, \nabla x\right)$
$|\cdot|_N$	p. 107	Lebesgue measure on $\mathbb{R}^N$
$R(x)$	p. 110	Rayleigh quotient
λ_2^*	p. 118	second Lusternik-Schnirelmann eigenvalue
λ_2	p. 118	second eigenvalue
$m^{\varphi}(y)$	p. 124	$\inf\left\{\, \|y^*\|_{Y^*} :\; y^* \in \partial\varphi(y) \right\}$

of φ on a set C with critical value c

$(K_c^\varphi)_\delta$	p. 151	$(K_c^\varphi)_\delta \stackrel{df}{=} \left\{ x \in C : \ d_X(x, K_c^\varphi) < \delta \right\}$
$E_{c,\delta,\varepsilon}^\varphi$	p. 151	$E_{c,\delta,\varepsilon}^\varphi \stackrel{df}{=} \left\{ x \in C : \ \left| \varphi(x) - c \right| \leq \varepsilon \right\} \cap (K_c^\varphi)_\delta^c$
φ^c	p. 151	$\varphi^c \stackrel{df}{=} \left\{ x \in C : \ \varphi(x) \leq c \right\}$
i_C	p. 157	indicator function of a set C
PS_c-condition	p. 160	generalized nonsmooth Palais-Smale condition at level c
PS-condition	p. 160	generalized nonsmooth Palais-Smale condition
d_∞	p. 167	sup-metric on $C(B; X)$
$R_{a,b}$	p. 176	ring centered at the origin of radiuses a and b
$\mathrm{PS}_{c,+}^*$	p. 184	Palais-Smale-type condition
$\mathrm{PS}_{c,-}^*$	p. 185	Palais-Smale-type condition
PS_c^*	p. 185	Palais-Smale-type condition
K_c^F	p. 198	$K_c^F \stackrel{df}{=} \left\{ x \in X : \ \left| dF \right|(x, c) = 0 \right\}$, the set of critical points of F at level c
m^F	p. 198	$m^F(y) \stackrel{df}{=} \min \left\{ b \in \mathbb{R} : \ b \in F(y) \right\}$
$L^\infty(T)_+$	p. 209	positive cone in $L^\infty(T)$ (for the pointwise ordering)
$C(T; \mathbb{R}^N)$	p. 210	space of continuous functions from T into $\mathbb{R}^N$
G_u	p. 210	generalized mean value map
$C^1(T; \mathbb{R}^N)$	p. 212	space of C^1-functions from T into $\mathbb{R}^N$
$\langle \cdot, \cdot \rangle_p$	p. 216	duality brackets for the pair of spaces $\left(L^{p'}(T; \mathbb{R}^N), L^p(T; \mathbb{R}^N) \right)$
$\widehat{A}$	p. 223	lifting (realization) operator
$\mathcal{L}$	p. 225	Lebesgue σ-field of T
i_C	p. 232	indicator function of a set C
$D(f, B_{R_0}, 0)$	p. 237	Brouwer degree
$\langle \cdot, \cdot \rangle_0$	p. 270	duality brackets for the pair of spaces $\left(W_0^{1,p}(T; \mathbb{R}^N), W^{-1,p'}(T; \mathbb{R}^N) \right)$
$L^\infty(T)_+$	p. 383	positive cone in $L^\infty(\Omega)$
$\|\cdot\|_{l^\infty}$	p. 391	l^∞-norm in $\mathbb{R}^N$

J	p. 414	standard symplectic matrix
S^{2N-1}	p. 420	unit sphere in $\mathbb{R}^{2N-1}$, i.e. $S^{2N-1} = \{y \in \mathbb{R}^{2N} : \|y\|_{\mathbb{R}^{2N}} = 1\}$
S^1	p. 426	unit circle in $\mathbb{R}^2$, i.e. $S^1 = \{e^{i\vartheta} : \vartheta \in [0, 2\pi]\} = \mathbb{R}/2\pi\mathbb{Z}$
P_n	p. 431	radial projection
$\{T_\vartheta\}_{\vartheta \in S^1}$	p. 436	representation of the group S^1 over Z
$M_N(A; \mathbb{R})$	p. 437	space of all continuous maps $\eta \colon A \longrightarrow \mathbb{R}^{2N} \setminus \{0\}$ equivariant with respect to (T, R)
Z^0	p. 438	space of fixed points of a representation T of the group S^1 over Z
$F(p)$	p. 512	$F(p) \overset{df}{=} \{x \in W^{1,p}(\Omega) : \int_\Omega m(z)\lvert x(z)\rvert^p \, dz \leq 0\}$
$BV(\Omega)$	p. 537	space of functions of bounded variation
$TV(\Omega)$	p. 537	space of functions with finite total variation
ψ^∞	p. 540	recession function of ψ
$\mathcal{M}(\Omega; \mathbb{R}^N)$	p. 540	space of all vector measures on Ω with values in $\mathbb{R}^N$ which are of bounded variation
$m \prec\prec \mu$	p. 540	m is an absolutely continuous measure with respect to μ
$m \perp \mu$	p. 540	m is singular with respect to μ
$u.\mu$	p. 540	the element of $\mathcal{M}(\Omega; \mathbb{R}^N)$, defined by $(u.\mu)(C) \overset{df}{=} \int_C u(z) \, d\mu(z)$ for all $C \in \mathcal{B}(\Omega)$
$\frac{dm}{d\mu}$	p. 541	Radon-Nikodym derivative of m with respect to μ
$\lvert D\varphi\rvert(x)$	p. 551	strong slope of $\varphi \colon X \longrightarrow \overline{\mathbb{R}}$ at $x \in \operatorname{dom} \varphi$
$\partial^- \varphi(x)$	p. 551	extended subdifferential of $\varphi \in \Gamma_0(X)$ at $x \in \operatorname{dom} \varphi$
$C_0^1(\overline{\Omega})$	p. 654	$C_0^1(\overline{\Omega}) \overset{df}{=} \{x \in C^1(\overline{\Omega}) : x(z) = 0 \text{ for all } z \in \partial\Omega\}$
$C_0^1(\overline{\Omega})_+$	p. 654	$C_0^1(\overline{\Omega})_+ \overset{df}{=} \{x \in C_0^1(\overline{\Omega}) : x(z) \geq 0 \text{ for all } z \in \overline{\Omega}\}$

References

Adams, R. (1975). *Sobolev Spaces*. Vol. 65 of *Pure and Applied Mathematics*. Academic Press. New York/London.

Addou, I. (1999). Multiplicity of solutions for quasilinear elliptic boundary-value problems. *Electron. J. Differential Equations*.

Adly, S. & Goeleven, D. (1995). Periodic solutions for a class of hemivariational inequalities. *Comm. Appl. Nonlinear Anal.* **2**, 45–57.

Adly, S., Buttazzo, G. & Théra, M. (1998). Critical points for nonsmooth energy functions and applications. *Nonlinear Anal.* **32**, 711–718.

Agarwal, R., Lü, H. & O'Regan, D. (2002). Eigenvalues and the one-dimensional p-Laplacian. *J. Math. Anal. Appl.* **266**, 383–400.

Ahmad, S. & Lazer, A. (1984). Critical point theory and a theorem of Amaral and Pera. *Boll. Un. Mat. Ital. B (6)* **3**, 583–598.

Ahmad, S., Lazer, A. & Paul, J. (1976). Elementary critical point theory and perturbations of elliptic boundary value problems at resonance. *Indiana Univ. Math. J.* **25**, 933 944.

Akhiezer, N. & Glazman, I. (1961). *Theory of Linear Operators in Hilbert Spaces. Volume I.* Frederick Ungar Publishing Co. New York.

Akhiezer, N. & Glazman, I. (1963). *Theory of Linear Operators in Hilbert Spaces. Volume II.* Frederick Ungar Publisher Co. New York.

Allegretto, W. & Huang, Y. (1995). Eigenvalues of the indefinite-weight p-Laplacian in weighted spaces. *Funkcial. Ekvac.* **38**, 233–242.

Allegretto, W. & Huang, Y. (1998). A Picone's identity for the p-Laplacian and applications. *Nonlinear Anal.* **32**, 819–830.

Allegretto, W. & Huang, Y. (1999). Principal eigenvalues and Sturm comparison via Picone's identity. *J. Differential Equations* **156**, 427–438.

Ambrosetti, A. (1992). Critical points and nonlinear variational problems. *Mém. Soc. Math. France (N.S.)*.

Ambrosetti, A. & Badiale, M. (1989). The dual variational problem and elliptic problems with discontinuous nonlinearities. *J. Math. Anal. Appl.* **140**, 363–373.

Ambrosetti, A. & Lupo, D. (1984). On a class of nonlinear Dirichlet problems with multiple solutions. *Nonlinear Anal.* **8**, 1145–1150.

Ambrosetti, A. & Mancini, G. (1978a). Existence and multiplicity results for nonlinear elliptic problems with linear part at resonance. The case of simple eigenvalue. *J. Differential Equations* **28**, 220–245.

Ambrosetti, A. & Mancini, G. (1978b). Theorems of existence and multiplicity for nonlinear elliptic problems with noninvertible linear part. *Ann. Scuola Norm. Sup. Pisa Cl. Sci. (4)* **5**, 15–28.

Ambrosetti, A. & Mancini, G. (1982). On a theorem by Ekeland and Lasry concerning the number of periodic Hamiltonian trajectories. *J. Differential Equations* **43**, 249–256.

Ambrosetti, A. & Rabinowitz, P. (1973). Dual variational methods in critical point theory and applications. *J. Funct. Anal.* **14**, 349–381.

Ambrosetti, A. & Turner, R. (1988). Some discontinuous variational problems. *Differential Integral Equations* **1**, 341–349.

Ambrosetti, A., Brézis, H. & Cerami, G. (1994). Combined effects of concave and convex nonlinearities in some elliptic problems. *J. Funct. Anal.* **122**, 519–543.

Ambrosetti, A., Garcia Azorero, J. & Peral, I. (1996). Multiplicity results for some nonlinear elliptic equations. *J. Funct. Anal.* **137**, 219–242.

Anane, A. (1987). Simplicité et islolation de la première valeur du p-laplacien avec poids. *C. R. Acad. Sci. Paris Sér. I Math.* **305**, 725–728.

Anane, A. & Gossez, J.-P. (1990). Strongly nonlinear elliptic problems near resonance: A variational approach. *Comm. Partial Differential Equations* **15**, 1141–1159.

Anane, A. & Tsouli, N. (1996). On the second eigenvalue of the p-Laplacian. *in* A. Benkirane & J.-P. Gossez, eds, 'Nonlinear Partial Differential Equations (Fès, 1994)'. Vol. 343 of *Pitman Res. Notes in Math. Ser.* Logman. Harlow. pp. 1–9.

Ang, D., Schmitt, K. & Le, V. (1990). Noncoercive variational inequalities: Some applications. *Nonlinear Anal.* **15**, 497–512.

Anzellotti, G., Buttazzo, G. & Dal Maso, G. (1986). Dirichlet problems for demi-continuous functionals. *Nonlinear Anal.* **10**, 603–613.

Arcoya, D. & Orsina, L. (1997). Landesmann-Lazer conditions and quasilinear elliptic equations. *Nonlinear Anal.* **28**, 1623–1632.

Aubin, J.-P. & Clarke, F. (1979). Shadow prices and duality for a class of optimal control problems. *SIAM J. Control Optim.* **17**, 567–586.

Aubin, J.-P. & Ekeland, I. (1984). *Applied Nonlinear Analysis*. Pure and Applied Mathematics. Wiley. New York.

Aubin, J.-P. & Frankowska, H. (1990). *Set-Valued Analysis*. Vol. 2 of *Systems & Control: Foundations & Applications*. Birkhäuser Verlag. Boston, MA.

Bader, R. (2001). A topological fixed-point index theory for evolution inclusions. *Z. Anal. Anwendungen* **20**, 3–15.

Bader, R. & Papageorgiou, N. (2002). Nonlinear boundary value problems for differential inclusions. *Math. Nachr.* **244**, 5–25.

Badiale, M. (1995). Semilinear elliptic problems in $\mathbb{R}^N$ with discontinuous nonlinearities. *Atti Sem. Mat. Fis. Univ. Modena* **43**, 293–305.

Badiale, M. & Tarantello, G. (1997). Existence and multiplicity results for elliptic problems with critical growth and discontinuous nonlinearities. *Nonlinear Anal.* **29**, 639–677.

Baiocchi, C. & Capelo, A. (1984). *Variational and Quasivariational Inequalities*. Wiley. New York.

Bandle, C., Coffman, C. & Marcus, M. (1987). Nonlinear elliptic problems in annular domain. *J. Differential Equations* **69**, 322–345.

Barbu, V. (1976). *Nonlinear Semigroups and Differential Equations in Banach Spaces*. Noordhoff International Publishing. Leiden.

Barbu, V. (1984). *Optimal Control of Variational Inequalities*. Vol. 100 of *Research Notes in Mathematics*. Pitman. Boston, MA.

Barbu, V. & Precupanu, T. (1986). *Convexity and Optimization in Banach Spaces*. Vol. 10 of *Mathematics and its Applications (East European Series)*. D. Reidel Publishing Co. Dordrecht.

Bartolo, P., Benci, V. & Fortunato, D. (1983). Abstract critical point theorems and applications to some nonlinear problems with "strong" resonance at infinity. *Nonlinear Anal.* **7**, 981–1012.

Benci, V. (1981). A geometrical index for the group S^1 and some applications to the study of periodic solutions of ordinary differential equations. *Comm. Pure Appl. Math.* **34**, 393–432.

Berberian, S. (1976). *Introduction to Hilbert Spaces*. Chelsea Publishing Co. New York.

Berestycki, H. & de Figueiredo, D. (1981). Double resonance in semilinear elliptic problems. *Comm. Partial Differential Equations* **6**, 91–120.

Berestycki, H., Lasry, J.-M., Mancini, G. & Ruf, B. (1985). Existence of multiple periodic orbits on star-shaped Hamiltonian surfaces. *Comm. Pure Appl. Math.* **38**, 253–289.

Bertsch, M. & Klaver, M. (1991). On positive solutions of $\Delta u + \lambda f(u) = 0$ with f discontinuous. *J. Math. Anal. Appl.* **157**, 417–446.

Binding, P., Drábek, P. & Huang, Y. (1997). On Neumann boundary value problems for some quasilinear elliptic equations. *Electron. J. Differential Equations*.

Boccardo, L., Drábek, P., Giachetti, D. & Kučera, M. (1986). Generalization of Fredholm alternative for nonlinear differential operators. *Nonlinear Anal.* **10**, 1083–1103.

Bocea, M. (1997). Multiple solutions for a class of eigenvalue problems involving a nonlinear monotone operator in hemivariational inequalities. *Appl. Anal.* **65**, 395–407.

Bocea, M., Motreanu, D. & Panagiotopoulos, P. (2000). Multiple solutions for a double eigenvalue hemivariational inequality on a sphere-like type manifold. *Nonlinear Anal.* **42**, 737–749.

Bocea, M., Panagiotopoulos, P. & Rădulescu, V. (1999). A perturbation result for a double eigenvalue hemivariational inequality with constraints and applications. *J. Global Optim.* **14**, 137–156.

Bouchala, J. & Drábek, P. (2000). Strong resonance for some quasilinear elliptic equations. *J. Math. Anal. Appl.* **245**, 7–19.

Bouguima, M. (1995). A quasilinear elliptic problem with a discontinuous nonlinearity. *Nonlinear Anal.* **25**, 1115–1121.

Brézis, H. (1983). *Analyse Fonctionnelle. Théorie et Applications.* Masson. Paris.

Brézis, H. & Browder, F. (1978). Strongly nonlinear elliptic boundary value problems. *Ann. Scuola Norm. Sup. Pisa Cl. Sci. (4)* **5**, 587–603.

Brézis, H. & Browder, F. (1982). Some properties of higher order Sobolev spaces. *J. Math. Pures Appl. (9)* **61**, 245–259.

Brézis, H. & Nirenberg, L. (1991). Remarks on finding critical points. *Comm. Pure Appl. Math.* **44**, 939–963.

Browder, F. (1965). Nonlinear elliptic boundary value problems. II. *Trans. Amer. Math. Soc.* **117**, 530–550.

Browder, F. (1976). *Nonlinear Operators and Nonlinear Equations of Evolution in Banach Spaces.* AMS. Providence, RI.

Browder, F. (1977). Pseudo-monotone operators and nonlinear elliptic boundary value problems on unbounded domain. *Proc. Nat. Acad. Sci. U.S.A.* **74**, 2659–2661.

Browder, F. & Hess, P. (1972). Nonlinear mappings of monotone type in Banach spaces. *J. Funct. Anal.* **11**, 251–294.

Cabada, A. & Nieto, J. (1990). Extremal solutions of second order nonlinear periodic boundary value problems. *Appl. Math. Comput.* **40**, 135–145.

Các, N. (1988). On an elliptic boundary value problem at double resonance. *J. Math. Anal. Appl.* **132**, 473–483.

Caklović, L., Li, S. & Willem, M. (1990). A note on the Palais-Smale condition and coercivity. *Differential Integral Equations* **3**, 799–800.

Calvert, B. & Gupta, C. (1978). Nonlinear elliptic boundary value problems in L^p-spaces and sums of ranges of accretive operators. *Nonlinear Anal.* **2**, 1–26.

Canino, A. & Degiovanni, M. (1995). Nonsmooth critical point theory and quasilinear elliptic equations. *in* 'Topological Methods in Differential Equations and Inclusions'. Vol. 472 of *NATO Adv. Sci. Inst. Ser. C Math. Phys. Sci.* Kluwer. Dordrecht. pp. 1–50.

Carl, S. (1992). A combined variational-monotone iterative method for elliptic boundary value problems with discontinuous nonlinearity. *Appl. Anal.* **43**, 21–45.

Carl, S. (1997). Quasilinear elliptic equations with discontinuous nonlinearities in $\mathbb{R}^N$. *Nonlinear Anal.* **30**, 1743–1751. Proceedings of the Second World Congress of Nonlinear Analysts, Part 3 (Athens, 1996).

Carl, S. (2001). Existence of extremal solutions of boundary hemivariational inequalities. *J. Differential Equations* **171**, 370–396.

Carl, S. & Dietrich, H. (1995). The weak upper-lower solution method for quasilinear elliptic equations with generalized subdifferentiable perturbations. *Appl. Anal.* **56**, 263–278.

Carl, S. & Heikkilä, S. (1992). An existence result for elliptic differential inclusions with discontinuous nonlinearity. *Nonlinear Anal.* **18**, 471–479.

Carl, S. & Heikkilä, S. (1998). Elliptic equations with discontinuous nonlinearities in $\mathbb{R}^N$. *Nonlinear Anal.* **31**, 217–227.

Casas, E. & Fernández, L. (1989). A Green's formula for quasilinear elliptic operators. *J. Math. Anal. Appl.* **142**, 62–73.

Castro, A. & Lazer, A. (1979). Critical point theory and the number of solutions of a nonlinear Dirichlet problem. *Ann. Mat. Pura Appl.* **70**, 113–137.

Castro, A., Maya, C. & Shivaji, R. (2003). Nonlinear eigenvalue problems with semipositone structure. *Electron. J. Differ. Equ. Conf.* **5**, 33–49. Proceedings of the Conference on Nonlinear Differential Equations (Coral Gables, FL, 1999).

Cerami, G. (1978). Un criterio di esistenza per i punti critici su varieta' illimitate. *Ist. Lombardo Accad. Sci. Lett. Rend. A* **112**, 332–336.

Cerami, G. (1983). Soluzioni positive di problemi con parte non lineare discontinua e applicazione a un problema di frontiera libera. *Boll. Un. Mat. Ital. B (6)* **2**, 321–338.

Chang, K.-C. (1978). On the multiple solutions of the elliptic differential equations with discontinuous nonlinear terms. *Sci. Sinica* **21**, 139–158.

Chang, K.-C. (1980). The obstacle problem and partial differential equations with discontinuous nonlinearities. *Comm. Pure Appl. Math.* **33**, 117–146.

Chang, K.-C. (1981). Variational methods for nondifferentiable functionals and their applications to partial differential equations. *J. Math. Anal. Appl.* **80**, 102–129.

Chang, K.-C. (1993). *Infinite-Dimensional Morse Theory and Multiple Solution Problems*. Vol. 6 of *Progress in Nonlinear Differential Equations and Their Applications*. Birkhäuser Verlag. Boston, MA.

Chen, J. & Li, S. (2002). On multiple solutions of a singular quasilinear equation on unbounded domain. *J. Math. Anal. Appl.* **275**, 733–746.

Chiappinelli, R. & de Figueiredo, D. (1993). Bifurcation from infinity and multiple solutions for an elliptic system. *Differential Integral Equations* **6**, 757–771.

Chiappinelli, R., Mawhin, J. & Nugari, R. (1992). Bifurcation from infinity and multiple solutions for some Dirichlet problems with unbounded nonlinearities. *Nonlinear Anal.* **18**, 1099–1112.

Cîrstea, F. & Rădulescu, V. (2000). Multiplicity of solutions for a class of nonsymmetric eigenvalue hemivariational inequalities. *J. Global Optim.* **17**, 43–54.

Clarke, F. (1981). Periodic solutions of Hamiltonian inclusions. *J. Differential Equations* **40**, 1–6.

Clarke, F. (1983). *Optimization and Nonsmooth Analysis*. Wiley. New York.

Clarke, F. & Ekeland, I. (1980). Hamiltonian trajectories having prescribed minimal period. *Comm. Pure Appl. Math.* **33**, 103–116.

Coffman, C. (1969). A minimum-maximum principle for a class of non-linear integral equations. *J. Analyse Math.* **22**, 391–419.

Coffman, C. & Marcus, M. (1989). Existence and uniqueness results for semi-linear Dirichlet problems in annuli. *Arch. Rational Mech. Anal.* **108**, 293–307.

Corvellec, J.-N. (1995). Morse theory for continuous functionals. *J. Math. Anal. Appl.* **196**, 1050–1072.

Corvellec, J.-N. & Degiovanni, M. (1997). Nontrivial solutions of quasilinear equations via nonsmooth Morse theory. *J. Differential Equations* **136**, 268–293.

Corvellec, J.-N., Degiovanni, M. & Marzocchi, M. (1993). Deformation properties for continuous functionals and critical point theory. *Topol. Methods Nonlinear Anal.* **1**, 151–171.

Costa, D. & Gonçalves, J. (1990). Critical point theory for nondifferentiable functionals and applications. *J. Math. Anal. Appl.* **153**, 470–485.

Costa, D. & Magalhães, C. (1995). Existence results for perturbations of the p-Laplacian. *Nonlinear Anal.* **24**, 409–418.

Costa, D. & Silva, E. (1991). The Palais-Smale condition versus coercivity. *Nonlinear Anal.* **16**, 371–381.

Costa, D. & Silva, E. (1993). Existence of solutions for a class of resonant elliptic problems. *J. Math. Anal. Appl.* **175**, 411–424.

Costa, D. & Silva, E. (1995). On a class of resonant problems at higher eigenvalues. *Differential Integral Equations* **8**, 663–671.

Damascelli, L. (1998). Comparison theorems for some quasilinear degenerate elliptic operators and applications to symmetry and monotonicity results. *Ann. Inst. H. Poincaré Anal. Non Linéaire* **15**, 493–516.

Dancer, E. & Sweers, G. (1989). On the existence of a maximal weak solution for a semilinear elliptic equation. *Differential Integral Equations* **2**, 533–540.

Dang, H. & Oppenheimer, S. (1996). Existence and uniqueness results for some nonlinear boundary value problems. *J. Math. Anal. Appl.* **198**, 35–48.

De Blasi, F. & Pianigiani, G. (1985). The Baire category method in existence problems for a class of multivalued differential equations with nonconvex right-hand side. *Funkcial. Ekvac.* **28**, 139–156.

De Blasi, F. & Pianigiani, G. (1991). Non-convex-valued differential inclusions in Banach spaces. *J. Math. Anal. Appl.* **157**, 469–494.

De Blasi, F. & Pianigiani, G. (1992). On the density of extremal solutions of differential inclusions. *Ann. Polon. Math.* **56**, 133–142.

De Blasi, F. & Pianigiani, G. (1993). Topological properties of nonconvex differential inclusions. *Nonlinear Anal.* **20**, 871–894.

De Blasi, F., Górniewicz, L. & Pianigiani, G. (1999). Topological degree and periodic solutions of differential inclusions. *Nonlinear Anal.* **37**, 217–243.

De Coster, C. (1995). Pairs of positive solutions for the one-dimensional p-Laplacian. *Nonlinear Anal.* **23**, 669–681.

de Figueiredo, D. (1982). Positive solutions of semilinear elliptic problems. *in* 'Differential Equations (San Paulo, 1981)'. Vol. 957 of *Lecture Notes in Math.* Springer-Verlag. New York pp. 34–85.

de Figueiredo, D. (1989). *Lectures on the Ekeland Variational Principle with Applications and Detours.* Vol. 81 of *Tata Institute of Fundamental Research Lectures on Mathematics and Physics.* Springer-Verlag. Berlin.

De Giorgi, E., Marino, A. & Tosques, M. (1980). Problemi di evoluzione in spazi metrici e curve di massima pendenza. *Atti Accad. Naz. Lincei Rend. Cl. Sci. Fis. Mat. Natur. (8)* **68**, 180–187.

de Thélin, F. (1986). Sur l'espace propre associé à la première valeur propre du pseudo-laplacien. *C. R. Acad. Sci. Paris Sér. I Math.* **303**, 355–358.

Degiovanni, M. (1997). Nonsmooth critical point theory and applications. *Nonlinear Anal.* **30**, 89–99. Proceedings of the Second World Congress of Nonlinear Analysts, Part 1 (Athens, 1996).

Degiovanni, M. & Marzocchi, M. (1994). A critical point theory for nonsmooth functionals. *Ann. Mat. Pura Appl. (4)* **167**, 73–100.

Degiovanni, M., Marzocchi, M. & Rădulescu, V. (2000). Multiple solutions of hemivariational inequalities with area-type term. *Calc. Var. Partial Differential Equations* **10**, 355–387.

del Pino, M. & Manásevich, R. (1991). Multiple solutions for the p-Laplacian under global nonresonance. *Proc. Amer. Math. Soc.* **112**, 131–138.

del Pino, M., Elgueta, M. & Manásevich, R. (1988). A homotopic deformation along p of a Leray-Schauder degree result and existence for $\left(|u'|^{p-2}u'\right)' + f(t,u) = 0$, $u(0) = u(I) = 0$, $1 < p$. *J. Differential Equations* **80**, 1–13.

del Pino, M., Elgueta, M. & Manásevich, R. (1991). Generalizing Hartman's oscillation result for $\left(|x'|^{p-2}x'\right)' + c(t)|x|^{p-2}x = 0$, $p > 1$. *Houston J. Math.* **17**, 63–70.

del Pino, M., Manásevich, R. & Murúa, A. (1992). Existence and multiplicity of solutions with prescribed period for a second order quasilinear ODE. *Nonlinear Anal.* **18**, 79–92.

Delgado, M. & Suárez, A. (2000). Weak solutions for some quasilinear elliptic equations by the sub-supersolution method. *Nonlinear Anal.* **42**, 995–1002.

Denkowski, Z., Migórski, S. & Papageorgiou, N. (2003a). *An Introduction to Nonlinear Analysis: Theory.* Kluwer/Plenum. New York.

Denkowski, Z., Migórski, S. & Papageorgiou, N. (2003b). *An Introduction to Nonlinear Analysis: Applications.* Kluwer/Plenum. New York.

Di Benedetto, E. (1983). $C^{1+\alpha}$ local regularity of weak solutions of degenerate elliptic equations. *Nonlinear Anal.* **7**, 827–850.

Di Benedetto, E. (1995). *Partial Differential Equations*. Birkhäuser Verlag. Boston, MA.

Ding, Y. (1994). A remark on the linking theorem with applications. *Nonlinear Anal.* **22**, 237–250.

Dontchev, A. & Zolezzi, T. (1994). *Well-Posed Optimization Problems*. Vol. 1404 of *Lecture Notes in Math.* Springer-Verlag. Berlin.

Douka, P. & Papageorgiou, N. (submitted). Extremal solutions for nonlinear second order differential inclusions. submitted.

Drábek, P. & Robinson, S. (1999). Resonance problems for the p-Laplacian. *J. Funct. Anal.* **169**, 189–200.

Du, Y. (1993). Critical point theorems with relaxed boundary conditions and applications. *Bull. Austral. Math. Soc.* **47**, 101–118.

Dugundji, J. & Granas, A. (1982). *Fixed Point Theory. I.* Państwowe Wydawnictwo Naukowe. Warszawa.

Dunford, N. & Schwartz, J. (1958). *Linear Operators. I.* Wiley. New York.

Ekeland, I. (1974). On the variational principle. *J. Math. Anal. Appl.* **47**, 324–353.

Ekeland, I. (1979). Nonconvex minimization problems. *Bull. Amer. Math. Soc. (N.S.)* **1**, 443–474.

Ekeland, I. & Lasry, J.-M. (1980). On the number of periodic trajectories for a Hamiltonian flow on a convex energy surface. *Ann. of Math. (2)* **112**, 283–319.

Ekeland, I. & Temam, R. (1976). *Convex Analysis and Variational Problems*. North-Holland Publishing Co. Amsterdam-Oxford.

El Hachimi, A. & Gossez, J.-P. (1994). A note on a nonresonance condition for a quasilinear elliptic problem. *Nonlinear Anal.* **2**, 229–236.

Erbe, L. (1982). Existence of solutions to boundary value problems for second order differential equations. *Nonlinear Anal.* **6**, 1155–1162.

Erbe, L. & Krawcewicz, W. (1991*a*). Boundary value problems for differential inclusions. *in* 'Differential Equations (Colorado Springs, CO, 1989)'. Vol. 127 of *Lecture Notes in Pure and Appl. Math.* Marcel-Dekker. New York. pp. 115–135.

Erbe, L. & Krawcewicz, W. (1991*b*). Nonlinear boundary value problems for differential inclusions $y'' \in F(t, y, y')$. *Ann. Polon. Math.* **54**, 195–226.

Erbe, L. & Wang, H. (1994). On the existence of positive solutions of ordinary differential equations. *Proc. Amer. Math. Soc.* **120**, 743–748.

Erbe, L., Hu, S. & Wang, H. (1994). Multiple positive solutions of some boundary value problems. *J. Math. Anal. Appl.* **184**, 640–648.

Evans, L. & Gariepy, R. (1992). *Measure Theory and Fine Properties of Functions.* CRC Press. Boca Raton, FL.

Fabry, C. & Fayyad, D. (1992). Periodic solutions of second order differential equations with a p-Laplacian and asymmetric nonlinearities. *Rend. Ist. Mat. Univ. Trieste* **24**, 207–227.

Fabry, C. & Habets, P. (1986). Upper and lower solutions for second-order boundary value problems with nonlinear boundary conditions. *Nonlinear Anal.* **10**, 985–1007.

Fan, X. (1992). Existence of multiple periodic orbits on star-shaped Lipschitz-Hamiltonian surfaces. *J. Differential Equations* **98**, 91–110.

Faraci, F. (2003). Multiplicity results for a Neumann problem involving the p-Laplacian. *J. Math. Anal. Appl.* **277**, 180–189.

Filippakis, M., Gasiński, L. & Papageorgiou, N. (2004*a*). Existence theorems for periodic differential inclusions in $\mathbb{R}^n$. *Acta Math. Appl. Sinica* **20**, 33–344.

Filippakis, M., Gasiński, L. & Papageorgiou, N. (2004*b*). Quasilinear hemivariational inequalities with strong resonance at infinity. *Nonlinear Anal.* **56**, 331–345.

Filippakis, M., Gasiński, L. & Papageorgiou, N. (to appear*a*). Nonsmooth generalized guiding functions for periodic differential inclusions. *NoDEA Nonlinear Differential Equations Appl.*

Filippakis, M., Gasiński, L. & Papageorgiou, N. (to appear*b*). Positive solutions for second order multivalued boundary value problems. *in* R. Agerwal & D. O'Regan, eds, 'Nonlinear Analysis and Applications'. Kluwer. pp. 531–548.

Fink, A. & Gatica, J. (1993). Positive solutions of second order systems of boundary value problems. *J. Math. Anal. Appl.* **180**, 93–108.

Fonda, A. & Lupo, D. (1989). Periodic solutions of second order ordinary differential equations. *Boll. Un. Mat. Ital. B (7)* **3**, 291–299.

Fonseca, I. & Gangbo, W. (1995). *Degree Theory in Analysis and Applications.* Vol. 2 of *Oxford Lecture Series in Mathematics and its Applications.* The Clarendon Press. New York.

Frigon, M. (1990). *Application de la théorie de la transversalité topologique à des problèmes non linéaires pour des équations différentielles ordinaires.* Vol. 296 of *Dissertationes Math. (Rozprawy Mat.).*

Frigon, M. (1991). Problémes aux limites pour des inclusions différentielles sans condition de croissance. *Ann. Polon. Math.* **54**, 69–83.

Frigon, M. (1995). Théorème d'existence de solutions d'inclusions differentielles. *in* 'Topological Methods in Differential Equations and Inclusions (Montreal, PQ, 1994)'. Vol. 472 of *NATO Adv. Sci. Inst. Ser. C Math. Phys. Sci.* Kluwer. Dordrecht. pp. 51–87.

Frigon, M. (1998). On a critical point theory for multivalued functionals and applications to partial differential inclusions. *Nonlinear Anal.* **31**, 735–753.

Frigon, M. & Granas, A. (1991). Problèmes aux limites pour des inclusions différentielles de type semicontinues inférieurement. *Riv. Mat. Univ. Parma (4)* **17**, 87–97.

Gaines, R. & Mawhin, J. (1977). *Coincidence Degree and Nonlinear Differential Equations.* Vol. 568 of *Lecture Notes in Math.* Springer-Verlag. Berlin-New York.

Gao, W. & Wang, J. (1995). On a nonlinear second order periodic boundary value problem with Carathéodory functions. *Ann. Polon. Math.* **62**, 283–291.

Garaizar, X. (1987). Existence of positive radial solutions for semilinear elliptic equations in the annulus. *J. Differential Equations* **70**, 69–92.

García Huidobro, M., Manásevich, R. & Zanolin, F. (1993). Strongly nonlinear second-order ODEs with unilateral conditions. *Differential Integral Equations* **6**, 1057–1078.

García-Melián, J. & Sabina de Lis, J. (1998). Maximum and comparison principles for operators involving the p-Laplacian. *J. Math. Anal. Appl.* **218**, 49–65.

Gasiński, L. & Papageorgiou, N. (1999). Nonlinear hemivariational inequalities at resonance. *Bull. Austral. Math. Soc.* **60**, 353–364.

Gasiński, L. & Papageorgiou, N. (2000). Multiple solutions for nonlinear hemivariational inequalities near resonance. *Funkcial. Ekvac.* **43**, 271–284.

Gasiński, L. & Papageorgiou, N. (2001*a*). Existence of solutions and of multiple solutions for eigenvalue problems of hemivariational inequalities. *Adv. Math. Sci. Appl.* **11**, 437–464.

Gasiński, L. & Papageorgiou, N. (2001*b*). An existence theorem for nonlinear hemivariational inequalities at resonance. *Bull. Austral. Math. Soc.* **63**, 1–14.

Gasiński, L. & Papageorgiou, N. (2001*c*). Multiple solutions for semilinear hemivariational inequalities at resonance. *Publ. Math. Debrecen* **59**, 121–146.

Gasiński, L. & Papageorgiou, N. (2001d). Semilinear hemivariational inequalities at resonance. *Rend. Circ. Mat. Palermo (2)* **50**, 217–238.

Gasiński, L. & Papageorgiou, N. (2001e). Solutions and multiple solutions for quasilinear hemivariational inequalities at resonance. *Proc. Roy. Soc. Edinburgh Sect. A* **131**, 1091–1111.

Gasiński, L. & Papageorgiou, N. (2002a). A multiplicity result for nonlinear second order periodic equations with nonsmooth potential. *Bull. Belg. Math. Soc. Simon Stevin* **9**, 245–258.

Gasiński, L. & Papageorgiou, N. (2002b). Strongly resonant semilinear and quasilinear hemivariational inequalities. *Acta Sci. Math. (Szeged)* **68**, 727–750.

Gasiński, L. & Papageorgiou, N. (2003a). Nonlinear second-order multivalued boundary value problems. *Proc. Indian Acad. Sci. Math. Sci.* **113**, 293–319.

Gasiński, L. & Papageorgiou, N. (2003b). On the existence of multiple periodic solutions for equations driven by the p-Laplacian and with a non-smooth potential. *Proc. Edinburgh Math. Soc. (2)* **46**, 229–249.

Gasiński, L. & Papageorgiou, N. (2003c). Strongly nonlinear multivalued boundary value problems. *Nonlinear Anal.* **52**, 1219–1238.

Gasiński, L. & Papageorgiou, N. (2003d). Two bounded solutions of opposite sign for nonlinear hemivariational inequalities at resonance. *Publ. Math. Debrecen* **63**, 29–49.

Gasiński, L. & Papageorgiou, N. (to appear). On nonlinear elliptic hemivariational inequalities of second order. *Acta Math. Sci.*

Gazzola, F. & Rădulescu, V. (2000). A nonsmooth critical point theory approach to some nonlinear elliptic equations in $\mathbb{R}^n$. *Differential Integral Equations* **13**, 47–60.

Ghoussoub, N. (1993a). *Duality and Perturbation Methods in Critical Point Theory.* Cambridge University Press. Cambridge.

Ghoussoub, N. (1993b). A min-max principle with a relaxed boundary condition. *Proc. Amer. Math. Soc.* **117**, 439–447.

Gilbarg, D. & Trudinger, N. (2001). *Elliptic Partial Differential Equations of Second Order.* Springer-Verlag. Berlin.

Giusti, E. (1984). *Minimal Surfaces and Functions of Bounded Variation.* Vol. 80 of *Monographs in Mathematics.* Birkhäuser Verlag. Basel.

Godoy, T., Gossez, J.-P. & Paczka, S. (2002). On the antimaximum principle for the p-Laplacian with indefinite weight. *Nonlinear Anal.* **51**, 449–467.

Goeleven, D., Motreanu, D. & Panagiotopoulos, P. (1997). Multiple solutions for a class of eigenvalue problems in hemivariational inequalities. *Nonlinear Anal.* **29**, 9–26.

Goeleven, D., Motreanu, D. & Panagiotopoulos, P. (1998). Eigenvalue problems for variational-hemivariational inequalities at resonance. *Nonlinear Anal.* **33**, 161–180.

Gohberg, I. & Goldberg, S. (1981). *Basic Operator Theory.* Birkhäuser Verlag. Boston, MA.

Gonçalves, J. & Miyagaki, O. (1992). Multiple nontrivial solutions of semilinear strongly resonant elliptic equations. *Nonlinear Anal.* **19**, 43–52.

Gonçalves, J. & Miyagaki, O. (1995). Three solutions for a strongly resonant elliptic problem. *Nonlinear Anal.* **24**, 265–272.

Granas, A., Guenther, R. & Lee, J. (1985). *Nonlinear Boundary Value Problems for Ordinary Differential Equations.* Vol. 244 of *Dissertationes Math. (Rozprawy Mat.).*

Guedda, M. & Véron, L. (1989). Quasilinear elliptic equations involving critical Sobolev exponents. *Nonlinear Anal.* **13**, 879–902.

Guo, D. & Lakshmikantham, V. (1988). *Nonlinear Problems in Abstract Cones.* Vol. 5 of *Notes and Reports in Mathematics in Science and Engineering.* Academic Press. Boston, MA.

Guo, D., Sun, J. & Qi, G. (1988). Some extensions of the mountain pass lemma. *Differential Integral Equations* **1**, 351–358.

Guo, Z. (1993). Boundary value problems for a class of quasilinear ordinary differential equations. *Differential Integral Equations* **6**, 705–719.

Guo, Z. (1996). On the number of positive solutions for quasilinear elliptic problems. *Nonlinear Anal.* **27**, 229–247.

Gupta, C. & Hess, P. (1976). Existence theorems for nonlinear noncoercive operator equations and nonlinear elliptic boundary value problems. *J. Differential Equations* **22**, 305–313.

Halidias, N. & Papageorgiou, N. (1997/1998*b*). On a class of discontinuous nonlinear elliptic equations. *Ann. Univ. Sci. Budapest. Eötvös Sect. Math.* **40**, 149–154.

Halidias, N. & Papageorgiou, N. (1998*a*). Existence and relaxation results for nonlinear second-order multivalued boundary value problems in $\mathbb{R}^N$. *J. Differential Equations* **147**, 123–154.

Halidias, N. & Papageorgiou, N. (1998*c*). Second order multivalued boundary value problems. *Arch. Math. (Basel)* **34**, 267–284.

Halidias, N. & Papageorgiou, N. (2000*a*). Existence of solutions for quasilinear second order differential inclusions with nonlinear boundary conditions. *J. Comput Appl. Math.* **113**, 51–64. Special Issue "Fixed Point Theory with Applications in Nonlinear Analysis".

Halidias, N. & Papageorgiou, N. (2000*b*). Quasilinear elliptic problems with multivalued terms. *Czechoslovak Math. J.* **50(125)**, 803–823.

Halmos, P. (1998). *Introduction to Hilbert Space and the Theory of Spectral Multiplicity*. AMS Chelsea Publishing. Providence, RI.

Hartman, P. (1960). On boundary value problems for systems of ordinary nonlinear second order differential equations. *Trans. Amer. Math. Soc.* **96**, 493–509.

Hartman, P. (2002). *Ordinary Differential Equations*. Vol. 38 of *Classics in Applied Mathematics*. SIAM. Philadelphia, PA.

Heikkilä, S. (1990). On an elliptic boundary value problem with discontinuous nonlinearity. *Appl. Anal.* **37**, 183–189.

Heikkilä, S. & Hu, S. (1993). On fixed points of multifunctions in ordered spaces. *Applicable Anal.* **51**, 115–127.

Heikkilä, S. & Lakshmikantham, V. (1994). *Monotone Iterative Techniques for Discontinuous Nonlinear Differential Equations*. Vol. 181 of *Monographs and Textbooks in Pure and Applied Mathematics*. Marcel Dekker. New York.

Hess, P. (1973*a*). On nonlinear mappings of monotone type with respect to two Banach spaces. *J. Math. Pures Appl. (9)* **52**, 13–26.

Hess, P. (1973*b*). A strongly nonlinear elliptic boundary value problem. *J. Math. Anal. Appl.* **43**, 241–249.

Hess, P. (1976). On strongly nonlinear elliptic problems. *in* 'Functional Analysis'. Vol. 18 of *Lecture Notes in Pure and Appl. Math.* Marcel Dekker. New York. pp. 91–109. Proc. Brazilian Math. Soc. Sympos., Inst. Mat. Univ. Estad. Campinas, San Paulo, 1974.

Hess, P. (1978). Nonlinear perturbations of linear elliptic and parabolic problems at resonance: Existence of multiple solutions. *Ann. Scuola Norm. Sup. Pisa Cl. Sci.* **5**, 527–537.

Hess, P. (1980). On nontrivial solutions of a nonlinear elliptic boundary value problems. *in* 'Confer. Sem. Mat. Univ. Bari'. Vol. 173.

Hirano, N. & Nishimura, T. (1993). Multiplicity results for semilinear elliptic problems at resonance with jumping nonlinearities. *J. Math. Anal. Appl.* **180**, 566–586.

Hiriart Urruty, J.-B. & Lemaréchal, C. (1993). *Convex Analysis and Minimization Algorithms II*. Springer-Verlag. Berlin.

Hu, J. (2001). The existence of homoclinic orbits in Hamiltonian inclusions. *Nonlinear Anal.* **46**, 169–180.

Hu, S. & Papageorgiou, N. (1995). On the existence of periodic solutions for nonconvex-valued differential inclusions in $\mathbb{R}^N$. *Proc. Amer. Math. Soc.* **123**, 3043–3050.

Hu, S. & Papageorgiou, N. (1997). *Handbook of Multivalued Analysis. Volume I: Theory*. Vol. 419 of *Theory. Mathematics and its Applications*. Kluwer. Dordrecht.

Hu, S. & Papageorgiou, N. (2000). *Handbook of Multivalued Analysis. Volume II: Applications*. Vol. 500 of *Theory. Mathematics and its Applications*. Kluwer. Dordrecht.

Hu, S. & Papageorgiou, N. (2001). Nonlinear elliptic problems of Neumann-type. *Rend. Circ. Mat. Palermo (2)* **50**, 47–66.

Hu, S. & Papageorgiou, N. (submitted). Positive periodic and homoclinic solutions for nonlinear differential equations with nonsmooth potential. submitted.

Hu, S., Kandilakis, D. & Papageorgiou, N. (1999). Periodic solutions for non-convex differential inclusions. *Proc. Amer. Math. Soc.* **127**, 89–94.

Hu, S., Kourogenis, N. & Papageorgiou, N. (1999). Nonlinear elliptic eigen-value problems with discontinuities. *J. Math. Anal. Appl.* **233**, 406–424.

Hu, S., Matzakos, N. & Papageorgiou, N. (2001). Neumann problems for elliptic equations with discontinuities. *Atti Sem. Mat. Fis. Univ. Modena* **49**, 183–204.

Huang, Y. (1990). On eigenvalue problems of p-Laplacian with Neumann boundary conditions. *Proc. Amer. Math. Soc.* **109**, 177–184.

Iannacci, R. & Nkashama, M. (1995). Nonlinear elliptic partial differential equations at resonance: Higher eigenvalues. *Nonlinear Anal.* **25**, 455–471.

Ioffe, A. (1978). Survey of measurable selection theorems: Russian literature supplement. *SIAM J. Control Optim.* **16**, 728–732.

Ioffe, A. & Schwartzman, E. (1996). Metric critical point theory. I. Morse regularity and homotopic stability of a minimum. *J. Math. Pures Appl. (9)* **75**, 125–153.

Ioffe, A. & Schwartzman, E. (1997). Metric critical point theory. II. deformation techniques. *in* 'New Results in Operator Theory and its Applicatons'. Vol. 98 of *Oper. Theory Adv. Appl.*. Birkhäuser Verlag. Basel. pp. 131–144.

Ioffe, A. & Tihomirov, V. (1979). *Theory of Extremal Problems.* Vol. 6 of *Studies in Mathematics and its Applications.* North-Holland Publishing Co. Amsterdam.

Kandilakis, D. & Papageorgiou, N. (1996). Existence theorems for nonlinear boundary value problems for second order differential inclusions. *J. Differential Equations* **132**, 107–125.

Kandilakis, D. & Papageorgiou, N. (1999). Neumann problem for a class of quasilinear differential equations. *Atti Sem. Mat. Fis. Univ. Modena* **48**, 163–177.

Kandilakis, D., Kourogenis, N. & Papageorgiou, N. (submitted). Two nontrivial critical points for nonsmooth functionals via local linking and applications. submitted.

Kato, T. (1976). *Perturbation Theory for Linear Operators.* Classics in Mathematics. Springer-Verlag. Berlin.

Kenmochi, N. (1975). Pseudomonotone operators and nonlinear elliptic boundary value problems. *J. Math. Soc. Japan* **27**, 121–149.

Kesavan, S. (1989). *Topics in Functional Analysis and Applications.* Wiley. New York.

Kisielewicz, M. (1991). *Differential Inclusions and Optimal Control.* Vol. 44 of *Mathematics and its Applications (East European Series).* Kluwer. Dordrecht.

Klein, E. & Thompson, A. (1984). *Theory of Correspondences.* Series of Monographs and Advanced Texts. Wiley. New York.

Knobloch, H.-W. (1971). On the existence of periodic solutions for second order vector differential equations. *J. Differential Equations* **9**, 67–85.

Korman, P. & Lazer, A. (1994). Homoclinic orbits for a class of symmetric Hamiltonian systems. *Electron. J. Differential Equations.*

Kourogenis, N. (2002). Strongly nonlinear second order differential inclusions with generalized boundary conditions. *J. Math. Anal. Appl.* **287**, 348–364.

Kourogenis, N. & Papageorgiou, N. (1998*a*). Discontinuous quasilinear elliptic problems at resonance. *Colloq. Math.* **78**, 213–223.

Kourogenis, N. & Papageorgiou, N. (1998*b*). On a class of quasilinear differential equations: The Neumann problem. *Methods Appl. Anal.* **5**, 273–282.

Kourogenis, N. & Papageorgiou, N. (2000*a*). Nonsmooth critical point theory and nonlinear elliptic equations at resonance. *J. Austral. Math. Soc. Ser. A* **69**, 245–271.

Kourogenis, N. & Papageorgiou, N. (2000*b*). A weak nonsmooth Palais-Smale condition and coercivity. *Rend. Circ. Mat. Palermo (2)* **49**, 521–526.

Kourogenis, N. & Papageorgiou, N. (2001). Multiple solutions for discontinuous strongly resonant elliptic problems. *J. Math. Soc. Japan* **53**, 17–34.

Kourogenis, N. & Papageorgiou, N. (2003). Nonlinear hemivariational inequalities of second order using the method of upper-lower solutions. *Proc. Amer. Math. Soc.* **131**, 2359–2369.

Kourogenis, N., Papadrianos, J. & Papageorgiou, N. (2002). Extensions of nonsmooth critical point theory and applications. *Atti Sem. Mat. Fis. Univ. Modena* **50**, 381–414.

Krasnoselskii, M. (1964). *Topological Methods in the Theory of Nonlinear Integral Equations*. The Macmillan Co. New York.

Kufner, A., John, O. & Fučik, S. (1977). *Function Spaces*. Monographs and Textbooks on Mechanics of Solids and Fluids; Mechanics: Analysis. Noordhoff International Publishing. Leyden.

Kuratowski, K. (1966). *Topology*. Academic Press. New York.

Kuzin, I. & Pohozaev, S. (1997). *Entire Solutions of Semilinear Elliptic Equations*. Vol. 33 of *Progress in Nonlinear Differential Equations and Their Applications*. Birkhäuser Verlag. Basel.

Kyritsi, S. & Papageorgiou, N. (2001). Hemivariational inequalities with the potential crossing the first eigenvalue. *Bull. Austral. Math. Soc.* **64**, 381–393.

Kyritsi, S. & Papageorgiou, N. (2004). Multiple solutions of constant sign for nonlinear nonsmooth eigenvalue problems near resonance. *Calc. Var. Partial Differential Equations* **20**, 1–24.

Kyritsi, S. & Papageorgiou, N. (preprint). Minimizers of functionals on manifolds and nonlinear eigenvalue problems.

Kyritsi, S. & Papageorgiou, N. (to appear*a*). Nonsmooth critical point theory on closed convex sets and nonlinear hemivariational inequalities. *Nonlinear Anal.*

Kyritsi, S. & Papageorgiou, N. (to appear*b*). Solvability of semilinear hemivariational inequalities at resonance using generalized Landesman-Lazer conditions. *Monatsh. Math.*

Kyritsi, S., Matzakos, N. & Papageorgiou, N. (2002). Periodic problems for strongly nonlinear second-order differential inclusions. *J. Differential Equations* **183**, 279–302.

Ladyzhenskaya, O. & Uraltseva, N. (1968). *Linear and Quasilinear Elliptic Equations*. Academic Press. New York.

Landes, R. (1980). On Galerkin's method in the existence theory of quasilinear elliptic equations. *J. Funct. Anal.* **39**, 123–148.

Landesman, E. & Lazer, A. (1969/1970). Nonlinear perturbations of linear elliptic boundary value problems at resonance. *J. Math. Mech.* **19**, 609–623.

Landesman, E., Robinson, S. & Rumbos, A. (1995). Multiple solutions of semilinear elliptic problems at resonance. *Nonlinear Anal.* **24**, 1049–1059.

Lazer, A. & Leach, D. (1969). Bounded perturbations of forced harmonic oscillations at resonance. *Ann. Mat. Pura Appl. (4)* **82**, 49–68.

Le, V. (1997). On global bifurcation of variational inequalities and applications. *J. Differential Equations* **141**, 254–294.

Le, V. (2000). Existence of positive solutions of variational inequalities by subsolution-supersolution approach. *J. Math. Anal. Appl.* **252**, 65–90.

Le, V. (2001). Subsolution-supersolution method in variational inequalities. *Nonlinear Anal.* **45**, 775–800.

Le, V. & Schmitt, K. (1995). Minimization problems for noncoercive functionals subject to constraints. *Trans. Amer. Math. Soc.* **347**, 4485–4513.

Lebourg, G. (1975). Valeur moyenne pour gradient généralisé. *C. R. Acad. Sci. Paris Sér. A-B* **281**, 795–797.

Li, S. & Willem, M. (1995). Applications of local linking to critical point theory. *J. Math. Anal. Appl.* **189**, 6–32.

Li, W. & Zhen, H. (1995). The applications of sums of ranges of accretive operators to nonlinear equations involving the p-Laplacian operator. *Nonlinear Anal.* **24**, 185–193.

Lieberman, G. (1988). Boundary regularity for solutions of degenerate elliptic equations. *Nonlinear Anal.* **12**, 1203–1219.

Lindqvist, P. (1990). On the equation $\mathrm{div}(\|\nabla x\|^{p-2}\nabla x) + \lambda|x|^{p-2}x = 0$. *Proc. Amer. Math. Soc.* **109**, 157–164.

Lindqvist, P. (1992). Addendum to "On the equation $\mathrm{div}(\|\nabla x\|^{p-2}\nabla x) + \lambda|x|^{p-2}x = 0$". *Proc. Amer. Math. Soc.* **116**, 583–584.

Lions, P.-L. (1982). On the existence of positive solutions of semilinear elliptic equations. *SIAM Rev.* **24**, 441–467.

Liu, J. & Li, S. (1984). An existence theorem of multiple critical points and its applications. *Kexue Tongbao (Chinese)* **29**, 1025–1027.

Liu, Z. (1999). On quasilinear elliptic hemivariational inequalities of higher order. *Acta Math. Hungar.* **85**, 1–8.

Ma, J. (2000). Multiple nonnegative solutions of second-order systems of boundary value problems. *Nonlinear Anal.* **42**, 1003–1010.

Ma, J. & Tang, C.-L. (2002). Periodic solutions for some nonautonomous second-order systems. *J. Math. Anal. Appl.* **275**, 482–494.

Manásevich, R. & Mawhin, J. (1998). Periodic solutions for nonlinear systems with p-Laplacian-like operators. *J. Differential Equations* **145**, 367–393.

Manásevich, R. & Mawhin, J. (2000). Boundary value problems for nonlinear perturbations of vector p-Laplacian-like operators. *J. Korean Math. Soc.* **37**, 665–685.

Manes, A. & Micheletti, A. (1973). Un'estensione della teoria variazionale classica degli autovalori per operatori ellittici del secondo ordine. *Boll. Un. Mat. Ital.* **7**, 285–301.

Marano, S. (1995). Elliptic boundary-value problems with discontinuous nonlinearities. *Set-Valued Anal.* **3**, 167–180.

Marano, S. & Motreanu, D. (2002*a*). Infinitely many critical points for nondifferentiable functions and applications to a Neumann-type problem involving the p-Laplacian. *J. Differential Equations* **182**, 108–120.

Marano, S. & Motreanu, D. (2002*b*). On a three critical points theorem for non-differentiable functions and applications to nonlinear boundary value problems. *Nonlinear Anal.* **48**, 37–52.

Marcus, M. & Mizel, V. (1972). Absolute continuity on tracks and mappings of Sobolev spaces. *Arch. Rational Mech. Anal.* **45**, 294–320.

Marino, A. (1989). The calculus of variations and some semilinear variational inequalities of elliptic and parabolic type. *in* 'Partial Differential Equations and the Calculus of Variations. Vol. II'. Vol. 2 of *Progress in Nonlinear Differential Equations and Their Applications*. Birkhäuser Verlag. Boston, MA. pp. 787–822.

Marzocchi, M. (1995). Multiple solutions of quasilinear equations involving an area-type term. *J. Math. Anal. Appl.* **196**, 1093–1104.

Massabó, I. (1980). Elliptic boundary value problems at resonance with discontinuous nonlinearities. *Boll. Un. Mat. Ital. B (5)* **17**, 1308–1320.

Massabó, I. & Stuart, C. (1978). Elliptic eigenvalue problems with discontinuous nonlinearities. *J. Math. Anal. Appl.* **6**, 261–281.

Matzakos, N. & Papageorgiou, N. (2001). Boundary value problems for strongly nonlinear second order differential inclusions. *Nonlinear Funct. Anal. Appl.* **6**, 229–246.

Mawhin, J. (1979). *Topological Degree Methods in Nonlinear Boundary Value Problems*. Vol. 40 of *CBMS Regional Conference Series in Mathematics*. AMS. Providence, RI.

Mawhin, J. (1984). Remarks on: "Critical point theory and a theorem of Amaral and Pera" [*Boll. Un. Math. Ital. B (6)* **3** (1984), no.3, 583–598] by Ahmad and Lazer. *Boll. Un. Mat. Ital. A (6)* **3**, 229–238

Mawhin, J. (1993). Topological degree and boundary value problems for nonlinear differential equations. *in* M. Furi & P. Zecca, eds, 'Topological Methods for Ordinary Differential Equations (Montecatini Terma, 1991)'. Vol. 1537 of *Lecture Notes in Math.* Springer Verlag. Berlin. pp. 74–142.

Mawhin, J. (1995). Continuation theorems and periodic solutions of ordinary differential quations. *in* 'Topological Methods in Differential Equations and Inclusions (Montreal, PQ, 1994)'. Vol. 472 of *NATO Adv. Sci. Inst. Ser. C Math. Phys. Sci.* Kluwer. Dordrecht. pp. 291–375.

Mawhin, J. (2000). Some boundary value problems for Hartman-type perturbations of the ordinary vector p-Laplacian. *Nonlinear Anal.* **40**, 497–503.

Mawhin, J. (2001). Periodic solutions of systems with p-Laplacian-like operators. *in* 'Nonlinear Analysis and its Applications to Differential Equations (Lisbon 1998)'. Vol. 43 of *Progress in Nonlinear Differential Equations and Their Applications.* Birkhäuser Verlag. Boston, MA. pp. 37–63.

Mawhin, J. & Schmitt, K. (1988). Landesman-Lazer type problems at an eigenvalue of odd multiplicity. *Results Math.* **14**, 139–146.

Mawhin, J. & Schmitt, K. (1990). Nonlinear eigenvalue problems with the parameter near resonance. *Ann. Polon. Math.* **51**, 241–248.

Mawhin, J. & Ureña, A. (2002). A Hartman-Nagumo inequality for the vector ordinary p-Laplacian and applications to nonlinear boundary value problems. *J. Inequal. Appl.* **7**, 701–725.

Mawhin, J. & Willem, M. (1989). *Critical Point Theory and Hamiltonian Systems.* Vol. 74 of *Applied Mathematical Sciences.* Springer-Verlag. New York.

Mizoguchi, N. (1991). Existence of nontrivial solutions of partial differential equations with discontinuous nonlinearities. *Nonlinear Anal.* **16**, 1025–1034.

Moreau, J. (1968). La notion de sur-potentiel et les liaisons unilatérales en élastostaique. *C. R. Acad. Sci. Paris Sér. A-B* **267**, 954–957.

Motreanu, D. (1995). Existence of critical points in a general setting. *Set-Valued Anal.* **3**, 295–305.

Motreanu, D. (2001). Eigenvalue problems for variational-hemivariational inequalities in the sense of P.D. Panagiotopoulos. *Nonlinear Anal.* **47**, 5101–5112. Proceedings of the Third World Congress of Nonlinear Analysts, Part 8 (Catania, 2000).

Motreanu, D. & Panagiotopoulos, P. (1995*a*). An eigenvalue problem for a hemivariational inequality involving a nonlinear compact operators. *Set-Valued Anal.* **3**, 157–166.

Motreanu, D. & Panagiotopoulos, P. (1995*b*). A minimax approach to the eigenvalue problem of hemivariational inequalities and applications. *Appl. Anal.* **58**, 53–76.

Motreanu, D. & Panagiotopoulos, P. (1995*c*). Nonconvex energy functions, related eigenvalue hemivariational inequalities on the sphere and applications. *J. Global Optim.* **6**, 163–177.

Motreanu, D. & Panagiotopoulos, P. (1996). On the eigenvalue problem for hemivariational inequalities: Existence and multiplicity of solutions. *J. Math. Anal. Appl.* **197**, 75–89.

Motreanu, D. & Panagiotopoulos, P. (1997). Double eigenvalue problems for hemivariational inequalities. *Arch. Rational Mech. Anal.* **140**, 225–251.

Motreanu, D. & Rădulescu, V. (2000). Existence results for inequality problems with lack of convexity. *Numer. Funct. Anal. Optim.* **21**, 869–884.

Nagumo, M. (1942). Über das Randwertproblem der nicht linearen gewöhnlichen Differentialgleichung zweiter Ordnung. *Proc. Phys.-Math. Soc. Japan (3)* **24**, 845–851.

Naniewicz, Z. (1994). Hemivariational inequality approach to constrained problems for star-shaped admissible sets. *J. Optim. Theory Appl.* **83**, 97–112.

Naniewicz, Z. (1995). Hemivariational inequalities with functionals which are not locally Lipschitz. *Nonlinear Anal.* **25**, 1307–1330.

Naniewicz, Z. (1997). Hemivariational inequalities as necessary conditions for optimality for a class of nonsmooth nonconvex functionals. *Nonlinear World* **4**, 117–133.

Naniewicz, Z. & Panagiotopoulos, P. (1995). *Mathematical Theory of Hemivariational Inequalities and Applications.* Vol. 188 of *Monographs and Textbooks in Pure and Applied Mathematics.* Marcel Dekker. New York.

Nieto, J. (1989). Nonlinear second order periodic boundary value problems with Carathéodory functions. *Appl. Anal.* **34**, 111–128.

Omari, P. (1986). A monotone method for constructing extremal solutions of second order scalar boundary value problems. *Appl. Math. Comput.* **18**, 257–275.

O'Regan, D. (1993). Some general existence principles and results for $(\varphi(y'))' = qf(t,y,y')$, $0 < t < 1$. *SIAM J. Math. Anal.* **24**, 648–668.

Ôtani, M. (1984a). On certain second order ordinary differential equations associated with Sobolev-Poincaré-type inequalities. *Nonlinear Anal.* **8**, 1255–1270.

Ôtani, M. (1984b). A remark on certain nonlinear elliptic equations. *Proc. Fac. Sci. Tokai Univ.* **19**, 23–28.

Palais, R. & Smale, S. (1964). A generalized Morse theory. *Bull. Amer. Math. Soc.* **70**, 165–172.

Palmucci, M. & Papalini, F. (2002). A nonlinear multivalued problem with nonlinear boundary conditions. *in* 'Set Valued Mappings with Applications to Nonlinear Analysis'. Vol. 4 of *Ser. Math. Anal. Appl.* Taylor & Francis. London. pp. 383–402.

Panagiotopoulos, P. (1985). *Inequality Problems in Mechanics and Applications.* Birkhäuser Verlag. Boston, MA.

Panagiotopoulos, P. (1988). *Hemivariational Inequalities and Their Applications.* Birkhäuser Verlag. Boston, MA.

Panagiotopoulos, P. (1991). Coercive and semicoercive hemivariational inequalities. *Nonlinear Anal.* **16**, 209–231.

Panagiotopoulos, P. (1993). *Hemivariational Inequalities. Applications to Mechanics and Engineering.* Springer-Verlag. Berlin.

Panagiotopoulos, P., Fundo, M. & Rădulescu, V. (1999). Existence theorems of Hartman-Stampacchia-type for hemivariational inequalities and applications. *J. Global Optim.* **15**, 41–54.

Papageorgiou, E. & Papageorgiou, N. (2004a). Strongly nonlinear multivalued periodic problems with maximal monotone terms. *Differential Integral Equations* **17**, 443–480.

Papageorgiou, E. & Papageorgiou, N. (2004b). Two nontrivial solutions for quasilinear periodic problems. *Proc. Amer. Math. Soc.* **132**, 429–434.

Papageorgiou, E. & Papageorgiou, N. (submitted). Nonlinear boundary value problems involving the *p*-Laplacian-like operators. submitted.

Papageorgiou, E. & Papageorgiou, N. (to appear). Nonlinear second order periodic systems with nonsmooth potential. *Czechoslovak Math. J.*

Papageorgiou, N. & Papalini, F. (2001). Periodic and boundary value problems for second order differential equations. *Proc. Indian Acad. Sci. Math. Sci.* **111**, 107–125.

Papageorgiou, N. & Smyrlis, G. (2003). *On Nonlinear Hemivariational Inequalities.* Vol. 419 of *Dissertationes Math. (Rozprawy Mat.).*

Papageorgiou, N. & Yannakakis, N. (1999). Nonlinear boundary value problems. *Proc. Indian Acad. Sci. Math. Sci.* **109**, 211–230.

Papageorgiou, N. & Yannakakis, N. (2004). Periodic solutions for second order equations with the scalar p-Laplacian and nonsmooth potential. *Funkcial. Ekvac.* **47**, 107–117.

Papalini, F. (2002). Nonlinear eigenvalue Neumann problems with discontinuities. *J. Math. Anal. Appl.* **273**, 137–152.

Papalini, F. (2003). A quasilinear Neumann problem with discontinuous nonlinearity. *Math. Nachr.* **250**, 82–97.

Pascali, D. & Sburlanu, S. (1978). *Nonlinear Mappings of Monotone Type.* Sijthoff and Noordhoff International Publishers. Alpen aan den Rijn.

Perron, O. (1923). Eine neue Behandlung der ersten Randwertaufgabe für $\delta u = 0$. *Math. Z.* **18**, 42–54.

Phelps, R. (1984). *Convex Functions, Monotone Operators and Differentiability.* Vol. 1364 of *Lecture Notes in Math.* Springer-Verlag. Berlin.

Protter, M. & Weinberger, H. (1967). *Minimax Principles in Differential Equations.* Prentice-Hall. New York.

Pruszko, T. (1984). *Some Applications of the Topological Degree Theory to Multivalued Boundary Value Problems.* Vol. 229 of *Dissertationes Math. (Rozprawy Mat.).*

Pucci, P. & Serrin, J. (1984). Extension of the mountain pass theorem. *J. Funct. Anal.* **59**, 185–210.

Quittner, P. (1989). Solvability and multiplicity results for variational inequalities. *Comment. Math. Univ. Carolin.* **30**, 281–302.

Rabinowitz, P. (1973). Some aspects of nonlinear eigenvalue problems. *Rocky Mountain J. Math.* **3**, 161–202.

Rabinowitz, P. (1978a). Periodic solutions of Hamiltonian systems. *Comm. Pure Appl. Math.* **31**, 157–184.

Rabinowitz, P. (1978b). Some critical points theorems and applications to semilinear elliptic partial differential equations. *Ann. Scuola Norm. Sup. Pisa Cl. Sci. (4)* **5**, 215–223.

Rabinowitz, P. (1978c). Some minimax theorems and applications to nonlinear partial differential equations. *in* L. Cesari, R. Kannan & H. Weinberg, eds, 'Nonlinear Analysis'. Academic Press. New York. pp. 161–177.

Rabinowitz, P. (1986). *Minimax Methods in Critical Point Theory with Applications to Differential Equations.* Vol. 65 of *CBMS Regional Conference Series in Mathematics.* AMS. Providence, RI.

Rabinowitz, P. (1990). Homoclinic orbits for a class of Hamiltonian systems. *Proc. Roy. Soc. Edinburgh Sect. A* **114**, 33–38.

Rachunková, I. (1999). Upper and lower solutions and topological degree. *J. Math. Anal. Appl.* **234**, 311–327.

Ramos, M. & Sánchez, L. (1997). A variational approach to multiplicity in elliptic problems near resonance. *Proc. Roy. Soc. Edinburgh Sect. A* **127**, 385–394.

Ribarska, N., Tsachev, J. & Krastanov, M. (2001). A note on "On a critical point theory for multivalued functionals and applications to partial differential inclusions" [*Nonlinear Anal.* **31** (1998), no. 5-6, 735–753] by M. Frigon]. *Nonlinear Anal.* **43**, 153–158.

Robinson, S. (1993). Double resonance in semilinear elliptic boundary value problem over bounded and unbounded domain. *Nonlinear Anal.* **21**, 407–424.

Rockafellar, R. (1970). *Convex Analysis.* Vol. 28 of *Princeton Mathematical Series.* Princeton Univ. Press. Princeton, NJ.

Rockafellar, R. (1971). Convex integral functionals and duality. *in* 'Contributions to Nonlinear Functional Analysis (Proc. Sympos., Math. Res. Center, Univ. Wisconsin, Madison, Wis., 1971)'. Academic Press. New York. pp. 215–236.

Rădulescu, V. (1993). Mountain pass theorems for nondifferentiable functions and applications. *Proc. Japan Acad. Ser. A Math. Sci.* **69**, 193–198.

Rădulescu, V. & Panagiotopoulos, P. (1998). Perturbations of hemivariational inequalities with constraints and applications. *J. Global Optim.* **12**, 285–297.

Rudd, M. & Schmitt, K. (2002). Variational inequalities of elliptic and parabolic type. *Taiwanese J. Math.* **6**, 287–322.

Schoenenberger-Deuel, J. & Hess, P. (1974/75). A criterion for the existence of solutions of non-linear elliptic boundary value problems. *Proc. Roy. Soc. Edinburgh Sect. A* **74**, 49–54.

Showalter, R. (1977). *Hilbert Space Methods for Partial Differential Equations.* Vol. 1 of *Monographs and Studies in Mathematics.* Pitman. London.

Showalter, R. (1997). *Monotone Operators in Banach Spaces and Nonlinear Partial Differential Equations.* Vol. 49 of *Mathematical Surveys and Monographs.* AMS. Providence, RI.

Silva, E. (1991). Linking theorems and applications to semilinear elliptic problems at resonance. *Nonlinear Anal.* **16**, 455–477.

Simon, L. (1988). On strongly nonlinear elliptic variational inequalities. *Acta Math. Hungar.* **52**, 147–164.

Simon, L. (1990). On uniqueness, regularity and stability of strongly nonlinear elliptic variational inequalities. *Acta Math. Hungar.* **55**, 372–392.

Stampacchia, G. (1965). Le problème de Dirichlet pour les équations elliptiques du second order à coefficients discontinus. *Ann. Inst. Fourier (Grenoble)* **15**, 189–258.

Struwe, M. (1982). A note on a result of Ambrosetti and Mancini. *Ann. Mat. Pura Appl.* **131**, 107–115.

Struwe, M. (1990). *Variational Methods. Applications to Nonlinear Partial Differential Equations and Hamiltonian Systems.* Springer-Verlag. Berlin.

Stuart, C. (1976). *Boundary-Value Problems with Discontinuous Non-Linearities.* Vol. 564 of *Lecture Notes in Math.* Springer-Verlag. Berlin.

Stuart, C. (1976/1977). Differential equations with discontinuous nonlinearities. *Arch. Rational Mech. Anal.* **63**, 59–75.

Stuart, C. (1978). Maximal and minimal solutions of elliptic differential equations with discontinuous nonlinearities. *Math. Z.* **163**, 239–249.

Stuart, C. & Toland, J. (1980). A variational method for boundary value problems with discontinuous nonlinearities. *J. London Math. Soc. (2)* **21**, 319–328.

Su, J. (2002). Semilinear elliptic boundary value problems with double resonance between two consecutive eigenvalues. *Nonlinear Anal.* **48**, 881–895.

Su, J. & Tang, C.-L. (2001). Multiplicity results for semilinear elliptic equations with resonance at higher eigenvalues. *Nonlinear Anal.* **44**, 311–321.

Szulkin, A. (1985). Positive solutions of variational inequalities: A degree-theoretic approach. *J. Differential Equations* **57**, 90–111.

Szulkin, A. (1986). Minimax principles for lower semicontinuous functions and applications to nonlinear boundary value problems. *Ann. Inst. H. Poincaré Anal. Non Linéaire* **3**, 77–109.

Tang, C.-L. (1998a). Existence and multiplicity of periodic solutions for nonautonomous second order systems. *Nonlinear Anal.* **32**, 299–304.

Tang, C.-L. (1998b). Periodic solutions for nonautonomous second order systems with sublinear nonlinearity. *Proc. Amer. Math. Soc.* **126**, 3269–3270.

Tang, C.-L. (1998c). Solvability of the forced Duffing equation at resonance. *J. Math. Anal. Appl.* **219**, 110–124.

Tang, C.-L. (1999). Multiplicity of periodic solutions for second-order systems with a small forcing term. *Nonlinear Anal.* **38**, 471–479.

Tang, C.-L. & Wu, X.-P. (2001a). Existence and multiplicity of solutions semilinear elliptic equations. *J. Math. Anal. Appl.* **256**, 1–12.

Tang, C. L. & Wu, X.-P. (2001b). Periodic solutions for second order systems with not uniformly coercive potential. *J. Math. Anal. Appl.* **259**, 386–397.

Tehrani, H. (1996). H^1 versus C^1 local minimizers on manifolds. *Nonlinear Anal.* **26**, 1491–1509.

Thews, K. (1980). Nontrivial solutions of elliptic equations at resonance. *Proc. Roy. Soc. Edinburgh Sect. A* **85**, 119–129.

Thompson, H. (1996). Second order ordinary differential equations with fully nonlinear two point boundary conditions. *Pacific J. Math.* **172**, 255–297.

Tolksdorf, P. (1984). Regularity for a more general class of quasilinear elliptic equations. *J. Differential Equations* **51**, 126–150.

Ton, B. (1971). Pseudo-monotone operators in Banach spaces and nonlinear elliptic equations. *Math. Z.* **121**, 243–252.

Ubilla, P. (1995a). Homoclinic orbits for a quasi-linear Hamiltonian system. *J. Math. Anal. Appl.* **193**, 573–587.

Ubilla, P. (1995b). Multiplicity results for the 1-dimensional generalized p-Laplacian. *J. Math. Anal. Appl.* **190**, 611–623.

Vaĭnberg, M. (1973). *Variational Method and Method of Monotone Operators in the Theory of Nonlinear Equations.* Halsted Press. New York.

Vázquez, J. (1984). A strong maximum principle for some quasilinear elliptic equations. *Appl. Math. Optim.* **12**, 191–202.

Wagner, D. (1977). Survey of measurable selection theorems. *SIAM J. Control Optimization* **15**, 859–903.

Wagner, D. (1980). Survey of measurable selection theorems: An update. *in* 'Measure Theory in Oberwolfach 1979 (Proc. Conf., Oberwolfach, 1979)'. Vol. 794 of *Lecture Notes in Math.* Springer-Verlag. Berlin. pp. 176–219.

Wang, H. (1994). On the existence of positive solutions for semilinear elliptic equations in the annulus. *J. Differential Equations* **109**, 1–7.

Wang, H. (2003). On the number of positive solutions of nonlinear systems. *J. Math. Anal. Appl.* **281**, 287–306.

Wang, J., Gao, W. & Lin, Z. (1995). Boundary value problems for general second order equations and similarity solutions to the Rayleigh problem. *Tôhoku Math. J. (2)* **47**, 327–344.

Wang, M., Cabada, A. & Nieto, J. (1993). Monotone method for nonlinear second order periodic boundary value problems with Carathéodory functions. *Ann. Polon. Math.* **58**, 221–235.

Ward, J. (1984). Applications of critical point theory to weakly nonlinear boundary value problems at resonance. *Houston J. Math.* **10**, 291–305.

Webb, J. (1980). Boundary value problems for strongly nonlinear elliptic equations. *J. London Math. Soc. (2)* **21**, 123–132.

Wei, Z. & Wu, X. (1992). A multiplicity result for quasilinear elliptic equations involving critical Sobolev exponents. *Nonlinear Anal.* **18**, 559–567.

Weinstein, A. (1978). Periodic orbits for convex Hamiltonian systems. *Ann. of Math. (2)* **108**, 507–518.

Willem, M. (1996). *Minimax Theorems.* Vol. 24 of *Progress in Nonlinear Differential Equations and Their Applications.* Birkhäuser Verlag. Boston, MA.

Wu, X.-P. & Tang, C.-L. (1999). Periodic solutions of a class of non-autonomous second-order systems. *J. Math. Anal. Appl.* **236**, 227–235.

Wu, X.-P. & Tang, C.-L. (2001). Some existence theorems for elliptic resonance problems. *J. Math. Anal. Appl.* **264**, 133–146.

Yang, Z. & Guo, Z. (1995). On the structure of positive solutions for quasilinear ordinary differential equations. *Appl. Anal.* **58**, 31–51.

Yosida, K. (1978). *Functional Analysis.* Springer-Verlag. Berlin/New York.

Zeidler, E. (1985). *Nonlinear Functional Analysis and its Applications III. Variational Methods and Optimization.* Springer-Verlag. New York.

Zeidler, E. (1990a). *Nonlinear Functional Analysis and its Applications II/A. Linear Monotone Operators.* Springer-Verlag. New York.

Zeidler, E. (1990b). *Nonlinear Functional Analysis and its Applications II/B. Nonlinear Monotone Operators.* Springer-Verlag. New York.

Zhong, C.-K. (1997). On Ekeland's variational principle and a minimax theorem. *J. Math. Anal. Appl.* **205**, 239–250.

Ziemer, W. (1989). *Weakly Differentiable Functions. Sobolev Spaces and Functions of Bounded Variation.* Springer-Verlag. New York.

Zou, W. & Liu, J. (2001). Multiple solutions for resonant elliptic equations via local linking theory and Morse theory. *J. Differential Equations* **170**, 68–95.

Index